AVIATION MAINTENANCE TECHNICIAN HANDBOOK

GENERAL

2008

U.S. DEPARTMENT OF TRANSPORTATION
FEDERAL AVIATION ADMINISTRATION
Flight Standards Service

Published by Books Express Publishing
ISBN 978-1-78266-086-6

Books Express publications are available from all good retail and online booksellers. For publishing proposals and direct ordering please contact us at: info@books-express.com

PREFACE

The *Aviation Maintenance Technician Handbook—General* was developed as one of a series of three handbooks for persons preparing for mechanic certification with airframe or powerplant ratings, or both. It is intended that this handbook will provide basic information on principles, fundamentals, and technical procedures in the subject matter areas common to both the airframe and powerplant ratings. Emphasis in this volume is on theory and methods of application.

The handbook is designed to aid students enrolled in a formal course of instruction preparing for FAA certification as a maintenance technician, as well as for current technicians who wish to improve their knowledge.

This volume contains information on mathematics, aircraft drawings, weight and balance, aircraft materials, processes and tools, physics, electricity, inspection, ground operations, and FAA regulations governing the certification and work of maintenance technicians. New to this volume is a section addressing how successful aviation maintenance technicians incorporate knowledge and awareness of ethics, professionalism, and human factors in the field.

Because there are so many different types of airframes and powerplants in use today, it is reasonable to expect that differences exist in the components and systems of each. To avoid undue repetition, the practice of using representative systems and units is implemented throughout the handbook. Subject matter treatment is from a generalized point of view, and should be supplemented by reference to manufacturers' manuals or other textbooks if more detail is desired. This handbook is not intended to replace, substitute for, or supersede official regulations or manufacturer instructions.

The companion handbooks to *Aviation Maintenance Technician Handbook—General* are the *Airframe and Powerplant Mechanics Powerplant Handbook*, AC 65-12A, and the *Airframe and Powerplant Mechanics Airframe Handbook*, AC 65-15A.

This handbook may be purchased from the Superintendent of Documents, U.S. Government Printing Office (GPO), Washington, DC 20402-9325, or from the GPO website:

Aviation Maintenance Technician Handbook—General was produced by the Federal Aviation Administration (FAA) with the assistance of Aviation Supplies and Academics, Inc., and the following authors: Ann Riley, Bob Aardema, Pete Vosbury, Mary Ann Eiff, H.G. Frautschy, Ron Serkenburg, Don Shaffer, Tom Wild, and Terry Michmerhuizen.

Grateful acknowledgment is extended to the contributors of photography for this handbook. Contained throughout the book are photographs by H.G. Frautschy, Sam Chui, Jeremy Irish, Pete Vosbury, Matt Wallman, Fred Gay, Jim Groom, Emmanuel Perez, John Martin, Ad Vercruijsse, Marc Michel, Bill Blanchard, Wim Houquet, Jean-Luc Altherr, Andrew Doughty, Erwin van Dijck, Kathleen and Walter Hudson, Paul Lintott, Michael Davis, and Dave Baldwin.

Contents

Chapter 7 Fluid Lines & Fittings 7-1

Chapter 8 Inspection Fundamentals 8-1

Chapter 9
Hand Tools & Measuring Devices 9-1

Chapter 10
Basic Electricity 10-1

AVIATION MAINTENANCE TECHNICIAN HANDBOOK

GENERAL

2008

U.S. DEPARTMENT OF TRANSPORTATION
FEDERAL AVIATION ADMINISTRATION
Flight Standards Service

Chapter 12 Publications, Forms, & Records 12-1

Chapter 13 The Mechanic Certificate 13-1

Chapter 14

Mathematics

Mathematics in Aviation Maintenance

Mathematics is woven into many areas of everyday life. Performing mathematical calculations with success requires an understanding of the correct methods and procedures, and practice and review of these principles. Mathematics may be thought of as a set of tools. The aviation mechanic will need these tools to successfully complete the maintenance, repair, installation, or certification of aircraft equipment.

Many examples of using mathematical principles by the aviation mechanic are available. Tolerances in turbine engine components are critical, making it necessary to measure within a ten-thousandth of an inch. Because of these close tolerances, it is important that the aviation mechanic be able to make accurate measurements and mathematical calculations. An aviation mechanic working on aircraft fuel systems will also use mathematical principles to calculate volumes and capacities of fuel tanks. The use of fractions and surface area calculations are required to perform sheet metal repair on aircraft structures.

Whole Numbers

Whole numbers are the numbers 0, 1, 2, 3, 4, 5, and so on.

Addition of Whole Numbers

Addition is the process in which the value of one number is added to the value of another. The result is called the sum. When working with whole numbers, it is important to understand the principle of the place value. The place value in a whole number is the value of the position of the digit within the number. For example, in the number 512, the 5 is in the hundreds column, the 1 is in the tens column, and the 2 is in the ones column. The place values of three whole numbers are shown in Figure 1-1.

	Place Value				
	Ten Thousands	Thousands	Hundreds	Tens	Ones
35 shown as				3	5
269 shown as			2	6	9
12,749 shown as	1	2	7	4	9

Figure 1-1. Example of place values of whole numbers.

When adding several whole numbers, such as 4,314, 122, 93,132, and 10, align them into columns according to place value and then add.

$$\begin{array}{r} 4{,}314 \\ 122 \\ 93{,}132 \\ +\ 10 \\ \hline 97{,}578 \end{array}$$

This is the sum of the four whole numbers.

Subtraction of Whole Numbers

Subtraction is the process in which the value of one number is taken from the value of another. The answer is called the difference. When subtracting two whole numbers, such as 3,461 from 97,564, align them into columns according to place value and then subtract.

$$\begin{array}{r} 97{,}564 \\ -3{,}461 \\ \hline 94{,}103 \end{array}$$

This is the difference of the two whole numbers.

Multiplication of Whole Numbers

Multiplication is the process of repeated addition. For example, 4×3 is the same as $4 + 4 + 4$. The result is called the product.

Example: How many hydraulic system filters are in the supply room if there are 35 cartons and each carton contains 18 filters?

$$\begin{array}{r} 18 \\ \times\ 35 \\ \hline 90 \\ 54 \\ \hline 630 \end{array}$$

Therefore, there are 630 filters in the supply room.

Division of Whole Numbers

Division is the process of finding how many times one number (called the divisor) is contained in another number (called the dividend). The result is the quotient, and any amount left over is called the remainder.

$$\text{divisor}\overline{\big)\,\text{dividend}}^{\;\text{quotient}}$$

Example: 218 landing gear bolts need to be divided between 7 aircraft. How many bolts will each aircraft receive?

$$\begin{array}{r} 31 \\ 7\overline{)218} \\ -21 \\ \hline 8 \\ -7 \\ \hline 1 \end{array}$$

The solution is 31 bolts per aircraft with a remainder of 1 bolt left over.

Fractions

A fraction is a number written in the form $^N/_D$ where N is called the numerator and D is called the denominator. The fraction bar between the numerator and denominator shows that division is taking place.

Some examples of fractions are: $\frac{17}{18}\quad\frac{2}{3}\quad\frac{5}{8}$

The denominator of a fraction cannot be a zero. For example, the fraction $^2/_0$ is not allowed. An improper fraction is a fraction in which the numerator is equal to or larger than the denominator. For example, $^4/_4$ or $^{15}/_8$ are examples of improper fractions.

Finding the Least Common Denominator

To add or subtract fractions, they must have a common denominator. In math, the least common denominator (LCD) is commonly used. One way to find the LCD is to list the multiples of each denominator and then choose the smallest one that they have in common.

Example: Add $^1/_5 + {}^1/_{10}$ by finding the least common denominator.

Multiples of 5 are: 5, 10, 15, 20, 25, and on. Multiples of 10 are: 10, 20, 30, 40, and on. Notice that 10, 20, and 30 are in both lists, but 10 is the smallest or least common denominator (LCD). The advantage of finding the LCD is that the final answer is more likely to be in lowest terms.

A common denominator can also be found for any group of fractions by multiplying all of the denominators together. This number will not always be the LCD, but it can still be used to add or subtract fractions.

Example: Add $^2/_3 + {}^3/_5 + {}^4/_7$ by finding a common denominator.

A common denominator can be found by multiplying the denominators $3 \times 5 \times 7$ to get 105.

$$\left(\frac{2}{3}+\frac{3}{5}+\frac{4}{7}\right)=\left(\frac{70}{105}+\frac{63}{105}+\frac{60}{105}\right)=\frac{193}{105}=1\frac{88}{105}$$

Addition of Fractions

In order to add fractions, the denominators must be the same number. This is referred to as having "common denominators."

Example: Add $^1/_7$ to $^3/_7$

$$\frac{1}{7}+\frac{3}{7}=\frac{1+3}{7}=\frac{4}{7}$$

If the fractions do not have the same denominator, then one or all of the denominators must be changed so that every fraction has a common denominator.

Example: Find the total thickness of a panel made from $^3/_{32}$-inch thick aluminum, which has a paint coating that is $^1/_{64}$-inch thick. To add these fractions, determine a common denominator. The least common denominator for this example is 1, so only the first fraction must be changed since the denominator of the second fraction is already 64.

$$\left(\frac{3}{32}+\frac{1}{64}\right)=\left(\frac{3\times 2}{32\times 2}+\frac{1}{64}\right)=\left(\frac{6}{64}+\frac{1}{64}\right)=\frac{6+1}{64}=\frac{7}{64}$$

Therefore, $^7/_{64}$ is the total thickness.

Subtraction of Fractions

In order to subtract fractions, they must have a common denominator.

Example: Subtract $^2/_{17}$ from $^{10}/_{17}$

$$\frac{10}{17}-\frac{2}{17}=\frac{10-2}{17}=\frac{8}{17}$$

If the fractions do not have the same denominator, then one or all of the denominators must be changed so that every fraction has a common denominator.

Example: The tolerance for rigging the aileron droop of an airplane is $\frac{7}{8}$ inch ± $\frac{1}{5}$ inch. What is the minimum droop to which the aileron can be rigged? To subtract these fractions, first change both to common denominators. The common denominator in this example is 40. Change both fractions to $\frac{1}{40}$, as shown, then subtract.

$$\left(\frac{7}{8}-\frac{1}{5}\right)=\left(\frac{7\times 5}{8\times 5}-\frac{1\times 8}{5\times 8}\right)=\left(\frac{35}{40}-\frac{8}{40}\right)=\left(\frac{35-8}{40}\right)=\frac{27}{40}$$

Therefore, $\frac{27}{40}$ is the minimum droop.

Multiplication of Fractions

Multiplication of fractions does not require a common denominator. To multiply fractions, first multiply the numerators. Then, multiply the denominators.

Example:

$$\frac{3}{5}\times\frac{7}{8}\times\frac{1}{2}=\frac{3\times 7\times 1}{5\times 8\times 2}=\frac{21}{80}$$

The use of cancellation when multiplying fractions is a helpful technique which divides out or cancels all common factors that exist between the numerators and denominators. When all common factors are cancelled before the multiplication, the final product will be in lowest terms.

Example:

$$\left(\frac{14}{15}\times\frac{3}{7}\right)=\left(\frac{\overset{2}{\cancel{14}}}{\underset{5}{\cancel{15}}}\times\frac{\overset{1}{\cancel{3}}}{\underset{1}{\cancel{7}}}\right)=\left(\frac{2\times 1}{5\times 1}\right)=\frac{2}{5}$$

Division of Fractions

Division of fractions does not require a common denominator. To divide fractions, first change the division symbol to multiplication. Next, invert the second fraction. Then, multiply the fractions.

Example: Divide $\frac{7}{8}$ by $\frac{4}{3}$

$$\left(\frac{7}{8}\div\frac{4}{3}\right)=\left(\frac{7}{8}\times\frac{3}{4}\right)=\left(\frac{7\times 3}{8\times 4}\right)=\frac{21}{32}$$

Example: In Figure 1-2, the center of the hole is in the center of the plate. Find the distance that the center of the hole is from the edges of the plate. To find the answer, the length and width of the plate should each be divided in half. First, change the mixed numbers to improper fractions:

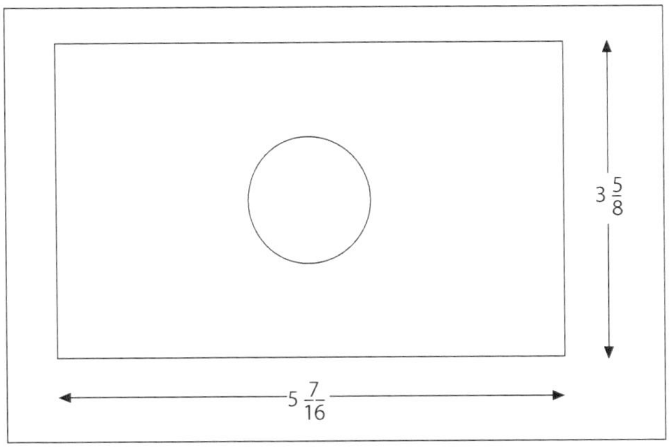

Figure 1-2. Center hole of the plate.

$$5\tfrac{7}{16}\text{ inches} = \tfrac{87}{16}\text{ inches}$$
$$3\tfrac{5}{8}\text{ inches} = \tfrac{29}{8}\text{ inches}$$

Then, divide each improper fraction by 2 to find the center of the plate.

$$\frac{87}{16}\div\frac{2}{1}=\frac{87}{16}\times\frac{1}{2}=\frac{87}{32}\text{ inches}$$

$$\frac{29}{8}\div\frac{2}{1}=\frac{29}{8}\times\frac{1}{2}=\frac{29}{16}\text{ inches}$$

Finally, convert each improper fraction to a mixed number:

$$\frac{87}{32}=87\div 32=2\frac{23}{32}\text{ inches}$$

$$\frac{29}{16}=29\div 16=1\frac{13}{16}\text{ inches}$$

Therefore, the distance to the center of the hole from each of the plate edges is $2\frac{23}{32}$ inches and $1\frac{13}{16}$ inches.

Reducing Fractions

A fraction needs to be reduced when it is not in "lowest terms." Lowest terms means that the numerator and denominator do not have any factors in common. That is, they cannot be divided by the same number (or factor). To reduce a fraction, determine what the common factor(s) are and divide these out of the numerator and denominator. For example when both the numerator and denominator are even numbers, they can both be divided by 2.

Example: The total travel of a jackscrew is $\frac{13}{16}$ inch. If the travel in one direction from the neutral position is $\frac{7}{16}$ inch, what is the travel in the opposite direction?

$$\frac{13}{16}-\frac{7}{16}=\frac{13-7}{16}=\frac{6}{16}$$

The fraction $\frac{6}{16}$ is not in lowest terms because the numerator (6) and the denominator (16) have a common factor of 2. To reduce $\frac{6}{16}$, divide the numerator and the denominator by 2. The final reduced fraction is $\frac{3}{8}$ as shown below.

$$\frac{6}{16} = \frac{6 \div 2}{16 \div 2} = \frac{3}{8}$$

Therefore, the travel in the opposite direction is $\frac{3}{8}$ inch.

Mixed Numbers

A mixed number is a combination of a whole number and a fraction.

Addition of Mixed Numbers

To add mixed numbers, add the whole numbers together. Then add the fractions together by finding a common denominator. The final step is to add the sum of the whole numbers to the sum of the fractions for the final result.

Example: The cargo area behind the rear seat of a small airplane can handle solids that are $4\frac{3}{4}$ feet long. If the rear seats are removed, then $2\frac{1}{3}$ feet is added to the cargo area. What is the total length of the cargo area when the rear seats are removed?

$$4\frac{3}{4} + 2\frac{1}{3} = (4+2) + \left(\frac{3}{4} + \frac{1}{3}\right) = 6 + \left(\frac{9}{12} + \frac{4}{12}\right) = 6\frac{13}{12} =$$

$7\frac{1}{12}$ feet of cargo room.

Subtraction of Mixed Numbers

To subtract mixed numbers, find a common denominator for the fractions. Subtract the fractions from each other (it may be necessary to borrow from the larger whole number when subtracting the fractions). Subtract the whole numbers from each other. The final step is to combine the final whole number with the final fraction.

Example: What is the length of the grip of the bolt shown in Figure 1-3? The overall length of the bolt is $3\frac{1}{2}$ inches, the shank length is $3\frac{1}{8}$ inches, and the threaded portion is $1\frac{5}{16}$ inches long. To find the grip, subtract the length of the threaded portion from the length of the shank.

$$3\tfrac{1}{8} \text{ inches} - 1\tfrac{5}{16} \text{ inches} = \text{grip length}$$

To subtract, start with the fractions. Borrowing will be necessary because $\frac{5}{16}$ is larger than $\frac{1}{8}$ (or $\frac{2}{16}$). From the whole number 3, borrow 1, which is actually $\frac{16}{16}$. After borrowing, the first mixed number will now be

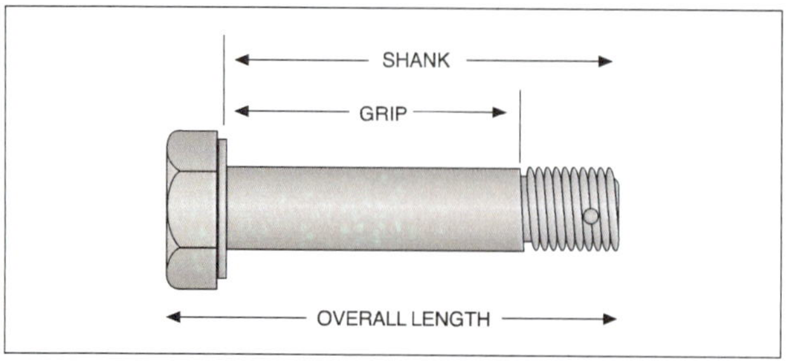

Figure 1-3. Bolt dimensions.

$2\frac{18}{16}$. (Because, $3\frac{1}{8} = 3\frac{2}{16} = 2 + 1 + \frac{2}{16} = 2 + \frac{16}{16} + \frac{2}{16} = 2\frac{18}{16}$.)

$$3\frac{1}{8} - 1\frac{5}{16} = 3\frac{2}{16} - 1\frac{5}{16} = 2\frac{18}{16} - 1\frac{5}{16} = 1\frac{13}{16}$$

Therefore, the grip length of the bolt is $1\frac{13}{16}$ inches.

(Note: The value for the overall length of the bolt was given in the example, but it was not needed to solve the problem. This type of information is sometimes referred to as a "distracter" because it distracts from the information needed to solve the problem.)

The Decimal Number System

The Origin and Definition

The number system that we use every day is called the decimal system. The prefix in the word decimal is a Latin root for the word "ten." The decimal system probably had its origin in the fact that we have ten fingers (or digits). The decimal system has ten digits: 0, 1, 2, 3, 4, 5, 6, 7, 8, and 9. It is a base 10 system and has been in use for over 5,000 years. A decimal is a number with a decimal point. For example, 0.515, .10, and 462.625 are all decimal numbers. Like whole numbers, decimal numbers also have place value. The place values are based on powers of 10, as shown in Figure 1-4.

	Place Value										
	Millions	Hundred Thousands	Ten Thousands	Thousands	Hundreds	Tens	Ones	Tenths	Hundredths	Thousandths	Ten Thousandths
1,623,051	1	6	2	3	0	5	1				
0.0531							0	0	5	3	1
32.4						3	2	4			

Figure 1-4. Place values.

Addition of Decimal Numbers

To add decimal numbers, they must first be arranged so that the decimal points are aligned vertically and according to place value. That is, adding tenths with tenths, ones with ones, hundreds with hundreds, and so forth.

Example: Find the total resistance for the circuit diagram shown in Figure 1-5. The total resistance of a series circuit is equal to the sum of the individual resistances. To find the total resistance, R_T, the individual resistances are added together.

$$R_T = 2.34 + 37.5 + .09$$

Arrange the resistance values in a vertical column so that the decimal points are aligned and then add.

$$\begin{array}{r} 2.34 \\ 37.5 \\ +\ .09 \\ \hline 39.93 \end{array}$$

Therefore, the total resistance, $R_T = 39.93$ ohms.

Subtraction of Decimal Numbers

To subtract decimal numbers, they must first be arranged so that the decimal points are aligned vertically and according to place value. That is, subtracting tenths from tenths, ones from ones, hundreds from hundreds, and so forth.

Example: A series circuit containing two resistors has a total resistance (R_T) of 37.272 ohms. One of the resistors (R_1) has a value of 14.88 ohms. What is the value of the other resistor (R_2)?

$$R_2 = R_T - R_1 = 37.272 - 14.88$$

Arrange the decimal numbers in a vertical column so that the decimal points are aligned and then subtract.

$$\begin{array}{r} 37.272 \\ -\,14.88 \\ \hline 22.392 \end{array}$$

Therefore, the second resistor, $R_2 = 22.392$ ohms.

Multiplication of Decimal Numbers

To multiply decimal numbers, vertical alignment of the decimal point is not required. Instead, align the numbers to the right in the same way as whole numbers are multiplied (with no regard to the decimal points or place values) and then multiply. The last step is to place the decimal point in the correct place in the answer. To do this, "count" the number of decimal places in each of the numbers, add the total, and then "give" that number of decimal places to the result.

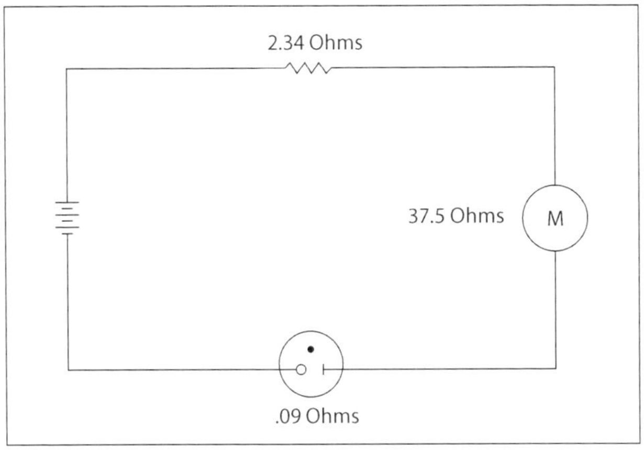

Figure 1-5. Circuit diagram.

Example: To multiply 0.2×6.03, arrange the numbers vertically and align them to the right. Multiply the numbers, ignoring the decimal points for now.

$$\begin{array}{r} 6.03 \\ \times\ 0.2 \\ \hline 1206 \end{array} \quad \text{(ignore the decimal points, for now)}$$

After multiplying the numbers, count the total number of decimal places in both numbers. For this example, 6.03 has 2 decimal places and 0.2 has 1 decimal place. Together there are a total of 3 decimal places. The decimal point for the answer will be placed 3 decimal places from the right. Therefore, the answer is 1.206.

$$\begin{array}{rl} 6.03 & \longleftarrow \text{2 decimal places} \\ \times\ 0.2 & \longleftarrow \text{1 decimal place} \\ \hline 1.206 & \longleftarrow \text{3 decimal places} \end{array}$$

Example: Using the formula Watts = Amperes × Voltage, what is the wattage of an electric drill that uses 9.45 amperes from a 120 volt source? Align the numbers to the right and multiply.

After multiplying the numbers, count the total number of decimal places in both numbers. For this example, 9.45 has 2 decimal places and 120 has no decimal place. Together there are 2 decimal places. The decimal point for the answer will be placed 2 decimal places from the right. Therefore, the answer is 1,134.00 watts, or 1,134 watts.

$$
\begin{array}{r l}
9.45 & \longleftarrow \text{2 decimal places} \\
\times\ 120 & \longleftarrow \text{no decimal places} \\
\hline
000 & \\
1890 & \\
945 & \\
\hline
1{,}134.00 & \longleftarrow \text{2 decimal places}
\end{array}
$$

Division of Decimal Numbers

Division of decimal numbers is performed the same way as whole numbers, unless the divisor is a decimal.

$$
\text{divisor}\overline{\smash{\big)}\,\text{dividend}}^{\ \text{quotient}}
$$

When the divisor is a decimal, it must be changed to a whole number before dividing. To do this, move the decimal in the divisor to the right until there are no decimal places. At the same time, move the decimal point in the dividend to the right the same number of places. Then divide. The decimal in the quotient will be placed directly above the decimal in the dividend.

Example: Divide 0.144 by 0.12

$$
.12\overline{\smash{\big)}\,0.144} \;=\; 12.\overline{\smash{\big)}\,14.4}^{\ 1.2}
\qquad
\begin{array}{r}
14.4 \\
12 \\
\hline
24 \\
24 \\
\hline
0
\end{array}
$$

Move the decimal in the divisor (0.12) two places to the right. Next move the decimal in the dividend (0.144) two places to the right. Then divide. The result is 1.2.

Example: The wing area of an airplane is 262.6 square feet and its span is 40.4 feet. Find the mean chord of its wing using the formula: Area ÷ span = mean chord.

$$
40.4\overline{\smash{\big)}\,262.6} \;=\; 404.\overline{\smash{\big)}\,2626.0}^{\ 6.5}
\qquad
\begin{array}{r}
2626.0 \\
2424 \\
\hline
2020 \\
2020 \\
\hline
0
\end{array}
$$

Move the decimal in the divisor (40.4) one place to the right. Next move the decimal in the dividend (262.6) one place to the right. Then divide. The mean chord length is 6.5 feet.

Rounding Off Decimal Numbers

Occasionally, it is necessary to round off a decimal number to some value that is practical to use.

For example, a measurement is calculated to be 29.4948 inches. To use this measurement, we can use the process of "rounding off." A decimal is "rounded off" by keeping the digits for a certain number of places and discarding the rest. The degree of accuracy desired determines the number of digits to be retained. When the digit immediately to the right of the last retained digit is 5 or greater, round up by 1. When the digit immediately to the right of the last retained digit is less than 5, leave the last retained digit unchanged.

Example: An actuator shaft is 2.1938 inches in diameter. Round to the nearest tenth.

The digit in the tenths column is a 1. The digit to the right of the 1 is a 9. Since 9 is greater than or equal to 5, "round up" the 1 to a 2. Therefore, 2.1938 rounded to the nearest tenth is 2.2.

Example: The outside diameter of a bearing is 2.1938 inches. Round to the nearest hundredth.

The digit in the hundredths column is a 9. The digit to the right of the 9 is a 3. Since 3 is less than 5, do not round up the 9. Therefore, 2.1938 to the nearest hundredth is 2.19.

Example: The length of a bushing is 2.1938 inches. Round to the nearest thousandth.

The digit in the thousandths column is a 3. The digit to the right of the 3 is an 8. Since 8 is greater than or equal to 5, "round up" the 3 to a 4. Therefore, 2.1938 to the nearest thousandth is 2.194.

Converting Decimal Numbers to Fractions

To change a decimal number to a fraction, "read" the decimal, and then write it into a fraction just as it is read as shown below.

Example: One oversized rivet has a diameter of 0.52 inches. Convert 0.52 to a fraction. The decimal 0.52 is read as "fifty-two hundredths."

Therefore,

$$
0.52 = \frac{52 \;\longleftarrow \text{"fifty-two"}}{100 \;\longleftarrow \text{"hundredths"}} \quad \text{which can be reduced to } \frac{13}{25}
$$

A dimension often appears in a maintenance manual or on a blueprint as a decimal instead of a fraction. In order to use the dimension, it may need to be converted to a fraction. An aviation mechanic frequently uses a steel rule that is calibrated in units of $^1/_{64}$ of an inch. To change a decimal to the nearest equivalent common fraction, multiply the decimal by 64. The product of the decimal and 64 will be the numerator of the fraction and 64 will be the denominator. Reduce the fraction, if needed.

Example: The width of a hex head bolt is 0.3123 inches. Convert the decimal 0.3123 to a common fraction to decide which socket would be the best fit for the bolt head. First, multiply the 0.3123 decimal by 64:

$$0.3123 \times 64 = 19.9872$$

Next, round the product to the nearest whole number: $19.98722 \approx 20$.

Use this whole number (20) as the numerator and 64 as the denominator: $^{20}/_{64}$.

Now, reduce $^{20}/_{64}$ to $^5/_{16}$.

Therefore, the correct socket would be the $^5/_{16}$ inch socket ($^{20}/_{64}$ reduced).

Example: When accurate holes of uniform diameter are required for aircraft structures, they are first drilled approximately $^1/_{64}$ inch undersized and then reamed to the final desired diameter. What size drill bit should be selected for the undersized hole if the final hole is reamed to a diameter of 0.763 inches? First, multiply the 0.763 decimal by 64.

$$0.763 \times 64 = 48.832$$

Next, round the product to the nearest whole number: $48.832 \approx 49$.

Use this number (49) as the numerator and 64 as the denominator: $^{49}/_{64}$ is the closest fraction to the final reaming diameter of 0.763 inches. To determine the drill size for the initial undersized hole, subtract $^1/_{64}$ inch from the finished hole size.

$$\frac{49}{64} - \frac{1}{64} = \frac{48}{64} = \frac{3}{4}$$

Therefore, a $^3/_4$-inch drill bit should be used for the initial undersized holes.

Converting Fractions to Decimals

To convert any fraction to a decimal, simply divide the top number (numerator) by the bottom number (denominator). Every fraction will have an approximate decimal equivalent.

Example:

$$\frac{1}{2} = 1 \div 2 = 2\overline{)1.0} = .5 \qquad \begin{array}{r} 1.0 \\ \hline 0 \end{array} \qquad \text{Therefore, } \frac{1}{2} = .5$$

$$\frac{3}{8} = 3 \div 8 = 8\overline{)3.000} = .375 \qquad \begin{array}{r} -24 \\ \hline 60 \\ -56 \\ \hline 40 \\ -40 \\ \hline 0 \end{array} \qquad \text{Therefore, } \frac{3}{8} = .375$$

Calculator tip: Numerator (top number) ÷ Denominator (bottom number) = the decimal equivalent of the fraction.

Some fractions when converted to decimals produce a repeating decimal.

Example:

$$\frac{1}{3} = 1 \div 3 = 3\overline{)1.00} = .33 \qquad \begin{array}{r} -9 \\ \hline 10 \\ -9 \\ \hline 1 \end{array} \qquad = .\overline{3} \text{ or } .33$$

This decimal can be represented with bar, or can be rounded. (A bar indicates that the number(s) beneath it are repeated to infinity.)

Other examples of repeating decimals:

$$.212121\ldots = .\overline{21}$$

$$.6666\ldots = .\overline{6} \text{ or } .67$$

$$.254254\ldots = .\overline{254}$$

Decimal Equivalent Chart

Figure 1-6 (on the next page) is a fraction to decimal to millimeter equivalency chart. Measurements starting at $^1/_{64}$ inch and up to 23 inches have been converted to decimal numbers and to millimeters.

Ratio

A ratio is the comparison of two numbers or quantities. A ratio may be expressed in three ways: as a fraction, with a colon, or with the word "to." For example, a gear ratio of 5:7 can be expressed as any of the following:

$^5/_7$ or 5:7 or 5 to 7

fraction	decimal	mm
1/64	0.015	0.396
1/32	0.031	0.793
3/64	0.046	1.190
1/16	0.062	1.587
5/64	0.078	1.984
3/32	0.093	2.381
7/64	0.109	2.778
1/8	0.125	3.175
9/64	0.140	3.571
5/32	0.156	3.968
11/64	0.171	4.365
5/16	0.187	4.762
13/64	0.203	5.159
7/32	0.218	5.556
15/64	0.234	5.953
1/4	0.25	6.35
17/64	0.265	6.746
9/32	0.281	7.143
19/64	0.296	7.540
5/16	0.312	7.937
21/64	0.328	8.334
11/32	0.343	8.731
23/64	0.359	9.128
3/8	0.375	9.525
25/64	0.390	9.921
13/32	0.406	10.318
27/64	0.421	10.715
7/16	0.437	11.112
29/64	0.453	11.509
15/32	0.468	11.906
31/64	0.484	12.303
1/2	0.5	12.7
33/64	0.515	13.096
17/32	0.531	13.493
35/64	0.546	13.890
39341	0.562	14.287
37/64	0.578	14.684
19/32	0.593	15.081
39/64	0.609	15.478
5/8	0.625	15.875
41/64	0.640	16.271
21/32	0.656	16.668
43/64	0.671	17.065
11/16	0.687	17.462
45/64	0.703	17.859
23/32	0.718	18.256
47/64	0.734	18.653
3/4	0.75	19.05
49/64	0.765	19.446
25/32	0.781	19.843
51/64	0.796	20.240
13/16	0.812	20.637
53/64	0.828	21.034
27/32	0.843	21.431
55/64	0.859	21.828
7/8	0.875	22.225
57/64	0.890	22.621
29/32	0.906	23.018
59/64	0.921	23.415
15/16	0.937	23.812
61/64	0.953	24.209
31/32	0.968	24.606
63/64	0.984	25.003
1	1	25.4

fraction	decimal	mm
1 1/64	1.015	25.796
1 1/32	1.031	26.193
1 3/64	1.046	26.590
1 1/16	1.062	26.987
1 5/64	1.078	27.384
1 3/32	1.093	27.781
1 7/64	1.109	28.178
1 1/8	1.125	28.575
1 9/64	1.140	28.971
1 5/32	1.156	29.368
1 11/64	1.171	29.765
1 3/16	1.187	30.162
1 13/64	1.203	30.559
1 7/32	1.218	30.956
1 15/64	1.234	31.353
1 1/4	1.25	31.75
1 17/64	1.265	32.146
1 9/32	1.281	32.543
1 19/64	1.296	32.940
1 5/16	1.312	33.337
1 21/64	1.328	33.734
1 11/32	1.343	34.131
1 23/64	1.359	34.528
1 3/8	1.375	34.925
1 25/64	1.390	35.321
1 13/32	1.406	35.718
1 27/64	1.421	36.115
1 7/16	1.437	36.512
1 29/64	1.453	36.909
1 15/32	1.468	37.306
1 31/64	1.484	37.703
1 1/2	1.5	38.1
1 33/64	1.515	38.496
1 17/32	1.531	38.893
1 35/64	1.546	39.290
1 9/16	1.562	39.687
1 37/64	1.578	40.084
1 19/32	1.593	40.481
1 39/64	1.609	40.878
1 5/8	1.625	41.275
1 41/64	1.640	41.671
1 21/32	1.656	42.068
1 43/64	1.671	42.465
1 11/16	1.687	42.862
1 45/64	1.703	43.259
1 23/32	1.718	43.656
1 47/64	1.734	44.053
1 3/4	1.75	44.45
1 49/64	1.765	44.846
1 25/32	1.781	45.243
1 51/64	1.796	45.640
1 13/16	1.812	46.037
1 53/64	1.828	46.434
1 27/32	1.843	46.831
1 55/64	1.859	47.228
1 7/8	1.875	47.625
1 57/64	1.890	48.021
1 29/32	1.906	48.418
1 59/64	1.921	48.815
1 15/16	1.937	49.212
1 61/64	1.953	49.609
1 31/32	1.968	50.006
1 63/64	1.984	50.403
2	2	50.8

fraction	decimal	mm
2 1/64	2.015	51.196
2 1/32	2.031	51.593
2 3/64	2.046	51.990
2 1/16	2.062	52.387
2 5/64	2.078	52.784
2 3/32	2.093	53.181
2 7/64	2.109	53.578
2 1/8	2.125	53.975
2 9/64	2.140	54.371
2 5/32	2.156	54.768
2 11/64	2.171	55.165
2 3/16	2.187	55.562
2 13/64	2.203	55.959
2 7/32	2.218	56.356
2 15/64	2.234	56.753
2 1/4	2.25	57.15
2 17/64	2.265	57.546
2 9/32	2.281	57.943
2 19/64	2.296	58.340
2 5/16	2.312	58.737
2 21/64	2.328	59.134
2 11/32	2.343	59.531
2 23/64	2.359	59.928
2 3/8	2.375	60.325
2 25/64	2.390	60.721
2 13/32	2.406	61.118
2 27/64	2.421	61.515
2 7/16	2.437	61.912
2 29/64	2.453	62.309
2 15/32	2.468	62.706
2 31/64	2.484	63.103
2 1/2	2.5	63.5
2 33/64	2.515	63.896
2 17/32	2.531	64.293
2 35/64	2.546	64.690
2 9/16	2.562	65.087
2 37/64	2.578	65.484
2 19/32	2.593	65.881
2 39/64	2.609	66.278
2 5/8	2.625	66.675
2 41/64	2.640	67.071
2 21/32	2.656	67.468
2 43/64	2.671	67.865
2 11/16	2.687	68.262
2 45/64	2.703	68.659
2 23/32	2.718	69.056
2 47/64	2.734	69.453
2 3/4	2.75	69.85
2 49/64	2.765	70.246
2 25/32	2.781	70.643
2 51/64	2.796	71.040
2 13/16	2.812	71.437
2 53/64	2.828	71.834
2 27/32	2.843	72.231
2 55/64	2.859	72.628
2 7/8	2.875	73.025
2 57/64	2.890	73.421
2 29/32	2.906	73.818
2 59/64	2.921	74.215
2 15/16	2.937	74.612
2 61/64	2.953	75.009
2 31/32	2.968	75.406
2 63/64	2.984	75.803
3	3	76.2

Figure 1-6. Fractions, decimals, and millimeters.

Aviation Applications

Ratios have widespread application in the field of aviation. For example:

Compression ratio on a reciprocating engine is the ratio of the volume of a cylinder with the piston at the bottom of its stroke to the volume of the cylinder with the piston at the top of its stroke. For example, a typical compression ratio might be 10:1 (or 10 to 1).

Aspect ratio is the ratio of the length (or span) of an airfoil to its width (or chord). A typical aspect ratio for a commercial airliner might be 7:1 (or 7 to 1).

Air-fuel ratio is the ratio of the weight of the air to the weight of fuel in the mixture being fed into the cylinders of a reciprocating engine. For example, a typical air-fuel ratio might be 14.3:1 (or 14.3 to 1).

Glide ratio is the ratio of the forward distance traveled to the vertical distance descended when an aircraft is operating without power. For example, if an aircraft descends 1,000 feet while it travels through the air for a distance of two linear miles (10,560 feet), it has a glide ratio of 10,560:1,000 which can be reduced to 10.56: 1 (or 10.56 to 1).

Gear Ratio is the number of teeth each gear represents when two gears are used in an aircraft component. In Figure 1-7, the pinion gear has 8 teeth and a spur gear has 28 teeth. The gear ratio is 8:28 or 2:7.

Figure 1-7. Gear ratio.

Speed Ratio. When two gears are used in an aircraft component, the rotational speed of each gear is represented as a speed ratio. As the number of teeth in a gear decreases, the rotational speed of that gear increases, and vice-versa. Therefore, the speed ratio of two gears is the inverse (or opposite) of the gear ratio. If two gears have a gear ratio of 2:9, then their speed ratio is 9:2.

Example: A pinion gear with 10 teeth is driving a spur gear with 40 teeth. The spur gear is rotating at 160 rpm. Determine the speed of the pinion gear.

$$\frac{\text{Teeth in Pinion Gear}}{\text{Teeth in Spur Gear}} = \frac{\text{Speed of Spur Gear}}{\text{Speed of Pinion Gear}}$$

$$\frac{10 \text{ teeth}}{40 \text{ teeth}} = \frac{160 \text{ rpm}}{S_P \text{ (speed of pinion gear)}}$$

To solve for S_P, multiply 40 × 160, then divide by 10. The speed of the pinion gear is 640 rpm.

Example: If the cruising speed of an airplane is 200 knots and its maximum speed is 250 knots, what is the ratio of cruising speed to maximum speed? First, express the cruising speed as the numerator of a fraction whose denominator is the maximum speed.

$$\text{Ratio} = \frac{200}{250}$$

Next, reduce the resulting fraction to its lowest terms.

$$\text{Ratio} = \frac{200}{250} = \frac{4}{5}$$

Therefore, the ratio of cruising speed to maximum speed is 4:5.

Another common use of ratios is to convert any given ratio to an equivalent ratio with a denominator of 1.

Example: Express the ratio 9:5 as a ratio with a denominator of 1.

$$R = \frac{9}{5} = \frac{?}{1} \quad \text{Since } 9 \div 5 = 1.8, \text{ then } \frac{9}{5} = \frac{1.8}{1}$$

Therefore, 9:5 is the same ratio as 1.8:1. In other words, 9 to 5 is the same ratio as 1.8 to 1.

Proportion

A proportion is a statement of equality between two or more ratios. For example,

$$\frac{3}{4} = \frac{6}{8} \text{ or } 3:4 = 6:8$$

This proportion is read as, "3 is to 4 as 6 is to 8."

Extremes and Means

The first and last terms of the proportion (the 3 and 8 in this example) are called the extremes. The second and third terms (the 4 and 6 in this example) are called the means. In any proportion, the product of the extremes is equal to the product of the means.

In the proportion 2:3 = 4:6, the product of the extremes, 2 × 6, is 12; the product of the means, 3 × 4, is also 12. An inspection of any proportion will show this to be true.

Solving Proportions

Normally when solving a proportion, three quantities will be known, and the fourth will be unknown. To solve for the unknown, multiply the two numbers along the diagonal and then divide by the third number.

Example: Solve for X in the proportion given below.

$$\frac{65}{80} = \frac{X}{100}$$

First, multiply 65 × 100: 65 × 100 = 6500
Next, divide by 80: 6500 ÷ 80 = 81.25
Therefore, X = 81.25.

Example: An airplane flying a distance of 300 miles used 24 gallons of gasoline. How many gallons will it need to travel 750 miles?

The ratio here is: "miles to gallons;" therefore, the proportion is set up as:

$$\begin{matrix}\text{Miles} \\ \text{Gallons}\end{matrix} \quad \frac{300}{24} = \frac{750}{G}$$

Solve for G: (750 × 24) ÷ 300 = 60

Therefore, to fly 750 miles, 60 gallons of gasoline will be required.

Percentage

Percentage means "parts out of one hundred." The percentage sign is "%". Ninety percent is expressed as 90% (= 90 parts out of 100). The decimal 0.90 equals $^{90}/_{100}$, or 90 out of 100, or 90%.

Expressing a Decimal Number as a Percentage

To express a decimal number in percent, move the decimal point two places to the right (adding zeros if necessary) and then affix the percent symbol.

Example: Express the following decimal numbers as a percent:

.90 = 90%
.5 = 50%
1.25 = 125%
.335 = 33.5%

Expressing a Percentage as a Decimal Number

Sometimes it may be necessary to express a percentage as a decimal number. To express a percentage as a decimal number, move the decimal point two places to the left and drop the % symbol.

For example: Express the following percentages as decimal numbers:

90% = .90
50% = .50
5% = .05
150% = 1.5

Expressing a Fraction as a Percentage

To express a fraction as a percentage, first change the fraction to a decimal number (by dividing the numerator by the denominator), and then convert the decimal number to a percentage as shown earlier.

Example: Express the fraction $^5/_8$ as a percentage.

$$\frac{5}{8} = 5 \div 8 = 0.625 = 62.5\%$$

Finding a Percentage of a Given Number

This is the most common type of percentage calculation. Here are two methods to solve percentage problems: using algebra or using proportions. Each method is shown below to find a percent of a given number.

Example: In a shipment of 80 wingtip lights, 15% of the lights were defective. How many of the lights were defective?

Algebra Method:

15% of 80 lights = N (number of defective lights)
0.15 × 80 = N
12 = N

Therefore, 12 defective lights were in the shipment.

Proportion Method:

$$\frac{N}{80} = \frac{15}{100}$$

To solve for N: N × 100 = 80 × 15
N × 100 = 1200
N = 1200 ÷ 100
N = 12
or
N = (80 × 15) ÷ 100
N = 12

Finding What Percentage One Number Is of Another

Example: A small engine rated at 12 horsepower is found to be delivering only 10.75 horsepower. What is the motor efficiency expressed as a percent?

Algebra Method:
N% of 12 rated horsepower = 10.75 actual horsepower
N% × 12 = 10.75
N% = 10.75 ÷ 12
N% = .8958
N = 89.58
Therefore, the motor efficiency is 89.58%.

Proportion Method:

$$\frac{10.75}{12} = \frac{N}{100}$$

To solve for N: N × 12 = 10.75 × 100
N × 12 = 1075
N = 1075 ÷ 12
N = 89.58
or
N = (1075 × 100) ÷ 12
N = 89.58
Therefore, the motor efficiency is 89.58%.

Finding a Number When a Percentage of It Is Known

Example: Eighty ohms represents 52% of a microphone's total resistance. Find the total resistance of this microphone.

Algebraic Method:
52% of N = 80 ohms
52% × N = 80
N = 80 ÷ .52
N = 153.846
The total resistance of the microphone is 153.846 ohms.

Proportion Method:

$$\frac{80}{N} = \frac{52}{100}$$

Solve for N: N × 52 = 80 × 100
N × 52 = 8,000
N = 8,000 ÷ 52
N = 153.846 ohms
or
N = (80 × 100) ÷ 52
N = 153.846 ohms

Positive and Negative Numbers (Signed Numbers)

Positive numbers are numbers that are greater than zero. Negative numbers are numbers less than zero. [Figure 1-8] Signed numbers are also called integers.

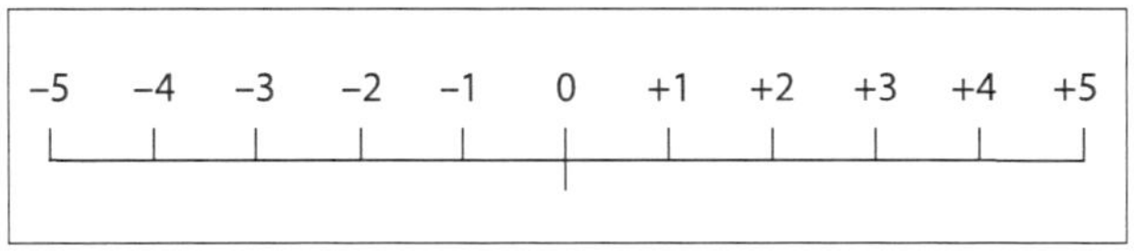

Figure 1-8. A scale of signed numbers.

Addition of Positive and Negative Numbers

The sum (addition) of two positive numbers is positive. The sum (addition) of two negative numbers is negative. The sum of a positive and a negative number can be positive or negative, depending on the values of the numbers. A good way to visualize a negative number is to think in terms of debt. If you are in debt by $100 (or, −100) and you add $45 to your account, you are now only $55 in debt (or −55).

Therefore: −100 + 45 = −55.

Example: The weight of an aircraft is 2,000 pounds. A radio rack weighing 3 pounds and a transceiver weighing 10 pounds are removed from the aircraft. What is the new weight? For weight and balance purposes, all weight removed from an aircraft is given a minus sign, and all weight added is given a plus sign.

2,000 + −3 + −10 = 2,000 + −13 = 1987
Therefore, the new weight is 1,987 pounds.

Subtraction of Positive and Negative Numbers

To subtract positive and negative numbers, first change the "–" (subtraction symbol) to a "+" (addition symbol), and change the sign of the second number to its opposite (that is, change a positive number to a negative number or vice versa). Finally, add the two numbers together.

Example: The daytime temperature in the city of Denver was 6° below zero (−6°). An airplane is cruising at 15,000 feet above Denver. The temperature at 15,000 feet is 20° colder than in the city of Denver. What is the temperature at 15,000 feet?

Subtract 20 from −6: −6 – 20 = −6 + −20 = −26

The temperature is −26°, or 26° below zero at 15,000 feet above the city.

Multiplication of Positive and Negative Numbers

The product of two positive numbers is always positive. The product of two negative numbers is always positive. The product of a positive and a negative number is always negative.

Examples:
$3 \times 6 = 18 \quad -3 \times 6 = -18 \quad -3 \times -6 = 18 \quad 3 \times -6 = -18$

Division of Positive and Negative Numbers

The quotient of two positive numbers is always positive. The quotient of two negative numbers is always positive. The quotient of a positive and negative number is always negative.

Examples:
$6 \div 3 = 2 \quad -6 \div 3 = -2 \quad -6 \div -3 = 2 \quad 6 \div -3 = -2$

Powers

The power (or exponent) of a number is a shorthand method of indicating how many times a number, called the base, is multiplied by itself. For example, 3^4 means "3 to the power of 4." That is, 3 multiplied by itself 4 times. The 3 is the base and 4 is the power.

Examples:
$2^3 = 2 \times 2 \times 2 = 8$.
Read "two to the third power equals 8."

$10^5 = 10 \times 10 \times 10 \times 10 \times 10 = 100{,}000$
Read "ten to the fifth power equals 100,000."

Special Powers

Squared. When a number has a power of 2, it is commonly referred to as "squared." For example, 7^2 is read as "seven squared" or "seven to the second power." To remember this, think about how a square has two dimensions: length and width.

Cubed. When a number has a power of 3, it is commonly referred to as "cubed." For example, 7^3 is read as "seven cubed" or "seven to the third power." To remember this, think about how a cube has three dimensions: length, width, and depth.

Power of Zero. Any non-zero number raised to the zero power always equals 1.

Example:
$7^0 = 1 \qquad 181^0 = 1 \qquad (-24)^0 = 1$

Negative Powers

A number with a negative power equals its reciprocal with the same power made positive.

Example: The number 2^{-3} is read as "2 to the negative 3rd power," and is calculated by:

$$2^{-3} = \frac{1}{2^3} = \frac{1}{2 \times 2 \times 2} = \frac{1}{8}$$

Powers of Ten	Expansion	Value
Positive Exponents		
10^6	10 × 10 × 10 × 10 × 10 × 10	1,000,000
10^5	10 × 10 × 10 × 10 × 10	100,000
10^4	10 × 10 × 10 × 10	10,000
10^3	10 × 10 × 10	1,000
10^2	10 × 10	100
10^1	10	10
10^0		1
Negative Exponents		
10^{-1}	1/10	1/10 = 0.1
10^{-2}	1/(10 × 10)	1/100 = 0.01
10^{-3}	1/(10 × 10 × 10)	1/1,000 = 0.001
10^{-4}	1/(10 × 10 × 10 × 10)	1/10,000 = 0.0001
10^{-5}	1/(10 × 10 × 10 × 10 × 10)	1/100,000 = 0.00001
10^{-6}	1/(10 × 10 × 10 × 10 × 10 × 10)	1/1,000,000 = 0.000001

Figure 1-9. Powers of ten.

When using a calculator to raise a negative number to a power, always place parentheses around the negative number (before raising it to a power) so that the entire number gets raised to the power.

Law of Exponents

When multiplying numbers with powers, the powers can be added as long as the bases are the same.

Example:

$$3^2 \times 3^4 = (3 \times 3) \times (3 \times 3 \times 3 \times 3) = 3 \times 3 \times 3 \times 3 \times 3 \times 3 = 3^6$$

$$\text{or } 3^2 \times 3^4 = 3^{(2+4)} = 3^6$$

When dividing numbers with powers, the powers can be subtracted as long as the bases are the same.

Example:

$$10^4 \div 10^2 = \frac{10 \times 10 \times 10 \times 10}{10 \times 10} = \frac{\cancel{10} \times \cancel{10} \times 10 \times 10}{\cancel{10} \times \cancel{10}} = 10 \times 10 = 10^2$$

$$\text{or } 10^4 \div 10^2 = 10^{(4-2)} = 10^2$$

Powers of Ten

Because we use the decimal system of numbers, powers of ten are frequently seen in everyday applications. For example, scientific notation uses powers of ten. Also, many aircraft drawings are scaled to powers of ten. Figure 1-9 gives more information on the powers of ten and their values.

Roots

A root is a number that when multiplied by itself a specified number of times will produce a given number.

The two most common roots are the square root and the cube root. For more examples of roots, see the chart in Figure 1-10, Functions of Numbers (on page 1-14).

Square Roots

The square root of 25, written as $\sqrt{25}$, equals 5. That is, when the number 5 is squared (multiplied by itself), it produces the number 25. The symbol $\sqrt{\ }$ is called a radical sign. Finding the square root of a number is the most common application of roots. The collection of numbers whose square roots are whole numbers are called perfect squares. The first ten perfect squares are: 1, 4, 9, 16, 25, 36, 49, 64, 81, and 100. The square root of each of these numbers is 1, 2, 3, 4, 5, 6, 7, 8, 9, and 10, respectively.

For example, $\sqrt{36} = 6$ and $\sqrt{81} = 9$

To find the square root of a number that is not a perfect square, use either a calculator or the estimation method. A longhand method does exist for finding square roots, but with the advent of calculators and because of its lengthy explanation, it is no longer included in this handbook. The estimation method uses the knowledge of perfect squares to approximate the square root of a number.

Example: Find the square root of 31. Since 31 falls between the two perfect roots 25 and 36, we know that $\sqrt{31}$ must be between $\sqrt{25}$ and $\sqrt{36}$. Therefore, $\sqrt{31}$ must be greater than 5 and less than 6 because $\sqrt{25} = 5$ and $\sqrt{36} = 6$. If you estimate the square root of 31 at 5.5, you are close to the correct answer. The square root of 31 is actually 5.568.

Cube Roots

The cube root of 125, written as $\sqrt[3]{125}$, equals 5. That is, when the number 5 is cubed (5 multiplied by itself then multiplying the product (25) by 5 again), it produces the number 125. It is common to confuse the "cube" of a number with the "cube root" of a number. For clarification, the cube of 27 = $27^3 = 27 \times 27 \times 27 = 19{,}683$. However, the cube root of 27 = $\sqrt[3]{27} = 3$.

Fractional Powers

Another way to write a root is to use a fraction as the power (or exponent) instead of the radical sign. The square root of a number is written with a $\frac{1}{2}$ as the exponent instead of a radical sign. The cube root of a number is written with an exponent of $\frac{1}{3}$ and the fourth root with an exponent of $\frac{1}{4}$ and so on.

Example: $\sqrt{31} = 31^{1/2}$ $\sqrt[3]{125} = 125^{1/3}$ $\sqrt[4]{16} = 16^{1/4}$

Functions of Numbers Chart

The Functions of Numbers chart [Figure 1-10] is included in this chapter for convenience in making computations. Each column in the chart is listed below, with new concepts explained.

- Number, (N)
- N squared, (N^2)
- N cubed, (N^3)
- Square root of N, ($\sqrt{N}$)
- Cube root of N, ($\sqrt[3]{N}$)
- Circumference of a circle with diameter = N. Circumference is the linear measurement of the distance around a circle. The circumference is calculated by multiplying the diameter of the circle by 3.1416 (3.1416 is the number referred to as pi, which has the symbol π). If the diameter of a circle is 10 inches, then the circumference would be 31.416 inches because $10 \times 3.1416 = 31.4160$.
- Area of a circle with diameter = N. Area of a circle is the number of square units of measurement contained in the circle with a diameter of N. The area of a circle equals π multiplied by the radius squared. This is calculated by the formula: $A = \pi \times r^2$. Remember that the radius is equal to one-half of the diameter.

  Example: A cockpit instrument gauge has a round face that is 3 inches in diameter. What is the area of the face of the gauge? From Figure 1-10 for N = 3, the answer is 7.0686 square inches. This is calculated by:

  If the diameter of the gauge is 3 inches, then the radius = $D/2 = 3/2 = 1.5$ inches.

  Area = $\pi \times r^2 = 3.1416 \times 1.5^2 = 3.1416 \times 2.25 = 7.0686$ square inches.

Scientific Notation

Scientific notation is used as a type of shorthand to express very large or very small numbers. It is a way to write numbers so that they do not take up as much space on the page. The format of a number written in scientific notation has two parts. The first part is a number greater than or equal to 1 and less than 10 (for example, 2.35). The second part is a power of 10 (for example, 10^6). The number 2,350,000 is expressed in scientific notation as 2.35×10^6. It is important that the decimal point is always placed to the right of the first digit. Notice that very large numbers always have

Number	Square	Cube	Square Root	Cube Root	Circumference	Area
Number (N)	N Squared (N^2)	N Cubed (N^3)	Square Root of N ($\sqrt{N}$)	Cube Root of N ($\sqrt[3]{N}$)	Circumference of a circle with diameter = N	Area of a circle with diameter = N
1	1	1	1.000	1.000	3.142	0.785
2	4	8	1.414	1.260	6.283	3.142
3	9	27	1.732	1.442	9.425	7.069
4	16	64	2.000	1.587	12.566	12.566
5	25	125	2.236	1.710	15.708	19.635
6	36	216	2.449	1.817	18.850	28.274
7	49	343	2.646	1.913	21.991	38.484
8	64	512	2.828	2.000	25.133	50.265
9	81	729	3.000	2.080	28.274	63.617
10	100	1,000	3.162	2.154	31.416	78.540
11	121	1,331	3.317	2.224	34.558	95.033
12	144	1,728	3.464	2.289	37.699	113.01
13	169	2,197	3.606	2.351	40.841	132.73
14	196	2,744	3.742	2.410	43.982	153.94
15	225	3,375	3.873	2.466	47.124	176.71
16	256	4,096	4.000	2.520	50.265	201.06
17	289	4,913	4.123	2.571	53.407	226.98
18	324	5,832	4.243	2.621	56.549	254.47
19	361	6,859	4.359	2.668	59.690	283.53
20	400	8,000	4.472	2.714	62.832	314.16
21	441	9,261	4.583	2.759	65.973	346.36
22	484	10,648	4.690	2.802	69.115	380.13
23	529	12,167	4.796	2.844	72.257	415.48
24	576	13,824	4.899	2.885	75.398	452.39
25	625	15,625	5.000	2.924	78.540	490.87
26	676	17,576	5.099	2.963	81.681	530.93
27	729	19,683	5.196	3.000	84.823	572.55
28	784	21,952	5.292	3.037	87.965	615.75
29	841	24,389	5.385	3.072	91.106	660.52
30	900	27,000	5.477	3.107	94.248	706.86
31	961	29,791	5.568	3.141	97.389	754.77
32	1,024	32,768	5.657	3.175	100.531	804.25
33	1,089	35,937	5.745	3.208	103.672	855.30
34	1,156	39,304	5.831	3.240	106.814	907.92
35	1,225	42,875	5.916	3.271	109.956	962.11
36	1,296	46,656	6.000	3.302	113.097	1017.88
37	1,369	50,653	6.083	3.332	116.239	1075.21
38	1,444	54,872	6.164	3.362	119.380	1134.11
39	1,521	59,319	6.245	3.391	122.522	1194.59
40	1,600	64,000	6.325	3.420	125.664	1256.64
41	1,681	68,921	6.403	3.448	128.805	1320.25
42	1,764	74,088	6.481	3.476	131.947	1385.44
43	1,849	79,507	6.557	3.503	135.088	1452.20
44	1,936	85,184	6.633	3.530	138.230	1520.53
45	2,025	91,125	6.708	3.557	141.372	1590.43
46	2,116	97,336	6.782	3.583	144.513	1661.90
47	2,209	103,823	6.856	3.609	147.655	1734.94
48	2,304	110,592	6.928	3.634	150.796	1809.56
49	2,401	117,649	7.000	3.659	153.938	1885.74
50	2,500	125,000	7.071	3.684	157.080	1963.49

Figure 1-10. Functions of numbers.

Number	Square	Cube	Square Root	Cube Root	Circumference	Area
Number (N)	N Squared (N^2)	N Cubed (N^3)	Square Root of N ($\sqrt{N}$)	Cube Root of N ($\sqrt[3]{N}$)	Circumference of a circle with diameter = N	Area of a circle with diameter = N
51	2,601	132,651	7.141	3.708	160.221	2042.82
52	2,704	140,608	7.211	3.733	163.363	2123.71
53	2,809	148,877	7.280	3.756	166.504	2206.18
54	2,916	157,464	7.348	3.780	169.646	2290.22
55	3,025	166,375	7.416	3.803	172.787	2375.83
56	3,136	175,616	7.483	3.826	175.929	2463.01
57	3,249	185,193	7.550	3.849	179.071	2551.76
58	3,364	195,112	7.616	3.871	182.212	2642.08
59	3,481	205,379	7.681	3.893	185.354	2733.97
60	3,600	216,000	7.746	3.915	188.495	2827.43
61	3,721	226,981	7.810	3.937	191.637	2922.46
62	3,844	238,328	7.874	3.958	194.779	3019.07
63	3,969	250,047	7.937	3.979	197.920	3117.24
64	4,096	262,144	8.000	4.000	201.062	3216.99
65	4,225	274,625	8.062	4.021	204.203	3318.30
66	4,356	287,496	8.124	4.041	207.345	3421.19
67	4,489	300,763	8.185	4.062	210.487	3525.65
68	4,624	314,432	8.246	4.082	213.628	3631.68
69	4,761	328,509	8.307	4.102	216.770	3739.28
70	4,900	343,000	8.367	4.121	219.911	3848.45
71	5,041	357,911	8.426	4.141	223.053	3959.19
72	5,184	373,248	8.485	4.160	226.194	4071.50
73	5,329	389,017	8.544	4.179	229.336	4185.38
74	5,476	405,224	8.602	4.198	232.478	4300.84
75	5,625	421,875	8.660	4.217	235.619	4417.86
76	5,776	438,976	8.718	4.236	238.761	4536.46
77	5,929	456,533	8.775	4.254	241.902	4656.62
78	6,084	474,552	8.832	4.273	245.044	4778.36
79	6,241	493,039	8.888	4.291	248.186	4901.67
80	6,400	512,000	8.944	4.309	251.327	5026.54
81	6,561	531,441	9.000	4.327	254.469	5152.99
82	6,724	551,368	9.055	4.344	257.610	5281.01
83	6,889	571,787	9.110	4.362	260.752	5410.60
84	7,056	592,704	9.165	4.380	263.894	5541.76
85	7,225	614,125	9.220	4.397	267.035	5674.50
86	7,396	636,056	9.274	4.414	270.177	5808.80
87	7,569	658,503	9.327	4.431	273.318	5944.67
88	7,744	681,472	9.381	4.448	276.460	6082.12
89	7,921	704,969	9.434	4.465	279.602	6221.13
90	8,100	729,000	9.487	4.481	282.743	6361.72
91	8,281	753,571	9.539	4.498	285.885	6503.88
92	8,464	778,688	9.592	4.514	289.026	6647.60
93	8,649	804,357	9.644	4.531	292.168	6792.90
94	8,836	830,584	9.695	4.547	295.309	6939.77
95	9,025	857,375	9.747	4.563	298.451	7088.21
96	9,216	884,736	9.798	4.579	301.593	7238.22
97	9,409	912,673	9.849	4.595	304.734	7389.81
98	9,604	941,192	9.900	4.610	307.876	7542.96
99	9,801	970,299	9.950	4.626	311.017	7697.68
100	10,000	1,000,000	10.000	4.642	314.159	7853.98

Figure 1-10. Functions of numbers. (continued)

a positive power of 10 and very small numbers always have a negative power of 10.

Example: The velocity of the speed of light is over 186,000,000 mph. This can be expressed as 1.86×10^8 mph in scientific notation. The mass of an electron is approximately 0.000,000,000,000,000,000,000,000 ,000,911 grams. This can be expressed in scientific notation as 9.11×10^{-28} grams.

Converting Numbers from Standard Notation to Scientific Notation

Example: Convert 1,244,000,000,000 to scientific notation as follows. First, note that the decimal point is to the right of the last zero. (Even though it is not usually written, it is assumed to be there.)

1,244,000,000,000 = 1,244,000,000,000

To change to the format of scientific notation, the decimal point must be moved to the position between the first and second digits, which in this case is between the 1 and the 2. Since the decimal point must be moved 12 places to the left to get there, the power of 10 will be 12. Remember that large numbers always have a positive exponent. Therefore, 1,244,000,000,000 = 1.244×10^{12} when written in scientific notation.

Example: Convert 0.000000457 from standard notation to scientific notation. To change to the format of scientific notation, the decimal point must be moved to the position between the first and second numbers, which in this case is between the 4 and the 5. Since the decimal point must be moved 7 places to the right to get there, the power of 10 will be −7. Remember that small numbers (those less than one) will have a negative exponent. Therefore, 0.000000457 = 4.57×10^{-7} when written in scientific notation.

Converting Numbers from Scientific Notation to Standard Notation

Example: Convert 3.68×10^7 from scientific notation to standard notation, as follows. To convert from scientific notation to standard notation, move the decimal place 7 places to the right. $3.68 \times 10^7 = 36800000 =$ 36,800,000. Another way to think about the conversion is $3.68 \times 10^7 = 3.68 \times 10{,}000{,}000 = 36{,}800{,}000$.

Example: Convert 7.1543×10^{-10} from scientific notation to standard notation. Move the decimal place 10 places to the left: $7.1543 \times 10^{-10} = .00000000071543$. Another way to think about the conversion is $7.1543 \times 10^{-10} = 7.1543 \times .0000000001 = .00000000071543$

Conversion	Large numbers with positive powers of 10	Small numbers with negative powers of 10
From standard notation to scientific notation	Move decimal place to the left	Move decimal place to the right
From scientific notation to standard notation	Move decimal place to the right	Move decimal place to the left

Figure 1-11. Converting between scientific and standard notation.

When converting, remember that large numbers always have positive powers of ten and small numbers always have negative powers of ten. Refer to Figure 1-11 to determine which direction to move the decimal point.

Addition, Subtraction, Multiplication, and Division of Scientific Numbers

To add, subtract, multiply, or divide numbers in scientific notation, change the scientific notation number back to standard notation. Then add, subtract, multiply or divide the standard notation numbers. After the computation, change the final standard notation number back to scientific notation.

Algebra

Algebra is the branch of mathematics that uses letters or symbols to represent variables in formulas and equations.

For example, in the equation $D = V \times T$, where Distance = Velocity × Time, the variables are: D, V, and T.

Equations

Algebraic equations are frequently used in aviation to show the relationship between two or more variables. Equations normally have an equals sign (=) in the expression.

Example: The formula $A = \pi \times r^2$ shows the relationship between the area of a circle (A) and the length of the radius (r) of the circle. The area of a circle is equal to π (3.1416) times the radius squared. Therefore, the larger the radius, the larger the area of the circle.

Algebraic Rules

When solving for a variable in an equation, you can add, subtract, multiply or divide the terms in the equation, you do the same to both sides of the equals sign.

Examples: Solve the following equations for the value N.

$3N = 21$
To solve for N, divide both sides by 3.
$3N \div 3 = 21 \div 3$
$N = 7$

$N + 17 = 59$
To solve for N, subtract 17 from both sides.
$N + 17 - 17 = 59 - 17$
$N = 42$

$N - 22 = 100$
To solve for N, add 22 to both sides.
$N - 22 + 22 = 100 + 22$
$N = 122$

$\frac{N}{5} = 50$
To solve for N, multiply both sides by 5.
$\frac{N}{5} \times 5 = 50 \times 5$
$N = 250$

Solving for a Variable

Another application of algebra is to solve an equation for a given variable.

Example: Using the formula given in Figure 1-12, find the total capacitance (C_T) of the series circuit containing three capacitors with

$$C_1 = .1 \text{ microfarad}$$
$$C_2 = .015 \text{ microfarad}$$
$$C_3 = .05 \text{ microfarad}$$

First, substitute the given values into the formula:

$$C_T = \frac{1}{\frac{1}{C_1} + \frac{1}{C_2} + \frac{1}{C_3}}$$

Figure 1-12. Total capacitance in a series circuit.

$$C_T = \frac{1}{\frac{1}{C_1} + \frac{1}{C_2} + \frac{1}{C_3}} = \frac{1}{\frac{1}{0.1} + \frac{1}{0.015} + \frac{1}{0.05}} = \frac{1}{10 + 66.66 + 20}$$

Therefore, $C_T = \frac{1}{96.66} = .01034$ microfarad. The microfarad (10^{-6} farad) is a unit of measurement of capacitance. This will be discussed in greater length beginning on page 10-51 in chapter 10, Electricity.

Use of Parentheses

In algebraic equations, parentheses are used to group numbers or symbols together. The use of parentheses helps us to identify the order in which we should apply mathematical operations. The operations inside the parentheses are always performed first in algebraic equations.

Example: Solve the algebraic equation $N = (4 + 3)^2$. First, perform the operation inside the parentheses. That is, $4 + 3 = 7$. Then complete the exponent calculation $N = (7)^2 = 7 \times 7 = 49$.

When using more complex equations, which may combine several terms and use multiple operations, grouping the terms together helps organize the equation. Parentheses, ( ), are most commonly used in grouping, but you may also see brackets, [ ]. When a term or expression is inside one of these grouping symbols, it means that any operation indicated to be done on the group is done to the entire term or expression.

Example:
Solve the equation $N = 2 \times [(9 \div 3) + (4 + 3)^2]$. Start with the operations inside the parentheses ( ), then perform the operations inside the brackets [ ].

$N = 2 \times [(9 \div 3) + (4 + 3)^2]$
$N = 2 \times [3 + (7)^2]$ First, complete the operations inside the parentheses ( ).
$N = 2 \times [3 + 49]$
$N = 2 \times [52]$ Second, complete the operations inside the brackets [ ].
$N = 104$

Order of Operation

In algebra, rules have been set for the order in which operations are evaluated. These same universally accepted rules are also used when programming algebraic equations in calculators. When solving the following equation, the order of operation is given below:

$$N = (62 - 54)^2 + 6^2 - 4 + 3 \times [8 + (10 \div 2)] + \sqrt{25} + (42 \times 2) \div 4 + \tfrac{3}{4}$$

1. Parentheses. First, do everything in parentheses, ( ). Starting from the innermost parentheses. If the expression has a set of brackets, [ ], treat these exactly like parentheses. If you are working with a fraction, treat the top as if it were in parentheses and the denominator as if it were in parentheses, even if there are none shown. From the equation

above, completing the calculation in parentheses gives the following:

$N = (8)^2 + 6^2 - 4 + 3 \times [8 + (5)] + \sqrt{25} + (84) \div 4 + 3/4$,

then

$N = (8)^2 + 6^2 - 4 + 3 \times [13] + \sqrt{25} + 84 \div 4 + 3/4$

2. Exponents. Next, clear any exponents. Treat any roots (square roots, cube roots, and so forth) as exponents. Completing the exponents and roots in the equation gives the following:

   $N = 64 + 36 - 4 + 3 \times 13 + 5 + 84 \div 4 + 3/4$

3. Multiplication and Division. Evaluate all of the multiplications and divisions from left to right. Multiply and divide from left to right in one step. A common error is to use two steps for this (that is, to clear all of the multiplication signs and then clear all of the division signs), but this is not the correct method. Treat fractions as division. Completing the multiplication and division in the equation gives the following:

   $N = 64 + 36 - 4 + 39 + 5 + 21 + 3/4$

4. Addition and Subtraction. Evaluate the additions and subtractions from left to right. Like above, addition and subtraction are computed left to right in one step. Completing the addition and subtraction in the equation gives the following:

   $X = 161\,3/4$

Order of Operation for Algebraic Equations

1. Parentheses
2. Exponents
3. Multiplication and Division
4. Addition and Subtraction

Use the acronym PEMDAS to remember the order of operation in algebra. PEMDAS is an acronym for parentheses, exponents, multiplication, division, addition, and subtraction. To remember it, many use the sentence, "Please Excuse My Dear Aunt Sally." Always remember, however, to multiply/divide or add/subtract in one sweep from left to right, not separately.

Computing Area of Two-dimensional Solids

Area is a measurement of the amount of surface of an object. Area is usually expressed in such units as square inches or square centimeters for small surfaces or in square feet or square meters for larger surfaces.

Rectangle

A rectangle is a four-sided figure with opposite sides of equal length and parallel. [Figure 1-13] All of the angles are right angles. A right angle is a 90° angle. The rectangle is a very familiar shape in mechanics. The formula for the area of a rectangle is:

$$\text{Area} = \text{Length} \times \text{Width} = L \times W$$

Example: An aircraft floor panel is in the form of a rectangle having a length of 24 inches and a width of 12 inches. What is the area of the panel expressed in square inches? First, determine the known values and substitute them in the formula.

$$A = L \times W = 24 \text{ inches} \times 12 \text{ inches} = 288 \text{ square inches}$$

Square

A square is a four-sided figure with all sides of equal length and parallel. [Figure 1-14] All angles are right angles. The formula for the area of a square is:

$$\text{Area} = \text{Length} \times \text{Width} = L \times W$$

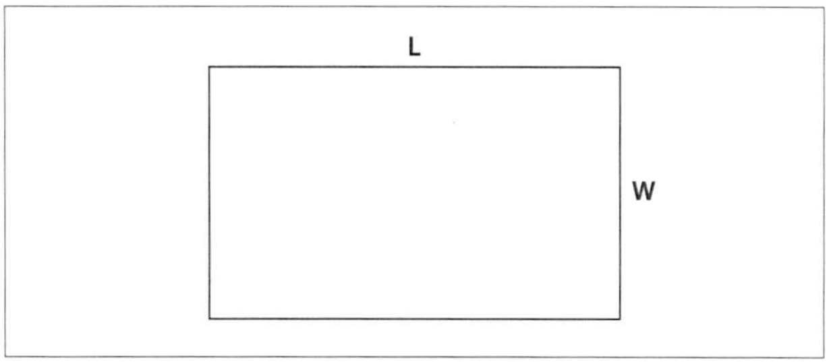

Figure 1-13. Rectangle.

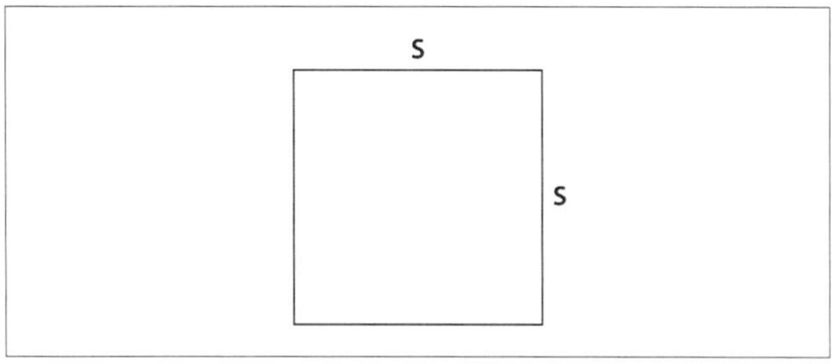

Figure 1-14. Square.

Since the length and the width of a square are the same value, the formula for the area of a square can also be written as:

$$\text{Area} = \text{Side} \times \text{Side} = S^2$$

Example: What is the area of a square access plate whose side measures 25 inches? First, determine the known value and substitute it in the formula.

$A = L \times W = 25 \text{ inches} \times 25 \text{ inches} = 625 \text{ square inches}$

Triangle

A triangle is a three-sided figure. The sum of the three angles in a triangle is always equal to 180°. Triangles are often classified by their sides. An equilateral triangle has 3 sides of equal length. An isosceles triangle has 2 sides of equal length. A scalene triangle has three sides of differing length. Triangles can also be classified by their angles: An acute triangle has all three angles less than 90°. A right triangle has one right angle (a 90° angle). An obtuse triangle has one angle greater than 90°. Each of these types of triangles is shown in Figure 1-15.

The formula for the area of a triangle is

$$\text{Area} = \tfrac{1}{2} \times (\text{Base} \times \text{Height}) = \tfrac{1}{2} \times (B \times H)$$

Example: Find the area of the obtuse triangle shown in Figure 1-16. First, substitute the known values in the area formula.

$$A = \tfrac{1}{2} \times (B \times H) = \tfrac{1}{2} \times (2'6'' \times 3'2'')$$

Next, convert all dimensions to inches:

$$2'6'' = (2 \times 12'') + 6'' = (24 + 6) = 30 \text{ inches}$$
$$3'2'' = (3 \times 12'') + 2'' = (36 + 2) = 38 \text{ inches}$$

Now, solve the formula for the unknown value:

$A = \tfrac{1}{2} \times (30 \text{ inches} \times 38 \text{ inches}) = 570 \text{ square inches}$

Parallelogram

A parallelogram is a four-sided figure with two pairs of parallel sides. [Figure 1-17] Parallelograms do not necessarily have four right angles. The formula for the area of a parallelogram is:

$$\text{Area} = \text{Length} \times \text{Height} = L \times H$$

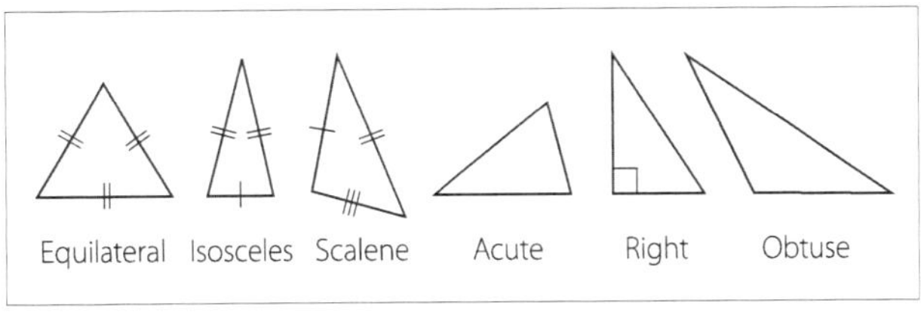

Figure 1-15. Types of triangles.

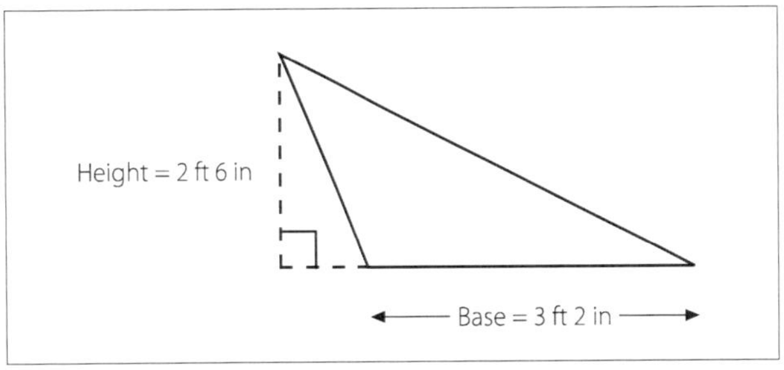

Figure 1-16. Obtuse triangle.

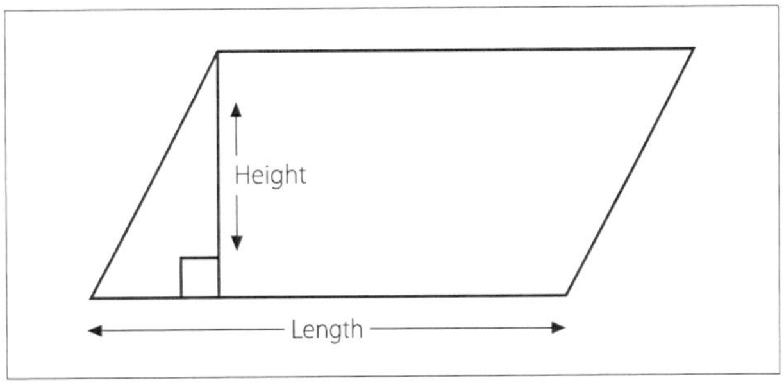

Figure 1-17. Parallelogram.

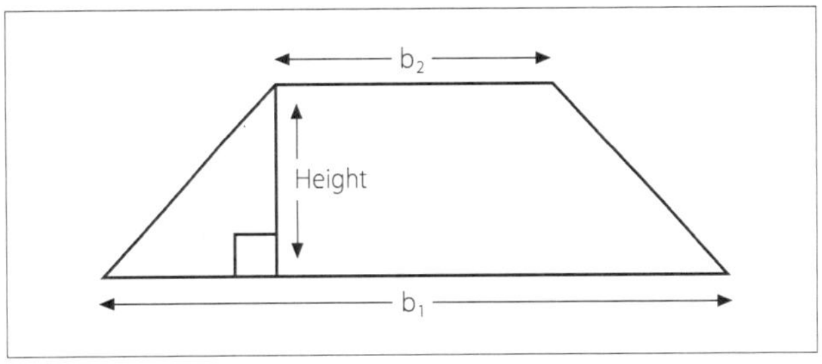

Figure 1-18. Trapezoid.

Trapezoid

A trapezoid is a four-sided figure with one pair of parallel sides. [Figure 1-18] The formula for the area of a trapezoid is:

$$\text{Area} = \tfrac{1}{2}(\text{Base}_1 + \text{Base}_2) \times \text{Height}$$

Example: What is the area of a trapezoid in Figure 1-19 whose bases are 14 inches and 10 inches, and whose height (or altitude) is 6 inches? First, substitute the known values in the formula.

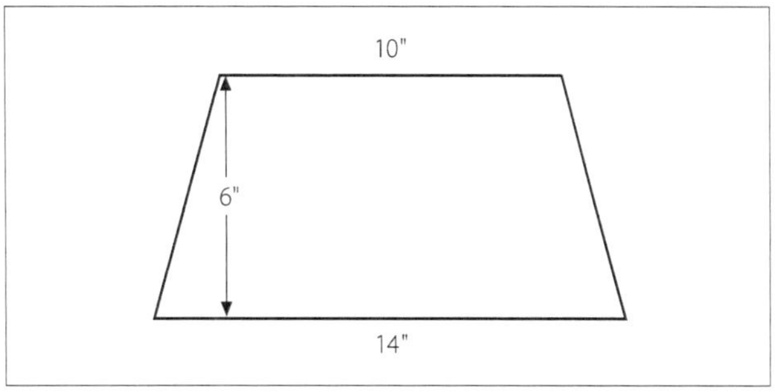

Figure 1-19. Trapezoid, with dimensions.

$A = \frac{1}{2}(b_1 + b_2) \times H$
$= \frac{1}{2}(14 \text{ inches} + 10 \text{ inches}) \times 6 \text{ inches}$

$A = \frac{1}{2}(24 \text{ inches}) \times 6 \text{ inches}$
$= 12 \text{ inches} \times 6 \text{ inches} = 72 \text{ square inches}.$

Circle

A circle is a closed, curved, plane figure. [Figure 1-20] Every point on the circle is an equal distance from the center of the circle. The diameter is the distance across the circle (through the center). The radius is the distance from the center to the edge of the circle. The diameter is always twice the length of the radius. The circumference, or distance around, a circle is equal to the diameter times π.

$$\text{Circumference} = C = d\,\pi$$

The formula for the area of a circle is:

$$\text{Area} = \pi \times \text{radius}^2 = \pi \times r^2$$

Example: The bore, or "inside diameter," of a certain aircraft engine cylinder is 5 inches. Find the area of the cross section of the cylinder.

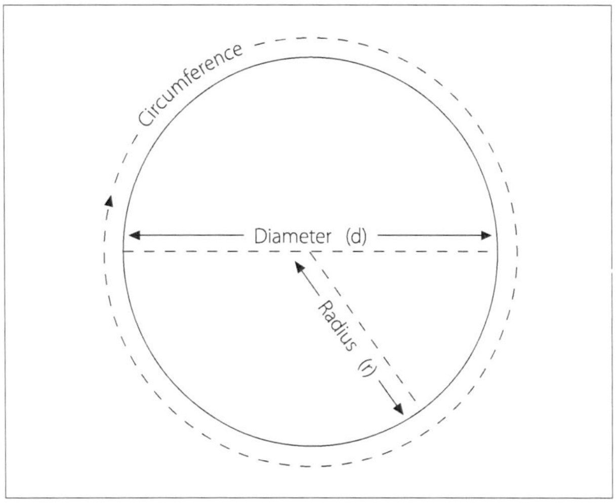

Figure 1-20. Circle.

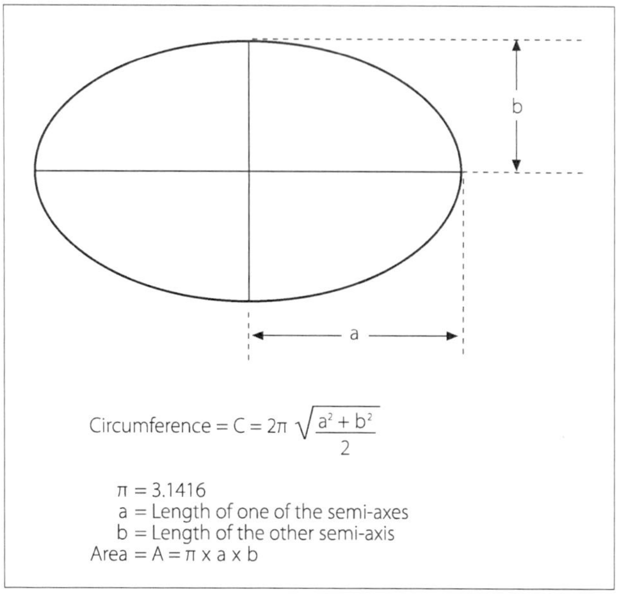

Figure 1-21. Ellipse.

First, substitute the known values in the formula:

$$A = \pi \times r^2.$$

The diameter is 5 inches, so the radius is 2.5 inches. (diameter = radius × 2)

$A = 3.1416 \times (2.5 \text{ inches})^2 = 3.1416 \times 6.25$ square inches = 19.635 square inches

Ellipse

An ellipse is a closed, curved, plane figure and is commonly called an oval. [Figure 1-21] In a radial engine, the articulating rods connect to the hub by pins, which travel in the pattern of an ellipse (i.e., an elliptical or obital path).

Wing Area

To describe the shape of a wing [Figure 1-23], several terms are required. To calculate wing area, it will be necessary to know the meaning of the terms "span" and "chord." The wingspan, S, is the length of the wing from wingtip to wingtip. The chord is the average width

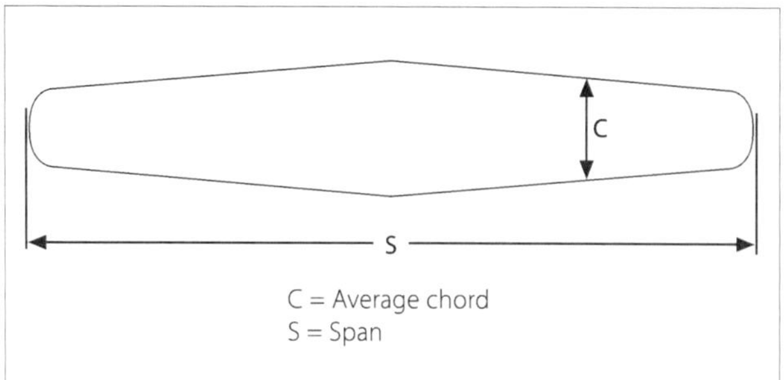

Figure 1-22. Wing planform.

Object	Area	Formula	Figure
Rectangle	Length × Width	$A = L \times W$	1-13
Square	Length × Width or Side × Side	$A = L \times W$ or $A = S^2$	1-14
Triangle	½ × (Length × Height) or ½ × (Base × Height) or (Base × Height) ÷ 2	$A = ½ (L \times H)$ or $A = ½ (B \times H)$ or $A = (B \times H) \div 2$	1-15
Parallelogram	Length × Height	$A = L \times H$	1-17
Trapezoid	½ ($base_1$ + $base_2$) × Height	$A = ½ (b_1 + b_2) \times H$	1-18
Circle	π × $radius^2$	$A = \pi \times r^2$	1-20
Ellipse	π × semi-axis A × semi-axis B	$A = \pi \times A \times B$	1-21

Figure 1-23. Formulas to compute area.

of the wing from leading edge to trailing edge. If the wing is a tapered wing, the average width, known as the mean chord (C), must be known to find the area. The formula for calculating wing area is:

Area of a wing = Span × Mean Chord

Example: Find the area of a tapered wing whose span is 50 feet and whose mean chord is 6'8". First, substitute the known values in the formula.

A = S × C
= 50 feet × 6 feet 8 inches
(Note: 8 inches = 8⁄12 feet = .67 feet)
= 50 feet × 6.67 feet
= 333.5 square feet

Units of Area

A square foot measures 1 foot by 1 foot. It also measures 12 inches by 12 inches. Therefore, one square foot also equals 144 square inches (that is, 12 × 12 = 144). To convert square feet to square inches, multiply by 144. To convert square inches to square feet, divide by 144.

A square yard measures 1 yard by 1 yard. It also measures 3 feet by 3 feet. Therefore, one square yard also equals 9 square feet (that is, 3 × 3 = 9). To convert square yards to square feet, multiply by 9. To convert square feet to square yards, divide by 9. Refer to Figure 1-37, Applied Mathematics Formula Sheet, at the end of the chapter for a comparison of different units of area.

Figure 1-23 summarizes the formulas for computing the area of two-dimensional solids.

Computing Volume of Three-Dimensional Solids

Three-dimensional solids have length, width, and height. There are many three-dimensional solids, but the most common are rectangular solids, cubes, cylinders, spheres, and cones. Volume is the amount of space within a solid. Volume is expressed in cubic units. Cubic inches or cubic centimeters are used for small spaces and cubic feet or cubic meters for larger spaces.

Rectangular Solid

A rectangular solid is a three-dimensional solid with six rectangle-shaped sides. [Figure 1-24] The volume is the number of cubic units within the rectangular solid. The formula for the volume of a rectangular solid is:

Volume = Length × Width × Height = L × W × H

In Figure 1-24, the rectangular solid is 3 feet by 2 feet by 2 feet.

The volume of the solid in Figure 1-24 is = 3 ft × 2 ft × 2 ft = 12 cubic feet.

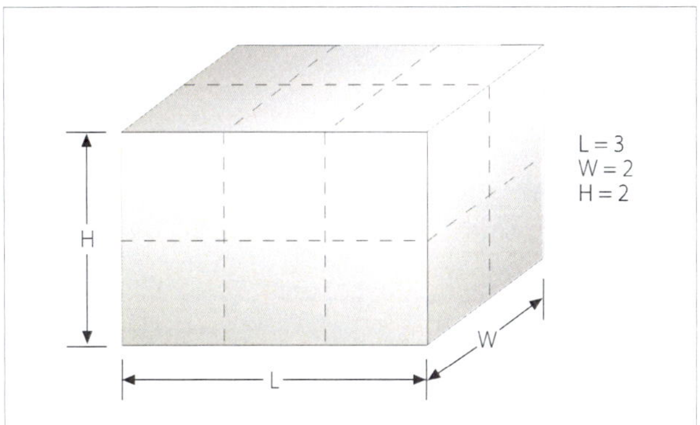

Figure 1-24. Rectangular solid.

Example: A rectangular baggage compartment measures 5 feet 6 inches in length, 3 feet 4 inches in width, and 2 feet 3 inches in height. How many cubic feet of baggage will it hold? First, substitute the known values into the formula.

$$\begin{aligned} V &= L \times W \times H \\ &= 5'6'' \times 3'4'' \times 2'3'' \\ &= 5.5 \text{ ft} \times 3.33 \text{ ft} \times 2.25 \text{ ft} \\ &= 41.25 \text{ cubic feet} \end{aligned}$$

Cube

A cube is a solid with six square sides. [Figure 1-25] A cube is just a special type of rectangular solid. It has the same formula for volume as does the rectangular solid which is Volume = Length × Width × Height = L × W × H. Because all of the sides of a cube are equal, the volume formula for a cube can also be written as:

$$\text{Volume} = \text{Side} \times \text{Side} \times \text{Side} = S^3$$

Example: A large, cube-shaped carton contains a shipment of smaller boxes inside of it. Each of the smaller boxes is 1 ft × 1 ft × 1 ft. The measurement of the large carton is 3 ft × 3 ft × 3 ft. How many of the smaller boxes are in the large carton? First, substitute the known values into the formula.

$$\begin{aligned} V &= L \times W \times H \\ &= 3 \text{ ft} \times 3 \text{ ft} \times 3 \text{ ft} \\ &= 27 \text{ cubic feet of volume in the large carton} \end{aligned}$$

Since each of the smaller boxes has a volume of 1 cubic foot, the large carton will hold 27 boxes.

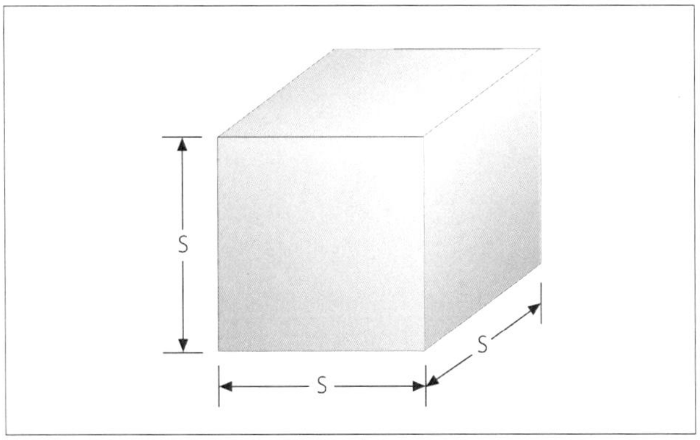

Figure 1-25. Cube.

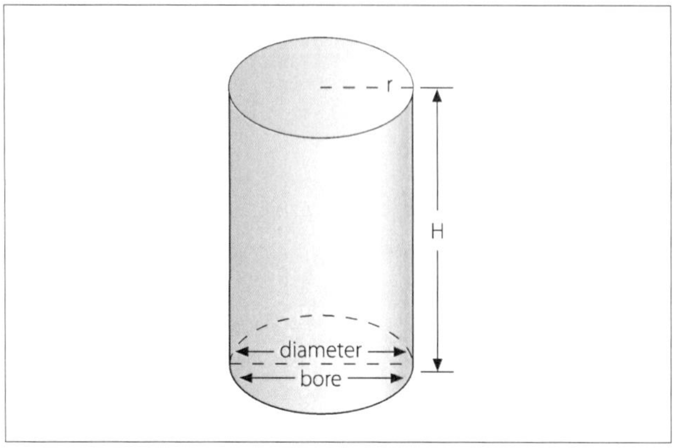

Figure 1-26. Cylinder.

Cylinder

A solid having the shape of a can, or a length of pipe, or a barrel is called a cylinder. [Figure 1-26] The ends of a cylinder are identical circles. The formula for the volume of a cylinder is:

$$\text{Volume} = \pi \times \text{radius}^2 \times \text{height of the cylinder} = \pi r^2 \times H$$

One of the most important applications of the volume of a cylinder is finding the piston displacement of a cylinder in a reciprocating engine. Piston displacement is the total volume (in cubic inches, cubic centimeters, or liters) swept by all of the pistons of a reciprocating engine as they move in one revolution of the crankshaft. The formula for piston displacement is given as:

$$\text{Piston Displacement} = \pi \times (\text{bore divided by } 2)^2 \times \text{stroke} \times (\#\text{ cylinders})$$

The bore of an engine is the inside diameter of the cylinder. The stroke of the engine is the length the piston travels inside the cylinder. [Figure 1-27]

Example: Find the piston displacement of one cylinder in a multi-cylinder aircraft engine. The engine has a cylinder bore of 5.5 inches and a stroke of 5.4 inches. First, substitute the known values in the formula.

$$V = \pi \times r^2 \times h = (3.1416) \times (5.5 \div 2)^2 \times (5.4)$$

$$V = 23.758 \times 5.4 = 128.29 \text{ cubic inches}$$

The piston displacement of one cylinder is 128.29 cubic inches. For an eight cylinder engine, then the total engine displacement would be:

$$\text{Total Displacement for 8 cylinders} = 8 \times 128.29 = 1026.32 \text{ cubic inches of displacement}$$

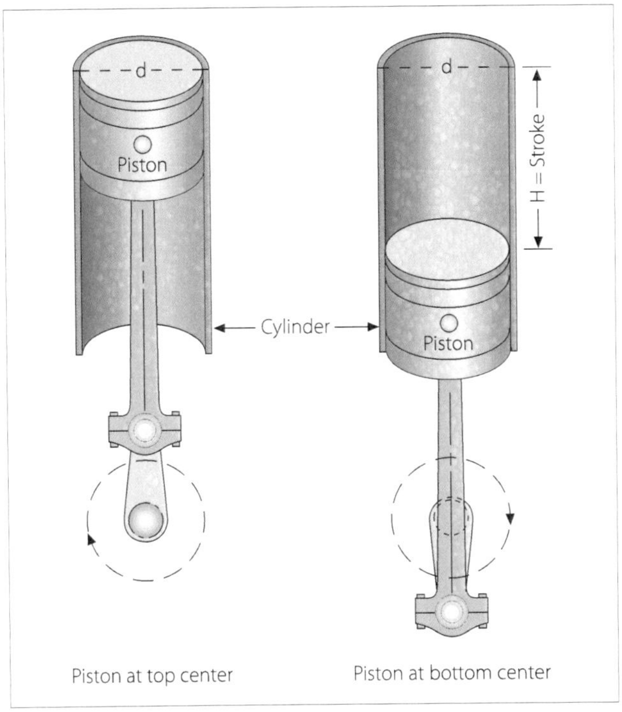

Figure 1-27. Cylinder displacement.

Sphere

A solid having the shape of a ball is called a sphere. [Figure 1-28] A sphere has a constant diameter. The radius (r) of a sphere is one-half of the diameter (D). The formula for the volume of a sphere is given as:

$$V = \frac{4}{3} \times \pi \times \text{radius}^3 = \frac{4}{3} \times \pi \times r^3 \quad \text{or} \quad V = \frac{1}{6} \times \pi D^3$$

Example: A pressure tank inside the fuselage of a cargo aircraft is in the shape of a sphere with a diameter of 34 inches. What is the volume of the pressure tank?

$$V = \frac{4}{3} \times \pi \times \text{radius}^3 = \frac{4}{3} \times (3.1416) \times (\frac{34}{2})^3$$
$$= 1.33 \times 3.1416 \times 17^3 = 1.33 \times 3.1416 \times 4913$$
$$V = 20{,}528.125 \text{ cubic inches}$$

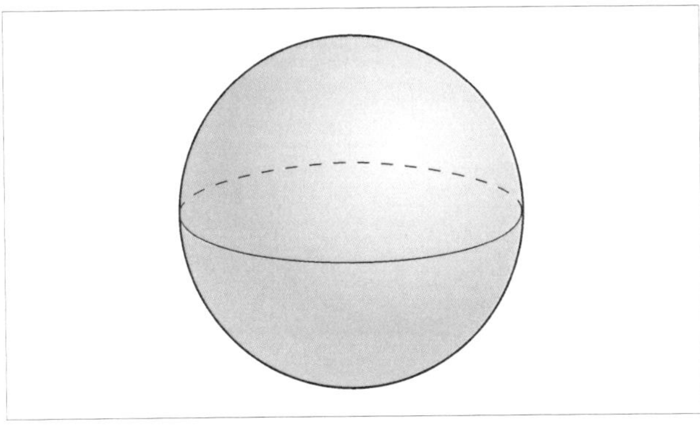

Figure 1-28. Sphere.

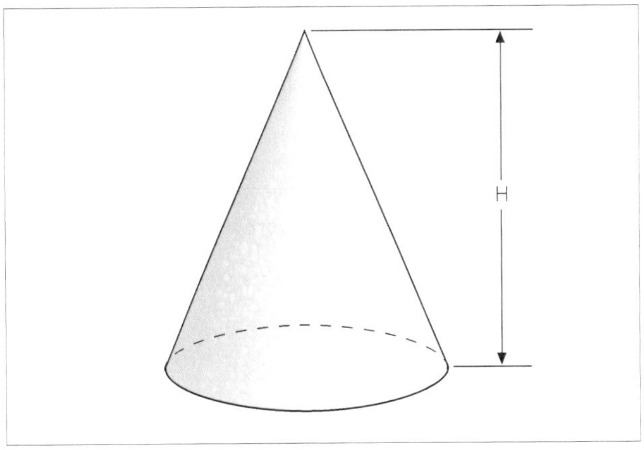

Figure 1-29. Cone.

Cone

A solid with a circle as a base and with sides that gradually taper to a point is called a cone. [Figure 1-29] The formula for the volume of a cone is given as:

$$V = \frac{1}{3} \times \pi \times \text{radius}^2 \times \text{height} = \frac{1}{3} \times \pi \times r^2 \times H$$

Units of Volume

Since all volumes are not measured in the same units, it is necessary to know all the common units of volume and how they are related to each other. For example, the mechanic may know the volume of a tank in cubic feet or cubic inches, but when the tank is full of gasoline, he or she will be interested in how many gallons it contains. Refer to Figure 1-37, Applied Mathematics Formula Sheet, at the end of the chapter for a comparison of different units of volume.

Computing Surface Area of Three-dimensional Solids

The surface area of a three-dimensional solid is the sum of the areas of the faces of the solid. Surface area is a different concept from that of volume. For example, surface area is the amount of sheet metal needed to build a rectangular fuel tank while volume is the amount of fuel that the tank can contain.

Rectangular Solid

The formula for the surface area of a rectangular solid [Figure 1-24] is given as:

Surface Area =

2 × [(Width × Length) + (Width × Height) + (Length × Height)]

= 2 × [(W × L) + (W × H) + (L × H)]

Cube

The formula for the surface area of a cube [Figure 1-25] is given as:

$$\text{Surface Area} = 6 \times (\text{Side} \times \text{Side}) = 6 \times S^2$$

Example: What is the surface area of a cube with a side measure of 8 inches?

$$\begin{aligned}\text{Surface Area} &= 6 \times (\text{Side} \times \text{Side})\\ &= 6 \times S^2 = 6 \times 8^2 = 6 \times 64\\ &= 384 \text{ square inches}\end{aligned}$$

Cylinder

The formula for the surface area of a cylinder [Figure 1-26] is given as:

$$\begin{aligned}\text{Surface Area} &= 2 \times \pi \times \text{radius}^2 + \pi \times \text{diameter} \times \text{height}\\ &= 2 \times \pi \times r^2 + \pi \times D \times H\end{aligned}$$

Sphere

The formula for the surface area of a sphere [Figure 1-28] is given as:

$$\text{Surface Area} = 4 \times \pi \times \text{radius}^2 = 4 \times \pi \times r^2$$

Cone

The formula for the surface area of a right circular cone [Figure 1-29] is given as:

$$\begin{aligned}\text{Surface Area} &= \pi \times \text{radius} \times [\text{radius} + (\text{radius}^2 + \text{height}^2)^{1/2}]\\ &= \pi \times r \times [r + (r^2 + H^2)^{1/2}]\end{aligned}$$

Figure 1-30 summarizes the formulas for computing the volume and surface area of three-dimensional solids.

Solid	Volume	Surface Area	Figure
Rectangular Solid	$L \times W \times H$	$2 \times [(W \times L) + (W \times H) + (L \times H)]$	1-23
Cube	S^3	$6 \times S^2$	1-24
Cylinder	$\pi \times r^2 \times H$	$2 \times \pi \times r^2 + \pi \times D \times H$	1-25
Sphere	$\frac{4}{3} \times \pi \times r^3$	$4 \times \pi \times r^2$	1-27
Cone	$\frac{1}{3} \times \pi \times r^2 \times H$	$\pi \times r \times [r + (r^2 + H^2)^{1/2}]$	1-28

Figure 1-30. Formulas to compute volume and surface area.

Trigonometric Functions

Trigonometry is the study of the relationship between the angles and sides of a triangle. The word trigonometry comes from the Greek *trigonon*, which means three angles, and *metro*, which means measure.

Right Triangle, Sides and Angles

In Figure 1-31, notice that each angle is labeled with a capital letter. Across from each angle is a corresponding side, each labeled with a lower case letter. This triangle is a right triangle because angle C is a 90° angle. Side a is opposite from angle A, and is sometimes referred to as the "opposite side." Side b is next to, or adjacent to, angle A and is therefore referred to as the "adjacent side." Side c is always across from the right angle and is referred to as the "hypotenuse."

Sine, Cosine, and Tangent

The three primary trigonometric functions and their abbreviations are: sine (sin), cosine (cos), and tangent (tan). These three functions can be found on most scientific calculators. The three trigonometric functions are actually ratios comparing two of the sides of the triangle as follows:

$$\text{Sine (sin) of angle A} = \frac{\text{opposite side (side a)}}{\text{hypotenuse (side c)}}$$

$$\text{Cosine (cos) of angle A} = \frac{\text{adjacent side (side b)}}{\text{hypotenuse (side c)}}$$

$$\text{Tangent (tan) of angle A} = \frac{\text{opposite side (side a)}}{\text{adjacent side (side b)}}$$

Example: Find the sine of a 30° angle.

Calculator Method:
Using a calculator, select the "sin" feature, enter the number 30, and press "enter." The calculator should display the answer as 0.5. This means that when angle

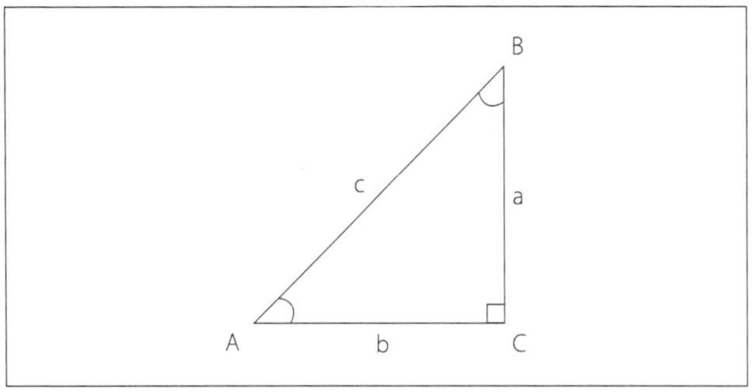

Figure 1-31. Right triangle.

A equals 30°, then the ratio of the opposite side (a) to the hypotenuse (c) equals 0.5 to 1, so the hypotenuse is twice as long as the opposite side for a 30° angle. Therefore, sin 30° = 0.5.

Trigonometric Table Method:
When using a trigonometry table, find 30° in the first column. Next, find the value for sin 30° under the second column marked "sine" or "sin." The value for sin 30° should be 0.5.

Pythagorean Theorem

The Pythagorean Theorem is named after the ancient Greek mathematician, Pythagoras (~500 B.C.). This theorem is used to find the third side of any right triangle when two sides are known. The Pythagorean Theorem states that $a^2 + b^2 = c^2$. [Figure 1-32] Where c = the hypotenuse of a right triangle, a is one side of the triangle and b is the other side of the triangle.

Example: What is the length of the longest side of a right triangle, given the other sides are 7 inches and 9 inches? The longest side of a right triangle is always side c, the hypotenuse. Use the Pythagorean Theorem to solve for the length of side c as follows:

$$a^2 + b^2 = c^2$$

$$7^2 + 9^2 = c^2$$

$$49 + 81 = c^2$$

$$130 = c^2$$

If $c^2 = 130$ then $c = \sqrt{130} = 11.4$ inches
Therefore, side c = 11.4 inches.

Example: The cargo door opening in a military airplane is a rectangle that is 5 ½ feet tall by 7 feet wide. A section of square steel plate that is 8 feet wide by 8 feet tall by 1 inch thick must fit inside the airplane. Can the square section of steel plate fit through the cargo door? It is obvious that the square steel plate will not fit horizontally through the cargo door. The steel plate is 8 feet wide and the cargo door is only 7 feet wide. However, if the steel plate is tilted diagonally, will it fit through the cargo door opening?

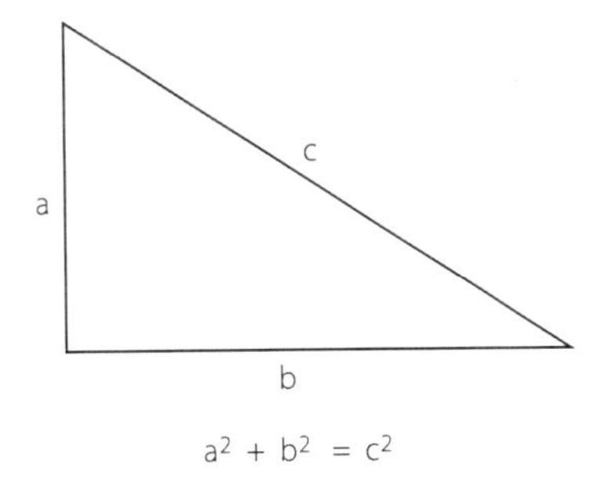

Figure 1-32. Pythagorean Theorem.

The diagonal distance across the cargo door opening can be calculated using the Pythagorean Theorem where "a" is the cargo door height, "b" is the cargo door width, and "c" is the diagonal distance across the cargo door opening.

$$a^2 + b^2 = c^2$$

$$(5.5\text{ ft})^2 + (7\text{ ft})^2 = c^2$$

$$30.25 + 49 = c^2$$

$$79.25 = c^2 \quad c = 8.9\text{ ft}$$

The diagonal distance across the cargo door opening is 8.9 feet, so the 8-foot wide square steel plate will fit diagonally through the cargo door opening and into the airplane.

Measurement Systems

Conventional (U.S. or English) System

Our conventional (U.S. or English) system of measurement is part of our cultural heritage from the days when the thirteen colonies were under British rule. It started as a collection of Anglo-Saxon, Roman, and Norman-French weights and measures. For example, the inch represents the width of the thumb and the foot is from the length of the human foot. Tradition holds that King Henry I decreed that the yard should be the distance from the tip of his nose to the end of his thumb. Since medieval times, commissions appointed by various English monarchs have reduced the chaos of measurement by setting specific standards for some of the most important units. Some of the conventional units of measure are: inches, feet, yards, miles, ounces, pints, gallons, and pounds. Because the conventional system was not set up systematically, it contains a random collection of conversions. For example, 1 mile = 5,280 feet and 1 foot = 12 inches.

Metric System

The metric system, also known as the International System of Units (SI), is the dominant language of measurement used today. Its standardization and decimal features make it well-suited for engineering and aviation work.

The metric system was first envisioned by Gabriel Mouton, Vicar of St. Paul's Church in Lyons, France. The meter is the unit of length in the metric system, and it is equal to one ten-millionth of the distance from the equator to the North Pole. The liter is the unit of volume and is equal to one cubic decimeter. The gram is the unit of mass and is equal to one cubic centimeter of water.

All of the metric units follow a consistent naming scheme, which consists of attaching a prefix to the unit. For example, since kilo stands for 1,000 one kilometer equals 1,000 meters. Centi is the prefix for one hundredth, so one meter equals one hundred centimeters. Milli is the prefix for one thousandths and one gram equals one thousand milligrams. Refer to Figure 1-33 for the names and definitions of metric prefixes.

Measurement Systems and Conversions

The United States primarily uses the conventional (U.S. or English) system, although it is slowly integrating the metric system (SI). A recommendation to transition to the metric system within ten years was initiated in the 1970s. However, this movement lost momentum, and the United States continues to use both measurement systems. Therefore, information to convert between the conventional (U.S., or English) system and the metric (SI) system has been included in Figure 1-37, Applied Mathematics Formula Sheet, at the end of this chapter. Examples of its use are as follows:

To convert inches to millimeters, multiply the number of inches by 25.4.

Example: 20 inches = 20 × 25.4 = 508 mm

Prefix	Means
tera (10^{12})	One trillion times
giga (10^{9})	One billion times
mega (10^{6})	One million times
kilo (10^{3})	One thousand times
hecto (10^{2})	One hundred times
deca (10^{1})	Ten times
deci (10^{-1})	One tenth of
centi (10^{-2})	One hundredth of
milli (10^{-3})	One thousandth of
micro (10^{-6})	One millionth of
nano (10^{-9})	One billionth of
pico (10^{-12})	One trillionth of

Figure 1-33. Names and definitions of metric prefixes.

To convert ounces to grams, multiply the number of ounces by 28.35.

Example: 12 ounces = 12 × 28.35 = 340.2 grams

The Binary Number System

The binary number system has only two digits: 0 and 1. The prefix in the word "binary" is a Latin root for the word "two" and its use was first published in the late 1700s. The use of the binary number system is based on the fact that switches or valves have two states: open or closed (on/off).

Currently, one of the primary uses of the binary number system is in computer applications. Information is stored as a series of 0s and 1s, forming strings of binary numbers. An early electronic computer, ENIAC (Electronic Numerical Integrator And Calculator), was built in 1946 at the University of Pennsylvania and contained 17,000 vacuum tubes, along with 70,000 resistors, 10,000 capacitors, 1,500 relays, 6,000 manual switches and 5 million soldered joints. Computers obviously have changed a great deal since then, but are still based on the same binary number system. The binary number system is also useful when working with digital electronics because the two basic conditions of electricity, on and off, can be represented by the two digits of the binary number system. When the system is on, it is represented by the digit 1, and when it is off, it is represented by the digit zero.

Place Values

The binary number system is a base-2 system. That is, the place values in the binary number system are based on powers of 2. An 8-bit binary number system is shown in Figure 1-34 on the next page.

Converting Binary Numbers to Decimal Numbers

To convert a binary number to a decimal number, add up the place values that have a 1 (place values that have a zero do not contribute to the decimal number conversion).

Example: Convert the binary number 10110011 to a decimal number. Using the Place Value chart shown in Figure 1-35, add up the place values of the '1s' in the binary number (ignore the place values with a zero in the binary number).

The binary number 10110011
= 128 + 0 + 32 + 16 + 0 + 0 + 2 + 1
= 179 in the decimal number system

	Place Value								
	2^7 or 128	2^6 or 64	2^5 or 32	2^4 or 16	2^3 or 8	2^2 or 4	2^1 or 2	2^0 or 1	
10011001 shown as	1	0	0	1	1	0	0	1	= 153
00101011 shown as	0	0	1	0	1	0	1	1	= 43

Figure 1-34. Binary system.

	Place Value								
	2^7 or 128	2^6 or 64	2^5 or 32	2^4 or 16	2^3 or 8	2^2 or 4	2^1 or 2	2^0 or 1	
10110011 shown as	1	0	1	1	0	0	1	1	
	128 +	0 +	32 +	16 +	0 +	0 +	2 +	1	= 179

Figure 1-35. Conversion from binary number to decimal number.

	Place Value							
	2^7 or 128	2^6 or 64	2^5 or 32	2^4 or 16	2^3 or 8	2^2 or 4	2^1 or 2	2^0 or 1
35 shown as	0	0	1	0	0	0	0	1
124 shown as	0	1	1	1	1	1	0	0
96 shown as	0	1	1	0	0	0	0	0
255 shown as	1	1	1	1	1	1	1	1
233 shown as	1	1	1	0	1	0	0	1

Figure 1-36. Conversion from decimal number to binary number.

Converting Decimal Numbers to Binary Numbers

To convert a decimal number to a binary number, the place values in the binary system are used to create a sum of numbers that equal the value of the decimal number being converted. Start with the largest binary place value and subtract from the decimal number. Continue this process until all of the binary digits are determined.

Example: Convert the decimal number 233 to a binary number.

Start by subtracting 128 (the largest place value from the 8-bit binary number) from 233.

$$233 - 128 = 105$$

A "1" is placed in the first binary digit space: 1XXXXXXX.

Continue the process of subtracting the binary number place values:

$$105 - 64 = 41$$

A "1" is placed in the second binary digit space: 11XXXXXX.

$$41 - 32 = 9$$

A "1" is placed in the third binary digit space: 111XXXXX.

Since 9 is less than 16 (the next binary place value), a "0" is placed in the fourth binary digit space: 1110XXXX.

$$9 - 8 = 1$$

A "1" is placed in the fifth binary digit space: 11101XXX

Since 1 is less than 4 (the next binary place value), a 0 is placed in the sixth binary digit space: 111010XX.

Since 1 is less than 2 (the next binary place value), a 0 is placed in the seventh binary digit space: 1110100X.

$$1 - 1 = 0$$

A "1" is placed in the eighth binary digit space: 11101001.

The decimal number 233 is equivalent to the binary number 11101001, as shown in Figure 1-36.

Additional decimal number to binary number conversions are shown in Figure 1-36.

Conversion Factors

Length		
1 inch	2.54 centimeters	25.4 millimeters
1 foot	12 inches	30.48 centimeters
1 yard	3 feet	0.9144 meters
1 mile	5,280 feet	1,760 yards
1 millimeter	0.0394 inches	
1 kilometer	0.62 miles	

Area		
1 square inch	6.45 square centimeters	
1 square foot	144 square inches	0.093 square meters
1 square yard	9 square feet	0.836 square meters
1 acre	43,560 square feet	
1 square mile	640 acres	2.59 square kilometers
1 square centimeter	0.155 square inches	
1 square meter	1.195 square yards	
1 square kilometer	0.384 square miles	

Volume					
1 fluid ounce	29.57 cubic centimeters				
1 cup	8 fluid ounce				
1 pint	2 cups	16 fluid ounces	0.473 liters		
1 quart	2 pints	4 cups	32 fluid ounces	0.9463 liters	
1 gallon	4 quarts	8 pints	16 cups	128 ounces	3.785 liters
1 gallon	231 cubic inches				
1 liter	0.264 gallons	1.057 quarts			
1 cubic foot	1,728 cubic inches				
1 cubic foot	7.5 gallons				
1 cubic yard	27 cubic feet				
1 board foot	1 inch by 12 inches by 12 inches				

Weight			
1 ounce	28.350 grams		
1 pound	16 ounces	453.592 grams	0.4536 kilograms
1 ton	2,000 pounds		
1 milligram	0.001 grams		
1 kilogram	1,000 grams	2.2 pounds	
1 gram	0.0353 ounces		

Temperature	
degrees Fahrenheit to degrees Celsius	Celsius = $\frac{5}{9}$ × (degrees Fahrenheit – 32)
degrees Celsius to degrees Fahrenheit	Fahrenheit = $\frac{9}{5}$ × (degrees Celsius) + 32

Formulas for Area of Two-Dimensional Objects

Object	Area	Formula	Figure
Rectangle	Length × Width	A = L × W	1-13
Square	Length × Width or Side × Side	A = L × W or A = S^2	1-14
Triangle	½ × (Length × Height) or ½ × (Base × Height) or (Base × Height) ÷ 2	A = ½ (L × H) or A = ½ (B × H) or A = (B × H) ÷ 2	1-15
Parallelogram	Length × Height	A = L × H	1-17
Trapezoid	½ ($base_1$ + $base_2$) × Height	A = ½ ($b_1 + b_2$) × H	1-18
Circle	π × $radius^2$	A = π × r^2	1-20

Figure 1-37. Applied Mathematics Formula Sheet.

Order of Operation for Algebraic Equations

1. Parentheses
2. Exponents
3. Multiplication and Division
4. Addition and Subtraction

Use the acronym PEMDAS to remember the order of operation in algebra. PEMDAS is an acronym for parentheses, exponents, multiplication, division, addition, and subtraction. To remember it, many use the sentence, "Please Excuse My Dear Aunt Sally."

Trigonometric Equations

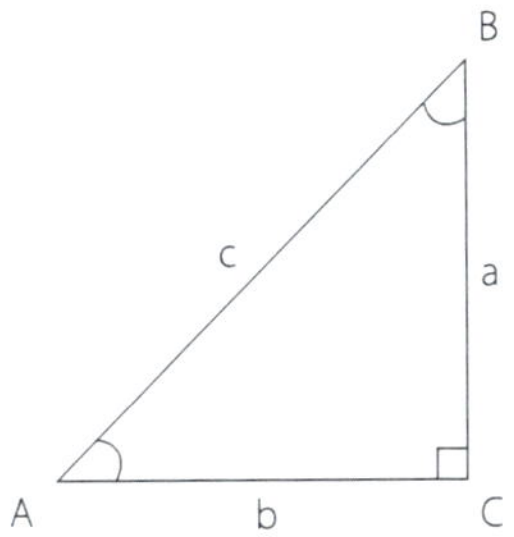

$$\text{Sine (sin) of angle A} = \frac{\text{opposite side (side a)}}{\text{hypotenuse (side c)}}$$

$$\text{Cosine (cos) of angle A} = \frac{\text{adjacent side (side b)}}{\text{hypotenuse (side c)}}$$

$$\text{Tangent (tan) of angle A} = \frac{\text{opposite side (side a)}}{\text{adjacent side (side b)}}$$

Pythagorean Theorem

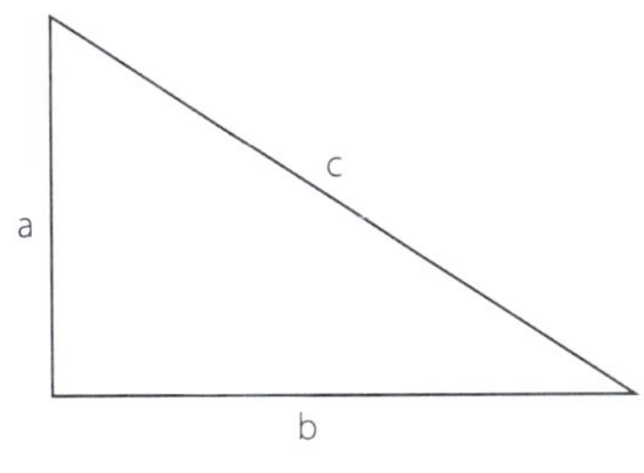

$$a^2 + b^2 = c^2$$

Circumference of a Circle

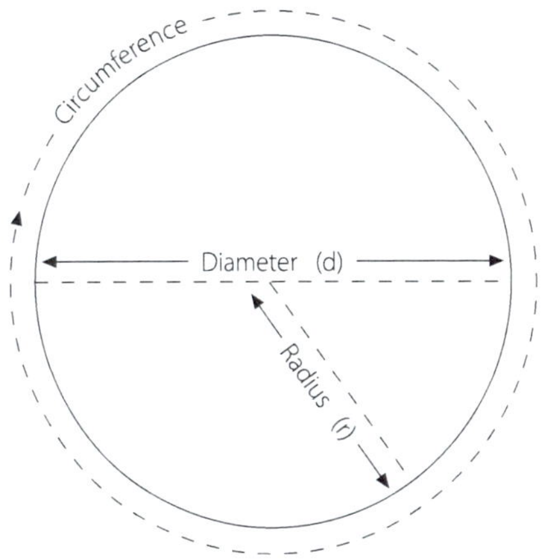

Circumference of an Ellipse

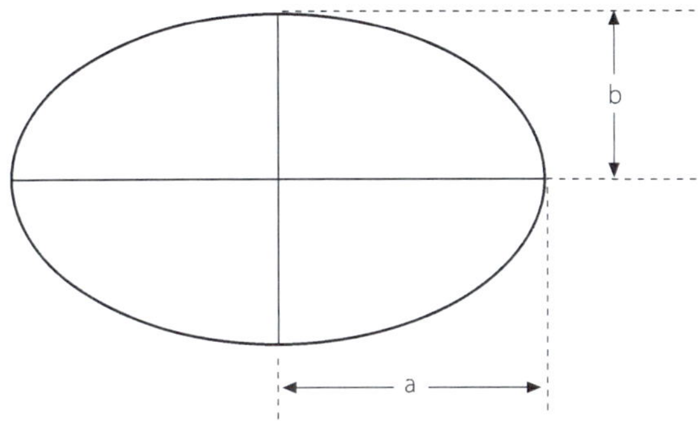

$$\text{Circumference} = C = 2\pi\sqrt{\frac{a^2 + b^2}{2}}$$

$\pi = 3.1416$
a = Length of one of the semi-axes
b = Length of the other semi-axis
Area = $A = \pi \times a \times b$

Formulas for Surface Area and Volume of Three-Dimensional Solids

Solid	Volume	Surface Area	Figure
Rectangular Solid	$L \times W \times H$	$2 \times [(W \times L) + (W \times H) + (L \times H)]$	1-23
Cube	S^3	$6 \times S^2$	1-24
Cylinder	$\pi \times r^2 \times H$	$2 \times \pi \times r^2 + \pi \times D \times H$	1-25
Sphere	$\frac{4}{3} \times \pi \times r^3$	$4 \times \pi \times r^2$	1-27
Cone	$\frac{1}{3} \times \pi \times r^2 \times H$	$\pi \times r \times [r + (r^2 + H^2)^{1/2}]$	1-28

Figure 1-37. Applied Mathematics Formula Sheet. (continued)

Names and Symbols for Metric Prefixes

Prefix	Means
tera (10^{12})	One trillion times
giga (10^{9})	One billion times
mega (10^{6})	One million times
kilo (10^{3})	One thousand times
hecto (10^{2})	One hundred times
deca (10^{1})	Ten times
deci (10^{-1})	One tenth of
centi (10^{-2})	One hundredth of
milli (10^{-3})	One thousandth of
micro (10^{-6})	One millionth of
nano (10^{-9})	One billionth of
pico (10^{-12})	One trillionth of

Powers of Ten

Powers of Ten	Expansion	Value
Positive Exponents		
10^{6}	10 × 10 × 10 × 10 × 10 × 10	1,000,000
10^{5}	10 × 10 × 10 × 10 × 10	100,000
10^{4}	10 × 10 × 10 × 10	10,000
10^{3}	10 × 10 × 10	1,000
10^{2}	10 × 10	100
10^{1}	10	10
10^{0}		1
Negative Exponents		
10^{-1}	1/10	1/10 = 0.1
10^{-2}	1/(10 × 10)	1/100 = 0.01
10^{-3}	1/(10 × 10 × 10)	1/1,000 = 0.001
10^{-4}	1/(10 × 10 × 10 × 10)	1/10,000 = 0.0001
10^{-5}	1/(10 × 10 × 10 × 10 × 10)	1/100,000 = 0.00001
10^{-6}	1/(10 × 10 × 10 × 10 × 10 × 10)	1/1,000,000 = 0.000001

Figure 1-37. Applied Mathematics Formula Sheet. (continued)

Chapter 2 Aircraft Drawings

The exchange of ideas is essential to everyone, regardless of his or her vocation or position. Usually, this exchange is carried on by the oral or written word; but under some conditions, the use of these alone is impractical. Industry discovered that it could not depend entirely upon written or spoken words for the exchange of ideas because misunderstanding and misinterpretation arose frequently. A written description of an object can be changed in meaning just by misplacing a comma; the meaning of an oral description can be completely changed by the use of a wrong word. To avoid these possible errors, industry uses drawings to describe objects. For this reason, drawing is the draftsman's language.

Drawing, as we use it, is a method of conveying ideas concerning the construction or assembly of objects. This is done with the help of lines, notes, abbreviations, and symbols. It is very important that the aviation mechanic who is to make or assemble the object understand the meaning of the different lines, notes, abbreviations, and symbols that are used in a drawing. (See especially the "Lines and Their Meanings" section of this chapter.)

Computer Graphics

From the early days of aviation, development of aircraft, aircraft engines, and other components relied heavily on aircraft drawings. For most of the 20th century, drawings were created on a drawing "board" with pen or pencil and paper. However, with the introduction and advancement of computers in the later decades of the 20th century, the way drawings are created changed dramatically. Computers were used not only to create drawings, but they were being used to show items in "virtual reality," from any possible viewing angle. Further development saw computer software programs with the capability of assembling separately created parts to check for proper fit and possible interferences. Additionally, with nearly instantaneous information sharing capability through computer networking and the Internet, it became much easier for designers to share their work with other designers and manufacturers virtually anytime, anywhere in the world. Using new computer controlled manufacturing techniques, it literally became possible to design a part and have it precisely manufactured without ever having it shown on paper. New terms and acronyms became commonplace. The more common of these terms are:

- Computer Graphics—drawing with the use of a computer,
- Computer Aided Design Drafting (CADD)—where a computer is used in the design and drafting process,
- Computer Aided Design (CAD)—where a computer is used in the design of a product,
- Computer Aided Manufacturing (CAM)—where a computer is used in the manufacturing of a product, and
- Computer Aided Engineering (CAE)—where a computer is used in the engineering of a product.

As computer hardware and software continue to evolve, there continues to be a greater amount of CAE done in less time at lower cost. In addition to product design, some of the other uses of CAE are product analysis, assembly, simulations and maintenance information. [Figure 2-1]

Purpose and Function of Aircraft Drawings

Drawings and prints are the link between the engineers who design an aircraft and the workers who build, maintain, and repair it. A print may be a copy of a working drawing for an aircraft part or group of parts, or for a design of a system or group of systems. They are made by placing a tracing of the drawing over a sheet of chemically treated paper and exposing it to a strong light for a short period of time. When the

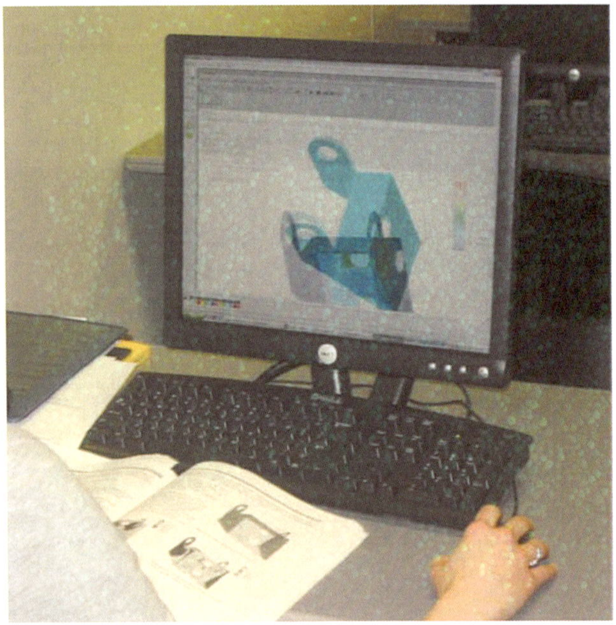

Figure 2-1. Computer graphics work station.

exposed paper is developed, it turns blue where the light has penetrated the transparent tracing. The inked lines of the tracing, having blocked out the light, show as white lines on a blue background. Other types of sensitized paper have been developed; prints may have a white background with colored lines or a colored background with white lines.

Drawings created using computers may be viewed as they appear on the computer monitor, or they may be printed out in "hard copy" by use of an ink jet or laser printer. Larger drawings may be printed by use of a plotter or large format printer. Large printers can print drawings up to 42 inches high with widths up to 600 inches by use of continuous roll paper. [Figure 2-2]

Figure 2-2. Large format printer.

Care and Use of Drawings

Drawings are both expensive and valuable; consequently, they should be handled carefully. Open drawings slowly and carefully to prevent tearing the paper. When the drawing is open, smooth out the fold lines instead of bending them backward.

To protect drawings from damage, never spread them on the floor or lay them on a surface covered with tools or other objects that may make holes in the paper. Hands should be free of oil, grease, or other unclean matter that can soil or smudge the print.

Never make notes or marks on a print as they may confuse other persons and lead to incorrect work. Only authorized persons are permitted to make notes or changes on prints, and they must sign and date any changes they make.

When finished with a drawing, fold and return it to its proper place. Prints are folded originally in a proper size for filing, and care should be taken so that the original folds are always used.

Types of Drawings

Drawings must give such information as size and shape of the object and all of its parts, specifications for material to be used, how the material is to be finished, how the parts are to be assembled, and any other information essential to making and assembling the particular object.

Drawings may be divided into three classes: (1) detail, (2) assembly, and (3) installation. [Figure 2-3]

Detail Drawing

A detail drawing is a description of a single part, describing by lines, notes, and symbols the specifications for size, shape, material, and methods of manufacture to be used in making the part. Detail drawings are usually rather simple; and, when single parts are small, several detail drawings may be shown on the same sheet or print. (See detail drawing at the top of Figure 2-3.)

Assembly Drawing

An assembly drawing is a description of an object made up of two or more parts. Examine the assembly drawing in the center of Figure 2-3. It describes the object by stating, in a general way, size and shape. Its primary purpose is to show the relationship of the various parts. An assembly drawing is usually more complex than a detail drawing, and is often accompanied by detail drawings of various parts.

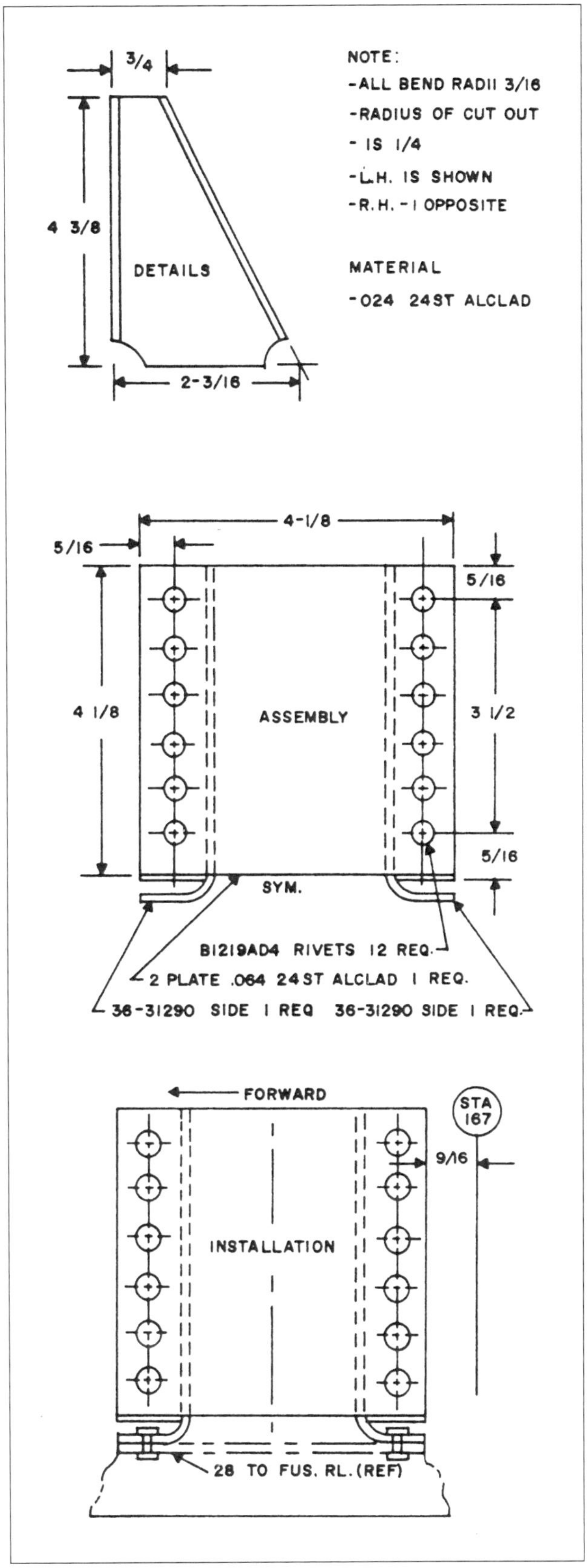

Figure 2-3. Types of drawings.

Installation Drawing

An installation drawing is one which includes all necessary information for a part or an assembly in the final installed position in the aircraft. It shows the dimensions necessary for the location of specific parts with relation to the other parts and reference dimensions that are helpful in later work in the shop. (See installation drawing at the bottom of Figure 2-3.)

Sectional View Drawings

A section or sectional view is obtained by cutting away part of an object to show the shape and construction at the cutting plane. The part or parts cut away are shown by the use of section (crosshatching) lines. Types of sections are described in the following paragraphs.

Full Section

A full section view is used when the interior construction or hidden features of an object cannot be shown clearly by exterior views. For example, Figure 2-4, a sectional view of a coaxial cable connector, shows the internal construction of the connector.

Half Section

In a half section, the cutting plane extends only halfway across the object, leaving the other half of the object as an exterior view.

Half sections are used to advantage with symmetrical objects to show both the interior and exterior.

Figure 2-5 is a half sectional view of a quick disconnect used in aircraft fluid systems.

Revolved Section

A revolved section drawn directly on the exterior view shows the shape of the cross section of a part, such as the spoke of a wheel. An example of a revolved section is shown in Figure 2-6.

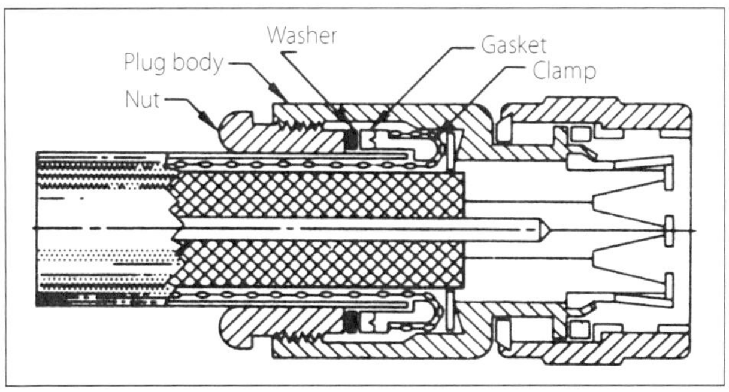

Figure 2-4. Sectional view of a cable connector.

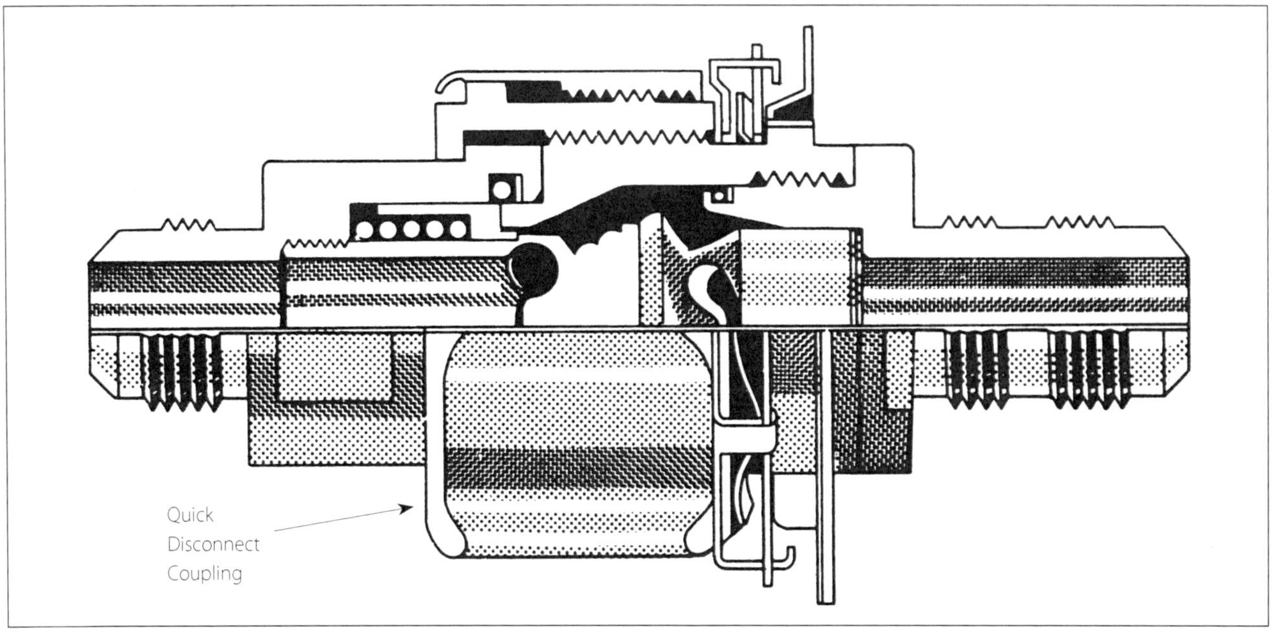

Figure 2-5. Half section.

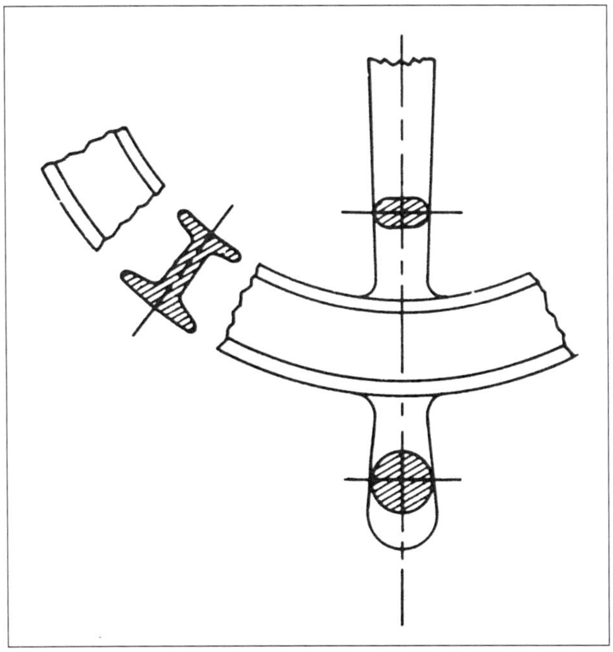

Figure 2-6. Revolved sections.

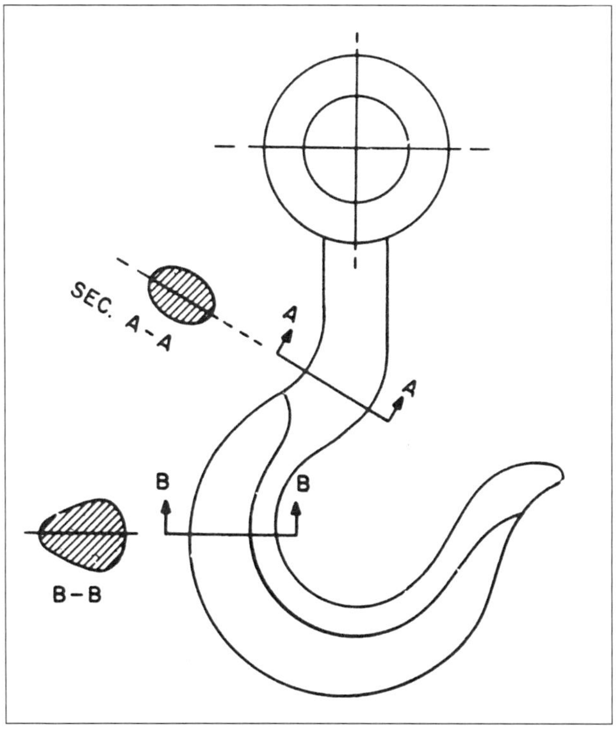

Figure 2-7. Removed sections.

Removed Section

A removed section illustrates particular parts of an object. It is drawn like revolved sections, except it is placed at one side and, to bring out pertinent details, often drawn to a larger scale than the view on which it is indicated.

Figure 2-7 is an illustration of removed sections. Section A-A shows the cross-sectional shape of the object at cutting plane line A-A. Section B-B shows the cross-sectional shape at cutting plane line B-B. These sectional views are drawn to the same scale as the principal view. Note that they are often drawn to a larger scale to bring out pertinent details.

Title Blocks

Every print must have some means of identification. This is provided by a title block. [Figure 2-8] The title block consists of a drawing number and certain

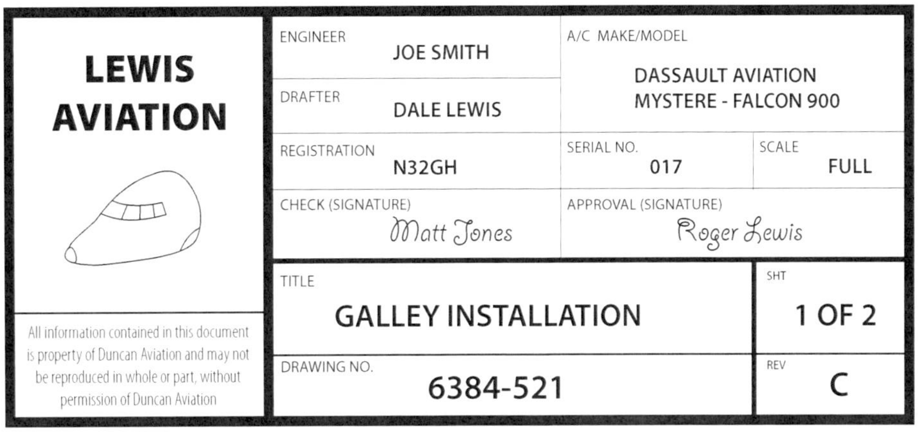

Figure 2-8. Title block.

other data concerning the drawing and the object it represents. This information is grouped in a prominent place on the print, usually in the lower right-hand corner. Sometimes the title block is in the form of a strip extending almost the entire distance across the bottom of the sheet.

Although title blocks do not follow a standard form insofar as layout is concerned, all of them present essentially the following information:

1. A drawing number to identify the print for filing purposes and to prevent confusing it with any other print.
2. The name of the part or assembly.
3. The scale to which it is drawn.
4. The date.
5. The name of the firm.
6. The name of the draftsmen, the checker, and the person approving the drawing.

Drawing or Print Numbers

All prints are identified by a number, which appears in a number block in the lower right-hand corner of the title block. It may also be shown in other places—such as near the top border line, in the upper right-hand corner, or on the reverse side of the print at both ends—so that the number will show when the print is folded or rolled. The purpose of the number is quick identification of a print. If a print has more than one sheet and each sheet has the same number, this information is included in the number block, indicating the sheet number and the number of sheets in the series.

Reference and Dash Numbers

Reference numbers that appear in the title block refer you to the numbers of other prints. When more than one detail is shown on a drawing, dash numbers are used. Both parts would have the same drawing number plus an individual number, such as 40267-1 and 40267-2.

In addition to appearing in the title block, dash numbers may appear on the face of the drawing near the parts they identify. Dash numbers are also used to identify right-hand and left-hand parts.

In aircraft, many parts on the left side are like the corresponding parts on the right side but in reverse. The left-hand part is always shown in the drawing. The right-hand part is called for in the title block. Above the title block a notation is found, such as: 470204-1LH shown; 470204-2RH opposite. Both parts carry the same number, but the part called for is distinguished by a dash number. Some prints have odd numbers for left-hand parts and even numbers for right-hand parts.

Universal Numbering System

The universal numbering system provides a means of identifying standard drawing sizes. In the universal numbering system, each drawing number consists of six or seven digits. The first digit is always 1, 2, 4, or 5, and indicates the size of the drawing. The remaining digits identify the drawing. Many firms have modified this basic system to conform to their particular needs. Letters may be used instead of numbers. The letter or number depicting the standard drawing size may be prefixed to the number, separated from it by a dash. Other numbering systems provide a separate box preceding the drawing number for the drawing size identifier. In another modification of this system, the part number of the depicted assembly is assigned as the drawing number.

Bill of Material

A list of the materials and parts necessary for the fabrication or assembly of a component or system is often included on the drawing. The list is usually in ruled columns in which are listed the part number, name of the part, material from which the part is to be constructed, the quantity required, and the source of the part or material. A typical bill of material is shown in Figure 2-9. On drawings that do not have a bill of material, the data may be indicated directly on the drawing.

On assembly drawings, each item is identified by a number in a circle or square. An arrow connecting the number with the item assists in locating it in the bill of material.

BILL OF MATERIAL			
ITEM	PART NO.	REQUIRED	SOURCE
CONNECTOR	UG-21D/U	2	STOCK

Figure 2-9. A bill of material.

Other Drawing Data

Revision Block

Revisions to a drawing are necessitated by changes in dimensions, design, or materials. The changes are usually listed in ruled columns either adjacent to the title block or at one corner of the drawing. All changes to approved drawings must be carefully noted on all existing prints of the drawing.

When drawings contain such corrections, attention is directed to the changes by lettering or numbering them and listing those changes against the symbol in a revision block. [Figure 2-10] The revision block contains the identification symbol, the date, the nature of the revision, the authority for the change, and the name of the draftsman who made the change.

To distinguish the corrected drawing from its previous version, many firms are including, as part of the title block, a space for entering the appropriate symbol to designate that the drawing has been changed or revised.

Notes

Notes are added to drawings for various reasons. Some of these notes refer to methods of attachment or construction. Others give alternatives, so that the drawing can be used for different styles of the same object. Still others list modifications that are available. Notes may be found alongside the item to which they refer. If the notes are lengthy, they may be placed elsewhere on the drawing and identified by letters or numbers. Notes are used only when the information cannot be conveyed in the conventional manner or when it is desirable to avoid crowding the drawing. Figure 2-3 illustrates one method of depicting notes.

When the note refers to a specific part, a light line with an arrowhead leads from the note to the part. If it applies to more than one part, the note is so worded to eliminate ambiguity as to the parts to which it pertains. If there are several notes, they are generally grouped together and numbered consecutively.

Zone Numbers

Zone numbers on drawings are similar to the numbers and letters printed on the borders of a map. They help locate a particular point. To find a point, mentally draw horizontal and vertical lines from the letters and numerals specified; the point where these lines intersect is the area sought.

Use the same method to locate parts, sections, and views on large drawings, particularly assembly drawings. Parts numbered in the title block can be located on the drawing by finding the numbers in squares along the lower border. Zone numbers read from right to left.

REV	ZONE	REVISION DESCRIPTION	DATE	APPR
A	ALL SHTS	INITIAL RELEASE	12/05/05	RL
B	PG2 C-2	ADDED ADDITIONAL MOUNTING POINTS	12/05/05	RL
C	PG2 A-1	ADDED ACCESS PANEL IN BULKHEAD	01/02/06	RL

Figure 2-10. Revision block.

Station Numbers and Location Identification on Aircraft

A numbering system is used on large assemblies for aircraft to locate stations such as fuselage stations. Fuselage station 185 indicates a location that is 185 inches from the datum of the aircraft. The measurement is usually taken from the nose or zero station, but in some instances it may be taken from the firewall or some other point chosen by the manufacturer. Just as forward and aft locations on aircraft are made by reference to the datum, locations left and right of the aircraft's longitudinal axis are made by reference to the buttock line and are called butt stations. Vertical locations on an airplane are made in reference to the waterline.

The same station numbering system is used for wing and stabilizer frames. The measurement is taken from the centerline or zero station of the aircraft. Figure 2-11 (on page 2-8) shows use of the fuselage stations (FS), waterline locations (WL), and left and right buttock line locations (RBL and LBL).

Allowances and Tolerances

When a given dimension on a print shows an allowable variation, the plus (+) figure indicates the maximum, and the minus (−) figure the minimum allowable variation. The sum of the plus and minus allowance figures is called tolerance. For example, using 0.225 + 0.0025 − 0.0005, the plus and minus figures indicate the part will be acceptable if it is not more than 0.0025 larger than the 0.225 given dimension, or not more than 0.0005 smaller than the 0.225 dimension. Tolerance in this example is 0.0030 (0.0025 max plus 0.0005 min).

If the plus and minus allowances are the same, you will find them presented as 0.224 ± 0.0025. The tolerance would then be 0.0050. Allowance can be indicated in either fractional or decimal form. When very accurate dimensions are necessary, decimal allowances are used. Fractional allowances are sufficient when precise tolerances are not required. Standard tolerances of −0.010 or −1/32 may be given in the title block of many drawings, to apply throughout the drawing.

Finish Marks

Finish marks are used to indicate the surface that must be machine finished. Such finished surfaces have a better appearance and allow a closer fit with adjoining parts. During the finishing process, the required limits and tolerances must be observed. Do not confuse machined finishes with those of paint, enamel, chromium plating, and similar coating.

Scale

Some drawings are made exactly the same size as the drawn part; they have a scale of 1:1. Other scales may be used. However, when drawings are made on a computer, drawing sizes may be easily increased (zoom in) or decreased (zoom out). Some electronic printers have the same capability. Furthermore, when a 1:1 copy of a print is made, the copy size may differ slightly from that of the original. For accurate information, refer to the dimensions shown on the drawing.

Application

When shown near the title block, application may refer to the aircraft, assembly, sub-assembly or next installation on which the part would be used.

Methods of Illustration

Applied Geometry

Geometry is the branch of mathematics that deals with lines, angles, figures and certain assumed properties in space. Applied geometry, as used in drawing, makes use of these properties to accurately and correctly represent objects graphically. In the past, draftsmen utilized a variety of instruments with various scales, shapes and curves to make their drawings. Today, computer software graphics programs showing drawings provide nearly any scale, shape and curve imaginable, outdating the need for additional instruments.

A number of methods are used to illustrate objects graphically. The most common are orthographic projections, pictorial drawings, diagrams, and flowcharts.

Orthographic Projection Drawings

In order to show the exact size and shape of all the parts of complex objects, a number of views are necessary. This is the system used in orthographic projection.

In orthographic projection, there are six possible views of an object, because all objects have six sides—front, top, bottom, rear, right side, and left side. Figure 2-12(a) shows an object placed in a transparent box, hinged at the edges. The projections on the sides of the box are the views as seen looking straight at the object through each side. If the outlines of the object are drawn on each surface and the box opened as shown in (b), then laid flat as shown in (c), the result is a six-view orthographic projection.

It is seldom necessary to show all six views to portray an object clearly; therefore, only those views necessary to illustrate the required characteristics of the object

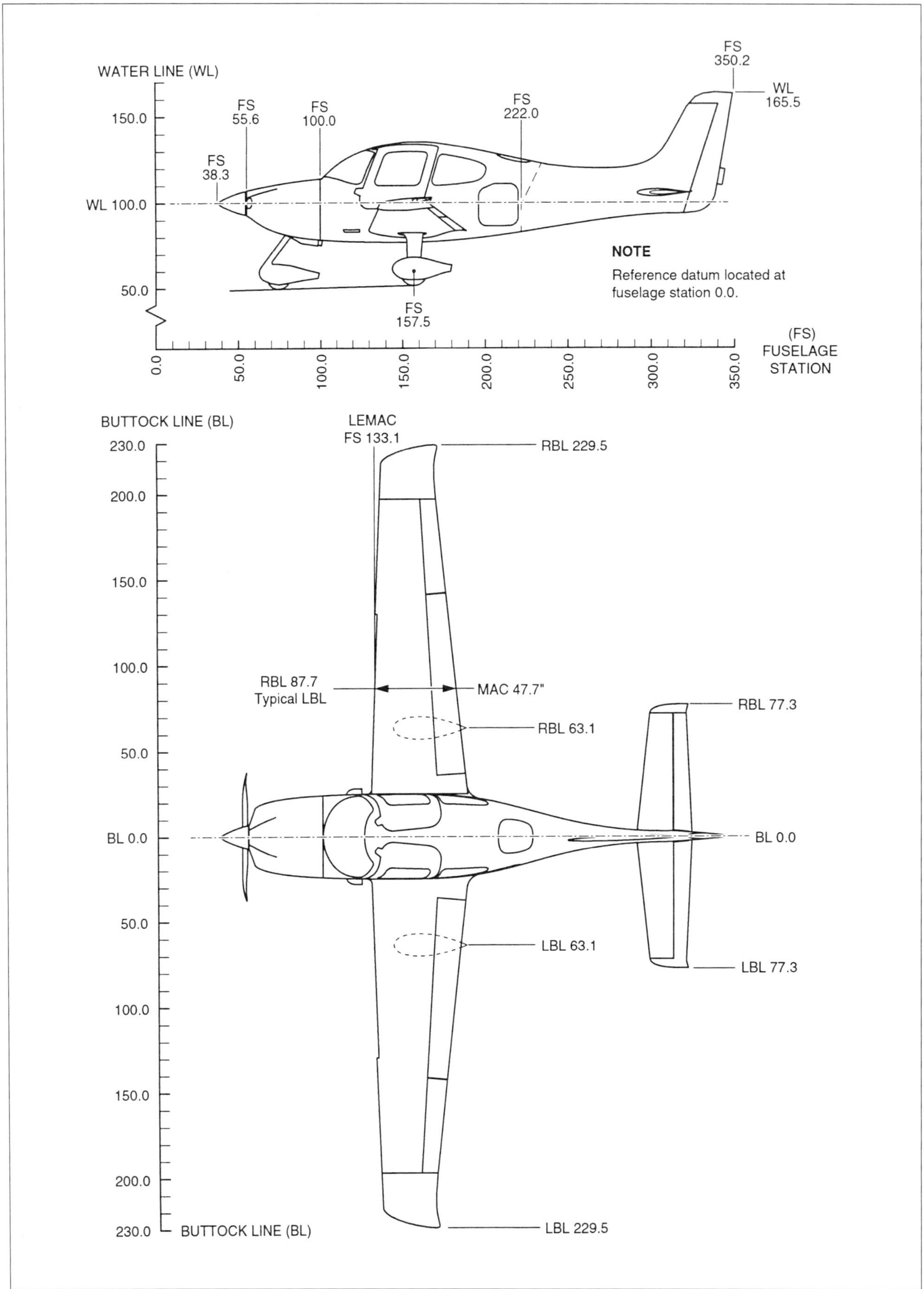

Figure 2-11. Station numbers and location identification on aircraft.

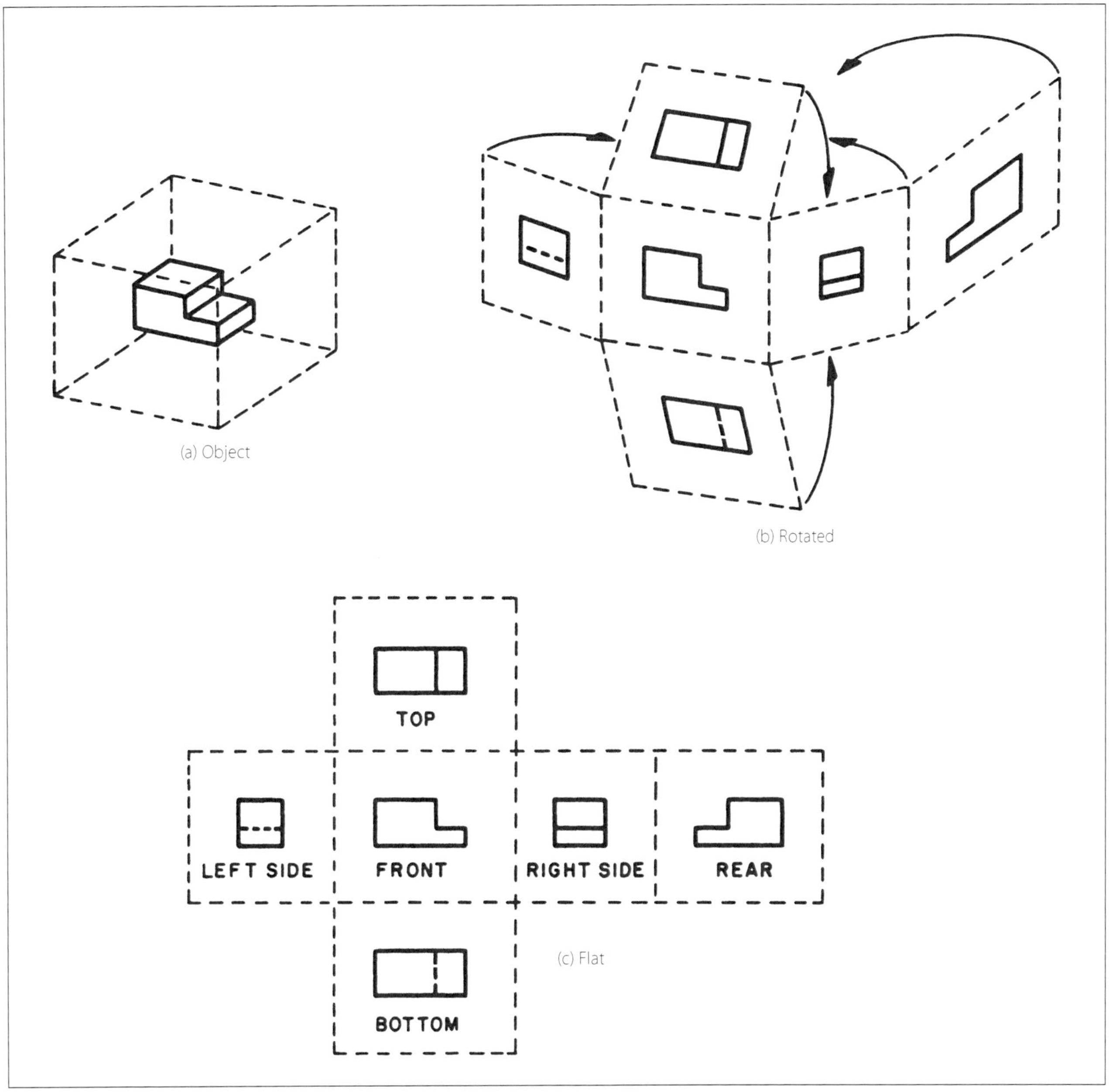

Figure 2-12. Orthographic projection.

are drawn. One-, two-, and three-view drawings are the most common. Regardless of the number of views used, the arrangement is generally as shown in Figure 2-12, with the front view as principal view. If the right side view is shown, it will be to the right of the front view. If the left side view is shown, it will be to the left of the front view. The top and bottom views, if included, will be shown in their respective positions relative to the front view.

One-view drawings are commonly used for objects of uniform thickness such as gaskets, shims, and plates. A dimensional note gives the thickness as shown in Figure 2-13. One-view drawings are also commonly used for cylindrical, spherical, or square parts if all the necessary dimensions can be properly shown in one view.

When space is limited and two views must be shown, symmetrical objects are often represented by half views, as illustrated in Figure 2-14.

Aircraft drawings seldom show more than two principal or complete views of an object. Instead, there will be usually one complete view and one or more detail views or sectional views.

Detail View

A detail view shows only a part of the object but in greater detail and to a larger scale than the principal

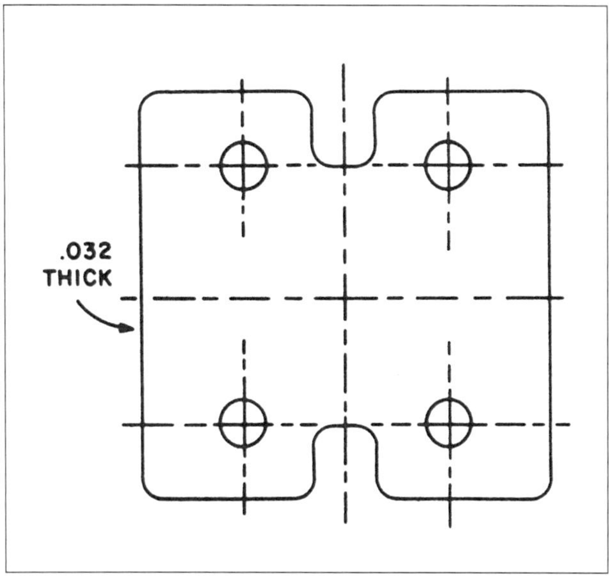

Figure 2-13. One view drawing.

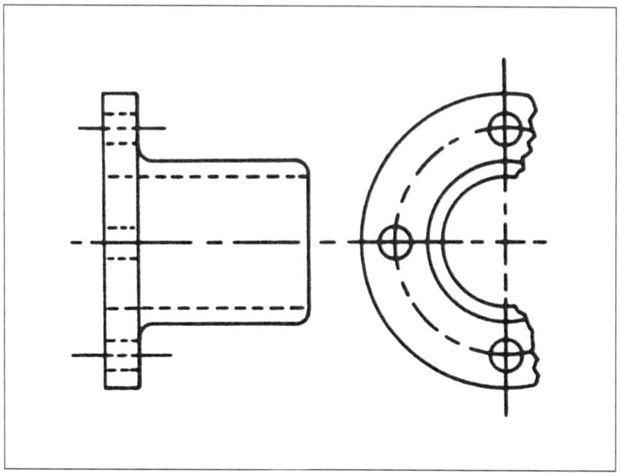

Figure 2-14. Symmetrical object with exterior half view.

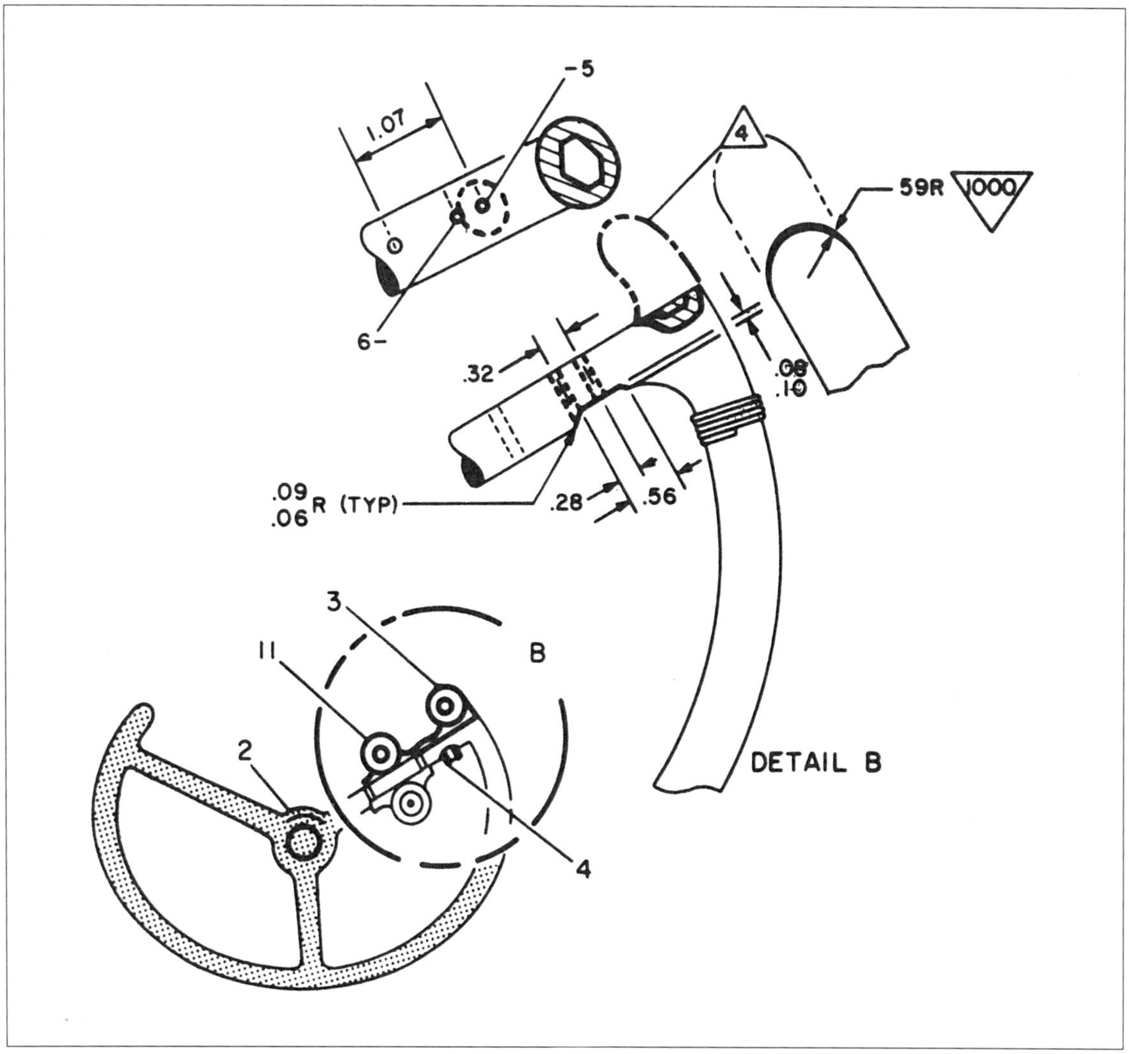

Figure 2-15. Detail view.

view. The part that is shown in detail elsewhere on the drawing is usually encircled by a heavy line on the principal view. Figure 2-15 is an example of the use of detail views.

The principal view shows the complete control wheel, while the detail view is an enlarged drawing of a portion of the control wheel.

Pictorial Drawings

A pictorial drawing [Figure 2-16] is similar to a photograph. It shows an object as it appears to the eye, but it is not satisfactory for showing complex forms and shapes. Pictorial drawings are useful in showing the general appearance of an object and are used extensively with orthographic projection drawings. Pictorial drawings are used in maintenance, overhaul, and part numbers. Three types of pictorial drawings are used frequently by aircraft engineers and technicians: (1) perspective, (2) isometric, and (3) oblique.

Perspective Drawings

A perspective view [Figure 2-17(a)] shows an object as it appears to an observer. It most closely resembles the way an object would look in a photograph. Because of perspective, some of the lines of an object are not parallel and therefore the actual angles and dimensions are not accurate.

Isometric Drawings

An isometric view [Figure 2-17(b)] uses a combination of the views of an orthographic projection and tilts the object forward so that portions of all three views can be seen in one view. This provides the observer with a three-dimensional view of the object. Unlike a perspective drawing where lines converge and dimensions are not true, lines in an isometric drawing are parallel and dimensioned as they are in an orthographic projection.

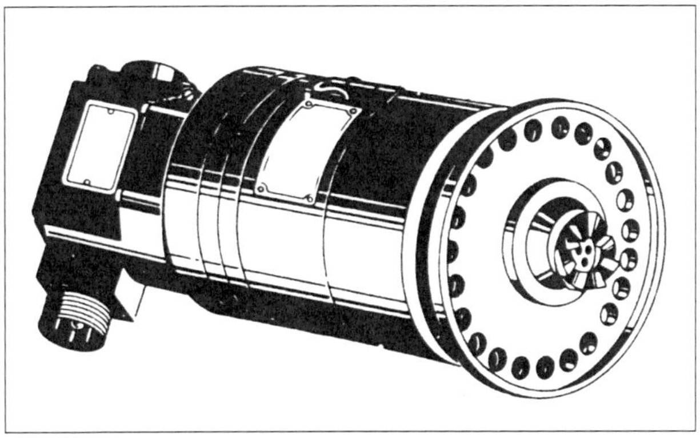

Figure 2-16. Pictorial drawing.

Oblique Drawings

An oblique view [Figure 2-17(c)] is similar to an isometric view except for one distinct difference. In an oblique drawing, two of the three drawing axes are always at right angles to each other.

Exploded View Drawings

An exploded view drawing is a pictorial drawing of two or more parts that fit together as an assembly. The view shows the individual parts and their relative position to the other parts before they are assembled.

Diagrams

A diagram may be defined as a graphic representation of an assembly or system, indicating the various parts and expressing the methods or principles of operation.

There are many types of diagrams; however, those with which the aviation mechanic will be concerned during the performance of his or her job may be grouped into four classes or types: (1) installation, (2) schematic, (3) block, and (4) wiring diagrams.

Installation Diagrams

Figure 2-18 is an example of an installation diagram. This is a diagram of the installation of the flight guid-

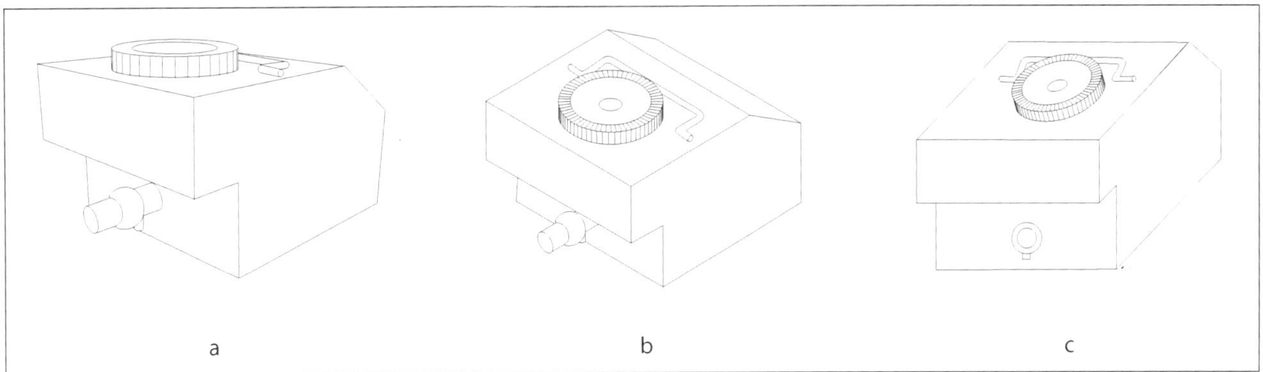

Figure 2-17. (a) Perspective, (b) isometric, and (c) oblique drawings.

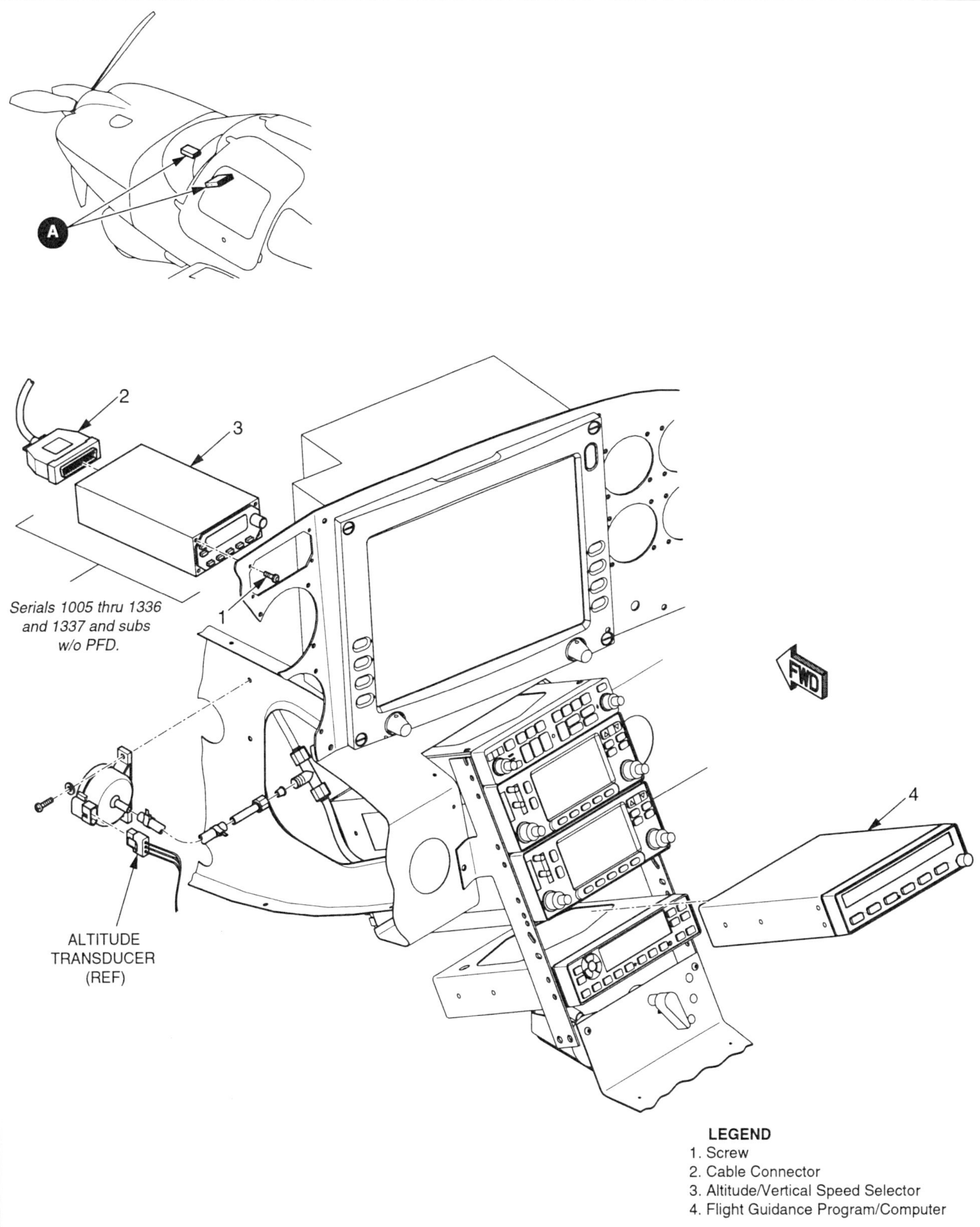

Figure 2-18. Example of installation diagram (flight guidance components).

ance control components of an aircraft. It identifies each of the components in the systems and shows their location in the aircraft. Each number (1, 2, 3, and 4) on the detail shows the location of the individual flight guidance system components within the cockpit of the aircraft. Installation diagrams are used extensively in aircraft maintenance and repair manuals, and are invaluable in identifying and locating components and understanding the operation of various systems.

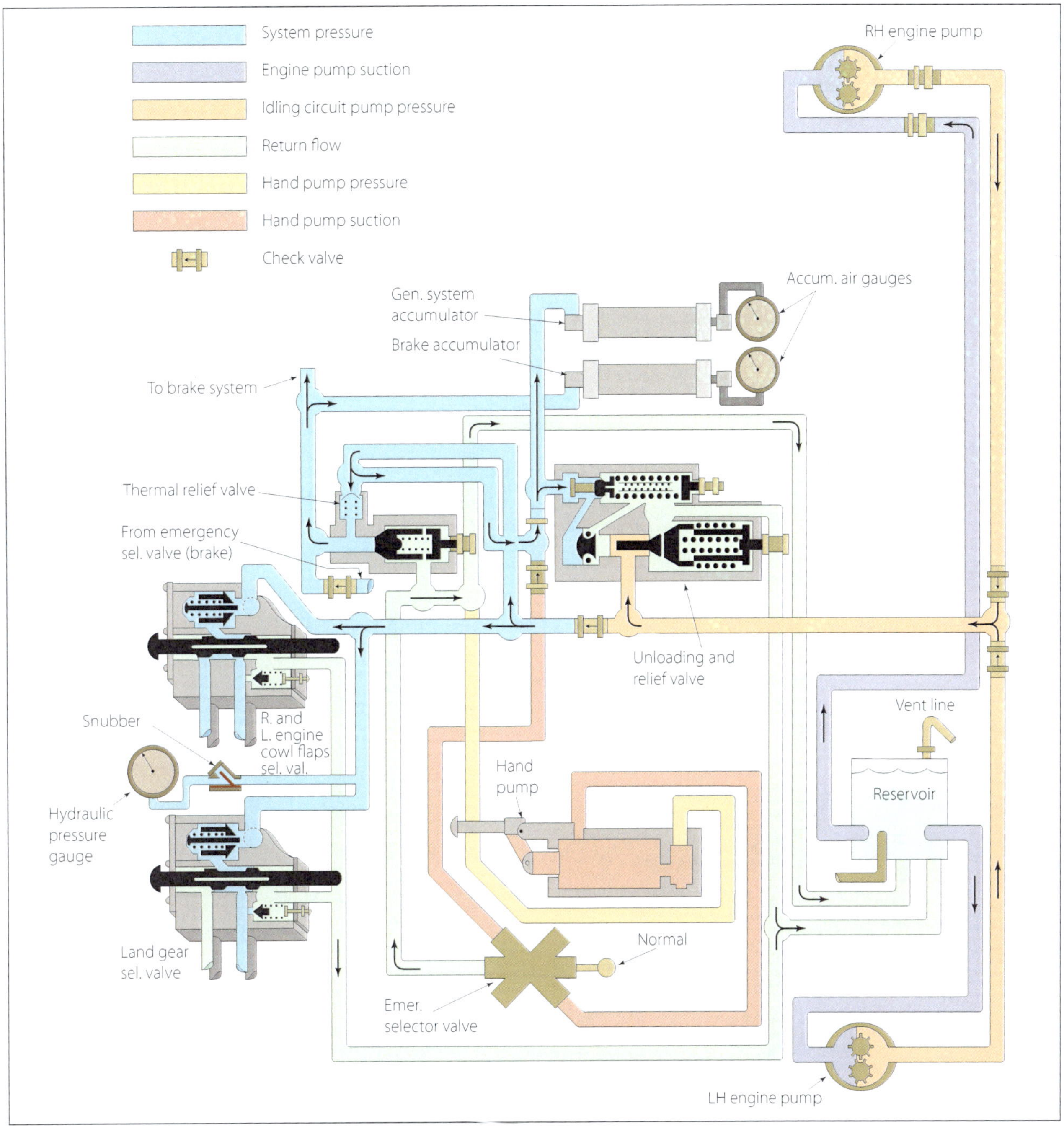

Figure 2-19. Aircraft hydraulic system schematic.

Schematic Diagrams

Schematic diagrams do not indicate the location of individual components in the aircraft, but locate components with respect to each other within the system. Figure 2-19 illustrates a schematic diagram of an aircraft hydraulic system. The hydraulic pressure gauge is not necessarily located above the landing gear selector valve in the aircraft. It is, however, connected to the pressure line that leads to the selector valve.

Schematic diagrams of this type are used mainly in troubleshooting. Note that each line is coded for ease of reading and tracing the flow. Each component is identified by name, and its location within the system can be ascertained by noting the lines that lead into and out of the unit.

Schematic diagrams and installation diagrams are used extensively in aircraft manuals.

Block Diagrams

Block diagrams [Figure 2-20] are used to show a simplified relationship of a more complex system of components. Individual components are drawn as

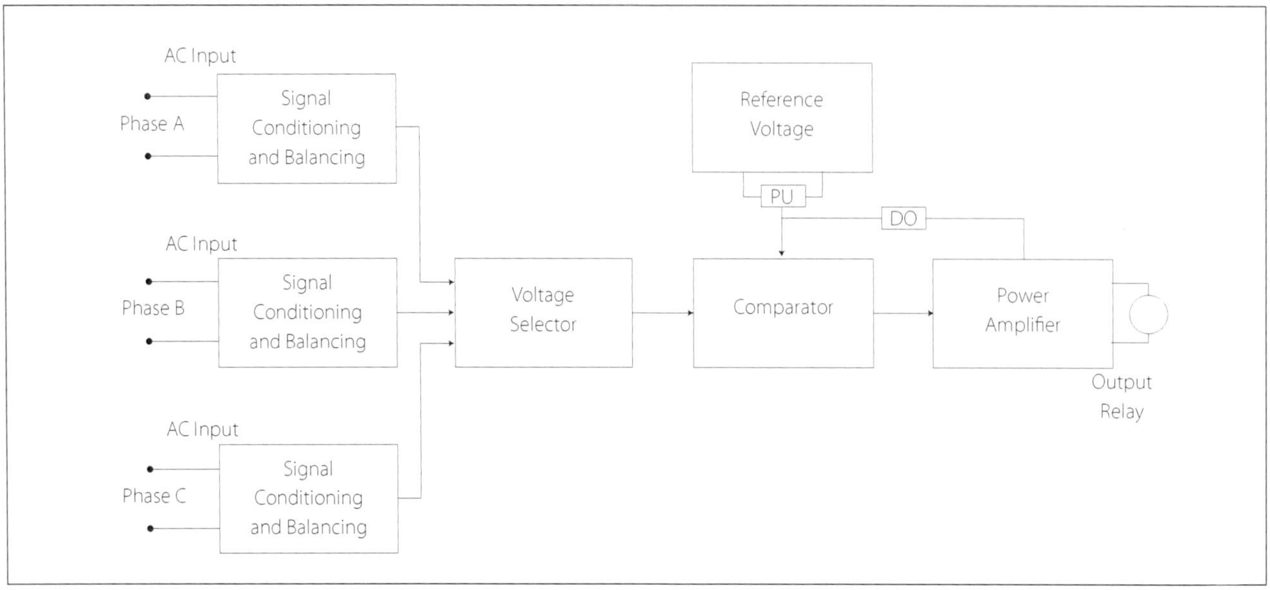

Figure 2-20. Block diagram.

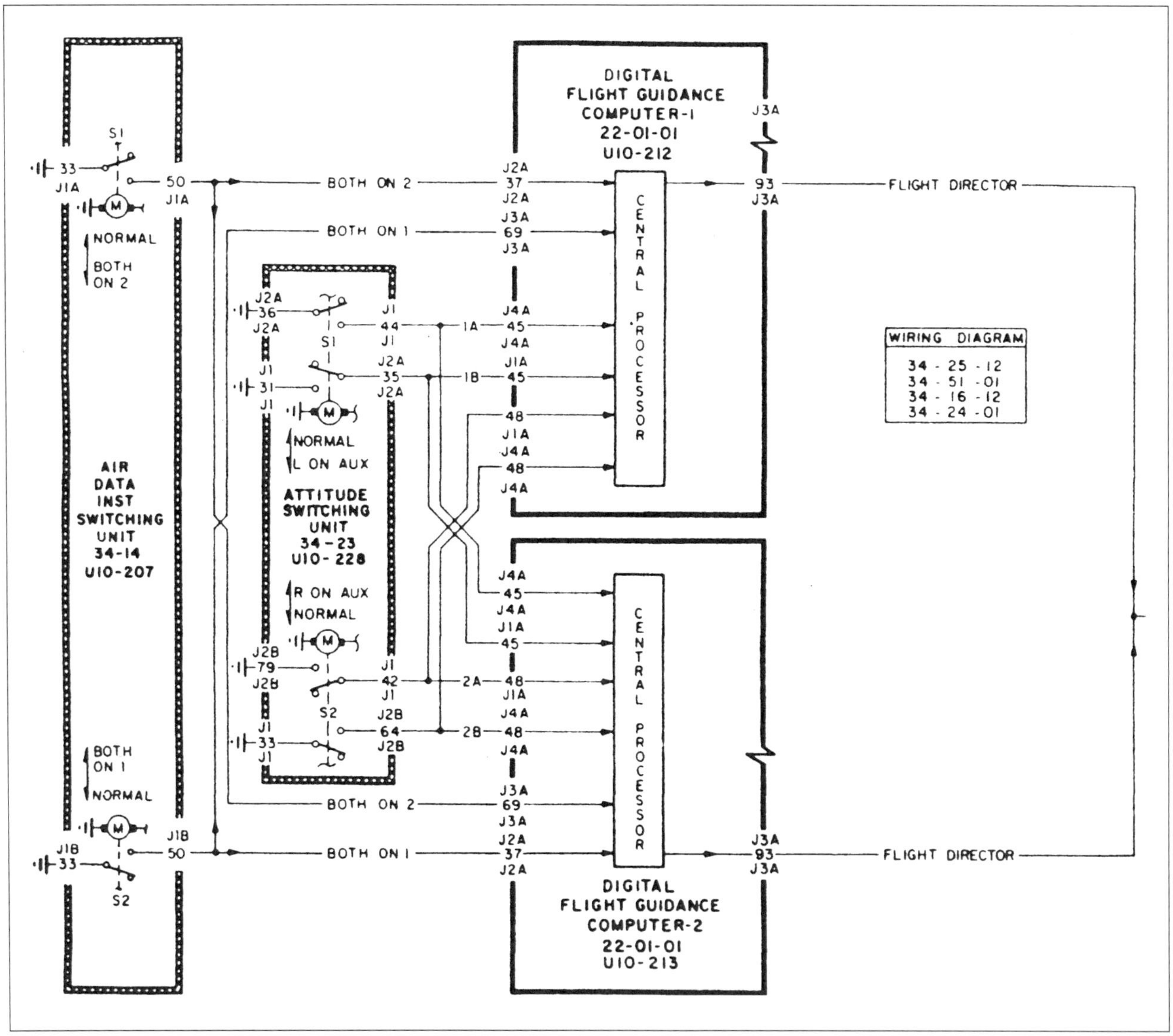

Figure 2-21. Wiring diagram.

a rectangle (block) with lines connecting it to other components (blocks) that it interfaces with during operation.

Wiring Diagrams

Wiring diagrams [Figure 2-21] show the electrical wiring and circuitry, coded for identification, of all the electrical appliances and devices used on aircraft. These diagrams, even for relatively simple circuits, can be quite complicated. For technicians involved with electrical repairs and installations, a thorough knowledge of wiring diagrams and electrical schematics is essential.

Flowcharts

Flowcharts are used to illustrate a particular sequence, or flow of events.

Troubleshooting Flowchart

Troubleshooting flowcharts are frequently used for the detection of faulty components. They often consist of a series of yes or no questions. If the answer to a question is yes, one course of action is followed. If the answer is no, a different course of action is followed. In this simple manner, a logical solution to a particular problem may be achieved. Another type of flowchart, developed specifically for analysis of digitally controlled components and systems, is the logic flowchart.

Logic Flowchart

A logic flowchart [Figure 2-22] uses standardized symbols to indicate specific types of logic gates and their relationship to other digital devices in a system. Since digital systems make use of binary mathematics consisting of 1s and 0s, voltage or no voltage, a light pulse or no light pulse, and so forth, logic flowcharts consist of individual components that take an input and provide an output which is either the same as the input or opposite. By analyzing the input or multiple inputs, it is possible to determine the digital output or outputs.

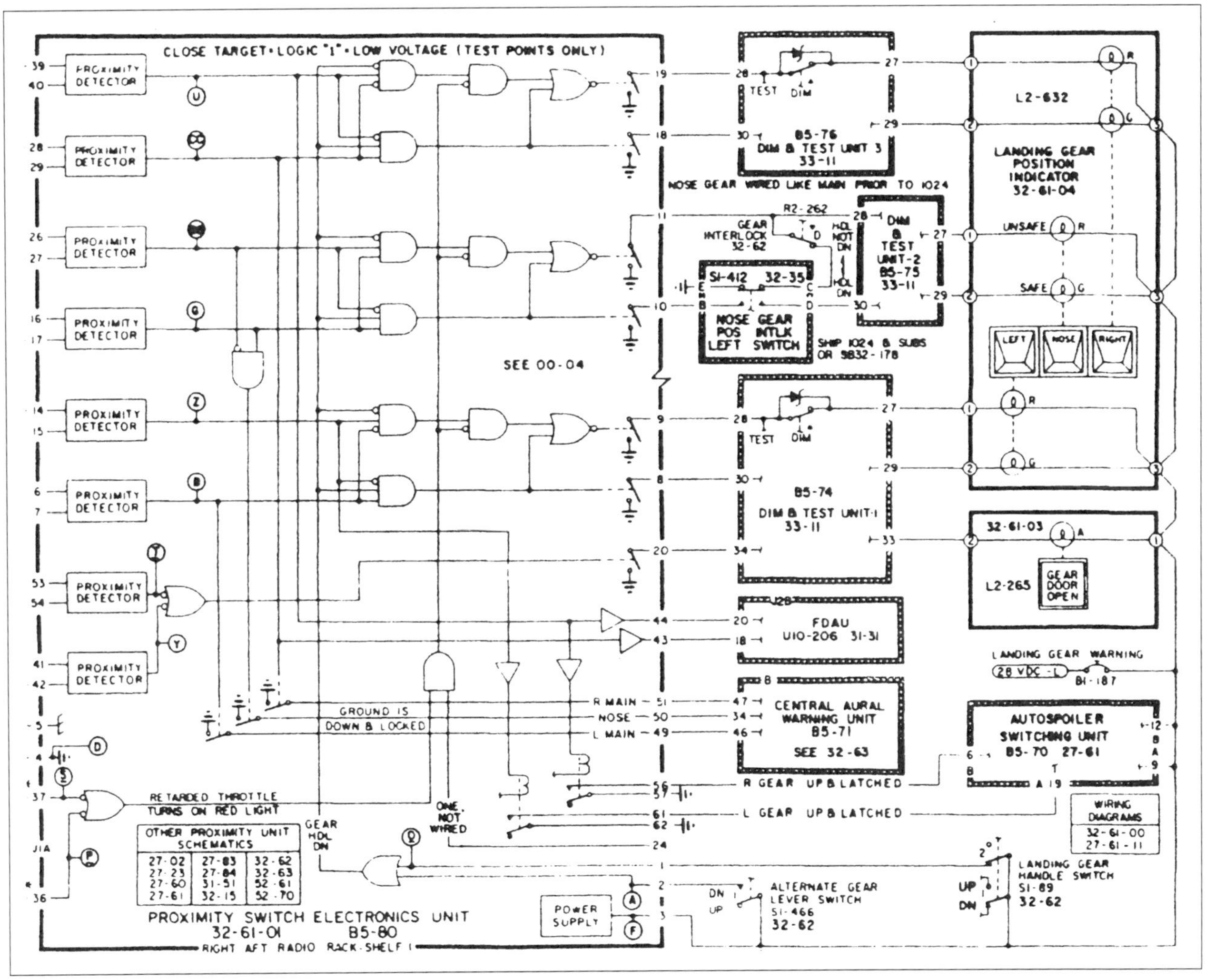

Figure 2-22. Logic flowchart.

Lines and Their Meanings

Every drawing is composed of lines. Lines mark the boundaries, edges, and intersection of surfaces. Lines are used to show dimensions and hidden surfaces and to indicate centers. Obviously, if the same kind of line is used to show all of these variations, a drawing becomes a meaningless collection of lines. For this reason, various kinds of standardized lines are used on aircraft drawings. These are illustrated in Figure 2-23, and their correct uses are shown in Figure 2-24.

Most drawings use three widths, or intensities, of lines: thin, medium, or thick. These lines may vary somewhat on different drawings, but there will always be a noticeable difference between a thin and a thick line, with the width of the medium line somewhere between the two.

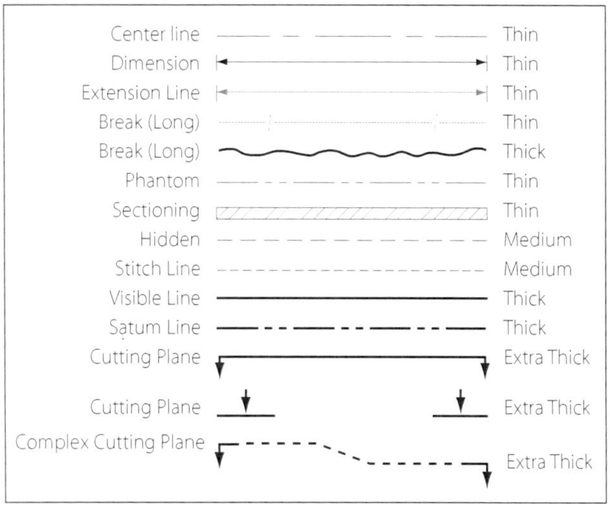

Figure 2-23. The meaning of lines.

Centerlines

Centerlines are made up of alternate long and short dashes. They indicate the center of an object or part of an object. Where centerlines cross, the short dashes intersect symmetrically. In the case of very small circles, the centerlines may be shown unbroken.

Dimension Lines

A dimension line is a light solid line, broken at the midpoint for insertion of measurement indications, and having opposite pointing arrowheads at each end to show origin and termination of a measurement. They are generally parallel to the line for which the dimension is given, and are usually placed outside the outline of the object and between views if more than one view is shown.

All dimensions and lettering are placed so that they will read from left to right. The dimension of an angle is indicated by placing the degree of the angle in its arc. The dimensions of circular parts are always given in terms of the diameter of the circle and are usually marked with the letter D or the abbreviation DIA following the dimension. The dimension of an arc is given in terms of its radius and is marked with the letter R following the dimension. Parallel dimensions are placed so that the longest dimension is farthest from the outline and the shortest dimension is closest to the outline of the object. On a drawing showing several views, the dimensions will be placed upon each view to show its details to the best advantage.

In dimensioning distances between holes in an object, dimensions are usually given from center to center rather than from outside to outside of the holes. When a

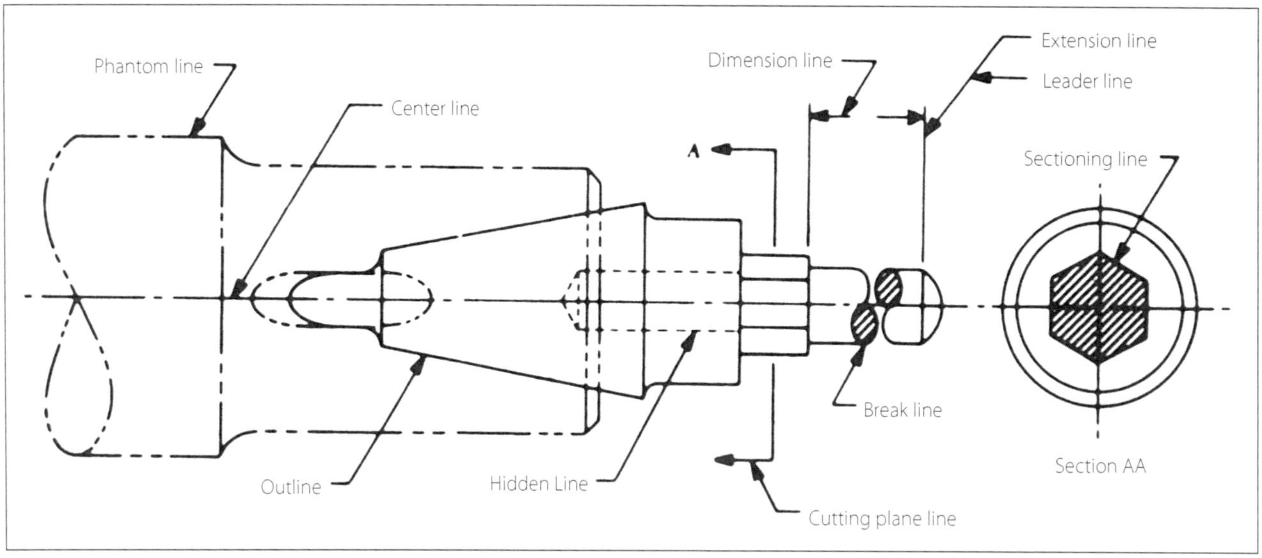

Figure 2-24. Correct use of lines.

number of holes of various sizes are shown, the desired diameters are given on a leader followed by notes indicating the machining operations for each hole. If a part is to have three holes of equal size, equally spaced, this information is explicitly stated. For precision work, sizes are given in decimals. Diameters and depths are given for counterbored holes. For countersunk holes, the angle of countersinking and the diameters are given. Study the examples shown in Figure 2-25.

The dimensions given for tolerances signifies the amount of clearance allowable between moving parts. A positive allowance is indicated for a part that is to slide or revolve upon another part. A negative allowance is one given for a force fit. Whenever possible, the tolerance and allowances for desired fits conform to those set up in the American Standard for Tolerances, Allowances, and Gauges for Metal Fits. The classes of fits specified in the standard may be indicated on assembly drawings.

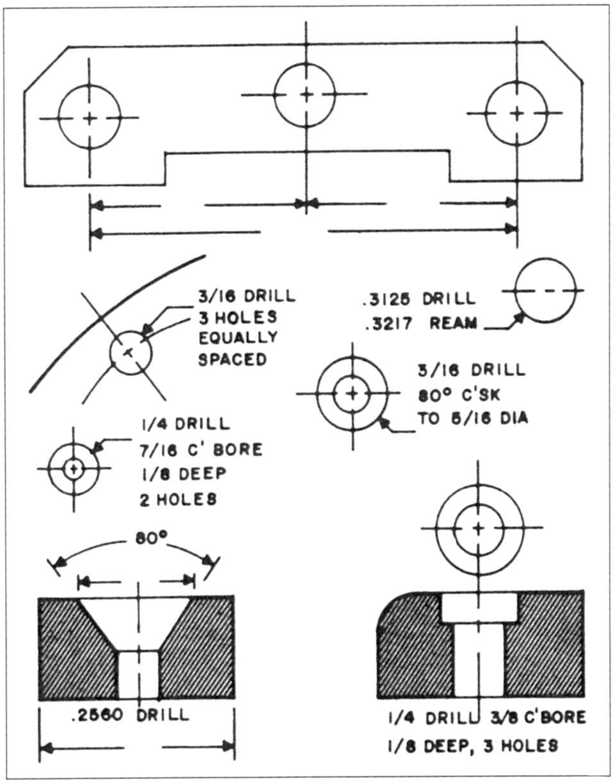

Figure 2-25. Dimensioning holes.

Extension Lines

Extensions are used to extend the line showing the side or edge of a figure for the purpose of placing a dimension to that side or edge. They are very narrow and have a short break where they extend from the object and extend a short distance past the arrow of the dimensioning line.

Sectioning Lines

Sectioning lines indicate the exposed surfaces of an object in sectional view. They are generally thin full lines but may vary with the kind of material shown in section.

Phantom Lines

Phantom lines, composed of one long and two short evenly spaced dashes, indicate the alternate position of parts of the object or the relative position of a missing part.

Break Lines

Break lines indicate that a portion of the object is not shown on the drawing. Short breaks are made by solid, freehand lines. For long breaks, solid ruled lines with zigzags are used. Shafts, rods, tubes, and other such parts which have a portion of their length broken out have the ends of the break drawn as indicated in Figure 2-24.

Leader Lines

Leader lines are solid lines with one arrowhead and indicate a part or portion to which a note, number, or other reference applies.

Hidden Lines

Hidden lines indicate invisible edges or contours. Hidden lines consist of short dashes evenly spaced and are frequently referred to as dash lines.

Outline or Visible Lines

The outline or visible line is used for all lines on the drawing representing visible lines on the object.

Stitch Lines

Stitch lines indicate stitching or sewing lines and consist of a series of evenly spaced dashes.

Cutting Plane and Viewing Plane Lines

Cutting plane lines indicate the plane in which a sectional view of the object is taken. In Figure 2-24, plane line A-A indicates the plane in which section A-A is taken.

Viewing plane lines indicate the plane from which a surface is viewed.

Drawing Symbols

The drawings for a component are composed largely of symbols and conventions representing its shape and material. Symbols are the shorthand of drawing. They graphically portray the characteristics of a component with a minimal amount of drawing.

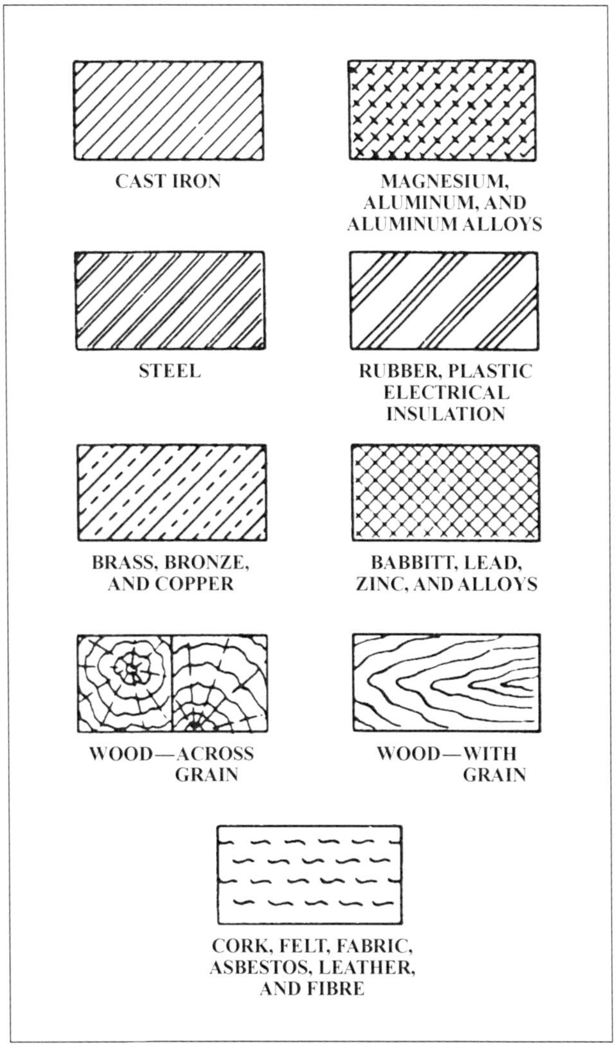

Figure 2-26. Standard material symbols.

Material Symbols

Section line symbols show the kind of material from which the part is to be constructed. The material may not be indicated symbolically if its exact specification is shown elsewhere on the drawing. In this case, the more easily drawn symbol for cast iron is used for the sectioning, and the material specification is listed in the bill of materials or indicated in a note. Figure 2-26 illustrates a few standard material symbols.

Shape Symbols

Symbols can be used to excellent advantage when needed to show the shape of an object. Typical shape symbols used on aircraft drawings are shown in Figure 2-27. Shape symbols are usually shown on a drawing as a revolved or removed section.

Electrical Symbols

Electrical symbols [Figure 2-28] represent various electrical devices rather than an actual drawing of the units. Having learned what the various symbols indicate, it becomes relatively simple to look at an electrical diagram and determine what each unit is, what function it serves, and how it is connected in the system.

Reading and Interpreting Drawings

Aircraft technicians do not necessarily need to be accomplished in making drawings. However, they must have a working knowledge of the information that is to be conveyed to them. They most frequently encounter drawings for construction and assembly of

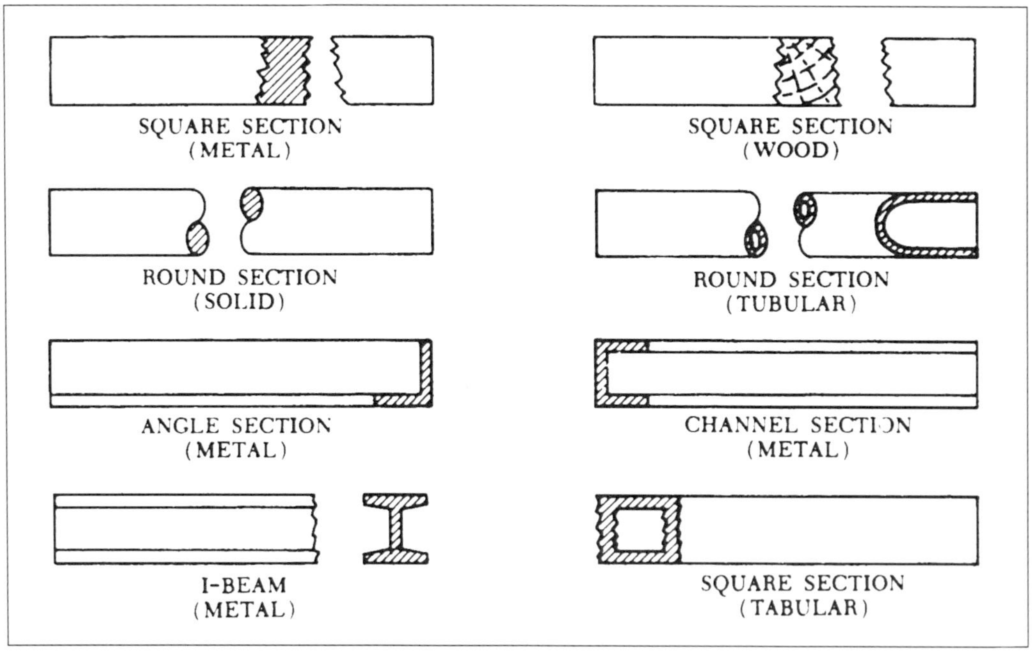

Figure 2-27. Shape symbols.

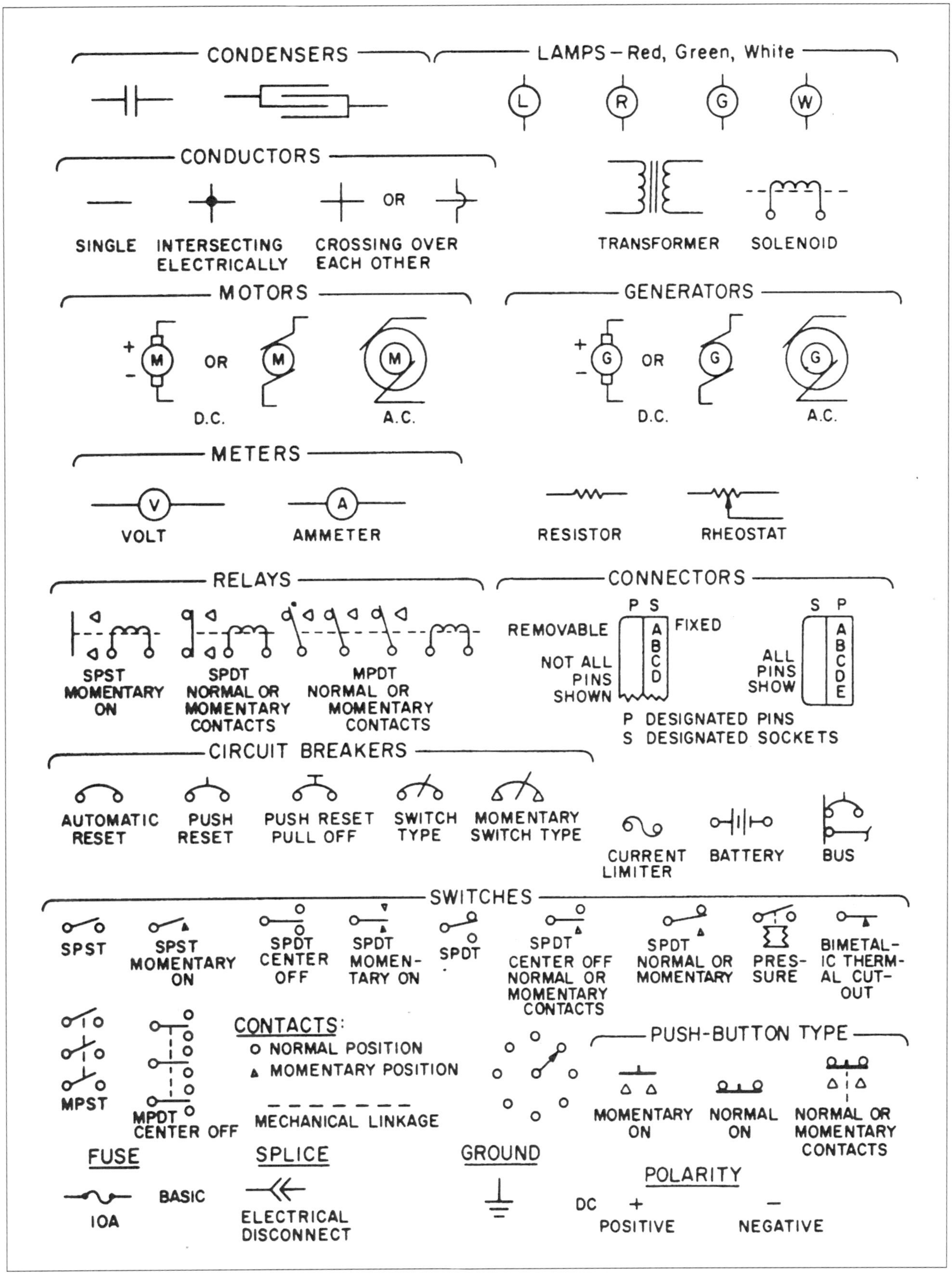

Figure 2-28. Electrical symbols.

new aircraft and components, during modifications, and for making repairs.

A drawing cannot be read all at once any more than a whole page of print can be read at a glance. Both must be read a line at a time. To read a drawing effectively, follow a systematic procedure.

Upon opening a drawing, read the drawing number and the description of the article. Next, check the model affected, the latest change letter, and the next assembly listed. Having determined that the drawing is the correct one, proceed to read the illustration(s).

In reading a multiview drawing, first get a general idea of the shape of the object by scanning all the views; then select one view for a more careful study. By referring back and forth to the adjacent view, it will be possible to determine what each line represents.

Each line on a view represents a change in the direction of a surface but another view must be consulted to determine what the change is. For example, a circle on one view may mean either a hole or a protruding boss, as in the top view of the object in Figure 2-29. Looking at the top view, we see two circles; however, the other view must be consulted to determine what each circle represents.

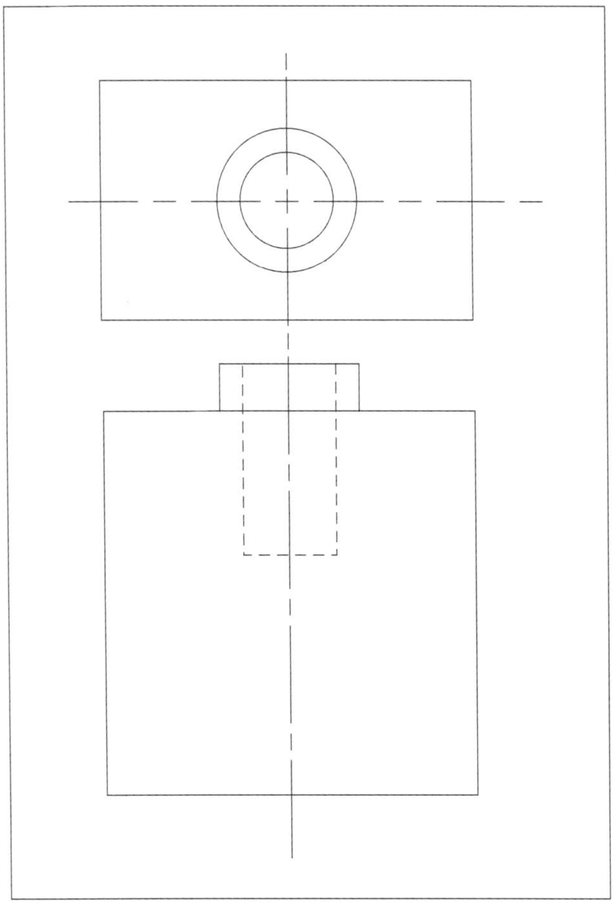

Figure 2-29. Reading views.

A glance at the other view tells us that the smaller circle represents a hole, and the larger circle represents a protruding boss. In the same way, the top view must be consulted to determine the shape of the hole and the protruding boss.

It can be seen from this example that one cannot read a print by looking at a single view when more than one view is given. Two views will not always describe an object and when three views are given, all three must be consulted to be sure the shape has been read correctly.

After determining the shape of an object, determine its size. Information on dimensions and tolerances is given so that certain design requirements may be met. Dimensions are indicated by figures either with or without the inch mark. If no inch mark is used, the dimension is in inches. It is customary to give part dimensions and an overall dimension that gives the greatest length of the part. If the overall dimension is missing, it can be determined by adding the separate part dimensions.

Drawings may be dimensioned in decimals or fractions. This is especially true in reference to tolerances. Instead of using plus and minus signs for tolerances, many figures give the complete dimension for both tolerances. For example, if a dimension is 2 inches with a plus or minus tolerance of 0.01, the drawing would show the total dimensions as:

2.01
1.99

A print tolerance (usually found in the title block) is a general tolerance that can be applied to parts where the dimensions are noncritical. Where a tolerance is not shown on a dimension line, the print tolerance applies.

To complete the reading of a drawing, read the general notes and the contents of the material block, check and find the various changes incorporated, and read the special information given in or near views and sections.

Drawing Sketches

A sketch is a simple rough drawing that is made rapidly and without much detail. Sketches may take many forms—from a simple pictorial presentation to a multi-view orthographic projection.

Just as aircraft technicians need not be highly skilled in making drawings, they need not be accomplished artists. However, in many situations, they will need to prepare a drawing to present an idea for a new design, a modification, or a repair method. The medium of sketching is an excellent way of accomplishing this.

The rules and conventional practices for making mechanical drawings are followed to the extent that all views needed to portray an object accurately are shown in their proper relationship. It is also necessary to observe the rules for correct line use [Figures 2-23 and 2-24] and dimensioning.

Sketching Techniques

To make a sketch, first determine what views are necessary to portray the object; then block in the views, using light construction lines. Next, complete the details, darken the object outline, and sketch extension and dimension lines. Complete the drawing by adding notes, dimensions, title, date, and when necessary, the sketcher's name. The steps in making a sketch of an object are illustrated in Figure 2-30.

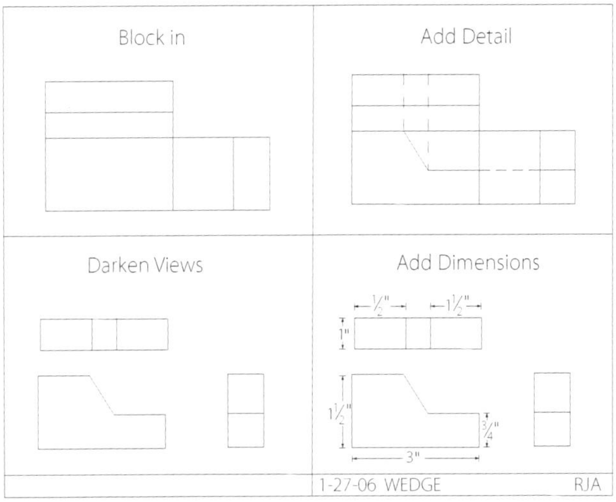

Figure 2-30. Steps in sketching.

Basic Shapes

Depending on the complexity of the sketch, basic shapes such as circles and rectangles may be drawn in freehand or by use of templates. If the sketch is quite complicated or the technician is required to make frequent sketches, use of a variety of templates and other drafting tools is highly recommended.

Repair Sketches

A sketch is frequently drawn for repairs or for use in manufacturing a replacement part. Such a sketch must provide all necessary information to those persons who must make the repair or manufacture the part.

The degree to which a sketch is complete will depend on its intended use. Obviously, a sketch used only to represent an object pictorially need not be dimensioned. If a part is to be manufactured from the sketch, it should show all the necessary construction details.

Care of Drafting Instruments

Good drawing instruments are expensive precision tools. Reasonable care given to them during their use and storage will prolong their service life.

T-squares, triangles, and scales should not be used or placed where their surfaces or edges may be damaged. Use a drawing board only for its intended purpose and not in a manner that will mar the working surface.

Compasses, dividers, and pens will provide better results with less annoyance, if they are correctly shaped and sharpened and are not damaged by careless handling.

Store drawing instruments in a place where they are not likely to be damaged by contact with other tools or equipment. Protect compass and divider points by inserting them into a piece of soft rubber or similar material. Never store ink pens without first cleaning and drying them thoroughly.

Graphs and Charts

Graphs and charts are frequently used to convey information graphically or information given certain conditions. They often utilize values shown on the x and y axes that can be projected up and across to arrive at a specific result. Also, when data is entered into a computer database, software programs can create a variety of different bar graphs, pie charts, and so forth, to graphically represent that data.

Reading and Interpreting Graphs and Charts

When interpreting information shown on graphs and charts, it is extremely important that all the notes and legend information be carefully understood in order to eliminate any misinterpretation of the information presented.

Nomograms

A nomogram is a graph that usually consists of three sets of data. Knowledge of any two sets of data enables the interpreter to obtain the value for the third unknown corresponding value. One type of nomogram consists of three parallel scales graduated for different variables so that when a straight edge connects any two values, the third can be read directly. Other types may use values on the x and y axes of a graph with the third corresponding value determined by the intersection of the x and y values with one of a series of curved lines. Figure 2-31 is an example of a nomogram that shows the relationship between aviation fuels, specific weight, and temperature.

Microfilm and Microfiche

The practice of recording drawings, parts catalogs, and maintenance and overhaul manuals on microfilms was utilized extensively in the past. Microfilm is available as regular 16 mm or 35 mm film. Since 35 mm film is larger, it provides a better reproduction of drawings. Microfiche is a card with pages laid out in a grid format. Microfilm and microfiche require use of special devices for both reading and printing the information.

Most modern aircraft manufacturers have replaced microfilm and microfiche with digital storage methods utilizing CDs, DVDs and other data storage devices. A great deal of service and repair information for older aircraft has been transferred to digital storage devices. However, there may still be a need to access information using the old methods. A well-equipped shop should have available, both the old microfilm and microfiche equipment, as well as new computer equipment.

Digital Images

Though not a drawing, a digital image created by a digital camera can be extremely helpful to aviation maintenance technicians in evaluating and sharing information concerning the airworthiness or other information about aircraft. Digital images can be rap-

Density Variation of Aviation Fuel
Based on Average Specific Gravity

Fuel	Average Specific Gravity ar 15 °C (59 °F)
Aviation Kerosene Jet A and Jet A1	.812
Jet B (JP-4)	.785
AV Gas Grade 100/130	.703

Note: The fuel quantity indicator is calibrated for correct indication when using Aviation Kerosene Jet A and Jet A1. When using other fuels, multiply the indicated fuel quantity in pounds by .99 for Jet B (JP-4) or by .98 for Aviation Gasoline (100/130) to obtain actual fuel quantity in pounds.

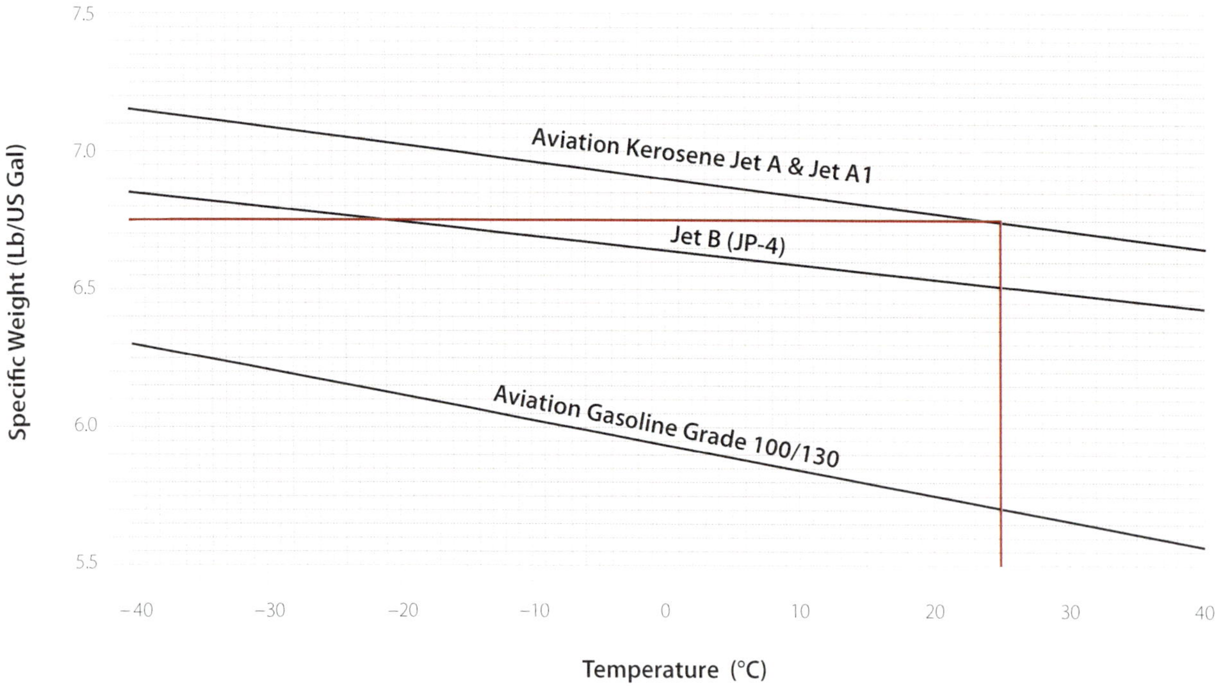

Figure 2-31. Nomogram.

Figure 2-32. Digital image of impact damage.

idly transmitted over the World Wide Web as attachments to e-mail messages. Images of structural fatigue cracks, failed parts, or other flaws, as well as desired design and paint schemes, are just a few examples of the types of digital images that might be shared by any number of users over the Internet. Figure 2-32 is a digital image of impact damage to a composite structure taken with a simple digital camera. To provide information about the extent of the damage, a measurement scale, or other object, such as a coin, can be placed near the area of concern before the picture is taken. Also, within the text of the e-mail, the technician should state the exact location of the damage, referenced to fuselage station, wing station, and so forth.

Physical science, or physics as it is most often called, is a very interesting and exciting topic. For the individual who likes technical things and is a hands-on type person, it is an invaluable tool. Physics allows us to explain how engines work, both piston and gas turbine; how airplanes and helicopters fly; and countless other things related to the field of aviation and aerospace. In addition to allowing us to explain the operation of the things around us, it also allows us to quantify them. For example, through the use of physics we can explain what the concept of thrust means for a jet engine, and then follow it up by mathematically calculating the pounds of thrust being created.

Physics is the term applied to that area of knowledge regarding the basic and fundamental nature of matter and energy. It does not attempt to determine why matter and energy behave as they do in their relation to physical phenomena, but rather how they behave. The people who maintain and repair aircraft should have a knowledge of basic physics, which is sometimes called the science of matter and energy.

Matter

Matter is the foundation or the building blocks for any discussion of physics. According to the dictionary, matter is what all things are made of; whatever occupies space, has mass, and is perceptible to the senses in some way. According to the Law of Conservation, matter cannot be created or destroyed, but it is possible to change its physical state. When liquid gasoline vaporizes and mixes with air, and then burns, it might seem that this piece of matter has disappeared and no longer exists. Although it no longer exists in the state of liquid gasoline, the matter still exists in the form of the gases given off by the burning fuel.

Characteristics of Matter

Mass and Weight

Mass is a measure of the quantity of matter in an object. In other words, how many molecules are in the object, or how many atoms, or to be more specific, how many protons, neutrons, and electrons. The mass of an object does not change regardless of where you take it in the universe, and it does not change with a change of state. The only way to change the mass of an object is to add or take away atoms. Mathematically, mass can be stated as follows:

$$\text{Mass} = \text{Weight} \div \text{Acceleration due to gravity}$$

The acceleration due to gravity here on earth is 32.2 feet per second per second (32.2 fps/s). An object weighing 32.2 pounds (lb) here on earth is said to have a mass of 1 slug. A slug is a quantity of mass that will accelerate at a rate of 1 ft/s^2 when a force of 1 pound is applied. In other words, under standard atmospheric conditions (gravity equal to 32.2) a mass of one slug is equal to 32.2 lb.

Weight is a measure of the pull of gravity acting on the mass of an object. The more mass an object has, the more it will weigh under the earth's force of gravity. Because it is not possible for the mass of an object to go away, the only way for an object to be weightless is for gravity to go away. We view astronauts on the space shuttle and it appears that they are weightless. Even though the shuttle is quite a few miles above the surface of the earth, the force of gravity has not gone away, and the astronauts are not weightless. The astronauts and the space shuttle are in a state of free fall, so relative to the shuttle the astronauts appear to be weightless. Mathematically, weight can be stated as follows:

$$\text{Weight} = \text{Mass} \times \text{Gravity}$$

Attraction

Attraction is the force acting mutually between particles of matter, tending to draw them together. Sir Isaac Newton called this the "Law of Universal Gravitation." He showed how each particle of matter attracts every other particle, how people are bound to the earth, and how the planets are attracted in the solar system.

Porosity

Porosity means having pores or spaces where smaller particles may fit when a mixture takes place. This is sometimes referred to as granular—consisting or appearing to consist of small grains or granules.

Impenetrability

Impenetrability means that no two objects can occupy the same place at the same time. Thus, two portions of matter cannot at the same time occupy the same space.

Density

The density of a substance is its weight per unit volume. The unit volume selected for use in the English system of measurement is 1 cubic foot (ft^3). In the metric system, it is 1 cubic centimeter (cm^3). Therefore, density is expressed in pounds per cubic foot ($^{lb}/_{ft^3}$) or in grams per cubic centimeter ($^{g}/_{cm^3}$).

To find the density of a substance, its weight and volume must be known. Its weight is then divided by its volume to find the weight per unit volume. For example, the liquid which fills a certain container weighs 1,497.6 lb. The container is 4 ft long, 3 ft wide, and 2 ft deep. Its volume is 24 ft^3 (4 ft × 3 ft × 2 ft). If 24 ft^3 of liquid weighs 1,497.6 lb, then 1 ft^3 weighs 1,497.6 ÷ 24, or 62.4 lb. Therefore, the density of the liquid is 62.4 lb/ft^3. This is the density of water at 4°C (Centigrade) and is usually used as the standard for comparing densities of other substances. In the metric system, the density of water is $1\,^{g}/_{cm^3}$. The standard temperature of 4°C is used when measuring the density of liquids and solids. Changes in temperature will not change the weight of a substance, but will change the volume of the substance by expansion or contraction, thus changing its weight per unit volume.

The procedure for finding density applies to all substances; however, it is necessary to consider the pressure when finding the density of gases. Pressure is more critical when measuring the density of gases than it is for other substances. The density of a gas increases in direct proportion to the pressure exerted on it. Standard conditions for the measurement of the densities of gases have been established at 0°C for temperature and a pressure of 76 cm of mercury (Hg). (This is the average pressure of the atmosphere at sea level.) Density is computed based on these conditions for all gases.

Specific Gravity

It is often necessary to compare the density of one substance with that of another. For this purpose, a standard is needed. Water is the standard that physicists have chosen to use when comparing the densities of all liquids and solids. For gases, air is most commonly used. However, hydrogen is sometimes used as a standard for gases. In physics, the word "specific" implies a ratio. Thus, specific gravity is calculated by comparing the weight of a definite volume of the given substance with the weight of an equal volume of water. The terms "specific weight" or "specific density" are sometimes used to express this ratio.

The following formulas are used to find the specific gravity of liquids and solids.

$$\text{Specific Gravity} = \frac{\text{Weight of the substance}}{\text{Weight of an equal volume of water}}$$

OR

$$\text{Specific Gravity} = \frac{\text{Density of the substance}}{\text{Density of water}}$$

The same formulas are used to find the density of gases by substituting air or hydrogen for water.

Specific gravity is not expressed in units, but as pure numbers. For example, if a certain hydraulic fluid has a specific gravity of 0.8, 1 ft^3 of the liquid weighs 0.8 times as much as 1 ft^3 of water: 62.4 times 0.8, or 49.92 lb.

Specific gravity and density are independent of the size of the sample under consideration and depend only upon the substance of which it is made. See Figure 3-1 for typical values of specific gravity for various substances.

A device called a hydrometer is used for measuring specific gravity of liquids. This device consists of a tubular glass float contained in a larger glass tube. [Figure 3-2] The larger glass tube provides the container for the liquid. A rubber suction bulb draws the liquid up into the container. There must be enough liquid to raise the float and prevent it from touching the bottom. The float is weighted and has a vertically graduated scale. To determine specific gravity, the scale is read at the surface of the liquid in which the float is immersed. An indication of 1000 is read when the float is immersed in pure water. When immersed in a liquid of greater density, the float rises, indicating a greater specific gravity. For liquids of lesser density the float sinks, indicating a lower specific gravity.

An example of the use of the hydrometer is to determine the specific gravity of the electrolyte (battery

Liquid	Specific Gravity	Solid	Specific Gravity	Gas	Specific Gravity
Gasoline	0.72	Ice	0.917	Hydrogen	0.0695
Jet Fuel Jp-4	0.785	Aluminum	2.7	Helium	0.138
Ethyl Alcohol	0.789	Titanium	4.4	Acetylene	0.898
Jet Fuel Jp-5	0.82	Zinc	7.1	Nitrogen	0.967
Kerosene	0.82	Iron	7.9	Air	1.000
Lube Oil	0.89	Brass	8.4	Oxygen	1.105
Synthetic Oil	0.928	Copper	8.9	Carbon Dioxide	1.528
Water	1.000	Lead	11.4		
Sulfuric Acid	1.84	Gold	19.3		
Mercury	13.6	Platinum	21.5		

Figure 3-1. Specific gravity of various substances.

liquid) in an aircraft battery. When a battery is discharged, the calibrated float immersed in the electrolyte will indicate approximately 1150. The indication of a charged battery is between 1275 and 1310. The values 1150, 1275, and 1310 actually represent 1.150, 1.275, and 1.310. The electrolyte in a discharged battery is 1.15 times denser than water, and in a charged battery 1.275 to 1.31 times denser than water.

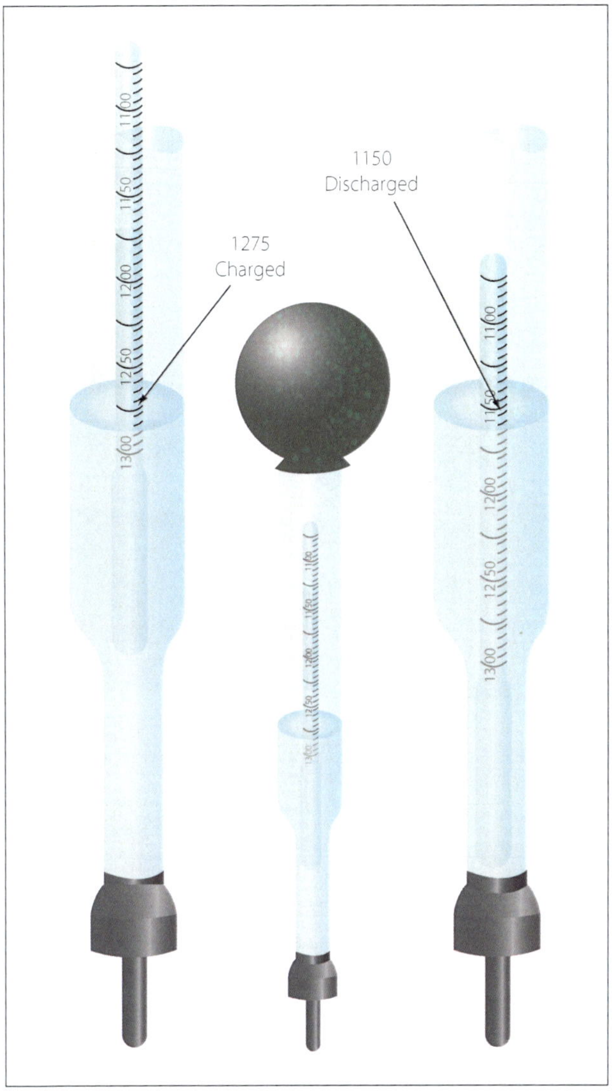

Figure 3-2. Hydrometer for checking battery specific gravity.

Energy

Energy is typically defined as something that gives us the capacity to perform work. As individuals, saying that we feel full of energy is probably indicating that we can perform a lot of work. Energy can be classified as one of two types: either potential or kinetic.

Potential Energy

Potential energy is defined as being energy at rest, or energy that is stored. Potential energy may be classified into three groups: (1) that due to position, (2) that due to distortion of an elastic body, and (3) that which produces work through chemical action. Water in an elevated reservoir, and an airplane raised off the ground sitting on jacks are examples of the first group; a stretched bungee chord on a Piper Tri-Pacer or compressed spring are examples of the second group; and energy in aviation gasoline, food, and storage batteries are examples of the third group.

To calculate the potential energy of an object due to its position, as in height, the following formula is used:

$$\text{Potential Energy} = \text{Weight} \times \text{Height}$$

A calculation based on this formula will produce an answer that has units of foot-pounds (ft-lb) or inch-pounds (in-lb), which are the same units that apply to work. Work, which is covered later in this chapter, is described as a force being applied over a measured distance, with the force being pounds and the distance being feet or inches. It can be seen that potential energy and work have a lot in common.

Example: *A Boeing 747 weighing 450,000 pounds needs to be raised 4 feet in the air so maintenance can be done on the landing gear. How much potential energy does the airplane possess because of this raised position?*

Potential Energy = Weight × Height
PE = 450,000 lb × 4 ft
PE = 1,800,000 ft-lb

As mentioned previously, aviation gasoline possesses potential energy because of its chemical nature. Gasoline has the potential to release heat energy, based on its British thermal unit (BTU) content. One pound of aviation gas contains 18,900 BTU of heat energy, and each BTU is capable of 778 ft-lb of work. So if we multiply 778 by 18,900, we find that one pound of aviation gas is capable of 14,704,200 ft-lb of work. Imagine the potential energy in the completely serviced fuel tanks of an airplane.

Kinetic Energy

Kinetic energy is defined as being energy in motion. An airplane rolling down the runway or a rotating flywheel on an engine are both examples of kinetic energy. Kinetic energy has the same units as potential energy, namely foot-pounds or inch-pounds. To calculate the kinetic energy for something in motion, the following formula is used:

$$\text{Kinetic Energy} = \tfrac{1}{2}\,\text{Mass} \times \text{Velocity}^2$$

To use the formula, we will show the mass as weight ÷ gravity and the velocity of the object will be in feet per second. This is necessary to end up with units in foot-pounds.

Example: *A Boeing 777 weighing 600,000 lb is moving down the runway on its takeoff roll with a velocity of 200 fps. How many foot-pounds of kinetic energy does the airplane possess? [Figure 3-3]*

$$\text{Kinetic Energy} = \tfrac{1}{2}\,\text{Mass} \times \text{Velocity}^2$$
$$\text{Kinetic Energy} = \tfrac{1}{2} \times 600{,}000 \div 32.2 \times 200^2$$
$$\text{KE} = 372{,}670{,}000 \text{ ft-lb}$$

Figure 3-3. Kinetic energy (Boeing 777 taking off).

Force, Work, Power, and Torque

Force

Before the concept of work, power, or torque can be discussed, we must understand what force means. According to the dictionary, force is the intensity of an impetus, or the intensity of an input. For example, if we apply a force to an object, the tendency will be for the object to move. Another way to look at it is that for work, power, or torque to exist, there has to be a force that initiates the process.

The unit for force in the English system of measurement is pounds, and in the metric system it is newtons. One pound of force is equal to 4.448 newtons. When we calculate the thrust of a turbine engine, we use the formula "Force = Mass × Acceleration," and the thrust of the engine is expressed in pounds. The GE90-115 turbofan engine (powerplant for the Boeing 777-300), for example, has 115,000 pounds of thrust.

Work

The study of machines, both simple and complex, is in one sense a study of the energy of mechanical work. This is true because all machines transfer input energy, or the work done on the machine, to output energy, or the work done by the machine.

Work, in the mechanical sense of the term, is done when a resistance is overcome by a force acting through a measurable distance. Two factors are involved: (1) force and (2) movement through a distance. As an example, suppose a small aircraft is stuck in the snow. Two men push against it for a period of time, but the aircraft does not move. According to the technical definition, no work was done in pushing against the aircraft. By definition, work is accomplished only when an object is displaced some distance against a resistive force. To calculate work, the following formula is used:

$$\text{Work} = \text{Force (F)} \times \text{distance (d)}$$

In the English system, the force will be identified in pounds and the distance either in feet or inches, so the units will be foot-pounds or inch-pounds. Notice

Figure 3-4. Airbus A-320 being jacked.

these are the same units that were used for potential and kinetic energy.

In the metric system, the force is identified in newtons (N) and the distance in meters, with the resultant units being joules. One pound of force is equal to 4.448 N and one meter is equal to 3.28 feet. One joule is equal to 1.36 ft-lb.

Example: *How much work is accomplished by jacking a 150,000-lb Airbus A-320 airplane a vertical height of 3 ft? [Figure 3-4]*

Work = Force × Distance
= 150,000 lb × 4 ft
= 600,000 ft-lb

Example: *How much work is accomplished when a tow tractor is hooked up to a tow bar and a Boeing 737-800 airplane weighing 130,000 lb is pushed 80 ft into the hangar? The force on the tow bar is 5,000 lb.*

Work = Force × Distance
= 5,000 lb × 80 ft
= 400,000 ft-lb

In this last example, notice the force does not equal the weight of the airplane. This is because the airplane is being moved horizontally and not lifted vertically. In virtually all cases, it takes less work to move something horizontally than it does to lift it vertically. Most people can push their car a short distance if it runs out of gas, but they cannot get under their car and lift it off the ground.

Friction and Work

In calculating work done, the actual resistance overcome is measured. This is not necessarily the weight of the object being moved. [Figure 3-5] A 900-lb load is being pulled a distance of 200 ft. This does not mean that the work done (force × distance) is 180,000 ft-lb (900 lb × 200 ft). This is because the person pulling the load is not working against the total weight of the load, but rather against the rolling friction of the cart, which may be no more than 90 lb.

Friction is an important aspect of work. Without friction it would be impossible to walk. One would have to shove oneself from place to place, and would have to bump against some obstacle to stop at a destination. Yet friction is a liability as well as an asset, and

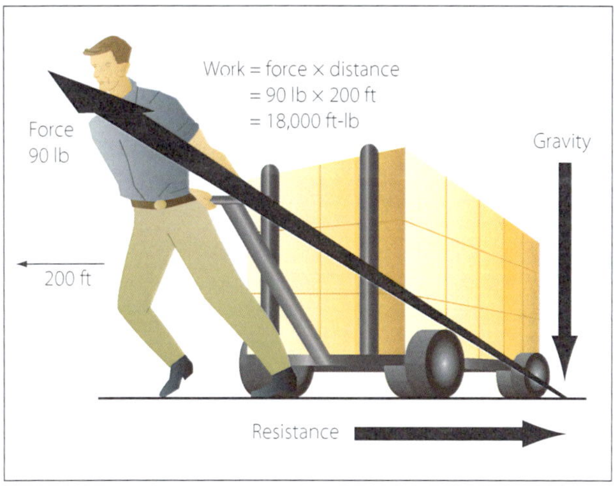

Figure 3-5. The effect of friction on work.

requires consideration when dealing with any moving mechanism.

In experiments relating to friction, measurement of the applied forces reveals that there are three kinds of friction. One force is required to start a body moving, while another is required to keep the body moving at constant speed. Also, after a body is in motion, a definitely larger force is required to keep it sliding than to keep it rolling.

Thus, the three kinds of friction may be classified as: (1) starting (static) friction, (2) sliding friction, and (3) rolling friction.

Static Friction

When an attempt is made to slide a heavy object along a surface, the object must first be broken loose or started. Once in motion, it slides more easily. The "breaking loose" force is, of course, proportional to the weight of the body. The force necessary to start the body moving slowly is designated "F," and "F'" is the normal force pressing the body against the surface (usually its weight). Since the nature of the surfaces rubbing against each other is important, they must be considered. The nature of the surfaces is indicated by the coefficient of starting friction which is designated by the letter "k." This coefficient can be established for various materials and is often published in tabular form. Thus, when the load (weight of the object) is known, starting friction can be calculated by using the following formula:

$$F = kF'$$

For example, if the coefficient of sliding friction of a smooth iron block on a smooth, horizontal surface is 0.3, the force required to start a 10 lb block would be 3 lb; a 40-lb block, 12 lb.

Starting friction for objects equipped with wheels and roller bearings is much smaller than that for sliding objects. Nevertheless, a locomotive would have difficulty getting a long train of cars in motion all at one time. Therefore, the couples between the cars are purposely made to have a few inches of play. When starting the train, the engineer backs the engine until all the cars are pushed together. Then, with a quick start forward the first car is set in motion. This technique is employed to overcome the static friction of each wheel (as well as the inertia of each car). It would be impossible for the engine to start all of the cars at the same instant, for static friction, which is the resistance of being set in motion, would be greater than the force exerted by the engine. Once the cars are in motion, however, static friction is greatly reduced and a smaller force is required to keep the train in motion than was required to start it.

Sliding Friction

Sliding friction is the resistance to motion offered by an object sliding over a surface. It pertains to friction produced after the object has been set in motion, and is always less than starting friction. The amount of sliding resistance is dependent on the nature of the surface of the object, the surface over which it slides, and the normal force between the object and the surface. This resistive force may be computed by using the following formula.

$$F = mN$$

In the formula above, "F" is the resistive force due to friction expressed in pounds; "N" is the force exerted on or by the object perpendicular (normal) to the surface over which it slides; and "m" (mu) is the coefficient of sliding friction. On a horizontal surface, N is equal to the weight of the object in pounds. The area of the sliding object exposed to the sliding surface has no effect on the results. A block of wood, for example, will not slide any easier on one of the broad sides than it will on a narrow side, (assuming all sides have the same smoothness). Therefore, area does not enter into the equation above.

Rolling Friction

Resistance to motion is greatly reduced if an object is mounted on wheels or rollers. The force of friction for objects mounted on wheels or rollers is called rolling friction. This force may be computed by the same equation used in computing sliding friction, but the values of "m" will be much smaller. For example, the value of "m" for rubber tires on concrete or macadam is about 0.02. The value of "m" for roller bearings is very small, usually ranging from 0.001 to 0.003 and is often disregarded.

Example: *An aircraft with a gross weight of 79,600 lb is towed over a concrete ramp. What force must be exerted by the towing vehicle to keep the airplane rolling after once set in motion?*

$$\begin{aligned} F &= mN \\ &= 0.02 \text{ mu} \times 79{,}600 \text{ lb} \\ &= 1{,}592 \text{ lb} \end{aligned}$$

Power

The concept of power involves the previously discussed topic of work, which was a force being applied over a measured distance, but adds one more consideration—time. In other words, how long does it take to accomplish the work. If someone asked the average person if he or she could lift one million pounds 5 feet off the ground, the answer most assuredly would be no. This person would probably assume that he or she is to lift it all at once. What if he or she is given 365 days to lift it, and could lift small amounts of weight at a time? The work involved would be the same, regardless of how long it took to lift the weight, but the power required is different. If the weight is to be lifted in a shorter period of time, it will take more power. The formula for power is as follows:

$$\text{Power} = \text{Force} \times \text{distance} \div \text{time}$$

The units for power will be foot-pounds per minute, foot-pounds per second, inch-pounds per minute or second, and possibly mile-pounds per hour. The units depend on how distance and time are measured.

Many years ago there was a desire to compare the power of the newly evolving steam engine to that of horses. People wanted to know how many horses the steam engine was equivalent to. Because of this, the value we currently know as one horsepower (hp) was developed, and it is equal to 550 foot-pounds per second (ft-lb/s). It was found that the average horse could lift a weight of 550 lb, one foot off the ground, in one second. The values we use today, in order to convert power to horsepower, are as follows:

1 hp = 550 ft-lb/s
1 hp = 33,000 ft-lb/min.
1 hp = 375 mile pounds per hour (mi-lb/hr)
1 hp = 746 watts (electricity conversion)

To convert power to horsepower, divide the power by the appropriate conversion based on the units being used.

Example: *What power would be needed, and also horsepower, to raise the GE-90 turbofan engine into position to install it on a Boeing 777-300 airplane? The engine weighs 19,000 lb, and it must be lifted 4 ft in 2 minutes.*

$$\begin{aligned}\text{Power} &= \text{Force} \times \text{distance} \div \text{time}\\ &= 19{,}000 \text{ lb} \times 4 \text{ ft} \div 2 \text{ min.}\\ &= 38{,}000 \text{ ft-lb/min.}\end{aligned}$$

$$\text{Hp} = 38{,}000 \text{ ft-lb/min.} \div 33{,}000 \text{ ft-lb/min.}$$
$$\text{Hp} = 1.15$$

The hoist that will be used to raise this engine into position will need to be powered by an electric motor because the average person will not be able to generate 1.15 hp in their arms for the necessary 2 minutes.

Torque

Torque is a very interesting concept and occurrence, and it is definitely something that needs to be discussed in conjunction with work and power. Whereas work is described as a force acting through a distance, torque is described as a force acting along a distance. Torque is something that creates twisting and tries to make something rotate.

If we push on an object with a force of 10 lb and it moves 10 inches in a straight line, we have done 100 in-lb of work. By comparison, if we have a wrench 10 inches long that is on a bolt, and we push down on it with a force of 10 lb, a torque of 100 lb-in is applied to the bolt. If the bolt was already tight and did not move as we pushed down on the wrench, the torque of 100 lb-in would still exist. The formula for torque is:

$$\text{Torque} = \text{Force} \times \text{distance}$$

Even though the formula looks the same as the one for calculating work, recognize that the distance value in this formula is not the linear distance an object moves, but rather the distance along which the force is applied.

Notice that with torque nothing had to move, because the force is being applied along a distance and not through a distance. Notice also that although the units of work and torque appear to be the same, they are not. The units of work were inch-pounds and the units of torque were pound-inches, and that is what differentiates the two.

Torque is very important when thinking about how engines work, both piston engines and gas turbine engines. Both types of engines create torque in advance of being able to create work or power. With a piston engine, a force in pounds pushes down on the top of the piston and tries to make it move. The piston is attached to the connecting rod, which is attached to the crankshaft at an offset. That offset would be like the length of the wrench discussed earlier, and the force acting along that length is what creates torque. [Figure 3-6]

For the cylinder in Figure 3-6, there is a force of 500 lb pushing down on the top of the piston. The connecting rod attaches to the crankshaft at an offset distance of 4 in. The product of the force and the offset distance is the torque, in this case 2,000 lb-in.

In a turbine engine, the turbine blades at the back of the engine extract energy from the high velocity exhaust gases. The energy extracted becomes a force in pounds pushing on the turbine blades, which happen to be a certain number of inches from the center of the shaft they are trying to make rotate. The number of inches from the turbine blades to the center of the shaft would be like the length of the wrench discussed earlier.

Mathematically, there is a relationship between the horsepower of an engine and the torque of an engine. The formula that shows this relationship is as follows:

$$\text{Torque} = \text{Horsepower} \times 5{,}252 \div \text{rpm}$$

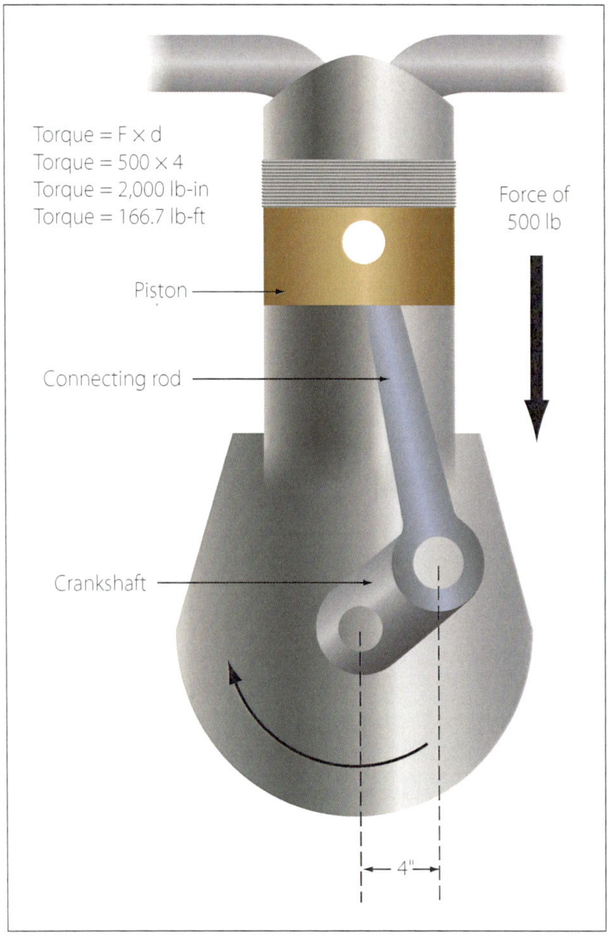

Figure 3-6. Piston engine and torque.

Example: *A Cessna 172R has a Lycoming IO-360 engine that creates 180 horsepower at 2,700 rpm. How many pound-feet of torque is the engine producing?*

$$\begin{aligned}\text{Torque} &= 180 \times 5{,}252 \div 2{,}700\\ &= 350 \text{ lb-ft}\end{aligned}$$

Simple Machines

A machine is any device with which work may be accomplished. In application, machines can be used for any of the following purposes, or combinations of these purposes.

1. Machines are used to transform energy, as in the case of a generator transforming mechanical energy into electrical energy.
2. Machines are used to transfer energy from one place to another, as in the examples of the connecting rods, crankshaft, and reduction gears transferring energy from an aircraft's engine to its propeller.
3. Machines are used to multiply force; for example, a system of pulleys may be used to lift a heavy load. The pulley system enables the load to be raised by exerting a force that is smaller than the weight of the load.
4. Machines can be used to multiply speed. A good example is the bicycle, by which speed can be gained by exerting a greater force.
5. Machines can be used to change the direction of a force. An example of this use is the flag hoist. A downward force on one side of the rope exerts an upward force on the other side, raising the flag toward the top of the pole.

There are only six simple machines. They are the lever, the pulley, the wheel and axle, the inclined plane, the screw, and the gear. Physicists, however, recognize only two basic principles in machines: the lever and the inclined plane. The pulley (block and tackle), the wheel and axle, and gears operate on the machine principle of the lever. The wedge and the screw use the principle of the inclined plane.

An understanding of the principles of simple machines provides a necessary foundation for the study of compound machines, which are combinations of two or more simple machines.

Mechanical Advantage of Machines

As identified in statements 3 and 4 under simple machines, a machine can be used to multiply force or to multiply speed. It cannot, however, multiply force and speed at the same time. In order to gain one, it must lose the other. To do otherwise would mean the machine has more power going out than coming in, and that is not possible.

In reference to machines, mechanical advantage is a comparison of the output force to the input force, or the output distance to the input distance. If there is a mechanical advantage in terms of force, there will be a fractional disadvantage in terms of distance. The following formulas can be used to calculate mechanical advantage.

Mechanical Advantage = Force Out ÷ Force In

Or

Mechanical Advantage = Distance Out ÷ Distance In

The Lever

The simplest machine, and perhaps the most familiar one, is the lever. A seesaw is a familiar example of a lever, with two people sitting on either end of a board and a pivoting point in the middle. There are three basic parts in all levers. They are the fulcrum "F," a force or effort "E," and a resistance "R." Shown in Figure 3-7 are the pivot point "F" (fulcrum), the effort "E" which is applied at a distance "L" from the fulcrum, and a resistance "R" which acts at a distance "l" from the fulcrum. Distances "L" and "l" are the lever arms.

The concept of torque was discussed earlier in this chapter, and torque is very much involved in the operation of a lever. When a person sits on one end of a seesaw, that person applies a downward force in pounds which acts along the distance to the center of the seesaw. This combination of force and distance creates torque, which tries to cause rotation.

First Class Lever

In the first class lever, the fulcrum is located between the effort and the resistance. As mentioned earlier, the seesaw is a good example of a lever, and it happens to be a first class lever. The amount of weight and the distance from the fulcrum can be varied to suit the need. Increasing the distance from the applied effort to the fulcrum, compared to the distance from the fulcrum to the weight being moved, increases the advantage provided by the lever. Crowbars, shears, and pliers are common examples of this class of lever. The proper balance of an airplane is also a good example, with the center of lift on the wing being the pivot point (fulcrum) and the weight fore and aft of this point being the effort and the resistance.

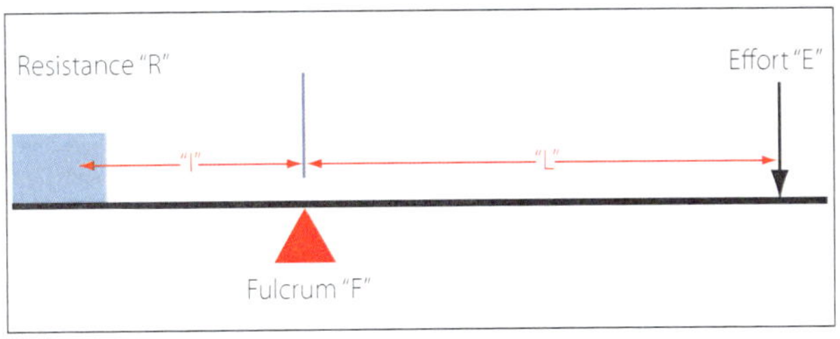

Figure 3-7. First class lever.

When calculating how much effort is required to lift a specific weight, or how much weight can be lifted by a specific effort, the following formula can be used.

Effort (E) × Effort Arm (L) = Resistance (R) × Resistance Arm (l)

What this formula really shows is the input torque (effort times effort arm) equals the output torque (resistance times resistance arm). This formula and concept apply to all three classes of levers, and to all simple machines in general.

Example: *A first class lever is to be used to lift a 500-lb weight. The distance from the weight to the fulcrum is 12 inches and from the fulcrum to the applied effort is 60 inches. How much force is required to lift the weight?*

Effort (E) × Effort Arm (L) = Resistance (R) × Resistance Arm (l)

E × 60 in = 500 lb × 12 in

E = 500 lb × 12 in ÷ 60 in

E = 100 lb

The mechanical advantage of the lever in this example would be:

Mechanical Advantage = Force Out ÷ Force In

= 500 lb ÷ 100 lb

= 5, or 5 to 1

An interesting thing to note with this example lever is if the applied effort moved down 10 inches, the weight on the other end would only move up 2 inches. The weight being lifted would only move one-fifth as far. The reason for this is the concept of work. Because a lever cannot have more work output than input, if it allows you to lift 5 times more weight, you will only move it $\frac{1}{5}$ as far as you move the effort.

Second Class Lever

The second class lever has the fulcrum at one end and the effort is applied at the other end. The resistance is somewhere between these points. A wheelbarrow is a good example of a second class lever, with the wheel at one end being the fulcrum, the handles at the opposite end being the applied effort, and the bucket

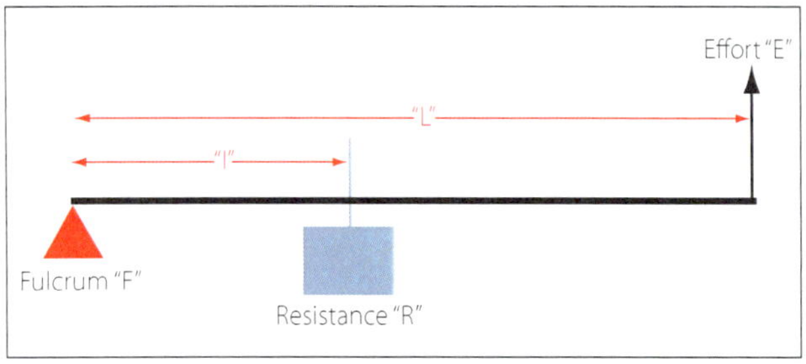

Figure 3-8. Second class lever.

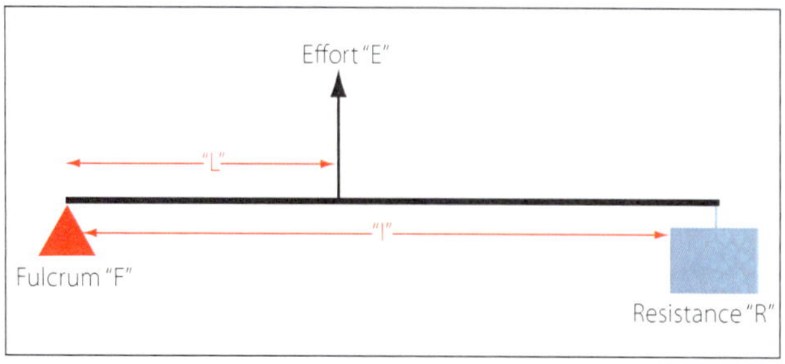

Figure 3-9. Third class lever.

in the middle being where the weight or resistance is placed. [Figure 3-8]

Both first and second class levers are commonly used to help in overcoming big resistances with a relatively small effort. The first class lever, however, is more versatile. Depending on how close or how far away the weight is placed from the fulcrum, the first class lever can be made to gain force or gain distance, but not both at the same time. The second class lever can only be made to gain force.

Example: *The distance from the center of the wheel to the handles on a wheelbarrow is 60 inches. The weight in the bucket is 18 inches from the center of the wheel. If 300 lb is placed in the bucket, how much force must be applied at the handles to lift the wheelbarrow?*

Effort (E) × Effort Arm (L) = Resistance (R) × Resistance Arm (l)
E × 60 inches = 300 lb × 18 in
E = 300 lb × 18 in ÷ 60 in
E = 90 lb

The mechanical advantage of the lever in this example would be:

Mechanical Advantage = Force Out ÷ Force In
= 300 lb ÷ 90 lb
= 3.33, or 3.33 to 1

Third Class Lever

There are occasions when it is desirable to speed up the movement of the resistance even though a large amount of effort must be used. Levers that help accomplish this are third class levers. As shown in Figure 3-9, the fulcrum is at one end of the lever and the weight or resistance to be overcome is at the other end, with the effort applied at some point between. Third class levers are easily recognized because the effort is applied between the fulcrum and the resistance. The retractable main landing gear on an airplane is a good example of a third class lever. The top of the landing gear, where it attaches to the airplane, is the pivot point. The wheel and brake assembly at the bottom of the landing gear is the resistance. The hydraulic actuator that makes the gear retract is attached somewhere in the middle, and that is the applied effort.

The Pulley

Pulleys are simple machines in the form of a wheel mounted on a fixed axis and supported by a frame. The wheel, or disk, is normally grooved to accommodate a rope. The wheel is sometimes referred to as a "sheave" (sometimes "sheaf"). The frame that supports the wheel is called a block. A block and tackle consists of a pair of blocks. Each block contains one or more pulleys and a rope connecting the pulley(s) of each block.

Single Fixed Pulley

A single fixed pulley is really a first class lever with equal arms. In Figure 3-10, the arm from point "R" to point "F" is equal to the arm from point "F" to point "E" (both distances being equal to the radius of the pulley). When a first class lever has equal arms, the mechanical advantage is 1. Thus, the force of the pull on the rope must be equal to the weight of the object being lifted. The only advantage of a single fixed pulley is to change the direction of the force, or pull on the rope.

Single Movable Pulley

A single pulley can be used to magnify the force exerted. In Figure 3-11, the pulley is movable, and both ropes extending up from the pulley are sharing in the support of the weight. This single movable pulley acts like a second class lever, with the effort arm (EF) being the diameter of the pulley and the resistance arm (FR) being the radius of the pulley. This type of pulley would have a mechanical advantage of two because the diameter of the pulley is double the radius of the pulley. In use, if someone pulled in 4 ft of the effort rope, the weight would only rise off the floor 2 ft. If the weight was 100 lb, the effort applied would only need to be 50 lb. With this type of pulley, the effort will always be one-half of the weight being lifted.

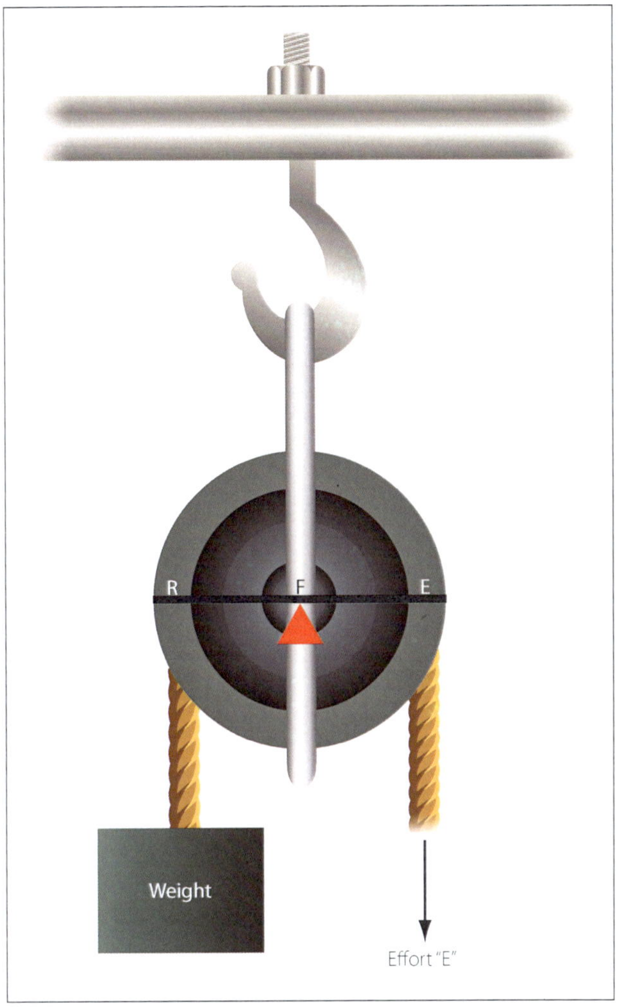

Figure 3-10. Single fixed pulley.

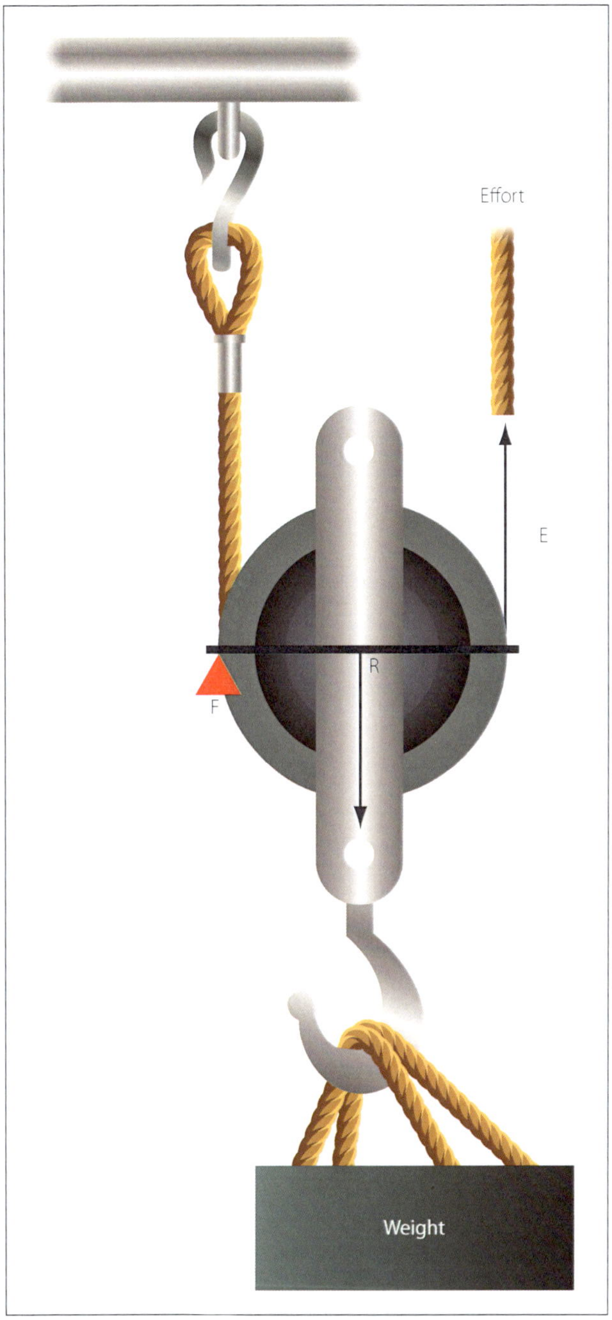

Figure 3-11. Single movable pulley.

Block and Tackle

A block and tackle is made up of multiple pulleys, some of them fixed and some movable. In Figure 3-12, the block and tackle is made up of four pulleys, the top two being fixed and the bottom two being movable. Viewing the figure from right to left, notice there are four ropes supporting the weight and a fifth rope where the effort is applied. The number of weight supporting ropes determines the mechanical advantage of a block and tackle, so in this case the mechanical advantage is four. If the weight was 200 lb, it would require a 50 lb effort to lift it.

The Gear

Two gears with teeth on their outer edges, as shown in Figure 3-13, act like a first class lever when one gear drives the other. The gear with the input force is called the drive gear, and the other is called the driven gear. The effort arm is the diameter of the driven gear, and the resistance arm is the diameter of the drive gear.

Notice that the two gears turn in opposite directions (the bottom one clockwise and the top one counter-clockwise). The gear on top (yellow) is 9 inches in diameter and has 45 teeth, and the gear on the bottom (blue) is 12 inches in diameter and has 60 teeth.

Imagine that the blue gear is driving the yellow one (blue is the drive, yellow is the driven). The mechanical advantage in terms of force would be the effort arm divided by the resistance arm, or 9 ÷ 12, which is 0.75. This would actually be called a fractional disadvantage, because there would be less force out than force in.

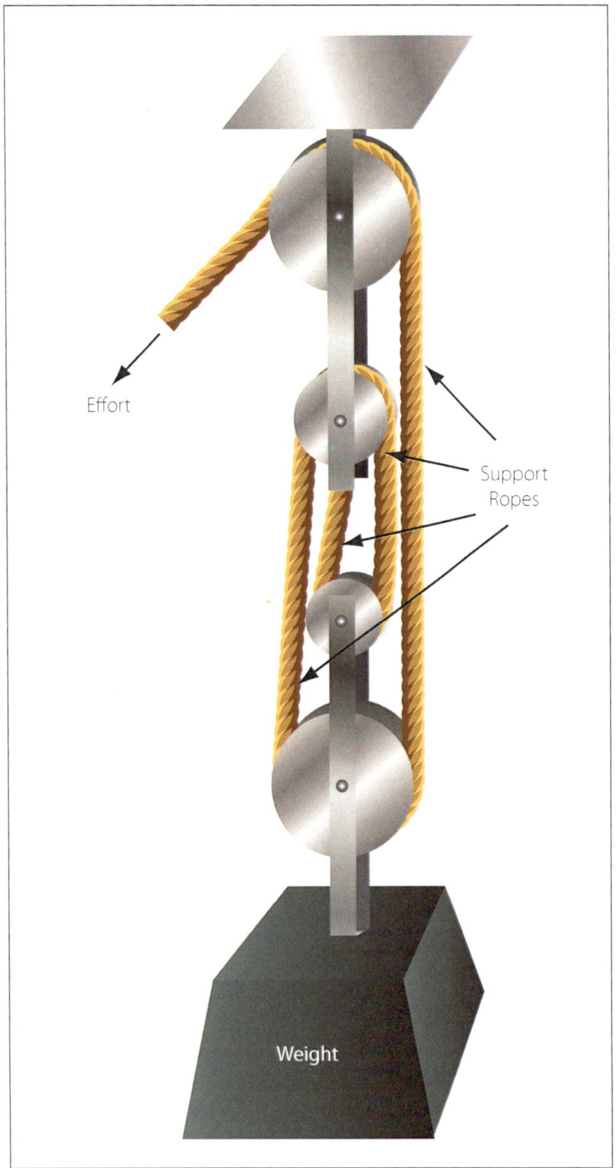

Figure 3-12. Block and tackle.

The mechanical advantage in terms of distance (rpm in this case), would be 12 ÷ 9, or 1.33.

This analysis tells us that when a large gear drives a small one, the small one turns faster and has less available force. In order to be a force gaining machine, the small gear needs to turn the large one. When the terminology reduction gearbox is used, such as a propeller reduction gearbox, it means that there is more rpm going in than is coming out. The end result is an increase in force, and ultimately torque.

Bevel gears are used to change the plane of rotation, so that a shaft turning horizontally can make a vertical shaft rotate. The size of the gears and their number of teeth determine the mechanical advantage, and whether force is being increased or rpm is being increased. If each gear

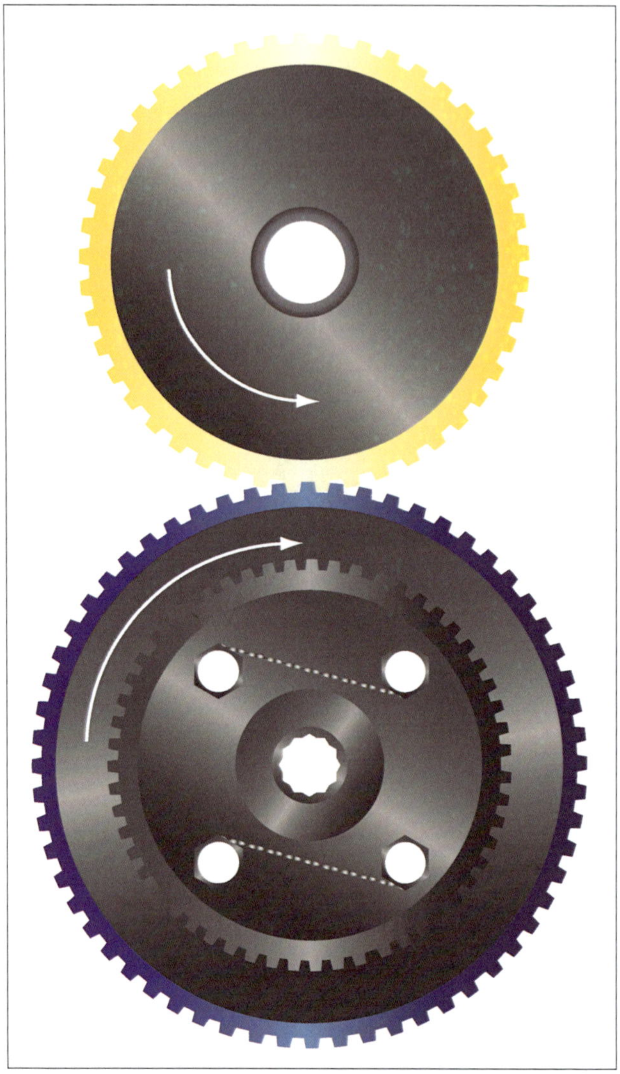

Figure 3-13. Spur gears.

has the same number of teeth, there would be no change in force or rpm. [Figure 3-14]

The worm gear has an extremely high mechanical advantage. The input force goes into the spiral worm gear, which drives the spur gear. One complete revolution of the worm gear only makes the spur gear turn an amount equal to one tooth. The mechanical advantage is equal to the number of teeth on the spur gear, which in this case is 25. This is a force gaining machine, to the tune of 25 times more output force. [Figure 3-15]

The planetary sun gear system is typical of what would be found in a propeller reduction gearbox. The power output shaft of the engine would drive the sun gear in the middle, which rotates the planetary gears and ultimately the ring gear. In this example, the sun gear has 28 teeth, each planet gear has 22 teeth, and the ring gear has 82 teeth. To figure how much gear reduction is taking place, the number of teeth on the

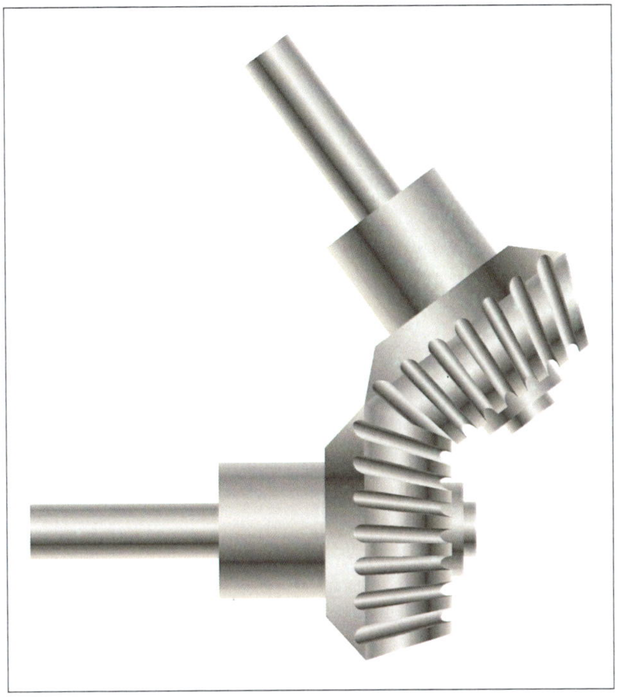

Figure 3-14. Bevel gears.

Figure 3-15. Worm gear.

ring gear is divided by the number of teeth on the sun gear. In this case, the gear reduction is 2.93, meaning the engine has an rpm 2.93 times greater than the propeller. [Figure 3-16]

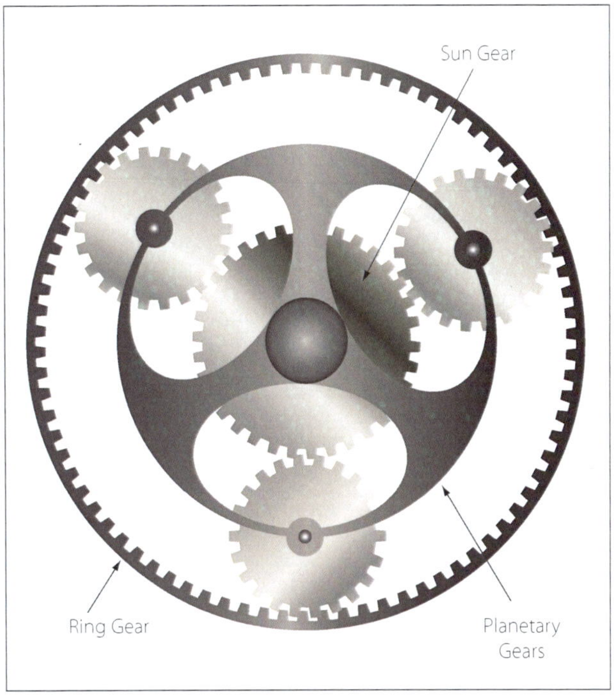

Figure 3-16. Planetary sun gear.

Inclined Plane

The inclined plane is a simple machine that facilitates the raising or lowering of heavy objects by application of a small force over a relatively long distance. Some familiar examples of the inclined plane are mountain highways and a loading ramp on the back of a moving truck. When weighing a small airplane, like a Cessna 172, an inclined plane (ramp) can be used to get the airplane on the scales by pushing it, rather than jacking it. A ramp can be seen in Figure 3-17, where a Cessna 172 right main gear is sitting on an electronic scale. The airplane was pushed up the ramps to get it on the scales.

With an inclined plane, the length of the incline is the effort arm and the vertical height of the incline is the resistance arm. If the length of the incline is five times greater than the height, there will be a force advantage, or mechanical advantage, of five. The Cessna 172 in Figure 3-17 weighed 1,600 lb on the day of the weighing. The ramp it is sitting on is 6 inches tall (resistance

Figure 3-17. Ramp in use with a Cessna 172.

Figure 3-18. A bolt and nut as an inclined plane.

arm) and the length of the ramp is 24 inches (effort arm). To calculate the force needed to push the airplane up the ramps, use the same formula introduced earlier when levers were discussed, as follows:

Effort (E) × Effort Arm (L) = Resistance (R) × Resistance Arm (l)
E × 24 in = 1,600 lb × 6 in
E = 1,600 lb × 6 in ÷ 24 in
E = 400 lb

Bolts, screws, and wedges are also examples of devices that operate on the principle of the inclined plane. A bolt, for example, has a spiral thread that runs around its circumference. As the thread winds around the bolt's circumference, it moves a vertical distance equal to the space between the threads. The circumference of the bolt is the effort arm and the distance between the threads is the resistance arm. [Figure 3-18] Based on this analysis, it can be seen that a fine threaded bolt (more threads per inch) has a greater mechanical advantage than a coarse threaded bolt.

A chisel is a good example of a wedge. A chisel might be 8 inches long and only ½ inch wide, with a sharp tip and tapered sides. The 8-inch length is the effort arm and the ½-inch width is the resistance arm. This chisel would provide a force advantage (mechanical advantage) of 16.

Stress

Whenever a machine is in operation, be it a simple machine like a lever or a screw, or a more complex machine like an aircraft piston engine or a hydraulically operated landing gear, the parts and pieces of that machine will experience something called stress. Whenever an external force is applied to an object, like a weight pushing on the end of a lever, a reaction will occur inside the object which is known as stress. Stress is typically measured in pounds per square foot or pounds per square inch (psi).

An external force acting on an object causes the stress to manifest itself in one of five forms, or combination of those five. The five forms are tension, compression, torsion, bending, and shear.

Tension

Tension is a force that tries to pull an object apart. In the block and tackle system discussed earlier in this chapter, the upper block that housed the two fixed pulleys was secured to an overhead beam. The movable lower block and its two pulleys were hanging by ropes, and the weight was hanging below the entire assembly. The weight being lifted would cause the ropes and the blocks to be under tension. The weight is literally trying to pull the rope apart, and ultimately would cause the rope to break if the weight was too great.

Compression

Compression is a force that tries to crush an object. An excellent example of compression is when a sheet metal airplane is assembled using the fastener known as a rivet. The rivet passes through a hole drilled in the pieces of aluminum, and then a rivet gun on one side and a bucking bar on the other apply a force. This applied force tries to crush the rivet and makes it expand to fill the hole and securely hold the aluminum pieces together. [Figure 3-19]

Torsion

Torsion is the stress an object experiences when it is twisted, which is what happens when torque is applied to a shaft. Torsion is actually made up of two other stresses: tension and compression. When a shaft is twisted, tension is experienced at a diagonal to the shaft and compression acts 90 degrees to the tension. [Figure 3-20]

The turbine shaft on a turbofan engine, which connects to the compressor in order to drive it, is under a torsion stress. The turbine blades extract energy from the high velocity air as a force in pounds. This force in pounds acts along the length from the blades to the center of the shaft, and creates the torque that causes rotation. [Figure 3-21]

Bending

An airplane in flight experiences a bending force on the wing as aerodynamic lift tries to raise the wing. This force of lift actually causes the skin on the top of the

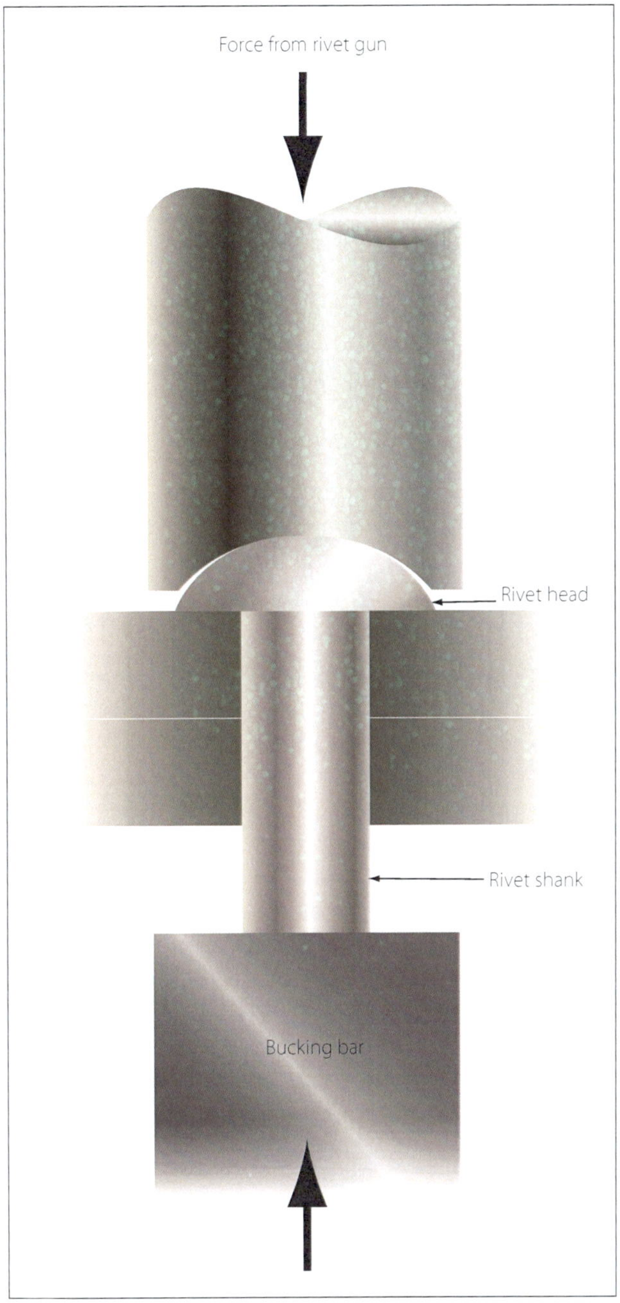

Figure 3-19. A rivet fastener and compression.

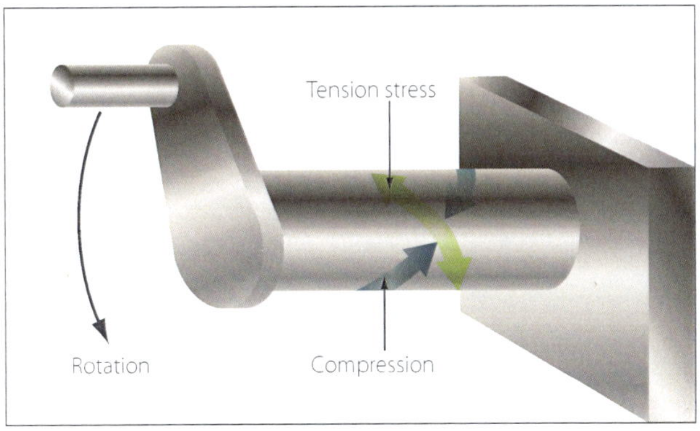

Figure 3-20. Torsion on a rotating shaft, made up of tension and compression.

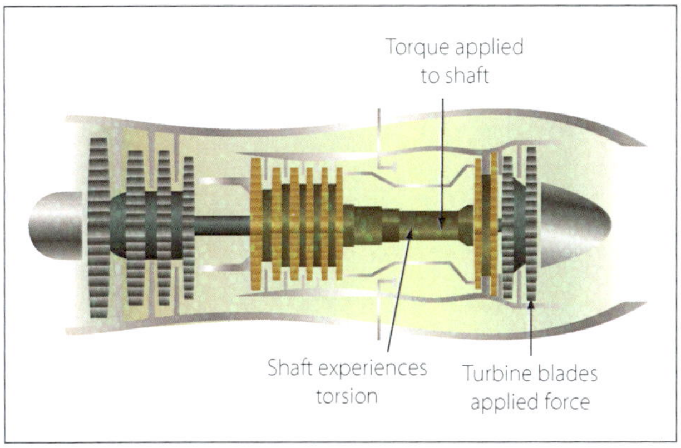

Figure 3-21. Turbofan engine, torque creating torsion in the shaft.

wing to compress and the skin on the bottom of the wing to be under tension. When the airplane is on the ground sitting on its landing gear, the force of gravity tries to bend the wing downward, subjecting the bottom of the wing to compression and the top of the wing to tension. [Figure 3-22] During the testing that occurs prior to FAA certification, an airplane manufacturer intentionally bends the wing up and down to make sure it can take the stress without failing.

Shear

When a shear stress is applied to an object, the force tries to cut or slice through, like a knife cutting through butter. A clevis bolt, which is often used to secure a cable to a part of the airframe, has a shear stress acting on it. As shown in Figure 3-23, a fork fitting is secured

Figure 3-22. Airplane on the ground, wing under tension and compression.

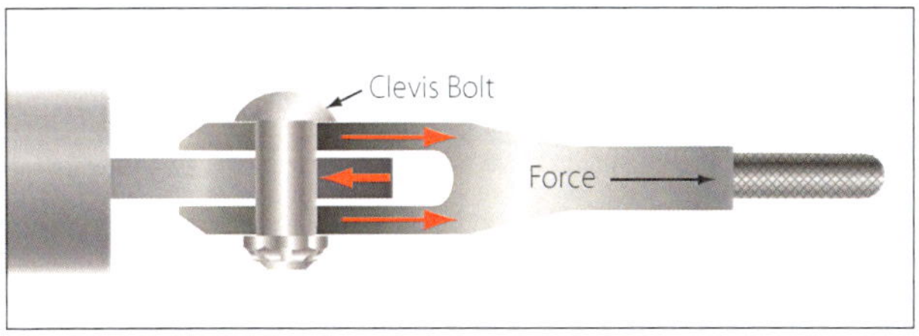

Figure 3-23. Clevis bolt, red arrows show opposing forces trying to shear the bolt.

to the end of the cable, and the fork attaches to an eye on the airframe with the clevis bolt. When the cable is put under tension, the fork tries to slide off the eye by cutting through the clevis bolt. This bolt would be designed to take very high shear loads.

Strain

If the stress acting on an object is great enough, it can cause the object to change its shape or to become distorted. One characteristic of matter is that it tends to be elastic, meaning it can be forced out of shape when a force is applied, and then return to its original shape when the force is removed. When an object becomes distorted by an applied force, the object is said to be strained.

On turbine engine test cells, the thrust of the engine is typically measured by what are called strain gages. When the force (thrust) of the engine is pulling out against the strain gages, the amount of distortion is measured and then translated into the appropriate thrust reading.

A deflecting beam style of torque wrench uses the strain on the drive end of the wrench and the resulting distortion of the beam to indicate the amount of torque on a bolt or nut. [Figure 3-24]

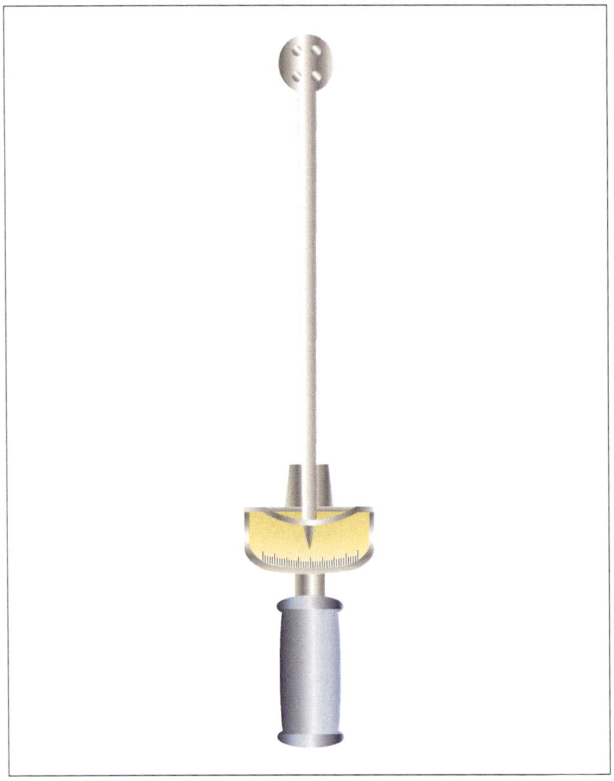

Figure 3-24. Deflecting beam torque wrench, measures strain by distortion.

Motion

The study of the relationship between the motion of bodies or objects and the forces acting on them is often called the study of "force and motion." In a more specific sense, the relationship between velocity, acceleration, and distance is known as kinematics.

Uniform Motion

Motion may be defined as a continuing change of position or place, or as the process in which a body undergoes displacement. When an object is at different points in space at different times, that object is said to be in motion, and if the distance the object moves remains the same for a given period of time, the motion may be described as uniform. Thus, an object in uniform motion always has a constant speed.

Speed and Velocity

In everyday conversation, speed and velocity are often used as if they mean the same thing. In physics they have definite and distinct meanings. Speed refers to how fast an object is moving, or how far the object will travel in a specific time. The speed of an object tells nothing about the direction an object is moving. For example, if the information is supplied that an airplane leaves New York City and travels 8 hours at a speed of 150 mph, this information tells nothing about the direction in which the airplane is moving. At the end of 8 hours, it might be in Kansas City, or if it traveled in a circular route, it could be back in New York City.

Velocity is that quantity in physics which denotes both the speed of an object and the direction in which the object moves. Velocity can be defined as the rate of motion in a particular direction. Velocity is also described as being a vector quantity, a vector being a line of specific length, having an arrow on one end or the other. The length of the line indicates the number value and the arrow indicates the direction in which that number is acting.

Two velocity vectors, such as one representing the velocity of an airplane and one representing the velocity of the wind, can be added together in what is called vector analysis. Figure 3-25 demonstrates this, with vectors "A" and "B" representing the velocity of the airplane and the wind, and vector "C" being the resultant. With no wind, the speed and direction of the airplane would be that shown by vector "A." When accounting for the wind direction and speed, the airplane ends up flying at the speed and direction shown by vector "C."

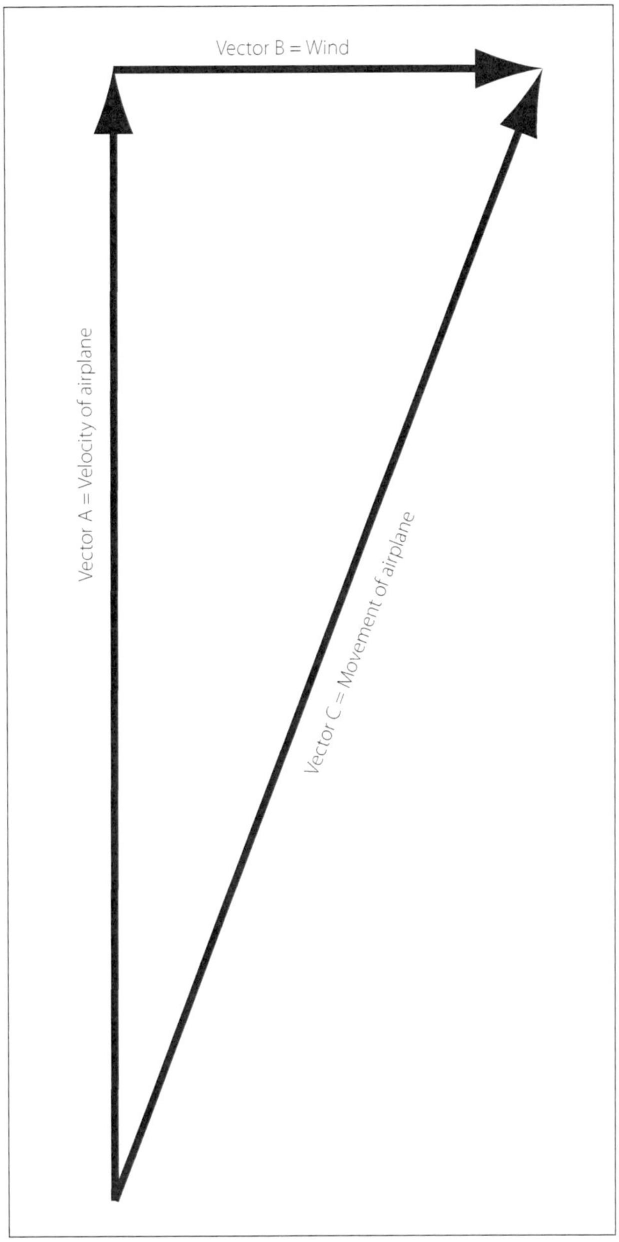

Figure 3-25. Vector analysis for airplane velocity and wind velocity.

Imagine that an airplane is flying in a circular pattern at a constant speed. Because of the circular pattern, the airplane is constantly changing direction, which means the airplane is constantly changing velocity. The reason for this is the fact that velocity includes direction.

To calculate the speed of an object, the distance it travels is divided by the elapsed time. If the distance is measured in miles and the time in hours, the units of speed will be miles per hour (mph). If the distance is measured in feet and the time in seconds, the units of speed will be feet per second (fps). To convert mph to fps, multiply by 1.467. Velocity is calculated the same way, the only difference being it must be recalculated every time the direction changes.

Acceleration

Acceleration is defined as the rate of change of velocity. If the velocity of an object is increased from 20 mph to 30 mph, the object has been accelerated. If the increase in velocity is 10 mph in 5 seconds, the rate of change in velocity is 10 mph in 5 seconds, or 2 mph per second. If this were multiplied by 1.467, it could also be expressed as an acceleration of 2.93 feet per second per second (fps/s). By comparison, the acceleration due to gravity is 32.2 fps/s.

To calculate acceleration, the following formula is used.

$$\text{Acceleration (A)} = \frac{\text{Velocity Final (Vf)} - \text{Velocity Initial (Vi)}}{\text{Time (t)}}$$

Example: *An Air Force F-15 fighter is cruising at 400 mph. The pilot advances the throttles to full afterburner and accelerates to 1,200 mph in 20 seconds. What is the average acceleration in mph/s and fps/s?*

$$A = \frac{Vf - Vi}{t}$$

$$A = \frac{1200 - 400}{20}$$

$A = 40\ ^{mph}/_{s}$, or by multiplying by 1.467, $58.7\ ^{fps}/_{s}$

In the example just shown, the acceleration was found to be 58.7 fps/s. Since 32.2 fps/s is equal to the acceleration due to gravity, divide the F-15's acceleration by 32.2 to find out how many G forces the pilot is experiencing. In this case, it would be 1.82 Gs.

Newton's Law of Motion

First Law

When a magician snatches a tablecloth from a table and leaves a full setting of dishes undisturbed, he is not displaying a mystic art; he is demonstrating the principle of inertia. Inertia is responsible for the discomfort felt when an airplane is brought to a sudden halt in the parking area and the passengers are thrown forward in their seats. Inertia is a property of matter. This property of matter is described by Newton's first law of motion, which states:

Objects at rest tend to remain at rest and objects in motion tend to remain in motion at the same speed and in the same direction, unless acted on by an external force.

Second Law

Bodies in motion have the property called momentum. A body that has great momentum has a strong tendency to remain in motion and is therefore hard to stop. For example, a train moving at even low velocity is difficult to stop because of its large mass. Newton's second law applies to this property. It states:

When a force acts upon a body, the momentum of that body is changed. The rate of change of momentum is proportional to the applied force. Based on Newton's second law, the formula for calculating thrust is derived, which states that force equals mass times acceleration (F = MA). Earlier in this chapter, it was determined that mass equals weight divided by gravity, and acceleration equals velocity final minus velocity initial divided by time. Putting all these concepts together, the formula for thrust is:

$$\text{Force} = \frac{\text{Weight (Velocity final} - \text{Velocity initial)}}{\text{Gravity (Time)}}$$

$$\text{Force} = \frac{W\,(Vf - Vi)}{Gt}$$

Example: *A turbojet engine is moving 150 lb of air per second through the engine. The air enters going 100 fps and leaves going 1,200 fps. How much thrust, in pounds, is the engine creating?*

$$F = \frac{W\,(Vf - Vi)}{Gt}$$

$$F = \frac{150\,(1200 - 100)}{32.2\,(1)}$$

$$F = 5{,}124 \text{ lb of thrust}$$

Third Law

Newton's third law of motion is often called the law of action and reaction. It states that for every action there is an equal and opposite reaction. This means that if a force is applied to an object, the object will supply a resistive force exactly equal to and in the opposite direction of the force applied. It is easy to see how this might apply to objects at rest. For example, as a man stands on the floor, the floor exerts a force against his feet exactly equal to his weight. But this law is also applicable when a force is applied to an object in motion.

Forces always occur in pairs. The term "acting force" means the force one body exerts on a second body, and reacting force means the force the second body exerts on the first.

When an aircraft propeller pushes a stream of air backward with a force of 500 lb, the air pushes the blades forward with a force of 500 lb. This forward force causes the aircraft to move forward. A turbofan engine exerts a force on the air entering the inlet duct, causing it to accelerate out the fan duct and the tailpipe. The air accelerating to the rear is the action, and the force inside the engine that makes it happen is the reaction, also called thrust.

Circular Motion

Circular motion is the motion of an object along a curved path that has a constant radius. For example, if one end of a string is tied to an object and the other end is held in the hand, the object can be swung in a circle. The object is constantly deflected from a straight (linear) path by the pull exerted on the string, as shown in Figure 3-26. When the weight is at point A, due to inertia it wants to keep moving in a straight line and end up at point B. Because of the force being exerted on the string, it is forced to move in a circular path and end up at point C.

The string exerts a centripetal force on the object, and the object exerts an equal but opposite force on the string, obeying Newton's third law of motion. The

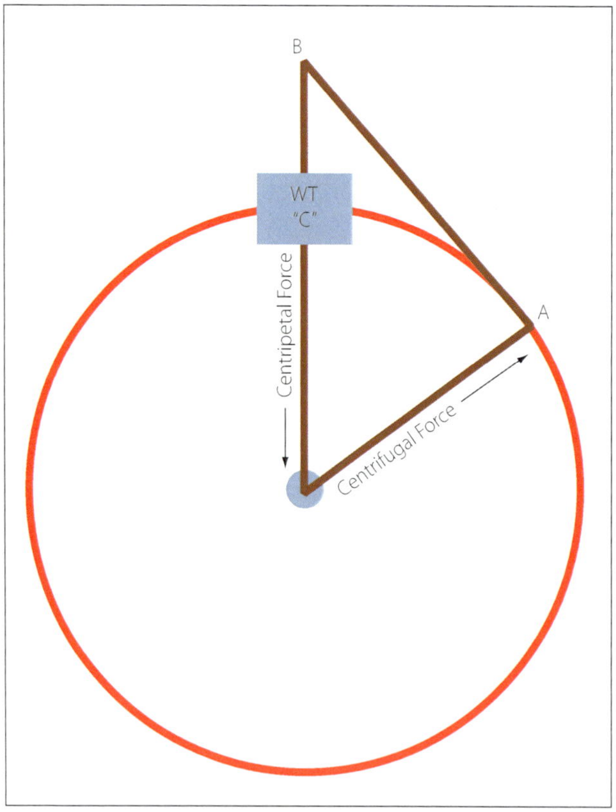

Figure 3-26. Circular motion.

force that is equal to centripetal force, but acting in an opposite direction, is called centrifugal force.

Centripetal force is always directly proportional to the mass of the object in circular motion. Thus, if the mass of the object in Figure 3-26 is doubled, the pull on the string must be doubled to keep the object in its circular path, provided the speed of the object remains constant.

Centripetal force is inversely proportional to the radius of the circle in which an object travels. If the string in Figure 3-26 is shortened and the speed remains constant, the pull on the string must be increased since the radius is decreased, and the string must pull the object from its linear path more rapidly. Using the same reasoning, the pull on the string must be increased if the object is swung more rapidly in its orbit. Centripetal force is thus directly proportional to the square of the velocity of the object. The formula for centripetal force is:

$$\text{Centripetal Force} = \text{Mass } (\text{Velocity}^2) \div \text{Radius}$$

For the formula above, mass would typically be converted to weight divided by gravity, velocity would be in feet per second, and the radius would be in feet.

Example: *What would the centripetal force be if a 10 pound weight was moving in a 3-ft radius circular path at a velocity of 500 fps?*

$$\text{Centripetal Force} = \text{Mass } (\text{Velocity}^2) \div \text{Radius}$$
$$\text{Centripetal Force} = 10\,(500^2) \div 32.2\,(3)$$
$$= 25{,}880 \text{ lb}$$

In the condition identified in the example, the object acts like it weighs 2,588 times more than it actually does. It can also be said that the object is experiencing 2,588 Gs (force of gravity). The fan blades in a large turbofan engine, when the engine is operating at maximum rpm, are experiencing many thousands of Gs for the same reason.

Heat

Heat is a form of energy. It is produced only by the conversion of one of the other forms of energy. Heat may also be defined as the total kinetic energy of the molecules of any substance.

Some forms of energy which can be converted into heat energy are as follows:

- *Mechanical Energy.* This includes all methods of producing increased motion of molecules such as friction, impact of bodies, or compression of gases.
- *Electrical Energy.* Electrical energy is converted to heat energy when an electric current flows through any form of resistance such as an electric iron, electric light, or an electric blanket.
- *Chemical Energy.* Most forms of chemical reaction convert stored potential energy into heat. Some examples are the explosive effects of gunpowder, the burning of oil or wood, and the combining of oxygen and grease.
- *Radiant Energy.* Electromagnetic waves of certain frequencies produce heat when they are absorbed by the bodies they strike such as x-rays, light rays, and infrared rays.
- *Nuclear Energy.* Energy stored in the nucleus of atoms is released during the process of nuclear fission in a nuclear reactor or atomic explosion.
- *The Sun.* All heat energy can be directly or indirectly traced to the nuclear reactions occurring in the sun.

When a gas is compressed, work is done and the gas becomes warm or hot. Conversely, when a gas under high pressure is allowed to expand, the expanding gas becomes cool. In the first case, work was converted into energy in the form of heat; in the second case heat energy was expended. Since heat is given off or absorbed, there must be a relationship between heat energy and work. Also, when two surfaces are rubbed together, the friction develops heat. However, work was required to cause the heat, and by experimentation, it has been shown that the work required and the amount of heat produced by friction are proportional. Thus, heat can be regarded as a form of energy.

According to this theory of heat as a form of energy, the molecules, atoms, and electrons in all bodies are in a continual state of motion. In a hot body, these small particles possess relatively large amounts of kinetic energy, but in cooler bodies they have less. Because the small particles are given motion, and hence kinetic energy, work must be done to slide one body over the other. Mechanical energy apparently is transformed, and what we know as heat is really kinetic energy of the small molecular subdivisions of matter.

Heat Energy Units

Two different units are used to express quantities of heat energy. They are the calorie and the BTU. One calorie is equal to the amount of heat required to change the temperature of 1 gram of water 1 degree Centigrade.

This term "calorie" (spelled with a lower case c) is 1/1,000 of the Calorie (spelled with a capital C) used in the measurement of the heat energy in foods. One BTU is defined as the amount of heat required to change the temperature of 1 lb of water 1 degree Fahrenheit (1°F). The calorie and the gram are seldom used in discussing aviation maintenance. The BTU, however, is commonly referred to in discussions of engine thermal efficiencies and the heat content of aviation fuel.

A device known as the calorimeter is used to measure quantities of heat energy. For example, it may be used to determine the quantity of heat energy available in 1 pound of aviation gasoline. A given weight of the fuel is burned in the calorimeter, and the heat energy is absorbed by a large quantity of water. From the weight of the water and the increase in its temperature, it is possible to compute the heat yield of the fuel. A definite relationship exists between heat and mechanical energy. This relationship has been established and verified by many experiments which show that:

One BTU of heat energy = 778 ft-lb of work

As discussed earlier in this chapter under the topic "Potential Energy," one pound of aviation gasoline contains 18,900 BTU of heat energy. Since each BTU is capable of 778 ft-lb of work, 1 lb of aviation gasoline is capable of 14,704,200 ft-lb of work.

Heat Energy and Thermal Efficiency

Thermal efficiency is the relationship between the potential for power contained in a specific heat source, and how much usable power is actually created when that heat source is used. The formula for calculating thermal efficiency is:

Thermal Efficiency =
Horsepower Produced ÷ Potential Horsepower in Fuel

For example, consider the piston engine used in a small general aviation airplane, which typically consumes 0.5 lb of fuel per hour for each horsepower it creates. Imagine that the engine is creating 200 hp. If we multiply 0.5 by the horsepower of 200, we find the engine is consuming 100 lb of fuel per hour, or 1.67 lb per minute. Earlier in this chapter, one horsepower was found to be 33,000 ft-lb of work per minute. The potential horsepower in the fuel burned for this example engine would be:

$$\text{Hp} = \frac{1.67 \text{ lb per minute} \times 18{,}900 \text{ BTU per lb} \times 778 \text{ ft lb per BTU}}{33{,}000 \text{ ft-lb/min}}$$

$$\text{Hp} = 744$$

The example engine is burning enough fuel that it has the potential to create 744 horsepower, but it is only creating 200. The thermal efficiency of the engine would be:

$$\begin{aligned}\text{Thermal Efficiency} &= \text{Hp Produced} \div \text{Hp in Fuel}\\ &= 200 \div 744\\ &= .2688 \text{ or } 26.88\%\end{aligned}$$

More than 70 percent of the energy in the fuel is not being used to create usable horsepower. The wasted energy is in the form of friction and heat. A tremendous amount of heat is given up to the atmosphere and not used inside the engine to create power.

Heat Transfer

There are three methods by which heat is transferred from one location to another or from one substance to another. These three methods are conduction, convection, and radiation.

Conduction

Heat transfer always takes place by areas of high heat energy migrating to areas of low heat energy. Heat transfer by conduction requires that there be physical contact between an object that has a large amount of heat energy and one that has a smaller amount of heat energy.

Everyone knows from experience that the metal handle of a heated pan can burn the hand. A plastic or wood handle, however, remains relatively cool even though it is in direct contact with the pan. The metal transmits the heat more easily than the wood because it is a better conductor of heat. Different materials conduct heat at different rates. Some metals are much better conductors of heat than others. Aluminum and copper are used in pots and pans because they conduct heat very rapidly. Woods and plastics are used for handles because they conduct heat very slowly.

Figure 3-27 illustrates the different rates of conduction of various metals. Of those listed, silver is the best conductor and lead is the poorest. As previously mentioned, copper and aluminum are used in pots and pans because they are good conductors. It is interesting to note that silver, copper, and aluminum are also excellent conductors of electricity.

Liquids are poorer conductors of heat than metals. Notice that the ice in the test tube shown in Figure 3-28 is not melting rapidly even though the water at the top is boiling. The water conducts heat so poorly that not enough heat reaches the ice to melt it.

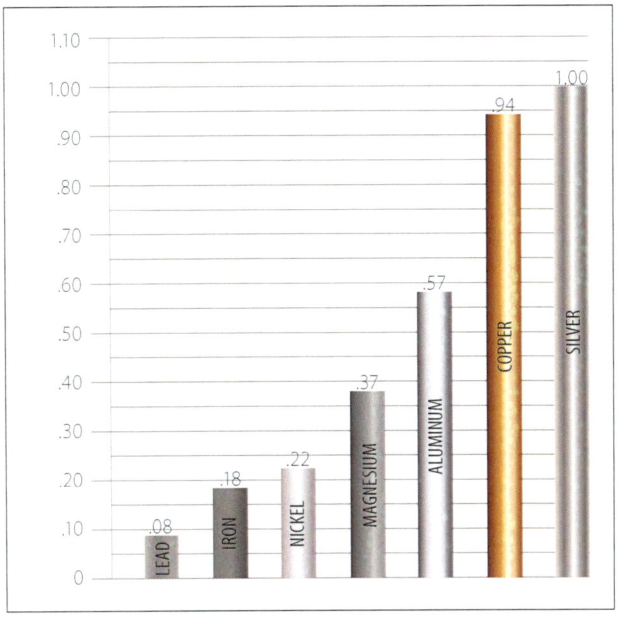

Figure 3-27. Conductivity of various metals.

Gases are even poorer conductors of heat than liquids. It is possible to stand quite close to a stove without being burned because air is such a poor conductor. Since conduction is a process whereby the increase in molecular energy is passed along by actual contact, gases are poor conductors.

At the point of application of the heat source, the molecules become violently agitated. These molecules strike adjacent molecules causing them to become agitated. This process continues until the heat energy is distributed evenly throughout the substance. Because molecules are farther apart in gases than in solids, the gases are much poorer conductors of heat.

Materials that are poor conductors are used to prevent the transfer of heat and are called heat insulators. A wooden handle on a pot or a soldering iron serves as a heat insulator. Certain materials, such as finely spun glass or asbestos, are particularly poor heat conductors. These materials are therefore used for many types of insulation.

Convection

Convection is the process by which heat is transferred by movement of a heated fluid (gas or liquid). For example, an incandescent light bulb will, when heated, become increasingly hotter until the air surrounding it begins to move. The motion of the air is upward. This upward motion of the heated air carries the heat away from the hot light bulb by convection. Transfer of heat by convection may be hastened by using a ventilating fan to move the air surrounding a hot object. The rate of cooling of a hot electronics component, such as the CPU in a computer, can be increased if it is provided with copper fins that conduct heat away from the hot surface. The fins provide large surfaces against which cool air can be blown.

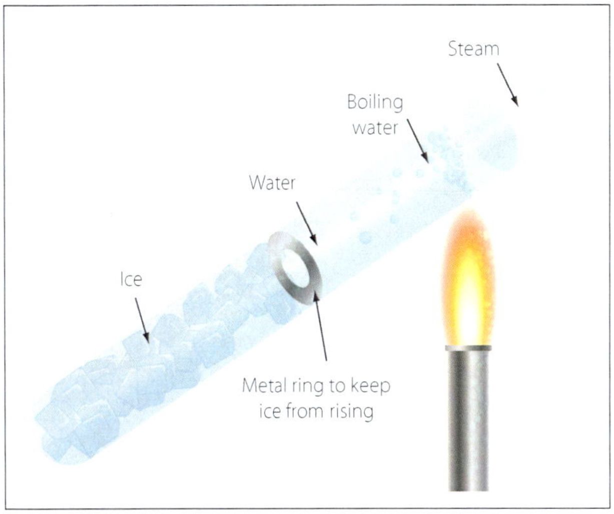

Figure 3-28. Water as a poor conductor.

A convection process may take place in a liquid as well as in a gas. A good example of this is a pan of water sitting on the stove. The bottom of the pan becomes hot because it conducts heat from the surface it is in contact with. The water on the bottom of the pan also heats up because of conduction. As the heated water starts to rise and cooler water moves in to take its place, the convection process begins.

When the circulation of gas or liquid is not rapid enough to remove sufficient heat, fans or pumps are used to accelerate the motion of the cooling material. In some installations, pumps are used to circulate water or oil to help cool large equipment. In airborne installations, electric fans and blowers are used to aid convection.

An aircraft air-cooled piston engine is a good example of convection being used to transfer heat. The engine shown in Figure 3-29 is a Continental IO-520, with six heavily finned air-cooled cylinders. This engine does not depend on natural convection for cooling, but rather forced air convection coming from the propeller on the engine. The heat generated inside the engine finds its way to the cylinder cooling fins by conduction, meaning transfer within the metal of the cylinder. Once the heat gets to the fins, forced air flowing around the cylinders carries the heat away.

Figure 3-29. Aircraft piston engine cooled by convection.

Radiation

Conduction and convection cannot wholly account for some of the phenomena associated with heat transfer. For example, the heat one feels when sitting in front of an open fire cannot be transferred by convection because the air currents are moving toward the fire. It cannot be transferred through conduction because the conductivity of the air is very small, and the cooler currents of air moving toward the fire would more than overcome the transfer of heat outward. Therefore, there must be some way for heat to travel across space other than by conduction and convection.

The existence of another process of heat transfer is still more evident when the heat from the sun is considered. Since conduction and convection take place only through some medium, such as a gas or a liquid, heat from the sun must reach the earth by another method, since space is an almost perfect vacuum. Radiation is the name given to this third method of heat transfer.

The term "radiation" refers to the continual emission of energy from the surface of all bodies. This energy is known as radiant energy. It is in the form of electromagnetic waves, radio waves, or x-rays, which are all alike except for a difference in wave length. These waves travel at the velocity of light and are transmitted through a vacuum more easily than through air because air absorbs some of them. Most forms of energy can be traced back to the energy of sunlight. Sunlight is a form of radiant heat energy that travels through space to reach the earth. These electromagnetic heat waves are absorbed when they come in contact with non-transparent bodies. The result is that the motion of the molecules in the body is increased as indicated by an increase in the temperature of the body.

The differences between conduction, convection, and radiation may now be considered. First, although conduction and convection are extremely slow, radiation takes place at the speed of light. This fact is evident at the time of an eclipse of the sun when the shutting off of the heat from the sun takes place at the same time as the shutting off of the light. Second, radiant heat may pass through a medium without heating it. For example, the air inside a greenhouse may be much warmer than the glass through which the sun's rays pass. Third, although heat transfer by conduction or convection may travel in roundabout routes, radiant heat always travels in a straight line. For example, radiation can be cut off with a screen placed between the source of heat and the body to be protected.

Specific Heat

One important way in which substances differ is in the requirement of different quantities of heat to produce the same temperature change in a given mass of the substance. Each substance requires a quantity of heat, called its specific heat capacity, to increase the temperature of a unit of its mass 1°C. The specific heat of a substance is the ratio of its specific heat capacity to the specific heat capacity of water. Specific heat is expressed as a number which, because it is a ratio, has no units and applies to both the English and the metric systems.

It is fortunate that water has a high specific heat capacity. The larger bodies of water on the earth keep the air and solid matter on or near the surface of the earth at a fairly constant temperature. A great quantity of heat is required to change the temperature of a large lake or river. Therefore, when the temperature falls below that of such bodies of water, they give off large quantities of heat. This process keeps the atmospheric temperature at the surface of the earth from changing rapidly.

The specific heat values of some common materials are listed in Figure 3-30.

Temperature

Temperature is a dominant factor affecting the physical properties of fluids. It is of particular concern when calculating changes in the state of gases.

The four temperature scales used extensively are the Centigrade, the Fahrenheit, the absolute or Kelvin, and the Rankine scales. The Centigrade scale is constructed by using the freezing and boiling points of water, under standard conditions, as fixed points of zero and 100, respectively, with 100 equal divisions between. The Fahrenheit scale uses 32° as the freezing point of water and 212° as the boiling point, and has 180 equal divisions between. The absolute or Kelvin scale is constructed with its zero point established as minus 273°C, meaning 273° below the freezing point of water. The relationships of the other fixed points of the scales are shown in Figure 3-31.

When working with temperatures, always make sure which system of measurement is being used and know how to convert from one to another. The conversion formulas are as follows:

$$\text{Degrees Fahrenheit} = (1.8 \times \text{Degrees Celsius}) + 32$$
$$\text{Degrees Celsius} = (\text{Degrees Fahrenheit} - 32) \times \tfrac{5}{9}$$
$$\text{Degrees Kelvin} = \text{Degrees Celsius} + 273$$
$$\text{Degrees Rankine} = \text{Degrees Fahrenheit} + 460$$

For purposes of calculations, the Rankine scale is commonly used to convert Fahrenheit to absolute. For Fahrenheit readings above zero, 460° is added. Thus, 72°F equals 460° plus 72°, or 532° absolute. If the Fahrenheit reading is below zero, it is subtracted from 460°. Thus −40°F equals 460° minus 40°, or 420° absolute. It should be stressed that the Rankine scale does not indicate absolute temperature readings in accordance with the Kelvin scale, but these conversions may be used for the calculations of changes in the state of gases.

The Kelvin and Centigrade scales are used more extensively in scientific work; therefore, some technical manuals may use these scales in giving directions and operating instructions. The Fahrenheit scale is commonly used in the United States, and most people are familiar with it. Therefore, the Fahrenheit scale is used in most areas of this book.

Material	Specific Heat
Lead	0.031
Mercury	0.033
Brass	0.094
Copper	0.095
Iron or Steel	0.113
Glass	0.195
Alcohol	0.547
Aluminum	0.712
Water	1.000

Figure 3-30. Specific heat value for various substances.

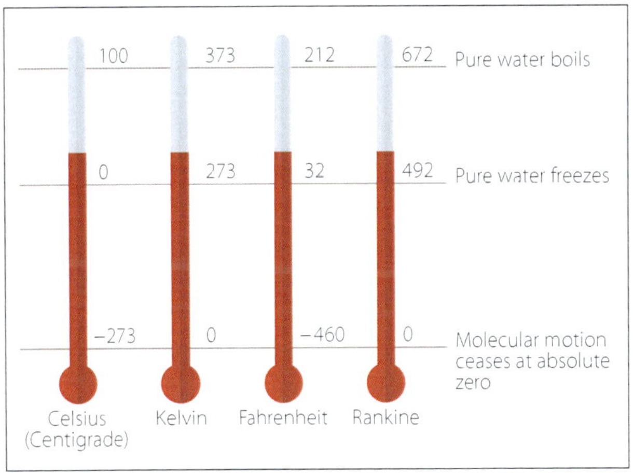

Figure 3-31. Comparison of temperature scales.

Thermal Expansion/Contraction

Thermal expansion takes place in solids, liquids, and gases when they are heated. With few exceptions, solids will expand when heated and contract when cooled. Because the molecules of solids are much closer together and are more strongly attracted to each other, the expansion of solids when heated is very slight in comparison to the expansion in liquids and gases. The expansion of fluids is discussed in the study of Boyle's law. Thermal expansion in solids must be explained in some detail because of its close relationship to aircraft metals and materials.

Because some substances expand more than others, it is necessary to measure experimentally the exact rate of expansion of each one. The amount that a unit length of any substance expands for a one degree rise in temperature is known as the coefficient of linear expansion for that substance. The coefficient of linear expansion for various materials is shown in Figure 3-32.

To estimate the expansion of any object, such as a steel rail, it is necessary to know three things about it: its length, the rise in temperature to which it is subjected, and its coefficient of expansion. This relationship is expressed by the equation:

Expansion = (coefficient) × (length) × (rise in temperature)

If a steel rod measures exactly 9 ft at 21°C, what is its length at 55°C? The coefficient of expansion for steel is 11×10^{-6}.

$$\text{Expansion} = (11 \times 10^{-6}) \times (9 \text{ feet}) \times 34°$$
$$\text{Expansion} = 0.003366 \text{ feet}$$

Substance	Coefficient of Expansion Per Degree Centigrade
Aluminum	25×10^{-6}
Brass or Bronze	19×10^{-6}
Brick	9×10^{-6}
Copper	17×10^{-6}
Glass (Plate)	9×10^{-6}
Glass (Pyrex)	3×10^{-6}
Ice	51×10^{-6}
Iron or Steel	11×10^{-6}
Lead	29×10^{-6}
Quartz	0.4×10^{-6}
Silver	19×10^{-6}

Figure 3-32. Coefficient of expansion for various materials.

This amount, when added to the original length of the rod, makes the rod 9.003366 ft long. Its length has only increased by $^{4}/_{100}$ of an inch.

The increase in the length of the rod is relatively small, but if the rod were placed where it could not expand freely, there would be a tremendous force exerted due to thermal expansion. Thus, thermal expansion must be taken into consideration when designing airframes, powerplants, or related equipment.

Pressure

Pressure is the amount of force acting on a specific amount of surface area. The force is typically measured in pounds and the surface area in square inches, making the units of pressure pounds per square inch or psi. If a 100-lb weight was placed on top of a block with a surface area of 10 in^2, the average weight distribution would be 10 lb for each of the square inches (100 ÷ 10), or 10 psi.

When atmospheric pressure is being measured, in addition to psi, other means of pressure measurement can be used. These include inches or millimeters of mercury, and millibars. Standard day atmospheric pressure is equal to 14.7 psi, 29.92 inches of mercury ("Hg), 760 millimeters of mercury (mm Hg), or 1013.2 millibars. The relationship between these units of measure is as follows:

1 psi = 2.04 "Hg
1 psi = 51.7 mm Hg
1 psi = 68.9 millibars

The concept behind measuring pressure in inches of mercury involves filling a test tube with the liquid mercury and then covering the top. The test tube is then turned upside down and placed in an open container of mercury, and the top is uncovered. Gravity acting on the mercury in the test tube will try to make the mercury run out. Atmospheric pressure pushing down on the mercury in the open container tries to make the mercury stay in the test tube. At some point these two forces (gravity and atmospheric pressure) will equal out and the mercury will stabilize at a certain height in the test tube. Under standard day atmospheric conditions, the air in a 1-in^2 column extending all the way to the top of the atmosphere would weigh 14.7 lb. A 1 in^2 column of mercury, 29.92 inches tall, would also weigh 14.7 lb. That is why 14.7 psi is equal to 29.92 "Hg. Figure 3-33 demonstrates this point.

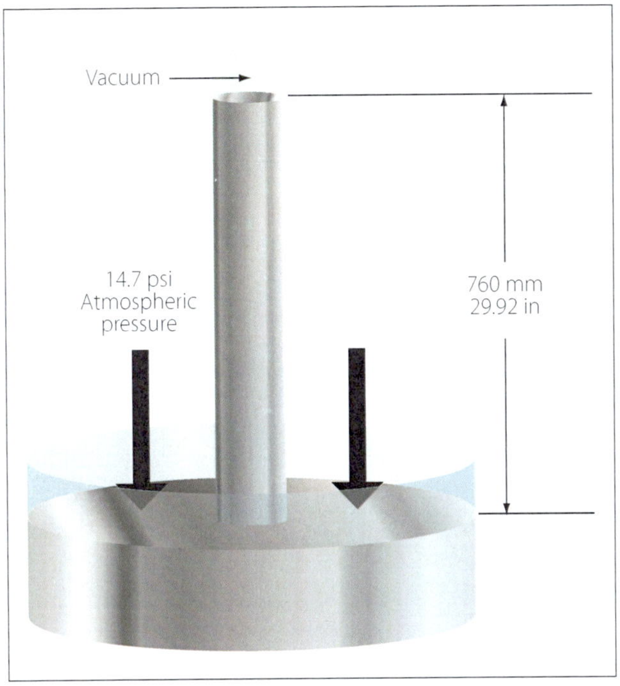

Figure 3-33. Atmospheric pressure as inches of mercury.

Gauge Pressure

When an instrument, such as an oil pressure gauge, fuel pressure gauge, or hydraulic system pressure gauge, displays pressure which is over and above ambient, the reading is referred to as gauge pressure (psig). This can be seen on the fuel pressure gauge shown in Figure 3-34. When the oil, fuel, or hydraulic pump is not turning, and there is no pressure being created, the gauge will read zero.

Absolute Pressure

A gauge that includes atmospheric pressure in its reading is measuring what is known as absolute pressure, or psia. Absolute pressure is equal to gauge pressure plus atmospheric pressure. If someone hooked up a psia indicating instrument to an engine's oil system, the gauge would read atmospheric pressure when the engine was not running. Since this would not make good sense to the typical operator, psia gauges are not used in this type of application. For the manifold pressure on a piston engine, a psia gauge does make good sense. Manifold pressure on a piston engine can read anywhere from less than atmospheric pressure if the engine is not supercharged, to more than atmospheric if it is supercharged. The only gauge that has the flexibility to show this variety of readings is the absolute pressure gauge. Figure 3-35 shows a manifold pressure gauge, with a readout that ranges from 10 "Hg to 35 "Hg. Remember that 29.92 "Hg is standard day atmospheric.

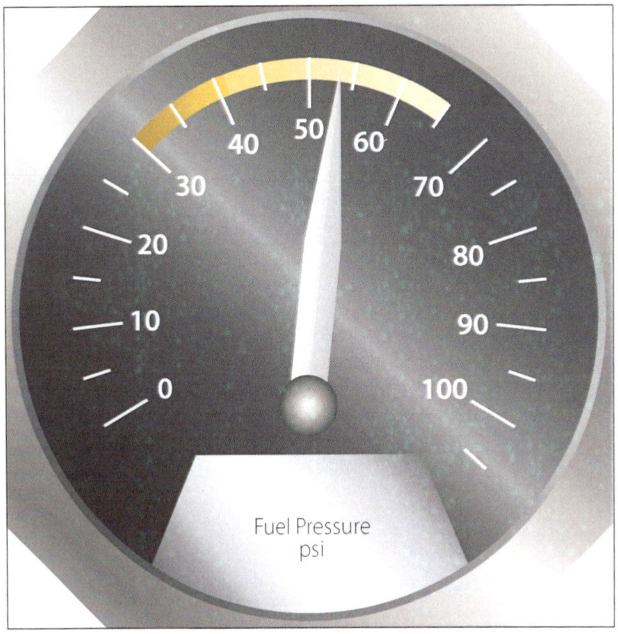

Figure 3-34. Psig read on a fuel pressure gauge.

Figure 3-35. Manifold pressure gauge indicating absolute pressure.

Differential Pressure

Differential pressure, or psid, is the difference between pressures being read at two different locations within a system. For example, in a turbine engine oil system the pressure is read as it enters the oil filter, and also as it leaves the filter. These two readings are sent to a transmitter which powers a light located on the flight deck. Across anything that poses a resistance to flow, like an oil filter, there will be a drop in pressure. If the filter starts to clog, the pressure drop will become

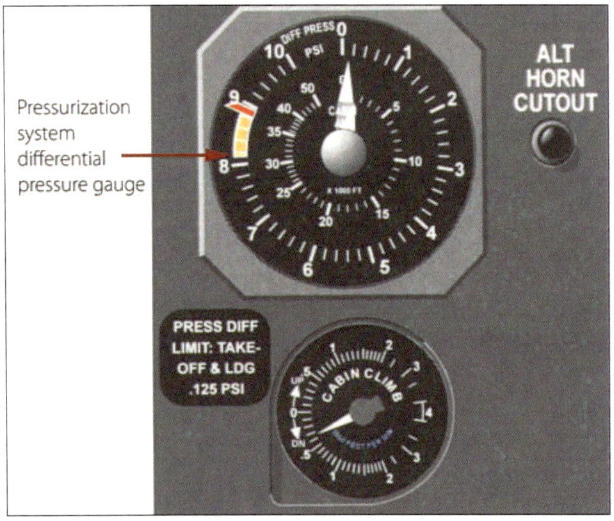

Figure 3-36. Differential pressure gauge.

greater, eventually causing the advisory light on the flight deck to come on.

Figure 3-36 shows a differential pressure gauge for the pressurization system on a Boeing 737. In this case, the difference in pressure is between the inside and the outside of the airplane. If the pressure difference becomes too great, the structure of the airplane could become overstressed.

Gas Laws

The simple structure of gases makes them readily adaptable to mathematical analysis from which has evolved a detailed theory of the behavior of gases. This is called the kinetic theory of gases. The theory assumes that a body of gas is composed of identical molecules which behave like minute elastic spheres, spaced relatively far apart and continuously in motion.

The degree of molecular motion is dependent upon the temperature of the gas. Since the molecules are continuously striking against each other and against the walls of the container, an increase in temperature with the resulting increase in molecular motion causes a corresponding increase in the number of collisions between the molecules. The increased number of collisions results in an increase in pressure because a greater number of molecules strike against the walls of the container in a given unit of time.

If the container were an open vessel, the gas would expand and overflow from the container. However, if the container is sealed and possesses elasticity (such as a rubber balloon), the increased pressure causes the container to expand. For instance, when making a long drive on a hot day, the pressure in the tires of an automobile increases, and a tire which appeared to be somewhat "soft" in cool morning temperature may appear normal at a higher midday temperature.

Such phenomena as these have been explained and set forth in the form of laws pertaining to gases and tend to support the kinetic theory.

Boyle's Law

As previously stated, compressibility is an outstanding characteristic of gases. The English scientist, Robert Boyle, was among the first to study this characteristic that he called the "springiness of air." By direct measurement he discovered that when the temperature of a combined sample of gas was kept constant and the absolute pressure doubled, the volume was reduced to half the former value. As the applied absolute pressure was decreased, the resulting volume increased. From these observations, he concluded that for a constant temperature the product of the volume and absolute pressure of an enclosed gas remains constant. Boyle's law is normally stated: "The volume of an enclosed dry gas varies inversely with its absolute pressure, provided the temperature remains constant." The following formula is used for Boyle's law calculations. Remember, pressure needs to be in the absolute.

$$\text{Volume 1} \times \text{Pressure 1} = \text{Volume 2} \times \text{Pressure 2}$$
Or
$$V1P1 = V2P2$$

Example: *10 ft^3 of nitrogen is under a pressure of 500 psia. If the volume is reduced to 7 ft^3, what will the new pressure be? [Figure 3-37]*

$$V1P1 = V2P2$$
$$10\ (500) = 7\ (P2)$$
$$10\ (500) \div 7 = P2$$
$$P2 = 714.29 \text{ psia}$$

The useful applications of Boyle's law are many and varied. Some applications more common to aviation are: (1) the carbon dioxide (CO_2) bottle used to inflate life rafts and life vests; (2) the compressed oxygen and the acetylene tanks used in welding; (3) the compressed air brakes and shock absorbers; and (4) the use of oxygen tanks for high altitude flying and emergency use.

Charles' Law

The French scientist, Jacques Charles, provided much of the foundation for the modern kinetic theory of gases. He found that all gases expand and contract in direct proportion to the change in the absolute tem-

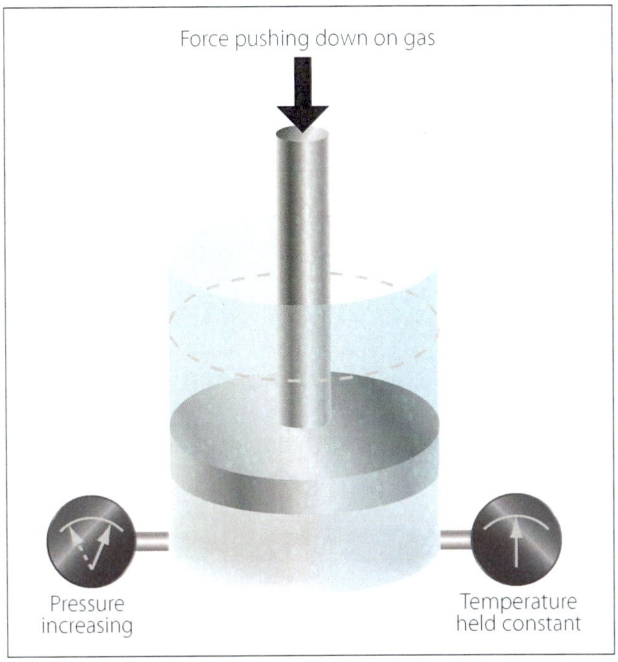

Figure 3-37. Boyle's law example.

perature, provided the pressure is held constant. As a formula, this law is shown as follows:

$$\text{Volume 1} \times \text{Absolute Temperature 2} = \text{Volume 2} \times \text{Absolute Temperature 1}$$

Or

$$V1T2 = V2T1$$

Charles' law also works if the volume is held constant, and pressure and temperature are the variables. In this case, the formula would be as follows:

$$P1T2 = P2T1$$

For this second formula, pressure and temperature must be in the absolute.

Example: *A 15-ft^3 cylinder of oxygen is at a temperature of 70°F and a pressure of 750 psig. The cylinder is placed in the sun and the temperature of the oxygen increases to 140°F. What would be the new pressure in psig?*

70 degrees Fahrenheit = 530 degrees Rankine
140 degrees Fahrenheit = 600 degrees Rankine
750 psig + 14.7 = 764.7 psia

$$P1T2 = P2T1$$
$$764.7\ (600) = P2\ (530)$$
$$P2 = 764.7\ (600) \div 530$$
$$P2 = 865.7 \text{ psia}$$
$$P2 = 851 \text{ psig}$$

General Gas Law

By combining Boyle's and Charles' laws, a single expression can be derived which states all the information contained in both. The formula which is used to express the general gas law is as follows:

$$\frac{\text{Pressure 1 (Volume 1)}}{\text{Temperature 1}} = \frac{\text{Pressure 2 (Volume 2)}}{\text{Temperature 2}}$$

Or

$$P1\ (V1)\ (T2) = P2\ (V2)\ (T1)$$

When using the general gas law formula, temperature and pressure must be in the absolute.

Example: *20 ft^3 of the gas argon is compressed to 15 ft^3. The gas starts out at a temperature of 60°F and a pressure of 1,000 psig. After being compressed, its temperature is 90°F. What would its new pressure be in psig?*

60 degrees Fahrenheit = 520 degrees Rankine
90 degrees Fahrenheit = 550 degrees Rankine
1,000 psig + 14.7 = 1,014.7 psia

$$P1\ (V1)\ (T2) = P2\ (V2)\ (T1)$$
$$1{,}014.7\ (20)\ (550) = P2\ (15)\ (520)$$
$$P2 = 1{,}431 \text{ psia}$$
$$P2 = 1{,}416.3 \text{ psig}$$

Dalton's Law

If a mixture of two or more gases that do not combine chemically is placed in a container, each gas expands throughout the total space and the absolute pressure of each gas is reduced to a lower value, called its partial pressure. This reduction is in accordance with Boyle's law. The pressure of the mixed gases is equal to the sum of the partial pressures. This fact was discovered by Dalton, an English physicist, and is set forth in Dalton's law: "A mixture of several gases which do not react chemically exerts a pressure equal to the sum of the pressures which the several gases would exert separately if each were allowed to occupy the entire space alone at the given temperature."

Fluid Mechanics

A fluid, by definition, is any substance that is able to flow if it is not in some way confined or restricted. Liquids and gases are both classified as fluids, and often act in a very similar way. One significant difference comes into play when a force is applied to these fluids. In this case, liquids tend to be incompressible and gases are highly compressible. Many of the principles that aviation is based on, such as the theory of lift on a wing and the force generated by a hydraulic system, can be explained and quantified by using the laws of fluid mechanics.

Buoyancy

A solid body submerged in a liquid or a gas weighs less than when weighed in free space. This is because of the upward force, called buoyant force, which any fluid exerts on a body submerged in it. An object will float if this upward force of the fluid is greater than the weight of the object. Objects denser than the fluid, even though they sink readily, appear to lose a part of their weight when submerged. A person can lift a larger weight under water than he or she can possibly lift in the air.

The following experiment is illustrated in Figure 3-38. The overflow can is filled to the spout with water. The heavy metal cube is first weighed in still air and weighs 10 lb. It is then weighed while completely submerged in the water and it weighs 3 lb. The difference between the two weights is the buoyant force of the water. As the cube is lowered into the overflow can, the water is caught in the catch bucket. The volume of water which overflows equals the volume of the cube. (The volume of irregular shaped objects can be measured by this method.) If this experiment is performed carefully, the weight of the water displaced by the metal cube exactly equals the buoyant force of the water, which the scale shows to be 7 lb.

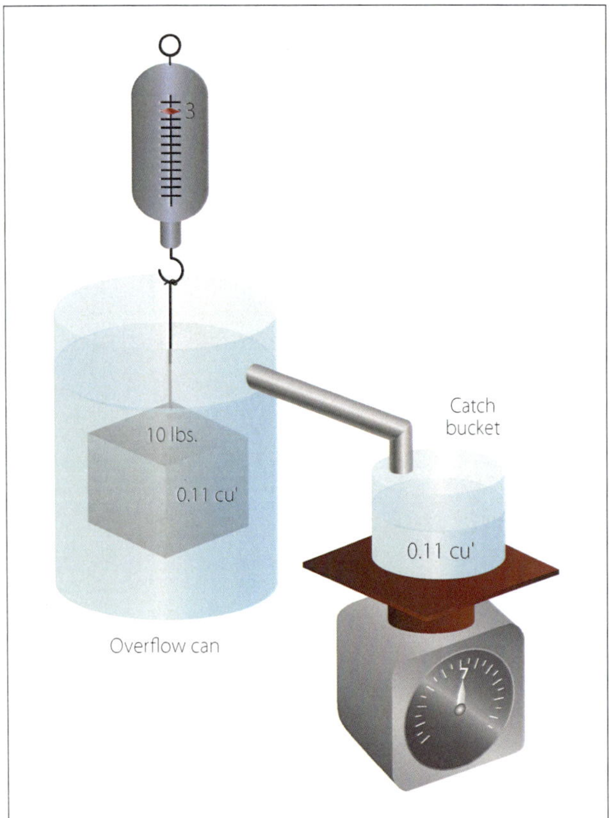

Figure 3-38. Example of buoyancy.

Archimedes (287–212 B.C.) performed similar experiments. As a result, he discovered that the buoyant force which a fluid exerts upon a submerged body is equal to the weight of the fluid the body displaces. This statement is referred to as Archimedes' principle. This principle applies to all fluids, gases as well as liquids. Just as water exerts a buoyant force on submerged objects, air exerts a buoyant force on objects submerged in it.

The amount of buoyant force available to an object can be calculated by using the following formula:

Buoyant Force = Volume of Object × Density of Fluid Displaced

If the buoyant force is more than the object weighs, the object will float. If the buoyant force is less than the object weighs, the object will sink. For the object that sinks, its measurable weight will be less by the weight of the displaced fluid.

Example: *A 10-ft^3 object weighing 700 lb is placed in pure water. Will the object float? If the object sinks, what is its measurable weight in the submerged condition? If the object floats, how many cubic feet of its volume is below the water line?*

Buoyant Force = Volume of Object × Density of Fluid Displace
= 10 (62.4)
= 624 lb

Because the buoyant force is less than the object weighs, the object will sink. The difference between the buoyant force and the object's weight will be its measurable weight, or 76 lb.

Two good examples of buoyancy are a helium filled airship and a seaplane on floats. An airship is able to float in the atmosphere and a seaplane is able to float on water. That means both have more buoyant force than weight. Figure 3-39 is a DeHavilland Twin Otter seaplane, with a gross takeoff weight of 12,500 lb. At a minimum, the floats on this airplane must be large enough to displace a weight in water equal to the airplane's weight. According to Title 14 of the Code of Federal Regulations (14 CFR) part 23, the floats must be 80 percent larger than the minimum needed to support the airplane. For this airplane, the necessary size of the floats would be calculated as follows:

Divide the airplane weight by the density of water.
12,500 ÷ 62.4 = 200.3 ft^3

Multiply this volume by 80%.
200.3 × 80% = 160.2 ft^3

Add the two volumes together to get the total volume of the floats.
200.3 + 160.2 = 360.5 ft^3

By looking at the Twin Otter in Figure 3-39, it is obvious that much of the volume of the floats is out of the water. This is accomplished by making sure the floats have at least 80 percent more volume than the minimum necessary.

Figure 3-39. De Havilland Twin Otter seaplane.

Some of the large Goodyear airships have a volume of 230,000 ft^3. Since the fluid they are submerged in is air, to find the buoyant force of the airship, the volume of the airship is multiplied by the density of air (.07651 $^{lb}/_{ft^3}$). For this Goodyear airship, the buoyant force is 17,597 lb. Figure 3-40 shows an inside view of the Goodyear airship.

The ballonets, items 2 and 4 in the picture, are air chambers within the airship. Through the air scoop, item 9, air can be pumped into the ballonets or evacuated from the ballonets in order to control the weight of the airship. Controlling the weight of the airship controls how much positive or negative lift it has. Although the airship is classified as a lighter-than-air aircraft, it is in fact flown in a condition slightly heavier than air.

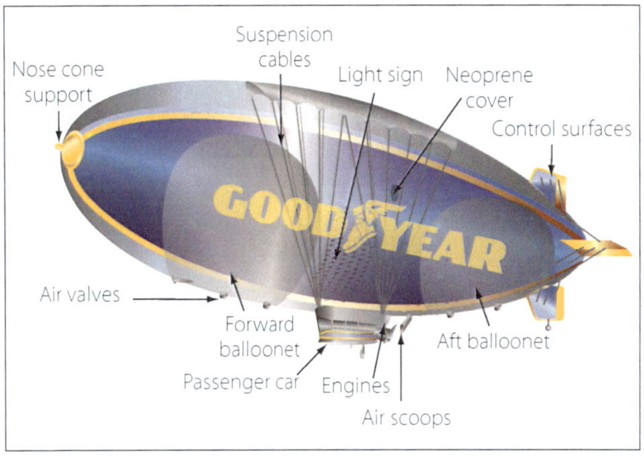

Figure 3-40. The Goodyear Airship and buoyancy.

Fluid Pressure

The pressure exerted on the bottom of a container by a liquid is determined by the height of the liquid and not by the shape of the container. This can be seen in Figure 3-41, where three different shapes and sizes of containers are full of colored water. Even though they are different shapes and have different volumes of liquid, each one has a height of 231 inches. Because of this height, each one would exert a pressure on the bottom of 8.34 psi. The container on the left, with a surface area of 1 in^2, contains a volume of 231 in^3 (one gallon). One gallon of water weighs 8.34 lb, which is why the pressure on the bottom is 8.34 psi.

Still thinking about Figure 3-41, if the pressure was measured half way down, it would be half of 8.34, or 4.17 psi. In other words, the pressure is adjustable by varying the height of the column. Pressure based on the column height of a fluid is known as static pressure. With liquids, such as gasoline, it is sometimes referred to as a head of pressure. For example, if a carburetor needs to have 2 psi supplied to its inlet (head of pressure), this could be accomplished by having the

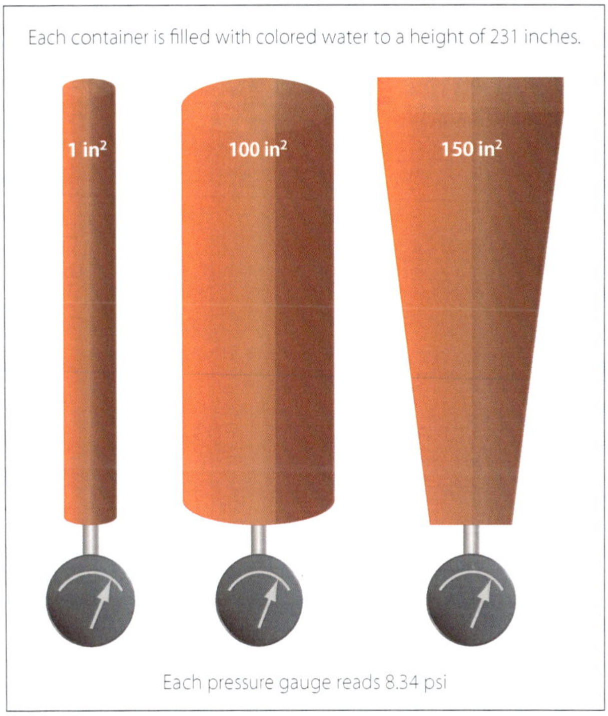

Figure 3-41. Fluid pressure based on column height.

fuel tank positioned the appropriate number of inches higher than the carburetor.

As identified in the previous paragraph, pressure due to the height of a fluid column is known as static pressure. When a fluid is in motion, and its velocity is converted to pressure, that pressure is known as ram. When ram pressure and static pressure are added together, the result is known as total pressure. In the inlet of a gas turbine engine, for example, total pressure is often measured to provide a signal to the fuel metering device or to provide a signal to a gauge on the flight deck.

Pascal's Law

The foundations of modern hydraulics and pneumatics were established in 1653 when Pascal discovered that pressure set up in a fluid acts equally in all directions. This pressure acts at right angles to containing surfaces. When the pressure in the fluid is caused solely by the fluid's height, the pressure against the walls of the container is equal at any given level, but it is not equal if the pressure at the bottom is compared to the pressure half way down. The concept of the pressure set up in a fluid, and how it relates to the force acting on the fluid and the surface area through which it acts, is Pascal's law.

In Figure 3-41, if a piston is placed at the top of the cylinder and an external force pushes down on the piston, additional pressure will be created in the liquid. If the additional pressure is 100 psi, this 100 psi will act equally and undiminished from the top of the cylinder all the way to the bottom. The gauge at the bottom will now read 108.34 psi, and if a gauge were positioned half way down the cylinder, it would read 104.17 psi (100 plus half of 8.34).

Pascal's law, when dealing with the variables of force, pressure, and area, is dealt with by way of the following formula.

$$\text{Force} = \text{Pressure} \times \text{Area}$$

In this formula, the force is in units of pounds, the pressure is in pounds per square inch (psi), and the area is in square inches. By transposing the original formula, we have two additional formulas, as follows:

$$\text{Pressure} = \text{Force} \div \text{Area}$$

And

$$\text{Area} = \text{Force} \div \text{Pressure}$$

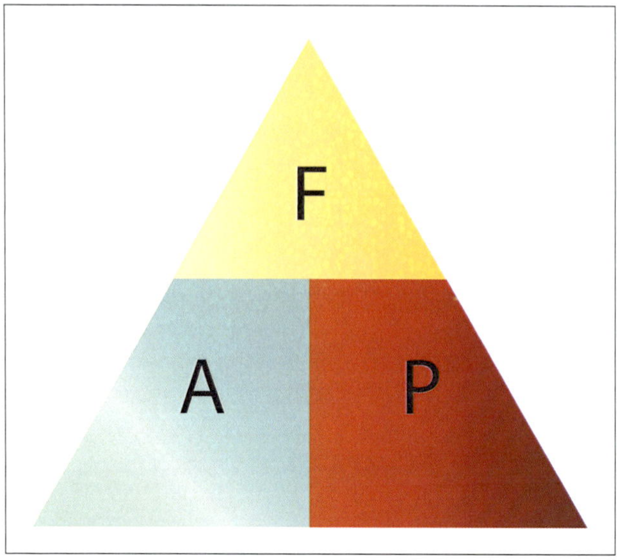

Figure 3-42. Force, area, pressure relationship.

An easy and convenient way to remember the formulas for Pascal's law, and the relationship between the variables, is with the triangle shown in Figure 3-42. If the variable we want to solve for is covered up, the position of the remaining two variables shows the proper math relationship. For example, if the "A" (area) is covered up, what remains is the "F" on the top and the "P" on the bottom, meaning force divided by pressure.

The simple hydraulic system in Figure 3-43 has a 5-lb force acting on a piston with a $\frac{1}{2}$-in^2 surface area. Based on Pascal's law, the pressure in the system would be equal to the force applied divided by the area of the piston, or 10 psi. As shown in Figure 3-43, the pressure of 10 psi is present everywhere in the fluid.

The hydraulic system in Figure 3-44 is a little more complex than the one in Figure 3-43. In Figure 3-44, the input force of 5 lb is acting on a $\frac{1}{2}$-in^2 piston, creating a pressure of 10 psi. The input cylinder and piston is connected to a second cylinder, which contains a 5-in^2 piston. The pressure of 10 psi created by the

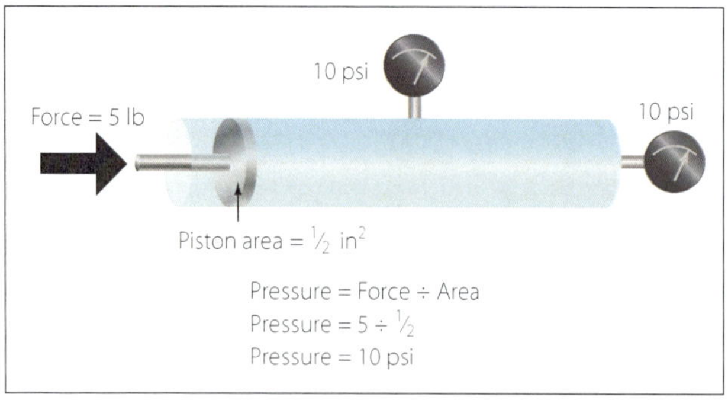

Figure 3-43. Pressure created in a hydraulic system.

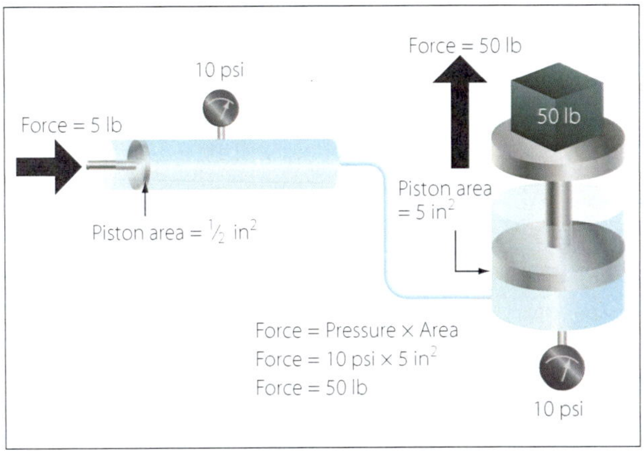

Figure 3-44. Output force created in a hydraulic system.

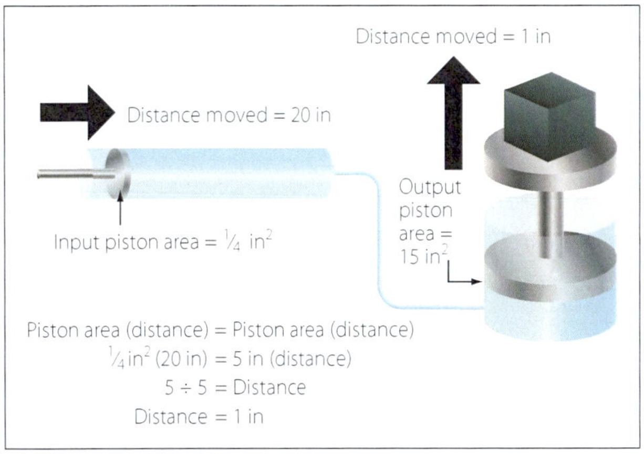

Figure 3-45. Piston movement in a hydraulic system.

input piston pushes on the piston in the second cylinder, creating an output force of 50 pounds.

More often than not, the purpose of a hydraulic system is to generate a large output force, with the input force being much less. In Figure 3-44, the input force is 5 lb and the output force is 50 lb, or 10 times greater. The relationship between the output force and the input force, as discussed earlier in this chapter, is known as mechanical advantage. The mechanical advantage in Figure 3-44 would be 50 divided by 5, or 10. The following formulas can be used to calculate mechanical advantage.

Mechanical Advantage = Force Out ÷ Force In
Or
Mechanical Advantage = Distance Out ÷ Distance In

Earlier in this chapter when simple machines, such as levers and gears were discussed, it was identified that no machine allows us to gain work. The same statement holds true for a hydraulic system, that we get no more work out of a hydraulic system than we put in. Since work is equal to force times distance, if we gain force with a hydraulic system, we must lose distance. We only get the same work out, if the system is 100 percent efficient.

In order to think about the distance that the output piston will move in response to the movement of the input piston, the volume of fluid displaced must be considered. In the study of geometry, one learns that the volume of a cylinder is equal to the cylinder's surface area multiplied by its height. So when a piston of 2 in^2 moves down in a cylinder a distance of 10 in, it displaces a volume of fluid equal to 20 in^3 (2 in^2 × 10 in). The 20 in^3 displaced by the first piston is what moves over to the second cylinder and causes its piston to move. In a simple two-piston hydraulic system, the relationship between the piston area and the distance moved is shown by the following formula.

Input Piston Area (Distance Moved) =
Output Piston Area (Distance Moved)

In essence, this formula shows that the volume in is equal to the volume out. This concept is shown in Figure 3-45, where a small input piston moves a distance of 20 inches, and the larger output piston only moves a distance of 1 inch.

Example: *A two-piston hydraulic system, like that shown in Figure 3-45, has an input piston with an area of $^1/_4$ in^2 and an output piston with an area of 15 in^2. An input force of 50 lb is applied, and the input piston moves 30 inches. What is the pressure in the system, how much force is generated by the output piston, how far would the output piston move, and what is the mechanical advantage?*

Pressure = Force ÷ Area
= 50 ÷ ¼
= 200 psi

Force = Pressure × Area
= 200 × 15
= 3,000 lb

Mechanical Advantage = Force Out ÷ Force In
= 3,000 ÷ 50
= 60

Input Piston Area
(Distance Moved) = Output Piston Area
(Distance Moved)
¼ (30) = 15 (Distance Moved)
¼ (30) ÷ 15 = Distance Moved
Distance Moved = ½ in

Part of understanding Pascal's law and hydraulics involves utilizing formulas, and recognizing the relationship between the individual variables. Before the numbers are plugged into the formulas, it is often possible to analyze the variables in the system and come to a realization about what is happening. For example, look at the variables in Figure 3-45 and notice that the output piston is 20 times larger than the input piston (5 in^2 compared to ¼ in^2). That comparison tells us that the output force will be 20 times greater than the input force, and also that the output piston will only move 1/20 as far. Without doing any formula based calculations, we can conclude that the hydraulic system in question has a mechanical advantage of 20.

Bernoulli's Principle

Bernoulli's principle was originally stated to explain the action of a liquid flowing through the varying cross-sectional areas of tubes. In Figure 3-46 a tube is shown in which the cross-sectional area gradually decreases to a minimum diameter in its center section. A tube constructed in this manner is called a "venturi," or "venturi tube." Where the cross-sectional area is decreasing, the passageway is referred to as a converging duct. As the passageway starts to spread out, it is referred to as a diverging duct.

As a liquid (fluid) flows through the venturi tube, the gauges at points "A," "B," and "C" are positioned to register the velocity and the static pressure of the liquid. The venturi in Figure 3-46 can be used to illustrate Bernoulli's principle, which states that: The static pressure of a fluid (liquid or gas) decreases at points where the velocity of the fluid increases, provided no energy is added to nor taken away from the fluid. The velocity of the air is kinetic energy and the static pressure of the air is potential energy.

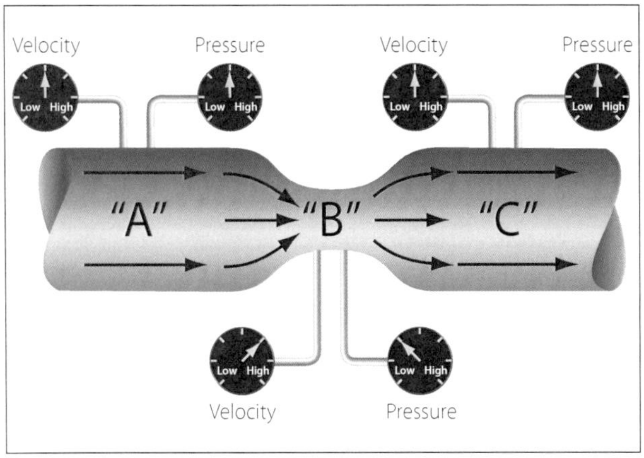

Figure 3-46. Bernoulli's principle and a venturi.

In the wide section of the venturi (points A and C of Figure 3-46), the liquid moves at low velocity, producing a high static pressure, as indicated by the pressure gauge. As the tube narrows in the center, it must contain the same volume of fluid as the two end areas. In this narrow section, the liquid moves at a higher velocity, producing a lower pressure than that at points A and C, as indicated by the velocity gauge reading high and the pressure gauge reading low. A good application for the use of the venturi principle is in a float-type carburetor. As the air flows through the carburetor on its way to the engine, it goes through a venturi, where the static pressure is reduced. The fuel in the carburetor, which is under a higher pressure, flows into the lower pressure venturi area and mixes with the air.

Bernoulli's principle is extremely important in understanding how some of the systems used in aviation work, including how the wing of an airplane generates lift or why the inlet duct of a turbine engine on a subsonic airplane is diverging in shape. The wing on a slow moving airplane has a curved top surface and a relatively flat bottom surface. The curved top surface acts like half of the converging shaped middle of a venturi. As the air flows over the top of the wing, the air speeds up, and its static pressure decreases. The static pressure on the bottom of the wing is now greater than the pressure on the top, and this pressure difference creates the lift on the wing. Bernoulli's principle and the concept of lift on a wing is covered in greater depth in "Aircraft Theory of Flight" located in this chapter.

Sound

Sound has been defined as a series of disturbances in matter that the human ear can detect. This definition can also be applied to disturbances which are beyond the range of human hearing. There are three elements which are necessary for the transmission and reception of sound. These are the source, a medium for carrying the sound, and the detector. Anything which moves back and forth (vibrates) and disturbs the medium around it may be considered a sound source.

An example of the production and transmission of sound is the ring of a bell. When the bell is struck and begins to vibrate, the particles of the medium (the surrounding air) in contact with the bell also vibrate. The vibrational disturbance is transmitted from one particle of the medium to the next, and the vibrations travel in a "wave" through the medium until they reach the ear. The eardrum, acting as detector, is set in motion by the vibrating particles of air, and the brain interprets the eardrum's vibrations as the characteristic sound associated with a bell.

Wave Motion

Since sound is a wave motion in matter, it can best be understood by first considering water waves. When an object is thrown into a pool, a series of circular waves travel away from the disturbance. In Figure 3-47 such waves are seen from a top perspective, with the waves traveling out from the center. In the cross-section perspective in Figure 3-47, notice that the water waves are a succession of crests and troughs. The wavelength is the distance from the crest of one wave to the crest of the next. Water waves are known as transverse waves because the motion of the water molecules is up and down, or at right angles to the direction in which the waves are traveling. This can be seen by observing a cork on the water, bobbing up and down as the waves pass by.

Sound travels through matter in the form of longitudinal wave motions. These waves are called longitudinal waves because the particles of the medium vibrate back and forth longitudinally in the direction of propagation. [Figure 3-48] When the tine of a tuning fork moves in an outward direction, the air immediately in front of the tine is compressed so that its momentary pressure is raised above that at other points in the surrounding medium. Because air is elastic, this disturbance is transmitted progressively in an outward direction from the tine in the form of a compression wave.

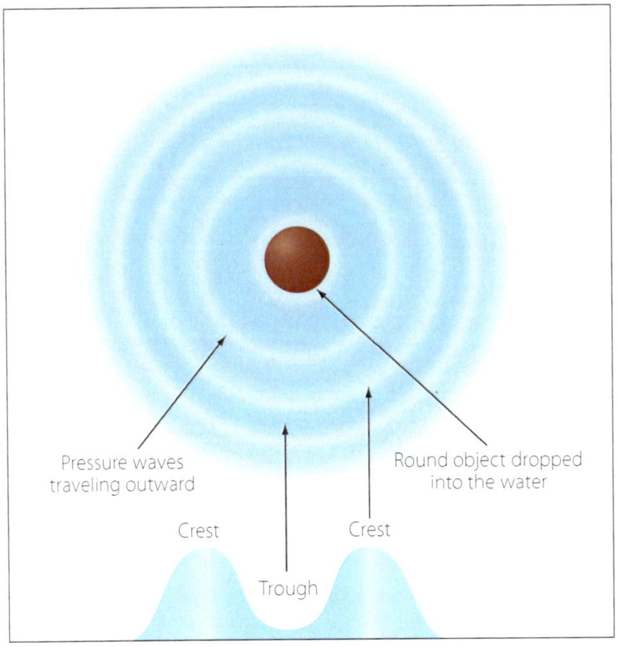

Figure 3-47. Relationship between sound and waves in water.

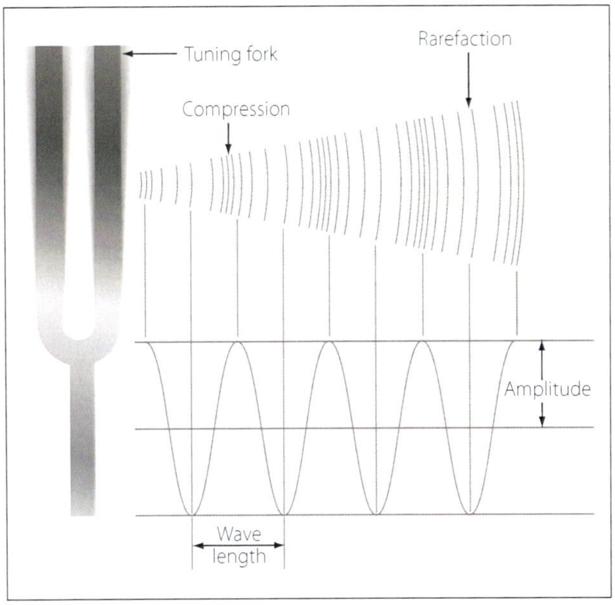

Figure 3-48. Sound propagation by a tuning fork.

When the tine returns and moves in an inward direction, the air in front of the tine is rarefied so that its momentary pressure is reduced below that at other points in the surrounding medium. This disturbance is transmitted in the form of a rarefaction (expansion) wave and follows the compression wave through the medium. The progress of any wave involves two distinct motions: (1) The wave itself moves forward with constant speed, and (2) simultaneously, the particles of the medium that convey the wave vibrate harmonically. Examples of harmonic motion are the motion of a clock pendulum, the balance wheel in a watch, and the piston in a reciprocating engine.

Speed of Sound

In any uniform medium, under given physical conditions, sound travels at a definite speed. In some substances, the velocity of sound is higher than in others. Even in the same medium under different conditions of temperature, pressure, and so forth, the velocity of sound varies. Density and elasticity of a medium are the two basic physical properties which govern the velocity of sound.

In general, a difference in density between two substances is sufficient to indicate which one will be the faster transmission medium for sound. For example, sound travels faster through water than it does through air at the same temperature. However, there are some surprising exceptions to this rule of thumb. An outstanding example among these exceptions involves comparison of the speed of sound in lead and aluminum at the same temperature. Sound travels at 16,700 fps in aluminum at 20°C, and only 4,030 fps in lead at 20°C, despite the fact that lead is much more dense than aluminum. The reason for such exceptions is found in the fact, mentioned above, that sound velocity depends on elasticity as well as density.

Using density as a rough indication of the speed of sound in a given substance, it can be stated as a general rule that sound travels fastest in solid materials, slower in liquids, and slowest in gases. The velocity of sound in air at 0°C (32°F) is 1,087 fps and increases by 2 fps for each Centigrade degree of temperature rise (1.1 fps for each degree Fahrenheit).

Mach Number

In the study of aircraft that fly at supersonic speeds, it is customary to discuss aircraft speed in relation to the velocity of sound (approximately 760 miles per hour (mph) at 59°F). The term "Mach number" has been given to the ratio of the speed of an aircraft to the speed of sound, in honor of Ernst Mach, an Austrian scientist. If the speed of sound at sea level is 760 mph, an aircraft flying at a Mach number of 1.2 at sea level would be traveling at a speed of 760 mph × 1.2 = 912 mph.

Frequency of Sound

The term "pitch" is used to describe the frequency of a sound. The outstanding recognizable difference between the tones produced by two different keys on a piano is a difference in pitch. The pitch of a tone is proportional to the number of compressions and rarefactions received per second, which in turn, is determined by the vibration frequency of the sounding source. A good example of frequency is the noise generated by a turbofan engine on a commercial airliner. The high tip speeds of the fan in the front of the engine creates a high frequency sound, and the hot exhaust creates a low frequency sound.

Loudness

When a bell rings, the sound waves spread out in all directions and the sound is heard in all directions. When a bell is struck lightly, the vibrations are of small amplitude and the sound is weak. A stronger blow produces vibrations of greater amplitude in the bell, and the sound is louder. It is evident that the amplitude of the air vibrations is greater when the amplitude of the vibrations of the source is increased. Hence, the loudness of the sound depends on the amplitude of the vibrations of the sound waves. As the distance from the source increases, the energy in each wave spreads out, and the sound becomes weaker.

As the sound wave advances, variations in pressure occur at all points in the transmitting medium. The greater the pressure variations, the more intense the sound wave. The intensity is proportional to the square of the pressure variation regardless of the frequency. Thus, by measuring pressure changes, the intensities of sounds having different frequencies can be compared directly.

Measurement of Sound Intensity

Sound intensity is measured in decibels, with a decibel being the ratio of one sound to another. One decibel (dB) is the smallest change in sound intensity the human ear can detect. A faint whisper would have an intensity of 20 dB, and a pneumatic drill would be 80 dB. The engine on a modern jetliner, at takeoff thrust, would have a sound intensity of 90 dB when heard by someone standing 150 ft away. A 110 dB noise, by comparison, would sound twice as loud as the jetliner's engine. Figure 3-49 shows the sound intensity from a variety of different sources.

Doppler Effect

When sound is coming from a moving object, the object's forward motion adds to the frequency as sensed from the front and takes away from the frequency as sensed from the rear. This change in frequency is known as the Doppler effect, and it explains

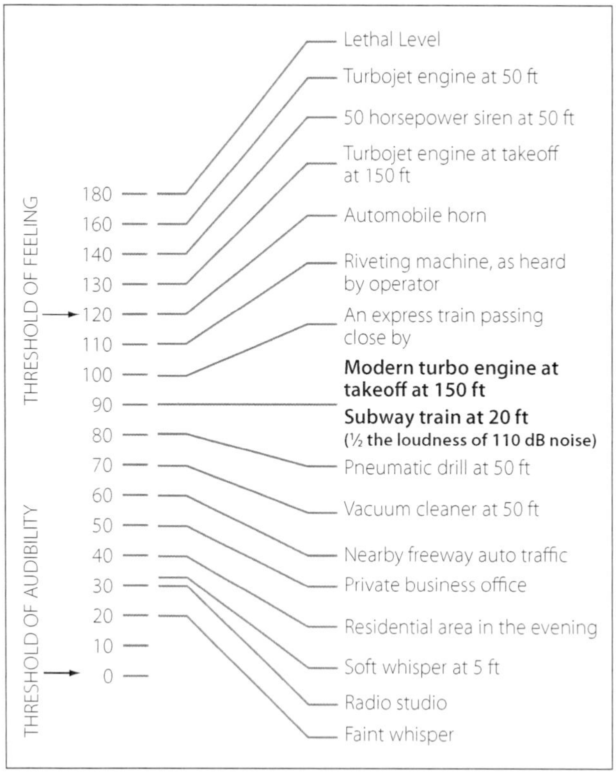

Figure 3-49. Sound intensity from different sources.

why the sound from an airplane seems different as it approaches compared to how it sounds as it flies overhead. As it approaches, it becomes both louder and higher pitched. As it flies away, the loudness and pitch both decrease noticeably.

If an airplane is flying at or higher than the speed of sound, the sound energy cannot travel out ahead of the airplane, because the airplane catches up to it the instant it tries to leave. The sound energy being created by the airplane piles up, and attaches itself to the structure of the airplane. As the airplane approaches, a person standing on the ground will not be able to hear it until it gets past their position, because the sound energy is actually trailing behind the airplane. When the sound of the airplane is heard, it will be in the form of what is called a sonic boom.

Resonance

All types of matter, regardless of whether it is a solid, liquid, or gas, have a natural frequency at which the atoms within that matter vibrate. If two pieces of matter have the same natural frequency, and one of them starts to vibrate, it can transfer its wave energy to the other one and cause it to vibrate. This transfer of energy is known as resonance. Some piston engine powered airplanes have an rpm range that they are placarded to avoid, because spinning the prop at that rpm can cause vibration problems. The difficulty lies in the natural frequency of the metal in the prop, and the frequency of vibration that will be set up with a particular tip speed for the prop. At that particular rpm, stresses can be set up that could lead to the propeller coming apart.

The Atmosphere

Aviation is so dependent upon that category of fluids called gases and the effect of forces and pressures acting upon gases, that a discussion of the subject of the atmosphere is important to the persons maintaining and repairing aircraft.

Data available about the atmosphere may determine whether a flight will succeed, or whether it will even become airborne. The various components of the air around the earth, the changes in temperatures and pressures at different levels above the earth, the properties of weather encountered by aircraft in flight, and many other detailed data are considered in the preparation of flight plans.

Pascan and Torricelli have been credited with developing the barometer, an instrument for measuring atmospheric pressure. The results of their experiments are still used today with very little improvement in design or knowledge. They determined that air has weight which changes as altitude is changed with respect to sea level. Today scientists are also interested in how the atmosphere affects the performance of the aircraft and its equipment.

Composition of the Atmosphere

The atmosphere is a complex and ever changing mixture. Its ingredients vary from place to place and from day to day. In addition to a number of gases, it contains quantities of foreign matter such as pollen, dust, bacteria, soot, volcanic ash, spores, and dust from outer space. The composition of the air remains almost constant from sea level up to its highest level, but its density diminishes rapidly with altitude. Six miles up, for example, it is too thin to support respiration, and 12 miles up, there is not enough oxygen to support combustion, except in some specially designed turbine engine powered airplanes. At a point several hundred miles above the earth, some gas particles spray out into space, some are dragged by gravity and fall back into the ocean of air below, while others never return. Physicists disagree as to the boundaries of the outer fringes of the atmosphere. Some think it begins 240 miles above the earth and extends to 400 miles; others place its lower edge at 600 miles and its upper boundary at 6,000 miles.

There are also certain nonconformities at various levels. Between 12 and 30 miles, high solar ultraviolet radiation reacts with oxygen molecules to produce a thin curtain of ozone, a very poisonous gas without which life on earth could not exist. This ozone filters out a portion of the sun's lethal ultraviolet rays, allowing only enough to come through to give us sunburn, kill bacteria, and prevent rickets. At 50 to 65 miles up, most of the oxygen molecules begin to break down under solar radiation into free atoms, and to form hydroxy ions (OH) from water vapor. Also in this region, all the atoms become ionized.

Studies of the atmosphere have revealed that the temperature does not decrease uniformly with increasing altitude; instead it gets steadily colder up to a height of about 7 miles, where the rate of temperature change slows down abruptly and remains almost constant at −55° Centigrade (218° Kelvin) up to about 20 miles. Then the temperature begins to rise to a peak value of 77° Centigrade (350° Kelvin) at the 55 mile level. Thereafter it climbs steadily, reaching 2,270° Centigrade (2,543° Kelvin) at a height of 250 to 400 miles. From the 50 mile level upward, a man or any other living creature, without the protective cover of the atmosphere, would be broiled on the side facing the sun and frozen on the other.

The atmosphere is divided into concentric layers or levels. Transition through these layers is gradual and without sharply defined boundaries. However, one boundary, the tropopause, exists between the first and second layer. The tropopause is defined as the point in the atmosphere at which the decrease in temperature (with increasing altitude) abruptly ceases. The four atmosphere layers are the troposphere, stratosphere, ionosphere, and the exosphere. The upper portion of the stratosphere is often called the chemosphere or ozonosphere, and the exosphere is also known as the mesosphere.

The troposphere extends from the earth's surface to about 35,000 ft at middle latitudes, but varies from 28,000 ft at the poles to about 54,000 ft at the equator. The troposphere is characterized by large changes in temperature and humidity and by generally turbulent conditions. Nearly all cloud formations are within the troposphere. Approximately three-fourths of the total weight of the atmosphere is within the troposphere. The stratosphere extends from the upper limits of the troposphere (and the tropopause) to an average altitude of 60 miles.

The ionosphere ranges from the 50 mile level to a level of 300 to 600 miles. Little is known about the characteristics of the ionosphere, but it is thought that many electrical phenomena occur there. Basically, this layer is characterized by the presence of ions and free electrons, and the ionization seems to increase with altitude and in successive layers.

The exosphere (or mesosphere) is the outer layer of the atmosphere. It begins at an altitude of 600 miles and extends to the limits of the atmosphere. In this layer, the temperature is fairly constant at 2,500° Kelvin, and propagation of sound is thought to be impossible due to lack of molecular substance.

Atmospheric Pressure

The human body is under pressure, since it exists at the bottom of a sea of air. This pressure is due to the weight of the atmosphere. On a standard day at sea level, if a 1-in^2 column of air extending to the top of the atmosphere was weighed, it would weigh 14.7 lb. That is why standard day atmospheric pressure is said to be 14.7 pounds per square inch (14.7 psi).

Since atmospheric pressure at any altitude is due to the weight of air above it, pressure decreases with increased altitude. Obviously, the total weight of air above an area at 15,000 ft would be less than the total weight of the air above an area at 10,000 ft.

Atmospheric pressure is often measured by a mercury barometer. A glass tube somewhat over 30 inches in length is sealed at one end and then filled with mercury. It is then inverted and the open end placed in a dish of mercury. Immediately, the mercury level in the inverted tube will drop a short distance, leaving a small volume of mercury vapor at nearly zero absolute pressure in the tube just above the top of the liquid mercury column. Gravity acting on the mercury in the tube will try to make the mercury run out. Atmospheric pressure pushing down on the mercury in the open container tries to make the mercury stay in the tube. At some point these two forces (gravity and atmospheric pressure) will equilibrate and the mercury will stabilize at a certain height in the tube. Under standard day atmospheric conditions, the air in a 1-in^2 column extending to the top of the atmosphere would weigh 14.7 lb. A 1-in^2 column of mercury, 29.92 inches tall, would also weigh 14.7 lb. That is why 14.7 psi is equal to 29.92 "Hg. Figure 3-50 demonstrates this point.

A second means of measuring atmospheric pressure is with an aneroid barometer. This mechanical instrument is a much better choice than a mercury barometer for use on airplanes. Aneroid barometers (altimeters) are used to indicate altitude in flight. The calibrations are made in thousands of feet rather than in psi or inches of mercury. For example, the standard pressure at sea

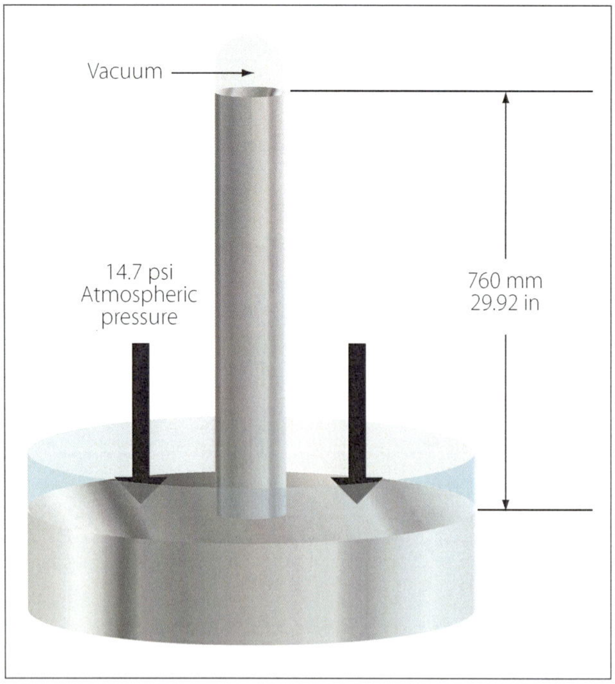

Figure 3-50. Atmospheric pressure as inches of mercury.

Figure 3-51. An airplane's altimeter is an aneroid barometer.

level is 29.92 "Hg, or 14.7 psi. At 10,000 feet above sea level, standard pressure is 20.58 "Hg, or 10.10 psi. Altimeters are calibrated so that if the pressure exerted by the atmosphere is 10.10 psi, the altimeter will point to 10,000 ft. [Figure 3-51]

Atmospheric Density

Since both temperature and pressure decrease with altitude, it might appear that the density of the atmosphere would remain fairly constant with increased altitude. This is not true, however, because pressure drops more rapidly with increased altitude than does the temperature. The result is that density decreases with increased altitude.

By use of the general gas law, studied earlier, it can be shown that for a particular gas, pressure and temperature determine the density. Since standard pressure and temperatures have been associated with each altitude, the density of the air at these standard temperatures and pressures must also be considered standard. Thus, a particular atmospheric density is associated with each altitude. This gives rise to the expression "density altitude," symbolized "Hd." A density altitude of 15,000 ft is the altitude at which the density is the same as that considered standard for 15,000 ft. Remember, however, that density altitude is not necessarily true altitude. For example, on a day when the atmospheric pressure is higher than standard and the temperature is lower than standard, the density which is standard at 10,000 ft might occur at 12,000 ft. In this case, at an actual altitude of 12,000 ft, we have air that has the same density as standard air at 10,000 ft. Density altitude is a calculated altitude obtained by correcting pressure altitude for temperature.

Water Content of the Atmosphere

In the troposphere, the air is rarely completely dry. It contains water vapor in one of two forms: (1) fog or (2) water vapor. Fog consists of minute droplets of water held in suspension by the air. Clouds are composed of fog. The height to which some clouds extend is a good indication of the presence of water in the atmosphere almost up to the stratosphere. The presence of water vapor in the air is quite evident in Figure 3-52, with a military F-18 doing a high-speed fly-by at nearly Mach 1. The temperature and pressure changes that occur as the airplane approaches supersonic flight cause the water vapor in the air to condense and form the vapor cloud that is visible.

As a result of evaporation, the atmosphere always contains some moisture in the form of water vapor. The moisture in the air is called the humidity of the air. Moisture does not consist of tiny particles of liquid held in suspension in the air as in the case of fog, but is an invisible vapor truly as gaseous as the air with which it mixes. Fog and humidity both affect the performance of an aircraft. In flight, at cruising power, the effects are small and receive no consideration. During takeoff, however, humidity has important effects. Two things are done to compensate for the effects of humidity on takeoff performance. Since humid air is less dense than dry air, the allowable takeoff gross weight of an aircraft is generally reduced for operation in areas that are

Figure 3-52. F-18 high-speed fly-by and a vapor cloud.

consistently humid. Second, because the power output of reciprocating engines is decreased by humidity, the manifold pressure may need to be increased above that recommended for takeoff in dry air in order to obtain the same power output.

Engine power output is calculated on dry air. Since water vapor is incombustible, its pressure in the atmosphere is a total loss as far as contributing to power output. The mixture of water vapor and air is drawn through the carburetor, and fuel is metered into it as though it were all air. This mixture of water vapor, air, and fuel enters the combustion chamber where it is ignited. Since the water vapor will not burn, the effective fuel/air ratio is enriched and the engine operates as though it were on an excessively rich mixture. The resulting horsepower loss under humid conditions can therefore be attributed to the loss in volumetric efficiency due to displaced air, and the incomplete combustion due to an excessively rich fuel/air mixture.

The reduction in power that can be expected from humidity is usually given in charts in the flight manual. There are several types of charts in use. Some merely show the expected reduction in power due to humidity; others show the boost in manifold pressure necessary to restore full takeoff power.

The effect of fog on the performance of an engine is very noticeable, particularly on engines with high compression ratios. Normally, some detonation will occur during acceleration, due to the high BMEP (brake mean effective pressures) developed. However, on a foggy day it is difficult to cause detonation to occur. The explanation of this lies in the fact that fog consists of particles of water that have not vaporized. When these particles enter the cylinders, they absorb a tremendous amount of heat energy in the process of vaporizing. The temperature is thus lowered, and the decrease is sufficient to prevent detonation.

Fog will generally cause a decrease in horsepower output. However, with a supercharged engine, it will be possible to use higher manifold pressures without danger of detonation.

Absolute Humidity

Absolute humidity is the actual amount of the water vapor in a mixture of air and water. It is expressed either in grams per cubic meter or pounds per cubic foot. The amount of water vapor that can be present in the air is dependent upon the temperature and pressure. The higher the temperatures, the more water vapor the air is capable of holding, assuming constant pressure. When air has all the water vapor it can hold at the prevailing temperature and pressure, it is said to be saturated.

Relative Humidity

Relative humidity is the ratio of the amount of water vapor actually present in the atmosphere to the amount that would be present if the air were saturated at the prevailing temperature and pressure. This ratio is usually multiplied by 100 and expressed as a percentage. Suppose, for example, that a weather report includes the information that the temperature is 75°F and the relative humidity is 56 percent. This indicates that the air holds 56 percent of the water vapor required to saturate it at 75°F. If the temperature drops and the absolute humidity remains constant, the relative humidity will increase. This is because less water vapor is required to saturate the air at the lower temperature.

Dew Point

The dew point is the temperature to which humid air must be cooled at constant pressure to become saturated. If the temperature drops below the dew point, condensation occurs. People who wear eyeglasses have experience going from cold outside air into a warm room and having moisture collect quickly on their glasses. This happens because the glasses were below the dew point temperature of the air in the room. The air immediately in contact with the glasses was cooled below its dew point temperature, and some of the water vapor was condensed out. This principle is applied in determining the dew point. A vessel is cooled until water vapor begins to condense on its surface. The temperature at which this occurs is the dew point.

Vapor Pressure

Vapor pressure is the portion of atmospheric pressure that is exerted by the moisture in the air (expressed in tenths of an inch of mercury). The dew point for a given condition depends on the amount of water pressure present; thus, a direct relationship exists between the vapor pressure and the dew point.

Standard Atmosphere

If the performance of an aircraft is computed, either through flight tests or wind tunnel tests, some standard reference condition must be determined first in order to compare results with those of similar tests. The conditions in the atmosphere vary continuously, and it is generally not possible to obtain exactly the same set of conditions on two different days or even on two successive flights. For this reason, a set group of standards must be used as a point of reference. The set of standard conditions presently used in the United States is known as the U.S. Standard Atmosphere.

The standard atmosphere approximates the average conditions existing at 40° latitude, and is determined on the basis of the following assumptions. The standard sea level conditions are:

Pressure at 0 altitude (P0) = 29.92 "Hg
Temperature at 0 altitude (T0) = 15°C or 59°F
Gravity at 0 altitude (G0) = 32.174 fps/s

The U.S. Standard Atmosphere is in agreement with the International Civil Aviation Organization (ICAO) Standard Atmosphere over their common altitude range. The ICAO Standard Atmosphere has been adopted as standard by most of the principal nations of the world.

Aircraft Theory of Flight

Before a technician can consider performing maintenance on an aircraft, it is necessary to understand the pieces that make up the aircraft. Names like fuselage, empennage, wing, and so many others, come into play when describing what an airplane is and how it operates. For helicopters, names like main rotor, anti-torque rotor, and autorotation come to mind as a small portion of what needs to be understood about rotorcraft. The study of physics, which includes basic aerodynamics, is a necessary part of understanding why aircraft operate the way they do.

Four Forces of Flight

During flight, there are four forces acting on an airplane. These forces are lift, weight, thrust, and drag. [Figure 3-53] Lift is the upward force created by the wing, weight is the pull of gravity on the airplane's mass, thrust is the force created by the airplane's propeller or turbine engine, and drag is the friction caused by the air flowing around the airplane.

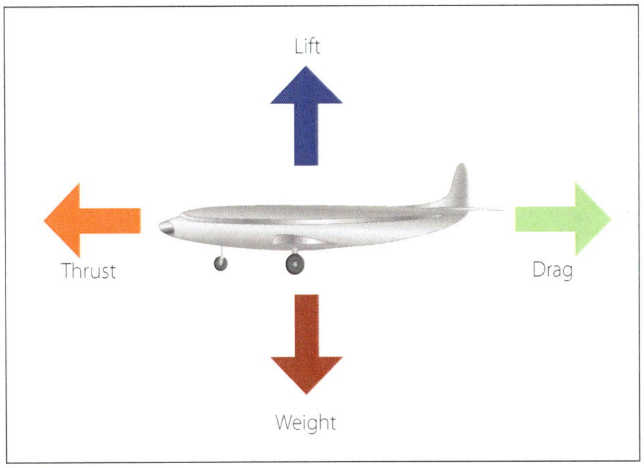

Figure 3-53. Four forces acting on an airplane.

All four of these forces are measured in pounds. Any time the forces are not in balance, something about the airplane's condition is changing. The possibilities are as follows:

1. When an airplane is accelerating, it has more thrust than drag.
2. When an airplane is decelerating, it has less thrust than drag.
3. When an airplane is at a constant velocity, thrust and drag are equal.
4. When an airplane is climbing, it has more lift than weight.
5. When an airplane is descending, it has more weight than lift.
6. When an airplane is at a constant altitude, lift and weight are equal.

Bernoulli's Principle and Subsonic Flow

The basic concept of subsonic airflow and the resulting pressure differentials was discovered by Daniel Bernoulli, a Swiss physicist. Bernoulli's principle, as we refer to it today, states that "as the velocity of a fluid increases, the static pressure of that fluid will decrease, provided there is no energy added or energy taken away." A direct application of Bernoulli's principle is the study of air as it flows through either a converging or a diverging passage, and to relate the findings to some aviation concepts.

A converging shape is one whose cross-sectional area gets progressively smaller from entry to exit. A diverging shape is just the opposite, with the cross-sectional area getting larger from entry to exit. Figure 3-54 shows a converging shaped duct, with the air entering on the left at subsonic velocity and exiting on the right. Looking at the pressure and velocity gauges, and the indicated velocity and pressure, notice that the air exits at an increased velocity and a decreased static pressure. Because a unit of air must exit the duct when another unit enters, the unit leaving must increase its velocity as it flows into a smaller space.

In a diverging duct, just the opposite would happen. From the entry point to the exit point, the duct is spreading out and the area is getting larger. [Figure 3-55] With the increase in cross-sectional area, the velocity of the air decreases and the static pressure increases. The total energy in the air has not changed. What has

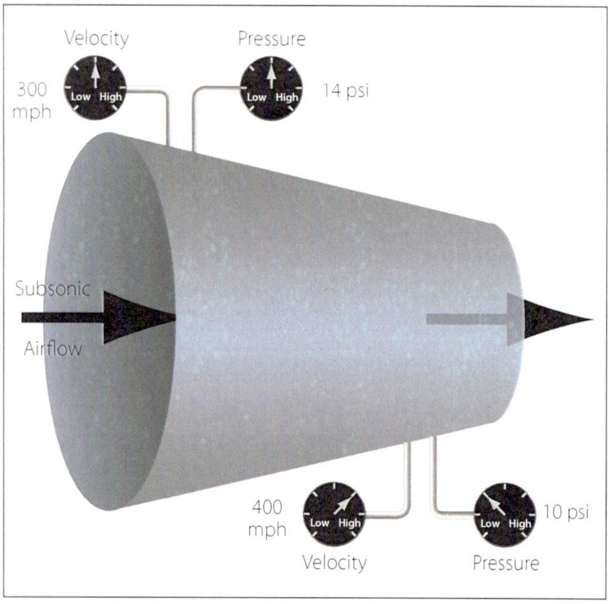

Figure 3-54. Bernoulli's principle and a converging duct.

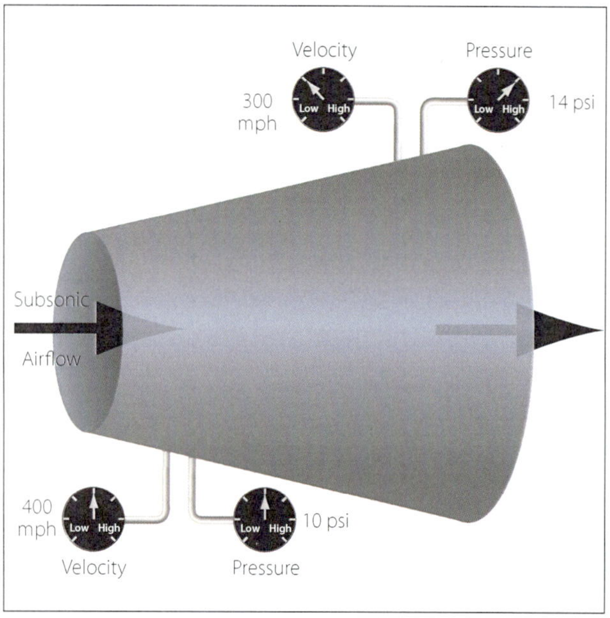

Figure 3-55. Bernoulli's principle and a diverging duct.

been lost in velocity (kinetic energy) is gained in static pressure (potential energy).

In the discussion of Bernoulli's principle earlier in this chapter, a venturi was shown in Figure 3-46. In Figure 3-56, a venturi is shown again, only this time a wing is shown tucked up into the recess where the venturi's converging shape is. There are two arrows showing airflow. The large arrow shows airflow within the venturi, and the small arrow shows airflow on the outside heading toward the leading edge of the wing.

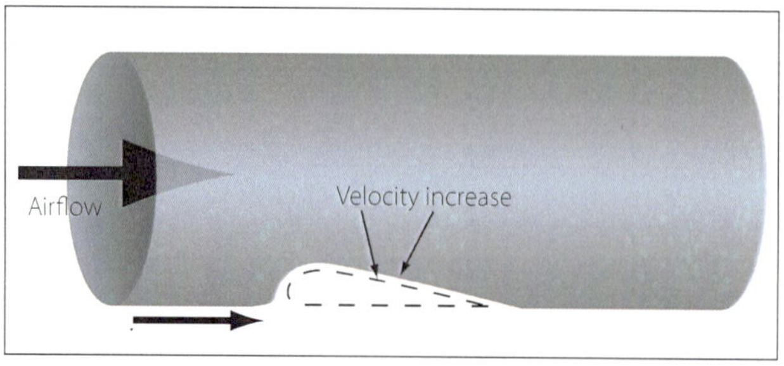

Figure 3-56. Venturi with a superimposed wing.

In the converging part of the venturi, velocity would increase and static pressure would decrease. The same thing would happen to the air flowing around the wing, with the velocity over the top increasing and static pressure decreasing.

In Figure 3-56, the air reaching the leading edge of the wing separates into two separate flows. Some of the air goes over the top of the wing and some travels along the bottom. The air going over the top, because of the curvature, has farther to travel. With a greater distance to travel, the air going over the top must move at a greater velocity. The higher velocity on the top causes the static pressure on the top to be less than it is on the bottom, and this difference in static pressures is what creates lift.

For the wing shown in Figure 3-56, imagine it is 5 ft wide and 15 ft long, for a surface area of 75 ft^2 (10,800 in^2). If the difference in static pressure between the top and bottom is 0.1 psi, there will be $\frac{1}{10}$ lb of lift for each square inch of surface area. Since there are 10,800 in^2 of surface area, there would be 1,080 lb of lift (0.1 × 10,800).

Lift and Newton's Third Law

Newton's third law identifies that for every force there is an equal and opposite reacting force. In addition to Bernoulli's principle, Newton's third law can also be used to explain the lift being created by a wing. As the air travels around a wing and leaves the trailing edge, the air is forced to move in a downward direction. Since a force is required to make something change direction, there must be an equal and opposite reacting force. In this case, the reacting force is what we call lift. In order to calculate lift based on Newton's third law, Newton's second law and the formula "Force = Mass × Acceleration" would be used. The mass would be the weight of air flowing over the wing every second, and the acceleration would be the change in velocity the wing imparts to the air.

The lift on the wing as described by Bernoulli's principle, and lift on the wing as described by Newton's third law, are not separate or independent of each other. They are just two different ways to describe the same thing, namely the lift on a wing.

Airfoils

An airfoil is any device that creates a force, based on Bernoulli's principles or Newton's laws, when air is caused to flow over the surface of the device. An airfoil can be the wing of an airplane, the blade of a propeller, the rotor blade of a helicopter, or the fan blade of a turbofan engine. The wing of an airplane moves through the air because the airplane is in motion, and generates lift by the process previously described. By comparison, a propeller blade, helicopter rotor blade, or turbofan engine fan blade rotates through the air. These rotating blades could be referred to as rotating wings, as is common with helicopters when they are called rotary wing aircraft. The rotating wing can be viewed as a device that creates lift, or just as correctly, it can be viewed as a device that creates thrust.

In Figure 3-57 an airfoil (wing) is shown, with some of the terminology that is used to describe a wing. The terms and their meaning are as follows:

Camber

The camber of a wing is the curvature which is present on top and bottom surfaces. The camber on the top is much more pronounced, unless the wing is a symmetrical airfoil, which has the same camber top and bottom. The bottom of the wing, more often than not, is relatively flat. The increased camber on top is what causes the velocity of the air to increase and the static pressure to decrease. The bottom of the wing has less velocity and more static pressure, which is why the wing generates lift.

Chord Line

The chord line is an imaginary straight line running from the wing's leading edge to its trailing edge. The angle between the chord line and the longitudinal axis of the airplane is known as the angle of incidence.

Relative Wind

Whatever direction the airplane is flying, the relative wind is in the opposite direction. If the airplane is flying due north, and someone in the airplane is not shielded from the elements, that person will feel like the wind is coming directly from the south.

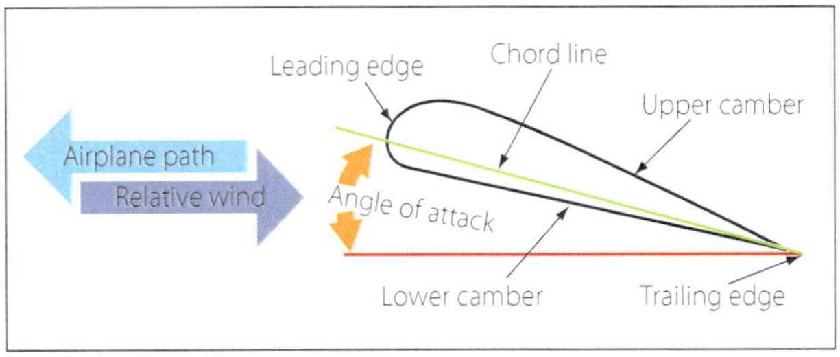

Figure 3-57. Wing terminology.

Angle of Attack

The angle between the chord line and the relative wind is the angle of attack. As the angle of attack increases, the lift on the wing increases. If the angle of attack becomes too great, the airflow can separate from the wing and the lift will be destroyed. When this occurs, a condition known as a stall takes place.

There are a number of different shapes, known as planforms, that a wing can have. A wing in the shape of a rectangle is very common on small general aviation airplanes. An elliptical shape or tapered wing can also be used, but these do not have as desirable a stall characteristic. For airplanes that operate at high subsonic speeds, sweptback wings are common, and for supersonic flight, a delta shape might be used.

The aspect ratio of a wing is the relationship between its span (wingtip to wingtip measurement) and the chord of the wing. If a wing has a long span and a very narrow chord, it is said to have a high aspect ratio. A higher aspect ratio produces less drag for a given flight speed, and is typically found on glider type aircraft.

The angle of incidence of a wing is the angle formed by the intersection of the wing chord line and the horizontal plane passing through the longitudinal axis of the aircraft. Many airplanes are designed with a greater angle of incidence at the root of the wing than at the tip, and this is referred to as washout. This feature causes the inboard part of the wing to stall before the outboard part, which helps maintain aileron control during the initial stages of a wing stall.

Boundary Layer Airflow

The boundary layer is a very thin layer of air lying over the surface of the wing and, for that matter, all other surfaces of the airplane. Because air has viscosity, this layer of air tends to adhere to the wing. As the wing moves forward through the air, the boundary layer at first flows smoothly over the streamlined shape of the airfoil. Here the flow is called the laminar layer.

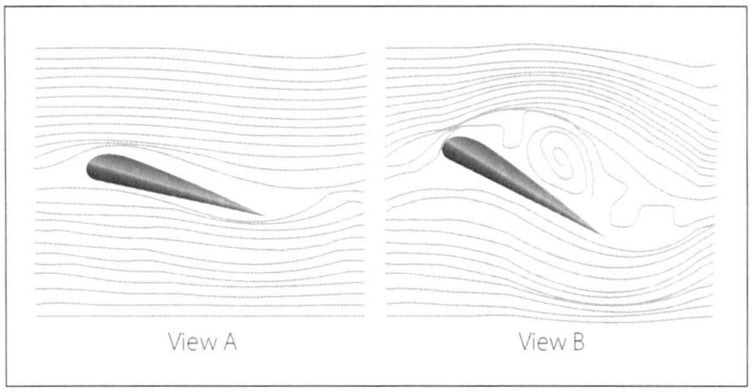

Figure 3-58. Wing boundary layer separation.

As the boundary layer approaches the center of the wing, it begins to lose speed due to skin friction and it becomes thicker and turbulent. Here it is called the turbulent layer. The point at which the boundary layer changes from laminar to turbulent is called the transition point. Where the boundary layer becomes turbulent, drag due to skin friction is relatively high. As speed increases, the transition point tends to move forward. As the angle of attack increases, the transition point also tends to move forward. With higher angles of attack and further thickening of the boundary layer, the turbulence becomes so great the air breaks away from the surface of the wing. At this point, the lift of the wing is destroyed and a condition known as a stall has occurred. In Figure 3-58, view A shows a normal angle of attack and the airflow staying in contact with the wing. View B shows an extreme angle of attack and the airflow separating and becoming turbulent on the top of the wing. In view B, the wing is in a stall.

Boundary Layer Control

One way of keeping the boundary layer air under control, or lessening its negative effect, is to make the wing's surface as smooth as possible and to keep it free of dirt and debris. As the friction between the air and the surface of the wing increases, the boundary layer thickens and becomes more turbulent and eventually a wing stall occurs. With a smooth and clean wing surface, the onset of a stall is delayed and the wing can operate at a higher angle of attack. One of the reasons ice forming on a wing can be such a serious problem is because of its effect on boundary layer air. On a high-speed airplane, even a few bugs splattered on the wing's leading edge can negatively affect boundary layer air.

Other methods of controlling boundary layer air include wing leading edge slots, air suction through small holes on the wing's upper surface, and the use of devices called vortex generators.

A wing leading edge slot is a duct that allows air to flow from the bottom of the wing, through the duct, to the top of the wing. As the air flows to the top of the wing, it is directed along the wing's surface at a high velocity and helps keep the boundary layer from becoming turbulent and separating from the wing's surface.

Another way of controlling boundary layer air is to create a suction on the top of the wing through a large number of small holes. The suction on the top of the wing draws away the slow-moving turbulent air, and helps keep the remainder of the airflow in contact with the wing.

Vortex generators are used on airplanes that fly at high subsonic speed, where the velocity of the air on the top of the wing can reach Mach 1. As the air reaches Mach 1 velocity, a shock wave forms on the top of the wing, and the subsequent shock wave causes the air to separate from the wing's upper surface. Vortex generators are short airfoils, arranged in pairs, located on the wing's upper surface. They are positioned such that they pull high-energy air down into the boundary layer region and prevent airflow separation.

Wingtip Vortices

Wingtip vortices are caused by the air beneath the wing, which is at the higher pressure, flowing over the wingtip and up toward the top of the wing. The end result is a spiral or vortex that trails behind the wingtip anytime lift is being produced. This vortex is also referred to as wake turbulence, and is a significant factor in determining how closely one airplane can follow behind another on approach to land. The wake turbulence of a large airplane can cause a smaller airplane, if it is following too closely, to be thrown out of control. Vortices from the wing and from the horizontal stabilizer are quite visible on the MD-11 shown in Figure 3-59.

Upwash and downwash refer to the effect an airfoil has on the free airstream. Upwash is the deflection of the oncoming airstream, causing it to flow up and over the wing. Downwash is the downward deflection of the airstream after it has passed over the wing and is leaving the trailing edge. This downward deflection is what creates the action and reaction described under lift and Newton's third law.

Axes of an Aircraft

An airplane in flight is controlled around one or more of three axes of rotation. These axes of rotation are the longitudinal, lateral, and vertical. On the airplane, all three axes intersect at the center of gravity (CG). As the airplane pivots on one of these axes, it is in essence

Figure 3-59. Wing and horizontal stabilizer vortices on an MD-11.

pivoting around the center of gravity (CG). The center of gravity is also referred to as the center of rotation.

On the brightly colored airplane shown in Figure 3-60, the three axes are shown in the colors red (vertical axis), blue (longitudinal axis), and orange (lateral axis). The flight control that makes the airplane move around the axis is shown in a matching color.

The rudder, in red, causes the airplane to move around the vertical axis and this movement is described as being a yaw. The elevator, in orange, causes the airplane to move around the lateral axis and this movement is described as being a pitch. The ailerons, in blue, cause the airplane to move around the longitudinal axis and this movement is described as being a roll.

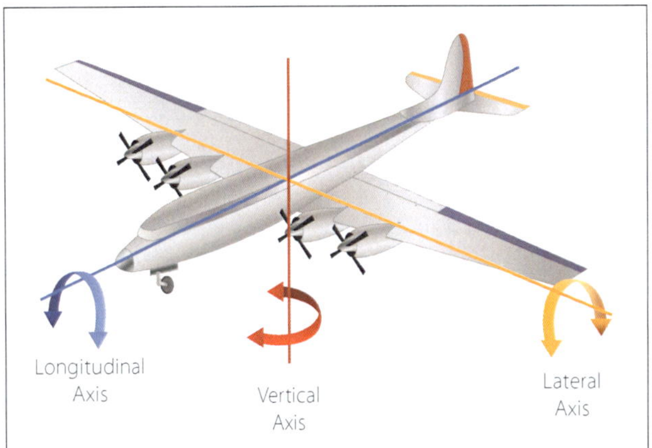

Figure 3-60. The three axes of an airplane.

Aircraft Stability

When an airplane is in straight-and-level flight at a constant velocity, all the forces acting on the airplane are in equilibrium. If that straight-and-level flight is disrupted by a disturbance in the air, such as wake turbulence, the airplane might pitch up or down, yaw left or right, or go into a roll. If the airplane has what is characterized as stability, once the disturbance goes away, the airplane will return to a state of equilibrium.

Static Stability

The initial response that an airplane displays after its equilibrium is disrupted is referred to as its static stability. If the static stability is positive, the airplane will tend to return to its original position after the disruptive force is removed. If the static stability is negative, the airplane will continue to move away from its original position after the disruptive force is removed. If an airplane with negative static stability has the nose pitch up because of wake turbulence, the tendency will be for the nose to continue to pitch up even after the turbulence goes away. If an airplane tends to remain in a displaced position after the force is removed, but does not continue to move toward even greater displacement, its static stability is described as being neutral.

Dynamic Stability

The dynamic stability of an airplane involves the amount of time it takes for it to react to its static stability after it has been displaced from a condition of equilibrium. Dynamic stability involves the oscillations that typically occur as the airplane tries to return to its original position or attitude. Even though an airplane may have positive static stability, it may have dynamic stability which is positive, neutral, or negative.

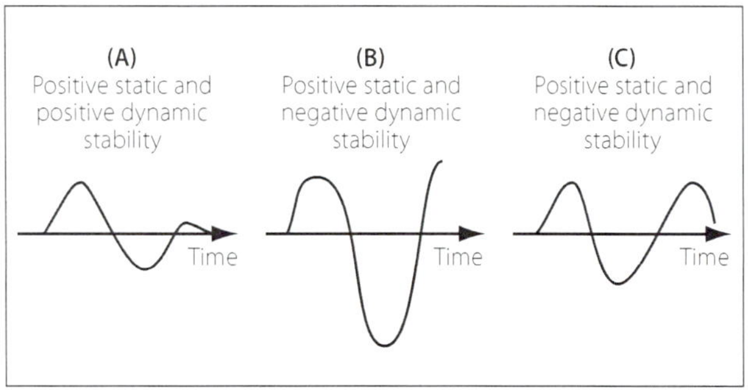

Figure 3-61. Dynamic stability.

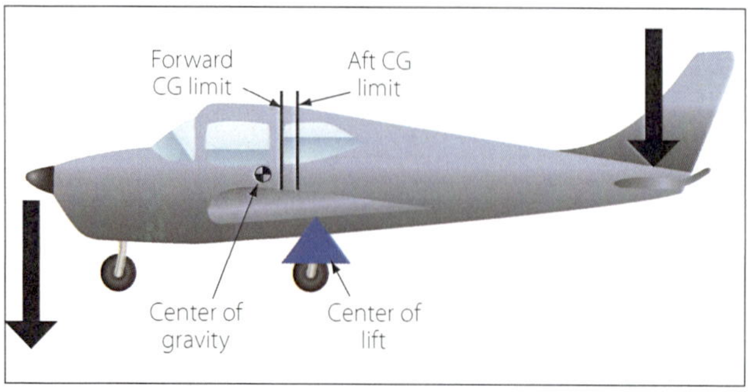

Figure 3-62. Longitudinal stability and balance.

Imagine that an airplane in straight-and-level flight is disturbed and pitches nose up. If the airplane has positive static stability, the nose will pitch back down after the disturbance is removed. If it immediately returns to straight-and-level flight, it is also said to have positive dynamic stability. The airplane, however, may pass through level flight and remain pitched down, and then continue the recovery process by pitching back up. This pitching up and then down is known as an oscillation. If the oscillations lessen over time, the airplane is still classified as having positive dynamic stability. If the oscillations increase over time, the airplane is classified as having negative dynamic stability. If the oscillations remain the same over time, the airplane is classified as having neutral dynamic stability.

Figure 3-61 shows the concept of dynamic stability. In view A, the displacement from equilibrium goes through three oscillations and then returns to equilibrium. In view B, the displacement from equilibrium is increasing after two oscillations, and will not return to equilibrium. In view C, the displacement from equilibrium is staying the same with each oscillation.

Longitudinal Stability

Longitudinal stability for an airplane involves the tendency for the nose to pitch up or pitch down, rotating around the lateral axis (wingtip to wingtip). If an airplane is longitudinally stable, it will return to a properly trimmed angle of attack after the force that upset its flightpath is removed.

The weight and balance of an airplane, which is based on both the design characteristics of the airplane and the way it is loaded, is a major factor in determining longitudinal stability. There is a point on the wing of an airplane, called the center of pressure or center of lift, where all the lifting forces concentrate. In flight, the airplane acts like it is being lifted from or supported by this point. This center of lift runs from wingtip to wingtip. There is also a point on the airplane, called the center of gravity, where the mass or weight of the airplane is concentrated. For an airplane to have good longitudinal stability, the center of gravity is typically located forward of the center of lift. This gives the airplane a nose-down pitching tendency, which is balanced out by the force generated at the horizontal stabilizer and elevator. The center of gravity has limits within which it must fall. If it is too far forward, the forces at the tail might not be able to compensate and it may not be possible to keep the nose of the airplane from pitching down.

In Figure 3-62, the center of lift, center of gravity, and center of gravity limits are shown. It can be seen that the center of gravity is not only forward of the center of lift, it is also forward of the center of gravity limit. At the back of the airplane, the elevator trailing edge is deflected upward to create a downward force on the tail, to try and keep the nose of the airplane up. This airplane would be highly unstable longitudinally, especially at low speed when trying to land. It is especially dangerous if the center of gravity is behind the aft limit. The airplane will now have a tendency to pitch nose up, which can lead to the wing stalling and possible loss of control of the airplane.

Lateral Stability

Lateral stability of an airplane takes place around the longitudinal axis, which is from the airplane's nose to its tail. If one wing is lower than the other, good lateral stability will tend to bring the wings back to a level flight attitude. One design characteristic that tends to give an airplane good lateral stability is called dihedral. Dihedral is an upward angle for the wings with respect to the horizontal, and it is usually just a few degrees.

Imagine a low wing airplane with a few degrees of dihedral experiencing a disruption of its flightpath such that the left wing drops. When the left wing drops, this will cause the airplane to experience a sideslip toward the low wing. The sideslip causes the low wing to experience a higher angle of attack, which increases

Figure 3-63. The dihedral of a wing.

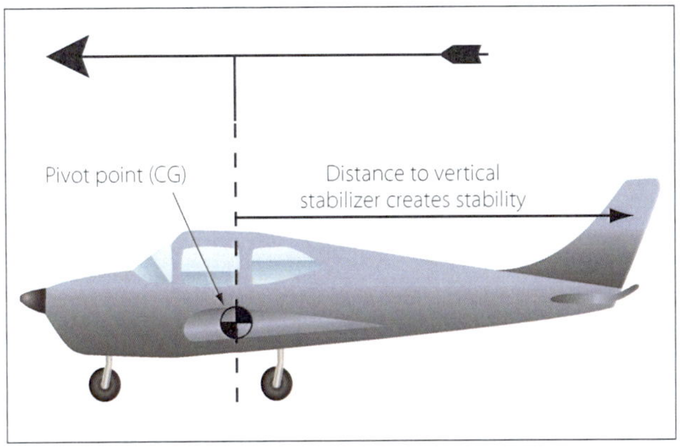

Figure 3-64. Directional stability caused by distance to vertical stabilizer.

its lift and raises it back to a level flight attitude. The dihedral on a wing is shown in Figure 3-63.

Directional Stability

Movement of the airplane around its vertical axis, and the airplane's ability to not be adversely affected by a force creating a yaw type of motion, is called directional stability. The vertical fin gives the airplane this stability, causing the airplane to align with the relative wind. In flight, the airplane acts like the weather vane we use around our home to show the direction the wind is blowing. The distance from the pivot point on a weather vane to its tail is greater than the distance from its pivot point to the nose. So when the wind blows, it creates a greater torque force on the tail and forces it to align with the wind. On an airplane, the same is true. With the CG being the pivot point, it is a greater distance from the CG to the vertical stabilizer than it is from the CG to the nose. [Figure 3-64]

Dutch Roll

The dihedral of the wing tries to roll the airplane in the opposite direction of how it is slipping, and the vertical fin will try to yaw the airplane in the direction of the slip. These two events combine in a way that affects lateral and directional stability. If the wing dihedral has the greatest effect, the airplane will have a tendency to experience a Dutch roll. A Dutch roll is a small amount of oscillation around both the longitudinal and vertical axes. Although this condition is not considered dangerous, it can produce an uncomfortable feeling for passengers. Commercial airliners typically have yaw dampers that sense a Dutch roll condition and cancel it out.

Flight Control Surfaces

The purpose of flight controls is to allow the pilot to maneuver the airplane, and to control it from the time it starts the takeoff roll until it lands and safely comes to a halt. Flight controls are typically associated with the wing and the vertical and horizontal stabilizers, because these are the parts of the airplane that flight controls most often attach to. In flight, and to some extent on the ground, flight controls provide the airplane with the ability to move around one or more of the three axes. Flight controls function by changing the shape or aerodynamic characteristics of the surface they are attached to.

Flight Controls and the Lateral Axis

The lateral axis of an airplane is a line that runs below the wing, from wingtip to wingtip, passing through the airplane's center of gravity. Movement around this axis is called pitch, and control around this axis is called longitudinal control. The flight control that handles this job is the elevator attached to the horizontal stabilizer, a fully moving horizontal stabilizer, or on a v-tail configured airplane, it is called ruddervators. An elevator on a Cessna 182 can be seen in Figure 3-65. In Figure 3-66, a fully moving horizontal stabilizer, known as a stabilator, can be seen on a Piper Cherokee Arrow, and Figure 3-67 shows a ruddervator on a Beechcraft Bonanza.

Depending on the airplane being discussed, movement around the lateral axis happens as a result of the pilot moving the control wheel or yoke, the control stick,

Figure 3-65. Elevator on a Cessna 182 provides pitch control.

Figure 3-66. Moving horizontal stabilizer, known as a stabilator, on a Piper Cherokee Arrow provides pitch control.

Figure 3-67. Ruddervators on a Beechcraft Bonanza provide pitch control.

Figure 3-68. Cessna 182 control wheel and rudder pedals.

or on some airplanes, a side stick. On the airplanes shown in Figures 3-65, 3-66, and 3-67, a control wheel or yoke is used.

On the Cessna 182 shown in Figure 3-65, pulling back on the control wheel causes the trailing edge of the elevator to deflect upward, causing an increased downward force that raises the nose of the airplane. Movement of the elevator causes the nose of the airplane to pitch up or pitch down by rotating around the lateral axis. The Cessna 182 control wheel can be seen in Figure 3-68.

On the Piper Cherokee Arrow shown in Figure 3-66, pulling back on the control wheel causes the entire horizontal surface (stabilator) to move, with the trailing edge deflecting upward. The anti-servo tab seen on the Cherokee provides a control feel similar to what would be experienced by moving an elevator. Without this tab, the stabilator might be too easy to move and a pilot could overcontrol the airplane.

The ruddervators shown on the Beechcraft Bonanza in Figure 3-67 are also moved by the control wheel, with their trailing edges deflecting upward when the control wheel is pulled back. As the name implies, these surfaces also act as the rudder for this airplane.

Flight Controls and the Longitudinal Axis

The longitudinal axis of the airplane runs through the middle of the airplane, from nose to tail, passing through the center of gravity. Movement around this axis is known as roll, and control around this axis is called lateral control. Movement around this axis is controlled by the ailerons, and on jet transport airplanes, it is aided by surfaces on the wing known as spoilers.

The ailerons move as a result of the pilot rotating the control wheel to the left or to the right, much the same as turning the steering wheel on an automobile. [Figure 3-68] When a pilot turns the control wheel to the left, the airplane is being asked to turn or bank to the left. Turning the control wheel to the left causes the trailing edge of the aileron on the left wing to rise up into the airstream, and the aileron on the right wing lowers down into the airstream. This increases the lift on the right wing and decreases the lift on the left wing, causing the right wing to move up and the airplane to bank to the left.

In Figure 3-69, an Air Force F-15 can be seen doing an aileron roll. Notice that the left aileron is up and the right aileron is down, which would cause the airplane to roll around the longitudinal axis in a counterclockwise direction.

Flight Controls and the Vertical Axis

The vertical axis of an airplane runs from top to bottom through the middle of the airplane, passing through the center of gravity. Movement around this axis is known as yaw, and control around this axis is called directional control. Movement around this axis is controlled by the rudder, or in the case of the Beechcraft Bonanza in Figure 3-67, by the ruddervators.

Figure 3-69. Air Force F-15 doing an aileron roll.

The feet of the pilot are on the rudder pedals, and pushing on the left or right rudder pedal makes the rudder move left or right. When the right rudder pedal is pushed, the trailing edge of the rudder moves to the right, and the nose of the airplane yaws to the right. The rudder pedals of a Cessna 182 can be seen in Figure 3-68.

Even though the rudder of the airplane will make the nose yaw to the left or the right, the rudder is not what turns the airplane. For what is called a coordinated turn to occur, both the ailerons and rudder come into play. Let's say we want to turn the airplane to the right. We start by turning the control wheel to the right, which raises the right aileron and lowers the left aileron and initiates the banking turn. The increased lift on the left wing also increases the induced drag on the left wing, which tries to make the nose of the airplane yaw to the left. To counteract this, when the control wheel is moved to the right, a small amount of right rudder is used to keep the nose of the airplane from yawing to the left. Once the nose of the airplane is pointing in the right direction, pressure on the rudder is no longer needed. The rudder of a Piper Cherokee Arrow can be seen in Figure 3-70.

Figure 3-70. Rudder on a Piper Cherokee Arrow.

Tabs

Trim Tabs

Trim tabs are small movable surfaces that attach to the trailing edge of flight controls. These tabs can be controlled from the flight deck, and their purpose is to create an aerodynamic force that keeps the flight control in a deflected position. Trim tabs can be installed on any of the primary flight controls.

A very common flight control to find fitted with a trim tab is the elevator. In order to be stable in flight, most airplanes have the center of gravity located forward of the center of lift on the wing. This causes a nose heavy condition, which needs to be balanced out by having the elevator deflect upwards and create a downward force. To relieve the pilot of the need to hold back pressure on the control wheel, a trim tab on the elevator can be adjusted to hold the elevator in a slightly deflected position. An elevator trim tab for a Cessna 182 is shown in Figure 3-71.

Anti-servo Tab

Some airplanes, like a Piper Cherokee Arrow, do not have a fixed horizontal stabilizer and movable elevator. The Cherokee uses a moving horizontal surface known as a stabilator. Because of the location of the pivot point for this movable surface, it has a tendency to be extremely sensitive to pilot input. To reduce the sensitivity, a full length anti-servo tab is installed on the trailing edge of the stabilator. As the trailing edge of the stabilator moves down, the anti-servo tab moves down and creates a force trying to raise the trailing edge. With this force acting against the movement of the stabilator, it reduces the sensitivity to pilot input. The anti-servo tab on a Piper Cherokee Arrow is shown in Figure 3-70.

Figure 3-71. Elevator trim tab on a Cessna 182.

Balance Tab

On some airplanes, the force needed to move the flight controls can be excessive. In these cases, a balance tab can be used to generate a force that assists in the movement of the flight control. Just the opposite of anti-servo tabs, balance tabs move in the opposite direction of the flight control's trailing edge, providing a force that helps the flight control move.

Servo Tab

On large airplanes, because the force needed to move the flight controls is beyond the capability of the pilot, hydraulic actuators are used to provide the necessary force. In the event of a hydraulic system malfunction or failure, some of these airplanes have servo tabs on the trailing edge of the primary flight controls. When the control wheel is pulled back in an attempt to move the elevator, the servo tab moves and creates enough aerodynamic force to move the elevator. The servo tab is acting like a balance tab, but rather than assisting the normal force that moves the elevator, it becomes the sole force that makes the elevator move. Like the balance tab, the servo tab moves in the opposite direction of the flight control's trailing edge. The Boeing 727 has servo tabs that back up the hydraulic system in the event of a failure. During normal flight, the servo tabs act like balance tabs. [Figure 3-73]

Supplemental Lift-Modifying Devices

If the wing of an airplane was designed to produce the maximum lift possible at low airspeed, to accommodate takeoffs and landings, it would not be suited for higher speed flight because of the enormous amount of drag it would produce. To give the wing the ability to produce maximum low speed lift without being drag prohibitive, retractable high lift devices, such as flaps and slats, are utilized.

Flaps

The most often used lift-modifying device, for small airplanes and large, is the wing flap. Flaps can be installed on the leading edge or trailing edge, with the leading edge versions used only on larger airplanes. Flaps change the camber of the wing, and they increase both the lift and the drag for any given angle of attack. The four different types of flaps in use are the plain, split, slotted, and Fowler. [Figure 3-72]

Plain flaps attach to the trailing edge of the wing, inboard of the ailerons, and form part of the wing's overall surface. When deployed downward, they increase the effective camber of the wing and the wing's chord line. Both of these factors cause the wing to create more lift and more drag.

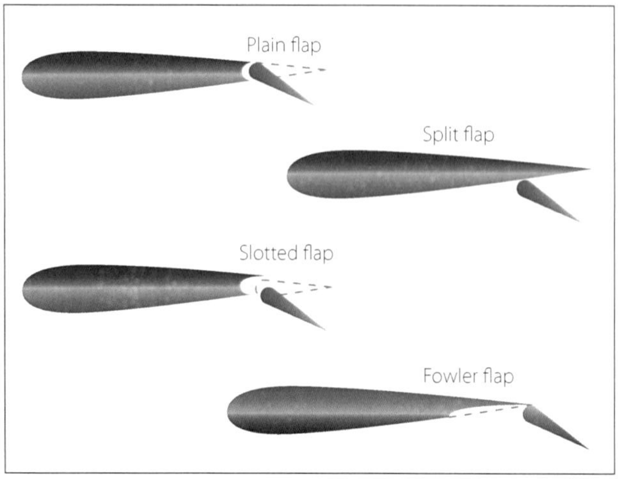

Figure 3-72. Four types of wing flaps.

The split flap attaches to the bottom of the wing, and deploys downward without changing the top surface of the wing. This type of flap creates more drag than the plain flap because of the increase in turbulence.

The slotted flap is similar to the plain flap, except when it deploys, the leading edge drops down a small amount. By having the leading edge drop down slightly, a slot opens up, which lets some of the high pressure air on the bottom of the wing flow over the top of the flap. This additional airflow over the top of the flap produces additional lift.

The Fowler flap attaches to the back of the wing using a track and roller system. When it deploys, it moves aft in addition to deflecting downward. This increases the total wing area, in addition to increasing the wing camber and chord line. This type of flap is the most effective of the four types, and it is the type used on commercial airliners and business jets.

Leading Edge Slots

Leading edge slots are ducts or passages in the leading edge of a wing that allow high pressure air from the bottom of the wing to flow to the top of the wing. This ducted air flows over the top of the wing at a high velocity and helps keep the boundary layer air from becoming turbulent and separating from the wing. Slots are often placed on the part of the wing ahead of the ailerons, so during a wing stall, the inboard part of the wing stalls first and the ailerons remain effective.

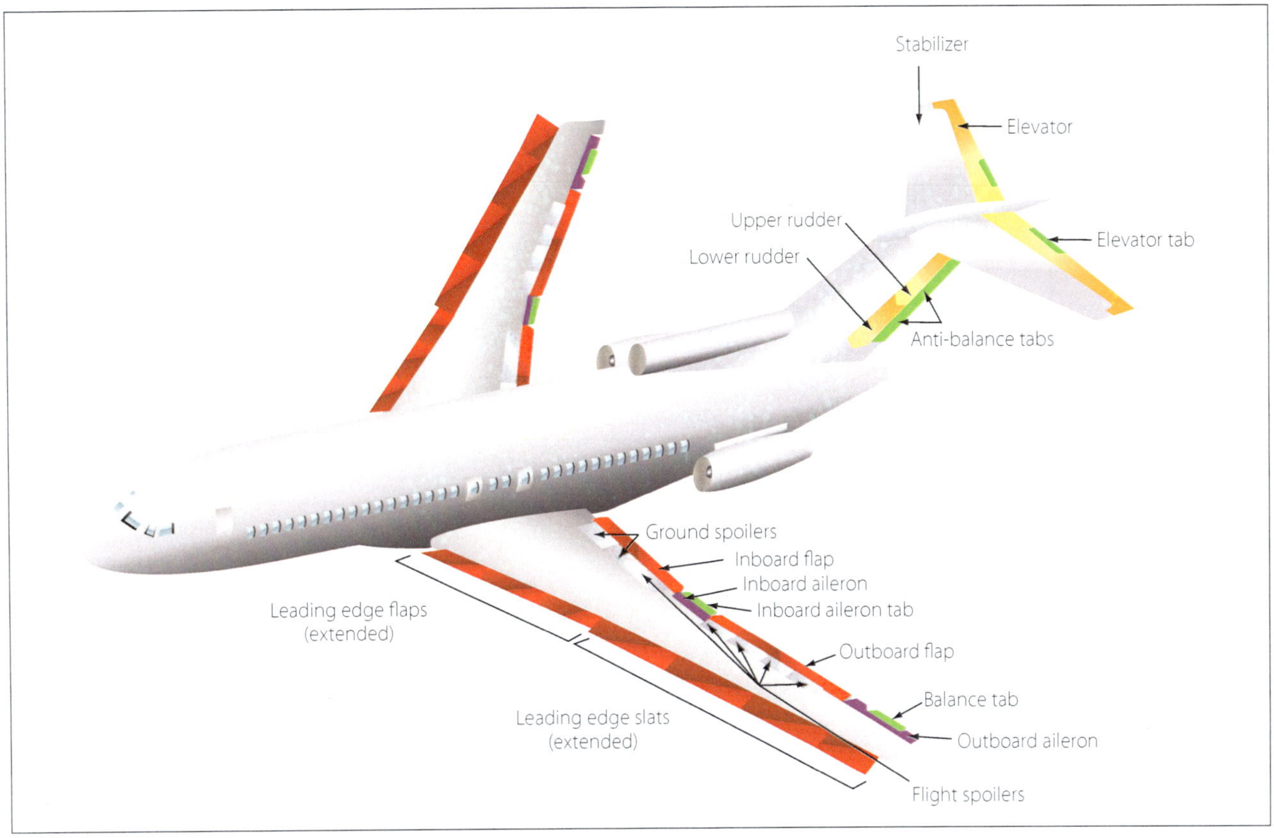

Figure 3-73. Boeing 727 flight controls.

Leading Edge Slats

Leading edge slats serve the same purpose as slots, the difference being that slats are movable and can be retracted when not needed. On some airplanes, leading edge slats have been automatic in operation, deploying in response to the aerodynamic forces that come into play during a high angle of attack. On most of today's commercial airliners, the leading edge slats deploy when the trailing edge flaps are lowered.

The flight controls of a large commercial airliner are shown in Figure 3-73. The controls by color are as follows:

1. All aerodynamic tabs are shown in green.
2. All leading and trailing edge high lift devices are shown in red (leading edge flaps and slats, trailing edge inboard and outboard flaps).
3. The tail mounted primary flight controls are in yellow (rudder and elevator).
4. The wing mounted primary flight controls are in purple (inboard and outboard aileron).

High-Speed Aerodynamics

Compressibility Effects

When air is flowing at subsonic speed, it acts like an incompressible fluid. As discussed earlier in this chapter, when air at subsonic speed flows through a diverging shaped passage, the velocity decreases and the static pressure rises, but the density of the air does not change. In a converging shaped passage, subsonic air speeds up and its static pressure decreases. When supersonic air flows through a converging passage, its velocity decreases and its pressure and density both increase. [Figure 3-74] At supersonic flow, air acts like a compressible fluid. Because air behaves differently

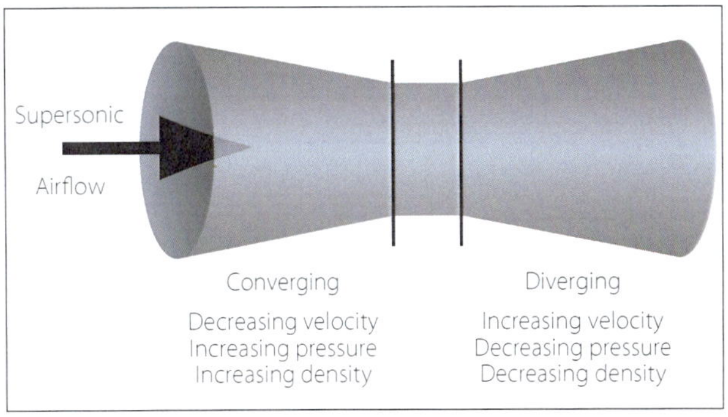

Figure 3-74. Supersonic airflow through a venturi.

when flowing at supersonic velocity, airplanes that fly supersonic must have wings with a different shape.

The Speed of Sound

Sound, in reference to airplanes and their movement through the air, is nothing more than pressure disturbances in the air. As discussed earlier in this chapter, it is like dropping a rock in the water and watching the waves flow out from the center. As an airplane flies through the air, every point on the airplane that causes a disturbance creates sound energy in the form of pressure waves. These pressure waves flow away from the airplane at the speed of sound, which at standard day temperature of 59°F, is 761 mph. The speed of sound in air changes with temperature, increasing as temperature increases. Figure 3-75 shows how the speed of sound changes with altitude.

Altitude in Feet	Temperature (°F)	Speed of Sound (mph)
0	59.00	761
1,000	55.43	758
2,000	51.87	756
3,000	48.30	753
4,000	44.74	750
5,000	41.17	748
6,000	37.60	745
7,000	34.04	742
8,000	30.47	740
9,000	26.90	737
10,000	23.34	734
15,000	5.51	721
20,000	−12.32	707
25,000	−30.15	692
30,000	−47.98	678
35,000	−65.82	663
*36,089	−69.70	660
40,000	−69.70	660
45,000	−69.70	660
50,000	−69.70	660
55,000	−69.70	660
60,000	−69.70	660
65,000	−69.70	660
70,000	−69.70	660
75,000	−69.70	660
80,000	−69.70	660
85,000	−64.80	664
90,000	−56.57	671
95,000	−48.34	678
100,000	−40.11	684

* Altitude at which temperature stops decreasing

Figure 3-75. Altitude and temperature versus speed of sound.

Subsonic, Transonic, and Supersonic Flight

When an airplane is flying at subsonic speed, all of the air flowing around the airplane is at a velocity of less than the speed of sound (known as Mach 1). Keep in mind that the air accelerates when it flows over certain parts of the airplane, like the top of the wing, so an airplane flying at 500 mph could have air over the top of the wing reach a speed of 600 mph. How fast an airplane can fly and still be considered in subsonic flight varies with the design of the wing, but as a Mach number, it will typically be just over Mach 0.8.

When an airplane is flying at transonic speed, part of the airplane is experiencing subsonic airflow and part is experiencing supersonic airflow. Over the top of the wing, probably about halfway back, the velocity of the air will reach Mach 1 and a shock wave will form. The shock wave forms 90 degrees to the airflow and is known as a normal shock wave. Stability problems can be encountered during transonic flight, because the shock wave can cause the airflow to separate from the wing. The shock wave also causes the center of lift to shift aft, causing the nose to pitch down. The speed at which the shock wave forms is known as the critical Mach number. Transonic speed is typically between Mach 0.80 and 1.20.

When an airplane is flying at supersonic speed, the entire airplane is experiencing supersonic airflow. At this speed, the shock wave which formed on top of the wing during transonic flight has moved all the way aft and has attached itself to the wing trailing edge. Supersonic speed is from Mach 1.20 to 5.0. If an airplane flies faster than Mach 5, it is said to be in hypersonic flight.

Shock Waves

Sound coming from an airplane is the result of the air being disturbed as the airplane moves through it, and the resulting pressure waves that radiate out from the source of the disturbance. For a slow moving airplane, the pressure waves travel out ahead of the airplane, traveling at the speed of sound. When the speed of the airplane reaches the speed of sound, however, the pressure waves (sound energy) cannot get away from the airplane. At this point the sound energy starts to pile up, initially on the top of the wing, and eventually attaching itself to the wing leading and trailing edges. This piling up of sound energy is called a shock wave. If the shock waves reach the ground, and cross the path of a person, they will be heard as a sonic boom. Figure 3-76A shows a wing in slow speed flight, with many

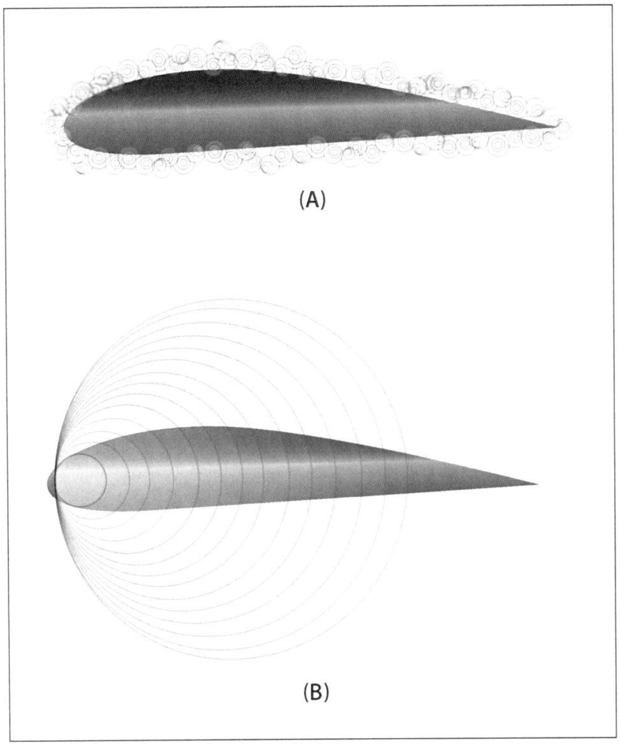

Figure 3-76. Sound energy in subsonic and supersonic flight.

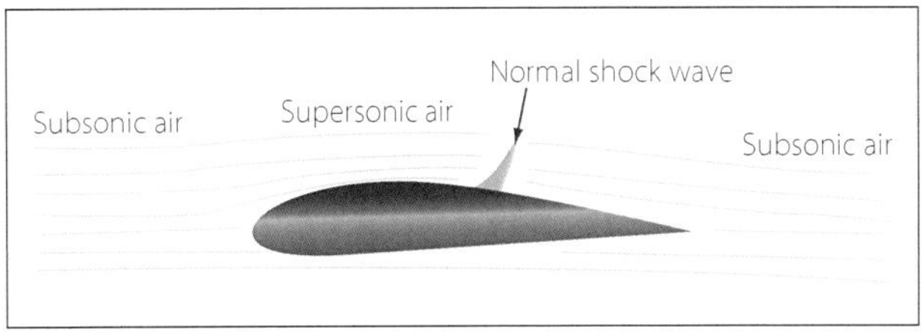

Figure 3-77. Normal shock wave.

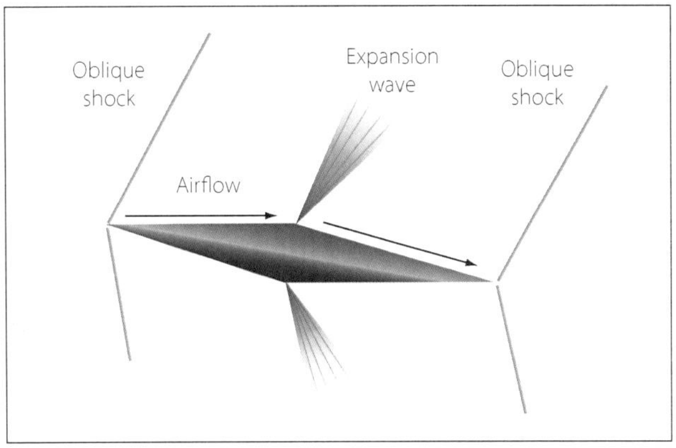

Figure 3-78. Supersonic airfoil with oblique shock waves and expansion waves.

disturbances on the wing generating sound pressure waves that are radiating outward. View B is the wing of an airplane in supersonic flight, with the sound pressure waves piling up toward the wing leading edge.

Normal Shock Wave

When an airplane is in transonic flight, the shock wave that forms on top of the wing, and eventually on the bottom of the wing, is called a normal shock wave. If the leading edge of the wing is blunted, instead of being rounded or sharp, a normal shock wave will also form in front of the wing during supersonic flight. Normal shock waves form perpendicular to the airstream. The velocity of the air behind a normal shock wave is subsonic, and the static pressure and density of the air are higher. Figure 3-77 shows a normal shock wave forming on the top of a wing.

Oblique Shock Wave

An airplane that is designed to fly supersonic will have very sharp edged surfaces, in order to have the least amount of drag. When the airplane is in supersonic flight, the sharp leading edge and trailing edge of the wing will have shock waves attach to them. These shock waves are known as oblique shock waves. Behind an oblique shock wave the velocity of the air is lower, but still supersonic, and the static pressure and density are higher. Figure 3-78 shows an oblique shock wave on the leading and trailing edges of a supersonic airfoil.

Expansion Wave

Earlier in the discussion of high-speed aerodynamics, it was stated that air at supersonic speed acts like a compressible fluid. For this reason, supersonic air, when given the opportunity, wants to expand outward. When supersonic air is flowing over the top of a wing, and the wing surface turns away from the direction of flow, the air will expand and follow the new direction. At the point where the direction of flow changes, an expansion wave will occur. Behind the expansion wave the velocity increases, and the static pressure and density decrease. An expansion wave is not a shock wave. Figure 3-78 shows an expansion wave on a supersonic airfoil.

High-Speed Airfoils

Transonic flight is the most difficult flight regime for an airplane, because part of the wing is experiencing subsonic airflow and part is experiencing supersonic airflow. For a subsonic airfoil, the aerodynamic center (the point of support) is approximately 25 percent of the way back from the wing leading edge. In supersonic flight, the aerodynamic center moves back to 50 percent of the wing's chord, causing some significant changes in the airplane's control and stability.

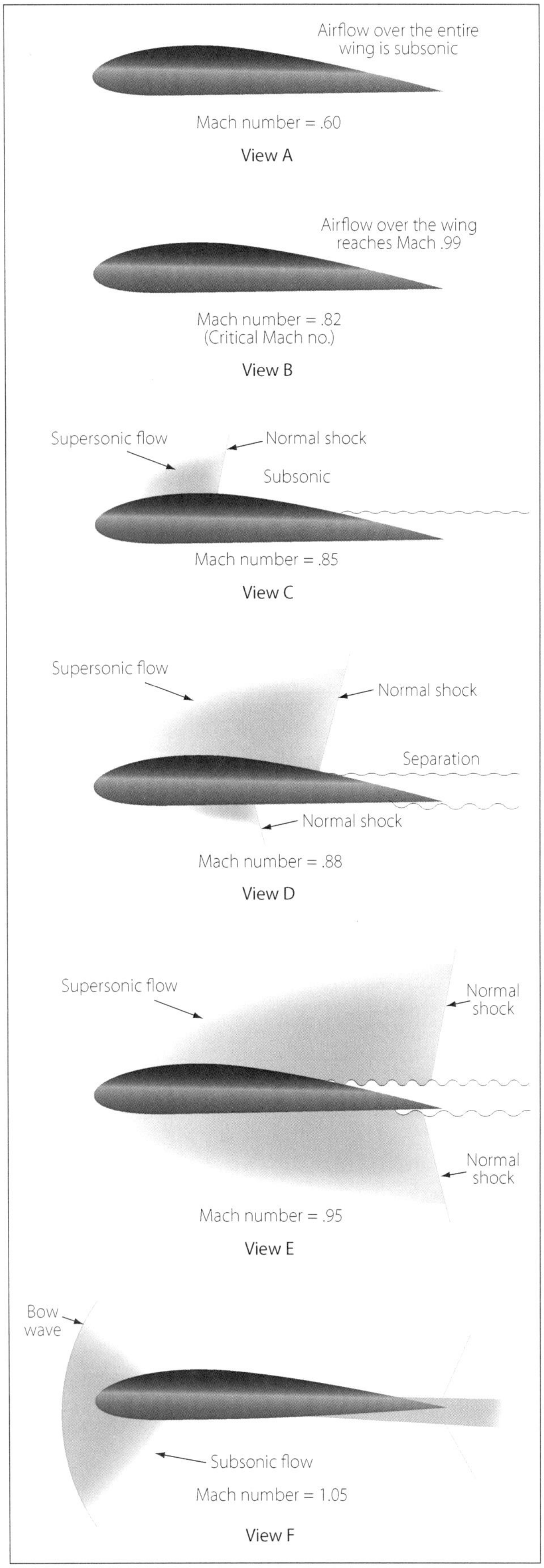

Figure 3-79. Airflow with progressively greater Mach numbers.

If an airplane designed to fly subsonic, perhaps at a Mach number of 0.80, flies too fast and enters transonic flight, some noticeable changes will take place with respect to the airflow over the wing. Figure 3-79 shows six views of a wing, with each view showing the Mach number getting higher. The scenario for the six views is as follows:

A. The Mach number is fairly low, and the entire wing is experiencing subsonic airflow.

B. The velocity has reached the critical Mach number, where the airflow over the top of the wing is reaching Mach 1 velocity.

C. The velocity has surpassed the critical Mach number, and a normal shock wave has formed on the top of the wing. Some airflow separation starts to occur behind the shock wave.

D. The velocity has continued to increase beyond the critical Mach number, and the normal shock wave has moved far enough aft that serious airflow separation is occurring. A normal shock wave is now forming on the bottom of the wing as well. Behind the normal shock waves, the velocity of the air is subsonic and the static pressure has increased.

E. The velocity has increased to the point that both shock waves on the wing (top and bottom) have moved to the back of the wing and attached to the trailing edge. Some airflow separation is still occurring.

F. The forward velocity of the airfoil is greater than Mach 1, and a new shock wave has formed just forward of the leading edge of the wing. If the wing has a sharp leading edge, the shock wave will attach itself to the sharp edge.

The airfoil shown in Figure 3-79 is not properly designed to handle supersonic airflow. The bow wave in front of the wing leading edge of view F would be attached to the leading edge, if the wing was a double wedge or biconvex design. These two wing designs are shown in Figure 3-80.

Aerodynamic Heating

One of the problems with airplanes and high-speed flight is the heat that builds up on the airplane's surface because of air friction. When the SR-71 Blackbird airplane is cruising at Mach 3.5, skin temperatures on its surface range from 450°F to over 1,000°F. To withstand this high temperature, the airplane was constructed of titanium alloy, instead of the traditional alu-

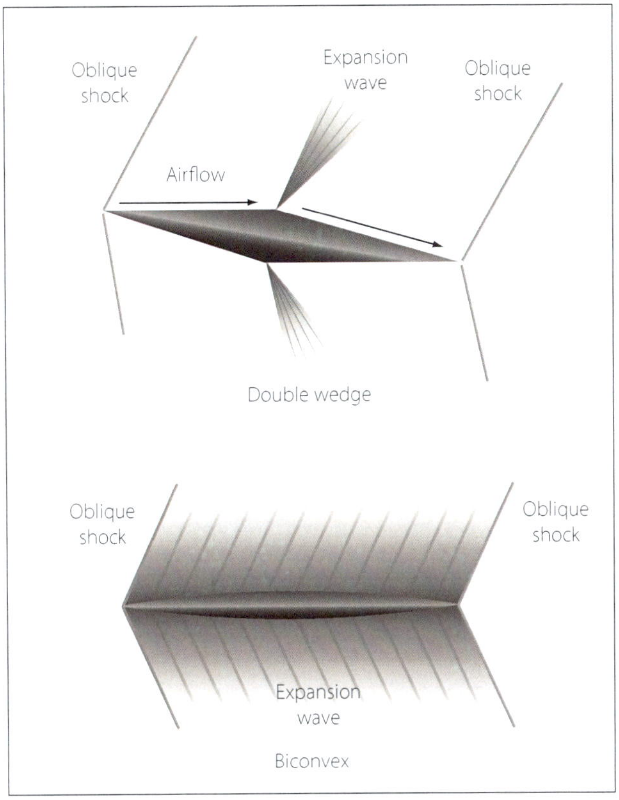

Figure 3-80. Double wedge and biconvex supersonic wing design.

minum alloy. The supersonic transport Concorde was originally designed to cruise at Mach 2.2, but its cruise speed was reduced to Mach 2.0 because of structural problems that started to occur because of aerodynamic heating. If airplanes capable of hypersonic flight are going to be built in the future, one of the obstacles that will have to be overcome is the stress on the airplane's structure caused by heat.

Helicopter Aerodynamics

The helicopter, as we know it today, falls under the classification known as rotorcraft. Rotorcraft are also known as rotary wing aircraft, because instead of their wing being fixed like it is on an airplane, the wing rotates. The rotating wing of a rotorcraft can be thought of as a lift producing device, like the wing of an airplane, or as a thrust producing device, like the propeller on a piston engine.

Helicopter Structures and Airfoils

The main parts that make up a helicopter are the cabin, landing gear, tail boom, powerplant, transmission, main rotor, and tail rotor. [Figure 3-81]

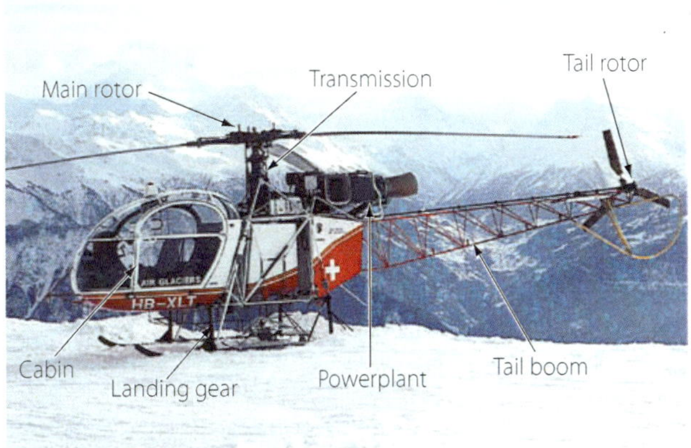

Figure 3-81. Main components of a helicopter.

Main Rotor Systems

The classification of main rotor systems is based on how the blades move relative to the main rotor hub. The principal classifications are known as fully articulated, semi-rigid, and rigid.

In the fully articulated rotor system, the blades are attached to the hub with multiple hinges. The blades are hinged in a way that allows them to move up and down and fore and aft, and bearings provide for motion around the pitch change axis. Rotor systems using this type of arrangement typically have three or more blades. The hinge that allows the blades to move up and down is called the flap hinge, and movement around this hinge is called flap. The hinge that allows the blades to move fore and aft is called a drag or lag hinge. Movement around this hinge is called dragging, lead/lag, or hunting. These hinges and their associated movement are shown in Figure 3-82. The main rotor head of a Eurocopter model 725 is shown in Figure 3-83, with the drag hinge and pitch change rods visible.

The semi-rigid rotor system is used with a two blade main rotor. The blades are rigidly attached to the hub, with the hub and blades able to teeter like a seesaw. The teetering action allows the blades to flap, with one blade dropping down while the other blade rises. The blades are able to change pitch independently of each other. Figure 3-84 shows a Bell Jet Ranger helicopter in flight. This helicopter uses a semi-rigid rotor system, which is evident because of the way the rotor is tilted forward when the helicopter is in forward flight.

With a rigid rotor system, the blades are not hinged for movement up and down (flapping) or for movement fore and aft (drag). The blades are able to move around the pitch change axis, with each blade being able to independently change its blade angle. The rigid

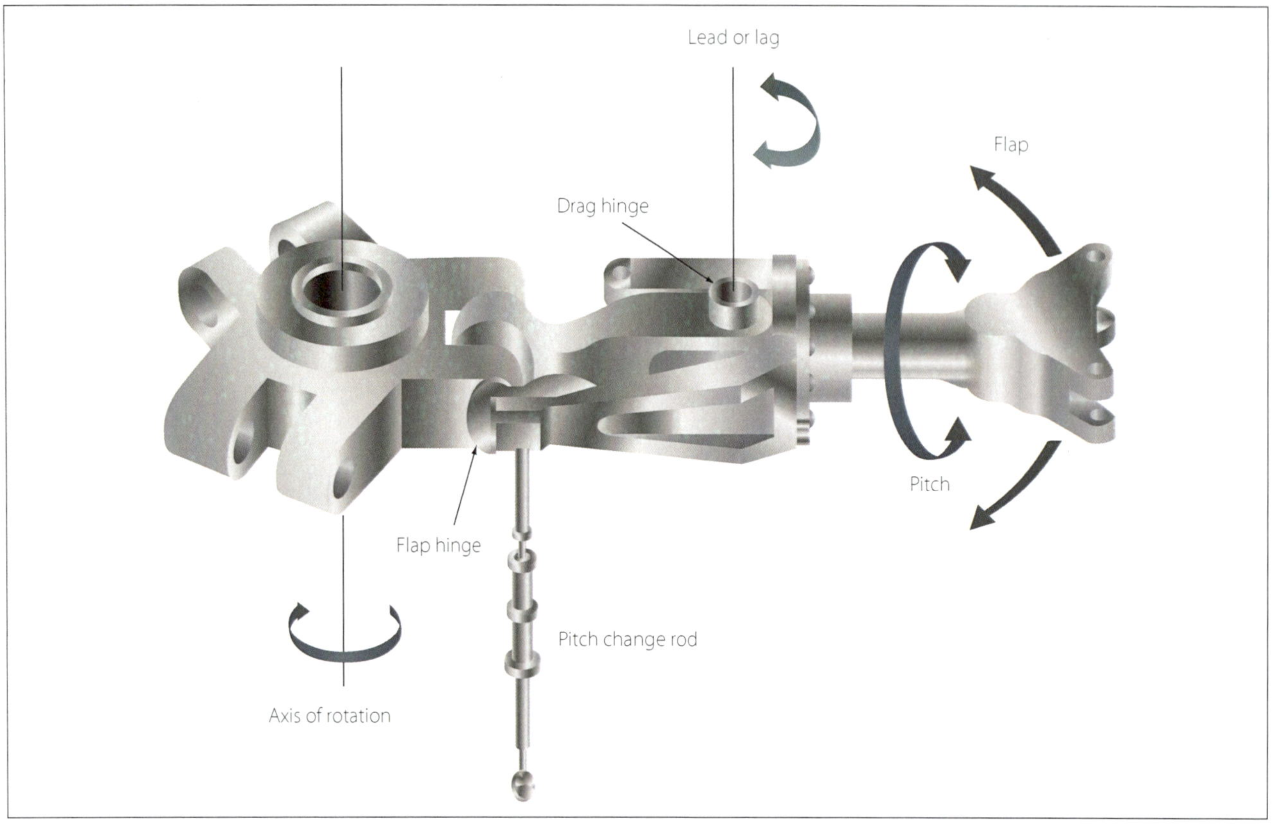

Figure 3-82. Fully articulated main rotor head.

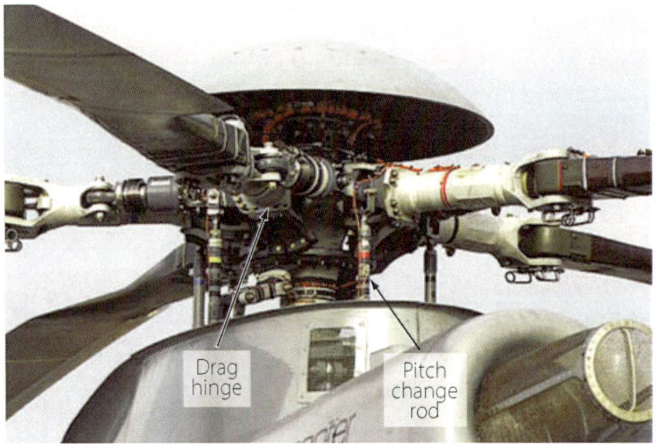

Figure 3-83. Eurocopter 725 main rotor head.

Figure 3-84. Bell Jet Ranger with semi-rigid main rotor.

rotor system uses blades that are very strong and yet flexible. They are flexible enough to bend when they need to, without the use of hinges or a teetering rotor, to compensate for the uneven lift that occurs in forward flight. The Eurocopter model 135 uses a rigid rotor system. [Figure 3-85]

Anti-Torque Systems

Any time a force is applied to make an object rotate, there will be an equal force acting in the opposite direction. If the helicopter's main rotor system rotates clockwise when viewed from the top, the helicopter will try to rotate counterclockwise. Earlier in this chapter, it was discovered that torque is what tries to make something rotate. For this reason, a helicopter uses what is called an anti-torque system to counteract the force trying to make it rotate.

One method that is used on a helicopter to counteract torque is to place a spinning set of blades at the end of the tail boom. These blades are called a tail rotor or anti-torque rotor, and their purpose is to create a force (thrust) that acts in the opposite direction of the way the helicopter is trying to rotate. The tail rotor force, in pounds, multiplied by the distance from the tail rotor to the main rotor, in feet, creates a torque in pound-feet that counteracts the main rotor torque.

Figure 3-85. Eurocopter Model 135 rigid rotor system.

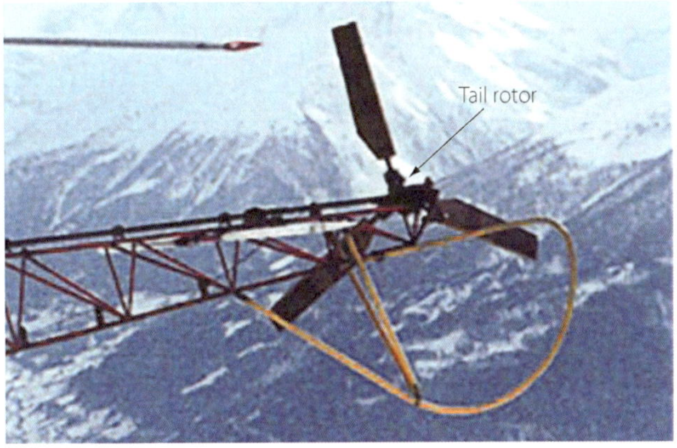

Figure 3-86. Aerospatiale helicopter tail rotor.

Figure 3-86 shows a three bladed tail rotor on an Aerospatiale AS-315B helicopter. This tail rotor has open tipped blades that are variable pitch, and the helicopter's anti-torque pedals (positioned like rudder pedals on an airplane) control the amount of thrust they create. Some potential problems with this tail rotor system are as follows:

- The spinning blades are deadly if someone walks into them.
- When the helicopter is in forward flight and a vertical fin may be in use to counteract torque, the tail rotor robs engine power and creates drag.

An alternative to the tail rotor seen in Figure 3-86 is a type of anti-torque rotor known as a fenestron, or "fan-in-tail" design as seen in Figure 3-87. Because the rotating blades in this design are enclosed in a shroud, they present less of a hazard to personnel on the ground and they create less drag in flight.

A third method of counteracting the torque of the helicopter's main rotor is a technique called the "no tail rotor" system, or NOTAR. This system uses a high

Figure 3-87. Fenestron on a Eurocopter Model 135.

volume of air at low pressure, which comes from a fan driven by the helicopter's engine. The fan forces air into the tail boom, where a portion of it exits out of slots on the right side of the boom and, in conjunction with the main rotor downwash, creates a phenomenon called the "Coanda effect." The air coming out of the slots on the right side of the boom causes a higher velocity, and therefore lower pressure, on that side of the boom. The higher pressure on the left side of the boom creates the primary force that counteracts the torque of the main rotor. The remainder of the air travels back to a controllable rotating nozzle in the helicopter's tail. The air exits the nozzle at a high velocity, and creates an additional force (thrust) that helps counteracts the torque of the main rotor. A NOTAR system is shown in Figures 3-88 and 3-89.

For helicopters with two main rotors, such as the Chinook that has a main rotor at each end, no anti-torque

Figure 3-88. McDonnell Douglas 520 NOTAR.

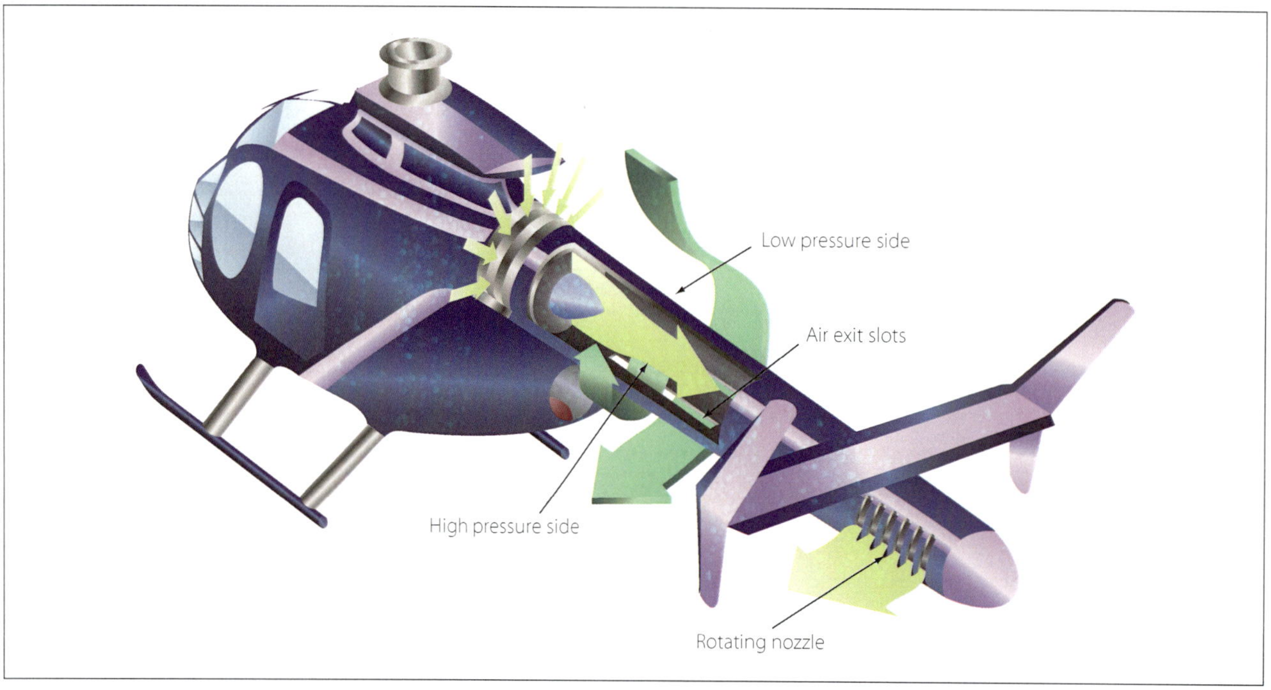

Figure 3-89. Airflow for a NOTAR.

rotor is needed. For this type of helicopter, the two main rotors turn in opposite directions, and each one cancels out the torque of the other.

Helicopter Axes of Flight

Helicopters, like airplanes, have a vertical, lateral, and longitudinal axis that passes through the helicopter's center of gravity. Helicopters yaw around the vertical axis, pitch around the lateral axis, and rotate around the longitudinal axis. Figure 3-90 shows the three axes of a helicopter and how they relate to the helicopter's movement. All three axes will intersect at the helicopter's center of gravity, and the helicopter pivots around this point. Notice in the figure that the vertical axis passes almost through the center of the main rotor, because the helicopter's center of gravity needs to be very close to this point.

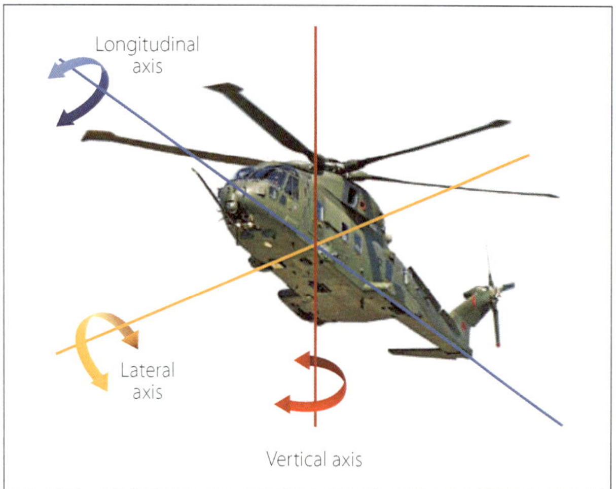

Figure 3-90. Three axes of rotation for a helicopter.

Control Around the Vertical Axis

For a single main rotor helicopter, control around the vertical axis is handled by the anti-torque rotor (tail rotor) or from the fan's airflow on a NOTAR type helicopter. Like in an airplane, rotation around this axis is known as yaw. The pilot controls yaw by pushing on the anti-torque pedals located on the cockpit floor, in the same way the airplane pilot controls yaw by pushing on the rudder pedals. To make the nose of the helicopter yaw to the right, the pilot pushes on the right anti-torque pedal. When viewed from the top, if the helicopter tries to spin in a counterclockwise direction because of the torque of the main rotor, the pilot will also push on the right anti-torque pedal to counteract the main rotor torque. By using the anti-torque pedals, the pilot can intentionally make the helicopter rotate in either direction around the vertical axis. The anti-torque pedals can be seen in Figure 3-91.

Some helicopters have a vertical stabilizer, such as those shown in Figures 3-90 and 3-92. In forward flight, the vertical stabilizer creates a force that helps counteract the torque of the main rotor, thereby reducing the power needed to drive the anti-torque system located at the end of the tail boom.

Figure 3-91. Helicopter cockpit controls.

Control Around the Longitudinal and Lateral Axes

Movement around the longitudinal and lateral axes is handled by the helicopter's main rotor. In the cockpit, there are two levers that control the main rotor, known as the collective and cyclic pitch controls. The collective pitch lever is on the side of the pilot's seat, and the cyclic pitch lever is at the front of the seat in the middle. [Figure 3-91]

When the collective pitch control lever is raised, the blade angle of all the rotor blades increases uniformly and they create the lift that allows the helicopter to take off vertically. The grip on the end of the collective pitch control is the throttle for the engine, which is rotated to increase engine power as the lever is raised. On many helicopters, the throttle automatically rotates and increases engine power as the collective lever is raised. The collective pitch lever may have adjustable friction built into it, so the pilot does not have to hold upward pressure on it during flight.

The cyclic pitch control lever, like the yoke of an airplane, can be pulled back or pushed forward, and can be moved left and right. When the cyclic pitch lever is pushed forward, the rotor blades create more lift as they pass through the back half of their rotation and less lift as they pass through the front half. The difference in lift is caused by changing the blade angle (pitch) of the rotor blades. The pitch change rods that were seen earlier, in Figures 3-82 and 3-83, are controlled by the cyclic pitch lever and they are what change the pitch of the rotor blades. The increased lift in the back either causes the main rotor to tilt forward, the nose of the helicopter to tilt downward, or both. The end result is the helicopter moves in the forward direction. If the cyclic pitch lever is pulled back, the rotor blade lift will be greater in the front and the helicopter will back up.

If the cyclic pitch lever is moved to the left or the right, the helicopter will bank left or bank right. For the helicopter to bank to the right, the main rotor blades must create more lift as they pass by the left side of the helicopter. Just the opposite is true if the helicopter is banking to the left. By creating more lift in the back than in the front, and more lift on the left than on the right, the helicopter can be in forward flight and banking to the right. In Figure 3-92, an Agusta A-109 can be seen in forward flight and banking to the right. The

Figure 3-92. Agusta A-109 banking to the right.

Figure 3-93. Air Force CH-53 in a hover.

rotor blade in the rear and the one on the left are both in an upward raised position, meaning they have both experienced the condition called flap.

Some helicopters use a horizontal stabilizer, similar to what is seen on an airplane, to help provide additional stability around the lateral axis. A horizontal stabilizer can be seen on the Agusta A-109 in Figure 3-92.

Helicopters in Flight

Hovering

For a helicopter, hovering means that it is in flight at a constant altitude, with no forward, aft, or sideways movement. In order to hover, a helicopter must be producing enough lift in its main rotor blades to equal the weight of the aircraft. The engine of the helicopter must be producing enough power to drive the main rotor, and also to drive whatever type of anti-torque system is being used. The ability of a helicopter to hover is affected by many things, including whether or not it is in ground effect, the density altitude of the air, the available power from the engine, and how heavily loaded it is.

For a helicopter to experience ground effect, it typically needs to be no higher off the ground than one half of its main rotor system diameter. If a helicopter has a main rotor diameter of 40 ft, it will be in ground effect up to an altitude of approximately 20 ft. Being close to the ground affects the velocity of the air through the rotor blades, causing the effective angle of attack of the blades to increase and the lift to increase. So, if a helicopter is in ground effect, it can hover at a higher gross weight than it can when out of ground effect. On a windy day, the positive influence of ground effect is lessened, and at a forward speed of 5 to 10 mph the positive influence becomes less. In Figure 3-93, an Air Force CH-53 is seen in a hover, with all the rotor blades flapping up as a result of creating equal lift.

Forward Flight

In the early days of helicopter development, the ability to hover was mastered before there was success in attaining forward flight. The early attempts at forward flight resulted in the helicopter rolling over when it tried to depart from the hover and move in any direction. The cause of the rollover is what we now refer to as dissymmetry of lift.

When a helicopter is in a hover, all the rotor blades are experiencing the same velocity of airflow. When the helicopter starts to move, the velocity of airflow seen by the rotor blades changes. For helicopters built in the United States, the main rotor blades turn in a counterclockwise direction when viewed from the top. Viewed from the top, as the blades move around the right side of the helicopter, they are moving toward the nose; as they move around the left side of the helicopter, they are moving toward the tail. When the helicopter starts moving forward, the blade on the right side is moving toward the relative wind, and the blade on the left side is moving away from the relative wind. This causes the blade on the right side to create more lift and the blade on the left side to create less lift. Figure 3-94 shows how this occurs.

In Figure 3-94, blade number 2 would be called the advancing blade, and blade number 1 would be called the retreating blade. The advancing blade is moving toward the relative wind, and therefore experiences a greater velocity of airflow. The increased lift created by the blade on the right side will try to roll the helicopter to the left. If this condition is allowed to exist, it will ultimately lead to the helicopter crashing.

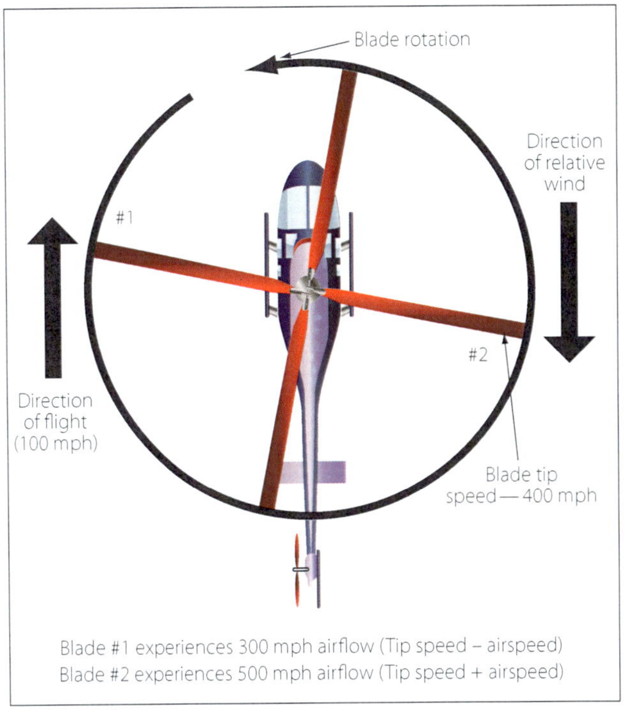

Figure 3-94. Dissymmetry of lift for rotor blades.

Blade Flapping

To solve the problem of dissymmetry of lift, helicopter designers came up with a hinged design that allows the rotor blade to flap up when it experiences increased lift, and to flap down when it experiences decreased lift. When a rotor blade advances toward the front of the helicopter and experiences an increased velocity of airflow, the increase in lift causes the blade to flap up. This upward motion of the blade changes the direction of the relative wind in relation to the chord line of the blade, and causes the angle of attack to decrease. The decrease in the angle of attack decreases the lift on the blade. The retreating blade experiences a reduced velocity of airflow and reduced lift, and flaps down. By flapping down, the retreating blade ends up with an increased angle of attack and an increase in lift. The end result is the lift on the blades is equalized, and the tendency for the helicopter to roll never materializes.

The semi-rigid and fully articulated rotor systems have flapping hinges that automatically allow the blades to move up or down with changes in lift. The rigid type of rotor system has blades that are flexible enough to bend up or down with changes in lift.

Advancing Blade and Retreating Blade Problems

As a helicopter flies forward at higher and higher speeds, the blade advancing toward the relative wind sees the airflow at an ever increasing velocity. Eventually, the velocity of the air over the rotor blade will reach sonic velocity, much like the critical Mach number for the wing of an airplane. When this happens, a shock wave will form and the air will separate from the rotor blade, resulting in a high-speed stall.

As the helicopter's forward speed increases, the relative wind over the retreating blade decreases, resulting in a loss of lift. The loss of lift causes the blade to flap down and the effective angle of attack to increase. At a high enough forward speed, the angle of attack will increase to a point that the rotor blade stalls. The tip of the blade stalls first, and then progresses in toward the blade root.

When approximately 25 percent of the rotor system is stalled, due to the problems with the advancing and retreating blades, control of the helicopter will be lost. Conditions that will lead to the rotor blades stalling include high forward speed, heavy gross weight, turbulent air, high-density altitude, and steep or abrupt turns.

Autorotation

The engine on a helicopter drives the main rotor system by way of a clutch and a transmission. The clutch allows the engine to be running and the rotor system not to be turning, while the helicopter is on the ground, and it also allows the rotor system to disconnect from the engine while in flight, if the engine fails. Having the rotor system disconnect from the engine in the event of an engine failure is necessary if the helicopter is to be capable of a flight condition called autorotation.

Autorotation is a flight condition where the main rotor blades are driven by the force of the relative wind passing through the blades, rather than by the engine. This flight condition is similar to an airplane gliding if its engine fails while in flight. As long as the helicopter maintains forward airspeed, while decreasing altitude, and the pilot lowers the blade angle on the blades with the collective pitch, the rotor blades will continue to rotate. The altitude of the helicopter, which equals potential energy, is given up in order to have enough energy (kinetic energy) to keep the rotor blades turning. As the helicopter nears the ground, the cyclic pitch control is used to slow the forward speed and to flare the helicopter for landing. With the airspeed bled off, and the helicopter now close to the ground, the final step is to use the collective pitch control to cushion the landing. The airflow through the rotor blades in normal forward flight and in an autorotation flight condition are shown in Figure 3-95. In Figure 3-96, a Bell Jet Ranger is shown approaching the ground in the final stage of an autorotation.

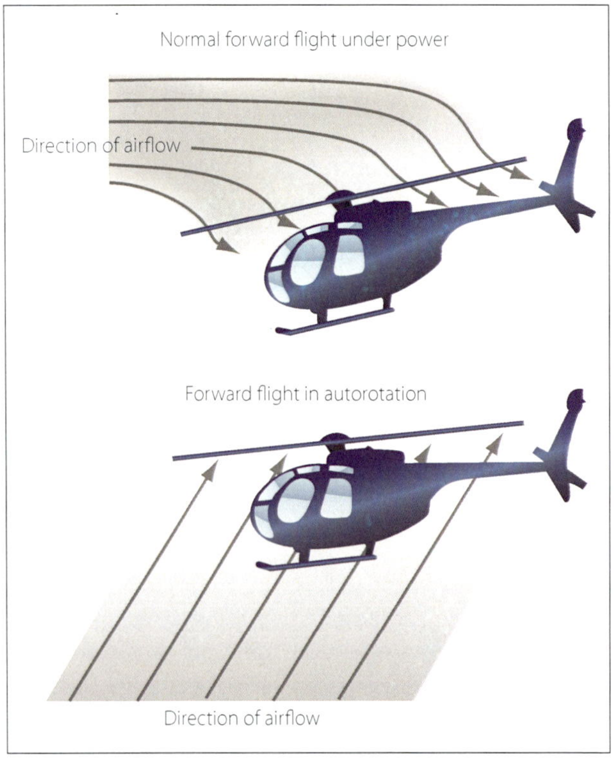

Figure 3-95. Rotor blade airflow during normal flight and during autorotation.

Figure 3-96. Bell Jet Ranger in final stage of autorotation.

Weight-Shift Control, Flexible Wing Aircraft Aerodynamics

A weight-shift control, flexible wing type aircraft consists of a fabric-covered wing, often referred to as the sail, attached to a tubular structure that has wheels, seats, and an engine and propeller. The wing structure is also tubular, with the fabric covering creating the airfoil shape. The shape of the wing varies among the different models of weight-shift control aircraft being produced, but a delta shaped wing is a very popular design. Within the weight-shift control aircraft community, these aircraft are typically referred to as trikes. [Figure 3-97]

Figure 3-97. Weight-shift control aircraft in level flight.

In Figure 3-97, the trike's mast is attached to the wing at the hang point on the keel of the wing with a hang point bolt and safety cable. There is also a support tube, known as a king post, extending up from the top of the wing, with cables running down and secured to the tubular wing structure. The cables running down from the king post as part of the upper rigging are there to support the wing when the aircraft is on the ground, and to handle negative loads when in flight. The lines that run from the king post to the trailing edge of the wing are known as reflex cables. These cables maintain the shape of the wing when it is in a stalled state by holding the trailing edge of the wing up which helps raise the nose during recovery from the stall. If the aircraft goes into an inadvertent stall, having the trailing edge of the wing in a slightly raised position helps raise the nose of the aircraft and get it out of the stall. The passenger seat is centered under the wing's aerodynamic center, with the weight of the pilot being forward of this point and the weight of the engine and propeller being aft.

Unlike a traditional airplane, the trike does not have a rudder, elevator, or ailerons. Instead, it has a wing that can be pivoted forward or aft, and left or right. In Figure 3-98, the pilot's hand is on a control bar that is connected to a pivot point just forward of where the wing attaches. There are cables attached to the ends of the bar that extend up to the wing's leading and

Figure 3-98. Weight-shift control aircraft getting ready for flight.

trailing edge, and to the left and right side of the cross bar. Running from the wing leading edge to trailing edge are support pieces known as battens. The battens fit into pockets, and they give the wing its cambered shape. The names of some of the primary parts of the trike are shown in Figure 3-98, and these parts will be referred to when the flight characteristics of the trike are described in the paragraphs that follow.

In order to fly the trike, engine power is applied to get the aircraft moving. As the groundspeed of the aircraft reaches a point where flight is possible, the pilot pushes forward on the control bar, which causes the wing to pivot where it attaches to the mast and the leading edge of the wing tilts up. When the leading edge of the wing tilts up, the angle of attack and the lift of the wing increase. With sufficient lift, the trike rotates and starts climbing. Pulling back on the bar reduces the angle of attack, and allows the aircraft to stop climbing and to fly straight and level. Once the trike is in level flight, airspeed can be increased or decreased by adding or taking away engine power by use of the throttle.

Stability in flight along the longitudinal axis (nose to tail), for a typical airplane, is achieved by having the horizontal stabilizer and elevator generate a force that balances out the airplane's nose heavy tendency. Because the trike does not have a horizontal stabilizer or elevator, it must create stability along the longitudinal axis in a different way. The trike has a sweptback delta wing, with the trailing edge of the wingtips located well aft of the aircraft center of gravity. Pressure acting on the tips of the delta wing creates the force that balances out the nose heavy tendency.

The wings of weight-shift control aircraft are designed in a way that allows them to change their shape when subjected to an external force. This is possible because the frame leading edges and the sail are flexible, which is why they are sometimes referred to as flexible wing aircraft. This produces somewhat different aerodynamic effects when compared with a normal fixed-wing aircraft. A traditional small airplane, like a Cessna 172, turns or banks by using the ailerons, effectively altering the camber of the wing and thereby generating differential lift. By comparison, weight shift on a trike actually causes the wing to twist, which changes the angle of attack on the wing and causes the differential lift to exist that banks the trike. The cross-bar (wing spreader) of the wing frame is allowed to float slightly with respect to the keel, and this, along with some other geometric considerations allows the sail to "billow shift." Billow shift can be demonstrated on the ground by grabbing the trailing edge of one end of the

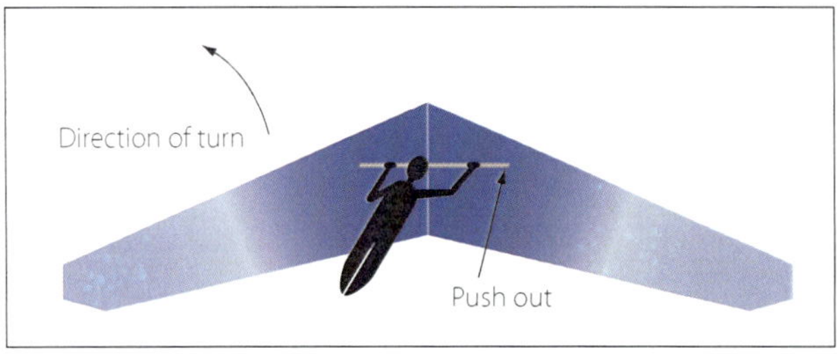

Figure 3-99. Direction of turn based on weight shift.

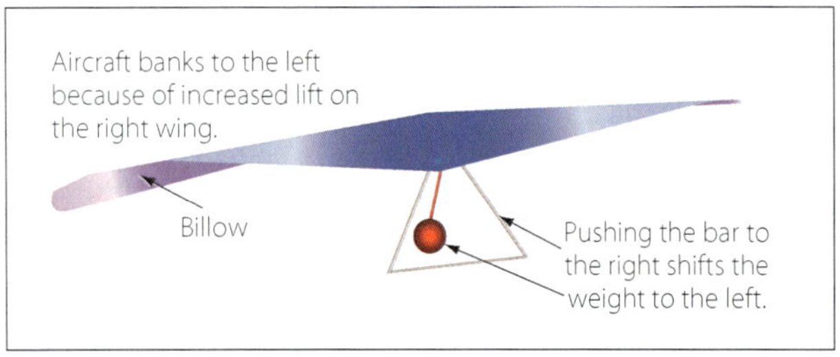

Figure 3-100. Weight shift to the left causing a left-hand turn.

wing and lifting up on it. If this was done, the fabric on the other end of the wing would become slightly flatter and tighter, and the wing's angle of attack would increase.

If the pilot pushes the bar to the right, the wing pivots with the left wingtip dropping down and the right wingtip rising up, causing the aircraft to bank to the left. This motion is depicted in Figure 3-99, showing a hang glider as an example. The shift in weight to the left increases the wing loading on the left, and lessens it on the right. The increased loading on the left wing increases its washout and reduces its angle of attack and lift. The increased load on the left wing causes the left wing to billow, which causes the fabric to tighten on the right wing and the angle of attack and lift to increase. The change in lift is what banks the aircraft to the left. Billow on the left wing is depicted in Figure 3-100.

Shifting weight to the right causes the aircraft to bank right. The weight of the trike and its occupants acts like a pendulum, and helps keep the aircraft stable in flight. Pushing or pulling on the bar while in flight causes the weight hanging below the wing to shift its position relative to the wing, which is why the trike is referred to as a weight-shift aircraft.

Once the trike is in flight and flying straight and level, the pilot only needs to keep light pressure on the bar that controls the wing. If the trike is properly balanced and there is no air turbulence, the aircraft will remain stable even if the pilot's hands are removed from the bar. The same as with any airplane, increasing engine power will make the aircraft climb and decreasing power will make it descend. The throttle is typically controlled with a foot pedal, like a gas pedal in an automobile.

A trike lands in a manner very similar to an airplane. When it is time to land, the pilot reduces engine power with the foot-operated throttle, causing airspeed and wing lift to decrease. As the trike descends, the rate of descent can be controlled by pushing forward or pulling back on the bar, and varying engine power. When the trike is almost to the point of touchdown, the engine power will be reduced and the angle of attack of the wing will be increased, to cushion the descent and provide a smooth landing. If the aircraft is trying to land in a very strong crosswind, the landing may not be so smooth. When landing in a cross wind, the pilot will land in a crab to maintain direction down the runway. Touchdown is done with the back wheels first, then letting the front wheel down.

A trike getting ready to touch down can be seen in Figure 3-101. The control cables coming off the control bar can be seen, and the support mast and the cables on top of the wing, including the luff lines, can also be seen.

Powered Parachute Aerodynamics

A powered parachute has a carriage very similar to the weight-shift control aircraft. Its wing, however, has no support structure or rigidity and only takes on the shape of an airfoil when it is inflated by the blast of air from the propeller and the forward speed of the aircraft. In Figure 3-102, a powered parachute is on its approach to land with the wing fully inflated and rising up above the aircraft. Each colored section of the inflated wing is made up of cells that are open in the front to allow air to ram in, and closed in the back to keep the air trapped inside. In between all the cells there are holes that allow the air to flow from one cell to the next, in order to equalize the pressure within the inflated wing. The wing is attached to the carriage of the aircraft by a large number of nylon or kevlar lines that run from the tips of the wing all the way to the center. The weight of the aircraft acting on these lines and their individual lengths cause the inflated wing to take its shape. The lines attach to the carriage of the aircraft at a location very close to where the center of gravity is located, and this attachment point is adjustable to account for balance changes with occupants of varying weights.

As in weight-shift control aircraft, the powered parachute does not have the traditional flight controls of a fixed-wing airplane. When the wing of the aircraft

Figure 3-101. Weight-shift control aircraft landing.

is inflated and the aircraft starts moving forward, the wing starts generating lift. Once the groundspeed is sufficient for the wing's lift greater than the weight of the aircraft, the aircraft lifts off the ground. Unlike an airplane, where the pilot has a lot of control over when the airplane rotates by deciding when to pull back on the yoke, the powered parachute will not take off until it reaches a specific airspeed. The powered parachute will typically lift off the ground at a speed somewhere between 28 and 30 mph, and will have an airspeed in flight of approximately 30 mph.

Figure 3-102. Powered parachute with the wing inflated.

Once the powered parachute is in flight, control over climbing and descending is handled with engine power. Advancing the throttle makes the aircraft climb, and retarding the throttle makes it descend. The inflated wing creates a lot of drag in flight, so reducing the engine power creates a very controllable descent of the aircraft. The throttle, for controlling engine power, is typically located on the right-hand side of the pilot. [Figure 3-103]

Turning of the powered parachute in flight is handled by foot-operated pedals (steering bars) located at the front of the aircraft. These bars can be seen in Figure 3-103. Each foot-operated pedal controls a set of lines, usually made out of nylon, that runs up to the trailing edge of each wingtip. When the right foot pedal is pushed, the line pulls down on the trailing edge of the right wingtip. As the trailing edge of the right wing drops down, drag is increased on the right side and the aircraft turns right. When pressure is taken off the foot

Figure 3-103. Two seat powered parachute.

Figure 3-104. Powered parachute wing trailing edge.

pedal, the drag on the entire airfoil equalizes and the aircraft resumes its straight-and-level flight.

To land a powered parachute, the first action the pilot takes is to reduce engine power and allow the aircraft to descend. With the power reduced to idle, the aircraft will descend at a rate of approximately 5 to 10 fps. As the aircraft approaches the ground, the descent rate can be lessened by increasing the engine power. Just before touchdown, the pilot pushes on both foot-operated pedals to drop the trailing edges on both sides of the wing. This action increases the drag on the wing uniformly, causing the wing to pivot aft, which raises the wing leading edge and increases the angle of attack and lift.

In Figure 3-104, the pilot is pushing on both foot pedals and the left and right wing trailing edges are deflected downward. The aircraft has just touched down and the wing is trailing behind the aircraft, caused by the high angle of attack and the additional drag on the wing. The increase in lift reduces the descent rate to almost nothing, and provides for a gentle landing. If the pilot pushes on the foot pedals too soon, the wing may pivot too far aft before touchdown, resulting in an unacceptable descent rate. In that case, it might be a relatively hard landing.

Chapter 4: Aircraft Weight & Balance

The weight of an aircraft and its balance are extremely important for operating an aircraft in a safe and efficient manner. When a manufacturer designs an aircraft and the Federal Aviation Administration (FAA) certifies it, the specifications identify the aircraft's maximum weight and the limits within which it must balance.

The maximum allowable weight is based on the surface area of the wing, and how much lift it will generate at a safe and appropriate airspeed. If a small general aviation airplane, for example, required a takeoff speed of 200 miles per hour (mph) to generate enough lift to support its weight, that would not be safe. Taking off and landing at lower airspeeds is certainly safer than doing so at higher speeds.

Where an aircraft balances is also a significant factor in determining if the aircraft is safe to operate. An aircraft that does not have good balance can exhibit poor maneuverability and controllability, making it difficult or impossible to fly. This could result in an accident, causing damage to the aircraft and injury to the people on board. Safety is the primary reason for concern about an aircraft's weight and balance.

A secondary reason for concern about weight and balance, but also a very important one, is the efficiency of the aircraft. Improper loading reduces the efficiency of an aircraft from the standpoint of ceiling, maneuverability, rate of climb, speed, and fuel consumption. If an airplane is loaded in such a way that it is extremely nose heavy, higher than normal forces will need to be exerted at the tail to keep the airplane in level flight. The higher than normal forces at the tail will create additional drag, which will require additional engine power and therefore additional fuel flow in order to maintain airspeed.

The most efficient condition for an aircraft is to have the point where it balances fall very close to, or perhaps exactly at, the aircraft's center of lift. If this were the case, little or no flight control force would be needed to keep the aircraft flying straight and level. In terms of stability and safety, however, this perfectly balanced condition might not be desirable. All of the factors that affect aircraft safety and efficiency, in terms of its weight and balance, are discussed in detail in this chapter.

Need and Requirements for Aircraft Weighing

Every aircraft type certificated by the FAA, before leaving the factory for delivery to its new owner, receives a weight and balance report as part of its required aircraft records. The weight and balance report identifies the empty weight of the aircraft and the location at which the aircraft balances, known as the center of gravity. If the manufacturer chooses to do so, it can weigh every aircraft it produces and issue the weight and balance report based on that weighing. As an alternative, the manufacturer is permitted to weigh an agreed upon percentage of a particular model of aircraft produced, perhaps 10 to 20 percent, and apply the average to all the aircraft.

After the aircraft leaves the factory and is delivered to its owner, the need or requirement for placing the aircraft on scales and reweighing it varies depending on the type of aircraft and how it is used. For a small general aviation airplane being used privately, such as a Cessna 172, there is no FAA requirement that it be periodically reweighed. There is, however, an FAA requirement that the airplane always have a current and accurate weight and balance report. If the weight and balance report for an aircraft is lost, the aircraft must be weighed and a new report must be created. If the airplane has new equipment installed, such as a radio or a global positioning system, a new weight and balance report must be created. If the installer of the equipment wants to place the airplane on scales and weigh it after the installation, that is a perfectly acceptable way of creating the new report. If the installer knows the exact weight and location of the new equipment, it is also possible to create a new report by doing a series of mathematical calculations.

Over a period of time, almost all aircraft have a tendency to gain weight. Examples of how this can happen include an airplane being repainted without the old paint being removed, and the accumulation of dirt, grease, and oil in parts of the aircraft that are not easily accessible for cleaning. When new equipment is installed, and its weight and location are mathematically accounted for, some miscellaneous weight might be overlooked, such as wire and hardware. For this reason, even if the FAA does not require it, it is a good practice to periodically place an aircraft on scales and confirm its actual empty weight and empty weight center of gravity.

Some aircraft are required to be weighed and have their center of gravity calculated on a periodic basis, typically every 3 years. Examples of aircraft that fall under this requirement are:

1. Air taxi and charter twin-engine airplanes operating under Title 14 of the Code of Federal Regulations (14 CFR) part 135, section (§)135.185(a).
2. Airplanes with a seating capacity of 20 or more passengers or a maximum payload of 6,000 pounds or more, as identified in 14 CFR part 125, §125.91(b). This paragraph applies to most airplanes operated by the airlines, both main line and regional, and to many of the privately operated business jets.

Weight and Balance Terminology

Datum

The datum is an imaginary vertical plane from which all horizontal measurements are taken for balance purposes, with the aircraft in level flight attitude. If the datum was viewed on a drawing or photograph of an aircraft, it would appear as a vertical line which is perpendicular (90 degrees) to the aircraft's horizontal axis. For each aircraft make and model, the location of all items is identified in reference to the datum. For example, the fuel in a tank might be 60 inches (60") behind the datum, and a radio on the flight deck might be 90" forward of the datum.

There is no fixed rule for the location of the datum, except that it must be a location that will not change during the life of the aircraft. For example, it would not be a good idea to have the datum be the tip of the propeller spinner or the front edge of a seat, because changing to a new design of spinner or moving the seat would cause the datum to change. It might be located at or near the nose of the aircraft, a specific number of inches forward of the nose, at the engine firewall,

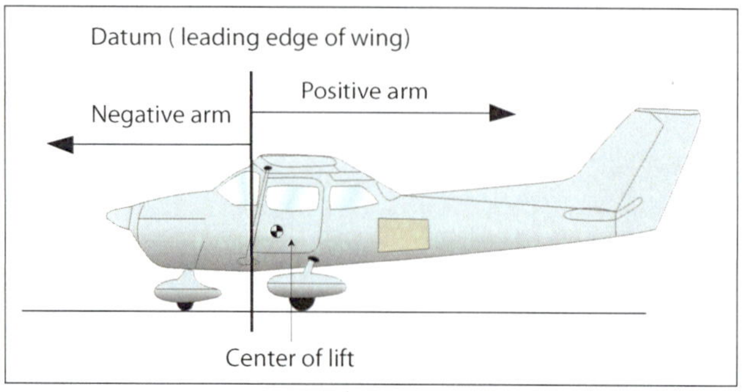

Figure 4-1. Datum location and its effect on positive and negative arms.

at the center of the main rotor shaft of a helicopter, or any place that can be imagined. The manufacturer has the choice of locating the datum where it is most convenient for measurement, equipment location, and weight and balance computation. Figure 4-1 shows an aircraft with the leading edge of the wing being the datum.

The location of the datum is identified in the Aircraft Specifications or Type Certificate Data Sheet. Aircraft certified prior to 1958 fell under the Civil Aeronautics Administration, and had their weight and balance information contained in a document known as Aircraft Specifications. Aircraft certified since 1958 fall under the FAA and have their weight and balance information contained in a document known as a Type Certificate Data Sheet. The Aircraft Specifications typically included the aircraft equipment list. For aircraft with a Type Certificate Data Sheet, the equipment list is a separate document.

Arm

The arm is the horizontal distance that a part of the aircraft or a piece of equipment is located from the datum. The arm's distance is always given or measured in inches, and, except for a location which might be exactly on the datum, it is preceded by the algebraic sign for positive (+) or negative (−). The positive sign indicates an item is located aft of the datum and the negative sign indicates an item is located forward of the datum. If the manufacturer chooses a datum that is at the most forward location on an aircraft (or some distance forward of the aircraft), all the arms will be positive numbers. Location of the datum at any other point on the aircraft will result in some arms being positive numbers, or aft of the datum, and some arms being negative numbers, or forward of the datum. Figure 4-1 shows an aircraft where the datum is the leading edge of the wing. For this aircraft, any item (fuel, seat, radio, and so forth) located forward of the wing leading edge will have a negative arm, and any item located aft of the wing leading edge will have a positive arm. If an item was located exactly at the wing leading edge, its arm would be zero, and mathematically it would not matter whether its arm was considered to be positive or negative.

The arm of each item is usually included in parentheses immediately after the item's name or weight in the Aircraft Specifications, Type Certificate Data Sheet, or equipment list for the aircraft. In a Type Certificate Data Sheet, for example, the fuel quantity might be identified as 150 gallons (gal) (+138) and the nose baggage limit as 200 pounds (lb) (−55). These numbers indicate that the fuel is located 138" aft of the datum and the nose baggage is located 55" forward of the datum. If the arm for a particular piece of equipment is not known, its exact location must be accurately measured. When the arm for a piece of equipment is being determined, the measurement is taken from the datum to the piece of equipment's own center of gravity.

Moment

A moment is the product of a weight multiplied by its arm. The moment for a piece of equipment is in fact a torque value, measured in units of inch-pounds (in-lb). To obtain the moment of an item with respect to the datum, multiply the weight of the item by its horizontal distance from the datum. Likewise, the moment of an item with respect to the center of gravity (CG) of an aircraft can be computed by multiplying its weight by the horizontal distance from the CG.

A 5 lb radio located 80" from the datum would have a moment of 400 inch-pounds (in-lb) (5 lb × 8"). Whether the value of 400 in-lb is preceded by a positive (+) or negative (−) sign depends on whether the moment is the result of a weight being removed or added and its location in relation to the datum. This situation is shown in Figure 4-2, where the moment ends up being a positive number because the weight and arm are both positive.

The algebraic sign of the moment, based on the datum location and whether weight is being installed or removed, would be as follows:

- Weight being added aft of the datum produces a positive moment (+weight, +arm).
- Weight being added forward of the datum produces a negative moment (+weight, −arm).
- Weight being removed aft of the datum produces a negative moment (−weight, +arm).
- Weight being removed forward of the datum produces a positive moment (−weight, −arm)

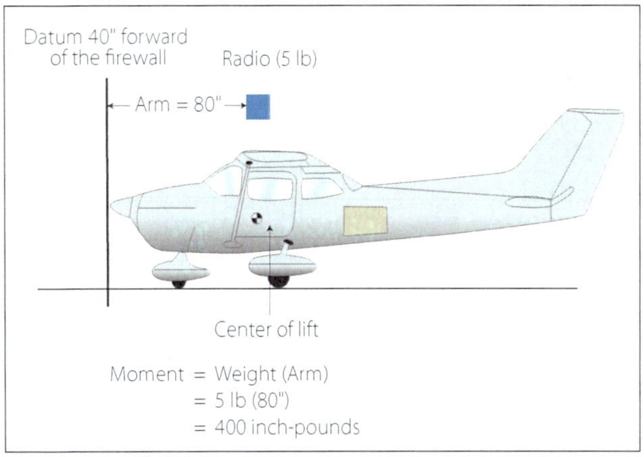

Figure 4-2. Moment of a radio located aft of the datum.

When dealing with positive and negative numbers, remember that the product of like signs produces a positive answer and the product of unlike signs produces a negative answer.

Center of Gravity

The center of gravity (CG) of an aircraft is a point about which the nose heavy and tail heavy moments are exactly equal in magnitude. It is the balance point for the aircraft. An aircraft suspended from this point would have no tendency to rotate in either a nose-up or nose-down attitude. It is the point about which the weight of an airplane or any object is concentrated.

Figure 4-3 shows a first class lever with the pivot point (fulcrum) located at the center of gravity for the lever. Even though the weights on either side of the fulcrum are not equal, and the distances from each weight to the fulcrum are not equal, the product of the weights and arms (moments) are equal, and that is what produces a balanced condition.

Maximum Weight

The maximum weight is the maximum authorized weight of the aircraft and its contents, and is indicated in the Aircraft Specifications or Type Certificate Data Sheet. For many aircraft, there are variations to the maximum allowable weight, depending on the pur-

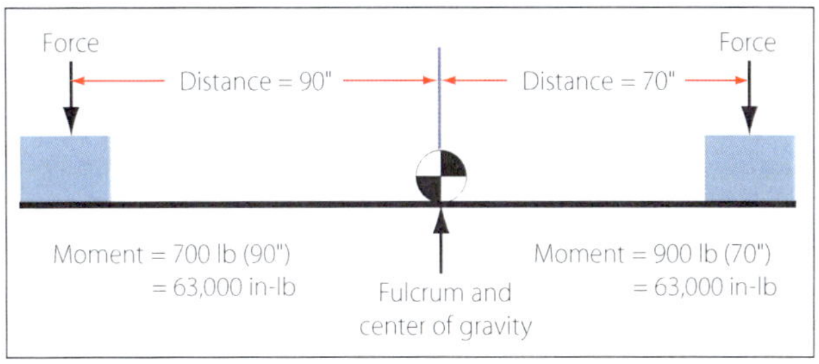

Figure 4-3. Center of gravity and a first class lever.

pose and conditions under which the aircraft is to be flown. For example, a certain aircraft may be allowed a maximum gross weight of 2,750 lb when flown in the normal category, but when flown in the utility category, which allows for limited aerobatics, the same aircraft's maximum allowable gross weight might only be 2,175 lb. There are other variations when dealing with the concept of maximum weight, as follows:

- **Maximum Ramp Weight**—the heaviest weight to which an aircraft can be loaded while it is sitting on the ground. This is sometimes referred to as the maximum taxi weight.
- **Maximum Takeoff Weight**—the heaviest weight an aircraft can have when it starts the takeoff roll. The difference between this weight and the maximum ramp weight would equal the weight of the fuel that would be consumed prior to takeoff.
- **Maximum Landing Weight**—the heaviest weight an aircraft can have when it lands. For large wide body commercial airplanes, it can be 100,000 lb less than maximum takeoff weight, or even more.
- **Maximum Zero Fuel Weight**—the heaviest weight an aircraft can be loaded to without having any usable fuel in the fuel tanks. Any weight loaded above this value must be in the form of fuel.

Empty Weight

The empty weight of an aircraft includes all operating equipment that has a fixed location and is actually installed in the aircraft. It includes the weight of the airframe, powerplant, required equipment, optional or special equipment, fixed ballast, hydraulic fluid, and residual fuel and oil. Residual fuel and oil are the fluids that will not normally drain out because they are trapped in the fuel lines, oil lines, and tanks. They must be included in the aircraft's empty weight. For most aircraft certified after 1978, the full capacity of the engine oil system is also included in the empty weight. Information regarding residual fluids in aircraft systems that must be included in the empty weight, and whether or not full oil is included, will be indicated in the Aircraft Specifications or Type Certificate Data Sheet.

Other terms that are sometimes used when describing empty weight include basic empty weight, licensed empty weight, and standard empty weight. The term "basic empty weight" typically applies when the full capacity of the engine oil system is included in the value. The term "licensed empty weight" typically applies when only the weight of residual oil is included in the value, so it generally involves only aircraft certified prior to 1978. Standard empty weight would be a value supplied by the aircraft manufacturer, and it would not include any optional equipment that might be installed in a particular aircraft. For most people working in the aviation maintenance field, the basic empty weight of the aircraft is the most important one.

Empty Weight Center of Gravity

The empty weight center of gravity for an aircraft is the point at which it balances when it is in an empty weight condition. The concepts of empty weight and center of gravity were discussed earlier in this chapter, and now they are being combined into a single concept.

One of the most important reasons for weighing an aircraft is to determine its empty weight center of gravity. All other weight and balance calculations, including loading the aircraft for flight, performing an equipment change calculation, and performing an adverse condition check, begin with knowing the empty weight and empty weight center of gravity. This crucial information is part of what is contained in the aircraft weight and balance report.

Useful Load

To determine the useful load of an aircraft, subtract the empty weight from the maximum allowable gross weight. For aircraft certificated in both normal and utility categories, there may be two useful loads listed in the aircraft weight and balance records. An aircraft with an empty weight of 900 lb will have a useful load of 850 lb, if the normal category maximum weight is listed as 1,750 lb. When the aircraft is operated in the utility category, the maximum gross weight may be reduced to 1,500 lb, with a corresponding decrease in the useful load to 600 lb. Some aircraft have the same useful load regardless of the category in which they are certificated.

The useful load consists of fuel, any other fluids that are not part of empty weight, passengers, baggage, pilot, copilot, and crewmembers. Whether or not the weight of engine oil is considered to be a part of useful load depends on when the aircraft was certified, and can be determined by looking at the Aircraft Specifications or Type Certificate Data Sheet. The payload of an aircraft is similar to the useful load, except it does not include fuel.

A reduction in the weight of an item, where possible, may be necessary to remain within the maximum weight allowed for the category in which an aircraft is operating. Determining the distribution of these weights is called a weight check.

Minimum Fuel

There are times when an aircraft will have a weight and balance calculation done, known as an extreme condition check. This is a pencil and paper check in which the aircraft is loaded in as nose heavy or tail heavy a condition as possible to see if the center of gravity will be out of limits in that situation. In a forward adverse check, for example, all useful load in front of the forward CG limit is loaded, and all useful load behind this limit is left empty. An exception to leaving it empty is the fuel tank. If the fuel tank is located behind the forward CG limit, it cannot be left empty because the aircraft cannot fly without fuel. In this case, an amount of fuel is accounted for, which is known as minimum fuel. Minimum fuel is typically that amount needed for 30 minutes of flight at cruise power.

For a piston engine powered aircraft, minimum fuel is calculated based on the METO (maximum except take-off) horsepower of the engine. For each METO horsepower of the engine, one-half pound of fuel is used. This amount of fuel is based on the assumption that the piston engine in cruise flight will burn 1 lb of fuel per hour for each horsepower, or ½ lb for 30 minutes. The piston engines currently used in small general aviation aircraft are actually more efficient than that, but the standard for minimum fuel has remained the same.

Minimum fuel is calculated as follows:

Minimum Fuel (pounds) =
Engine METO Horsepower ÷ 2

For example, if a forward adverse condition check was being done on a piston engine powered twin, with each engine having a METO horsepower of 500, the minimum fuel would be 250 lb (500 METO Hp ÷ 2).

For turbine engine powered aircraft, minimum fuel is not based on engine horsepower. If an adverse condition check is being performed on a turbine engine powered aircraft, the aircraft manufacturer would need to supply information on minimum fuel.

Tare Weight

When aircraft are placed on scales and weighed, it is sometimes necessary to use support equipment to aid in the weighing process. For example, to weigh a tail dragger airplane, it is necessary to raise the tail in order to get the airplane level. To level the airplane, a jack might be placed on the scale and used to raise the tail. Unfortunately, the scale is now absorbing the weight of the jack in addition to the weight of the airplane. This extra weight is known as tare weight, and must be subtracted from the scale reading. Other examples of tare weight are wheel chocks placed on the scales and ground locks left in place on retractable landing gear.

Procedures for Weighing an Aircraft

General Concepts

The most important reason for weighing an aircraft is to find out its empty weight (basic empty weight), and to find out where it balances in the empty weight condition. When an aircraft is to be flown, the pilot in command must know what the loaded weight of the aircraft is, and where its loaded center of gravity is. In order for the loaded weight and center of gravity to be calculated, the pilot or dispatcher handling the flight must first know the empty weight and empty weight center of gravity.

Earlier in this chapter it was identified that the center of gravity for an object is the point about which the nose heavy and tail heavy moments are equal. One method that could be used to find this point would involve lifting an object off the ground twice, first suspending it from a point near the front, and on the second lift suspending it from a point near the back. With each lift, a perpendicular line (90 degrees) would be drawn from the suspension point to the ground. The two perpendicular lines would intersect somewhere in the object, and the point of intersection would be the center of gravity. This concept is shown in Figure 4-4, where an irregular shaped object is suspended from two different points. The perpendicular line from the first suspension point is shown in red, and the new

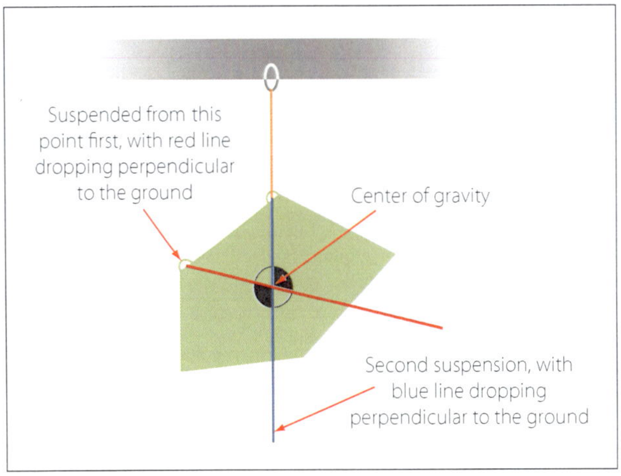

Figure 4-4. Center of gravity determined by two suspension points.

suspension point line is shown in blue. Where the red and blue lines intersect is the center of gravity.

If an airplane were suspended from two points, one at the nose and one at the tail, the perpendicular drop lines would intersect at the center of gravity the same way they do for the object in Figure 4-4. Suspending an airplane from the ceiling by two hooks, however, is clearly not realistic. Even if it could be done, determining where in the airplane the lines intersect would not be possible.

A more realistic way to find the center of gravity for an object, especially an airplane, is to place it on a minimum of two scales and to calculate the moment value for each scale reading. In Figure 4-5, there is a plank that is 200" long, with the left end being the datum (zero arm), and 6 weights placed at various locations along the length of the plank. The purpose of Figure 4-5 is to show how the center of gravity can be calculated when the arms and weights for an object are known.

To calculate the center of gravity for the object in Figure 4-5, the moments for all the weights need to be calculated and then summed, and the weights need to be summed. In the four column table in Figure 4-6, the item, weight, and arm are listed in the first three columns, with the information coming from Figure 4-5. The moment value in the fourth column is the product of the weight and arm. The weight and moment columns are summed, with the center of gravity being equal to the total moment divided by the total weight. The arm column is not summed. The number appearing at the bottom of that column is the center of gravity. The calculation would be as shown in Figure 4-6.

For the calculation shown in Figure 4-6, the total moment is 52,900 in-lb, and the total weight is 495 lb. The center of gravity is calculated as follows:

Center of Gravity = Total Moment ÷ Total Weight
= 52,900 in-lb ÷ 495 lb
= 106.9" (106.87 rounded off to tenths)

An interesting characteristic exists for the problem presented in Figure 4-5, and the table showing the center of gravity calculation. If the datum (zero arm) for the object was in the middle of the 200" long plank, with 100" of negative arm to the left and 100" of positive arm to the right, the solution would show the center of gravity to be in the same location. The arm for the center of gravity would not be the same number, but its physical location would be the same. Figure 4-7 and Figure 4-8 show the new calculation.

Center of Gravity = Total Moment ÷ Total Weight
= 3,400 in-lb ÷ 495 lb
= 6.9" (6.87 rounded off to tenths)

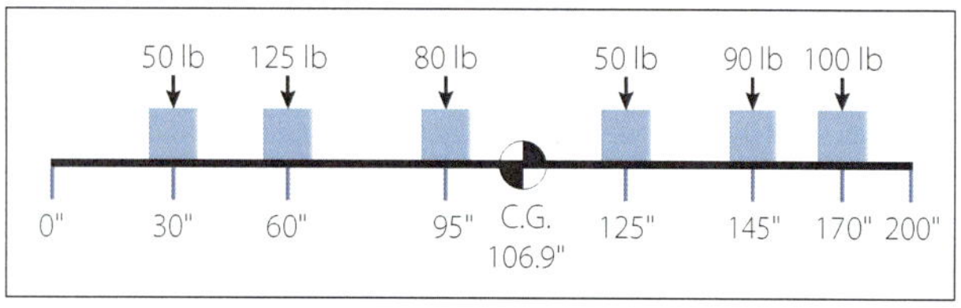

Figure 4-5. Center of gravity for weights on a plank, datum at one end.

Item	Weight (lb)	Arm (inches)	Moment (in-lb)
50 pound weight	50	+30	1,500
125 pound weight	125	+60	7,500
80 pound weight	80	+95	7,600
50 pound weight	50	+125	6,250
90 pound weight	90	+145	13,050
100 pound weight	100	+170	17,000
Total	495	+106.9	52,900

Figure 4-6. Center of gravity calculation for weights on a plank with datum at one end.

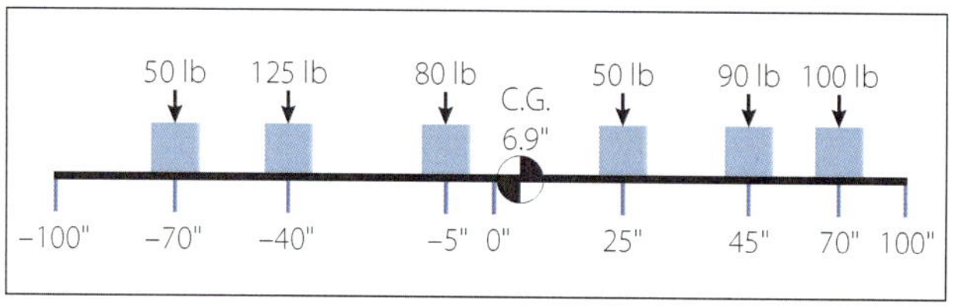

Figure 4-7. Center of gravity for weights on a plank, datum in the middle.

Item	Weight (lb)	Arm (inches)	Moment (in-lb)
50 pound weight	50	-70	-3,500
125 pound weight	125	-40	-5,000
80 pound weight	80	-5	-400
50 pound weight	50	+25	1,250
90 pound weight	90	+45	4,050
100 pound weight	100	+70	7,000
Total	495	+6.9	3,400

Figure 4-8. Center of gravity calculation for weights on a plank with datum in the middle.

In Figure 4-7, the center of gravity is 6.9" to the right of the plank's center. Even though the arm is not the same number, in Figure 4-5 the center of gravity is also 6.9" to the right of center (CG location of 106.9 with the center being 100). Because both problems are the same in these two figures, except for the datum location, the center of gravity must be the same.

The definition for center of gravity states that it is the point about which all the moments are equal. We can prove that the center of gravity for the object in Figure 4-7 is correct by showing that the total moments on either side of this point are equal. Using 6.87 as the CG location for slightly greater accuracy, instead of the rounded off 6.9 number, the moments to the left of the CG would be as shown in Figure 4-9.

The moments to the right of the CG, as shown in Figure 4-7, would be as shown in Figure 4-10.

Disregarding the slightly different decimal value, the moment in both of the previous calculations is 10,651 in-lb. Showing that the moments are equal is a good way of proving that the center of gravity has been properly calculated.

Weight and Balance Data

In order to weigh an aircraft and calculate its empty weight and empty weight center of gravity, a technician must have access to weight and balance information about the aircraft. Possible sources of weight and balance data are as follows:

- **Aircraft Specifications**—applies primarily to aircraft certified under the Civil Aeronautics Administration, when the specifications also included a list of equipment with weights and arms.
- **Aircraft Operating Limitations**—supplied by the aircraft manufacturer.
- **Aircraft Flight Manual**—supplied by the aircraft manufacturer.
- **Aircraft Weight and Balance Report**—supplied by the aircraft manufacturer when the aircraft is new, and by the technician when an aircraft is reweighed in the field.
- **Aircraft Type Certificate Data Sheet**—applies primarily to aircraft certified under the FAA and the Federal Aviation Regulations, where the equipment list with weights and arms is a separate document.

The document in Figure 4-11 is a Type Certificate Data Sheet (TCDS) for a Piper twin-engine airplane known as the Seneca (PA-34-200). The main headings for the information typically contained in a TCDS are included, but much of the information contained under these headings has been removed if it did not directly pertain to weight and balance. Information on only one model of Seneca is shown, because to show all the different models would make the document excessively long. The portion of the TCDS that has the most direct application to weight and balance is highlighted in yellow.

Some of the important weight and balance information found in a Type Certificate Data Sheet is as follows:

1. Center of gravity range
2. Maximum weight
3. Leveling means
4. Number of seats and location
5. Baggage capacity
6. Fuel capacity
7. Datum location
8. Engine horsepower
9. Oil capacity
10. Amount of fuel in empty weight
11. Amount of oil in empty weight

Item	Weight (lb)	Arm (inches)	Moment (in-lb)
50 pound weight	50	76.87	3,843.50
125 pound weight	125	46.87	5,858.75
80 pound weight	80	11.87	949.60
Total	255	135.61	10,651.85

Figure 4-9. Moments to the left of the center of gravity.

Item	Weight (lb)	Arm (inches)	Moment (in-lb)
50 pound weight	50	18.13	906.50
90 pound weight	90	38.13	3,431.70
100 pound weight	100	63.13	6,313.00
Total	240	119.39	10,651.25

Figure 4-10. Moments to the right of the center of gravity.

DEPARTMENT OF TRANSPORTATION
FEDERAL AVIATION ADMINISTRATION

TYPE CERTIFICATE DATA SHEET NO. A7SO – REVISION 14

PIPER
PA-34-200
PA-34-200T
PA-34-220T

June 1, 2001

This data sheet, which is a part of type certificate No. A7SO, prescribes conditions and limitations under which the product for which the type certificate was issued meets the airworthiness requirements of the Federal Aviation Regulations.

Type Certificate Holder:
The New Piper Aircraft, Inc.
2926 Piper Drive
Vero Beach, Florida 32960

I. Model PA-34-200 (Seneca), 7 PCLM (Normal Category), Approved 7 May 1971.

Engines S/N 34-E4, 34-7250001 through 34-7250214:
1 Lycoming LIO-360-C1E6 with fuel injector,
Lycoming P/N LW-10409 or LW-12586 (right side); and
1 Lycoming IO-360-C1E6 with fuel injector,
Lycoming P/N LW-10409 or LW 12586 (left side).
S/N 34-7250215 through 34-7450220:
1 Lycoming LIO-360-C1E6 with fuel injector,
Lycoming P/N LW-12586 (right side); and
1 Lycoming IO-360-C1E6 with fuel injector,
Lycoming P/N LW-12586 (left side).

Fuel 100/130 minimum grade aviation gasoline

Engine Limits For all operations, 2,700 RPM (200 hp)

Propeller and Propeller Limits

Left Engine
1 Hartzell, Hub Model HC-C2YK-2 ( ) E, Blade Model C7666A-0;
1 Hartzell, Hub Model HC-C2YK-2 ( ) EU, Blade Model C7666A-0;
1 Hartzell, Hub Model HC-C2YK-2 ( ) EF, Blade Model FC7666A-0;
1 Hartzell, Hub Model HC-C2YK-2 ( ) EFU, Blade Model FC7666A-0;
1 Hartzell, Hub Model HC-C2YK-2CG (F), Blade Model (F) C7666A
(This model includes the Hartzell damper); or
1 Hartzell, Hub Model HC-C2YK-2CGU (F), Blade Model (F) C7666A
(This model includes the Hartzell damper).
Note: HC-( )2YK-( ) may be substituted for HC-( )2YR-( ) per Hartzell Service Advisory 61.

Right Engine
1 Hartzell, Hub Model HC-C2YK-2 ( ) LE, Blade Model JC7666A-0;
1 Hartzell, Hub Model HC-C2YK-2 ( ) LEU, Blade Model JC7666A-0;
1 Hartzell, Hub Model HC-C2YK-2 ( ) LEF, Blade Model FJC7666A-0;
1 Hartzell, Hub Model HC-C2YK-2 ( ) LEFU, Blade Model FJC7666A-0;
1 Hartzell, Hub Model HC-C2YK-2CLG (F), Blade Model (F) JC7666A
(This model includes the Hartzell damper); or

Figure 4-11. Type Certificate Data Sheet.

1 Hartzell, Hub Model HC-C2YK-2CLGU (F), Blade Model (F) JC7666A
(This model includes the Hartzell damper.)
Note: HC-( )2YK-( ) may be substituted for HC-( )2YR-( ) per Hartzell Service Advisory 61.
Pitch setting: High 79° to 81°, Low 13.5° at 30" station.
Diameter: Not over 76", not under 74". No further reduction permitted.

Spinner: Piper P/N 96388 Spinner Assembly and P/N 96836 Cap Assembly, or P/N 78359-0 Spinner Assembly and P/N 96836-2 Cap Assembly (See NOTE 4)

Governor Assembly:
1 Hartzell hydraulic governor, Model F-6-18AL (Right);
1 Hartzell hydraulic governor, Model F-6-18A (Left).
Avoid continuous operation between 2,200 and 2,400 RPM unless aircraft is equipped with Hartzell propellers which incorporate a Hartzell damper on both left and right engine as noted above.

Airspeed Limits	VNE (Never exceed)	217 mph	(188 knots)
	VNO (Maximum structural cruise)	190 mph	(165 knots)
	VA (Maneuvering, 4,200 lb)	146 mph	(127 knots)
	VA (Maneuvering, 4,000 lb)	146 mph	(127 knots)
	VA (Maneuvering, 2,743 lb)	133 mph	(115 knots)
	VFE (Flaps extended)	125 mph	(109 knots)
	VLO (Landing gear operating)		
	Extension	150 mph	(130 knots)
	Retract	125 mph	(109 knots)
	VLE (Landing gear extended)	150 mph	(130 knots)
	VMC (Minimum control speed)	80 mph	(69 knots)

CG Range (Gear Extended)

S/N 34-E4, 34-7250001 through 34-7250214 (See NOTE 3):
(+86.4) to (+94.6) at 4,000 lb
(+82.0) to (+94.6) at 3,400 lb
(+80.7) to (+94.6) at 2,780 lb
S/N 34-7250215 through 34-7450220:
(+87.9) to (+94.6) at 4,200 lb
(+82.0) to (+94.6) at 3,400 lb
(+80.7) to (+94.6) at 2,780 lb
Straight line variation between points given.
Moment change due to gear retracting landing gear (-32 in-lb)

Empty Weight CG Range None

Maximum Weight

S/N 34-E4, 34-7250001 through 34-7250214:
4,000 lb – Takeoff
4,000 lb – Landing
See NOTE 3.

Maximum Weight

S/N 34-7250215 through 34-7450220:
4,200 lb – Takeoff
4,000 lb – Landing

No. of Seats 7 (2 at +85.5, 3 at +118.1, 2 at +155.7)

Maximum Baggage 200 lb (100 lb at +22.5, 100 lb at +178.7)

Figure 4-11. Type Certificate Data Sheet. (continued)

Fuel Capacity	98 gallons (2 wing tanks) at (+93.6) (93 gallons usable) See NOTE 1 for data on system fuel.
Oil Capacity	8 qt per engine (6 qt per engine usable) See NOTE 1 for data on system oil.

Control Surface Movements	Ailerons	(±2°)	Up 30°	Down 15°
	Stabilator		Up 12.5° (+0,-1°)	Down 7.5° (±1°)
Rudder	(±1°)	Left 35°		Right 35°
	Stabilator Trim Tab (Stabilator neutral)	(±1°)	Down 10.5°	Up 6.5°
	Wing Flaps	(±2°)	Up 0°	Down 40°
	Rudder Trim Tab (Rudder neutral)	(±1°)	Left 17°	Right 22°
	Nosewheel Travel	S/N 34-E4, 34-7250001 through 34-7350353: (±1°)	Left 21°	Right 21°
	Nosewheel Travel	S/N 34-7450001 through 34-7450220: (±1°)	Left 27°	Right 27°

Manufacturer's Serial Numbers 3449001 and up.

DATA PERTINENT TO ALL MODELS

Datum 78.4" forward of wing leading edge from the inboard edge of the inboard fuel tank.

Leveling Means Two screws left side fuselage below window.

Certification Basis Type Certificate No. A7SO issued May 7, 1971, obtained by the manufacturer under the delegation option authorization. Date of Type Certificate application July 23, 1968.

Model PA-34-200 (Seneca I):
14 CFR part 23 as amended by Amendment 23-6 effective August 1, 1967; 14 CFR part 23.959 as amended by Amendment 23-7 effective September 14, 1969; and 14 CFR part 23.1557(c)(1) as amended by Amendment 23-18 effective May 2, 1977. Compliance with 14 CFR part 23.1419 as amended by Amendment 23-14 effective December 20, 1973, has been established with optional ice protection provisions.

Production Basis Production Certificate No. 206. Production Limitation Record issued and the manufacturer is authorized to issue an airworthiness certificate under the delegation option provisions of 14 CFR part 21.

Equipment The basic required equipment as prescribed in the applicable airworthiness regulations (see Certification Basis) must be installed in the aircraft for certification. In addition, the following items of equipment are required:

Figure 4-11. Type Certificate Data Sheet. (continued)

MODEL	AFM/POH	REPORT #	APPROVED	SERIAL EFFECTIVITY
PA-34-200(Seneca)	AFM	VB-353	7/2/71	34-E4, 34-7250001 through 34-7250214
	AFM	VB-423	5/20/72	34-7250001 through 34-7250189 when Piper Kit 760-607 is installed; 34-7250190 through 34-7250214 when Piper Kit 760-611 is installed; and 34-7250215 through 34-7350353
	AFM	VB-563	5/14/73	34-7450001 through 34-7450220
	AFM Supp.	VB-588	7/20/73	34-7250001 through 34-7450039 when propeller with dampers are installed
	AFM Supp.	VB-601	11/9/73	34-7250001 through 34-745017 when ice protection system is installed

<u>NOTES</u>

NOTE 1 Current Weight and Balance Report, including list of equipment included in certificated empty weight, and loading instructions when necessary, must be provided for each aircraft at the time of original certification. The certificated empty weight and corresponding center of gravity locations must include undrainable system oil (not included in oil capacity) and unusable fuel as noted below:

Fuel: 30.0 lb at (+103.0) for PA-34 series, except Model PA-34-220T (Seneca V), S/N 3449001 and up
Fuel: 36.0 lb at (+103.0) for Model PA-34-220T (Seneca V), S/N 3449001 and up
Oil: 6.2 lb at (+ 39.6) for Model PA-34-200
Oil: 12.0 lb at (+ 43.7) for Models PA-34-200T and PA-34-220T

NOTE 2 All placards required in the approved Airplane Flight Manual or Pilot's Operating Handbook and approved Airplane Flight Manual or Pilot's Operating Handbook supplements must be installed in the appropriate location.

NOTE 3 The Model PA-34-200; S/N 34-E4, 34-7250001 through 34-7250189, may be operated at a maximum takeoff weight of 4,200 lb when Piper Kit 760-607 is installed. S/N 34-7250190 through 34-7250214 may be operated at a maximum takcoff weight of 4,200 lb when Piper Kit 760-611 is installed.

NOTE 4 The Model PA-34-200; S/N 34-E4, 34-7250001 through 34-7250189, may be operated without spinner domes or without spinner domes and rear bulkheads when Piper Kit 760-607 has been installed. ---
----

NOTE 5 The Model PA-34-200 may be operated in known icing conditions when equipped with spinner assembly and the following kits: --

NOTE 6 Model PA-34-200T; S/N 34-7570001 through 34-8170092, may be operated in known icing conditions when equipped with deicing equipment installed per Piper Drawing No. 37700 and spinner assembly.

Figure 4-11. Type Certificate Data Sheet. (continued)

NOTE 7 The following serial numbers are not eligible for import certification to the United States: --

NOTE 8 Model PA-34-200; S/N 34-E4, S/N 34-7250001 through 34-7450220, and Model PA-34-200T; S/N 34-7570001 through 34-8170092, and Model PA-34-220T may be operated subject to the limitations listed in the Airplane Flight Manual or Pilot's Operating Handbook with rear cabin and cargo door removed.

NOTE 9 In the following serial numbered aircraft, rear seat location is farther aft as shown and the center seats may be removed and replaced by CLUB SEAT INSTALLATION, which has a more aft CG location as shown in "No. of Seats," above:PA-34-200T: S/N 34-7770001 through 34-8170092.

NOTE 10 These propellers are eligible on Teledyne Continental L/TSIO-360-E only.

NOTE 11 With Piper Kit 764-048V installed weights are as follows: 4,407 lb – Takeoff 4,342 lb - Landing (All weight in excess of 4,000 lb must be fuel) Zero fuel weight may be increased to a maximum of 4,077.7 lb when approved wing options are installed (See POH VB-1140).

NOTE 12 With Piper Kit 764-099V installed, weights are as follows: 4,430 lb - Ramp 4,407 lb - Takeoff, Landing, and Zero Fuel (See POH VB-1150).

NOTE 13 With Piper Kit 766-203 installed, weights are as follows: 4,430 lb - Ramp 4,407 lb - Takeoff, Landing and Zero Fuel (See POH VB-1259).

NOTE 14 With Piper Kit 766-283 installed, weights are as follows: 4,430 lb - Ramp 4,407 lb - Takeoff, Landing and Zero Fuel (See POH VB-1558).

NOTE 15 With Piper Kit 766-608 installed, weights are as follows: 4,430 lb - Ramp 4,407 lb - Takeoff, Landing and Zero Fuel (See POH VB-1620).

NOTE 16 With Piper Kit 766-632 installed, weights are as follows: 4,430 lb - Ramp 4,407 lb - Takeoff, Landing and Zero Fuel (See POH VB-1649).

NOTE 17 The bolt and stack-up that connect the upper drag link to the nose gear trunnion are required to be replaced every 500 hours' time-in-service. ---
----

Figure 4-11. Type Certificate Data Sheet. (continued)

Weight and Balance Equipment

Scales

Two types of scales are typically used to weigh aircraft: those that operate mechanically with balance weights or springs, and those that operate electronically with what are called load cells. The balance weight type of mechanical scale, known as a beam scale, is similar to that found in a doctor's office, in which a bar rises up when weight is put on the scale. A sliding weight is then moved along the bar until the bar is centered between a top and bottom stop.

The sliding weight provides the capability to measure up to 50 lb, and the cup holds fixed weights that come in 50 lb equivalent units. As an example of this scale in use, let's say the nosewheel of a small airplane is placed on the scale with an applied weight of 580 lb. To find out what the applied weight is, a technician would place 550 lb of equivalent weight in the cup, and then slide the weight on the beam out to the 30 lb point. The 580 lb applied by the nosewheel would now be balanced by the 580 lb of equivalent weight, and the end of the beam would be centered between the top and bottom stop.

A mechanical scale based on springs is like the typical bathroom scale. When weight is applied to the scale, a spring compresses, which causes a wheel that displays the weight to rotate. It would be difficult to use this type of scale to weigh anything other than a very small aircraft, because these scales typically measure only up to 300 lb. The accuracy of this type of scale is also an issue.

Electronic scales that utilize load cells come in two varieties: the platform type and the type that mounts to the top of a jack. The platform type of electronic scale sits on the ground, with the tire of the airplane sitting on top of the platform. Built into the platform is an electronic load cell, which senses the weight being applied to it and generates a corresponding electrical signal. Inside the load cell there is an electronic grid that experiences a proportional change in electrical resistance as the weight being applied to it increases. An electrical cable runs from the platform scale to a display unit, which interprets the resistance change of the load cell and equates it to a specific number of pounds. A digital readout on the display typically shows the weight. In Figure 4-12, a Piper Archer is being weighed using platform scales that incorporate electronic load cells. In this case, the platform scales are secured to the hangar floor and stay permanently in place.

Figure 4-12. Weighing a Piper Archer using electronic platform scales.

In Figure 4-13, a Mooney M20 airplane is being weighed with portable electronic platform scales. Notice in the picture of the Mooney that its nose tire is deflated (close-up shown in the lower right corner of the photo). This was done to get the airplane in a level flight attitude. This type of scale is easy to transport and can be powered by household current or by a battery contained in the display unit. The display unit for these scales is shown in Figure 4-14.

The display unit for the portable scales is very simple to operate. [Figure 4-14] In the lower left corner is the power switch, and in the lower right is the switch for selecting pounds or kilograms. The red, green, and yellow knobs are potentiometers for zeroing the three scales, and next to them are the on/off switches for the scales. Before the weight of the airplane is placed on the scales, each scale switch is turned on and the potentiometer knob turned until the digital display reads zero. In Figure 4-14, the nose scale is turned on and the readout of 546 lb is for the Mooney airplane in Figure 4-13. If all three scale switches are turned on at the same time, the total weight of the airplane will be displayed.

The second type of electronic scale utilizes a load cell that attaches to the top of a jack. The top of the load cell has a concave shape that matches up with the jack pad on the aircraft, with the load cell absorbing all the weight of the aircraft at each jacking point. Each load cell has an electrical cable attached to it, which connects to the display unit that shows the weight being absorbed by each load cell. An important advantage of weighing an aircraft this way is that it allows the technician to level the aircraft. An aircraft needs to be in a flight level attitude when it is weighed. If an aircraft is sitting on floor scales, the only way to level the aircraft might be to deflate tires and landing gear

Figure 4-13. Mooney M20 being weighed with portable electronic platform scales.

struts. When an aircraft is weighed using load cells on jacks, leveling the aircraft is easy by simply adjusting the height with the jacks. Figure 4-15 shows a regional jet on jacks with the load cells in place.

Spirit Level

Before an aircraft can be weighed and reliable readings obtained, it must be in a level flight attitude. One method that can be used to check for a level condition is to use a spirit level, sometimes thought of as a carpenter's level, by placing it on or against a specified place on the aircraft. Spirit levels consist of a vial full of liquid, except for a small air bubble. When the air bubble is centered between the two black lines, a level condition is indicated.

In Figure 4-16, a spirit level is being used on a Mooney M20 to check for a flight level attitude. By looking in the Type Certificate Data Sheet, it is determined that the leveling means is two screws on the left side of the airplane fuselage, in line with the trailing edge of the wing.

Plumb Bob

A plumb bob is a heavy metal object, cylinder or cone shape, with a sharp point at one end and a string attached to the other end. If the string is attached to a given point on an aircraft, and the plumb bob is allowed to hang down so the tip just touches the ground, the point where the tip touches will be perpendicular to where the string is attached. An example of the use of

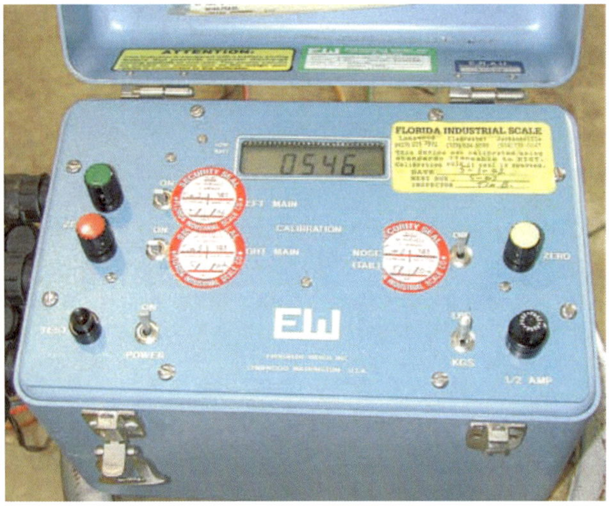

Figure 4-14. Display unit showing nosewheel weight for Mooney M20.

Figure 4-15. Airplane on jacks with load cells in use.

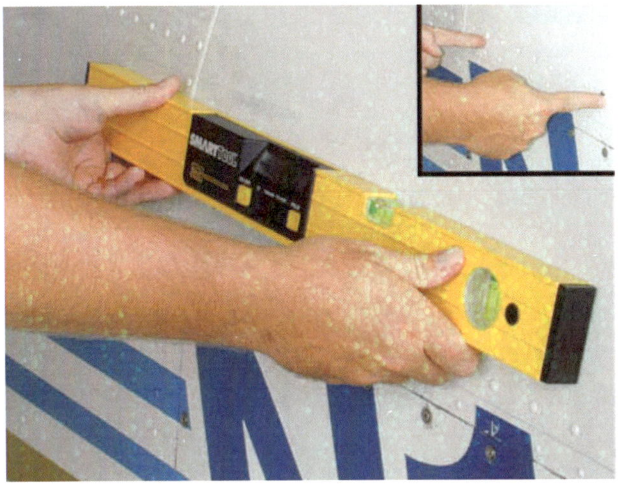

Figure 4-16. Spirit level being used on a Mooney M20.

Figure 4-17. Plumb bob dropped from a wing leading edge.

a plumb bob would be measuring the distance from an aircraft's datum to the center of the main landing gear axle. If the leading edge of the wing was the datum, a plumb bob could be dropped from the leading edge and a chalk mark made on the hangar floor. The plumb bob could also be dropped from the center of the axle on the main landing gear, and a chalk mark made on the floor. With a tape measure, the distance between the two chalk marks could be determined, and the arm for the main landing gear would be known. Plumb bobs can also be used to level an aircraft, described on page 4-27 of the Helicopter Weight and Balance section of this chapter. Figure 4-17 shows a plumb bob being dropped from the leading edge of an aircraft wing.

Hydrometer

When an aircraft is weighed with full fuel in the tanks, the weight of the fuel must be accounted for by mathematically subtracting it from the scale readings. To subtract it, its weight, arm, and moment must be known. Although the standard weight for aviation gasoline is 6.0 lb/gal and jet fuel is 6.7 lb/gal, these values are not exact for all conditions. On a hot day versus a cold day, these values can vary dramatically. On a hot summer day in the state of Florida, aviation gasoline checked with a hydrometer typically weighs between 5.85 and 5.9 lb/gal. If 100 gallons of fuel were involved in a calculation, using the actual weight versus the standard weight would make a difference of 10 to 15 lb.

When an aircraft is weighed with fuel in the tanks, the weight of fuel per gallon should be checked with a hydrometer. A hydrometer consists of a weighted glass tube which is sealed, with a graduated set of markings on the side of the tube. The graduated markings and their corresponding number values represent units of pounds per gallon. When placed in a flask with fuel in it, the glass tube floats at a level dependent on the density of the fuel. Where the fuel intersects the markings on the side of the tube indicates the pounds per gallon.

Preparing an Aircraft for Weighing

Weighing an aircraft is a very important and exacting phase of aircraft maintenance, and must be carried out with accuracy and good workmanship. Thoughtful preparation saves time and prevents mistakes.

To begin, assemble all the necessary equipment, such as:

1. Scales, hoisting equipment, jacks, and leveling equipment.
2. Blocks, chocks, or sandbags for holding the airplane on the scales.
3. Straightedge, spirit level, plumb bobs, chalk line, and a measuring tape.
4. Applicable Aircraft Specifications and weight and balance computation forms.

If possible, aircraft should be weighed in a closed building where there are no air currents to cause incorrect scale readings. An outside weighing is permissible if wind and moisture are negligible.

Fuel System

When weighing an aircraft to determine its empty weight, only the weight of residual (unusable) fuel should be included. To ensure that only residual fuel is accounted for, the aircraft should be weighed in one of the following three conditions.

1. Weigh the aircraft with absolutely no fuel in the aircraft tanks or fuel lines. If an aircraft is weighed

in this condition, the technician can mathematically add the proper amount of residual fuel to the aircraft, and account for its arm and moment. The proper amount of fuel can be determined by looking at the Aircraft Specifications or Type Certificate Data Sheet.

2. Weigh the aircraft with only residual fuel in the tanks and lines.
3. Weigh the aircraft with the fuel tanks completely full. If an aircraft is weighed in this condition, the technician can mathematically subtract the weight of usable fuel, and account for its arm and moment. A hydrometer can be used to determine the weight of each gallon of fuel, and the Aircraft Specifications or Type Certificate Data Sheet can be used to identify the fuel capacity. If an aircraft is to be weighed with load cells attached to jacks, the technician should check to make sure it is permissible to jack the aircraft with the fuel tanks full. It is possible that this may not be allowed because of stresses that would be placed on the aircraft.

Never weigh an aircraft with the fuel tanks partially full, because it will be impossible to determine exactly how much fuel to account for.

Oil System

For aircraft certified since 1978, full engine oil is typically included in an aircraft's empty weight. This can be confirmed by looking at the Type Certificate Data Sheet. If full oil is to be included, the oil level needs to be checked and the oil system serviced if it is less than full. If the Aircraft Specifications or Type Certificate Data Sheet specifies that only residual oil is part of empty weight, this can be accommodated by one of the following two methods.

1. Drain the engine oil system to the point that only residual oil remains.
2. Check the engine oil quantity, and mathematically subtract the weight of the oil that would leave only the residual amount. The standard weight for lubricating oil is 7.5 lb/gal (1.875 pounds per quart (lb/qt)), so if 7 qt of oil needed to be removed, the technician would subtract 13.125 lb at the appropriate arm.

Miscellaneous Fluids

Unless otherwise noted in the Aircraft Specifications or manufacturer's instructions, hydraulic reservoirs and systems should be filled, drinking and washing water reservoirs and lavatory tanks should be drained, and constant speed drive oil tanks should be filled.

Flight Controls

The position of such items as spoilers, slats, flaps, and helicopter rotor systems is an important factor when weighing an aircraft. Always refer to the manufacturer's instructions for the proper position of these items.

Other Considerations

Inspect the aircraft to see that all items included in the certificated empty weight are installed in the proper location. Remove items that are not regularly carried in flight. Also look in the baggage compartments to make sure they are empty. Replace all inspection plates, oil and fuel tank caps, junction box covers, cowling, doors, emergency exits, and other parts that have been removed. All doors, windows, and sliding canopies should be in their normal flight position. Remove excessive dirt, oil, grease, and moisture from the aircraft.

Some aircraft are not weighed with the wheels on the scales, but are weighed with the scales placed either at the jacking points or at special weighing points. Regardless of what provisions are made for placing the aircraft on the scales or jacks, be careful to prevent it from falling or rolling off, thereby damaging the aircraft and equipment. When weighing an aircraft with the wheels placed on the scales, release the brakes to reduce the possibility of incorrect readings caused by side loads on the scales.

All aircraft have leveling points or lugs, and care must be taken to level the aircraft, especially along the longitudinal axis. With light, fixed-wing airplanes, the lateral level is not as critical as it is with heavier airplanes. However, a reasonable effort should be made to level the light airplanes along the lateral axis. Helicopters must be level longitudinally and laterally when they are weighed. Accuracy in leveling all aircraft longitudinally cannot be overemphasized.

Weighing Points

When an aircraft is being weighed, the arms must be known for the points where the weight of the aircraft is being transferred to the scales. If a tricycle gear small airplane has its three wheels sitting on floor scales, the weight transfer to each scale happens through the center of the axle for each wheel. If an airplane is weighed while it is on jacks, the weight transfer happens through the center of the jack pad. For a helicopter with skids for landing gear, determining the arm for the weighing points can be difficult if the skids are sitting directly on floor scales. The problem is that the skid is in contact with the entire top portion of the scale, and it is impossible to know exactly where the center of weight transfer is occurring. In such a case, place a piece of pipe between the skid and the scale, and the center of the pipe will now be the known point of weight transfer.

The arm for each of the weighing points is the distance from the center of the weight transfer point to the aircraft's datum. If the arms are not known, based on previous weighing of the aircraft or some other source of data, they must be measured when the aircraft is weighed. This involves dropping a plumb bob from the center of each weighing point and from the aircraft datum, and putting a chalk mark on the hangar floor representing each point. The perpendicular distance between the datum and each of the weighing points can then be measured. In Figure 4-18, the distance from the nosewheel centerline to the datum is being measured on a Cessna 310 airplane. Notice the chalk lines on the hangar floor, which came as a result of dropping a plumb bob from the nosewheel axle centerline and from the datum. The nosewheel sitting on an electronic scale can be seen in the background.

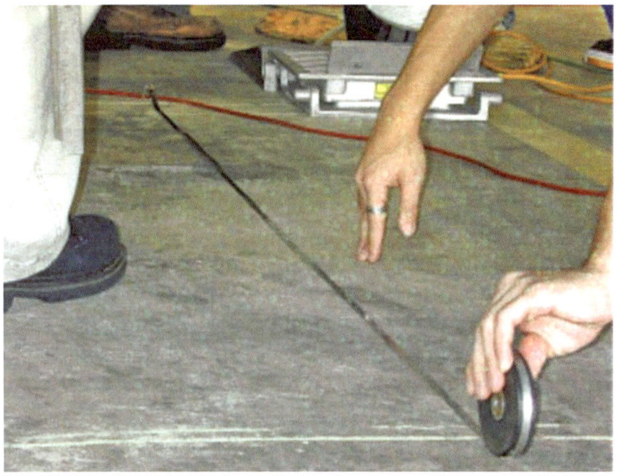

Figure 4-18. Measuring the nosewheel arm on a Cessna 310.

Center of Gravity Range

The center of gravity range for an aircraft is the limits within which the aircraft must balance. It is identified as a forward most limit (arm) and an aft most limit (arm). In the Type Certificate Data Sheet for the Piper Seneca airplane, shown earlier in this chapter, the range is given as follows:

CG Range: (Gear Extended)

S/N 34-E4, 34-7250001 through 34-7250214 (See NOTE 3):

(+86.4") to (+94.6") at 4,000 lb

(+82.0") to (+94.6") at 3,400 lb

(+80.7") to (+94.6") at 2,780 lb

Straight line variation between points given.

Moment change due to gear retracting landing gear (–32 in-lb)

Because the Piper Seneca is a retractable gear airplane, the specifications identify that the range applies when the landing gear is extended, and that the airplane's total moment will be decreased by 32 when the gear retracts. To know how much the center of gravity will change when the gear is retracted, the moment of 32 in-lb would need to be divided by the loaded weight of the airplane. For example, if the airplane weighed 3,500 lb, the center of gravity would move forward 0.009" (32 ÷ 3500).

Based on the numbers given, up to a loaded weight of 2,780 lb, the forward CG limit is +80.7" and the aft CG limit is +94.6". As the loaded weight of the airplane increases to 3,400 lb and eventually to the maximum of 4,000 lb, the forward CG limit moves aft. In other words, as the loaded weight of the airplane increases, the CG range gets smaller. The range gets smaller as a result of the forward limit moving back, while the aft limit stays in the same place.

The data sheet identifies that there is a straight line variation between the points given. The points being referred to are the forward and aft center of gravity limits. From a weight of 2,780 lb to a weight of 3,400 lb, the forward limit moves from +80.7" to +82.0", and if plotted on a graph, that change would form a straight line. From 3,400 lb to 4,000 lb, the forward limit moves from +82 to +86.4", again forming a straight line. Plotted on a graph, the CG limits would look like what is shown in Figure 4-19. When graphically plotted, the CG limits form what is known as the CG envelope.

In Figure 4-19, the red line represents the forward limit up to a weight of 2,780 lb. The blue and green lines represent the straight line variation that occurs for the forward limit as the weight increases up to a maximum of 4,000 lb. The yellow line represents the maximum weight for the airplane, and the purple line represents the aft limit.

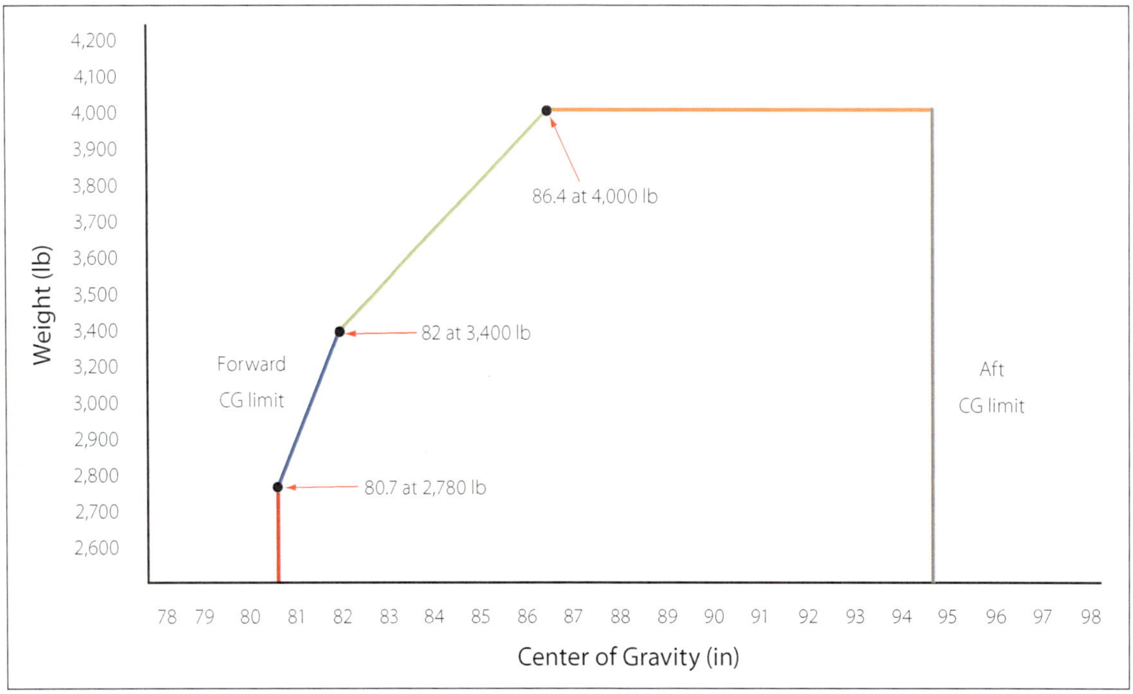

Figure 4-19. Center of gravity envelope for the Piper Seneca.

Empty Weight Center of Gravity Range

For some aircraft, a center of gravity range is given for the aircraft in the empty weight condition. This practice is not very common with airplanes, but is often done for helicopters. This range would only be listed for an airplane if it was very small and had limited positions for people and fuel. If the empty weight CG of an aircraft falls within the empty weight CG limits, it is known that the loaded CG of the aircraft will be within limits if standard loading is used. This information will be listed in the Aircraft Specifications or Type Certificate Data Sheet, and if it does not apply, it will be identified as "none."

Operating Center of Gravity Range

All aircraft will have center of gravity limits identified for the operational condition, with the aircraft loaded and ready for flight. If an aircraft can operate in more than one category, such as normal and utility, more than one set of limits might be listed. As shown earlier for the Piper Seneca airplane, the limits can change as the weight of the aircraft increases. In order to legally fly, the center of gravity for the aircraft must fall within the CG limits.

Standard Weights Used for Aircraft Weight and Balance

Unless the specific weight for an item is known, the standard weights used in aircraft weight and balance are as follows:

- Aviation gasoline — 6 lb/gal
- Turbine fuel — 6.7 lb/gal
- Lubricating oil — 7.5 lb/gal
- Water — 8.35 lb/gal
- Crew and passengers — 170 lb per person

Example Weighing of an Airplane

In Figure 4-20, a tricycle gear airplane is being weighed by using three floor scales. The specifications on the airplane and the weighing specific data are as follows:

- Aircraft Datum: Leading edge of the wing
- Leveling Means: Two screws, left side of fuselage below window
- Wheelbase: 100"
- Fuel Capacity: 30 gal aviation gasoline at +95"
- Unusable Fuel: 6 lb at +98"
- Oil Capacity: 8 qt at –38"
- Note 1: Empty weight includes unusable fuel and full oil
- Left Main Scale Reading: 650 lb
- Right Main Scale Reading: 640 lb
- Nose Scale Reading: 225 lb

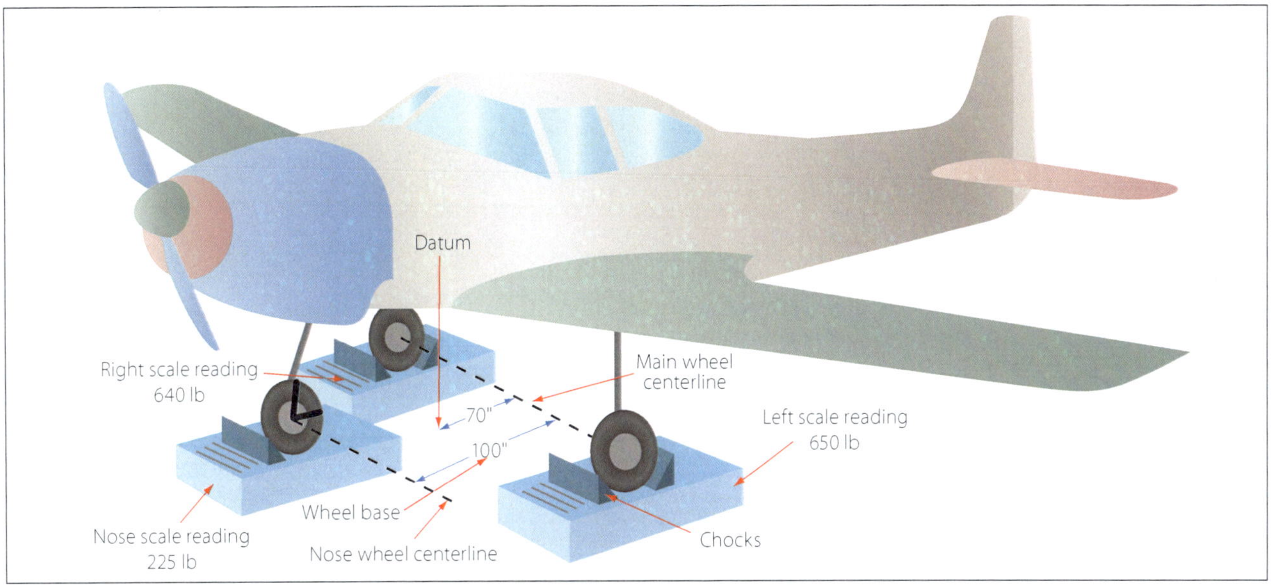

Figure 4-20. Example airplane being weighed.

- Tare Weight: 5 lb chocks on left main
  5 lb chocks on right main
  2.5 lb chock on nose
- During Weighing: Fuel tanks full and oil full
  Hydrometer check on
  fuel shows 5.9 lb/gal

By analyzing the data identified for the airplane being weighed in Figure 4-20, the following needed information is determined.

- Because the airplane was weighed with the fuel tanks full, the full weight of the fuel must be subtracted and the unusable fuel added back in. The weight of the fuel being subtracted is based on the pounds per gallon determined by the hydrometer check (5.9 lb/gal).
- Because wheel chocks are used to keep the airplane from rolling off the scales, their weight must be subtracted from the scale readings as tare weight.
- Because the main wheel centerline is 70" behind the datum, its arm is a +70".
- The arm for the nosewheel is the difference between the wheelbase (100") and the distance from the datum to the main wheel centerline (70"). Therefore, the arm for the nosewheel is −30".

To calculate the airplane's empty weight and empty weight center of gravity, a six column chart is used. Figure 4-21 shows the calculation for the airplane in Figure 4-20.

Item	Weight (lb)	Tare (lb)	Net Wt. (lb)	Arm (inches)	Moment (in-lb)
Nose	225	-2.5	222.5	-30	-6,675
Left Main	650	-5	645	+70	45,150
Right Main	640	-5	635	+70	44,450
Subtotal	1,515	-12.5	1,502.5		82,925
Fuel Total			-177	+95	-16,815
Fuel Unuse			+6	+98	588
Oil			Full		
Total			1,331.5	+50.1	66,698

Figure 4-21. Center of gravity calculation for airplane being weighed.

Based on the calculation shown in the chart, the center of gravity is at +50.1", which means it is 50.1" aft of the datum. This places the center of gravity forward of the main landing gear, which must be the case for a tricycle gear airplane. This number is the result of dividing the total moment of 66,698 in-lb by the total weight of 1,331.5 lb.

Loading an Aircraft for Flight

The ultimate test of whether or not there is a problem with an airplane's weight and balance is when it is loaded and ready to fly. The only real importance of an airplane's empty weight and empty weight center of gravity is how it affects the loaded weight and balance of the airplane, since an airplane doesn't fly when it is empty. The pilot in command is responsible for the weight and balance of the loaded airplane, and he or she makes the final decision on whether or not the airplane is safe to fly.

Example Loading of an Airplane

As an example of an airplane being loaded for flight, the Piper Seneca twin will be used. The Type Certificate Data Sheet for this airplane was shown earlier in this chapter, and its center of gravity range and CG envelope were also shown.

The information from the Type Certificate Data Sheet that pertains to this example loading is as follows:

CG Range (Gear Extended) S/N 34-7250215 through 34-7450220:

(+87.9") to (+94.6") at 4,200 lb

(+82.0") to (+94.6") at 3,400 lb

(+80.7") to (+94.6") at 2,780 lb

Straight line variation between points given.

−32 in-lb moment change due to gear retracting landing gear

Empty Weight CG Range None

Maximum Weight S/N 34-7250215 through 34-7450220:

4,200 lb—Takeoff

4,000 lb—Landing

No. of Seats 7 (2 at +85.5", 3 at +118.1", 2 at +155.7")

Maximum Baggage 200 lb (100 lb at +22.5, 100 lb at +178.7)

Fuel Capacity 98 gal (2 wing tanks) at (+93.6") (93 gal usable). See NOTE 1 for data on system fuel.

For the example loading of the airplane, the following information applies:

- Airplane Serial Number: 34-7250816
- Airplane Empty Weight: 2,650 lb
- Airplane Empty Weight CG: +86.8"

For today's flight, the following useful load items will be included:

- 1 pilot at 180 lb at an arm of +85.5"
- 1 passenger at 160 lb at an arm of +118.1"
- 1 passenger at 210 lb at an arm of +118.1"
- 1 passenger at 190 lb at an arm of +118.1"
- 1 passenger at 205 lb at an arm of +155.7"
- 50 lb of baggage at +22.5"
- 100 lb of baggage at +178.7"
- 80 gal of fuel at +93.6"

To calculate the loaded weight and CG of this airplane, a four column chart will be used, as shown in Figure 4-22.

Based on the information in the Type Certificate Data Sheet, the maximum takeoff weight of this airplane is 4,200 lb, and the aft-most CG limit is +94.6". The loaded airplane in the chart above is 25 lb too heavy, and the CG is 1.82" too far aft. To make the airplane

Item	Weight (lb)	Arm (inches)	Moment (in-lb)
Empty Weight	2,650	+86.80	230,020.0
Pilot	180	+85.50	15,390.0
Passenger	160	+118.10	18,896.0
Passenger	210	+118.10	24,801.0
Passenger	190	+118.10	22,439.0
Passenger	205	155.70	31,918.5
Baggage	50	+178.70	1,125.0
Baggage	100	+93.60	17,870.0
Fuel	480	11.87	44,928.0
Total	4,255	+96.42	407,387.5

Figure 4-22. Center of gravity calculation for Piper Seneca.

Item	Weight (lb)	Arm (inches)	Moment (in-lb)
Empty Weight	2,650	+86.8	230,020.0
Pilot	180	+85.5	15,390.0
Passenger	210	+85.5	17,955.0
Passenger	160	+118.1	24,801.0
Passenger	190	+118.1	22,439.0
Passenger	205	+118.1	24,210.5
Baggage	100	+22.5	2,250.0
Baggage	25	+178.7	4,467.5
Fuel	480	93.6	44,928.0
Total	4,200	+92.0	386,461.0

Figure 4-23. Center of gravity calculation for Piper Seneca with weights shifted.

safe to fly, the load needs to be reduced by 25 lb and some of the load needs to be shifted forward. For example, the baggage can be reduced by 25 lb, and a full 100 lb of it can be placed in the more forward compartment. One passenger can be moved to the forward seat next to the pilot, and the aft-most passenger can then be moved forward. If these changes are made, the four column calculation will be as shown in Figure 4-23.

With the changes made, the loaded weight is now at the maximum allowable of 4,200 lb, and the CG has moved forward 4.41". The airplane is now safe to fly.

Weight and Balance Extreme Conditions

A weight and balance extreme condition check, sometimes called an adverse condition check, involves loading the aircraft in as nose heavy or tail heavy a condition as possible, and seeing if the center of gravity falls outside the allowable limits. This check is done with pencil and paper. In other words, the aircraft is not actually loaded in an adverse way and an attempt made to fly it.

On what is called a forward extreme condition check, all useful load items in front of the forward CG limit are loaded, and all useful load items behind the forward CG limit are left empty. So if there are two seats and a baggage compartment located in front of the forward CG limit, two people weighing 170 lb each will be put in the seats, and the maximum allowable baggage will be put in the baggage compartment. Any seat or baggage compartment located behind the forward CG limit will be left empty. If the fuel is located behind the forward CG limit, minimum fuel will be shown in the tank. Minimum fuel is calculated by dividing the engine's METO horsepower by 2.

On an aft extreme condition check, all useful load items behind the aft CG limit are loaded, and all useful load items in front of the aft CG limit are left empty. Even though the pilot's seat will be located in front of the aft CG limit, the pilot's seat cannot be left empty. If the fuel tank is located forward of the aft CG limit, minimum fuel will be shown.

Example Forward and Aft Extreme Condition Checks

Using the stick airplane in Figure 4-24 as an example, adverse forward and aft checks will be calculated. Some of the data for the airplane is shown in the figure, such as seat, baggage, and fuel information. The center of gravity limits are shown, with arrows pointing in the direction where maximum and minimum weights will be loaded. On the forward check, any useful load item located in front of 89" will be loaded, and anything behind that location will be left empty. On the aft check, maximum weight will be added behind 99" and minimum weight in front of that location. For either of the checks, if fuel is not located in a maximum weight location, minimum fuel must be accounted for. Notice that the front seats show a location of 82" to 88", meaning they are adjustable fore and aft. In a forward check, the pilot's seat will be shown at 82", and in the aft check it will be at 88". Additional specifications for the airplane shown in Figure 4-24 are as follows:

- Airplane Empty Weight: 1,850 lb
- Empty Weight CG: +92.45"
- CG Limits: +89" to +99"
- Maximum Weight: 3,200 lb
- Fuel Capacity: 45 gal at +95" (44 usable)
  40 gal at +102" (39 usable)

In evaluating the two extreme condition checks, the following key points should be recognized.

- The total arm is the airplane center of gravity, and is found by dividing the total moment by the total weight.
- For the forward check, the only thing loaded behind the forward limit was minimum fuel.
- For the forward check, the pilot and passenger seats were shown at the forward position of 82".

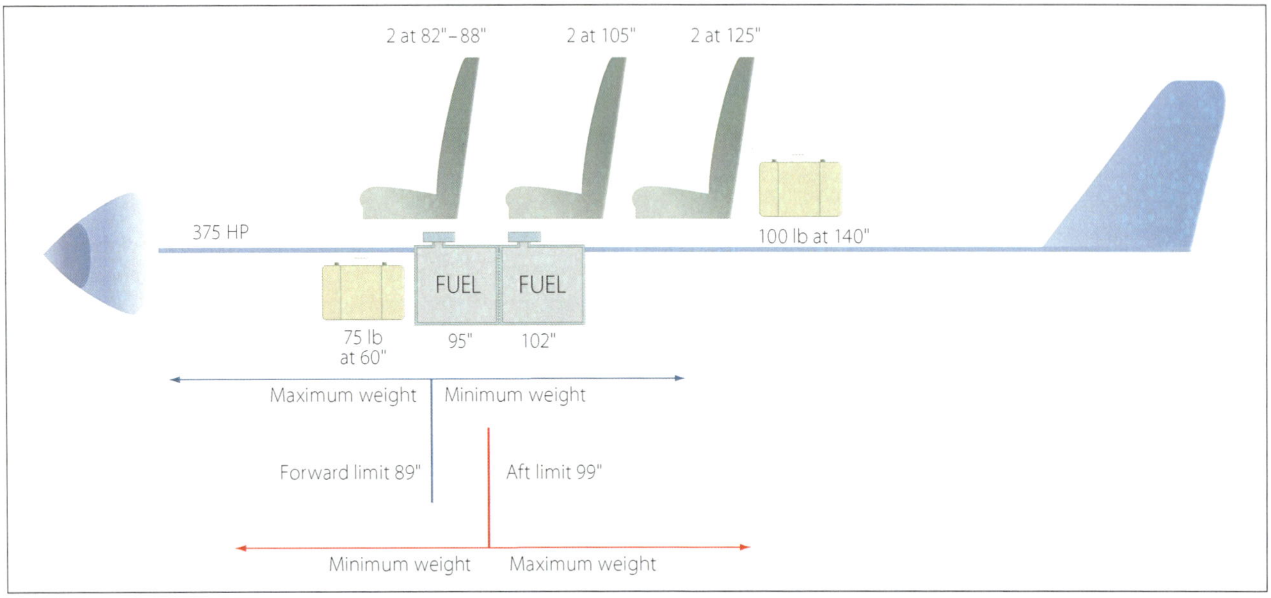

Figure 4-24. Example airplane for extreme condition checks.

- For the forward check, the CG was within limits, so the airplane could be flown this way.
- For the aft check, the only thing loaded in front of the aft limit was the pilot, at an arm of 88".

Extreme Condition Forward Check			
Item	**Weight (lb)**	**Arm (inches)**	**Moment (in-lb)**
Empty Weight	1,850.0	+92.45	171,032.5
Pilot	170.0	+82.00	13,940.0
Passenger	170.0	+82.00	13,940.0
Baggage	75.0	+60.00	4,500.0
Fuel	187.5	+95.00	17,812.5
Total	2,452.5	+90.20	221,225.0

Extreme Condition Aft Check			
Item	**Weight (lb)**	**Arm (inches)**	**Moment (in-lb)**
Empty Weight	1,850	+92.45	171,032.5
Pilot	170	+88.00	14,960.0
2 Passengers	340	+105.00	35,700.0
2 Passengers	340	+125.00	42,500.0
Baggage	100	+140.00	14,000.0
Fuel	234	+102.00	23,868.0
Total	3,034	+99.60	302,060.5

Figure 4-25. Center of gravity extreme condition checks.

- For the aft check, the fuel tank at 102" was filled, which more than accounted for the required minimum fuel.
- For the aft check, the CG was out of limits by 0.6", so the airplane should not be flown this way.

Equipment Change and Aircraft Alteration

When the equipment in an aircraft is changed, such as the installation of a new radar system or ground proximity warning system, or the removal of a radio or seat, the weight and balance of an aircraft will change. An alteration performed on an aircraft, such as a cargo door being installed or a reinforcing plate being attached to the spar of a wing, will also change the weight and balance of an aircraft. Any time the equipment is changed or an alteration is performed, the new empty weight and empty weight center of gravity must be determined. This can be accomplished by placing the aircraft on scales and weighing it, or by mathematically calculating the new weight and balance. The mathematical calculation is acceptable if the exact weight and arm of all the changes are known.

Example Calculation After an Equipment Change

A small twin-engine airplane has some new equipment installed, and some of its existing equipment removed. The details of the equipment changes are as follows:

- Airplane Empty Weight: 2,350 lb
- Airplane Empty Weight CG: +24.7"
- Airplane Datum: Leading edge of the wing
- Radio Installed: 5.8 lb at an arm of –28"
- Global Positioning System Installed: 7.3 lb at an arm of –26"
- Emergency Locater Transmitter Installed: 2.8 lb at an arm of +105"
- Strobe Light Removed: 1.4 lb at an arm of +75"
- Automatic Direction Finder (ADF) removed: 3 lb at an arm of –28"
- Seat Removed: 34 lb at an arm of +60"

To calculate the new empty weight and empty weight center of gravity, a four column chart is used. The calculation would be as shown in Figure 4-26.

In evaluating the weight and balance calculation shown in Figure 4-26, the following key points should be recognized.

- The weight of the equipment needs to be identified with a plus or minus to signify whether it is being installed or removed.
- The sign of the moment (plus or minus) is determined by the signs of the weight and arm.
- The strobe and the ADF are both being removed (negative weight), but only the strobe has a negative moment. This is because the arm for the ADF is also negative, and two negatives multiplied together produce a positive result.

Item	Weight (lb)	Arm (inches)	Moment (in-lb)
Empty Weight	2,350.0	+24.70	58,045.0
Radio Install	+5.8	−28.00	−162.4
GPS Install	+7.3	−26.00	−189.8
ELT Install	+2.8	+105.00	294.0
Strobe Remove	−1.4	+75.00	−105.0
ADF Remove	−3.0	−28.00	84.0
Seat Remove	−34.0	+60.00	−2,040.0
Total	2,327.5	24.03	55,925.8

Figure 4-26. Center of gravity calculation after equipment change.

- The total arm is the airplane's center of gravity, and is found by dividing the total moment by the total weight.
- The result of the equipment change is that the airplane's weight was reduced by 22.5 lb and the center of gravity has moved forward 0.67".

The Use of Ballast

Ballast is used in an aircraft to attain the desired CG balance, when the center of gravity is not within limits or is not at the location desired by the operator. It is usually located as far aft or as far forward as possible to bring the CG within limits, while using a minimum amount of weight. Ballast that is installed to compensate for the removal or installation of equipment items and that is to remain in the aircraft for long periods is called permanent ballast. It is generally lead bars or plates bolted to the aircraft structure. It may be painted red and placarded: PERMANENT BALLAST—DO NOT REMOVE. The installation of permanent ballast results in an increase in the aircraft empty weight, and it reduces the useful load.

Temporary ballast, or removable ballast, is used to meet certain loading conditions that may vary from time to time. It generally takes the form of lead shot bags, sand bags, or other weight items that are not permanently installed. Temporary ballast should be placarded: BALLAST, XX LB. REMOVAL REQUIRES WEIGHT AND BALANCE CHECK. The baggage compartment is usually the most convenient location for temporary ballast.

Whenever permanent or temporary ballast is installed, it must be placed in an approved location and secured in an appropriate manner. If permanent ballast is being bolted to the structure of the aircraft, the location must be one that was previously approved and designed for the installation, or it must be approved by the FAA as a major alteration before the aircraft is returned to service. When temporary ballast is placed in a baggage compartment, it must be secured in a way that prevents it from becoming a projectile if the aircraft encounters turbulence or an unusual flight attitude.

To calculate how much ballast is needed to bring the center of gravity within limits, the following formula is used.

$$\text{Ballast Needed} = \frac{\text{Loaded weight of aircraft (distance CG is out of limits)}}{\text{Arm from ballast location to affected limit}}$$

Item	Weight (lb)	Arm (inches)	Moment (in-lb)
Empty Weight	1,850	+92.45	171,032.5
Pilot	170	+88.00	14,960.0
2 Passengers	340	+105.00	35,700.0
2 Passengers	340	+125.00	42,500.0
Baggage	100	+140.00	14,000.0
Fuel	234	+102.00	23,868.0
Total	3,034	+99.60	302,060.5

Figure 4-27. Extreme condition check.

Figure 4-24 on page 4-22 and Figure 4-27 show an aft extreme condition check being performed on an airplane. In this previously shown example, the airplane's center of gravity was out of limits by 0.6". If there were a need or a desire to fly the airplane loaded this way, one way to make it possible would be the installation of temporary ballast in the front of the airplane. The logical choice for placement of this ballast is the forward baggage compartment.

The center of gravity for this airplane is 0.6" too far aft. If the forward baggage compartment is used as a temporary ballast location, the ballast calculation will be as follows:

Ballast Needed =

$$\frac{\text{Loaded weight of aircraft (distance CG is out of limits)}}{\text{Arm from ballast location to affected limit}}$$

$$= \frac{3{,}034 \text{ lb } (0.6")}{39"}$$

$$= 46.68 \text{ lb}$$

When ballast is calculated, the answer should always be rounded up to the next higher whole pound, or in this case, 47 lb of ballast would be used. To ensure the ballast calculation is correct, the weight of the ballast should be plugged back into the four column calculation and a new center of gravity calculated.

Item	Weight (lb)	Arm (inches)	Moment (in-lb)
Loaded Weight	3,034	+99.60	302,060.5
Ballast	47	+60.00	2,820.0
Total	3,081	+98.96	304,880.5

Figure 4-28. Ballast calculation.

The aft limit for the airplane was 99", and the new CG is at 98.96", which puts it within acceptable limits. The new CG did not fall exactly at 99" because the amount of needed ballast was rounded up to the next higher whole pound. If the ballast could have been placed farther forward, such as being bolted to the engine firewall, less ballast would have been needed. That is why ballast is always placed as far away from the affected limit as possible.

In evaluating the ballast calculation shown above, the following key points should be recognized.

- The loaded weight of the aircraft, as identified in the formula, is what the airplane weighed when the CG was out of limits.
- The distance the CG is out of limits is the difference between the CG location and the CG limit, in this case 99.6" minus 99".
- The affected limit identified in the formula is the CG limit which has been exceeded. If the CG is too far aft, it is the aft limit that has been exceeded.
- The aft limit for this example airplane is 99", and the ballast is being placed in the baggage compartment at an arm of 60". The difference between the two is 39", the quantity divided by in the formula.

Viewed as a first class lever problem, Figure 4-29 shows what this ballast calculation would look like. A ballast weight of 46.68 lb on the left side of the lever multiplied by the arm of 39" (99 minus 60) would equal the aircraft weight of 3,034 lb multiplied by the distance the CG is out of limits, which is 0.6" (99.6 minus 99).

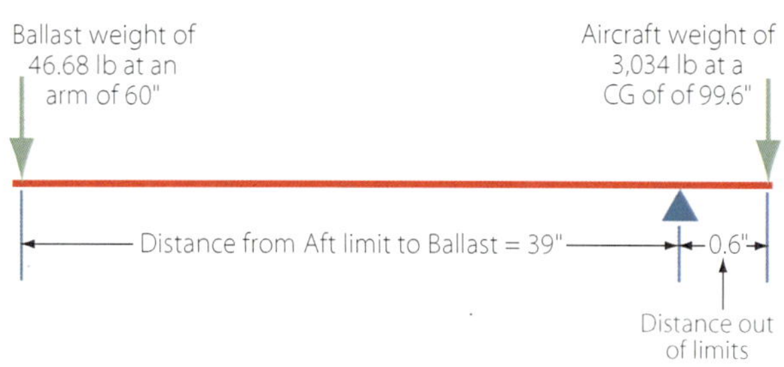

In order to balance at the aft limit of 99", the moment to the left of the fulcrum must equal the moment to the right of the fulcrum. The moment to the right is the weight of the airplane multiplied by 0.6". The moment to the left is the ballast weight multiplied by 39".

Figure 4-29. Ballast calculation as a first class lever.

Loading Graphs and CG Envelopes

The weight and balance computation system, commonly called the loading graph and CG envelope system, is an excellent and rapid method for determining the CG location for various loading arrangements. This method can be applied to any make and model of aircraft, but is more often seen with small general aviation aircraft.

Aircraft manufacturers using this method of weight and balance computation prepare graphs similar to those shown in Figures 4-30 and 4-31 for each make and model aircraft at the time of original certification. The graphs become a permanent part of the aircraft records, and are typically found in the Airplane Flight Manual or Pilot's Operating Handbook (AFM/POH). These graphs, used in conjunction with the empty weight and empty weight CG data found in the weight and balance report, allow the pilot to plot the CG for the loaded aircraft.

The loading graph illustrated in Figure 4-30 is used to determine the index number (moment value) of any item or weight that may be involved in loading the aircraft. To use this graph, find the point on the verti-

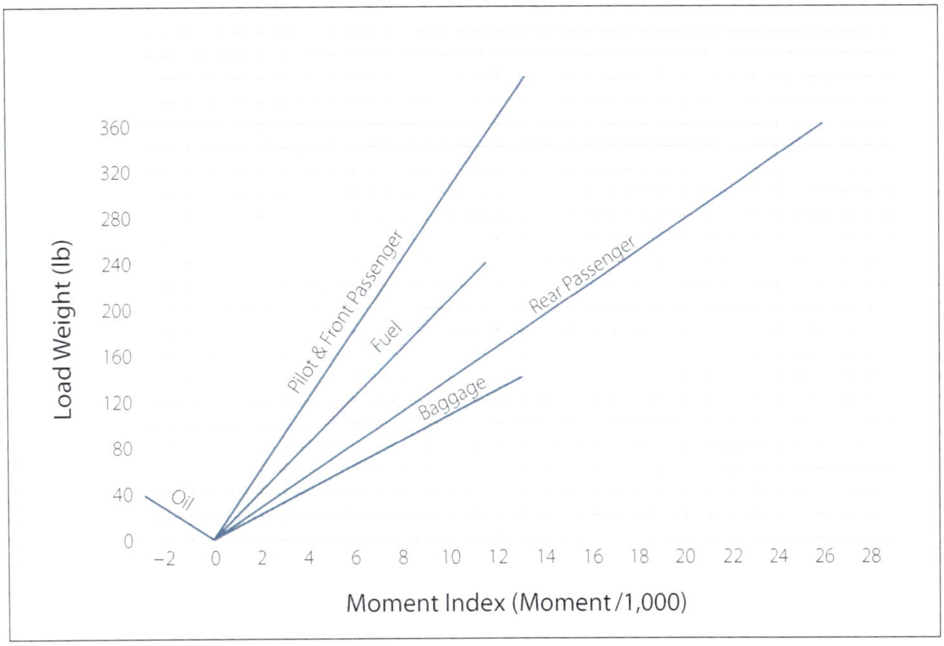

Figure 4-30. Aircraft loading graph.

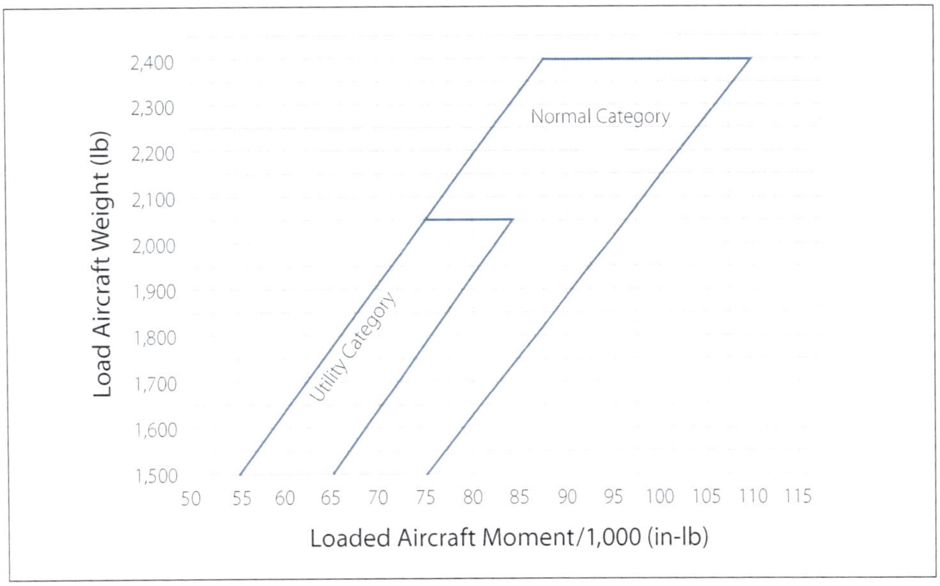

Figure 4-31. CG envelope.

cal scale that represents the known weight. Project a horizontal line to the point where it intersects the proper diagonal weight line (i.e., pilot, copilot, baggage). Where the horizontal line intersects the diagonal, project a vertical line downward to determine the loaded moment (index number) for the weight being added.

After the moment for each item of weight has been determined, all weights are added and all moments are added. The total weight and moment is then plotted on the CG envelope. [Figure 4-31] The total weight is plotted on the vertical scale of the graph, with a horizontal line projected out from that point. The total moment is plotted on the horizontal scale of the graph, with a vertical line projected up from that point. Where the horizontal and vertical plot lines intersect on the graph is the center of gravity for the loaded aircraft. If the point where the plot lines intersect falls inside the CG envelope, the aircraft CG is within limits. In Figure 4-31, there are actually two CG envelopes, one for the aircraft in the Normal Category and one for the aircraft in the Utility Category.

Item	Weight (lb)	Moment (in-lb)
Aircraft Empty Wt.	1,400	53,900
Pilot	180	6,000
Front Passengers	140	4,500
Rear Passengers	210	15,000
Baggage	100	9,200
Fuel	228	10,800
Total	2,258	99,400

Figure 4-32. Aircraft load chart.

The loading graph and CG envelope shown in Figures 4-30 and 4-31 are for an airplane with the following specifications and weight and balance data.

- Number of Seats: 4
- Fuel Capacity (Usable): 38 gal of Av Gas
- Oil Capacity: 8 qt (included in empty weight)
- Baggage: 120 lb
- Empty Weight: 1,400 lb
- Empty Weight CG: 38.5"
- Empty Weight Moment: 53,900 in-lb

An example of loading the airplane for flight and calculating the total loaded weight and the total loaded moment is shown in Figures 4-32 and 4-33. The use of the loading graph to determine the moment for each of the useful load items is shown in Figure 4-33. The color used for each useful load item in Figure 4-32 matches the color used for the plot on the loading graph.

The total loaded weight of the airplane is 2,258 pounds and the total loaded moment is 99,400 in-lb. These two numbers can now be plotted on the CG envelope to see if the airplane is within CG limits. Figure 4-34 shows the CG envelope, with the loaded weight and moment of the airplane plotted. The CG location shown falls

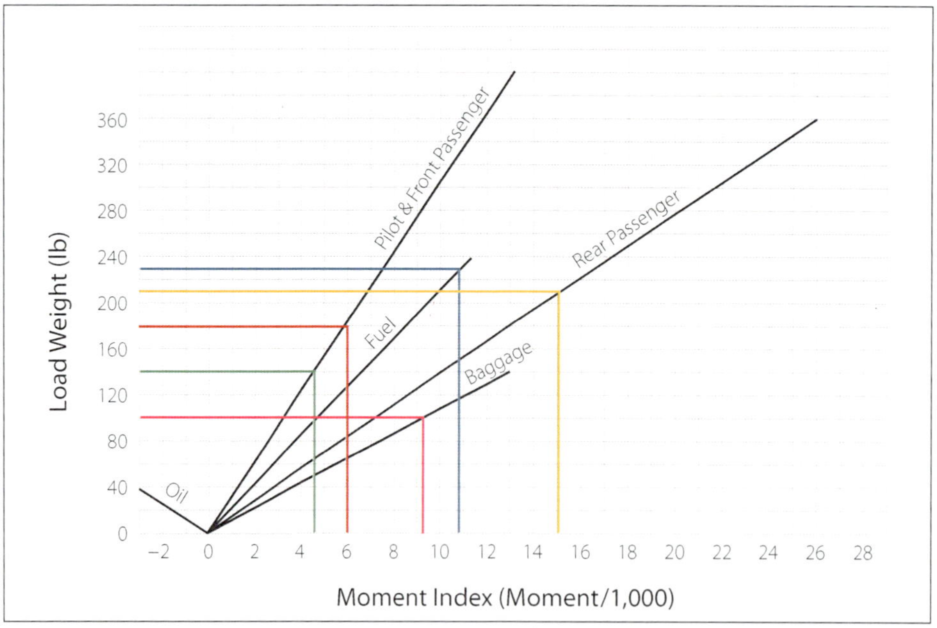

Figure 4-33. Example plots on a loading graph.

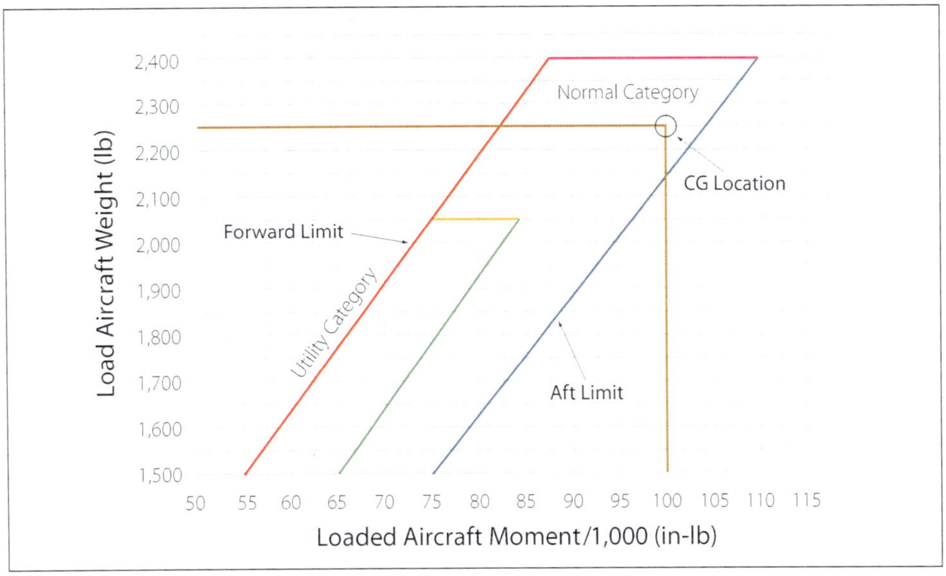

Figure 4-34. CG envelope example plot.

within the normal category envelope, so the airplane is within CG limits for this category.

It is interesting to note that the lines that form the CG envelope are actually graphic plots of the forward and aft CG limits. In Figure 4-34, the red line is a graphic plot of the forward limit, and the blue and green lines are graphic plots of the aft limit for the two different categories.

Helicopter Weight and Balance

General Concepts

All of the terminology and concepts that apply to airplane weight and balance also apply generally to helicopter weight and balance. There are some specific differences, however, which need to be identified.

Most helicopters have a much more restricted CG range than do airplanes. In some cases, this range is less than 3". The exact location and length of the CG range is specified for each helicopter, and usually extends a short distance fore and aft of the main rotor mast or centered between the main rotors of a dual rotor system. Whereas airplanes have a center of gravity range only along the longitudinal axis, helicopters have both longitudinal and lateral center of gravity ranges. Because the wings extend outward from the center of gravity, airplanes tend to have a great deal of lateral stability. A helicopter, on the other hand, acts like a pendulum, with the weight of the helicopter hanging from the main rotor shaft.

Ideally, the helicopter should have such perfect balance that the fuselage remains horizontal while in a hover. If the helicopter is too nose heavy or tail heavy while it is hovering, the cyclic pitch control will be used to keep the fuselage horizontal. If the CG location is too extreme, it may not be possible to keep the fuselage horizontal or maintain control of the helicopter.

Helicopter Weighing

When a helicopter is being weighed, the location of both longitudinal and lateral weighing points must be known to determine its empty weight and empty weight CG. This is because helicopters have longitudinal and lateral CG limits. As with the airplane, the longitudinal arms are measured from the datum, with locations behind the datum being positive arms and locations in front of the datum being negative arms. Laterally, the arms are measured from the butt line, which is a line from the nose to the tail running through the middle of the helicopter. When facing forward, arms to the right of the butt line are positive; to the left they are negative.

Before a helicopter is weighed, it must be leveled longitudinally and laterally. This can be done with a spirit level, but more often than not it is done with a plumb bob. For example, the Bell JetRanger has a location inside the aft cabin where a plumb can be attached, and allowed to hang down to the cabin floor. On the cabin floor is a plate bearing cross hairs, with the cross hairs corresponding to the horizontal and lateral axis of the helicopter. When the point of the plumb bob falls in the middle of the cross hairs, the helicopter is level along both axes. If the tip of the plumb bob falls forward of this point, the nose of the helicopter is too low; if it falls to the left of this point, the left side of the helicopter is

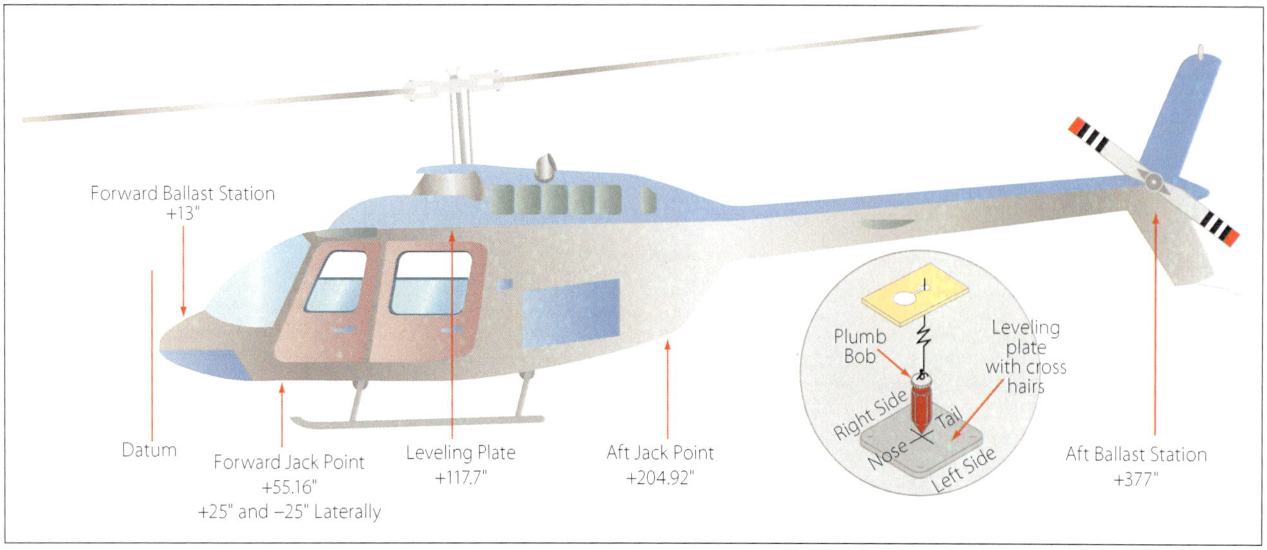

Figure 4-35. Bell JetRanger.

too low. In other words, the tip of the plumb bob will always move toward the low point.

A Bell JetRanger helicopter is shown in Figure 4-35, with the leveling plate depicted on the bottom right of the figure. The helicopter has three jack pads, two at the front and one in the back. To weigh this helicopter, three jacks would be placed on floor scales, and the helicopter would be raised off the hangar floor. To level the helicopter, the jacks would be adjusted until the plumb bob point falls exactly in the middle of the cross hairs.

As an example of weighing a helicopter, consider the Bell JetRanger in Figure 4-35, and the following specifications and weighing data.

- Datum: 55.16" forward of the front jack point centerline
- Leveling Means: Plumb line from ceiling left rear cabin to index plate on floor
- Longitudinal CG Limits:
  +106" to +111.4" at 3,200 lb
  +106" to +112.1" at 3,000 lb
  +106" to +112.4" at 2,900 lb
  +106" to +113.4" at 2,600 lb
  +106" to +114.2" at 2,350 lb
  +106" to +114.2" at 2,100 lb
  Straight line variation between points given
- Lateral CG Limits:
  2.3" left to 3.0" right at longitudinal CG +106.0"
  3.0" left to 4.0" right at longitudinal CG +108" to +114.2"
  Straight line variation between points given
- Fuel and Oil: Empty weight includes unusable fuel and unusable oil
- Left Front Scale Reading: 650 lb
- Left Front Jack Point: Longitudinal arm of +55.16"
  Lateral arm of –25"
- Right Front Scale Reading: 625 lb
- Right Front Jack Point: Longitudinal arm of +55.16"
  Lateral arm of +25"
- Aft Scale Reading: 710 lb
- Aft Jack Point: Longitudinal arm of +204.92"
  Lateral arm of 0.0"
- Notes: The helicopter was weighed with unusable fuel and oil. Electronic scales were used, which were zeroed with the jacks in place, so no tare weight needs to be accounted for.

Longitudinal CG Calculation					
Item	Scale (lb)	Tare Wt. (lb)	Nt. Wt. (lb)	Arm (inches)	Moment (in-lb)
Left Front	650	0	650	+55.16	35,854.0
Right Front	625	0	625	+55.16	34,475.0
Aft	710	0	710	+204.92	145,493.2
Total	1,985		1,985	+108.73	215,822.2

Lateral CG Calculation					
Item	Scale (lb)	Tare Wt. (lb)	Nt. Wt. (lb)	Arm (inches)	Moment (in-lb)
Left Front	650	0	650	–25	–16,250
Right Front	625	0	625	+25	+15,625
Aft	710	0	710	0	0
Total	1,985		1,985	+.31	–625

Figure 4-36. Center of gravity calculation for Bell JetRanger.

Using six column charts for the calculations, the empty weight and the longitudinal and lateral center of gravity for the helicopter would be as shown in Figure 4-36.

Based on the calculations in Figure 4-36, it has been determined that the empty weight of the helicopter is 1,985 lb, the longitudinal CG is at +108.73", and the lateral CG is at –0.31".

Weight and Balance—Weight-Shift Control Aircraft and Powered Parachutes

The terminology, theory, and concepts of weight and balance that applies to airplanes also applies to weight-shift aircraft and powered parachutes. Weight is still weight, and the balance point is still the balance point. There are, however, a few differences that need to be discussed. Before reading about the specifics of weight and balance on weight-shift control and powered parachute aircraft, be sure to read about their aerodynamic characteristics in Chapter 3, Physics.

Weight-shift control aircraft and powered parachutes do not fall under the same Code of Federal Regulations that govern certified airplanes and helicopters and, therefore, do not have Type Certificate Data Sheets or the same type of FAA mandated weight and balance reports. Weight and balance information and guidelines are left to the individual owners and the companies with which they work in acquiring this type of aircraft. Overall, the industry that is supplying these aircraft is regulating itself well, and the safety record is good for those aircraft being operated by experienced pilots.

The FAA has recently (2005) accepted a new classification of aircraft, known as Light Sport Aircraft (LSA). A new set of standards is being developed which will have an impact on the weight-shift control and powered parachute aircraft and how their weight and balance is handled.

Weight-Shift Control Aircraft

Weight-shift control aircraft, commonly known by the name "trikes," have very few options for loading because they have very few places to put useful load items. Some trikes have only one seat and a fuel tank, so the only variables for a flight are amount of fuel and weight of the pilot. Some trikes have two seats and a small storage bin, in addition to the fuel tank.

The most significant factor affecting the weight and balance of a trike is the weight of the pilot; if the aircraft has two seats, the weight of the passenger must be considered. The trike acts somewhat like a single main rotor helicopter because the weight of the aircraft is hanging like a pendulum under the wing. Figure 4-37 shows a two-place trike, in which the mast and the nose strut come together slightly below the wing attach point. When the trike is in flight, the weight of the aircraft is hanging from the wing attach point. The weight of the engine and fuel is behind this point, the passenger is almost directly below this point, and the pilot is forward of this point. The balance of the aircraft is determined by how all these weights compare.

The wing attach point, with respect to the wing keel, is an adjustable location. The attach point can be loosened

Figure 4-37. Weight and balance for a weight-shift aircraft.

and moved slightly forward or slightly aft, depending on the weight of the occupants. For example, if the aircraft is flown by a heavy person, the attach point can be moved a little farther aft, bringing the wing forward, to compensate for the change in center of gravity.

Figure 4-38 shows a close-up of the wing attach point, and the small amount of forward and aft movement that is available.

Powered Parachutes

Powered parachutes have many of the same characteristics as weight-shift control aircraft when it comes to weight and balance. They have the same limited loading, with only one or two seats and a fuel tank. They also act like a pendulum, with the weight of the aircraft hanging beneath the inflated wing (parachute).

The point at which the inflated wing attaches to the structure of the aircraft is adjustable to compensate for pilots and passengers of varying weights. With a very heavy pilot, the wing attach point would be moved forward to prevent the aircraft from being too nose heavy. Figure 4-39 illustrates the structure of a powered parachute with the adjustable wing attach points.

Weight and Balance for Large Airplanes

Weight and balance for large airplanes is almost identical to what it is for small airplanes, on a much larger scale. If a technician can weigh a small airplane and calculate its empty weight and empty weight center of gravity, that same technician should be able to do it for

Figure 4-38. Wing attach point for a weight-shift aircraft.

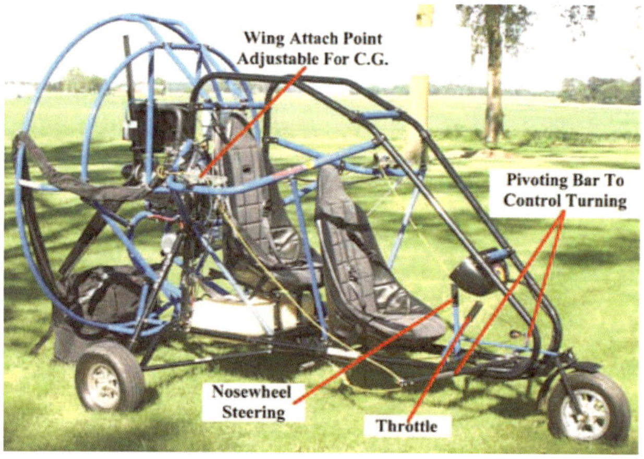

Figure 4-39. Powered parachute structure with wing attach points.

a large airplane. The jacks and scales will be larger, and it may take more personnel to handle the equipment, but the concepts and processes are the same.

Built-In Electronic Weighing

One difference that may be found with large airplanes is the incorporation of electronic load cells in the aircraft's landing gear. With this type of system, the airplane is capable of weighing itself as it sits on the tarmac. The load cells are built into the axles of the landing gear, or the landing gear strut, and they work in the same manner as load cells used with jacks. This system is currently in use on the Boeing 747-400, Boeing 777, Boeing 787, McDonnell Douglas MD-11, and the wide body Airbus airplanes like the A-330, A-340, and A-380.

The Boeing 777, utilizes two independent systems that provide information to the airplane's flight management computer. If the two systems agree on the weight and center of gravity of the airplane, the data being provided are considered accurate and the airplane can be dispatched based on that information. The flight crew has access to the information on the flight deck by accessing the flight management computer and bringing up the weight and balance page.

Mean Aerodynamic Chord

On small airplanes and on all helicopters, the center of gravity location is identified as being a specific number of inches from the datum. The center of gravity range is identified the same way. On larger airplanes, from private business jets to large jumbo jets, the center of gravity and its range are typically identified in relation to the width of the wing.

The width of the wing on an airplane is known as the chord. If the leading edge and trailing edge of a wing are parallel to each other, the chord of the wing is the same along the wing's length. Business jets and commercial transport airplanes have wings that are tapered and that are swept back, so the width of their wings is different along their entire length. The width is greatest where the wing meets the fuselage and progressively decreases toward the tip. In relation to the aerodynamics of the wing, the average length of the chord on these tapered swept-back wings is known as the mean aerodynamic chord (MAC).

On these larger airplanes, the CG is identified as being at a location that is a specific percent of the mean aerodynamic chord (% MAC). For example, imagine that the MAC on a particular airplane is 100", and the CG falls 20" behind the leading edge of the MAC. That means it falls one-fifth of the way back, or at 20% of the MAC.

Figure 4-40 shows a large twin-engine commercial transport airplane. The datum is forward of the nose of the airplane, and all the arms shown in the figure are being measured from that point. The center of gravity for the airplane is shown as an arm measured in inches. In the lower left corner of the figure, a cross section of the wing is shown, with the same center of gravity information being presented.

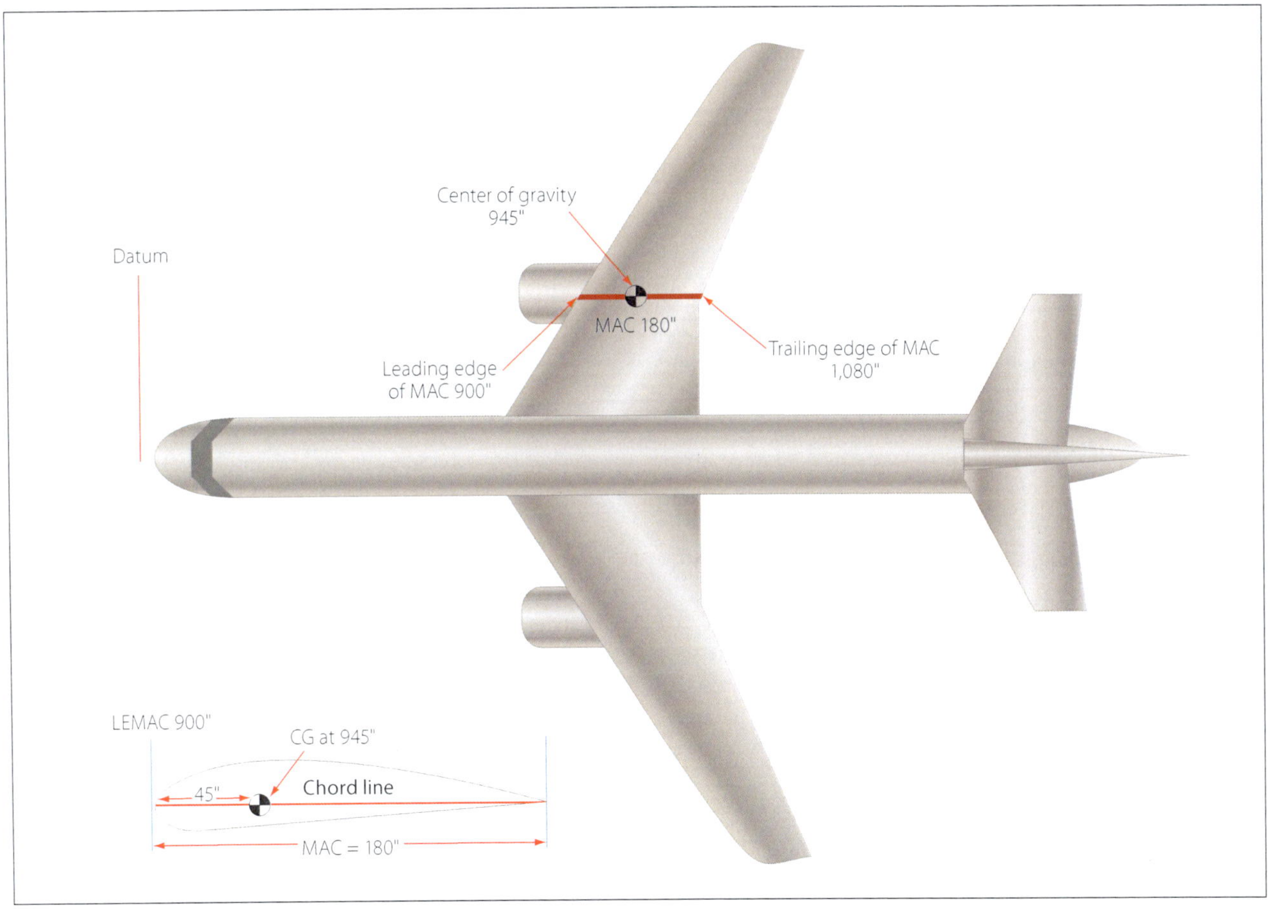

Figure 4-40. Center of gravity location on a large commercial transport.

To convert the center of gravity location from inches to a percent of MAC, for the airplane shown in Figure 4-40, the steps are as follows:

1. Identify the center of gravity location, in inches from the datum.
2. Identify the leading edge of the MAC (LEMAC), in inches from the datum.
3. Subtract LEMAC from the CG location.
4. Divide the difference by the length of the MAC.
5. Convert the result in decimals to a percentage by multiplying by 100.

As a formula, the solution to solve for the percent of MAC would be:

$$\textbf{Percent of MAC} = \frac{\textbf{CG} - \textbf{LEMAC}}{\textbf{MAC}} \times \textbf{100}$$

The result using the numbers shown in Figure 4-40 would be:

$$\begin{aligned}\text{Percent of MAC} &= \frac{\text{CG} - \text{LEMAC}}{\text{MAC}} \times 100 \\ &= \frac{945 - 900}{180} \times 100 \\ &= 25\%\end{aligned}$$

If the center of gravity is known in percent of MAC, and there is a need to know the CG location in inches from the datum, the conversion would be done as follows:

1. Convert the percent of MAC to a decimal by dividing by 100.
2. Multiply the decimal by the length of the MAC.
3. Add this number to LEMAC.

As a formula, the solution to convert a percent of MAC to an inch value would be:

CG in inches = MAC % ÷ 100 × MAC + LEMAC

For the airplane in Figure 4-40, if the CG was at 32.5% of the MAC, the solution would be:

$$\begin{aligned}\text{CG in inches} &= \text{MAC } \% \div 100 \times \text{MAC} + \text{LEMAC} \\ &= 32.5 \div 100 \times 180 + 900 \\ &= 958.5\end{aligned}$$

Weight and Balance Records

When a technician gets involved with the weight and balance of an aircraft, it almost always involves a calculation of the aircraft's empty weight and empty weight center of gravity. Only on rare occasions will the technician be involved in calculating extreme conditions, how much ballast is needed, or the loaded weight and balance of the aircraft. Calculating the empty weight and empty weight CG might involve putting the aircraft on scales and weighing it, or a pencil and paper exercise after installing a new piece of equipment.

The FAA requires that a current and accurate empty weight and empty weight center of gravity be known for an aircraft. This information must be included in the weight and balance report, which is a part of the aircraft permanent records. The weight and balance report must be in the aircraft when it is being flown.

There is no required format for this report, but Figure 4-41 is a good example of recording the data obtained from weighing an aircraft. As it is currently laid out, the form would accommodate either a tricycle gear or tail dragger airplane. Depending on the gear type, either the nose or the tail row would be used. If an airplane is being weighed using jacks and load cells, or if a helicopter is being weighed, the item names must be changed to reflect the weight locations.

If an equipment change is being done on an aircraft, and the new weight and balance is calculated mathematically instead of weighing the aircraft, the same type of form shown in Figure 4-41 can be used. The only change would be the use of a four column solution, instead of six columns, and there would be no tare weight or involvement with fuel and oil.

Aircraft Weight and Balance Report

Results of Aircraft Weighing

Make ______________________ Model ______________________

Serial # ______________________ N# ______________________

Datum Location ____

Leveling Means ____

Scale Arms: Nose __________ Tail __________ Left Main __________ Right Main __________

Scale Weights: Nose __________ Tail __________ Left Main __________ Right Main __________

Tare Weights: Nose __________ Tail __________ Left Main __________ Right Main __________

Weight and Balance Calculation

Item	Scale (lb)	Tare Wt. (lb)	Net Wt. (lb)	Arm (inches)	Moment (in-lb)
Nose					
Tail					
Left Main					
Right Main					
Subtotal					
Fuel					
Oil					
Misc.					
Total					

Aircraft Current Empty Weight: ______________________________

Aircraft Current Empty Weight CG: ______________________________

Aircraft Maximum Weight: ______________________________

Aircraft Useful Load: ______________________________

Computed By: ______________________________ (print name)

______________________________ (signature)

Certificate #: ______________________ (A&P, Repair Station, etc.)

Date: ______________

Figure 4-41. Aircraft weight and balance report.

Chapter 5 Aircraft Materials, Processes, & Hardware

Aircraft Metals

Knowledge and understanding of the uses, strengths, limitations, and other characteristics of structural metals is vital to properly construct and maintain any equipment, especially airframes. In aircraft maintenance and repair, even a slight deviation from design specification, or the substitution of inferior materials, may result in the loss of both lives and equipment. The use of unsuitable materials can readily erase the finest craftsmanship. The selection of the correct material for a specific repair job demands familiarity with the most common physical properties of various metals.

Properties of Metals

Of primary concern in aircraft maintenance are such general properties of metals and their alloys as hardness, malleability, ductility, elasticity, toughness, density, brittleness, fusibility, conductivity contraction and expansion, and so forth. These terms are explained to establish a basis for further discussion of structural metals.

Hardness

Hardness refers to the ability of a material to resist abrasion, penetration, cutting action, or permanent distortion. Hardness may be increased by cold working the metal and, in the case of steel and certain aluminum alloys, by heat treatment. Structural parts are often formed from metals in their soft state and are then heat treated to harden them so that the finished shape will be retained. Hardness and strength are closely associated properties of metals.

Strength

One of the most important properties of a material is strength. Strength is the ability of a material to resist deformation. Strength is also the ability of a material to resist stress without breaking. The type of load or stress on the material affects the strength it exhibits.

Density

Density is the weight of a unit volume of a material. In aircraft work, the specified weight of a material per cubic inch is preferred since this figure can be used in determining the weight of a part before actual manufacture. Density is an important consideration when choosing a material to be used in the design of a part in order to maintain the proper weight and balance of the aircraft.

Malleability

A metal which can be hammered, rolled, or pressed into various shapes without cracking, breaking, or leaving some other detrimental effect, is said to be malleable. This property is necessary in sheet metal that is worked into curved shapes, such as cowlings, fairings, or wingtips. Copper is an example of a malleable metal.

Ductility

Ductility is the property of a metal which permits it to be permanently drawn, bent, or twisted into various shapes without breaking. This property is essential for metals used in making wire and tubing. Ductile metals are greatly preferred for aircraft use because of their ease of forming and resistance to failure under shock loads. For this reason, aluminum alloys are used for cowl rings, fuselage and wing skin, and formed or extruded parts, such as ribs, spars, and bulkheads. Chrome molybdenum steel is also easily formed into desired shapes. Ductility is similar to malleability.

Elasticity

Elasticity is that property that enables a metal to return to its original size and shape when the force which causes the change of shape is removed. This property is extremely valuable because it would be highly undesirable to have a part permanently distorted after an applied load was removed. Each metal has a point known as the elastic limit, beyond which it cannot be loaded without causing permanent distortion. In aircraft construction, members and parts are so designed that the maximum loads to which they are subjected will not stress them beyond their elastic limits. This desirable property is present in spring steel.

Toughness

A material which possesses toughness will withstand tearing or shearing and may be stretched or otherwise deformed without breaking. Toughness is a desirable property in aircraft metals.

Brittleness

Brittleness is the property of a metal which allows little bending or deformation without shattering. A brittle metal is apt to break or crack without change of shape. Because structural metals are often subjected to shock loads, brittleness is not a very desirable property. Cast iron, cast aluminum, and very hard steel are examples of brittle metals.

Fusibility

Fusibility is the ability of a metal to become liquid by the application of heat. Metals are fused in welding. Steels fuse around 2,600 °F and aluminum alloys at approximately 1,100 °F.

Conductivity

Conductivity is the property which enables a metal to carry heat or electricity. The heat conductivity of a metal is especially important in welding because it governs the amount of heat that will be required for proper fusion. Conductivity of the metal, to a certain extent, determines the type of jig to be used to control expansion and contraction. In aircraft, electrical conductivity must also be considered in conjunction with bonding, to eliminate radio interference.

Thermal Expansion

Thermal expansion refers to contraction and expansion that are reactions produced in metals as the result of heating or cooling. Heat applied to a metal will cause it to expand or become larger. Cooling and heating affect the design of welding jigs, castings, and tolerances necessary for hot rolled material.

Ferrous Aircraft Metals

Many different metals are required in the repair of aircraft. This is a result of the varying needs with respect to strength, weight, durability, and resistance to deterioration of specific structures or parts. In addition, the particular shape or form of the material plays an important role. In selecting materials for aircraft repair, these factors plus many others are considered in relation to the mechanical and physical properties. Among the common materials used are ferrous metals. The term "ferrous" applies to the group of metals having iron as their principal constituent.

Iron

If carbon is added to iron, in percentages ranging up to approximately 1 percent, the product is vastly superior to iron alone and is classified as carbon steel. Carbon steel forms the base of those alloy steels produced by combining carbon steel with other elements known to improve the properties of steel. A base metal (such as iron) to which small quantities of other metals have been added is called an alloy. The addition of other metals changes or improves the chemical or physical properties of the base metal for a particular use.

Steel and Steel Alloys

To facilitate the discussion of steels, some familiarity with their nomenclature is desirable. A numerical index, sponsored by the Society of Automotive Engineers (SAE) and the American Iron and Steel Institute (AISI), is used to identify the chemical compositions of the structural steels. In this system, a four-numeral series is used to designate the plain carbon and alloy steels; five numerals are used to designate certain types of alloy steels. The first two digits indicate the type of steel, the second digit also generally (but not always) gives the approximate amount of the major alloying element, and the last two (or three) digits are intended to indicate the approximate middle of the carbon range. However, a deviation from the rule of indicating the carbon range is sometimes necessary.

Small quantities of certain elements are present in alloy steels that are not specified as required. These elements are considered as incidental and may be present to the maximum amounts as follows: copper, 0.35 percent; nickel, 0.25 percent; chromium, 0.20 percent; molybdenum, 0.06 percent.

The list of standard steels is altered from time to time to accommodate steels of proven merit and to provide for changes in the metallurgical and engineering requirements of industry. [Figure 5-1]

Metal stock is manufactured in several forms and shapes, including sheets, bars, rods, tubing, extrusions, forgings, and castings. Sheet metal is made in a number of sizes and thicknesses. Specifications designate thicknesses in thousandths of an inch. Bars and rods are supplied in a variety of shapes, such as round, square, rectangular, hexagonal, and octagonal. Tubing can be obtained in round, oval, rectangular, or streamlined

Series Designation	Types
100xx	Nonsulphurized carbon steels
11xx	Resulphurised carbon steels (free machining)
12xx	Rephosphorized and resulphurised carbon steels (free machining)
13xx	Manganese 1.75%
*23xx	Nickel 3.50%
*25xx	Nickel 5.00%
31xx	Nickel 1.25%, chromium 0.65%
33xx	Nickel 3.50%, chromium 1.55%
40xx	Molybdenum 0.20 or 0.25%
41xx	Chromium 0.50% or 0.95%, molybdenum 0.12 or 0.20%
43xx	Nickel 1.80%, chromium 0.5 or 0.80%, molybdenum 0.25%
44xx	Molybdenum 0.40%
45xx	Molybdenum 0.52%
46xx	Nickel 1.80%, molybdenum 0.25%
47xx	Nickel 1.05% chromium 0.45%, molybdenum 0.20 or 0.35%
48xx	Nickel 3.50%, molybdenum 0.25%
50xx	Chromium 0.25, or 0.40 or 0.50%
50xxx	Carbon 1.00%, chromium 0.50%
51xx	Chromium 0.80, 0.90, 0.95 or 1.00%
51xxx	Carbon 1.00%, chromium 1.05%
52xxx	Carbon 1.00%, chromium 1.45%
61xx	Chromium 0.60, 0.80, 0.95%, vanadium 0.12%, 0.10% min., or 0.15% min.
81xx	Nickel 0.30%, chromium 0.40%, molybdenum 0.12%
86xx	Nickel 0.55%, chromium 0.50%, molybdenum 0.20%
87xx	Nickel 0.55%, chromium 0.05%, molybdenum 0.25%
88xx	Nickel 0.55%, chromium 0.05%, molybdenum 0.35%
92xx	Manganese 0.85%, silicon 2.00%, chromium 0 or 0.35%
93xx	Nickel 3.25%, chromium 1.20%, molybdenum 0.12%
94xx	Nickel 0.45%, chromium 0.40%, molybdenum 0.12%
98xx	Nickel 1.00%, chromium 0.80%, molybdenum 0.25%

*Not included in the current list of standard steels

Figure 5-1. SAE numerical index.

shapes. The size of tubing is generally specified by outside diameter and wall thickness.

The sheet metal is usually formed cold in such machines as presses, bending brakes, drawbenches, or rolls. Forgings are shaped or formed by pressing or hammering heated metal in dies. Castings are produced by pouring molten metal into molds. The casting is finished by machining.

Spark testing is a common means of identifying various ferrous metals. In this test the piece of iron or steel is held against a revolving grinding stone and the metal is identified by the sparks thrown off. Each ferrous metal has its own peculiar spark characteristics. The spark streams vary from a few tiny shafts to a shower of sparks several feet in length. (Few nonferrous metals give off sparks when touched to a grinding stone. Therefore, these metals cannot be successfully identified by the spark test.)

Identification by spark testing is often inexact unless performed by an experienced person, or the test pieces differ greatly in their carbon content and alloying constituents.

Wrought iron produces long shafts that are straw colored as they leave the stone and white at the end. Cast iron sparks are red as they leave the stone and turn to a straw color. Low carbon steels give off long, straight shafts having a few white sprigs. As the carbon content of the steel increases, the number of sprigs along each shaft increases and the stream becomes whiter in color. Nickel steel causes the spark stream to contain small white blocks of light within the main burst.

Types, Characteristics, and Uses of Alloyed Steels

Steel containing carbon in percentages ranging from 0.10 to 0.30 percent is classed as low carbon steel. The equivalent SAE numbers range from 1010 to 1030. Steels of this grade are used for making such items as safety wire, certain nuts, cable bushings, or threaded rod ends. This steel in sheet form is used for secondary structural parts and clamps, and in tubular form for moderately stressed structural parts.

Steel containing carbon in percentages ranging from 0.30 to 0.50 percent is classed as medium carbon steel. This steel is especially adaptable for machining or forging, and where surface hardness is desirable. Certain rod ends and light forgings are made from SAE 1035 steel.

Steel containing carbon in percentages ranging from 0.50 to 1.05 percent is classed as high carbon steel. The addition of other elements in varying quantities adds to the hardness of this steel. In the fully heat-treated condition it is very hard, will withstand high shear and wear, and will have little deformation. It has limited use in aircraft. SAE 1095 in sheet form is used for making flat springs and in wire form for making coil springs.

The various nickel steels are produced by combining nickel with carbon steel. Steels containing from 3 to 3.75 percent nickel are commonly used. Nickel increases the hardness, tensile strength, and elastic limit of steel without appreciably decreasing the ductility. It also intensifies the hardening effect of heat treatment. SAE 2330 steel is used extensively for aircraft parts, such as bolts, terminals, keys, clevises, and pins.

Chromium steel is high in hardness, strength, and corrosion resistant properties, and is particularly adaptable for heat-treated forgings which require greater toughness and strength than may be obtained in plain carbon steel. It can be used for such articles as the balls and rollers of antifriction bearings.

Chrome-nickel or stainless steels are the corrosion resistant metals. The anticorrosive degree of this steel is determined by the surface condition of the metal as well as by the composition, temperature, and concentration of the corrosive agent. The principal alloy of stainless steel is chromium. The corrosion resistant steel most often used in aircraft construction is known as 18-8 steel because of its content of 18 percent chromium and 8 percent nickel. One of the distinctive features of 18-8 steel is that its strength may be increased by cold working.

Stainless steel may be rolled, drawn, bent, or formed to any shape. Because these steels expand about 50 percent more than mild steel and conduct heat only about 40 percent as rapidly, they are more difficult to weld. Stainless steel can be used for almost any part of an aircraft. Some of its common applications are in the fabrication of exhaust collectors, stacks and manifolds, structural and machined parts, springs, castings, tie rods, and control cables.

The chrome-vanadium steels are made of approximately 18 percent vanadium and about 1 percent chromium. When heat treated, they have strength, toughness, and resistance to wear and fatigue. A special grade of this steel in sheet form can be cold formed into intricate shapes. It can be folded and flattened without signs of breaking or failure. SAE 6150 is used for making springs; chrome-vanadium with high carbon content, SAE 6195, is used for ball and roller bearings.

Molybdenum in small percentages is used in combination with chromium to form chrome-molybdenum steel, which has various uses in aircraft. Molybdenum is a strong alloying element. It raises the ultimate strength of steel without affecting ductility or workability. Molybdenum steels are tough and wear resistant, and they harden throughout when heat treated. They are especially adaptable for welding and, for this reason, are used principally for welded structural parts and assemblies. This type steel has practically replaced carbon steel in the fabrication of fuselage tubing, engine mounts, landing gears, and other structural parts. For example, a heat-treated SAE X4130 tube is approximately four times as strong as an SAE 1025 tube of the same weight and size.

A series of chrome-molybdenum steel most used in aircraft construction is that series containing 0.25 to 0.55 percent carbon, 0.15 to 0.25 percent molybdenum, and 0.50 to 1.10 percent chromium. These steels, when suitably heat treated, are deep hardening, easily machined, readily welded by either gas or electric methods, and are especially adapted to high temperature service.

Inconel is a nickel-chromium-iron alloy closely resembling stainless steel (corrosion resistant steel, CRES) in appearance. Aircraft exhaust systems use both alloys interchangeably. Because the two alloys look very much alike, a distinguishing test is often necessary. One method of identification is to use an electrochemical technique, as described in the following paragraph, to identify the nickel (Ni) content of the alloy. Inconel has a nickel content greater than 50 percent, and the electrochemical test detects nickel.

The tensile strength of Inconel is 100,000 psi annealed, and 125,000 psi when hard rolled. It is highly resistant to salt water and is able to withstand temperatures as high as 1,600 °F. Inconel welds readily and has working qualities quite similar to those of corrosion resistant steels.

Electrochemical Test

Prepare a wiring assembly as shown in Figure 5-2, and prepare the two reagents (ammonium fluoride and dimethylglyoxime solutions) placing them in separate dedicated dropper solution bottles. Before testing, you must thoroughly clean the metal in order for the electrolytic deposit to take place. You may use nonmetallic hand scrubbing pads or 320 to 600 grit "crocus cloth" to remove deposits and corrosion products (thermal oxide).

Connect the alligator clip of the wiring assembly to the bare metal being tested. Place one drop of a 0.05 percent reagent grade ammonium fluoride solution in deionized water on the center of a 1 inch × 1 inch sheet of filter paper. Lay the moistened filter paper over the bare metal alloy being tested. Firmly press the end of the aluminum rod over the center of the moist paper. Maintain connection for 10 seconds while rocking the aluminum rod on the filter paper. Ensure that the light emitting diode (LED) remains lit (indicating good electrical contact and current flow) during this period. Disconnect the wiring assembly and set it aside. Remove the filter paper and examine it to determine that a light spot appears where the connection was made.

Deposit one drop of 1.0 percent solution of reagent grade dimethylglyoxime in ethyl alcohol on the filter paper (same side that was in contact with the test metal). A bright, distinctly pink spot will appear within seconds on the filter paper if the metal being tested is Inconel. A brown spot will appear if the test metal is stainless steel. Some stainless steel alloys may leave a very light pink color. However, the shade and depth of color will be far less than would appear for Inconel. For flat surfaces, the test spot will be circular while for curved surfaces, such as the outside of a tube or pipe, the test spot may appear as a streak. (Refer to Figure 5-3 for sample test results.) This procedure should not be used in the heat affected zone of weldments or on nickel coated surfaces.

Nonferrous Aircraft Metals

The term "nonferrous" refers to all metals which have elements other than iron as their base or principal constituent. This group includes such metals as alu-

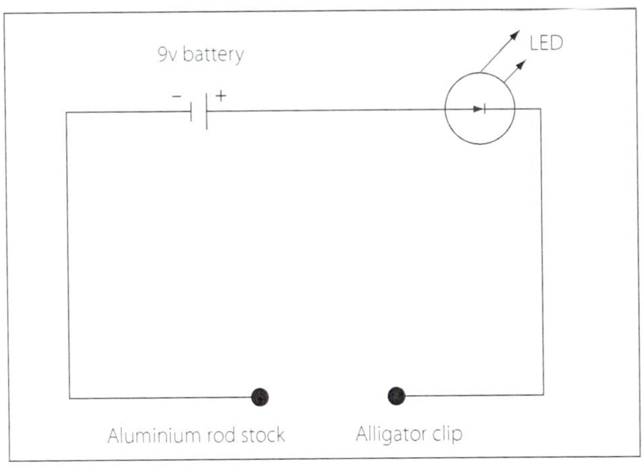

Figure 5-2. Wiring assembly schematic.

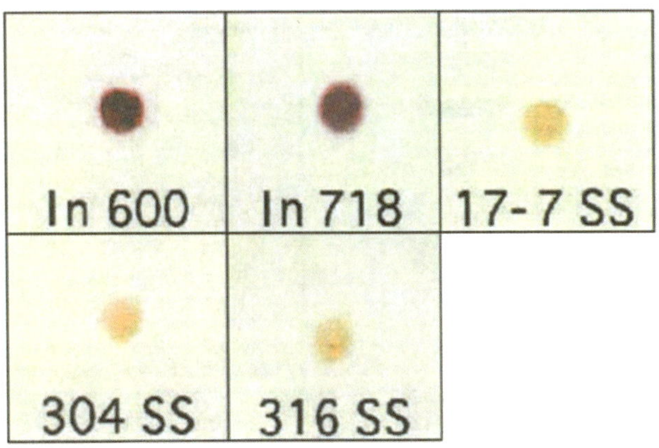

Figure 5-3. Electrochemical test results, Inconel and stainless steel alloys.

minum, titanium, copper, and magnesium, as well as such alloyed metals as Monel and babbit.

Aluminum and Aluminum Alloys

Commercially pure aluminum is a white lustrous metal which stands second in the scale of malleability, sixth in ductility, and ranks high in its resistance to corrosion. Aluminum combined with various percentages of other metals forms alloys which are used in aircraft construction.

Aluminum alloys in which the principal alloying ingredients are manganese, chromium, or magnesium and silicon show little attack in corrosive environments. Alloys in which substantial percentages of copper are used are more susceptible to corrosive action. The total percentage of alloying elements is seldom more than 6 or 7 percent in the wrought alloys.

Aluminum is one of the most widely used metals in modern aircraft construction. It is vital to the aviation industry because of its high strength to weight ratio and its comparative ease of fabrication. The outstanding characteristic of aluminum is its light weight. Aluminum melts at the comparatively low temperature of 1,250 °F. It is nonmagnetic and is an excellent conductor.

Commercially pure aluminum has a tensile strength of about 13,000 psi, but its strength may be approximately doubled by rolling or other cold working processes. By alloying with other metals, or by using heat-treating processes, the tensile strength may be raised to as high as 65,000 psi or to within the strength range of structural steel.

Aluminum alloys, although strong, are easily worked because they are malleable and ductile. They may be rolled into sheets as thin as 0.0017 inch or drawn into wire 0.004 inch in diameter. Most aluminum alloy sheet stock used in aircraft construction range from 0.016 to 0.096 inch in thickness; however, some of the larger aircraft use sheet stock which may be as thick as 0.356 inch.

The various types of aluminum may be divided into two general classes: (1) casting alloys (those suitable for casting in sand, permanent mold, or die castings) and (2) wrought alloys (those which may be shaped by rolling, drawing, or forging). Of these two, the wrought alloys are the most widely used in aircraft construction, being used for stringers, bulkheads, skin, rivets, and extruded sections.

Aluminum casting alloys are divided into two basic groups. In one, the physical properties of the alloys are determined by the alloying elements and cannot be changed after the metal is cast. In the other, the alloying elements make it possible to heat treat the casting to produce the desired physical properties.

The casting alloys are identified by a letter preceding the alloy number. When a letter precedes a number, it indicates a slight variation in the composition of the original alloy. This variation in composition is simply to impart some desirable quality. In casting alloy 214, for example, the addition of zinc to improve its pouring qualities is indicated by the letter A in front of the number, thus creating the designation A214.

When castings have been heat treated, the heat treatment and the composition of the casting is indicated by the letter T, followed by an alloying number. An example of this is the sand casting alloy 355, which has several different compositions and tempers and is designated by 355-T6, 355-T51, or C355-T51.

Aluminum alloy castings are produced by one of three basic methods: (1) sand mold, (2) permanent mold, or (3) die cast. In casting aluminum, it must be remembered that in most cases different types of alloys must be used for different types of castings. Sand castings and die castings require different types of alloys than those used in permanent molds.

Sand and permanent mold castings are parts produced by pouring molten metal into a previously prepared mold, allowing the metal to solidify or freeze, and then removing the part. If the mold is made of sand, the part is a sand casting; if it is a metallic mold (usually cast iron) the part is a permanent mold casting. Sand and permanent castings are produced by pouring liquid metal into the mold, the metal flowing under the force of gravity alone.

The two principal types of sand casting alloys are 112 and 212. Little difference exists between the two metals from a mechanical properties standpoint, since both are adaptable to a wide range of products.

The permanent mold process is a later development of the sand casting process, the major difference being in the material from which the molds are made. The advantage of this process is that there are fewer openings (called porosity) than in sand castings. The sand and the binder, which is mixed with the sand to hold it together, give off a certain amount of gas which causes porosity in a sand casting.

Permanent mold castings are used to obtain higher mechanical properties, better surfaces, or more accurate dimensions. There are two specific types of permanent mold castings: (1) permanent metal mold with metal cores, and (2) semipermanent types containing sand cores. Because finer grain structure is produced in alloys subjected to the rapid cooling of metal molds, they are far superior to the sand type castings. Alloys 122, A132, and 142 are commonly used in permanent mold castings, the principal uses of which are in internal combustion engines.

Die castings used in aircraft are usually aluminum or magnesium alloy. If weight is of primary importance, magnesium alloy is used because it is lighter than aluminum alloy. However, aluminum alloy is frequently used because it is stronger than most magnesium alloys.

A die casting is produced by forcing molten metal under pressure into a metallic die and allowing it to solidify; then the die is opened and the part removed. The basic difference between permanent mold casting and die casting is that in the permanent mold process the metal flows into the die under gravity. In the die casting operation, the metal is forced under great pressure.

Die castings are used where relatively large production of a given part is involved. Remember, any shape which can be forged can be cast.

Wrought aluminum and wrought aluminum alloys are divided into two general classes: non-heat-treatable alloys and heat-treatable alloys.

Non-heat-treatable alloys are those in which the mechanical properties are determined by the amount of cold work introduced after the final annealing operation. The mechanical properties obtained by cold working are destroyed by any subsequent heating and cannot be restored except by additional cold working, which is not always possible. The "full hard" temper is produced by the maximum amount of cold work that is commercially practicable. Metal in the "as fabricated" condition is produced from the ingot without any subsequent controlled amount of cold working or thermal treatment. There is, consequently, a variable amount of strain hardening, depending upon the thickness of the section.

For heat-treatable aluminum alloys, the mechanical properties are obtained by heat treating to a suitable temperature, holding at that temperature long enough to allow the alloying constituent to enter into solid solution, and then quenching to hold the constituent in solution. The metal is left in a supersaturated, unstable state and is then age hardened either by natural aging at room temperature or by artificial aging at some elevated temperature.

Wrought Aluminum

Wrought aluminum and wrought aluminum alloys are designated by a four digit index system. The system is broken into three distinct groups: 1xxx group, 2xxx through 8xxx group, and 9xxx group (which is currently unused).

The first digit of a designation identifies the alloy type. The second digit indicates specific alloy modifications. Should the second number be zero, it would indicate no special control over individual impurities. Digits 1 through 9, however, when assigned consecutively as needed for the second number in this group, indicate the number of controls over individual impurities in the metal.

The last two digits of the 1xxx group are used to indicate the hundredths of 1 percent above the original 99 percent designated by the first digit. Thus, if the last two digits were 30, the alloy would contain 99 percent plus 0.30 percent of pure aluminum, or a total of 99.30 percent pure aluminum. Examples of alloys in this group are:

- 1100—99.00 percent pure aluminum with one control over individual impurities.
- 1130—99.30 percent pure aluminum with one control over individual impurities.
- 1275—99.75 percent pure aluminum with two controls over individual impurities.

In the 2xxx through 8xxx groups, the first digit indicates the major alloying element used in the formation of the alloy as follows:

- 2xxx—copper
- 3xxx—manganese
- 4xxx—silicon
- 5xxx—magnesium
- 6xxx—magnesium and silicon
- 7xxx—zinc
- 8xxx—other elements

In the 2xxx through 8xxx alloy groups, the second digit in the alloy designation indicates alloy modifications. If the second digit is zero, it indicates the original alloy, while digits 1 through 9 indicate alloy modifications.

The last two of the four digits in the designation identify the different alloys in the group. [Figure 5-4]

Alloy	Percentage of alloying elements (aluminum and normal impurities constitute remainder)								
	Copper	Silicon	Manganese	Magnesium	Zinc	Nickel	Chromium	Lead	Bismuth
1100	—	—	—	—	—	—	—	—	—
3003	—	—	1.2	—	—	—	—	—	—
2011	5.5	—	—	—	—	—	—	0.5	0.5
2014	4.4	0.8	0.8	0.4	—	—	—	—	—
2017	4.0	—	0.5	0.5	—	—	—	—	—
2117	2.5	—	—	0.3	—	—	—	—	—
2018	4.0	—	—	0.5	—	2.0	—	—	—
2024	4.5	—	0.6	1.5	—	—	—	—	—
2025	4.5	0.8	0.8	—	—	—	—	—	—
4032	0.9	12.5	—	1.0	—	0.9	—	—	—
6151	—	1.0	—	0.6	—	—	0.25	—	—
5052	—	—	—	2.5	—	—	0.25	—	—
6053	—	0.7	—	1.3	—	—	0.25	—	—
6061	0.25	0.6	—	1.0	—	—	0.25	—	—
7075	1.6	—	—	2.5	5.6	—	0.3	—	—

Figure 5-4. Nominal composition of wrought aluminum alloys.

Effect of Alloying Element

1000 series. 99 percent aluminum or higher, excellent corrosion resistance, high thermal and electrical conductivity, low mechanical properties, excellent workability. Iron and silicon are major impurities.

2000 series. Copper is the principal alloying element. Solution heat treatment, optimum properties equal to mild steel, poor corrosion resistance unclad. It is usually clad with 6000 or high purity alloy. Its best known alloy is 2024.

3000 series. Manganese is the principal alloying element of this group which is generally non-heat treatable. The percentage of manganese which will be alloy effective is 1.5 percent. The most popular is 3003, which is of moderate strength and has good working characteristics.

4000 series. Silicon is the principal alloying element of this group, and lowers melting temperature. Its primary use is in welding and brazing. When used in welding heat-treatable alloys, this group will respond to a limited amount of heat treatment.

5000 series. Magnesium is the principal alloying element. It has good welding and corrosion resistant characteristics. High temperatures (over 150°F) or excessive cold working will increase susceptibility to corrosion.

6000 series. Silicon and magnesium form magnesium silicide which makes alloys heat treatable. It is of medium strength, good forming qualities, and has corrosion resistant characteristics.

7000 series. Zinc is the principal alloying element. The most popular alloy of the series is 6061. When coupled with magnesium, it results in heat-treatable alloys of very high strength. It usually has copper and chromium added. The principal alloy of this group is 7075.

Hardness Identification

Where used, the temper designation follows the alloy designation and is separated from it by a dash: i.e., 7075-T6, 2024-T4, and so forth. The temper designation consists of a letter indicating the basic temper which may be more specifically defined by the addition of one or more digits. These designations are as follows:

- F—as fabricated
- O—annealed, recrystallized (wrought products only)
- H—strain hardened
- H1 (plus one or more digits)—strain hardened only
- H2 (plus one or more digits)—strain hardened and partially annealed
- H3 (plus one or more digits)—strain hardened and stabilized

The digit following the designations H1, H2, and H3 indicates the degree of strain hardening, number 8 representing the ultimate tensile strength equal to that achieved by a cold reduction of approximately 75 percent following a full anneal, 0 representing the annealed state.

Magnesium and Magnesium Alloys

Magnesium, the world's lightest structural metal, is a silvery white material weighing only two-thirds as much as aluminum. Magnesium does not possess sufficient strength in its pure state for structural uses, but when alloyed with zinc, aluminum, and manganese it produces an alloy having the highest strength to weight ratio of any of the commonly used metals.

Magnesium is probably more widely distributed in nature than any other metal. It can be obtained from such ores as dolomite and magnesite, and from sea water, underground brines, and waste solutions of potash. With about 10 million pounds of magnesium in 1 cubic mile of sea water, there is no danger of a dwindling supply.

Some of today's aircraft require in excess of one-half ton of this metal for use in hundreds of vital spots. Some wing panels are fabricated entirely from magnesium alloys, weigh 18 percent less than standard aluminum panels, and have flown hundreds of satisfactory hours. Among the aircraft parts that have been made from magnesium with a substantial savings in weight are nosewheel doors, flap cover skin, aileron cover skin, oil tanks, floorings, fuselage parts, wingtips, engine nacelles, instrument panels, radio masts, hydraulic fluid tanks, oxygen bottle cases, ducts, and seats.

Magnesium alloys possess good casting characteristics. Their properties compare favorably with those of cast aluminum. In forging, hydraulic presses are ordinarily used, although, under certain conditions, forging can be accomplished in mechanical presses or with drop hammers.

Magnesium alloys are subject to such treatments as annealing, quenching, solution heat treatment, aging, and stabilizing. Sheet and plate magnesium are annealed at the rolling mill. The solution heat treatment is used to put as much of the alloying ingredients as possible into solid solution, which results in high tensile strength and maximum ductility. Aging is applied to castings following heat treatment where maximum hardness and yield strength are desired.

Magnesium embodies fire hazards of an unpredictable nature. When in large sections, its high thermal conductivity makes it difficult to ignite and prevents it from burning. It will not burn until the melting point of 1,204 °F is reached. However, magnesium dust and fine chips are ignited easily. Precautions must be taken to avoid this if possible. Should a fire occur, it can be extinguished with an extinguishing powder, such as soapstone or graphite. Water or any standard liquid or foam fire extinguisher cause magnesium to burn more rapidly and can cause explosions.

Magnesium alloys produced in the United States consist of magnesium alloyed with varying proportions of aluminum, manganese, and zinc. These alloys are designated by a letter of the alphabet, with the number 1 indicating high purity and maximum corrosion resistance.

Many of the magnesium alloys manufactured in the United States are produced by the Dow Chemical Company and have been given the trade name of Dowmetal™ alloys. To distinguish between these alloys, each is assigned a letter. Thus, we have Dowmetal J, Dowmetal M, and so forth.

Another manufacturer of magnesium alloys is the American Magnesium Corporation, a subsidiary of the Aluminum Company of America. This company uses an identification system similar to that used for aluminum alloys, with the exception that magnesium alloy numbers are preceded with the letters AM. Thus, AM240C is a cast alloy, and AM240C4 is the same alloy in the heat-treated state. AM3S0 is an annealed wrought alloy, and AM3SRT is the same alloy rolled after heat treatment.

Titanium and Titanium Alloys

Titanium was discovered by an English priest named Gregot. A crude separation of titanium ore was accomplished in 1825. In 1906 a sufficient amount of pure titanium was isolated in metallic form to permit a study. Following this study, in 1932, an extraction process was developed which became the first commercial method for producing titanium. The United States Bureau of Mines began making titanium sponge in 1946, and 4 years later the melting process began.

The use of titanium is widespread. It is used in many commercial enterprises and is in constant demand for such items as pumps, screens, and other tools and fixtures where corrosion attack is prevalent. In aircraft construction and repair, titanium is used for fuselage skins, engine shrouds, firewalls, longerons, frames, fittings, air ducts, and fasteners.

Titanium is used for making compressor disks, spacer rings, compressor blades and vanes, through bolts, turbine housings and liners, and miscellaneous hardware for turbine engines.

Titanium, in appearance, is similar to stainless steel. One quick method used to identify titanium is the spark test. Titanium gives off a brilliant white trace ending in a brilliant white burst. Also, identification can be accomplished by moistening the titanium and using it to draw a line on a piece of glass. This will leave a dark line similar in appearance to a pencil mark.

Titanium falls between aluminum and stainless steel in terms of elasticity, density, and elevated temperature strength. It has a melting point of from 2,730 °F to 3,155 °F, low thermal conductivity, and a low coefficient of expansion. It is light, strong, and resistant to stress corrosion cracking. Titanium is approximately 60 percent heavier than aluminum and about 50 percent lighter than stainless steel.

Because of the high melting point of titanium, high temperature properties are disappointing. The ultimate yield strength of titanium drops rapidly above 800 °F. The absorption of oxygen and nitrogen from the air at temperatures above 1,000 °F makes the metal so brittle on long exposure that it soon becomes worthless. However, titanium does have some merit for short time exposure up to 3,000 °F where strength is not important. Aircraft firewalls demand this requirement.

Titanium is nonmagnetic and has an electrical resistance comparable to that of stainless steel. Some of the base alloys of titanium are quite hard. Heat treating and alloying do not develop the hardness of titanium to the high levels of some of the heat-treated alloys of steel. It was only recently that a heat-treatable titanium alloy was developed. Prior to the development of this alloy, heating and rolling was the only method of forming that could be accomplished. However, it is possible to form the new alloy in the soft condition and heat treat it for hardness.

Iron, molybdenum, and chromium are used to stabilize titanium and produce alloys that will quench harden and age harden. The addition of these metals also adds ductility. The fatigue resistance of titanium is greater than that of aluminum or steel.

Titanium becomes softer as the degree of purity is increased. It is not practical to distinguish between the various grades of commercially pure or unalloyed titanium by chemical analysis; therefore, the grades are determined by mechanical properties.

Titanium Designations

The A-B-C classification of titanium alloys was established to provide a convenient and simple means of describing all titanium alloys. Titanium and titanium alloys possess three basic types of crystals: A (alpha), B (beta), and C (combined alpha and beta). Their characteristics are:

- A (alpha)—all around performance; good weldability; tough and strong both cold and hot, and resistant to oxidation.
- B (beta)—bendability; excellent bend ductility; strong both cold and hot, but vulnerable to contamination.
- C (combined alpha and beta for compromise performances)—strong when cold and warm, but weak when hot; good bendability; moderate contamination resistance; excellent forgeability.

Titanium is manufactured for commercial use in two basic compositions: commercially pure titanium and alloyed titanium. A-55 is an example of a commercially pure titanium. It has a yield strength of 55,000 to 80,000 psi and is a general purpose grade for moderate to severe forming. It is sometimes used for nonstructural aircraft parts and for all types of corrosion resistant applications, such as tubing. Type A-70 titanium is closely related to type A-55 but has a yield strength of 70,000 to 95,000 psi. It is used where higher strength is required, and it is specified for many moderately stressed aircraft parts. For many corrosion applications, it is used interchangeably with type A-55. Both type A-55 and type A-70 are weldable.

One of the widely used titanium base alloys is designated as C-110M. It is used for primary structural members and aircraft skin, has 110,000 psi minimum yield strength, and contains 8 percent manganese.

Type A-110AT is a titanium alloy which contains 5 percent aluminum and 2.5 percent tin. It also has a high minimum yield strength at elevated temperatures with the excellent welding characteristics inherent in alpha-type titanium alloys.

Corrosion Characteristics

The corrosion resistance of titanium deserves special mention. The resistance of the metal to corrosion is caused by the formation of a protective surface film of stable oxide or chemi-absorbed oxygen. Film is often produced by the presence of oxygen and oxidizing agents.

Corrosion of titanium is uniform. There is little evidence of pitting or other serious forms of localized attack. Normally, it is not subject to stress corrosion, corrosion fatigue, intergranular corrosion, or galvanic corrosion. Its corrosion resistance is equal or superior to 18-8 stainless steel.

Laboratory tests with acid and saline solutions show titanium polarizes readily. The net effect, in general, is to decrease current flow in galvanic and corrosion cells. Corrosion currents on the surface of titanium and metallic couples are naturally restricted. This partly accounts for good resistance to many chemicals; also, the material may be used with some dissimilar metals with no harmful galvanic effect on either.

Copper and Copper Alloys

Copper is one of the most widely distributed metals. It is the only reddish colored metal and is second only to silver in electrical conductivity. Its use as a structural material is limited because of its great weight. However, some of its outstanding characteristics, such as its high electrical and heat conductivity, in many cases overbalance the weight factor.

Because it is very malleable and ductile, copper is ideal for making wire. It is corroded by salt water but is not affected by fresh water. The ultimate tensile strength of copper varies greatly. For cast copper, the tensile strength is about 25,000 psi, and when cold rolled or cold drawn its tensile strength increases to a range of 40,000 to 67,000 psi.

In aircraft, copper is used primarily in the electrical system for bus bars, bonding, and as lockwire.

Beryllium copper is one of the most successful of all the copper base alloys. It is a recently developed alloy containing about 97 percent copper, 2 percent beryllium, and sufficient nickel to increase the percentage of elongation. The most valuable feature of this metal is that the physical properties can be greatly stepped up by heat treatment, the tensile strength rising from 70,000 psi in the annealed state to 200,000 psi in the heat-treated state. The resistance of beryllium copper to fatigue and wear makes it suitable for diaphragms, precision bearings and bushings, ball cages, and spring washers.

Brass is a copper alloy containing zinc and small amounts of aluminum, iron, lead, manganese, magnesium, nickel, phosphorous, and tin. Brass with a zinc content of 30 to 35 percent is very ductile, but that containing 45 percent has relatively high strength.

Muntz metal is a brass composed of 60 percent copper and 40 percent zinc. It has excellent corrosion resistant qualities in salt water. Its strength can be increased by heat treatment. As cast, this metal has an ultimate tensile strength of 50,000 psi, and it can be elongated 18 percent. It is used in making bolts and nuts, as well as parts that come in contact with salt water.

Red brass, sometimes termed "bronze" because of its tin content, is used in fuel and oil line fittings. This metal has good casting and finishing properties and machines freely.

Bronzes are copper alloys containing tin. The true bronzes have up to 25 percent tin, but those with less than 11 percent are most useful, especially for such items as tube fittings in aircraft.

Among the copper alloys are the copper aluminum alloys, of which the aluminum bronzes rank very high in aircraft usage. They would find greater usefulness in structures if it were not for their strength to weight ratio as compared with alloy steels. Wrought aluminum bronzes are almost as strong and ductile as medium carbon steel, and they possess a high degree of resistance to corrosion by air, salt water, and chemicals. They are readily forged, hot or cold rolled, and many react to heat treatment.

These copper base alloys contain up to 16 percent of aluminum (usually 5 to 11 percent), to which other metals, such as iron, nickel, or manganese, may be added. Aluminum bronzes have good tearing qualities, great strength, hardness, and resistance to both shock and fatigue. Because of these properties, they are used for diaphragms, gears, and pumps. Aluminum bronzes are available in rods, bars, plates, sheets, strips, and forgings.

Cast aluminum bronzes, using about 89 percent copper, 9 percent aluminum, and 2 percent of other elements, have high strength combined with ductility, and are resistant to corrosion, shock, and fatigue. Because of these properties, cast aluminum bronze is used in bearings and pump parts. These alloys are useful in areas exposed to salt water and corrosive gases.

Manganese bronze is an exceptionally high strength, tough, corrosion resistant copper zinc alloy containing aluminum, manganese, iron and, occasionally, nickel or tin. This metal can be formed, extruded, drawn, or rolled to any desired shape. In rod form, it is generally used for machined parts, for aircraft landing gears and brackets.

Silicon bronze is a more recent development composed of about 95 percent copper, 3 percent silicon, and 2 percent manganese, zinc, iron, tin, and aluminum. Although not a bronze in the true sense because of its small tin content, silicon bronze has high strength and great corrosion resistance.

Monel

Monel, the leading high nickel alloy, combines the properties of high strength and excellent corrosion resistance. This metal consists of 68 percent nickel, 29 percent copper, 0.2 percent iron, 1 percent manganese, and 1.8 percent of other elements. It cannot be hardened by heat treatment.

Monel, adaptable to casting and hot or cold working, can be successfully welded. It has working properties similar to those of steel. When forged and annealed, it has a tensile strength of 80,000 psi. This can be increased by cold working to 125,000 psi, sufficient for classification among the tough alloys.

Monel has been successfully used for gears and chains to operate retractable landing gears, and for structural parts subject to corrosion. In aircraft, Monel is used for parts demanding both strength and high resistance to corrosion, such as exhaust manifolds and carburetor needle valves and sleeves.

K-Monel

K-Monel is a nonferrous alloy containing mainly nickel, copper, and aluminum. It is produced by adding a small amount of aluminum to the Monel formula. It is corrosion resistant and capable of being hardened by heat treatment.

K-Monel has been successfully used for gears, and structural members in aircraft which are subjected to corrosive attacks. This alloy is nonmagnetic at all temperatures. K-Monel sheet has been successfully welded by both oxyacetylene and electric arc welding.

Nickel and Nickel Alloys

There are basically two nickel alloys used in aircraft. They are Monel and Inconel. Monel contains about 68 percent nickel and 29 percent copper, plus small amounts of iron and manganese. Nickel alloys can be welded or easily machined. Some of the nickel Monel, especially the nickel Monels containing small amounts of aluminum, are heat-treatable to similar tensile strengths of steel. Nickel Monel is used in gears and parts that require high strength and toughness, such as exhaust systems that require high strength and corrosion resistance at elevated temperatures.

Inconel alloys of nickel produce a high strength, high temperature alloy containing approximately 80 percent nickel, 14 percent chromium, and small amounts of iron and other elements. The nickel Inconel alloys are frequently used in turbine engines because of their ability to maintain their strength and corrosion resistance under extremely high temperature conditions.

Inconel and stainless steel are similar in appearance and are frequently found in the same areas of the engine. Sometimes it is important to identify the difference between the metal samples. A common test is to apply one drop of cupric chloride and hydrochloric acid solution to the unknown metal and allow it to remain for 2 minutes. At the end of the soak period, a shiny spot indicates the material is nickel Inconel, and a copper colored spot indicates stainless steel.

Substitution of Aircraft Metals

In selecting substitute metals for the repair and maintenance of aircraft, it is very important to check the appropriate structural repair manual. Aircraft manufacturers design structural members to meet a specific load requirement for a particular aircraft. The methods of repairing these members, apparently similar in construction, will thus vary with different aircraft.

Four requirements must be kept in mind when selecting substitute metals. The first and most important of these is maintaining the original strength of the structure. The other three are: (1) maintaining contour or aerodynamic smoothness, (2) maintaining original weight, if possible, or keeping added weight to a minimum, and (3) maintaining the original corrosion resistant properties of the metal.

Metalworking Processes

There are three methods of metalworking: (1) hot working, (2) cold working, and (3) extruding. The method used will depend on the metal involved and the part required, although in some instances both hot and cold working methods may be used to make a single part.

Hot Working

Almost all steel is hot worked from the ingot into some form from which it is either hot or cold worked to the finished shape. When an ingot is stripped from its mold, its surface is solid, but the interior is still molten. The ingot is then placed in a soaking pit which retards loss of heat, and the molten interior gradually solidifies. After soaking, the temperature is equalized throughout the ingot, then it is reduced to intermediate size by rolling, making it more readily handled.

The rolled shape is called a bloom when its section dimensions are 6 inches × 6 inches or larger and approximately square. The section is called a billet when it is approximately square and less than 6 inches × 6 inches. Rectangular sections which have a width greater than twice their thickness are called slabs. The slab is the intermediate shape from which sheets are rolled.

Blooms, billets, or slabs are heated above the critical range and rolled into a variety of shapes of uniform cross section. Common rolled shapes are sheet, bar, channel, angle, and I-beam. As discussed later in this chapter, hot rolled material is frequently finished by cold rolling or drawing to obtain accurate finish dimensions and a bright, smooth surface.

Complicated sections which cannot be rolled, or sections of which only a small quantity is required, are usually forged. Forging of steel is a mechanical working at temperatures above the critical range to shape the metal as desired. Forging is done either by pressing or hammering the heated steel until the desired shape is obtained.

Pressing is used when the parts to be forged are large and heavy; this process also replaces hammering where high grade steel is required. Since a press is slow acting, its force is uniformly transmitted to the center of the section, thus affecting the interior grain structure as well as the exterior to give the best possible structure throughout.

Hammering can be used only on relatively small pieces. Since hammering transmits its force almost instantly, its effect is limited to a small depth. Thus, it is necessary to use a very heavy hammer or to subject the part to repeated blows to ensure complete working of the section. If the force applied is too weak to reach the center, the finished forged surface will be concave. If the center was properly worked, the surface will be convex or bulged. The advantage of hammering is that the operator has control over both the amount of pressure applied and the finishing temperature, and is able to produce small parts of the highest grade. This type of forging is usually referred to as smith forging. It is used extensively where only a small number of parts are needed. Considerable machining time and material are saved when a part is smith forged to approximately the finished shape.

Steel is often harder than necessary and too brittle for most practical uses when put under severe internal strain. To relieve such strain and reduce brittleness, it is tempered after being hardened. This consists of heating the steel in a furnace to a specified temperature and then cooling it in air, oil, water, or a special solution. Temper condition refers to the condition of metal or metal alloys with respect to hardness or toughness. Rolling, hammering, or bending these alloys, or heat treating and aging them, causes them to become tougher and harder. At times these alloys become too hard for forming and have to be re-heat treated or annealed.

Metals are annealed to relieve internal stresses, soften the metal, make it more ductile, and refine the grain structure. Annealing consists of heating the metal to a prescribed temperature, holding it there for a specified length of time, and then cooling the metal back to room temperature. To produce maximum softness, the metal must be cooled very slowly. Some metals must be furnace cooled; others may be cooled in air.

Normalizing applies to iron base metals only. Normalizing consists of heating the part to the proper temperature, holding it at that temperature until it is uniformly heated, and then cooling it in still air. Normalizing is used to relieve stresses in metals.

Strength, weight, and reliability are three factors which determine the requirements to be met by any material used in airframe construction and repair. Airframes must be strong and yet as light weight as possible. There are very definite limits to which increases in strength can be accompanied by increases in weight. An airframe so heavy that it could not support a few hundred pounds of additional weight would be of little use.

All metals, in addition to having a good strength/weight ratio, must be thoroughly reliable, thus minimizing the possibility of dangerous and unexpected failures. In addition to these general properties, the material selected for a definite application must possess specific qualities suitable for the purpose.

The material must possess the strength required by the dimensions, weight, and use. The five basic stresses which metals may be required to withstand are tension, compression, shear, bending, and torsion.

The tensile strength of a material is its resistance to a force which tends to pull it apart. Tensile strength is measured in pounds per square inch (psi) and is calculated by dividing the load in pounds required to pull the material apart by its cross-sectional area in square inches.

The compression strength of a material is its resistance to a crushing force which is the opposite of tensile strength. Compression strength is also measured in psi. When a piece of metal is cut, the material is subjected, as it comes in contact with the cutting edge, to a force known as shear. Shear is the tendency on the part of parallel members to slide in opposite directions. It is like placing a cord or thread between the blades of a pair of scissors (shears). The shear strength is the shear force in psi at which a material fails. It is the load divided by the shear area.

Bending can be described as the deflection or curving of a member due to forces acting upon it. The bending strength of material is the resistance it offers to deflecting forces. Torsion is a twisting force. Such action would occur in a member fixed at one end and twisted at the other. The torsional strength of material is its resistance to twisting.

The relationship between the strength of a material and its weight per cubic inch, expressed as a ratio, is known as the strength/weight ratio. This ratio forms the basis for comparing the desirability of various materials for use in airframe construction and repair. Neither strength nor weight alone can be used as a means of true comparison. In some applications, such as the skin of monocoque structures, thickness is more important than strength, and, in this instance, the material with the lightest weight for a given thickness or gauge is best. Thickness or bulk is necessary to prevent bucking or damage caused by careless handling.

Corrosion is the eating away or pitting of the surface or the internal structure of metals. Because of the thin sections and the safety factors used in aircraft design and construction, it would be dangerous to select a material possessing poor corrosion resistant characteristics.

Another significant factor to consider in maintenance and repair is the ability of a material to be formed, bent, or machined to required shapes. The hardening of metals by cold working or forming is termed work hardening. If a piece of metal is formed (shaped or bent) while cold, it is said to be cold worked. Practically all the work an aviation mechanic does on metal is cold work. While this is convenient, it causes the metal to become harder and more brittle.

If the metal is cold worked too much, that is, if it is bent back and forth or hammered at the same place too often, it will crack or break. Usually, the more malleable and ductile a metal is, the more cold working it can stand. Any process which involves controlled heating and cooling of metals to develop certain desirable characteristics (such as hardness, softness, ductility, tensile strength, or refined grain structure) is called heat treatment or heat treating. With steels the term "heat treating" has a broad meaning and includes such processes as annealing, normalizing, hardening, and tempering.

In the heat treatment of aluminum alloys, only two processes are included: (1) the hardening and toughening process, and (2) the softening process. The hardening and toughening process is called heat treating, and the softening process is called annealing. Aircraft metals are subjected to both shock and fatigue (vibrational) stresses. Fatigue occurs in materials which are exposed to frequent reversals of loading or repeatedly applied loads, if the fatigue limit is reached or exceeded. Repeated vibration or bending will ultimately cause a minute crack to occur at the weakest point. As vibration or bending continues, the crack lengthens until the part completely fails. This is termed shock and fatigue failure. Resistance to this condition is known as shock and fatigue resistance. It is essential that materials used for critical parts be resistant to these stresses.

Heat treatment is a series of operations involving the heating and cooling of metals in the solid state. Its purpose is to change a mechanical property or combination of mechanical properties so that the metal will be more useful, serviceable, and safe for a definite purpose. By heat treating, a metal can be made harder, stronger, and more resistant to impact. Heat treating can also make a metal softer and more ductile. No one heat treating operation can produce all of these characteristics. In fact, some properties are often improved at the expense of others. In being hardened, for example, a metal may become brittle.

The various heat-treating processes are similar in that they all involve the heating and cooling of metals. They differ, however, in the temperatures to which the metal is heated, the rate at which it is cooled, and, of course, in the final result.

The most common forms of heat treatment for ferrous metals are hardening, tempering, normalizing, annealing, and casehardening. Most nonferrous metals can be annealed and many of them can be hardened by heat treatment. However, there is only one nonferrous metal, titanium, that can be casehardened, and none can be tempered or normalized.

Internal Structure of Metals

The results obtained by heat treatment depend to a great extent on the structure of the metal and on the manner in which the structure changes when the metal is heated and cooled. A pure metal cannot be hardened by heat treatment because there is little change in its structure when heated. On the other hand, most alloys respond to heat treatment since their structures change with heating and cooling.

An alloy may be in the form of a solid solution, a mechanical mixture, or a combination of a solid solution and a mechanical mixture. When an alloy is in the form of a solid solution, the elements and compounds which form the alloy are absorbed, one into the other, in much the same way that salt is dissolved in a glass of water, and the constituents cannot be identified even under a microscope.

When two or more elements or compounds are mixed but can be identified by microscopic examination, a mechanical mixture is formed. A mechanical mixture can be compared to the mixture of sand and gravel in concrete. The sand and gravel are both visible. Just as the sand and gravel are held together and kept in place by the matrix of cement, the other constituents of an alloy are embedded in the matrix formed by the base metal.

An alloy in the form of a mechanical mixture at ordinary temperatures may change to a solid solution when heated. When cooled back to normal temperature, the alloy may return to its original structure. On the other hand, it may remain a solid solution or form a combination of a solid solution and mechanical mixture. An alloy which consists of a combination of solid solution and mechanical mixture at normal temperatures may change to a solid solution when heated. When cooled, the alloy may remain a solid solution, return to its original structure, or form a complex solution.

Heat-Treating Equipment

Successful heat treating requires close control over all factors affecting the heating and cooling of metals. Such control is possible only when the proper equipment is available and the equipment is selected to fit the particular job. Thus, the furnace must be of the proper size and type and must be so controlled that temperatures are kept within the limits prescribed for each operation. Even the atmosphere within the furnace affects the condition of the part being heat treated. Further, the quenching equipment and the quenching medium must be selected to fit the metal and the heat-treating operation. Finally, there must be equipment for handling parts and materials, for cleaning metals, and for straightening parts.

Furnaces and Salt Baths

There are many different types and sizes of furnaces used in heat treatment. As a general rule, furnaces are designed to operate in certain specific temperature ranges and attempted use in other ranges frequently results in work of inferior quality.

In addition, using a furnace beyond its rated maximum temperature shortens its life and may necessitate costly and time consuming repairs.

Fuel fired furnaces (gas or oil) require air for proper combustion and an air compressor or blower is therefore necessary. These furnaces are usually of the muffler type; that is, the combustion of the fuel takes place outside of and around the chamber in which the work is placed. If an open muffler is used, the furnace should be designed to prevent the direct impingement of flame on the work.

In furnaces heated by electricity, the heating elements are generally in the form of wire or ribbon. Good design requires incorporation of additional heating elements at locations where maximum heat loss may be expected. Such furnaces commonly operate at up to a maximum temperature of about 2,000 °F. Furnaces operating at temperatures up to about 2,500 °F usually employ resistor bars of sintered carbides.

Temperature Measurement and Control

Temperature in the heat-treating furnace is measured by a thermoelectric instrument known as a pyrometer. This instrument measures the electrical effect of a thermocouple and, hence, the temperature of the metal being treated. A complete pyrometer consists of three parts—a thermocouple, extension leads, and meter.

Furnaces intended primarily for tempering may be heated by gas or electricity and are frequently equipped with a fan for circulating the hot air.

Salt baths are available for operating at either tempering or hardening temperatures. Depending on the composition of the salt bath, heating can be conducted at temperatures as low as 325 °F to as high as 2,450 °F. Lead baths can be used in the temperature range of 650 °F to 1,700 °F. The rate of heating in lead or salt baths is much faster in furnaces.

Heat-treating furnaces differ in size, shape, capacity, construction, operation, and control. They may be circular or rectangular and may rest on pedestals or directly on the floor. There are also pit-type furnaces, which are below the surface of the floor. When metal is to be heated in a bath of molten salt or lead, the furnace must contain a pot or crucible for the molten bath.

The size and capacity of a heat-treating furnace depends on the intended use. A furnace must be capable of heating rapidly and uniformly, regardless of the desired maximum temperature or the mass of the charge. An oven-type furnace should have a working space (hearth) about twice as long and three times as wide as any part that will be heated in the furnace.

Accurate temperature measurement is essential to good heat treating. The usual method is by means of thermocouples: the most common base metal couples are copper-constantan (up to about 700 °F), iron-constantan (up to about 1,400 °F), and chromel-alumel (up to about 2,200 °F). The most common noble metal couples (which can be used up to about 2,800 °F) are platinum coupled with either the alloy 87 percent platinum (13 percent rhodium) or the alloy 90 percent platinum (10 percent rhodium). The temperatures quoted are for continuous operation.

The life of thermocouples is affected by the maximum temperature (which may frequently exceed those given above) and by the furnace atmosphere. Iron-constantan is more suited for use in reducing and chromel-alumel in oxidizing atmospheres. Thermocouples are usually encased in metallic or ceramic tubes closed at the hot end to protect them from the furnace gases. A necessary attachment is an instrument, such as a millivoltmeter or potentiometer, for measuring the electromotive force generated by the thermocouple. In the interest of accurate control, place the hot junction of the thermocouple as close to the work as possible. The use of an automatic controller is valuable in controlling the temperature at the desired value.

Pyrometers may have meters either of the indicating type or recording type. Indicating pyrometers give direct reading of the furnace temperature. The recording type produces a permanent record of the temperature range throughout the heating operation by means of an inked stylus attached to an arm which traces a line on a sheet of calibrated paper or temperature chart.

Pyrometer installations on all modern furnaces provide automatic regulation of the temperature at any desired setting. Instruments of this type are called controlling potentiometer pyrometers. They include a current regulator and an operating mechanism, such as a relay.

Heating

The object in heating is to transform pearlite (a mixture of alternate strips of ferrite and iron carbide in a single grain) to austenite as the steel is heated through the critical range. Since this transition takes time, a relatively slow rate of heating must be used. Ordinarily, the cold steel is inserted when the temperature in the furnace is from 300 °F to 500 °F below the hardening temperature. In this way, too rapid heating through the critical range is prevented.

If temperature measuring equipment is not available, it becomes necessary to estimate temperatures by some other means. An inexpensive, yet fairly accurate method involves the use of commercial crayons, pellets, or paints that melt at various temperatures within the range of 125 °F to 1,600 °F. The least accurate method of temperature estimation is by observation of the color of the hot hearth of the furnace or of the work. The heat colors observed are affected by many factors, such as the conditions of artificial or natural light, the character of the scale on the work, and so forth. Steel begins to appear dull red at about 1,000 °F, and as the temperature increases, the color changes gradually through various shades of red to orange, to yellow, and finally to white. A rough approximation of the correspondence between color and temperature is indicated in Figure 5-5.

It is also possible to secure some idea of the temperature of a piece of carbon or low alloy steel, in the low temperature range used for tempering, from the color of the thin oxide film that forms on the cleaned surface of the steel when heated in this range. The approximate temperature/color relationship is indicated on the lower portion of the scale in Figure 5-5.

It is often necessary or desirable to protect steel or cast iron from surface oxidation (scaling) and loss of carbon from the surface layers (decarburization). Commercial furnaces, therefore, are generally equipped with some means of atmosphere control. This usually is in the form of a burner for burning controlled amounts of gas and air and directing the products of combustion into the furnace muffle. Water vapor, a product of this combustion, is detrimental and many furnaces are equipped with a means for eliminating it. For furnaces not equipped with atmosphere control, a variety of external atmosphere generators are available. The gas so generated is piped into the furnace and one generator may supply several furnaces. If no method of atmosphere control is available, some degree of protection may be secured by covering the work with cast iron borings or chips.

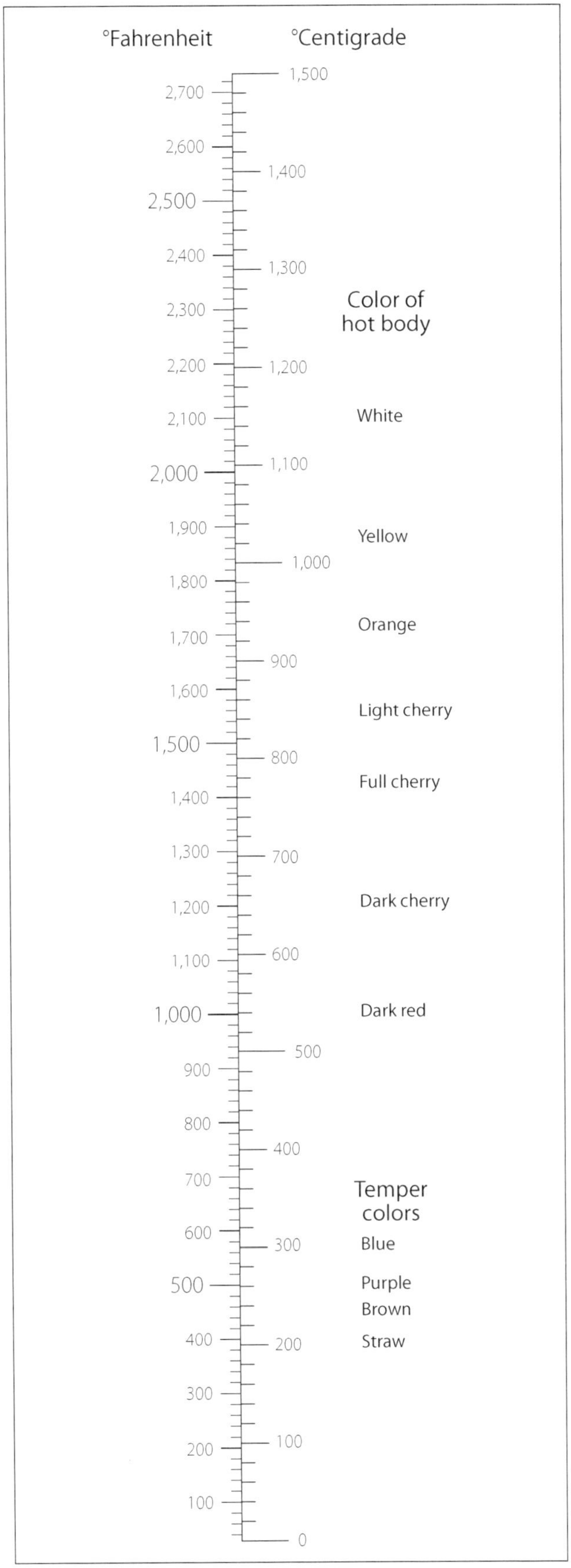

Figure 5-5. Temperature chart indicating conversion of Centigrade to Fahrenheit or visa versa, color temperature scale for hardening temperature range, and tempering temperature range.

Since the work in salt or lead baths is surrounded by the liquid heating medium, the problem of preventing scaling or decarburization is simplified.

Vacuum furnaces also are used for annealing steels, especially when a bright nonoxidized surface is a prime consideration.

Soaking

The temperature of the furnace must be held constant during the soaking period, since it is during this period that rearrangement of the internal structure of the steel takes place. Soaking temperatures for various types of steel are specified in ranges varying as much as 100 °F. [Figure 5-6] Small parts are soaked in the lower part of the specified range and heavy parts in the upper part of the specified range. The length of the soaking period depends upon the type of steel and the size of the part. Naturally, heavier parts require longer soaking to ensure equal heating throughout. As a general rule, a soaking period of 30 minutes to 1 hour is sufficient for the average heat-treating operation.

Cooling

The rate of cooling through the critical range determines the form that the steel will retain. Various rates of cooling are used to produce the desired results. Still air is a slow cooling medium, but is much faster than furnace cooling. Liquids are the fastest cooling media and are therefore used in hardening steels.

There are three commonly used quenching liquids—brine, water, and oil. Brine is the strongest quenching medium, water is next, and oil is the least. Generally, an oil quench is used for alloy steels, and brine or water for carbon steels.

Quenching Media

Quenching solutions act only through their ability to cool the steel. They have no beneficial chemical action on the quenched steel and in themselves impart no unusual properties. Most requirements for quenching media are met satisfactorily by water or aqueous solutions of inorganic salts, such as table salt or caustic soda, or by some type of oil. The rate of cooling is relatively rapid during quenching in brine, somewhat less rapid in water, and slow in oil.

Brine usually is made of a 5 to 10 percent solution of salt (sodium chloride) in water. In addition to its greater cooling speed, brine has the ability to "throw" the scale from steel during quenching. The cooling ability of both water and brine, particularly water, is considerably affected by their temperature. Both

Steel No.	Temperatures			Quenching medium (n)	Tempering (drawing) temperature for tensile strength (psi)				
	Normalizing air cool (°F)	Annealing (°F)	Hardening (°F)		100,000 (°F)	125,000 (°F)	150,000 (°F)	180,000 (°F)	200,000 (°F)
1020	1,650–1,750	1,600–1,700	1,575–1,675	Water	—	—	—	—	—
1022 (x1020)	1,650–1,750	1,600–1,700	1,575–1,675	Water	—	—	—	—	—
1025	1,600–1,700	1,575–1,650	1,575–1,675	Water	(a)	—	—	—	—
1035	1,575–1,650	1,575–1,625	1,525–1,600	Water	875	—	—	—	—
1045	1,550–1,600	1,550–1,600	1,475–1,550	Oil or water	1,150	—	—	(n)	—
1095	1,475–1,550	1,450–1,500	1,425–1,500	Oil	(b)	—	1,100	850	750
2330	1,475–1,525	1,425–1,475	1,450–1,500	Oil or water	1,100	950	800	—	—
3135	1,600–1,650	1,500–1,550	1,475–1,525	Oil	1,250	1,050	900	750	650
3140	1,600–1,650	1,500–1,550	1,475–1,525	Oil	1,325	1,075	925	775	700
4037	1,600	1,525–1,575	1,525–1,575	Oil or water	1,225	1,100	975	—	—
4130 (x4130)	1,600–1,700	1,525–1,575	1,525–1,625	Oil (c)	(d)	1,050	900	700	575
4140	1,600–1,650	1,525–1,575	1,525–1,575	Oil	1,350	1,100	1,025	825	675
4150	1,550–1,600	1,475–1,525	1,550–1,550	Oil	—	1,275	1,175	1,050	950
4340 (x4340)	1,550–1,625	1,525–1,575	1,475–1,550	Oil	—	1,200	1,050	950	850
4640	1,675–1,700	1,525–1,575	1,500–1,550	Oil	—	1,200	1,050	750	625
6135	1,600–1,700	1,550–1,600	1,575–1,625	Oil	1,300	1,075	925	800	750
6150	1,600–1,650	1,525–1,575	1,550–1,625	Oil	(d)(e)	1,200	1,000	900	800
6195	1,600–1,650	1,525–1,575	1,500–1,550	Oil	(f)	—	—	—	—
NE8620	—	—	1,525–1,575	Oil	—	1,000	—	—	—
NE8630	1,650	1,525–1,575	1,525–1,575	Oil	—	1,125	975	775	675
NE8735	1,650	1,525–1,575	1,525–1,575	Oil	—	1,175	1,025	875	775
NE8740	1,625	1,500–1,550	1,500–1,550	Oil	—	1,200	1,075	925	850
30905	—	(g)(h)	(i)	—	—	—	—	—	—
51210	1,525–1,575	1,525–1,575	1,775–1,825 (j)	Oil	1,200	1,100	(k)	750	—
51335	—	1,525–1,575	1,775–1,850	Oil	—	—	—	—	—
52100	1,625–1,700	1,400–1,450	1,525–1,550	Oil	(f)	—	—	—	—
Corrosion resisting (16-2)(1)	—	—	—	—	(m)	—	—	—	—
Silicon Chromium (for springs)	—	—	1,700–1,725	Oil	—	—	—	—	—

Figure 5-6. Heat treatment procedures for steels.

Notes:

(a) Draw at 1,150°F for tensile strength of 70,000 psi.

(b) For spring temper draw at 800–900 °F. Rockwell hardness C-40–45.

(c) Bars or forgings may be quenched in water from 1,500–1,600 °F.

(d) Air cooling from the normalizing temperature will produce a tensile strength of approximately 90,000 psi.

(e) For spring temper draw at 850–950 °F. Rockwell hardness C-40–45.

(f) Draw at 350–450 °F to remove quenching strains. Rockwell hardness C-60–65.

(g) Anneal at 1,600–1,700 °F to remove residual stresses due to welding or cold work. May be applied only to steel containing titanium or columbium.

(h) Anneal at 1,900–2,100 °F to produce maximum softness and corrosion resistance. Cool in air or quench in water.

(i) Harden by cold work only.

(j) Lower side of range for sheet 0.06 inch and under. Middle of range for sheet and wire 0.125 inch. Upper side of range for forgings.

(k) Not recommended for intermediate tensile strengths because of low impact.

(l) AN-QQ-S-770—It is recommended that, prior to tempering, corrosion-resisting (16 Cr-2 Ni) steel be quenched in oil from a temperature of 1,875–1,900 °F, after a soaking period of 30 minutes at this temperature. To obtain a tensile strength at 115,000 psi, the tempering temperature should be approximately 525 °F. A holding time at these temperatures of about 2 hours is recommended. Tempering temperatures between 700 °F and 1,100 °F will not be approved.

(m) Draw at approximately 800 °F and cool in air for Rockwell hardness of C-50.

(n) Water used for quenching shall be within the temperature range of 80–150 °F.

Figure 5-6. Heat-treatment procedures for steels. (continued)

should be kept cold—well below 60 °F. If the volume of steel being quenched tends to raise the temperature of the bath appreciably, add ice or use some means of refrigeration to cool the quenching bath.

There are many specially prepared quenching oils on the market; their cooling rates do not vary widely. A straight mineral oil with a Saybolt viscosity of about 100 at 100 °F is generally used. Unlike brine and water, the oils have the greatest cooling velocity at a slightly elevated temperature—about 100–140 °F—because of their decreased viscosity at these temperatures.

When steel is quenched, the liquid in immediate contact with the hot surface vaporizes; this vapor reduces the rate of heat abstraction markedly. Vigorous agitation of the steel or the use of a pressure spray quench is necessary to dislodge these vapor films and thus permit the desired rate of cooling.

The tendency of steel to warp and crack during the quenching process is difficult to overcome because certain parts of the article cool more rapidly than others. The following recommendations will greatly reduce the warping tendency.

1. Never throw a part into the quenching bath. By permitting it to lie on the bottom of the bath, it is apt to cool faster on the top side than on the bottom side, thus causing it to warp or crack.
2. Agitate the part slightly to destroy the coating of vapor that could prevent it from cooling evenly and rapidly. This allows the bath to dissipate its heat to the atmosphere.
3. Immerse irregular shaped parts so that the heavy end enters the bath first.

Quenching Equipment

The quenching tank should be of the proper size to handle the material being quenched. Use circulating pumps and coolers to maintain approximately constant temperatures when doing a large amount of quenching. To avoid building up a high concentration of salt in the quenching tank, make provisions for adding fresh water to the quench tank used for molten salt baths.

Tank location in reference to the heat-treating furnace is very important. Situate the tank to permit rapid transfer of the part from the furnace to the quenching medium. A delay of more than a few seconds will, in many instances, prove detrimental to the effectiveness of the heat treatment. When heat treating material of thin section, employ guard sheets to retard the loss of heat during transfer to the quench tank. Provide a rinse tank to remove all salt from the material after quenching if the salt is not adequately removed in the quenching tank.

Heat Treatment of Ferrous Metals

The first important consideration in the heat treatment of a steel part is to know its chemical composition. This, in turn, determines its upper critical point. When the upper critical point is known, the next consideration is the rate of heating and cooling to be used. Carrying out these operations involves the use of uniform heating furnaces, proper temperature controls, and suitable quenching mediums.

Behavior of Steel During Heating and Cooling

Changing the internal structure of a ferrous metal is accomplished by heating to a temperature above its upper critical point, holding it at that temperature for a time sufficient to permit certain internal changes to occur, and then cooling to atmospheric temperature under predetermined, controlled conditions.

At ordinary temperatures, the carbon in steel exists in the form of particles of iron carbide scattered throughout an iron matrix known as "ferrite." The number, size, and distribution of these particles determine the hardness of the steel. At elevated temperatures, the carbon is dissolved in the iron matrix in the form of a solid solution called "austenite," and the carbide particles appear only after the steel has been cooled. If the cooling is slow, the carbide particles are relatively coarse and few. In this condition, the steel is soft. If the cooling is rapid, as by quenching in oil or water, the carbon precipitates as a cloud of very fine carbide particles, and the steel is hard. The fact that the carbide particles can be dissolved in austenite is the basis of the heat treatment of steel. The temperatures at which this transformation takes place are called the critical points and vary with the composition of the steel. The percentage of carbon in the steel has the greatest influence on the critical points of heat treatment.

Hardening

Pure iron, wrought iron, and extremely low carbon steels cannot be appreciably hardened by heat treatment, since they contain no hardening element. Cast iron can be hardened, but its heat treatment is limited. When cast iron is cooled rapidly, it forms white iron, which is hard and brittle. When cooled slowly, it forms gray iron, which is soft but brittle under impact.

In plain carbon steel, the maximum hardness depends almost entirely on the carbon content of the steel. As the carbon content increases, the ability of the steel to be hardened increases. However, this increase in the ability to harden with an increase in carbon content continues only to a certain point. In practice, that point is 0.85 percent carbon content. When the carbon content is increased beyond 0.85 percent, there is no increase in wear resistance.

For most steels, the hardening treatment consists of heating the steel to a temperature just above the upper critical point, soaking or holding for the required length of time, and then cooling it rapidly by plunging the hot steel into oil, water, or brine. Although most steels must be cooled rapidly for hardening, a few may be cooled in still air. Hardening increases the hardness and strength of the steel but makes it less ductile.

When hardening carbon steel, it must be cooled to below 1,000 °F in less than 1 second. Should the time required for the temperature to drop to 1,000 °F exceed 1 second, the austenite begins to transform into fine pearlite. This pearlite varies in hardness, but is much harder than the pearlite formed by annealing and much softer than the martensite desired. After the 1,000 °F temperature is reached, the rapid cooling must continue if the final structure is to be all martensite.

When alloys are added to steel, the time limit for the temperature drop to 1,000 °F increases above the 1 second limit for carbon steels. Therefore, a slower quenching medium will produce hardness in alloy steels.

Because of the high internal stresses in the "as quenched" condition, steel must be tempered just before it becomes cold. The part should be removed from the quenching bath at a temperature of approximately 200 °F, since the temperature range from 200 °F down to room temperature is the cracking range.

Hardening temperatures and quenching mediums for the various types of steel are listed in Figure 5-6.

Hardening Precautions

A variety of different shapes and sizes of tongs for handling hot steels is necessary. It should be remembered that cooling of the area contacted by the tongs is retarded and that such areas may not harden, particularly if the steel being treated is very shallow hardening. Small parts may be wired together or quenched in baskets made of wire mesh.

Special quenching jigs and fixtures are frequently used to hold steels during quenching in a manner to restrain distortion.

When selective hardening is desired, portions of the steel may be protected by covering with alundum cement or some other insulating material. Selective hardening may be accomplished also by the use of water or oil jets designed to direct quenching medium on the areas to be hardened. This also is accomplished by the induction and flame hardening procedures previously described, particularly on large production jobs.

Shallow hardening steels, such as plain carbon and certain varieties of alloy steels, have such a high critical cooling rate that they must be quenched in brine or water to effect hardening. In general, intricately shaped sections should not be made of shallow hardening steels because of the tendency of these steels to warp and crack during hardening. Such items should be made of deeper hardening steels capable of being hardened by quenching in oil or air.

Tempering

Tempering reduces the brittleness imparted by hardening and produces definite physical properties within the steel. Tempering always follows, never precedes, the hardening operation. In addition to reducing brittleness, tempering softens the steel.

Tempering is always conducted at temperatures below the low critical point of the steel. In this respect, tempering differs from annealing, normalizing, or hardening, all of which require temperatures above the upper critical point. When hardened steel is reheated, tempering begins at 212 °F and continues as the temperature increases toward the low critical point. By selecting a definite tempering temperature, the resulting hardness and strength can be predetermined. Approximate temperatures for various tensile strengths are listed in Figure 5-6. The minimum time at the tempering temperature should be 1 hour. If the part is over 1 inch in thickness, increase the time by 1 hour for each additional inch of thickness. Tempered steels used in aircraft work have from 125,000 to 200,000 psi ultimate tensile strength.

Generally, the rate of cooling from the tempering temperature has no effect on the resulting structure; therefore, the steel is usually cooled in still air after being removed from the furnace.

Annealing

Annealing of steel produces a fine grained, soft, ductile metal without internal stresses or strains. In the annealed state, steel has its lowest strength. In general, annealing is the opposite of hardening.

Annealing of steel is accomplished by heating the metal to just above the upper critical point, soaking at that temperature, and cooling very slowly in the furnace. (Refer to Figure 5-6 for recommended temperatures.) Soaking time is approximately 1 hour per inch of thickness of the material. To produce maximum softness in steel, the metal must be cooled very slowly. Slow cooling is obtained by shutting off the heat and allowing the furnace and metal to cool together to 900 °F or lower, then removing the metal from the furnace and cooling in still air. Another method is to bury the heated steel in ashes, sand, or other substance that does not conduct heat readily.

Normalizing

The normalizing of steel removes the internal stresses set up by heat treating, welding, casting, forming, or machining. Stress, if not controlled, will lead to failure. Because of the better physical properties, aircraft steels are often used in the normalized state, but seldom, if ever, in the annealed state.

One of the most important uses of normalizing in aircraft work is in welded parts. Welding causes strains to be set up in the adjacent material. In addition, the weld itself is a cast structure as opposed to the wrought structure of the rest of the material. These two types of structures have different grain sizes, and to refine the grain as well as to relieve the internal stresses, all welded parts should be normalized after fabrication.

Normalizing is accomplished by heating the steel above the upper critical point and cooling in still air. The more rapid quenching obtained by air cooling, as compared to furnace cooling, results in a harder and stronger material than that obtained by annealing. Recommended normalizing temperatures for the various types of aircraft steels are listed in Figure 5-6.

Casehardening

Casehardening produces a hard wear-resistant surface or case over a strong, tough core. Casehardening is ideal for parts which require a wear-resistant surface and, at the same time, must be tough enough internally to withstand the applied loads. The steels best suited to casehardening are the low carbon and low alloy steels. If high carbon steel is casehardened, the hardness penetrates the core and causes brittleness. In casehardening, the surface of the metal is changed chemically by introducing a high carbide or nitride content. The core is unaffected chemically.

When heat treated, the surface responds to hardening while the core toughens. The common forms of casehardening are *carburizing, cyaniding,* and *nitriding.* Since cyaniding is not used in aircraft work, only carburizing and nitriding are discussed in this section.

Carburizing

Carburizing is a casehardening process in which carbon is added to the surface of low carbon steel. Thus, a carburized steel has a high carbon surface and a low carbon interior. When the carburized steel is heat treated, the case is hardened while the core remains soft and tough.

A common method of carburizing is called "pack carburizing." When carburizing is to be done by this method, the steel parts are packed in a container with charcoal or some other material rich in carbon. The container is then sealed with fire clay, placed in a furnace, heated to approximately 1,700 °F, and soaked at that temperature for several hours. As the temperature increases, carbon monoxide gas forms inside the container and, being unable to escape, combines with the gamma iron in the surface of the steel. The depth to which the carbon penetrates depends on the length of the soaking period. For example, when carbon steel is soaked for 8 hours, the carbon penetrates to a depth of about 0.062 inch.

In another method of carburizing, called "gas carburizing," a material rich in carbon is introduced into the furnace atmosphere. The carburizing atmosphere is produced by the use of various gases or by the burning of oil, wood, or other materials. When the steel parts are heated in this atmosphere, carbon monoxide combines with the gamma iron to produce practically the same results as those described under the pack carburizing process.

A third method of carburizing is that of "liquid carburizing." In this method, the steel is placed in a molten salt bath that contains the chemicals required to produce a case comparable with one resulting from pack or gas carburizing.

Alloy steels with low carbon content as well as low carbon steels may be carburized by any of the three processes. However, some alloys, such as nickel, tend to retard the absorption of carbon. As a result, the time required to produce a given thickness of case varies with the composition of the metal.

Nitriding

Nitriding is unlike other casehardening processes in that, before nitriding, the part is heat treated to produce definite physical properties. Thus, parts are hardened and tempered before being nitrided. Most steels can be nitrided, but special alloys are required for best results. These special alloys contain aluminum as one of the alloying elements and are called "nitralloys."

In nitriding, the part is placed in a special nitriding furnace and heated to a temperature of approximately 1,000 °F. With the part at this temperature, ammonia gas is circulated within the specially constructed furnace chamber. The high temperature cracks the ammonia gas into nitrogen and hydrogen. The ammonia which does not break down is caught in a water trap below the regions of the other two gases. The nitrogen reacts with the iron to form nitride. The iron nitride is dispersed in minute particles at the surface and works inward. The depth of penetration depends on the length of the treatment. In nitriding, soaking periods as long as 72 hours are frequently required to produce the desired thickness of case.

Nitriding can be accomplished with a minimum of distortion, because of the low temperature at which parts are casehardened and because no quenching is required after exposure to the ammonia gas.

Heat Treatment of Nonferrous Metals

Aluminum Alloys

In the wrought form, commercially pure aluminum is known as 1100. It has a high degree of resistance to corrosion and is easily formed into intricate shapes. It is relatively low in strength and does not have the properties required for structural aircraft parts. High strengths are generally obtained by the process of alloying. The resulting alloys are less easily formed and, with some exceptions, have lower resistance to corrosion than 1100 aluminum.

Alloying is not the only method of increasing the strength of aluminum. Like other materials, aluminum becomes stronger and harder as it is rolled, formed, or otherwise cold worked. Since the hardness depends on the amount of cold working done, 1100 and some wrought aluminum alloys are available in several strain hardened tempers. The soft or annealed condition is designated O. If the material is strain hardened, it is said to be in the H condition.

The most widely used alloys in aircraft construction are hardened by heat treatment rather than by cold work. These alloys are designated by a somewhat different set of symbols: T4 and W indicate solution heat treated and quenched but not aged, and T6 indicates an alloy in the heat treated hardened condition.

- W—Solution heat treated, unstable temper
- T—Treated to produce stable tempers other than F, O, or H
- T2—Annealed (cast products only)
- T3—Solution heat treated and then cold worked
- T4—Solution heat treated
- T5—Artificially aged only
- T6—Solution heat treated and then artificially aged
- T7—Solution heat treated and then stabilized
- T8—Solution heat treated, cold worked, and then artificially aged

- T9—Solution heat treated, artificially aged, and then cold worked
- T10—Artificially aged and then cold worked

Additional digits may be added to T1 through T10 to indicate a variation in treatment which significantly alters the characteristics of the product.

Aluminum alloy sheets are marked with the specification number on approximately every square foot of material. If for any reason this identification is not on the material, it is possible to separate the heat-treatable alloys from the non-heat-treatable alloys by immersing a sample of the material in a 10 percent solution of caustic soda (sodium hydroxide). The heat-treatable alloys will turn black due to the copper content, whereas the others will remain bright. In the case of clad material, the surface will remain bright, but there will be a dark area in the middle when viewed from the edge.

Alclad Aluminum

The terms "Alclad and Pureclad" are used to designate sheets that consist of an aluminum alloy core coated with a layer of pure aluminum to a depth of approximately 5 ½ percent on each side. The pure aluminum coating affords a dual protection for the core, preventing contact with any corrosive agents, and protecting the core electrolytically by preventing any attack caused by scratching or from other abrasions.

There are two types of heat treatments applicable to aluminum alloys. One is called solution heat treatment, and the other is known as precipitation heat treatment. Some alloys, such as 2017 and 2024, develop their full properties as a result of solution heat treatment followed by about 4 days of aging at room temperature. Other alloys, such as 2014 and 7075, require both heat treatments.

The alloys that require precipitation heat treatment (artificial aging) to develop their full strength also age to a limited extent at room temperature; the rate and amount of strengthening depends upon the alloy. Some reach their maximum natural or room temperature aging strength in a few days, and are designated as -T4 or -T3 temper. Others continue to age appreciably over a long period of time.

Because of this natural aging, the -W designation is specified only when the period of aging is indicated, for example, 7075-W (½ hour). Thus, there is considerable difference in the mechanical and physical properties of freshly quenched (-W) material and material that is in the -T3 or -T4 temper.

The hardening of an aluminum alloy by heat treatment consists of four distinct steps:

1. Heating to a predetermined temperature.
2. Soaking at temperature for a specified length of time.
3. Rapidly quenching to a relatively low temperature.
4. Aging or precipitation hardening either spontaneously at room temperature, or as a result of a low temperature thermal treatment.

The first three steps above are known as solution heat treatment, although it has become common practice to use the shorter term, "heat treatment." Room temperature hardening is known as natural aging, while hardening done at moderate temperatures is called artificial aging, or precipitation heat treatment.

Solution Heat Treatment

Temperature

The temperatures used for solution heat treating vary with different alloys and range from 825 °F to 980 °F. As a rule, they must be controlled within a very narrow range (±10 °F) to obtain specified properties.

If the temperature is too low, maximum strength will not be obtained. When excessive temperatures are used, there is danger of melting the low melting constituents of some alloys with consequent lowering of the physical properties of the alloy. Even if melting does not occur, the use of higher than recommended temperatures promotes discoloration and increases quenching strains.

Time at Temperature

The time at temperature, referred to as soaking time, is measured from the time the coldest metal reaches the minimum limit of the desired temperature range. The soaking time varies, depending upon the alloy and thickness, from 10 minutes for thin sheets to approximately 12 hours for heavy forgings. For the heavy sections, the nominal soaking time is approximately 1 hour for each inch of cross-sectional thickness. [Figure 5-7]

Choose the minimum soaking time necessary to develop the required physical properties. The effect of an abbreviated soaking time is obvious. An excessive soaking period aggravates high temperature oxidation. With clad material, prolonged heating results in exces-

Thickness (inch)	Time (minutes)
Up to .032	30
.032 to 1/8	30
1/8 to 1/4	40
Over 1/4	60

Note: Soaking time starts when the metal (or the molten bath) reaches a temperature within the range specified above.

Figure 5-7. Typical soaking times for heat treatment.

sive diffusion of copper and other soluble constituents into the protective cladding and may defeat the purpose of cladding.

Quenching

After the soluble constituents are in solid solution, the material is quenched to prevent or retard immediate reprecipitation. Three distinct quenching methods are employed. The one to be used in any particular instance depends upon the part, the alloy, and the properties desired.

Cold Water Quenching

Parts produced from sheet, extrusions, tubing, small forgings, and similar type material are generally quenched in a cold water bath. The temperature of the water before quenching should not exceed 85 °F.

Using a sufficient quantity of water keeps the temperature rise under 20 °F. Such a drastic quench ensures maximum resistance to corrosion. This is particularly important when working with such alloys as 2017, 2024, and 7075. This is the reason a drastic quench is preferred, even though a slower quench may produce the required mechanical properties.

Hot Water Quenching

Large forgings and heavy sections can be quenched in hot or boiling water. This type of quench minimizes distortion and alleviates cracking which may be produced by the unequal temperatures obtained during the quench. The use of a hot water quench is permitted with these parts because the temperature of the quench water does not critically affect the resistance to corrosion of the forging alloys. In addition, the resistance to corrosion of heavy sections is not as critical a factor as for thin sections.

Spray Quenching

High velocity water sprays are useful for parts formed from clad sheet and for large sections of almost all alloys. This type of quench also minimizes distortion and alleviates quench cracking. However, many specifications forbid the use of spray quenching for bare 2017 and 2024 sheet materials because of the effect on their resistance to corrosion.

Lag Between Soaking and Quenching

The time interval between the removal of the material from the furnace and quenching is critical for some alloys and should be held to a minimum. When solution heat treating 2017 or 2024 sheet material, the elapsed time must not exceed 10 seconds. The allowable time for heavy sections may be slightly greater.

Allowing the metal to cool slightly before quenching promotes reprecipitation from the solid solution. The precipitation occurs along grain boundaries and in certain slip planes causing poorer formability. In the case of 2017, 2024, and 7075 alloys, their resistance to intergranular corrosion is adversely affected.

Reheat Treatment

The treatment of material which has been previously heat treated is considered a reheat treatment. The unclad heat-treatable alloys can be solution heat treated repeatedly without harmful effects.

The number of solution heat treatments allowed for clad sheet is limited due to increased diffusion of core and cladding with each reheating. Existing specifications allow one to three reheat treatments of clad sheet depending upon cladding thickness.

Straightening After Solution Heat Treatment

Some warping occurs during solution heat treatment, producing kinks, buckles, waves, and twists. These imperfections are generally removed by straightening and flattening operations.

Where the straightening operations produce an appreciable increase in the tensile and yield strengths and a slight decrease in the percent of elongation, the material is designated -T3 temper. When the above values are not materially affected, the material is designated -T4 temper.

Precipitation Heat Treating

As previously stated, the aluminum alloys are in a comparatively soft state immediately after quenching from a solution heat-treating temperature. To obtain their maximum strengths, they must be either naturally aged or precipitation hardened.

During this hardening and strengthening operation, precipitation of the soluble constituents from the supersaturated solid solution takes place. As precipitation progresses, the strength of the material increases, often by a series of peaks, until a maximum is reached. Further aging (overaging) causes the strength to steadily decline until a somewhat stable condition is obtained. The submicroscopic particles that are precipitated provide the keys or locks within the grain structure and between the grains to resist internal slippage and distortion when a load of any type is applied. In this manner, the strength and hardness of the alloy are increased.

Precipitation hardening produces a great increase in the strength and hardness of the material with corresponding decreases in the ductile properties. The process used to obtain the desired increase in strength is therefore known as aging, or precipitation hardening.

The strengthening of the heat-treatable alloys by aging is not due merely to the presence of a precipitate. The strength is due to both the uniform distribution of a finely dispersed submicroscopic precipitate and its effects upon the crystal structure of the alloy.

The aging practices used depend upon many properties other than strength. As a rule, the artificially aged alloys are slightly overaged to increase their resistance to corrosion. This is especially true with the artificially aged high copper content alloys that are susceptible to intergranular corrosion when inadequately aged.

The heat-treatable aluminum alloys are subdivided into two classes: those that obtain their full strength at room temperature and those that require artificial aging.

The alloys that obtain their full strength after 4 or 5 days at room temperature are known as natural aging alloys. Precipitation from the supersaturated solid solution starts soon after quenching, with 90 percent of the maximum strength generally being obtained in 24 hours. Alloys 2017 and 2024 are natural aging alloys.

The alloys that require precipitation thermal treatment to develop their full strength are artificially aged alloys. However, these alloys also age a limited amount at room temperature, the rate and extent of the strengthening depending upon the alloys.

Many of the artificially aged alloys reach their maximum natural or room temperature aging strengths after a few days. These can be stocked for fabrication in the -T4 or -T3 temper. High zinc content alloys such as 7075 continue to age appreciably over a long period of time, their mechanical property changes being sufficient to reduce their formability.

The advantage of -W temper formability can be utilized, however, in the same manner as with natural aging alloys; that is, by fabricating shortly after solution heat treatment, or retaining formability by the use of refrigeration.

Refrigeration retards the rate of natural aging. At 32 °F, the beginning of the aging process is delayed for several hours, while dry ice (−50 °F to −100 °F) retards aging for an extended period of time.

Precipitation Practices

The temperatures used for precipitation hardening depend upon the alloy and the properties desired, ranging from 250 °F to 375 °F. They should be controlled within a very narrow range (±5 °F) to obtain best results. [Figure 5-8]

The time at temperature is dependent upon the temperature used, the properties desired, and the alloy. It ranges from 8 to 96 hours. Increasing the aging tem-

Alloy	Solution heat treatment			Precipitation heat treatment		
	Temperature (°F)	Quench	Temperature designation	Temperature (°F)	Time of aging	Temperature designation
2017	930–950	Cold water	T4			T
2117	930–950	Cold water	T4			T
2024	910–930	Cold water	T4			T
6053	960–980	Water	T4	445–455 345–355	1–2 hours or 8 hours	T5 T6
6061	960–980	Water	T4	315–325 345–355	18 hours or 8 hours	T6 T6
7075	870	Water		250	24 hours	T6

Figure 5-8. Conditions for heat treatment of aluminum alloys.

perature decreases the soaking period necessary for proper aging. However, a closer control of both time and temperature is necessary when using the higher temperatures.

After receiving the thermal precipitation treatment, the material should be air cooled to room temperature. Water quenching, while not necessary, produces no ill effects. Furnace cooling has a tendency to produce overaging.

Annealing of Aluminum Alloys

The annealing procedure for aluminum alloys consists of heating the alloys to an elevated temperature, holding or soaking them at this temperature for a length of time depending upon the mass of the metal, and then cooling in still air. Annealing leaves the metal in the best condition for cold working. However, when prolonged forming operations are involved, the metal will take on a condition known as "mechanical hardness" and will resist further working. It may be necessary to anneal a part several times during the forming process to avoid cracking. Aluminum alloys should not be used in the annealed state for parts or fittings.

Clad parts should be heated as quickly and carefully as possible, since long exposure to heat tends to cause some of the constituents of the core to diffuse into the cladding. This reduces the corrosion resistance of the cladding.

Heat Treatment of Aluminum Alloy Rivets

Aluminum alloy rivets are furnished in the following compositions: Alloys 1100, 5056, 2117, 2017, and 2024.

Alloy 1100 rivets are used in the "as fabricated" condition for riveting aluminum alloy sheets where a low strength rivet is suitable. Alloy 5056 rivets are used in the "as fabricated" condition for riveting magnesium alloy sheets.

Alloy 2117 rivets have moderately high strength and are suitable for riveting aluminum alloy sheets. These rivets receive only one heat treatment, which is performed by the manufacturer, and are anodized after being heat treated. They require no further heat treatment before they are used. Alloy 2117 rivets retain their characteristics indefinitely after heat treatment and can be driven anytime. Rivets made of this alloy are the most widely used in aircraft construction.

Alloy 2017 and 2024 rivets are high strength rivets suitable for use with aluminum alloy structures. They are purchased from the manufacturer in the heat-treated condition. Since the aging characteristics of these alloys at room temperatures are such that the rivets are unfit for driving, they must be reheat treated just before they are to be used. Alloy 2017 rivets become too hard for driving in approximately 1 hour after quenching. Alloy 2024 rivets become hardened in 10 minutes after quenching. Both of these alloys may be reheat treated as often as required; however, they must be anodized before the first reheat treatment to prevent intergranular oxidation of the material. If these rivets are stored in a refrigerator at a temperature lower than 32 °F immediately after quenching, they will remain soft enough to be usable for several days.

Rivets requiring heat treatment are heated either in tubular containers in a salt bath, or in small screen wire baskets in an air furnace. The heat treatment of alloy 2017 rivets consists of subjecting the rivets to a temperature between 930 °F to 950 °F for approximately 30 minutes, and immediately quenching in cold water. These rivets reach maximum strength in about 9 days after being driven. Alloy 2024 rivets should be heated to a temperature of 910 °F to 930 °F and immediately quenched in cold water. These rivets develop a greater shear strength than 2017 rivets and are used in locations where extra strength is required. Alloy 2024 rivets develop their maximum shear strength in 1 day after being driven.

The 2017 rivet should be driven within approximately 1 hour and the 2024 rivet within 10 to 20 minutes after heat treating or removal from refrigeration. If not used within these times, the rivets should be re-heat treated before being refrigerated.

Heat Treatment of Magnesium Alloys

Magnesium alloy castings respond readily to heat treatment, and about 95 percent of the magnesium used in aircraft construction is in the cast form. The heat treatment of magnesium alloy castings is similar to the heat treatment of aluminum alloys in that there are two types of heat treatment: (1) solution heat treatment and (2) precipitation (aging) heat treatment. Magnesium, however, develops a negligible change in its properties when allowed to age naturally at room temperatures.

Solution Heat Treatment

Magnesium alloy castings are solution heat treated to improve tensile strength, ductility, and shock resistance. This heat-treatment condition is indicated by using the symbol -T4 following the alloy designation. Solution heat treatment plus artificial aging is designated -T6. Artificial aging is necessary to develop the full properties of the metal.

Solution heat-treatment temperatures for magnesium alloy castings range from 730 °F to 780 °F, the exact range depending upon the type of alloy. The temperature range for each type of alloy is listed in Specification MIL-H-6857. The upper limit of each range listed in the specification is the maximum temperature to which the alloy may be heated without danger of melting the metal.

The soaking time ranges from 10 to 18 hours, the exact time depending upon the type of alloy as well as the thickness of the part. Soaking periods longer than 18 hours may be necessary for castings over 2 inches in thickness. **NEVER** heat magnesium alloys in a salt bath as this may result in an explosion.

A serious potential fire hazard exists in the heat treatment of magnesium alloys. If through oversight or malfunctioning of equipment, the maximum temperatures are exceeded, the casting may ignite and burn freely. For this reason, the furnace used should be equipped with a safety cutoff that will turn off the power to the heating elements and blowers if the regular control equipment malfunctions or fails. Some magnesium alloys require a protective atmosphere of sulfur dioxide gas during solution heat treatment. This aids in preventing the start of a fire even if the temperature limits are slightly exceeded.

Air quenching is used after solution heat treatment of magnesium alloys since there appears to be no advantage in liquid cooling.

Precipitation Heat Treatment

After solution treatment, magnesium alloys may be given an aging treatment to increase hardness and yield strength. Generally, the aging treatments are used merely to relieve stress and stabilize the alloys in order to prevent dimensional changes later, especially during or after machining. Both yield strength and hardness are improved somewhat by this treatment at the expense of a slight amount of ductility. The corrosion resistance is also improved, making it closer to the "as cast" alloy.

Precipitation heat treatment temperatures are considerably lower than solution heat-treatment temperatures and range from 325 °F to 500 °F. Soaking time ranges from 4 to 18 hours.

Heat Treatment of Titanium

Titanium is heat treated for the following purposes:

- Relief of stresses set up during cold forming or machining.
- Annealing after hot working or cold working, or to provide maximum ductility for subsequent cold working.
- Thermal hardening to improve strength.

Stress Relieving

Stress relieving is generally used to remove stress concentrations resulting from forming of titanium sheet. It is performed at temperatures ranging from 650 °F to 1,000 °F. The time at temperature varies from a few minutes for a very thin sheet to an hour or more for heavier sections. A typical stress relieving treatment is 900 °F for 30 minutes, followed by an air cool.

The discoloration or scale which forms on the surface of the metal during stress relieving is easily removed by pickling in acid solutions. The recommended solution contains 10 to 20 percent nitric acid and 1 to 3 percent hydrofluoric acid. The solution should be at room temperature or slightly above.

Full Annealing

The annealing of titanium and titanium alloys provides toughness, ductility at room temperature, dimensional and structural stability at elevated temperatures, and improved machinability.

The full anneal is usually called for as preparation for further working. It is performed at 1,200–1,650 °F. The time at temperature varies from 16 minutes to several hours, depending on the thickness of the material and the amount of cold work to be performed. The usual treatment for the commonly used alloys is 1,300 °F for 1 hour, followed by an air cool. A full anneal generally results in sufficient scale formation to require the use of caustic descaling, such as sodium hydride salt bath.

Thermal Hardening

Unalloyed titanium cannot be heat treated, but the alloys commonly used in aircraft construction can be strengthened by thermal treatment, usually at some sacrifice in ductility. For best results, a water quench from 1,450 °F, followed by reheating to 900 °F for 8 hours is recommended.

Casehardening

The chemical activity of titanium and its rapid absorption of oxygen, nitrogen, and carbon at relatively low temperatures make casehardening advantageous for special applications. Nitriding, carburizing, or carbonitriding can be used to produce a wear-resistant case of 0.0001 to 0.0002 inch in depth.

Hardness Testing

Hardness testing is a method of determining the results of heat treatment as well as the state of a metal prior to heat treatment. Since hardness values can be tied in with tensile strength values and, in part, with wear resistance, hardness tests are a valuable check of heat-treat control and of material properties.

Practically all hardness testing equipment now uses the resistance to penetration as a measure of hardness. Included among the better known hardness testers are the Brinell and Rockwell, both of which are described and illustrated in this section. Also included is a popular portable-type hardness tester currently in use.

Brinell Tester

The Brinell hardness tester [Figure 5-9] uses a hardened spherical ball, which is forced into the surface of the metal. This ball is 10 millimeters (0.3937 inch) in diameter. A pressure of 3,000 kilograms is used for ferrous metals and 500 kilograms for nonferrous metals. The pressure must be maintained at least 10 seconds for ferrous metals and at least 30 seconds for nonferrous metals. The load is applied by hydraulic pressure. The hydraulic pressure is built up by a hand pump or an electric motor, depending on the model of tester. A pressure gauge indicates the amount of pressure. There is a release mechanism for relieving the pressure after the test has been made, and a calibrated microscope is provided for measuring the diameter of the impression in millimeters. The machine has various shaped anvils for supporting the specimen and an elevating screw for bringing the specimen in contact with the ball penetrator. These are attachments for special tests.

To determine the Brinell hardness number for a metal, measure the diameter of the impression, using the calibrated microscope furnished with the tester. Then convert the measurement into the Brinell hardness number on the conversion table furnished with the tester.

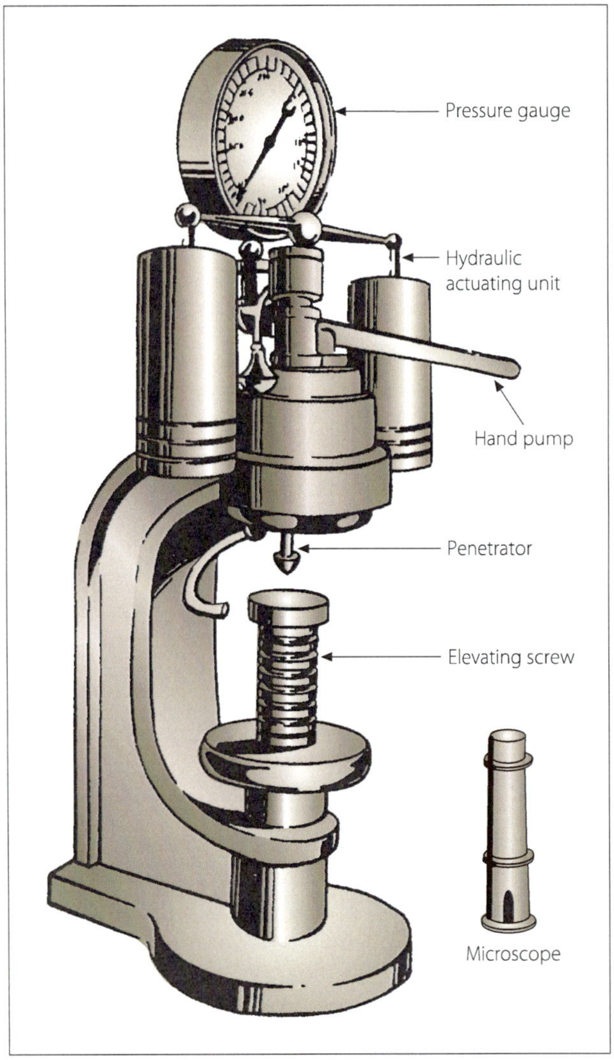

Figure 5-9. Brinell hardness tester.

Rockwell Tester

The Rockwell hardness tester [Figure 5-10] measures the resistance to penetration, as does the Brinell tester. Instead of measuring the diameter of the impres-

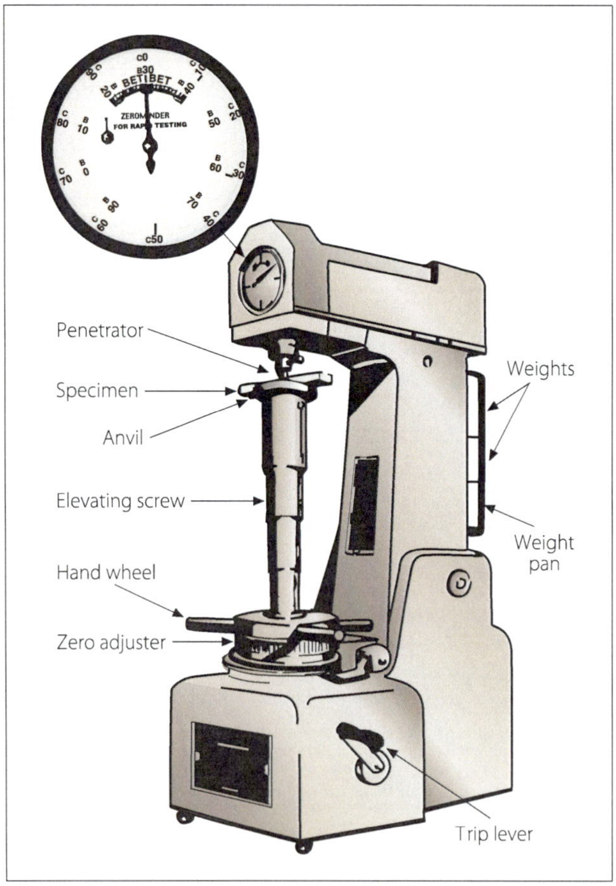

Figure 5-10. Rockwell hardness tester.

sion, the Rockwell tester measures the depth, and the hardness is indicated directly on a dial attached to the machine. The dial numbers in the outer circle are black, and the inner numbers are red. Rockwell hardness numbers are based on the difference between the depth of penetration at major and minor loads. The greater this difference, the lower the hardness number and the softer the material.

Two types of penetrators are used with the Rockwell tester: a diamond cone and a hardened steel ball. The load which forces the penetrator into the metal is called the major load and is measured in kilograms. The results of each penetrator and load combination are reported on separate scales, designated by letters. The penetrator, the major load, and the scale vary with the kind of metal being tested.

For hardened steels, the diamond penetrator is used; the major load is 150 kilograms; and the hardness is read on the "C" scale. When this reading is recorded, the letter "C" must precede the number indicated by the pointer. The C-scale setup is used for testing metals ranging in hardness from C-20 to the hardest steel (usually about C-70). If the metal is softer than C-20, the B-scale setup is used. With this setup, the 1⁄16-inch ball is used as a penetrator; the major load is 100 kilograms; and the hardness is read on the B-scale.

In addition to the "C" and "B" scales, there are other setups for special testing. The scales, penetrators, major loads, and dial numbers to be read are listed in Figure 5-11.

The Rockwell tester is equipped with a weight pan, and two weights are supplied with the machine. One weight is marked in red. The other weight is marked in black. With no weight in the weight pan, the machine applies a major load of 60 kilograms. If the scale setup calls for a 100 kilogram load, the red weight is placed in the pan. For a 150 kilogram load, the black weight is added to the red weight. The black weight is always used with the red weight; it is never used alone.

Practically all testing is done with either the B-scale setup or the C-scale setup. For these scales, the colors may be used as a guide in selecting the weight (or weights) and in reading the dial. For the B-scale test, use the red weight and read the red numbers. For a C-scale test, add the black weight to the red weight and read the black numbers.

In setting up the Rockwell machine, use the diamond penetrator for testing materials known to be hard. If the hardness is unknown, try the diamond, since the steel ball may be deformed if used for testing hard materials. If the metal tests below C-22, then change to the steel ball.

Scale Symbol	Penetrator	Major Load (kg)	Dial Color/Number
A	Diamond	60	Black
B	1⁄16-inch ball	100	Red
C	Diamond	150	Black
D	Diamond	100	Black
E	1⁄8-inch ball	100	Red
F	1⁄16-inch ball	60	Red
G	1⁄16-inch ball	150	Red
H	1⁄8-inch ball	60	Red
K	1⁄8-inch ball	150	Red

Figure 5-11. Standard Rockwell hardness scales.

Use the steel ball for all soft materials, those testing less than B-100. Should an overlap occur at the top of the B-scale and the bottom of the C-scale, use the C-scale setup.

Before the major load is applied, securely lock the test specimen in place to prevent slipping and to seat the anvil and penetrator properly. To do this, apply a load of 10 kilograms before the lever is tripped. This preliminary load is called the minor load. The minor load is 10 kilograms regardless of the scale setup.

The metal to be tested in the Rockwell tester must be ground smooth on two opposite sides and be free of scratches and foreign matter. The surface should be perpendicular to the axis of penetration, and the two opposite ground surfaces should be parallel. If the specimen is tapered, the amount of error will depend on the taper. A curved surface will also cause a slight error in the hardness test. The amount of error depends on the curvature; i.e., the smaller the radius of curvature, the greater the error. To eliminate such error, a small flat should be ground on the curved surface if possible.

Clad aluminum alloy sheets cannot be tested directly with any accuracy with a Rockwell hardness tester. If the hardness value of the base metal is desired, the pure aluminum coating must be removed from the area to be checked prior to testing.

Barcol Tester

The Barcol tester [Figure 5-12] is a portable unit designed for testing aluminum alloys, copper, brass, or other relatively soft materials. It should not be used on

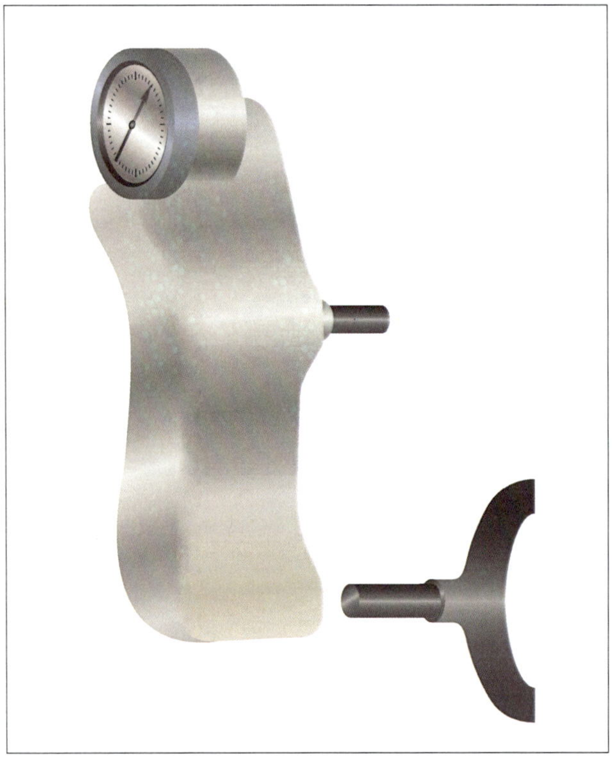

Figure 5-12. Barcol portable hardness tester.

aircraft steels. Approximate range of the tester is 25 to 100 Brinell. The unit can be used in any position and in any space that will allow for the operator's hand. It is of great value in the hardness testing of assembled or installed parts, especially to check for proper heat treatment. The hardness is indicated on a dial conveniently divided into 100 graduations.

The design of the Barcol tester is such that operating experience is not necessary. It is only necessary to exert a light pressure against the instrument to drive the spring loaded indenter into the material to be tested. The hardness reading is instantly indicated on the dial.

Several typical readings for aluminum alloys are listed in Figure 5-13. Note that the harder the material is, the higher the Barcol number will be. To prevent damage to the point, avoid sliding or scraping when it is in contact with the material being tested. If the point should become damaged, it must be replaced with a new one. Do not attempt to grind the point.

Each tester is supplied with a test disk for checking the condition of the point. To check the point, press the instrument down on the test disk. When the downward pressure brings the end of the lower plunger guide against the surface of the disk, the indicator reading should be within the range shown on the test disk.

Alloy and Temper	Barcol Number
1100-O	35
3003-O	42
3003-H14	56
2024-O	60
5052-O	62
5052-H34	75
6061-T	78
2024-T	85

Figure 5-13. Typical Barcol readings for aluminum alloy.

Forging

Forging is the process of forming a product by hammering or pressing. When the material is forged below the recrystallization temperature, it is called cold forged. When worked above the recrystallization temperature, it is referred to as hot forged. Drop forging is a hammering process that uses a hot ingot that is placed between a pair of formed dies in a machine called a drop hammer and a weight of several tons is dropped on the upper die. This results in the hot metal being forced to take the form of the dies. Because the process is very rapid, the grain structure of the metal is altered, resulting in a significant increase in the strength of the finished part.

Casting

Casting is formed by melting the metal and pouring it into a mold of the desired shape. Since plastic deformation of the metal does not occur, no alteration of the grain shape or orientation is possible. The gain size of the metal can be controlled by the cooling rate, the alloys of the metal, and the thermal treatment. Castings are normally lower in strength and are more brittle than a wrought product of the same material. For intricate shapes or items with internal passages, such as turbine blades, casting may be the most economical process. Except for engine parts, most metal components found on an aircraft are wrought instead of cast.

All metal products start in the form of casting. Wrought metals are converted from cast ingots by plastic deformation. For high strength aluminum alloys, an 80 to 90 percent reduction (dimensional change in thickness) of the material is required to obtain the high mechanical properties of a fully wrought structure.

Both iron and aluminum alloys are cast for aircraft uses. Cast iron contains 6 to 8 percent carbon and silicon.

Cast iron is a hard unmalleable pig iron made by casting or pouring into a mold. Cast aluminum alloy has been heated to its molten state and poured into a mold to give it the desired shape.

Extruding

The extrusion process involves the forcing of metal through an opening in a die, thus causing the metal to take the shape of the die opening. The shape of the die will be the cross section of an angle, channel, tube, or some other shape. Some metals such as lead, tin, and aluminum may be extruded cold; however, most metals are heated before extrusion. The main advantage of the extrusion process is its flexibility. For example, because of its workability, aluminum can be economically extruded to more intricate shapes and larger sizes than is practical with other metals.

Extruded shapes are produced in very simple as well as extremely complex sections. In this process a cylinder of aluminum, for instance, is heated to 750–850 °F and is then forced through the opening of a die by a hydraulic ram. The opening is the shape desired for the cross section of the finished extrusion.

Many structural parts, such as channels, angles, T-sections, and Z-sections, are formed by the extrusion process.

Aluminum is the most extruded metal used in aircraft. Aluminum is extruded at a temperature of 700–900 °F (371–482 °C) and requires pressure of up to 80,000 psi (552 MPa). After extrusion, the product frequently will be subjected to both thermal and mechanical processes to obtain the desired properties. Extrusion processes are limited to the more ductile materials.

Cold Working/Hardening

Cold working applies to mechanical working performed at temperatures below the critical range. It results in a strain hardening of the metal. In fact, the metal often becomes so hard that it is difficult to continue the forming process without softening the metal by annealing.

Since the errors attending shrinkage are eliminated in cold working, a much more compact and better metal is obtained. The strength and hardness, as well as the elastic limit, are increased; but the ductility decreases. Since this makes the metal more brittle, it must be heated from time to time during certain operations to remove the undesirable effects of the working.

While there are several cold working processes, the two with which the aviation mechanic will be principally concerned are cold rolling and cold drawing. These processes give the metals desirable qualities which cannot be obtained by hot working.

Cold rolling usually refers to the working of metal at room temperature. In this operation, the materials that have been rolled to approximate sizes are pickled to remove the scale, after which they are passed through chilled finishing rolls. This gives a smooth surface and also brings the pieces to accurate dimensions. The principal forms of cold rolled stocks are sheets, bars, and rods.

Cold drawing is used in making seamless tubing, wire, streamlined tie rods, and other forms of stock. Wire is made from hot rolled rods of various diameters. These rods are pickled in acid to remove scale, dipped in lime water, and then dried in a steam room where they remain until ready for drawing. The lime coating adhering to the metal serves as a lubricant during the drawing operation.

The size of the rod used for drawing depends upon the diameter wanted in the finished wire. To reduce the rod to the desired size, it is drawn cold through a die. One end of the rod is filed or hammered to a point and slipped through the die opening. Here it is gripped by the jaws of the drawing block and pulled through the die. This series of operations is done by a mechanism known as a drawbench.

To reduce the rod gradually to the desired size, it is necessary to draw the wire through successively smaller dies. Because each of these drawings reduces the ductility of the wire, it must be annealed from time to time before further drawings can be accomplished. Although cold working reduces the ductility, it increases the tensile strength of the wire.

In making seamless steel aircraft tubing, the tubing is cold drawn through a ring shaped die with a mandrel or metal bar inside the tubing to support it while the drawing operations are being performed. This forces the metal to flow between the die and the mandrel and affords a means of controlling the wall thickness and the inside and outside diameters.

Nonmetallic Aircraft Materials

The use of magnesium, plastic, fabric, and wood in aircraft construction has nearly disappeared since the mid-1950s. Aluminum has also greatly diminished in use, from 80 percent of airframes in 1950 to about 15 percent aluminum and aluminum alloys today for airframe construction. Replacing those materials are nonmetallic aircraft materials, such as reinforced plastics and advanced composites.

Wood

The earliest aircraft were constructed of wood and cloth. Today, except for restorations and some homebuilt aircraft, very little wood is used in aircraft construction.

Plastics

Plastics are used in many applications throughout modern aircraft. These applications range from structural components of thermosetting plastics reinforced with fiberglass to decorative trim of thermoplastic materials to windows.

Transparent Plastics

Transparent plastic materials used in aircraft canopies, windshields, windows and other similar transparent enclosures may be divided into two major classes or groups. These plastics are classified according to their reaction to heat. The two classes are: thermoplastic and thermosetting.

Thermoplastic materials will soften when heated and harden when cooled. These materials can be heated until soft, and then formed into the desired shape. When cooled, they will retain this shape. The same piece of plastic can be reheated and reshaped any number of times without changing the chemical composition of the materials.

Thermosetting plastics harden upon heating, and reheating has no softening effect. These plastics cannot be reshaped once being fully cured by the application of heat.

In addition to the above classes, transparent plastics are manufactured in two forms: monolithic (solid) and laminated. Laminated transparent plastics are made from transparent plastic face sheets bonded by an inner layer material, usually polyvinyl butyryl. Because of its shatter resistant qualities, laminated plastic is superior to solid plastics and is used in many pressurized aircraft.

Most of the transparent sheet used in aviation is manufactured in accordance with various military specifications. A new development in transparent plastics is stretched acrylic. Stretched acrylic is a type of plastic which, before being shaped, is pulled in both directions to rearrange its molecular structure. Stretched acrylic panels have a greater resistance to impact and are less subject to shatter; its chemical resistance is greater, edging is simpler, and crazing and scratches are less detrimental.

Individual sheets of plastic are covered with a heavy masking paper to which a pressure sensitive adhesive has been added. This paper helps to prevent accidental scratching during storage and handling. Be careful to avoid scratches and gouges which may be caused by sliding sheets against one another or across rough or dirty tables.

If possible, store sheets in bins which are tilted at approximately 10° from vertical. If they must be stored horizontally, piles should not be over 18 inches high, and small sheets should be stacked on the larger ones to avoid unsupported overhang. Store in a cool, dry place away from solvent fumes, heating coils, radiators, and steam pipes. The temperature in the storage room should not exceed 120 °F.

While direct sunlight does not harm acrylic plastic, it will cause drying and hardening of the masking adhesive, making removal of the paper difficult. If the paper will not roll off easily, place the sheet in an oven at 250 °F for 1 minute, maximum. The heat will soften the masking adhesive for easy removal of the paper.

If an oven is not available, remove hardened masking paper by softening the adhesive with aliphatic naphtha. Rub the masking paper with a cloth saturated with naphtha. This will soften the adhesive and free the paper from the plastic. Sheets so treated must be washed immediately with clean water, taking care not to scratch the surfaces.

Note: Aliphatic naphtha is not to be confused with aromatic naphtha and other dry cleaning solvents which have harmful effects on plastic. However, aliphatic naphtha is flammable and all precautions regarding the use of flammable liquids must be observed.

Composite Materials

In the 1940s, the aircraft industry began to develop synthetic fibers to enhance aircraft design. Since that time, composite materials have been used more and more. When composites are mentioned, most people think of only fiberglass, or maybe graphite or aramids (Kevlar). Composites began in aviation, but now are being embraced by many other industries, including auto racing, sporting goods, and boating, as well as defense industry uses.

A "composite" material is defined as a mixture of different materials or things. This definition is so general that it could refer to metal alloys made from several different metals to enhance the strength, ductility, conductivity or whatever characteristics are desired. Likewise, the composition of composite materials is a combination of reinforcement, such as a fiber, whisker, or particle, surrounded and held in place by a resin, forming a structure. Separately, the reinforcement and the resin are very different from their combined state. Even in their combined state, they can still be individually identified and mechanically separated. One composite, concrete, is composed of cement (resin) and gravel or reinforcement rods for the reinforcement to create the concrete.

Advantages/Disadvantages of Composites

Some of the many advantages for using composite materials are:

- High strength to weight ratio
- Fiber-to-fiber transfer of stress allowed by chemical bonding
- Modulus (stiffness to density ratio) 3.5 to 5 times that of steel or aluminum
- Longer life than metals
- Higher corrosion resistance
- Tensile strength 4 to 6 times that of steel or aluminum
- Greater design flexibility
- Bonded construction eliminates joints and fasteners
- Easily repairable

The disadvantages of composites include:

- Inspection methods difficult to conduct, especially delamination detection (Advancements in technology will eventually correct this problem.)
- Lack of long term design database, relatively new technology methods
- Cost
- Very expensive processing equipment
- Lack of standardized system of methodology
- Great variety of materials, processes, and techniques
- General lack of repair knowledge and expertise
- Products often toxic and hazardous
- Lack of standardized methodology for construction and repairs

The increased strength and the ability to design for the performance needs of the product makes composites much superior to the traditional materials used in today's aircraft. As more and more composites are used, the costs, design, inspection ease, and information about strength to weight advantages will help composites become the material of choice for aircraft construction.

Composite Safety

Composite products can be very harmful to the skin, eyes, and lungs. In the long or short term, people can become sensitized to the materials with serious irritation and health issues. Personal protection is often uncomfortable, hot, and difficult to wear; however, a little discomfort while working with the composite materials can prevent serious health issues or even death.

Respirator particle protection is very important to protecting the lungs from permanent damage from tiny glass bubbles and fiber pieces. At a minimum, a dust mask approved for fiberglass is a necessity. The best protection is a respirator with dust filters. The proper fit of a respirator or dust mask is very important because if the air around the seal is breathed, the mask cannot protect the wearer's lungs. When working with resins, it is important to use vapor protection. Charcoal filters in a respirator will remove the vapors for a period of time. If you can smell the resin vapors after placing the mask back on after a break, replace the filters immediately. Sometimes, charcoal filters last less than 4 hours. Store the respirator in a sealed bag when not in use. If working with toxic materials for an extended period of time, a supplied air mask and hood are recommended.

Avoid skin contact with the fibers and other particles by wearing long pants and long sleeves along with gloves or barrier creams. The eyes must be protected using leak-proof goggles (no vent holes) when working with resins or solvents because chemical damage to the eyes is usually irreversible.

Fiber Reinforced Materials

The purpose of reinforcement in reinforced plastics is to provide most of the strength. The three main forms of fiber reinforcements are particles, whiskers, and fibers.

A particle is a square piece of material. Glass bubbles (Q-cell) are hollow glass spheres, and since their dimensions are equal on all axes, they are called a particle.

A whisker is a piece of material that is longer than it is wide. Whiskers are usually single crystals. They are very strong and used to reinforce ceramics and metals.

Fibers are single filaments that are much longer than they are wide. Fibers can be made of almost any material, and are not crystalline like whiskers. Fibers are the base for most composites. Fibers are smaller than the finest human hair and are normally woven into cloth-like materials.

Laminated Structures

Composites can be made with or without an inner core of material. Laminated structure with a core center is called a sandwich structure. Laminate construction is strong and stiff, but heavy. The sandwich laminate is equal in strength, and its weight is much less; less weight is very important to aerospace products.

The core of a laminate can be made from nearly anything. The decision is normally based on use, strength, and fabricating methods to be used.

Various types of cores for laminated structures include rigid foam, wood, metal, or the aerospace preference of honeycomb made from paper, Nomex, carbon, fiberglass or metal. Figure 5-14 shows a typical sandwich structure. It is very important to follow proper techniques to construct or repair laminated structures to ensure the strength is not compromised. A sandwich assembly is made by taking a high-density laminate or solid face and backplate and sandwiching a core in the middle. The selection of materials for the face and backplate are decided by the design engineer, depending on the intended application of the part. It is important to follow manufacturers' maintenance manual specific instructions regarding testing and repair procedures as they apply to a particular aircraft.

Reinforced Plastic

Reinforced plastic is a thermosetting material used in the manufacture of radomes, antenna covers, and wingtips, and as insulation for various pieces of electrical equipment and fuel cells. It has excellent dielectric characteristics which make it ideal for radomes; however, its high strength-to-weight ratio, resistance to mildew, rust, and rot, and ease of fabrication make it equally suited for other parts of the aircraft.

Reinforced plastic components of aircraft are formed of either solid laminates or sandwich-type laminates. Resins used to impregnate glass cloths are of the contact pressure type (requiring little or no pressure during cure). These resins are supplied as a liquid which can vary in viscosity from a waterlike consistency to a thick syrup. Cure or polymerization is effected by the use of a catalyst, usually benzoyl peroxide.

Solid laminates are constructed of three or more layers of resin impregnated cloths "wet laminated" together to form a solid sheet facing or molded shape.

Sandwich-type laminates are constructed of two or more solid sheet facings or a molded shape enclosing a fiberglass honeycomb or foam-type core. Honeycomb cores are made of glass cloths impregnated with a polyester or a combination of nylon and phenolic resins. The specific density and cell size of honeycomb cores varies over a considerable latitude. Honeycomb cores are normally fabricated in blocks that are later cut to the desired thickness on a bandsaw.

Foam-type cores are formulated from combinations of alkyd resins and metatoluene di-isocyanate. Sandwich-type fiberglass components filled with foam-type cores

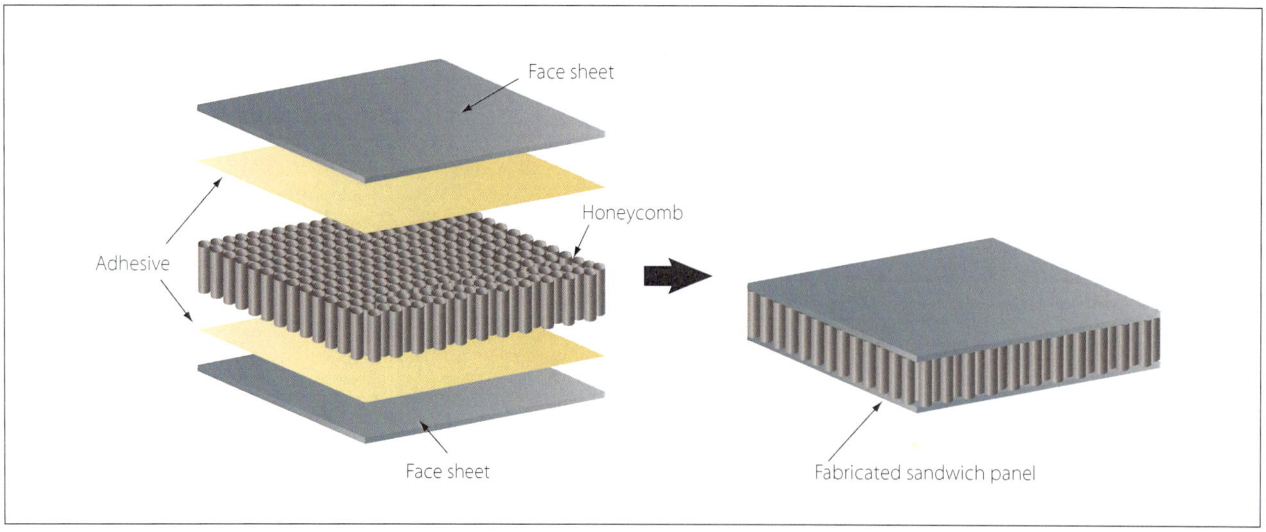

Figure 5-14. Sandwich structure.

are manufactured to exceedingly close tolerances on overall thickness of the molded facing and core material. To achieve this accuracy, the resin is poured into a close tolerance, molded shape. The resin formulation immediately foams up to fill the void in the molded shape and forms a bond between the facing and the core.

Rubber

Rubber is used to prevent the entrance of dirt, water, or air, and to prevent the loss of fluids, gases, or air. It is also used to absorb vibration, reduce noise, and cushion impact loads.

The term "rubber" is as all inclusive as the term "metal." It is used to include not only natural rubber, but all synthetic and silicone rubbers.

Natural Rubber

Natural rubber has better processing and physical properties than synthetic or silicone rubber. These properties include: flexibility, elasticity, tensile strength, tear strength, and low heat buildup due to flexing (hysteresis). Natural rubber is a general purpose product; however, its suitability for aircraft use is somewhat limited because of its inferior resistance to most influences that cause deterioration. Although it provides an excellent seal for many applications, it swells and often softens in all aircraft fuels and in many solvents (naphthas, and so forth). Natural rubber deteriorates more rapidly than synthetic rubber. It is used as a sealing material for water/methanol systems.

Synthetic Rubber

Synthetic rubber is available in several types, each of which is compounded of different materials to give the desired properties. The most widely used are the butyls, Bunas, and neoprene.

Butyl is a hydrocarbon rubber with superior resistance to gas permeation. It is also resistant to deterioration; however, its comparative physical properties are significantly less than those of natural rubber. Butyl will resist oxygen, vegetable oils, animal fats, alkalies, ozone, and weathering.

Like natural rubber, butyl will swell in petroleum or coal tar solvents. It has a low water absorption rate and good resistance to heat and low temperature. Depending on the grade, it is suitable for use in temperatures ranging from −65 °F to 300 °F. Butyl is used with phosphate ester hydraulic fluids (Skydrol), silicone fluids, gases, ketones, and acetones.

Buna-S rubber resembles natural rubber both in processing and performance characteristics. Buna-S is as water resistant as natural rubber, but has somewhat better aging characteristics. It has good resistance to heat, but only in the absence of severe flexing. Generally, Buna-S has poor resistance to gasoline, oil, concentrated acids, and solvents. Buna-S is normally used for tires and tubes as a substitute for natural rubber.

Buna-N is outstanding in its resistance to hydrocarbons and other solvents; however, it has poor resilience in solvents at low temperature. Buna-N compounds have good resistance to temperatures up to 300 °F, and may be procured for low temperature applications down to −75 °F. Buna-N has fair tear, sunlight, and ozone resistance. It has good abrasion resistance and good breakaway properties when used in contact with metal. When used as a seal on a hydraulic piston, it will not stick to the cylinder wall. Buna-N is used for oil and gasoline hose, tank linings, gaskets, and seals.

Neoprene can take more punishment than natural rubber and has better low temperature characteristics. It possesses exceptional resistance to ozone, sunlight, heat, and aging. Neoprene looks and feels like rubber. Neoprene, however, is less like rubber in some of its characteristics than butyl or Buna. The physical characteristics of neoprene, such as tensile strength and elongation, are not equal to natural rubber but do have a definite similarity. Its tear resistance as well as its abrasion resistance is slightly less than that of natural rubber. Although its distortion recovery is complete, it is not as rapid as natural rubber.

Neoprene has superior resistance to oil. Although it is good material for use in nonaromatic gasoline systems, it has poor resistance to aromatic gasolines. Neoprene is used primarily for weather seals, window channels, bumper pads, oil resistant hose, and carburetor diaphragms. It is also recommended for use with Freons™ and silicate ester lubricants.

Thiokol, known also as polysulfide rubber, has the highest resistance to deterioration but ranks the lowest in physical properties. Thiokol, in general, is not seriously affected by petroleum, hydrocarbons, esters, alcohols, gasoline, or water. Thiokols are ranked low in such physical properties as compression set, tensile strength, elasticity, and tear abrasion resistance. Thiokol is used for oil hose, tank linings for aromatic aviation gasolines, gaskets, and seals.

Silicone rubbers are a group of plastic rubber materials made from silicon, oxygen, hydrogen, and carbon. The silicons have excellent heat stability and very low temperature flexibility. They are suitable for gaskets, seals, or other applications where elevated temperatures up to 600 °F are prevalent. Silicone rubbers are also resistant to temperatures down to −150 °F. Throughout this temperature range, silicone rubber remains extremely flexible and useful with no hardness or gumminess. Although this material has good resistance to oils, it reacts unfavorably to both aromatic and nonaromatic gasolines.

Silastic, one of the best known silicones, is used to insulate electrical and electronic equipment. Because of its dielectric properties over a wide range of temperatures, it remains flexible and free from crazing and cracking. Silastic is also used for gaskets and seals in certain oil systems.

Shock Absorber Cord

Shock absorber cord is made from natural rubber strands encased in a braided cover of woven cotton cords treated to resist oxidation and wear. Great tension and elongation are obtained by weaving the jacket upon the bundle of rubber strands while they are stretched about three times their original length.

There are two types of elastic shock absorbing cord. Type I is a straight cord, and type II is a continuous ring, known as a "bungee." The advantages of the type II cord are that it is easily and quickly replaced and does not need to be secured by stretching and whipping. Shock cord is available in standard diameters from ¼ inch to 13⁄16 inch.

Three colored threads are braided into the outer cover for the entire length of the cord. Two of these threads are of the same color and represent the year of manufacture; the third thread, a different color, represents the quarter of the year in which the cord was made. The code covers a 5-year period and then repeats itself. This makes it easy to figure forward or backward from the years shown in Figure 5-15.

Year	Threads	Color
2000	2	black
2001	2	green
2002	2	red
2003	2	blue
2004	2	yellow
2005	2	black
2006	2	green
2007	2	red
2008	2	blue
2009	2	yellow
2010	2	black

Quarter Marking		
Quarter	Threads	Color
Jan., Feb., Mar.	1	red
Apr., May, June	1	blue
July, Aug., Sept.	1	green
Oct., Nov., Dec.	1	yellow

Figure 5-15. Shock absorber cord color coding.

Seals

Seals are used to prevent fluid from passing a certain point, as well as to keep air and dirt out of the system in which they are used. The increased use of hydraulics and pneumatics in aircraft systems has created a need for packings and gaskets of varying characteristics and design to meet the many variations of operating speeds and temperatures to which they are subjected. No one style or type of seal is satisfactory for all installations. Some of the reasons for this are: (1) pressure at which the system operates, (2) the type fluid used in the system, (3) the metal finish and the clearance between adjacent parts, and (4) the type motion (rotary or reciprocating), if any. Seals are divided into three main classes: (1) packings, (2) gaskets, and (3) wipers.

Packings

Packings are made of synthetic or natural rubber. They are generally used as "running seals," that is, in units that contain moving parts, such as actuating cylinders, pumps, selector valves, and so forth. Packings are made in the form of O-rings, V-rings, and U-rings, each designed for a specific purpose. [Figure 5-16]

O-Ring Packings

O-ring packings are used to prevent both internal and external leakage. This type of packing ring seals effectively in both directions and is the type most commonly used. In installations subject to pressures above 1,500 psi, backup rings are used with O-rings to prevent extrusion.

When an O-ring packing is subjected to pressure from both sides, as in actuating cylinders, two backup rings must be used (one on either side of the O-ring). When an O-ring is subject to pressure on only one side, a

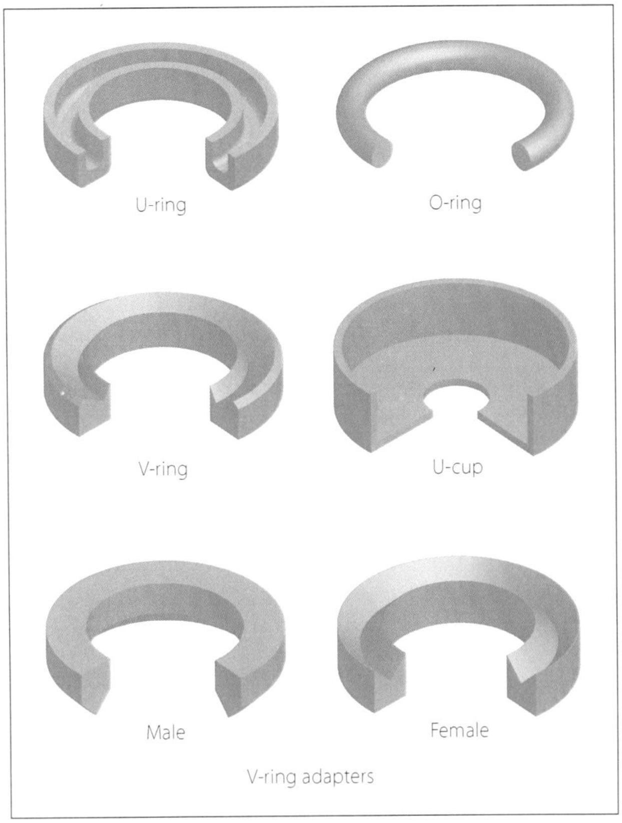

Figure 5-16. Packing rings.

single backup ring is generally used. In this case, the backup ring is always placed on the side of the O-ring away from the pressure.

The materials from which O-rings are manufactured have been compounded for various operating conditions, temperatures, and fluids. An O-ring designed specifically for use as a static (stationary) seal probably will not do the job when installed on a moving part, such as a hydraulic piston. Most O-rings are similar in appearance and texture, but their characteristics may differ widely. An O-ring is useless if it is not compatible with the system fluid and operating temperature.

Advances in aircraft design have necessitated new O-ring compositions to meet changed operating conditions. Hydraulic O-rings were originally established under AN specification numbers (6227, 6230, and 6290) for use in MIL-H-5606 fluid at temperatures ranging from −65 °F to +160 °F. When new designs raised operating temperatures to a possible 275 °F, more compounds were developed and perfected.

Recently, a compound was developed that offered improved low temperature performance without sacrificing high temperature performance, rendering the other series obsolete. This superior material was adopted in the MS28775 series. This series is now the standard for MIL-H-5606 systems in which the temperature may vary from −65 °F to +275 °F.

Manufacturers provide color coding on some O-rings, but this is not a reliable or complete means of identification. The color coding system does not identify sizes, but only system fluid or vapor compatibility and in some cases the manufacturer. Color codes on O-rings that are compatible with MIL-H-5606 fluid will always contain blue, but may also contain red or other colors. Packings and gaskets suitable for use with Skydrol fluid will always be coded with a green stripe, but may also have a blue, grey, red, green, or yellow dot as a part of the color code. Color codes on O-rings that are compatible with hydrocarbon fluid will always contain red, but will never contain blue. A colored stripe around the circumference indicates that the O-ring is a boss gasket seal. The color of the stripe indicates fluid compatibility: red for fuel, blue for hydraulic fluid.

The coding on some rings is not permanent. On others it may be omitted due to manufacturing difficulties or interference with operation. Furthermore, the color coding system provides no means to establish the age of the O-ring or its temperature limitations.

Because of the difficulties with color coding, O-rings are available in individual hermetically sealed envelopes, labeled with all pertinent data. When selecting an O-ring for installation, the basic part number on the sealed envelope provides the most reliable compound identification.

Although an O-ring may appear perfect at first glance, slight surface flaws may exist. These flaws are often capable of preventing satisfactory O-ring performance under the variable operating pressures of aircraft systems; therefore, O-rings should be rejected for flaws that will affect their performance. Such flaws are difficult to detect, and one aircraft manufacturer recommends using a 4 power magnifying glass with adequate lighting to inspect each ring before it is installed.

By rolling the ring on an inspection cone or dowel, the inner diameter surface can also be checked for small cracks, particles of foreign material, or other irregularities that will cause leakage or shorten the life of the O-ring. The slight stretching of the ring when it is rolled inside out will help to reveal some defects not otherwise visible.

Backup Rings

Backup rings (MS28782) made of Teflon™ do not deteriorate with age, are unaffected by any system fluid or vapor, and can tolerate temperature extremes in excess of those encountered in high pressure hydraulic systems. Their dash numbers indicate not only their size but also relate directly to the dash number of the O-ring for which they are dimensionally suited. They are procurable under a number of basic part numbers, but they are interchangeable; that is, any Teflon™ backup ring may be used to replace any other Teflon™ backup ring if it is of proper overall dimension to support the applicable O-ring. Backup rings are not color coded or otherwise marked and must be identified from package labels.

The inspection of backup rings should include a check to ensure that surfaces are free from irregularities, that the edges are clean cut and sharp, and that scarf cuts are parallel. When checking Teflon™ spiral backup rings, make sure that the coils do not separate more than ¼ inch when unrestrained.

V-Ring Packings

V-ring packings (AN6225) are one-way seals and are always installed with the open end of the "V" facing the pressure. V-ring packings must have a male and female adapter to hold them in the proper position after installation. It is also necessary to torque the seal retainer to the value specified by the manufacturer of the component being serviced, or the seal may not give satisfactory service. An installation using V-rings is shown in Figure 5-17.

U-Ring Packings

U-ring packings (AN6226) and U-cup packings are used in brake assemblies and brake master cylinders. The U-ring and U-cup will seal pressure in only one direction; therefore, the lip of the packings must face toward the pressure. U-ring packings are primarily low pressure packings to be used with pressures of less than 1,000 psi.

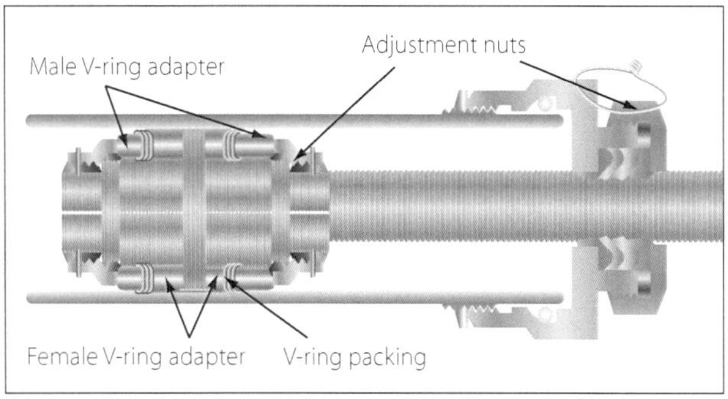

Figure 5-17. V-ring installation.

Gaskets

Gaskets are used as static (stationary) seals between two flat surfaces. Some of the more common gasket materials are asbestos, copper, cork, and rubber. Asbestos sheeting is used wherever a heat resistant gasket is needed. It is used extensively for exhaust system gaskets. Most asbestos exhaust gaskets have a thin sheet of copper edging to prolong their life.

A solid copper washer is used for spark plug gaskets where it is essential to have a noncompressible, yet semisoft gasket.

Cork gaskets can be used as an oil seal between the engine crankcase and accessories, and where a gasket is required that is capable of occupying an uneven or varying space caused by a rough surface or expansion and contraction.

Rubber sheeting can be used where there is a need for a compressible gasket. It should not be used in any place where it may come in contact with gasoline or oil because the rubber will deteriorate very rapidly when exposed to these substances. Gaskets are used in fluid systems around the end caps of actuating cylinders, valves, and other units. The gasket generally used for this purpose is in the shape of an O-ring, similar to O-ring packings.

Wipers

Wipers are used to clean and lubricate the exposed portions of piston shafts. They prevent dirt from entering the system and help protect the piston shaft against scoring.

Wipers may be either metallic or felt. They are sometimes used together, a felt wiper installed behind a metallic wiper.

Sealing Compounds

Certain areas of all aircraft are sealed to withstand pressurization by air, to prevent leakage of fuel, to prevent passage of fumes, or to prevent corrosion by sealing against the weather. Most sealants consist of two or more ingredients properly proportioned and compounded to obtain the best results. Some materials are ready for use as packaged, but others will require mixing before application.

One Part Sealants

One part sealants are prepared by the manufacturer and are ready for application as packaged. However, the consistency of some of these compounds may be altered to satisfy a particular method of application. If thinning is desired, use the thinner recommended by the sealant manufacturer.

Two Part Sealants

Two part sealants are compounds requiring separate packaging to prevent cure prior to application and are identified as the base sealing compound and the accelerator. Any alteration of the prescribed ratios will reduce the quality of the material. Generally, two-part sealants are mixed by combining equal portions (by weight) of base compound and accelerator.

All sealant material should be carefully weighed in accordance with the sealant manufacturer's recommendations. Sealant material is usually weighed with a balance scale equipped with weights specially prepared for various quantities of sealant and accelerator.

Before weighing the sealant materials, thoroughly stir both the base sealant compound and the accelerator. Do not use accelerator which is dried out, lumpy, or flaky. Preweighed sealant kits do not require weighing of the sealant and accelerator before mixing when the entire quantity is to be mixed.

After determining the proper amount of base sealant compound and accelerator, add the accelerator to the base sealant compound. Immediately after adding the accelerator, thoroughly mix the two parts by stirring or folding, depending on the consistency of the material. Carefully mix the material to prevent entrapment of air in the mixture. Overly rapid or prolonged stirring will build up heat in the mixture and shorten the normal application time (working life) of the mixed sealant.

To ensure a well mixed compound, test by smearing a small portion on a clean, flat metal or glass surface. If flecks or lumps are found, continue mixing. If the flecks or lumps cannot be eliminated, reject the batch.

The working life of mixed sealant is from ½ hour to 4 hours (depending upon the class of sealant); therefore, apply mixed sealant as soon as possible or place in refrigerated storage. Figure 5-18 presents general information concerning various sealants.

The curing rate of mixed sealants varies with changes in temperature and humidity. Curing of sealants will be extremely slow if the temperature is below 60 °F. A temperature of 77 °F with 50 percent relative humidity is the ideal condition for curing most sealants.

Curing may be accelerated by increasing the temperature, but the temperature should never be allowed to exceed 120 °F at any time in the curing cycle. Heat may be applied by using infrared lamps or heated air. If heated air is used, it must be properly filtered to remove moisture and dirt.

Heat should not be applied to any faying surface sealant installation until all work is completed. All faying surface applications must have all attachments, permanent or temporary, completed within the application limitations of the sealant.

Sealant must be cured to a tack-free condition before applying brush top coatings. (Tack-free consistency is the point at which a sheet of cellophane pressed onto the sealant will no longer adhere.)

Aircraft Hardware

Aircraft hardware is the term used to describe the various types of fasteners and miscellaneous small items used in the manufacture and repair of aircraft. The importance of aircraft hardware is often overlooked because of its small size; however, the safe and efficient operation of any aircraft is greatly dependent upon the correct selection and use of aircraft hardware.

An aircraft, even though made of the best materials and strongest parts, would be of doubtful value unless those parts were firmly held together. Several methods are used to hold metal parts together; they include riveting, bolting, brazing, and welding. The process used must produce a union that will be as strong as the parts that are joined.

Identification

Most items of aircraft hardware are identified by their specification number or trade name. Threaded fasteners and rivets are usually identified by AN (Air Force-Navy), NAS (National Aircraft Standard), or MS (Military Standard) numbers. Quick-release fasteners are usually identified by factory trade names and size designations.

Threaded Fasteners

Various types of fastening devices allow quick dismantling or replacement of aircraft parts that must be taken apart and put back together at frequent intervals. Riveting or welding these parts each time they are serviced would soon weaken or ruin the joint. Furthermore, some joints require greater tensile strength and stiffness than rivets can provide. Bolts and screws are two types of fastening devices which give the required security of attachment and rigidity. Generally, bolts are used where great strength is required, and screws are used where strength is not the deciding factor. Bolts

Sealant Base	Accelerator (Catalyst)	Mixing Ratio by Weight	Application Life (Work)	Storage (Shelf) Life After Mixing	Storage (Shelf) Life Unmixed	Temperature Range	Application and Limitations
EC-801(black) MIL-S-7502A Class B-2	EC-807	12 parts of EC-807 to 100 parts of EC-801	2–4 hours	5 days at −20 °F after flash freeze at −65 °F	6 months	−65 °F to 200 °F	Faying surfaces, fillet seals, and packing gaps
EC-800 (red)	None	Use as is	8–12 hours	Not applicable	6–9 months	−65 °F to 200 °F	Coating rivet
EC-612 P (pink) MIL-P-20628	None	Use as is	Indefinite non-drying	Not applicable	6–9 months	−40 °F to 200 °F	Packing voids up to ¼ inch
PR-1302HT (red) MIL-S-8784	PR-1302HT-A	10 parts of PR-1302HT-A to 100 parts of PR-1302HT	2–4 hours	5 days at −20 °F after flash freeze at −65 °F	6 months	−65 °F to 200 °F	Sealing access door gaskets
PR-727 potting compound MIL-S-8516B	PR-727A	12 parts of PR-727A to 100 parts of PR-727	1½ hours minimum	5 days at −20 °F after flash freeze at −65 °F	6 months	−65 °F to 200 °F	Potting electrical connections and bulkhead seals
HT-3 (greygreen)	None	Use as is	Solvent release, sets up in 2–4 hours	Not applicable	6–9 months	−60 °F to 850 °F	Sealing hot air ducts passing through bulkheads
EC-776 (clear amber) MIL-S-4383B	None	Use as is	8–12 hours	Not applicable	Indefinite in airtight containers	−65 °F to 250 °F	Top coating

Figure 5-18. General sealant information.

and screws are similar in many ways. They are both used for fastening or holding, and each has a head on one end and screw threads on the other. Regardless of these similarities, there are several distinct differences between the two types of fasteners. The threaded end of a bolt is always blunt while that of a screw may be either blunt or pointed.

The threaded end of a bolt usually has a nut screwed onto it to complete the assembly. The threaded end of a screw may fit into a female receptacle, or it may fit directly into the material being secured. A bolt has a fairly short threaded section and a comparatively long grip length or unthreaded portion; whereas a screw has a longer threaded section and may have no clearly defined grip length. A bolt assembly is generally tightened by turning the nut on the bolt; the head of the bolt may or may not be designed for turning. A screw is always tightened by turning its head.

When it becomes necessary to replace aircraft fasteners, a duplicate of the original fastener should be used if at all possible. If duplicate fasteners are not available, extreme care and caution must be used in selecting substitutes.

Classification of Threads

Aircraft bolts, screws, and nuts are threaded in the NC (American National Coarse) thread series, the NF (American National Fine) thread series, the UNC (American Standard Unified Coarse) thread series, or the UNF (American Standard Unified Fine) thread series. There is one difference between the American National series and the American Standard Unified series that should be pointed out. In the 1-inch diameter size, the NF thread specifies14 threads per inch (1-14 NF), while the UNF thread specifies 12 threads per inch (1-12 UNF). Both types of threads are designated by the number of times the incline (threads) rotates around a 1-inch length of a given diameter bolt or screw. For example, a 4-28 thread indicates that a ¼-inch (4⁄16 inch) diameter bolt has 28 threads in 1 inch of its threaded length.

Threads are also designated by Class of fit. The Class of a thread indicates the tolerance allowed in manufacturing. Class 1 is a loose fit, Class 2 is a free fit, Class 3 is a medium fit, and Class 4 is a close fit. Aircraft bolts are almost always manufactured in the Class 3, medium fit.

A Class 4 fit requires a wrench to turn the nut onto a bolt, whereas a Class 1 fit can easily be turned with the fingers. Generally, aircraft screws are manufactured with a Class 2 thread fit for ease of assembly.

Bolts and nuts are also produced with right-hand and left-hand threads. A right-hand thread tightens when turned clockwise; a left-hand thread tightens when turned counterclockwise.

Aircraft Bolts

Aircraft bolts are fabricated from cadmium- or zinc-plated corrosion resistant steel, unplated corrosion resistant steel, or anodized aluminum alloys. Most bolts used in aircraft structures are either general purpose, AN bolts, or NAS internal wrenching or close tolerance bolts, or MS bolts. In certain cases, aircraft manufacturers make bolts of different dimensions or greater strength than the standard types. Such bolts are made for a particular application, and it is of extreme importance to use like bolts in replacement. Special bolts are usually identified by the letter "S" stamped on the head.

AN bolts come in three head styles—hex head, clevis, and eyebolt. [Figure 5-19] NAS bolts are available in hex head, internal wrenching, and countersunk head styles. MS bolts come in hex head and internal wrenching styles.

General Purpose Bolts

The hex head aircraft bolt (AN-3 through AN-20) is an all-purpose structural bolt used for general applications involving tension or shear loads where a light drive fit is permissible (0.006-inch clearance for a ⅝-inch hole, and other sizes in proportion).

Alloy steel bolts smaller than No. 10-32 and aluminum alloy bolts smaller than ¼ inch in diameter are not used in primary structures. Aluminum alloy bolts and nuts are not used where they will be repeatedly removed for purposes of maintenance and inspection. Aluminum alloy nuts may be used with cadmium-plated steel bolts loaded in shear on land airplanes, but are not used on seaplanes due to the increased possibility of dissimilar metal corrosion.

The AN-73 drilled head bolt is similar to the standard hex bolt, but has a deeper head which is drilled to receive wire for safetying. The AN-3 and the AN-73 series bolts are interchangeable, for all practical purposes, from the standpoint of tension and shear strengths.

Close Tolerance Bolts

This type of bolt is machined more accurately than the general purpose bolt. Close tolerance bolts may be hex headed (AN-173 through AN-186) or have a 100° countersunk head (NAS-80 through NAS-86). They are used in applications where a tight drive fit is required. (The bolt will move into position only when struck with a 12- to 14-ounce hammer.)

Internal Wrenching Bolts

These bolts, (MS-20004 through MS-20024 or NAS-495) are fabricated from high-strength steel and are suitable for use in both tension and shear applications. When they are used in steel parts, the bolt hole must be slightly countersunk to seat the large corner radius of the shank at the head. In Dural material, a special heat-treated washer must be used to provide an adequate bearing surface for the head. The head of the internal wrenching bolt is recessed to allow the insertion of an internal wrench when installing or removing the bolt. Special high-strength nuts are used on these bolts. Replace an internal wrenching bolt with another internal wrenching bolt. Standard AN hex head bolts and washers cannot be substituted for them as they do not have the required strength.

Identification and Coding

Bolts are manufactured in many shapes and varieties. A clear-cut method of classification is difficult. Bolts can be identified by the shape of the head, method of securing, material used in fabrication, or the expected usage.

AN-type aircraft bolts can be identified by the code markings on the bolt heads. The markings generally denote the bolt manufacturer, the material of which the bolt is made, and whether the bolt is a standard AN-type or a special purpose bolt. AN standard steel

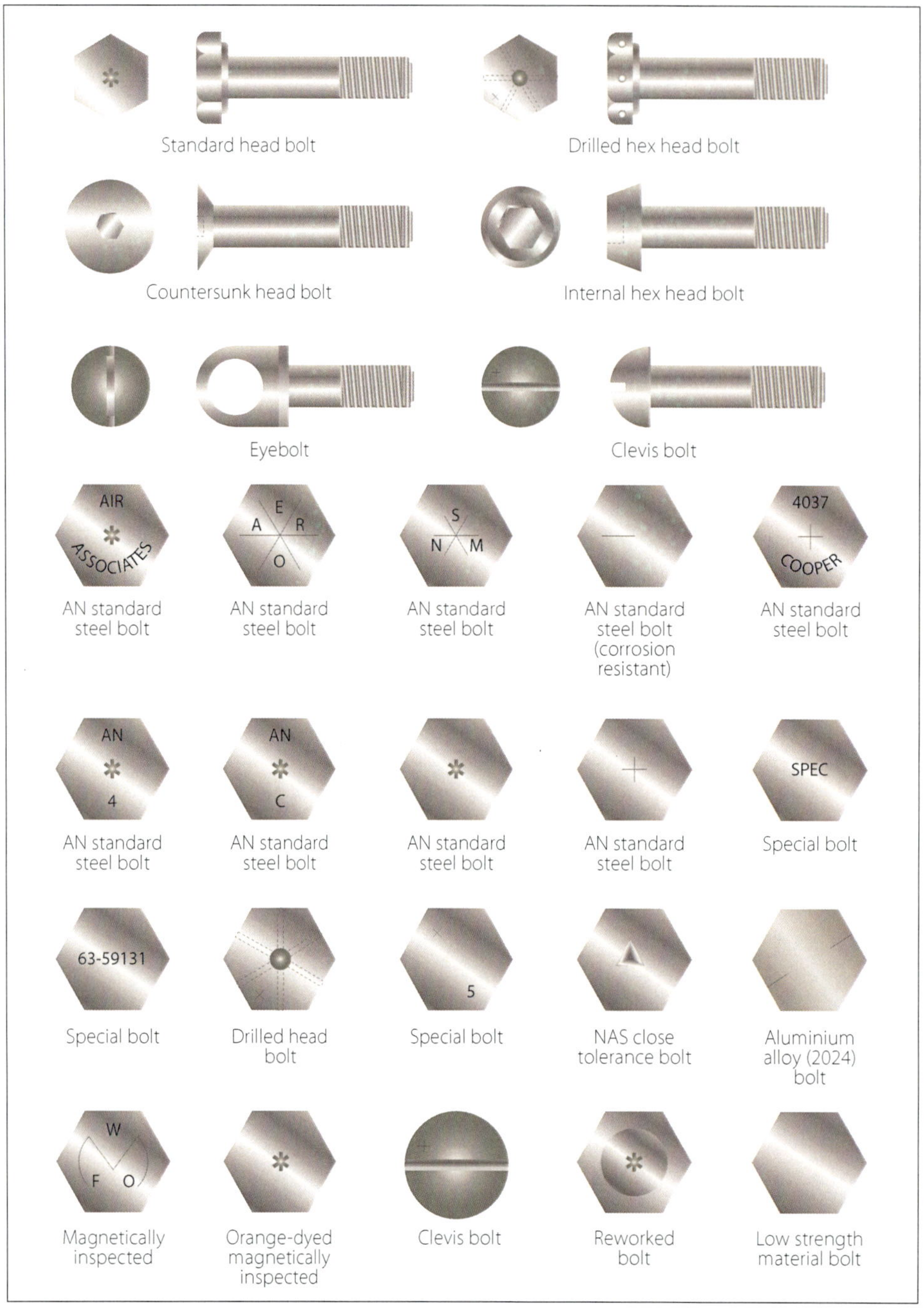

Figure 5-19. Aircraft bolt identification.

bolts are marked with either a raised dash or asterisk; corrosion resistant steel is indicated by a single raised dash; and AN aluminum alloy bolts are marked with two raised dashes. Additional information, such as bolt diameter, bolt length, and grip length may be obtained from the bolt part number.

For example, in the bolt part number AN3DD5A, the "AN" designates that it is an Air Force-Navy Standard bolt, the "3" indicates the diameter in sixteenths of an inch ($^3/_{16}$), the "DD" indicates the material is 2024 aluminum alloy. The letter "C" in place of the "DD" would indicate corrosion resistant steel, and the absence of the letters would indicate cadmium plated steel. The "5" indicates the length in eighths of an inch ($^5/_8$), and the "A" indicates that the shank is undrilled. If the letter "H" preceded the "5" in addition to the "A" following it, the head would be drilled for safetying.

Close tolerance NAS bolts are marked with either a raised or recessed triangle. The material markings for NAS bolts are the same as for AN bolts, except that they may be either raised or recessed. Bolts inspected magnetically (Magnaflux) or by fluorescent means

(Zyglo) are identified by means of colored lacquer, or a head marking of a distinctive type.

Special-Purpose Bolts

Bolts designed for a particular application or use are classified as special-purpose bolts. Clevis bolts, eyebolts, Jo-bolts, and lockbolts are special-purpose bolts.

Clevis Bolts

The head of a clevis bolt is round and is either slotted to receive a common screwdriver or recessed to receive a crosspoint screwdriver. This type of bolt is used only where shear loads occur and never in tension. It is often inserted as a mechanical pin in a control system.

Eyebolt

This type of special purpose bolt is used where external tension loads are to be applied. The eyebolt is designed for the attachment of such devices as the fork of a turnbuckle, a clevis, or a cable shackle. The threaded end may or may not be drilled for safetying.

Jo-Bolt

Jo-bolt is a trade name for an internally threaded three-piece rivet. The Jo-bolt consists of three parts—a threaded steel alloy bolt, a threaded steel nut, and an expandable stainless steel sleeve. The parts are factory preassembled. As the Jo-bolt is installed, the bolt is turned while the nut is held. This causes the sleeve to expand over the end of the nut, forming the blind head and clamping against the work. When driving is complete, a portion of the bolt breaks off. The high shear and tensile strength of the Jo-bolt makes it suitable for use in cases of high stresses where some of the other blind fasteners would not be practical. Jo-bolts are often a part of the permanent structure of late model aircraft. They are used in areas which are not often subjected to replacement or servicing. (Because it is a three-part fastener, it should not be used where any part, in becoming loose, could be drawn into the engine air intake.) Other advantages of using Jo-bolts are their excellent resistance to vibration, weight saving, and fast installation by one person.

Presently, Jo-bolts are available in four diameters: The 200 series, approximately $\frac{3}{16}$ inch in diameter; the 260 series, approximately $\frac{1}{4}$ inch in diameter; the 312 series, approximately $\frac{5}{16}$ inch in diameter; and the 375 series, approximately $\frac{3}{8}$ inch in diameter. Jo-bolts are available in three head styles which are: F (flush), P (hex head), and FA (flush millable).

Lockbolts

Lockbolts are used to attach two materials permanently. They are lightweight and are equal in strength to standard bolts. Lockbolts are manufactured by several companies and conform to Military Standards. Military Standards specify the size of a lockbolt's head in relation to the shank diameter, plus the alloy used in its construction. The only drawback to lockbolt installations is that they are not easily removable compared to nuts and bolts.

The lockbolt combines the features of a high-strength bolt and rivet, but it has advantages over both. The lockbolt is generally used in wing splice fittings, landing gear fittings, fuel cell fittings, longerons, beams, skin splice plates, and other major structural attachments. It is more easily and quickly installed than the conventional rivet or bolt and eliminates the use of lockwashers, cotter pins, and special nuts. Like the rivet, the lockbolt requires a pneumatic hammer or "pull gun" for installation; when installed, it is rigidly and permanently locked in place. Three types of lockbolts are commonly used: the pull type, the stump type, and the blind type. [Figure 5-20]

Pull type. Pull-type lockbolts are used mainly in aircraft primary and secondary structures. They are installed

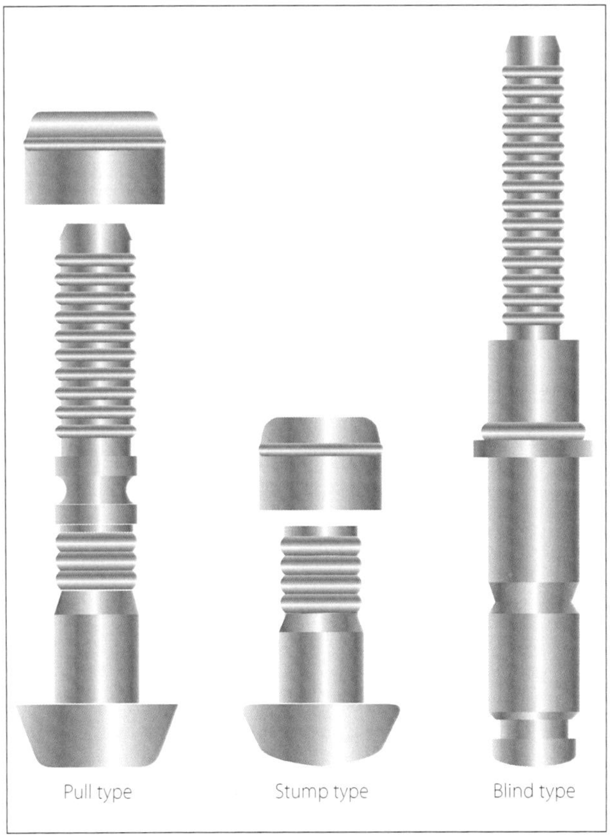

Figure 5-20. Lockbolt types.

very rapidly and have approximately one-half the weight of equivalent AN steel bolts and nuts. A special pneumatic "pull gun" is required to install this type of lockbolt. Installation can be accomplished by one person since bucking is not required.

Stump type. Stump-type lockbolts, although they do not have the extended stem with pull grooves, are companion fasteners to pull-type lockbolts. They are used primarily where clearance will not permit installation of the pull-type lockbolt. A standard pneumatic riveting hammer (with a hammer set attached for swaging the collar into the pin locking grooves) and a bucking bar are tools necessary for the installation of stump-type lockbolts.

Blind type. Blind-type lockbolts come as complete units or assemblies. They have exceptional strength and sheet pull-together characteristics. Blind lockbolts are used where only one side of the work is accessible and, generally, where it is difficult to drive a conventional rivet. This type of lockbolt is installed in the same manner as the pull-type lockbolt.

Common features. Common features of the three types of lockbolts are the annular locking grooves on the pin and the locking collar which is swaged into the pin's lock grooves to lock the pin in tension. The pins of the pull- and blind-type lockbolts are extended for pull installation. The extension is provided with pulling grooves and a tension breakoff groove.

Composition. The pins of pull- and stump-type lockbolts are made of heat-treated alloy steel or high strength aluminum alloy. Companion collars are made of aluminum alloy or mild steel. The blind lockbolt consists of a heat-treated alloy steel pin, blind sleeve and filler sleeve, mild steel collar, and carbon steel washer.

Substitution. Alloy steel lockbolts may be used to replace steel high-shear rivets, solid steel rivets, or AN bolts of the same diameter and head type. Aluminum alloy lockbolts may be used to replace solid aluminum alloy rivets of the same diameter and head type. Steel and aluminum alloy lockbolts may also be used to replace steel and 2024T aluminum alloy bolts, respectively, of the same diameter. Blind lockbolts may be used to replace solid aluminum alloy rivets, stainless steel rivets, or all blind rivets of the same diameter.

Numbering system. The numbering systems for the various types of lockbolts are explained by the breakouts in Figure 5-23.

Grip No.	Grip Range		Grip No.	Grip Range	
	Min.	Max.		Min.	Max.
1	.031	.094	17	1.031	1.094
2	.094	.156	18	1.094	1.156
3	.156	.219	19	1.156	1.219
4	.219	.281	20	1.219	1.281
5	.281	.344	21	1.281	1.344
6	.344	.406	22	1.344	1.406
7	.406	.469	23	1.406	1.469
8	.469	.531	24	1.469	1.531
9	.531	.594	25	1.531	1.594
10	.594	.656	26	1.594	1.656
11	.656	.718	27	1.656	1.718
12	.718	.781	28	1.718	1.781
13	.781	.843	29	1.781	1.843
14	.843	.906	30	1.843	1.906
15	.906	.968	31	1.906	1.968
16	.968	1.031	32	1.968	2.031
			33	2.031	2.094

Figure 5-21. Pull- and stump-type lockbolt grip ranges.

Grip Range. To determine the bolt grip range required for any application, measure the thickness of the material with a hook scale inserted through the hole. Once this measurement is determined, select the correct grip range by referring to the charts provided by the rivet manufacturer. Examples of grip range charts are shown in Figures 5-21 and 5-24.

When installed, the lockbolt collar should be swaged substantially throughout the complete length of the collar. The tolerance of the broken end of the pin relative to the top of the collar must be within the dimensions given in Figure 5-22.

Pin diameter	Tolerance Below		Above
3⁄16	0.079	to	0.032
1⁄4	0.079	to	0.050
5⁄16	0.079	to	0.050
3⁄8	0.079	to	0.060

Figure 5-22. Pin tolerance ranges.

Pull-type lockbolt

ALPP H T 8 8

ALPP | Head type
- ACT509 = close tolerance AN-509 C-sink head
- ALPP = pan head
- ALPB = brazier head
- ALP509 = standard AN-509 C-sink head
- ALP426 = standard AN-426 C-sink head

H | Class fit
- H = hole filling (interference fit)
- N = non-hole filling (clearance fit)

T | Pin Materials
- E = 75S-T6 aluminum alloy
- T = heat-treated alloy steel

8 | Body diameter in 32nds of an inch

8 | Grip length in 16ths of an inch

Blind-type lockbolt

BL 8 4

BL | Blind Lockbolt

8 | Diameter in 32nds of an inch

4 | Grip length in 16ths of an inch, ± 1/32 inch

Lockbolt collar

LC C C

LC | Lockbolt collar

C | Material
- C = 24ST aluminum alloy (green color). Use with heat-treated alloy lockbolts only.
- F = 61ST aluminum alloy (plain color). Use with 75ST aluminum alloy lockbolts only.
- R = mild steel (cadmium plated). Use with heat-treated alloy steel lockbolts for high temperature applications only.

C | Diameter of a pin in 32nds of an inch

Stump-type lockbolt

ALSF E 8 8

ALSF | Head type
- ASCT509 = close tolerance AN-509 C-sink head
- ALSF = flathead type.
- ALS509 = standard AN-509 C-sink head
- ALS426 = standard AN-426 C-sink head

E | Pin materials
- E = 75S-T6 aluminum alloy
- T = heat-treated alloy steel

8 | Body diameter in 32nds of an inch

8 | Grip length in 16ths of an inch

Figure 5-23. Lockbolt numbering system.

When removal of a lockbolt becomes necessary, remove the collar by splitting it axially with a sharp, cold chisel. Be careful not to break out or deform the hole. The use of a backup bar on the opposite side of the collar being split is recommended. The pin may then be driven out with a drift punch.

Aircraft Nuts

Aircraft nuts are made in a variety of shapes and sizes. They are made of cadmium plated carbon steel, stainless steel, or anodized 2024T aluminum alloy, and may be obtained with either right- or left-hand threads. No identifying marking or lettering appears on nuts. They can be identified only by the characteristic metallic luster or color of the aluminum, brass, or the insert when the nut is of the self-locking type. They can be further identified by their construction.

Aircraft nuts can be divided into two general groups: non-self-locking and self-locking nuts. Non-self-locking nuts are those that must be safetied by external locking devices, such as cotter pins, safety wire, or locknuts. Self-locking nuts contain the locking feature as an integral part.

Non-Self-Locking Nuts

Most of the familiar types of nuts, including the plain nut, the castle nut, the castellated shear nut, the plain

¼-inch Diameter			5/16-inch Diameter		
Grip No.	Grip Range		Grip No.	Grip Range	
	Min.	Max.		Min.	Max.
1	.031	.094	2	.094	.156
2	.094	.156	3	.156	.219
3	.156	.219	4	.219	.281
4	.219	.281	5	.281	.344
5	.281	.344	6	.344	.406
6	.344	.406	7	.406	.469
7	.406	.469	8	.469	.531
8	.469	.531	9	.531	.594
9	.531	.594	10	.594	.656
10	.594	.656	11	.656	.718
11	.656	.718	12	.718	.781
12	.718	.781	13	.781	.843
13	.781	.843	14	.843	.906
14	.843	.906	15	.906	.968
15	.906	.968	16	.968	1.031
16	.968	1.031	17	1.031	1.094
17	1.031	1.094	18	1.094	1.156
18	1.094	1.156	19	1.156	1.219
19	1.156	1.219	20	1.219	1.281
20	1.219	1.281	21	1.281	1.343
21	1.281	1.343	22	1.343	1.406
22	1.344	1.406	23	1.406	1.469
23	1.406	1.469	24	1.460	1.531
24	1.469	1.531			
25	1.531	1.594			

Figure 5-24. Blind-type lockbolt grip ranges.

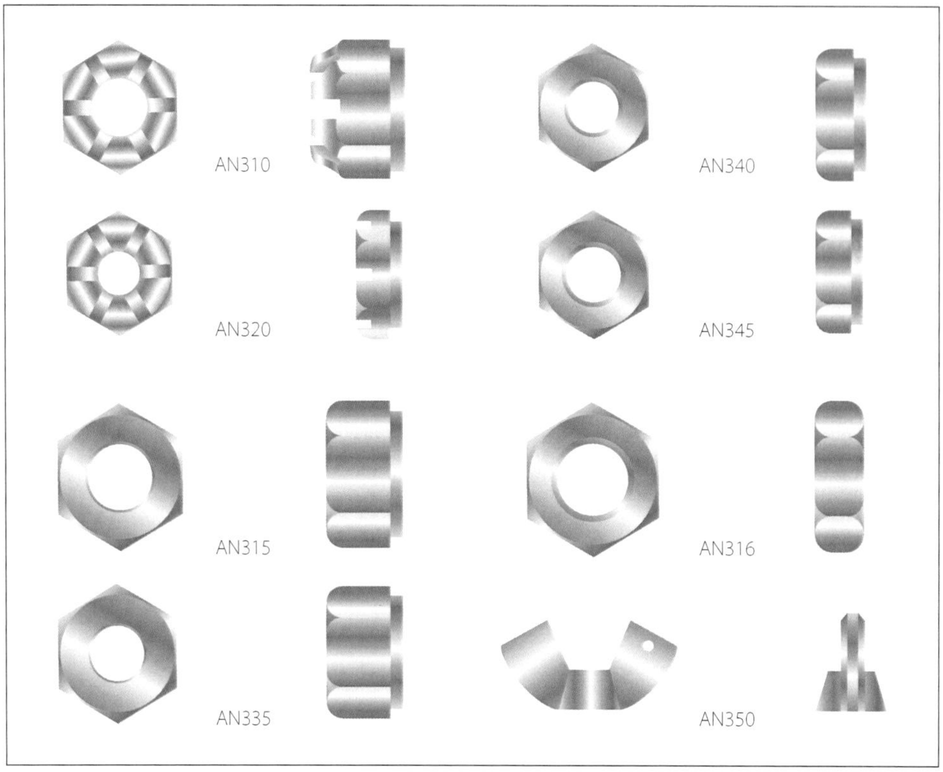

Figure 5-25. Non-self-locking nuts.

hex nut, the light hex nut, and the plain checknut are the non-self-locking type. [Figure 5-25]

The castle nut, AN310, is used with drilled shank AN hex head bolts, clevis bolts, eyebolts, drilled head bolts, or studs. It is fairly rugged and can withstand large tensional loads. Slots (called castellations) in the nut are designed to accommodate a cotter pin or lockwire for safety.

The castellated shear nut, AN320, is designed for use with devices (such as drilled clevis bolts and threaded taper pins) which are normally subjected to shearing stress only. Like the castle nut, it is castellated for safetying. Note, however, that the nut is not as deep or as strong as the castle nut; also that the castellations are not as deep as those in the castle nut.

The plain hex nut, AN315 and AN335 (fine and coarse thread), is of rugged construction. This makes it suitable for carrying large tensional loads. However, since it requires an auxiliary locking device, such as a checknut or lockwasher, its use on aircraft structures is somewhat limited.

The light hex nut, AN340 and AN345 (fine and coarse thread), is a much lighter nut than the plain hex nut and must be locked by an auxiliary device. It is used for miscellaneous light tension requirements.

The plain checknut, AN316, is employed as a locking device for plain nuts, set screws, threaded rod ends, and other devices.

The wing nut, AN350, is intended for use where the desired tightness can be obtained with the fingers and where the assembly is frequently removed.

Self-Locking Nuts

As their name implies, self-locking nuts need no auxiliary means of safetying but have a safetying feature included as an integral part of their construction. Many types of self-locking nuts have been designed and their use has become quite widespread. Common applications are: (1) attachment of antifriction bearings and control pulleys; (2) attachment of accessories, anchor nuts around inspection holes and small tank installation openings; and (3) attachment of rocker box covers and exhaust stacks. Self-locking nuts are acceptable for use on certificated aircraft subject to the restrictions of the manufacturer.

Self-locking nuts are used on aircraft to provide tight connections which will not shake loose under severe vibration. Do not use self-locking nuts at joints which subject either the nut or bolt to rotation. They may be used with antifriction bearings and control pulleys, provided the inner race of the bearing is clamped to the supporting structure by the nut and bolt. Plates must be attached to the structure in a positive manner to eliminate rotation or misalignment when tightening the bolts or screws.

The two general types of self-locking nuts currently in use are the all-metal type and the fiber lock type. For the sake of simplicity, only three typical kinds of self-locking nuts are considered in this handbook: the Boots self-locking and the stainless steel self-locking nuts, representing the all-metal types; and the elastic stop nut, representing the fiber insert type.

Boots Self-Locking Nut

The Boots self-locking nut is of one piece, all-metal construction, designed to hold tight in spite of severe vibration. Note in Figure 5-26 that it has two sections and is essentially two nuts in one, a locking nut and a load-carrying nut. The two sections are connected with a spring which is an integral part of the nut.

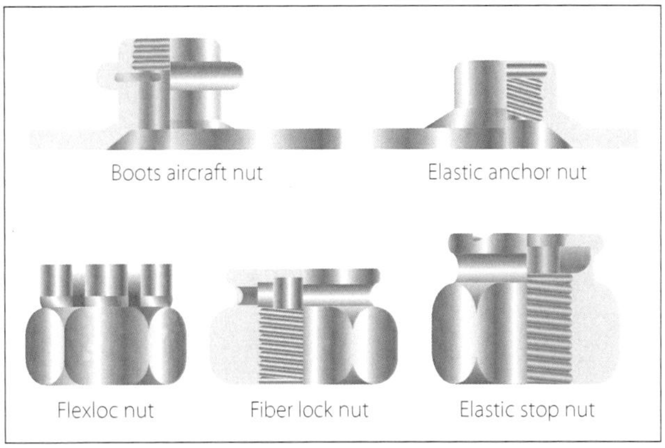

Figure 5-26. Self-locking nuts.

The spring keeps the locking and load-carrying sections such a distance apart that the two sets of threads are out of phase; that is, so spaced that a bolt which has been screwed through the load carrying section must push the locking section outward against the force of the spring to engage the threads of the locking section properly.

Thus, the spring, through the medium of the locking section, exerts a constant locking force on the bolt in the same direction as a force that would tighten the nut. In this nut, the load-carrying section has the thread strength of a standard nut of comparable size, while the locking section presses against the threads of the bolt and locks the nut firmly in position. Only a wrench applied to the nut will loosen it. The nut can be removed and reused without impairing its efficiency.

Boots self-locking nuts are made with three different spring styles and in various shapes and sizes. The wing type, which is the most common, ranges in size for No. 6 up to $\frac{1}{4}$ inch, the Rol-top ranges from $\frac{1}{4}$ inch to $\frac{1}{6}$ inch, and the bellows type ranges in size from No. 8 up to $\frac{3}{8}$ inch. Wing-type nuts are made of anodized aluminum alloy, cadmium-plated carbon steel, or stainless steel. The Rol-top nut is cadmium-plated steel, and the bellows type is made of aluminum alloy only.

Stainless Steel Self-Locking Nut

The stainless steel self-locking nut may be spun on and off with the fingers, as its locking action takes place only when the nut is seated against a solid surface and tightened. The nut consists of two parts: a case with a beveled locking shoulder and key, and a threaded insert with a locking shoulder and slotted keyway. Until the nut is tightened, it spins on the bolt easily because the threaded insert is the proper size for the bolt. However, when the nut is seated against a solid surface and tightened, the locking shoulder of the insert is pulled downward and wedged against the locking shoulder of the case. This action compresses the threaded insert and causes it to clench the bolt tightly. The cross-sectional view in Figure 5-27 shows how the key of the case fits into the slotted keyway of the insert so that when the case is turned, the threaded insert is turned with it. Note that the slot is wider than the key. This permits the slot to be narrowed and the insert to be compressed when the nut is tightened.

Elastic Stop Nut

The elastic stop nut is a standard nut with the height increased to accommodate a fiber locking collar. This

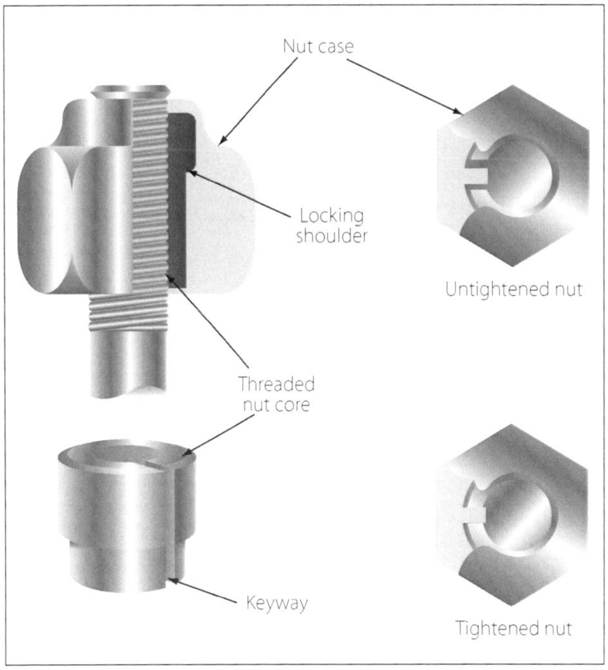

Figure 5-27. Stainless steel self-locking nut.

fiber collar is very tough and durable and is unaffected by immersion in hot or cold water or ordinary solvents, such as ether, carbon tetrachloride, oils, and gasoline. It will not damage bolt threads or plating.

As shown in Figure 5-28, the fiber locking collar is not threaded and its inside diameter is smaller than the largest diameter of the threaded portion or the outside diameter of a corresponding bolt. When the nut is screwed onto a bolt, it acts as an ordinary nut until the bolt reaches the fiber collar. When the bolt is screwed into the fiber collar, however, friction (or drag) causes the fiber to be pushed upward. This creates a heavy downward pressure on the load carrying part and automatically throws the load carrying sides of the nut and bolt threads into positive contact. After the bolt has been forced all the way through the fiber collar, the downward pressure remains constant. This pressure locks and holds the nut securely in place even under severe vibration.

Nearly all elastic stop nuts are steel or aluminum alloy. However, such nuts are available in practically any kind of metal. Aluminum alloy elastic stop nuts are supplied with an anodized finish. Steel nuts are cadmium plated.

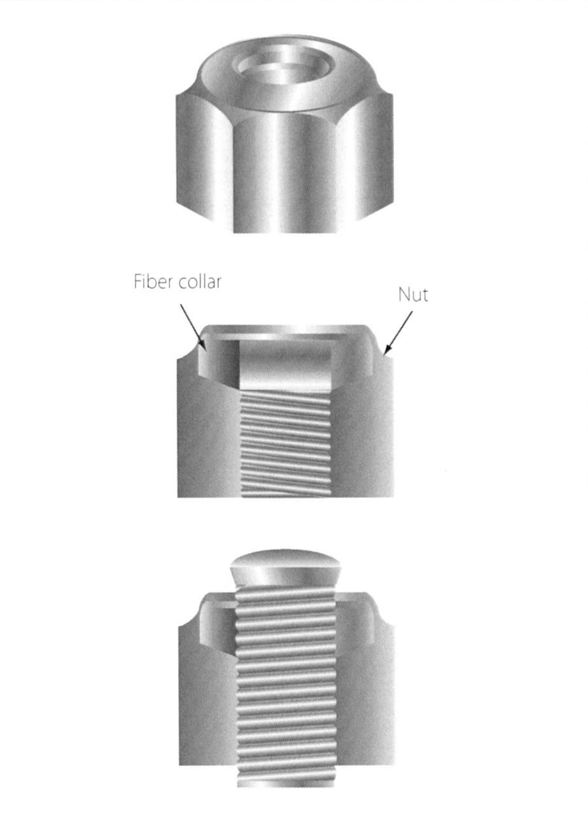

Figure 5-28. Elastic stop nut.

Normally, elastic stop nuts can be used many times with complete safety and without detriment to their locking efficiency. When reusing elastic stop nuts, be sure the fiber has not lost its locking friction or become brittle. If a nut can be turned with the fingers, replace it.

After the nut has been tightened, make sure the rounded or chamfered end of the bolts, studs, or screws extends at least the full round or chamfer through the nut. Flat end bolts, studs, or screws should extend at least $\frac{1}{32}$ inch through the nut. Bolts of $\frac{5}{16}$-inch diameter and over with cotter pin holes may be used with self-locking nuts, but only if free from burrs around the holes. Bolts with damaged threads and rough ends are not acceptable. Do not tap the fiber locking insert. The self-locking action of the elastic stop nut is the result of having the bolt threads impress themselves into the untapped fiber.

Do not install elastic stop nuts in places where the temperature is higher than 250 °F, because the effectiveness of the self-locking action is reduced beyond this point. Self-locking nuts may be used on aircraft engines and accessories when their use is specified by the engine manufacturer.

Self-locking nut bases are made in a number of forms and materials for riveting and welding to aircraft structure or parts. [Figure 5-29] Certain applications require the installation of self-locking nuts in channels, an arrangement which permits the attachment of many nuts with only a few rivets. These channels are track-like bases with regularly spaced nuts which are either removable or nonremovable. The removable type carries a floating nut, which can be snapped in or out of the channel, thus making possible the easy removal of damaged nuts. Nuts, such as the clinch-type and spline-type, which depend on friction for their anchorage, are not acceptable for use in aircraft structures.

Sheet Spring Nuts

Sheet spring nuts, such as speed nuts, are used with standard and sheet metal self-tapping screws in nonstructural locations. They find various uses in supporting line clamps, conduit clamps, electrical equipment, access doors, and the like, and are available in several types. Speed nuts are made from spring steel and are arched prior to tightening. This arched spring lock prevents the screw from working loose. These nuts should be used only where originally used in the fabrication of the aircraft.

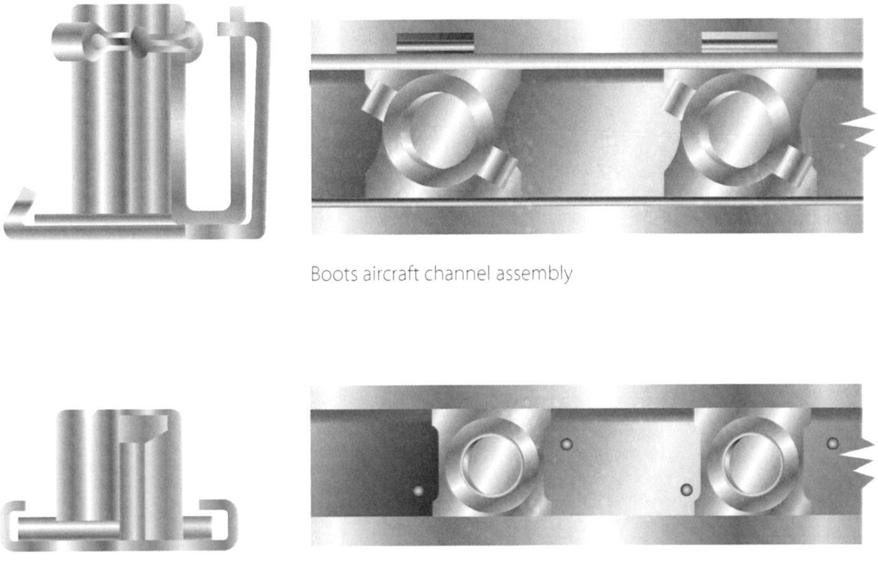

Figure 5-29. Self-locking nut bases.

Internal and External Wrenching Nuts

Two commercial types of high strength internal or external wrenching nuts are available; they are the internal and external wrenching elastic stop nut and the *Unbrako* internal and external wrenching nut. Both are of the self-locking type, are heat treated, and are capable of carrying high strength bolt tension loads.

Identification and Coding

Part numbers designate the type of nut. The common types and their respective part numbers are: Plain, AN315 and AN335; castle AN310; plain check, AN316; light hex, AN340 and AN345; and castellated shear, AN320. The patented self-locking types are assigned part numbers ranging from MS20363 through MS20367. The Boots, the Flexloc, the fiber locknut, the elastic stop nut, and the self-locking nut belong to this group. Part number AN350 is assigned to the wing nut.

Letters and digits following the part number indicate such items as material, size, threads per inch, and whether the thread is right or left hand. The letter "B" following the part number indicates the nut material to be brass, a "D" indicates 2017-T aluminum alloy, a "DD" indicates 2024-T aluminum alloy, a "C" indicates stainless steel, and a dash in place of a letter indicates cadmium-plated carbon steel.

The digit (or two digits) following the dash or the material code letter is the dash number of the nut, and it indicates the size of the shank and threads per inch of the bolt on which the nut will fit. The dash number corresponds to the first figure appearing in the part number coding of general purpose bolts. A dash and the number 3, for example, indicates that the nut will fit an AN3 bolt (10-32); a dash and the number 4 means it will fit an AN4 bolt (1/4-28); a dash and the number 5, an AN5 bolt (5/16-24); and so on.

The code numbers for self-locking nuts end in three or four digit numbers. The last two digits refer to threads per inch, and the one or two preceding digits stand for the nut size in 16ths of an inch.

Some other common nuts and their code numbers are:

Code Number AN310D5R:
AN310 = aircraft castle nut
D = 2024-T aluminum alloy
5 = $^5/_{16}$ inch diameter
R = right-hand thread (usually 24 threads per inch)

Code Number AN320-10:
AN320 = aircraft castellated shear nut, cadmium-plated carbon steel
10 = $^5/_8$ inch diameter, 18 threads per inch (this nut is usually right-hand thread)

Code Number AN350B1032:
AN350 = aircraft wing nut
B = brass
10 = number 10 bolt
32 = threads per inch

Aircraft Washers

Aircraft washers used in airframe repair are either plain, lock, or special type washers.

Plain Washers

Plain washers [Figure 5-30], both the AN960 and AN970, are used under hex nuts. They provide a smooth bearing surface and act as a shim in obtaining correct grip length for a bolt and nut assembly. They are used to adjust the position of castellated nuts in respect to drilled cotter pin holes in bolts. Use plain washers under lockwashers to prevent damage to the surface material.

Aluminum and aluminum alloy washers may be used under bolt heads or nuts on aluminum alloy or magnesium structures where corrosion caused by dissimilar metals is a factor. When used in this manner, any electric current flow will be between the washer and the steel bolt. However, it is common practice to use a cadmium plated steel washer under a nut bearing directly against a structure as this washer will resist the cutting action of a nut better than an aluminum alloy washer.

The AN970 steel washer provides a greater bearing area than the AN960 washer and is used on wooden structures under both the head and the nut of a bolt to prevent crushing the surface.

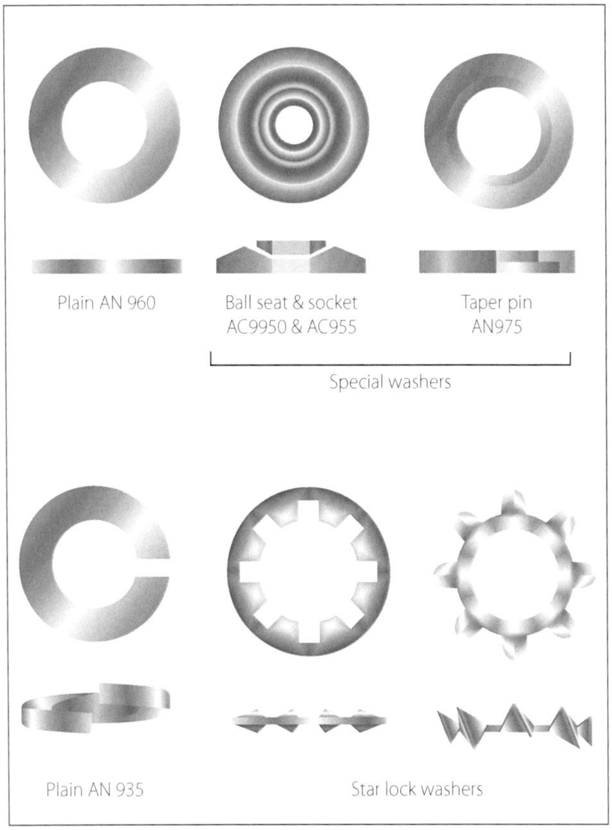

Figure 5-30. Various types of washers.

Lockwashers

Lockwashers, both the AN935 and AN936, are used with machine screws or bolts where the self-locking or castellated-type nut is not appropriate. The spring action of the washer (AN935) provides enough friction to prevent loosening of the nut from vibration. [Figure 5-30]

Lockwashers should never be used under the following conditions:

- With fasteners to primary or secondary structures
- With fasteners on any part of the aircraft where failure might result in damage or danger to the aircraft or personnel
- Where failure would permit the opening of a joint to the airflow
- Where the screw is subject to frequent removal
- Where the washers are exposed to the airflow
- Where the washers are subject to corrosive conditions
- Where the washer is against soft material without a plain washer underneath to prevent gouging the surface

Shakeproof Lockwashers

Shakeproof lockwashers are round washers designed with tabs or lips that are bent upward across the sides of a hex nut or bolt to lock the nut in place. There are various methods of securing the lockwasher to prevent it from turning, such as an external tab bent downward 90° into a small hole in the face of the unit, or an internal tab which fits a keyed bolt.

Shakeproof lockwashers can withstand higher heat than other methods of safetying and can be used under high vibration conditions safely. They should be used only once because the tabs tend to break when bent a second time.

Special Washers

The ball socket and seat washers, AC950 and AC955, are special washers used where a bolt is installed at an angle to a surface, or where perfect alignment with a surface is required. These washers are used together. [Figure 5-30]

The NAS143 and MS20002 washers are used for internal wrenching bolts of the NAS144 through NAS158 series. This washer is either plain or countersunk. The countersunk washer (designated as NAS143C and MS20002C) is used to seat the bolt head shank radius, and the plain washer is used under the nut.

Installation of Nuts, Washers, and Bolts

Bolt and Hole Sizes

Slight clearances in bolt holes are permissible wherever bolts are used in tension and are not subject to reversal of load. A few of the applications in which clearance of holes may be permitted are in pulley brackets, conduit boxes, lining trim, and miscellaneous supports and brackets.

Bolt holes are to be normal to the surface involved to provide full bearing surface for the bolt head and nut, and must not be oversized or elongated. A bolt in such a hole will carry none of its shear load until parts have yielded or deformed enough to allow the bearing surface of the oversized hole to contact the bolt. In this respect, remember that bolts do not become swaged to fill up the holes as do rivets.

In cases of oversized or elongated holes in critical members, obtain advice from the aircraft or engine manufacturer before drilling or reaming the hole to take the next larger bolt. Usually, such factors as edge distance, clearance, or load factor must be considered. Oversized or elongated holes in noncritical members can usually be drilled or reamed to the next larger size.

Many bolt holes, particularly those in primary connecting elements, have close tolerances. Generally, it is permissible to use the first lettered drill size larger than the normal bolt diameter, except where the AN hexagon bolts are used in light drive fit (reamed) applications and where NAS close tolerance bolts or AN clevis bolts are used.

Light drive fits for bolts (specified on the repair drawings as 0.0015 inch maximum clearance between bolt and hole) are required in places where bolts are used in repair, or where they are placed in the original structure.

The fit of holes and bolts cannot be defined in terms of shaft and hole diameters; it is defined in terms of the friction between bolt and hole when sliding the bolt into place. A tight drive fit, for example, is one in which a sharp blow of a 12- or 14-ounce hammer is required to move the bolt. A bolt that requires a hard blow and sounds tight is considered to fit too tightly. A light drive fit is one in which a bolt will move when a hammer handle is held against its head and pressed by the weight of the body.

Installation Practices

Examine the markings on the bolt head to determine that each bolt is of the correct material. It is of extreme importance to use like bolts in replacement. In every case, refer to the applicable Maintenance Instructions Manual and Illustrated Parts Breakdown.

Be sure that washers are used under both the heads of bolts and nuts unless their omission is specified. A washer guards against mechanical damage to the material being bolted and prevents corrosion of the structural members. An aluminum alloy washer should be used under the head and nut of a steel bolt securing aluminum alloy or magnesium alloy members. Any corrosion that occurs then attacks the washer rather than the members. Steel washers should be used when joining steel members with steel bolts.

Whenever possible, place the bolt with the head on top or in the forward position. This positioning tends to prevent the bolt from slipping out if the nut is accidentally lost.

Be certain that the bolt grip length is correct. Grip length is the length of the unthreaded portion of the bolt shank. Generally speaking, the grip length should equal the thickness of the material being bolted together. However, bolts of slightly greater grip length may be used if washers are placed under the nut or the bolt head. In the case of plate nuts, add shims under the plate.

Safetying of Bolts and Nuts

It is very important that all bolts or nuts, except the self-locking type, be safetied after installation. This prevents them from loosening in flight due to vibration. Methods of safetying are discussed later in this chapter.

Repair of Damaged Internal Threads

Installation or replacement of bolts is simple when compared to the installation or replacement of studs. Bolt heads and nuts are cut in the open, whereas studs are installed into internal threads in a casting or built-up assembly. Damaged threads on bolts or nuts can be seen and only require replacement of the defective part. If internal threads are damaged, two alternatives are apparent: the part may be replaced or the threads repaired or replaced. Correction of the thread problem is usually cheaper and more convenient. Two methods of repairing are by replacement bushings or helicoils.

Replacement Bushings

Bushings are usually special material (steel or brass spark plug bushings into aluminum cylinder heads). A material that will resist wear is used where removal and replacement is frequent. The external threads on the bushing are usually coarse. The bushing is installed, a thread lock compound may or may not be used, and staked to prevent loosening. Many bushings have left-hand threads external and right-hand threads internal. With this installation, removal of the bolt or stud (right-hand threads) tends to tighten the bushing.

Bushings for common installations, such as spark plugs, may be up to 0.040 oversize (in increments of 0.005). Original installation and overhaul shop replacements are shrunk fit: a heated cylinder head and a frozen bushing.

Helicoils

Helicoils are precision formed screw thread coils of 18-8 stainless steel wire having a diamond shaped cross section. [Figure 5-31] They form unified coarse or unified fine thread classes 2-band 3B when assembled into (helicoil) threaded holes. The assembled insert accommodates UNJ (controlled radius root) male threaded members. Each insert has a driving tang with a notch to facilitate removal of the tang after the insert is screwed into a helicoil tapped hole.

They are used as screw thread bushings. In addition to being used to restore damaged threads, they are used in the original design of missiles, aircraft engines, and all types of mechanical equipment and accessories to protect and strengthen tapped threads in light materials, metals, and plastics, particularly in locations which require frequent assembly and disassembly, and/or where a screw locking action is desired.

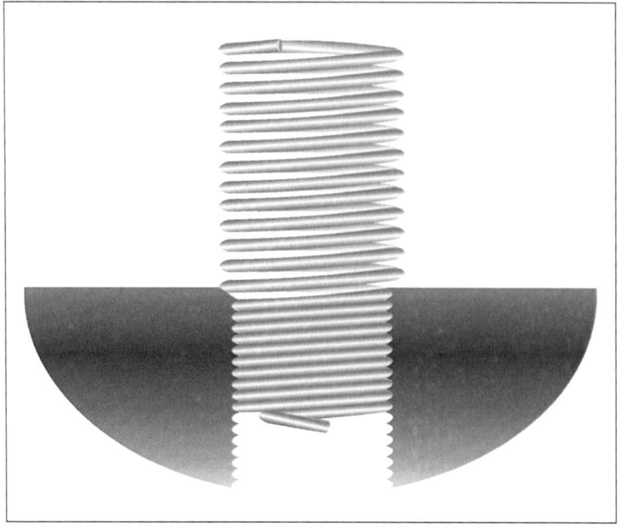

Figure 5-31. Helicoil insert.

Helicoil installation [Figure 5-32] is a 5 or 6 step operation, depending upon how the last step is classed.

Step 1: Determine what threads are damaged.

Step 2: (a) New installation of helicoil. Drill out damaged threads to minimum depth specified.

(b) Previously installed helicoil. Using proper size extracting tool, place edge of blade in 90° from the edge of the insert. Tap with hammer to seat tool. Turn to left, applying pressure, until insert backs out. Threads are not damaged if insert is properly removed.

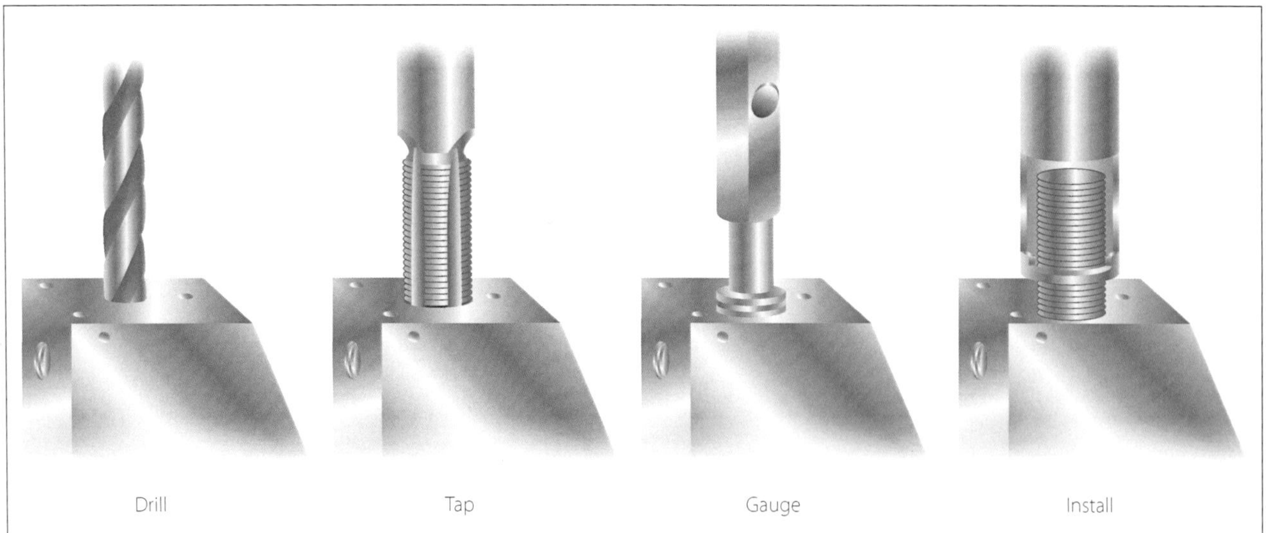

Figure 5-32. Helicoil installation.

Step 3: Tap. Use the tap of required nominal thread size. The tapping procedure is the same as standard thread tapping. Tap length must be equal to or exceed the requirement.

Step 4: Gauge. Threads may be checked with a helicoil thread gauge.

Step 5: Insert assembly. Using proper tool, install insert to a depth that puts end of top coil $\frac{1}{4}$ to $\frac{1}{2}$ turn below the top surface of the tapped hole.

Step 6: Tang breakoff. Select proper breakoff tool. Tangs should be removed from all drilled through holes. In blind holes, the tangs may be removed when necessary if enough hole depth is provided below the tang of the installed insert.

These are not to be considered specific instructions on helicoil installation. The manufacturer's instruction must be followed when making an installation.

Helicoils are available for the following threads: unified coarse, unified fine, metric, spark plug, and national taper pipe threads.

Fastener Torque

Torque and Torque Wrenches

As the speed of an aircraft increases, each structural member becomes more highly stressed. It is therefore extremely important that each member carry no more and no less than the load for which it was designed. To distribute the loads safely throughout a structure, it is necessary that proper torque be applied to all nuts, bolts, studs, and screws. Using the proper torque allows the structure to develop its designed strength and greatly reduces the possibility of failure due to fatigue.

Torque wrenches. The three most commonly used torque wrenches are the flexible beam, rigid frame, and the ratchet types. [Figure 5-33] When using the flexible beam and the rigid frame torque wrenches, the torque value is read visually on a dial or scale mounted on the handle of the wrench.

To use the ratchet type, unlock the grip and adjust the handle to the desired setting on the micrometer type scale, then relock the grip. Install the required socket or adapter to the square drive of the handle. Place the wrench assembly on the nut or bolt and pull the wrench assembly on the nut or bolt and pull in a clockwise direction with a smooth, steady motion. (A fast or jerky motion will result in an improperly torqued unit.) When the applied torque reaches the torque value indicated on

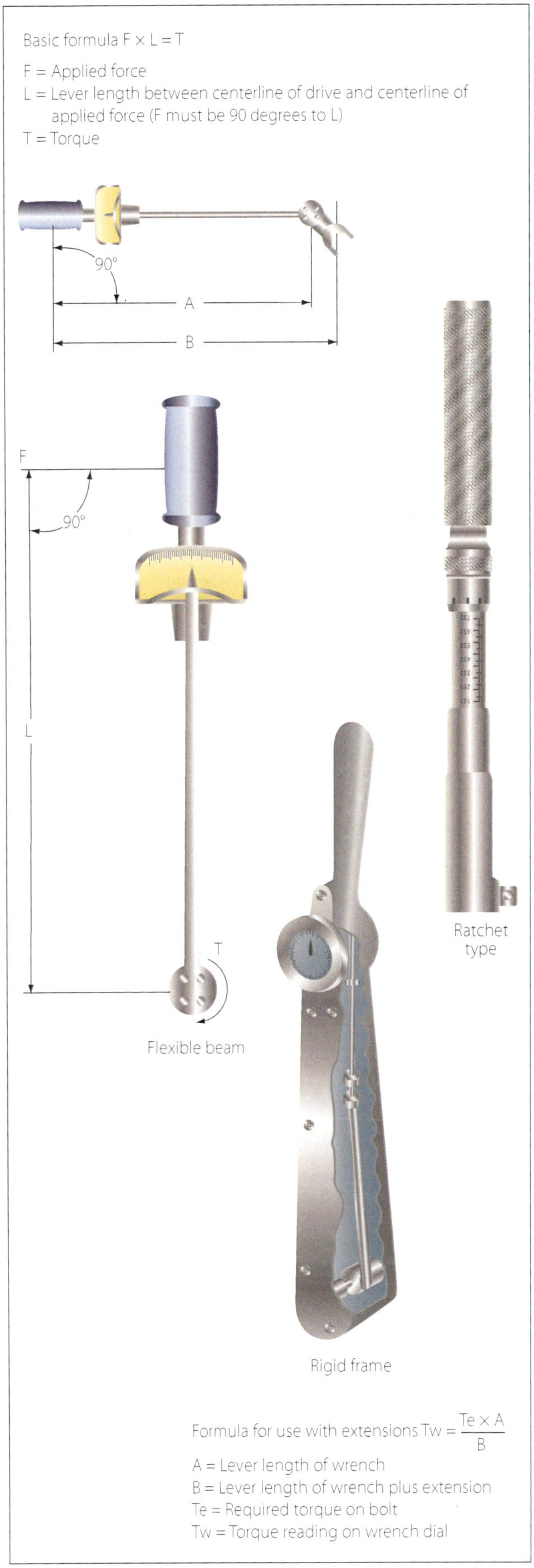

Figure 5-33. Common torque wrenches.

the handle setting, the handle will automatically release or "break" and move freely for a short distance.

The release and free travel is easily felt, so there is no doubt about when the torquing process is completed.

To assure getting the correct amount of torque on the fasteners, all torque wrenches must be tested at least once a month or more often if necessary.

Note: It is not advisable to use a handle length extension on a flexible beam type torque wrench at any time. A handle extension alone has no effect on the reading of the other types. The use of a drive end extension on any type of torque wrench makes the use of the formula mandatory. When applying the formula, force must be applied to the handle of the torque wrench at the point from which the measurements were taken. If this is not done, the torque obtained will be incorrect.

Torque Tables. Use the standard torque table as a guide in tightening nuts, studs, bolts, and screws whenever specific torque values are not called out in maintenance procedures. The following rules apply for correct use of the torque table: [Figure 5-34]

1. To obtain values in foot-pounds, divide inch-pounds by 12.
2. Do not lubricate nuts or bolts except for corrosion-resistant steel parts or where specifically instructed to do so.
3. Always tighten by rotating the nut first if possible. When space considerations make it necessary to tighten by rotating the bolt head, approach the high side of the indicated torque range. Do not exceed the maximum allowable torque value.
4. Maximum torque ranges should be used only when materials and surfaces being joined are of sufficient thickness, area, and strength to resist breaking, warping, or other damage.
5. For corrosion resisting steel nuts, use torque values given for shear type nuts.
6. The use of any type of drive end extension on a torque wrench changes the dial reading required to obtain the actual values indicated in the standard torque range tables. When using a drive end extension, the torque wrench reading must be computed by use of the proper formula, which is included in the handbook accompanying the torque wrench.

Cotter Pin Hole Line Up

When tightening castellated nuts on bolts, the cotter pin holes may not line up with the slots in the nuts for the range of recommended values. Except in cases of highly stressed engine parts, the nut may not be over torque. Remove hardware and realign the holes. The torque loads specified may be used for all unlubricated cadmium-plated steel nuts of the fine or coarse thread series which have approximately equal number of threads and equal face bearing areas. These values do not apply where special torque requirements are specified in the maintenance manual.

If the head end, rather than the nut, must be turned in the tightening operation, maximum torque values may be increased by an amount equal to shank friction, provided the latter is first measured by a torque wrench.

Aircraft Rivets

Sheets of metal must be fastened together to form the aircraft structure, and this is usually done with solid aluminum alloy rivets. A rivet is a metal pin with a formed head on one end when the rivet is manufactured. The shank of the rivet is inserted into a drilled hole, and its shank is then upset (deformed) by a hand or pneumatic tool. The second head, formed either by hand or by pneumatic equipment, is called a "shop head." The shop head functions in the same manner as a nut on a bolt. In addition to their use for joining aircraft skin sections, rivets are also used for joining spar sections, for holding rib sections in place, for securing fittings to various parts of the aircraft, and for fastening innumerable bracing members and other parts together. The rivet creates a bond that is at least as strong as the material being joined.

Two of the major types of rivets used in aircraft are the common solid shank type, which must be driven using a bucking bar, and the special (blind) rivets, which may be installed where it is impossible to use a bucking bar.

Aircraft rivets are not hardware store rivets. Rivets purchased at a hardware store should never be used as a substitute for aircraft quality rivets. The rivets may be made from very different materials, the strength of the rivets differs greatly, and their shear strength qualities are very different. The countersunk heads on hardware store rivets are 78°, whereas countersunk aircraft rivets have 100° angle heads for more surface contact to hold it in place.

Bolt, Stud or Screw Size		Torque Values in Inch-Pounds for Tightening Nuts			
		On standard bolts, studs and screws having a tensile strength of 125,000 to 140,000 psi		On bolts, studs, and screws having a tensile strength of 140,000 to 160,000 psi	On high-strength bolts, studs, and screws having a tensile strength of 160,000 psi and over
		Shear type nuts (AN320, AN364 or equivalent)	Tension type nuts and threaded machine parts (AN-310, AN365 or equivalent)	Any nut, except shear type	Any nut, except shear type
8–32	8–36	7–9	12–15	14–17	15–18
10–24	10–32	12–15	20–25	23–30	25–35
¼–20		25–30	40–50	45–49	50–68
	¼–28	30–40	50–70	60–80	70–90
5⁄16–18		48–55	80–90	85–117	90–144
	5⁄16–24	60–85	100–140	120–172	140–203
⅜–16		95–110	160–185	173–217	185–248
	⅜–24	95–110	160–190	175–271	190–351
7⁄16–14		140–155	235–255	245–342	255–428
	7⁄16–20	270–300	450–500	475–628	500–756
½–13		240–290	400–480	440–636	480–792
	½–20	290–410	480–690	585–840	690–990
9⁄16–12		300–420	500–700	600–845	700–990
	9⁄16–18	480–600	800–1000	900–1,220	1,000–1,440
⅝–11		420–540	700–900	800–1,125	900–1,350
	⅝–18	660–780	1,100–1,300	1,200–1,730	1,300–2,160
¾–10		700–950	1,150–1,600	1,380–1,925	1,600–2,250
	¾–16	1,300–1,500	2,300–2,500	2,400–3,500	2,500–4,500
⅞–9		1,300–1,800	2,200–3,000	2,600–3,570	3,000–4,140
	⅞–14	1,500–1,800	2,500–3,000	2,750–4,650	3,000–6,300
1"–8		2,200–3,000	3,700–5,000	4,350–5,920	5,000–6,840
	1"–14	2,200–3,300	3,700–5,500	4,600–7,250	5,500–9,000
1⅛–8		3,300–4,000	5,500–6,500	6,000–8,650	6,500–10,800
	1⅛–12	3,000–4,200	5,000–7,000	6,000–10,250	7,000–13,500
1¼–8		4,000–5,000	6,500–8,000	7,250–11,000	8,000–14,000
	1¼–12	5,400–6,600	9,000–11,000	10,000–16,750	11,000–22,500

Figure 5-34. Standard torque table (inch-pounds).

Standards and Specifications

The FAA requires that the structural strength and integrity of type-certificated aircraft conform to all airworthiness requirements. These requirements apply to performance, structural strength, and integrity as well flight characteristics. To meet these requirements, each aircraft must meet the same standards. To accomplish standardization, all materials and hardware must be manufactured to a standard of quality. Specifications and standards for aircraft hardware are usually identified by the organization that originated them. Some of the common standardizing organizations include:

AMS	Aeronautical Material Specifications
AN	Air Force-Navy
AND	Air Force-Navy Design
AS	Aeronautical Standard
ASA	American Standards Association
ASTM	American Society for Testing Materials
MS	Military Standard
NAF	Naval Aircraft Factory
NAS	National Aerospace Standard
SAE	Society of Automotive Engineers

When a MS20426-AD4-6 rivet is required, the specifications have already been written for it in the Military Standard (MS) specifications. That information is available to the aircraft manufacturers and to the rivet manufacturers as well as to the mechanic. The specifications designate the material to be used as well as the head type, diameter, and length of the rivet. The use of standardized materials in the production of aircraft makes each aircraft exactly the same as the previous one and makes them less expensive to build.

Aircraft rivets are manufactured to much higher standards and specifications than rivets manufactured for general use. When aircraft manufacturers started building all-metal aircraft in the 1930s, different manufacturers had different rivet head designs. Brazier heads, modified brazier heads, button heads, mushroom heads, flatheads, and 78° countersunk heads were used. As aircraft standardized, four rivet head designs almost completely replaced all of the others. Rivets exposed to the airflow over the top of the structure are usually either universal head MS20470 or 100° countersunk head MS20426 rivets. For rivets used in internal structures, the roundhead MS20430 and the flathead MS20442 are generally used.

Solid Shank Rivets

Solid shank rivets are generally used in repair work. They are identified by the kind of material of which they are made, their head type, size of shank, and their temper condition. The designation of the solid shank rivet head type, such as universal head, roundhead, flathead, countersunk head, and brazier head, depends on the cross-sectional shape of the head. [Figure 5-37] The temper designation and strength are indicated by special markings on the head of the rivet.

The material used for the majority of aircraft solid shank rivets is aluminum alloy. The strength and temper conditions of aluminum alloy rivets are identified by digits and letters similar to those adopted for the identification of strength and temper conditions of aluminum and aluminum alloy stock. The 1100, 2017-T, 2024-T, 2117-T, and 5056 rivets are the five grades usually available.

The 1100 rivet, which is composed of 99.45 percent pure aluminum, is very soft. It is for riveting the softer aluminum alloys, such as 1100, 3003, and 5052, which are used for nonstructural parts (all parts where strength is not a factor). The riveting of map cases is a good example of where a rivet of 1100 aluminum alloy may be used.

The 2117-T rivet, known as the field rivet, is used more than any other for riveting aluminum alloy structures. The field rivet is in wide demand because it is ready for use as received and needs no further heat treating or annealing. It also has a high resistance to corrosion.

The 2017-T and 2024-T rivets are used in aluminum alloy structures where more strength is needed than is obtainable with the same size 2217-T rivet. These rivets are known as "ice box rivets," are annealed, and must be kept refrigerated until they are to be driven. The 2017-T rivet should be driven within approximately 1 hour and the 2024-T rivet within 10 to 20 minutes after removal from refrigeration.

The 5056 rivet is used for riveting magnesium alloy structures because of its corrosion-resistant qualities in combination with magnesium.

Mild steel rivets are used for riveting steel parts. The corrosion-resistant steel rivets are for riveting corrosion-resistant steels in firewalls, exhaust stack brackets, and similar structures.

Monel rivets are used for riveting nickel-steel alloys. They can be substituted for those made of corrosion-resistant steel in some cases.

The use of copper rivets in aircraft repair is limited. Copper rivets can be used only on copper alloys or nonmetallic materials such as leather.

Metal temper is an important factor in the riveting process, especially with aluminum alloy rivets. Aluminum alloy rivets have the same heat-treating characteristics as aluminum alloy stock. They can be hardened and annealed in the same manner as aluminum. The rivet must be soft, or comparatively soft, before a good head can be formed. The 2017-T and 2024-T rivets are annealed before being driven. They harden with age.

The process of heat treating (annealing) rivets is much the same as that for stock. Either an electric air furnace, a salt bath, or a hot oil bath is needed. The heat-treating range, depending on the alloy, is 625 °F to 950 °F. For convenient handling, rivets are heated in a tray or a wire basket. They are quenched in cold water (70 °F) immediately after heat treating.

The 2017-T and 2024-T rivets, which are heat-treatable rivets, begin to age harden within a few minutes after being exposed to room temperature. Therefore, they must be used immediately after quenching or else be placed in cold storage.

The most commonly used means for holding heat-treatable rivets at low temperature (below 32 °F) is to keep them in a refrigerator. They are referred to as "icebox" rivets. Under this storage condition, they will remain soft enough for driving for periods up to 2 weeks. Any rivets not used within that time should be removed for reheat treating.

Icebox rivets attain about one-half their maximum strength in approximately 1 hour after driving and full strength in about 4 days. When 2017-T rivets are exposed to room temperature for 1 hour or longer, they must be subject to reheat treatment. This also applies to 2024-T rivets exposed to room temperature for a period exceeding 10 minutes.

Once an icebox rivet has been taken from the refrigerator, it should not be mixed with the rivets still in cold storage. If more rivets are removed from the refrigerator than can be used in 15 minutes, they should be placed in a separate container and stored for reheat treatment. Heat treatment of rivets may be repeated a number of times if done properly. Proper heating times and temperatures are shown in Figure 5-35.

Heating Time — Air Furnace		
Rivet Alloy	Time at Temperature	Heat Treating Temperature
2024	1 hour	910 °F–930 °F
2017	1 hour	925 °F–950 °F
Heating Time — Salt Bath		
Rivet Alloy	Time at Temperature	Heat Treating Temperature
2024	30 minutes	910 °F–930 °F
2017	30 minutes	925 °F–950 °F

Figure 5-35. Rivet heating times and temperatures.

Most metals, and therefore aircraft rivet stock, are subject to corrosion. Corrosion may be the result of local climatic conditions or the fabrication process used. It is reduced to a minimum by using metals which are highly resistant to corrosion and possess the correct strength-to-weight ratio.

Ferrous metals placed in contact with moist salt air will rust if not properly protected. Nonferrous metals, those without an iron base, do not rust, but a similar process known as corrosion takes place. The salt in moist air (found in the coastal areas) attacks the aluminum alloys. It is a common experience to inspect the rivets of an aircraft which has been operated near salt water and find them badly corroded.

If a copper rivet is inserted into an aluminum alloy structure, two dissimilar metals are brought in contact with each other. Remember, all metals possess a small electrical potential. Dissimilar metals in contact with each other in the presence of moisture cause an electrical current to flow between them and chemical byproducts to be formed. Principally, this results in the deterioration of one of the metals.

Certain aluminum alloys react to each other and, therefore, must be thought of as dissimilar metals. The commonly used aluminum alloys may be divided into the two groups shown in Figure 5-36.

Group A	Group B
1100	2117
3003	2017
5052	2124
6053	7075

Figure 5-36. Aluminum groupings.

Material	Head Marking	AN Material Code	AN425 78° Counter-sunk Head	AN426 100° Counter-sunk Head MS20426*	AN427 100° Counter-sunk Head MS20427*	AN430 Round Head MS20470*	AN435 Round Head MS20613* MS20615*	AN441 Flat Head	AN442 Flat Head MS20470*	AN455 Brazier Head MS20470*	AN456 Brazier Head MS20470*	AN470 Universal Head MS20470*	Heat Treat Before Use	Shear Strength psi	Bearing Strength psi
1100	Plain	A	X	X		X			X	X	X	X	No	10,000	25,000
2117T	Recessed Dot	AD	X	X		X			X	X	X	X	No	30,000	100,000
2017T	Raised Dot	D	X	X		X			X	X	X	X	Yes	34,000	113,000
2017T-HD	Raised Dot	D	X	X		X			X	X	X	X	No	38,000	126,000
2024T	Raised Double Dash	DD	X	X		X			X	X	X	X	Yes	41,000	136,000
5056T	Raised Cross	B		X		X			X	X	X	X	No	27,000	90,000
7075-T73	Three Raised Dashes		X	X		X			X	X	X	X	No		
Carbon Steel	Recessed Triangle				X		X MS20613*	X					No	35,000	90,000
Corrosion Resistant Steel	Recessed Dash	F			X		X MS20613*						No	65,000	90,000
Copper	Plain	C			X		X	X					No	23,000	
Monel	Plain	M			X			X					No	49,000	
Monel (Nickel-Copper Alloy)	Recessed Double Dots	C					X MS20615*						No	49,000	
Brass	Plain						X MS20615*						No		
Titanium	Recessed Large and Small Dot			MS20426									No	95,000	

* New specifications are for design purposes.

Figure 5-37. Rivet identification chart.

Members within either group A or group B can be considered as similar to each other and will not react to others within the same group. A corroding action will take place, however, if any metal of group A comes in contact with a metal in group B in the presence of moisture.

Avoid the use of dissimilar metals whenever possible. Their incompatibility is a factor which was considered when the AN Standards were adopted. To comply with AN Standards, the manufacturers must put a protective surface coating on the rivets. This may be zinc chromate, metal spray, or an anodized finish.

The protective coating on a rivet is identified by its color. A rivet coated with zinc chromate is yellow, an anodized surface is pearl gray, and the metal sprayed rivet is identified by a silvery gray color. If a situation arises in which a protective coating must be applied on the job, paint the rivet with zinc chromate before it is used and again after it is driven.

Identification

Markings on the heads of rivets are used to classify their characteristics. These markings may be either a raised teat, two raised teats, a dimple, a pair of raised dashes, a raised cross, a single triangle, or a raised dash; some other heads have no markings.

The different markings indicate the composition of the rivet stock. As explained previously, the rivets have different colors to identify the manufacturers' protective surface coating.

Roundhead rivets are used in the interior of the aircraft, except where clearance is required for adjacent members. The roundhead rivet has a deep, rounded top surface. The head is large enough to strengthen the sheet around the hole and, at the same time, offer resistance to tension.

The flathead rivet, like the roundhead rivet, is used on interior structures. It is used where maximum strength is needed and where there isn't sufficient clearance to use a roundhead rivet. It is seldom, if ever, used on external surfaces. The brazier head rivet has a head of large diameter, which makes it particularly adaptable for riveting thin sheet stock (skin). The brazier head rivet offers only slight resistance to the airflow, and because of this factor, it is frequently used for riveting skin on exterior surfaces, especially on aft sections of the fuselage and empennage. It is used for riveting thin sheets exposed to the slipstream. A modified brazier head rivet is also manufactured; it is simply a brazier head of reduced diameter.

The universal head rivet is a combination of the roundhead, flathead, and brazier head. It is used in aircraft construction and repair in both interior and exterior locations. When replacement is necessary for protruding head rivets—roundhead, flathead, or brazier head—they can be replaced by universal head rivets.

The countersunk head rivet is flat topped and beveled toward the shank so that it fits into a countersunk or dimpled hole and is flush with the material's surface. The angle at which the head slopes may vary from 78° to 120°. The 100° rivet is the most commonly used type. These rivets are used to fasten sheets over which other sheets must fit. They are also used on exterior surfaces of the aircraft because they offer only slight resistance to the slipstream and help to minimize turbulent airflow.

The markings on the heads of rivets indicate the material of which they are made and, therefore, their strength. Figure 5-37 identifies the rivet head markings and the materials indicated by them. Although there are three materials indicated by a plain head, it is possible to distinguish their difference by color. The 1100 is aluminum color; the mild steel is a typical steel color; and the copper rivet is a copper color. Any head marking can appear on any head style of the same material.

Each type of rivet is identified by a part number so that the user can select the correct rivet for the job. The type of rivet head is identified by AN or MS standard numbers. The numbers selected are in series and each series represents a particular type of head. [Figure 5-37]

The most common numbers and the types of heads they represent are:

AN426 or MS20426—countersunk head rivets (100°)

AN430 or MS20430—roundhead rivets

AN441—flathead rivets

AN456—brazier head rivets

AN470 or MS20470—universal head rivets

There are also letters and numbers added to a part number. The letters designate alloy content; the numbers designate rivet diameter and length. The letters in common use for alloy designation are:

A—aluminum alloy, 1100 or 3003 composition

AD—aluminum alloy, 2117-T composition

D—aluminum alloy, 2017-T composition

DD—aluminum alloy, 2024-T composition

B—aluminum alloy, 5056 composition

C—copper

M—monel

The absence of a letter following the AN standard number indicates a rivet manufactured from mild steel.

The first number following the material composition letters expresses the diameter of the rivet shank in 32nds of an inch (Examples: 3 indicates $\frac{3}{32}$, 5 indicates $\frac{5}{32}$, and so forth). [Figure 5-38]

The last number(s), separated by a dash from the preceding number, expresses the length of the rivet shank in 16ths of an inch (Examples: 3 indicates $\frac{3}{16}$, 7 indicates $\frac{7}{16}$, 11 indicates $\frac{11}{16}$, and so forth). [Figure 5-38]

An example of identification marking of a rivet is:

AN470AD3-5—complete part number

AN—Air Force-Navy standard number

470—universal head rivet

AD—2117-T aluminum alloy

3—$\frac{3}{32}$ in diameter

5—$\frac{5}{16}$ in length

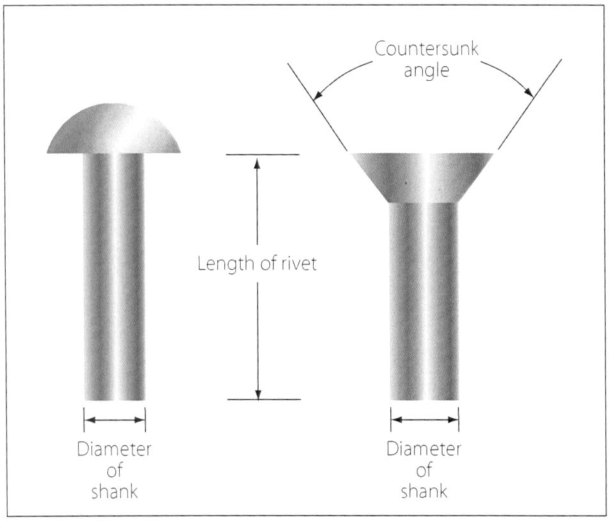

Figure 5-38. Methods of measuring rivets.

Blind Rivets

There are many places on an aircraft where access to both sides of a riveted structure or structural part is impossible, or where limited space will not permit the use of a bucking bar. Also, in the attachment of many nonstructural parts, such as aircraft interior furnishings, flooring, deicing boots, and the like, the full strength of solid shank rivets is not necessary.

For use in such places, special rivets have been designed which can be bucked from the front. They are sometimes lighter than solid shank rivets, yet amply strong for their intended use. These rivets are produced by several manufacturers and have unique characteristics that require special installation tools, special installation procedures and special removal procedures. That is why they are called special rivets. Because these rivets are often inserted in locations where one head (usually the shop head) cannot be seen, they are also called blind rivets.

Mechanically Expanded Rivets

Two classes of mechanically expanded rivets are discussed here:

(1) Nonstructural.

(a) Self-plugging (friction lock) rivets.

(b) Pull-thru rivets.

(2) Mechanical lock, flush fracturing, self plugging rivets.

Self-Plugging Rivets (friction lock)

The self-plugging (friction lock) blind rivets are manufactured by several companies; the same general basic information about their fabrication, composition, uses, selection, installation, inspection, and removal procedures apply to all of them.

Self-plugging (friction lock) rivets are fabricated in two parts: a rivet head with a hollow shank or sleeve, and a stem that extends through the hollow shank. Figure 5-39 illustrates a protruding head and a countersunk head self-plugging rivet produced by one manufacturer.

Several events, in their proper sequence, occur when a pulling force is applied to the stem of the rivet: (1) the stem is pulled into the rivet shank; (2) the mandrel portion of the stem forces the rivet shank to expand; and (3) when friction (or pulling action pressure) becomes great enough, it will cause the stem to snap at a breakoff groove on the stem. The plug portion (bottom end of the stem) is retained in the shank of the rivet giving

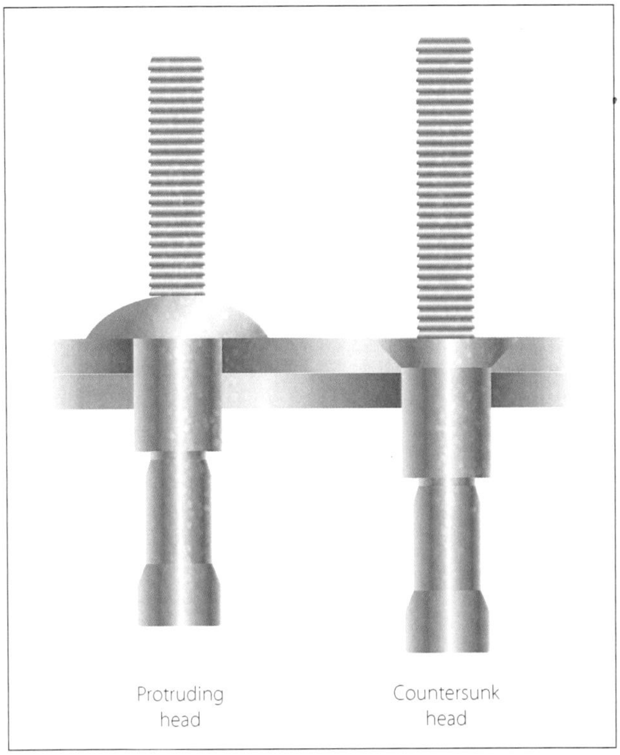

Figure 5-39. Self-plugging (friction lock) rivets.

the rivet much greater shear strength than could be obtained from a hollow rivet.

Self-plugging (friction lock) rivets are fabricated in two common head styles: (1) a protruding head similar to the MS20470 or universal head, and (2) a 100° countersunk head. Other head styles are available from some manufacturers.

The stem of the self-plugging (friction lock) rivet may have a knot or knob on the upper portion, or it may have a serrated portion. [Figure 5-39]

Self-plugging (friction lock) rivets are fabricated from several materials. Rivets are available in the following material combinations: stem 2017 aluminum alloy and sleeve 2117 aluminum alloy; stem 2017 aluminum alloy and sleeve 5056 aluminum alloy; and stem steel and sleeve steel.

Self-plugging (friction lock) rivets are designed so that installation requires only one person; it is not necessary to have the work accessible from both sides. The pulling strength of the rivet stem is such that a uniform job can always be assured. Because it is not necessary to have access to the opposite side of the work, self-plugging (friction lock) rivets can be used to attach assemblies to hollow tubes, corrugated sheet, hollow boxes, and so forth. Because a hammering force is not necessary to install the rivet, it can be used to attach assemblies to plywood or plastics.

Factors to consider in the selection of the correct rivet for installation are: (1) installation location, (2) composition of the material being riveted, (3) thickness of the material being riveted, and (4) strength desired.

If the rivet is to be installed on an aerodynamically smooth surface, or if clearance for an assembly is needed, countersunk head rivets should be selected. In other areas where clearance or smoothness is not a factor, the protruding head type rivet may be utilized.

Material composition of the rivet shank depends upon the type of material being riveted. Aluminum alloy 2117 shank rivets can be used on most aluminum alloys. Aluminum alloy 5056 shank rivets should be used when the material being riveted is magnesium. Steel rivets should always be selected for riveting assemblies fabricated from steel.

The thickness of the material being riveted determines the overall length of the shank of the rivet. As a general rule, the shank of the rivet should extend beyond the material thickness approximately $^{3}/_{64}$ inch to $^{1}/_{8}$ inch before the stem is pulled. [Figure 5-40]

Pull-Thru Rivets

The pull-thru blind rivets are manufactured by several companies; the same general basic information about their fabrication, composition, uses, selection, installation, inspection, and removal procedures apply to all of them.

Pull-thru rivets are fabricated in two parts: a rivet head with a hollow shank or sleeve and a stem that extends

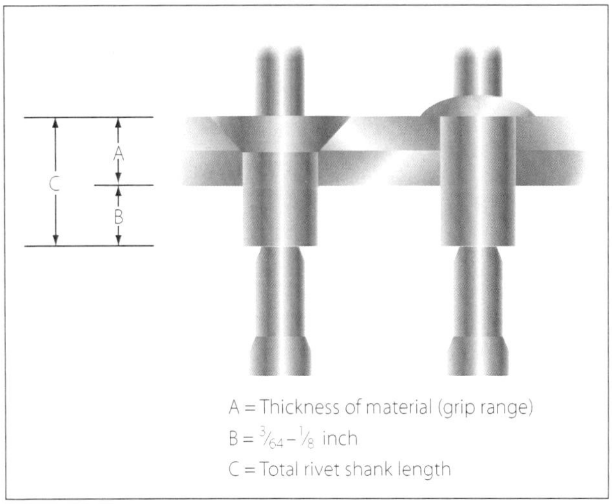

Figure 5-40. Determining length of friction lock rivets.

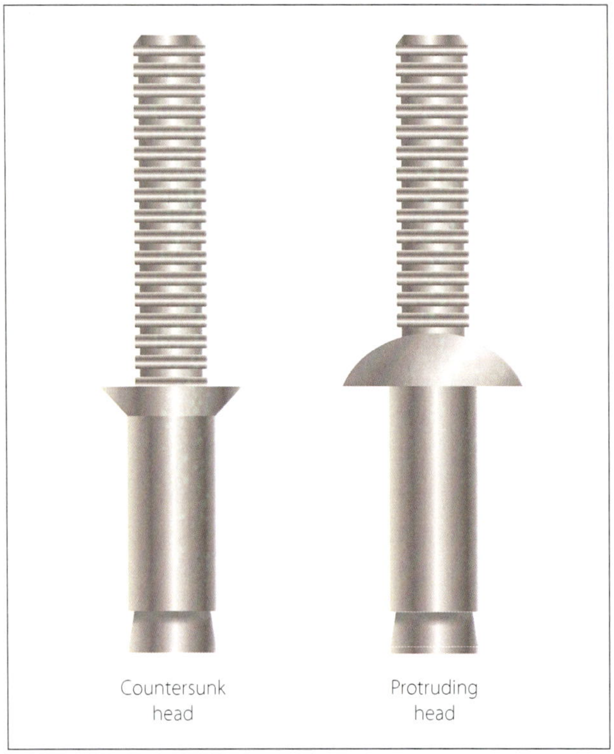

Figure 5-41. Pull-thru rivets.

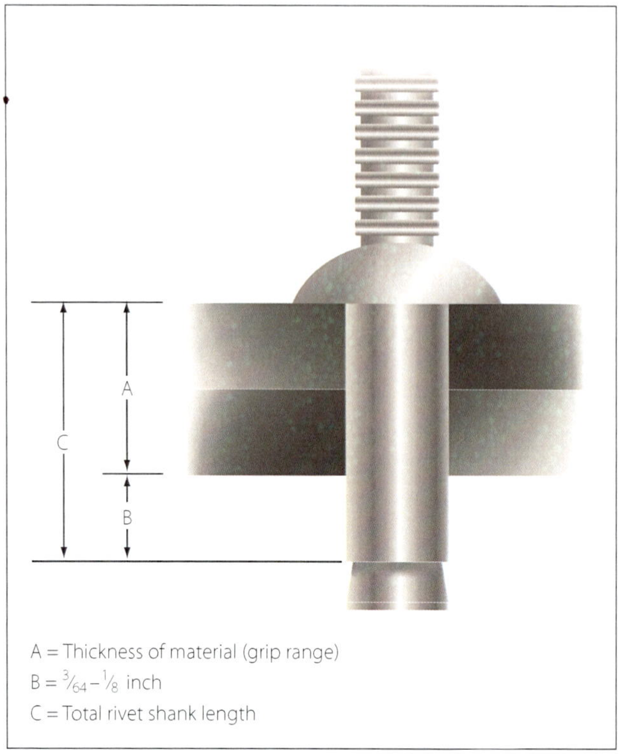

Figure 5-42. Determining length of pull-thru rivets.

through the hollow shank. Figure 5-41 illustrates a protruding head and a countersunk head pull-thru rivet.

Several events, in their proper sequence, occur when a pulling force is applied to the stem of the rivet: (1) The stem is pulled through the rivet shank; (2) the mandrel portion of the stem forces the shank to expand forming the blind head and filling the hole.

Pull-thru rivets are fabricated in two common head styles: (1) protruding head similar to the MS20470 or universal head, and (2) a 100° countersunk head. Other head styles are available from some manufacturers.

Pull-thru rivets are fabricated from several materials. Following are the most commonly used: 2117-T4 aluminum alloy, 5056 aluminum alloy, Monel.

Pull-thru rivets are designed so that installation requires only one person; it is not necessary to have the work accessible from both sides.

Factors to consider in the selection of the correct rivet for installation are: (1) installation location, (2) composition of the material being riveted, (3) thickness of the material being riveted, and (4) strength desired.

The thickness of the material being riveted determines the overall length of the shank of the rivet. As a general rule, the shank of the rivet should extend beyond the material thickness approximately $^{3}/_{64}$ inch to $^{1}/_{8}$ inch before the stem is pulled. [Figure 5-42]

Each company that manufactures pull-thru rivets has a code number to help users obtain correct rivet for the grip range of a particular installation. In addition, MS numbers are used for identification purposes. Numbers are similar to those shown on the preceding pages.

Self-Plugging Rivets (Mechanical Lock)

Self-plugging (mechanical lock) rivets are similar to self-plugging (friction lock) rivets, except for the manner in which the stem is retained in the rivet sleeve. This type of rivet has a positive mechanical locking collar to resist vibrations that cause the friction lock rivets to loosen and possibly fall out. [Figure 5-45] Also, the mechanical locking type rivet stem breaks off flush with the head and usually does not require further stem trimming when properly installed. Self-plugging (mechanical lock) rivets display all the strength characteristics of solid shank rivets and in most cases can be substituted rivet for rivet.

Bulbed Cherrylock Rivets

The large blind head of this fastener introduced the word "bulb" to blind rivet terminology. In conjunction with the unique residual preload developed by the high stem break load, its proven fatigue strength makes it

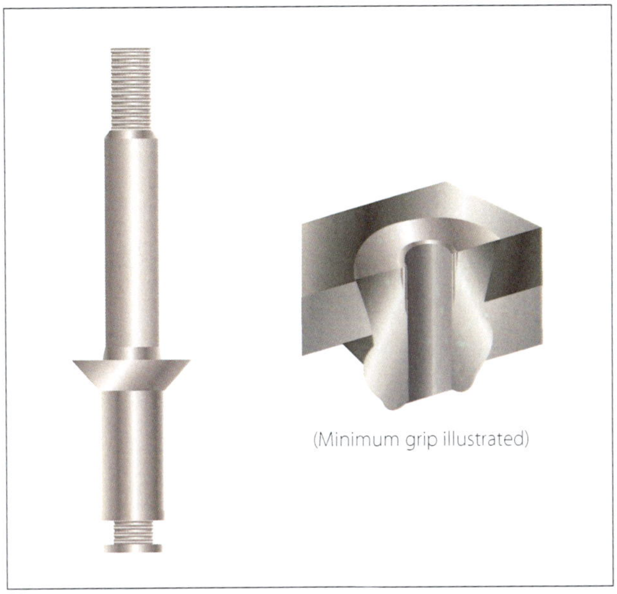

Figure 5-43. Bulbed cherrylock rivet.

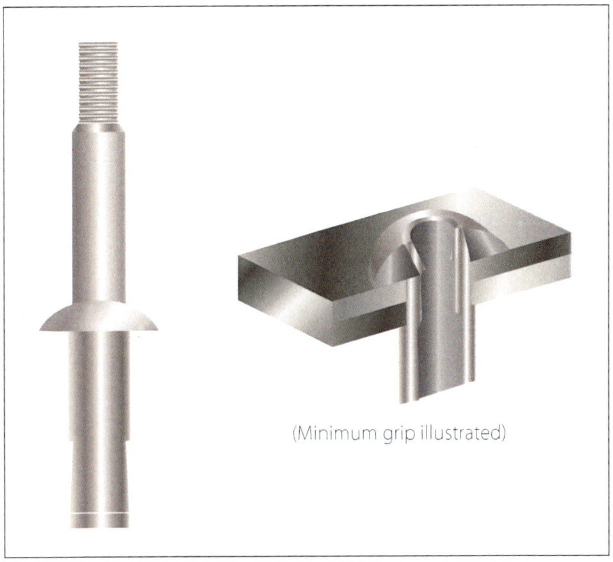

Figure 5-44. Wiredraw cherrylock rivet.

the only blind rivet interchangeable structurally with solid rivets. [Figure 5-43]

Wiredraw Cherrylock Rivets

There is a wide range of sizes, materials, and strength levels from which to select. This fastener is especially suited for sealing applications and joints requiring an excessive amount of sheet takeup. [Figure 5-44]

Huck Mechanical Locked Rivets

Self-plugging (mechanical lock) rivets are fabricated in two sections: a head and shank (including a conical recess and locking collar in the head), and a serrated stem that extends through the shank. Unlike the friction lock rivet, the mechanical lock rivet has a locking collar that forms a positive lock for retention of the stem in the shank of the rivet. This collar is seated in position during the installation of the rivet.

Material

Self-plugging (mechanical lock) rivets are fabricated with sleeves (rivet shanks) of 2017 and 5056 aluminum alloys, Monel, or stainless steel.

The mechanical lock type of self-plugging rivet can be used in the same applications as the friction lock type of rivet. In addition, because of its greater stem retention characteristic, installation in areas subject to considerable vibration is recommended.

The same general requirements must be met in the selection of the mechanical lock rivet as for the friction lock rivet. Composition of the material being joined together determines the composition of the rivet sleeve; for example, 2017 aluminum alloy rivets for most aluminum alloys and 5056 aluminum rivets for magnesium.

Figure 5-46 depicts the sequences of a typical mechanically locked blind rivet. The form and function may vary slightly between blind rivet styles and specifics should be obtained from manufacturers.

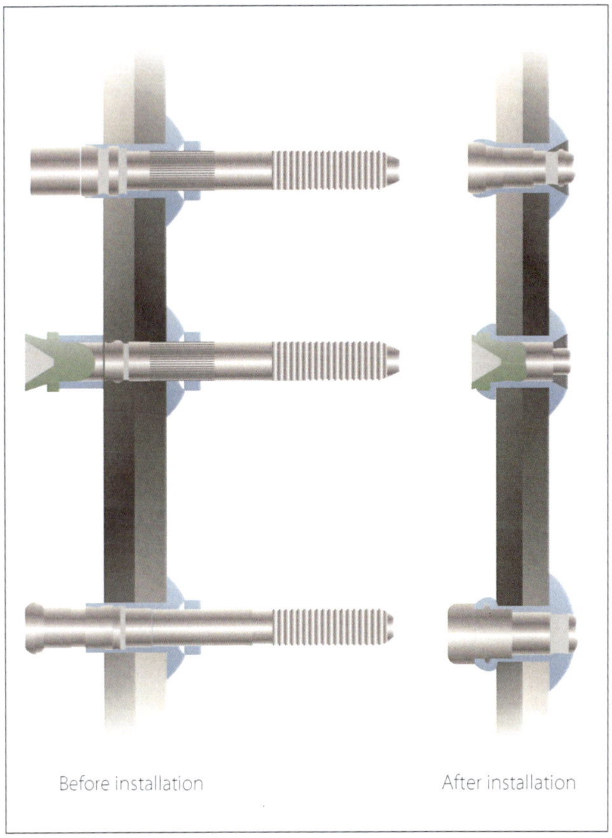

Figure 5-45. Self-plugging (mechanical lock) rivets.

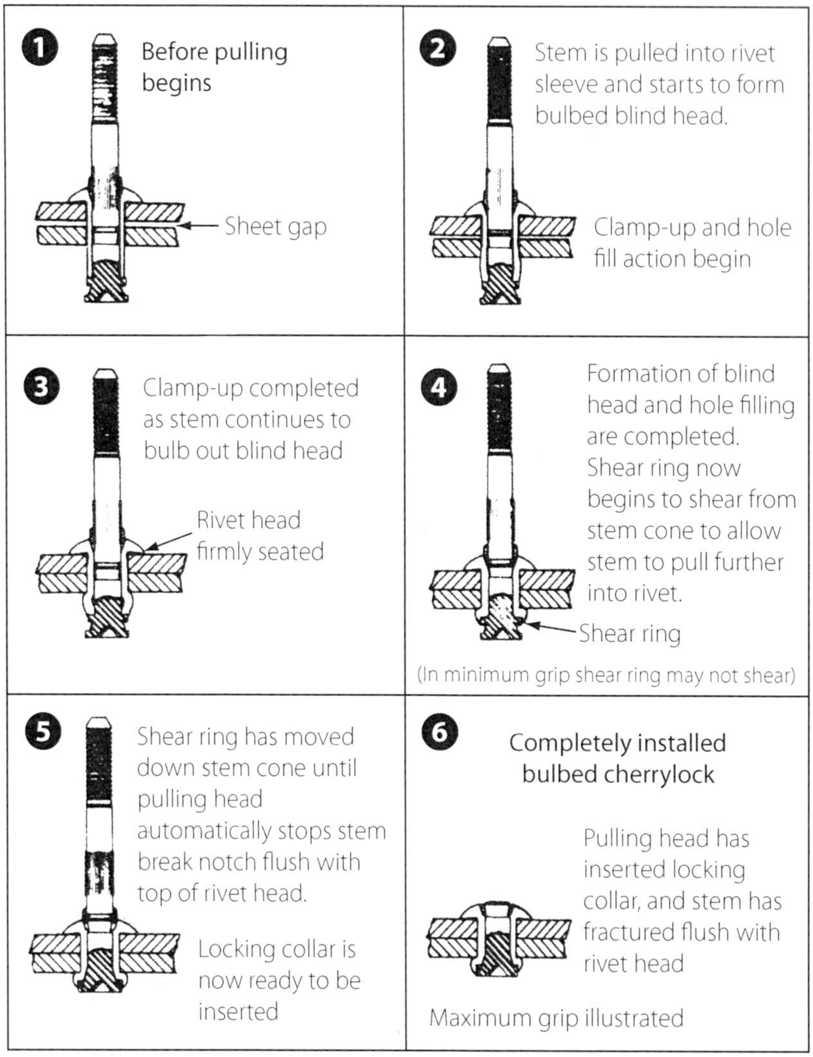

Figure 5-46. Cherrylock rivet installation.

Head Styles

Self-plugging mechanical locked blind rivets are available in several head styles depending on the installation requirements. [Figure 5-47]

Diameters

Shank diameters are measured in 1/32-inch increments and are generally identified by the first dash number: -3 indicates $\frac{3}{32}$ inch diameter, -4 indicates $\frac{1}{8}$ inch diameter, and so forth.

Both nominal and $\frac{1}{64}$-inch oversize diameters are available.

Grip Length

Grip length refers to the maximum total sheet thickness to be riveted and is measured in $\frac{1}{6}$ of an inch. This is generally identified by the second dash number. Unless otherwise noted, most blind rivets have their grip lengths (maximum grip) marked on the rivet head and have a total grip range of $\frac{1}{16}$ inch. [Figure 5-48]

To determine the proper grip rivet to use, measure the material thickness with a grip selection gauge (available from blind rivet manufacturers). The proper use of a grip selector gauge is shown in Figure 5-49.

The thickness of the material being riveted determines the overall length of the shank of the rivet. As a general rule, the shank of the rivet should extend beyond the material thickness approximately $\frac{3}{64}$ inch to $\frac{1}{8}$ inch before the stem is pulled. [Figure 5-50]

Rivet Identification

Each company that manufactures self-plugging (friction lock) rivets has a code number to help users obtain the correct rivet for the grip range or material thickness of a particular installation. In addition, MS numbers are used for identification purposes. Figures

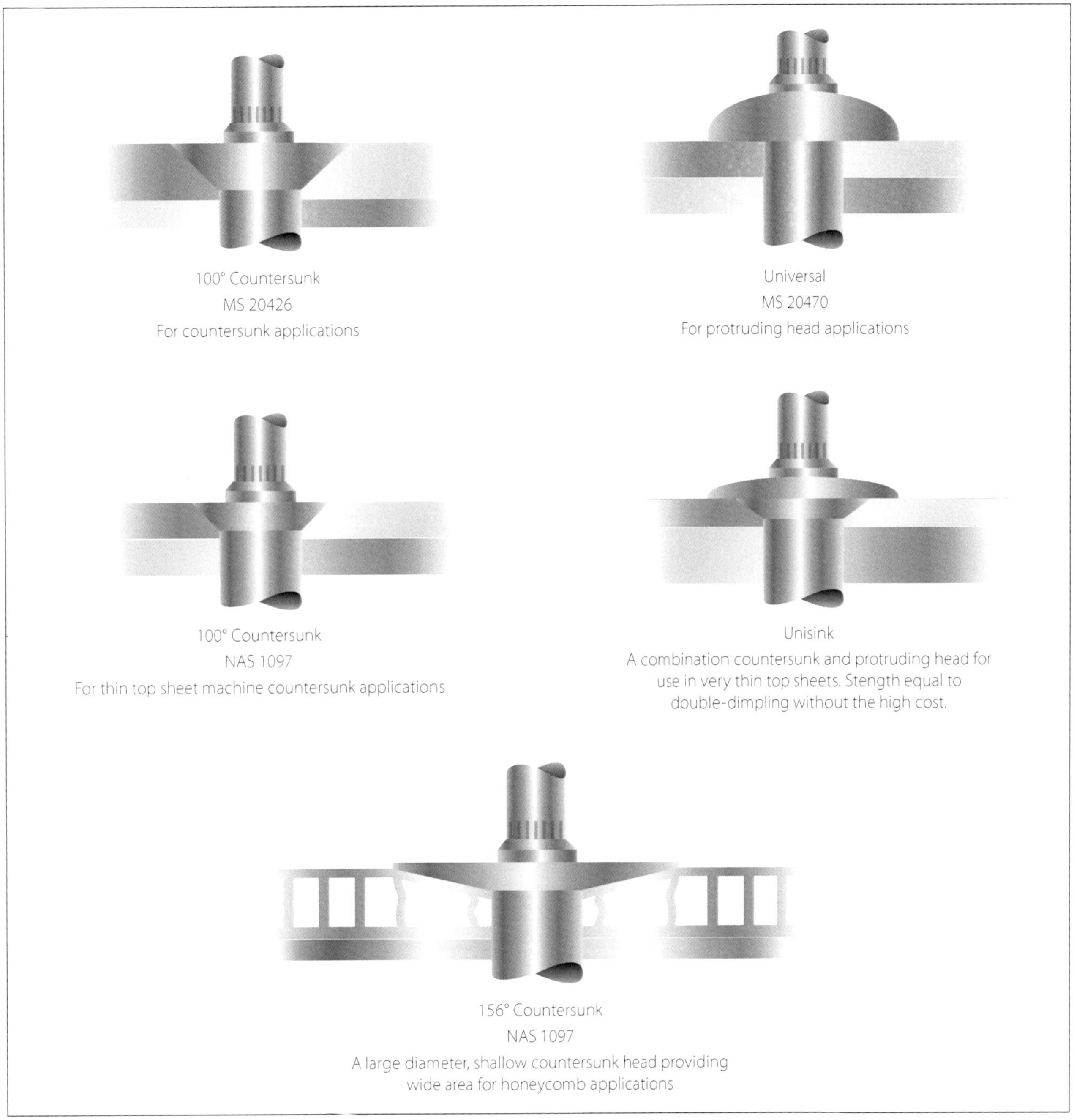

Figure 5-47. Cherrylock rivet heads.

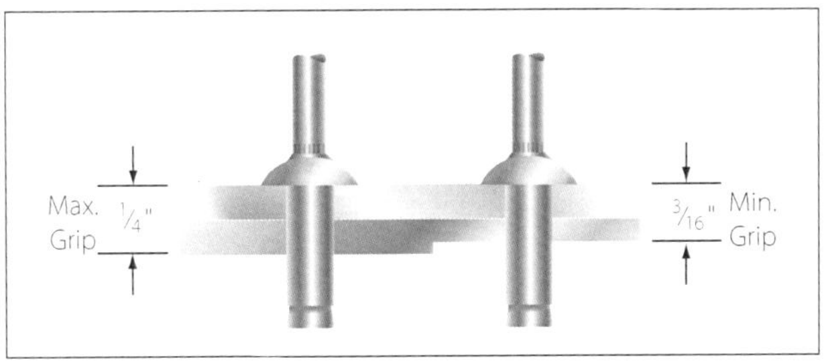

Figure 5-48. Typical grip length.

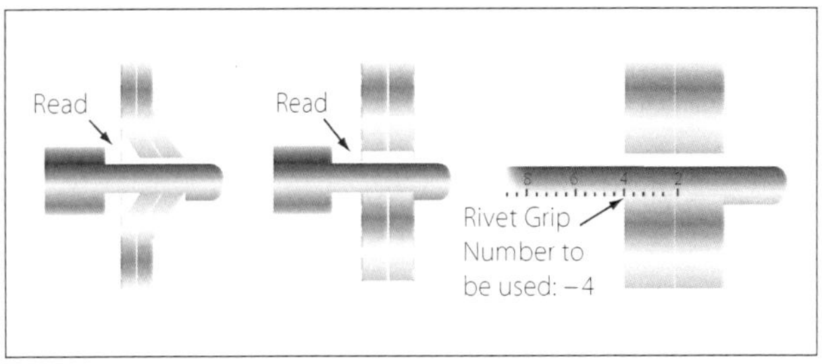

Figure 5-49. Grip gauge use.

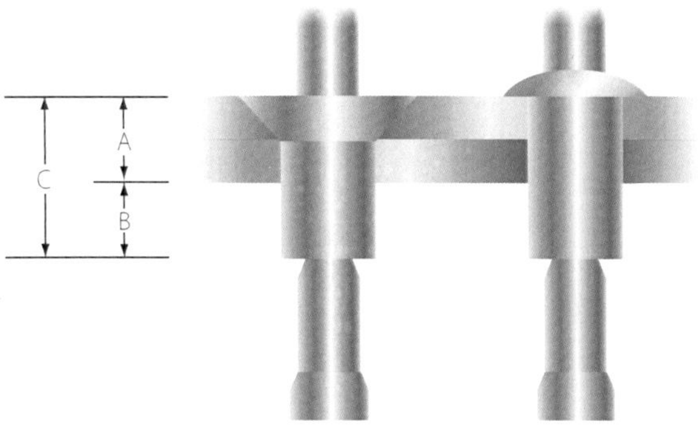

Figure 5-50. Determining rivet length.

5-51 through 5-54 contain examples of part numbers for self-plugging (friction lock) rivets that are representative of each.

Special Shear and Bearing Load Fasteners

Many special fasteners produce high strength with light weight and can be used in place of conventional AN bolts and nuts. When AN bolts are tightened with the nut, the bolt stretches, narrowing the diameter and then the bolt is no longer tight in the hole. Special fasteners eliminate this loose fit because they are held in place by a collar that is squeezed into position. These fasteners are not under the same tensile loads as a bolt during installation. Special fasteners are also used extensively for light sport aircraft (LSA). Always follow the aircraft manufacturer's recommendations.

Huck Manufacturing Comany

9SP-B A 6 3

9SP-B | **Head Style**
9SP-B = brazier or universal head
9SP-100 = 100° countersunk head

A | **Material composition of shank**
A = 2017 aluminium alloy
B = 5056 aluminium alloy
R = mild steel

6 | **Shank diameter in 32nds of an inch:**
4 = $\frac{1}{8}$ inch 6 = $\frac{3}{16}$ inch
5 = $\frac{5}{32}$ inch 8 = $\frac{1}{4}$ inch

3 | Grip range (material thickness) in 16ths of an inch

Figure 5-51. Huck Manufacturing Company codes.

Olympic Screw and Rivet Corporation

RV 2 0 0 4 2

RV | **Manufacturer**
Olympic Screw and Rivet Corporation

2 | **Rivet type**
2 = self plugging (friction lock)
5 = holow pull thru

0 | **Material composition of shank**
0 = 2017 aluminium alloy
5 = 5056 aluminium alloy
7 = mild steel

0 | **Head style**
0 = universal head
1 = 100° countersunk

4 | **Shank diameter in 32nds of an inch:**
4 = $\frac{1}{8}$ inch 6 = $\frac{3}{16}$ inch
5 = $\frac{5}{32}$ inch 8 = $\frac{1}{4}$ inch

2 | Grip range in 16ths of an inch

Figure 5-52. Olympic Screw and Rivet Corporation codes.

Townsend Company, Cherry Rivet Division

CR 163 6 6

CR | **Cherry rivet**

163 | **Series number**
Designates rivet material, type of rivet, and head style (163 = 2117 aluminium alloy, self-plugging (friction lock) rivet, protruding head)

6 | **Shank diameter in 32nds of an inch:**
4 = $\frac{1}{8}$ inch 6 = $\frac{3}{16}$ inch
5 = $\frac{5}{32}$ inch 8 = $\frac{1}{4}$ inch

6 | **Grip range (material thickness):**
knob stem in 32nds of an inch; serrated stem in 16ths of an inch

Figure 5-53. Townsend Company, Cherry Rivet Division codes.

Military Standard Number

MS 20600 B 4 K 2

MS | **Military Standard**

20600 | **Type of rivet and head style:**
20600 = self-plugging (friction lock) protruding head
20600 = self-plugging (friction lock) 100° countersunk head

B | **Material composition of sleeve:**
AD = 2117 aluminium alloy
B = 5056 aluminium alloy

4 | **Shank diameter in 32nds of an inch:**
4 = $\frac{1}{8}$ inch 6 = $\frac{3}{16}$ inch
5 = $\frac{5}{32}$ inch 8 = $\frac{1}{4}$ inch

K | **Type of stem:**
K = knot head stem
W = serrated stem

2 | **Grip range (material thickness) in 16ths of an inch**

Figure 5-54. Military Standard Numbers.

Pin Rivets

Pin (Hi-Shear) rivets are classified as special rivets but are not of the blind type. Access to both sides of the material is required to install this type of rivet. Pin rivets have the same shear strength as bolts of equal diameters, are about 40 percent of the weight of a bolt, and require only about one-fifth as much time for installation as a bolt, nut, and washer combination. They are approximately three times as strong as solid shank rivets.

Pin rivets are essentially threadless bolts. The pin is headed at one end and is grooved about the circumference at the other. A metal collar is swaged onto the grooved end effecting a firm, tight fit. [Figure 5-55] Pin rivets are fabricated in a variety of materials but should be used only in shear applications. They should never be used where the grip length is less than the shank diameter.

Part numbers for pin rivets can be interpreted to give the diameter and grip length of the individual rivets. A typical part number breakdown would be:

NAS177-14-17

NAS = National Aircraft Standard

177 = 100° countersunk head rivet

OR 178 = flathead rivet

14 = Nominal diameter in 32nds of an inch

17 = Maximum grip length in 16ths of an inch

Figure 5-55. Pin (Hi-Shear) rivet.

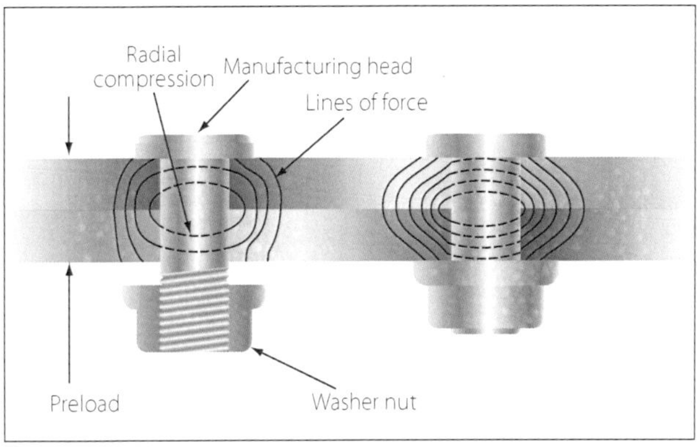

Figure 5-56. Taper-Lok special fasteners.

Taper-Lok

Taper-Loks are the strongest special fasteners used in aircraft construction. The Taper-Lok exerts a force on the walls of the hole because of its tapered shape. The Taper-Lok is designed to completely fill the hole, but unlike the rivet, it fills the hole without deforming the shank. Instead, the washer head nut squeezes the metal with tremendous force against the tapered walls of the hole. This creates radial compression around the shank and vertical compression lines as the metals are squeezed together. The combination of these forces generates strength unequaled by any other fastener. [Figure 5-56]

Hi-Tigue

The Hi-Tigue special fastener has a bead that encircles the bottom of its shank. The bead preloads the hole it fills, resulting in increased joint strength. At installation, the bead presses against the sidewall of the hole, exerting radial force that strengthens the surrounding area. Because it is preloaded, the joint is not subjected to the constant cyclic action that normally causes a joint to become cold worked and eventually fail.

Hi-Tigue fasteners are made of aluminum, titanium, and stainless steel alloys. The collars are composed of compatible metal alloys and come in two types: sealing and non-sealing. Just like the Hi-Loks, they can be installed using an Allen wrench and a box-end wrench. [Figure 5-57]

Captive Fasteners

Captive fasteners are used for quick removal of engine nacelles, inspection panels, and areas where fast and easy access is important. A captive fastener has the ability to turn in the body in which it is mounted, but which will not drop out when it is unscrewed from the part it is holding.

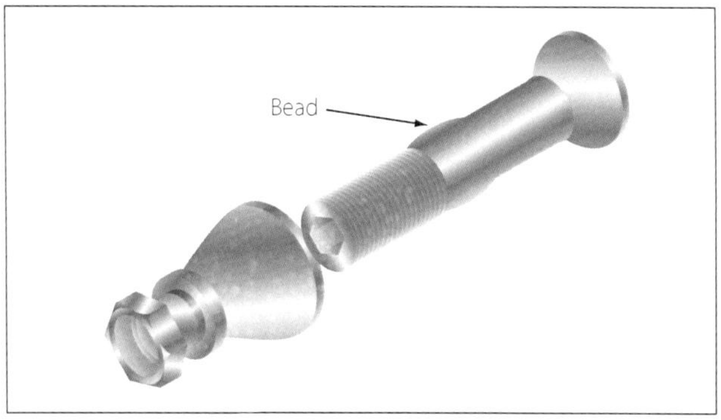

Figure 5-57. Hi-Tigue special fasteners.

Turnlock Fasteners

Turnlock fasteners are used to secure inspection plates, doors, and other removable panels on aircraft. Turnlock fasteners are also referred to by such terms as quick opening, quick action, and stressed panel fasteners. The most desirable feature of these fasteners is that they permit quick and easy removal of access panels for inspection and servicing purposes.

Turnlock fasteners are manufactured and supplied by a number of manufacturers under various trade names. Some of the most commonly used are the Dzus, Camloc, and Airloc.

Dzus Fasteners

The Dzus turnlock fastener consists of a stud, grommet, and receptacle. Figure 5-58 illustrates an installed Dzus fastener and the various parts.

The grommet is made of aluminum or aluminum alloy material. It acts as a holding device for the stud. Grommets can be fabricated from 1100 aluminum tubing, if none are available from normal sources.

The spring is made of steel, cadmium plated to prevent corrosion. The spring supplies the force that locks or secures the stud in place when two assemblies are joined.

The studs are fabricated from steel and are cadmium plated. They are available in three head styles: wing, flush, and oval. Body diameter, length, and head type may be identified or determined by the markings found on the head of the stud. [Figure 5-59] The diameter is always measured in sixteenths of an inch. Stud length is measured in hundredths of an inch and is the dis-

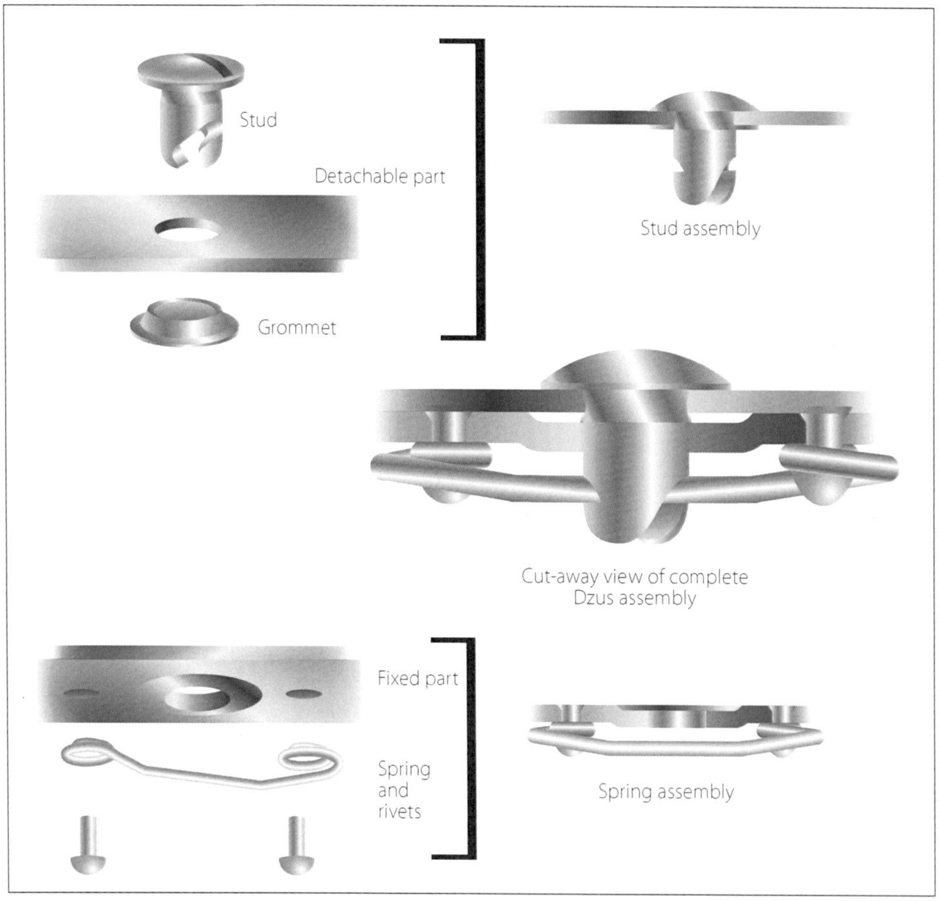

Figure 5-58. Dzus fastener.

Figure 5-59. Dzus identification.

tance from the head of the stud to the bottom of the spring hole.

A quarter of a turn of the stud (clockwise) locks the fastener. The fastener may be unlocked only by turning the stud counterclockwise. A Dzus key or a specially ground screwdriver locks or unlocks the fastener.

Camloc Fasteners

Camloc fasteners are made in a variety of styles and designs. Included among the most commonly used are the 2600, 2700, 40S51, and 4002 series in the regular line, and the stressed panel fastener in the heavy duty line. The latter is used in stressed panels which carry structural loads.

The Camloc fastener is used to secure aircraft cowlings and fairings. It consists of three parts: a stud assembly, a grommet, and a receptacle. Two types of receptacles are available: rigid and floating. [Figure 5-60]

The stud and grommet are installed in the removable portion; the receptacle is riveted to the structure of the aircraft. The stud and grommet are installed in either a plain, dimpled, countersunk, or counterbored hole, depending upon the location and thickness of the material involved.

A quarter turn (clockwise) of the stud locks the fastener. The fastener can be unlocked only by turning the stud counterclockwise.

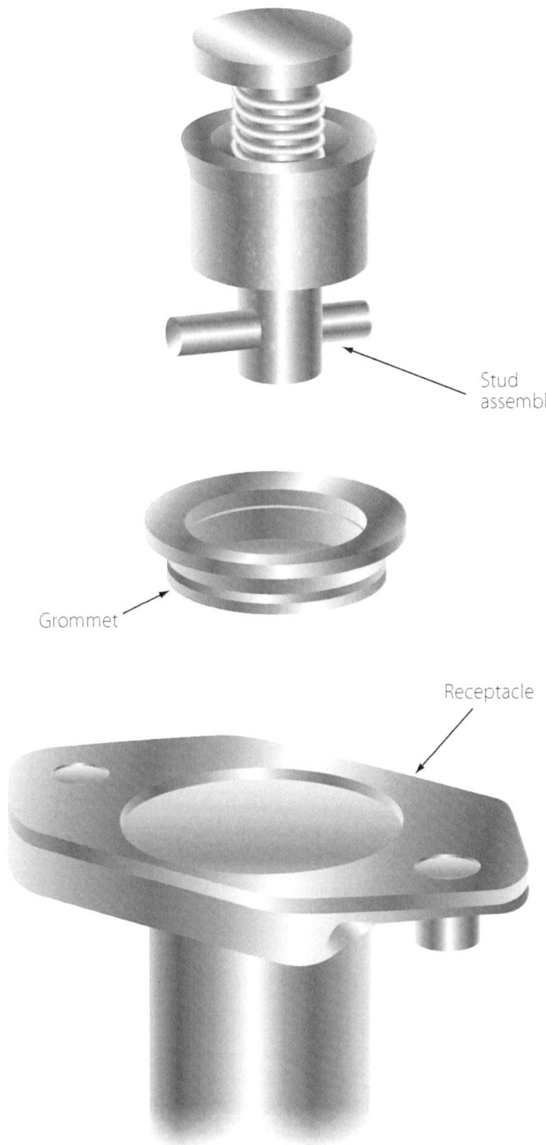

Figure 5-60. Camloc fastener.

Airloc Fasteners

The Airloc fastener consists of three parts: a stud, a cross pin, and a stud receptacle. [Figure 5-61] The studs are manufactured from steel and casehardened to prevent excessive wear. The stud hole is reamed for a press fit of the cross pin.

The total amount of material thickness to be secured with the Airloc fastener must be known before the correct length of stud can be selected for installation. The total thickness of material that each stud will satisfactorily lock together is stamped on the head of the stud in thousandths of an inch (0.040, 0.070, 0.190, and so forth). Studs are manufactured in three head styles: flush, oval, and wing.

The cross pin [Figure 5-61] is manufactured from chrome-vanadium steel and heat treated to provide maximum strength, wear, and holding power. It should never be used the second time; once removed from the stud, replace it with a new pin.

Receptacles for Airloc fasteners are manufactured in two types: rigid and floating. Sizes are classified by number—No. 2, No. 5, and No. 7. They are also classified by the center-to-center distance between the rivet holes of the receptacle: No. 2 is ¾ inch; No. 5 is 1 inch; and No. 7 is 1⅜ inch. Receptacles are fabricated from

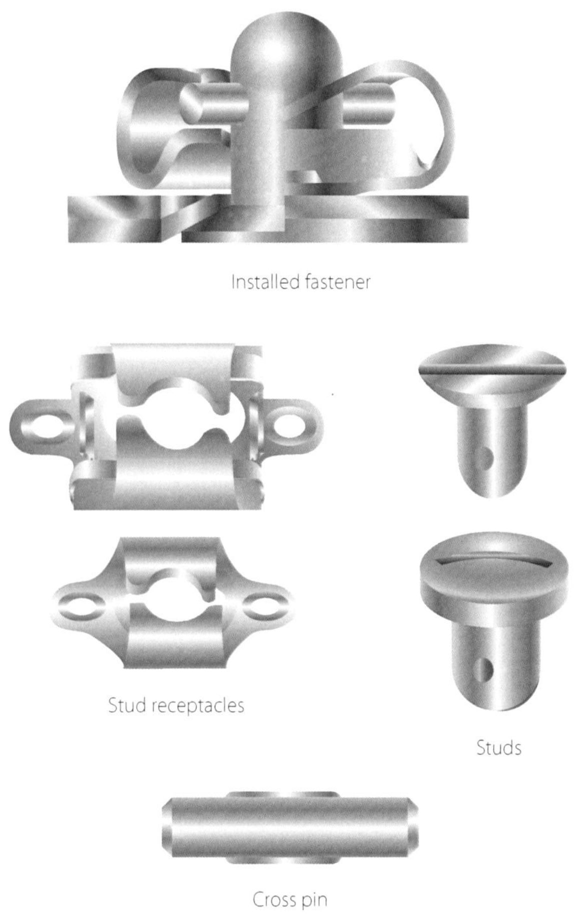

Figure 5-61. Airloc fastener.

high-carbon, heat-treated steel. An upper wing assures ejection of the stud when unlocked and enables the cross pin to be held in a locked position between the upper wing, cam, stop, and wing detent, regardless of the tension to which the receptacle is subjected.

Screws

Screws are the most commonly used threaded fastening devices on aircraft. They differ from bolts inasmuch as they are generally made of lower strength materials. They can be installed with a loose fitting thread, and the head shapes are made to engage a screwdriver or wrench. Some screws have a clearly defined grip or unthreaded portion while others are threaded along their entire length.

Several types of structural screws differ from the standard structural bolts only in head style. The material in them is the same, and a definite grip length is provided. The AN525 washer head screw and the NAS220 through NAS227 series are such screws.

Commonly used screws are classified in three groups: (1) structural screws, which have the same strength as equal size bolts; (2) machine screws, which include the majority of types used for general repair; and (3) self-tapping screws, which are used for attaching lighter parts. A fourth group, drive screws, are not actually screws but nails. They are driven into metal parts with a mallet or hammer and their heads are not slotted or recessed.

Structural Screws

Structural screws are made of alloy steel, are properly heat treated, and can be used as structural bolts. These screws are found in the NAS204 through NAS235 and AN509 and AN525 series. They have a definite grip and the same shear strength as a bolt of the same size. Shank tolerances are similar to AN hex head bolts, and the threads are National Fine. Structural screws are available with round, brazier, or countersunk heads. The recessed head screws are driven by either a Phillips or a Reed & Prince screwdriver.

The AN509 (100°) flathead screw is used in countersunk holes where a flush surface is necessary.

The AN525 washer head structural screw is used where raised heads are not objectionable. The washer head screw provides a large contact area.

Machine Screws

Machine screws are usually of the flathead (countersunk), roundhead, or washer head types. These are general purpose screws and are available in low carbon steel, brass, corrosion-resistant steel, and aluminum alloy.

Roundhead screws, AN515 and AN520, have either slotted or recessed heads. The AN515 screw has coarse threads, and the AN520 has fine threads.

Countersunk machine screws are listed as AN505 and AN510 for 82°, and AN507 for 100°. The AN505 and AN510 correspond to the AN515 and AN520 roundhead in material and usage.

The fillister head screw, AN500 through AN503, is a general purpose screw and is used as a capscrew in light mechanisms. This could include attachments of cast aluminum parts such as gearbox cover plates.

The AN500 and AN501 screws are available in low carbon steel, corrosion-resistant steel, and brass. The AN500 has coarse threads, while the AN501 has fine threads. They have no clearly defined grip length.

Screws larger than No. 6 have a hole drilled through the head for safetying purposes.

The AN502 and AN503 fillister head screws are made of heat-treated alloy steel, have a small grip, and are available in fine and coarse threads. These screws are used as capscrews where great strength is required. The coarse threaded screws are commonly used as capscrews in tapped aluminum alloy and magnesium castings because of the softness of the metal.

Self-Tapping Screws

Machine *self-tapping screws* are listed as AN504 and AN506. The AN504 screw has a roundhead, and the AN506 is 82° countersunk. These screws are used for attaching removable parts, such as nameplates, to castings and parts in which the screw cuts its own threads.

AN530 and AN531 self-tapping sheet metal screws, such as the Parker-Kalon Z-type sheet metal screw, are blunt on the end. They are used in the temporary attachment of metal for riveting, and in the permanent assembly of nonstructural assemblies. Self-tapping screws should not be used to replace standard screws, nuts, bolts, or rivets.

Drive Screws

Drive screws, AN535, correspond to the Parker-Kalon U-type. They are plain head self-tapping screws used as capscrews for attaching nameplates in castings and for sealing drain holes in corrosion proofing tubular structures. They are not intended to be removed after installation.

Identification and Coding for Screws

The coding system used to identify screws is similar to that used for bolts. There are AN and NAS screws. NAS screws are structural screws. Part numbers 510, 515, 550, and so on, catalog screws into classes, such as roundhead, flathead, washer head, and so forth. Letters and digits indicate their material composition, length, and thickness. Examples of AN and NAS code numbers follow.

AN501B-416-7

AN = Air Force-Navy standard

501 = fillister head, fine thread

B = brass

416 = $^4/_{16}$-inch diameter

7 = $^7/_{16}$-inch length

The letter "D" in place of the "B" would indicate that the material is 2017-T aluminum alloy. The letter "C" would designate corrosion resistant steel. An "A" placed before the material code letter would indicate that the head is drilled for safetying.

NAS144DH-22

NAS = National Aircraft Standard

144 = head style; diameter and thread—1/4-28 bolt, internal wrenching

DH = drilled head

22 = screw length in 16ths of an inch—$1^3/_8$ inches long

The basic NAS number identifies the part. The suffix letters and dash numbers separate different sizes, plating material, drilling specifications, and so forth. The dash numbers and suffix letters do not have standard meanings. It is necessary to refer to a specific NAS page in the Standards book for the legend.

Riveted and Rivetless Nutplates

When access to the back of a screw or bolt installation is impractical, riveted or *rivetless nutplates* are used to secure the connection of panels. One example in aircraft this technique is especially useful is to secure the floorboards to the stringers and to each other.

Nutplates

Nuts that are made to be riveted in place in aircraft are called nutplates. Their purpose is to allow bolts and screws to be inserted without having to hold the nut. They are permanently mounted to enable inspection panels and access doors to be easily removed and installed. When many screws are used on a panel, to make installation easier, normally floating anchor nuts are used. The floating anchor nut fits into a small bracket which is riveted to the aircraft skin. The nut is free to move, which makes it much easier to align it with the screw. For production ease, sometimes ganged anchor nuts are used for inspection panels. Ganged anchor nuts allow the nuts to float in a channel, making alignment with the screw easy.

Self-locking nutplates are made under several standards and come in several shapes and sizes. Figure 5-62 shows an MS21078 two-lug nutplate with a nonmetallic insert, and an MS21047 lightweight, all-metal, 450 °F (232 °C) nutplate. Nutplates can also have three riveting points if the added strength is required.

Rivnuts

This is the trade name of a hollow, blind rivet made of 6053 aluminum alloy, counterbored and threaded on the inside. Rivnuts can be installed by one person using

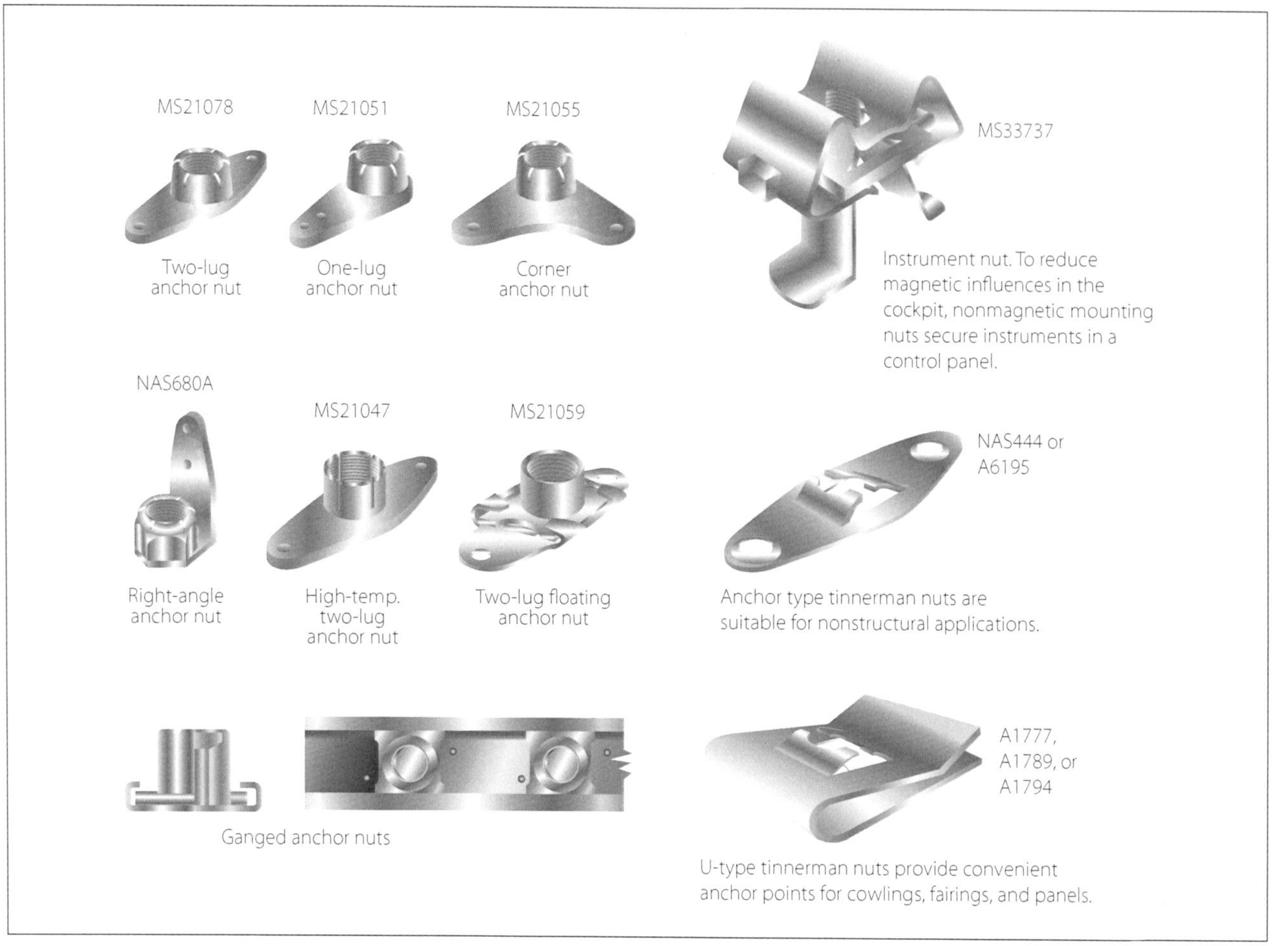

Figure 5-62. Various nutplates.

a special tool which heads the rivet on the blind side of the material. The Rivnut is threaded on the mandrel of the heading tool and inserted in the rivet hole. The heading tool is held at right angles to the material, the handle is squeezed, and the mandrel crank is turned clockwise after each stroke. Continue squeezing the handle and turning the mandrel crank of the heading tool until a solid resistance is felt, which indicates that the rivet is set.

The Rivnut is used primarily as a nut plate and in the attachment of deicer boots to the leading edges of wings. It may be used as a rivet in secondary structures or for the attachment of accessories such as brackets, fairings, instruments, or soundproofing materials.

Rivnuts are manufactured in two head types, each with two ends: the flathead with open or closed end, and the countersunk head with open or closed end. All Rivnuts, except the thin head countersunk type, are available with or without small projections (keys) attached to the head to keep the Rivnut from turning. Keyed Rivnuts are used as a nut plate, while those without keys are used for straight blind riveting repairs where no torque loads are imposed. A keyway cutter is needed when installing Rivnuts which have keys.

The countersunk style Rivnut is made with two different head angles: the 100° with 0.048 and 0.063 inch head thickness, and the 115° with 0.063 inch head thickness. Each of these head styles is made in three sizes: 6-32, 8-32, and 10-32. These numbers represent the machine screw size of the threads on the inside of the Rivnut. The actual outside diameters of the shanks are $^{3}/_{16}$ inch for the 6-32 size, $^{7}/_{32}$ inch for the 8-32 size, and $^{1}/_{4}$ inch for the 10-32 size.

Open end Rivnuts are the most widely used and are recommended in preference to the closed end type wherever possible. However, closed end Rivnuts must be used in pressurized compartments.

Rivnuts are manufactured in six grip ranges. The minimum grip length is indicated by a plain head, and the next higher grip length by one radial dash mark on the head. Each succeeding grip range is indicated by an additional radial dash mark until five marks indicate the maximum range.

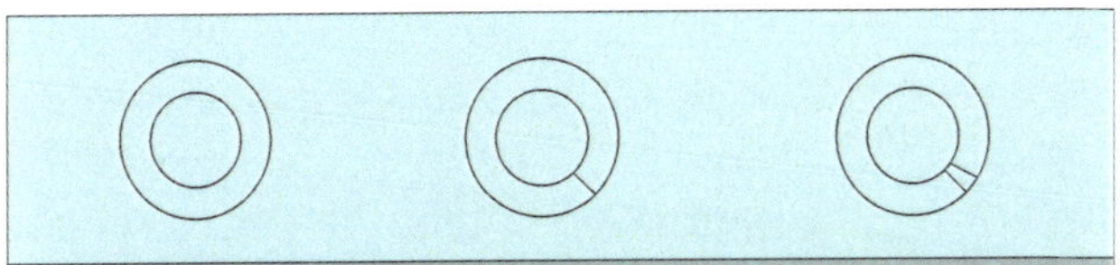

Flat—0.32 Head Thickness		
6-45	6-75	6-100
8-45	8-75	8-100
10-45	10-75	10-100
6B45	6B75	6B100
8B45	8B75	8B100
10B45	10B75	10B100
6K45	6K75	6K100
8K45	8K75	8K100
10K45	10K75	10K100
6KB45	6KB75	6KB100
8KB45	8KB75	8KB100
10KB45	10KB75	10KB100
100°—0.48 Head Thickness		
6-91	6-121	6-146
8-91	8-121	8-146
10-91	10-121	10-146
6B91	6B121	6B146
8B91	8B121	8B146
10B91	10B121	10B146
100°—0.63 Head Thickness		
6-106	6-136	6-161
8-106	8-136	8-161
10-106	10-136	10-161
6B106	6B136	6B161
8B106	8B136	8B161
10B106	10B136	10B161
6K106	6K136	6K161
8K106	8K136	8K161
10K106	10K136	10K161
6KB106	6KB136	6KB161
8KB106	8KB136	8KB161
10KB106	10KB136	10KB161

Figure 5-63. Rivnut data chart.

Notice in Figure 5-63 that some part number codes consist of a "6," an "8," or a "10," a "dash," and two or three more numbers. In some, the dash is replaced by the letters "K" or "KB." The first number indicates the machine screw size of the thread, and the last two or three numbers indicate the maximum grip length in thousandths of an inch. A dash between the figures indicates that the Rivnut has an open end and is keyless; a "B" in place of the dash means it has a closed end and is keyless; a "K" means it has an open end and has a key; and a "KB" indicates that it has a closed end and a key. If the last two or three numbers are divisible by five, the Rivnut has a flathead; if they are not divisible by five, the Rivnut has a countersunk head.

An example of a part number code is:

10KB106

10 = Grip length

KB = Closed end and key

106 = Screw and thread size

Dill Lok-Skrus and Dill Lok-Rivets

Dill "Lok-Skru" and "Lok-Rivet" are trade names for internally threaded rivets. They are used for blind attachment of such accessories as fairings, fillets, access door covers, door and window frames, floor panels, and the like. Lok-Skrus and Lok-Rivets are similar to the Rivnut in appearance and application; however, they come in two parts and require more clearance on the blind side than the Rivnut to accommodate the barrel. [Figure 5-64]

The Lok-Rivet and the Lok-Skru are alike in construction, except the Lok-Skru is tapped internally for fastening an accessory by using an attaching screw, whereas the Lok-Rivet is not tapped and can be used only as a rivet. Since both Lok-Skrus and Lok-Rivets

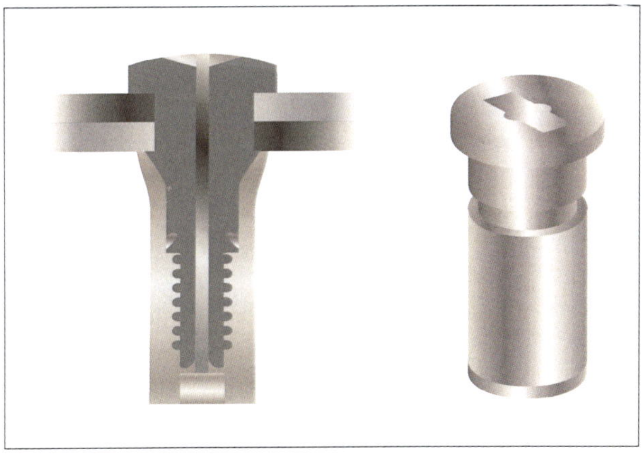

Figure 5-64. Internally threaded rivet (Rivnut).

are installed in the same manner, the following discussions for the Lok-Skru also applies to the Lok-Rivet.

The main parts of a Lok-Skru are the barrel, the head, and an attachment screw. The barrel is made of aluminum alloy and comes in either closed or open ends. The head is either aluminum alloy or steel, and the attachment screw is made of steel. All of the steel parts are cadmium plated, and all of aluminum parts are anodized to resist corrosion. When installed, the barrel screws up over the head and grips the metal on the blind side. The attaching screw is then inserted if needed. There are two head types: the flathead and the countersunk head. The Lok-Skru is tapped for 7-32, 8-32, 10-32, or 10-24 screws, and the diameters vary from 0.230 inch for 6-32 screws, to 0.292 inch for 10-32 screws. Grip ranges vary from 0.010 inch to 0.225 inch.

Deutsch Rivets

This rivet is a high strength blind rivet used on late model aircraft. It has a minimum shear strength of 75,000 psi, and can be installed by one person. The Deutsch rivet consists of two parts: the stainless steel sleeve and the hardened steel drive pin. [Figure 5-65] The pin and sleeve are coated with a lubricant and a corrosion inhibitor.

The Deutsch rivet is available in diameters of $^{3}/_{16}$, $^{1}/_{4}$, or $^{3}/_{8}$ inch. Grip lengths for this rivet range from $^{3}/_{16}$ to 1 inch. Some variation is allowed in grip length when installing the rivet; for example, a rivet with a grip length of $^{3}/_{16}$ inch can be used where the total thickness of materials is between 0.198 and 0.228 inch.

When driving a Deutsch rivet, an ordinary hammer or a pneumatic rivet gun and a flathead set are used. The rivet is seated in the previously drilled hole and then the pin is driven into the sleeve. The driving action causes the pin to exert pressure against the sleeve and forces the sides of the sleeve out. This stretching forms a shop head on the end of the rivet and provides positive fastening. The ridge on the top of the rivet head locks the pin into the rivet as the last few blows are struck.

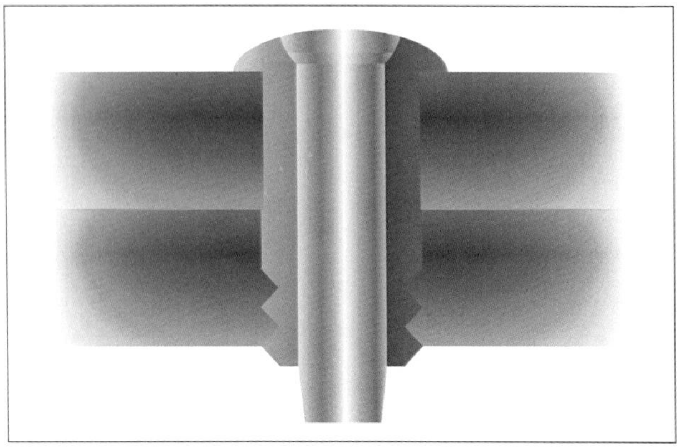

Figure 5-65. Deutsch rivet.

Sealing Nutplates

When securing nutplates in pressurized aircraft and in fuel cells, a sealing nutplate is used instead of the open ended variety previously described. Care must be taken to use exactly the correct length of bolt or screw. If a bolt or screw is too short, there will not be enough threads to hold the device in place. If the bolt or screw is too long, it will penetrate the back side of the nutplate and compromise the seal. Normally, a sealant is also used to ensure complete sealing of the nutplate. Check the manufacturer's specifications for the acceptable sealant to be used for sealing nutplates.

Hole Repair and Hole Repair Hardware

Many of the blind fasteners are manufactured in oversized diameters to accommodate slightly enlarged holes resulting from drilling out the original fastener. When using rivets or even bolts, care must be taken to ensure the hole is not elongated or slanted.

To reduce the chances of an incorrectly drilled rivet or bolt hole, use a slightly smaller drill bit first, then enlarge to the correct diameter. The last step to prepare the hole for the fastener is to deburr the hole using either a very large drill bit or a special deburring tool. This practice also works well when drilling out a previously attached fastener. If the drill bit does not exactly find the center of the rivet or bolt or screw, the hole can easily be elongated, but when using a smaller drill bit, drill the head only off the fastener, then the ring and stem that is left can be pushed out with a pin punch of the appropriate diameter. If an incorrectly drilled hole is found, the options are to redrill the hole to the next larger diameter for an acceptable fastener or repair the hole using an Acres fastener sleeve.

Repair of Damaged Holes with Acres Fastener Sleeves

Acres fastener sleeves are thin-wall tubular elements with a flared end. The sleeves are installed in holes to accept standard bolts and rivet-type fasteners. The existing fastener holes are drilled $^{1}/_{64}$ inch oversize for installation of the sleeves. The sleeves are manufactured in 1-inch increments. Along their length, grooves provide a place to break or cut off excess length to match fastener grip range. The grooves also provide a place to hold adhesive or sealing agents when bonding the sleeve into the hole.

Advantages and Limitations

The sleeves are used in holes which must be drilled $^{1}/_{64}$ inch oversize to clean up corrosion or other damage. The oversize hole with the sleeve installed allows the use of the original diameter fastener in the repaired hole. The sleeves can be used in areas of high galvanic corrosion where the corrosion must be confined to a readily replaceable part. Oversizing of holes reduces the net cross-sectional area of a part and should not be done unless absolutely required.

Consult the manufacturer of the aircraft, aircraft engine or aircraft component prior to repair of damaged holes with Acres sleeves.

Identification

The sleeve is identified by a standard code number [Figure 5-66A] which represents the type and style of sleeve, a material code, the fastener shank diameter, surface finish code letter and grip tang for the sleeve. The type and material of the sleeve is represented by the basic code number. The first dash number represents the diameter of the sleeve for the fastener installed and the second dash represents the grip length of the sleeve. The required length of the sleeve is determined on installation and the excess is broken off of the sleeve. A JK5512A-05N-10 is a 100° low profile head sleeve of aluminum alloy. The diameter is for a $^{5}/_{32}$-inch fastener with no surface finish and is $^{5}/_{8}$ inch in length.

Hole Preparation

Refer to Figure 5-66B for drill number for standard or close fit holes. Inspect hole after drilling to assure all corrosion is removed before installing the sleeve. The hole must also be the correct shape and free from burrs. The countersink must be enlarged to receive the flare of the sleeve so the sleeve is flush with the surrounding surface.

Installation

After selecting the correct type and diameter sleeve, use the 6501 sleeve breakoff tool for final installation length. Refer to Figure 5-66B for the sleeve breakoff procedure. The sleeve may be installed with or without being bonded in the hole. When bonding the sleeve in a hole, use MIL-S-8802A$^{1}/_{2}$ sealant. Reinstall original size fastener and torque as required.

Sleeve Removal

Sleeves not bonded in the hole may be removed by either driving them out with a drift pin of the same diameter as the outside diameter of the sleeve or they may be deformed and removed with a pointed tool. Bonded sleeves may be removed by this method, but care should be used not to damage the structure hole. If this method cannot be used, drill the sleeves out with a drill 0.004 to 0.008 inch smaller than the installation drill size. The remaining portion of the sleeve after drilling can be removed using a pointed tool and applying an adhesive solvent to the sealant.

Control Cables and Terminals

Cables are the most widely used linkage in primary flight control systems. Cable-type linkage is also used in engine controls, emergency extension systems for the landing gear, and various other systems throughout the aircraft.

Cable-type linkage has several advantages over the other types. It is strong and light weight, and its flexibility makes it easy to route through the aircraft. An aircraft cable has a high mechanical efficiency and can be set up without backlash, which is very important for precise control.

Cable linkage also has some disadvantages. Tension must be adjusted frequently due to stretching and temperature changes.

Aircraft control cables are fabricated from carbon steel or stainless steel.

Cable Construction

The basic component of a cable is a wire. The diameter of the wire determines the total diameter of the cable. A number of wires are preformed into a helical or spiral shape and then formed into a strand. These preformed strands are laid around a straight center strand to form a cable.

Cable designations are based on the number of strands and the number of wires in each strand. The most common aircraft cables are the 7 × 7 and 7 × 19.

The 7 × 7 cable consists of seven strands of seven wires each. Six of these strands are laid around the center strand. [Figure 5-67] This is a cable of medium flexibility and is used for trim tab controls, engine controls, and indicator controls.

The 7 × 19 cable is made up of seven strands of 19 wires each. Six of these strands are laid around the center strand. [Figure 5-67] This cable is extra flexible and is used in primary control systems and in other places where operation over pulleys is frequent.

Aircraft control cables vary in diameter, ranging from $^{1}/_{16}$ to $^{3}/_{8}$ inch. The diameter is measured as shown in Figure 5-67.

Acres Sleeve	Type	Basic Part Number
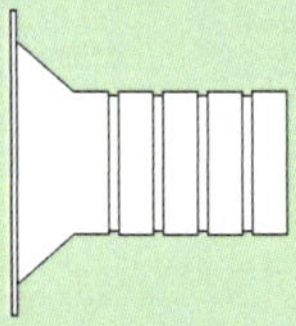	100° Tension head plus flange (509 type)	JK5610
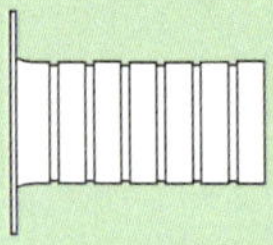	Protruding head (shear)	JK5511
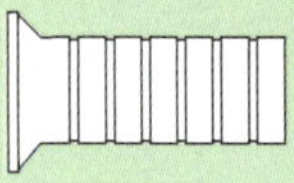	100° Low profile head	JK5512
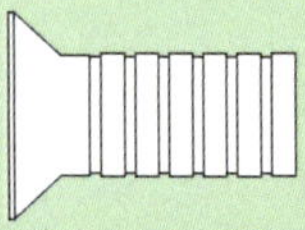	100° Standard profile head (509 type)	JK5516
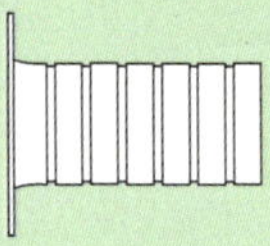	Protruding head (tension)	JK5517
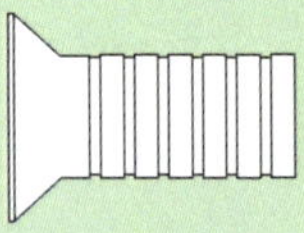	100° Oversize tension head ($^{1}/_{64}$ oversize bolt)	JK5533

Part Number Breakdown

JK5511 A 04 N 08 L

- JK5511 | Basic part number
- A | Material Code [1]
- 04 | Fastener shank diameter in 32nds
- N | Surface finish
  N = No finish
  C = Chemical film per MIL-C-554
- 08 | Length in sixteenth-inch increments
  (required installation length by breaking off at proper groove)
- L | "L" at end of part number indicates cetyl alcohol lubricant

Material	Material Code
5052 Aluminium alloy $^{1}/_{2}$ hard	A
6061 Aluminium alloy (T6 condition)	B
A286 Stainless steel (passivate)	C

Sleeve Part No.	Bolt Size	[2] Sleeve Length
JK5511( )04( )( ) JK5512( )04( )( ) JK5516( )04( )( ) JK5517( )04( )( )	$^{1}/_{8}$	8
JK5511( )45( )( ) JK5512 JK5516( )45( )( ) JK5517( )45( )( )	#6	8
JK5511( )05( )( ) JK5512( )05( )( ) JK5516( )05( )( ) JK5517( )05( )( )	$^{5}/_{32}$	10
JK5511( )55( )( ) JK5512( )55( )( ) JK5516( )55( )( ) JK5517( )55( )( ) JK5610( )55( )( )	#8	10
JK5511( )06( )( ) JK5512( )06( )( ) JK5516( )06( )( ) JK5517( )06( )( ) JK5610( )06( )( )	#10	12
JK5511( )08( )( ) JK5512( )08( )( ) JK5516( )08( )( ) JK5517( )08( )( ) JK5610( )08( )( )	$^{1}/_{4}$	16
JK5511( )10( )( ) JK5512( )10( )( ) JK5516( )10( )( ) JK5517( )10( )( ) JK5610( )10( )( )	$^{5}/_{16}$	16
JK5511( )12( )( ) JK5512( )12( )( ) JK5516( )12( )( ) JK5517( )12( )( ) JK5610( )12( )( )	$^{3}/_{8}$	16

Acres Sleeve for $^{1}/_{64}$ Oversize Bolt

[1] Sleeve Part No.	Bolt Size	[2] Sleeve Length
JK5533( )06( )( )	$^{13}/_{64}$	12
JK5533( )08( )( )	$^{17}/_{64}$	16
JK5533( )10( )( )	$^{21}/_{64}$	16
JK5533( )12( )( )	$^{25}/_{64}$	16

Notes:

1. Acres sleeve, JK5533 $^{1}/_{64}$ oversize available in A286 steel only
2. Acres sleeve length in sixteenth-inch increments

Figure 5-66A. Acres sleeve identification.

Hole Preparation for 1⁄64 Oversize Bolt

Bolt Size	Drill No.	Drill Dia.
13⁄64	7⁄32	0.2187
17⁄64	9⁄32	0.2812
21⁄64	11⁄32	0.3437
25⁄64	13⁄32	0.4062

Hole Preparation

Bolt Size	Standard Fit		Close Fit	
	Drill No.	Drill Dia.	Drill No.	Drill Dia.
1⁄8	9⁄64	0.1406	28	0.1405
#6	23	0.1540	24	0.1520
5⁄32	11⁄64	0.1719	18	0.1695
#8	15	0.1800	16	0.1770
#10	5	0.2055	6	0.2040
1⁄4	14	0.2660	17⁄64	0.2656
5⁄16	21⁄64	0.3281		
3⁄8	25⁄64	0.3908		

Installation Procedure

A. Drill out corrosion or damage to existing hole to 1⁄64 oversize.
B. Select proper type and length acres sleeve for existing fastener.
C. Bond sleeve in structure hole with MIL-S-8802 class A 1⁄2 sealant.

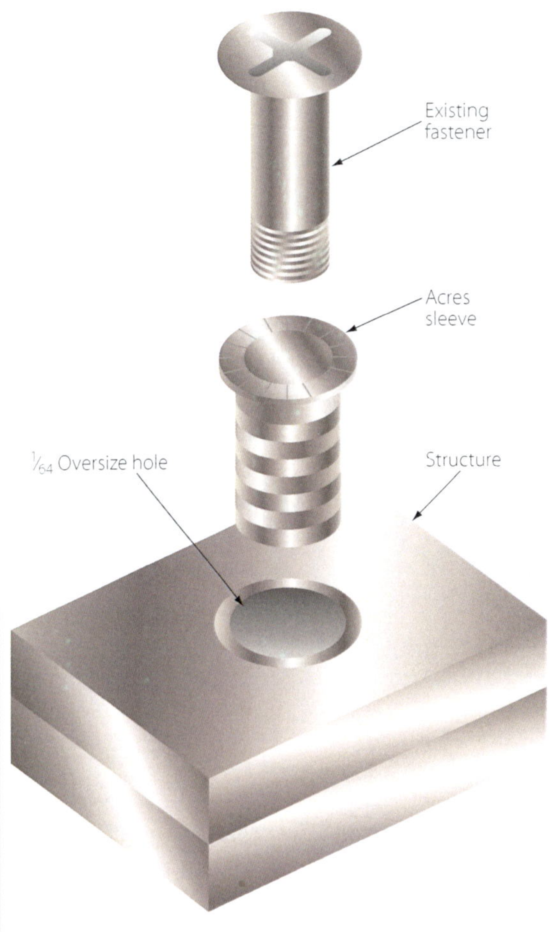

Acres sleeve installation

Figure 5-66B. Acres sleeve identification, installation, and breakoff procedure.

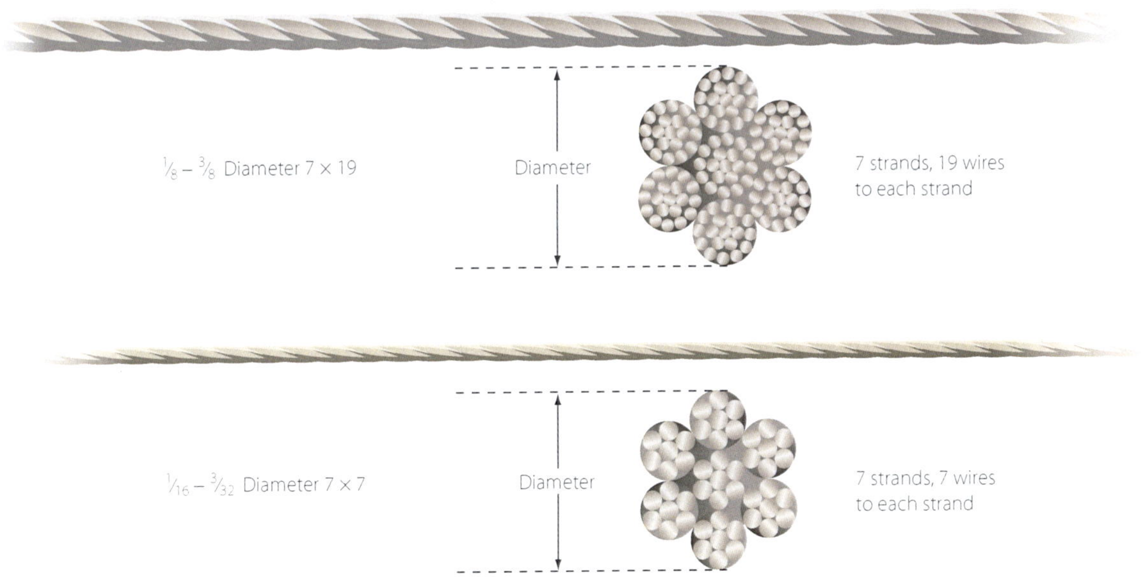

Figure 5-67. Cable cross sections.

Cable Fittings

Cables may be equipped with several different types of fittings, such as terminals, thimbles, bushings, and shackles.

Terminal fittings are generally of the swaged type. They are available in the threaded end, fork end, eye end, single shank ball end, and double shank ball end. The threaded end, fork end, and eye end terminals are used to connect the cable to a turnbuckle, bellcrank, or other linkage in the system. The ball end terminals are used for attaching cables to quadrants and special connections where space is limited. Figure 5-68 illustrates the various types of terminal fittings.

The thimble, bushing, and shackle fittings may be used in place of some types of terminal fittings when facilities and supplies are limited and immediate replacement of the cable is necessary.

Turnbuckles

A turnbuckle assembly is a mechanical screw device consisting of two threaded terminals and a threaded barrel. [Figure 5-69]

Turnbuckles are fitted in the cable assembly for the purpose of making minor adjustments in cable length and for adjusting cable tension. One of the terminals has right-hand threads and the other has left-hand threads. The barrel has matching right- and left-hand internal threads. The end of the barrel with the left-hand threads can usually be identified by a groove or knurl around that end of the barrel.

When installing a turnbuckle in a control system, it is necessary to screw both of the terminals an equal number of turns into the barrel. It is also essential that all turnbuckle terminals be screwed into the barrel until not more than three threads are exposed on either side of the turnbuckle barrel.

After a turnbuckle is properly adjusted, it must be safetied. The methods of safetying turnbuckles are discussed later in this chapter.

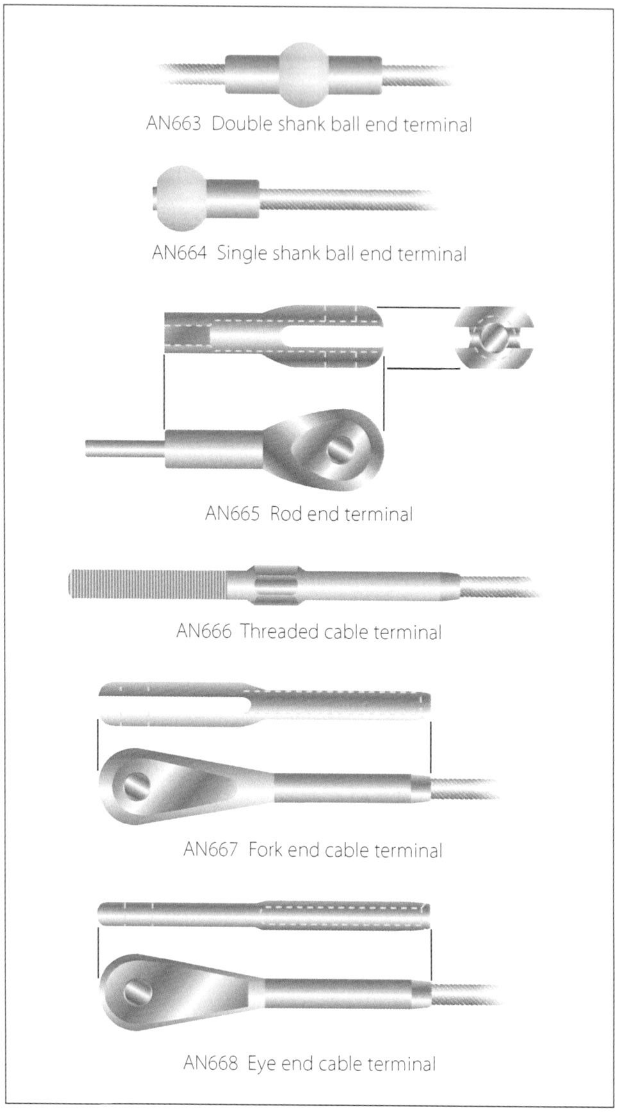

Figure 5-68. Types of terminal fittings.

Push-Pull Tube Linkage

Push-pull tubes are used as linkage in various types of mechanically operated systems. This type linkage eliminates the problem of varying tension and permits the transfer of either compression or tension stress through a single tube.

A push-pull tube assembly consists of a hollow aluminum alloy or steel tube with an adjustable end fitting

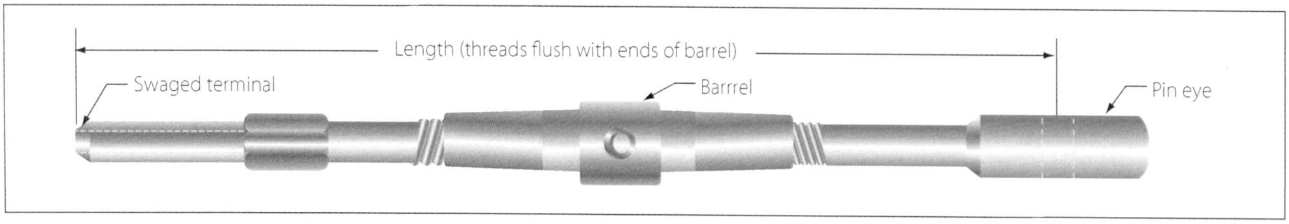

Figure 5-69. Typical turnbuckle assembly.

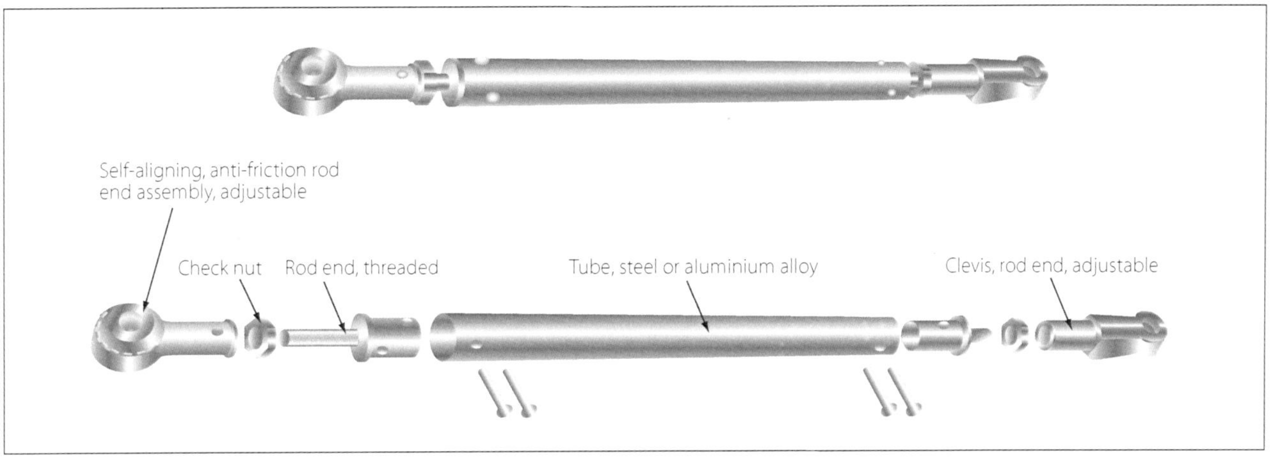

Figure 5-70. Push-pull tube assembly.

and a checknut at either end. [Figure 5-70] The checknuts secure the end fittings after the tube assembly has been adjusted to its correct length. Push-pull tubes are generally made in short lengths to prevent vibration and bending under compression loads.

Safetying Methods

To ensure fasteners do not separate from their nuts or holding ends, various safetying methods are used in aircraft from heavy aircraft to gliders to recreational aircraft.

Safetying is the process of securing all aircraft, bolts, nuts, screws, pins, and other fasteners so that they do not work loose due to vibration. A familiarity with the various methods and means of safetying equipment on an aircraft is necessary in order to perform maintenance and inspection.

There are various methods of safetying aircraft parts. The most widely used methods are safety wire, cotter pins, lockwashers, snaprings, and special nuts, such as self-locking nuts, pal nuts, and jamnuts. Some of these nuts and washers have been previously described in this chapter.

Pins

The three main types of pins used in aircraft structures are the taper pin, flathead pin, and cotter pin. Pins are used in shear applications and for safetying. Roll pins are finding increasing uses in aircraft construction.

Taper Pins

Plain and threaded taper pins (AN385 and AN386) are used in joints which carry shear loads and where absence of play is essential. The plain taper pin is drilled and usually safetied with wire. The threaded taper pin is used with a taper pin washer (AN975) and shear nut (safetied with a cotter pin or safety clip) or self-locking nut.

Flathead Pin

Commonly called a clevis pin, the flathead pin (MS20392) is used with tie rod terminals and in secondary controls which are not subject to continuous operation. The pin is customarily installed with the head up so that if the cotter pin fails or works out, the pin will remain in place.

Cotter Pins

The AN380 cadmium plated, low carbon steel cotter pin is used for safetying bolts, screws, nuts, other pins, and in various applications where such safetying is necessary. The AN381 corrosion resistant steel cotter pin is used in locations where nonmagnetic material is required, or in locations where resistance to corrosion is desired.

Roll Pins

The roll pin is a pressed fit pin with chamfered ends. It is tubular in shape and is slotted the full length of the tube. The pin is inserted with hand tools and is compressed as it is driven into place. Pressure exerted by the roll pin against the hole walls keeps it in place, until deliberately removed with a drift punch or pin punch.

Safety Wiring

Safety wiring is the most positive and satisfactory method of safetying capscrews, studs, nuts, bolt heads, and turnbuckle barrels which cannot be safetied by any other practical means. It is a method of wiring together two or more units in such a manner that any tendency of one to loosen is counteracted by the tightening of the wire.

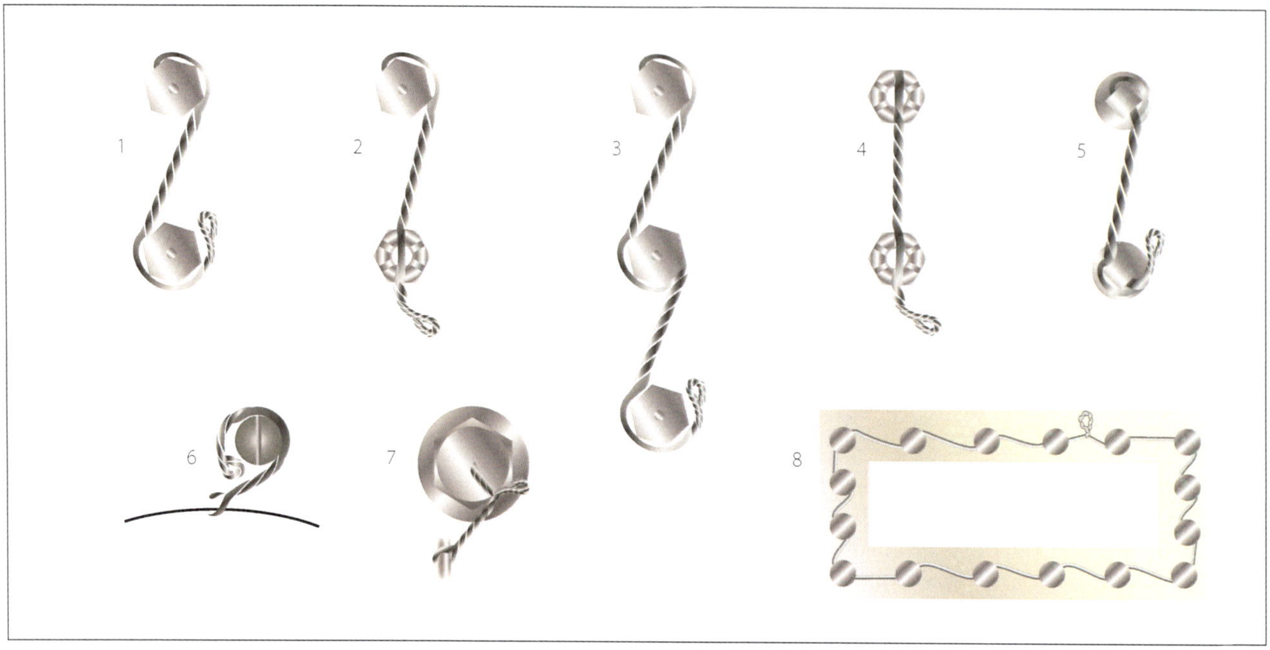

Figure 5-71. Safety wiring methods.

Nuts, Bolts, and Screws

Nuts, bolts, and screws are safety wired by the single wire or double twist method. The double twist method is the most common method of safety wiring. The single wire method may be used on small screws in a closely spaced closed geometrical pattern, on parts in electrical systems, and in places that are extremely difficult to reach. Safety wiring should always be per conventional methods or as required by the manufacturer, especially for Light Sport Aircraft (LSA).

Figure 5-71 is an illustration of various methods which are commonly used in safety wiring nuts, bolts, and screws. Careful study of Figure 5-71 shows that:

- Examples 1, 2, and 5 illustrate the proper method of safety wiring bolts, screws, squarehead plugs, and similar parts when wired in pairs.
- Example 3 illustrates several components wired in series.
- Example 4 illustrates the proper method of wiring castellated nuts and studs. (Note that there is no loop around the nut.)
- Examples 6 and 7 illustrate a single threaded component wired to a housing or lug.
- Example 8 illustrates several components in a closely spaced closed geometrical pattern, using a single wire method.

When drilled head bolts, screws, or other parts are grouped together, they are more conveniently safety wired to each other in a series rather than individually. The number of nuts, bolts, or screws that may be safety wired together is dependent on the application. For instance, when safety wiring widely spaced bolts by the double twist method, a group of three should be the maximum number in a series.

When safety wiring closely spaced bolts, the number that can be safety wired by a 24-inch length of wire is the maximum in a series. The wire is arranged so that if the bolt or screw begins to loosen, the force applied to the wire is in the tightening direction.

Parts being safety wired should be torqued to recommend values and the holes aligned before attempting the safetying operation. Never overtorque or loosen a torqued nut to align safety wire holes.

Oil Caps, Drain Cocks, and Valves

These units are safety wired as shown in Figure 5-72. In the case of the oil cap, the wire is anchored to an adjacent fillister head screw.

This system applies to any other unit which must be safety wired individually. Ordinarily, anchorage lips are conveniently located near these individual parts. When such provision is not made, the safety wire is fastened to some adjacent part of the assembly.

Electrical Connectors

Under conditions of severe vibration, the coupling nut of a connector may vibrate loose, and with sufficient vibration the connector may come apart. When this

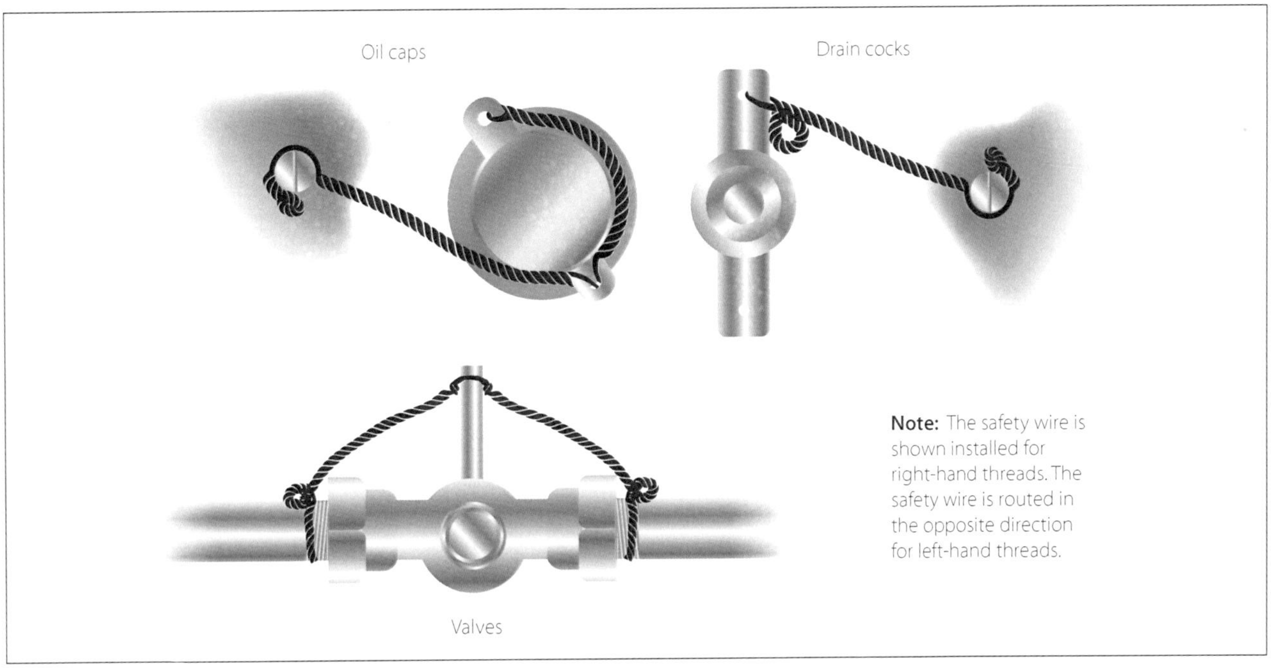

Figure 5-72. Safety wiring attachment for plug connectors.

occurs, the circuit carried by the cable opens. The proper protective measure to prevent this occurrence is by safety wiring as shown in Figure 5-73. The safety wire should be as short as practicable and must be installed in such a manner that the pull on the wire is in the direction which tightens the nut on the plug.

Turnbuckles

After a turnbuckle has been properly adjusted, it must be safetied. There are several methods of safetying turnbuckles; however, only two methods will be discussed in this section. These methods are illustrated in Figure 5-74(A) and Figure 5-74(B). The clip locking method is used only on the most modern aircraft. The older type aircraft still use the type turnbuckles that require the wire wrapping method.

Double Wrap Method. Of the methods using safety wire for safetying turnbuckles, the double wrap method is preferred, although the single wrap methods described

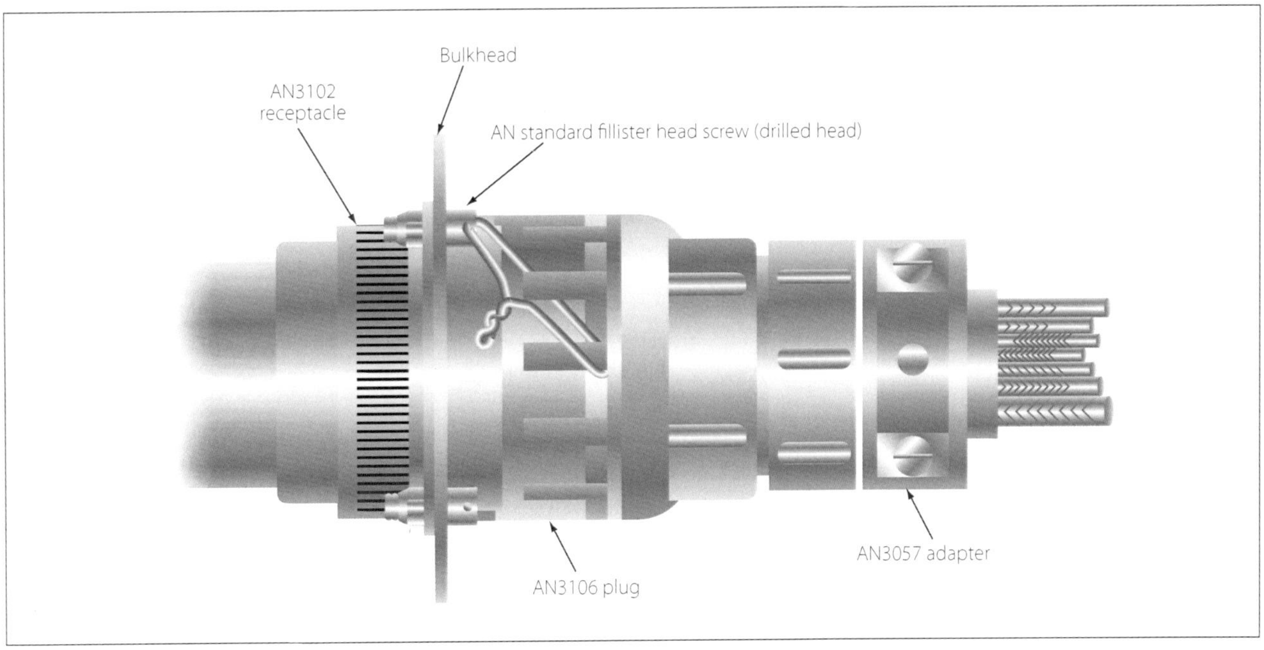

Figure 5-73. Safety wiring attachment for plug connectors.

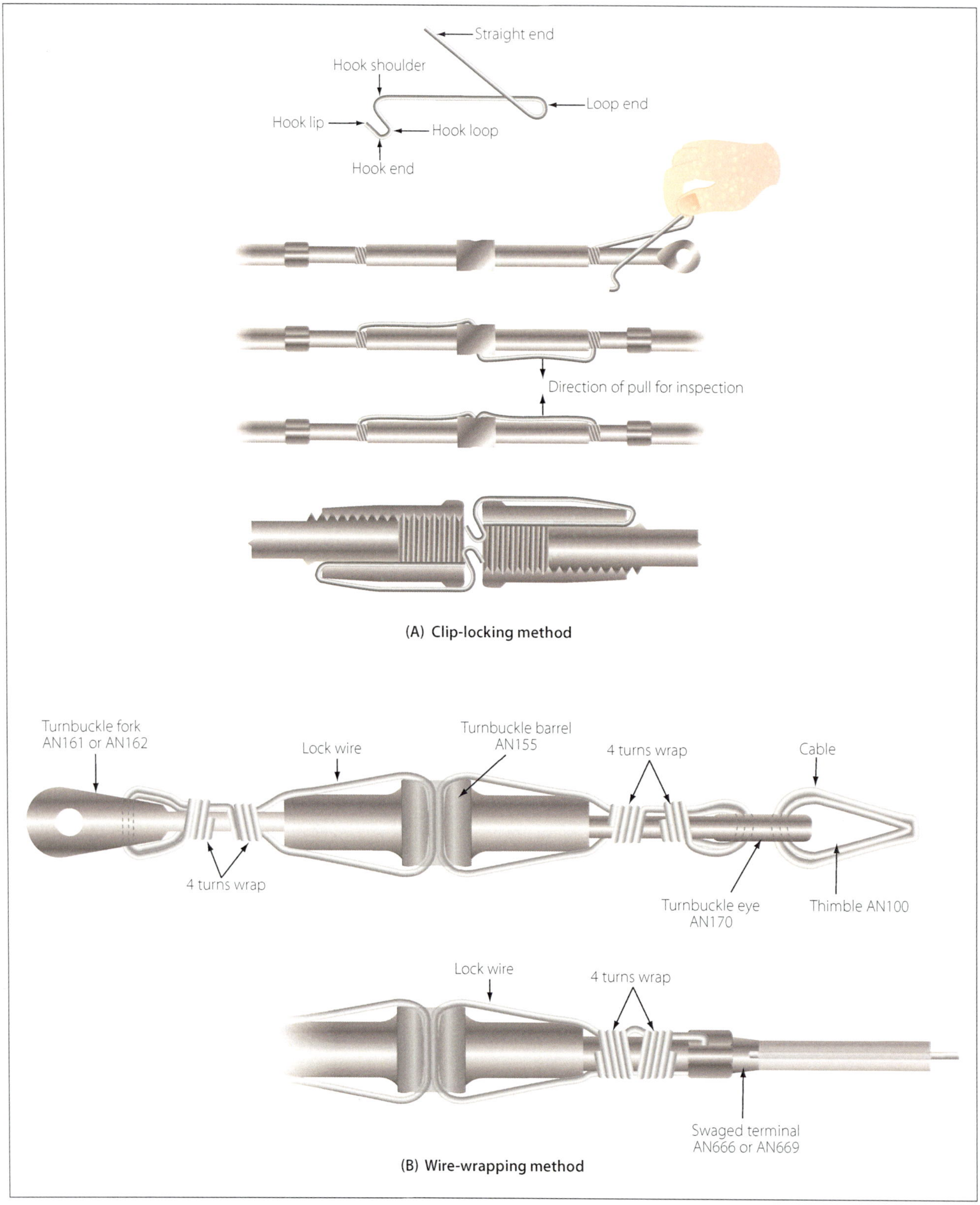

Figure 5-74. Safetying turnbuckles: (A) clip-locking method and (B) wire-wrapping method.

are satisfactory. The method of double wrap safetying is shown in Figure 5-74(B). Use two separate lengths of the proper wire as shown in Figure 5-75.

Run one end of the wire through the hole in the barrel of the turnbuckle and bend the ends of the wire toward opposite ends of the turnbuckle. Then pass the second length of the wire into the hole in the barrel and bend the ends along the barrel on the side opposite the first. Then pass the wires at the end of the turnbuckle in opposite directions through the holes in the turnbuckle

Cable size (in)	Type of Wrap	Diameter of Safety Wire (in)	Material (Annealed Condition)
1⁄16	Single	.020	Stainless steel
3⁄32	Single	.040	Copper, brass[1]
1⁄8	Single	.040	Stainless steel
1⁄8	Double	.040	Copper, brass[1]
1⁄8	Single	.057 min	Copper, brass[1]
5⁄32 and greater	Single	.057	Stainless steel

[1] Galvanized or tinned steel, or soft iron wires are also acceptable.

Figure 5-75. Turnbuckle safetying guide.

eyes or between the jaws of the turnbuckle fork, as applicable. Bend the laid wires in place before cutting off the wrapped wire. Wrap the remaining length of safety wire at least four turns around the shank and cut it off. Repeat the procedure at the opposite end of the turnbuckle.

When a swaged terminal is being safetied, pass the ends of both wires, if possible, through the hole provided in the terminal for this purpose and wrap both ends around the shank as described above.

If the hole is not large enough to allow passage of both wires, pass the wire through the hole and loop it over the free end of the other wire, and then wrap both ends around the shank as described.

Single Wrap Method. The single wrap safetying methods described in the following paragraphs are acceptable but are not the equal of the double wrap methods.

Pass a single length of wire through the cable eye or fork, or through the hole in the swaged terminal at either end of the turnbuckle assembly. Spiral each of the wire ends in opposite directions around the first half of the turnbuckle barrel so that the wires cross each other twice. Thread both wire ends through the hole in the middle of the barrel so that the third crossing of the wire ends is in the hole. Again, spiral the two wire ends in opposite directions around the remaining half of the turnbuckle, crossing them twice. Then, pass one wire end through the cable eye or fork, or through the hole in the swaged terminal. In the manner described above, wrap both wire ends around the shank for at least four turns each, cutting off the excess wire.

An alternate to the above method is to pass one length of wire through the center hole of the turnbuckle and bend the wire ends toward opposite ends of the turnbuckle. Then pass each wire end through the cable eye or fork, or through the hole in the swaged terminal and wrap each wire end around the shank for at least four turns, cutting off the excess wire. After safetying, no more than three threads of the turnbuckle threaded terminal should be exposed.

General Safety Wiring Rules

When using the safety wire method of safetying, the following general rules should be followed:

1. A pigtail of 1⁄4 to 1⁄2 inch (three to six twists) should be made at the end of the wiring. This pigtail must be bent back or under to prevent it from becoming a snag.
2. The safety wire must be new upon each application.
3. When castellated nuts are to be secured with safety wire, tighten the nut to the low side of the selected torque range, unless otherwise specified, and if necessary, continue tightening until a slot aligns with the hole.
4. All safety wires must be tight after installation, but not under such tension that normal handling or vibration will break the wire.
5. The wire must be applied so that all pull exerted by the wire tends to tighten the nut.
6. Twists should be tight and even, and the wire between the nuts as taut as possible without overtwisting.
7. The safety wire should always be installed and twisted so that the loop around the head stays down and does not tend to come up over the bolt head, causing a slack loop.

Cotter Pin Safetying

Cotter pin installation is shown in Figure 5-76. Castellated nuts are used with bolts that have been drilled for cotter pins. The cotter pin should fit neatly into the hole, with very little sideplay. The following general rules apply to cotter pin safetying:

1. The prong bent over the bolt end should not extend beyond the bolt diameter. (Cut it off if necessary.)
2. The prong bent down should not rest against the surface of the washer. (Again, cut it off if necessary.)
3. If the optional wraparound method is used, the prongs should not extend outward from the sides of the nut.
4. All prongs should be bent over a reasonable radius. Sharp angled bends invite breakage. Tapping lightly with a mallet is the best method of bending the prongs.

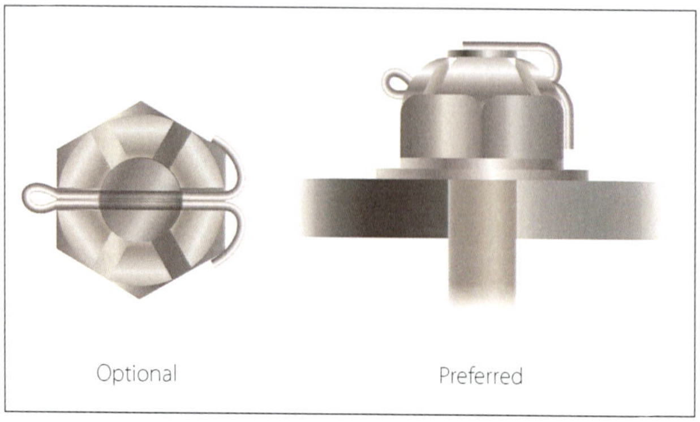

Figure 5-76. Cotter pin installations.

Snaprings

A snapring is a ring of metal, either round or flat in cross section, which is tempered to have springlike action. This springlike action will hold the snapring firmly seated in a groove. The external types are designed to fit in a groove around the outside of a shaft or cylinder, and may be safety wired. Safety wiring of an external type snapring is shown in Figure 5-77. The internal types fit in a groove inside a cylinder, and are never safetied. A special type of pliers is designed to install each type of snapring. Snaprings can be reused as long as they retain their shape and springlike action.

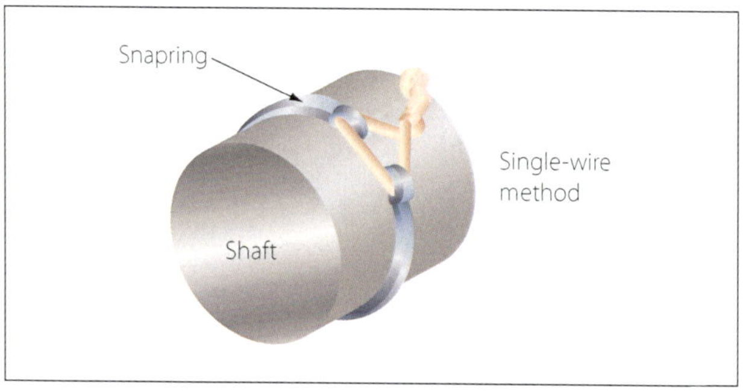

Figure 5-77. External type snapring with safety wire installation.

Chapter 6: Aircraft Cleaning & Corrosion Control

Corrosion Control

Many aircraft structures are made of metal, and the most insidious form of damage to those structures is corrosion. From the moment the metal is manufactured, it must be protected from the deleterious effects of the environment that surrounds it. This protection can be the introduction of certain elements into the base metal, creating a corrosion resistant alloy, or the addition of a surface coating of a chemical conversion coating, metal or paint. While in use, additional moisture barriers, such as viscous lubricants and protectants may be added to the surface.

The introduction of airframes built primarily of composite components has not eliminated the need for careful monitoring of aircraft with regard to corrosion. While the airframe itself may not be subject to corrosion, the use of metal components and accessories within the airframe means the aircraft maintenance technician must be on the alert for the evidence of corrosion when inspecting any aircraft.

This chapter provides an overview to the problems associated with aircraft corrosion. For more in-depth information on the subject, refer to the latest edition of FAA Advisory Circular (AC) 43-4A, Corrosion Control for Aircraft. The advisory circular is an extensive handbook, which deals with the sources of corrosion particular to aircraft structures, as well as steps the aircraft maintenance technician can take in the course of maintaining aircraft that have been attacked by corrosion.

Metal corrosion is the deterioration of the metal by chemical or electrochemical attack. This type of damage can take place internally as well as on the surface. As in the rotting of wood, this deterioration may change the smooth surface, weaken the interior, or damage or loosen adjacent parts.

Water or water vapor containing salt combines with oxygen in the atmosphere to produce the main source of corrosion in aircraft. Aircraft operating in a marine environment, or in areas where the atmosphere contains industrial fumes that are corrosive, are particularly susceptible to corrosive attacks. [Figure 6-1]

If left unchecked, corrosion can cause eventual structural failure. The appearance of corrosion varies with the metal. On the surface of aluminum alloys and magnesium, it appears as pitting and etching, and is

Figure 6-1. Seaplane operations.

often combined with a gray or white powdery deposit. On copper and copper alloys, the corrosion forms a greenish film; on steel, a reddish corrosion byproduct commonly referred to as rust. When the gray, white, green, or reddish deposits are removed, each of the surfaces may appear etched and pitted, depending upon the length of exposure and severity of attack. If these surface pits are not too deep, they may not significantly alter the strength of the metal; however, the pits may become sites for crack development, particularly if the part is highly stressed. Some types of corrosion burrow between the inside of surface coatings and the metal surface, and can spread until the part fails.

Types of Corrosion

There are two general classifications of corrosion that cover most of the specific forms: direct chemical attack and electrochemical attack. In both types of corrosion, the metal is converted into a metallic compound such as an oxide, hydroxide, or sulfate. The corrosion process always involves two simultaneous changes: The metal that is attacked or oxidized suffers what may be called anodic change, and the corrosive agent is reduced and may be considered as undergoing cathodic change.

Figure 6-2. Direct chemical attack in a battery compartment.

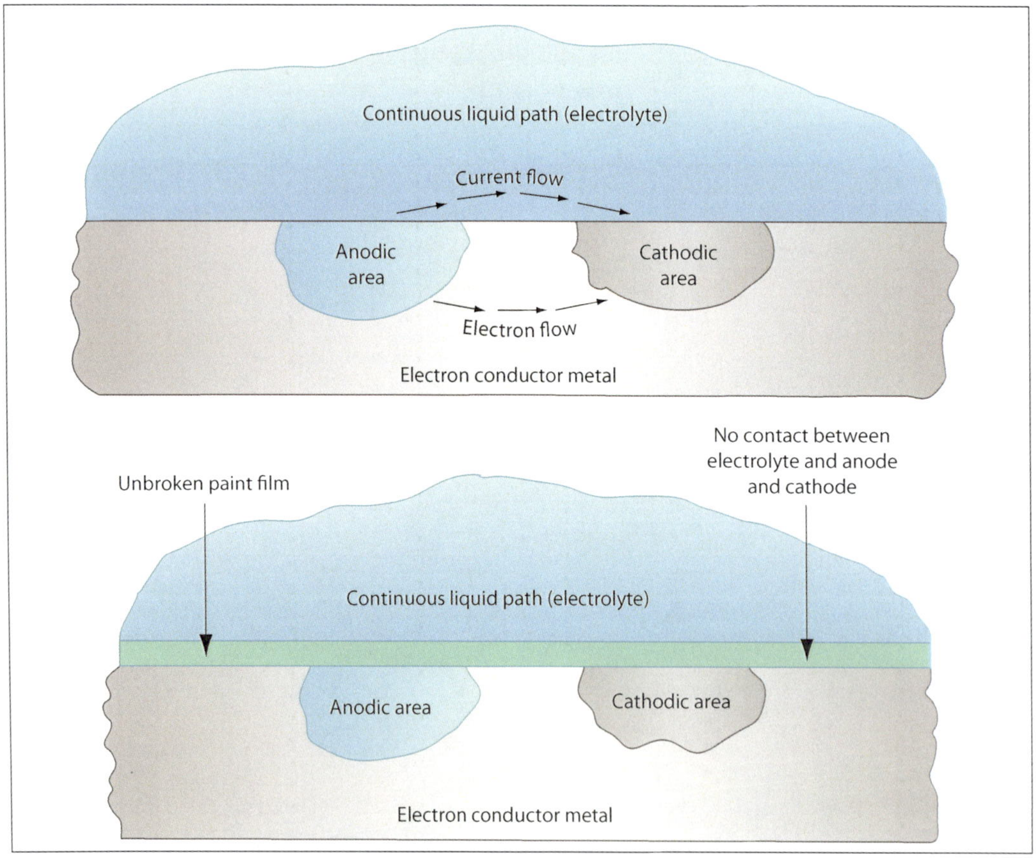

Figure 6-3. Electrochemical attack.

Direct Chemical Attack

Direct chemical attack, or pure chemical corrosion, is an attack resulting from a direct exposure of a bare surface to caustic liquid or gaseous agents. Unlike electrochemical attack where the anodic and cathodic changes may be taking place a measurable distance apart, the changes in direct chemical attack are occurring simultaneously at the same point. The most common agents causing direct chemical attack on aircraft are: (1) spilled battery acid or fumes from batteries; (2) residual flux deposits resulting from inadequately cleaned, welded, brazed, or soldered joints; and (3) entrapped caustic cleaning solutions. [Figure 6-2]

With the introduction of sealed lead-acid batteries and the use of nickel-cadmium batteries, spilled battery acid is becoming less of a problem. The use of these closed units lessens the hazards of acid spillage and battery fumes.

Many types of fluxes used in brazing, soldering, and welding are corrosive, and they chemically attack the metals or alloys with which they are used. Therefore, it is important to remove residual flux from the metal surface immediately after the joining operation. Flux residues are hygroscopic in nature; that is, they absorb moisture, and unless carefully removed, tend to cause severe pitting.

Caustic cleaning solutions in concentrated form should be kept tightly capped and as far from aircraft as possible. Some cleaning solutions used in corrosion removal are, in themselves, potentially corrosive agents; therefore, particular attention should be directed toward their complete removal after use on aircraft. Where entrapment of the cleaning solution is likely to occur, use a noncorrosive cleaning agent, even though it is less efficient.

Electrochemical Attack

An electrochemical attack may be likened chemically to the electrolytic reaction that takes place in electroplating, anodizing, or in a dry cell battery. The reaction in this corrosive attack requires a medium, usually water, which is capable of conducting a tiny current of electricity. When a metal comes in contact with a corrosive agent and is also connected by a liquid or gaseous path through which electrons may flow, corrosion begins as the metal decays by oxidation. [Figure 6-3] During the attack, the quantity of corrosive agent is reduced and, if not renewed or removed, may completely react with the metal, becoming neutralized. Different areas of the same metal surface have varying levels of electrical potential and, if connected by a conductor, such as salt water, will set up a series of corrosion cells and corrosion will commence.

All metals and alloys are electrically active and have a specific electrical potential in a given chemical environment. This potential is commonly referred to as the metal's "nobility." [Figure 6-4] The less noble a metal is, the more easily it can be corroded. The metals chosen for use in aircraft structures are a studied compromise with strength, weight, corrosion resistance, workability, and cost balanced against the structure's needs.

The constituents in an alloy also have specific electrical potentials that are generally different from each other. Exposure of the alloy surface to a conductive, corrosive medium causes the more active metal to

+ Corroded End (anodic, or least noble)
Magnesium Magnesium alloy Zinc
Aluminum (1100) Cadmium Aluminum 2024-T4 Steel or Iron Cast Iron Chromium-Iron (active) Ni-Resist Cast Iron
Type 304 Stainless steel (active) Type 316 Stainless steel (active)
Lead-Tin solder Lead Tin
Nickel (active) Inconel nickel-chromium alloy (active) Hastelloy Alloy C (active)
Brass Copper Bronze Copper-nickel alloy Monel nickel-copper alloy
Silver Solder Nickel (passive) Inconel nickel-chromium alloy (passive)
Chromium-Iron (passive) Type 304 Stainless steel (passive) Type 316 Stainless steel (passive) Hastelloy Alloy C (passive)
Silver Titanium Graphite Gold Platinum
– Protected End (cathodic, or most noble)

Figure 6-4. The galvanic series of metals and alloys.

become anodic and the less active metal to become cathodic, thereby establishing conditions for corrosion. These are called local cells. The greater the difference in electrical potential between the two metals, the greater will be the severity of a corrosive attack, if the proper conditions are allowed to develop.

The conditions for these corrosion reactions are the presence of a conductive fluid and metals having a difference in potential. If, by regular cleaning and surface refinishing, the medium is removed and the minute electrical circuit eliminated, corrosion cannot occur. This is the basis for effective corrosion control. The electrochemical attack is responsible for most forms of corrosion on aircraft structure and component parts.

Forms of Corrosion

There are many forms of corrosion. The form of corrosion depends on the metal involved, its size and shape, its specific function, atmospheric conditions, and the corrosion producing agents present. Those described in this section are the more common forms found on airframe structures.

Surface Corrosion

Surface corrosion appears as a general roughening, etching, or pitting of the surface of a metal, frequently accompanied by a powdery deposit of corrosion products. Surface corrosion may be caused by either direct chemical or electrochemical attack. Sometimes corrosion will spread under the surface coating and cannot be recognized by either the roughening of the surface or the powdery deposit. Instead, closer inspection will reveal the paint or plating is lifted off the surface in small blisters which result from the pressure of the underlying accumulation of corrosion products. [Figure 6-5]

Filiform corrosion gives the appearance of a series of small worms under the paint surface. It is often seen on surfaces that have been improperly chemically treated prior to painting. [Figure 6-6]

Dissimilar Metal Corrosion

Extensive pitting damage may result from contact between dissimilar metal parts in the presence of a conductor. While surface corrosion may or may not be taking place, a galvanic action, not unlike electroplating, occurs at the points or areas of contact where the insulation between the surfaces has broken down or been omitted. This electrochemical attack can be very serious because in many instances the action is taking place out of sight, and the only way to detect it prior to structural failure is by disassembly and inspection. [Figure 6-7]

The contamination of a metal's surface by mechanical means can also induce dissimilar metal corrosion. The improper use of steel cleaning products, such as steel wool or a steel wire brush on aluminum or magnesium, can force small pieces of steel into the metal being cleaned, which will then further corrode and

Figure 6-5. Surface corrosion.

Figure 6-6. Filiform corrosion.

Figure 6-7. Dissimilar metal corrosion.

ruin the adjoining surface. Carefully monitor the use of nonwoven abrasive pads, so that pads used on one type of metal are not used again on a different metal surface.

Intergranular Corrosion

This type of corrosion is an attack along the grain boundaries of an alloy and commonly results from a lack of uniformity in the alloy structure. Aluminum alloys and some stainless steels are particularly susceptible to this form of electrochemical attack. [Figure 6-8] The lack of uniformity is caused by changes that occur in the alloy during heating and cooling during the material's manufacturing process. Intergranular corrosion may exist without visible surface evidence. Very severe intergranular corrosion may sometimes cause the surface of a metal to "exfoliate." [Figure 6-9] This is a lifting or flaking of the metal at the

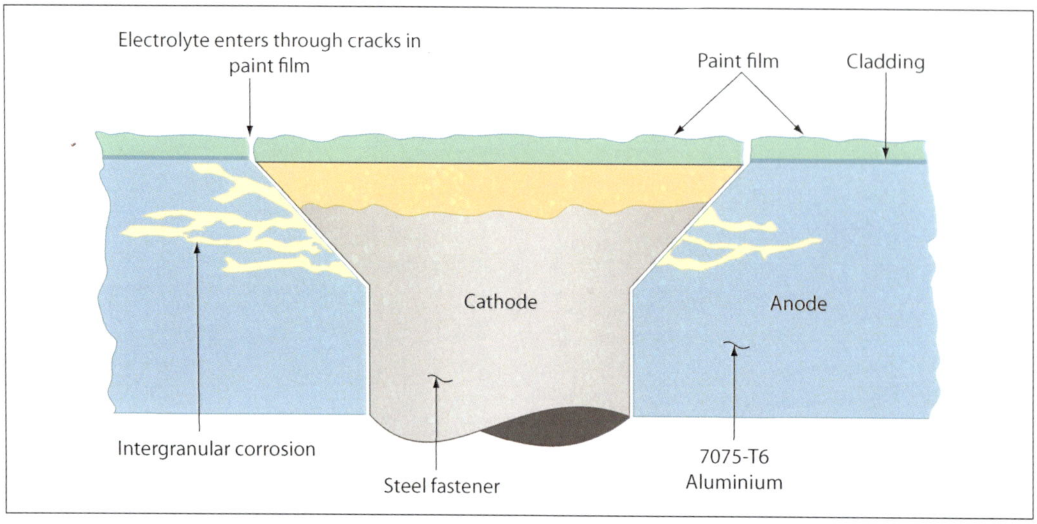

Figure 6-8. Intergranular corrosion of 7075-T6 aluminum adjacent to steel fastener.

surface due to delamination of the grain boundaries caused by the pressure of corrosion residual product buildup. This type of corrosion is difficult to detect in its initial stage. Extruded components such as spars can be subject to this type of corrosion. Ultrasonic and eddy current inspection methods are being used with a great deal of success.

Stress Corrosion

Stress corrosion occurs as the result of the combined effect of sustained tensile stresses and a corrosive environment. Stress corrosion cracking is found in most metal systems; however, it is particularly characteristic of aluminum, copper, certain stainless steels, and high strength alloy steels (over 240,000 psi). It usually occurs along lines of cold working and may be transgranular or intergranular in nature. Aluminum alloy bellcranks with pressed in bushings, landing gear shock struts with pipe thread type grease fittings, clevis pin joints, shrink fits, and overstressed tubing B-nuts are examples of parts which are susceptible to stress corrosion cracking.

Fretting Corrosion

Fretting corrosion is a particularly damaging form of corrosive attack that occurs when two mating surfaces, normally at rest with respect to one another, are subject to slight relative motion. It is characterized by pitting of the surfaces and the generation of considerable quantities of finely divided debris. Since the restricted movements of the two surfaces prevent the debris from escaping very easily, an extremely local-

Figure 6-9. Exfoliation.

ized abrasion occurs. [Figure 6-10] The presence of water vapor greatly increases this type of deterioration. If the contact areas are small and sharp, deep grooves resembling brinell markings or pressure indentations may be worn in the rubbing surface. As a result, this type of corrosion (on bearing surfaces) has also been called false brinelling.

Factors Affecting Corrosion

Many factors affect the type, speed, cause, and seriousness of metal corrosion. Some of these factors can be controlled and some cannot.

Climate

The environmental conditions under which an aircraft is maintained and operated greatly affect corrosion characteristics. In a predominately marine environment (with exposure to sea water and salt air), moisture-laden air is considerably more detrimental to an aircraft than it would be if all operations were conducted in a dry climate. Temperature considerations are important because the speed of electrochemical attack is increased in a hot, moist climate.

Foreign Material

Among the controllable factors which affect the onset and spread of corrosive attack is foreign material that adheres to the metal surfaces. Such foreign material includes:

- Soil and atmospheric dust.
- Oil, grease, and engine exhaust residues.
- Salt water and salt moisture condensation.
- Spilled battery acids and caustic cleaning solutions.
- Welding and brazing flux residues.

It is important that aircraft be kept clean. How often and to what extent an aircraft should be cleaned depends on several factors, including geographic location, model of aircraft, and type of operation.

Preventive Maintenance

Much has been done to improve the corrosion resistance of aircraft: improvements in materials, surface treatments, insulation, and in particular, modern protective finishes. All of these have been aimed at reducing the overall maintenance effort, as well as improving reliability. In spite of these improvements, corrosion and its control is a very real problem that requires continuous preventive maintenance. During any corrosion control maintenance, consult the Material Safety Data Sheet (MSDS) for information on any chemicals used in the process.

Corrosion preventive maintenance includes the following specific functions:

1. Adequate cleaning
2. Thorough periodic lubrication

Figure 6-10. Fretting corrosion.

3. Detailed inspection for corrosion and failure of protective systems
4. Prompt treatment of corrosion and touchup of damaged paint areas
5. Keeping drain holes free of obstructions
6. Daily draining of fuel cell sumps
7. Daily wipe down of exposed critical areas
8. Sealing of aircraft against water during foul weather and proper ventilation on warm, sunny days
9. Maximum use of protective covers on parked aircraft

After any period during which regular corrosion preventive maintenance is interrupted, the amount of maintenance required to repair accumulated corrosion damage and bring the aircraft back up to standard will usually be quite high.

Inspection

Inspection for corrosion is a continuing problem and should be handled on a daily basis. Overemphasizing a particular corrosion problem when it is discovered and then forgetting about corrosion until the next crisis is an unsafe, costly, and troublesome practice. Most scheduled maintenance checklists are complete enough to cover all parts of the aircraft or engine, and no part of the aircraft should go uninspected. Use these checklists as a general guide when an area is to be inspected for corrosion. Through experience it will be learned that most aircraft have trouble areas where, despite routine inspection and maintenance, corrosion will set in.

In addition to routine maintenance inspections, amphibians or seaplanes should be checked daily and critical areas cleaned or treated, as necessary.

Corrosion Prone Areas

Discussed briefly in this section are most of the trouble areas common to all aircraft. However, this coverage is not necessarily complete and may be amplified and expanded to cover the special characteristics of the particular aircraft model involved by referring to the applicable maintenance manual.

Exhaust Trail Areas

Both jet and reciprocating engine exhaust deposits are very corrosive and give particular trouble where gaps, seams, hinges, and fairings are located downstream from the exhaust pipes or nozzles. [Figure 6-11]

Figure 6-11. Exhaust nozzle area.

Deposits may be trapped and not reached by normal cleaning methods. Pay special attention to areas around rivet heads and in skin lap joints and other crevices. Remove and inspect fairings and access plates in the exhaust areas. Do not overlook exhaust deposit buildup in remote areas, such as the empennage surfaces. Buildup in these areas will be slower and may not be noticed until corrosive damage has begun.

Battery Compartments and Battery Vent Openings

Despite improvements in protective paint finishes and in methods of sealing and venting, battery compartments continue to be corrosion prone areas. Fumes from overheated electrolyte are difficult to contain and will spread to adjacent cavities and cause a rapid corrosive attack on all unprotected metal surfaces. Battery vent openings on the aircraft skin should be included in the battery compartment inspection and maintenance procedure. Regular cleaning and neutralization of acid deposits will minimize corrosion from this cause.

Bilge Areas

These are natural sumps for waste hydraulic fluids, water, dirt, and odds and ends of debris. Residual oil quite often masks small quantities of water that settle to the bottom and set up a hidden chemical cell.

Instead of using chemical treatments for the bilge water, current float manufacturers recommend the diligent maintenance of the internal coatings applied to the float's interior during manufacture. In addition to chemical conversion coatings applied to the surface of the sheet metal and other structural components, and to sealants installed in lap joints during construction, the interior compartments are painted to protect the bilge areas. When seaplane structures are repaired or restored, this level of corrosion protection must be maintained.

Inspection procedures should include particular attention paid to areas located under galleys and lavatories and to human waste disposal openings on the aircraft exteriors. Human waste products and the chemicals used in lavatories are very corrosive to common aircraft metals. Clean these areas frequently and keep the paint touched up.

Wheel Well and Landing Gear

More than any other area on the aircraft, this area probably receives more punishment due to mud, water, salt, gravel, and other flying debris.

Because of the many complicated shapes, assemblies, and fittings, complete area paint film coverage is difficult to attain and maintain. A partially applied preservative tends to mask corrosion rather than prevent it. Due to heat generated by braking action, preservatives cannot be used on some main landing gear wheels. During inspection of this area, pay particular attention to the following trouble spots:

1. Magnesium wheels, especially around bolt heads, lugs, and wheel web areas, particularly for the presence of entrapped water or its effects
2. Exposed rigid tubing, especially at B-nuts and ferrules, under clamps and tubing identification tapes
3. Exposed position indicator switches and other electrical equipment
4. Crevices between stiffeners, ribs, and lower skin surfaces, which are typical water and debris traps

Water Entrapment Areas

Design specifications require that aircraft have drains installed in all areas where water may collect. Daily inspection of low point drains should be a standard requirement. If this inspection is neglected, the drains may become ineffective because of accumulated debris, grease, or sealants.

Engine Frontal Areas and Cooling Air Vents

These areas are being constantly abraded with airborne dirt and dust, bits of gravel from runways, and rain erosion, which tends to remove the protective finish. Inspection of these areas should include all sections in the cooling air path, with special attention to places where salt deposits may be built up during marine operations. It is imperative that incipient corrosion be inhibited and that paint touchup and hard film preservative coatings are maintained intact on seaplane and amphibian engine surfaces at all times.

Wing Flap and Spoiler Recesses

Dirt and water may collect in flap and spoiler recesses and go unnoticed because they are normally retracted. For this reason, these recesses are potential corrosion problem areas. Inspect these areas with the spoilers and/or flaps in the fully deployed position.

External Skin Areas

External aircraft surfaces are readily visible and accessible for inspection and maintenance. Even here, certain types of configurations or combinations of materials become troublesome under certain operating conditions and require special attention.

Relatively little corrosion trouble is experienced with magnesium skins if the original surface finish and insulation are adequately maintained. Trimming, drilling, and riveting destroy some of the original surface treatment, which is never completely restored by touchup procedures. Any inspection for corrosion should include all magnesium skin surfaces with special attention to edges, areas around fasteners, and cracked, chipped, or missing paint.

Piano-type hinges are prime spots for corrosion due to the dissimilar metal contact between the steel pin and aluminum hinge. They are also natural traps for dirt, salt, and moisture. Inspection of hinges should include lubrication and actuation through several cycles to ensure complete lubricant penetration. Use water-displacing lubricants when servicing piano hinges. [Figures 6-12 and 6-13]

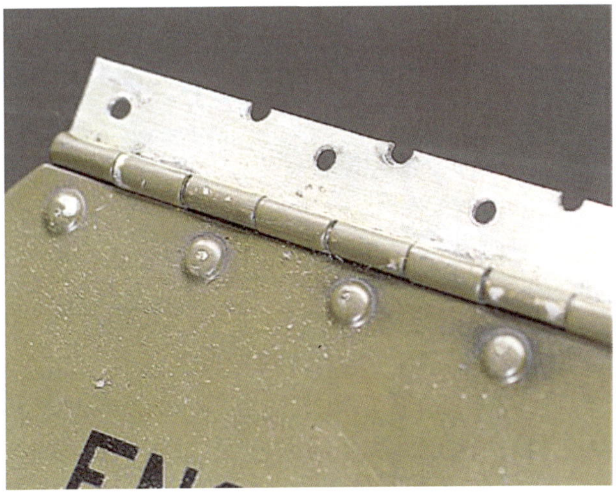

Figure 6-12. Piano hinge.

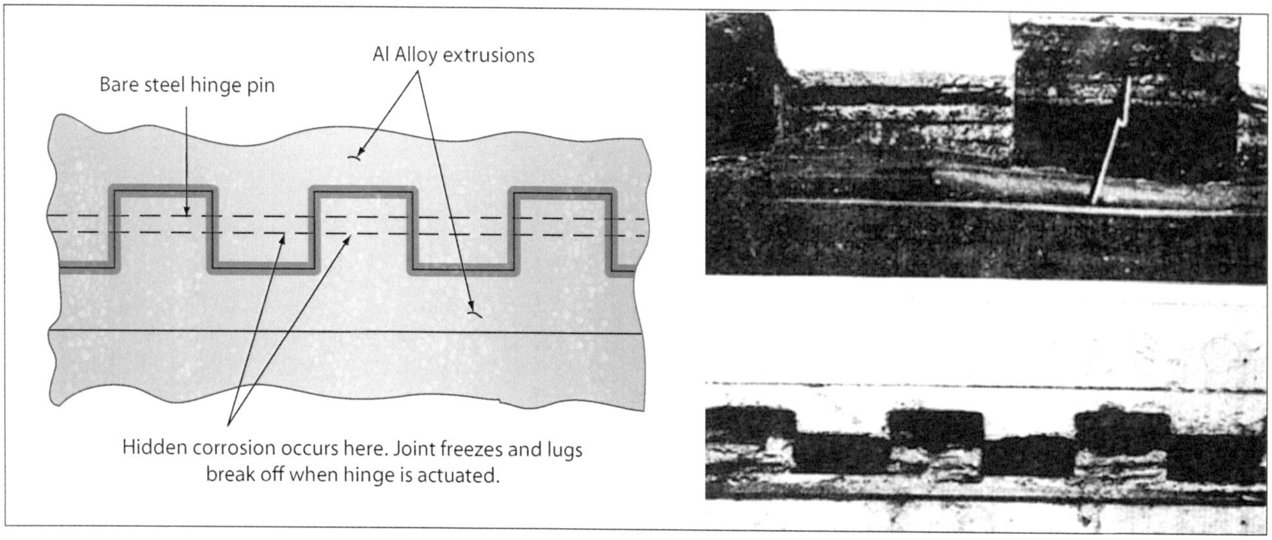

Figure 6-13. Hinge corrosion points.

Corrosion of metal skins joined by spot welding is the result of the entrance and entrapment of corrosive agents between the layers of metal. This type of corrosion is evidenced by corrosion products appearing at the crevices through which the corrosive agents enter. More advanced corrosive attack causes skin buckling and eventual spot weld fracture. Skin buckling in its early stages may be detected by sighting along spot welded seams or by using a straightedge. The only technique for preventing this condition is to keep potential moisture entry points, including seams and holes created by broken spot welds, filled with a sealant or a suitable preservative compound.

Miscellaneous Trouble Areas

Helicopter rotor heads and gearboxes, in addition to being constantly exposed to the elements, contain bare steel surfaces, many external working parts, and dissimilar metal contacts. Inspect these areas frequently for evidence of corrosion. The proper maintenance, lubrication, and the use of preservative coatings can prevent corrosion in these areas.

All control cables, whether plain carbon steel or corrosion resistant steel, should be inspected to determine their condition at each inspection period. In this process, inspect cables for corrosion by random cleaning of short sections with solvent soaked cloths. If external corrosion is evident, relieve tension and check the cable for internal corrosion. Replace cables that have internal corrosion. Remove light external corrosion with a nonwoven abrasive pad lightly soaked in oil or, alternatively, a steel wire brush. When corrosion products have been removed, recoat the cable with preservative.

Corrosion Removal

In general, any complete corrosion treatment involves the following: (1) cleaning and stripping of the corroded area, (2) removing as much of the corrosion products as practicable, (3) neutralizing any residual materials remaining in pits and crevices, (4) restoring protective surface films, and (5) applying temporary or permanent coatings or paint finishes.

The following paragraphs deal with the correction of corrosive attack on aircraft surface and components where deterioration has not progressed to the point requiring rework or structural repair of the part involved.

Surface Cleaning and Paint Removal

The removal of corrosion necessarily includes removal of surface finishes covering the attacked or suspected area. To assure maximum efficiency of the stripping compound, the area must be cleaned of grease, oil, dirt, or preservatives. This preliminary cleaning operation is also an aid in determining the extent of the spread of the corrosion, since the stripping operation will be held to the minimum consistent with full exposure of the corrosion damage. Extensive corrosion spread on any panel should be corrected by fully treating the entire section.

The selection of the type of materials to be used in cleaning will depend on the nature of the matter to be removed. Modern environmental standards encourage the use of water-based, non-toxic cleaning compounds whenever possible. In some locations, local or state laws may require the use of such products, and prohibit the use of solvents that contain volatile organic compounds (VOCs). Where permitted, dry cleaning solvent (P-D-680) may be used for removing oil, grease, or soft preservative compounds. For heavy-duty removal of thick or dried preservatives, other compounds of the solvent emulsion type are available.

The use of a general purpose, water rinsable stripper can be used for most applications. There are other methods for paint removal that have minimal impact upon the aircraft structure, and are considered "environmentally friendly."

Wherever practicable, chemical paint removal from any large area should be accomplished outside (in open air) and preferably in shaded areas. If inside removal is necessary, adequate ventilation must be assured. Synthetic rubber surfaces, including aircraft tires, fabric, and acrylics, must be thoroughly protected against possible contact with paint remover. Care must be exercised in using paint remover. Care must also be exercised in using paint remover around gas or watertight seam sealants, since the stripper will tend to soften and destroy the integrity of these sealants.

Mask off any opening that would permit the stripping compound to get into aircraft interiors or critical cavities. Paint stripper is toxic and contains ingredients harmful to both skin and eyes. Therefore, wear rubber gloves, aprons of acid repellent material, and goggle-type eyeglasses. The following is a general stripping procedure:

1. Brush the entire area to be stripped with a cover of stripper to a depth of 1⁄32 to 1⁄16 inch. Any paintbrush makes a satisfactory applicator, except that the bristles will be loosened by the effect of paint remover on the binder, and the brush should not be used for other purposes after being exposed to paint remover.
2. Allow the stripper to remain on the surface for a sufficient length of time to wrinkle and lift the paint. This may be from 10 minutes to several hours, depending on both the temperature and humidity, and the condition of the paint coat being removed. Scrub the surface with a bristle brush saturated with paint remover to further loosen finish that may still be adhering to the metal.
3. Reapply the stripper as necessary in areas where the paint remains tightly adhered or where the stripper has dried, and repeat the above process. Only nonmetallic scrapers should be used to assist in removing persistent paint finishes. Nonwoven abrasive pads intended for paint stripping may also prove to be useful in removing the loosened paint.
4. Remove the loosened paint and residual stripper by washing and scrubbing the surface with water and a broom, brush or fresh nonwoven abrasive pad. If water spray is available, use a low to medium pressure stream of water directly on the area being scrubbed. If steam-cleaning equipment is available and the area is sufficiently large, cleaning may be accomplished using this equipment together with a solution of steam-cleaning compound. On small areas, any method may be used that will assure complete rinsing of the cleaned area. Use care to dispose of the stripped residue in accordance with environmental laws.

Corrosion of Ferrous Metals

One of the most familiar types of corrosion is ferrous oxide (rust), generally resulting from atmospheric oxidation of steel surfaces. Some metal oxides protect the underlying base metal, but rust is not a protective coating in any sense of the word. Its presence actually promotes additional attack by attracting moisture from the air and acting as a catalyst for additional corrosion. If complete control of the corrosive attack is to be realized, all rust must be removed from steel surfaces.

Rust first appears on bolt heads, hold-down nuts, or other unprotected aircraft hardware. [Figure 6-14] Its presence in these areas is generally not dangerous and has no immediate effect on the structural strength of any major components. The residue from the rust may also contaminate other ferrous components, promoting corrosion of those parts. The rust is indicative of a need for maintenance and of possible corrosive attack in more critical areas. It is also a factor in the general appearance of the equipment. When paint failures occur or mechanical damage exposes highly stressed steel surfaces to the atmosphere, even the smallest amount of rusting is potentially dangerous in these areas and must be removed and controlled.

Rust removal from structural components, followed by an inspection and damage assessment, must be done as soon as feasible. [Figure 6-15]

Mechanical Removal of Iron Rust

The most practicable means of controlling the corrosion of steel is the complete removal of corrosion products by mechanical means and restoring corrosion preventive coatings. Except on highly stressed

Figure 6-14. Rust.

Figure 6-15. Rust on structural components.

steel surfaces, the use of abrasive papers and compounds, small power buffers and buffing compounds, hand wire brushing, or steel wool are all acceptable cleanup procedures. However, it should be recognized that in any such use of abrasives, residual rust usually remains in the bottom of small pits and other crevices. It is practically impossible to remove all corrosion products by abrasive or polishing methods alone. As a result, once a part cleaned in such a manner has rusted, it usually corrodes again more easily than it did the first time.

The introduction of variations of the nonwoven abrasive pad has also increased the options available for the removal of surface rust. [Figure 6-16] Flap wheels, pads intended for use with rotary or oscillating power tools, and hand-held nonwoven abrasive pads all can be used alone or with light oils to remove corrosion from ferrous components.

Figure 6-16. Nonwoven abrasive pads.

Chemical Removal of Rust

As environmental concerns have been addressed in recent years, interest in noncaustic chemical rust removal has increased. A variety of commercial products, which actively remove the iron oxide without chemically etching the base metal, are available and should be considered for use. Generally speaking, if at all possible, the steel part should be removed from the airframe for treatment, as it can be nearly impossible to remove all residues. The use of any caustic rust removal product will require the isolation of the part from any nonferrous metals during treatment, and will probably require inspection for proper dimensions.

Chemical Surface Treatment of Steel

There are approved methods for converting active rust to phosphates and other protective coatings. Other commercial preparations are effective rust converters where tolerances are not critical and where thorough rinsing and neutralizing of residual acid is possible. These situations are generally not applicable to assembled aircraft, and the use of chemical inhibitors on installed steel parts is not only undesirable but also very dangerous. The danger of entrapment of corrosive solutions and the resulting uncontrolled attack, which could occur when such materials are used under field conditions, outweigh any advantages to be gained from their use.

Removal of Corrosion from Highly Stressed Steel Parts

Any corrosion on the surface of a highly stressed steel part is potentially dangerous, and the careful removal of corrosion products is required. Surface scratches or change in surface structure from overheating can also cause sudden failure of these parts. Corrosion products must be removed by careful processing, using mild abrasive papers such as rouge or fine grit aluminum oxide, or fine buffing compounds on cloth buffing wheels. Nonwoven abrasive pads can also be used. It is essential that steel surfaces not be overheated during buffing. After careful removal of surface corrosion, reapply protective paint finishes immediately.

Corrosion of Aluminum and Aluminum Alloys

Corrosion on aluminum surfaces is usually quite obvious, since the products of corrosion are white and generally more voluminous than the original base metal. Even in its early stages, aluminum corrosion is evident as general etching, pitting, or roughness of the aluminum surfaces.

NOTE: Aluminum alloys commonly form a smooth surface oxidation that is from 0.001 to 0.0025 inch thick. This is not considered detrimental; the coating provides a hard shell barrier to the introduction of corrosive elements. Such oxidation is not to be confused with the severe corrosion discussed in this paragraph.

General surface attack of aluminum penetrates relatively slowly, but is speeded up in the presence of dissolved salts. Considerable attack can usually take place before serious loss of structural strength develops.

At least three forms of attack on aluminum alloys are particularly serious: (1) the penetrating pit-type corrosion through the walls of aluminum tubing, (2) stress corrosion cracking of materials under sustained stress, and (3) intergranular corrosion which is characteristic of certain improperly heat-treated aluminum alloys.

In general, corrosion of aluminum can be more effectively treated in place compared to corrosion occurring on other structural materials used in aircraft. Treatment includes the mechanical removal of as much of the corrosion products as practicable, and the inhibition of residual materials by chemical means, followed by the restoration of permanent surface coatings.

Treatment of Unpainted Aluminum Surfaces

Relatively pure aluminum has considerably more corrosion resistance compared with the stronger aluminum alloys. To take advantage of this characteristic, a thin coating of relatively pure aluminum is applied over the base aluminum alloy. The protection obtained is good, and the pure-aluminum clad surface (commonly called "Alclad") can be maintained in a polished condition. In cleaning such surfaces, however, care must be taken to prevent staining and marring of the exposed aluminum and, more important from a protection standpoint, to avoid unnecessary mechanical removal of the protective Alclad layer and the exposure of the more susceptible aluminum alloy base material. A typical aluminum corrosion treatment sequence follows:

1. Remove oil and surface dirt from the aluminum surface using any suitable mild cleaner. Use caution when choosing a cleaner; many commercial consumer products are actually caustic enough to induce corrosion if trapped between aluminum lap joints. Choose a neutral Ph product.
2. Hand polish the corroded areas with fine abrasives or with metal polish. Metal polish intended for use on clad aluminum aircraft surfaces must not be used on anodized aluminum since it is abrasive enough to actually remove the protective anodized film. It effectively removes stains and produces a highly polished, lasting surface on unpainted Alclad. If a surface is particularly difficult to clean, a cleaner and brightener compound for aluminum can be used before polishing to shorten the time and lessen the effort necessary to get a clean surface.
3. Treat any superficial corrosion present, using an inhibitive wipe down material. An alternate treatment is processing with a solution of sodium dichromate and chromium trioxide. Allow these solutions to remain on the corroded area for 5 to 20 minutes, and then remove the excess by rinsing and wiping the surface dry with a clean cloth.
4. Overcoat the polished surfaces with waterproof wax.

Aluminum surfaces that are to be subsequently painted can be exposed to more severe cleaning procedures and can also be given more thorough corrective treatment prior to painting. The following sequence is generally used:

1. Thoroughly clean the affected surfaces of all soil and grease residues prior to processing. Any general aircraft cleaning procedure may be used.
2. If residual paint films remain, strip the area to be treated. Procedures for the use of paint removers and the precautions to observe were previously mentioned in this chapter under "Surface Cleaning and Paint Removal."
3. Treat superficially corroded areas with a 10 percent solution of chromic acid and sulfuric acid. Apply the solution by swab or brush. Scrub the corroded area with the brush while it is still damp. While chromic acid is a good inhibitor for aluminum alloys, even when corrosion products have not been completely removed, it is important that the solution penetrate to the bottom of all pits and underneath any corrosion that may be present. Thorough brushing with a stiff fiber brush should loosen or remove most existing corrosion and assure complete penetration of the inhibitor into crevices and pits. Allow the chromic acid to remain in place for at least 5 minutes, and then remove the excess by flushing with water or wiping with a wet cloth. There are several commercial chemical surface treatment compounds, similar to the type described above, which may also be used.
4. Dry the treated surface and restore recommended permanent protective coatings as required in accordance with the aircraft manufacturer's procedures. Restoration of paint coatings should immediately follow any surface treatment performed. In any case, make sure that corrosion treatment is accomplished or is reapplied on the same day that paint refinishing is scheduled.

Treatment of Anodized Surfaces

As previously stated, anodizing is a common surface treatment of aluminum alloys. When this coating is damaged in service, it can only be partially restored by chemical surface treatment. Therefore, any corrosion correction of anodized surfaces should avoid destruction of the oxide film in the unaffected area. Do not use steel wool or steel wire brushes. Do not use severe abrasive materials.

Nonwoven abrasive pads have generally replaced aluminum wool, aluminum wire brushes, or fiber bristle brushes as the tools used for cleaning corroded anodized surfaces. Care must be exercised in any cleaning process to avoid unnecessary breaking of the adjacent protective film. Take every precaution to maintain as much of the protective coating as practicable. Otherwise, treat anodized surfaces in the same manner as other aluminum finishes. Chromic acid and other inhibitive treatments can be used to restore the oxide film.

Treatment of Intergranular Corrosion in Heat-Treated Aluminum Alloy Surfaces

As previously described, intergranular corrosion is an attack along grain boundaries of improperly or inadequately heat-treated alloys, resulting from precipitation of dissimilar constituents following heat treatment. In its most severe form, actual lifting of metal layers (exfoliation, *see* Figure 6-9) occurs.

More severe cleaning is a must when intergranular corrosion is present. The mechanical removal of all corrosion products and visible delaminated metal layers must be accomplished to determine the extent of the destruction and to evaluate the remaining structural strength of the component. Corrosion depth and removal limits have been established for some aircraft. Any loss of structural strength should be evaluated prior to repair or replacement of the part. If the manufacturer's limits do not adequately address the damage, a designated engineering representative (DER) can be brought in to assess the damage.

Corrosion of Magnesium Alloys

Magnesium is the most chemically active of the metals used in aircraft construction and is the most difficult to protect. When a failure in the protective coating does occur, the prompt and complete correction of the coating failure is imperative if serious structural damage is to be avoided. Magnesium attack is probably the easiest type of corrosion to detect in its early stages, since magnesium corrosion products occupy several times the volume of the original magnesium metal destroyed. The beginning of attack shows as a lifting of the paint films and white spots on the magnesium surface. These rapidly develop into snow-like mounds or even "white whiskers." [Figure 6-17] Reprotection involves the removal of corrosion products, the partial restoration of surface coatings by chemical treatment, and a reapplication of protective coatings.

Figure 6-17. Magnesium corrosion.

Treatment of Wrought Magnesium Sheet and Forgings

Magnesium skin corrosion usually occurs around edges of skin panels, underneath washers, or in areas physically damaged by shearing, drilling, abrasion, or impact. If the skin section can be removed easily, this should be done to assure complete inhibition and treatment. If insulating washers are involved, screws should at least be sufficiently loosened, to permit brush treatment of the magnesium under the insulating washer. Complete mechanical removal of corrosion products should be practiced insofar as practicable. Limit such mechanical cleaning to the use of stiff, hog bristle brushes, and similar nonmetallic cleaning tools (including nonwoven abrasive pads), particularly if treatment is to be performed under field conditions. Like aluminum, under no circumstances are steel or aluminum tools, steel, bronze or aluminum wool or other cleaning abrasive pads used on different metal surfaces to be used in cleaning magnesium. Any entrapment of particles from steel wire brushes or steel tools, or contamination of treated surfaces by dirty abrasives, can cause more trouble than the initial corrosive attack.

Corroded magnesium may generally be treated as follows:

1. Clean and strip the paint from the area to be treated. (Paint stripping procedures were discussed earlier in this chapter, and are also addressed in FAA Advisory Circular (AC) 43.13-1B, *Acceptable Methods, Techniques, and Practices—Aircraft Inspection and Repair.*)
2. Using a stiff, hog bristle brush or nonwoven abrasive pad, break loose and remove as much of the corrosion products as practicable. Steel wire brushes, carborundum abrasives, or steel cutting tools must not be used.
3. Treat the corroded area liberally with a chromic acid solution, to which has been added sulfuric acid, and work into pits and crevices by brushing the area while still wet with chromic acid, again using a nonmetallic brush.
4. Allow the chromic acid to remain in place for 5 to 20 minutes before wiping up the excess with a clean, damp cloth. Do not allow the excess solution to dry and remain on the surface, as paint lifting will be caused by such deposits.
5. As soon as the surfaces are dry, restore the original protective paint.

Treatment of Installed Magnesium Castings

Magnesium castings, in general, are more porous and more prone to penetrating attack than wrought magnesium skins. For all practical purposes, however, treatment is the same for all magnesium areas. Engine cases, bellcranks, fittings, numerous covers, plates, and handles are the most common magnesium castings.

When attack occurs on a casting, the earliest practicable treatment is required if dangerous corrosive penetration is to be avoided. In fact, engine cases submerged in saltwater overnight can be completely penetrated. If it is at all practicable, parting surfaces should be separated to effectively treat the existing attack and prevent its further progress. The same general treatment sequence in the preceding paragraph for magnesium skin should be followed.

If extensive removal of corrosion products from a structural casting is involved, a decision from the manufacturer may be necessary to evaluate the adequacy of structural strength remaining. Specific structural repair manuals usually include dimensional tolerance limits for critical structural members and should be referred to, if any question of safety is involved.

Treatment of Titanium and Titanium Alloys

Attack on titanium surfaces is generally difficult to detect. Titanium is, by nature, highly corrosion resistant, but it may show deterioration from the presence of salt deposits and metal impurities, particularly at high temperatures. Therefore, the use of steel wool, iron scrapers, or steel brushes for cleaning or for the removal of corrosion from titanium parts is prohibited.

If titanium surfaces require cleaning, hand polishing with aluminum polish or a mild abrasive is permissible, if fiber brushes only are used and if the surface is treated following cleaning with a suitable solution of sodium dichromate. Wipe the treated surface with dry cloths to remove excess solution, but do not use a water rinse.

Protection of Dissimilar Metal Contacts

Certain metals are subject to corrosion when placed in contact with other metals. This is commonly referred to as electrolytic or dissimilar metals corrosion. Contact of different bare metals creates an electrolytic action when moisture is present. If this moisture is salt water, the electrolytic action is accelerated. The result of dissimilar metal contact is oxidation (decomposition) of one or both metals. The chart shown in Figure 6-18 lists the metal combinations requiring a protective separator. The separating materials may be metal primer, aluminum tape, washers, grease, or sealant, depending on the metals involved.

Contacts Not Involving Magnesium

All dissimilar joints not involving magnesium are protected by the application of a minimum of two coats of zinc chromate or, preferably, epoxy primer in addition to normal primer requirements. Primer is applied by brush or spray and allowed to air dry 6 hours between coats.

Contacts Involving Magnesium

To prevent corrosion between dissimilar metal joints in which magnesium alloy is involved, each surface is insulated as follows:

At least two coats of zinc chromate or, preferably, epoxy primer are applied to each surface. Next, a layer of pressure sensitive vinyl tape 0.003 inch thick is applied smoothly and firmly enough to prevent air bubbles and wrinkles. To avoid creep back, the tape is not stretched during application. When the thickness of the tape interferes with the assembly of parts, where relative motion exists between parts, or when service temperatures above 250 °F are anticipated, the use of tape is eliminated and extra coats (minimum of three) of primer are applied.

Corrosion Limits

Corrosion, however slight, is damage. Therefore, corrosion damage is classified under the four standard types, as is any other damage. These types are: (1) negligible damage, (2) damage repairable by patching, (3) damage repairable by insertion, and (4) damage necessitating replacement of parts.

The term "negligible," as used here, does not imply that little or nothing should be done. The corroded surface should be cleaned, treated, and painted as appropriate. Negligible damage, generally, is corrosion that has scarred or eaten away the surface protective coats and begun to etch the metal. Corrosion damage extending to classifications of "repairable by patching" and "repairable by insertion" should be repaired in accordance with the applicable structural repair manual. When corrosion damage exceeds the damage limits to the extent that repair is not possible, the component or structure should be replaced.

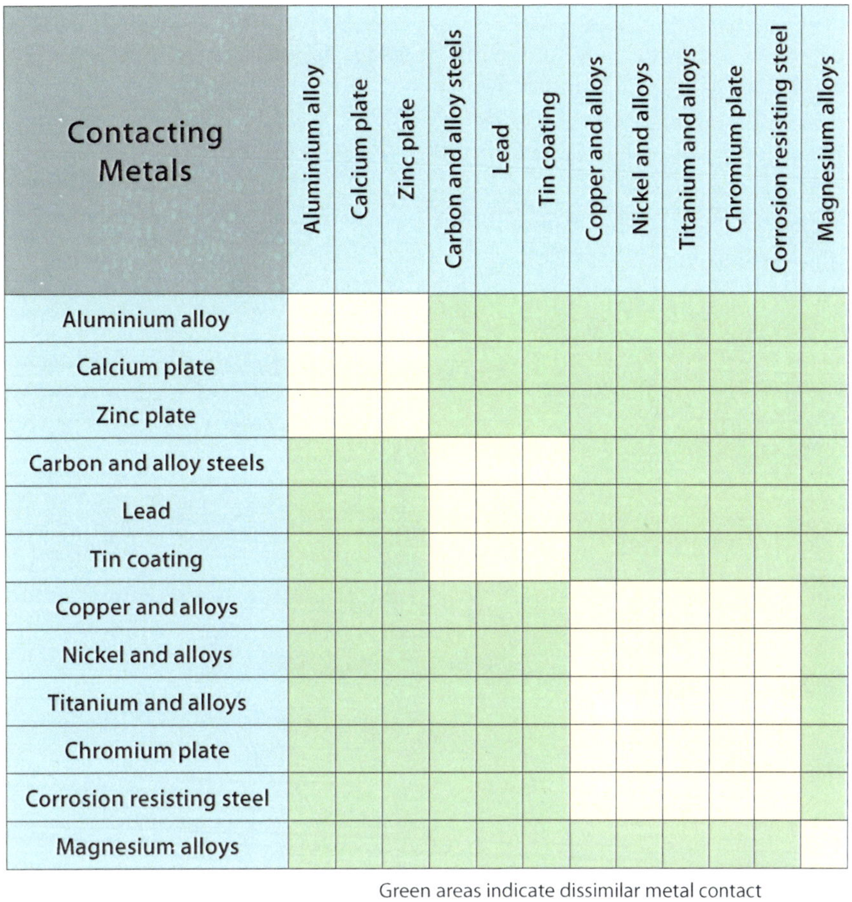

Figure 6-18. Dissimilar metal contacts that will result in electrolytic corrosion.

Processes and Materials Used in Corrosion Control

Metal Finishing

Aircraft parts are almost always given some type of surface finish by the manufacturer. The main purpose is to provide corrosion resistance; however, surface finishes may also be applied to increase wear resistance or to provide a suitable base for paint.

In most instances, the original finishes described in the following paragraphs cannot be restored in the field due to unavailable equipment or other limitations. However, an understanding of the various types of metal finishes is necessary if they are to be properly maintained in the field and if the partial restoration techniques used in corrosion control are to be effective.

Surface Preparation

Original surface treatments for steel parts usually include a cleaning treatment to remove all traces of dirt, oil, grease, oxides, and moisture. This is necessary to provide an effective bond between the metal surface and the final finish. The cleaning process may be either mechanical or chemical. In mechanical cleaning, the following methods are employed: wire brush, steel wool, emery cloth, sandblasting, or vapor blasting.

Chemical cleaning is preferred over mechanical since none of the base metal is removed by cleaning. There are various chemical processes now in use, and the type used will depend on the material being cleaned and the type of foreign matter being removed.

Steel parts are pickled to remove scale, rust, or other foreign matter, particularly before plating. The pickling solution can be either muriatic (hydrochloric) or sulfuric acid. Cost wise, sulfuric acid is preferable, but muriatic acid is more effective in removing certain types of scale.

The pickling solution is kept in a stoneware tank and is usually heated by means of a steam coil. Parts not to be electroplated after pickling are immersed in a lime bath to neutralize the acid from the pickling solution.

Electrocleaning is another type of chemical cleaning used to remove grease, oil, or organic matter. In this cleaning process, the metal is suspended in a hot alkaline solution containing special wetting agents, inhibitors, and materials to provide the necessary electrical conductivity. An electric current is then passed through the solution in a manner similar to that used in electroplating.

Aluminum and magnesium parts are also cleaned by using some of the foregoing methods. Blast cleaning using abrasive media is not applicable to thin aluminum sheets, particularly Alclad. Steel grits are not used on aluminum or corrosion resistant metals.

Polishing, buffing, and coloring of metal surfaces play a very important part in the finishing of metal surfaces. Polishing and buffing operations are sometimes used when preparing a metal surface for electroplating, and all three operations are used when the metal surface requires a high luster finish.

Chemical Treatments

Anodizing

Anodizing is the most common surface treatment of nonclad aluminum alloy surfaces. It is typically done in specialized facilities in accordance with Mil-C-5541E or AMS-C-5541. The aluminum alloy sheet or casting is the positive pole in an electrolytic bath in which chromic acid or other oxidizing agent produces an aluminum oxide film on the metal surface. Aluminum oxide is naturally protective, and anodizing merely increases the thickness and density of the natural oxide film. When this coating is damaged in service, it can only be partially restored by chemical surface treatments. Therefore, when an anodized surface is cleaned including corrosion removal, the technician should avoid unnecessary destruction of the oxide film.

The anodized coating provides excellent resistance to corrosion. The coating is soft and easily scratched, making it necessary to use extreme caution when handling it prior to coating it with primer.

Aluminum wool, nylon webbing impregnated with aluminum oxide abrasive, fine grade nonwoven abrasive pads or fiber bristle brushes are the approved tools for cleaning anodized surfaces. The use of steel wool, steel wire brushes, or harsh abrasive materials on any aluminum surfaces is prohibited. Producing a buffed or wire brush finish by any means is also prohibited. Otherwise, anodized surfaces are treated in much the same manner as other aluminum finishes.

In addition to its corrosion resistant qualities, the anodic coating is also an excellent bond for paint. In most cases, parts are primed and painted as soon as possible after anodizing. The anodic coating is a poor conductor of electricity; therefore, if parts require bonding, the coating is removed where the bonding wire is to be attached. Alclad surfaces that are to be left unpainted require no anodic treatment; however, if the Alclad surface is to be painted, it is usually anodized to provide a bond for the paint.

Alodizing

Alodizing is a simple chemical treatment for all aluminum alloys to increase their corrosion resistance and to improve their paint bonding qualities. Because of its simplicity, it is rapidly replacing anodizing in aircraft work.

The process consists of precleaning with an acidic or alkaline metal cleaner that is applied by either dipping or spraying. The parts are then rinsed with fresh water under pressure for 10 to 15 seconds. After thorough rinsing, Alodine® is applied by dipping, spraying, or brushing. A thin, hard coating results which ranges in color from light, bluish green with a slight iridescence on copper free alloys to an olive green on copper bearing alloys. The Alodine is first rinsed with clear, cold or warm water for a period of 15 to 30 seconds. An additional 10 to 15 second rinse is then given in a Deoxylyte® bath. This bath is to counteract alkaline material and to make the alodined aluminum surface slightly acid on drying.

Chemical Surface Treatment and Inhibitors

As previously described, aluminum and magnesium alloys in particular are protected originally by a variety of surface treatments. Steels may have been treated on the surface during manufacture. Most of these coatings can only be restored by processes that are completely impractical in the field. But, corroded areas where such protective films have been destroyed require some type of treatment prior to refinishing.

The labels on the containers of surface treatment chemicals will provide warnings if a material is toxic or flammable. However, the label might not be large enough to accommodate a list of all the possible hazards which may ensue if the materials are mixed with incompatible substances. The Material Safety Data Sheet (MSDS) should also be consulted for information. For example, some chemicals used in surface treatments will react violently if inadvertently mixed with paint thinners. Chemical surface treatment materials must be handled with extreme care and mixed exactly according to directions.

Chromic Acid Inhibitor

A 10 percent solution by weight of chromic acid, activated by a small amount of sulfuric acid, is particularly effective in treating exposed or corroded aluminum surfaces. It may also be used to treat corroded magnesium.

This treatment tends to restore the protective oxide coating on the metal surface. Such treatment must be followed by regular paint finishes as soon as practicable, and never later than the same day as the latest chromic acid treatment. Chromium trioxide flake is a powerful oxidizing agent and a fairly strong acid. It must be stored away from organic solvents and other combustibles. Either thoroughly rinse or dispose of wiping cloths used in chromic acid pickup.

Sodium Dichromate Solution

A less active chemical mixture for surface treatment of aluminum is a solution of sodium dichromate and chromic acid. Entrapped solutions of this mixture are less likely to corrode metal surfaces than chromic acid inhibitor solutions.

Chemical Surface Treatments

Several commercial, activated chromate acid mixtures are available under Specification MIL-C-5541 for field treatment of damaged or corroded aluminum surfaces. Take precautions to make sure that sponges or cloths used are thoroughly rinsed to avoid a possible fire hazard after drying.

Protective Paint Finishes

A good, intact paint finish is the most effective barrier between metal surfaces and corrosive media. The most common finishes include catalyzed polyurethane enamel, waterborne polyurethane enamel, and two-part epoxy paint. As new regulations regarding the emission of volatile organic compounds (VOCs) are put into effect, the use of waterborne paint systems have increased in popularity. Also still available are nitrate and butyrate dope finishes for fabric-covered aircraft. In addition, high visibility fluorescent materials may also be used, along with a variety of miscellaneous combinations of special materials. There may also be rain erosion resistant coatings on metal leading edges, and several different baked enamel finishes on engine cases and wheels.

Aircraft Cleaning

Cleaning an aircraft and keeping it clean are extremely important. From an aircraft maintenance technician's viewpoint, it should be considered a regular part of aircraft maintenance. Keeping the aircraft clean can mean more accurate inspection results, and may even allow a flight crewmember to spot an impending component failure. A cracked landing gear fitting covered with mud and grease may be easily overlooked. Dirt can hide cracks in the skin. Dust and grit cause hinge fittings to wear excessively. If left on the aircraft's outer surface, a film of dirt reduces flying speed and adds extra weight. Dirt or trash blowing or bouncing around the inside of the aircraft is annoying and dangerous. Small pieces of dirt blown into the eyes of the pilot at a critical moment can cause an accident. A coating of dirt and grease on moving parts makes a grinding compound that can cause excessive wear. Salt water has a serious corroding effect on exposed metal parts of the aircraft, and should be washed off immediately.

There are many different kinds of cleaning agents approved for use in cleaning aircraft. It is impractical to cover each of the various types of cleaning agents since their use varies under different conditions, such as the type of material to be removed, the aircraft finish, and whether the cleaning is internal or external.

In general, the types of cleaning agents used on aircraft are solvents, emulsion cleaners, soaps, and synthetic detergents. Their use must be in accordance with the applicable maintenance manual. The types of cleaning agents named above are also classed as light or heavy duty cleaners. The soap and synthetic detergent type cleaners are used for light duty cleaning, while the solvent and emulsion type cleaners are used for heavy duty cleaning. The light duty cleaners, which are nontoxic and nonflammable, should be used whenever possible. As mentioned previously, cleaners that can be effectively rinsed and neutralized must be used, or an alkaline cleaner may cause corrosion within the lap joints of riveted or spot-welded sheet metal components.

Exterior Cleaning

There are three methods of cleaning the aircraft exterior: (1) wet wash, (2) dry wash, and (3) polishing. Polishing can be further broken down into hand polishing and mechanical polishing. The type and extent of soiling and the final desired appearance determine the cleaning method to be used.

Wet wash removes oil, grease, or carbon deposits and most soils, with the exception of corrosion and oxide films. The cleaning compounds used are usually applied by spray or mop, after which high pressure running water is used as a rinse. Either alkaline or emulsion cleaners can be used in the wet wash method.

Dry wash is used to remove airport film, dust, and small accumulations of dirt and soil when the use of liquids is neither desirable nor practical. This method is not suitable for removing heavy deposits of carbon, grease, or oil, especially in the engine exhaust areas. Dry wash materials are applied with spray, mops, or cloths, and removed by dry mopping or wiping with clean, dry cloths.

Polishing restores the luster to painted and unpainted surfaces of the aircraft, and is usually performed after the surfaces have been cleaned. Polishing is also used to remove oxidation and corrosion. Polishing materials are available in various forms and degrees of abrasiveness. It is important that the aircraft manufacturer's instructions be used in specific applications.

The washing of aircraft should be performed in the shade whenever possible as cleaning compounds tend to streak the surface if applied to hot metal, or are permitted to dry on the area. Install covers over all openings where water or cleaners might enter and cause damage. Pay particular attention to instrument system components such as pitot-static fittings and ports.

Various areas of aircraft, such as the sections housing radar and the area forward of the cockpit that are finished with a flat-finish paint, should not be cleaned more than necessary and should never be scrubbed with stiff brushes or coarse rags. A soft sponge or cheesecloth with a minimum of manual rubbing is advisable. Any oil or exhaust stains on the surface should first be removed with a solvent such as kerosene or other petroleum base solvent. Rinse the surfaces immediately after cleaning to prevent the compound from drying on the surface.

Before applying soap and water to plastic surfaces, flush the plastic surfaces with fresh water to dissolve salt deposits and wash away dust particles. Plastic surfaces should be washed with soap and water, preferably by hand.

Rinse with fresh water and dry with a chamois, synthetic wipes designed for use on plastic windshields, or absorbent cotton. In view of the soft surface, do not rub plastic with a dry cloth since this is not only likely to cause scratches, but it also builds up an electrostatic charge that attracts dust particles to the surface. The charge, as well as the dust, may be removed by patting or gently blotting with a clean, damp chamois. Do not use scouring powder or other material that can mar the plastic surface. Remove oil and grease by rubbing gently with a cloth wet with soap and water. Do not use acetone, benzene, carbon tetrachloride, lacquer thinners, window cleaning sprays, gasoline, fire extinguisher or deicer fluid on plastics because they soften the plastic and will cause crazing. Finish cleaning the plastic by coating with a plastic polish intended for aircraft windows and windshields. These polishes can minimize small surface scratches and will also help keep static charges from building up on the surface of the windows.

Surface oil, hydraulic fluid, grease, or fuel can be removed from aircraft tires by washing with a mild soap solution. After cleaning, lubricate all grease fittings, hinges, and so forth, where removal, contamination, or dilution of the grease is suspected during washing of the aircraft.

Interior Cleaning

Keeping the interior of the aircraft clean is just as important as maintaining a clean exterior surface. Corrosion can establish itself on the inside structure to a greater degree because it is difficult to reach some areas for cleaning. Nuts, bolts, bits of wire, or other metal objects carelessly dropped and neglected, combined with moisture and dissimilar metal contact, can cause electrolytic corrosion.

When performing structural work inside the aircraft, clean up all metal particles and other debris as soon as possible. To make cleaning easier and prevent the metal particles and debris from getting into inaccessible areas, use a drop cloth in the work area to catch this debris.

A vacuum cleaner can be used to pick up dust and dirt from the interior of the cockpit and cabin.

Aircraft interior present certain problems during cleaning operations. The following is taken from The National Fire Protection Association (NFPA) Bulletin #410F, Aircraft Cabin Cleaning Operation.

"Basic to an understanding of the problem is the fact that aircraft cabin compartments constitute relatively small enclosures as measured by their cubic footage. This presents the possibility of restricted ventilation and the quick buildup of flammable vapor/air mixtures where there is any indiscriminate use of flammable cleaning agents or solvents. Within the same volume there may also exist the possibility of an ignition source in the form of an electrical fault, a friction or static spark, an open flame device, or some other potential introduced by concurrent maintenance work."

Wherever possible, use nonflammable agents in these operations to reduce to the minimum the fire and explosion hazards.

Types of Cleaning Operations

The principal areas of aircraft cabins which may need periodic cleaning are:

1. Aircraft passenger cabin areas (seats, carpets, side panels, headliners, overhead racks, curtains, ash trays, windows, doors, decorative panels of plastic, wood or similar materials).
2. Aircraft flight station areas (similar materials to those found in passenger cabin areas plus instrument panels, control pedestals, glare shields, flooring materials, metallic surfaces of instruments and flight control equipment, electrical cables and contacts, and so forth).
3. Lavatories and buffets (similar materials to those found in passenger cabin areas plus toilet facilities, metal fixtures and trim, trash containers, cabinets, wash and sink basins, mirrors, ovens, and so forth).

Nonflammable Aircraft Cabin Cleaning Agents and Solvents

1. Detergents and soaps. These have widespread application for most aircraft cleaning operations involving fabrics, headliners, rugs, windows, and similar surfaces that are not damageable by water solutions since they are colorfast and nonshrinkable. Care is frequently needed to prevent leaching of water-soluble fire retardant salts which may have been used to treat such materials in order to reduce their flame spread characteristics. Allowing water laced with fire retardant salts to come in contact with the aluminum framework of seats and seat rails can induce corrosion. Be careful to ensure only the necessary amount of water is applied to the seat materials when cleaning.
2. Alkaline cleaners. Most of these agents are water soluble and thus have no fire hazard properties. They can be used on fabrics, headliners, rugs, and similar surfaces in the same manner as detergent and soap solutions with only minor added limitations resulting from their inherent caustic character. This may increase their efficiency as cleaning agents but results in somewhat greater deteriorating effects on certain fabrics and plastics.
3. Acid solutions. A number of proprietary acid solutions are available for use as cleaning agents. They are normally mild solutions designed primarily to remove carbon smut or corrosive stains. As water-based solutions, they have no flash point but may require more careful and judicious use not only to prevent damage to fabrics, plastics, or other surfaces but also to protect the skin and clothing of those using the materials.
4. Deodorizing or disinfecting agents. A number of proprietary agents useful for aircraft cabin deodorizing or disinfecting are nonflammable. Most of these are designed for spray application (aerosol type) and have a nonflammable pressurizing agent, but it is best to check this carefully as some may contain a flammable compressed gas for pressurization.
5. Abrasives. Some proprietary nonflammable mild abrasive materials are available for rejuvenating painted or polished surfaces. They present no fire hazard.
6. Dry cleaning agents. Perchlorethylene and trichlorethylene as used at ambient temperatures are examples of nonflammable dry cleaning agents. These materials do have a toxicity hazard requiring care in their use, and in some locations, due to environmental laws, their use may be prohibited or severely restricted. In the same way, water-soluble agents can be detrimental. Fire retardant treated materials may be adversely affected by the application of these dry cleaning agents.

Flammable and Combustible Agents

1. High flash point solvents. Specially refined petroleum products, first developed as "Stoddard solvent" but now sold under a variety of trade names by different companies, have solvent properties approximating gasoline but have fire hazard properties similar to those of kerosene as commonly used (not heated). Most of these are stable products having a flash point from 100 °F to 140 °F with a comparatively low degree of toxicity.

2. Low flash point solvents. Class I (flash point at below 100 °F) flammable liquids should not be used for aircraft cleaning or refurbishing. Common materials falling into this "class" are acetone, aviation gasoline, methyl ethyl ketone, naphtha, and toluol. In cases where it is absolutely necessary to use a flammable liquid, use high flash point liquids (those having a flash point of 100 °F or more).
3. Mixed liquids. Some commercial solvents are mixtures of liquids with differing rates of evaporation, such as a mixture of one of the various naphthas and a chlorinated material. The different rates of evaporation may present problems from both the toxicity and fire hazard viewpoints. Such mixtures should not be used unless they are stored and handled with full knowledge of these hazards and appropriate precautions taken.

Container Controls

Flammable liquids should be handled only in approved containers or safety cans appropriately labeled.

Fire Prevention Precautions

During aircraft cleaning or refurbishing operations where flammable or combustible liquids are used, the following general safeguards are recommended:

1. Aircraft cabins should be provided with ventilation sufficient at all times to prevent the accumulation of flammable vapors. To accomplish this, doors to cabins shall be open to secure maximum advantage of natural ventilation. Where such natural ventilation is insufficient, approved mechanical ventilation equipment shall be provided and used. The accumulation of flammable vapors above 25 percent of the lower flammability limit of the particular vapor being used, measured at a point 5 feet from the location of use, shall result in emergency revisions of operations in progress.
2. All open flame and spark producing equipment or devices that might be brought within the vapor hazard area should be shut down and not operated during the period when flammable vapors may exist.
3. Electrical equipment of a hand portable nature used within an aircraft cabin shall be of the type approved for use in Class I, Group D, Hazardous Locations as defined by the National Electrical Code.
4. Switches to aircraft cabin lighting and to the aircraft electrical system components within the cabin area should not be worked on or switched on or off during cleaning operations.
5. Suitable warning signs should be placed in conspicuous locations at aircraft doors to indicate that flammable liquids are being or have been used in the cleaning or refurbishing operation in progress.

Fire Protection Recommendations

During aircraft cleaning or refurbishing operations where flammable liquids are used, the following general fire protection safeguards are recommended:

1. Aircraft undergoing such cleaning or refurbishing should preferably be located outside of the hangar buildings when weather conditions permit. This provides for added natural ventilation and normally assures easier access to the aircraft in the event of fire.
2. It is recommended that during such cleaning or refurbishing operations in an aircraft outside of the hangar that portable fire extinguishers be provided at cabin entrances having a minimum rating of 20-B and, at minimum, a booster hose line with an adjustable water spray nozzle being available capable of reaching the cabin area for use pending the arrival of airport fire equipment. As an alternate to the previous recommendations, a Class A fire extinguisher having a minimum rating of 4-A plus or a Class B fire extinguisher having a minimum rating of 20-B should be placed at aircraft cabin doors for immediate use if required.

   NOTE 1: All-purpose ABC (dry chemical) type extinguishers should not be used in situations where aluminum corrosion is a problem if the extinguisher is used.

   NOTE 2: Portable and semi-portable fire detection and extinguishing equipment has been developed, tested, and installed to provide protection to aircraft during construction and maintenance operations. Operators are urged to investigate the feasibility of utilizing such equipment during aircraft cabin cleaning and refurbishing operations.
3. Aircraft undergoing such cleaning or refurbishing where the work must be done under cover should be in hangars equipped with automatic fire protection equipment.

Powerplant Cleaning

Cleaning the powerplant is an important job and should be done thoroughly. Grease and dirt accumulations on an air-cooled engine provide an effective insulation against the cooling effect of air flowing over it. Such an accumulation can also cover up cracks or other defects.

When cleaning an engine, open or remove the cowling as much as possible. Beginning with the top, wash down the engine and accessories with a fine spray of kerosene or solvent. A bristle brush may be used to help clean some of the surfaces.

Fresh water and soap and approved cleaning solvents may be used for cleaning propeller and rotor blades. Except in the process of etching, caustic material should not be used on a propeller. Scrapers, power buffers, steel brushes, or any tool or substances that will mar or scratch the surface should not be used on propeller blades, except as recommended for etching and repair.

Water spray, rain, or other airborne abrasive material strikes a whirling propeller blade with such force that small pits are formed in the blade's leading edge. If preventive measures are not taken, corrosion causes these pits to rapidly grow larger. The pits may become so large that it is necessary to file the blade's leading edge until it is smooth.

Steel propeller blades have more resistance to abrasion and corrosion than aluminum alloy blades. Steel blades, if rubbed down with oil after each flight, retain a smooth surface for a long time.

Examine the propellers regularly because cracks in steel or aluminum alloy blades can become filled with oil, which tends to oxidize. This can readily be seen when the blade is inspected. Keeping the surface wiped with oil serves as a safety feature by helping to make cracks more obvious.

Propeller hubs must be inspected regularly for cracks and other defects. Unless the hubs are kept clean, defects may not be found. Clean steel hubs with soap and fresh water, or with an approved cleaning solvent. These cleaning solvents may be applied by cloths or brushes. Avoid tools and abrasives that scratch or otherwise damage the plating.

In special cases in which a high polish is desired, the use of a good grade of metal polish is recommended. Upon completion of the polishing, all traces of polish must be removed immediately, the blades cleaned, and then coated with clean engine oil. All cleaning substances must be removed immediately after completion of the cleaning of any propeller part. Soap in any form can be removed by rinsing repeatedly with fresh water. After rinsing, all surfaces should be dried and coated with clean engine oil. After cleaning the powerplant, all control arms, bellcranks, and moving parts should be lubricated according to instructions in the applicable maintenance manual.

Solvent Cleaners

In general, solvent cleaners used in aircraft cleaning should have a flashpoint of not less than 105 °F, if explosion proofing of equipment and other special precautions are to be avoided. Chlorinated solvents of all types meet the nonflammable requirements but are toxic, and safety precautions must be observed in their use. Use of carbon tetrachloride should be avoided. The Material Safety Data Sheet (MSDS) for each solvent should be consulted for handling and safety information.

AMT's should review the Material Safety Data Sheet (MSDS) available for any chemical, solvent or other materials they may come in contact with during the course of their maintenance activities. In particular, solvents and cleaning liquids, even those considered "environmentally friendly" can have varied detrimental effects on the skin, internal organs and/or nervous system. Active solvents such as methyl ethyl ketone (MEK) and acetone can be harmful or fatal if swallowed, and can be harmful when inhaled or absorbed through the skin in sufficient quantities.

Particular attention should be paid to recommended protective measures including gloves, respirators and face shields. A regular review of the MSDS will keep the AMT updated on any revisions that may be made by chemical manufacturers or government authorities.

Dry Cleaning Solvent

Stoddard solvent is the most common petroleum base solvent used in aircraft cleaning. Its flashpoint is slightly above 105 °F and can be used to remove grease, oils, or light soils. Dry cleaning solvent is preferable to kerosene for all cleaning purposes, but like kerosene, it leaves a slight residue upon evaporation, which may interfere with the application of some final paint films.

Aliphatic and Aromatic Naphtha

Aliphatic naphtha is recommended for wipe down of cleaned surfaces just before painting. This material can also be used for cleaning acrylics and rubber. It flashes at approximately 80 °F and must be used with care.

Aromatic naphtha should not be confused with the aliphatic material. It is toxic and attacks acrylics and rubber products, and must be used with adequate controls.

Safety Solvent

Safety solvent, trichloroethane (methyl chloroform), is used for general cleaning and grease removal. It is nonflammable under ordinary circumstances, and is used as a replacement for carbon tetrachloride. The use and safety precautions necessary when using chlorinated solvents must be observed. Prolonged use can cause dermatitis on some persons.

Methyl Ethyl Ketone (MEK)

MEK is also available as a solvent cleaner for metal surfaces and paint stripper for small areas. This is a very active solvent and metal cleaner, with a flashpoint of about 24 °F. It is toxic when inhaled, and safety precautions must be observed during its use. In most instances, it has been replaced with safer to handle and more environmentally friendly cleaning solvents.

Kerosene

Kerosene is mixed with solvent emulsion type cleaners for softening heavy preservative coatings. It is also used for general solvent cleaning, but its use should be followed by a coating or rinse with some other type of protective agent. Kerosene does not evaporate as rapidly as dry cleaning solvent and generally leaves an appreciable film on cleaned surfaces, which may actually be corrosive. Kerosene films may be removed with safety solvent, water emulsion cleaners, or detergent mixtures.

Cleaning Compound for Oxygen Systems

Cleaning compounds for use in the oxygen system are anhydrous (waterless) ethyl alcohol or isopropyl (anti-icing fluid) alcohol. These may be used to clean accessible components of the oxygen system such as crew masks and lines. Fluids should not be put into tanks or regulators.

Do not use any cleaning compounds which may leave an oily film when cleaning oxygen equipment. Instructions of the manufacturer of the oxygen equipment and cleaning compounds must be followed at all times.

Emulsion Cleaners

Solvent and water emulsion compounds are used in general aircraft cleaning. Solvent emulsions are particularly useful in the removal of heavy deposits, such as carbon, grease, oil, or tar. When used in accordance with instructions, these solvent emulsions do not affect good paint coatings or organic finishes.

Water Emulsion Cleaner

Material available under Specification MIL-C-22543A is a water emulsion cleaning compound intended for use on both painted and unpainted aircraft surfaces. This material is also acceptable for cleaning fluorescent painted surfaces and is safe for use on acrylics. However, these properties will vary with the material available, and a sample application should be checked carefully before general uncontrolled use.

Solvent Emulsion Cleaners

One type of solvent emulsion cleaner is nonphenolic and can be safely used on painted surfaces without softening the base paint. Repeated use may soften acrylic nitrocellulose lacquers. It is effective, however, in softening and lifting heavy preservative coatings. Persistent materials should be given a second or third treatment as necessary.

Another type of solvent emulsion cleaner has a phenolic base that is more effective for heavy duty application, but it also tends to soften paint coatings. It must be used with care around rubber, plastics, or other nonmetallic materials. Wear rubber gloves and goggles for protection when working with phenolic base cleaners.

Soaps and Detergent Cleaners

A number of materials are available for mild cleaning use. In this section, some of the more common materials are discussed.

Cleaning Compound, Aircraft Surfaces

Specification MIL-C-5410 Type I and II materials are used in general cleaning of painted and unpainted aircraft surfaces for the removal of light to medium soils, operational films, oils, or greases. They are safe to use on all surfaces, including fabrics, leather, and transparent plastics. Nonglare (flat) finishes should not be cleaned more than necessary and should never be scrubbed with stiff brushes.

Nonionic Detergent Cleaners

These materials may be either water soluble or oil soluble. The oil-soluble detergent cleaner is effective in a 3 to 5 percent solution in dry cleaning solvent for softening and removing heavy preservative coatings. This mixture's performance is similar to the emulsion cleaners mentioned previously.

Mechanical Cleaning Materials

Mechanical cleaning materials must be used with care and in accordance with directions given, if damage to finishes and surfaces is to be avoided.

Mild Abrasive Materials

No attempt is made in this section to furnish detailed instructions for using various materials listed. Some do's and don'ts are included as an aid in selecting materials for specific cleaning jobs.

The introduction of various grades of nonwoven abrasive pads (a common brand name produced by the 3M company is Scotch-Brite™) has given the aircraft maintenance technician a clean, inexpensive material for the removal of corrosion products and for other light abrasive needs. The pads can be used on most metals (although the same pad should not be used on different metals) and are generally the first choice when the situation arises. A very open form of this pad is also available for paint stripping, when used in conjunction with wet strippers.

Powdered pumice can be used for cleaning corroded aluminum surfaces. Similar mild abrasives may also be used.

Impregnated cotton wadding material is used for removal of exhaust gas stains and polishing corroded aluminum surfaces. It may also be used on other metal surfaces to produce a high reflectance.

Aluminum metal polish is used to produce a high luster, long lasting polish on unpainted aluminum clad surfaces. It should not be used on anodized surfaces because it will remove the oxide coat.

Three grades of aluminum wool, coarse, medium, and fine, are used for general cleaning of aluminum surfaces. Impregnated nylon webbing material is preferred over aluminum wool for the removal of corrosion products and stubborn paint films and for the scuffing of existing paint finishes prior to touchup.

Lacquer rubbing compound material can be used to remove engine exhaust residues and minor oxidation. Avoid heavy rubbing over rivet heads or edges where protective coatings may be worn thin.

Abrasive Papers

Abrasive papers used on aircraft surfaces should not contain sharp or needlelike abrasives which can imbed themselves in the base metal being cleaned or in the protective coating being maintained. The abrasives used should not corrode the material being cleaned. Aluminum oxide paper, 300 grit or finer, is available in several forms and is safe to use on most surfaces. Type I, Class 2 material under Federal Specification P-C-451 is available in $1\frac{1}{2}$ and 2 inch widths. Avoid the use of carborundum (silicon carbide) papers, particularly on aluminum or magnesium. The grain structure of carborundum is sharp, and the material is so hard that individual grains will penetrate and bury themselves even in steel surfaces. The use of emery paper or crocus cloth on aluminum or magnesium can cause serious corrosion of the metal by imbedded iron oxide.

Chemical Cleaners

Chemical cleaners must be used with great care in cleaning assembled aircraft. The danger of entrapping corrosive materials in faying surfaces and crevices counteracts any advantages in their speed and effectiveness. Any materials used must be relatively neutral and easy to remove. It is emphasized that all residues must be removed. Soluble salts from chemical surface treatments, such as chromic acid or dichromate treatment, will liquefy and promote blistering in the paint coatings.

Phosphoric-Citric Acid

A phosphoric-citric acid mixture (Type I) for cleaning aluminum surfaces is available and is ready to use as packaged. Type II is a concentrate that must be diluted with mineral spirits and water. Wear rubber gloves and goggles to avoid skin contact. Any acid burns may be neutralized by copious water washing, followed by treatment with a diluted solution of baking soda (sodium bicarbonate).

Baking Soda

Baking soda may be used to neutralize acid deposits in lead-acid battery compartments and to treat acid burns from chemical cleaners and inhibitors.

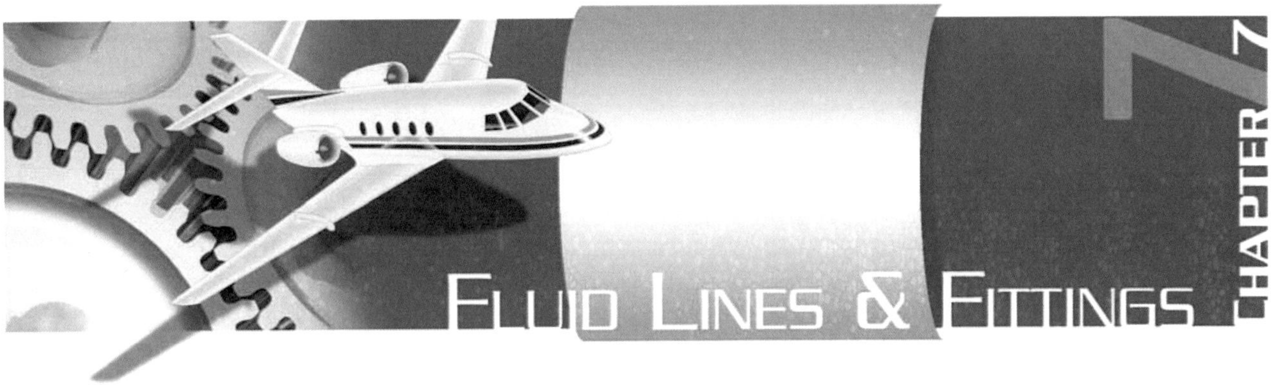

Chapter 7: Fluid Lines & Fittings

Aircraft fluid lines are usually made of metal tubing or flexible hose. Metal tubing (also called rigid fluid lines) is used in stationary applications and where long, relatively straight runs are possible. They are widely used in aircraft for fuel, oil, coolant, oxygen, instrument, and hydraulic lines. Flexible hose is generally used with moving parts or where the hose is subject to considerable vibration.

Occasionally, it may be necessary to repair or replace damaged aircraft fluid lines. Very often the repair can be made simply by replacing the tubing. However, if replacements are not available, the needed parts may have to be fabricated. Replacement tubing should be of the same size and material as the original tubing. All tubing is pressure tested prior to initial installation, and is designed to withstand several times the normal operating pressure to which it will be subjected. If a tube bursts or cracks, it is generally the result of excessive vibration, improper installation, or damage caused by collision with an object. All tubing failures should be carefully studied and the cause of the failure determined.

Rigid Fluid Lines

Tubing Materials

Copper

In the early days of aviation, copper tubing was used extensively in aviation fluid applications. In modern aircraft, aluminum alloy, corrosion resistant steel or titanium tubing have generally replaced copper tubing.

Aluminum Alloy Tubing

Tubing made from 1100 H14 (1/2-hard) or 3003 H14 (1/2-hard) is used for general purpose lines of low or negligible fluid pressures, such as instrument lines and ventilating conduits. Tubing made from 2024-T3, 5052-O, and 6061-T6 aluminum alloy materials is used in general purpose systems of low and medium pressures, such as hydraulic and pneumatic 1,000 to 1,500 psi systems, and fuel and oil lines.

Steel

Corrosion resistant steel tubing, either annealed CRES 304, CRES 321 or CRES 304-1/8-hard, is used extensively in high pressure hydraulic systems (3,000 psi or more) for the operation of landing gear, flaps, brakes, and in fire zones. Its higher tensile strength permits the use of tubing with thinner walls; consequently, the final installation weight is not much greater than that of the thicker wall aluminum alloy tubing. Steel lines are used where there is a risk of foreign object damage (FOD); i.e., the landing gear and wheel well areas. Although identification markings for steel tubing differ, each usually includes the manufacturer's name or trademark, the *Society of Automotive Engineers* (SAE) number, and the physical condition of the metal.

Titanium 3AL-2.5V

This type of tubing and fitting is used extensively in transport category and high performance aircraft hydraulic systems for pressures above 1,500 psi. Titanium is 30 percent stronger than steel and 50 percent lighter than steel. Cryofit fittings or swaged fittings are used with titanium tubing. Do not use titanium tubing and fittings in any oxygen system assembly. Titanium and titanium alloys are oxygen reactive. If a freshly formed titanium surface is exposed in gaseous oxygen, spontaneous combustion could occur at low pressures.

Material Identification

Before making repairs to any aircraft tubing, it is important to make accurate identification of tubing materials. Aluminum alloy, steel, or titanium tubing can be identified readily by sight where it is used as the basic tubing material. However, it is difficult to determine whether a material is carbon steel or stainless steel, or whether it is 1100, 3003, 5052-O, 6061-T6 or 2024-T3 aluminum alloy. To positively identify the material used in the original installation, compare code markings of the replacement tubing with the original markings on the tubing being replaced.

Aluminium Alloy Number	Color of Band
1100	White
3003	Green
2014	Gray
2024	Red
5052	Purple
6053	Black
6061	Blue and Yellow
7075	Brown and Yellow

Figure 7-1. Painted color codes used to identify aluminum alloy tubing.

On large aluminum alloy tubing, the alloy designation is stamped on the surface. On small aluminum tubing, the designation may be stamped on the surface; but more often it is shown by a color code, not more than 4" in width, painted at the two ends and approximately midway between the ends of some tubing. When the band consists of two colors, one-half the width is used for each color. [Figure 7-1]

If the code markings are hard or impossible to read, it may be necessary to test samples of the material for hardness by hardness testing.

Sizes

Metal tubing is sized by outside diameter (o.d.), which is measured fractionally in sixteenths of an inch. Thus, number 6 tubing is 6/16" (or $^{3}/_{8}$") and number 8 tubing is 8/16" (or $^{1}/_{2}$"), and so forth. The tube diameter is typically printed on all rigid tubing. In addition to other classifications or means of identification, tubing is manufactured in various wall thicknesses. Thus, it is important when installing tubing to know not only the material and outside diameter, but also the thickness of the wall. The wall thickness is typically printed on the tubing in thousands of an inch. To determine the inside diameter (i.d.) of the tube, subtract twice the wall thickness from the outside diameter.

For example, a number 10 piece of tubing with a wall thickness of 0.063" has an inside diameter of 0.625" – 2(0.063") = 0.499".

Fabrication of Metal Tube Lines

Damaged tubing and fluid lines should be repaired with new parts whenever possible. Unfortunately, sometimes replacement is impractical and repair is necessary. Scratches, abrasions, or minor corrosion on the outside of fluid lines may be considered negligible and can be smoothed out with a burnishing tool or aluminum wool. Limitations on the amount of damage that can be repaired in this manner are discussed in this chapter under "Rigid Tubing Inspection and Repair." If a fluid line assembly is to be replaced, the fittings can often be salvaged; then the repair will involve only tube forming and replacement.

Tube forming consists of four processes: Cutting, bending, flaring, and beading. If the tubing is small and made of soft material, the assembly can be formed by hand bending during installation. If the tube is $^{1}/_{4}$" diameter or larger, hand bending without the aid of tools is impractical.

Tube Cutting

When cutting tubing, it is important to produce a square end, free of burrs. Tubing may be cut with a tube cutter or a hacksaw. The cutter can be used with any soft metal tubing, such as copper, aluminum, or aluminum alloy. Correct use of the tube cutter is shown in Figure 7-2. Special chipless cutters are available for cutting aluminum 6061-T6, corrosion resistant steel and titanium tubing.

A new piece of tubing should be cut approximately 10 percent longer than the tube to be replaced to provide

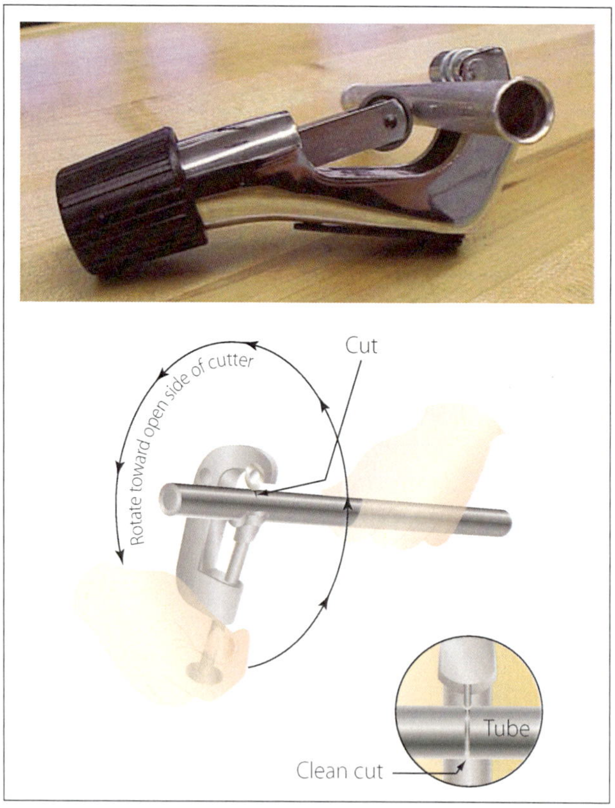

Figure 7-2. Tube cutting.

for minor variations in bending. Place the tubing in the cutting tool, with the cutting wheel at the point where the cut is to be made. Rotate the cutter around the tubing, applying a light pressure to the cutting wheel by intermittently twisting the thumbscrew. Too much pressure on the cutting wheel at one time could deform the tubing or cause excessive burring. After cutting the tubing, carefully remove any burrs from inside and outside the tube. Use a knife or the burring edge attached to the tube cutter. The deburring operation can be accomplished by the use of a deburring tool as shown in Figure 7-3. This tool is capable of removing both the inside and outside burrs by just turning the tool end for end.

When performing the deburring operation, use extreme care that the wall thickness of the end of the tubing is not reduced or fractured. Very slight damage of this type can lead to fractured flares or defective flares which will not seal properly. Use a fine-tooth file to file the end square and smooth.

If a tube cutter is not available, or if tubing of hard material is to be cut, use a fine-tooth hacksaw, preferably one having 32 teeth per inch. The use of a saw will decrease the amount of work hardening of the tubing during the cutting operation. After sawing, file the end of the tube square and smooth, removing all burrs.

An easy way to hold small diameter tubing, when cutting it, is to place the tube in a combination flaring tool and clamp the tool in a vise. Make the cut about one-half inch from the flaring tool. This procedure keeps sawing vibrations to a minimum and prevents damage to the tubing if it is accidentally hit with the hacksaw frame or file handle while cutting. Be sure all filings and cuttings are removed from the tube.

Tube Bending

The objective in tube bending is to obtain a smooth bend without flattening the tube. Tubing under 1/4" in diameter usually can be bent without the use of a bending tool. For larger sizes, either portable hand benders or production benders are usually used. Table 7-1 shows preferred methods and standard bend radii for bending tubing by tube size.

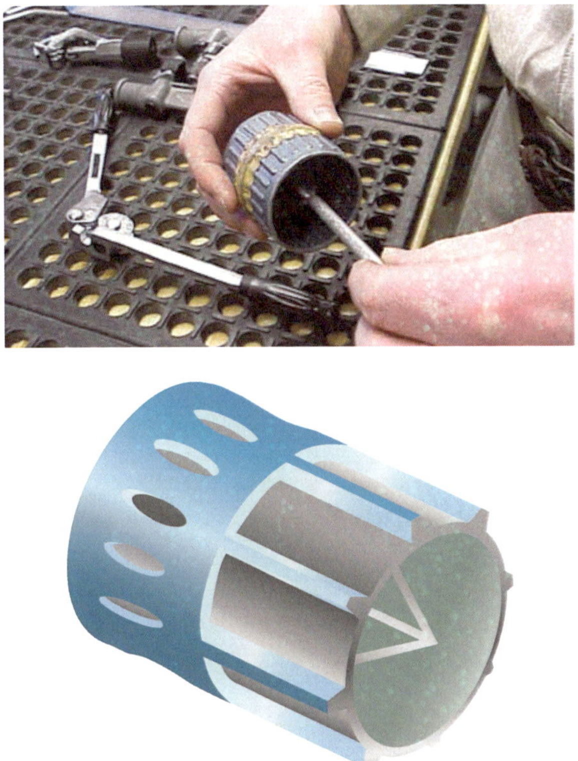

Figure 7-3. Deburring tool.

Using a hand bender, insert the tubing into the groove of the bender, so that the measured end is left of the form block. Align the two zeros and align the mark on the tubing with the L on the form handle. If the measured end is on the right side, then align the mark on the tubing with the R on the form handle. With a steady motion, pull the form handle till the zero mark

Type Bender	AB	AB	B	B	B	BC	B	BC	B	BC	C	BC	C
Tube od	1/8"	3/16"	1/4"	5/16"	3/8"	3/8"	7/16"	1/2"	1/2"	5/8"	5/8"	3/4"	3/4"
Standard Bend	3/8"	7/16"	9/16"	11/16"	11/16"	15/16"	1 3/8"	1 1/2"	1 1/4"	2"	1 1/2"	2 1/2"	1 3/4"

Type Bender	C	B	C	C	C	C	C	C	C	C	C	C	C
Tube od	7/8"	1"	1"	1 1/8"	1 1/4"	1 3/8"	1 3/8"	1 1/2"	1 1/2"	1 3/4"	2"	2 1/2"	3"
Standard Bend	2"	3 1/2"	3"	3 1/2"	3 3/4"	5"	6"	5"	6"	7"	8"	10"	12"

A – Hand   B – Portable hand benders   C – Production bender

Table 7-1. Standard bend radii to which standard bending tools will form the various sizes of tubes.

Figure 7-4. Tube bending.

on the form handle lines up with the desired angle of bend, as indicated on the radius block. [Figure 7-4]

Bend the tubing carefully to avoid excessive flattening, kinking, or wrinkling. A small amount of flattening in bends is acceptable, but the small diameter of the flattened portion must not be less than 75 percent of the original outside diameter. Tubing with flattened, wrinkled, or irregular bends should not be installed. Wrinkled bends usually result from trying to bend thin wall tubing without using a tube bender. Excessive flattening will cause fatigue failure of the tube. Examples of correct and incorrect tubing bends are shown in Figure 7-5.

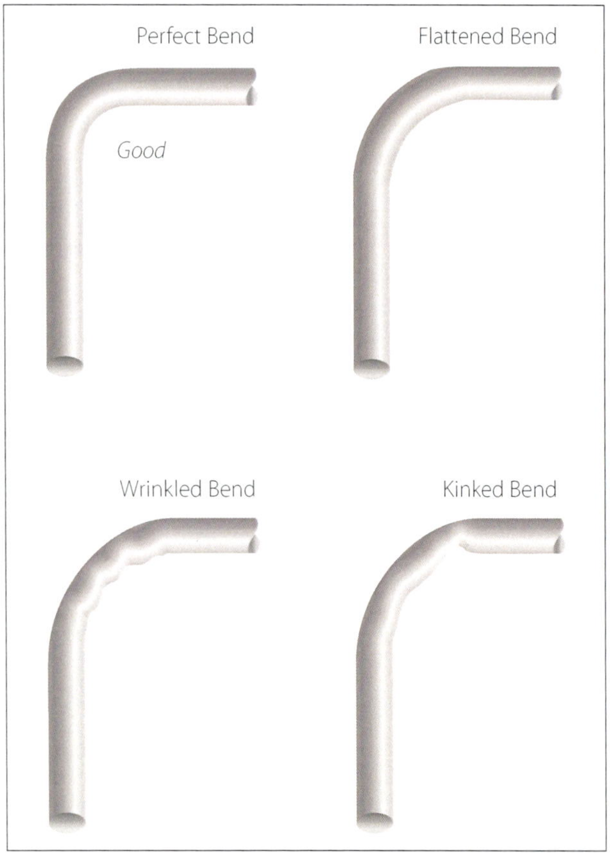

Figure 7-5. Correct and incorrect tubing bends.

Tube bending machines for all types of tubing are generally used in repair stations and large maintenance shops. With such equipment, proper bends can be made on large diameter tubing and on tubing made from hard material. The production CNC™ tube bender is an example of this type of machine. [Figure 7-6]

The ordinary production tube bender will accommodate tubing ranging from $\frac{1}{4}$" to $1\frac{1}{2}$" outside diameter. Benders for larger sizes are available, and the principle of their operation is similar to that of the hand tube bender. The radius blocks are so constructed that the radius of bend will vary with the tube diameter. The radius of bend is usually stamped on the block.

Alternative Bending Methods

When hand or production tube benders are not available or are not suitable for a particular bending operation, a filler of metallic composition or of dry sand may be used to facilitate bending. When using this method, cut the tube slightly longer than is required. The extra

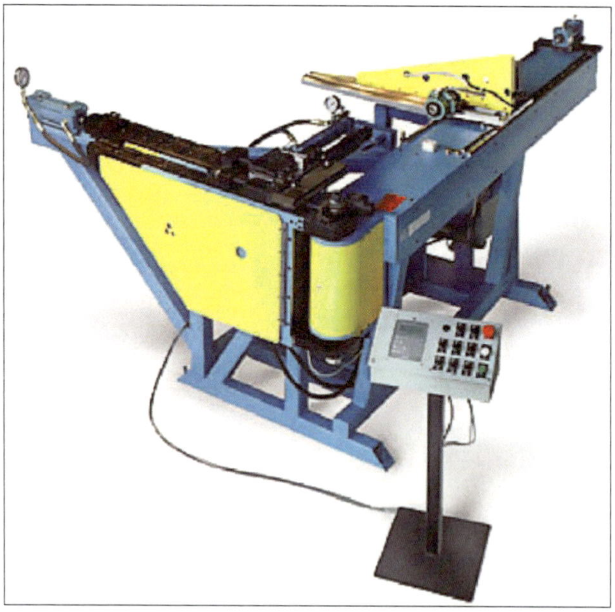

Figure 7-6. CNC tube bending machine.

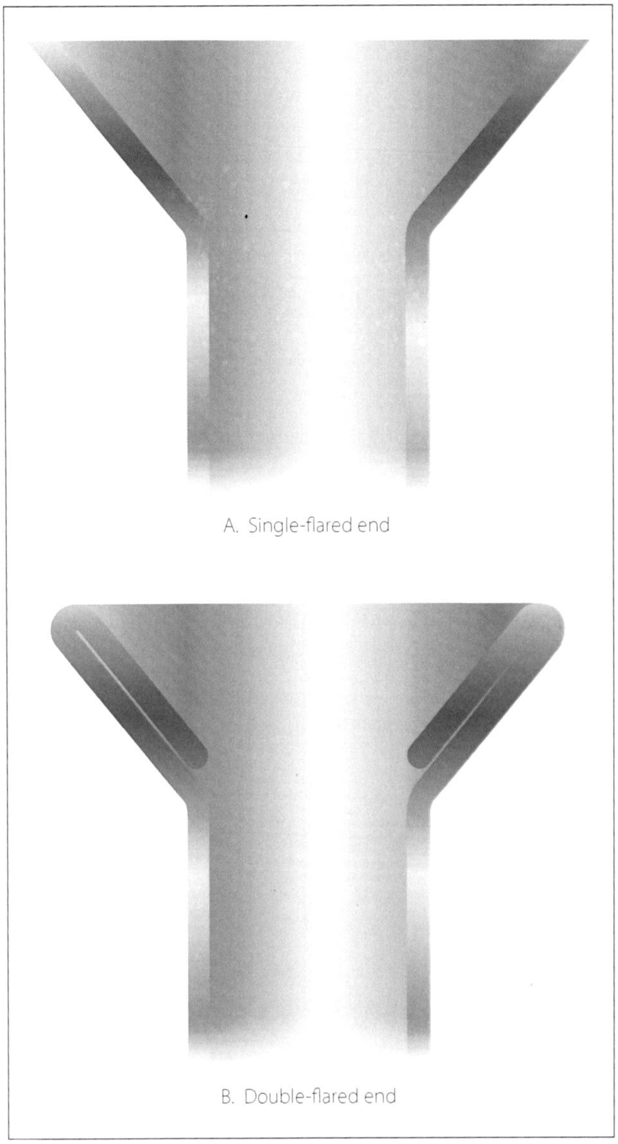

Figure 7-7. Cutaway view of single and double-flared tube ends.

length is for inserting a plug (which may be wooden) in each end. The tube can also be closed by flattening the ends or by soldering metal disks in them.

After plugging one end, fill and pack the tube with fine, dry sand and plug tightly. Both plugs must be tight so they will not be forced out when the bend is made. After the ends are closed, bend the tubing over a forming block shaped to the specified radius. In a modified version of the filler method, a fusible alloy is used instead of sand. In this method, the tube is filled under hot water with a fusible alloy that melts at 160 °F. The alloy-filled tubing is then removed from the water, allowed to cool, and bent slowly by hand around a forming block or with a tube bender. After the bend is made, the alloy is again melted under hot water and removed from the tubing. When using either filler methods, make certain that all particles of the filler are removed. Visually inspect with a borescope to make certain that no particles will be carried into the system in which the tubing is installed. Store the fusible alloy filler where it will be free from dust or dirt. It can be remelted and reused as often as desired. Never heat this filler in any other way than the prescribed method, as the alloy will stick to the inside of the tubing, making them both unusable.

Tube Flaring

Two kinds of flares are generally used in aircraft tubing: the single flare and the double flare. [Figure 7-7 (A and B)] Flares are frequently subjected to extremely high pressures; therefore, the flare on the tubing must be properly shaped or the connection will leak or fail. A flare made too small produces a weak joint, which may leak or pull apart; if made too large, it interferes with the proper engagement of the screw thread on the fitting and will cause leakage. A crooked flare is the result of the tubing not being cut squarely. If a flare is not made properly, flaws cannot be corrected by applying additional torque when tightening the fitting. The flare and tubing must be free from cracks, dents, nicks, scratches, or any other defects.

The flaring tool used for aircraft tubing has male and female dies ground to produce a flare of 35° to 37°. Under no circumstance is it permissible to use an automotive-type flaring tool which produces a flare of 45°. [Figure 7-8]

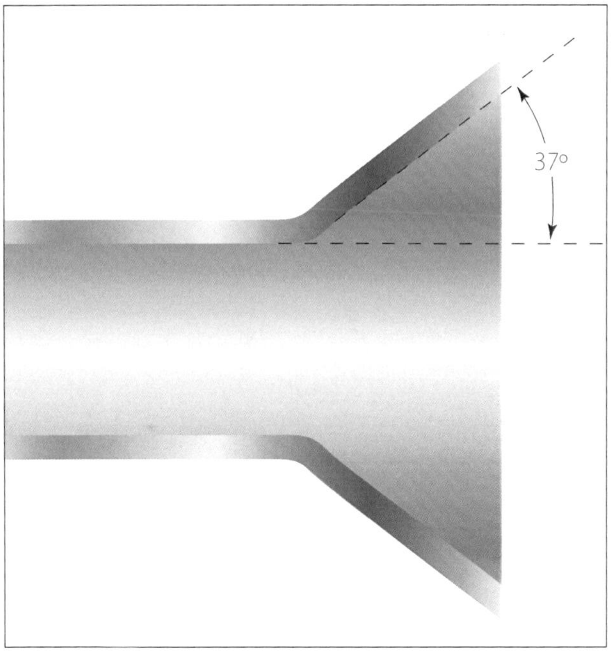

Figure 7-8. Flaring tool.

Figure 7-9. Hand flaring tool.

The single-flare hand flaring tool, similar to that shown in Figure 7-9, is used for flaring tubing. The tool consists of a flaring block or grip die, a yoke, and a flaring pin. The flaring block is a hinged double bar with holes corresponding to various sizes of tubing. These holes are countersunk on one end to form the outside support against which the flare is formed. The yoke is used to center the flaring pin over the end of the tube to be flared. Two types of flaring tools are used to make flares on tubing: the impact type and the rolling type.

Instructions for Rolling-Type Flaring Tools

Use these tools only to flare soft copper, aluminum, and brass tubing. Do not use with corrosion resistant steel or titanium. Cut the tube squarely and remove all burrs. Slip the fitting nut and sleeve on the tube. Loosen clamping screw used for locking the sliding segment in the die holder. This will permit their separation. The tools are self-gauging; the proper size flare is produced when tubing is clamped flush with the top of the die block. Insert tubing between the segments of the die block that correspond to the size of the tubing to be flared. Advance the clamp screw against the end segment and tighten firmly. Move the yoke down over the top of the die holder and twist it clockwise to lock it into position. Turn the feed screw down firmly, and continue until a slight resistance is felt. This indicates an accurate flare has been completed. Always read the tool manufacturer's instructions, because there are several different types of rolling-type flaring tools that use slightly different procedures.

Double Flaring

A double flare is used on soft aluminum alloy tubing $^3/_8$" outside diameter and under. This is necessary to prevent cutting off the flare and failure of the tube assembly under operating pressures. A double flare is smoother and more concentric than a single flare and therefore seals better. It is also more resistant to the shearing effect of torque.

Double Flaring Instructions

Deburr both the inside and outside of the tubing to be flared. Cut off the end of the tubing, if it appears damaged. Anneal brass, copper, and aluminum by heating to a dull red and cool rapidly in cold water. Open the flaring tool by unscrewing both clamping screws. Select the hole in the flaring bar that matches the tubing diameter and place the tubing with the end you have just prepared, extending above the top of the bar by a distance equal to the thickness of the shoulder of the adapter insert. Tighten clamping screws to hold tubing securely. Insert pilot of correctly sized adapter into tubing. Slip yoke over the flaring bars and center over adapter. Advance the cone downward until the shoulder of the adapter rests on the flaring bar. This bells out the end of the tubing. Next, back off the cone just enough to remove the adapter. After removing the adapter, advance the cone directly into the belled end of the tubing. This folds the tubing on itself and forms

Figure 7-10. Double flare tool.

an accurate double flare without cracking or splitting the tubing. To prevent thinning out of the flare wall, do not overtighten. [Figure 7-10]

Fittings

Rigid tubing may be joined to either an end item (such as a brake cylinder), another section of either rigid tubing, or to a flexible hose (such as a drain line). In the case of connection to an end item or another tube, fittings are required, which may or may not necessitate flaring of the tube. In the case of attachment to a hose, it may be necessary to bead the rigid tube so that a clamp can be used to hold the hose onto the tube.

Flareless Fittings

Although the use of flareless tube fittings eliminates all tube flaring, another operation, referred to as presetting, is necessary prior to installation of a new flareless tube assembly. *Flareless* tube assemblies should be preset with the proper size presetting tool or operation. Figure 7-11 (steps 1, 2, and 3) illustrates the presetting operation, which is performed as follows: Cut the tube to the correct length, with the ends perfectly square. Deburr the inside and outside of the tube. Slip the nut, then the sleeve, over the tube (step 1), lubricate the threads of the fitting and nut with hydraulic fluid. Place the fitting in a vise (step 2), and hold the tubing firmly and squarely on the seat in the fitting. (The tube must bottom firmly in the fitting.) Tighten the nut until the cutting edge of the sleeve grips the tube. To determine this point, slowly turn the tube back and forth while tightening the nut. When the tube no longer turns, the nut is ready for tightening. Final tightening depends upon the tubing (step 3). For aluminum alloy tubing up to and including ½" outside diameter, tighten the nut from 1 to 1⅙ turns. For steel tubing and aluminum alloy tubing over ½" outside diameter, tighten from 1⅙ to 1½ turns.

After presetting the sleeve, disconnect the tubing from the fitting and check the following points: The tube should extend $^{3}/_{32}$" to ⅛" beyond the sleeve pilot; otherwise, blowoff may occur. The sleeve pilot should contact the tube or have a maximum clearance of 0.005" for aluminum alloy tubing or 0.015" for steel tubing. A slight collapse of the tube at the sleeve cut is permissible. No movement of the sleeve pilot, except rotation, is permissible.

Beading

Tubing may be beaded with a hand beading tool, with machine beading rolls, or with grip dies. The method to be used depends on the diameter and wall thickness of the tube and the material from which it was made.

Figure 7-11. Presetting flareless tube assembly.

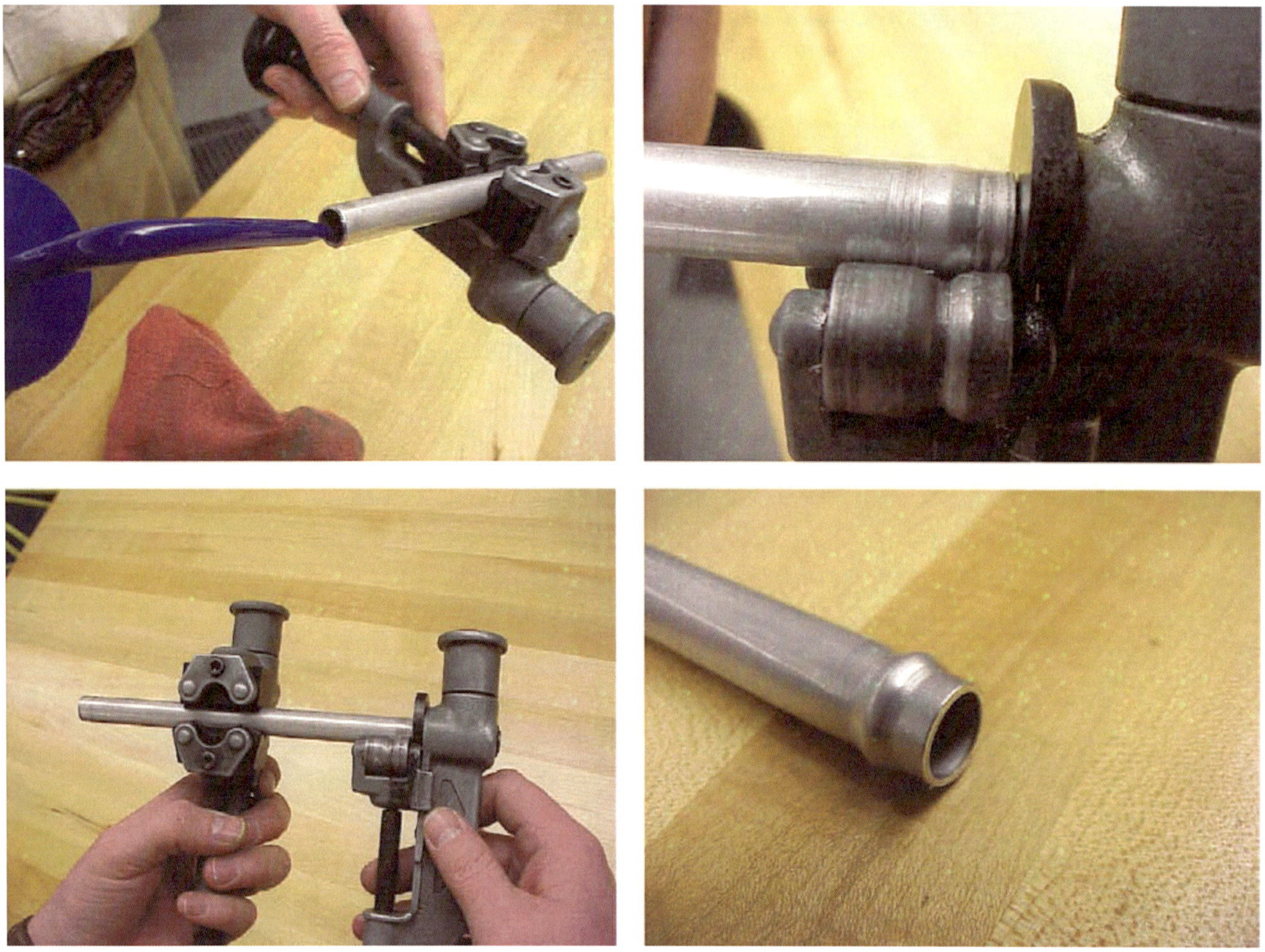

Figure 7-12. Hand beading tool.

The hand beading tool is used with tubing having $\frac{1}{4}$" to 1" outside diameter. The bead is formed by using the beader frame with the proper rollers attached. The inside and outside of the tube is lubricated with light oil to reduce the friction between the rollers during beading. The sizes, marked in sixteenths of an inch on the rollers, are for the outside diameter of the tubing that can be beaded with the rollers. [Figure 7-12]

Separate rollers are required for the inside of each tubing size, and care must be taken to use the correct parts when beading. The hand beading tool works somewhat like the tube cutter in that the roller is screwed down intermittently while rotating the beading tool around the tubing. In addition, a small vise (tube holder) is furnished with the kit.

Other methods and types of beading tools and machines are available, but the hand beading tool is used most often. As a rule, beading machines are limited to use with large diameter tubing, over $1\frac{15}{16}$", unless special rollers are supplied. The grip-die method of beading is confined to small tubing.

Fluid Line Identification

Fluid lines in aircraft are often identified by markers made up of color codes, words, and geometric symbols. These markers identify each line's function, content, and primary hazard. Figure 7-13 illustrates the various color codes and symbols used to designate the type of system and its contents.

Fluid lines are marked, in most instances with 1" tape or decals, as shown in Figure 7-14(A). On lines 4" in diameter (or larger), lines in oily environment, hot lines, and on some cold lines, steel tags may be used in place of tape or decals, as shown in Figure 7-14(B). Paint is used on lines in engine compartments, where there is the possibility of tapes, decals, or tags being drawn into the engine induction system.

In addition to the above-mentioned markings, certain lines may be further identified regarding specific function within a system; for example, drain, vent, pressure, or return. Lines conveying fuel may be marked FLAM; lines containing toxic materials are marked TOXIC in place of FLAM. Lines containing physically dangerous

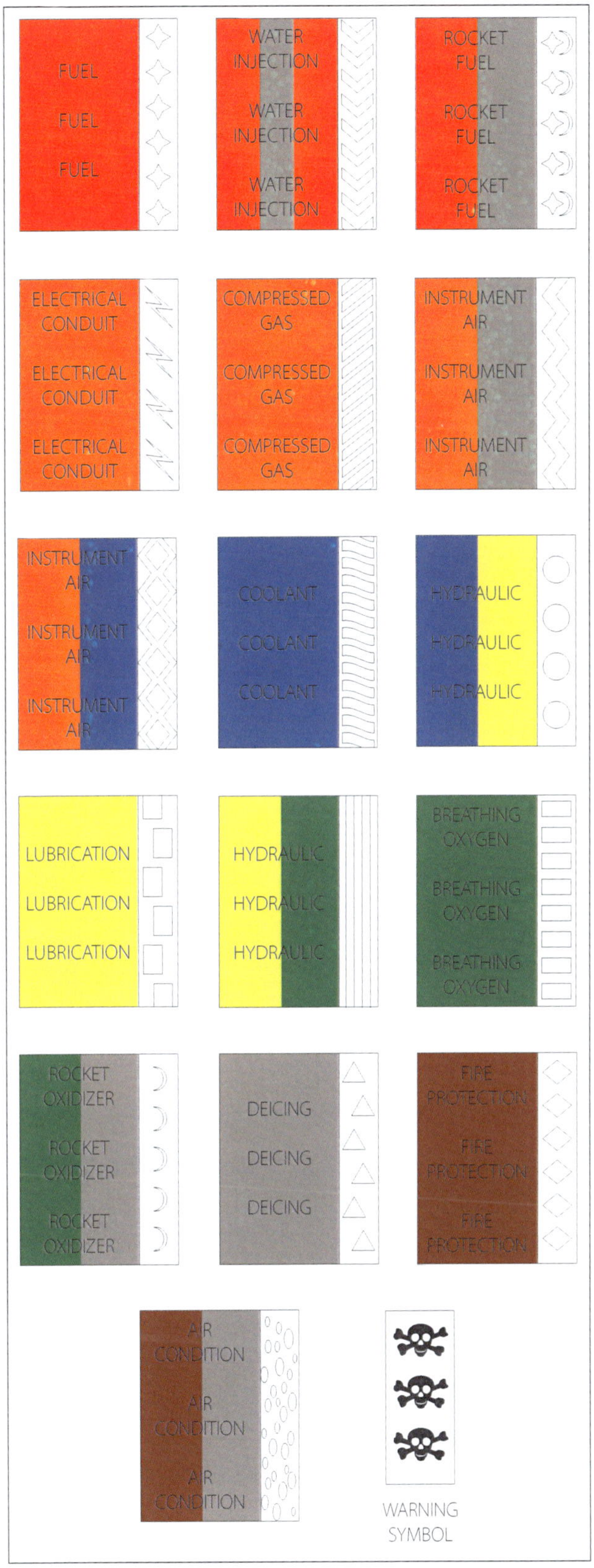

Figure 7-13. Identification of aircraft fluid lines.

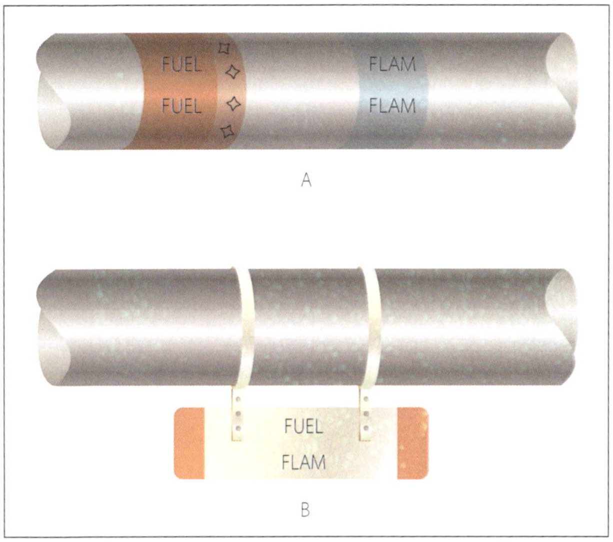

Figure 7-14. Fluid line identification using: (A) tape and decals and (B) metal tags.

materials, such as oxygen, nitrogen, or Freon™, may be marked PHDAN. [Figure 7-14]

Aircraft and engine manufacturers are responsible for the original installation of identification markers, but the aviation mechanic is responsible for their replacement when it becomes necessary. Tapes and decals are generally placed on both ends of a line and at least once in each compartment through which the line runs. In addition, identification markers are placed immediately adjacent to each valve, regulator, filter, or other accessory within a line. Where paint or tags are used, location requirements are the same as for tapes and decals.

Fluid Line End Fittings

Depending on the type and use, fittings will have either pipe threads or machine threads. Pipe threads are similar to those used in ordinary plumbing and are tapered, both internal and external. External threads are referred to as male threads and internal threads are female threads.

When two fittings are joined, a male into a female, the thread taper forms a seal. Some form of pipe thread lubricant approved for the particular fluid application should be used when joining pipe threads to prevent seizing and high-pressure leakage. Use care when applying thread lubricant so that the lubricant will not enter and contaminate the system. Do not use lubricants on oxygen lines. Oxygen will react with petroleum products and can ignite (special lubricants are available for oxygen systems).

Machine threads have no sealing capability and are similar to those used on common nuts and bolts. This type of fitting is used only to draw connections together or for attachment through bulkheads. A flared tube connection, a crush washer, or a synthetic seal is used to make the connection fluid tight. Machine threads have no taper and will not form a fluid-tight seal. The size of these fittings is given in dash numbers, which equal the nominal o.d. in sixteenths of an inch.

Universal Bulkhead Fittings

When a fluid line passes through a bulkhead, and it is desired to secure the line to the bulkhead, a bulkhead fitting should be used. The end of the fitting that passes through the bulkhead is longer than the other end(s), which allows a locknut to be installed, securing the fitting to the bulkhead.

Fittings attach one piece of tubing to another, or to system units. There are four types: (1) bead and clamp, (2) flared fittings, (3) flareless fittings, and (4) permanent fittings (Permaswage™, Permalite™, and Cyrofit™). The amount of pressure that the system carries and the material used are usually the deciding factors in selecting a connector.

The beaded type of fitting, which requires a bead and a section of hose and hose clamps, is used only in low- or medium-pressure systems, such as vacuum and coolant systems. The flared, flareless, or permanent-type fittings may be used as connectors in all systems, regardless of the pressure.

AN Flared Fittings

A flared tube fitting consists of a sleeve and a nut, as shown in Figure 7-15. The nut fits over the sleeve and, when tightened, draws the sleeve and tubing flare tightly against a male fitting to form a seal. Tubing used with this type of fitting must be flared before installation. The male fitting has a cone-shaped surface with the same angle as the inside of the flare. The sleeve supports the tube so that vibration does not concentrate at the edge of the flare, and distributes the shearing action over a wider area for added strength.

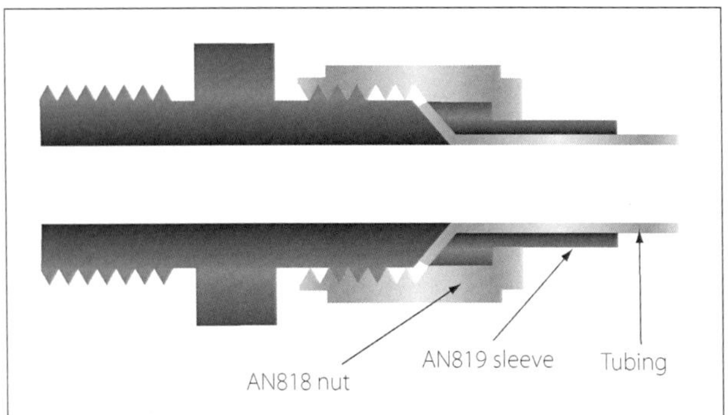

Figure 7-15. Flared tube fitting.

Fitting combinations composed of different alloys should be avoided to prevent dissimilar metal corrosion. As with all fitting combinations, ease of assembly, alignment, and proper lubrication should be assured when tightening fittings during installation.

Standard AN fittings are identified by their black or blue color. All AN steel fittings are colored black, all AN aluminum fittings are colored blue, and aluminum bronze fittings are cadmium plated and natural in appearance. A sampling of AN fittings is shown in Figure 7-16. Table 7-2 contains additional information on sizes, torques, and bend radii.

MS Flareless Fittings

MS flareless fittings are designed primarily for high-pressure (3,000 psi) hydraulic systems that may be subjected to severe vibration or fluctuating pressure. Using this type of fitting eliminates all tube flaring, yet provides a safe and strong, dependable tube connection. [Figure 7-17] The fitting consists of three parts: a body, a sleeve, and a nut. [Figure 7-18] The internal design of the body causes the sleeve to cut into the outside of the tube when the body and nut are joined. The counterbore shoulder within the body is designed with a reverse angle of 15° for steel connectors and 45° for aluminum fittings. This reverse angle prevents inward collapse of the tubing when tightened and provides a partial sealing force to be exerted against the periphery of the body counterbore.

Swaged Fittings

A popular repair system for connecting and repairing hydraulic lines on transport category aircraft is the use of Permaswage™ fittings. Swaged fittings create a permanent connection that is virtually maintenance free. Swaged fittings are used to join hydraulic lines in areas where routine disconnections are not required and are often used with titanium and corrosion resistant steel tubing. The fittings are installed with portable hydraulically powered tooling, which is compact enough to be used in tight spaces. [Figure 7-19] If the fittings need to be disconnected, cut the tubing with a tube cutter. Special installation tooling is available in portable kits. Always use the manufacturer's instructions to install swaged fittings. One of the latest developments is the Permalite™ fitting. Permalite™ is a tube fitting that is mechanically attached to the tube by axial swaging. The movement of the ring along the fitting body results in deformation of the tube with a leak-tight joint. [Figure 7-20]

— —AN744 to AN932— —

Material:

Aluminum alloy .... (code D)

Steel .... (code, absence of letter)

Brass .... (code B)

Aluminum bronze .... (code Z - for AN819 sleeve)

Size:

The dash number following the AN number indicates the size of the tubing (or hose) for which the fitting is made, in 16ths of an inch. This size measures the od of tubing and the id of hose. Fittings having pipe threads are coded by a dash number, indicating the pipe size in 8ths of an inch. The material code letter, as noted above, follows the dash number.

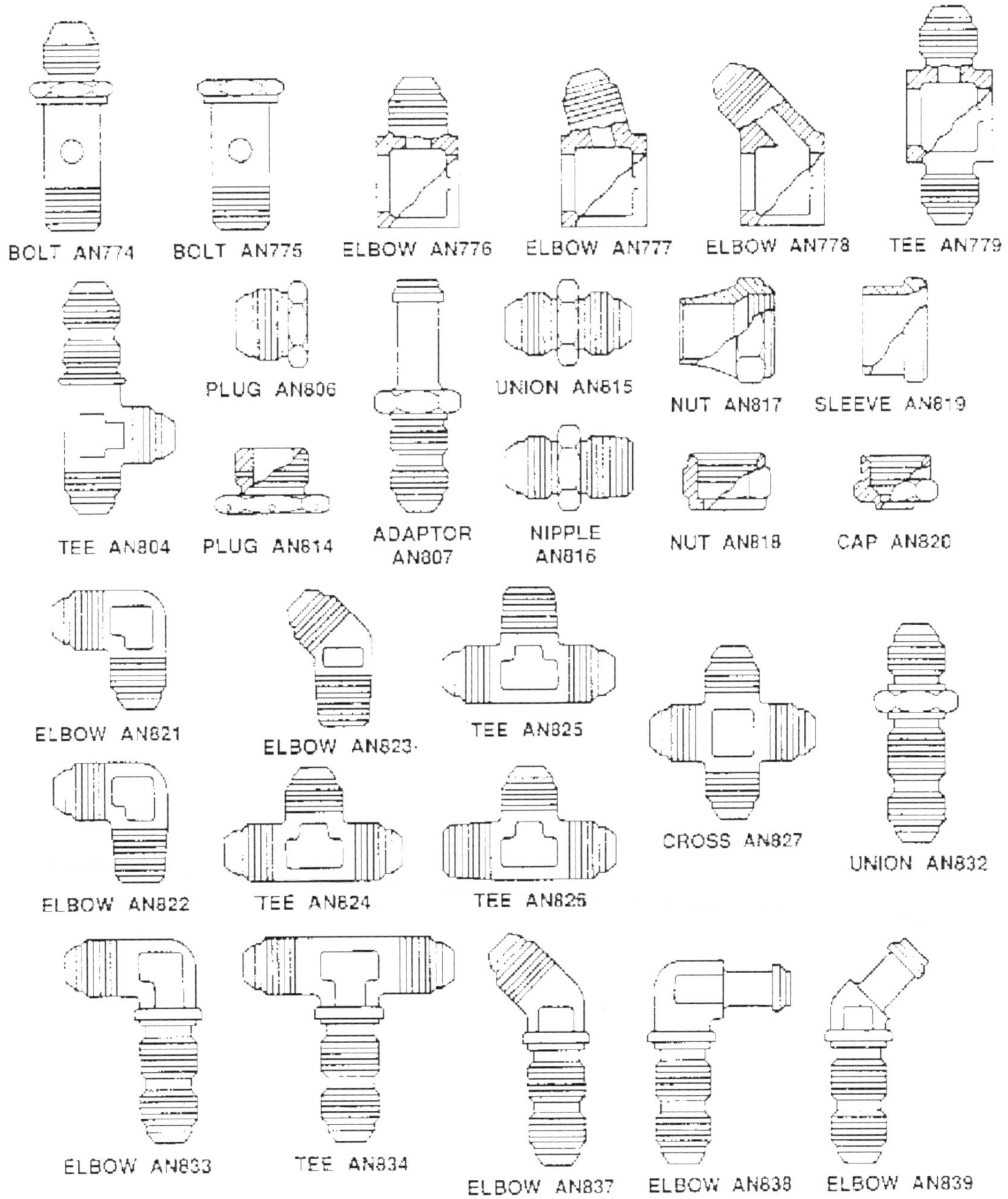

Figure 7-16. AN standard fittings.

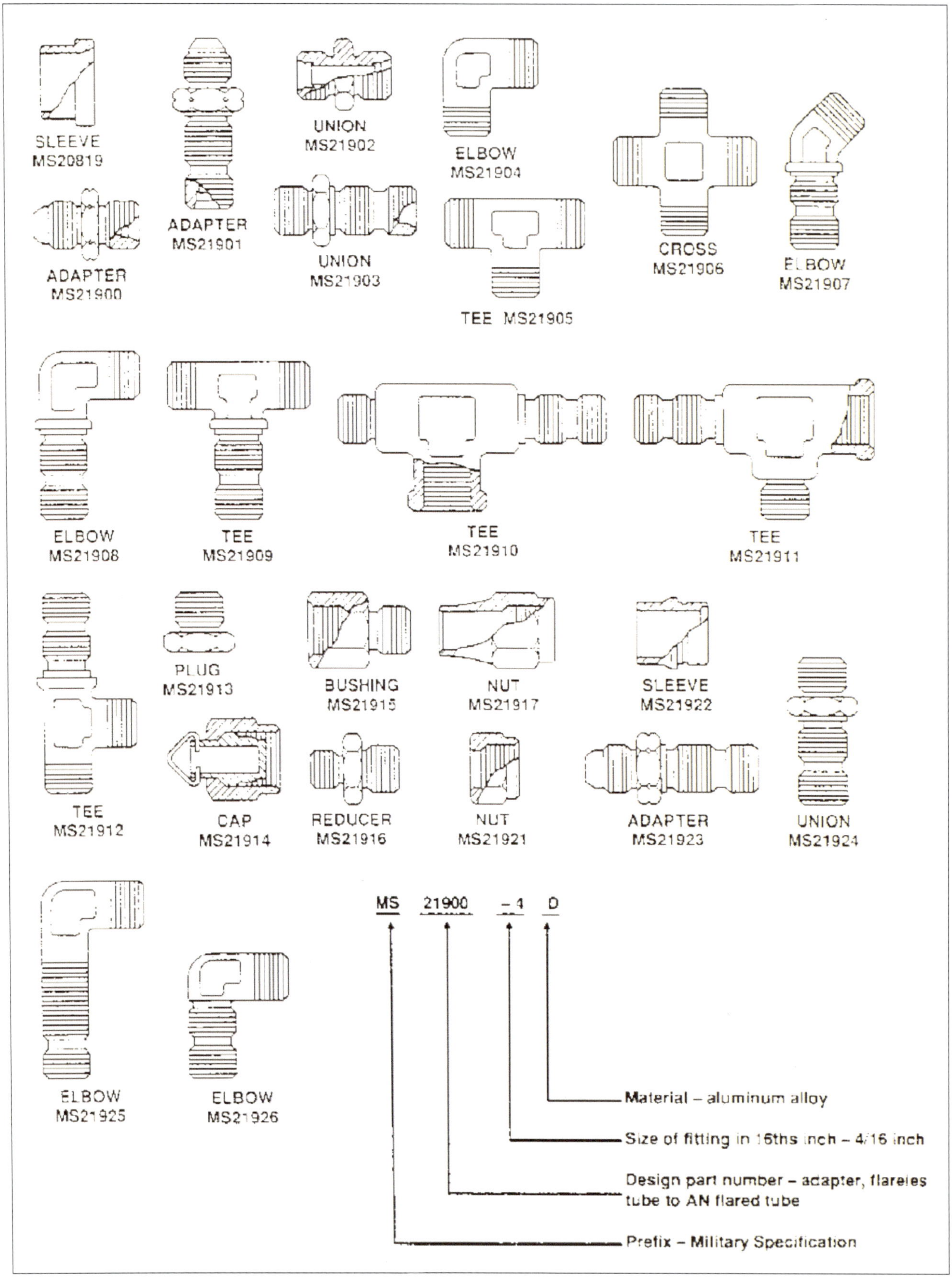

Figure 7-17. Typical MS flareless tube fittings.

Tubing o.d. (inches)	Fitting Bolt or Nut Size	Aluminium Alloy Tubing, Bolt, Fitting, or Nut Torque (in-lb)	Steel Tubing, Bolt Fitting, or Nut Torque (in-lb)	Hose End Fittings and Hose Assemblies MS28740 or Equivalent End Fitting		Minimum bend radii (inches)	
				Minimum	Maximum	Alum. alloy 1100-H14 5052-0	Steel
$1\frac{1}{8}$	-2	20–30				$\frac{3}{8}$	
$\frac{3}{16}$	-3	30–40	90–100	70	120	$\frac{7}{16}$	$\frac{21}{32}$
$\frac{1}{4}$	-4	40–65	135–150	100	250	$\frac{9}{16}$	$\frac{7}{8}$
$\frac{5}{16}$	-5	60–85	180–200	210	420	$\frac{3}{4}$	$1\frac{1}{8}$
$\frac{3}{8}$	-6	75–125	270–300	300	480	$\frac{15}{16}$	$1\frac{5}{16}$
$\frac{1}{2}$	-8	150–250	450–500	500	850	$1\frac{1}{4}$	$1\frac{3}{4}$
$\frac{5}{8}$	-10	200–350	650–700	700	1,150	$1\frac{1}{2}$	$2\frac{3}{16}$
$\frac{3}{4}$	-12	300–500	900–1,000			$1\frac{3}{4}$	$2\frac{5}{8}$
$\frac{7}{8}$	-14	500–600	1,000–1,100				
1	-16	500–700	1,200–1,400			3	$3\frac{1}{2}$
$1\frac{1}{4}$	-20	600–900	1,200–1,400			$3\frac{3}{4}$	$4\frac{3}{8}$
$1\frac{1}{2}$	-24	600–900	1,500–1,800			5	$5\frac{1}{4}$
$1\frac{3}{4}$	-28	850–1,050				7	$6\frac{1}{8}$
2	-32	950–1,150				8	7

Table 7-2. Flared fitting data.

Cryofit Fittings

Many transport category aircraft use Cryofit fittings to join hydraulic lines in areas where routine disconnections are not required. Cryofit fittings are standard fittings with a cryogenic sleeve. The sleeve is made of a shape memory alloy, Tinel™. The sleeve is manufactured 3 percent smaller, frozen in liquid nitrogen, and expanded to 5 percent larger than the line. During installation, the fitting is removed from the liquid nitrogen and inserted onto the tube. During a 10 to 15 second warming up period, the fitting contracts to its original size (3 percent smaller), biting down on the tube, forming a permanent seal. Cryofit fittings can only be removed by cutting the tube at the sleeve, though this leaves enough room to replace it with a swaged fitting without replacing the hydraulic line. It is frequently used with titanium tubing. The shape memory technology is also used for end fittings, flared fittings, and flareless fittings. [Figure 7-21]

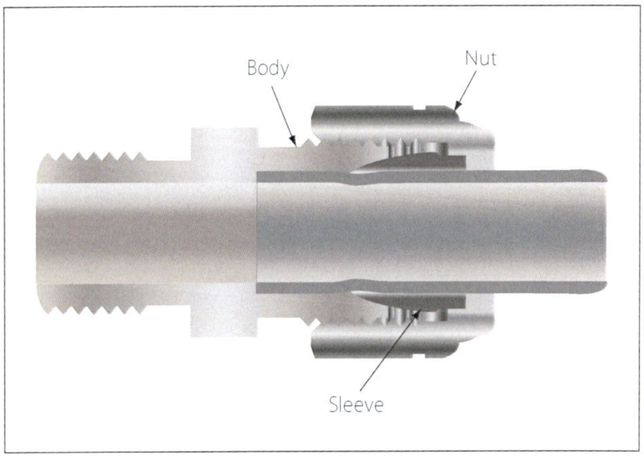

Figure 7-18. Flareless fitting.

Rigid Tubing Installation and Inspection

Before installing a line assembly in an aircraft, inspect the line carefully. Remove dents and scratches, and be sure all nuts and sleeves are snugly mated and securely fitted by proper flaring of the tubing. The line assembly should be clean and free of all foreign matter.

Connection and Torque

Never apply compound to the faces of the fitting or the flare, for it will destroy the metal-to-metal contact between the fitting and flare, a contact which is necessary to produce the seal. Be sure that the line assembly is properly aligned before tightening the fittings. Do

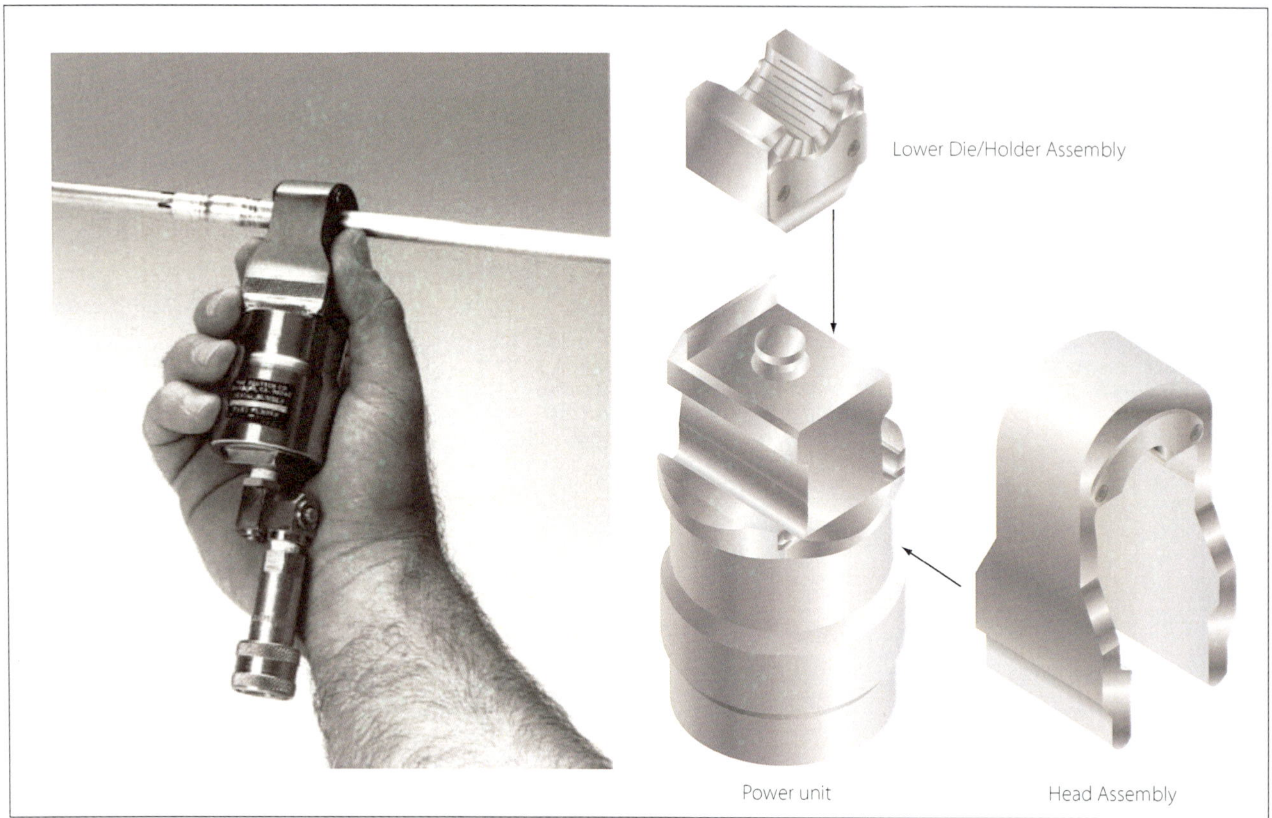

Figure 7-19. Swaged fitting tooling.

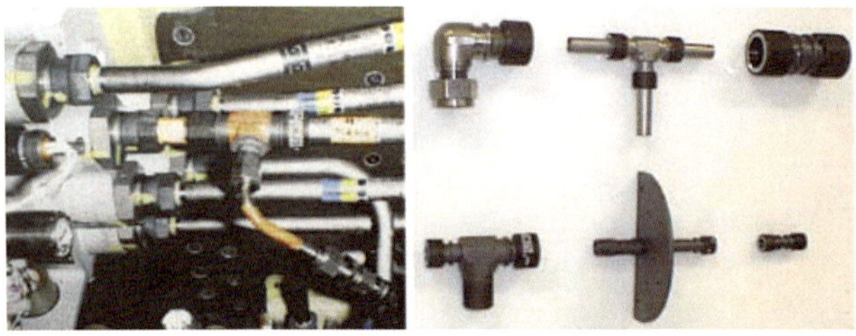

Figure 7-20. Permalite™ fittings.

Figure 7-21. Cryofit fittings.

not pull the installation into place with torque on the nut. Correct and incorrect methods of installing flared tube assemblies are illustrated in Figure 7-22. Proper torque values are given in Table 7-2. Remember that these torque values are for flared-type fittings only. Always tighten fittings to the correct torque value when installing a tube assembly. Overtightening a fitting may badly damage or completely cut off the tube flare, or it may ruin the sleeve or fitting nut. Failure to tighten sufficiently also may be serious, as this condition may allow the line to blow out of the assembly or to leak under system pressure. The use of torque wrenches and the prescribed torque values prevents overtightening or undertightening. If a tube fitting assembly is tightened properly, it may be removed and retightened many times before reflaring is necessary.

Flareless Tube Installation

Tighten the nut by hand until an increase in resistance to turning is encountered. Should it be impossible to run the nut down with the fingers, use a wrench, but be alert for the first signs of bottoming. It is important that the final tightening commence at the point where the nut just begins to bottom. Use a wrench and turn the nut one-sixth turn (one flat on a hex nut). Use a wrench on the connector to prevent it from turning while tightening the nut. After the tube assembly is

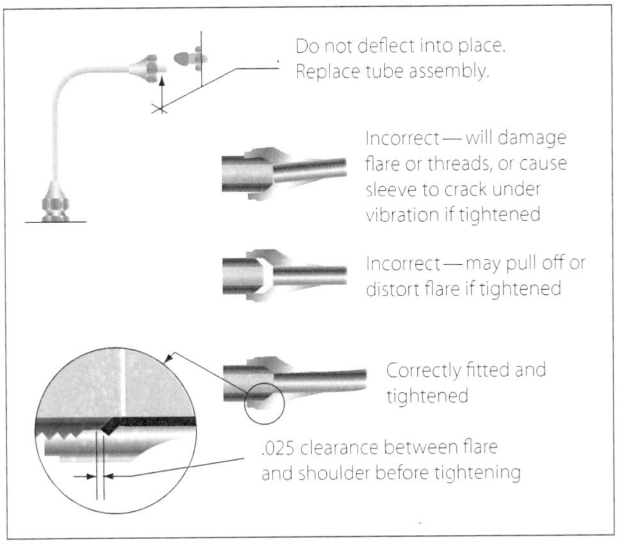

Figure 7-22. Correct and incorrect methods of tightening flared fittings.

installed, the system should be pressure tested. It is permissible to tighten the nut an additional one-sixth turn (making a total of one-third turn), should a connection leak. If leakage still occurs after tightening the nut a total of one-third turn, remove the assembly and inspect the components for scores, cracks, presence of foreign material, or damage from overtightening. Several aircraft manufacturers include torque values in their maintenance manuals to tighten the flareless fittings.

The following notes, cautions, and faults apply to the installation of rigid tubing.

Note: Overtightening a flareless tube nut drives the cutting edge of the sleeve deeply into the tube, causing the tube to be weakened to the point where normal in-flight vibration could cause the tube to shear. After inspection (if no discrepancies are found), reassemble the connections and repeat the pressure test procedures.

Caution: Never tighten the nut beyond one-third turn (two flats on the hex nut); this is the maximum the fitting may be tightened without the possibility of permanently damaging the sleeve and nut.

Common faults: Flare distorted into nut threads; sleeve cracked; flare cracked or split; flare out of round; inside of flare rough or scratched; and threads of nut or union dirty, damaged, or broken.

Rigid Tubing Inspection and Repair

Minor dents and scratches in tubing may be repaired. Scratches or nicks not deeper than 10 percent of the wall thickness in aluminum alloy tubing, which are not in the heel of a bend, may be repaired by burnishing with hand tools. The damage limits for hard, thin-walled corrosion-resistant steel and titanium tubing are considerably less than for aluminum tubing, and might depend on the aircraft manufacturer. Consult the aircraft maintenance manual for damage limits. Replace lines with severe die marks, seams, or splits in the tube. Any crack or deformity in a flare is unacceptable and is cause for rejection. A dent of less than 20 percent of the tube diameter is not objectionable, unless it is in the heel of a bend. To remove dents, draw a bullet of proper size through the tube by means of a length of cable, or push the bullet through a short straight tube by means of a dowel rod. In this case, a bullet is a ball bearing or slug normally made of steel or some other hard metal. In the case of soft aluminum tubing, a hard wood slug or dowel may even be used as a bullet. [Figure 7-23] A severely damaged line should be replaced. However, the line may be repaired by cutting out the damaged section and inserting a tube section of the same size and material. Flare both ends of the undamaged and replacement tube sections and make the connection by using standard unions, sleeves, and tube nuts. Aluminum 6061-T6, corrosion resistant steel 304-1/8h and Titanium 3AL-2.5V tubing can be repaired by swaged fittings. If the damaged portion is short enough, omit the insert tube and repair by using one repair union. [Figure 7-24] When repairing a damaged line, be very careful to remove all chips and burrs. Any open line that is to be left unattended for some time should be sealed, using metal, wood, rubber, or plastic plugs or caps.

When repairing a low-pressure line using a flexible fluid connection assembly, position the hose clamps carefully to prevent overhang of the clamp bands or chafing of the tightening screws on adjacent parts. If chafing can occur, the hose clamps should be repositioned on the hose. Figure 7-25 illustrates the design of a flexible fluid connection assembly and gives the maximum allowable angular and dimensional offset.

When replacing rigid tubing, ensure that the layout of the new line is the same as that of the line being replaced. Remove the damaged or worn assembly, taking care not to further damage or distort it, and use it

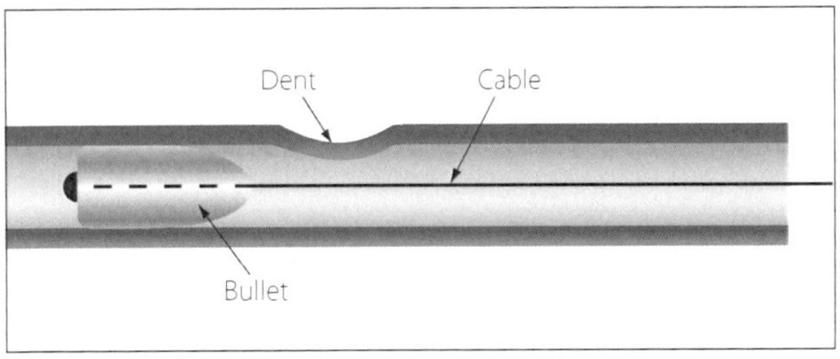

Figure 7-23. Dent removal using a bullet.

Type of Failure	Repair Method
1. Pin hole leak or circumferential crack in tubing. 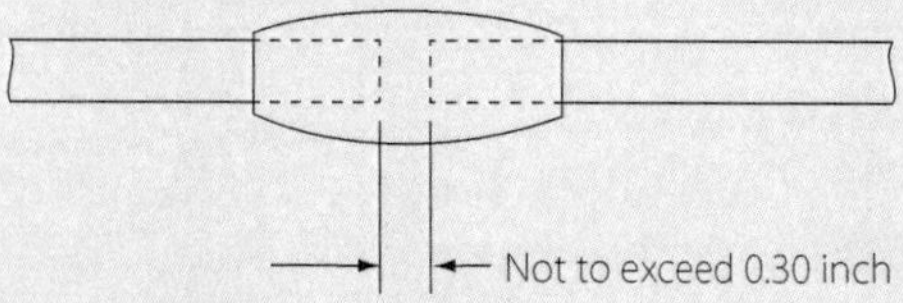	1. a. Make 1 or 2 cuts as necessary, to remove damaged section. If 2 cuts are required, the distance between them shall not exceed 0. 30". If distance is more than 0.30 inch, go to repair method 2. b. Swage 1 tube-to-tube union in tube section under repair.
2. Longitudinal crack in tubing (crack length in excess of 0.30"). 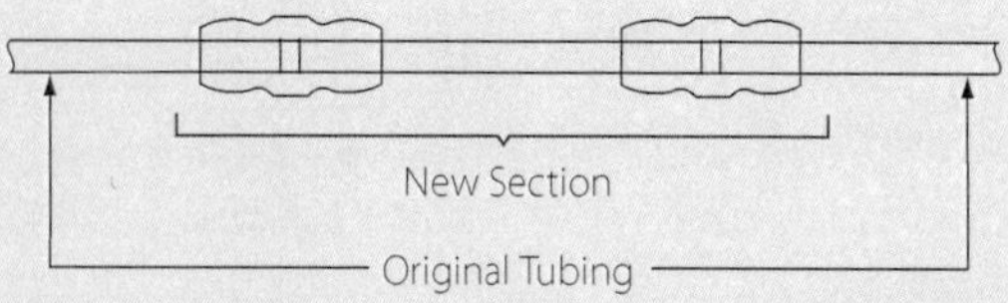	2. a. Make 2 cuts to enable removal of damaged section. b. Remove damaged section and duplicate. c. Swage replacement section into tubing under repair using 2 tube-to-tube unions.
3. Leaking tee or elbow (permanent tube connection type). 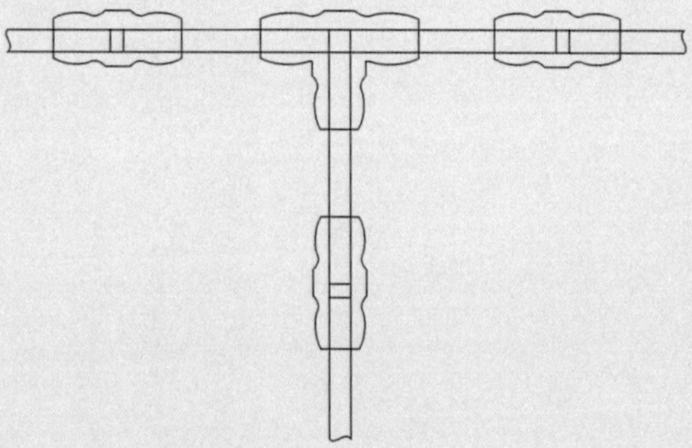	3. a. Cut out defective tee or elbow. b. Duplicate tubing sections for each branch. c. Swage splice sections to tee or elbow. d. Connect each splice section to tubing under repair using a tube to tube union.
4. Leaking flared, flareless, or lipseal end fittings. 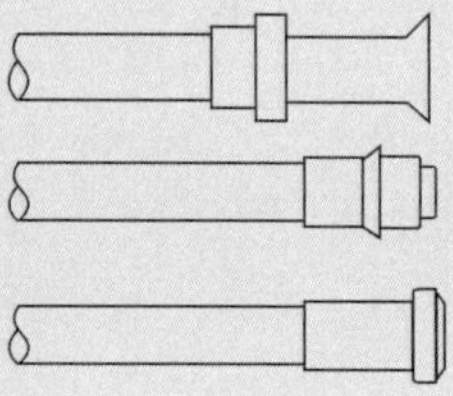	4. a. Cut tubing to remove defective fitting. b. Swage appropriate end fitting to tube end. c. Connect new end fitting to mating connection, torquing nut as required.

Figure 7-24. Permaswage™ repair.

as a forming template for the new part. If the old length of tubing cannot be used as a pattern, make a wire template, bending the pattern by hand as required for the new assembly. Then bend the tubing to match the wire pattern. Never select a path that does not require bends in the tubing. A tube cannot be cut or flared accurately enough so that it can be installed without bending and still be free from mechanical strain. Bends are also necessary to permit the tubing to expand or contract under temperature changes and to absorb vibration. If the tube is small (under $\frac{1}{4}$") and can be hand formed, casual bends may be made to allow for this. It the tube must be machine formed, definite bends must be made to avoid a straight assembly. Start all bends a reasonable distance from the fittings because the sleeves and nuts must be slipped back during the fabrication of flares and during inspections. In all cases, the new tube assembly should be so formed prior

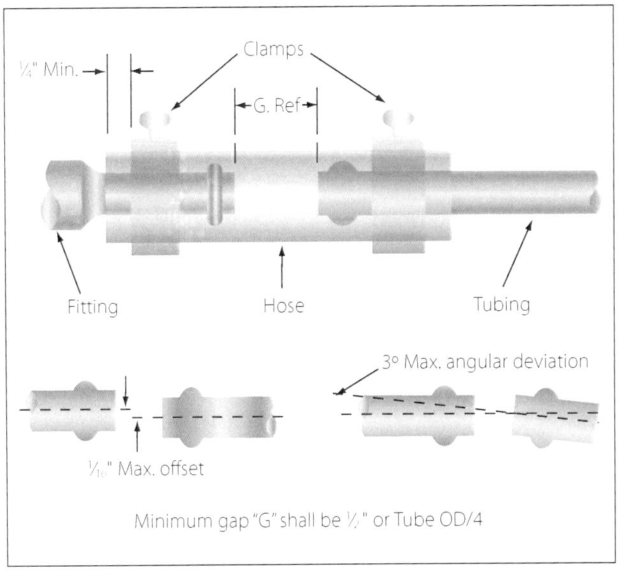

Figure 7-25. Flexible fluid connection assembly.

to installation that it will not be necessary to pull or deflect the assembly into alignment by means of the coupling nuts.

Flexible Hose Fluid Lines

Flexible hose is used in aircraft fluid systems to connect moving parts with stationary parts in locations subject to vibration or where a great amount of flexibility is needed. It can also serve as a connector in metal tubing systems.

Hose Materials and Construction

Pure rubber is never used in the construction of flexible fluid lines. To meet the requirements of strength, durability, and workability, among other factors, synthetics are used in place of pure rubber. Synthetic materials most commonly used in the manufacture of flexible hose are Buna-N, neoprene, butyl, ethylene propylene diene rubber (EPDM) and Teflon™. While Teflon™ is in a category of its own, the others are synthetic rubber.

Buna-N is a synthetic rubber compound which has excellent resistance to petroleum products. Do not confuse with Buna-S. Do not use for phosphate ester base hydraulic fluid (Skydrol).

Neoprene is a synthetic rubber compound which has an acetylene base. Its resistance to petroleum products is not as good as Buna-N, but it has better abrasive resistance. Do not use for phosphate ester base hydraulic fluid (Skydrol).

Butyl is a synthetic rubber compound made from petroleum raw materials. It is an excellent material to use with phosphate ester base hydraulic fluid (Skydrol). Do not use with petroleum products.

Flexible rubber hose consists of a seamless synthetic rubber inner tube covered with layers of cotton braid and wire braid and an outer layer of rubber-impregnated cotton braid. This type of hose is suitable for use in fuel, oil, coolant, and hydraulic systems. The types of hose are normally classified by the amount of pressure they are designed to withstand under normal operating conditions.

Low, Medium, and High Pressure Hoses

- Low pressure—below 250 psi. Fabric braid reinforcement.
- Medium pressure—up to 3,000 psi. One wire braid reinforcement. Smaller sizes carry up to 3,000 psi. Larger sizes carry pressure up to 1,500 psi.
- High pressure—all sizes up to 3,000 psi operating pressures.

Hose Identification

Lay lines and identification markings consisting of lines, letters, and numbers are printed on the hose. [Figure 7-26] Most hydraulic hose is marked to identify its type, the quarter and year of manufacture, and a 5-digit code identifying the manufacturer. These

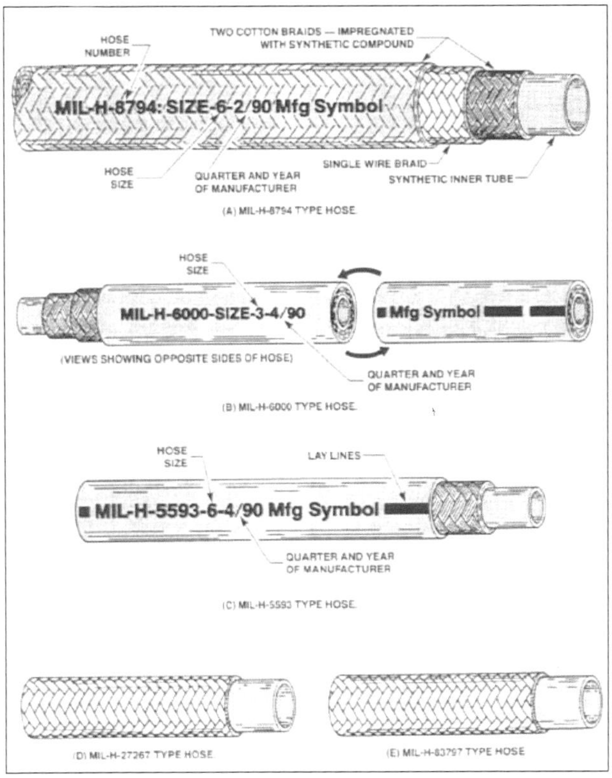

Figure 7-26. Hose identification markings.

markings are in contrasting colored letters and numerals which indicate the natural lay (no twist) of the hose and are repeated at intervals of not more than 9 inches along the length of the hose. Code markings assist in replacing a hose with one of the same specifications or a recommended substitute. Hose suitable for use with phosphate ester base hydraulic fluid will be marked Skydrol use. In some instances, several types of hose may be suitable for the same use. Therefore, to make the correct hose selection, always refer to the applicable aircraft maintenance or parts manual.

Teflon™ is the DuPont trade name for tetrafluoroethylene resin. It has a broad operating temperature range (−65 °F to +450 °F). It is compatible with nearly every substance or agent used. It offers little resistance to flow; sticky, viscous materials will not adhere to it. It has less volumetric expansion than rubber, and the shelf and service life is practically limitless. Teflon™ hose is flexible and designed to meet the requirements of higher operating temperatures and pressures in present aircraft systems. Generally, it may be used in the same manner as rubber hose. Teflon™ hose is processed and extruded into tube shape to a desired size. It is covered with stainless steel wire, which is braided over the tube for strength and protection. Teflon™ hose is unaffected by any known fuel, petroleum, or synthetic base oils, alcohol, coolants, or solvents commonly used in aircraft. Teflon™ hose has the distinct advantages of a practically unlimited storage time, greater operating temperature range, and broad usage (hydraulic, fuel, oil, coolant, water, alcohol, and pneumatic systems). Medium-pressure Teflon™ hose assemblies are sometimes preformed to clear obstructions and to make connections using the shortest possible hose length. Since preforming permits tighter bends that eliminate the need for special elbows, preformed hose assemblies save space and weight. Never straighten a preformed hose assembly. Use a support wire if the hose is to be removed for maintenance. [Figure 7-27]

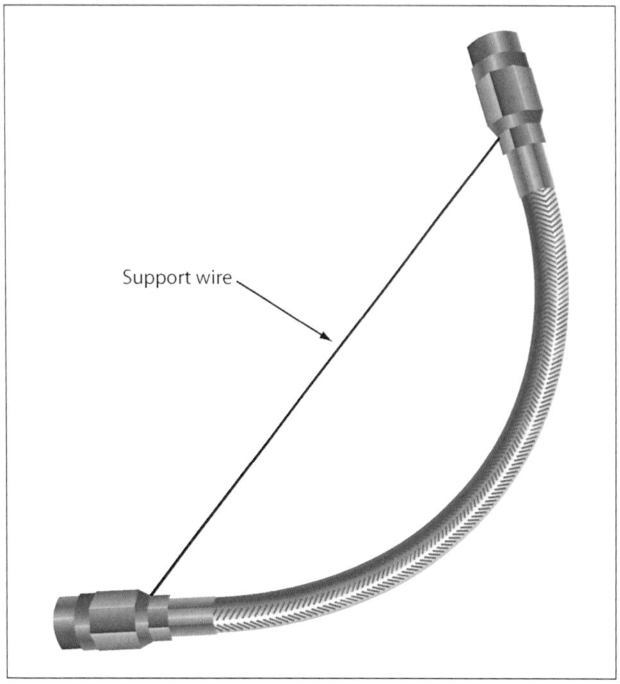

Figure 7-27. Suggested handling of preformed hose.

Flexible Hose Inspection

Check the hose and hose assemblies for deterioration at each inspection period. Leakage, separation of the cover or braid from the inner tube, cracks, hardening, lack of flexibility, or excessive "cold flow" are apparent signs of deterioration and reason for replacement. The term "cold flow" describes the deep, permanent impressions in the hose produced by the pressure of hose clamps or supports.

When failure occurs in a flexible hose equipped with swaged end fittings, the entire assembly must be replaced. Obtain a new hose assembly of the correct size and length, complete with factory installed end fittings. When failure occurs in hose equipped with reusable end fittings, a replacement line can be fabricated with the use of such tooling as may be necessary to comply with the assembly instructions of the manufacturer.

Fabrication and Replacement of Flexible Hose

To make a hose assembly, select the proper size hose and end fitting. [Figure 7-28] MS-type end fittings for flexible hose are detachable and may be reused if determined to be serviceable. The inside diameter of the fitting is the same as the inside diameter of the hose to which it is attached.

Flexible Hose Testing

All flexible hose must be proof-tested after assembly and applying pressure to the inside of the hose assembly. The proof-test medium may be a liquid or gas. For example, hydraulic, fuel, and oil lines are generally tested using hydraulic oil or water, whereas air or instrument lines are tested with dry, oil-free air or nitrogen. When testing with a liquid, all trapped air is bled from the assembly prior to tightening the cap or plug. Hose tests, using a gas, are conducted underwater. In all cases, follow the hose manufacturer's instructions for proof-test pressure and fluid to be used when testing a specific hose assembly. [Table 7-3]

When a flexible hose has been repaired or overhauled using existing hardware and new hose material, and

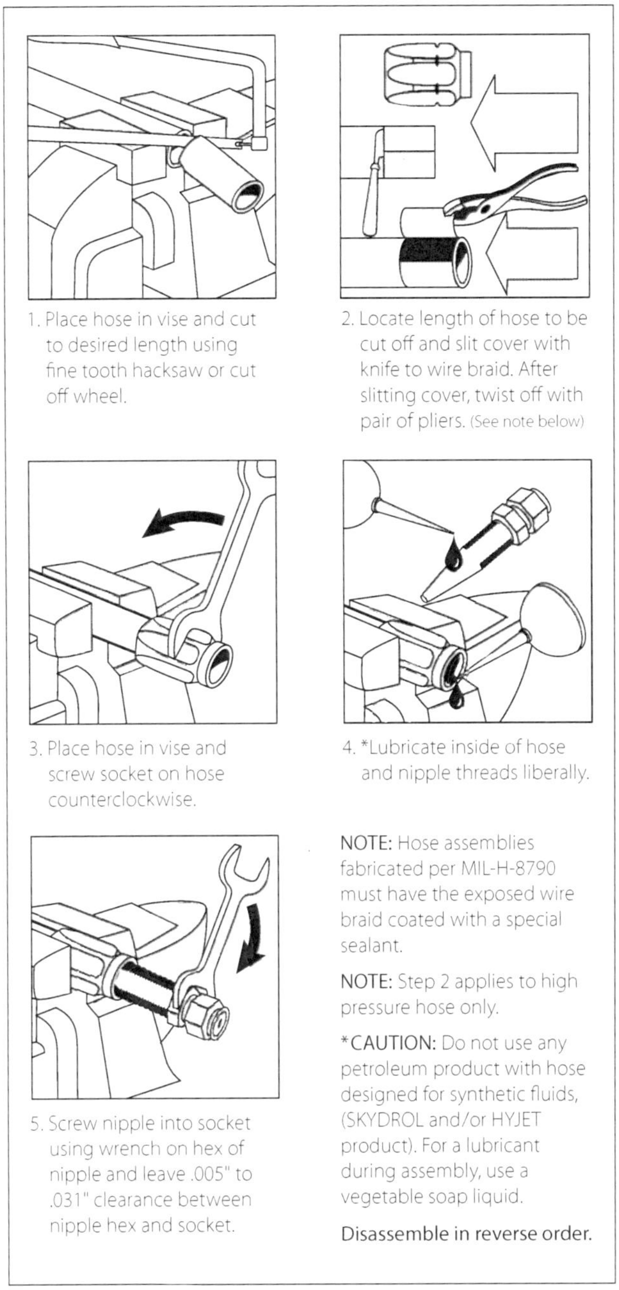

Figure 7-28. Assembly of MS fitting to flexible hose.

before the hose is installed on the aircraft, it is recommended that the hose be tested to at least 1.5 system pressure. A hydraulic hose burst test stand is used for testing flexible hose. [Figure 7-29] A new hose can be operationally checked after it is installed in the aircraft using system pressure.

Size Designations

Hose is also designated by a dash number, according to its size. The dash number is stenciled on the side of the hose and indicates the size tubing with which the hose is compatible. It does not denote inside or outside diameter. When the dash number of the hose

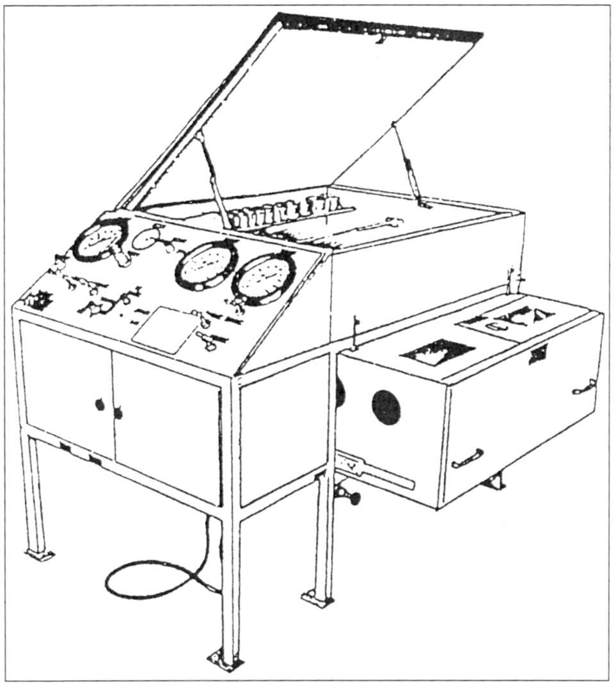

Figure 7-29. Hydraulic hose burst test stand.

corresponds with the dash number of the tubing, the proper size hose is being used. Dash numbers are shown in Figure 7-26.

Hose Fittings

Flexible hose may be equipped with either swaged fittings or detachable fittings, or they may be used with beads and hose clamps. Hoses equipped with swaged fittings are ordered by correct length from the manufacturer and ordinarily cannot be assembled by the mechanic. They are swaged and tested at the factory and are equipped with standard fittings. The detachable fittings used on flexible hoses may be detached and reused if they are not damaged; otherwise, new fittings must be used. [Figure 7-30]

Installation of Flexible Hose Assemblies

Slack—Hose assemblies must not be installed in a manner that will cause a mechanical load on the hose. When installing flexible hose, provide slack or bend in the hose line from 5 to 8 percent of its total length to provide for changes in length that will occur when pressure is applied. Flexible hose contracts in length and expands in diameter when pressurized. Protect all flexible hoses from excessive temperatures, either by locating the lines so they will not be affected or by installing shrouds around them.

Flex—When hose assemblies are subject to considerable vibration or flexing, sufficient slack must be left between rigid fittings. Install the hose so that flexure

SINGLE WIRE BRAID FABRIC COVERED

MIL. Part No.	Tube Size o.d. (inches)	Hose Size i.d. (inches)	Hose Size o.d. (inches)	Recomm. Operating Pressure (PSI)	Min. Burst Pressure (PSI)	Max. Proof Pressure (PSI)	Min. Bend Radius (inches)
MIL-H-8794-3-L	3/16	1/8	.45	3,000	12,000	6,000	3.00
MIL-H-8794-4-L	1/4	3/16	.52	3,000	12,000	6,000	3.00
MIL-H-8794-5-L	5/16	1/4	.58	3,000	10,000	5,000	3.38
MIL-H-8794-6-L	3/8	5/16	.67	2,000	9,000	4,500	4.00
MIL-H-8794-8-L	1/2	13/32	.77	2,000	8,000	4,000	4.63
MIL-H-8794-10-L	5/8	1/2	.92	1,750	7,000	3,500	5.50
MIL-H-8794-12-L	3/4	5/8	1.08	1,750	6,000	3,000	6.50
MIL-H-8794-16-L	1	7/8	1.23	800	3,200	1,600	7.38
MIL-H-8794-20-L	1 1/4	1 1/8	1.50	600	2,500	1,250	9.00
MIL-H-8794-24-L	1 1/2	1 3/8	1.75	500	2,000	1,000	11.00
MIL-H-8794-32-L	2	1 13/16	2.22	350	1,400	700	13.25
MIL-H-8794-40-L	2 1/2	2 3/8	2.88	200	1,000	300	24.00
MIL-H-8794-48-L	3	3	3.56	200	800	300	33.00

Construction: Seamless synthetic rubber inner tube reinforced with one fiber braid, one braid of high tensile steel wire and covered with an oil resistant rubber impregnated fiber braid.

Identification: Hose is identified by specification number, size number, quarter year and year, hose manufacturer's identification.

Uses: Hose is approved for use in aircraft hydraulic, pneumatic, coolant, fuel, and oil systems.

Operating Temperatures:
Sizes 3 through 12: Minus 65 °F to plus 250 °F
Sizes 16 through 48: Minus 40 °F to plus 275 °F

Note: Maximum temperatures and pressures should not be used simultaneously.

MULTIPLE WIRE BRAID RUBBER COVERED

MIL. Part No.	Tube Size o.d. (inches)	Hose Size i.d. (inches)	Hose Size o.d. (inches)	Recomm. Operating Pressure (PSI)	Min. Burst Pressure (PSI)	Max. Proof Pressure (PSI)	Min. Bend Radius (inches)
MIL-H-8788- 4-L	1/4	7/32	.63	3,000	16,000	8,000	3.00
MIL-H-8788- 5-L	5/16	9/32	.70	3,000	14,000	7,000	3.38
MIL-H-8788- 6-L	3/8	11/32	.77	3,000	14,000	7,000	5.00
MIL-H-8788- 8-L	1/2	7/16	.86	3,000	14,000	7,500	5.75
MIL-H-8788-10-L	5/8	9/16	1.03	3,000	12,000	6,000	6.50
MIL-H-8788-12-L	3/4	11/16	1.22	3,000	12,000	6,000	7.75
MIL-H-8788-16-L	1	7/8	1.50	3,000	10,000	5,000	9.63

Construction: Seamless synthetic rubber inner tube reinforced with one fiber braid, two or more steel wire braids, and covered with synthetic rubber cover (for gas applications request perforated cover).

Identification: Hose is identified by specification number, size number, quarter year and year, hose manufacturer's identification.

Uses: High pressure hydraulic, pneumatic, coolant, fuel and oil.

Operating Temperatures: Minus 65 °F to plus 200 °F

Table 7-3. Aircraft hose specifications.

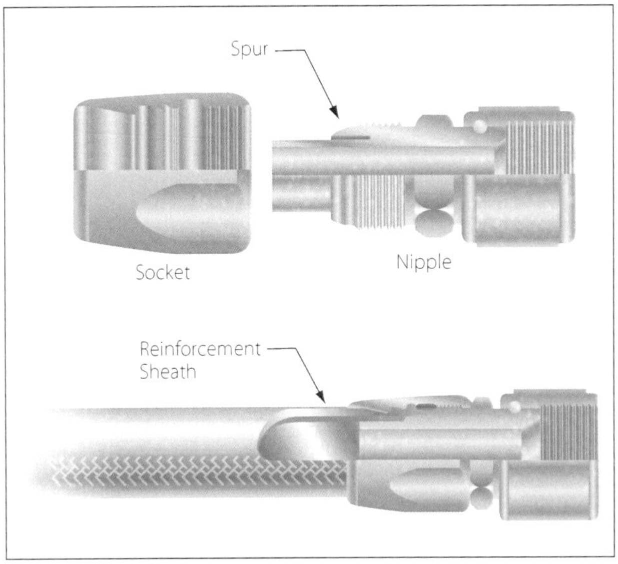

Figure 7-30. Reusable fittings for medium-pressure hose.

does not occur at the end fittings. The hose must remain straight for at least two hose diameters from the end fittings. Avoid clamp locations that will restrict or prevent hose flexure.

Twisting—Hoses must be installed without twisting to avoid possible rupture of the hose or loosening of the attaching nuts. Use of swivel connections at one or both ends will relieve twist stresses. Twisting of the hose can be determined from the identification stripe running along its length. This stripe should not spiral around the hose.

Bending—To avoid sharp bends in the hose assembly, use elbow fittings, hose with elbow-type end fittings, or the appropriate bend radii. Bends that are too sharp will reduce the bursting pressure of flexible hose considerably below its rated value. [Figure 7-31]

Clearance—The hose assembly must clear all other lines, equipment, and adjacent structure under every operating condition.

Flexible hose should be installed so that it will be subject to a minimum of flexing during operation. Although hose must be supported at least every 24 inches, closer supports are desirable. Flexible hose must never be stretched tightly between two fittings. If clamps do not seal at specified tightening, examine

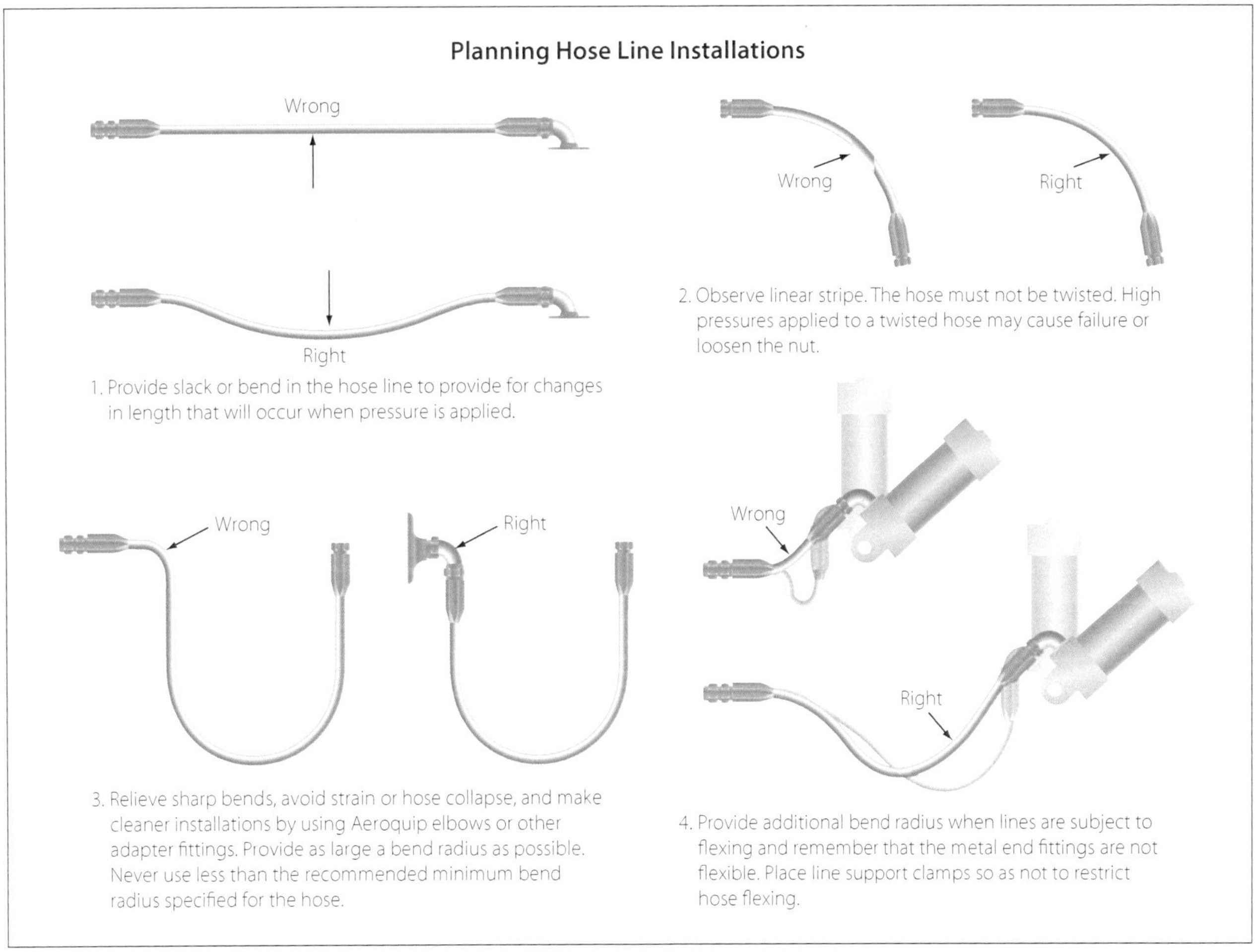

Figure 7-31. Flexible hose installation.

hose connections and replace parts as necessary. The above is for initial installation and should not be used for loose clamps.

For retightening loose hose clamps in service, proceed as follows: Non-self-sealing hose—if the clamp screw cannot be tightened with the fingers, do not disturb unless leakage is evident. If leakage is present, tighten one-fourth turn. Self-sealing hose—if looser than finger-tight, tighten to finger-tight and add one-fourth turn. [Table 7-4]

Hose Clamps

To ensure proper sealing of hose connections and to prevent breaking hose clamps or damaging the hose, follow the hose clamp tightening instructions carefully. When available, use the hose clamp torque-limiting wrench. These wrenches are available in calibrations of 15 and 25 in-lb limits. In the absence of torque-limiting wrenches, follow the finger-tight-plus-turns method. Because of the variations in hose clamp design and hose structure, the values given in Table 7-4 are approximate. Therefore, use good judgment when tightening hose clamps by this method. Since hose connections are subject to "cold flow" or a setting process, a follow-up tightening check should be made for several days after installation.

Support clamps are used to secure the various lines to the airframe or powerplant assemblies. Several types of support clamps are used for this purpose. The most commonly used clamps are the rubber-cushioned and plain. The rubber-cushioned clamp is used to secure lines subject to vibration; the cushioning prevents chafing of the tubing. [Figure 7-32] The plain clamp is used to secure lines in areas not subject to vibration.

A Teflon™-cushioned clamp is used in areas where the deteriorating effect of Skydrol, hydraulic fluid, or fuel is expected. However, because it is less resilient, it does not provide as good a vibration-damping effect as other cushion materials.

Initial Installation Only	Worm screw type clamp (10 threads per inch)	Clamps—radial and other type (28 threads per inch)
Self sealing hose approximately 15 in-lb	Finger-tight plus 2 complete turns	Finger-tight plus 2½ complete turns
All other aircraft hose approximately 25 in-lb	Finger-tight plus 1¼ complete turns	Finger-tight plus 2 complete turns

Table 7-4. Hose clamp tightening.

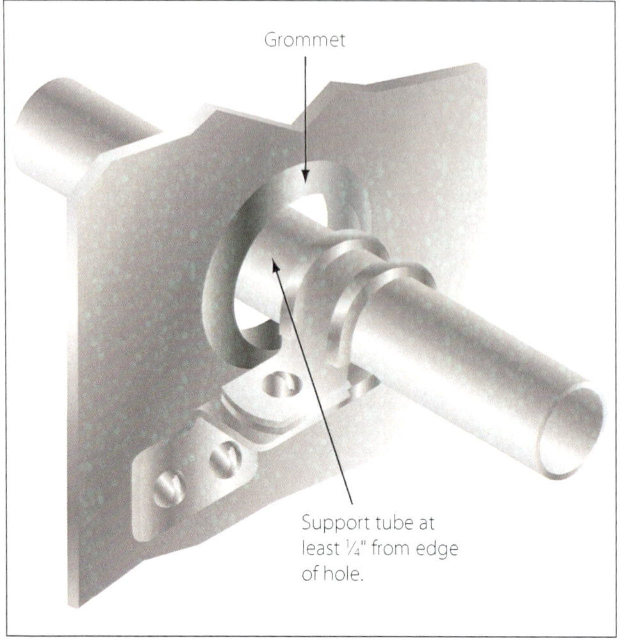

Figure 7-32. Rubber-cushioned clamp.

Use bonded clamps to secure metal hydraulic, fuel, or oil lines in place. Unbonded clamps should be used only for securing wiring. Remove any paint or anodizing from the portion of the tube at the bonding clamp location. Make certain that clamps are of the correct size. Clamps or supporting clips smaller than the outside diameter of the hose may restrict the flow of fluid through the hose. All fluid lines must be secured at specified intervals. The maximum distance between supports for rigid tubing is shown in Table 7-5.

Tube O. D. (in.)	Distance between supports (in.)	
	Aluminium Alloy	Steel
⅛	9½	11½
3/16	12	14
¼	13½	16
5/16	15	18
⅜	16½	20
½	19	23
⅝	22	25½
¾	24	27½
1	26½	30

Table 7-5. Maximum distance between supports for fluid tubing.

Chapter 8 Inspection Fundamentals

Inspections are visual examinations and manual checks to determine the condition of an aircraft or component. An aircraft inspection can range from a casual walk-around to a detailed inspection involving complete disassembly and the use of complex inspection aids.

An inspection system consists of several processes, including reports made by mechanics or the pilot or crew flying an aircraft and regularly scheduled inspections of an aircraft. An inspection system is designed to maintain an aircraft in the best possible condition. Thorough and repeated inspections must be considered the backbone of a good maintenance program. Irregular and haphazard inspection will invariably result in gradual and certain deterioration of an aircraft. The time spent in repairing an abused aircraft often totals far more than any time saved in hurrying through routine inspections and maintenance.

It has been proven that regularly scheduled inspections and preventive maintenance assure airworthiness. Operating failures and malfunctions of equipment are appreciably reduced if excessive wear or minor defects are detected and corrected early. The importance of inspections and the proper use of records concerning these inspections cannot be overemphasized.

Airframe and engine inspections may range from preflight inspections to detailed inspections. The time intervals for the inspection periods vary with the models of aircraft involved and the types of operations being conducted. The airframe and engine manufacturer's instructions should be consulted when establishing inspection intervals.

Aircraft may be inspected using flight hours as a basis for scheduling, or on a calendar inspection system. Under the calendar inspection system, the appropriate inspection is performed on the expiration of a specified number of calendar weeks. The calendar inspection system is an efficient system from a maintenance management standpoint. Scheduled replacement of components with stated hourly operating limitations is normally accomplished during the calendar inspection falling nearest the hourly limitation.

In some instances, a flight hour limitation is established to limit the number of hours that may be flown during the calendar interval.

Aircraft operating under the flight hour system are inspected when a specified number of flight hours are accumulated. Components with stated hourly operating limitations are normally replaced during the inspection that falls nearest the hourly limitation.

Basic Inspection Techniques/Practices

Before starting an inspection, be certain all plates, access doors, fairings, and cowling have been opened or removed and the structure cleaned. When opening inspection plates and cowling and before cleaning the area, take note of any oil or other evidence of fluid leakage.

Preparation

In order to conduct a thorough inspection, a great deal of paperwork and/or reference information must be accessed and studied before actually proceeding to the aircraft to conduct the inspection. The aircraft logbooks must be reviewed to provide background information and a maintenance history of the particular aircraft. The appropriate checklist or checklists must be utilized to ensure that no items will be forgotten or overlooked during the inspection. Also, many additional publications must be available, either in hard copy or in electronic format to assist in the inspections. These additional publications may include information provided by the aircraft and engine manufacturers, appliance manufacturers, parts venders, and the Federal Aviation Administration (FAA).

Aircraft Logs

"Aircraft logs," as used in this handbook, is an inclusive term which applies to the aircraft logbook and all supplemental records concerned with the aircraft. They may come in a variety of formats. For a small aircraft, the log may indeed be a small 5" × 8" logbook. For larger aircraft, the logbooks are often larger, in the form of a three-ring binder. Aircraft that have been in service for a long time are likely to have several logbooks.

The aircraft logbook is the record in which all data concerning the aircraft is recorded. Information gathered in this log is used to determine the aircraft condition, date of inspections, time on airframe, engines and propellers. It reflects a history of all significant events occurring to the aircraft, its components, and accessories, and provides a place for indicating compliance with FAA airworthiness directives or manufacturers' service bulletins. The more comprehensive the logbook, the easier it is to understand the aircraft's maintenance history.

When the inspections are completed, appropriate entries must be made in the aircraft logbook certifying that the aircraft is in an airworthy condition and may be returned to service. When making logbook entries, exercise special care to ensure that the entry can be clearly understood by anyone having a need to read it in the future. Also, if making a hand-written entry, use good penmanship and write legibly. To some degree, the organization, comprehensiveness, and appearance of the aircraft logbooks have an impact on the value of the aircraft. High quality logbooks can mean a higher value for the aircraft.

Checklists

Always use a checklist when performing an inspection. The checklist may be of your own design, one provided by the manufacturer of the equipment being inspected, or one obtained from some other source. The checklist should include the following:

1. Fuselage and hull group.
   a. Fabric and skin—for deterioration, distortion, other evidence of failure, and defective or insecure attachment of fittings.
   b. Systems and components—for proper installation, apparent defects, and satisfactory operation.
   c. Envelope gas bags, ballast tanks, and related parts—for condition.
2. Cabin and cockpit group.
   a. Generally—for cleanliness and loose equipment that should be secured.
   b. Seats and safety belts—for condition and security.
   c. Windows and windshields—for deterioration and breakage.
   d. Instruments—for condition, mounting, marking, and (where practicable) for proper operation.
   e. Flight and engine controls—for proper installation and operation.
   f. Batteries—for proper installation and charge.
   g. All systems—for proper installation, general condition, apparent defects, and security of attachment.
3. Engine and nacelle group.
   a. Engine section—for visual evidence of excessive oil, fuel, or hydraulic leaks, and sources of such leaks.
   b. Studs and nuts—for proper torquing and obvious defects.
   c. Internal engine—for cylinder compression and for metal particles or foreign matter on screens and sump drain plugs. If cylinder compression is weak, check for improper internal condition and improper internal tolerances.
   d. Engine mount—for cracks, looseness of mounting, and looseness of engine to mount.
   e. Flexible vibration dampeners—for condition and deterioration.
   f. Engine controls—for defects, proper travel, and proper safetying.
   g. Lines, hoses, and clamps—for leaks, condition, and looseness.
   h. Exhaust stacks—for cracks, defects, and proper attachment.
   i. Accessories—for apparent defects in security of mounting.
   j. All systems—for proper installation, general condition defects, and secure attachment.
   k. Cowling—for cracks and defects.
   l. Ground runup and functional check—check all powerplant controls and systems for correct response, all instruments for proper operation and indication.

4. Landing gear group.
   a. All units—for condition and security of attachment.
   b. Shock absorbing devices—for proper oleo fluid level.
   c. Linkage, trusses, and members—for undue or excessive wear, fatigue, and distortion.
   d. Retracting and locking mechanism—for proper operation.
   e. Hydraulic lines—for leakage.
   f. Electrical system—for chafing and proper operation of switches.
   g. Wheels—for cracks, defects, and condition of bearings.
   h. Tires—for wear and cuts.
   i. Brakes—for proper adjustment.
   j. Floats and skis—for security of attachment and obvious defects.
5. Wing and center section.
   a. All components—for condition and security.
   b. Fabric and skin—for deterioration, distortion, other evidence of failure, and security of attachment.
   c. Internal structure (spars, ribs compression members)—for cracks, bends, and security.
   d. Movable surfaces—for damage or obvious defects, unsatisfactory fabric or skin attachment and proper travel.
   e. Control mechanism—for freedom of movement, alignment, and security.
   f. Control cables—for proper tension, fraying, wear and proper routing through fairleads and pulleys.
6. Empennage group.
   a. Fixed surfaces—for damage or obvious defects, loose fasteners, and security of attachment.
   b. Movable control surfaces—for damage or obvious defects, loose fasteners, loose fabric, or skin distortion.
   c. Fabric or skin—for abrasion, tears, cuts or defects, distortion, and deterioration.
7. Propeller group.
   a. Propeller assembly—for cracks, nicks, bends, and oil leakage.
   b. Bolts—for proper torquing and safetying.
   c. Anti-icing devices—for proper operation and obvious defects.
   d. Control mechanisms—for proper operation, secure mounting, and travel.
8. Communication and navigation group.
   a. Radio and electronic equipment—for proper installation and secure mounting.
   b. Wiring and conduits—for proper routing, secure mounting, and obvious defects.
   c. Bonding and shielding—for proper installation and condition.
   d. Antennas—for condition, secure mounting, and proper operation.
9. Miscellaneous.
   a. Emergency and first aid equipment—for general condition and proper stowage.
   b. Parachutes, life rafts, flares, and so forth—inspect in accordance with the manufacturer's recommendations.
   c. Autopilot system—for general condition, security of attachment, and proper operation.

Publications

Aeronautical publications are the sources of information for guiding aviation mechanics in the operation and maintenance of aircraft and related equipment. The proper use of these publications will greatly aid in the efficient operation and maintenance of all aircraft. These include manufacturers' service bulletins, manuals, and catalogs; FAA regulations; airworthiness directives; advisory circulars; and aircraft, engine and propeller specifications.

Manufacturers' Service Bulletins/Instructions

Service bulletins or service instructions are two of several types of publications issued by airframe, engine, and component manufacturers.

The bulletins may include: (1) purpose for issuing the publication, (2) name of the applicable airframe, engine, or component, (3) detailed instructions for service, adjustment, modification or inspection, and source of parts, if required and (4) estimated number of manhours required to accomplish the job.

Maintenance Manual

The manufacturer's aircraft maintenance manual contains complete instructions for maintenance of all systems and components installed in the aircraft. It contains information for the mechanic who normally works on components, assemblies, and systems while they are installed in the aircraft, but not for the overhaul mechanic. A typical aircraft maintenance manual contains:

- A description of the systems (i.e., electrical, hydraulic, fuel, control)
- Lubrication instructions setting forth the frequency and the lubricants and fluids which are to be used in the various systems,
- Pressures and electrical loads applicable to the various systems,
- Tolerances and adjustments necessary to proper functioning of the airplane,
- Methods of leveling, raising, and towing,
- Methods of balancing control surfaces,
- Identification of primary and secondary structures,
- Frequency and extent of inspections necessary to the proper operation of the airplane,
- Special repair methods applicable to the airplane,
- Special inspection techniques requiring x-ray, ultrasonic, or magnetic particle inspection, and
- A list of special tools.

Overhaul Manual

The manufacturer's overhaul manual contains brief descriptive information and detailed step by step instructions covering work normally performed on a unit that has been removed from the aircraft. Simple, inexpensive items, such as switches and relays on which overhaul is uneconomical, are not covered in the overhaul manual.

Structural Repair Manual

This manual contains the manufacturer's information and specific instructions for repairing primary and secondary structures. Typical skin, frame, rib, and stringer repairs are covered in this manual. Also included are material and fastener substitutions and special repair techniques.

Illustrated Parts Catalog

This catalog presents component breakdowns of structure and equipment in disassembly sequence. Also included are exploded views or cutaway illustrations for all parts and equipment manufactured by the aircraft manufacturer.

Code of Federal Regulations (CFRs)

The CFRs were established by law to provide for the safe and orderly conduct of flight operations and to prescribe airmen privileges and limitations. A knowledge of the CFRs is necessary during the performance of maintenance, since all work done on aircraft must comply with CFR provisions.

Airworthiness Directives

A primary safety function of the FAA is to require correction of unsafe conditions found in an aircraft, aircraft engine, propeller, or appliance when such conditions exist and are likely to exist or develop in other products of the same design. The unsafe condition may exist because of a design defect, maintenance, or other causes. Title 14 of the Code of Federal Regulations (14 CFR) part 39, Airworthiness Directives, defines the authority and responsibility of the Administrator for requiring the necessary corrective action. The Airworthiness Directives (ADs) are published to notify aircraft owners and other interested persons of unsafe conditions and to prescribe the conditions under which the product may continue to be operated.

Airworthiness Directives are Federal Aviation Regulations and must be complied with unless specific exemption is granted.

Airworthiness Directives may be divided into two categories: (1) those of an emergency nature requiring immediate compliance upon receipt and (2) those of a less urgent nature requiring compliance within a relatively longer period of time. Also, ADs may be a one-time compliance item or a recurring item that requires future inspection on an hourly basis (accrued flight time since last compliance) or a calendar time basis.

The contents of ADs include the aircraft, engine, propeller, or appliance model and serial numbers affected. Also included are the compliance time or period, a description of the difficulty experienced, and the necessary corrective action.

Type Certificate Data Sheets

The type certificate data sheet (TCDS) describes the type design and sets forth the limitations prescribed by the applicable CFR part. It also includes any other limitations and information found necessary for type certification of a particular model aircraft.

Type certificate data sheets are numbered in the upper right-hand corner of each page. This number is the same as the type certificate number. The name of the type certificate holder, together with all of the approved models, appears immediately below the type certificate number. The issue date completes this group. This information is contained within a bordered text box to set it off.

The data sheet is separated into one or more sections. Each section is identified by a Roman numeral followed by the model designation of the aircraft to which the section pertains. The category or categories in which the aircraft can be certificated are shown in parentheses following the model number. Also included is the approval date shown on the type certificate.

The data sheet contains information regarding:

1. Model designation of all engines for which the aircraft manufacturer obtained approval for use with this model aircraft.
2. Minimum fuel grade to be used.
3. Maximum continuous and takeoff ratings of the approved engines, including manifold pressure (when used), rpm, and horsepower (hp).
4. Name of the manufacturer and model designation for each propeller for which the aircraft manufacturer obtained approval will be shown together with the propeller limits and any operating restrictions peculiar to the propeller or propeller engine combination.
5. Airspeed limits in both mph and knots.
6. Center of gravity range for the extreme loading conditions of the aircraft is given in inches from the datum. The range may also be stated in percent of MAC (Mean Aerodynamic Chord) for transport category aircraft.
7. Empty weight center of gravity (CG) range (when established) will be given as fore and aft limits in inches from the datum. If no range exists, the word "none" will be shown following the heading on the data sheet.
8. Location of the datum.
9. Means provided for leveling the aircraft.
10. All pertinent maximum weights.
11. Number of seats and their moment arms.
12. Oil and fuel capacity.
13. Control surface movements.
14. Required equipment.
15. Additional or special equipment found necessary for certification.
16. Information concerning required placards.

It is not within the scope of this handbook to list all the items that can be shown on the type certificate data sheets. Those items listed above serve only to acquaint aviation mechanics with the type of information generally included on the data sheets. Type certificate data sheets may be many pages in length. Figure 8-1 shows a typical TCDS.

When conducting a required or routine inspection, it is necessary to ensure that the aircraft and all the major items on it are as defined in the type certificate data sheets. This is called a conformity check, and verifies that the aircraft conforms to the specifications of the aircraft as it was originally certified. Sometimes alterations are made that are not specified or authorized in the TCDS. When that condition exists, a supplemental type certificate (STC) will be issued. STCs are considered a part of the permanent records of an aircraft, and should be maintained as part of that aircraft's logs.

Routine/Required Inspections

For the purpose of determining their overall condition, 14 CFR provides for the inspection of all civil aircraft at specific intervals, depending generally upon the type of operations in which they are engaged. The pilot in command of a civil aircraft is responsible for determining whether that aircraft is in condition for safe flight. Therefore, the aircraft must be inspected before each flight. More detailed inspections must be conducted by aviation maintenance technicians at least once each 12 calendar months, while inspection is required for others after each 100 hours of flight. In other instances, an aircraft may be inspected in accordance with a system set up to provide for total inspection of the aircraft over a calendar or flight time period.

To determine the specific inspection requirements and rules for the performance of inspections, refer to the CFR, which prescribes the requirements for the inspection and maintenance of aircraft in various types of operations.

Preflight/Postflight Inspections

Pilots are required to follow a checklist contained within the Pilot's Operating Handbook (POH) when operating aircraft. The first section of a checklist includes a section entitled Preflight Inspection. The preflight inspection checklist includes a "walk-around" section listing items that the pilot is to visually check for general condition as he or she walks around the airplane. Also, the pilot must ensure that fuel, oil and other items required for flight are at the proper levels

(Continued on page 8-12)

DEPARTMENT OF TRANSPORTATION
FEDERAL AVIATION ADMINISTRATION

A25CE
Revision 11
CESSNA
404
406
June 15, 1995

TYPE CERTIFICATE DATA SHEET NO. A25CE

This data sheet which is part of Type Certificate No. A25CE prescribes conditions and limitations under which the product for which the type certificate was issued meets the airworthiness requirements of the Federal Aviation Regulations.

Type Certificate Holder — Cessna Aircraft Company
P. O. Box 7704
Wichita, Kansas 67277

I - Model 404, Titan, (Normal Category), Approved July 21, 1976

Engines — Two Teledyne Continental GTSIO-520-M
Reduction gear ratio .667:1

Fuel — 100/130 or 100 low-lead minimum grade aviation gasoline
See NOTE 3 for optional anti-icing additive

Engine Limits — For all operations, 2235 propeller r.p.m. (375 hp.)
40.0 in. Hg. mp. up to critical altitude of 16,000 feet in standard atmosphere. Above 16,000 feet the following maximum mp. applies for maximum r.p.m.

Altitude (ft.)	Max. Allowable Mp. (in. Hg)
16,000	40.0
18,000	37.5
20,000	35.0
22,000	32.0
24,000	29.2
26,000	26.0
28,000	23.0
30,000	20.0

Propeller and Propeller Limits — Two McCauley full-feathering three-bladed propeller installations
(a) McCauley hub 3FF32C501 with 90UMB-0 blades
Diameter: not over 90.0 in., not under 88.5 in.
No further reduction permitted
Pitch settings at 30.0 in. sta.:
low 16.6°, ±0.2°, feathering 84.6°, ±0.3°

S/N 404-0001 through 404-0600
(b) Hydraulic governor McCauley DCF290D2/T6, DCFU290D2/T6, DCFS290D2/T6, DCFUS290D2/T6, DCF290D7/T6, DCFU290D7/T6 or DCFU290D13/T6, DCFS290D7/T6, DCFUS290D7/T6 or DCFUS290D13/T6

S/N 404-0601 and up
(b) Hydraulic governor McCauley DCF290D7/T6, DCFU290D7/T6 or DCFU290D13/T6, DCFS290D9/T6, DCFUS290D9/T6

(c) Propeller spinner and bulkhead assembly, McCauley D3534/D-4506

Page No.	1	2	3	4	5	6
Rev. No.	11	10	9	11	8	8

Figure 8-1. Type certificate data sheet.

Airspeed Limits (IAS)	Maneuvering	160 KIAS
	Maximum structural cruising	212 KIAS
	Never exceed	241 KIAS
	Landing gear operating	182 KIAS
	Landing gear extended	182 KIAS
	Flaps extended - takeoff	182 KIAS
	Flaps extended - landing	152 KIAS
	Minimum control	78 KIAS

C.G. Range (Landing Gear Extended)
(+170.31) to (+179.08) at 8400 lb.
(+165.6) at 6100 lb. or less
Straight line variation between points given
Landing gear retracted moment change: +1113 in./lb.

Empty Wt. C.G. Range
None

Leveling Means
Two screws located on W.L. 93.80 @ sta. 248.25 and sta. 272.65

Maximum Weight
S/N 404-0001 through 404-0200
Landing 8100 lb., takeoff 8400 lb.

S/N 404-0201 and up
Landing 8100 lb., ramp 8450 lb., takeoff 8400 lb.

No. of Seats
One through eleven
(2 at +137.0, 2 at +171.0, 2 at +199.0, 2 at +227.0, 1 at +255.0 and 2 at +296.0)
See manufacturer's equipment list for other seating arrangements

Maximum Baggage
250 lb. (+32.0), 350 lb. (+71.0), 400 lb. (+211.0), 400 lb. (+301.0) and 100 lb. (+317.0)

Fuel Capacity
S/N 404-0001 through 404-0200
348.0 gallons (2 wing tanks, 174.0 gallons each, 170.0 gallons usable at +181.2)

S/N 404-0201 and up
348.0 gallons (2 wing tanks, 174.0 gallons each, 172.0 gallons usable at +181.2)

See NOTE 1 for data on unusable fuel

Oil Capacity
26 quarts (13 quarts in each engine at +129.0; usable 7.0 quarts per engine)
See NOTE 1 for data on undrainable oil.

Maximum Operating Altitude
30,000 feet

Control Surface Movements

Wing flaps			Down 35°, ±1° (Inboard)
			Down 23°, +0°, -1° (Outboard)
Main surfaces			
Aileron	Up	25°, +1°, -0°	Down 15°, +1°, -0°
Elevator	Up	24°, +1°, -0°	Down 15°, +1°, -0°
Rudder	Right	32°, +1°, -0°	Left 32°, +1°, -0°
(Read degrees normal to rudder hinge line)			
Tab (main surface in neutral)			
Aileron	Up	15°, +1°, -0°	Down 15°, +1°, -0°
Elevator	Up	4°, +0.5°, -0.0°	Down 14°, +0.0°, -0.5°
Rudder	Right	9°, +0.0°, -0.5°	Left 9°, +1°, -0°
(Read degrees normal to rudder hinge line)			

Serial Nos. Eligible
404-0001 through 404-0859

Figure 8-1. Type certificate data sheet. (continued)

II - Model 406, Caravan II, (Normal Category), Approved June 27, 1986

Engines — Two Pratt and Whitney Aircraft of Canada, Ltd., PT6A-112 turboprops

Fuel — Aviation turbine fuel ASTM D-1655, Jet A, Jet A-1, or Jet B; MIL-T-5624, JP-4, JP-5; MIL-T-83133, JP-8. Anti-icing additive per MIL-I-27686D, MIL-I-27686E, or Phillips PFA55MB must be blended into the aircraft fuel in concentrations not less than 0.060% or more than 0.15% by volume. For emergency use of aviation gasoline and fueling procedures, refer to approved Airplane Flight Manual.

Engine Limits

	Operating Limits				
	Shaft Horsepower	Ng Gas Generator Speed (% r.p.m.)	Indicated Torque (ft.-lbs.)	Prop Shaft Speed (r.p.m.)	Maximum Permissible Interturbine Temp. (°C)
Takeoff static and max. continous	500	101.6	1382	1900	725
Starting (2 seconds)	-	-	-	-	1090
Maximum reverse	480	101.6	1382	1815	725

Propeller and Propeller Limits — Two McCauley three-bladed, full-feathering, reversible
Hub: 3GFR34C701
Blade: 93KB-0
Diameter: Not over 93 inches, not under 90-5/8 inches; no further reduction permitted
Pitch at 30-inch station:
Low pitch 18.5°, feathered 85.5°, reverse -13.5°

Airspeed Limits (IAS)

V_{MO} (Maximum operating)	
Sea level to 21,500 ft.	229 knots
M_{MO} Above 21,500 ft.	.52 mach
V_A (Maneuvering)	162 knots
V_{FE} (Flaps extended)	
30° (Landing)	180 knots
20° (Approach)	200 knots
10° (Takeoff)	200 knots
V_{MCA} (Air minimum control speed)	90 knots
V_{LO} (Landing gear operating)	180 knots
V_{LE} (Landing gear extended)	180 knots

C.G. Range (Landing Gear Extended) — (+166.99 in.) to (+180.28 in.) at 6,500 lb. or less (11% to 32% MAC)
(+172.42 in.) to (+180.28 in.) at 9,360 lb. (19.6% to 32% MAC)
Variation is linear between points
Landing gear retracting moment (+1346 in.-lb.)

Empty Wt. C.G. Range — None

Leveling Means — Two screws located on W.L. 93.80 @ sta. 248.25 and sta. 272.65

Maximum Weight

Takeoff	9,360 lb.
Landing	9,360 lb.
Zero fuel (with zero wing locker payload)	8,500 lb.
Ramp	9,435 lb.

Number of Seats — One through fourteen
(2 at +137.0, 2 at +166.0, 2 at +192.0, 2 at +218.0, 2 at +244.0, 2 at +270.0, and 2 at +296.0)
See manufacturer's equipment list for other seating arrangements

Figure 8-1. Type certificate data sheet. (continued)

Maximum Baggage	250 lb. (+32.0), 350 lb. (+71.0), 400 lb (+211.0), 400 lb. (+301.0), and 100 lb. (+317.0)
Fuel Capacity	3227 lb. (481.5 gal.) total in two wing tanks, 1613.5 lb. (240.75 gal.) each 3183 lb. (475 gal.) usable total, 1591.5 lb. (237.5 gal.) in each tank at sta. +181.9 Fuel weight based on 6.70 lb./gal. See NOTE 1 for data on unusable fuel
Oil Capacity	5.28 gal. total, 3.00 gal. usable (2.3 gal. in each engine-mounted tank at +142.1) See NOTE 1 for data on undrainable oil
Maximum Operating Altitude	30,000 ft.

Control Surface Movements

Elevator (horn faired)	Up	14°, +1°, -0°	Down	17°, +1°, -0°
Elevator trim tabs	Up	8°, +1°, -0°	Down	10°, +2°, -0°
Rudder (perpendicular to hinge 0° faired with fin)	Right	32°, +1°, -0°	Left	32°, +1°, -0°
Rudder trim tab (perpendicular to hinge)	Right	11°, +1°, -0°	Left	16°, +1°, -0°
Aileron	Up	25°, +1°, -0°	Down	14°, +1°, -0°
Aileron trim tab	Up	19°, +1°, -0°	Down	19°, +1°, -0°
Wing flap (inboard)			Down	30°, +1°, -0°
Wing flap (outboard)			Down	20°, +1°, -0°

Serial Nos. Eligible 406-0001 and on

Data Pertinent to All Models

Datum 100.0 inches forward of forward face of fuselage bulkhead forward of rudder pedals.

Certification Basis

Model 404:
Part 23 of the Federal Aviation Regulations effective February 1, 1965, as amended by 23-1 through 23-13 except Subpart B as amended through 23-14; and FAR 23.1385(c) as amended through 23-21; and 23.1327 as amended through 23-23. Findings of equivalent level of safety were made for FAR 23.1189(a), 23.1545, and 23.1583(a). In addition, effective S/N 404-0601, FAR 36 dated December 1, 1969, as amended by 36-1 through 36-4.

In addition to the above certification basis, compliance with ice protection has been demonstrated in accordance with FAR 23.1419 of Amendment 23-14 effective December 20, 1973, when ice protection equipment is installed in accordance with Cessna Drawing 5114400, Factory Kit (FK) No. 194, and Pilot's Operating Handbook and FAA Approved Airplane Flight Manual. Aircraft which have been modified in compliance with Accessory Kit (AK) No. 421-106 are considered to be equivalent to those with Factory Kit (FK) No. 194.

Model 406:
Part 23 of the Federal Aviation Regulations effective February 1, 1965, as amended by 23-1 through 23-13 except Subpart B as amended through 23-14; and 23.427, 23.929, 23.979, 23.1017, 23.1027, 23.1163, 23.1182, 23.1189, 23.1192 as amended through 23-14; 23.951, 23.997, 23.1013, 23.1015, 23.1019(a)(1), 23.1019(a)(2), 23.1019(a)(4), 23.1019(a)(5), 23.1019(b), 23.1183 as amended through 23-15; 23.933, 23.971, 23.977, 23.999, 23.1111, 23.1125, 23.1143, 23.1165, 23.1303 (a through d), 23.1385(c), 23.1549 as amended through 23-17; 23.901, 23.939, 23.943, 23.959, 23.967, 23.973, 23.975, 23.995, 23.1093, 23.1121, 23.1141, 23.1145, 23.1193, 23.1203, 23.1305 (a through u and w), 23.1337 as amended through 23-18; 23.1323, 23.1325, 23.1327, 23.1351, 23.1357, 23.1547 as amended through 23-20; 23.45, 23.49, 23.65, 23.67, 23.77, 23.161, 23.1043, 23.1353, 23.1521 as amended through 23-21; 23.1545 as amended through 23-23; 23.903, 23.1529 as amended through 23-26; SFAR 27 as amended by 27-1 through 27-4; Part 36 as amended by 36-1 through 36-12; SFAR 41C; and Exemption No. 4661 from exact compliance with the requirements of Section 23.207(c). Findings of equivalent level of safety were made for FAR 23.1189(a), and that design of

Figure 8-1. Type certificate data sheet. (continued)

the elevator tab control system provides the level of safety intended by the requirements of FAR 21.21(b)(2) by preventing an unsafe condition. Therefore, FAR 23.629(f), as amended by Amendment 23-23, is applicable to the elevator tab control system, in addition to other requirements in the cited certification basis.

In addition to the above certification basis, compliance with ice protection has been demonstrated in accordance with FAR 23.773 and 23.1419 of Amendment 23-14, FAR 23.1309 as amended through Amendment 23-17, and FAR 23.1416 of Amendment 23-23 when ice protection equipment is installed in accordance with Cessna Drawing 6015006, Factory Kit (FK) No. 194, and Pilot's Operating Handbook and FAA Approved Airplane Flight Manual. Aircraft which have been modified in accordance with Accessory Kit (AK) No. 421-106 are considered to be equivalent to those with Factory Kit (FK) No. 194.
Application for type certificate dated October 9, 1973. Type Certificate No. A25CE issued July 21, 1976, obtained by the manufacturer under delegation option procedures.

Production Basis — Production Certificate No. 312 issued and Delegation Option Manufacturer No. CE-3 authorized to issue airworthiness certificates under delegation option provisions of Part 21 of the Federal Aviation Regulations. Effective February 15, 1985, and on, Production Certificate No. 4 is applicable to all spares production for the Model 404 and to all production on the Model 406.

Equipment — The basic required equipment as prescribed in the applicable airworthiness regulations (see Certification Basis) must be installed in the aircraft for certification. In addition, the following item of equipment is required:

Stall Warning Indicator - Cessna Dwg. 5818008 (404), 5718030 (406) or
Angle-of-Attack Indicator System - Cessna Dwg. 0800302 (404)

NOTE 1. Current weight and balance report together with list of equipment included in certificated empty weight and loading instructions when necessary must be provided for each aircraft at the time of original certification.

The certified empty weight and corresponding center of gravity location must include undrainable oil (not included in Oil Capacity) and unusable fuel as follows:

(a) S/N 404-0001 through 404-0200
Fuel 48 lb. at (+177.6)

S/N 404-0201 and up
Fuel 28 lb. at (+177.6)

S/N 406-0001 and up
Fuel 44 lb. (6.5 gal.) at (+186.7)

(b) Oil 0.0 lb.

NOTE 2. The placards specified in the FAA Approved Airplane Flight Manual must be displayed.

NOTE 3. (404-0001 and up)
1%, by volume, isopropyl alcohol approved for use as fuel anti-icing additive when used as outlined in Cessna Service Letter ME73-25 dated November 2, 1973, or subsequent revisions.

NOTE 4. (S/N 404-0201 and up), (406-0001 and up)
An optional cargo configuration is available which excludes the passenger air distribution and seating. Such airplanes may be operated with passenger seats installed provided the operating rules for supplemental oxygen are complied with.

Figure 8-1. Type certificate data sheet. (continued)

NOTE 5. Aircraft operators must observe limitations and placards shown in the applicable Pilot's Operating Manual and FAA Approved Airplane Flight Manual, or later approved revisions as listed below:

Cessna P/N D1540-3-13:	Model 404 Serial 404-0001 through 404-0136 (1977 Model)
Cessna P/N D1563-1-13:	Model 404 Serial 404-0201 through 404-0246 (1978 Model)
Cessna P/N D1572-2-13PH:	Model 404 Serial 404-0401 through 404-0460 (1979 Model)
Cessna P/N D1583-3-13PH:	Model 404 Serial 404-0601 through 404-0695 (1980 Model)
Cessna P/N D1593-1-13PH:	Model 404 Serial 404-0801 through 404-0859 (1981 Model)
Cessna P/N D1624-13PH:	Model 406 Serial 406-0001 and on

NOTE 6. The Model 406 type design has been duplicated as Model F406 in Type Certificate No. A54EU. The type design file is common between Models 406 and F406 and duplicates of the type design file are maintained by the respective type certificate holders.

.....END.....

Figure 8-1. Type certificate data sheet. (continued)

and not contaminated. Additionally, it is the pilot's responsibility to review the airworthiness certificate, maintenance records, and other required paperwork to verify that the aircraft is indeed airworthy. After each flight, it is recommended that the pilot or mechanic conduct a postflight inspection to detect any problems that might require repair or servicing before the next flight.

Annual/100-Hour Inspections

Title 14 of the Code of Federal Regulations (14 CFR) part 91 discusses the basic requirements for annual and 100-hour inspections. With some exceptions, all aircraft must have a complete inspection annually. Aircraft that are used for commercial purposes and are likely to be used more frequently than noncommercial aircraft must have this complete inspection every 100 hours. The scope and detail of items to be included in annual and 100-hour inspections is included as appendix D of 14 CFR part 43 and shown as Figure 8-2. A properly written checklist, such as the one shown earlier in this chapter, will include all the items of appendix D. Although the scope and detail of annual and 100-hour inspections is identical, there are two significant differences. One difference involves persons authorized to conduct them. A certified airframe and powerplant maintenance technician can conduct a 100-hour inspection, whereas an annual inspection must be conducted by a certified airframe and powerplant maintenance technician with inspection authorization (IA). The other difference involves authorized overflight of the maximum 100 hours before inspection. An aircraft may be flown up to 10 hours beyond the 100-hour limit if necessary to fly to a destination where the inspection is to be conducted.

Progressive Inspections

Because the scope and detail of an annual inspection is very extensive and could keep an aircraft out of service for a considerable length of time, alternative

Appendix D to Part 43—Scope and Detail of Items (as Applicable to the Particular Aircraft) To Be Included in Annual and 100-Hour Inspections

(a) Each person performing an annual or 100-hour inspection shall, before that inspection, remove or open all necessary inspection plates, access doors, fairing, and cowling. He shall thoroughly clean the aircraft and aircraft engine.

(b) Each person performing an annual or 100-hour inspection shall inspect (where applicable) the following components of the fuselage and hull group:

(1) Fabric and skin—for deterioration, distortion, other evidence of failure, and defective or insecure attachment of fittings.

(2) Systems and components—for improper installation, apparent defects, and unsatisfactory operation.

(3) Envelope, gas bags, ballast tanks, and related parts—for poor condition.

(c) Each person performing an annual or 100-hour inspection shall inspect (where applicable) the following components of the cabin and cockpit group:

(1) Generally—for uncleanliness and loose equipment that might foul the controls.

(2) Seats and safety belts—for poor condition and apparent defects.

(3) Windows and windshields—for deterioration and breakage.

(4) Instruments—for poor condition, mounting, marking, and (where practicable) improper operation.

(5) Flight and engine controls—for improper installation and improper operation.

(6) Batteries—for improper installation and improper charge.

(7) All systems—for improper installation, poor general condition, apparent and obvious defects, and insecurity of attachment.

(d) Each person performing an annual or 100-hour inspection shall inspect (where applicable) components of the engine and nacelle group as follows:

(1) Engine section—for visual evidence of excessive oil, fuel, or hydraulic leaks, and sources of such leaks.

(2) Studs and nuts—for improper torquing and obvious defects.

(3) Internal engine—for cylinder compression and for metal particles or foreign matter on screens and sump drain plugs. If there is

Figure 8-2. Scope and detail of annual and 100-hour inspections.

Appendix D to Part 43—Scope and Detail of Items (as Applicable to the Particular Aircraft) To Be Included in Annual and 100-Hour Inspections (continued)

weak cylinder compression, for improper internal condition and improper internal tolerances.

(4) Engine mount—for cracks, looseness of mounting, and looseness of engine to mount.

(5) Flexible vibration dampeners—for poor condition and deterioration.

(6) Engine controls—for defects, improper travel, and improper safetying.

(7) Lines, hoses, and clamps—for leaks, improper condition, and looseness.

(8) Exhaust stacks—for cracks, defects, and improper attachment.

(9) Accessories—for apparent defects in security of mounting.

(10) All systems—for improper installation, poor general condition, defects, and insecure attachment.

(11) Cowling—for cracks and defects.

(e) Each person performing an annual or 100-hour inspection shall inspect (where applicable) the following components of the landing gear group:

(1) All units—for poor condition and insecurity of attachment.

(2) Shock absorbing devices—for improper oleo fluid level.

(3) Linkages, trusses, and members—for undue or excessive wear fatigue, and distortion.

(4) Retracting and locking mechanism—for improper operation.

(5) Hydraulic lines—for leakage.

(6) Electrical system—for chafing and improper operation of switches.

(7) Wheels—for cracks, defects, and condition of bearings.

(8) Tires—for wear and cuts.

(9) Brakes—for improper adjustment.

(10) Floats and skis—for insecure attachment and obvious or apparent defects.

(f) Each person performing an annual or 100-hour inspection shall inspect (where applicable) all components of the wing and center section assembly for poor general condition, fabric or skin deterioration, distortion, evidence of failure, and insecurity of attachment.

(g) Each person performing an annual or 100-hour inspection shall inspect (where applicable) all components and systems that make up the complete empennage assembly for poor general condition, fabric or skin deterioration, distortion, evidence of failure, insecure attachment, improper component installation, and improper component operation.

(h) Each person performing an annual or 100-hour inspection shall inspect (where applicable) the following components of the propeller group:

(1) Propeller assembly—for cracks, nicks, binds, and oil leakage.

(2) Bolts—for improper torquing and lack of safetying.

(3) Anti-icing devices—for improper operations and obvious defects.

(4) Control mechanisms—for improper operation, insecure mounting, and restricted travel.

(i) Each person performing an annual or 100-hour inspection shall inspect (where applicable) the following components of the radio group:

(1) Radio and electronic equipment—for improper installation and insecure mounting.

(2) Wiring and conduits—for improper routing, insecure mounting, and obvious defects.

(3) Bonding and shielding—for improper installation and poor condition.

(4) Antenna including trailing antenna—for poor condition, insecure mounting, and improper operation.

(j) Each person performing an annual or 100-hour inspection shall inspect (where applicable) each installed miscellaneous item that is not otherwise covered by this listing for improper installation and improper operation.

Figure 8-2. Scope and detail of annual and 100-hour inspections. (continued)

inspection programs designed to minimize down time may be utilized. A progressive inspection program allows an aircraft to be inspected progressively. The scope and detail of an annual inspection is essentially divided into segments or phases (typically four to six). Completion of all the phases completes a cycle that satisfies the requirements of an annual inspection. The advantage of such a program is that any required segment may be completed overnight and thus enable the aircraft to fly daily without missing any revenue earning potential. Progressive inspection programs include routine items such as engine oil changes and detailed items such as flight control cable inspection. Routine items are accomplished each time the aircraft comes in for a phase inspection and detailed items focus on detailed inspection of specific areas. Detailed inspections are typically done once each cycle. A cycle must be completed within 12 months. If all required phases are not completed within 12 months, the remaining phase inspections must be conducted before the end of the 12th month from when the first phase was completed.

Each registered owner or operator of an aircraft desiring to use a progressive inspection program must submit a written request to the FAA Flight Standards District Office (FSDO) having jurisdiction over the area in which the applicant is located. Title 14 of the Code of Federal Regulations (14 CFR) part 91, §91.409(d)

§91.409 Inspections.

(d) **Progressive inspection.** Each registered owner or operator of an aircraft desiring to use a progressive inspection program must submit a written request to the FAA Flight Standards district office having jurisdiction over the area in which the applicant is located, and shall provide—

- (1) A certificated mechanic holding an inspection authorization, a certificated airframe repair station, or the manufacturer of the aircraft to supervise or conduct the progressive inspection;
- (2) A current inspection procedures manual available and readily understandable to pilot and maintenance personnel containing, in detail—
  - (i) An explanation of the progressive inspection, including the continuity of inspection responsibility, the making of reports, and the keeping of records and technical reference material;
  - (ii) An inspection schedule, specifying the intervals in hours or days when routine and detailed inspections will be performed and including instructions for exceeding an inspection interval by not more than 10 hours while en route and for changing an inspection interval because of service experience;
  - (iii) Sample routine and detailed inspection forms and instructions for their use; and
  - (iv) Sample reports and records and instructions for their use;
- (3) Enough housing and equipment for necessary disassembly and proper inspection of the aircraft; and
- (4) Appropriate current technical information for the aircraft.

The frequency and detail of the progressive inspection shall provide for the complete inspection of the aircraft within each 12 calendar months and be consistent with the manufacturer's recommendations, field service experience, and the kind of operation in which the aircraft is engaged. The progressive inspection schedule must ensure that the aircraft, at all times, will be airworthy and will conform to all applicable FAA aircraft specifications, type certificate data sheets, airworthiness directives, and other approved data. If the progressive inspection is discontinued, the owner or operator shall immediately notify the local FAA Flight Standards district office, in writing, of the discontinuance. After the discontinuance, the first annual inspection under §91.409(a)(1) is due within 12 calendar months after the last complete inspection of the aircraft under the progressive inspection. The 100-hour inspection under §91.409(b) is due within 100 hours after that complete inspection. A complete inspection of the aircraft, for the purpose of determining when the annual and 100-hour inspections are due, requires a detailed inspection of the aircraft and all its components in accordance with the progressive inspection. A routine inspection of the aircraft and a detailed inspection of several components is not considered to be a complete inspection.

Figure 8-3. 14 CFR §91.409(d) Progressive inspection.

establishes procedures to be followed for progressive inspections and is shown in Figure 8-3.

Continuous Inspections

Continuous inspection programs are similar to progressive inspection programs, except that they apply to large or turbine-powered aircraft and are therefore more complicated.

Like progressive inspection programs, they require approval by the FAA Administrator. The approval may be sought based upon the type of operation and the CFR parts under which the aircraft will be operated. The maintenance program for commercially operated aircraft must be detailed in the approved operations specifications (OpSpecs) of the commercial certificate holder.

Airlines utilize a continuous maintenance program that includes both routine and detailed inspections. However, the detailed inspections may include different levels of detail. Often referred to as "checks," the A-check, B-check, C-check, and D-checks involve increasing levels of detail. A-checks are the least comprehensive and occur frequently. D-checks, on the other hand, are extremely comprehensive, involving major disassembly, removal, overhaul, and inspection of systems and components. They might occur only three to six times during the service life of an aircraft.

Altimeter and Transponder Inspections

Aircraft that are operated in controlled airspace under instrument flight rules (IFR) must have each altimeter and static system tested in accordance with procedures described in 14 CFR part 43, appendix E, within the preceding 24 calendar months. Aircraft having an air traffic control (ATC) transponder must also have each transponder checked within the preceding 24 months. All these checks must be conducted by appropriately certified individuals.

ATA iSpec 2200

In an effort to standardize the format for the way in which maintenance information is presented in aircraft maintenance manuals, the Air Transport Association of America (ATA) issued specifications for Manufacturers Technical Data. The original specification was called ATA Spec 100. Over the years, Spec 100 has been continuously revised and updated. Eventually, ATA Spec 2100 was developed for electronic documentation. These two specifications evolved into one document called ATA iSpec 2200. As a result of this standardization, maintenance technicians can always find information regarding a particular system in the same section of an aircraft maintenance manual, regardless of manufacturer. For example, if you are seeking information about the electrical system on any aircraft, you will always find that information in section (chapter) 24.

The ATA Specification 100 has the aircraft divided into systems, such as air conditioning, which covers the basic air conditioning system (ATA 21). Numbering in each major system provides an arrangement for breaking the system down into several subsystems. Late model aircraft, both over and under the 12,500 pound designation, have their parts manuals and maintenance manuals arranged according to the ATA coded system.

The following abbreviated table of ATA System, Subsystem, and Titles is included for familiarization purposes.

ATA Specification 100 Systems

Sys.	*Sub.*	*Title*
21		AIR CONDITIONING
21	00	General
21	10	Compression
21	20	Distribution
21	30	Pressurization Control
21	40	Heating
21	50	Cooling
21	60	Temperature Control
21	70	Moisture/Air Contaminate Control

The remainder of this list shows the systems and title with subsystems deleted in the interest of brevity. Consult specific aircraft maintenance manuals for a complete description of the subsystems used in them.

22	AUTO FLIGHT
23	COMMUNICATIONS
24	ELECTRICAL POWER
25	EQUIPMENT/FURNISHINGS
26	FIRE PROTECTION
27	FLIGHT CONTROLS
28	FUEL
29	HYDRAULIC POWER
30	ICE AND RAIN PROTECTION
31	INDICATING/RECORDING SYSTEMS
32	LANDING GEAR
33	LIGHTS
34	NAVIGATION

35 OXYGEN

36 PNEUMATIC

37 VACUUM/PRESSURE

38 WATER/WASTE

39 ELECTRICAL/ELECTRONIC PANELS AND MULTIPURPOSE COMPONENTS

49 AIRBORNE AUXILIARY POWER

51 STRUCTURES

52 DOORS

53 FUSELAGE

54 NACELLES/PYLONS

55 STABILIZERS

56 WINDOWS

57 WINGS

61 PROPELLERS

65 ROTORS

71 POWERPLANT

72 (T) TURBINE/TURBOPROP

72 (R) ENGINE RECIPROCATING

73 ENGINE FUEL AND CONTROL

74 IGNITION

75 BLEED AIR

76 ENGINE CONTROLS

77 ENGINE INDICATING

78 ENGINE EXHAUST

79 ENGINE OIL

80 STARTING

81 TURBINES (RECIPROCATING ENG)

82 WATER INJECTION

83 REMOTE GEAR BOXES (ENG DR)

Keep in mind that not all aircraft will have all these systems installed. Small and simple aircraft have far fewer systems than larger more complex aircraft.

Special Inspections

During the service life of an aircraft, occasions may arise when something out of the ordinary care and use of an aircraft might happen that could possibly affect its airworthiness. When these situations are encountered, special inspection procedures should be followed to determine if damage to the aircraft structure has occurred. The procedures outlined on the following pages are general in nature and are intended to acquaint the aviation mechanic with the areas which should be inspected. As such, they are not all inclusive. When performing any of these special inspections, always follow the detailed procedures in the aircraft maintenance manual. In situations where the manual does not adequately address the situation, seek advice from other maintenance technicians who are highly experienced with them.

Hard or Overweight Landing Inspection

The structural stress induced by a landing depends not only upon the gross weight at the time but also upon the severity of impact. However, because of the difficulty in estimating vertical velocity at the time of contact, it is hard to judge whether or not a landing has been sufficiently severe to cause structural damage. For this reason, a special inspection should be performed after a landing is made at a weight known to exceed the design landing weight or after a rough landing, even though the latter may have occurred when the aircraft did not exceed the design landing weight.

Wrinkled wing skin is the most easily detected sign of an excessive load having been imposed during a landing. Another indication which can be detected easily is fuel leakage along riveted seams. Other possible locations of damage are spar webs, bulkheads, nacelle skin and attachments, firewall skin, and wing and fuselage stringers. If none of these areas show adverse effects, it is reasonable to assume that no serious damage has occurred. If damage is detected, a more extensive inspection and alignment check may be necessary.

Severe Turbulence Inspection/Over "G"

When an aircraft encounters a gust condition, the airload on the wings exceeds the normal wingload supporting the aircraft weight. The gust tends to accelerate the aircraft while its inertia acts to resist this change. If the combination of gust velocity and airspeed is too severe, the induced stress can cause structural damage.

A special inspection should be performed after a flight through severe turbulence. Emphasis should be placed upon inspecting the upper and lower wing surfaces for excessive buckles or wrinkles with permanent set. Where wrinkles have occurred, remove a few rivets and examine the rivet shanks to determine if the rivets have sheared or were highly loaded in shear.

Through the inspection doors and other accessible openings, inspect all spar webs from the fuselage to the tip. Check for buckling, wrinkles, and sheared attachments. Inspect for buckling in the area around the nacelles and in the nacelle skin, particularly at the wing leading edge.

Check for fuel leaks. Any sizeable fuel leak is an indication that an area may have received overloads which have broken the sealant and opened the seams.

If the landing gear was lowered during a period of severe turbulence, inspect the surrounding surfaces carefully for loose rivets, cracks, or buckling. The interior of the wheel well may give further indications of excessive gust conditions. Inspect the top and bottom fuselage skin. An excessive bending moment may have left wrinkles of a diagonal nature in these areas.

Inspect the surface of the empennage for wrinkles, buckling, or sheared attachments. Also, inspect the area of attachment of the empennage to the fuselage. The above inspections cover the critical areas. If excessive damage is noted in any of the areas mentioned, the inspection should be continued until all damage is detected.

Lightning Strike

Although lightning strikes to aircraft are extremely rare, if a strike has occurred, the aircraft must be carefully inspected to determine the extent of any damage that might have occurred. When lightning strikes an aircraft, the electrical current must be conducted through the structure and be allowed to discharge or dissipate at controlled locations. These controlled locations are primarily the aircraft's static discharge wicks, or on more sophisticated aircraft, null field dischargers. When surges of high voltage electricity pass through good electrical conductors, such as aluminum or steel, damage is likely to be minimal or nonexistent. When surges of high voltage electricity pass through non-metallic structures, such as a fiberglass radome, engine cowl or fairing, glass or plastic window, or a composite structure that does not have built-in electrical bonding, burning and more serious damage to the structure could occur. Visual inspection of the structure is required. Look for evidence of degradation, burning or erosion of the composite resin at all affected structures, electrical bonding straps, static discharge wicks and null field dischargers.

Fire Damage

Inspection of aircraft structures that have been subjected to fire or intense heat can be relatively simple if visible damage is present. Visible damage requires repair or replacement. If there is no visible damage, the structural integrity of an aircraft may still have been compromised. Since most structural metallic components of an aircraft have undergone some sort of heat treatment process during manufacture, an exposure to high heat not encountered during normal operations could severely degrade the design strength of the structure. The strength and airworthiness of an aluminum structure that passes a visual inspection but is still suspect can be further determined by use of a conductivity tester. This is a device that uses eddy current and is discussed later in this chapter. Since strength of metals is related to hardness, possible damage to steel structures might be determined by use of a hardness tester such as a Rockwell C hardness tester.

Flood Damage

Like aircraft damaged by fire, aircraft damaged by water can range from minor to severe, depending on the level of the flood water, whether it was fresh or salt water and the elapsed time between the flood occurrence and when repairs were initiated. Any parts that were totally submerged should be completely disassembled, thoroughly cleaned, dried and treated with a corrosion inhibitor. Many parts might have to be replaced, particularly interior carpeting, seats, side panels, and instruments. Since water serves as an electrolyte that promotes corrosion, all traces of water and salt must be removed before the aircraft can again be considered airworthy.

Seaplanes

Because they operate in an environment that accelerates corrosion, seaplanes must be carefully inspected for corrosion and conditions that promote corrosion. Inspect bilge areas for waste hydraulic fluids, water, dirt, drill chips, and other debris. Additionally, since seaplanes often encounter excessive stress from the pounding of rough water at high speeds, inspect for loose rivets and other fasteners; stretched, bent or cracked skins; damage to the float attach fitting; and general wear and tear on the entire structure.

Aerial Application Aircraft

Two primary factors that make inspecting these aircraft different from other aircraft are the corrosive nature of some of the chemicals used and the typical flight profile. Damaging effects of corrosion may be detected in a much shorter period of time than normal use aircraft. Chemicals may soften the fabric or loosen the fabric tapes of fabric covered aircraft. Metal aircraft may need to have the paint stripped, cleaned, and repainted and corrosion treated annually. Leading edges of wings and other areas may require protective coatings or tapes. Hardware may require more frequent replacement.

During peak use, these aircraft may fly up to 50 cycles (takeoffs and landings) or more in a day, most likely from an unimproved or grass runway. This can greatly accelerate the failure of normal fatigue items. Landing gear and related items require frequent inspections. Because these aircraft operate almost continuously at very low altitudes, air filters tend to become obstructed more rapidly.

Special Flight Permits

For an aircraft that does not currently meet airworthiness requirements because of an overdue inspection, damage, expired replacement times for time-limited parts or other reasons, but is capable of safe flight, a special flight permit may be issued. Special flight permits, often referred to as ferry permits, are issued for the following purposes:

- Flying the aircraft to a base where repairs, alterations, or maintenance are to be performed, or to a point of storage.
- Delivering or exporting the aircraft.
- Production flight testing new production aircraft.
- Evacuating aircraft from areas of impending danger.
- Conducting customer demonstration flights in new production aircraft that have satisfactorily completed production flight tests.

Additional information about special flight permits may be found in 14 CFR part 21. Application forms for special flight permits may be requested from the nearest FAA Flight Standards District Office (FSDO).

Nondestructive Inspection/Testing

The preceding information in this chapter provided general information regarding aircraft inspection. The remainder of this chapter deals with several methods often used on specific components or areas on an aircraft when carrying out the more specific inspections. They are referred to as nondestructive inspection (NDI) or nondestructive testing (NDT). The objective of NDI and NDT is to determine the airworthiness of a component without damaging it, which would render it unairworthy. Some of these methods are simple, requiring little additional expertise, while others are highly sophisticated and require that the technician be highly trained and specially certified.

Additional information on NDI may be found by referring to chapter 5 of FAA Advisory Circular (AC) 43.13-1B, Acceptable Methods, Techniques, and Practices—Aircraft Inspection and Repair. Information regarding training, qualifications, and certification of NDI personnel may be found in FAA Advisory Circular (AC) 65-31A, Training, Qualification and Certification of Non-destructive Inspection (NDI) Personnel.

General Techniques

Before conducting NDI, it is necessary to follow preparatory steps in accordance with procedures specific to that type of inspection. Generally, the parts or areas must be thoroughly cleaned. Some parts must be removed from the aircraft or engine. Others might need to have any paint or protective coating stripped. A complete knowledge of the equipment and procedures is essential and if required, calibration and inspection of the equipment must be current.

Visual Inspection

Visual inspection can be enhanced by looking at the suspect area with a bright light, a magnifying glass, and a mirror (when required). Some defects might be so obvious that further inspection methods are not required. The lack of visible defects does not necessarily mean further inspection is unnecessary. Some defects may lie beneath the surface or may be so small that the human eye, even with the assistance of a magnifying glass, cannot detect them.

Borescope

Inspection by use of a borescope is essentially a visual inspection. A borescope is a device that enables the inspector to see inside areas that could not otherwise be inspected without disassembly. An example of an area that can be inspected with a borescope is the inside of a reciprocating engine cylinder. The borescope can be inserted into an open spark plug hole to detect damaged pistons, cylinder walls, or valves. Another example would be the hot section of a turbine engine to which access could be gained through the hole of a removed igniter or removed access plugs specifically installed for inspection purposes.

Borescopes are available in two basic configurations. The simpler of the two is a rigid type of small diameter telescope with a tiny mirror at the end that enables the user to see around corners. The other type uses fiber optics that enables greater flexibility. Many borescopes provide images that can be displayed on a computer or video monitor for better interpretation of what is being viewed and to record images for future reference. Most borescopes also include a light to illuminate the area being viewed.

Liquid Penetrant Inspection

Penetrant inspection is a nondestructive test for defects open to the surface in parts made of any nonporous material. It is used with equal success on such metals as aluminum, magnesium, brass, copper, cast iron, stainless steel, and titanium. It may also be used on ceramics, plastics, molded rubber, and glass.

Penetrant inspection will detect such defects as surface cracks or porosity. These defects may be caused by fatigue cracks, shrinkage cracks, shrinkage porosity, cold shuts, grinding and heat treat cracks, seams, forging laps, and bursts. Penetrant inspection will also indicate a lack of bond between joined metals.

The main disadvantage of penetrant inspection is that the defect must be open to the surface in order to let the penetrant get into the defect. For this reason, if the part in question is made of material which is magnetic, the use of magnetic particle inspection is generally recommended.

Penetrant inspection uses a penetrating liquid that enters a surface opening and remains there, making it clearly visible to the inspector. It calls for visual examination of the part after it has been processed, increasing the visibility of the defect so that it can be detected. Visibility of the penetrating material is increased by the addition of one of two types of dye, visible or fluorescent.

The visible penetrant kit consists of dye penetrant, dye remover emulsifier, and developer. The fluorescent penetrant inspection kit contains a black light assembly, as well as spray cans of penetrant, cleaner, and developer. The light assembly consists of a power transformer, a flexible power cable, and a hand-held lamp. Due to its size, the lamp may be used in almost any position or location.

Briefly, the steps for performing a penetrant inspection are:

1. Thorough cleaning of the metal surface.
2. Applying penetrant.
3. Removing penetrant with remover emulsifier or cleaner.
4. Drying the part.
5. Applying the developer.
6. Inspecting and interpreting results.

Interpretation of Results

The success and reliability of a penetrant inspection depends upon the thoroughness with which the part was prepared. Several basic principles applying to penetrant inspection are:

1. The penetrant must enter the defect in order to form an indication. It is important to allow sufficient time so the penetrant can fill the defect. The defect must be clean and free of contaminating materials so that the penetrant is free to enter.
2. If all penetrant is washed out of a defect, an indication cannot be formed. During the washing or rinsing operation, prior to development, it is possible that the penetrant will be removed from within the defect, as well as from the surface.
3. Clean cracks are usually easy to detect. Surface openings that are uncontaminated, regardless of how fine, are seldom difficult to detect with the penetrant inspection.
4. The smaller the defect, the longer the penetrating time. Fine crack-like apertures require a longer penetrating time than defects such as pores.
5. When the part to be inspected is made of a material susceptible to magnetism, it should be inspected by a magnetic particle inspection method if the equipment is available.
6. Visible penetrant-type developer, when applied to the surface of a part, will dry to a smooth, even, white coating. As the developer dries, bright red indications will appear where there are surface defects. If no red indications appear, there are no surface defects.
7. When conducting the fluorescent penetrant-type inspection, the defects will show up (under black light) as a brilliant yellow-green color and the sound areas will appear deep blue-violet.
8. It is possible to examine an indication of a defect and to determine its cause as well as its extent. Such an appraisal can be made if something is known about the manufacturing processes to which the part has been subjected.

The size of the indication, or accumulation of penetrant, will show the extent of the defect and the brilliance will be a measure of its depth. Deep cracks will hold more penetrant and will be broader and more brilliant. Very fine openings can hold only small amounts of penetrants and will appear as fine lines. Figure 8-4 shows some of the types of defects that can be located using dry penetrant.

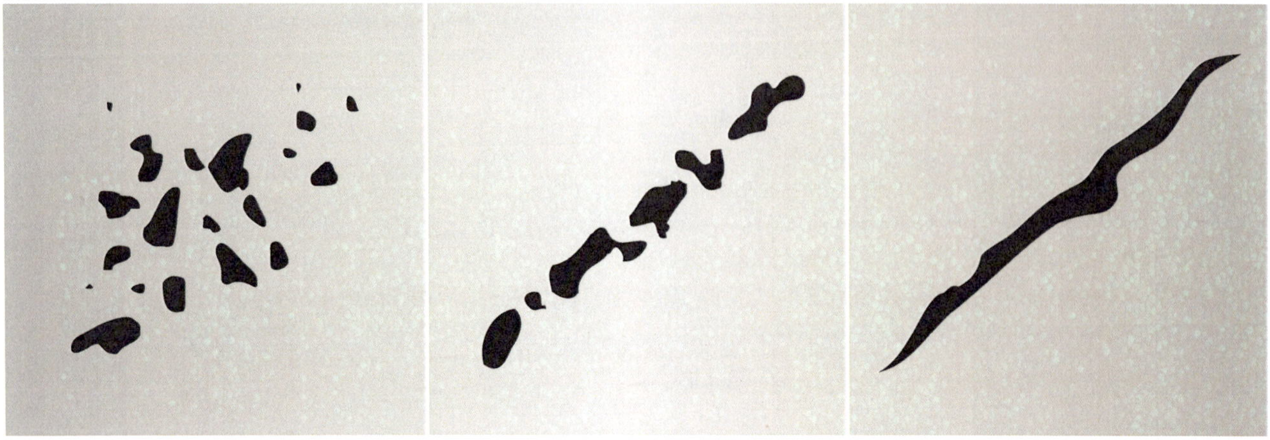

Figure 8-4. Types of defects.

False Indications

With the penetrant inspection, there are no false indications in the sense that they occur in the magnetic particle inspection. There are, however, two conditions which may create accumulations of penetrant that are sometimes confused with true surface cracks and discontinuities.

The first condition involves indications caused by poor washing. If all the surface penetrant is not removed in the washing or rinsing operation following the penetrant dwell time, the unremoved penetrant will be visible. Evidences of incomplete washing are usually easy to identify since the penetrant is in broad areas rather than in the sharp patterns found with true indications. When accumulations of unwashed penetrant are found on a part, the part should be completely reprocessed. Degreasing is recommended for removal of all traces of the penetrant.

False indications may also be created where parts press fit to each other. If a wheel is press fit onto a shaft, penetrant will show an indication at the fit line. This is perfectly normal since the two parts are not meant to be welded together. Indications of this type are easy to identify since they are regular in form and shape.

Eddy Current Inspection

Electromagnetic analysis is a term which describes the broad spectrum of electronic test methods involving the intersection of magnetic fields and circulatory currents. The most widely used technique is the eddy current.

Eddy currents are composed of free electrons under the influence of an induced electromagnetic field which are made to "drift" through metal.

Eddy current is used in aircraft maintenance to inspect jet engine turbine shafts and vanes, wing skins, wheels, bolt holes, and spark plug bores for cracks, heat or frame damage. Eddy current may also be used in repair of aluminum aircraft damaged by fire or excessive heat. Different meter readings will be seen when the same metal is in different hardness states. Readings in the affected area are compared with identical materials in known unaffected areas for comparison. A difference in readings indicates a difference in the hardness state of the affected area. In aircraft manufacturing plants, eddy current is used to inspect castings, stampings, machine parts, forgings, and extrusions. Figure 8-5 shows a technician performing an eddy current inspection on an aluminum wheel half.

Basic Principles

When an alternating current is passed through a coil, it develops a magnetic field around the coil, which in turn induces a voltage of opposite polarity in the coil and opposes the flow of original current. If this coil is placed in such a way that the magnetic field passes

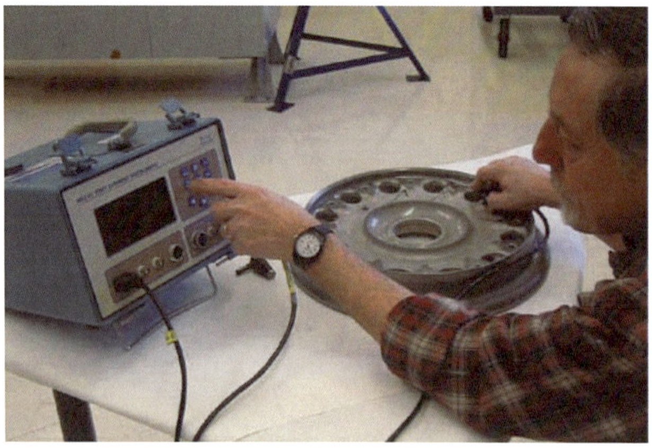
Figure 8-5. Eddy current inspection of wheel half.

through an electrically conducting specimen, eddy currents will be induced into the specimen. The eddy currents create their own field which varies the original field's opposition to the flow of original current. The specimen's susceptibility to eddy currents determines the current flow through the coil. [Figure 8-6]

The magnitude and phase of this counter field is dependent primarily upon the resistance and permeability of the specimen under consideration, and which enables us to make a qualitative determination of various physical properties of the test material. The interaction of the eddy current field with the original field results is a power change that can be measured by utilizing electronic circuitry similar to a Wheatstone bridge.

The specimen is either placed in or passed through the field of an electromagnetic induction coil, and its effect on the impedance of the coil or on the voltage output of one or more test coils is observed. The process, which involves electric fields made to explore a test piece for various conditions, involves the transmission of energy through the specimen much like the transmission of x-rays, heat, or ultrasound.

Eddy current inspection can frequently be performed without removing the surface coatings such as primer, paint, and anodized films. It can be effective in detecting surface and subsurface corrosion, pots and heat treat condition.

Ultrasonic Inspection

Ultrasonic detection equipment makes it possible to locate defects in all types of materials. Minute cracks, checks, and voids too small to be seen by x-ray can be located by ultrasonic inspection. An ultrasonic test instrument requires access to only one surface of the material to be inspected and can be used with either straight line or angle beam testing techniques.

Two basic methods are used for ultrasonic inspection. The first of these methods is immersion testing. In this method of inspection, the part under examination and the search unit are totally immersed in a liquid couplant, which may be water or any other suitable fluid.

The second method is called contact testing, which is readily adapted to field use and is the method discussed in this chapter. In this method, the part under examination and the search unit are coupled with a viscous material, liquid or a paste, which wets both the face of the search unit and the material under examination.

There are three basic ultrasonic inspection methods: (1) pulse echo; (2) through transmission; and (3) resonance.

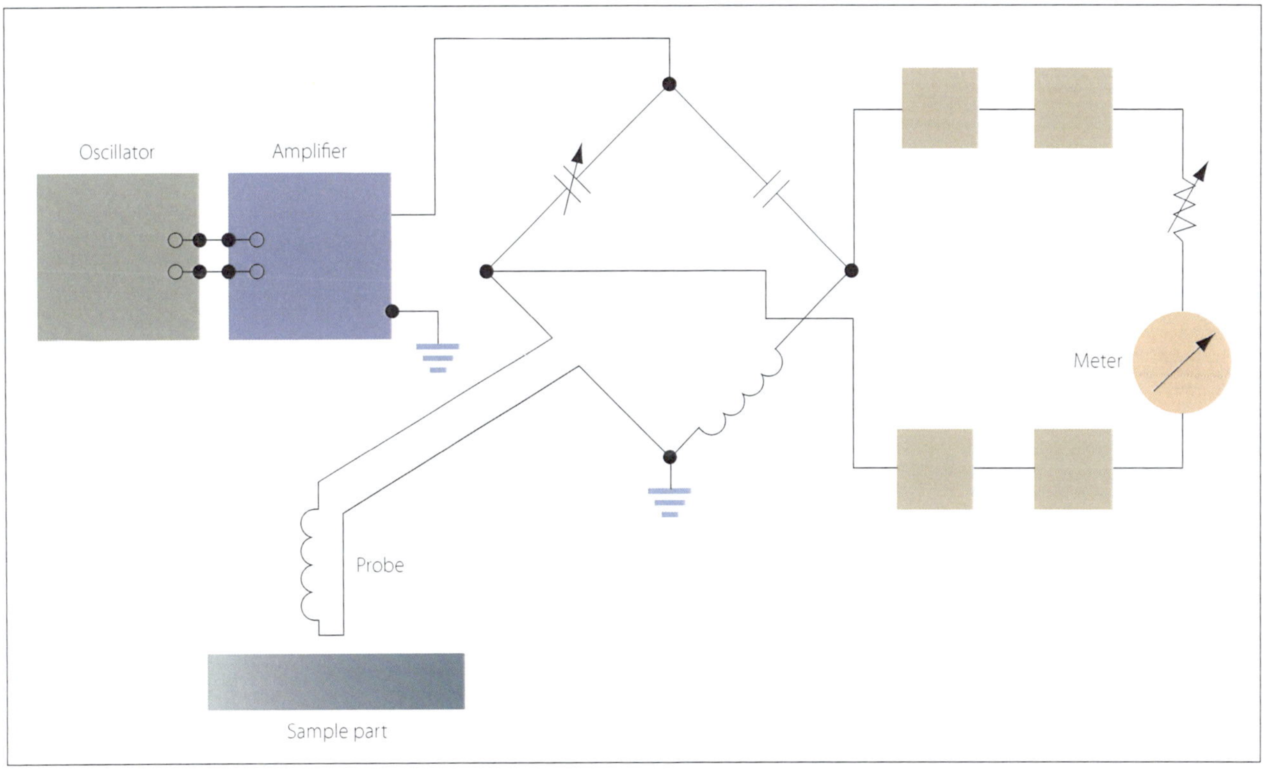

Figure 8-6. Eddy current inspection circuit.

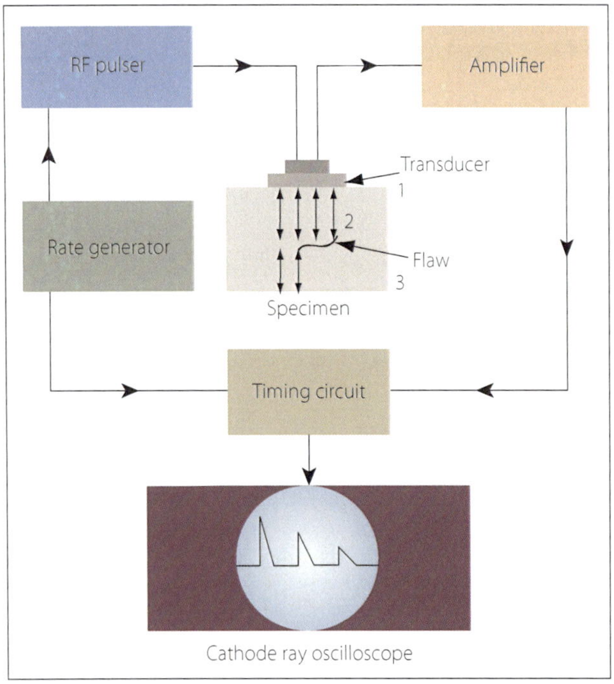

Figure 8-7. Block diagram of basic pulse-echo system.

Pulse Echo

Flaws are detected by measuring the amplitude of signals reflected and the time required for these signals to travel between specific surfaces and the discontinuity. [Figure 8-7]

The time base, which is triggered simultaneously with each transmission pulse, causes a spot to sweep across the screen of the cathode ray tube (CRT). The spot sweeps from left to right across the face of the scope 50 to 5,000 times per second, or higher if required for high speed automated scanning. Due to the speed of the cycle of transmitting and receiving, the picture on the oscilloscope appears to be stationary.

A few microseconds after the sweep is initiated, the rate generator electrically excites the pulser, and the pulser in turn emits an electrical pulse. The transducer converts this pulse into a short train of ultrasonic sound waves. If the interfaces of the transducer and the specimen are properly oriented, the ultrasound will be reflected back to the transducer when it reaches the internal flaw and the opposite surface of the specimen. The time interval between the transmission of the initial impulse and the reception of the signals from within the specimen are measured by the timing circuits. The reflected pulse received by the transducer is amplified, then transmitted to and displayed on the instrument screen. The pulse is displayed in the same relationship to the front and back pulses as the flaw is in relation to the front and back surfaces of the specimen. [Figure 8-8]

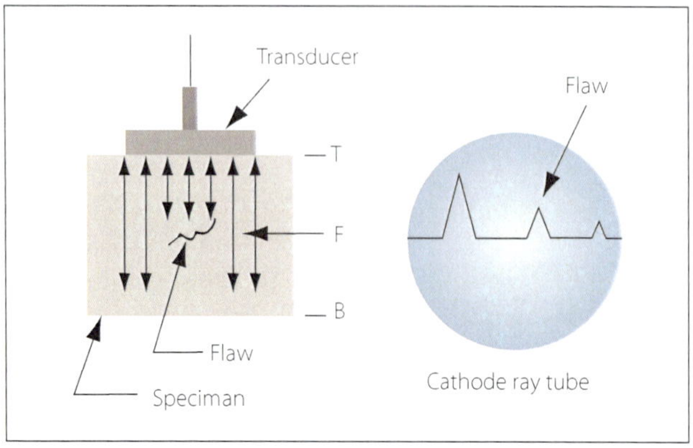

Figure 8-8. Pulse-echo display in relationship to flaw detection.

Pulse-echo instruments may also be used to detect flaws not directly underneath the probe by use of the angle-beam testing method. Angle beam testing differs from straight beam testing only in the manner in which the ultrasonic waves pass through the material being tested. As shown in Figure 8-9, the beam is projected into the material at an acute angle to the surface by means of a crystal cut at an angle and mounted in plastic. The beam or a portion thereof reflects successively from the surfaces of the material or any other discontinuity, including the edge of the piece. In straight beam testing, the horizontal distance on the screen between the initial pulse and the first back reflection represents the thickness of the piece; while in angle beam testing, this distance represents the width of the material between the searching unit and the opposite edge of the piece.

Through Transmission

Through transmission inspection uses two transducers, one to generate the pulse and another placed on the opposite surface to receive it. A disruption in the sound path will indicate a flaw and be displayed on the instrument screen. Through transmission is less sensitive to small defects than the pulse-echo method.

Resonance

This system differs from the pulse method in that the frequency of transmission may be continuously varied. The resonance method is used principally for thickness measurements when the two sides of the material being tested are smooth and parallel and the backside is inaccessible. The point at which the frequency matches the resonance point of the material being tested is the thickness determining factor.

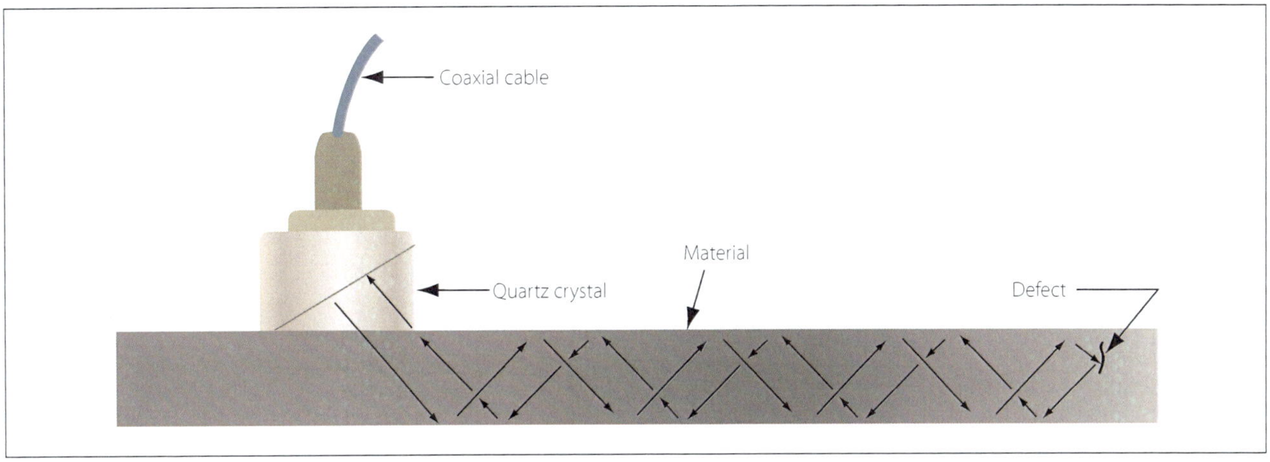

Figure 8-9. Pulse-echo angle beam testing.

It is necessary that the frequency of the ultrasonic waves corresponding to a particular dial setting be accurately known. Checks should be made with standard test blocks to guard against possible drift of frequency.

If the frequency of an ultrasonic wave is such that its wavelength is twice the thickness of a specimen (fundamental frequency), then the reflected wave will arrive back at the transducer in the same phase as the original transmission so that strengthening of the signal will occur. This results from constructive interference or a resonance and is shown as a high amplitude value on the indicating screen. If the frequency is increased such that three times the wavelength equals four times the thickness, the reflected signal will return completely out of phase with the transmitted signal and cancellation will occur. Further increase of the frequency causes the wavelength to be equal to the thickness again and gives a reflected signal in phase with the transmitted signal and a resonance once more.

By starting at the fundamental frequency and gradually increasing the frequency, the successive cancellations and resonances can be noted and the readings used to check the fundamental frequency reading. [Figure 8-10]

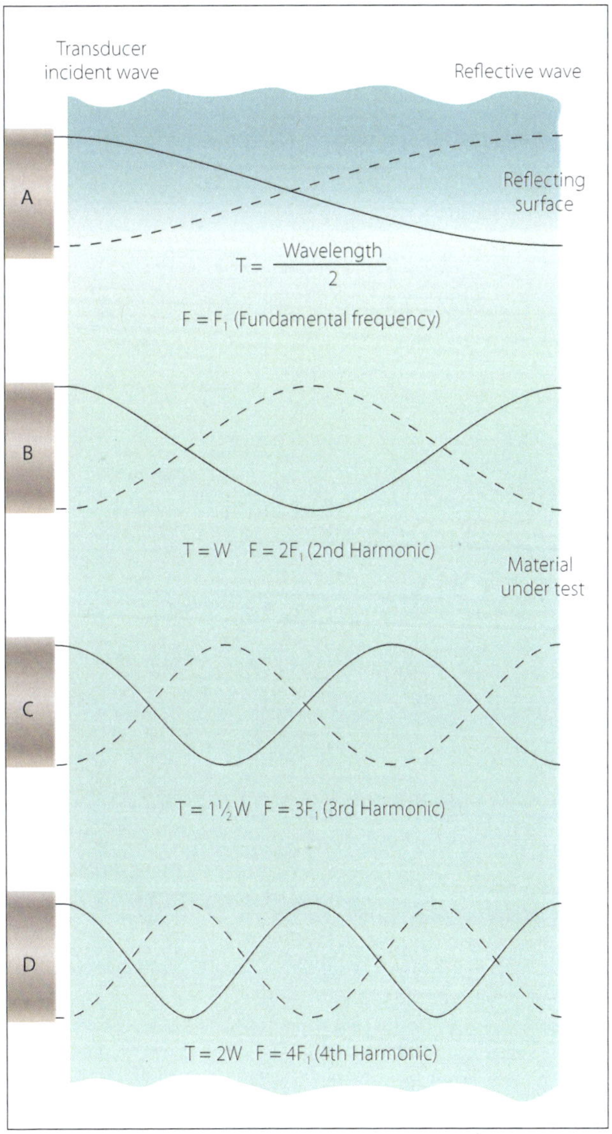

Figure 8-10. Conditions of ultrasonic resonance in a metal plate.

In some instruments, the oscillator circuit contains a motor driven capacitor which changes the frequency of the oscillator. [Figure 8-11] In other instruments, the frequency is changed by electronic means.

The change in frequency is synchronized with the horizontal sweep of a CRT. The horizontal axis thus represents a frequency range. If the frequency range contains resonances, the circuitry is arranged to present these vertically. Calibrated transparent scales are then placed in front of the tube, and the thickness can be read directly. The instruments normally operate

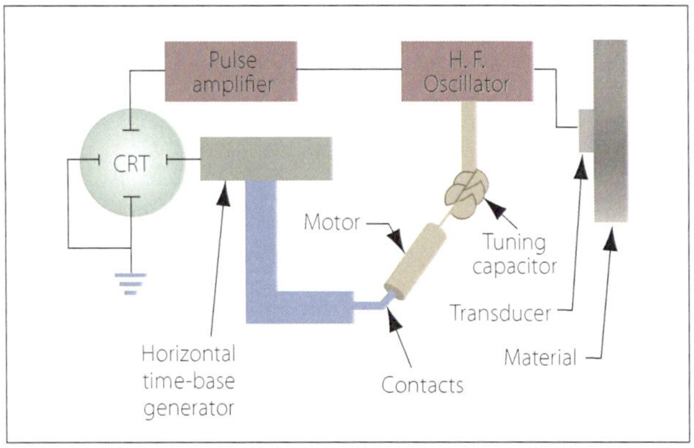

Figure 8-11. Block diagram of resonance thickness measuring system.

between 0.25 millicycle (mc) and 10 mc in four or five bands.

The resonance thickness instrument can be used to test the thickness of such metals as steel, cast iron, brass, nickel, copper, silver, lead, aluminum, and magnesium. In addition, areas of corrosion or wear on tanks, tubing, airplane wing skins, and other structures or products can be located and evaluated.

Direct reading dial-operated units are available that measure thickness between 0.025 inch and 3 inches with an accuracy of better than ±1 percent.

Ultrasonic inspection requires a skilled operator who is familiar with the equipment being used as well as the inspection method to be used for the many different parts being tested. [Figure 8-12]

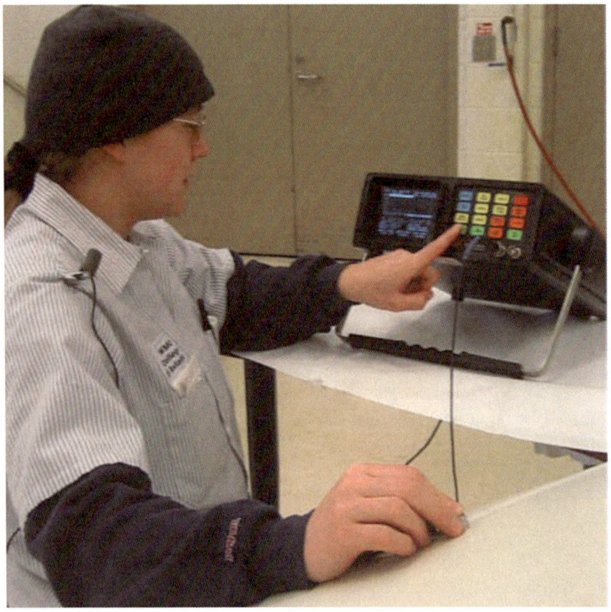

Figure 8-12. Ultrasonic inspection of a composite structure.

Acoustic Emission Inspection

Acoustic emission is an NDI technique that involves the placing of acoustic emission sensors at various locations on an aircraft structure and then applying a load or stress. The materials emit sound and stress waves that take the form of ultrasonic pulses. Cracks and areas of corrosion in the stressed airframe structure emit sound waves which are registered by the sensors. These acoustic emission bursts can be used to locate flaws and to evaluate their rate of growth as a function of applied stress. Acoustic emission testing has an advantage over other NDI methods in that it can detect and locate all of the activated flaws in a structure in one test. Because of the complexity of aircraft structures, application of acoustic emission testing to aircraft has required a new level of sophistication in testing technique and data interpretation.

Magnetic Particle Inspection

Magnetic particle inspection is a method of detecting invisible cracks and other defects in ferromagnetic materials such as iron and steel. It is not applicable to nonmagnetic materials.

In rapidly rotating, reciprocating, vibrating, and other highly stressed aircraft parts, small defects often develop to the point that they cause complete failure of the part. Magnetic particle inspection has proven extremely reliable for the rapid detection of such defects located on or near the surface. With this method of inspection, the location of the defect is indicated and the approximate size and shape are outlined.

The inspection process consists of magnetizing the part and then applying ferromagnetic particles to the surface area to be inspected. The ferromagnetic particles (indicating medium) may be held in suspension in a liquid that is flushed over the part; the part may be immersed in the suspension liquid; or the particles, in dry powder form, may be dusted over the surface of the part. The wet process is more commonly used in the inspection of aircraft parts.

If a discontinuity is present, the magnetic lines of force will be disturbed and opposite poles will exist on either side of the discontinuity. The magnetized particles thus form a pattern in the magnetic field between the opposite poles. This pattern, known as an "indication," assumes the approximate shape of the surface projection of the discontinuity. A discontinuity may be defined as an interruption in the normal physical structure or configuration of a part, such as a crack, forging lap, seam, inclusion, porosity, and the like.

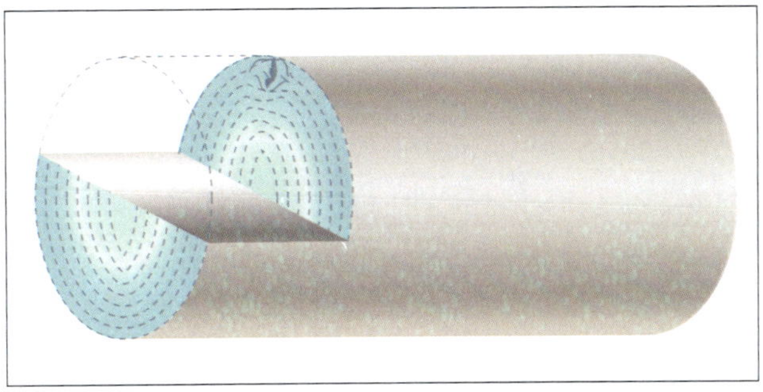

Figure 8-13. Flux leakage at transverse discontinuity.

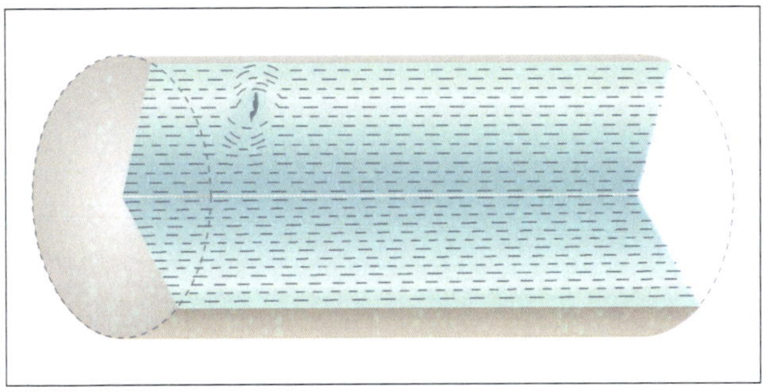

Figure 8-14. Flux leakage at longitudinal discontinuity.

A discontinuity may or may not affect the usefulness of a part.

Development of Indications

When a discontinuity in a magnetized material is open to the surface, and a magnetic substance (indicating medium) is available on the surface, the flux leakage at the discontinuity tends to form the indicating medium into a path of higher permeability. (Permeability is a term used to refer to the ease with which a magnetic flux can be established in a given magnetic circuit.) Because of the magnetism in the part and the adherence of the magnetic particles to each other, the indication remains on the surface of the part in the form of an approximate outline of the discontinuity that is immediately below it.

The same action takes place when the discontinuity is not open to the surface, but since the amount of flux leakage is less, fewer particles are held in place and a fainter and less sharply defined indication is obtained.

If the discontinuity is very far below the surface, there may be no flux leakage and no indication on the surface. The flux leakage at a transverse discontinuity is shown in Figure 8-13. The flux leakage at a longitudinal discontinuity is shown in Figure 8-14.

Types of Discontinuities Disclosed

The following types of discontinuities are normally detected by the magnetic particle test: cracks, laps, seams, cold shuts, inclusions, splits, tears, pipes, and voids. All of these may affect the reliability of parts in service.

Cracks, splits, bursts, tears, seams, voids, and pipes are formed by an actual parting or rupture of the solid metal. Cold shuts and laps are folds that have been formed in the metal, interrupting its continuity.

Inclusions are foreign material formed by impurities in the metal during the metal processing stages. They may consist, for example, of bits of furnace lining picked up during the melting of the basic metal or of other foreign constituents. Inclusions interrupt the continuity of the metal because they prevent the joining or welding of adjacent faces of the metal.

Preparation of Parts for Testing

Grease, oil, and dirt must be cleaned from all parts before they are tested. Cleaning is very important since any grease or other foreign material present can produce nonrelevant indications due to magnetic particles adhering to the foreign material as the suspension drains from the part.

Grease or foreign material in sufficient amount over a discontinuity may also prevent the formation of a pattern at the discontinuity. It is not advisable to depend upon the magnetic particle suspension to clean the part. Cleaning by suspension is not thorough and any foreign materials so removed from the part will contaminate the suspension, thereby reducing its effectiveness.

In the dry procedure, thorough cleaning is absolutely necessary. Grease or other foreign material will hold the magnetic powder, resulting in nonrelevant indications and making it impossible to distribute the indicating medium evenly over the part's surface.

All small openings and oil holes leading to internal passages or cavities should be plugged with paraffin or other suitable nonabrasive material.

Coatings of cadmium, copper, tin, and zinc do not interfere with the satisfactory performance of magnetic particle inspection, unless the coatings are unusually heavy or the discontinuities to be detected are unusually small.

Chromium and nickel plating generally will not interfere with indications of cracks open to the surface of the base metal but will prevent indications of fine discontinuities, such as inclusions.

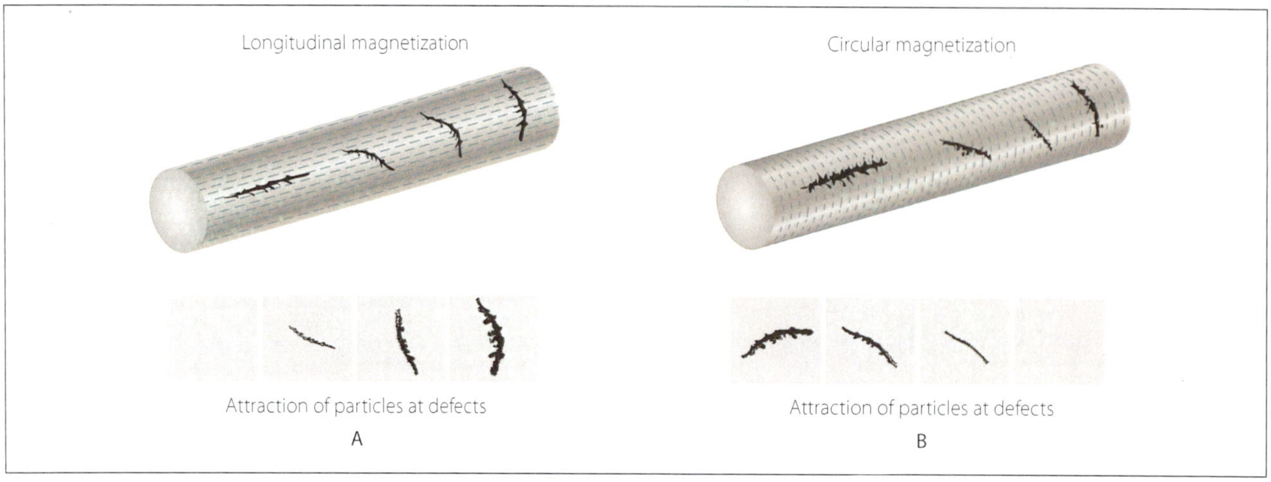

Figure 8-15. Effect of flux direction on strength of indication.

Because it is more strongly magnetic, nickel plating is more effective than chromium plating in preventing the formation of indications.

Effect of Flux Direction

To locate a defect in a part, it is essential that the magnetic lines of force pass approximately perpendicular to the defect. It is therefore necessary to induce magnetic flux in more than one direction since defects are likely to exist at any angle to the major axis of the part. This requires two separate magnetizing operations, referred to as circular magnetization and longitudinal magnetization. The effect of flux direction is illustrated in Figure 8-15.

Circular magnetization is the induction of a magnetic field consisting of concentric circles of force about and within the part which is achieved by passing electric current through the part. This type of magnetization will locate defects running approximately parallel to the axis of the part. Figure 8-16 illustrates circular magnetization of a camshaft.

Figure 8-16. Circular magnetization of a camshaft.

In longitudinal magnetization, the magnetic field is produced in a direction parallel to the long axis of the part. This is accomplished by placing the part in a solenoid excited by electric current. The metal part then becomes the core of an electromagnet and is magnetized by induction from the magnetic field created in the solenoid.

In longitudinal magnetization of long parts, the solenoid must be moved along the part in order to magnetize it. [Figure 8-17] This is necessary to ensure adequate field strength throughout the entire length of the part.

Solenoids produce effective magnetization for approximately 12 inches from each end of the coil, thus accommodating parts or sections approximately 30 inches in length. Longitudinal magnetization equivalent to that obtained by a solenoid may be accomplished by wrapping a flexible electrical conductor around the part. Although this method is not as convenient, it has

Figure 8-17. Longitudinal magnetization of crankshaft (solenoid method).

an advantage in that the coils conform more closely to the shape of the part, producing a somewhat more uniform magnetization.

The flexible coil method is also useful for large or irregularly shaped parts for which standard solenoids are not available.

Effect of Flux Density

The effectiveness of the magnetic particle inspection also depends on the flux density or field strength at the surface of the part when the indicating medium is applied. As the flux density in the part is increased, the sensitivity of the test increases because of the greater flux leakages at discontinuities and the resulting improved formation of magnetic particle patterns.

Excessively high flux densities may form nonrelevant indications; for example, patterns of the grain flow in the material. These indications will interfere with the detection of patterns resulting from significant discontinuities. It is therefore necessary to use a field strength high enough to reveal all possible harmful discontinuities but not strong enough to produce confusing nonrelevant indications.

Magnetizing Methods

When a part is magnetized, the field strength in the part increases to a maximum for the particular magnetizing force and remains at this maximum as long as the magnetizing force is maintained.

When the magnetizing force is removed, the field strength decreases to a lower residual value depending on the magnetic properties of the material and the shape of the part. These magnetic characteristics determine whether the continuous or residual method is used in magnetizing the part.

In the continuous inspection method, the part is magnetized and the indicating medium applied while the magnetizing force is maintained. The available flux density in the part is thus at a maximum. The maximum value of flux depends directly upon the magnetizing force and the permeability of the material of which the part is made.

The continuous method may be used in practically all circular and longitudinal magnetization procedures. The continuous procedure provides greater sensitivity than the residual procedure, particularly in locating subsurface discontinuities. The highly critical nature of aircraft parts and assemblies and the necessity for subsurface inspection in many applications have resulted in the continuous method being more widely used.

Inasmuch as the continuous procedure will reveal more nonsignificant discontinuities than the residual procedure, careful and intelligent interpretation and evaluation of discontinuities revealed by this procedure are necessary.

The residual inspection procedure involves magnetization of the part and application of the indicating medium after the magnetizing force has been removed. This procedure relies on the residual or permanent magnetism in the part and is more practical than the continuous procedure when magnetization is accomplished by flexible coils wrapped around the part.

In general, the residual procedure is used only with steels which have been heat treated for stressed applications.

Identification of Indications

The correct evaluation of the character of indications is extremely important but is sometimes difficult to make from observation of the indications alone. The principal distinguishing features of indications are shape, buildup, width, and sharpness of outline. These characteristics are more valuable in distinguishing between types of discontinuities than in determining their severity. Careful observation of the character of the magnetic particle pattern should always be included in the complete evaluation of the significance of an indicated discontinuity.

The most readily distinguished indications are those produced by cracks open to the surface. These discontinuities include fatigue cracks, heat treat cracks, shrink cracks in welds and castings, and grinding cracks. An example of a fatigue crack is shown in Figure 8-18.

Magnaglo Inspection

Magnaglo inspection is similar to the preceding method but differs in that a fluorescent particle solution is used and the inspection is made under black light. Efficiency of inspection is increased by the neon-like glow of defects allowing smaller flaw indications to be seen. This is an excellent method for use on gears, threaded parts, and aircraft engine components. The reddish brown liquid spray or bath that is used consists of Magnaglo paste mixed with a light oil at the ratio of 0.10 to 0.25 ounce of paste per gallon of oil.

After inspection, the part must be demagnetized and rinsed with a cleaning solvent.

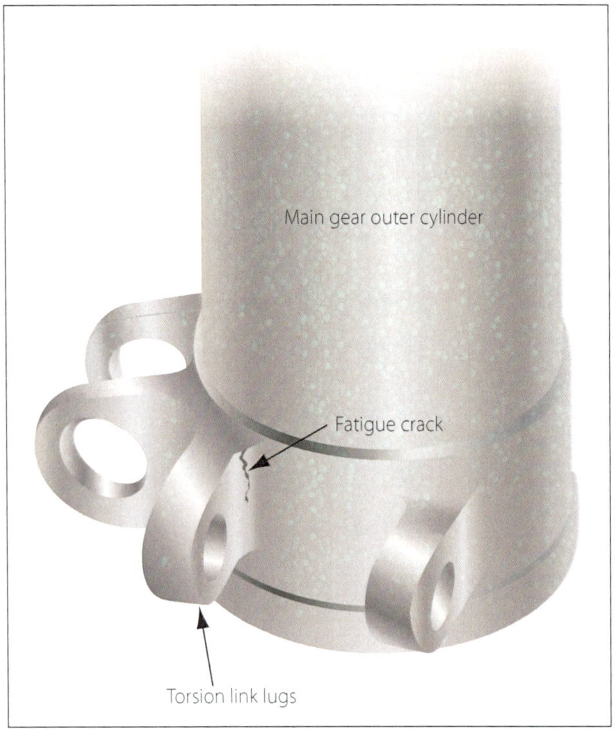

Figure 8-18. Fatigue crack in a landing gear.

Magnetizing Equipment

Fixed (Nonportable) General Purpose Unit

A fixed general purpose unit is shown in Figure 8-19. This unit provides direct current for wet continuous or residual magnetization procedures. Circular or longitudinal magnetization may be used and it may be powered with rectified alternating current (ac), as well as direct current (dc). The contact heads provide the electrical terminals for circular magnetization. One head is fixed in position with its contact plate mounted on a shaft surrounded by a pressure spring, so that the plate may be moved longitudinally. The plate is maintained in the extended position by the spring until pressure transmitted through the work from the movable head forces it back.

The motor driven movable head slides horizontally in longitudinal guides and is controlled by a switch. The spring allows sufficient overrun of the motor driven head to avoid jamming it and also provides pressure on the ends of the work to ensure good electrical contact.

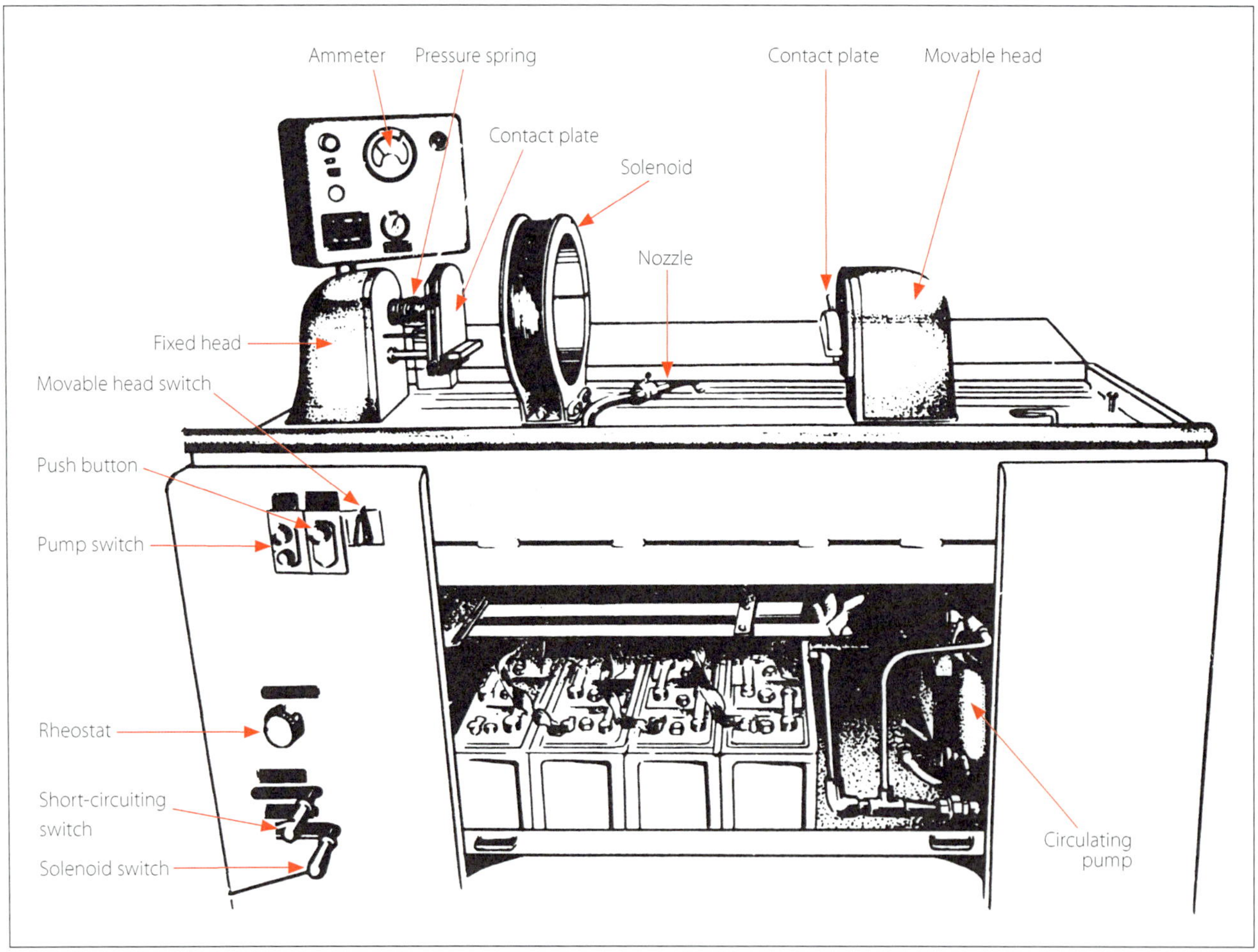

Figure 8-19. Fixed general-purpose magnetizing unit.

A plunger operated switch in the fixed head cuts out the forward motion circuit of the movable head motor when the spring has been properly compressed.

In some units the movable head is hand operated, and the contact plate is sometimes arranged for operation by an air ram. Both contact plates are fitted with various fixtures for supporting the work.

The magnetizing circuit is closed by depressing a pushbutton on the front of the unit. It is set to open automatically, usually after about one-half second.

The strength of the magnetizing current may be set manually to the desired value by means of the rheostat or increased to the capacity of the unit by the rheostat short circuiting switch. The current utilized is indicated on the ammeter.

Longitudinal magnetization is produced by the solenoid, which moves in the same guide rail as the movable head and is connected in the electrical circuit by means of a switch.

The suspension liquid is contained in a sump tank and is agitated and circulated by a pump. The suspension is applied to the work through a nozzle. The suspension drains from the work through a nonmetallic grill into a collecting pan that leads back to the sump. The circulating pump is operated by a pushbutton switch.

Portable General Purpose Unit

It is often necessary to perform the magnetic particle inspection at locations where fixed general purpose equipment is not available or to perform an inspection on members of aircraft structures without removing them from the aircraft. It is particularly useful for inspecting landing gear and engine mounts suspected of having developed cracks in service. Portable units supply both alternating current and direct current magnetization.

This unit is only a source of magnetizing and demagnetizing current and does not provide a means for supporting the work or applying the suspension. It operates on 200 volt, 60 cycle, alternating current and contains a rectifier for producing direct current when required. [Figure 8-20]

The magnetizing current is supplied through the flexible cables. The cable terminals may be fitted with prods, as shown in the illustration, or with contact clamps. Circular magnetization may be developed by using either the prods or clamps.

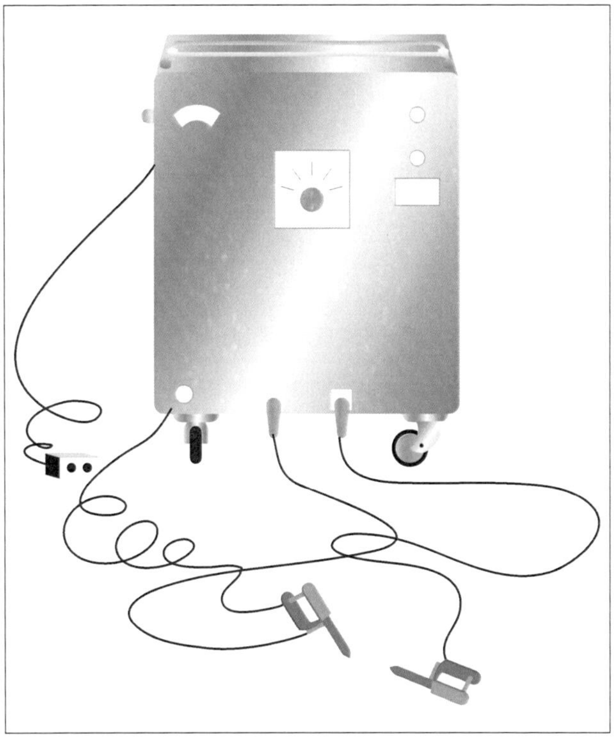

Figure 8-20. Portable general purpose unit.

Longitudinal magnetization is developed by wrapping the cable around the part.

The strength of the magnetizing current is controlled by an eight point tap switch, and the time duration for which it is applied is regulated by an automatic cutoff similar to that used in the fixed general purpose unit.

This portable unit also serves as a demagnetizer and supplies high amperage low voltage alternating current for this purpose. For demagnetization, the alternating current is passed through the part and gradually reduced by means of a current reducer.

In testing large structures with flat surfaces where current must be passed through the part, it is sometimes impossible to use contact clamps. In such cases, contact prods are used.

Prods can be used with the fixed general purpose unit as well as the portable unit. The part or assembly being tested may be held or secured above the standard unit and the suspension hosed onto the area; excess suspension drains into the tank. The dry procedure may also be used.

Prods should be held firmly against the surface being tested. There is a tendency for a high amperage current to cause burning at contact areas, but with proper care, such burning is usually slight. For applications where prod magnetization is acceptable, slight burning is normally acceptable.

Indicating Mediums

The various types of indicating mediums available for magnetic particle inspection may be divided into two general material types: wet and dry. The basic requirement for any indicating medium is that it produce acceptable indications of discontinuities in parts.

The contrast provided by a particular indicating medium on the background or part surface is particularly important. The colors most extensively used are black and red for the wet procedure and black, red, and gray for the dry procedure.

For acceptable operation, the indicating medium must be of high permeability and low retentivity. High permeability ensures that a minimum of magnetic energy will be required to attract the material to flux leakage caused by discontinuities. Low retentivity ensures that the mobility of the magnetic particles will not be hindered by the particles themselves becoming magnetized and attracting one another.

Demagnetizing

The permanent magnetism remaining after inspection must be removed by a demagnetization operation if the part is to be returned to service. Parts of operating mechanisms must be demagnetized to prevent magnetized parts from attracting filings, grindings, or chips inadvertently left in the system, or steel particles resulting from operational wear.

An accumulation of such particles on a magnetized part may cause scoring of bearings or other working parts. Parts of the airframe must be demagnetized so they will not affect instruments.

Demagnetization between successive magnetizing operations is not normally required unless experience indicates that omission of this operation results in decreased effectiveness for a particular application. Demagnetization may be accomplished in a number of different ways. A convenient procedure for aircraft parts involves subjecting the part to a magnetizing force that is continually reversing in direction and, at the same time, gradually decreasing in strength. As the decreasing magnetizing force is applied first in one direction and then the other, the magnetization of the part also decreases.

Standard Demagnetizing Practice

The simplest procedure for developing a reversing and gradually decreasing magnetizing force in a part involves the use of a solenoid coil energized by alternating current. As the part is moved away from the alternating field of the solenoid, the magnetism in the part gradually decreases.

A demagnetizer whose size approximates that of the work should be used. For maximum effectiveness, small parts should be held as close to the inner wall of the coil as possible.

Parts that do not readily lose their magnetism should be passed slowly in and out of the demagnetizer several times and, at the same time, tumbled or rotated in various directions. Allowing a part to remain in the demagnetizer with the current on accomplishes very little practical demagnetization.

The effective operation in the demagnetizing procedure is that of slowly moving the part out of the coil and away from the magnetizing field strength. As the part is withdrawn, it should be kept directly opposite the opening until it is 1 or 2 feet from the demagnetizer.

The demagnetizing current should not be cut off until the part is 1 or 2 feet from the opening as the part may be remagnetized if current is removed too soon.

Another procedure used with portable units is to pass alternating current through the part being demagnetized, while gradually reducing the current to zero.

Radiographic

X and gamma radiations, because of their unique ability to penetrate material and disclose discontinuities, have been applied to the radiographic (x-ray) inspection of metal fabrications and nonmetallic products.

The penetrating radiation is projected through the part to be inspected and produces an invisible or latent image in the film. When processed, the film becomes a radiograph or shadow picture of the object. This inspection medium and portable unit provides a fast and reliable means for checking the integrity of airframe structures and engines. [Figure 8-21]

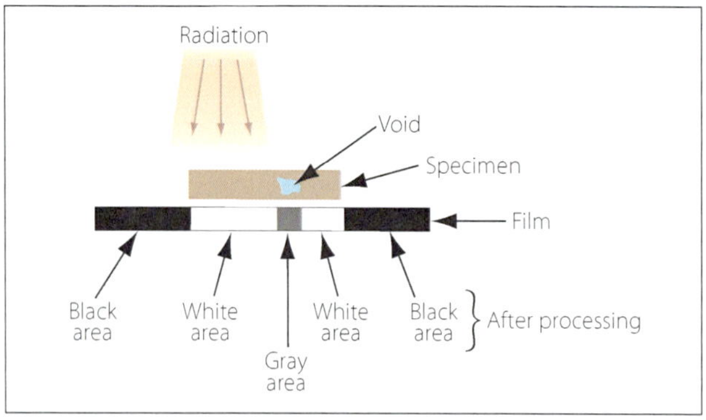

Figure 8-21. Radiograph.

Radiographic Inspection

Radiographic inspection techniques are used to locate defects or flaws in airframe structures or engines with little or no disassembly. This is in marked contrast to other types of nondestructive testing which usually require removal, disassembly, and stripping of paint from the suspected part before it can be inspected. Due to the radiation risks associated with x-ray, extensive training is required to become a qualified radiographer. Only qualified radiographers are allowed to operate the x-ray units.

Three major steps in the x-ray process discussed in subsequent paragraphs are: (1) exposure to radiation, including preparation, (2) processing of film, and (3) interpretation of the radiograph.

Preparation and Exposure

The factors of radiographic exposure are so interdependent that it is necessary to consider all factors for any particular radiographic exposure. These factors include but are not limited to the following:

- Material thickness and density
- Shape and size of the object
- Type of defect to be detected
- Characteristics of x-ray machine used
- The exposure distance
- The exposure angle
- Film characteristics
- Types of intensifying screen, if used

Knowledge of the x-ray unit's capabilities should form a background for the other exposure factors. In addition to the unit rating in kilovoltage, the size, portability, ease of manipulation, and exposure particulars of the available equipment should be thoroughly understood.

Previous experience on similar objects is also very helpful in the determination of the overall exposure techniques. A log or record of previous exposures will provide specific data as a guide for future radiographs.

Film Processing

After exposure to x-rays, the latent image on the film is made permanently visible by processing it successively through a developer chemical solution, an acid bath, and a fixing bath, followed by a clear water wash.

Radiographic Interpretation

From the standpoint of quality assurance, radiographic interpretation is the most important phase of radiography. It is during this phase that an error in judgment can produce disastrous consequences. The efforts of the whole radiographic process are centered in this phase; the part or structure is either accepted or rejected. Conditions of unsoundness or other defects which are overlooked, not understood, or improperly interpreted can destroy the purpose and efforts of radiography and can jeopardize the structural integrity of an entire aircraft. A particular danger is the false sense of security imparted by the acceptance of a part or structure based on improper interpretation.

As a first impression, radiographic interpretation may seem simple, but a closer analysis of the problem soon dispels this impression. The subject of interpretation is so varied and complex that it cannot be covered adequately in this type of document. Instead, this chapter gives only a brief review of basic requirements for radiographic interpretation, including some descriptions of common defects.

Experience has shown that, whenever possible, radiographic interpretation should be conducted close to the radiographic operation. When viewing radiographs, it is helpful to have access to the material being tested. The radiograph can thus be compared directly with the material being tested, and indications due to such things as surface condition or thickness variations can be immediately determined.

The following paragraphs present several factors which must be considered when analyzing a radiograph.

There are three basic categories of flaws: voids, inclusions, and dimensional irregularities. The last category, dimensional irregularities, is not pertinent to these discussions because its prime factor is one of degree, and radiography is not exact. Voids and inclusions may appear on the radiograph in a variety of forms ranging from a two-dimensional plane to a three-dimensional sphere. A crack, tear, or cold shut will most nearly resemble a two-dimensional plane, whereas a cavity will look like a three-dimensional sphere. Other types of flaws, such as shrink, oxide inclusions, porosity, and so forth, will fall somewhere between these two extremes of form.

It is important to analyze the geometry of a flaw, especially for items such as the sharpness of terminal points. For example, in a crack-like flaw the terminal points appear much sharper in a sphere-like flaw, such as a gas cavity. Also, material strength may be adversely affected by flaw shape. A flaw having sharp points could establish a source of localized stress concentration. Spherical flaws affect material strength to a far lesser degree than do sharp pointed flaws. Specifications and reference standards usually stipulate that sharp pointed flaws, such as cracks, cold shuts, and so forth, are cause for rejection.

Material strength is also affected by flaw size. A metallic component of a given area is designed to carry a certain load plus a safety factor. Reducing this area by including a large flaw weakens the part and reduces the safety factor. Some flaws are often permitted in components because of these safety factors; in this case, the interpreter must determine the degree of tolerance or imperfection specified by the design engineer. Both flaw size and flaw shape should be considered carefully, since small flaws with sharp points can be just as bad as large flaws with no sharp points.

Another important consideration in flaw analysis is flaw location. Metallic components are subjected to numerous and varied forces during their effective service life. Generally, the distribution of these forces is not equal in the component or part, and certain critical areas may be rather highly stressed. The interpreter must pay special attention to these areas. Another aspect of flaw location is that certain types of discontinuities close to one another may potentially serve as a source of stress concentrations creating a situation that should be closely scrutinized.

An inclusion is a type of flaw which contains entrapped material. Such flaws may be of greater or lesser density than the item being radiographed. The foregoing discussions on flaw shape, size, and location apply equally to inclusions and to voids. In addition, a flaw containing foreign material could become a source of corrosion.

Radiation Hazards

Radiation from x-ray units and radioisotope sources is destructive to living tissue. It is universally recognized that in the use of such equipment, adequate protection must be provided. Personnel must keep outside the primary x-ray beam at all times.

Radiation produces changes in all matter through which it passes. This is also true of living tissue. When radiation strikes the molecules of the body, the effect may be no more than to dislodge a few electrons, but an excess of these changes could cause irreparable harm. When a complex organism is exposed to radiation, the degree of damage, if any, depends on which of its body cells have been changed.

Vital organs in the center of the body that are penetrated by radiation are likely to be harmed the most. The skin usually absorbs most of the radiation and reacts earliest to radiation.

If the whole body is exposed to a very large dose of radiation, death could result. In general, the type and severity of the pathological effects of radiation depend on the amount of radiation received at one time and the percentage of the total body exposed. Smaller doses of radiation could cause blood and intestinal disorders in a short period of time. The more delayed effects are leukemia and other cancers. Skin damage and loss of hair are also possible results of exposure to radiation.

Inspection of Composites

Composite structures should be inspected for delamination, which is separation of the various plies, debonding of the skin from the core, and evidence of moisture and corrosion. Previously discussed methods including ultrasonic, acoustic emission, and radiographic inspections may be used as recommended by the aircraft manufacturer. The simplest method used in testing composite structures is the tap test.

Tap Testing

Tap testing, also referred to as the ring test or coin test, is widely used as a quick evaluation of any accessible surface to detect the presence of delamination or debonding. The testing procedure consists of lightly tapping the surface with a light hammer (maximum weight of 2 ounces), a coin or other suitable device. The acoustic response or "ring" is compared to that of a known good area. A "flat" or "dead" response indicates an area of concern. Tap testing is limited to finding defects in relatively thin skins, less than 0.080" thick. On honeycomb structures, both sides need to be tested. Tap testing on only one side would not detect debonding on the opposite side.

Electrical Conductivity

Composite structures are not inherently electrically conductive. Some aircraft, because of their relatively low speed and type of use, are not affected by electrical issues. Manufacturers of other aircraft, such as high-speed high-performance jets, are required to utilize various methods of incorporating aluminum into their structures to make them conductive. The aluminum is imbedded within the plies of the lay-ups either as a

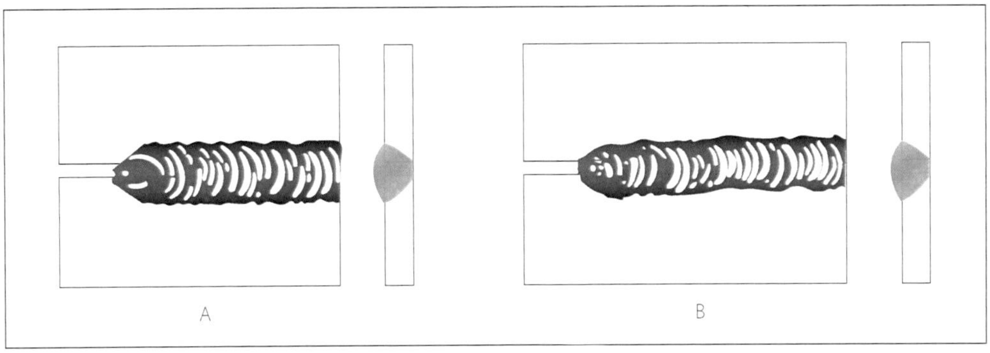

Figure 8-22. Examples of good welds.

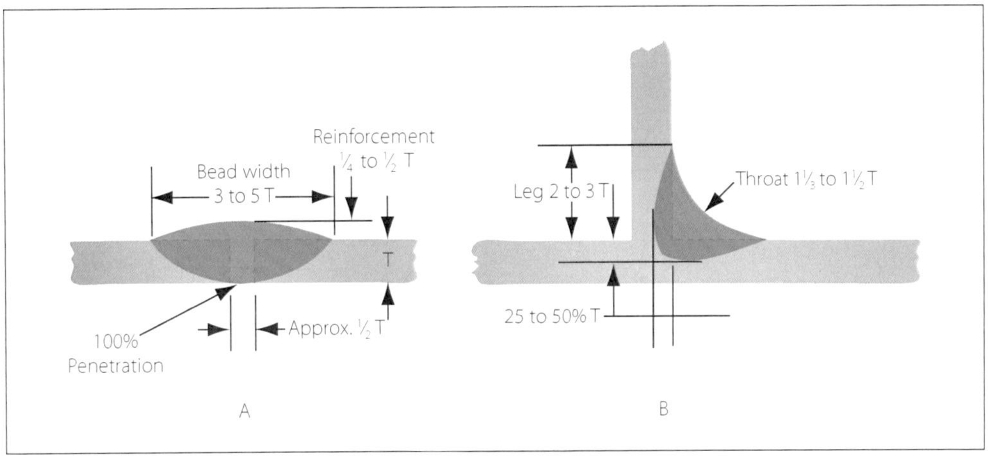

Figure 8-23. (A) Butt weld and (B) fillet weld, showing width and depth of bead.

thin wire mesh, screen, foil, or spray. When damaged sections of the structure are repaired, care must be taken to ensure that the conductive path be restored. Not only is it necessary to include the conductive material in the repair, but the continuity of the electrical path from the original conductive material to the replacement conductor and back to the original must be maintained. Electrical conductivity may be checked by use of an ohmmeter. Specific manufacturer's instructions must be carefully followed.

Inspection of Welds

A discussion of welds in this chapter will be confined to judging the quality of completed welds by visual means. Although the appearance of the completed weld is not a positive indication of quality, it provides a good clue about the care used in making it.

A properly designed joint weld is stronger than the base metal which it joins. The characteristics of a properly welded joint are discussed in the following paragraphs.

A good weld is uniform in width; the ripples are even and well feathered into the base metal, which shows no burn due to overheating. [Figure 8-22] The weld has good penetration and is free of gas pockets, porosity, or inclusions. The edges of the bead illustrated in Figure 8-22 (B) are not in a straight line, yet the weld is good since penetration is excellent.

Penetration is the depth of fusion in a weld. Thorough fusion is the most important characteristic contributing to a sound weld. Penetration is affected by the thickness of the material to be joined, the size of the filler rod, and how it is added. In a butt weld, the penetration should be 100 percent of the thickness of the base metal. On a fillet weld, the penetration requirements are 25 to 50 percent of the thickness of the base metal. The width and depth of bead for a butt weld and fillet weld are shown in Figure 8-23.

To assist further in determining the quality of a welded joint, several examples of incorrect welds are discussed in the following paragraphs.

The weld shown in Figure 8-24 (A) was made too rapidly. The long and pointed appearance of the

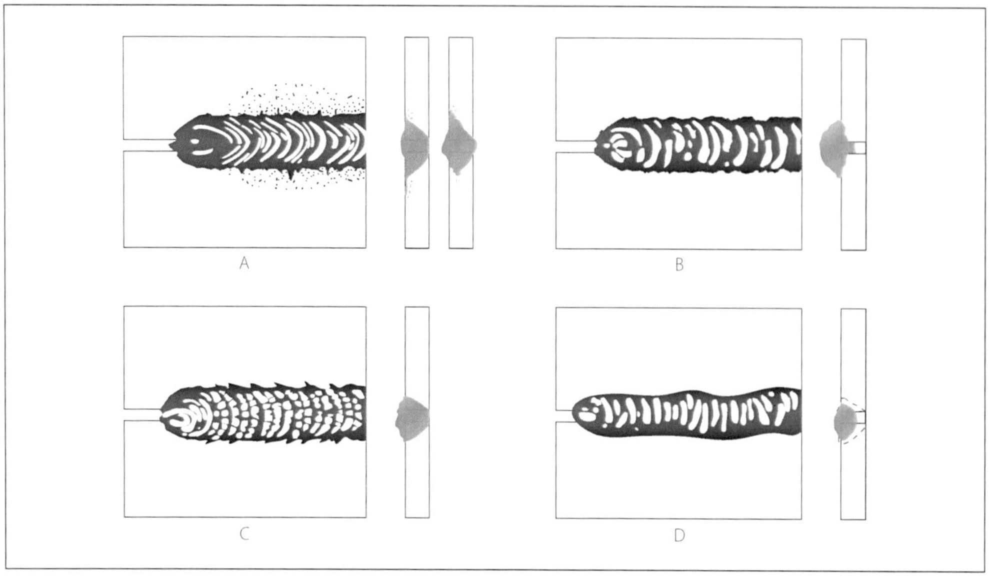

Figure 8-24. Examples of poor welds.

ripples was caused by an excessive amount of heat or an oxidizing flame. If the weld were cross-sectioned, it would probably disclose gas pockets, porosity, and slag inclusions.

Figure 8-24 (B) illustrates a weld that has improper penetration and cold laps caused by insufficient heat. It appears rough and irregular, and its edges are not feathered into the base metal.

The puddle has a tendency to boil during the welding operation if an excessive amount of acetylene is used. This often leaves slight bumps along the center and craters at the finish of the weld. Cross-checks are apparent if the body of the weld is sound. If the weld were cross-sectioned, pockets and porosity would be visible. Such a condition is shown in Figure 8-24 (C).

A bad weld with irregular edges and considerable variation in the depth of penetration is shown in D of Figure 8-24. It often has the appearance of a cold weld.

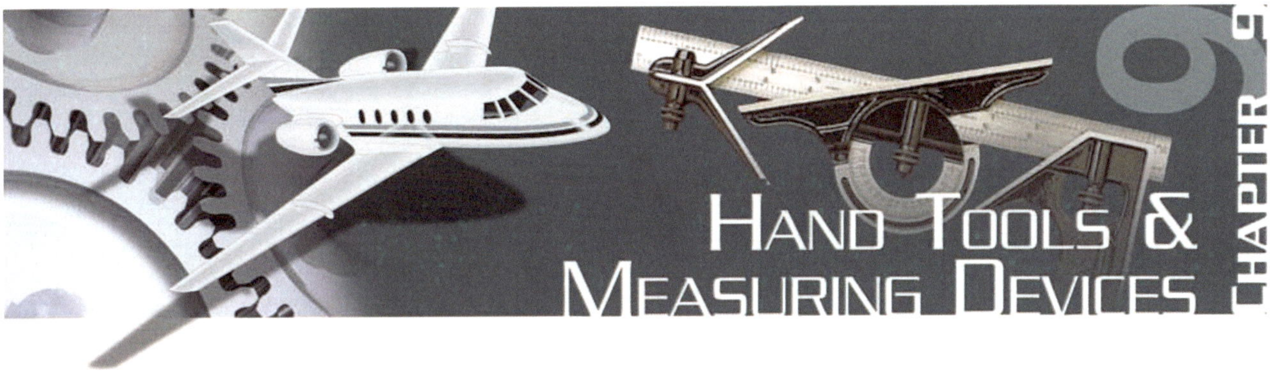

Chapter 9: Hand Tools & Measuring Devices

The aviation maintenance technician (AMT) spends a major portion of each day using a wide variety of hand tools to accomplish maintenance tasks. This chapter contains an overview of some of the hand tools an AMT can expect to use. An AMT encounters many special tools as their experience widens; large transport category aircraft have different maintenance tasks from those of a light airplane, and special hand tools are often required when working on complex aircraft.

This chapter outlines the basic knowledge required in using the most common hand tools and measuring instruments used in aircraft repair work. This information, however, cannot replace sound judgment on the part of the individual, nor additional training as the need arises. There are many times when ingenuity and resourcefulness can supplement the basic rules. Sound knowledge is required of these basic rules and of the situations in which they apply. The use of tools may vary, but good practices for safety, care, and storage of tools remain the same.

General Purpose Tools

Hammers and Mallets

Figure 9-1 shows some of the hammers that the aviation mechanic may be required to use. Metal head hammers are usually sized according to the weight of the head without the handle.

Occasionally it is necessary to use a soft-faced hammer, which has a striking surface made of wood, brass, lead, rawhide, hard rubber, or plastic. These hammers are intended for use in forming soft metals and striking surfaces that are easily damaged. Soft-faced hammers should not be used for striking punch heads, bolts, or nails, as using one in this fashion will quickly ruin this type of hammer.

A mallet is a hammer-like tool with a head made of hickory, rawhide, or rubber. It is handy for shaping thin metal parts without causing creases or dents with abrupt corners. Always use a wooden mallet when pounding a wood chisel or a gouge.

When using a hammer or mallet, choose the one best suited for the job. Ensure that the handle is tight. When striking a blow with the hammer, use the forearm as an extension of the handle. Swing the hammer by bending the elbow, not the wrist. Always strike the work squarely with the full face of the hammer. When striking a metal tool with a metal hammer, the use of safety glasses or goggles is strongly encouraged.

Always keep the faces of hammers and mallets smooth and free from dents, chips, or gouges to prevent marring the work.

Screwdrivers

The screwdriver can be classified by its shape, type of blade, and blade length. It is made for only one pur-

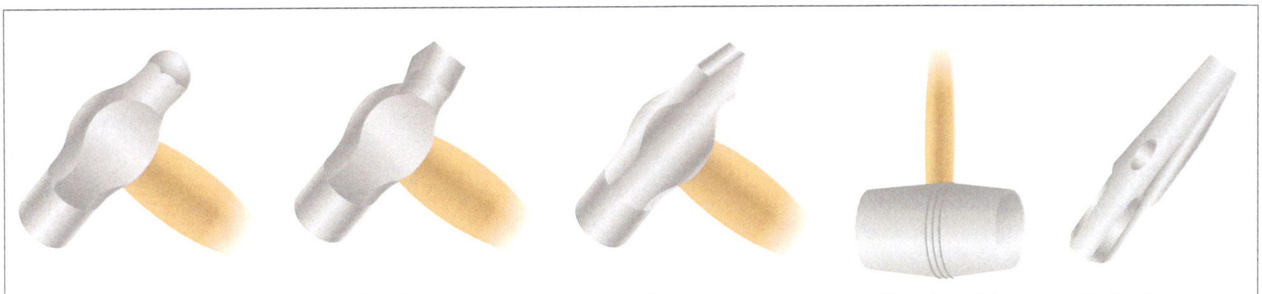

Figure 9-1. Hammers.

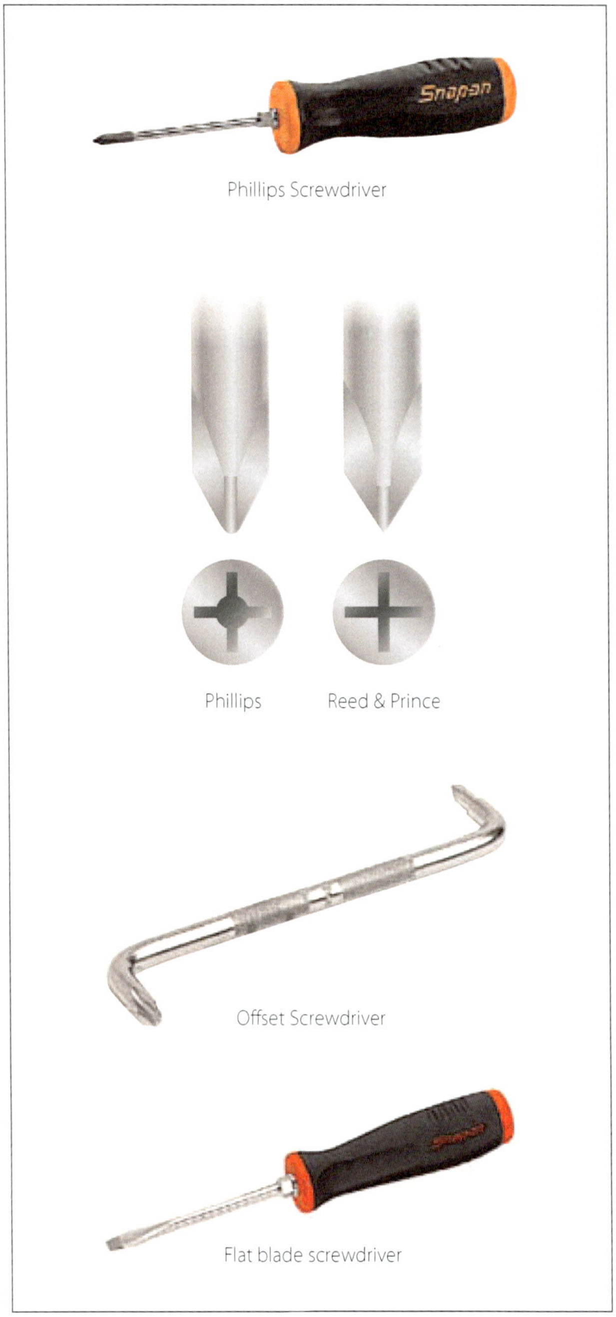

Figure 9-2. Typical screwdrivers.

pose, i.e., for loosening or tightening screws or screw head bolts. Figure 9-2 shows several different types of screwdrivers. When using the common screwdriver, select the largest screwdriver whose blade will make a good fit in the screw that is to be turned.

A common screwdriver must fill at least 75 percent of the screw slot. If the screwdriver is the wrong size, it cuts and burrs the screw slot, making it worthless. The damage may be so severe that the use of screw extractor may be required. A screwdriver with the wrong size blade may slip and damage adjacent parts of the structure.

The common screwdriver is used only where slotted head screws or fasteners are found on aircraft. An example of a fastener that requires the use of a common screwdriver is the camlock style fastener that is used to secure the cowling on some aircraft.

The two types of recessed head screws in common use are the Phillips and the Reed & Prince.

Both the Phillips and Reed & Prince recessed heads are optional on several types of screws. As shown in Figure 9-2, the Reed & Prince recessed head forms a perfect cross. The screwdriver used with this screw is pointed on the end. Since the Phillips screw has a slightly larger center in the cross, the Phillips screwdriver is blunt on the end. The Phillips screwdriver is not interchangeable with the Reed & Prince. The use of the wrong type screwdriver results in mutilation of the screwdriver and the screw head. When turning a recessed head screw, use only the proper recessed head screwdriver of the correct size. The most common crosspoint screwdrivers are the Number 1 and Number 2 Phillips.

An offset screwdriver may be used when vertical space is limited. Offset screwdrivers are constructed with both ends bent 90° to the shank handle. By using alternate ends, most screws can be seated or loosened even when the swinging space is limited. Offset screwdrivers are made for both standard and recessed head screws. Ratcheting right angle screwdrivers are also available, and often prove to be indispensable when working in close quarters.

A screwdriver should not be used for chiseling or prying. Do not use a screwdriver to check an electric circuit since an electric arc will burn the tip and make it useless. In some cases, an electric arc may fuse the blade to the unit being checked, creating a short circuit.

When using a screwdriver on a small part, always hold the part in the vise or rest it on a workbench. Do not hold the part in the hand, as the screwdriver may slip and cause serious personal injury.

Replaceable tip screwdrivers, commonly referred to as "10 in 1" screwdrivers, allow for the quick changing of a screwdriver tip, and economical replacement of the tip when it becomes worn. A wide variety of screwdriver tips, including flat, crosspoint (Phillips and Reed & Prince), Torx and square drive tips are available for use with the handles. [Figure 9-3]

The cordless hand-held power screwdriver has replaced most automatic or spiral screwdrivers for the removal of multiple screws from an airframe. Care must be

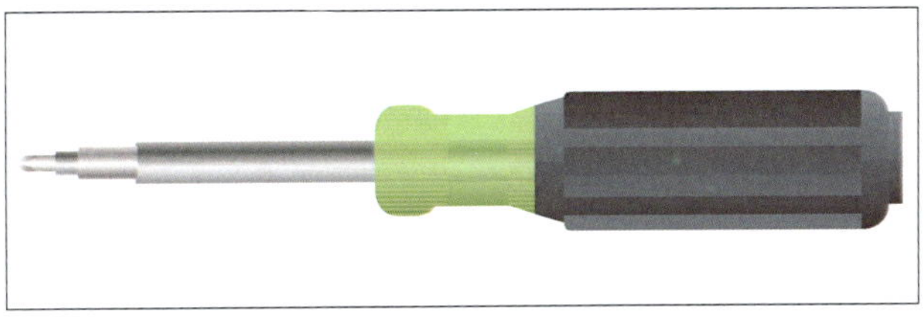

Figure 9-3. Replaceable tip screwdriver.

exercised when using a power screwdriver; if the slip clutch is set for too high a setting when installing a screw, the screwdriver tip will slip and rotate on top of the screw head, damaging it. The screw should be started by hand, to avoid driving the screw into the nut or nutplate in a cross-threaded manner. To avoid damaging the slot or receptacle in the head of the screw, the use of cordless power drills fitted with a removable tip driver to remove or install screws is not recommended, as the drill does not have a slip-clutch installed.

Pliers and Plier-Type Cutting Tools

As shown in Figure 9-4, the pliers used most frequently in aircraft repair work are the diagonal, needlenose, and duckbill. The size of pliers indicates their overall length, usually ranging from 5 to 12 inches.

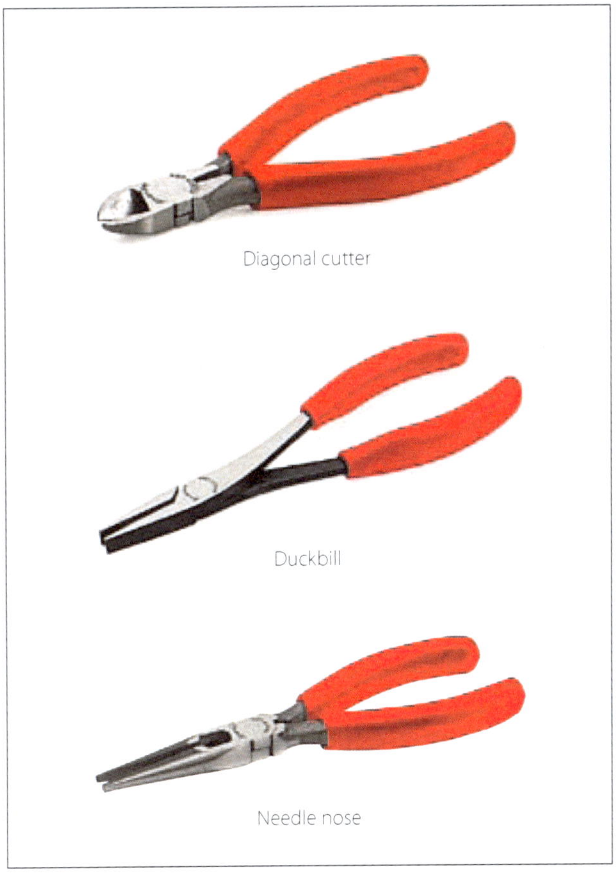

Figure 9-4. Pliers.

Roundnose pliers are used to crimp metal. They are not made for heavy work because too much pressure will spring the jaws, which are often wrapped to prevent scarring the metal.

Needlenose pliers have half round jaws of varying lengths. They are used to hold objects and make adjustments in tight places.

Duckbill pliers resemble a "duck's bill" in that the jaws are thin, flat, and shaped like a duck's bill. They are used exclusively for twisting safety wire.

Diagonal pliers are usually referred to as diagonals or "dikes." The diagonal is a short-jawed cutter with a blade set at a slight angle on each jaw. This tool can be used to cut wire, rivets, small screws, and cotter pins, besides being practically indispensable in removing or installing safety wire. The duckbill pliers and the diagonal cutting pliers are used extensively in aviation for the job of safety wiring.

Two important rules for using pliers are:

1. Do not make pliers work beyond their capacity. The long-nosed variety is especially delicate. It is easy to spring or break them, or nick the edges. If this occurs, they are practically useless.
2. Do not use pliers to turn nuts. In just a few seconds, a pair of pliers can damage a nut more than years of service.

Punches

Punches are used to locate centers for drawing circles, to start holes for drilling, to punch holes in sheet metal, to transfer location of holes in patterns, and to remove damaged rivets, pins or bolts.

Solid or hollow punches are the two types generally used. Solid punches are classified according to the shape of their points. Figure 9-5 shows several types of punches.

Prick punches are used to place reference marks on metal. This punch is often used to transfer dimensions from a paper pattern directly on the metal. To do this, first place the paper pattern directly on the metal. Then go over the outline of the pattern with the prick punch, tapping it lightly with a small hammer and making slight indentations on the metal at the major points on the drawing. These indentations can then be used as reference marks for cutting the metal. A prick punch should never be struck a heavy blow with a hammer because it may bend the punch or cause excessive damage to the material being worked.

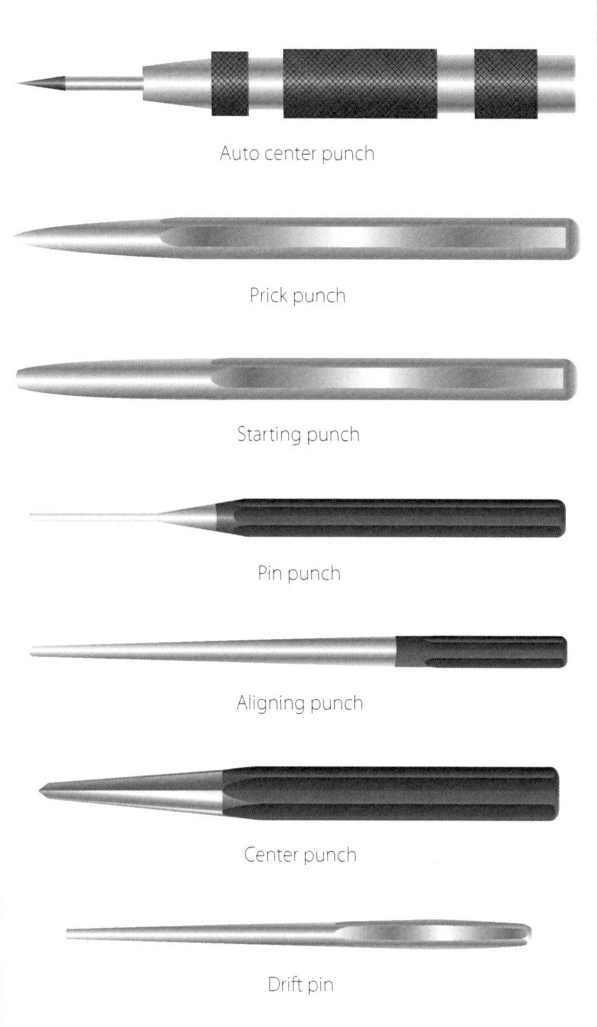

Figure 9-5. Punches.

Large indentations in metal, which are necessary to start a twist drill, are made with a center punch. It should never be struck with enough force to dimple the material around the indentation or to cause the metal to protrude through the other side of the sheet. A center punch has a heavier body than a prick punch and is ground to a point with an angle of about 60°.

The drive punch, which is often called a tapered punch, is used for driving out damaged rivets, pins, and bolts that sometimes bind in holes. The drive punch is therefore made with a flat face instead of a point. The size of the punch is determined by the width of the face, which is usually ⅛ inch to ¼ inch.

Pin punches, often called drift punches, are similar to drive punches and are used for the same purposes. The difference between the two is that the sides of a drive punch taper all the way to the face while the pin punch has a straight shank. Pin punches are sized by the diameter of the face, in thirty-seconds of an inch, and range from $^1/_{16}$ to ⅜ inch in diameter.

In general practice, a pin or bolt which is to be driven out is usually started and driven with a drive punch until the sides of the punch touch the side of the hole. A pin punch is then used to drive the pin or bolt the rest of the way out of the hole. Stubborn pins may be started by placing a thin piece of scrap copper, brass, or aluminum directly against the pin and then striking it with a hammer until the pin begins to move.

Never use a prick punch or center punch to remove objects from holes because the point of the punch will spread the object and cause it to bind even more.

The transfer punch is usually about 4 inches long. It has a point that tapers, and then turns straight for a short distance in order to fit a drill locating hole in a template. The tip has a point similar to that of a prick punch. As its name implies, the transfer punch is used to transfer the location of holes through the template or pattern to the material.

Wrenches

The wrenches most often used in aircraft maintenance are classified as open-end, box-end, socket, adjustable, ratcheting and special wrenches. The Allen wrench, although seldom used, is required on one special type of recessed screw. One of the most widely used metals for making wrenches is chrome-vanadium steel. Wrenches made of this metal are almost unbreakable. Solid, nonadjustable wrenches with open parallel jaws on one or both ends are known as open-end wrenches. These wrenches may have their jaws parallel to the handle or at an angle up to 90°; most are set at an angle of 15°. The wrenches are designed to fit a nut, bolt head, or other object, which makes it possible to exert a turning action.

Box-end wrenches are popular tools because of their usefulness in close quarters. They are called box wrenches since they box, or completely surround, the nut or bolt head. Practically all well-manufactured box-end wrenches are made with 12 points so they can be used in places having as little as 15° swing. In Figure 9-6, point A on the illustrated double broached hexagon wrench is nearer the centerline of the head and the wrench handle than point B, and also the centerline of nut C. If the wrench is inverted and installed on nut C, point A will be centered over side "Y" instead of side "X." The centerline of the handle will now be in the dotted line position. It is by reversing (turning the

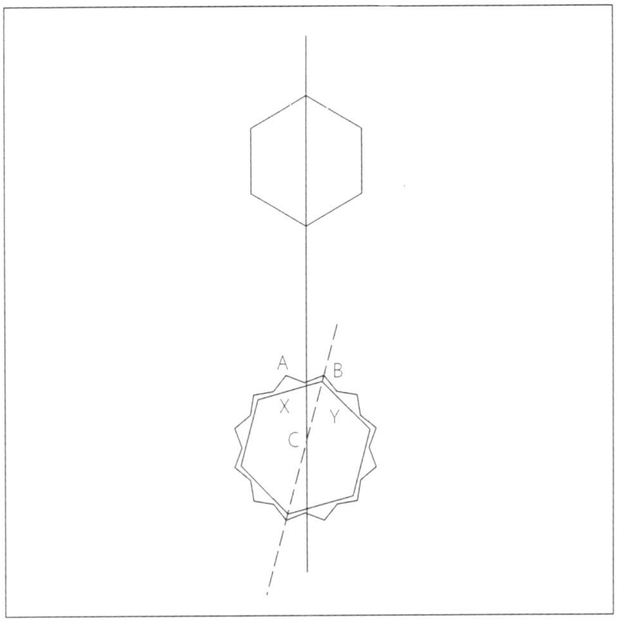

Figure 9-6. Box-end wrench use.

wrench over) the position of the wrench that a 15° arc may be made with the wrench handle.

Although box-end wrenches are ideal to break loose tight nuts or pull tight nuts tighter, time is lost turning the nut off the bolt once the nut is broken loose. Only when there is sufficient clearance to rotate the wrench in a complete circle can this tedious process be avoided.

After a tight nut is broken loose, it can be completely backed off or unscrewed more quickly with an open-end than with a box-end wrench. In this case, a combination wrench can be used; it has a box end on one end and an open-end wrench of the same size on the other.

Another option for removing a nut from a bolt is the ratcheting box-end wrench, which can be swung back and forth to remove the nut or bolt. The box-end, combination, and ratcheting wrenches are shown in Figure 9-7.

A socket wrench is made of two parts: (1) the socket, which is placed over the top of a nut or bolt head, and (2) a handle, which is attached to the socket. Many types of handles, extensions, and attachments are available to make it possible to use socket wrenches in almost any location or position. Sockets are made with either fixed or detachable handles. Socket wrenches with fixed handles are usually furnished as an accessory to a machine. They have either a four, six, or twelve-sided recess to fit a nut or bolt head that needs regular adjustment.

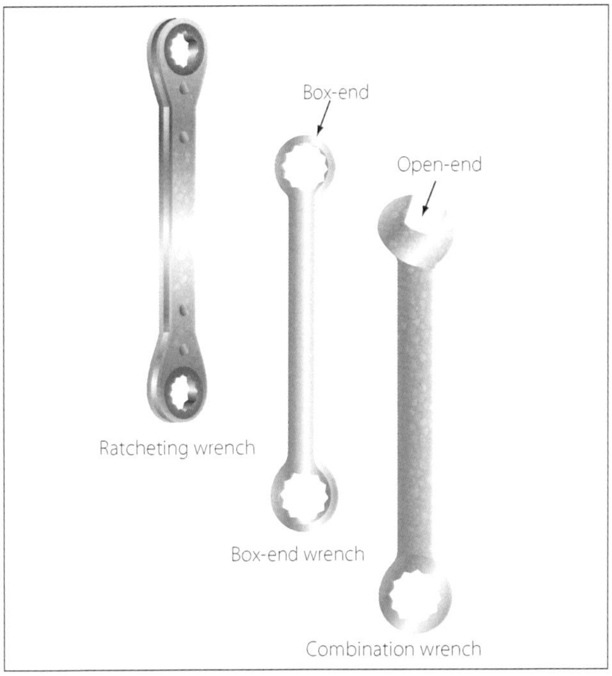

Figure 9-7. Box-end and combination wrenches.

Sockets with detachable handles usually come in sets and fit several types of handles, such as the T, ratchet, screwdriver grip, and speed handle. Socket wrench handles have a square lug on one end that fits into a square recess in the socket head. The two parts are held together by a light spring-loaded poppet. Two types of sockets, a set of handles, and an extension bar are shown in Figure 9-8.

The adjustable wrench is a handy utility tool that has smooth jaws and is designed as an open-end wrench. One jaw is fixed, but the other may be moved by a thumbscrew or spiral screwworm adjustment in the handle. The width of the jaws may be varied from 0 to ½ inch or more. The angle of the opening to the handle is 22½ degrees on an adjustable wrench. One adjustable wrench does the work of several open-end wrenches. Although versatile, they are not intended to replace the standard open-end, box-end, or socket wrenches. When using any adjustable wrench, always exert the pull on the side of the handle attached to the fixed jaw of the wrench. To minimize the possibility or rounding off the fastener, use care to fit the wrench to the bolt or nut to be turned.

Special Wrenches

The category of special wrenches includes the crowfoot, flare nut, spanner, torque, and Allen wrenches. [Figures 9-9 and 9-10]

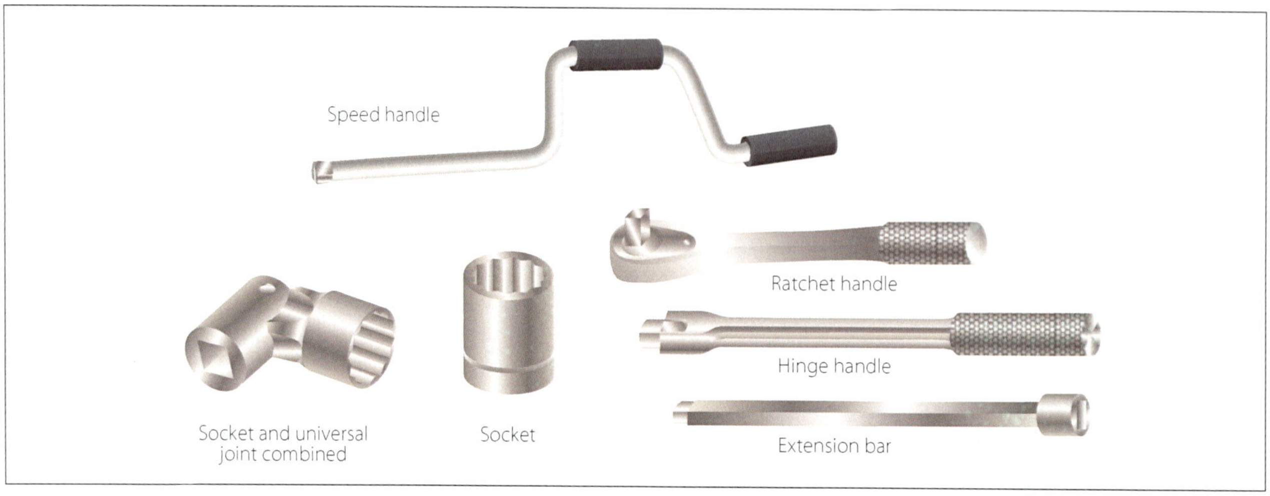

Figure 9-8. Socket wrench set.

The crowfoot wrench is normally used when accessing nuts that must be removed from studs or bolt that cannot be accessed using other tools.

The flare nut wrench has the appearance of a box-end wrench that has been cut open on one end. This opening allows the wrench to be used on the B-nut of a fuel, hydraulic, or oxygen line. Since it mounts using the standard square adapter, like the crowfoot wrench, it can be used in conjunction with a torque wrench.

The hook spanner is for a round nut with a series of notches cut in the outer edge. This wrench has a curved arm with a hook on the end that fits into one of the

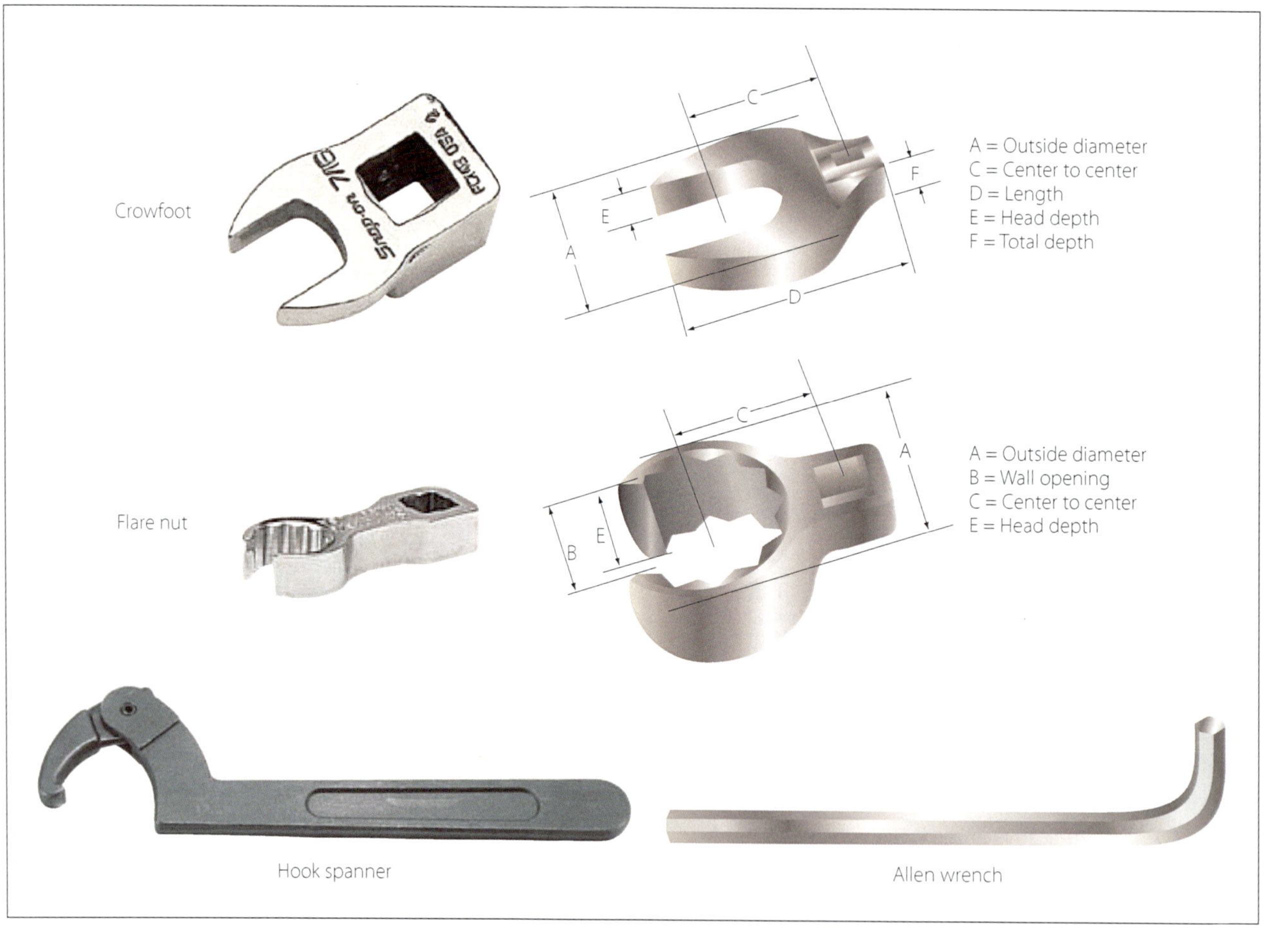

Figure 9-9. Special wrenches.

notches on the nut. The hook is placed in one of these notches with the handle pointing in the direction the nut is to be turned.

Some hook spanner wrenches are adjustable and will fit nuts of various diameters. U-shaped hook spanners have two lugs on the face of the wrench to fit notches cut in the face of the nut or screw plug. End spanners resemble a socket wrench, but have a series of lugs that fit into corresponding notches in a nut or plug. Pin spanners have a pin in place of a lug, and the pin fits into a round hole in the edge of a nut. Face pin spanners are similar to the U-shaped hook spanners except that they have pins instead of lugs.

Most headless setscrews are the hex-head Allen type and must be installed and removed with an Allen wrench. Allen wrenches are six-sided bars in the shape of an L, or they can be hex-shaped bars mounted in adapters for use with hand ratchets. They range in size from $\frac{3}{64}$ to $\frac{1}{2}$ inch and fit into a hexagonal recess in the setscrew.

Torque Wrench

There are times when definite pressure must be applied to a nut or bolt as it is installed. In such cases a torque wrench must be used. The torque wrench is a precision tool consisting of a torque indicating handle and appropriate adapter or attachments. It measures the amount of turning or twisting force applied to a nut, bolt, or screw.

Before each use, the torque wrench should be visually inspected for damage. If a bent pointer, cracked or broken glass (dial type), or signs of rough handling are found, the wrench must be tested. Torque wrenches must be tested at periodic intervals to ensure accuracy.

Commonly used torque wrenches include the deflecting beam (not shown), dial indicating, micrometer, and electronic setting types. [Figure 9-10]

When using the deflecting beam and the dial indicating torque wrenches, the torque is read visually on a dial or scale mounted on the handle of the wrench. The micrometer setting torque wrench is preset to the desired torque. When this torque is reached, the operator notices a sharp impulse or breakaway "click." For additional information on the installation of fasteners requiring the use of a torque wrench, refer to "Installation of Nuts, Washers, and Bolts" located in chapter 5.

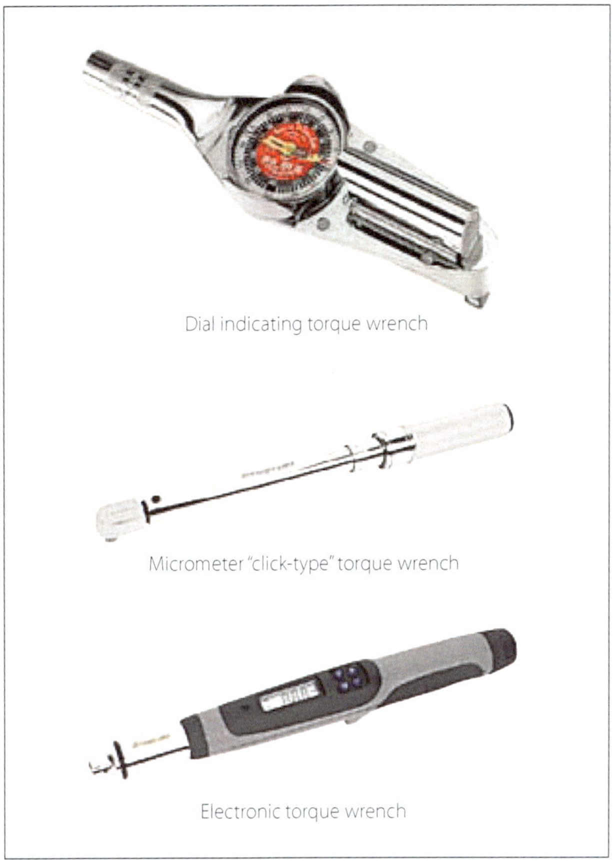

Figure 9-10. Torque wrenches.

Strap Wrenches

The strap wrench can prove to be an invaluable tool for the AMT. By their very nature, aircraft components such as tubing, pipes, small fittings, and round or irregularly shaped components are built to be as light as possible, while still retaining enough strength to function properly. The misuse of pliers or other gripping tools can quickly damage these parts. If it is necessary to grip a part to hold it in place, or to rotate it to facilitate removal, consider using a strap wrench that uses a plastic covered fabric strap to grip the part. [Figure 9-11]

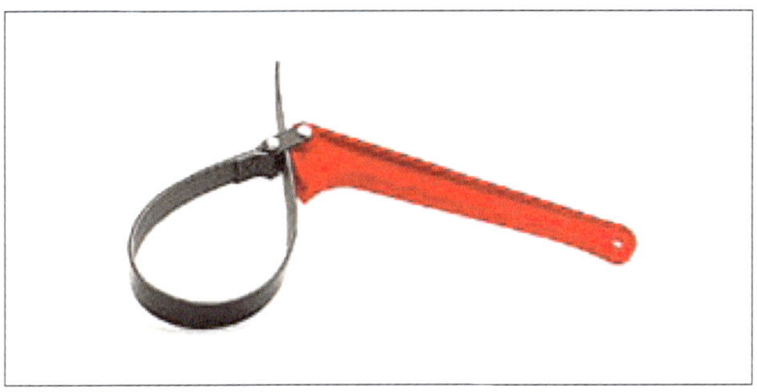

Figure 9-11. Strap wrench.

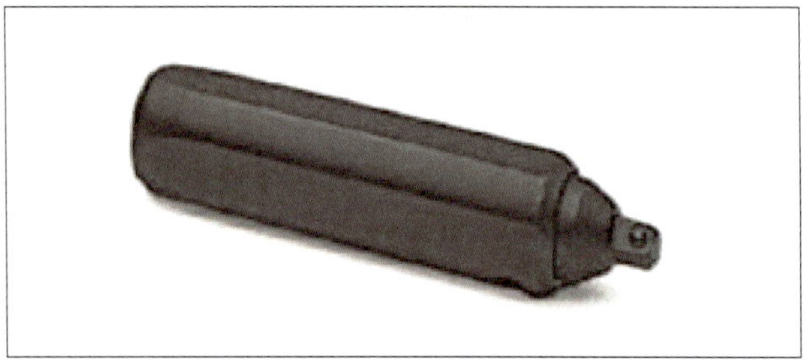

Figure 9-12. Impact driver.

Impact Drivers

In certain applications, the use of an impact driver may be required. Struck with a mallet, the impact driver uses cam action to impart a high amount of torque in a sharp impact to break loose a stubborn fastener. The drive portion of the impact driver can accept a number of different drive bits and sockets. The use of special bits and sockets specifically manufactured for use with an impact driver is required. [Figure 9-12]

Metal Cutting Tools

Hand Snips

There are several kinds of hand snips, each of which serves a different purpose. Straight, curved, hawksbill, and aviation snips are in common use. Straight snips are used for cutting straight lines when the distance is not great enough to use a squaring shear and for cutting the outside of a curve. The other types are used for cutting the inside of curves or radii. Snips should never be used to cut heavy sheet metal. [Figure 9-13]

Figure 9-13. Typical snips.

Aviation snips are designed especially for cutting heat-treated aluminum alloy and stainless steel. They are also adaptable for enlarging small holes. The blades have small teeth on the cutting edges and are shaped for cutting very small circles and irregular outlines. The handles are the compound leverage type, making it possible to cut material as thick as 0.051 inch. Aviation snips are available in two types, those which cut from right to left and those which cut from left to right.

Unlike the hacksaw, snips do not remove any material when the cut is made, but minute fractures often occur along the cut. Therefore, cuts should be made about $\frac{1}{32}$ inch from the layout line and finished by hand filing down to the line.

Hacksaws

The common hacksaw has a blade, a frame, and a handle. The handle can be obtained in two styles: pistol grip and straight. [Figure 9-14]

Hacksaw blades have holes in both ends; they are mounted on pins attached to the frame. When installing a blade in a hacksaw frame, mount the blade with the teeth pointing forward, away from the handle.

Blades are made of high-grade tool steel or tungsten steel and are available in sizes from 6 to 16 inches in length. The 10-inch blade is most commonly used. There are two types, the all-hard blade and the flexible blade. In flexible blades, only the teeth are hardened.

Selection of the best blade for the job involves finding the right type and pitch. An all-hard blade is best for sawing brass, tool steel, cast iron, and heavy cross-section materials. A flexible blade is usually best for sawing hollow shapes and metals having a thin cross section.

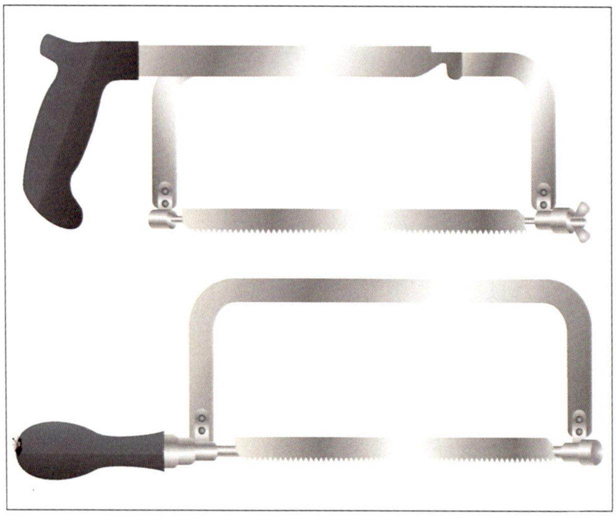

Figure 9-14. Hacksaws.

The pitch of a blade indicates the number of teeth per inch. Pitches of 14, 18, 24, and 32 teeth per inch are available. A blade with 14 teeth per inch is preferred when cutting machine steel, cold rolled steel, or structural steel. A blade with 18 teeth per inch is preferred for solid stock aluminum, bearing metal, tool steel, and cast iron. Use a blade with 24 teeth per inch when cutting thick-walled tubing, pipe, brass, copper, channel, and angle iron. Use the 32 teeth per inch blade for cutting thin-walled tubing and sheet metal. When using a hacksaw, observe the following procedures:

1. Select an appropriate saw blade for the job.
2. Assemble the blade in the frame so that the cutting edge of the teeth points away from the handle.
3. Adjust tension of the blade in the frame to prevent the saw from buckling and drifting.
4. Clamp the work in the vise in such a way that will provide as much bearing surface as possible and will engage the greatest number of teeth.
5. Indicate the starting point by nicking the surface with the edge of a file to break any sharp corner that might strip the teeth. This mark will also aid in starting the saw at the proper place.
6. Hold the saw at an angle that will keep at least two teeth in contact with the work at all times. Start the cut with a light, steady, forward stroke just outside the cutting line. At the end of the stroke, relieve the pressure and draw the blade back. (The cut is made on the forward stroke.)
7. After the first few strokes, make each stroke as long as the hacksaw frame will allow. This will prevent the blade from overheating. Apply just enough pressure on the forward stroke to cause each tooth to remove a small amount of metal. The strokes should be long and steady with a speed not more than 40 to 50 strokes per minute.
8. After completing the cut, remove chips from the blade, loosen tension on the blade, and return the hacksaw to its proper place.

Chisels

A chisel is a hard steel cutting tool that can be used for cutting and chipping any metal softer than the chisel itself. It can be used in restricted areas and for such work as shearing rivets, or splitting seized or damaged nuts from bolts. [Figure 9-15]

The size of a flat cold chisel is determined by the width of the cutting edge. Lengths will vary, but chisels are seldom under 5 inches or over 8 inches long.

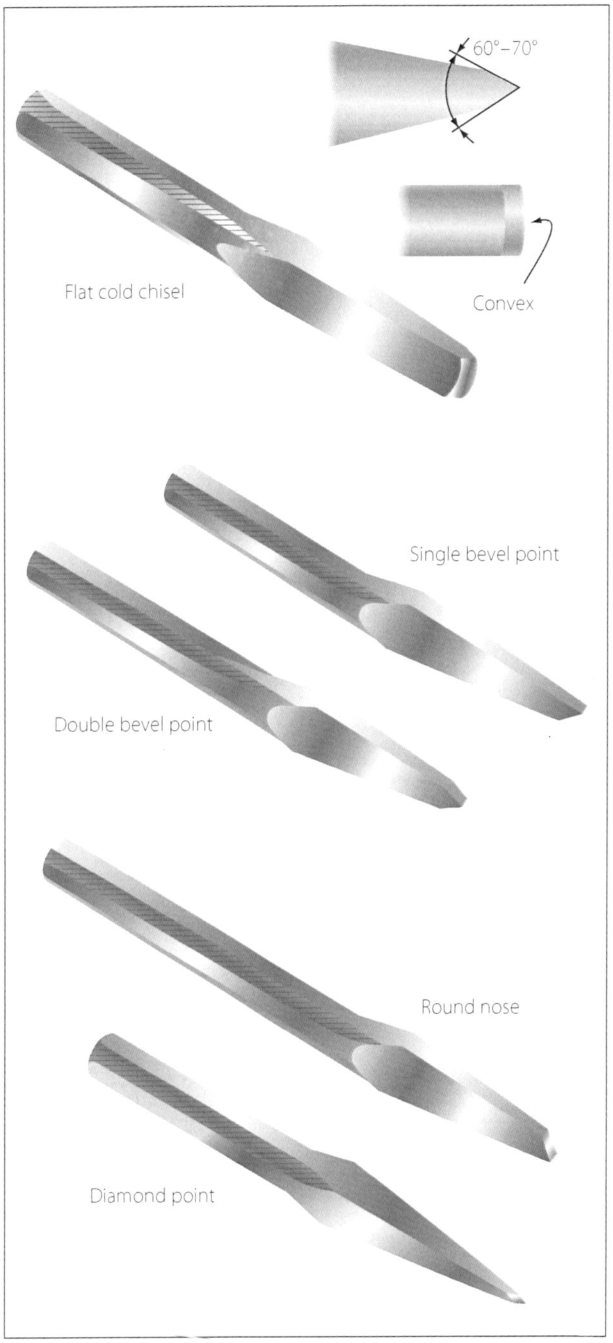

Figure 9-15. Chisels.

Chisels are usually made of eight-sided tool steel bar stock, carefully hardened and tempered. Since the cutting edge is slightly convex, the center portion receives the greatest shock when cutting, and the weaker corners are protected. The cutting angle should be 60° to 70° for general use, such as for cutting wire, strap iron, or small bars and rods.

When using a chisel, hold it firmly in one hand. With the other hand, strike the chisel head squarely with a ball peen hammer.

When cutting square corners or slots, a special cold chisel called a cape chisel should be used. It is like a flat chisel except the cutting edge is very narrow. It has the same cutting angle and is held and used in the same manner as any other chisel.

Rounded or semicircular grooves and corners that have fillets should be cut with a roundnose chisel. This chisel is also used to re-center a drill that has moved away from its intended center.

The diamond point chisel is tapered square at the cutting end, and then ground at an angle to provide the sharp diamond point. It is used for cutting B-grooves and inside sharp angles.

Files

Most files are made of high-grade tool steels that are hardened and tempered. Files are manufactured in a variety of shapes and sizes. They are known either by the cross section, the general shape, or by their particular use. The cuts of files must be considered when selecting them for various types of work and materials.

Files are used to square ends, file rounded corners, remove burrs and slivers from metal, straighten uneven edges, file holes and slots, and smooth rough edges.

Files have three distinguishing features: (1) their length, measured exclusive of the tang [Figure 9-16]; (2) their kind or name, which has reference to the relative coarseness of the teeth; and (3) their cut.

Files are usually made in two types of cuts: single cut and double cut. The single cut file has a single row of teeth extending across the face at an angle of 65° to 85° with the length of the file. The size of the cuts depends on the coarseness of the file. The double cut file has two rows of teeth that cross each other. For general work, the angle of the first row is 40° to 45°. The first row is generally referred to as "overcut," and the second row as "upcut;" the upcut is somewhat finer and not as deep as the overcut.

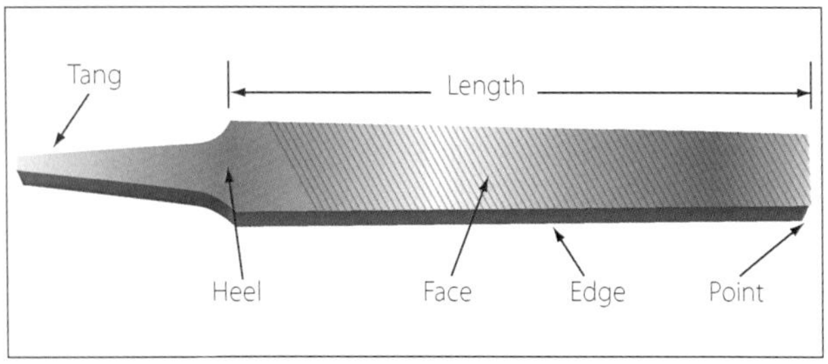

Figure 9-16. Hand file.

Files — Care and Use

Files and rasps are catalogued in three ways:

- *Length.* Measuring from the tip to the heel of the file. The tang is never included in the length.
- *Shape.* Refers to the physical configuration of the file (circular, rectangular, or triangular or a variation thereof).
- *Cut.* Refers to both the character of the teeth or the coarseness—rough, coarse, and bastard for use on heavier classes of work and second cut, smooth and dead smooth for finishing work.

Most Commonly Used Files [Figure 9-17]

Hand files—These are parallel in width and tapered in thickness. They have one safe edge (smooth edge) which permits filing in corners, and on other work where a safe edge is required. Hand files are double cut and used principally for finishing flat surfaces and similar work.

Flat files—These files are slightly tapered toward the point in both width and thickness. They cut on both edges as well as on the sides. They are the most common files in use. Flat files are double cut on both sides and single cut on both edges.

Mill files—These are usually tapered slightly in thickness and in width for about one-third of their length. The teeth are ordinarily single cut. These files are used for drawfiling and to some extent for filing soft metals.

Square files—These files may be tapered or blunt and are double cut. They are used principally for filing slots and key seats, and for surface filing.

Round or rattail files—These are circular in cross section and may be either tapered or blunt and single or double cut. They are used principally for filing circular openings or concave surfaces.

Triangular and three square files—These files are triangular in cross section. Triangular files are single cut and are used for filing the gullet between saw teeth. Three square files, which are double cut, may be used for filing internal angles, clearing out corners, and filing taps and cutters.

Half-round files—These files cut on both the flat and round sides. They may be single or double cut. Their shape permits them to be used where other files would be unsatisfactory.

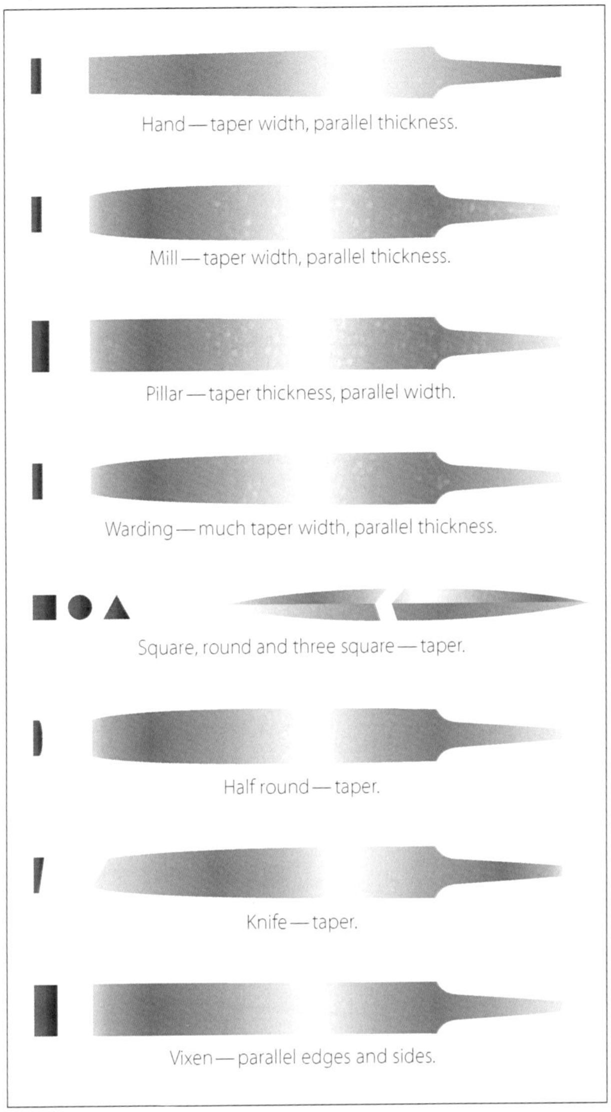

Figure 9-17. Types of files.

Lead float files—These are especially designed for use on soft metals. They are single cut and are made in various lengths.

Warding file—Rectangular in section and tapers to narrow point in width. This file is used for narrow space filing where other files cannot be used.

Knife file—Knife blade section. This file is used by tool and die makers on work having acute angles.

Wood file—Same section as flat and half-round files. This file has coarser teeth and is especially adaptable for use on wood.

Vixen (curved-tooth files)—Curved-tooth files are especially designed for rapid filing and smooth finish on soft metals and wood. The regular cut is adapted for tough work on cast iron, soft steel, copper, brass, aluminum, wood, slate, marble, fiber, rubber, and so forth. The fine cut gives excellent results on steel, cast iron, phosphor bronze, white brass, and all hard metals. The smooth cut is used where the amount of material to be removed is very slight, but where a superior finish is desired.

The following methods are recommended for using files:

1. *Crossfiling.* Before attempting to use a file, place a handle on the tang of the file. This is essential for proper guiding and safe use. In moving the file endwise across the work (commonly known as crossfiling), grasp the handle so that its end fits into and against the fleshy part of the palm with the thumb lying along the top of the handle in a lengthwise direction. Grasp the end of the file between the thumb and first two fingers. To prevent undue wear of the file, relieve the pressure during the return stroke.
2. *Drawfiling.* A file is sometimes used by grasping it at each end, crosswise to the work, then moving it lengthwise with the work. When done properly, work may be finished somewhat finer than when crossfiling with the same file. In drawfiling, the teeth of the file produce a shearing effect. To accomplish this shearing effect, the angle at which the file is held with respect to its line of movement varies with different files, depending on the angle at which the teeth are cut. Pressure should be relieved during the backstroke.
3. *Rounding corners.* The method used in filing a rounded surface depends upon its width and the radius of the rounded surface. If the surface is narrow or only a portion of a surface is to be rounded, start the forward stroke of the file with the point of the file inclined downward at approximately a 45° angle. Using a rocking chair motion, finish the stroke with the heel of the file near the curved surface. This method allows use of the full length of the file.
4. *Removing burred or slivered edges.* Practically every cutting operation on sheet metal produces burrs or slivers. These must be removed to avoid personal injury and to prevent scratching and marring of parts to be assembled. Burrs and slivers will prevent parts from fitting properly and should always be removed from the work as a matter of habit.

Lathe filing requires that the file be held against the work revolving in the lathe. The file should not be held rigid or stationary but should be stroked constantly with a slight gliding or lateral motion along the work. A standard mill file may be used for this operation, but the long angle lathe file provides a much cleaner shearing and self-clearing action. Use a file with "safe" edges to protect work with shoulders from being marred.

Care of Files

There are several precautions that any good craftsman will take in caring for files.

1. Choose the right file for the material and work to be performed.
2. Keep all files racked and separated so they do not bear against each other.
3. Keep the files in a dry place—rust will corrode the teeth points, dulling the file.
4. Keep files clean. Tap the end of the file against the bench after every few strokes, to loosen and clear the filings. Use the file card to keep files clean—a dirty file is a dull file. A dirty file can also contaminate different metals when the same file is used on multiple metal surfaces.

Particles of metal collect between the teeth of a file and may make deep scratches in the material being filed. When these particles of metal are lodged too firmly between the teeth and cannot be removed by tapping the edge of the file, remove them with a file card or wire brush. Draw the brush across the file so that the bristles pass down the gullet between the teeth. [Figure 9-18]

Figure 9-18. File card.

Drills

There are generally four types of portable drills used in aviation for holding and turning twist drills. Holes ¼ inch in diameter and under can be drilled using a hand drill. This drill is commonly called an "egg beater." The breast drill is designed to hold larger size twist drills than the hand drill. In addition, a breastplate is affixed at the upper end of the drill to permit the use of body weight to increase the cutting power of the drill. Electric and pneumatic power drills are available in various shapes and sizes to satisfy almost any requirement.

Pneumatic drills are preferred for use around flammable materials, since sparks from an electric drill are a fire or explosion hazard.

Twist Drills

A twist drill is a pointed tool that is rotated to cut holes in material. It is made of a cylindrical hardened steel bar having spiral flutes (grooves) running the length of the body, and a conical point with cutting edges formed by the ends of the flutes.

Twist drills are made of carbon steel or high-speed alloy steel. Carbon steel twist drills are satisfactory for the general run of work and are relatively inexpensive. The more expensive high-speed twist drills are used for the tough materials such as stainless steels. Twist drills have from one to four spiral flutes. Drills with two flutes are used for most drilling; those with three or four flutes are used principally to follow smaller drills or to enlarge holes.

The principal parts of a twist drill are the shank, the body, and the heel. [Figure 9-19] The drill shank is the end that fits into the chuck of a hand or power drill. The two shank shapes most commonly used in hand drills are the straight shank and the square or bit stock shank. The straight shank generally is used in hand, breast, and portable electric or pneumatic drills; the square shank is made to fit into a carpenter's brace. Tapered shanks generally are used in machine shop drill presses. [Figure 9-20]

The metal column forming the core of the drill is the body. The body clearance area lies just back of the margin; it is slightly smaller in diameter than the margin, to reduce the friction between the drill and the sides of the hole. The angle at which the drill point is ground is the lip clearance angle. On standard drills used to cut steel and cast iron, the angle should be 59° from the axis of the drill. For faster drilling of soft materials, sharper angles are used.

The diameter of a twist drill may be given in one of three ways: (1) by fractions, (2) letters, or (3) numbers. Fractionally, they are classified by sixteenths of an inch (from $\frac{1}{16}$ to $3\frac{1}{2}$ inch), by thirty-seconds (from $\frac{1}{32}$ to $2\frac{1}{2}$ inch), or by sixty-fourths (from $\frac{1}{64}$ to $1\frac{1}{4}$ inch). For a more exact measurement, a letter system is used with decimal equivalents: A (0.234 inch) to Z (0.413 inch). The number system of classification is

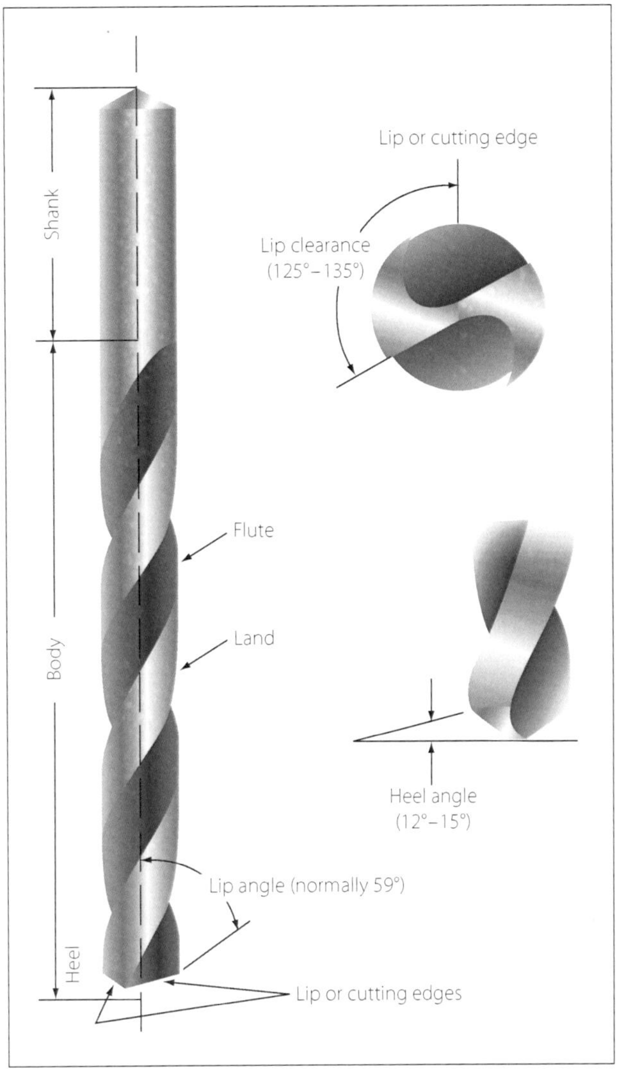

Figure 9-19. Twist drill.

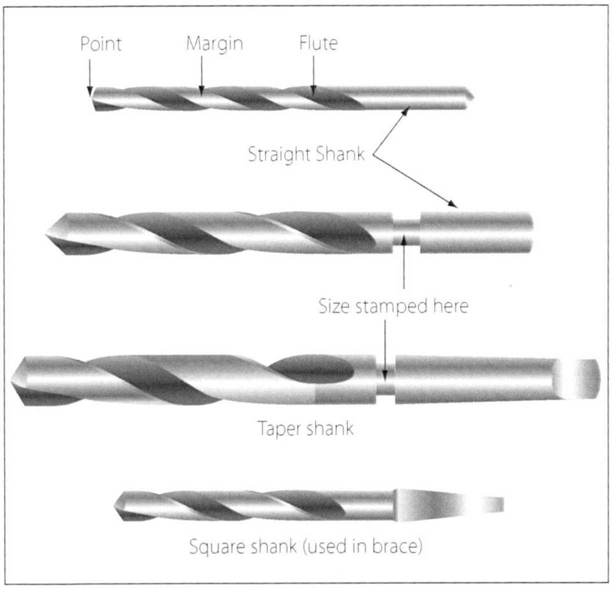

Figure 9-20. Drill types.

most accurate: No. 80 (0.0314 inch) to No. 1 (0.228 inch). Drill sizes and their decimal equivalents are shown in Figure 9-21.

The twist drill should be sharpened at the first sign of dullness. For most drilling, a twist drill with a cutting angle of 118° (59° on either side of center) will be sufficient; however, when drilling soft metals, a cutting angle of 90° may be more efficient.

Typical procedures for sharpening drills are as follows: [Figure 9-22]

1. Adjust the grinder tool rest to a convenient height for resting the back of the hand while grinding.
2. Hold the drill between the thumb and index finger of the right or left hand. Grasp the body of the drill near the shank with the other hand.
3. Place the hand on the tool rest with the centerline of the drill making a 59° angle with the cutting face of the grinding wheel. Lower the shank end of the drill slightly.
4. Slowly place the cutting edge of the drill against the grinding wheel. Gradually lower the shank of the drill as you twist the drill in a clockwise direction. Maintain pressure against the grinding surface only until you reach the heel of the drill.
5. Check the results of grinding with a gauge to determine whether or not the lips are the same length and at a 59° angle.

Alternatively, there are commercially available twist drill grinders available, as well as attachments for bench grinders which will ensure consistent, even sharpening of twist drills.

Reamers

Reamers are used to smooth and enlarge holes to exact size. Hand reamers have square end shanks so that they can be turned with a tap wrench or similar handle. The various types of reamers are illustrated in Figure 9-23.

A hole that is to be reamed to exact size must be drilled about 0.003 to 0.007 inch undersize. A cut that removes more than 0.007 inch places too much load on the reamer and should not be attempted.

Reamers are made of either carbon tool steel or high-speed steel. The cutting blades of a high-speed steel reamer lose their original keenness sooner than those

Millimeter	Decimal Equivalent	Fractional	Number	Millimeter	Decimal Equivalent	Fractional	Number	Millimeter	Decimal Equivalent	Fractional	Number	Millimeter	Decimal Equivalent	Fractional	Number
.1	.0039			—	.0410		59	2.2	.0866			—	.1470		26
.15	.0059			1.05	.0413			2.25	.0885			3.75	.1476		
.2	.0079			—	.0420		58	—	.0890		43	—	.1495		25
.25	.0098			—	.0430		57	2.3	.0905			3.8	.1496		
.3	.0118			1.1	.0433			2.35	.0925			—	.1520		24
—	.0135		80	1.15	.0452			—	.0935		42	3.9	.1535		
.35	.0138			—	.0465		56	2.38	.0937	3/32	—	—	.1540		23
—	.0145		79	1.19	.0469	3/64	—	2.4	.0945			3.97	.1562	5/32	—
.39	.0156	1/64	—	1.2	.0472			—	.0960		41	—	.1570		22
.4	.0157			1.25	.0492			2.45	.0964			4.0	.1575		
—	.0160		78	1.3	.0512			—	.0980		40	—	.1590		21
.45	.0177			—	.0520		55	2.5	.0984			—	.1610		20
—	.0180		77	1.35	.0531			—	.0995		39	4.1	.1614		
.5	.0197			—	.0550		54	—	.1015		38	4.2	.1654		
—	.0200		76	1.4	.0551			2.6	.1024			—	.1660		19
—	.0210		75	1.45	.0570			—	.1040		37	4.25	.1673		
.55	.0217			1.5	.0591			2.7	.1063			4.3	.1693		
—	.0225		74	—	.0595		53	—	.1065		36	—	.1695		18
.6	.0236			1.55	.0610			2.75	.1082			4.37	.1719	11/64	—
—	.0240		73	1.59	.0625	1/16	—	2.78	.1094	7/64	—	—	.1730		17
—	.0250		72	1.6	.0629			-	.1100		35	4.4	.1732		
.65	.0256			—	.0635		52	2.8	.1102			—	.1770		16
—	.0260		71	1.65	.0649			—	.1110		34	4.5	.1771		
—	.0280		70	1.7	.0669			—	.1130		33	—	.1800		15
.7	.0276			—	.0670		51	2.9	.1141			4.6	.1811		
—	.0292		69	1.75	.0689			—	.1160		32	—	.1820		14
.75	.0295			—	0.700		50	3.0	.1181			4.7	.1850		13
—	.0310		68	1.8	.0709			—	.1200		31	4.75	.1870		
.79	.0312	1/32	—	1.85	.0728			3.1	.1220			4.76	.1875	3/16	—
.8	.0315			—	.0730		49	3.18	.1250	1/8	—	4.8	.1890		12
—	.0320		67	1.9	.0748			3.2	.1260			—	1910		11
—	.0330		66	—	.0760		48	3.25	.1279			4.9	.1929		
.85	.0335			1.95	.0767			—	.1285		30	—	.1935		10
—	.0350		65	1.98	.0781	5/64	—	3.3	.1299			—	.1960		9
.9	.0354			—	.0785		47	3.4	.1338			5.0	.1968		
—	.0360		64	2.0	.0787			—	.1360		29	—	.1990		8
—	.0370		63	2.05	.0807			3.5	.1378			5.1	.2008		
.95	.0374			—	.0810		46	—	.1405		28	—	.2010		7
—	.0380		62	—	.0820		45	3.57	.1406	9/64		5.16	.2031	13/64	—
—	.0390		61	2.1	.0827			3.6	.1417			—	.2040		6
1.0	.0394			2.15	.0846			—	.1440		27	5.2	.2047		
—	.0400		60	—	.0860		44	3.7	.1457		—	—	.2055		5

Figure 9-21. Drill sizes.

Millimeter	Decimal Equivalent	Fractional	Number	Millimeter	Decimal Equivalent	Fractional	Number	Millimeter	Decimal Equivalent	Fractional	Number	Millimeter	Decimal Equivalent	Fractional	Number
5.25	.2067			7.25	.2854			9.5	.3740			16.5	.6496		
5.3	.2086			7.3	.2874			9.53	.3750	3/8	—	16.67	.6562	21/32	
—	.2090		4	—	2900		L	—	.3770		V	17.0	.6693		
5.4	.2126			7.4	.2913			9.6	.3780			17.06	.6719	43/64	
—	.2130			—	.2950		M	9.7	.3819			17.46	.6875	11/16	
5.5	.2165			7.5	.2953			9.75	.3838			17.5	.6890		
5.56	.2187	1/32	—	7.54	.2968	19/64	—	9.8	.3858			17.86	.7031	45/64	
5.6	.2205			7.6	.2992			—	.3860		W	18.0	7087		
—	.2210		2	—	.3020		N	9.9	.3898			18.26	.7187	23/32	
5.7	.2244			7.7	.3031			9.92	.3906	25/64	—	18.5	.7283		
5.75	.2263			7.75	.3051			10.0	.3937			18.65	.7344	47/64	
—	.2280		1	7.8	.3071				.3970		X	19.0	.7480		
5.8	.2283			7.9	.3110			—	.4040		Y	19.05	7500	3/4	
5.9	.2323			7.94	.3125	5/16	—	10.32	.4062	13/32	—	19.45	.7656	49/64	
—	.2340		A	8.0	.3150			—	.4130		Z	19.5	.7677		
5.95	.2344	15/64	—	—	.3160		O	10.5	.4134			19.84	.7812	25/32	
6.0	.2362			8.1	.3189			10.72	.4219	27/64		20.0	7874		
—	.2380		B	8.2	.3228			11.0	.4330			20.24	.7969	51/64	
6.1	.2401			—	.3230		P	11.11	.4375	7/16		20.5	.8071		
—	.2420		C	8.25	.3248			11.5	.4528			20.64	.8125	13/16	
6.2	.2441			8.3	.3268			11.51	.4531	29/64		21.0	.8268		
6.25	.2460		D	8.33	.3281	21/64	—	11.91	.4687	15/32		21.03	.8281	53/64	
6.3	.2480			8.4	.3307			12.0	.4724			21.43	.8437	27/32	
6.35	.2500	1/4	E	—	.3320		Q	12.30	.4843	31/64		21.5	.8465		
6.4	.2520			8.5	.3346			12.5	.4921			21.83	.8594	55/64	
6.5	.2559			8.6	.3386			12.7	.5000	1/2		22.0	.8661		
—	.2570		F	—	.3390		R	13.0	.5118			22.23	.8750	7/8	
6.6	.2598			8.7	.3425			13.10	.5156	33/64		22.5	.8858		
—	.2610		G	8.73	.3437	11/32	—	13.49	.5312	17/32		22.62	.8906	57/64	
6.7	.2638			8.75	.3445			13.5	.5315			23.0	.9055		
6.75	.2657	17/64	—	8.8	.3465			13.89	.5469	35/64		23.02	.9062	29/32	
6.75	.2657			—	.3480		S	14.0	.5512			23.42	.9219	59/64	
—	.2660		H	8.9	.3504			14.29	.5625	9/16		23.5	.9252		
6.8	.2677			9.0	.3543			14.5	.5709			23.81	.9375	15/16	
6.9	.2716			—	.3580		T	14.68	.5781	37/64		24.0	.9449		
—	.2720		I	9.1	.3583			15.0	.5906			24.21	.9531	61/64	
7.0	.2756			9.13	.3594	23/64	—	15.08	.5937	19/32		24.5	.9646		
—	.2770		J	9.2	.3622			15.48	.6094	39/32		24.61	.9687	31/32	
7.1	.2795			9.25	.3641			15.5	.6102			25.0	.9843		
—	.2811		K	9.3	.3661			15.88	.6250	5/8		25.03	.9844	63/64	
7.14	.2812	9/32	—	—	.3680		U	16.0	.6299			25.4	1.0000	1	
7.2	.2835			9.4	.3701			16.27	.6406	41/64					

Figure 9-21. Drill sizes (continued).

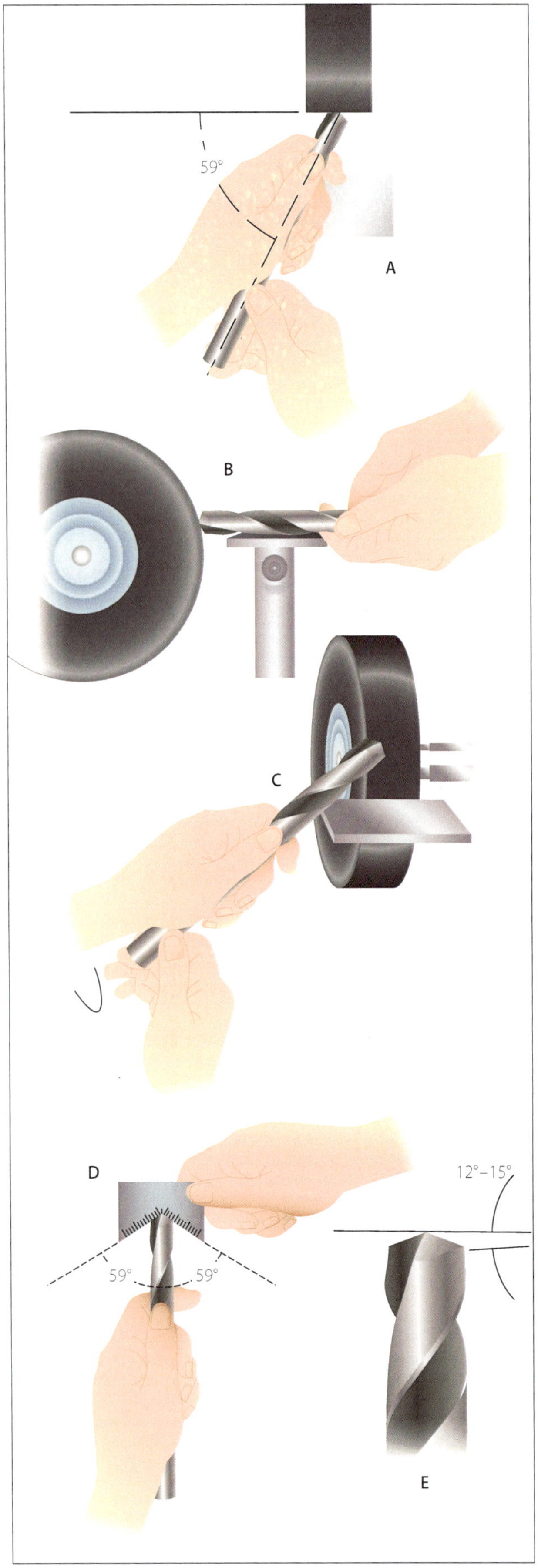

Figure 9-22. Drill sharpening procedures.

of a carbon steel reamer; however, after the first super keenness is gone, they are still serviceable. The high-speed reamer usually lasts much longer than the carbon steel type.

Reamer blades are hardened to the point of being brittle and must be handled carefully to avoid chipping them. When reaming a hole, rotate the reamer in the cutting direction only. Do not back a reamer out of a hole by rotating it opposite the cutting direction. Turn the reamer steadily and evenly to prevent chattering, or marking and scoring of the hole walls.

Reamers are available in any standard size. The straight fluted reamer is less expensive than the spiral fluted reamer, but the spiral type has less tendency to chatter. Both types are tapered for a short distance back of the end to aid in starting. Bottoming reamers have no taper and are used to complete the reaming of blind holes.

For general use, an expansion reamer is the most practical. This type is furnished in standard sizes from $\frac{1}{4}$ inch to 1 inch, increasing in diameter by $\frac{1}{32}$-inch increments.

Taper reamers, both hand and machine operated, are used to smooth and true taper holes and recesses.

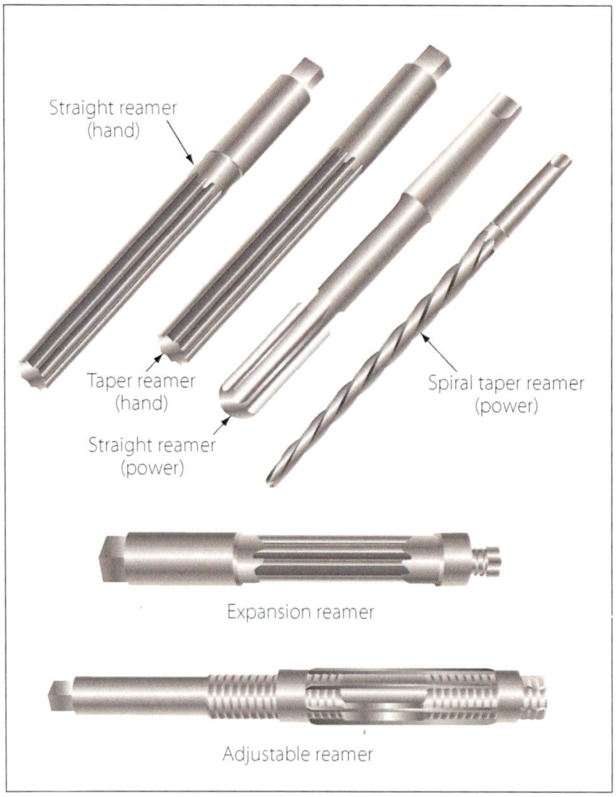

Figure 9-23. Reamers.

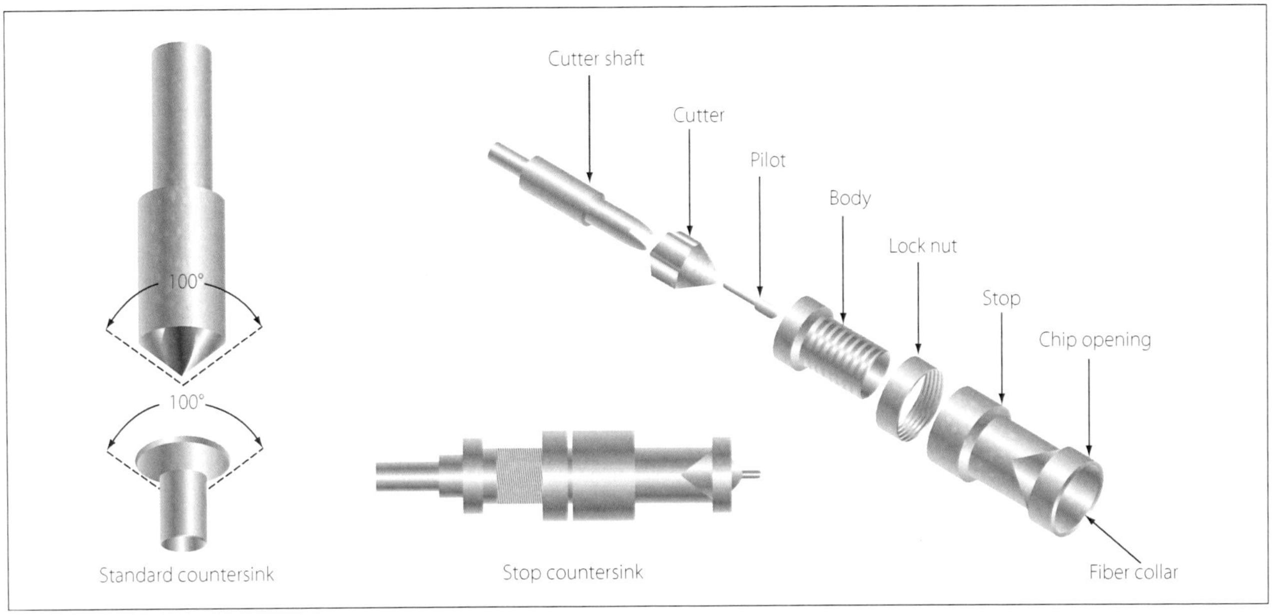

Figure 9-24. Countersinks.

Countersink

A countersink is a tool that cuts a cone shaped depression around the hole to allow a rivet or screw to set flush with the surface of the material. Countersinks are made with various angles to correspond to the various angles of the countersunk rivet and screwheads. The angle of the standard countersink shown in Figure 9-24 is 100°.

Special stop countersinks are available. Stop countersinks are adjustable to any desired depth, and the cutters are interchangeable so that holes of various countersunk angles may be made. Some stop countersinks have a micrometer set arrangement (in increments of 0.001 inch) for adjusting the cutting depths. [Figure 9-24]

When using a countersink, care must be taken not to remove an excessive amount of material, since this reduces the strength of flush joints.

Taps and Dies

A tap is used to cut threads on the inside of a hole, while a die is for cutting external threads on round stock. They are made of hard tempered steel and ground to an exact size. There are four types of threads that can be cut with standard taps and dies. They are: National Coarse, National Fine, National Extra Fine, and National Pipe.

Hand taps are usually provided in sets of three taps for each diameter and thread series. Each set contains a taper tap, a plug tap, and a bottoming tap. The taps in a set are identical in diameter and cross section; the only difference is the amount of taper. [Figure 9-25]

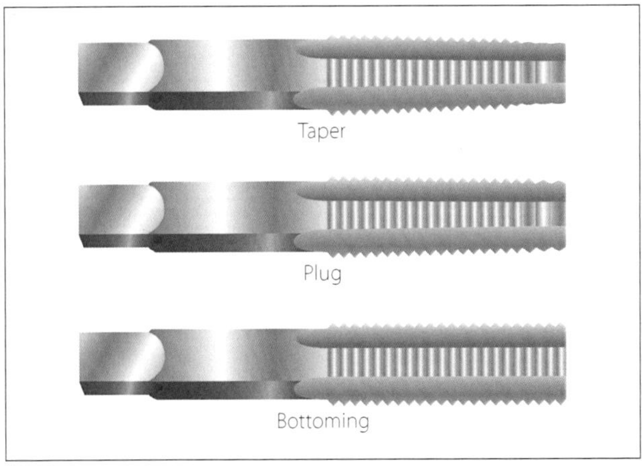

Figure 9-25. Hand taps.

The taper tap is used to begin the tapping process, because it is tapered back for 6 to 7 threads. This tap cuts a complete thread when it is cutting above the taper. It is the only tap needed when tapping holes that extend through thin sections. The plug tap supplements the taper tap for tapping holes in thick stock.

The bottoming tap is not tapered. It is used to cut full threads to the bottom of a blind hole.

Dies may be classified as adjustable round split die and plain round split die. The adjustable split die has an adjusting screw that can be tightened so that the die is

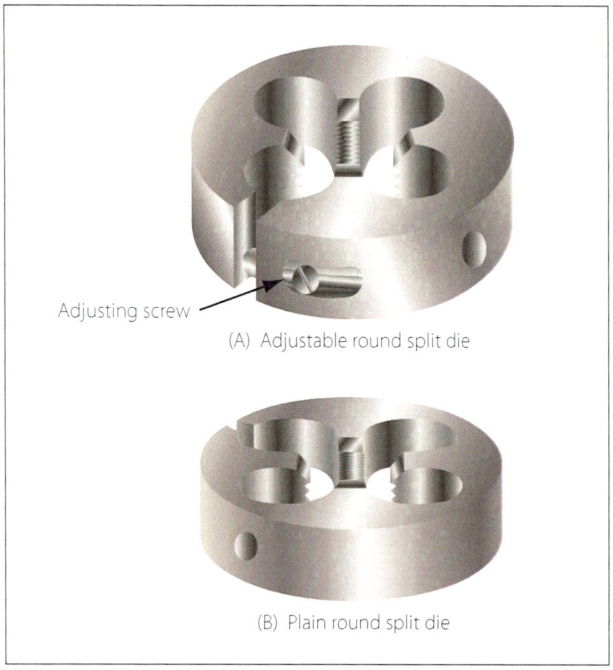

Figure 9-26. Types of dies.

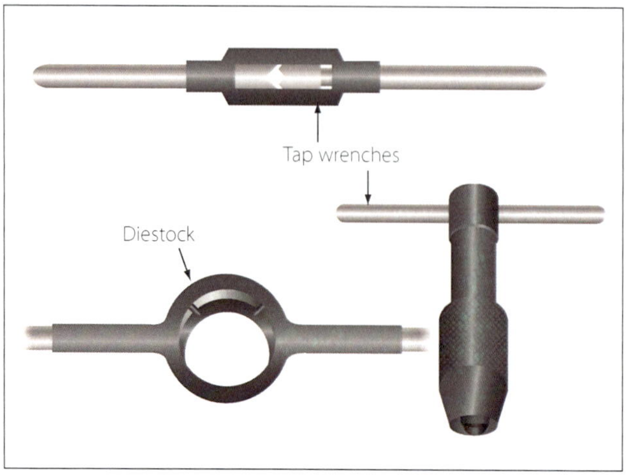

Figure 9-27. Diestock and tap wrenches.

spread slightly. By adjusting the die, the diameter and fit of the thread can be controlled. [Figure 9-26]

Solid dies are not adjustable; therefore, a variety of thread fits cannot be obtained with this type. There are many types of wrenches for turning taps, as well as turning dies. The T-handle, the adjustable tap wrench, and the diestock for round split dies shown in Figure 9-27 are a few of the more common types.

Information on thread sizes, fits, types, and so forth, is shown in Figures 9-28, 9-29, and 9-30.

Layout and Measuring Tools

Layout and measuring devices are precision tools. They are carefully machined, accurately marked and, in many cases, are made up of very delicate parts. When using these tools, be careful not to drop, bend, or scratch them. The finished product will be no more accurate than the measurements or the layout; therefore, it is very important to understand how to read, use, and care for these tools.

Rules

Rules are made of steel and are either rigid or flexible. The flexible steel rule will bend, but it should not be bent intentionally as it may be broken rather easily. In aircraft work, the unit of measure most commonly used is the inch. The inch may be divided into smaller parts by means of either common or decimal fraction divisions.

The fractional divisions for an inch are found by dividing the inch into equal parts: halves ($^1/_2$), quarters ($^1/_4$), eighths ($^1/_8$), sixteenths ($^1/_{16}$), thirty-seconds ($^1/_{32}$), and sixty-fourths ($^1/_{64}$).

The fractions of an inch may be expressed in decimals, called decimal equivalents of an inch; for example, $^1/_8$ inch is expressed as 0.0125 (one hundred twenty-five ten-thousandths of an inch).

Rules are manufactured in two basic styles—those divided or marked in common fractions and those divided or marked in decimals or divisions of one one-hundredth of an inch. A rule may be used either as a measuring tool or as a straightedge. [Figure 9-31]

Combination Sets

The combination set, as its name implies, is a tool that has several uses. It can be used for the same purposes as an ordinary tri-square, but it differs from the tri-square in that the head slides along the blade and can be clamped at any desired place. Combined with the square or stock head are a level and scriber. The head slides in a central groove on the blade or scale, which can be used separately as a rule. [Figure 9-32]

The spirit level in the stock head makes it convenient to square a piece of material with a surface and at the same time tell whether one or the other is plumb or level. The head can be used alone as a simple level.

The combination of square head and blade can also be used as a marking gauge to scribe lines at a 45° angle, as a depth gauge, or as a height gauge. A convenient scriber is held frictionally in the head by a small brass bushing.

National Coarse Thread Series Medium Fit Class 3 (NC)					National Fine Thread Series Medium Fit Class 3 (NF)				
Size and Threads	Diameter of body for thread	Body Drill	Tap Drill: Preferred diameter of hole	Tap Drill: Nearest standard drill size	Size and Threads	Diameter of body for thread	Body Drill	Tap Drill: Preferred diameter of hole	Tap Drill: Nearest standard drill size
					0-80	.060	52	.0472	3/64"
1-64	.073	47	.0575	#53	1-72	.073	47	.0591	#53
2-56	.086	42	.0682	#51	2-64	.086	42	.0700	#50
3-48	.099	37	.078	5/64	3-56	.099	37	.0810	#46
4-40	.112	31	.0866	#44	4-48	.112	31	.0911	#42
5-40	.125	29	.0995	#39	5-44	.125	25	1024	#38
6-32	.138	27	.1063	#36	6-40	.138	27	.113	#33
8-32	.164	18	.1324	#29	8-36	.164	18	.136	#29
10-24	.190	10	.1472	#26	10-32	.190	10	.159	#21
12-24	.216	2	.1732	#17	12-28	.216	2	.180	#15
1/4-20	.250	1/4	.1990	#8	1/4-28	.250	F	.213	#3
5/16-18	.3125	5/16	.2559	#F	5/16-24	.3125	5/16	.2703	I
3/8-16	.375	3/8	.3110	5/16"	3/8-24	.375	3/8	.332	Q
7/16-14	.4375	7/16	.3642	U	7/16-20	.4375	7/16	.386	W
1/2-13	.500	1/2	.4219	27/64"	1/2-20	.500	1/2	.449	7/16"
9/16-12	.5625	9/16	.4776	31/64"	9/16-18	.5625	9/16	.506	1/2"
5/8-11	.625	5/8	.5315	17/64"	5/8-18	.625	5/8	.568	9/16"
3/4-10	.750	3/4	.6480	41/64"	3/4-16	.750	3/4	.6688	11/16"
7/8-9	.875	7/8	.7307	49/64"	7/8-14	.875	7/8	.7822	51/64"
1-8	1.000	1.0	.8376	7/8"	1-14	1.000	1.0	.9072	49/64"

Figure 9-28. American (National) screw thread sizes.

Nominal size (inches)	Number of threads per inch	Pitch Diameter: A (inches)	Pitch Diameter: B (inches)	Size and Threads: L2 (inches)	Size and Threads: L1 (inches)	Pipe o.d. (inches)	Depth of thread (inches)	Tap drill for pipe threads: Minor diameter small end of pipe	Tap drill for pipe threads: Size drill
1/8	27	.36351	.37476	.2638	.180	.405	.02963	.33388	R
1/4	18	.47739	.48989	.4018	.200	.540	.04444	.43294	7/16
3/8	18	.61201	.62701	.4078	.240	.675	.04444	.56757	37/64
1/2	14	.75843	.77843	.5337	.320	.840	.05714	.70129	23/32
3/4	14	.96768	.98887	.5457	.339	1.050	.5714	.91054	59/64
1	11 1/2	1.21363	1.23863	.6828	.400	1.315	.06957	1.14407	1 5/32
1 1/4	11 1/2	1.55713	1.58338	.7068	.420	1.660	.06957	1.48757	1 1/2
1 1/2	11 1/2	1.79609	1.82234	.7235	.420	1.900	.06957	1.72652	1 47/64
2	11 1/2	2.26902	2.29627	.7565	.436	2.375	.06957	2.19946	1 7/32
2 1/2	8	2.71953	2.76216	1.1375	.682	2.875	.10000	2.61953	2 5/8
3	8	3.34062	3.8850	1.2000	.766	3.500	.10000	3.24063	3 1/4
3 1/2	8	3.83750	3.88881	1.2500	.821	4.000	.10000	3.73750	3 3/4
4	8	4.33438	4.38712	1.3000	.844	4.500	.10000	4.23438	4 1/4

Figure 9-29. American (National) pipe thread dimensions and tap drill sizes.

Diameter of drill	Soft metals 300 F.P.M.	Plastic and hard rubber 200 F.P.M.	Annealed cast iron 140 F.P.M.	Mild steel 100 F.P.M.	Malleable iron 90 F.P.M.	Hard cast iron 80 F.P.M.	Tool or hard steel 60 F.P.M.	Alloy steel cast steel 40 F.P.M.
1/16 (No. 53–80)	18320	12217	8554	6111	5500	4889	3667	2445
3/32 (No. 42–52)	12212	8142	5702	4071	3666	3258	2442	1649
1/8 (No. 31–41)	9160	6112	4278	3056	2750	2445	1833	1222
5/32 (No. 23–30)	7328	4888	3420	2444	2198	1954	1465	977
3/16 (No. 13–22)	6106	4075	2852	2037	1833	1630	1222	815
7/32 (No. 1–12)	5234	3490	444	1745	1575	1396	1047	698
1/4 (A–F)	4575	3055	2139	1527	1375	1222	917	611
9/32 (G–K)	4071	2715	1900	1356	1222	1084	814	542
9/16 (L, M, N)	3660	2445	1711	1222	1100	978	7333	489
11/32 (O–R)	3330	2220	1554	1110	1000	888	666	444
3/8 (S, T, U)	3050	2037	1426	1018	917	815	611	407
13/32 (V–Z)	2818	1878	1316	939	846	752	563	376
7/16	2614	1746	1222	873	786	698	524	349
15/32	2442	1628	1140	814	732	652	488	326
1/2	2287	1528	1070	764	688	611	458	306
9/16	2035	1357	950	678	611	543	407	271
3/8	1830	1222	856	611	550	489	367	244
11/16	1665	1110	777	555	500	444	333	222
3/4	1525	1018	713	509	458	407	306	204

Figure 9-30. Drill speeds.

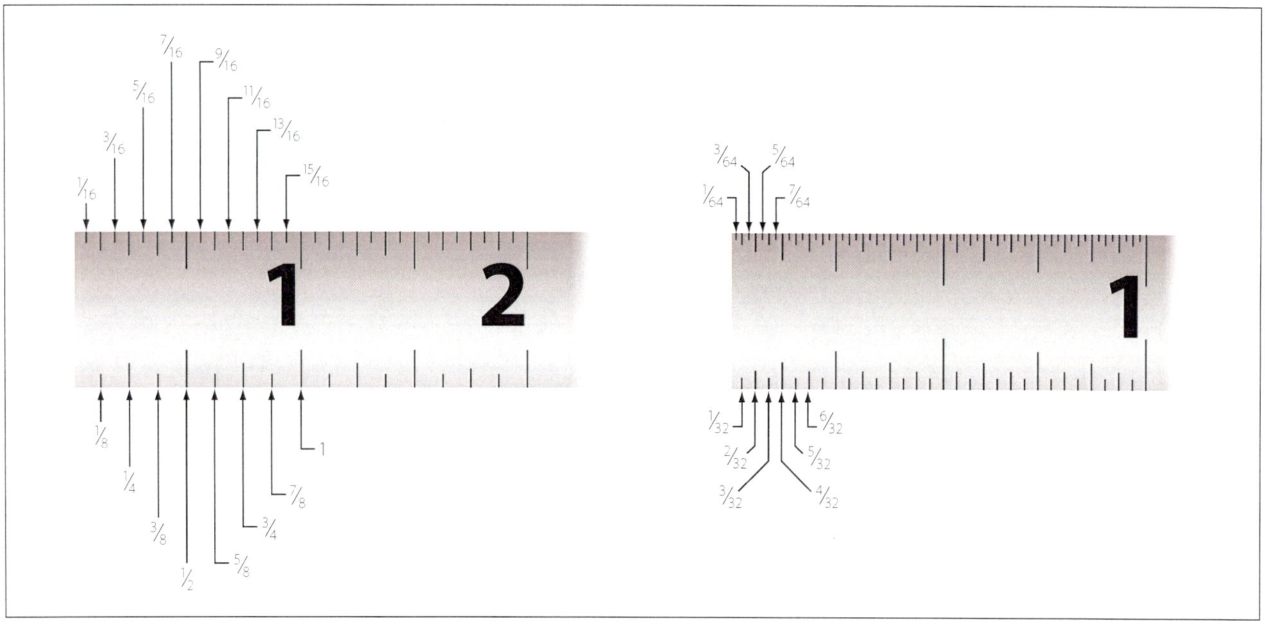

Figure 9-31. Rules.

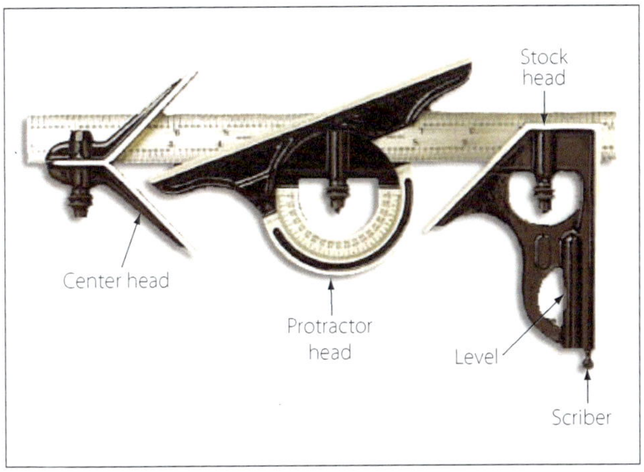

Figure 9-32. Combination set.

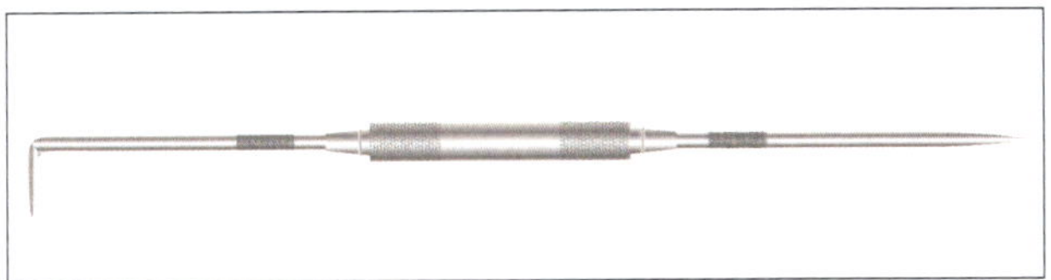

Figure 9-33. Scriber.

The center head is used to find the center of shafts or other cylindrical work. The protractor head can be used to check angles and also may be set at any desired angle to draw lines.

Scriber

The scriber is designed to serve the aviation mechanic in the same way a pencil or pen serves a writer. In general, it is used to scribe or mark lines on metal surfaces. The scriber is made of tool steel, 4 to 12 inches long, and has two needle pointed ends. One end is bent at a 90° angle for reaching and marking through holes. [Figure 9-33]

Before using a scriber, always inspect the points for sharpness. Be sure the straightedge is flat on the metal and in position for scribing. Tilt the scriber slightly in the direction toward which it will be moved, holding it like a pencil. Keep the scriber's point close to the guiding edge of the straightedge. The scribed line should be heavy enough to be visible, but no deeper than necessary to serve its purpose.

Dividers and Pencil Compasses

Dividers and pencil compasses have two legs joined at the top by a pivot. They are used to scribe circles and arcs and for transferring measurements from the rule to the work.

Pencil compasses have one leg tapered to a needle point; the other leg has a pencil or pencil lead inserted. Dividers have both legs tapered to needle points.

When using pencil compasses or dividers, the following procedures are suggested:

1. Inspect the points to make sure they are sharp.
2. To set the dividers or compasses, hold them with the point of one leg in the graduations on the rule. Turn the adjustment nut with the thumb and forefinger; adjust the dividers or compasses until the point of the other leg rests on the graduation of the rule that gives the required measurement.
3. To draw an arc or circle with either the pencil compasses or dividers, hold the thumb attachment on the top with the thumb and forefinger. With pressure exerted on both legs, swing the compass in a clockwise direction and draw the desired arc or circle.
4. The tendency for the legs to slip is avoided by inclining the compasses or dividers in the direction in which they are being rotated. In working on metals, the dividers are used only to scribe arcs or circles that will later be removed by cutting. All other arcs or circles are drawn with pencil compasses to avoid scratching the material.
5. On paper layouts, the pencil compasses are used for describing arcs and circles. Dividers should be used to transfer critical measurements because they are more accurate than a pencil compass.

Calipers

Calipers are used for measuring diameters and distances or for comparing distances and sizes. The three common types of calipers are inside, outside, and hermaphrodite calipers, such as gear tool calipers. [Figure 9-34]

Outside calipers are used for measuring outside dimensions—for example, the diameter of a piece of round stock. Inside calipers have outward curved legs for measuring inside diameters, such as diameters of holes, the distance between two surfaces, the width of slots, and other similar jobs. A hermaphrodite caliper is generally used as a marking gauge in layout work. It should not be used for precision measurement.

Micrometer Calipers

There are four types of micrometer calipers, each designed for a specific use: outside micrometer, inside micrometer, depth micrometer, and thread micrometer.

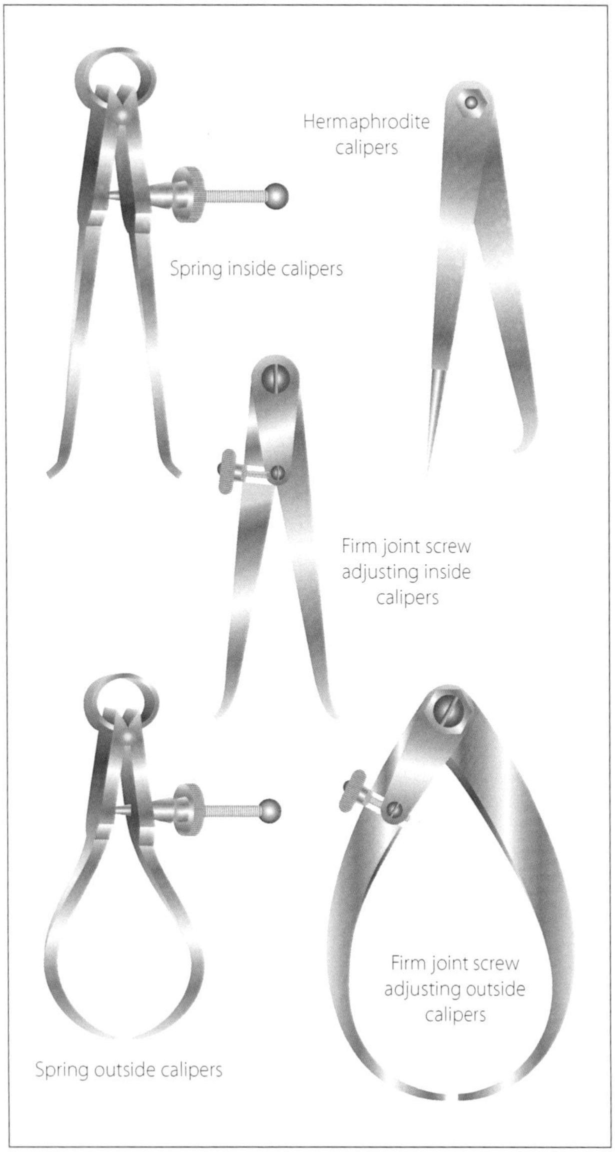

Figure 9-34. Calipers.

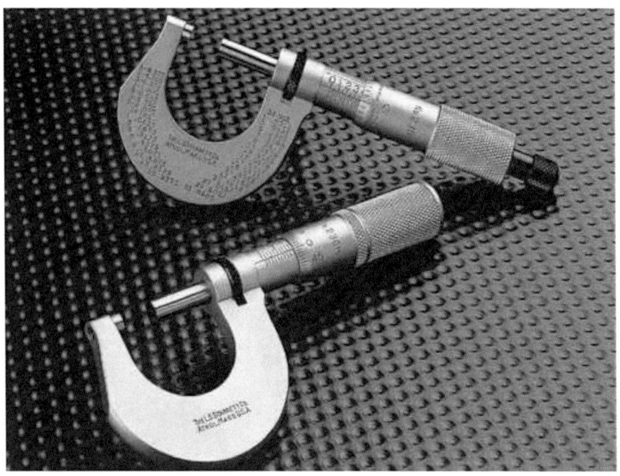

Figure 9-35. Outside micrometers.

Micrometers are available in a variety of sizes, either 0 to ½ inch, 0 to 1 inch, 1 to 2 inch, 2 to 3 inch, 3 to 4 inch, 4 to 5 inch, or 5 to 6 inch sizes. In addition to the micrometer inscribed with the measurement markings, micrometers equipped with electronic digital liquid crystal display (LCD) readouts are also in common use.

The AMT will use the outside micrometer more often than any other type. It may be used to measure the outside dimensions of shafts, thickness of sheet metal stock, the diameter of drills, and for many other applications. [Figure 9-35]

The smallest measurement which can be made with the use of the steel rule is one sixty-fourth of an inch in common fractions, and one one-hundredth of an inch in decimal fractions. To measure more closely than this (in thousandths and ten-thousandths of an inch), a micrometer is used. If a dimension given in a common fraction is to be measured with the micrometer, the fraction must be converted to its decimal equivalent.

All four types of micrometers are read in the same way. The method of reading an outside micrometer is discussed later in this chapter.

Micrometer Parts

The fixed parts of a micrometer are the frame, barrel, and anvil. The movable parts of a micrometer are the thimble and spindle. The thimble rotates the spindle which moves in the threaded portion inside the barrel. Turning the thimble provides an opening between the anvil and the end of the spindle where the work is measured. The size of the work is indicated by the graduations on the barrel and thimble. [Figure 9-36]

Reading a Micrometer

The lines on the barrel marked 1, 2, 3, 4, and so forth, indicate measurements of tenths, or 0.100 inch, 0.200 inch, 0.300 inch, 0.400 inch, respectively. [Figure 9-37]

Each of the sections between the tenths divisions (between 1, 2, 3, 4, and so forth) is divided into four parts of 0.025 inch each. One complete revolution of the thimble (from zero on the thimble around to the same zero) moves it one of these divisions (0.025 inch) along the barrel.

The bevel edge of the thimble is divided into 25 equal parts. Each of these parts represents one twenty-fifth of the distance the thimble travels along the barrel in moving from one of the 0.025 inch divisions to another. Thus, each division on the thimble represents one one-thousandth (0.001) of an inch.

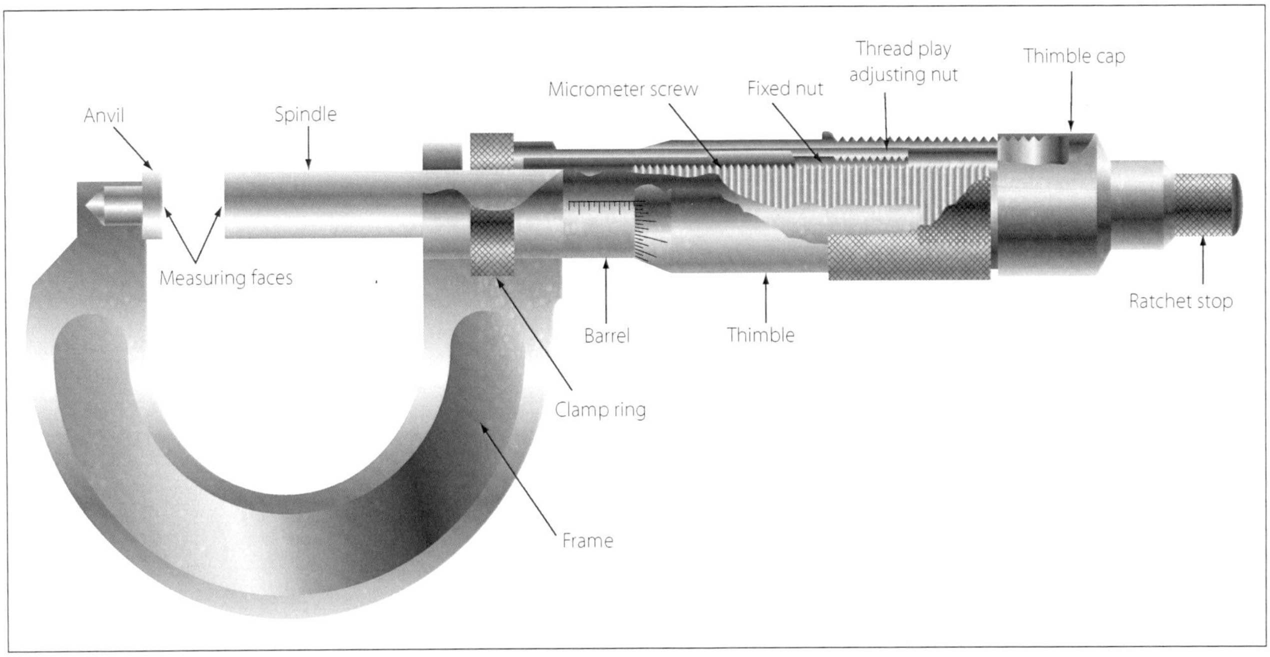

Figure 9-36. Outside micrometer parts.

These divisions are marked for convenience at every five spaces by 0, 5, 10, 15, and 20. When 25 of these graduations have passed the horizontal line on the barrel, the spindle (having made one revolution) has moved 0.025 inch.

The micrometer is read by first noting the last visible figure on the horizontal line of the barrel representing tenths of an inch. Add to this the length of barrel between the thimble and the previously noted number. (This is found by multiplying the number of graduations by 0.025 inch.) Add to this the number of divisions on the bevel edge of the thimble that coincides with the line of the graduation. The total of the three figures equals the measurement. (Figure 9-38 shows several sample readings.)

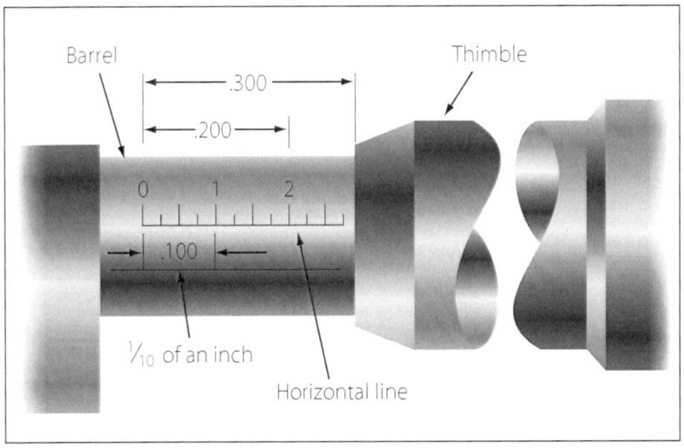

Figure 9-37. Micrometer measurements.

Vernier Scale

Some micrometers are equipped with a vernier scale that makes it possible to directly read the fraction of a division that is indicated on the thimble scale. Typical examples of the vernier scale as it applies to the micrometer are shown in Figure 9-39.

All three scales on a micrometer are not fully visible without turning the micrometer, but the examples shown in Figure 9-38 are drawn as though the barrel and thimble of the micrometer were laid out flat so that all three scales can be seen at the same time. The barrel scale is the lower horizontal scale, the thimble scale is vertical on the right, and the long horizontal lines (0 through 9 and 0) make up the vernier scale.

In reading a micrometer, an excellent way to remember the relative scale values is to remember that the 0.025 inch barrel scale graduations are established by the lead screw (40 threads per inch). Next, the thimble graduations divide the 0.025 inch into 25 parts, each equal to 0.001 inch. Then, the vernier graduations divide the 0.001 inch into 10 equal parts, each equal to 0.0001 inch. Remembering the values of the various scale graduations, the barrel scale reading is noted. The thimble scale reading is added to it; then the vernier scale reading is added to get the final reading. The vernier scale line to be read is always the one aligned exactly with any thimble graduation.

In the first example in Figure 9-39, the barrel reads 0.275 inch and the thimble reads more than 0.019 inch. The number 1 graduation on the thimble is aligned

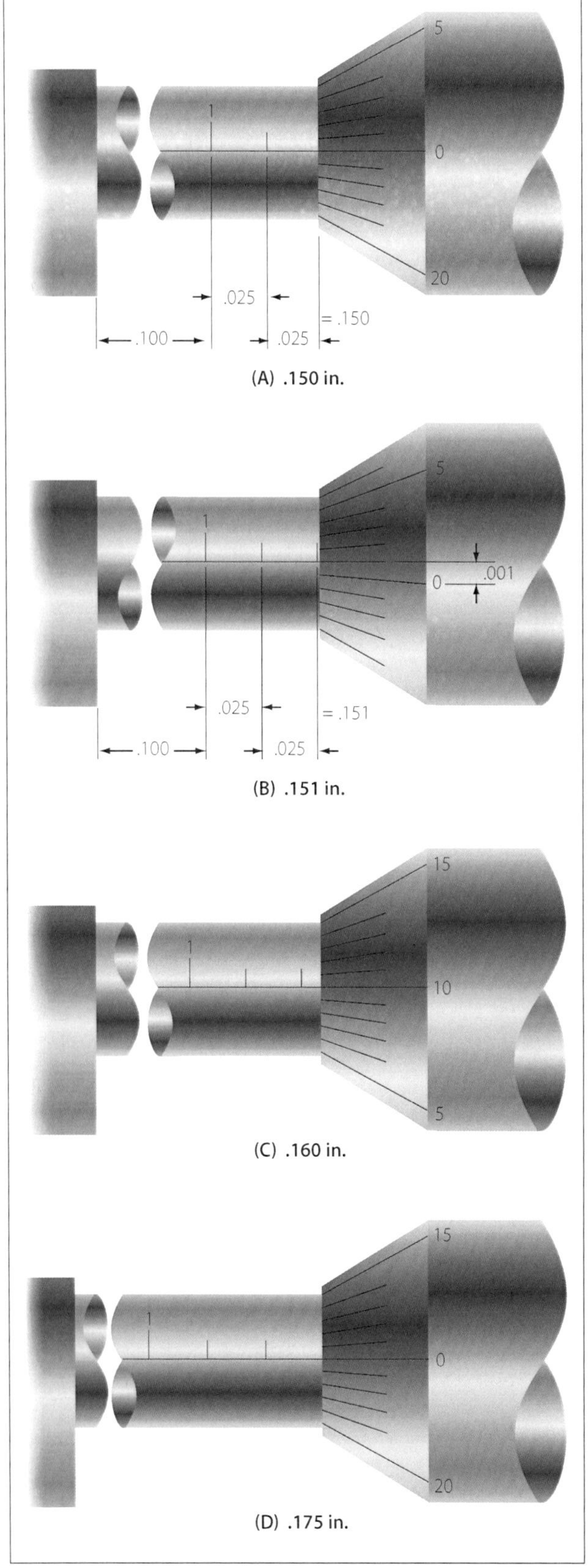

Figure 9-38. Reading a micrometer.

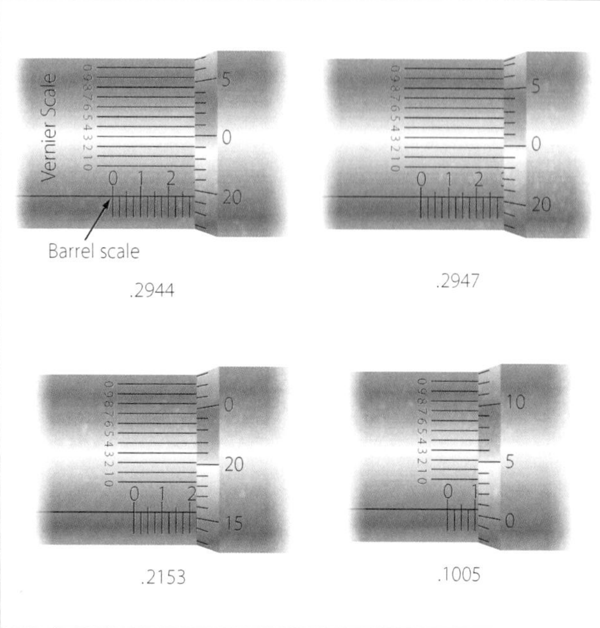

Figure 9-39. Vernier scale readings.

exactly with the number 4 graduation on the vernier scale. Thus, the final reading is 0.2944 inch.

In the second example in Figure 9-39, the barrel reads 0.275 inch, and the thimble reads more than 0.019 inch and less than 0.020 inch. On the vernier scale, the number 7 graduation coincides with a line on the thimble. This means that the thimble reading would be 0.0197 inch. Adding this to the barrel reading of 0.275 inch gives a total measurement of 0.2947 inch.

The third and fourth examples in Figure 9-39 are additional readings that would require use of the vernier scale for accurate readings to ten-thousandths of an inch.

Using a Micrometer

The micrometer must be handled carefully. If it is dropped, its accuracy may be permanently affected. Continually sliding work between the anvil and spindle may wear the surfaces. If the spindle is tightened too much, the frame may be sprung permanently and inaccurate readings will result.

To measure a piece of work with the micrometer, hold the frame of the micrometer in the palm of the hand with the little finger or third finger, whichever is more convenient. This allows the thumb and forefinger to be free to revolve the thimble for adjustment.

A variation of the micrometer is the dial indicator, which measures variations in a surface by using an accurately machined probe mechanically linked to a

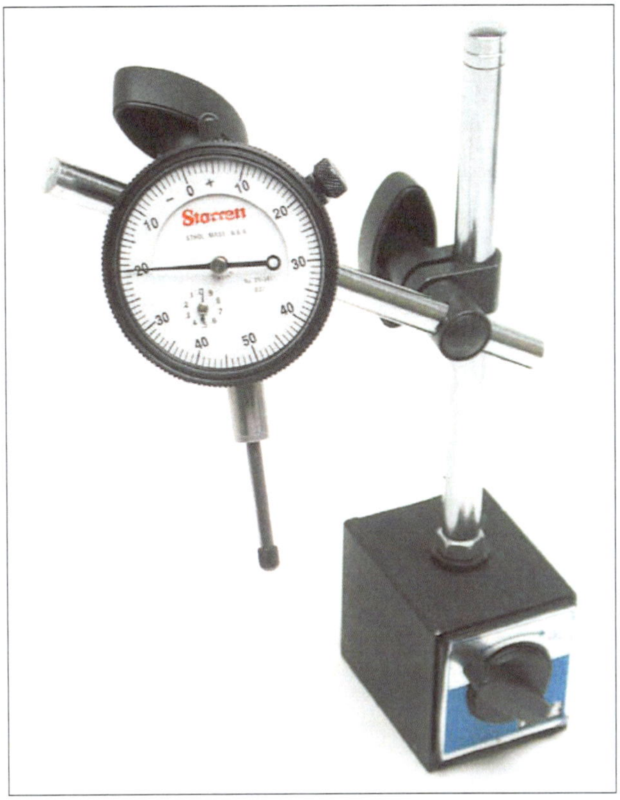

Figure 9-40. Dial Indicator.

circular hand whose movement indicates thousandths of an inch, or is displayed on a liquid crystal display (LCD) screen. [Figure 9-40]

A typical example would be using a dial indicator to measure the amount of runout, or bend, in a shaft. If a bend is suspected, the part can be rotated while resting between a pair of machined V-blocks. A dial indicator is then clamped to a machine table stand, and the probe of the indicator is positioned so it lightly contacts the surface. The outer portion of the dial is then rotated until the needle is pointed at zero. The part is then rotated, and the amount of bend, or runout, is displayed on the dial as the needle fluctuates. The total amount of the fluctuation is the runout.

Another common use for the dial indicator is to check for a warp in a rotating component such as a brake disc. In some cases, this can be done with the brake disc installed on the airplane, with the base clamped to a stationary portion of the structure.

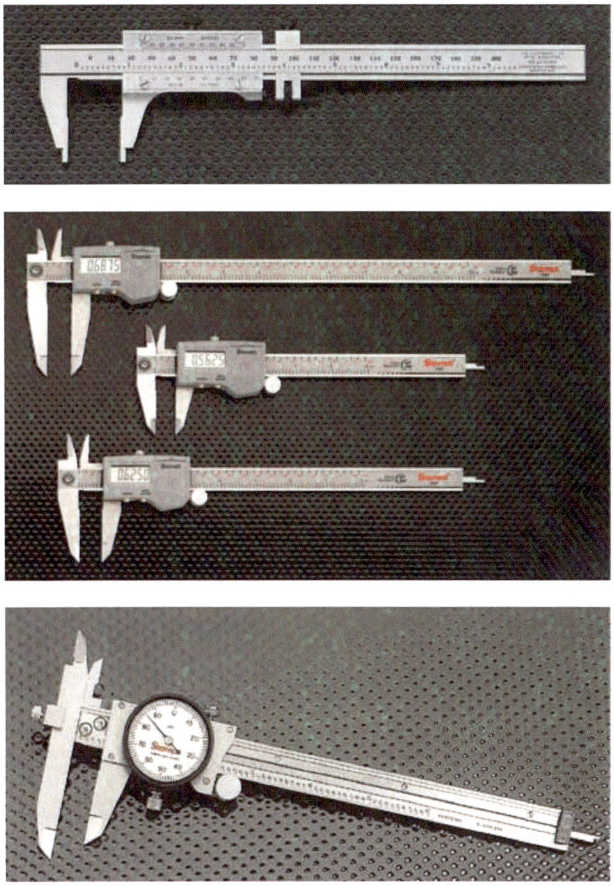

Figure 9-41. Electronic and dial indicator slide calipers.

In either case, it is imperative that the dial indicator be securely fastened so that movement of the indicator itself induces no errors in measurement.

Slide Calipers

Often used to measure the length of an object, the slide caliper provides greater accuracy than the ruler. It can, by virtue of its specially formed jaws, measure both inside and outside dimensions. As the tool's name implies, the slide caliper jaw is slid along a graduated scale, and its jaws then contact the inside or outside of the object to be measured. The measurement is then read on the scale located on the body of the caliper, or on the LCD screen. [Figure 9-41]

Some slide calipers also contain a depth gauge for measuring the depth of blind holes.

Chapter 10 Basic Electricity

Introduction to Electricity and Electronics

This chapter addresses the fundamental concepts that are the building blocks for advanced electrical knowledge and practical troubleshooting. Some of the questions addressed are: How does energy travel through a copper wire and through space? What is electric current, electromotive force, and what makes a landing light turn on or a hydraulic pump motor run? Each of these questions requires an understanding of many basic principles. By adding one basic idea on top of other basic ideas, it becomes possible to answer most of the interesting and practical questions about electricity or electronics.

Our understanding of electrical current must begin with the nature of matter. All matter is composed of molecules. All molecules are made up of atoms, which are themselves made up of electrons, protons, and neutrons.

General Composition of Matter

Matter

Matter can be defined as anything that has mass and has volume and is the substance of which physical objects are composed. Essentially, it is anything that can be touched. Mass is the amount of matter in a given object. Typically, the more matter there is in an object the more mass it will have. Weight is an indirect method of determining mass but not the same. The difference between mass and weight is that weight is determined by how much something or the fixed mass is pulled by gravity. Categories of matter are ordered by molecular activity. The four categories or states are: solids, liquids, gases, and plasma. For the purposes of the aircraft technician, only solids, liquids, and gases are considered.

Element

An element is a substance that cannot be reduced to a simpler form by chemical means. Iron, gold, silver, copper, and oxygen are examples of elements. Beyond this point of reduction, the element ceases to be what it is.

Compound

A compound is a chemical combination of two or more elements. Water is one of the most common compounds and is made up of two hydrogen atoms and one oxygen atom.

The Molecule

The smallest particle of matter that can exist and still retain its identity, such as water (H_2O), is called a molecule. A molecule of water is illustrated in Figure 10-1. Substances composed of only one type of atom are called elements. But most substances occur in nature as compounds, that is, combinations of two or more types of atoms. It would no longer retain the characteristics of water if it were compounded of one atom of hydrogen and two atoms of oxygen. If a drop of water is divided in two and then divided again and again until it cannot be divided any longer, it will still be water.

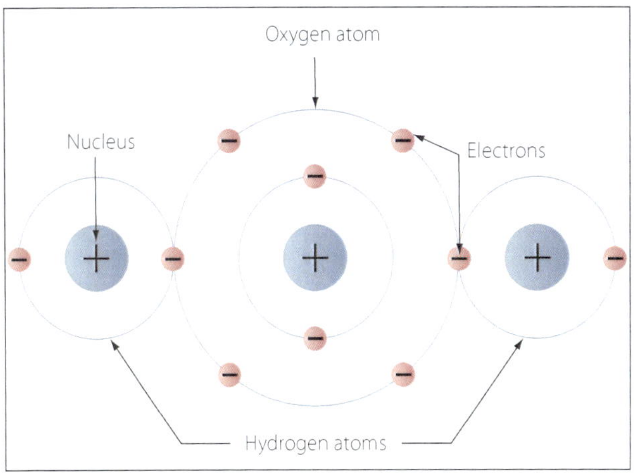

Figure 10-1. A water molecule.

The Atom

The atom is considered to be the most basic building block of all matter. Atoms are composed of three sub-atomic particles. These three sub-atomic particles are: protons, neutrons, and electrons. These three particles will determine the properties of the specific atoms. Elements are substances composed of the same atoms with specific properties. Oxygen is an example of this. The main property that defines each element is the number of neutrons, protons, and electrons. Hydrogen and helium are examples of elements. Both of these elements have neutrons, protons, and electrons but differ in the number of those items. This difference alone accounts for the variations in chemical and physical properties of these two different elements. There are over a 100 known elements in the periodic table, and they are categorized according to their properties on that table. The kinetic theory of matter also states that the particles that make up the matter are always moving. Thermal expansion is considered in the kinetic theory and explains why matter contracts when it is cool and expands when it is hot, with the exception of water/ice.

Electrons, Protons, and Neutrons

At the center of the atom is the nucleus, which contains the protons and neutrons. The protons are positively charged particles, and the neutrons are a neutrally charged particle. The neutron has approximately the same mass as the proton. The third particle of the atom is the electron that is a negatively charged particle with a very small mass compared to the proton. The proton's mass is approximately 1,837 times greater than the electron. Due to the proton and the neutron location in the central portion of the atom (nucleus) and the electron's position at the distant periphery of the atom, it is the electron that undergoes the change during chemical reactions. Since a proton weighs approximately 1,845 times as much as an electron, the number of protons and neutrons in its nucleus determines the overall weight of an atom. The weight of an electron is not considered in determining the weight of an atom. Indeed, the nature of electricity cannot be defined clearly because it is not certain whether the electron is a negative charge with no mass (weight) or a particle of matter with a negative charge.

Hydrogen represents the simplest form of an atom, as shown in Figure 10-2. At the nucleus of the hydrogen atom is one proton and at the outer shell is one orbiting electron. At a more complex level is the oxygen atom, as shown in Figure 10-3, which has eight electrons in two shells orbiting the nucleus with eight protons and eight neutrons. When the total positive charge of the protons in the nucleus equals the total negative charge of the electrons in orbit around the nucleus, the atom is said to have a neutral charge.

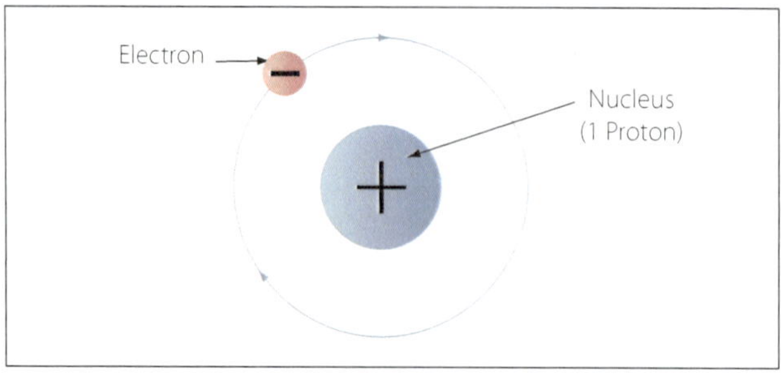

Figure 10-2. Hydrogen atom.

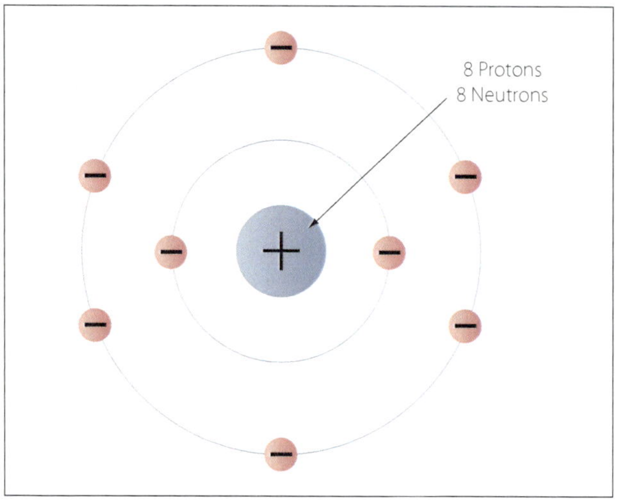

Figure 10-3. Oxygen atom.

Electron Shells and Energy Levels

Electrons require a certain amount of energy to stay in an orbit. This particular quantity is called the electron's energy level. By its motion alone, the electron possesses kinetic energy, while the electron's position in orbit determines its potential energy. The total energy of an electron is the main factor, which determines the radius of the electrons orbit.

Electrons of an atom will appear only at certain definite energy levels (shells). The spacing between energy levels is such that when the chemical properties of the various elements are cataloged, it is convenient to group several closely spaced permissible energy levels together into electron shells. The maximum number of electrons that can be contained in any shell or sub-shell is the same for all atoms and is defined as Electron Capacity = $2n^2$. In this equation n represents the energy level in question. The first shell can only contain two electrons; the second shell can only contain eight electrons; the third, 18 and so on until we reach the seventh shell for the heaviest atoms, which have six energy levels. Because the innermost shell is the lowest energy level, the shell begins to fill up from the shell closest to the nucleus and fill outward as the atomic number of the element increases. However, an energy level does not need to be completely filled before electrons begin to fill the next level. The Periodic Table of Elements should be checked to determine an element's electron configuration.

Valence Electrons

Valence is the number of chemical bonds an atom can form. Valence electrons are electrons that can participate in chemical bonds with other atoms. The number of electrons in the outermost shell of the atom is the determining factor in its valence. Therefore, the electrons contained in this shell are called valence electrons.

Ions

Ionization is the process by which an atom loses or gains electrons. Dislodging an electron from an atom will cause the atom to become positively charged. This net positively charged atom is called a positive ion or a cation. An atom that has gained an extra number of electrons is negatively charged and is called a negative ion or an anion. When atoms are neutral, the positively charged proton and the negatively charged electron are equal.

Free Electrons

Valence electrons are found drifting midway between two nuclei. Some electrons are more tightly bound to the nucleus of their atom than others and are positioned in a shell or sphere closer to the nucleus, while others are more loosely bound and orbit at a greater distance from the nucleus. These outermost electrons are called "free" electrons because they can be easily dislodged from the positive attraction of the protons in the nucleus. Once freed from the atom, the electron can then travel from atom to atom, becoming the flow of electrons commonly called current in a practical electrical circuit.

Electron Movement

The valence of an atom determines its ability to gain or lose an electron, which ultimately determines the chemical and electrical properties of the atom. These properties can be categorized as being a conductor, semiconductor or insulator, depending on the ability of the material to produce free electrons. When a material has a large number of free electrons available, a greater current can be conducted in the material.

Conductors

Elements such as gold, copper and silver possess many free electrons and make good conductors. The atoms in these materials have a few loosely bound electrons in their outer orbits. Energy in the form of heat can cause these electrons in the outer orbit to break loose and drift throughout the material. Copper and silver have one electron in their outer orbits. At room temperature, a piece of silver wire will have billions of free electrons.

Insulators

These are materials that do not conduct electrical current very well or not at all. Good examples of these are: glass, ceramic, and plastic. Under normal conditions, atoms in these materials do not produce free electrons. The absence of the free electrons means that electrical current cannot be conducted through the material. Only when the material is in an extremely strong electrical field will the outer electrons be dislodged. This action is called breakdown and usually causes physical damage to the insulator.

Semiconductors

This material falls in between the characteristics of conductors and insulators, in that they are not good at conducting or insulating. Silicon and germanium are the most widely used semiconductor materials. For a more detailed explanation on this topic refer to Page 10-101 in this chapter.

Metric Based Prefixes Used for Electrical Calculations

In any system of measurements, a single set of units is usually not sufficient for all the computations involved in electrical repair and maintenance. Small distances, for example, can usually be measured in inches, but larger distances are more meaningfully expressed in feet, yards, or miles. Since electrical values often vary from numbers that are a millionth part of a basic unit of measurement to very large values, it is often necessary to use a wide range of numbers to represent the values of such units as volts, amperes, or ohms. A series of prefixes which appear with the name of the unit have been devised for the various multiples or submultiples of the basic units. There are 12 of these prefixes, which are also known as conversion factors. Four of the most commonly used prefixes used in electrical work with a short definition of each are as follows:

Mega (M) means one million (1,000,000).
Kilo (k) means one thousand (1,000).
Milli (m) means one-thousandth ($^1/_{1,000}$).
Micro (μ) means one-millionth ($^1/_{1,000,000}$).

One of the most extensively used conversion factors, kilo, can be used to explain the use of prefixes with basic units of measurement. Kilo means 1,000, and when used with volts, is expressed as kilovolt, meaning 1,000 volts. The symbol for kilo is the letter "k". Thus, 1,000 volts is one kilovolt or 1kV. Conversely, one volt would equal one-thousandth of a kV, or $^1/_{1,000}$ kV. This could also be written 0.001 kV.

Similarly, the word "milli" means one-thousandth, and thus, 1 millivolt equals one-thousandth ($^1/_{1000}$) of a volt.

Figure 10-4 contains a complete list of the multiples used to express electrical quantities, together with the prefixes and symbols used to represent each number.

Static Electricity

Electricity is often described as being either static or dynamic. The difference between the two is based simply on whether the electrons are at rest (static) or in motion (dynamic). Static electricity is a build up of an electrical charge on the surface of an object. It is considered "static" due to the fact that there is no current flowing as in AC or DC electricity. Static electricity is usually caused when non-conductive materials such as rubber, plastic or glass are rubbed together, causing a transfer of electrons, which then results in an imbalance of charges between the two materials. The fact that there is an imbalance of charges between the two materials means that the objects will exhibit an attractive or repulsive force.

Number	Prefix	Symbol
1,000,000,000,000	tera	t
1,000,000,000	giga	g
1,000,000	mega	M
1,000	kilo	k
100	hecto	h
10	deka	dk
0.1	deci	d
0.01	centi	c
0.001	milli	m
0.000001	micro	μ
0.000000001	nano	n
0.000000000001	pico	p

Figure 10-4. Prefixes and symbols for multiples of basic quantities.

Attractive and Repulsive Forces

One of the most fundamental laws of static electricity, as well as magnetics, deals with attraction and repulsion. Like charges repel each other and unlike charges attract each other. All electrons possess a negative charge and as such will repel each other. Similarly, all protons possess a positive charge and as such will repel each other. Electrons (negative) and protons (positive) are opposite in their charge and will attract each other.

For example, if two pith balls are suspended, as shown in Figure 10-5, and each ball is touched with the charged glass rod, some of the charge from the rod

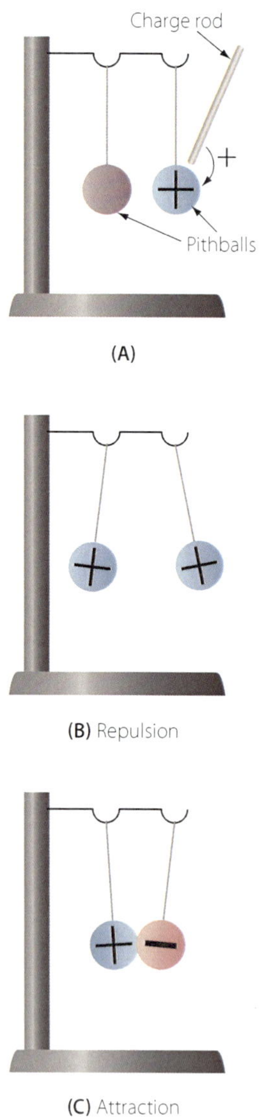

Figure 10-5. Reaction of like and unlike charges.

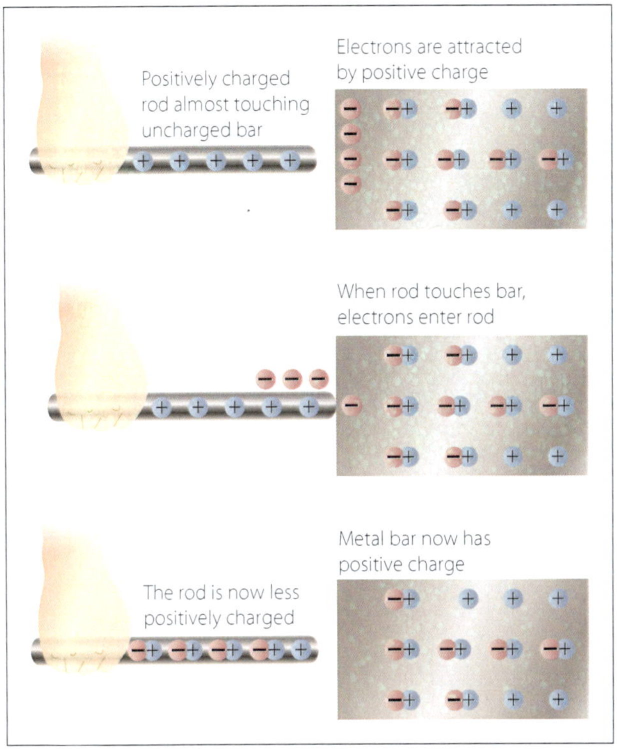

Figure 10-6. Charging by contact.

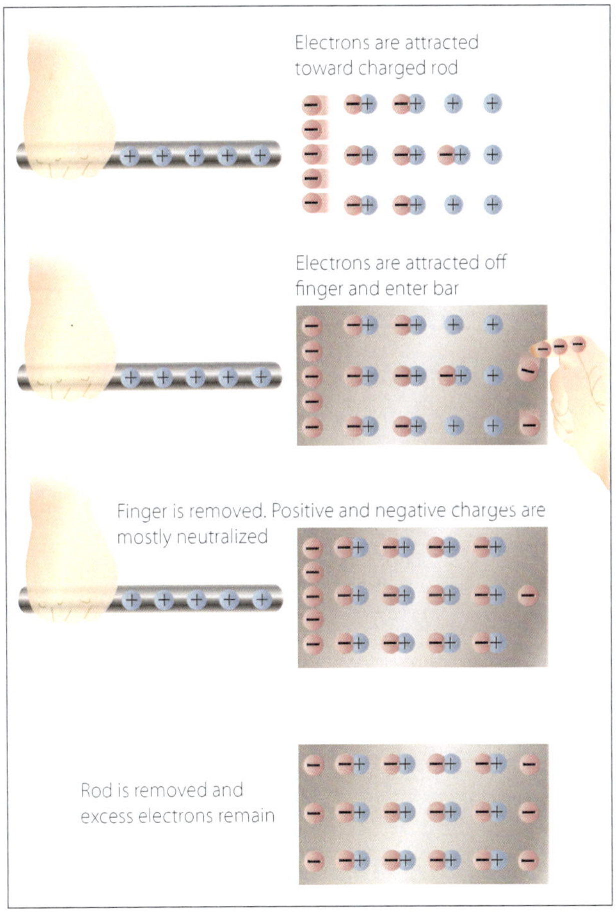

Figure 10-7. Charging a bar by induction.

is transferred to the balls. The balls now have similar charges and, consequently, repel each other as shown in part B of Figure 10-5. If a plastic rod is rubbed with fur, it becomes negatively charged and the fur is positively charged. By touching each ball with these differently charged sources, the balls obtain opposite charges and attract each other as shown in part C of Figure 10-5.

Although most objects become charged with static electricity by means of friction, a charged substance can also influence objects near it by contact. This is illustrated in Figure 10-6. If a positively charged rod touches an uncharged metal bar, it will draw electrons from the uncharged bar to the point of contact. Some electrons will enter the rod, leaving the metal bar with a deficiency of electrons (positively charged) and making the rod less positive than it was or, perhaps, even neutralizing its charge completely.

A method of charging a metal bar by induction is demonstrated in Figure 10-7. A positively charged rod is brought near, but does not touch, an uncharged metal bar. Electrons in the metal bar are attracted to the end of the bar nearest the positively charged rod, leaving a deficiency of electrons at the opposite end of the bar. If this positively charged end is touched by a neutral object, electrons will flow into the metal bar and neutralize the charge. The metal bar is left with an overall excess of electrons.

Electrostatic Field

A field of force exists around a charged body. This field is an electrostatic field (sometimes called a dielectric field) and is represented by lines extending in all directions from the charged body and terminating where there is an equal and opposite charge.

To explain the action of an electrostatic field, lines are used to represent the direction and intensity of the electric field of force. As illustrated in Figure 10-8, the intensity of the field is indicated by the number of lines per unit area, and the direction is shown by arrowheads

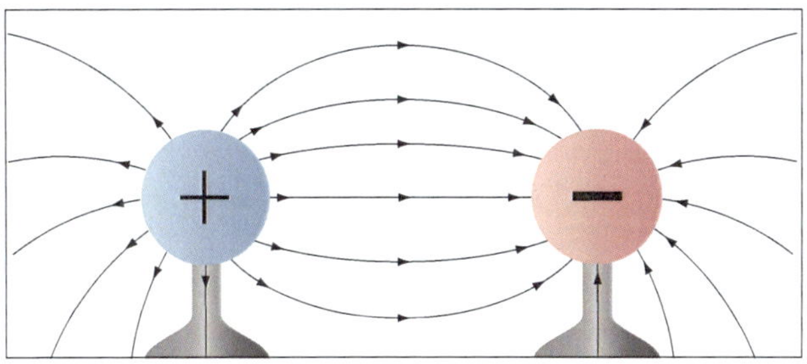

Figure 10-8. Direction of electric field around positive and negative charges.

on the lines pointing in the direction in which a small test charge would move or tend to move if acted upon by the field of force.

Either a positive or negative test charge can be used, but it has been arbitrarily agreed that a small positive charge will always be used in determining the direction of the field. Thus, the direction of the field around a positive charge is always away from the charge, as shown in Figure 10-8, because a positive test charge would be repelled. On the other hand, the direction of the lines about a negative charge is toward the charge, since a positive test charge is attracted toward it.

Figure 10-9 illustrates the field around bodies having like charges. Positive charges are shown, but regardless of the type of charge, the lines of force would repel each other if the charges were alike. The lines terminate on material objects and always extend from a positive charge to a negative charge. These lines are imaginary lines used to show the direction a real force takes.

It is important to know how a charge is distributed on an object. Figure 10-10 shows a small metal disk on which a concentrated negative charge has been placed. By using an electrostatic detector, it can be shown that the charge is spread evenly over the entire surface of the disk. Since the metal disk provides uniform resistance everywhere on its surface, the mutual repulsion of electrons will result in an even distribution over the entire surface.

Another example, shown in Figure 10-11, is the charge on a hollow sphere. Although the sphere is made of conducting material, the charge is evenly distributed over the outside surface. The inner surface is completely neutral. This phenomenon is used to safeguard operating personnel of the large Van de Graaff static generators used for atom smashing. The safest area for the operators is inside the large sphere, where millions of volts are being generated.

The distribution of the charge on an irregularly shaped object differs from that on a regularly shaped object. Figure 10-12 shows that the charge on such objects is not evenly distributed. The greatest charge is at the points, or areas of sharpest curvature, of the objects.

ESD Considerations

One of the most frequent causes of damage to a solid-state component or integrated circuits is the electro-

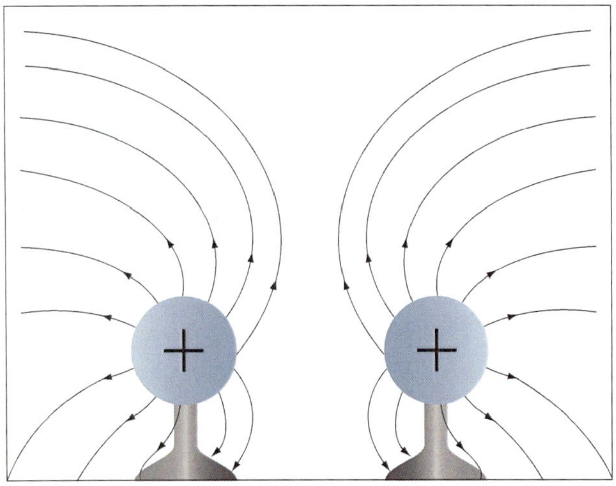

Figure 10-9. Field around two positively charged bodies.

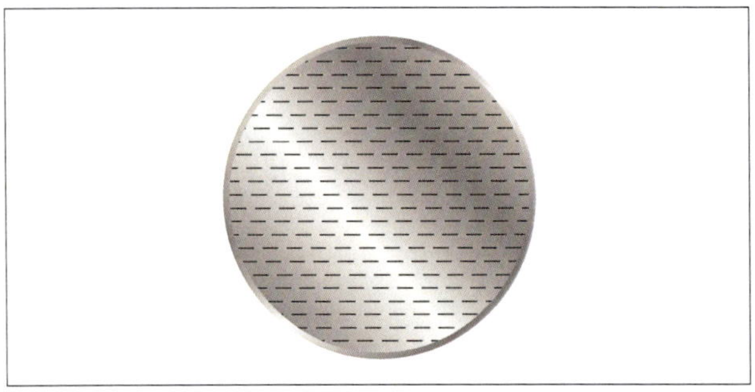

Figure 10-10. Even distribution of charge on metal disk.

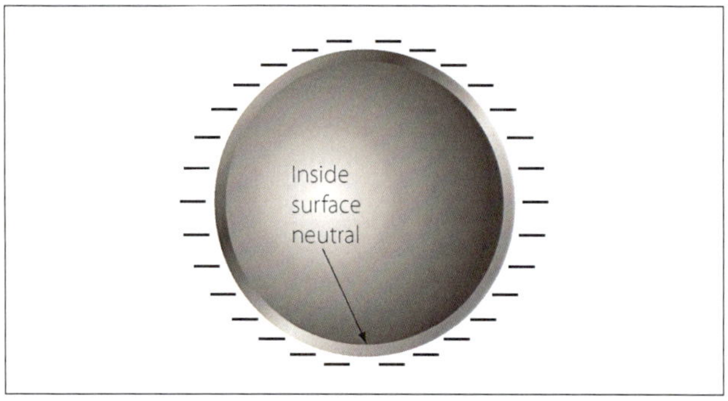

Figure10-11. Charge on a hollow sphere.

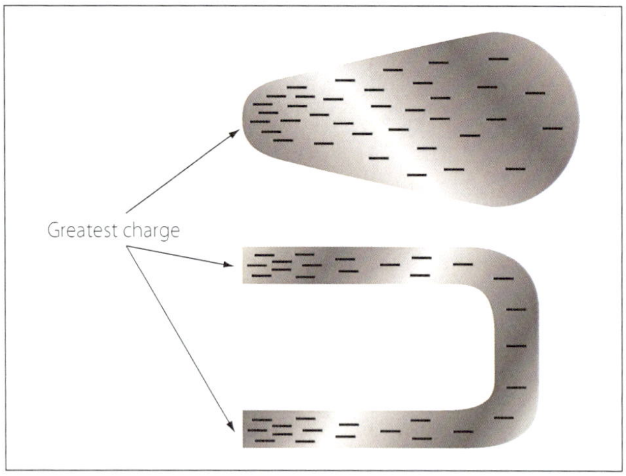

Figure 10-12. Charge on irregularly shaped objects.

static discharge (ESD) from the human body when one of these devices is handled. Careless handling of line replaceable units (LRUs), circuit cards, and discrete components can cause unnecessarily time consuming and expensive repairs. This damage can occur if a technician touches the mating pins for a card or box. Other sources for ESD can be the top of a toolbox that is covered with a carpet. Damage can be avoided by discharging the static electricity from your body by touching the chassis of the removed box, by wearing a grounding wrist strap, and exercising good professional handling of the components in the aircraft. This can include placing protective caps over open connectors and not placing an ESD sensitive component in an environment that will cause damage. Parts that are ESD sensitive are typically shipped in bags specially designed to protect components from electrostatic damage.

Other precautions that should be taken with working with electronic components are:

1. Always connect a ground between test equipment and circuit before attempting to inject or monitor a signal.
2. Ensure test voltages do not exceed maximum allowable voltage for the circuit components and transistors.
3. Ohmmeter ranges that require a current of more than one milliampere in the test circuit should not be used for testing transistors.
4. The heat applied to a diode or transistor, when soldering is required, should be kept to a minimum by using low-wattage soldering irons and heat-sinks.
5. Do not pry components off of a circuit board.
6. Power must be removed from a circuit before replacing a component.
7. When using test probes on equipment and the space between the test points is very close, keep the exposed portion of the leads as short as possible to prevent shorting.

Magnetism

Magnetism is defined as the property of an object to attract certain metallic substances. In general, these substances are ferrous materials; that is, materials composed of iron or iron alloys, such as soft iron, steel, and alnico. These materials, sometimes called magnetic materials, today include at least three nonferrous materials: nickel, cobalt, and gadolinium, which are magnetic to a limited degree. All other substances are considered nonmagnetic, and a few of these nonmagnetic substances can be classified as diamagnetic since they are repelled by both poles of a magnet.

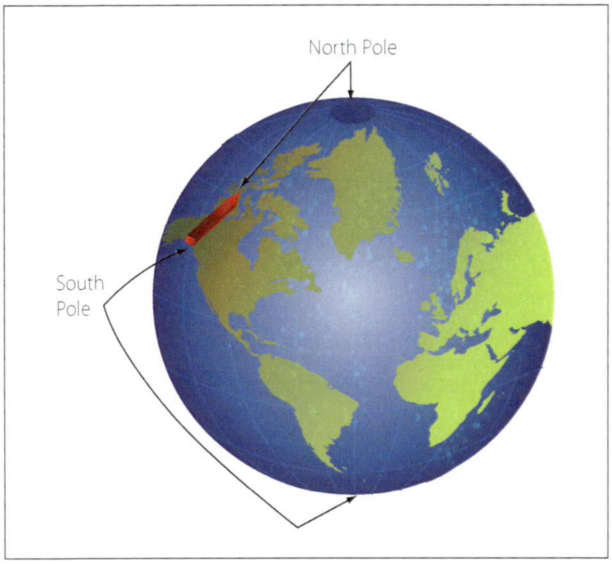

Figure 10-13. One end of magnetized strip points to the magnetic north pole.

Magnetism is an invisible force, the ultimate nature of which has not been fully determined. It can best be described by the effects it produces. Examination of a simple bar magnet similar to that illustrated in Figure 10-13 discloses some basic characteristics of all magnets. If the magnet is suspended to swing freely, it will align itself with the earth's magnetic poles. One end is labeled "N," meaning the north seeking end or pole of the magnet. If the "N" end of a compass or magnet is referred to as north seeking rather than north, there will be no conflict in referring to the pole it seeks, which is the north magnetic pole. The opposite end of the magnet, marked "S" is the south seeking end and points to the south magnetic pole. Since the earth is a giant magnet, its poles attract the ends of the magnet. These poles are not located at the geographic poles.

The somewhat mysterious and completely invisible force of a magnet depends on a magnetic field that surrounds the magnet as illustrated in Figure 10-14. This field always exists between the poles of a magnet, and will arrange itself to conform to the shape of any magnet.

The theory that explains the action of a magnet holds that each molecule making up the iron bar is itself a tiny magnet, with both north and south poles as illustrated in Figure 10-15A. These molecular magnets each possess a magnetic field, but in an unmagnetized state,

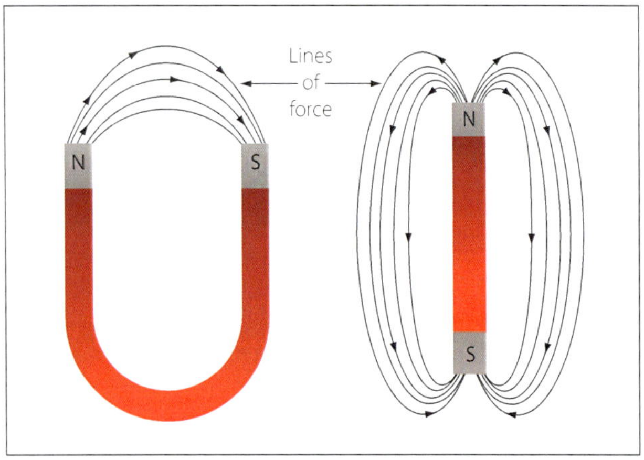

Figure 10-14. Magnetic field around magnets.

the molecules are arranged at random throughout the iron bar. If a magnetizing force, such as stroking with a lodestone, is applied to the unmagnetized bar, the molecular magnets rearrange themselves in line with the magnetic field of the lodestone, with all north ends of the magnets pointing in one direction and all south ends in the opposite direction. This is illustrated in Figure 10-15B. In such a configuration, the magnetic fields of the magnets combine to produce the total field of the magnetized bar.

When handling a magnet, avoid applying direct heat, or hammering or dropping it. Heating or sudden shock will cause misalignment of the molecules, causing the strength of a magnet to decrease. When a magnet is to be stored, devices known as "keeper bars" are installed to provide an easy path for flux lines from one pole to the other. This promotes the retention of the molecules in their north-south alignment.

The presence of the magnetic force or field around a magnet can best be demonstrated by the experiment illustrated in Figure 10-16. A sheet of transparent material, such as glass or Lucite™, is placed over a bar magnet and iron filings are sprinkled slowly on this transparent shield. If the glass or Lucite is tapped lightly, the iron filings will arrange themselves in a definite pattern around the bar, forming a series of lines from the north to south end of the bar to indicate the pattern of the magnetic field.

As shown, the field of a magnet is made up of many individual forces that appear as lines in the iron filing demonstration. Although they are not "lines" in the ordinary sense, this word is used to describe the individual nature of the separate forces making up the entire magnetic field. These lines of force are also referred to as magnetic flux.

They are separate and individual forces, since one line will never cross another; indeed, they actually repel one another. They remain parallel to one another and resemble stretched rubber bands, since they are held in place around the bar by the internal magnetizing force of the magnet.

The demonstration with iron filings further shows that the magnetic field of a magnet is concentrated at the ends of the magnet. These areas of concentrated flux are called the north and south poles of the magnet. There is a limit to the number of lines of force that can be crowded into a magnet of a given size. When a magnetizing force is applied to a piece of magnetic material, a point is reached where no more lines of force can be induced or introduced. The material is then said to be saturated.

The characteristics of the magnetic flux can be demonstrated by tracing the flux patterns of two bar magnets with like poles together, as shown in Figure 10-17. The two like poles repel one another because the lines of force will not cross each other. As the arrows on the individual lines indicate, the lines turn aside as the two like poles are brought near each other and travel in a

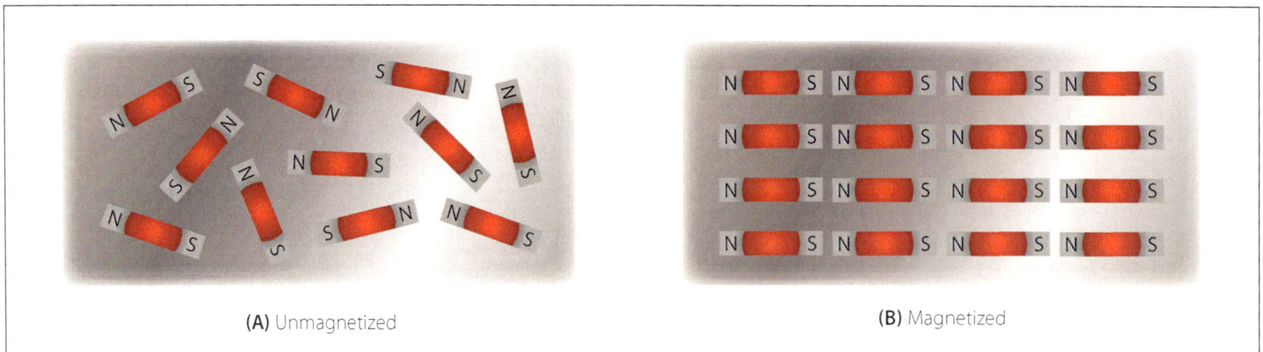

Figure 10-15. Arrangement of molecules in a piece of magnetic material.

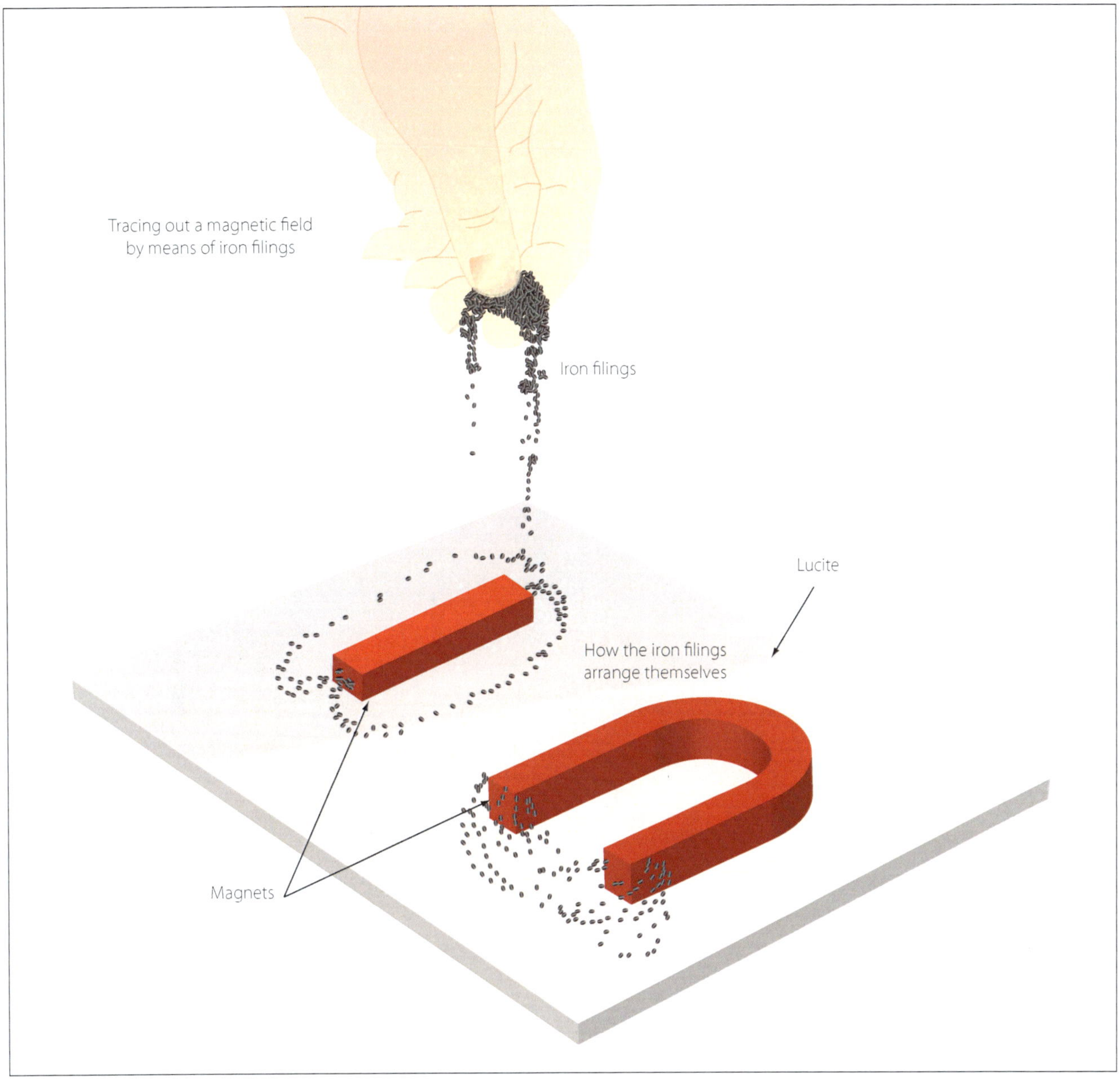

Figure 10-16. Tracing out a magnetic field with iron filings.

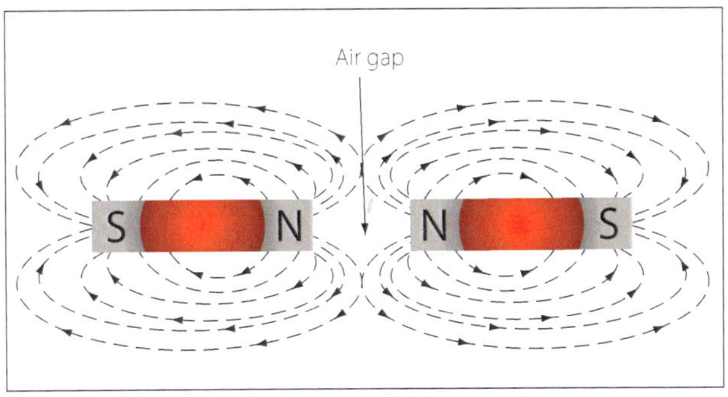

Figure 10-17. Like poles repel.

path parallel to each other. Lines moving in this manner repel each other, causing the magnets as a whole to repel each other.

By reversing the position of one of the magnets, the attraction of unlike poles can be demonstrated, as shown in Figure 10-18.

As the unlike poles are brought near each other, the lines of force rearrange their paths and most of the flux leaving the north pole of one magnet enters the south pole of the other. The tendency of lines of force to repel each other is indicated by the bulging of the flux in the air gap between the two magnets.

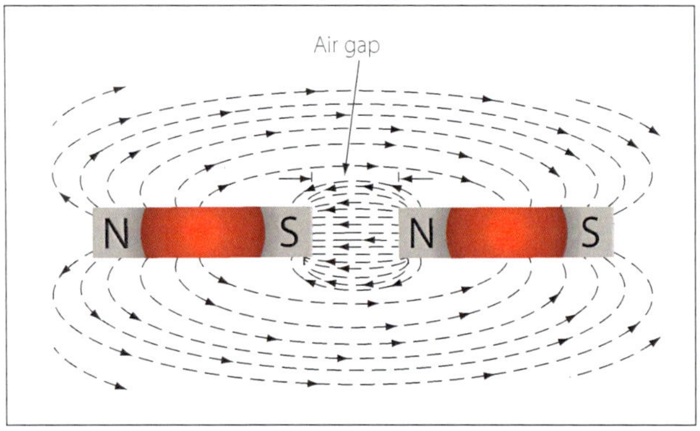

Figure 10-18. Unlike poles attract.

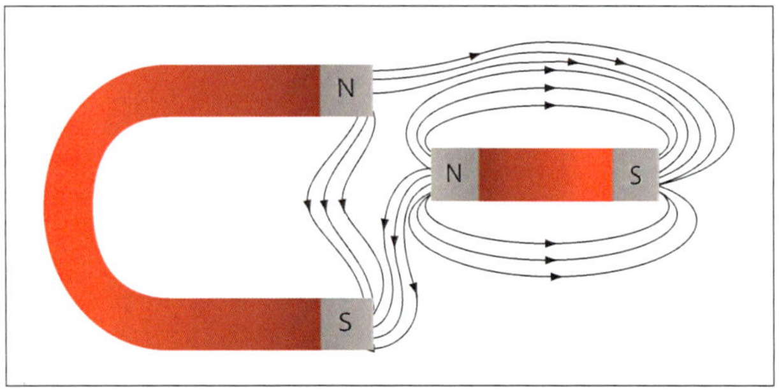

Figure 10-19. Bypassing flux lines.

To further demonstrate that lines of force will not cross one another, a bar magnet and a horseshoe magnet can be positioned to display a magnetic field similar to that of Figure 10-19. The magnetic fields of the two magnets do not combine, but are rearranged into a distorted flux pattern.

The two bar magnets may be held in the hands and the north poles brought near each other to demonstrate the force of repulsion between like poles. In a similar manner, the two south poles can demonstrate this force. The force of attraction between unlike poles can be felt by bringing a south and a north end together. These experiments are illustrated in Figure 10-20.

Figure 10-21 illustrates another characteristic of magnets. If the bar magnet is cut or broken into pieces,

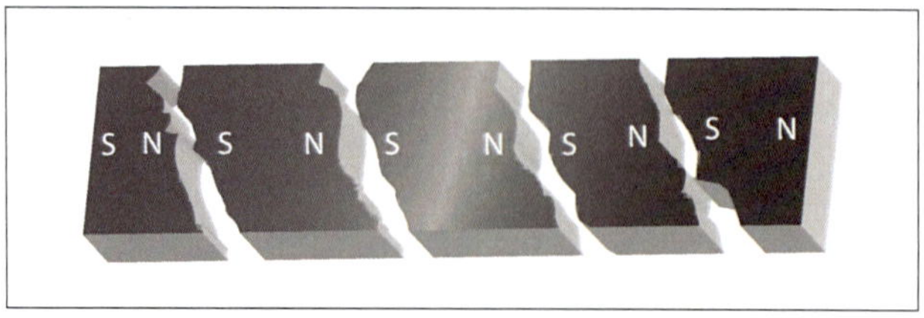

Figure 10-21. Magnetic poles in a broken magnet.

each piece immediately becomes a magnet itself, with a north and south pole. This feature supports the theory that each molecule is a magnet, since each successive division of the magnet produces still more magnets.

Since the magnetic lines of force form a continuous loop, they form a magnetic circuit. It is impossible to say where in the magnet they originate or start. Arbitrarily, it is assumed that all lines of force leave the north pole of any magnet and enter at the south pole.

There is no known insulator for magnetic flux, or lines of force, since they will pass through all materials. However, they will pass through some materials more easily than others.

Thus it is possible to shield items such as instruments from the effects of the flux by surrounding them with a material that offers an easier path for the lines of force. Figure 10-22 shows an instrument surrounded by a path of soft iron, which offers very little opposition to magnetic flux. The lines of force take the easier path, the path of greater permeability, and are guided away from the instrument.

Materials such as soft iron and other ferrous metals are said to have a high permeability, the measure of the ease with which magnetic flux can penetrate a material. The permeability scale is based on a perfect vacuum with a rating of one. Air and other nonmagnetic materials are so close to this that they are also considered to have a rating of one. The nonferrous metals with a permeability greater than one, such as nickel and cobalt, are called paramagnetic. The term

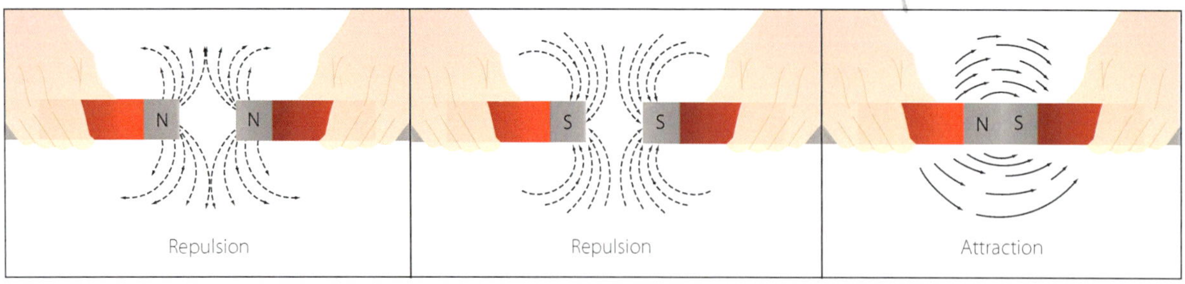

Figure 10-20. Repulsion and attraction of magnet poles.

Figure 10-22. Magnetic shield.

ferromagnetic is applied to iron and its alloys, which have by far the greatest permeability. Any substance, such as bismuth, having a permeability of less than one, is considered diamagnetic.

Reluctance, the measure of opposition to the lines of force through a material, can be compared to the resistance of an electrical circuit. The reluctance of soft iron, for instance, is much lower than that of air. Figure 10-23 demonstrates that a piece of soft iron placed near the field of a magnet can distort the lines of force, which follow the path of lowest reluctance through the soft iron.

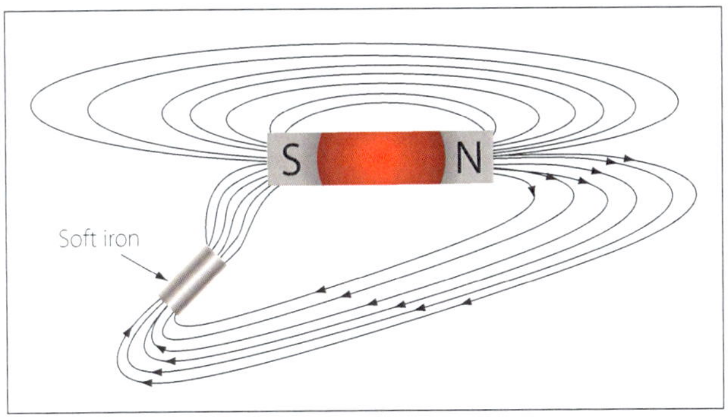

Figure 10-23. Effect of a magnetic substance in a magnetic field.

The magnetic circuit can be compared in many respects to an electrical circuit. The magnetomotive force, causing lines of force in the magnetic circuit, can be compared to the electromotive force or electrical pressure of an electrical circuit. The magnetomotive force is measured in gilberts, symbolized by the capital letter "F." The symbol for the intensity of the lines of force, or flux, is the Greek letter phi, and the unit of field intensity is the gauss. An individual line of force, called a maxwell, in an area of one square centimeter produces a field intensity of one gauss. Using reluctance rather than permeability, the law for magnetic circuits can be stated: a magnetomotive force of one gilbert will cause one maxwell, or line of force, to be set up in a material when the reluctance of the material is one.

Types of Magnets

Magnets are either natural or artificial. Since naturally occurring magnets or lodestones have no practical use, all magnets considered in this study are artificial or manmade. Artificial magnets can be further classified as permanent magnets, which retain their magnetism long after the magnetizing force has been removed, and temporary magnets, which quickly lose most of their magnetism when the external magnetizing force is removed.

Modern permanent magnets are made of special alloys that have been found through research to create increasingly better magnets. The most common categories of magnet materials are made out of Aluminum-Nickel-Cobalt (Alnicos), Strontium-Iron (Ferrites, also known as Ceramics), Neodymium-Iron-Boron (Neo magnets), and Samarium-Cobalt. Alnico, an alloy of iron, aluminum, nickel and cobalt, and is considered one of the very best. Others with excellent magnetic qualities are alloys such as Remalloy™ and Permendur™.

The ability of a magnet to hold its magnetism varies greatly with the type of metal and is known as retentivity. Magnets made of soft iron are very easily magnetized but quickly lose most of their magnetism when the external magnetizing force is removed. The small amount of magnetism remaining, called residual magnetism, is of great importance in such electrical applications as generator operation.

Horseshoe magnets are commonly manufactured in two forms. [Figure 10-24] The most common type is made from a long bar curved into a horseshoe shape, while a variation of this type consists of two bars connected by a third bar, or yoke.

Figure 10-24. Two forms of horseshoe magnets.

Magnets can be made in many different shapes, such as balls, cylinders, or disks. One special type of magnet is the ring magnet, or Gramme ring, often used in instruments. This is a closed loop magnet, similar to the type used in transformer cores, and is the only type that has no poles.

Sometimes special applications require that the field of force lie through the thickness rather than the length of a piece of metal. Such magnets are called flat magnets and are used as pole pieces in generators and motors.

Electromagnetism

In 1820, the Danish physicist, Hans Christian Oersted, discovered that the needle of a compass brought near a current carrying conductor would be deflected. When the current flow stopped, the compass needle returned to its original position. This important discovery demonstrated a relationship between electricity and magnetism that led to the electromagnet and to many of the inventions on which modern industry is based.

Oersted discovered that the magnetic field had no connection with the conductor in which the electrons were flowing, because the conductor was made of nonmagnetic copper. The electrons moving through the wire created the magnetic field around the conductor. Since a magnetic field accompanies a charged particle, the greater the current flow, and the greater the magnetic field. Figure 10-25 illustrates the magnetic field around

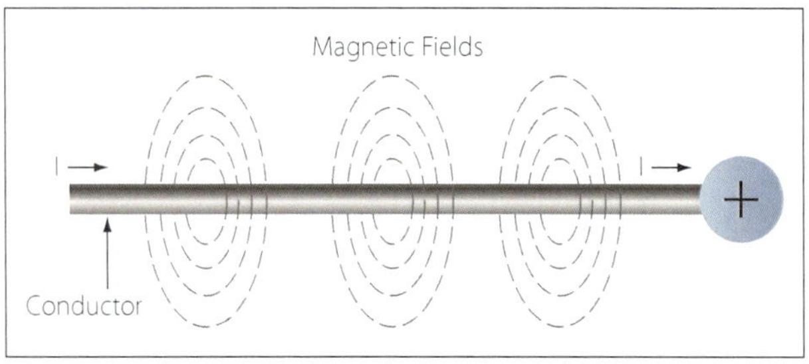

Figure 10-25. Magnetic field formed around a conductor in which current is flowing.

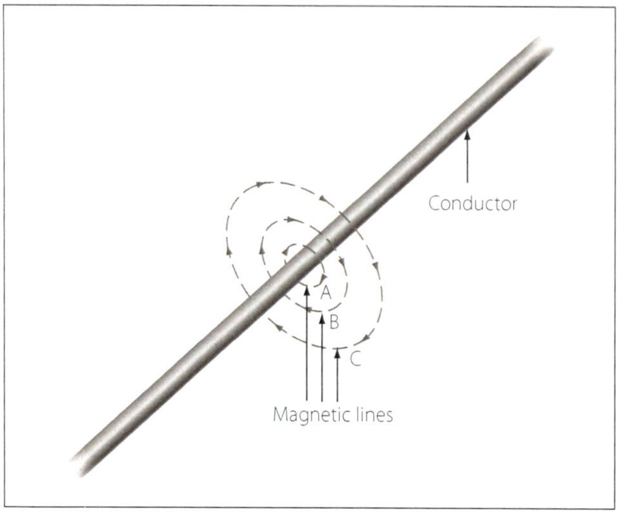

Figure 10-26. Expansion of magnetic field as current increases.

a current carrying wire. A series of concentric circles around the conductor represent the field, which if all the lines were shown would appear more as a continuous cylinder of such circles around the conductor.

As long as current flows in the conductor, the lines of force remain around it. [Figure 10-26] If a small current flows through the conductor, there will be a line of force extending out to circle A. If the current flow is increased, the line of force will increase in size to circle B, and a further increase in current will expand it to circle C. As the original line (circle) of force expands from circle A to B, a new line of force will appear at circle A. As the current flow increases, the number of circles of force increases, expanding the outer circles farther from the surface of the current carrying conductor.

If the current flow is a steady nonvarying direct current, the magnetic field remains stationary. When the current stops, the magnetic field collapses and the magnetism around the conductor disappears.

A compass needle is used to demonstrate the direction of the magnetic field around a current carrying conductor. Figure 10-27 View A shows a compass needle positioned at right angles to, and approximately one inch from, a current carrying conductor. If no current were flowing, the north seeking end of the compass needle would point toward the earth's magnetic pole. When current flows, the needle lines itself up at right angles to a radius drawn from the conductor. Since the compass needle is a small magnet, with lines of force extending from south to north inside the metal, it will turn until the direction of these lines agrees with the direction of the lines of force around the conductor. As

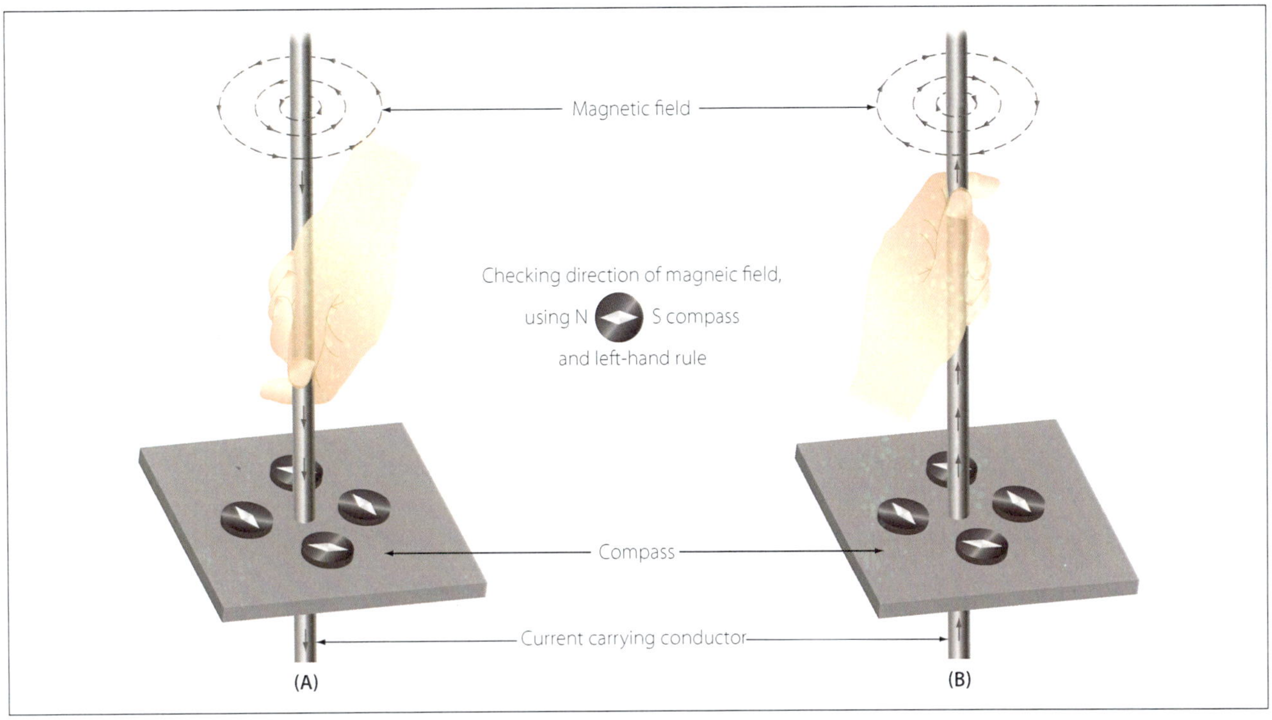

Figure 10-27. Magnetic field around a current-carrying conductor.

the compass needle is moved around the conductor, it will maintain itself in a position at right angles to the conductor, indicating that the magnetic field around a current carrying conductor is circular. As shown in View B of Figure 10-27, when the direction of current flow through the conductor is reversed, the compass needle will point in the opposite direction, indicating the magnetic field has reversed its direction.

A method used to determine the direction of the lines of force when the direction of the current flow is known, is shown in Figure 10-28. If the conductor is grasped in the left hand, with the thumb pointing in the direction of current flow, the fingers will be wrapped around the conductor in the same direction as the lines of the magnetic field. This is called the left-hand rule.

Although it has been stated that the lines of force have direction, this should not be construed to mean that the lines have motion in a circular direction around the conductor. Although the lines of force tend to act in a clockwise or counterclockwise direction, they are not revolving around the conductor.

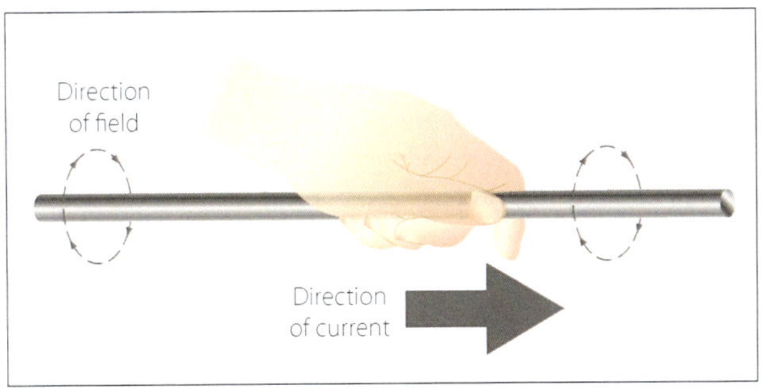

Figure 10-28. Left-hand rule.

Since current flows from negative to positive, many illustrations indicate current direction with a dot symbol on the end of the conductor when the electrons are flowing toward and a plus sign when the current is flowing away from the observer. [Figure 10-29]

When a wire is bent into a loop and an electric current flows through it, the left-hand rule remains valid. [Figure 10-30]

If the wire is coiled into two loops, many of the lines of force become large enough to include both loops. Lines of force go through the loops in the same direc-

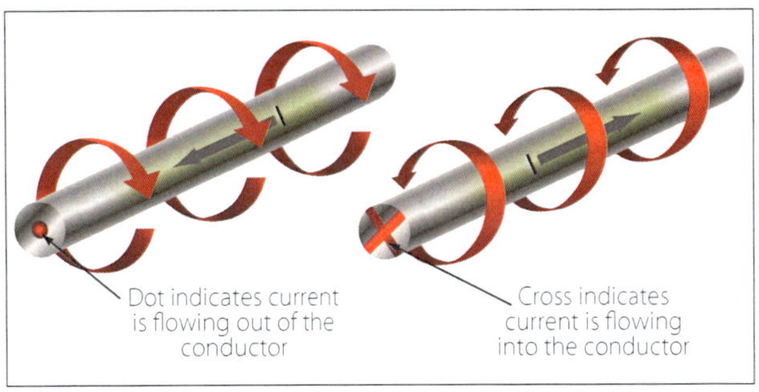

Figure 10-29. Direction of current flow in a conductor.

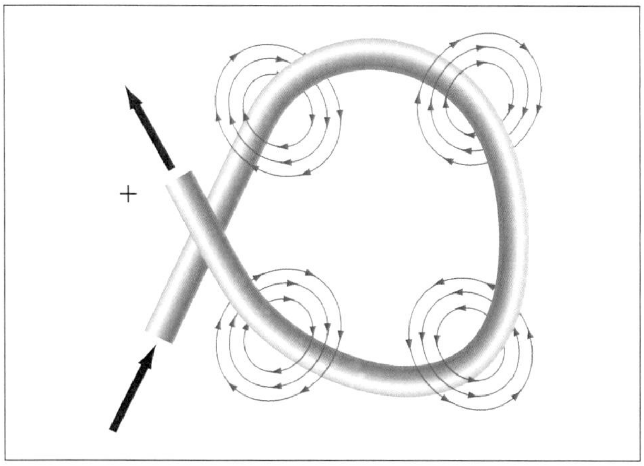

Figure 10-30. Magnetic field around a looped conductor.

tion, circle around the outside of the two coils, and come in at the opposite end. [Figure 10-31]

When a wire contains many such loops, it is called a coil. The lines of force form a pattern through all the loops, causing a high concentration of flux lines through the center of the coil. [Figure 10-32]

In a coil made from loops of a conductor, many of the lines of force are dissipated between the loops of the coil. By placing a soft iron bar inside the coil, the lines of force will be concentrated in the center of the coil, since soft iron has a greater permeability than air. [Figure 10-33] This combination of an iron core in a coil of

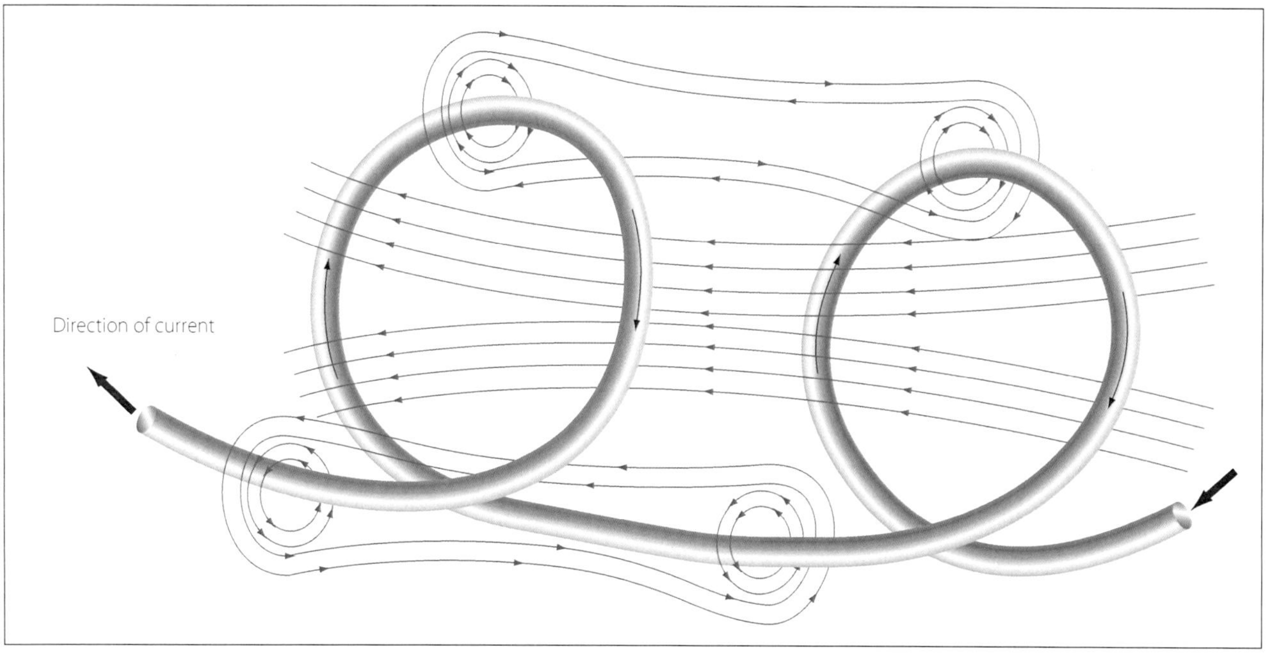

Figure 10-31. Magnetic field around a conductor with two loops.

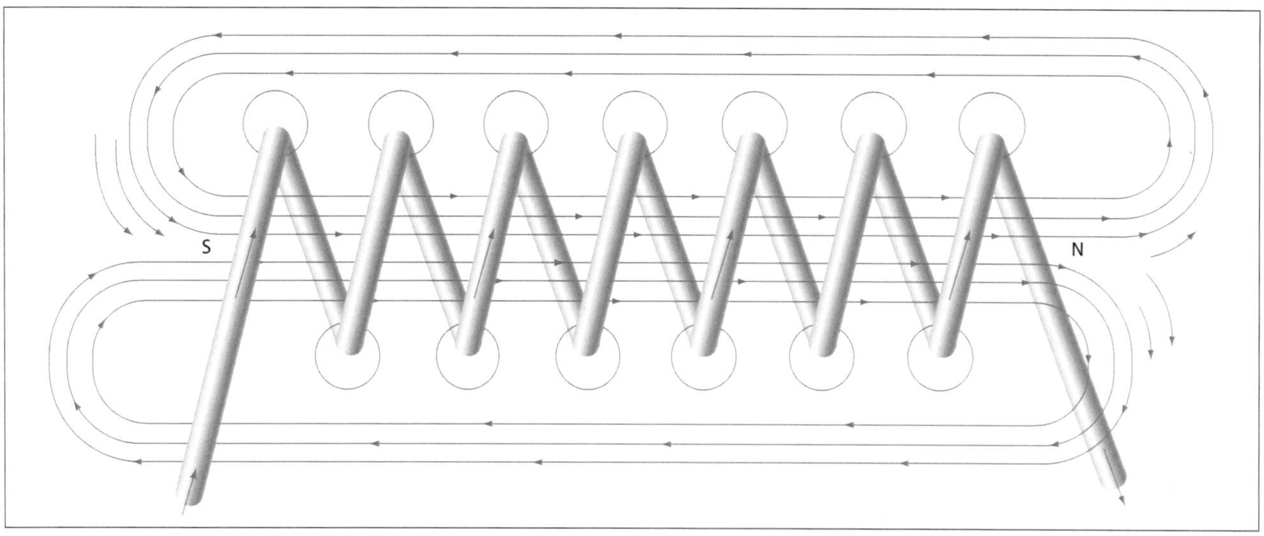

Figure 10-32. Magnetic field of a coil.

Figure 10-33. Electromagnet.

wire loops, or turns, is called an electromagnet, since the poles (ends) of the coil possess the characteristics of a bar magnet.

The addition of the soft iron core does two things for the current carrying coil. First, the magnetic flux is increased, and second, the flux lines are more highly concentrated.

When direct current flows through the coil, the core will become magnetized with the same polarity (location of north and south poles) as the coil would have without the core. If the current is reversed, the polarity will also be reversed.

The polarity of the electromagnet is determined by the left-hand rule in the same manner as the polarity of the coil without the core was determined. If the coil is grasped in the left hand in such a manner that the fingers curve around the coil in the direction of electron flow (minus to plus), the thumb will point in the direction of the north pole. [Figure 10-34]

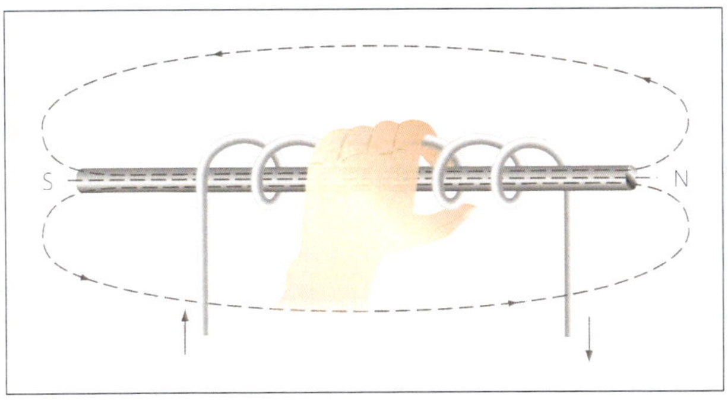

Figure 10-34. Left-hand rule applied to a coil.

The strength of the magnetic field of the electromagnet can be increased by either increasing the flow of current or the number of loops in the wire. Doubling the current flow approximately doubles the strength of the field, and in a similar manner, doubling the number of loops approximately doubles magnetic field strength. Finally, the type metal in the core is a factor in the field strength of the electromagnet.

A soft iron bar is attracted to either pole of a permanent magnet and, likewise, is attracted by a current carrying coil. The lines of force extend through the soft iron, magnetizing it by induction and pulling the iron bar toward the coil. If the bar is free to move, it will be drawn into the coil to a position near the center where the field is strongest. [Figure 10-35]

Electromagnets are used in electrical instruments, motors, generators, relays, and other devices. Some

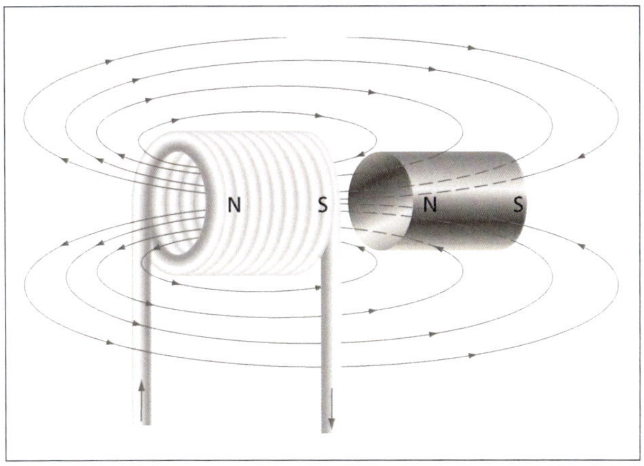

Figure 10-35. Solenoid with iron core.

electromagnetic devices operate on the principle that an iron core held away from the center of a coil will be rapidly pulled into a center position when the coil is energized. This principle is used in the solenoid, also called solenoid switch or relay, in which the iron core is spring-loaded off center and moves to complete a circuit when the coil is energized.

Conventional Flow and Electron Flow

Today's technician will find that there are two competing schools of thought and analytical practices regarding the flow of electricity. The two are called the conventional current theory and the electron theory.

Conventional Flow

Of the two, the conventional current theory was the first to be developed and, through many years of use, this method has become ingrained in electrical texts. The theory was initially advanced by Benjamin Franklin who reasoned that current flowed out of a positive source into a negative source or an area that lacked an abundance of charge. The notation assigned to the electric charges was positive (+) for the abundance of charge and negative (−) for a lack of charge. It then seemed natural to visualize the flow of current as being from the positive (+) to the negative (−).

Electron Flow

Later discoveries were made that proved that just the opposite is true. Electron flow is what actually happens where an abundance of electrons flow out of the negative (−) source to an area that lacks electrons or the positive (+) source.

Both conventional flow and electron flow are used in industry. Many textbooks in current use employ both electron flow and conventional flow methods. From the practical standpoint of the technician, troubleshooting a system, it makes little to no difference which way current is flowing as long as it is used consistently in the analysis.

Electromotive Force (Voltage)

Unlike current, which is easy to visualize as a flow, voltage is a variable that is determined between two points. Often we refer to voltage as a value across two points. It is the electromotive force (emf) or the push or pressure felt in a conductor that ultimately moves the electrons in a flow. The symbol for emf is the capital letter "E."

Across the terminals of the typical aircraft battery, voltage can be measured as the potential difference of 12 volts or 24 volts. That is to say that between the two terminal posts of the battery, there is an electromotive force of 12 or 24 volts available to push current through a circuit. Relatively free electrons in the negative terminal will move toward the excessive number of positive charges in the positive terminal. Recall from the discussion on static electricity that like charges repel each other but opposite charges attract each other. The net result is a flow or current through a conductor. There cannot be a flow in a conductor unless there is an applied voltage from a battery, generator, or ground power unit. The potential difference, or the voltage across any two points in an electrical system, can be determined by:

Where

$$E = \frac{\mathcal{E}}{Q}$$

E = potential difference in volts
$\mathcal{E}$ = energy expanded or absorbed in joules (J)
Q = Charge measured in coulombs

Figure 10-36 illustrates the flow of electrons of electric current. Two interconnected water tanks demonstrate that when a difference of pressure exists between the two tanks, water will flow until the two tanks are equalized. The illustration shows the level of water in tank A to be at a higher level, reading 10 psi (higher potential energy) than the water level in tank B, reading 2 psi (lower potential energy). Between the two tanks, there is 8-psi potential difference. If the valve in the interconnecting line between the tanks is opened, water will flow from tank A into tank B until the level of water (potential energy) of both tanks is equalized.

It is important to note that it was not the pressure in tank A that caused the water to flow; rather, it was the difference in pressure between tank A and tank B that caused the flow.

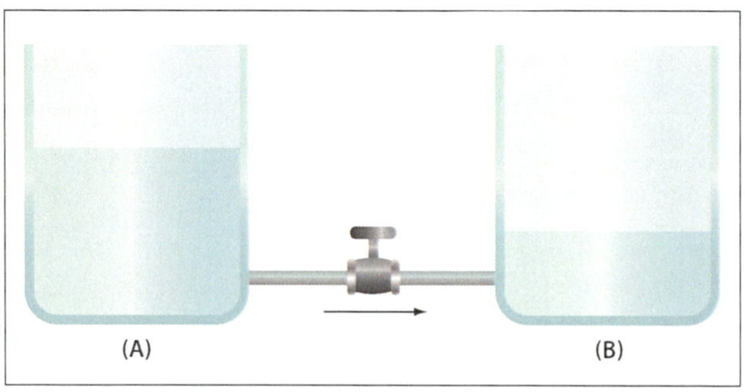

Figure 10-36. Difference of pressure.

This comparison illustrates the principle that electrons move, when a path is available, from a point of excess electrons (higher potential energy) to a point deficient in electrons (lower potential energy). The force that causes this movement is the potential difference in electrical energy between the two points. This force is called the electrical pressure or the potential difference or the electromotive force (electron moving force).

Current

Electrons in motion make up an electric current. This electric current is usually referred to as "current" or "current flow," no matter how many electrons are moving. Current is a measurement of a rate at which a charge flows through some region of space or a conductor. The moving charges are the free electrons found in conductors, such as copper, silver, aluminum, and gold. The term "free electron" describes a condition in some atoms where the outer electrons are loosely bound to their parent atom. These loosely bound electrons can be easily motivated to move in a given direction when an external source, such as a battery, is applied to the circuit. These electrons are attracted to the positive terminal of the battery, while the negative terminal is the source of the electrons. The greater amount of charge moving through the conductor in a given amount of time translates into a current.

$$\text{Current} = \frac{\text{Charge}}{\text{Time}}$$

Or

$$I = \frac{Q}{t}$$

Where:
I = Current in Amperes (A)
Q = Charge in Coulombs (C)
t = time

The System International unit for current is the Ampere (A), where

$$1\text{ A} = 1\frac{\text{C}}{\text{s}}$$

That is, 1 ampere (A) of current is equivalent to 1 coulomb (C) of charge passing through a conductor in 1 second(s). One coulomb of charge equals 6.28 billion billion electrons. The symbol used to indicate current in formulas or on schematics is the capital letter "I."

When current flow is one direction, it is called direct current (DC). Later in the text, we will discuss the form of current that periodically oscillates back and forth within the circuit. The present discussion will only be concerned with the use of direct current.

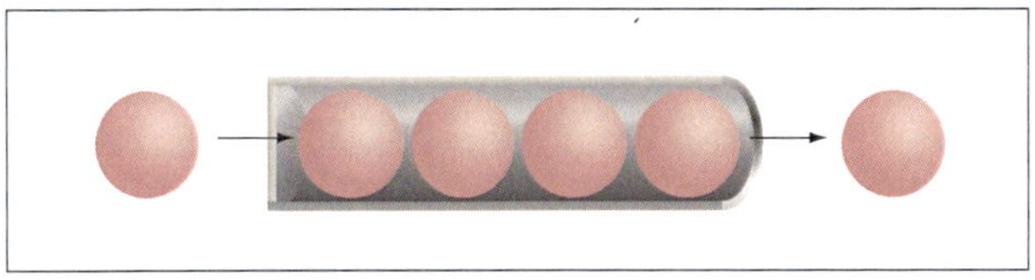

Figure 10-37. Electron movement.

The velocity of the charge is actually an average velocity and is called drift velocity. To understand the idea of drift velocity, think of a conductor in which the charge carriers are free electrons. These electrons are always in a state of random motion similar to that of gas molecules. When a voltage is applied across the conductor, an electromotive force creates an electric field within the conductor and a current is established. The electrons do not move in a straight direction but undergo repeated collisions with other nearby atoms. These collisions usually knock other free electrons from their atoms, and these electrons move on toward the positive end of the conductor with an average velocity called the drift velocity, which is relatively a slow speed. To understand the nearly instantaneous speed of the effect of the current, it is helpful to visualize a long tube filled with steel balls as shown in Figure 10-37. It can be seen that a ball introduced in one end of the tube, which represents the conductor, will immediately cause a ball to be emitted at the opposite end of the tube. Thus, electric current can be viewed as instantaneous, even though it is the result of a relatively slow drift of electrons.

Ohm's Law (Resistance)

The two fundamental properties of current and voltage are related by a third property known as resistance. In any electrical circuit, when voltage is applied to it, a current will result. The resistance of the conductor will determine the amount of current that flows under the given voltage. In most cases, the greater the circuit resistance, the less the current. If the resistance is reduced, then the current will increase. This relation is linear in nature and is known as Ohm's law.

By having a linearly proportional characteristic, it is meant that if one unit in the relationship increases or decreases by a certain percentage, the other variables in the relationship will increase or decrease by the same percentage. An example would be if the voltage across a resistor is doubled, then the current through the resistor doubles. It should be added that

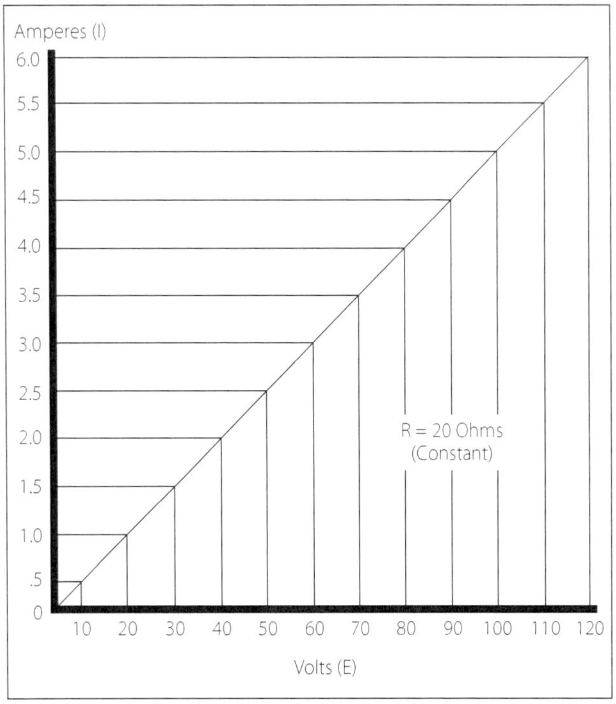

Figure 10-38. Voltage vs. current in a constant-resistance circuit.

this relationship is true only if the resistance in the circuit remains constant. For it can be seen that if the resistance changes, current also changes. A graph of this relationship is shown in Figure 10-38, which uses a constant resistance of 20Ω. The relationship between voltage and current in this example shows voltage plotted horizontally along the X axis in values from 0 to 120 volts, and the corresponding values of current are plotted vertically in values from 0 to 6.0 amperes along the Y axis. A straight line drawn through all the points where the voltage and current lines meet represents the equation $I = E/20$ and is called a linear relationship.

If $E = 10V$

Then $\frac{10V}{20\Omega} = 0.5A$

If $E = 60V$

Then $\frac{60V}{20\Omega} = 3A$

If $E = 120V$

Then $\frac{120V}{20\Omega} = 6A$

Ohm's law may be expressed as an equation, as follows:

Equation 1

$$I = \frac{E}{R}$$

I = Current in amperes (A)
E = Voltage (V)
R = Resistance (Ω)

Where I is current in amperes, E is the potential difference measured in volts, and R is the resistance measured in ohms. If any two of these circuit quantities are known, the third may be found by simple algebraic transposition. With this equation, we can calculate current in a circuit if the voltage and resistance are known. This same formula can be used to calculate voltage. By multiplying both sides of the equation 1 by R, we get an equivalent form of Ohm's law, which is:

Equation 2

$$E = I\,(R)$$

Finally, if we divide equation 2 by I, we will solve for resistance,

Equation 3

$$R = \frac{E}{I}$$

All three formulas presented in this section are equivalent to each other and are simply different ways of expressing Ohm's law.

The various equations, which may be derived by transposing the basic law, can be easily obtained by using the triangles in Figure 10-39.

The triangles containing E, R, and I are divided into two parts, with E above the line and I × R below it. To determine an unknown circuit quantity when the other two are known, cover the unknown quantity with a thumb. The location of the remaining uncovered letters in the triangle will indicate the mathematical operation to be performed. For example, to find I, refer to Figure 10-39A, and cover I with the thumb. The uncovered letters indicate that E is to be divided by R, or $I = E/R$. To find R, refer to Figure 10-39B, and cover R with the thumb. The result indicates that E is to be divided by I, or $R = E/I$. To find E, refer to Figure 10-39C, and cover E with the thumb. The result indicates I is to be multiplied by R, or $E = I \times R$.

This chart is useful when learning to use Ohm's law. It should be used to supplement the beginner's knowledge of the algebraic method.

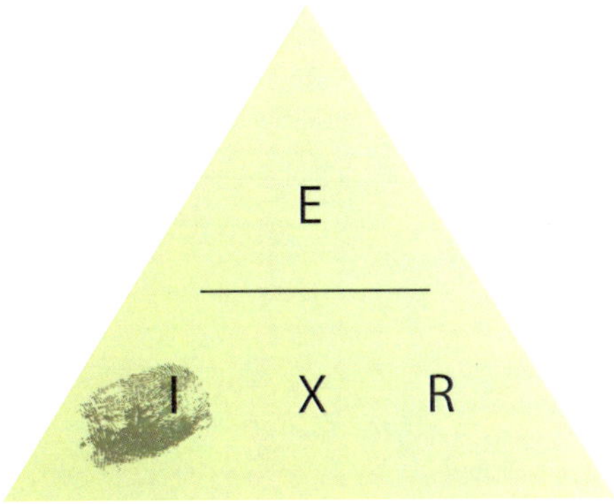

(A)
To find I (amperes), place thumb over I and divide E by R as indicated

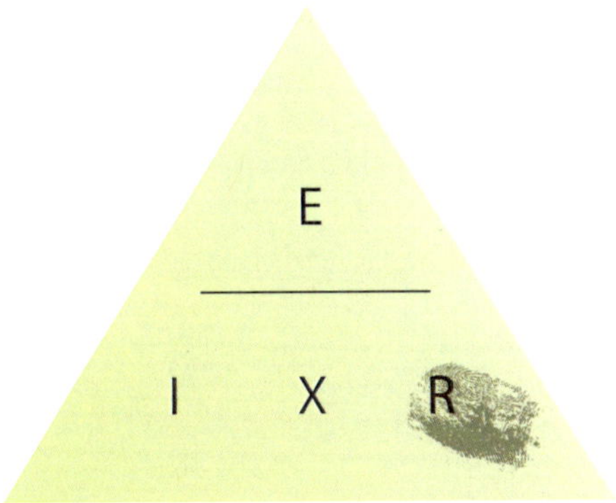

(B)
To find R (ohms), place thumb over R and divide as indicated

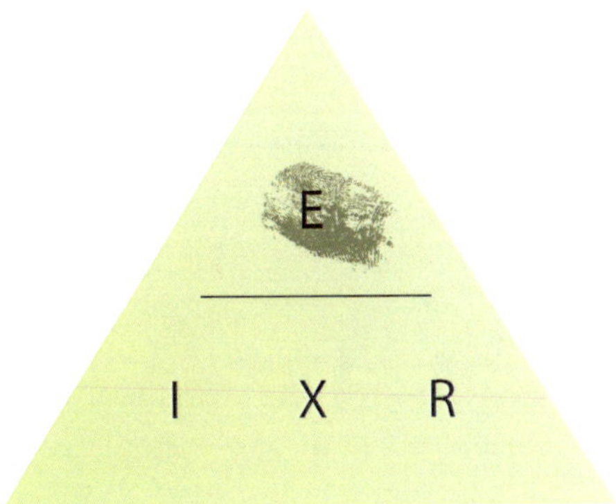

(C)
To find E (volts), place thumb over E multiply as indicated

Figure 10-39. Ohm's law chart.

Resistance of a Conductor

While wire of any size or resistance value may be used, the word "conductor" usually refers to materials that offer low resistance to current flow, and the word "insulator" describes materials that offer high resistance to current. There is no distinct dividing line between conductors and insulators; under the proper conditions, all types of material conduct some current. Materials offering a resistance to current flow midway between the best conductors and the poorest conductors (insulators) are sometimes referred to as "semiconductors," and find their greatest application in the field of transistors.

The best conductors are materials, chiefly metals, which possess a large number of free electrons; conversely, insulators are materials having few free electrons. The best conductors are silver, copper, gold, and aluminum; but some nonmetals, such as carbon and water, can be used as conductors. Materials such as rubber, glass, ceramics, and plastics are such poor conductors that they are usually used as insulators. The current flow in some of these materials is so low that it is usually considered zero. The unit used to measure resistance is called the ohm. The symbol for the ohm is the Greek letter omega (Ω). In mathematical formulas, the capital letter "R" refers to resistance. The resistance of a conductor and the voltage applied to it determine the number of amperes of current flowing through the conductor. Thus, 1 ohm of resistance will limit the current flow to 1 ampere in a conductor to which a voltage of 1 volt is applied.

Factors Affecting Resistance

1. The resistance of a metallic conductor is dependent on the type of conductor material. It has been pointed out that certain metals are commonly used as conductors because of the large number of free electrons in their outer orbits. Copper is usually considered the best available conductor material, since a copper wire of a particular diameter offers a lower resistance to current flow than an aluminum wire of the same diameter. However, aluminum is much lighter than copper, and for this reason as well as cost considerations, aluminum is often used when the weight factor is important.
2. The resistance of a metallic conductor is directly proportional to its length. The longer the length of a given size of wire, the greater the resistance. Figure 10-40 shows two wire conductors of different lengths. If 1 volt of electrical pressure is applied across the two ends of the conductor that is 1 foot in length and the resistance to the movement of free electrons is assumed to be 1 ohm, the current flow is limited to 1 ampere. If the same size conductor is doubled in length, the same electrons set in motion by the 1 volt applied now find twice the resistance;

Figure 10-40. Resistance varies with length of conductor.

consequently, the current flow will be reduced by one-half.

3. The resistance of a metallic conductor is inversely proportional to the cross-sectional area. This area may be triangular or even square, but is usually circular. If the cross-sectional area of a conductor is doubled, the resistance to current flow will be reduced in half. This is true because of the increased area in which an electron can move without collision or capture by an atom. Thus, the resistance varies inversely with the cross-sectional area of a conductor.
4. The fourth major factor influencing the resistance of a conductor is temperature. Although some substances, such as carbon, show a decrease in resistance as the ambient (surrounding) temperature increases, most materials used as conductors increase in resistance as temperature increases. The resistance of a few alloys, such as constantan and Manganin™, change very little as the temperature changes. The amount of increase in the resistance of a 1 ohm sample of a conductor, per degree rise in temperature above 0° Centigrade (C), the assumed standard, is called the temperature coefficient of resistance. For each metal, this is a different value; for example, for copper the value is approximately 0.00427 ohm. Thus, a copper wire having a resistance of 50 ohms at a temperature of 0 °C will have an increase in resistance of 50 × 0.00427, or 0.214 ohm, for each degree rise in temperature above 0 °C. The temperature coefficient of resistance must be considered where there is an appreciable change in temperature of a conductor during operation. Charts listing the temperature coefficient of resistance for different materials are available. Figure 10-41 shows a table for "resistivity" of some common electric conductors.

Conductor Material	Resistivity (Ohm meters @ 20 °C)
Silver	1.64×10^{-8}
Copper	1.72×10^{-8}
Aluminum	2.83×10^{-8}
Tungsten	5.50×10^{-8}
Nickel	7.80×10^{-8}
Iron	12.0×10^{-8}
Constantan	49.0×10^{-8}
Nichrome II	110×10^{-8}

Figure 10-41. Resistivity table.

The resistance of a material is determined by four properties: material, length, area, and temperature. The first three properties are related by the following equation at T = 20 °C (room temperature):

$$R = \frac{(\rho \times l)}{A}$$

Where
R = resistance in ohms
ρ = Resistivity of the material in circular mil-ohms per foot
l = Length of the sample in feet
A = area in circular mils

Resistance and Its Relation to Wire Sizing

Circular Conductors (Wires/Cables)

Because it is known that the resistance of a conductor is directly proportional to its length, and if we are given the resistance of the unit length of wire, we can readily calculate the resistance of any length of wire of that particular material having the same diameter. Also, because it is known that the resistance of a conductor is inversely proportional to its cross-sectional area, and if we are given the resistance of a length of wire with unit cross-sectional area, we can calculate the resistance of a similar length of wire of the same material with any cross-sectional area. Therefore, if we know the resistance of a given conductor, we can calculate the resistance for any conductor of the same material at the same temperature. From the relationship:

$$R = \frac{(\rho \times l)}{A}$$

It can also be written:

$$\frac{R_1}{R_2} = \frac{l_1}{l_2} = \frac{A_1}{A_2}$$

If we have a conductor that is 1 meter long with a cross-sectional area of 1 mm^2 and has a resistance of 0.017 ohm, what is the resistance of 50m of wire from the same material but with a cross-sectional area of 0.25 mm^2?

$$\frac{R_1}{R_2} = \frac{l_1}{l_2} = \frac{A_1}{A_2}$$

$$R^2 = 0.017\Omega \times \frac{50m}{1m} \times \frac{1mm^2}{0.25mm^2} = 3.4\Omega$$

While the System International (SI) units are commonly used in the analysis of electric circuits, electrical conductors in North America are still being manufactured using the foot as the unit length and the mil (one thousandth of an inch) as the unit of diameter. Before using the equation $R = {}^{(\rho \times l)}/_{A}$ to calculate the resistance of a conductor of a given AWG size, the cross-sectional area in square meters must be determined using the conversion factor 1 mil = 0.0254 mm. The most convenient unit of wire length is the foot. Using these standards, the unit of size is the mil-foot. Thus, a wire has unit size if it has a diameter of 1 mil and length of 1 foot.

In the case of using copper conductors, we are spared the task of tedious calculations by using a table as shown in Figure 10-42. Note that cross-sectional dimensions listed on the table are such that each decrease of one gauge number equals a 25 percent increase in the cross-sectional area. Because of this, a decrease of three gauge numbers represents an increase in cross-sectional area of approximately a 2:1 increase. Likewise, change of ten wire gauge numbers represents a 10:1 change in cross-sectional area—also, by doubling the cross-sectional area of the conductor, the resistance is cut in half. A decrease of three wire gauge numbers cuts the resistance of the conductor of a given length in half.

Rectangular Conductors (Bus Bars)

To compute the cross-sectional area of a conductor in square mils, the length in mils of one side is squared. In the case of a rectangular conductor, the length of one side is multiplied by the length of the other. For example, a common rectangular bus bar (large, special conductor) is 3/8 inch thick and 4 inches wide. The 3/8-inch thickness may be expressed as 0.375 inch. Since 1,000 mils equal 1 inch, the width in inches can be converted to 4,000 mils. The cross-sectional area of the rectangular conductor is found by converting 0.375 to mils (375 mils × 4,000 mils = 1,500,000 square mils).

AWG Number	Diameter in mils	Ohms per 1,000 ft.
0000	460.0	0.04901
000	409.6	0.06180
00	364.8	0.07793
0	324.9	0.09827
1	289.3	0.1239
2	257.6	0.1563
3	229.4	0.1970
4	204.3	0.2485
5	181.9	0.3133
6	162.0	0.3951
8	128.5	0.6282
10	101.9	0.9989
12	80.81	1.588
14	64.08	2.525
16	50.82	4.016
18	40.30	6.385
20	31.96	10.15
22	25.35	16.14
24	20.10	25.67
26	15.94	40.81
28	12.64	64.9
30	10.03	103.2

Figure 10-42. Conversion table when using copper conductors.

Power and Energy

Power in an Electrical Circuit

This section covers power in the DC circuit and energy consumption. Whether referring to mechanical or electrical systems, power is defined as the rate of energy consumption or conversion within that system—that is, the amount of energy used or converted in a given amount of time.

From the scientific discipline of physics, the fundamental expression for power is:

$$P = \frac{\mathcal{E}}{t}$$

Where

P = Power measured in Watts (W)

$\mathcal{E}$ = Energy ($\mathcal{E}$ is a script E) measured in Joules (J)

And

t = Time measured in Seconds (s)

The unit measurement for power is the watt (W), which refers to a rate of energy conversion of 1 joule/second. Therefore, the number of joules consumed in 1 second is equal to the number of watts. A simple example is given below.

Suppose 300 J of energy is consumed in 10 seconds. What would be the power in watts?

General formula $P = \frac{\text{energy}}{\text{time}}$

$$P = \frac{300\ J}{10\ s}$$

$$P = 30\ W$$

The watt is named for James Watt, the inventor of the steam engine. Watt devised an experiment to measure the power of a horse in order to find a means of measuring the mechanical power of his steam engine. One horsepower is required to move 33,000 pounds 1 foot in 1 minute. Since power is the rate of doing work, it is equivalent to the work divided by time. Stated as a formula, this is:

$$\text{Power} = \frac{33{,}000\ \text{ft-lb}}{60\ \text{sec}}$$

$$P = 550\ \text{ft-lb/sec}$$

Electrical power can be rated in a similar manner. For example, an electric motor rated as a 1 horsepower motor requires 746 watts of electrical energy.

Power Formulas Used in the Study of Electricity

When current flows through a resistive circuit, energy is dissipated in the form of heat. Recall that voltage can be expressed in the terms of energy and charge as given in the expression:

$$E = \frac{\mathcal{E}}{Q}$$

Where
E = potential difference in volts
W = energy expanded or absorbed in joules (J)
Q = Charge measured in coulombs

Current I, can also be expressed in terms of charge and time as given by the expression:

$$\text{Current} = \frac{\text{Charge}}{\text{Time}}$$

Or

$$I = \frac{Q}{t}$$

Where:
I = Current in Amperes (A)
Q = Charge in Coulombs (C)
t = time

When voltage $^W/_Q$ and current $^Q/_t$ are multiplied, the charge Q is divided out leaving the basic expression from physics:

$$E \times I = \frac{\mathcal{E}}{Q} \times \frac{Q}{t} = \frac{\mathcal{E}}{t} = \text{power}$$

For a simple DC electrical system, power dissipation can then be given by the equation:

General Power Formula P = I (E)
Where
P = Power
I = Current
E = Volts

If a circuit has a known voltage of 24 volts and a current of 2 amps, then the power in the circuit will be:

P = I (E)
P = 2A (24V)
P = 48 W

Now recall Ohm's laws which states that E = I(R). If we now substitute IR for E in the general formula, we get a formula that uses only current I and resistance R to determine the power in a circuit.

P = I (IR)

Second Form of Power Equation

$P = I^2R$

If a circuit has a known current of 2 amps and a resistance of 100 Ω, then the power in the circuit will be:

$P = I^2R$
$P = (2A)^2\ 100\ \Omega$
$P = 400\ W$

Using Ohm's law again, which can be stated as $I = {}^E/_R$, we can again make a substitution such that power can be determined by knowing only the voltage (E) and resistance (R) of the circuit.

$$P = \left(\frac{E}{R}\right)\left(E\right)$$

Third Form of Power Equation

$$P = \frac{E^2}{R}$$

If a circuit has a known voltage of 24 volts and a resistance of 20 Ω, then the power in the circuit will be:

$$P = \frac{E^2}{R}$$

$$P = \frac{(24V)^2}{20\ \Omega}$$

$$P = 28.8\ W$$

Power in a Series and Parallel Circuit

The total power dissipated in both a series and parallel circuit is equal to the sum of the power dissipated in each resistor in the circuit. Power is simply additive and can be stated as:

$$P_T = P_1 + P_2 + P_3 + \ldots\ldots P_N$$

Figure 10-43 provides a summary of all the possible transpositions of the Ohm's law formula and the power formula.

Energy in an Electrical Circuit

Energy is defined as the ability to do work. Because power is the rate of energy usage, power used over a span of time is actually energy consumption. If power and time are multiplied together, we will get energy.

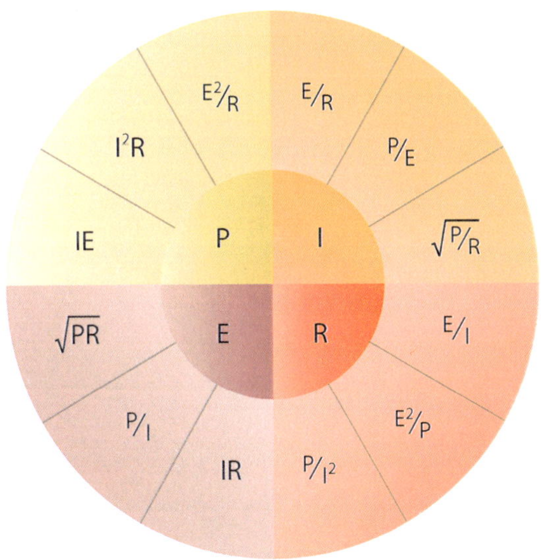

Figure 10-43. Ohm's law formula.

The joule is defined as a unit of energy. There is another unit of measure which is perhaps more familiar. Because power is expressed in watts and time in seconds, a unit of energy can be called a watt-second (Ws) or more recognizable from the electric bill, a kilowatt-hour (kWh). Refer to Page 3-3 for further discussion on energy.

Sources of Electricity

Electrical energy can be produced in a number of methods. The four most common are pressure, chemical, thermal, and light.

Pressure Source

This form of electrical generation is commonly known as piezoelectric (piezo or piez taken from Greek: to press; pressure; to squeeze) is a result of the application of mechanical pressure on a dielectric or nonconducting crystal. The most common piezoelectric materials used today are crystalline quartz and Rochelle salt. However, Rochelle salt is being superseded by other materials, such as barium titanate.

The application of a mechanical stress produces an electric polarization, which is proportional to this stress. This polarization establishes a voltage across the crystal. If a circuit is connected across the crystal a flow of current can be observed when the crystal is loaded (pressure is applied). An opposite condition can occur, where an application of a voltage between certain faces of the crystal can produce a mechanical distortion. This effect is commonly referred to as the piezoelectric effect.

Piezoelectric materials are used extensively in transducers for converting a mechanical strain into an electrical signal. Such devices include microphones, phonograph pickups and vibration-sensing elements. The opposite effect, in which a mechanical output is derived from an electrical signal input, is also widely used in headphones and loudspeakers.

Chemical Source

Chemical energy can be converted into electricity; the most common form of this is the battery. A primary battery produces electricity using two different metals in a chemical solution like alkaline electrolyte, where a chemical reaction between the metals and the chemicals frees more electrons in one metal than in the other. One terminal of the battery is attached to one of the metals such as zinc; the other terminal is attached to the other metal such as manganese oxide. The end that frees more electrons develops a positive charge and the other end develops a negative charge. If a wire is attached from one end of the battery to the other, electrons flow through the wire to balance the electrical charge.

Thermal Sources

The most common source of thermal electricity found in the aviation industry comes from thermocouples. Thermocouples are widely used as temperature sensors. They are cheap and interchangeable, have standard connectors, and can measure a wide range of temperatures. Thermocouples are pairs of dissimilar metal wires joined at least at one end, which generate a voltage between the two wires that is proportional to the temperature at the junction. This is called the Seebeck effect, in honor of Thomas Seebeck who first noticed the phenomena in 1821. It was also noticed that different metal combinations have a different voltage difference.

Thermocouples are usually pressed into service as ways to measure cylinder head temperatures and inter-turbine temperature.

Light Sources

A solar cell or a photovoltaic cell is a device that converts light energy into electricity. Fundamentally, the device contains certain chemical elements that when exposed to light energy, they release electrons.

Photons in sunlight are taken in by the solar panel or cell, where they are then absorbed by semiconducting materials, such as silicon. Electrons in the cell are broken loose from their atoms, allowing them to flow through the material to produce electricity. The complementary positive charges that are also created are called holes (absence of electron) and flow in the direction opposite of the electrons in a silicon solar panel.

Solar cells have many applications and have historically been used in earth orbiting satellites or space probes, handheld calculators, and wrist watches.

Schematic Representation of Electrical Components

The schematic is the most common place where the technician will find electronic symbols. The schematic is a diagram that depicts the interconnection and logic of an electronic or electrical circuit. Many symbols are employed for use in the schematic drawings, blueprints, and illustrations. This section briefly outlines some of the more common symbols and explains how to interpret them.

Conductors

The schematic depiction of a conductor is simple enough. This is generally shown as a solid line. However, the line types may vary depending on who drew the schematics and what exactly the line represents. While the solid line is used to depict the wire or conductor, schematics used for aircraft modifications can also use other line types such as a dashed to represent "existing" wires prior to modification and solid lines for "new" wires.

There are two methods employed to show wire crossovers and wire connections. Figure 10-44 shows the two methods of drawing wires that cross, version A and version B. Figure 10-45 shows the two methods for drawing wire that connect version A and version B. If version A in Figure 10-44 is used to depict crossovers, then version A for wire connections in Figure 10-45 will be used. The same can then be said about the use of version B methods. The technician will encounter both in common use.

Figure 10-46 shows a few examples of the more common wire types that the technician will encounter in schematics. They are the single wire, single shielded, shielded twisted pair or double and the shielded triple. This is not an exhaustive list of wire types but a fair

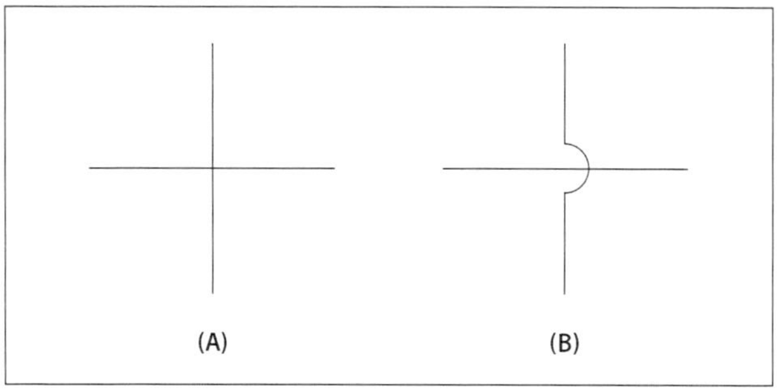

Figure 10-44. Unconnected crossover wires.

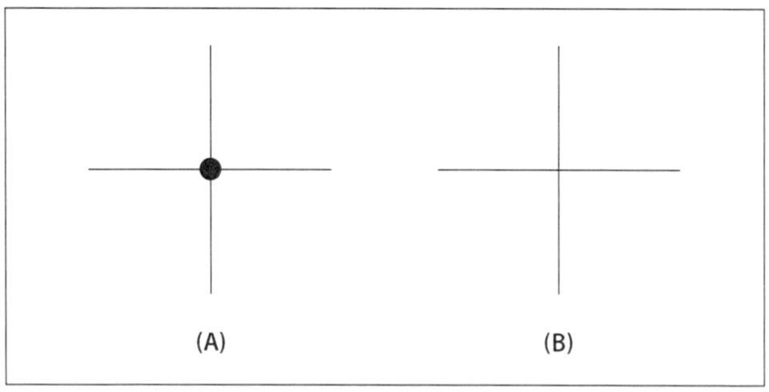

Figure 10-45. Connected wires.

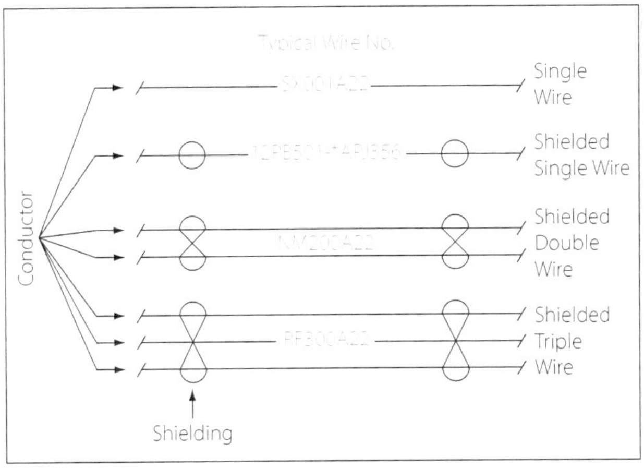

Figure 10-46. Common wire types.

representation of how they are depicted. Figure 10-46 also shows the wires having a wire number. These are shown for the sake of illustration and will vary from one installation agency to another. For further understanding of a wire numbering system, consult the appropriate wiring guide published by the agency that drew the prints. Regardless of the specifics of the wire numbering system, organizations that exercise professional wiring practices will have the installed wire marked in some manner. This will be an aid to the technician who has to troubleshoot or modify the system at a later date.

Types of Resistors

Fixed Resistor

Figure 10-47 is a schematic representation of a fixed resistor. Fixed resistors have built into the design a means of opposing current. The general use of a resistor in a circuit is to limit the amount of current flow. There are a number of methods used in construction and sizing of a resistor that control properties such as resistance value, the precision of the resistance value, and the ability to dissipate heat. While in some applications the purpose of the resistive element is used to generate heat, such as in propeller anti-ice boots, heat typically is the unwanted loss of energy.

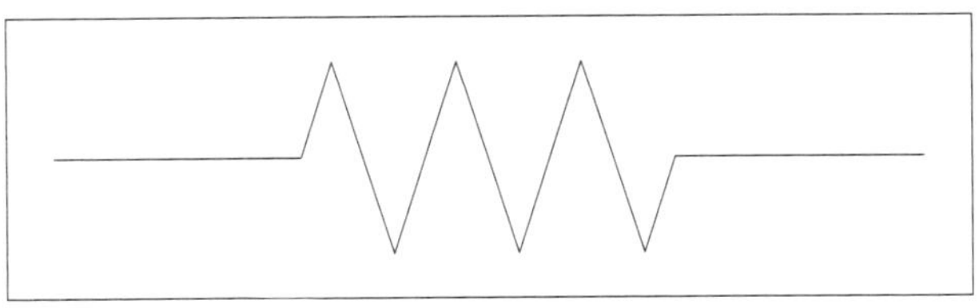

Figure 10-47. Fixed resistor schematic.

Carbon Composition

The carbon composed resistor is constructed from a mixture of finely grouped carbon/graphite, an insulation material for filler, and a substance for binding the material together. The amount of graphite in relation to the insulation material will determine the ohmic or resistive value of the resistor. This mixture is compressed into a rod, which is then fitted with axial leads or "pigtails." The finished product is then sealed in an insulating coating for isolation and physical protection.

There are other types of fixed resistors in common use. Included in this group are:

- Carbon film
- Metal oxide
- Metal film
- Metal glaze

The construction of a film resistor is accomplished by depositing a resistive material evenly on a ceramic rod. This resistive material can be graphite for the carbon film resistor, nickel chromium for the metal film resistor, metal and glass for the metal glaze resistor and last, metal and an insulating oxide for the metal oxide resistor.

Resistor Ratings

It is very difficult to manufacture a resistor to an exact standard of ohmic values. Fortunately, most circuit requirements are not extremely critical. For many uses, the actual resistance in ohms can be 20 percent higher or lower than the value marked on the resistor without causing difficulty. The percentage variation between the marked value and the actual value of a resistor is known as the "tolerance" of a resistor. A resistor coded for a 5 percent tolerance will not be more than 5 percent higher or lower than the value indicated by the color code.

The resistor color code is made up of a group of colors, numbers, and tolerance values. Each color is represented by a number, and in most cases, by a tolerance value. [Figure 10-48]

When the color code is used with the end-to-center band marking system, the resistor is normally marked with bands of color at one end of the resistor. The body or base color of the resistor has nothing to do with the color code, and in no way indicates a resistance value. To prevent confusion, this body will never be the same color as any of the bands indicating resistance value.

Resistor color code		
Color	**Number**	**Tolerance**
Black	0	—
Brown	1	1%
Red	2	2%
Orange	3	3%
Yellow	4	4%
Green	5	5%
Blue	6	6%
Violet	7	7%
Gray	8	8%
White	9	9%
Gold	—	5%
Silver	—	10%
No color	—	20%

Figure 10-48. Resistor color code.

When the end-to-center band marking system is used, either three or four bands will mark the resistor.

1. The first color band (nearest the end of the resistor) will indicate the first digit in the numerical resistance value. This band will never be gold or silver in color.
2. The second color band will always indicate the second digit of ohmic value. It will never be gold or silver in color. [Figure 10-49]
3. The third color band indicates the number of zeros to be added to the two digits derived from the first and second bands, except in the following two cases:
   (a) If the third band is gold in color, the first two digits must be multiplied by 10 percent.
   (b) If the third band is silver in color, the first two digits must be multiplied by 1 percent.
4. If there is a fourth color band, it is used as a multiplier for percentage of tolerance, as indicated in the color code chart in Figure 10-48. If there is no fourth band, the tolerance is understood to be 20 percent.

Figure 10-49 provides an example, which illustrates the rules for reading the resistance value of a resistor marked with the end-to-center band system. This resistor is marked with three bands of color, which must be read from the end toward the center.

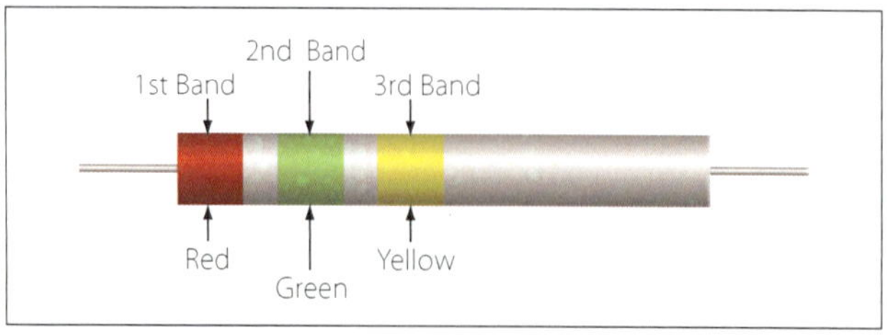

Color	Numerical Value	Significance
1st band—Red	2	1st digit
2nd band—Green	5	2nd digit
3rd band—Yellow	4	No. of zeroes to add

Figure 10-49. End-to-center band marking.

There is no fourth color band; therefore, the tolerance is understood to be 20 percent. 20 percent of 250,000 Ω, equals 50,000 Ω.

Since the 20 percent tolerance is plus or minus,

Maximum resistance
= 250,000 Ω + 50,000 Ω
= 300,000 Ω

Minimum resistance
= 250,000 Ω − 50,000 Ω
= 200,000 Ω

The following paragraphs provide a few extra examples of resistor color band decoding. Figure 10-50 contains a resistor with another set of colors. This resistor code should be read as follows:

The resistance of this resistor is 86,000 ± 10 percent ohms. The maximum resistance is 94,600 ohms, and the minimum resistance is 77,400 ohms.

As another example, the resistance of the resistor in Figure 10-51 is 960 ± 5 percent ohms. The maximum resistance is 1,008 ohms, and the minimum resistance is 912 ohms.

Sometimes circuit considerations dictate that the tolerance must be smaller than 20 percent. Figure 10-52 shows an example of a resistor with a 2 percent tolerance. The resistance value of this resistor is 2,500 ± 2 percent ohms. The maximum resistance is 2,550 ohms, and the minimum resistance is 2,450 ohms.

Figure 10-53 contains an example of a resistor with a black third color band. The color code value of black is zero, and the third band indicates the number of zeros to be added to the first two digits.

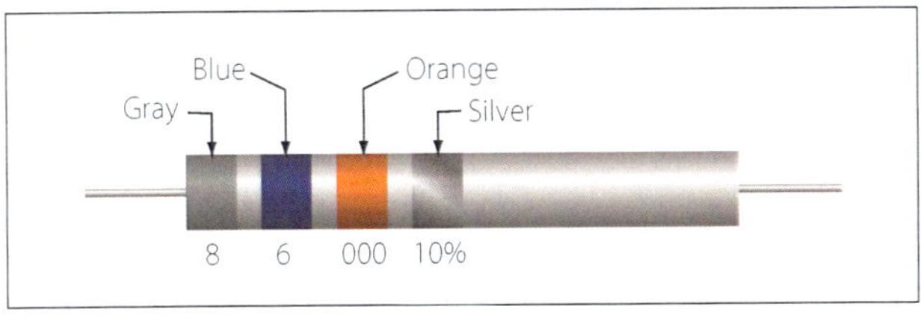

Figure 10-50. Resistor color code example.

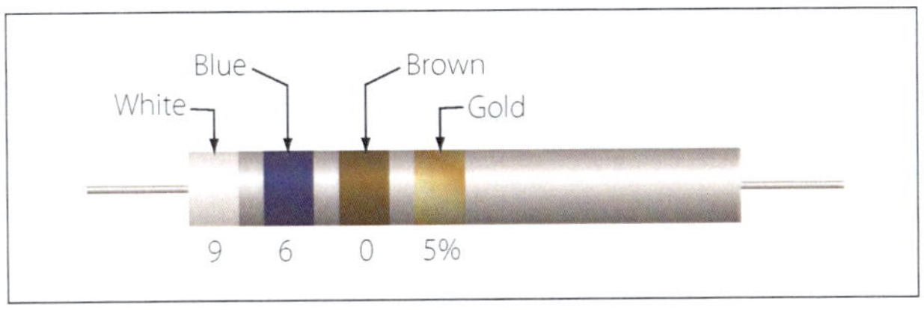

Figure 10-51. Resistor color code example.

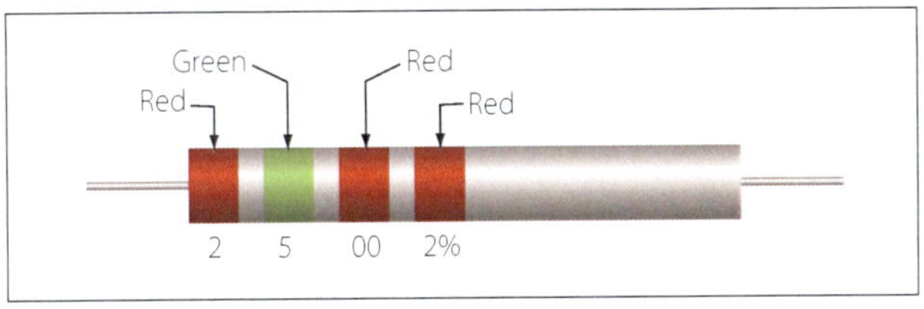

Figure 10-52. Resistor with two percent tolerance.

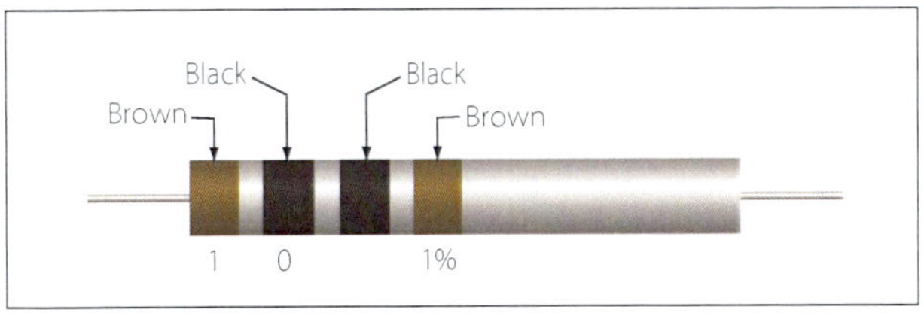

Figure 10-53. Resistor with black third color band.

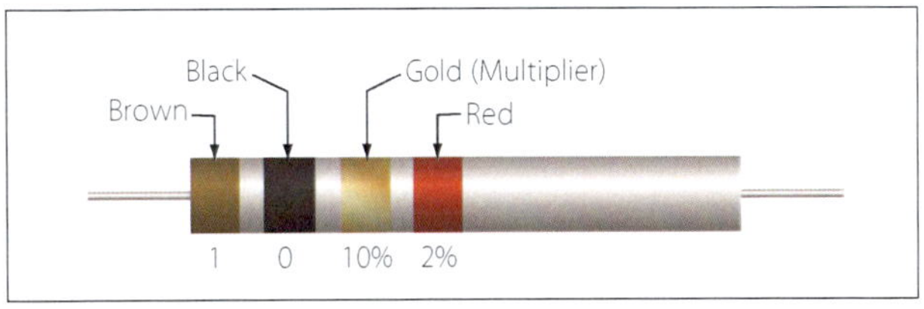

Figure 10-54. Resistor with gold third band.

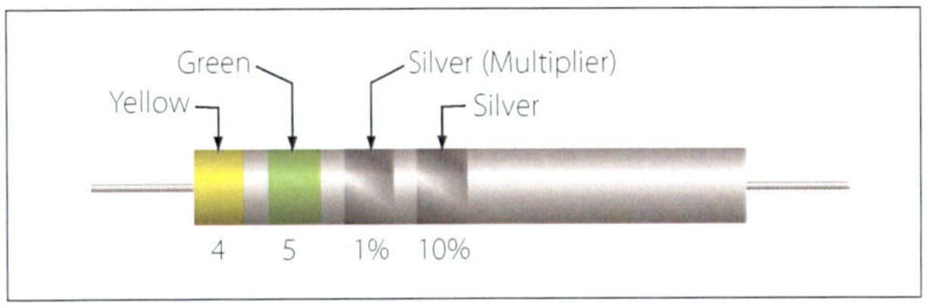

Figure 10-55. Resistor with a silver third band.

In this case, a zero number of zeros must be added to the first two digits; therefore, no zeros are added. Thus, the resistance value is 10 ± 1 percent ohms. The maximum resistance is 10.1 ohms, and the minimum resistance is 9.9 ohms. There are two exceptions to the rule stating the third color band indicates the number of zeros. The first of these exceptions is illustrated in Figure 10-54. When the third band is gold in color, it indicates that the first two digits must be multiplied by 10 percent. The value of this resistor in this case is:

$$10 \times 0.10 \pm 2\% = 1 = 0.02 \text{ ohms}$$

When the third band is silver, as is the case in Figure 10-55, the first two digits must be multiplied by 1 percent. The value of the resistor is 0.45 ± 10 percent ohms.

Wire-Wound

Wire-wound resistors typically control large amounts of current and have high power ratings. Resistors of this type are constructed by winding a resistance wire around an insulating rod, usually made of porcelain. The windings are then coated with an insulation material for physical protection and heat conduction. Both ends of the windings are then connected to terminals, which are used to connect the resistor to a circuit. [Figure 10-56]

A wire-wound resistor with tap is a special type of fixed resistor that can be adjusted. These adjustments can be made by moving a slide bar tap or by moving the tap to a preset incremental position. While the tap may be adjustable, the adjustments are usually set at the time of installation to a specific value and then operated in service as a fixed resistor. Another type of wire-wound resistor is that constructed of Manganin wire, used where high precision is needed.

Variable Resistors

Variable resistors are constructed so that the resistive value can be changed easily. This adjustment can be manual or automatic, and the adjustments can be made while the system that it is connected to is in operation. There are two basic types of manual adjustors. One is the rheostat and the second is the potentiometer.

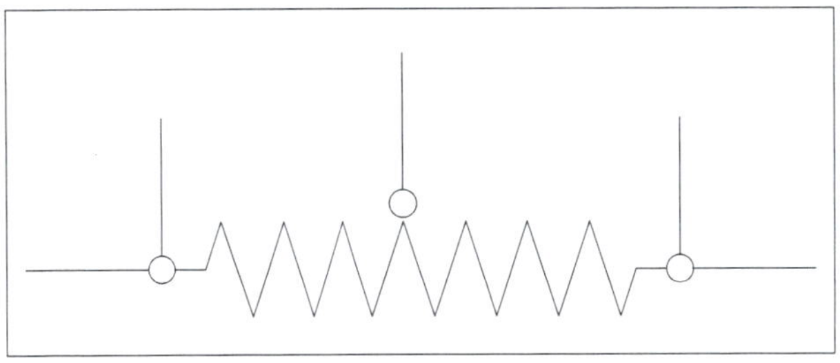

Figure 10-56. Wire-wound resistors.

Rheostat

The schematic symbol for the rheostat is shown in Figure 10-57. A rheostat is a variable resistor used to vary the amount of current flowing in a circuit. Figure 10-58 shows a rheostat connected in series with an ordinary resistance in a series circuit. As the slider arm moves from point A to B, the amount of rheostat resistance (AB) is increased. Since the rheostat resistance and the fixed resistance are in series, the total resistance in the circuit also increases, and the current in the circuit decreases. On the other hand, if the slider arm is moved toward point A, the total resistance decreases and the current in the circuit increases.

Potentiometer

The schematic symbol for the potentiometer is shown in Figure 10-59. The potentiometer is considered a three terminal device. As illustrated, terminals 1 and 2 have the entire value of the potentiometer resistance between them. Terminal 3 is the wiper or moving contact. Through this wiper, the resistance between terminals 1 and 3 or terminals 2 and 3 can be varied. While the rheostat is used to vary the current in a circuit, the potentiometer is used to vary the voltage in a circuit. A typical use for this component can be found in the volume controls on an audio panel and input devices for flight data recorders, among many other applications.

In Figure 10-60A, a potentiometer is used to obtain a variable voltage from a fixed voltage source to apply to an electrical load. The voltage applied to the load is the voltage between points 2 and 3. When the slider arm is moved to point 1, the entire voltage is applied to the electrical device (load); when the arm is moved to point 3, the voltage applied to the load is zero. The potentiometer makes possible the application of any voltage between zero and full voltage to the load.

The current flowing through the circuit of Figure 10-60 leaves the negative terminal electron flow of the battery and divides, one part flowing through the lower portion of the potentiometer (points 3 to 2) and the other part through the load. Both parts combine at point 2 and flow through the upper portion of the potentiometer (points 2 to 1) back to the positive terminal of the battery. In View B of Figure 10-60, a potentiometer and its schematic symbol are shown.

In choosing a potentiometer resistance, the amount of current drawn by the load should be considered as well

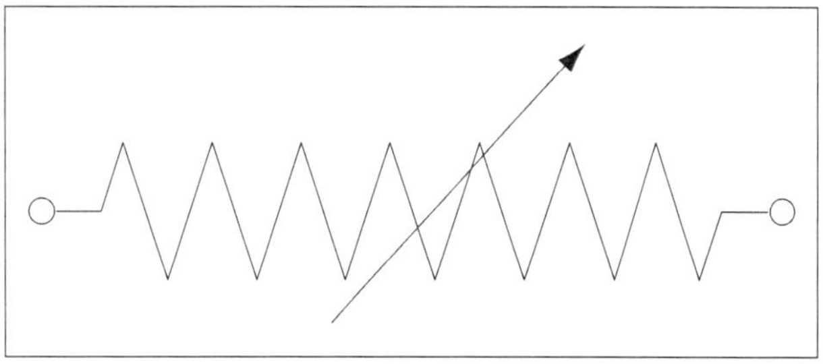

Figure 10-57. Rheostat schematic symbol.

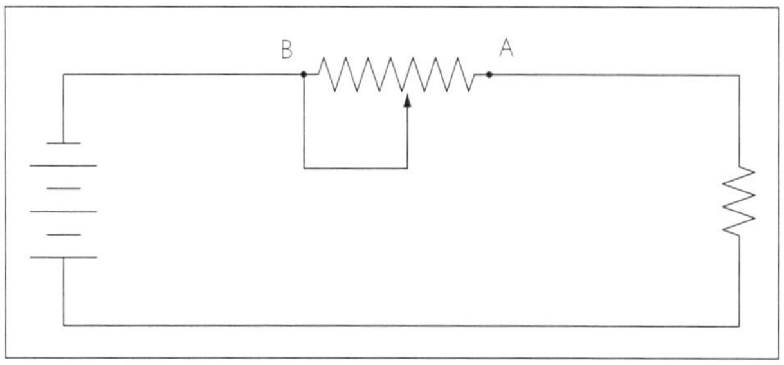

Figure 10-58. Rheostat connected in series.

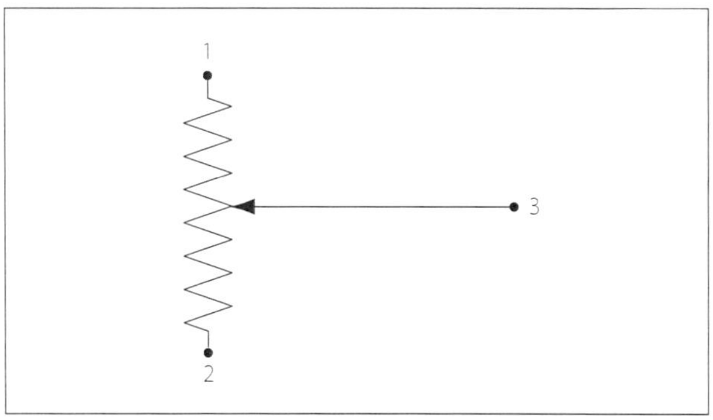

Figure 10-59. Potentiometer schematic symbol.

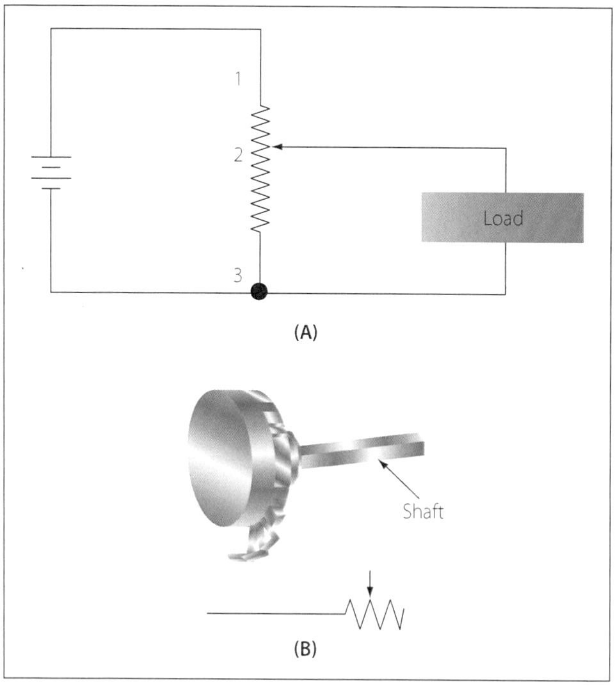

Figure 10-60. Potentiometer and schematic symbol.

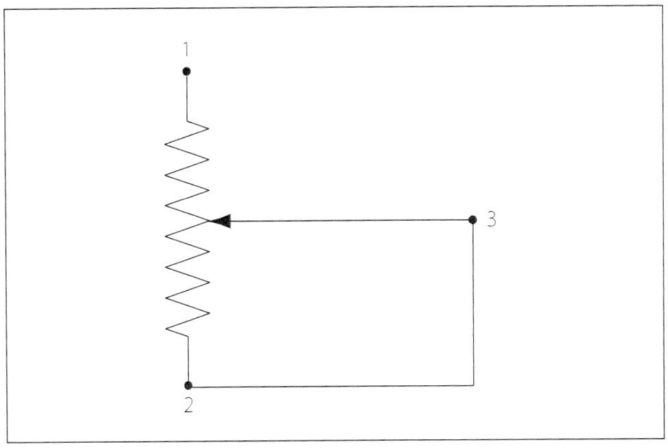

Figure 10-61. Potentiometer wired to function as rheostat.

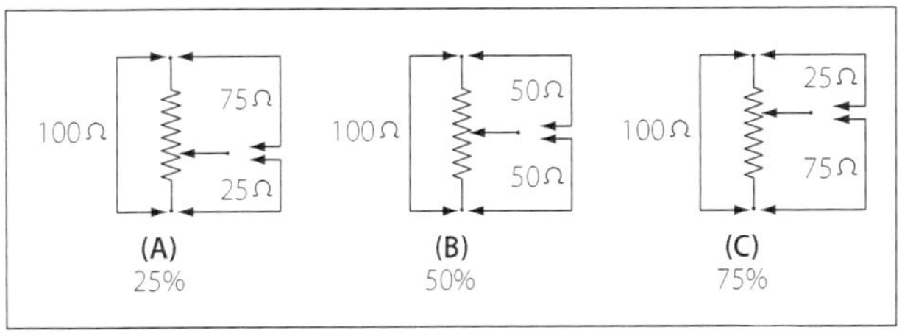

Figure 10-62. Linear potentiometer schematic.

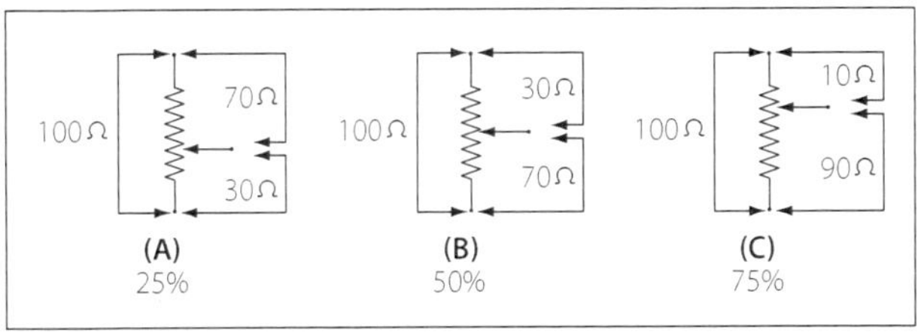

Figure 10-63. Tapered potentiometer.

as the current flow through the potentiometer at all settings of the slider arm. The energy of the current through the potentiometer is dissipated in the form of heat.

It is important to keep this wasted current as small as possible by making the resistance of the potentiometer as large as practicable. In most cases, the resistance of the potentiometer can be several times the resistance of the load. Figure 10-61 shows how a potentiometer can be wired to function as a rheostat.

Linear Potentiometers

In a linear potentiometer, the resistance between both terminal and the wiper varies linearly with the position of the wiper. To illustrate, one quarter of a turn on the potentiometer will result in one quarter of the total resistance. The same relationship exists when one-half or three-quarters of potentiometer movement. Figure 10-62 schematically depicts this.

Tapered Potentiometers

Resistance varies in a nonlinear manner in the case of the tapered potentiometer. Figure 10-63 illustrates this. Keep in mind that one-half of full potentiometer travel doesn't necessarily correspond to one-half the total resistance of the potentiometer.

Thermistors

Figure 10-64 shows the schematic symbol for the thermistor. The thermistor is a type of a variable resistor, which is temperature sensitive. This component has what is known as a negative temperature coefficient, which means that as the sensed temperature increases, the resistance of the thermistor decreases.

Photoconductive Cells

The photoconductive cell is similar to the thermistor. Like the thermistor, it has a negative temperature coefficient. Unlike the thermistor, the resistance is controlled by light intensity. This kind of component can be found in radio control heads where the intensity of the ambient light is sensed through the photoconductive cell resulting in the backlighting of the control heads to adjust to the cockpit lighting conditions. Figure 10-65 shows the schematic symbol component.

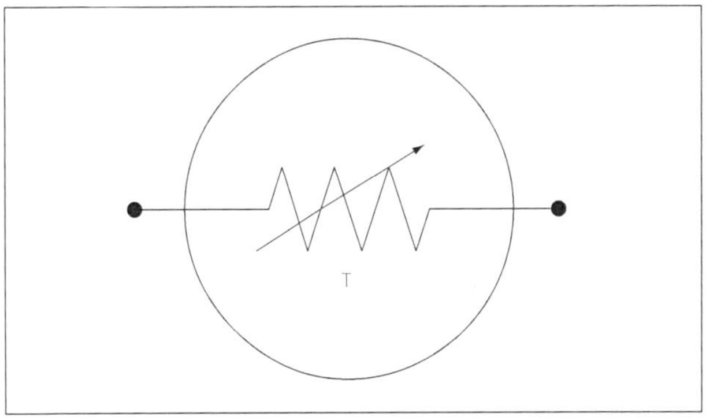

Figure 10-64. Schematic symbol for thermistor.

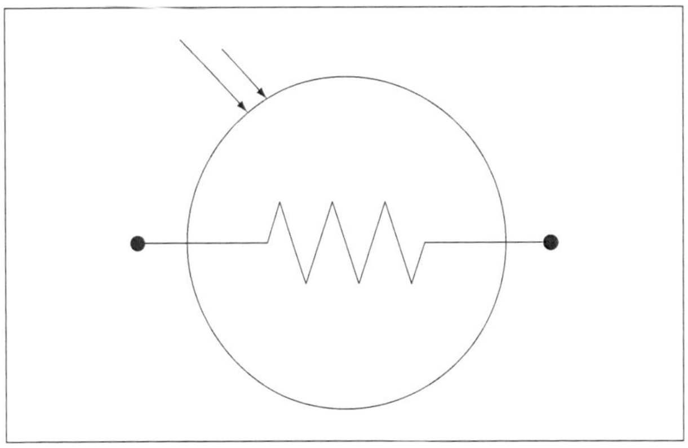

Figure 10-65. Photoconductive cell schematic symbol component.

Circuit Protection Devices

Perhaps the most serious trouble in a circuit is a direct short. The term, "direct short," describes a situation in which some point in the circuit, where full system voltage is present, comes in direct contact with the ground or return side of the circuit. This establishes a path for current flow that contains no resistance other than that present in the wires carrying the current, and these wires have very little resistance.

Most wires used in aircraft electrical circuits are small gauge, and their current carrying capacity is quite limited. The size of the wires used in any given circuit is determined by the amount of current the wires are expected to carry under normal operating conditions. Any current flow in excess of normal, such as the case of a direct short, would cause a rapid generation of heat. If the excessive current flow caused by the short is left unchecked, the heat in the wire will continue causing perhaps a portion of the wire to melt and at the very least, open the circuit.

To protect aircraft electrical systems from damage and failure caused by excessive current, several kinds of protective devices are installed in the systems. Fuses, circuit breakers, thermal protectors, and arc fault circuit breakers are used for this purpose.

Circuit protective devices, as the name implies, all have a common purpose—to protect the units and the wires in the circuit. Some are designed primarily to protect the wiring and to open the circuit in such a way as to stop the current flow when the current becomes greater than the wires can safely carry. Other devices are designed to protect a unit in the circuit by stopping the current flow to it when the unit becomes excessively warm.

Fuse

Figure 10-66 shows the schematic symbol for the fuse. Fuses are used to protect the circuit from over current conditions. The fuse is installed in the circuit so that all the current in the circuit passes through it. In most fuses, the strip of metal is made of an alloy of tin and bismuth, which will melt and open the circuit when the current exceeds the rated capacity of the fuse. For example, if a 5-amp fuse is placed into a circuit, the fuse will allow currents up to 5 amps to pass. Because the fuse is intended to protect the circuit, it is quite important that its capacity match the needs of the circuit in which it is used.

When replacing a fuse, consult the applicable manufacturer's instructions to be sure a fuse of the correct type and capacity is installed. Fuses are installed in two types of fuse holders in aircraft. "Plug-in holders" or in-line holders are used for small and low capacity fuses. "Clip" type holders are used for heavy high capacity fuses and current limiters.

Current Limiter

The current limiter is very much like the fuse. However, the current limiter link is usually made of copper and will stand a considerable overload for a short period of time. Like the fuse it will open up in an over current condition in heavy current circuits such as 30 amp or greater. These are used primarily to sectionalize an aircraft circuit or bus. Once the limiter is opened, it must be replaced. The schematic symbol for the current limiter is the same as that for the fuse.

Circuit Breaker

The circuit breaker is commonly used in place of a fuse and is designed to break the circuit and stop the current flow when the current exceeds a predetermined value. Unlike the fuse, the circuit breaker can be reset; whereas the fuse or current limiter must be replaced. Figure 10-67 shows the schematic symbol for a circuit breaker.

There are several types of circuit breakers in general use in aircraft systems. One is a magnetic type. When excessive current flows in the circuit, it makes an elec-

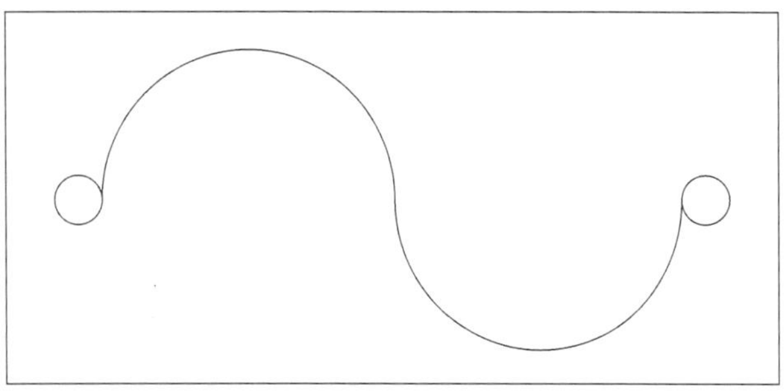

Figure 10-66. Schematic symbol for fuse.

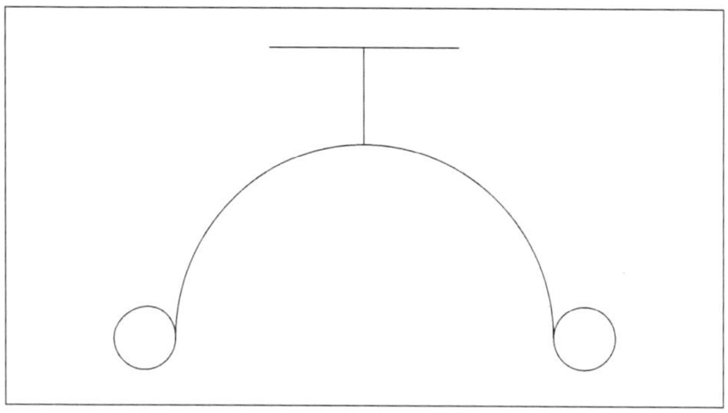

Figure 10-67. Schematic symbol for circuit breaker.

tromagnet strong enough to move a small armature, which trips the breaker. Another type is the thermal overload switch or breaker. This consists of a bimetallic strip which, when it becomes overheated from excessive current, bends away from a catch on the switch lever and permits the switch to trip open.

Most circuit breakers must be reset by hand. If the overload condition still exists, the circuit breaker will trip again to prevent damage to the circuit. At this point, it is usually not advisable to continue resetting the circuit breaker, but to initiate troubleshooting to determine the cause. Repeated resetting of a circuit breaker can lead to circuit or component damage or worse, the possibility of a fire or explosion.

Arc Fault Circuit Breaker

In recent years, the arc fault circuit breaker has begun to provide an additional layer of protection beyond that of the thermal protection already provided by conventional circuit breakers. The arc fault circuit breaker monitors the circuit for an electrical arcing signature, which can indicate possible wiring faults and unsafe conditions. These conditions can lead to fires or loss of power to critical systems. The arc fault circuit breaker is only beginning to make an appearance in the aircraft industry and is not widely used like the thermal type of circuit breaker.

Thermal Protectors

A thermal protector, or switch, is used to protect a motor. It is designed to open the circuit automatically whenever the temperature of the motor becomes excessively high. It has two positions—open and closed. The most common use for a thermal switch is to keep a motor from overheating. If a malfunction in the motor causes it to overheat, the thermal switch will break the circuit intermittently.

The thermal switch contains a bimetallic disk, or strip, which bends and breaks the circuit when it is heated. This occurs because one of the metals expands more than the other when they are subjected to the same temperature. When the strip or disk cools, the metals contract and the strip returns to its original position and closes the circuit.

Control Devices

Components in the electrical circuits are typically not all intended to operate continuously or automatically. Most of them are meant to operate at certain times, under certain conditions, to perform very definite functions. There must be some means of controlling their operation. Either a switch, or a relay, or both may be included in the circuit for this purpose.

Switches

Switches control the current flow in most aircraft electrical circuits. A switch is used to start, to stop, or to change the direction of the current flow in the circuit. The switch in each circuit must be able to carry the normal current of the circuit and must be insulated heavily enough for the voltage of the circuit.

An understanding of some basic definitions of the switch is necessary before any of the switch types are discussed. The number of poles, throws, and positions they have designates toggle switches, as well as some other type of switches.

Pole: the switch's movable blade or contactor. The number of poles is equal to the number of circuits, or paths for current flow, that can be completed through the switch at any one time.

Throw: indicates the number of circuits, or paths for current, that it is possible to complete through the switch with each pole or contactor.

Positions: indicates the number of places at which the operating device (toggle, plunger, and so forth) will come to rest and at the same time open or close one or more circuits.

Toggle Switch

Single-Pole, Single-Throw (SPST)

The single-pole, single-throw switch allows a connection between two contacts. One of two conditions will exist. Either the circuit is open in one position or closed in the other position. The schematic symbol for this switch is shown in Figure 10-68.

Single-Pole, Double-Throw (SPDT)

The single-pole, double-throw switch is shown in Figure 10-69. With this switch, contact between one contact can be made between one contact and the other.

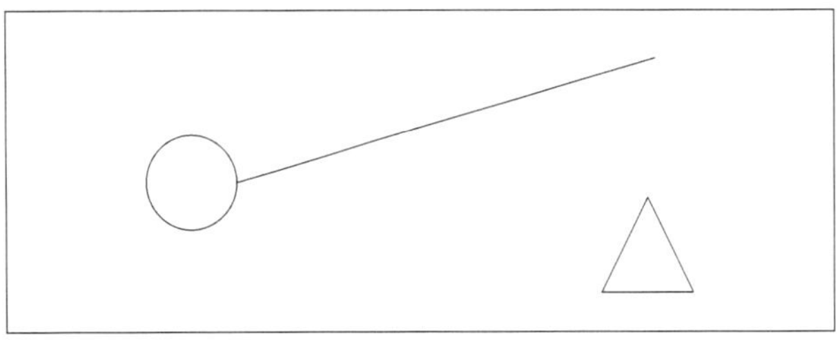

Figure 10-68. Single-pole, single-throw switch schematic symbol.

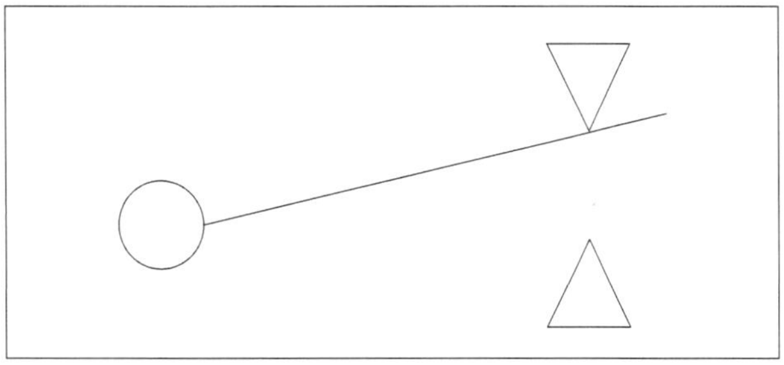

Figure 10-69. Single-pole, double-throw switch schematic symbol.

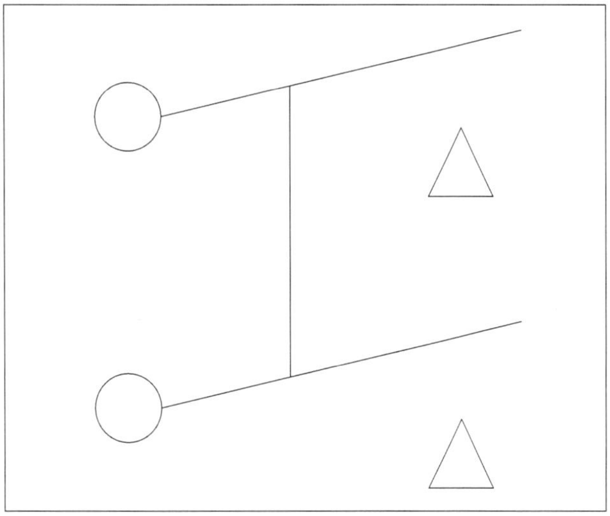

Figure 10-70. Double-pole, single-throw switch schematic symbol.

Double-Pole, Single-Throw (DPST)

The double-pole, single-throw switch connection can be made between one set of contacts and either of two other sets of contacts. The schematic symbol for this switch is shown in Figure 10-70.

Double-Pole, Double-Throw (DPDT)

The schematic symbol for the double-pole, double-throw switch is shown in Figure 10-71. This type of switch makes a connection from one set of contacts to either of two other sets of contacts.

A toggle switch that is spring-loaded to the OFF position and must be held in the ON position to complete the circuit is a momentary contact two-position switch. One that will come to rest at either of two positions, opening the circuit in one position and closing it in another, is a two-position switch. A toggle switch that will come to rest at any one of three positions is a three-position switch.

A switch that stays open, except when it is held in the closed position, is a normally open switch (usually identified as NO). One that stays closed, except when it is held in the open position is a normally closed switch (NC). Both kinds are spring loaded to their normal position and will return to that position as soon as they are released.

Locking toggles require the operator to pull out on the switch toggle before moving it in to another position. Once in the new position, the switch toggle is release back into a lock, which then prevents the switch from inadvertently being moved.

Pushbutton Switches

Pushbutton switches have one stationary contact and one movable contact. The movable contact is attached to the pushbutton. The pushbutton is either an insulator itself or is insulated from the contact. This switch is spring loaded and designed for momentary contact.

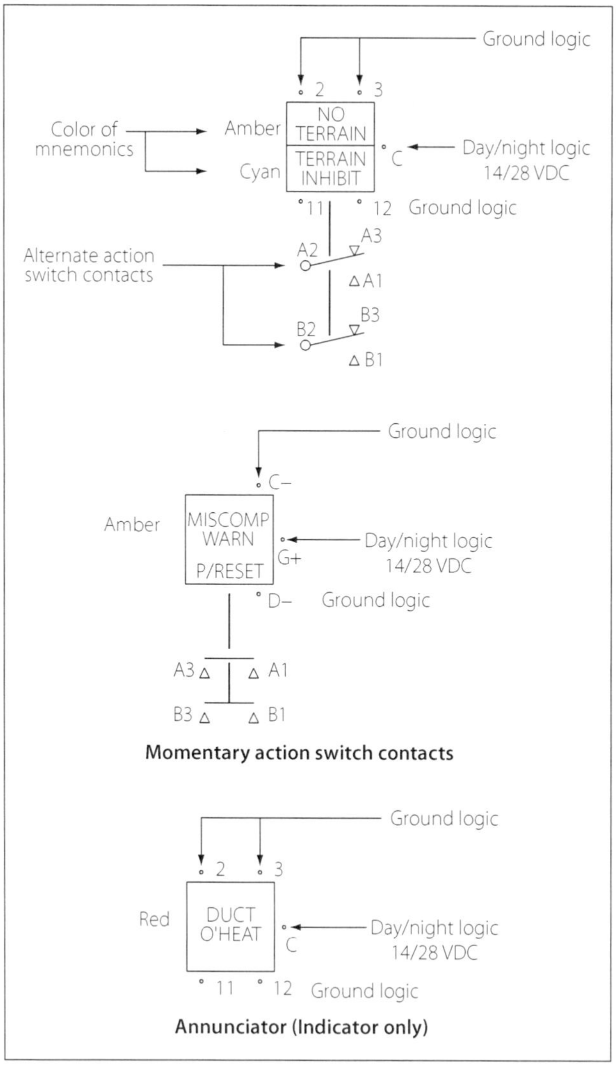

Figure 10-71. Double-pole, double-throw switch schematic symbol.

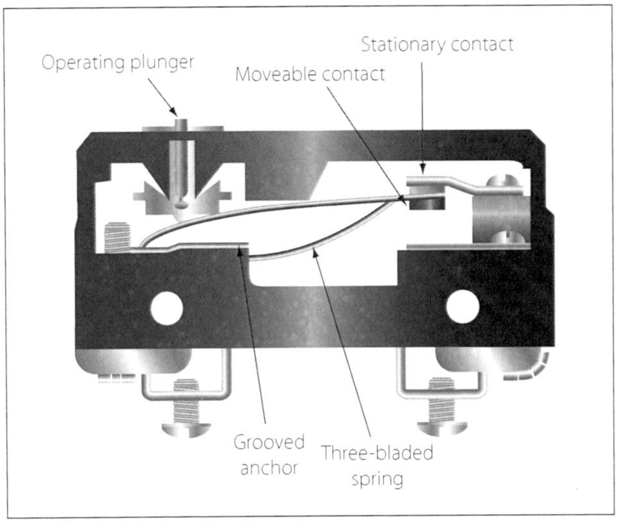

Figure 10-72. Cross-section of a microswitch.

Microswitches

A microswitch will open or close a circuit with a very small movement of the tripping device (1/16 inch or less). This is what gives the switch its name, since micro means small.

Microswitches are usually pushbutton switches. They are used primarily as limit switches to provide automatic control of landing gears, actuator motors, and the like. The diagram in Figure 10-72 shows a normally closed microswitch in cross-section and illustrates how these switches operate. When the operating plunger is pressed in, the spring and the movable contact are pushed, opening the contacts and the circuit.

Rotary Selector Switches

A rotary selector switch takes the place of several switches. When the knob of the switch is rotated, the switch opens one circuit and closes another. Ignition switches and voltmeter selector switches are typical examples of this kind of switch. [Figure 10-73]

Lighted Pushbutton Switches

Another more common switch found in today's aircraft is the lighted pushbutton switch. This type of switch takes the form of a 5/8-inch to 1-inch cube with incandescent or LED lights to indicate the function of the switch. Switch designs come in a number of configurations; the two most common are the alternate action and momentary action and will usually have a two-pole or four-pole switch body. Other less common switch actions are the alternate and momentary holding coil configurations. The less known holding or latching coil switch bodies are designed to have a magnetic coil inside the switch body that is energized through two contacts in the base of the switch. When the coil is energized and the switch is pressed, the switch contacts will remain latched until power is removed from the coil. This type of design allows for some degree of remote control over the switch body. Figure 10-71 illustrates a schematic representation of this switch design.

The display optics of the lighted pushbutton switch provide the crew with a clear message that is visible under a wide range of lighting conditions with very high luminance and wide viewing angles. While some displays are simply a transparent screen that is backlit by an incandescent light, the higher quality and more reliable switches are available in sunlight readable displays and night vision (NVIS) versions. Due to the sunlight environment of the cockpit, displays utilizing standard lighting techniques will "washout" when viewed in direct sunlight. Sunlight readable displays are designed to minimize this effect.

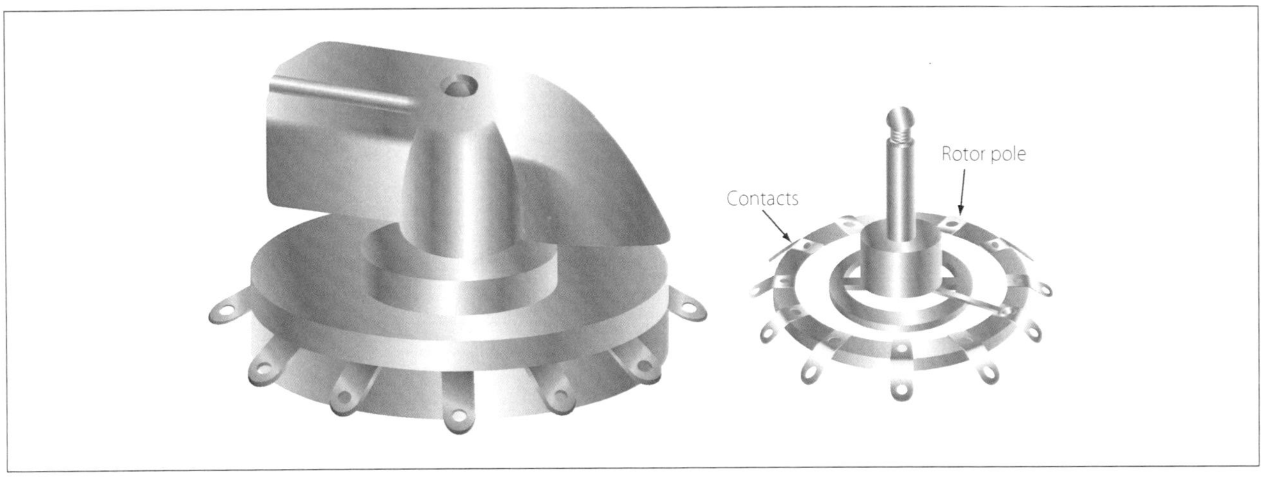

Figure 10-73. Rotary selector switch.

Lighted pushbutton switches can also be used in applications where a switch is not required and the optics are only for indications. This type of an indicator is commonly called an annunciator.

DIP Switches

The acronym "DIP" switch is defined as Dual In-Line Parallel switch in reference to the physical layout. DIP switches are commonly found in card cages, and line replaceable units (LRU) are used in most cases to adjust gains, control configurations, and so forth. Each one of the switches is generally an SPST slide or rocker switch. The technician may find this switch in packages ranging in size from DIP2 through DIP32. Some of the more common sizes are DIP4 and DIP8.

Switch Guards

Switch guards are covers that protect a switch from unintended operation. Prior to the operation of the switch, the guard is usually lifted. Switch guards are commonly found on systems such as fire suppression and override logics for various systems.

Relays

A relay is simply an electromechanical switch where a small amount of current can control a large amount of current. Figure 10-74 illustrates the basic relay in both schematic and pictorial format. When a voltage is applied to the coil of the relay, the electromagnet will be energized due to the current. When energized, an electromagnetic field will pull the common (C) or arm of the relay down. When the arm or common is pulled down, the circuit between the arm and the normally closed (NC) contacts is opened and the circuit between the arm and the normally open (NO) contacts are closed. When the energizing voltage is removed, the spring will return the arm contacts back to the normally closed (NC) contacts. The relay usually has two connections for the coil. The (+) side is designated as X1 and the ground-side of the coil is designated as X2.

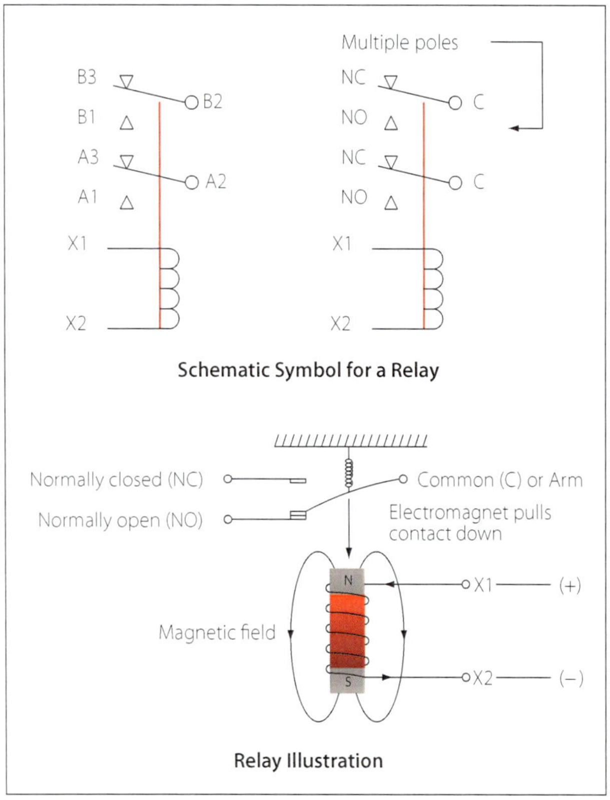

Figure 10-74. Basic relay.

Series DC Circuits

Introduction

The series circuit is the most basic electrical circuit and provides a good introduction to basic circuit analysis. The series circuit represents the first building block for all of the circuits to be studied and analyzed. Figure 10-75 shows this simple circuit with nothing more than a voltage source or battery, a conductor, and a resistor. This is classified as a series circuit because the components are connected end-to-end, so that the same current flows through each component equally. There is only one path for the current to take and the battery and resistor are in series with each other. Next is to make a few additions to the simple circuit in Figure 10-75.

Figure 10-76 shows an additional resistor and a little more detail regarding the values. With these values, we can now begin to learn more about the nature of the circuit. In this configuration, there is a 12-volt DC source in series with two resistors, R1 = 10 Ω and R2 = 30 Ω. For resistors in a series configuration, the total resistance of the circuit is equal to the sum of the individual resistors. The basic formula is:

$$R_T = R_1 + R_2 + R_3 + \ldots\ldots\ldots R_N$$

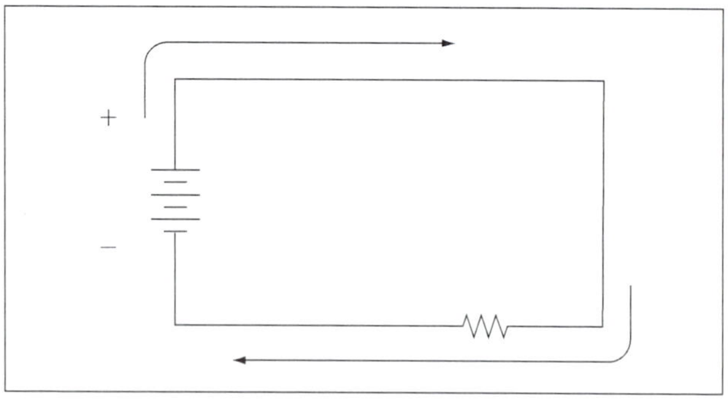

Figure 10-75. Simple DC circuit.

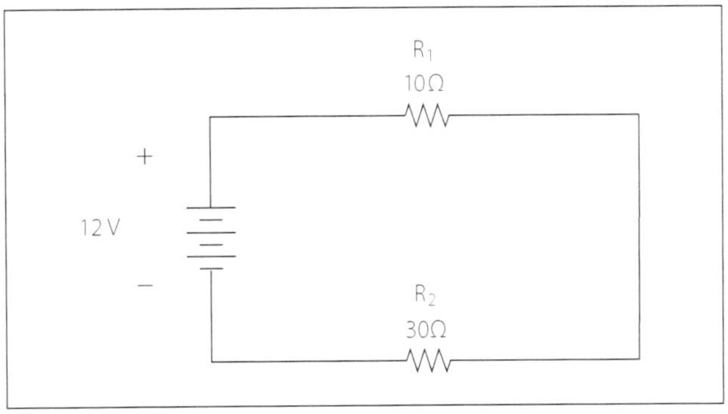

Figure 10-76. Inducing minimum voltage in an elementary generator.

For Figure 10-76, this will be:

$R_T = 10\Omega + 30\Omega$
$R_T = 40\ \Omega$

Now that the total resistance of the circuit is known, the current for the circuit can be determined. In a series circuit, the current cannot be different at different points within the circuit. The current through a series circuit will always be the same through each element and at any point. Therefore, the current in the simple circuit can now be determined using Ohm's law:

Formula, $E = I\ (R)$
Solve for current, $I = \dfrac{E}{R}$
The variables, $E = 12V$ and $R_T = 40\ \Omega$
Substitute variables, $I = \dfrac{12V}{40\ \Omega}$
Current in circuits, $I = 0.3A$

Ohm's law describes a relationship between the variables of voltage, current, and resistance that is linear and easy to illustrate with a few extra calculations.

First will be the act of changing the total resistance of the circuit while the other two remain constant. In this example, the R_T of the circuit in Figure 10-76 will be doubled. The effects on the total current in the circuit are:

Formula, $E = I\ (R)$
Solve for current, $I = \dfrac{E}{R}$
The variables, $E = 12V$ and $R_T = \mathbf{80\ \Omega}$
Substitute variables, $I = \dfrac{12V}{\mathbf{80\ \Omega}}$
Current in circuits, $I = \mathbf{0.15}A$

It can be seen quantitatively and intuitively that when the resistance of the circuit is doubled, the current is reduced by half the original value.

Next, reduce the R_T of the circuit in Figure 10-76 to half of its original value. The effects on the total current are:

Formula, $E = I\ (R)$
Solve for current, $I = \dfrac{E}{R}$
The variables, $E = 12V$ and $R_T = \mathbf{20\ \Omega}$
Substitute variables, $I = \dfrac{12V}{\mathbf{20\ \Omega}}$
Current in circuits, $I = \mathbf{0.6}A$

Voltage Drops and Further Application of Ohm's Law

The example circuit in Figure 10-77 will be used to illustrate the idea of voltage drop. It is important to differentiate between voltage and voltage drop when discussing series circuits. Voltage drop refers to the loss in electrical pressure or emf caused by forcing electrons through a resistor. Because there are two resistors in the example, there will be separate voltage drops. Each drop is associated with each individual resistor. The amount of electrical pressure required to force a given number of electrons through a resistance is proportional to the size of the resistor.

In Figure 10-77, the values used to illustrate the idea of voltage drop are:

Current, $I = 1mA$
$R_1 = 1\ k\ \Omega$
$R_2 = 3\ k\ \Omega$
$R_3 = 5\ k\ \Omega$

The voltage drop across each resistor will be calculated using Ohm's law. The drop for each resistor is the product of each resistance and the total current in the circuit. Keep in mind that the same current flows through series resistor.

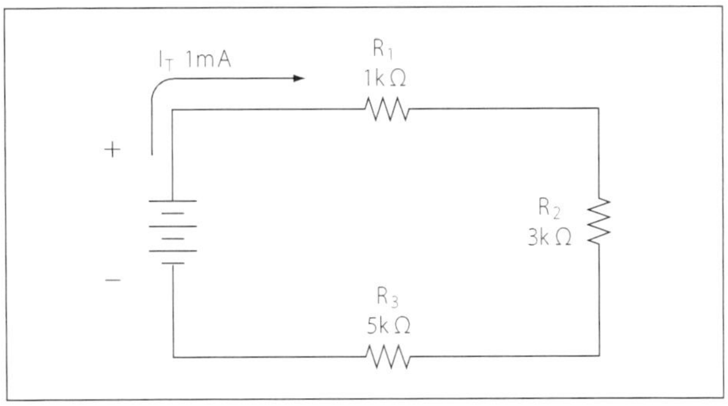

Figure 10-77. Example of three resistors in series.

Formula, $E = I(R)$

Voltage across R_1: $E_1 = I_T(R_1)$

$E_1 = 1 \text{ mA}(1 \text{ k}\Omega) = 1$ volt

Voltage across R_2: $E_2 = I_T(R_2)$

$E2 = 1 \text{ mA}(3 \text{ k}\Omega) = 3$ volt

Voltage across R_3: $E_3 = I_T(R_3)$

$E3 = 1 \text{ mA}(5 \text{ k}\Omega) = 5$ volt

The source voltage can now be determined, which can then be used to confirm the calculations for each voltage drop. Using Ohm's law:

Formula: $E = I(R)$

Source voltage = current times the total resistance

$E_S = I(R_T)$

$R_T = 1 \text{ k}\Omega + 3 \text{ k}\Omega + 5 \text{ k}\Omega$

$R_T = 9 \text{ k}\Omega$

Now: $E_S = I(R_T)$

Substitute $E_S = 1 \text{ mA}(9 \text{ k}\Omega)$

$E_S =$ **9** volts

Simple checks to confirm the calculation and to illustrate the concept of the voltage drop add up the individual values of the voltage drops and compare them to the results of the above calculation.

1 volt + 3 volts + 5 volts = **9** volts.

Voltage Sources in Series

A voltage source is an energy source that provides a constant voltage to a load. Two or more of these sources in series will equal the algebraic sum of all the sources connected in series. The significance of pointing out the algebraic sum is to indicate that the polarity of the sources must be considered when adding up the sources. The polarity will be indicated by a plus or minus sign depending on the source's position in the circuit.

In Figure 10-78 all of the sources are in the same direction in terms of their polarity. All of the voltages have the same sign when added up. In the case of Figure 10-78, three cells of a value of 1.5 volts are in series with the polarity in the same direction. The addition is simple enough:

$$E_T = 1.5\text{v} + 1.5\text{v} + 1.5\text{v} = +4.5 \text{ volts}$$

However, in Figure 10-79, one of the three sources has been turned around, and the polarity opposes the other two sources. Again the addition is simple:

$$E_T = +1.5\text{v} - 1.5\text{v} + 1.5\text{v} = +1.5 \text{ volts}$$

Kirchhoff's Voltage Law

A law of basic importance to the analysis of an electrical circuit is Kirchhoff's voltage law. This law simply states that the algebraic sum of all voltages around a closed path or loop is zero. Another way of saying it: The sum of all the voltage drops equals the total source voltage. A simplified formula showing this law is shown below:

With three resistors in the circuit:

$E_S - E_1 - E_2 - E_3 \ldots - E_N = 0$ volts

Notice that the sign of the source is opposite that of the individual voltage drops. Therefore, the algebraic sum equals zero. Written another way:

$E_S = E_1 + E_2 + E_3 \ldots + E_N$

The source voltage equals the sum of the voltage drops.

The polarity of the voltage drop is determined by the direction of the current flow. When going around the circuit, notice that the polarity of the resistor is opposite that of the source voltage. The positive on the resistor is facing the positive on the source and the negative towards the negative.

Figure 10-80 illustrates the very basic idea of Kirchhoff's voltage law. There are two resistors in this example. One has a drop of 14 volts and the other has a drop of 10 volts. The source voltage must equal the sum of the voltage drops around the circuit. By inspection it is easy to determine the source voltage as 24 volts.

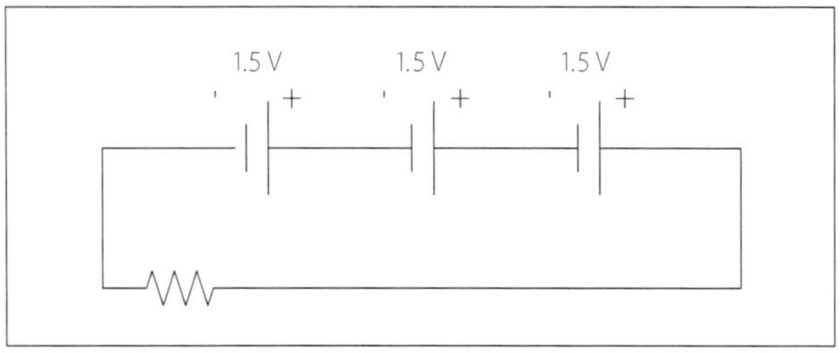

Figure 10-78. Voltage sources in series add algebraically.

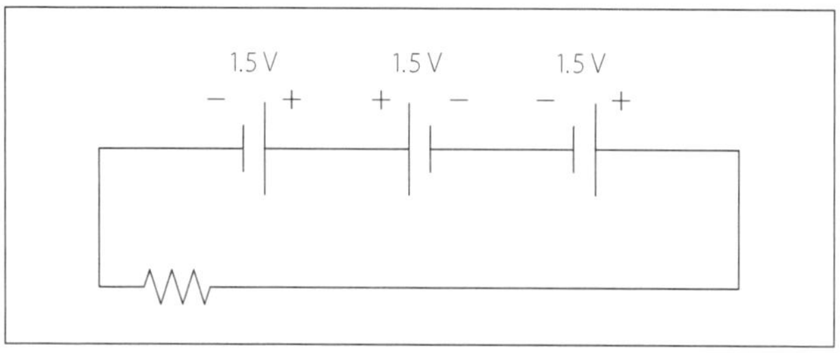

Figure 10-79. Voltage sources add algebraically; one source reversed.

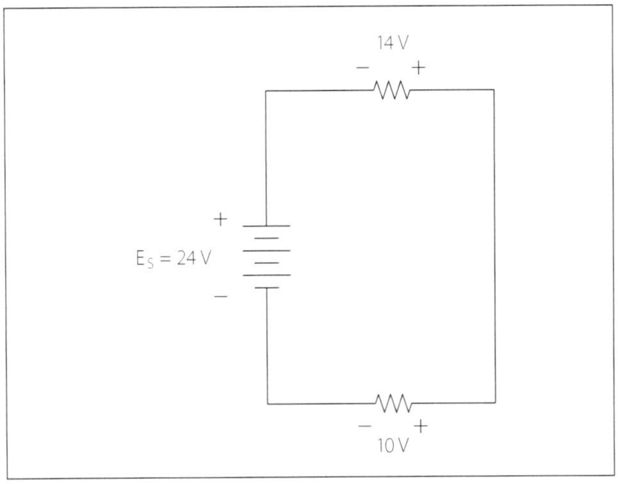

Figure 10-80. Kirchhoff's voltage law.

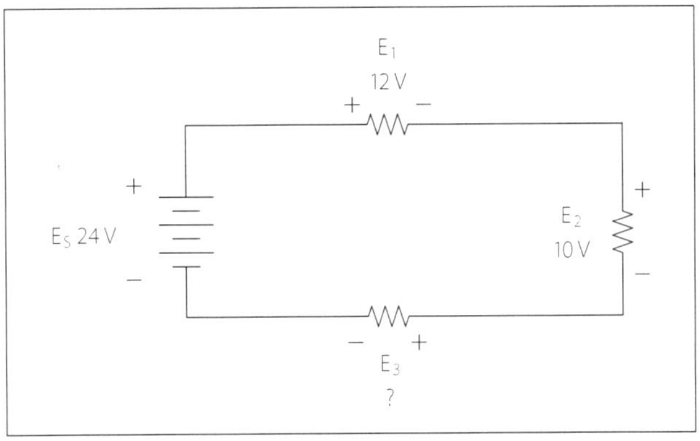

Figure 10-81. Determine the unknown voltage drop.

Figure 10-81 shows a series circuit with three voltage drops and one voltage source rated at 50 volts. Two of the voltage drops are known. However, the third is not known. Using Kirchhoff's voltage law, the third voltage drop can be determined.

With three resistors in the circuit:
$E_S - E_1 - E_2 - E_3 = 0$ volts

Substitute the known values:
$24v - 12v - 10v - E_3 = 0$

Collect known values: $2v - E_3 = 0$

Solve for the unknown: $E_3 = 2$ volts

Determine the value of E_4 in Figure 10-82. For this example, I = 200 mA.

First, the voltage drop across each of the individual resistors must be determined.

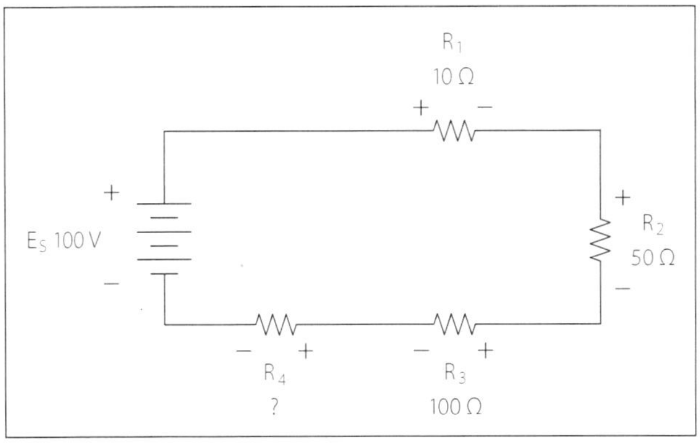

Figure 10-82. Determine the unknown voltage drop.

$E_1 = I\,(R_1)$
$E_1 = (200\text{ mA})\,(10\ \Omega)$
Voltage drop across R_1 $E_1 = 2$ volts

$E_2 = I\,(R_2)$
$E_2 = (200\text{ mA})\,(50\ \Omega)$
Voltage drop across R_2 $E_2 = 10$ volts

$E_3 = I\,(R_3)$
$E_3 = (200\text{ mA})\,(100\ \Omega)$
Voltage drop across R_3 $E_3 = 20$ volts

Kirchhoff's voltage law is now employed to determine the voltage drop across E_4.

With four resistors in the circuit:
$E_S - E_1 - E_2 - E_3 - E_4 = 0$ volts

Substituting values:
$100v - 2v - 10v - 20\ v - E_4 = 0$

Combine: $68\ v - E_4 = 0$

Solve for unknown: $E_4 = 68$ v

Using Ohm's law and substituting in E_4, the value for R_4 can now be determined.

Ohm's law: $R = \frac{E}{I}$

Specific application: $R_4 = \frac{E_4}{I}$

Substitute values: $R_4 = \frac{68V}{200\text{ mA}}$

Value for R_4: $R_4 = 340\ \Omega$

Voltage Dividers

Voltage dividers are devices that make it possible to obtain more than one voltage from a single power source. A voltage divider usually consists of a resistor, or resistors connected in series, with fixed or movable contacts and two fixed terminal contacts. As current flows through the resistor, different voltages can be obtained between the contacts.

Series circuits are used for voltage dividers. The voltage divider rule allows the technician to calculate the voltage across one or a combination of series resistors without having to first calculate the current in the circuit. Because the current flows through each resistor, the voltage drops are proportional to the ohmic values of the constituent resistors.

A typical voltage divider is shown in Figure 10-83.

To understand how a voltage divider works, examine Figure 10-84 carefully and observe the following:

Each load draws a given amount of current: I_1, I_2, I_3. In addition to the load currents, some bleeder current (I_B) flows. The current (I_T) is drawn from the power source and is equal to the sum of all currents.

The voltage at each point is measured with respect to a common point. Note that the common point is the point at which the total current (I_T) divides into separate currents (I_1, I_2, I_3).

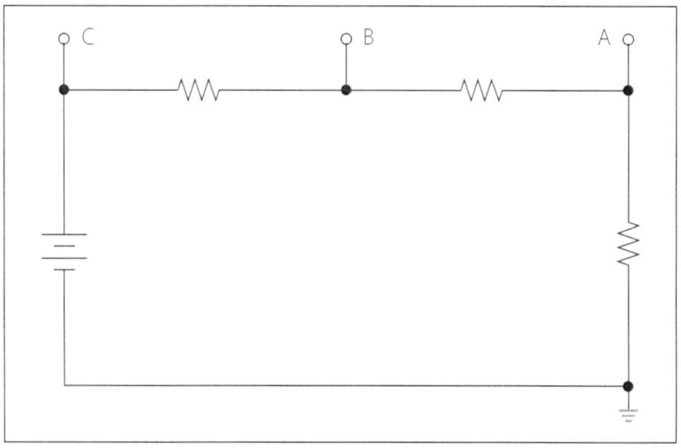

Figure 10-83. A voltage divider circuit.

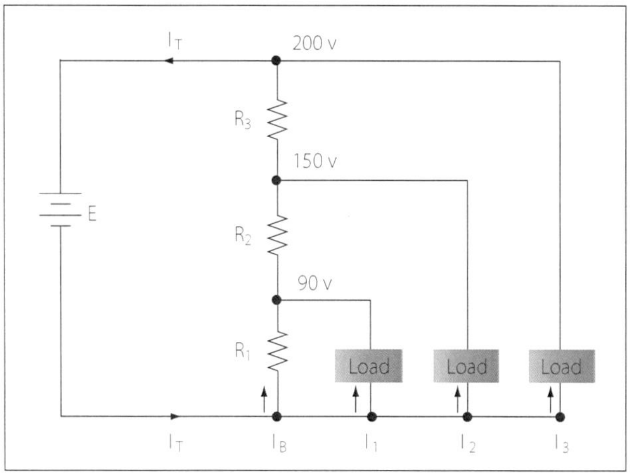

Figure 10-84. A typical voltage divider.

Each part of the voltage divider has a different current flowing in it. The current distribution is as follows:

Through R_1—bleeder current (I_B)
Through R_2—I_B plus I_1
Through R_3—I_B PIUS I_1, plus I_2

The voltage across each resistor of the voltage divider is:

90 volts across R_1
60 volts across R_2
50 volts across R_3

The voltage divider circuit discussed up to this point has had one side of the power supply (battery) at ground potential. In Figure 10-85 the common reference point (ground symbol) has been moved to a different point on the voltage divider. The voltage drop across R_1 is 20 volts; however, since tap A is connected to a point in the circuit that is at the same potential as the negative side of the battery, the voltage between tap A and the reference point is a negative (−) 20 volts. Since resistors R_2 and R_3 are connected to the positive side of the battery, the voltages between the reference point and tap B or C are positive.

The following rules provide a simple method of determining negative and positive voltages: (1) If current enters a resistance flowing away from the reference point, the voltage drop across that resistance is positive in respect to the reference point; (2) if current flows out of a resistance toward the reference point, the voltage drop across that resistance is negative in respect to the reference point. It is the location of the reference point that determines whether a voltage is negative or positive.

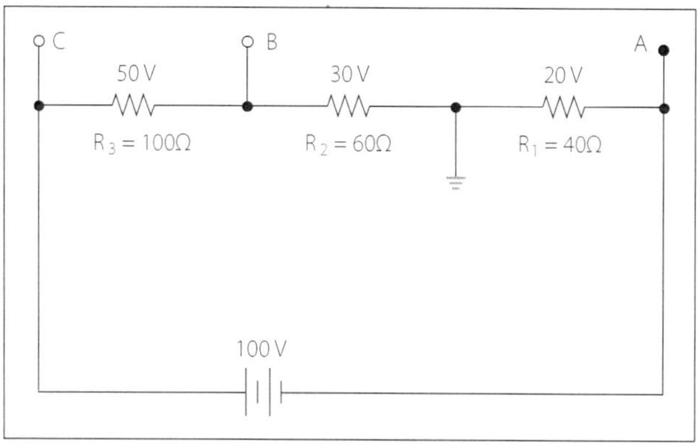

Figure 10-85. Positive and negative voltage on a voltage divider.

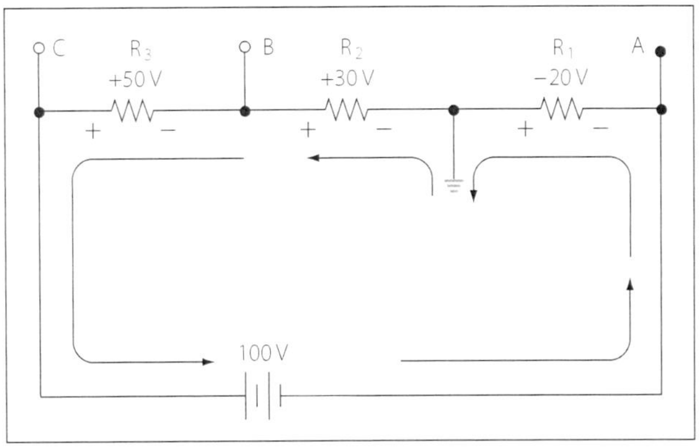

Figure 10-86. Current flow through a voltage divider.

Tracing the current flow provides a means for determining the voltage polarity. Figure 10-86 shows the same circuit with the polarities of the voltage drops and the direction of current flow indicated.

The current flows from the negative side of the battery to R_1. Tap A is at the same potential as the negative terminal of the battery since the slight voltage drop caused by the resistance of the conductor is disregarded; however, 20 volts of the source voltage are required to force the current through R_1 and this 20-volt drop has the polarity indicated. Stated another way, there are only 80 volts of electrical pressure left in the circuit on the ground side of R_1.

When the current reaches tap B, 30 more volts have been used to move the electrons through R_2, and in a similar manner the remaining 50 volts are used for R_3. But the voltages across R_2 and R_3 are positive voltages, since they are above ground potential.

Figure 10-87 shows the voltage divider used previously. The voltage drops across the resistances are the same; however, the reference point (ground) has been changed. The voltage between ground and tap A is now a negative 100 volts, or the applied voltage.

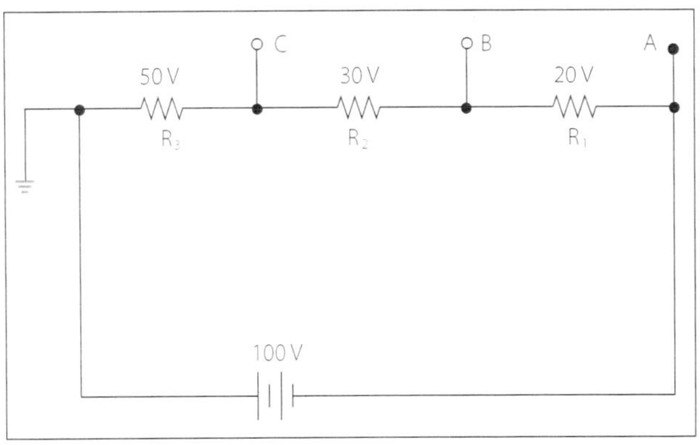

Figure 10-87. Voltage divider with changed ground.

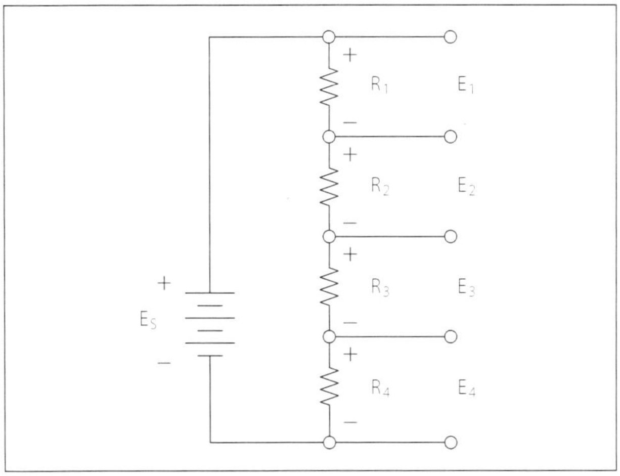

Figure 10-88. Four resistor voltage divider.

The voltage between ground and tap B is a negative 80 volts, and the voltage between ground and tap C is a negative 50 volts.

Determining the Voltage Divider Formula

Figure 10-88 shows the example network of four resistors and a voltage source. With a few simple calculations, a formula for determining the voltage divisions in a series circuit can be determined.

The voltage drop across any particular resistor shall be called E_X, where the subscript x is the value of a particular resistor (1, 2, 3, or 4). Using Ohm's law, the voltage drop across any resistor can be determined.

Ohm's law: $E_X = I\,(R_X)$

As seen earlier in the text, the current is equal to the source voltage divided by the total resistance of the series circuit.

Current: $I = \dfrac{E_S}{R_T}$

The current equation can now be substituted into the equation for Ohm's law.

Substitute: $E_X = \left(\dfrac{E_S}{R_T}\right)\left(R_X\right)$

Algebraic rearrange: $E_X = \left(\dfrac{R_X}{R_T}\right)\left(E_S\right)$

This equation is the general voltage divider formula. The explanation of this formula is that the voltage drop across any resistor or combination of resistors in a series circuit is equal to the ratio of the resistance value to the total resistance, divided by the value of the source voltage. Figure 10-89 illustrates this with a network of three resistors and one voltage source.

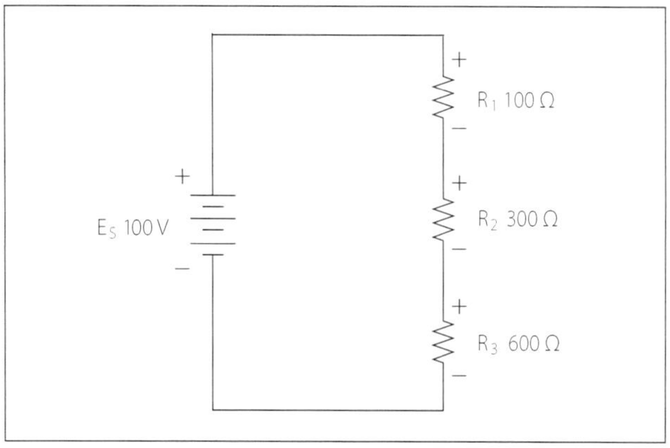

Figure 10-89. Network of three resistors and one voltage source.

$$E_X = \left(\frac{R_X}{R_T}\right) E_S$$

$R_T = 100\ \Omega + 300\ \Omega + 600\ \Omega = 1000\ \Omega$
$E_S = 100$ volts

Voltage drop over 100 Ω resistor is:

$$E_X = \left(\frac{100\ \Omega}{1000\ \Omega}\right) 100\ V$$

$E_{100\ \Omega} = 10\ V$

Voltage drop over 300 Ω resistor is:

$$E_X = \left(\frac{300\ \Omega}{1000\ \Omega}\right) 100\ V$$

$E_{100\ \Omega} = 30\ V$

Voltage drop over 600 Ω resistor is:

$$E_X = \left(\frac{600\ \Omega}{1000\ \Omega}\right) 100\ V$$

$E_{100\ \Omega} = 60\ V$

Checking work
$E_T = 10\ V + 30\ V + 60\ V = 100\ V$

Parallel DC Circuits

Overview

A circuit in which two of more electrical resistances or loads are connected across the same voltage source is called a parallel circuit. The primary difference between the series circuit and the parallel circuit is that more than one path is provided for the current in the parallel circuit. Each of these parallel paths is called a branch. The minimum requirements for a parallel circuit are the following:

- A power source.
- Conductors.

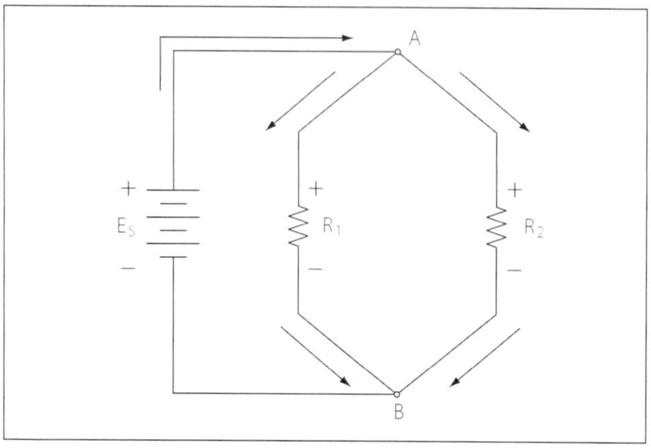

Figure 10-90. Basic parallel circuit.

- A resistance or load for each current path.
- Two or more paths for current flow.

Figure 10-90 depicts the most basic parallel circuit. Current flowing out of the source divides at point A in the diagram and goes through R_1 and R_2. As more branches are added to the circuit, more paths for the source current are provided.

Voltage Drops

The first point to understand is that the voltage across any branch is equal to the voltage across all of the other branches.

Total Parallel Resistance

The voltage across any branch is equal to the voltage across all of the other branches.

The parallel circuit consists of two or more resistors connected in such a way as to allow current flow to pass through all of the resistors at once. This eliminates the need for current to pass one resistor before passing through the next. When resistors are connected in parallel, the total resistance of the circuit decreases. The total resistance of a parallel combination is always less than the value of the smallest resistor in the circuit. In the series circuit, the current has to pass through the resistors one at a time. This gave a resistance to the current equal the sum of all the resistors. In the parallel circuit, the current has several resistors that it can pass through, actually reducing the total resistance of the circuit in relation to any one resistor value.

The amount of current passing through each resistor will vary according to its individual resistance. The total current of the circuit will then be the sum of the current in all branches. It can be determined by inspection that the total current will be greater than that of any given branch. Using Ohm's law to calculate the resistance based on the applied voltage and the total current, it can be determined that the total resistance is less than any individual branch.

An example of this is if there was a circuit with a 100 Ω resistor and a 5 Ω resistor; while the exact value must be calculated, it still can be said that the combined resistance between the two will be less than the 5 Ω.

Resistors in Parallel

The formula for the total parallel resistance is as follows:

$$\frac{1}{R_T} = \frac{1}{R_1} + \frac{1}{R_2} + \frac{1}{R_3} + \dots \frac{1}{R_N}$$

If the reciprocal of both sides is taken, then the general formula for the total parallel resistance is:

$$R_T = \frac{1}{\frac{1}{R_1} + \frac{1}{R_2} + \frac{1}{R_3} + \dots \frac{1}{R_N}}$$

Two Resistors in Parallel

Typically, it is more convenient to consider only two resistors at a time because this setup occurs in common practice. Any number of resistors in a circuit can be broken down into pairs. Therefore, the most common method is to use the formula for two resistors in parallel.

$$R_T = \frac{1}{\frac{1}{R_1} + \frac{1}{R_2}}$$

Combining the terms in the denominator and rewriting

$$R_T = \frac{R_1 R_2}{R_1 + R_2}$$

Put in words, this states that the total resistance for two resistors in parallel is equal to the product of both resistors divided by the sum of the two resistors. In the formula below, calculate the total resistance.

General formula $R_T = \frac{R_1 R_2}{R_1 + R_2}$

Known values $R_1 = 500\ \Omega$

$R_2 = 400\ \Omega$

$$R_T = \frac{500\ \Omega\ 400\ \Omega}{500\ \Omega + 400\ \Omega}$$

$$R_T = \frac{200000\ \Omega}{900\ \Omega}$$

$$R_T = 222.22\ \Omega$$

Current Source

A current source is an energy source that provides a constant value of current to a load even when the load changes in resistive value. The general rule to remember is that the total current produced by current sources in parallel is equal to the algebraic sum of the individual sources.

Kirchhoff's Current Law

Kirchhoff's current law can be stated as: The sum of the currents into a junction or node is equal to the sum of the currents flowing out of that same junction or node. A junction can be defined as a point in the circuit where two or more circuit paths come together. In the case of the parallel circuit, it is the point in the circuit where the individual branches join.

General formula $I_T = I_1 + I_2 + I_3$

Refer to Figure 10-91 for an illustration. Point A and point B represent two junctions or nodes in the circuit with three resistive branches in between. The voltage source provides a total current I_T into node A. At this point, the current must divide, flowing out of node A into each of the branches according to the resistive value of each branch. Kirchhoff's current law states that the current going in must equal that going out. Following the current through the three branches and back into node B, the total current I_T entering node B and leaving node B is the same as that which entered node A. The current then continues back to the voltage source.

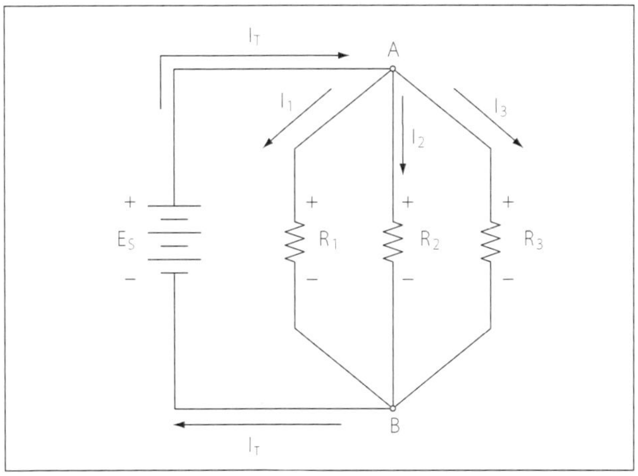

Figure 10-91. Kirchhoff's current law.

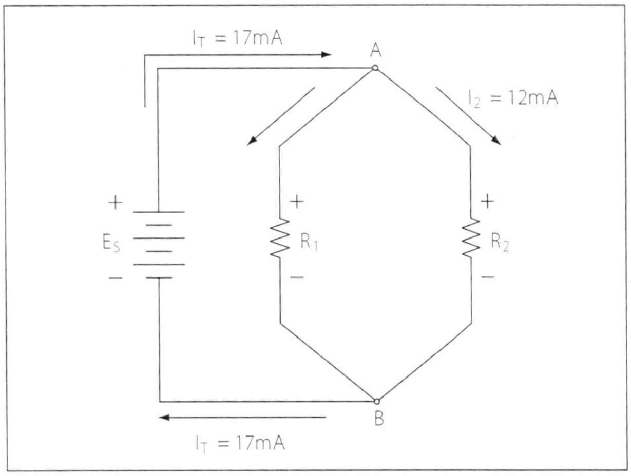

Figure 10-92. Individual branch currents.

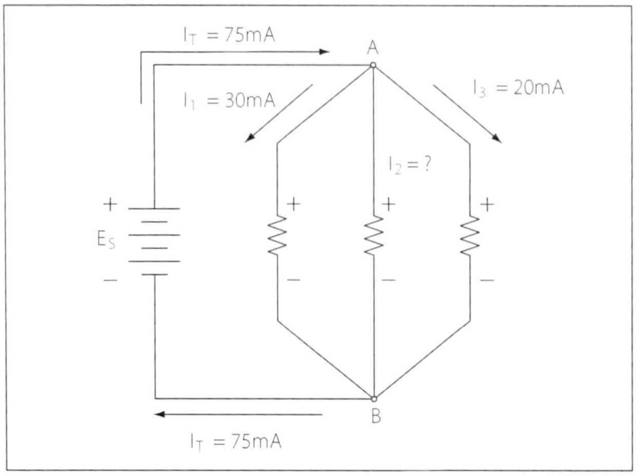

Figure 10-93. Determining an unknown circuit in branch 2.

Figure 10-92 shows that the individual branch currents are:

$I_1 = 5$ mA
$I_2 = 12$ mA

The total current flow into the node A equals the sum of the branch currents, which is:

$$I_T = I_1 + I_2$$

Substitute $I_T = 5mA + 12mA$

$$I_T = 17mA$$

The total current entering node B is also the same.

Figure 10-93 illustrates how to determine an unknown current in one branch. Note that the total current into a junction of the three branches is known. Two of the branch currents are known. By rearranging the general formula, the current in branch two can be determined.

General formula $I_T = I_1 + I_2 + I_3$
Substitute $75mA = 30mA + I_2 + 20mA$
Solve I_2 $I_2 = 75mA - 30mA - 20mA$
$I_2 = 25mA$

Current Dividers

It can now be easily seen that the parallel circuit is a current divider. As could be seen in Figure 10-90, there is a current through each of the two resistors. Because the same voltage is applied across both resistors in parallel, the branch currents are inversely proportional to the ohmic values of the resistors. Branches with higher resistance have less current than those with lower resistance. For example, if the resistive value of R_2 is twice as high as that of R_1, the current in R_2 will be half of that of R_1. All of this can be determined with Ohm's law.

By Ohm's law, the current through any one of the branches can be written as:

$$I_X = E_S / R_X$$

The voltage source appears across each of the parallel resistors and R_X represents any one the resistors. The source voltage is equal to the total current times the total parallel resistance.

$$E_S = I_T R_T$$

Substituting $I_T R_T$ for E_S $\quad I_X = \frac{I_T R_T}{R_X}$

Rearranging $\quad I_X = \left(\frac{R_T}{R_X}\right) I_T$

$$I_2 = \left(\frac{R_2}{R_T}\right) I_T$$

And $\quad I_1 = \left(\frac{R_1}{R_T}\right) I_T$

This formula is the general current divider formula. The current through any branch equals the total parallel resistance divided by the individual branch resistance, multiplied by the total current.

Series-Parallel DC Circuits

Overview

Most of the circuits that the technician will encounter will not be a simple series or parallel circuit. Circuits are usually a combination of both, known as series-parallel circuits, which are groups consisting of resistors in parallel and in series. An example of this type of circuit can be seen in Figure 10-94. While the series-parallel circuit can initially appear to be complex, the same rules that have been used for the series and parallel circuits can be applied to these circuits.

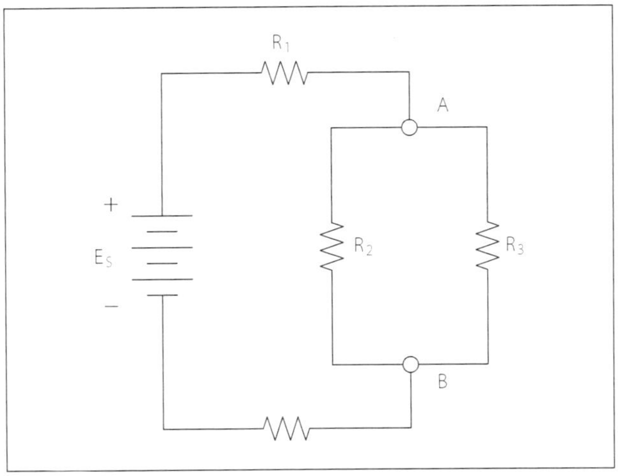

Figure 10-94. Series-parallel circuits.

The voltage source will provide a current out to resistor R_1, then to the group of resistors R_2 and R_3 and then to the next resistor R_4 before returning to the voltage source. The first step in the simplification process is to isolate the group R_2 and R_3 and recognize that they are a parallel network that can be reduced to an equivalent resistor. Using the formula for parallel resistance,

$$R_{23} = \frac{R_2 R_3}{R_2 + R_3}$$

R_2 and R_3 can be reduced to R_{23}. Figure 10-95 now shows an equivalent circuit with three series connected resistors. The total resistance of the circuit can now be simply determined by adding up the values of resistors R_1, R_{23}, and R_4.

Determining the Total Resistance

A more quantitative example for determining total resistance and the current in each branch in a combination circuit is shown in the following example. Also refer to Figure 10-96.

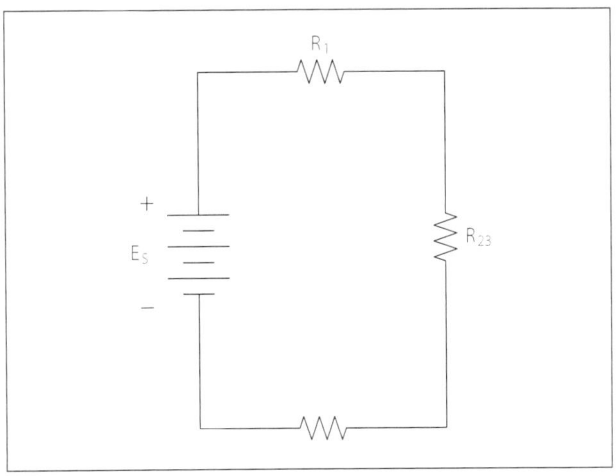

Figure 10-95. Equivalent circuit with three series connected resistors.

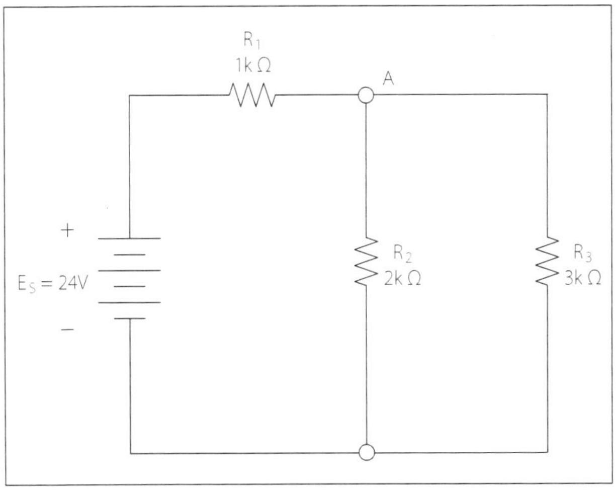

Figure 10-96. Determining total resistance.

The first step is to determine the current at junction A, leading into the parallel branch. To determine the I_T, the total resistance R_T of the entire circuit must be known. The total resistance of the circuit is given as:

$$
\begin{aligned}
& & R_T &= R_1 + R_{23} \\
&\text{Where} & R_{23} &= \left(\frac{R_2 R_3}{R_2 + R_3}\right) \text{ Parallel network} \\
&\text{Find } R_{EQ} & R_{23} &= \frac{2k\ \Omega\ 3k\ \Omega}{2k\ \Omega + 3k\ \Omega} \\
&\text{Solve for } R_{EQ} & R_{23} &= \frac{6000k\ \Omega}{5k\ \Omega} \\
& & R_{23} &= 1.2k\ \Omega \\
&\text{Solve for } R_T & R_T &= 1k\ \Omega + 1.2k\ \Omega \\
& & R_T &= 2.2k\ \Omega
\end{aligned}
$$

With the total resistance R_T now determined, the total I_T can be determined. Using Ohm's law:

$$
\begin{aligned}
& & I_T &= \frac{E_S}{R_T} \\
&\text{Substitute values} & I_T &= \frac{24V}{2.2k\ \Omega} \\
& & I_T &= 10.9mA
\end{aligned}
$$

The current through the parallel branches of R_2 and R_3 can be determined using the current divider rule discussed earlier in the text. Recall that:

Current divider rule $I_2 = \left(\frac{R_T}{R_2}\right)\left(I_T\right)$

And $I_3 = \left(\frac{R_T}{R_3}\right)\left(I_T\right)$

Substitute values for I_2

$$I_2 = \left(\frac{R_2 + R_3}{R_2}\right)\left(I_T\right)$$

$$I_2 = \left(\frac{2k\,\Omega + 3k\,\Omega}{2k\,\Omega}\right)\left(10.9mA\right)$$

$$I_2 = \left(\frac{5k\,\Omega}{2k\,\Omega}\right)\left(10.9mA\right)$$

$$I_2 = 2.5\,\Omega\,(10.9mA)$$

$$I_2 = 1.25mA$$

Now using Kirchhoff's current law, the current in branch with R_3 can be determined.

$I_T = I_2 + I_3$
$I_3 = I_T - I_2$
$I_3 = 10.9mA - 1.25mA$
$I_3 = 9.65mA$

Alternating Current and Voltage

Alternating current has largely replaced direct current in commercial power systems for a number of reasons. It can be transmitted over long distances more readily and more economically than direct current, since AC voltages can be increased or decreased by means of transformers.

Because more and more units are being operated electrically in airplanes, the power requirements are such that a number of advantages can be realized by using AC. Space and weight can be saved, since AC devices, especially motors, are smaller and simpler than DC devices. In most AC motors no brushes are required, and commutation trouble at high altitude is eliminated. Circuit breakers will operate satisfactorily under load at high altitudes in an AC system, whereas arcing is so excessive on DC systems that circuit breakers must be replaced frequently. Finally, most airplanes using a 24-volt DC system have special equipment that requires a certain amount of 400 cycle AC current.

AC and DC Compared

"AC" stands for Alternating Current. Many of the principles, characteristics, and effects of AC are similar to those of direct current. Similarly, there are a number of differences, which will be explained. Direct current flows constantly in only one direction with a constant polarity. It changes magnitude only when the circuit is opened or closed, as shown in the DC waveform in Figure 10-97. Alternating current changes direction at regular intervals, increases in value at a definite rate from zero to a maximum positive strength, and decreases back to zero; then it flows in the opposite direction, similarly increasing to a maximum negative value, and again decreasing to zero. DC and AC waveforms are compared in Figure 10-97.

Since alternating current constantly changes direction and intensity, the following two effects (to be discussed later) take place in AC circuits that do not occur in DC circuits:

1. Inductive reactance.
2. Capacitive reactance.

Generator Principles

After the discovery that an electric current flowing through a conductor creates a magnetic field around the conductor, there was considerable scientific speculation about whether a magnetic field could create a current flow in a conductor. In 1831, it was demonstrated this could be accomplished.

To show how an electric current can be created by a magnetic field, a demonstration similar to that illustrated in Figure 10-98 can be used. Several turns of

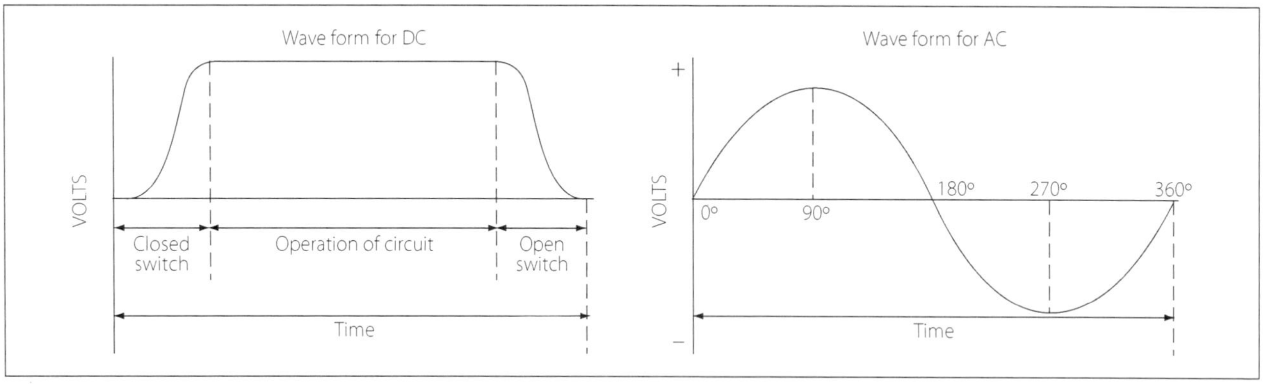

Figure 10-97. DC and AC voltage curves.

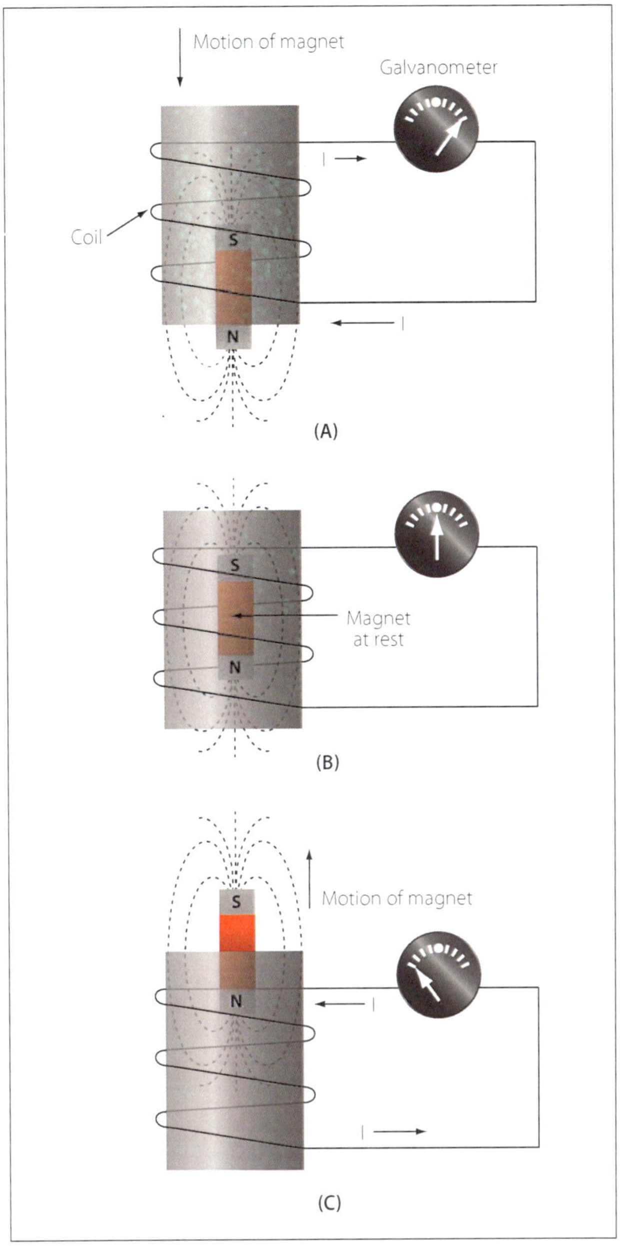

Figure 10-98. Inducing a current flow.

a conductor are wrapped around a cylindrical form, and the ends of the conductor are connected together to form a complete circuit, which includes a galvanometer. If a simple bar magnet is plunged into the cylinder, the galvanometer can be observed to deflect in one direction from its zero (center) position (Figure 10-98A).

When the magnet is at rest inside the cylinder, the galvanometer shows a reading of zero, indicating that no current is flowing (Figure 10-98B).

In Figure 10-98C, the galvanometer indicates a current flow in the opposite direction when the magnet is pulled from the cylinder.

The same results may be obtained by holding the magnet stationary and moving the cylinder over the magnet, indicating that a current flows when there is relative motion between the wire coil and the magnetic field. These results obey a law first stated by the German scientist, Heinrich Lenz. Lenz's law states:

The induced current caused by the relative motion of a conductor and a magnetic field always flows in such a direction that its magnetic field opposes the motion.

When a conductor is moved through a magnetic field, an electromotive force (emf) is induced in the conductor. [Figure 10-99] The direction (polarity) of the induced emf is determined by the magnetic lines of force and the direction the conductor is moved through the magnetic field. The generator left-hand rule (not to be confused with the left-hand rules used with a coil) can be used to determine the direction of the induced emf. [Figure 10-100] The left-hand rule is summed up as follows:

The first finger of the left hand is pointed in the direction of the magnetic lines of force (north to south), the

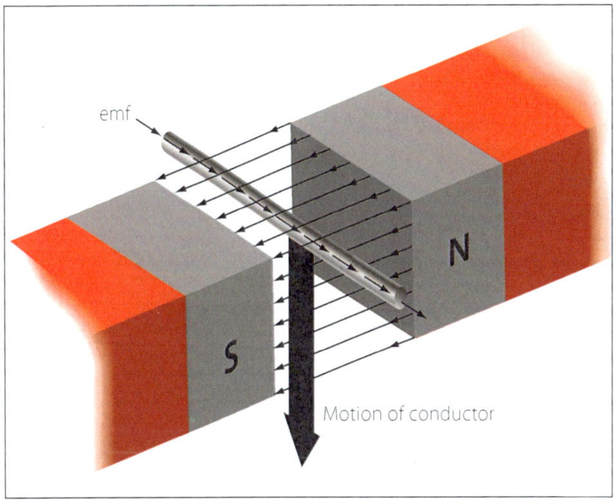

Figure 10-99. Inducing an emf in a conductor.

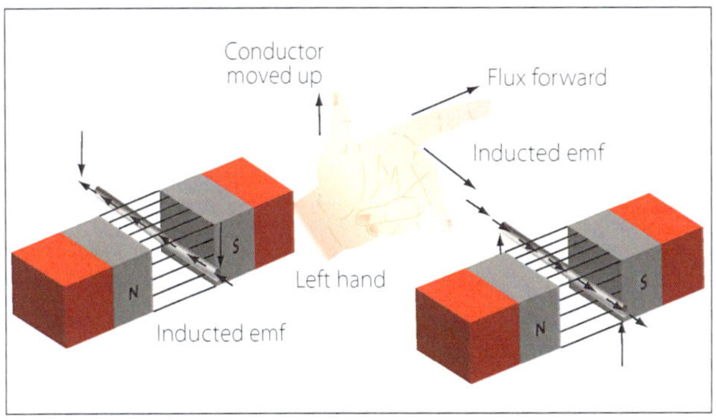

Figure 10-100. An application of the generator left-hand rule.

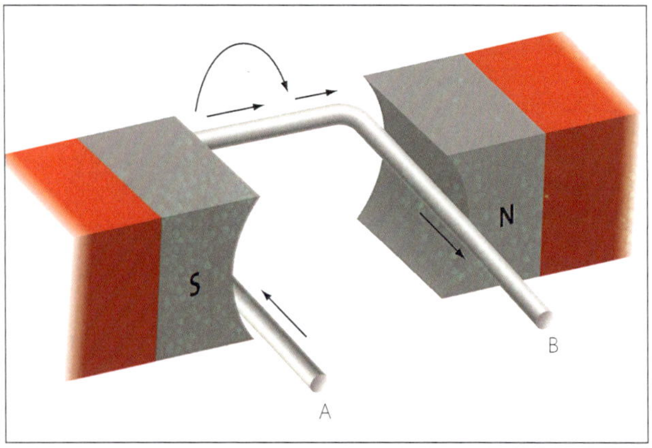

Figure 10-101. Voltage induced in a loop.

thumb is pointed in the direction of movement of the conductor through the magnetic field, and the second finger points in the direction of the induced emf.

When a loop conductor is rotated in a magnetic field, a voltage is induced in each side of the loop. [Figure 10-101] The two sides cut the magnetic field in opposite directions, and although the current flow is continuous, it moves in opposite directions with respect to the two sides of the loop. If sides A and B and the loop are rotated half a turn and the sides of the conductor have exchanged positions, the induced emf in each wire reverses its direction, since the wire formerly cutting the lines of force in an upward direction is now moving downward.

The value of an induced emf depends on three factors:

1. The number of wires moving through the magnetic field.
2. The strength of the magnetic field.
3. The speed of rotation.

Generators of Alternating Current

Generators used to produce an alternating current are called AC generators or alternators.

The simple generator constitutes one method of generating an alternating voltage. [Figure 10-102] It consists of a rotating loop, marked A and B, placed between two magnetic poles, N and S. The ends of the loop are connected to two metal slip rings (collector rings), C_1 and C_2. Current is taken from the collector rings by brushes. If the loop is considered as separate wires A and B, and the left-hand rule for generators is applied, then it can be observed that as wire A moves up across the field, a voltage is induced which causes the current to flow inward. As wire B moves down across the field, a voltage is induced which causes the current to flow outward. When the wires are formed into a loop, the voltages induced in the two sides of the loop are combined. Therefore, for explanatory purposes, the action of either conductor, A or B, while rotating in the magnetic field is similar to the action of the loop.

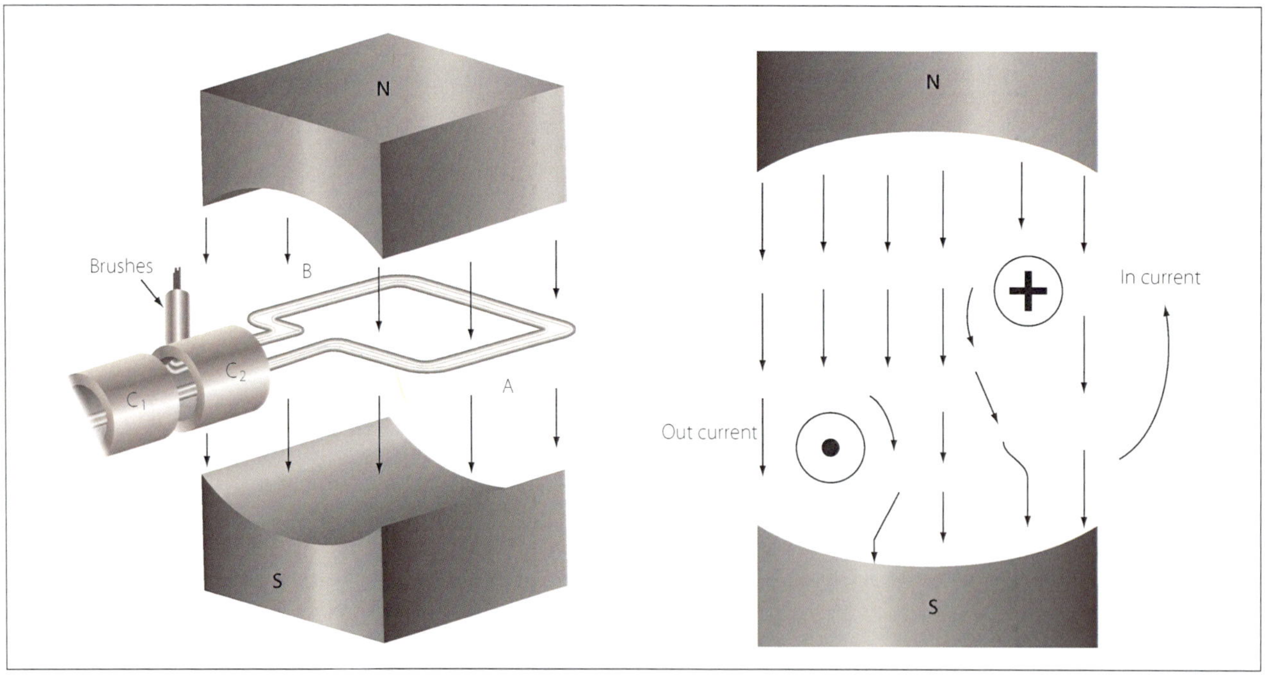

Figure 10-102. Simple generator.

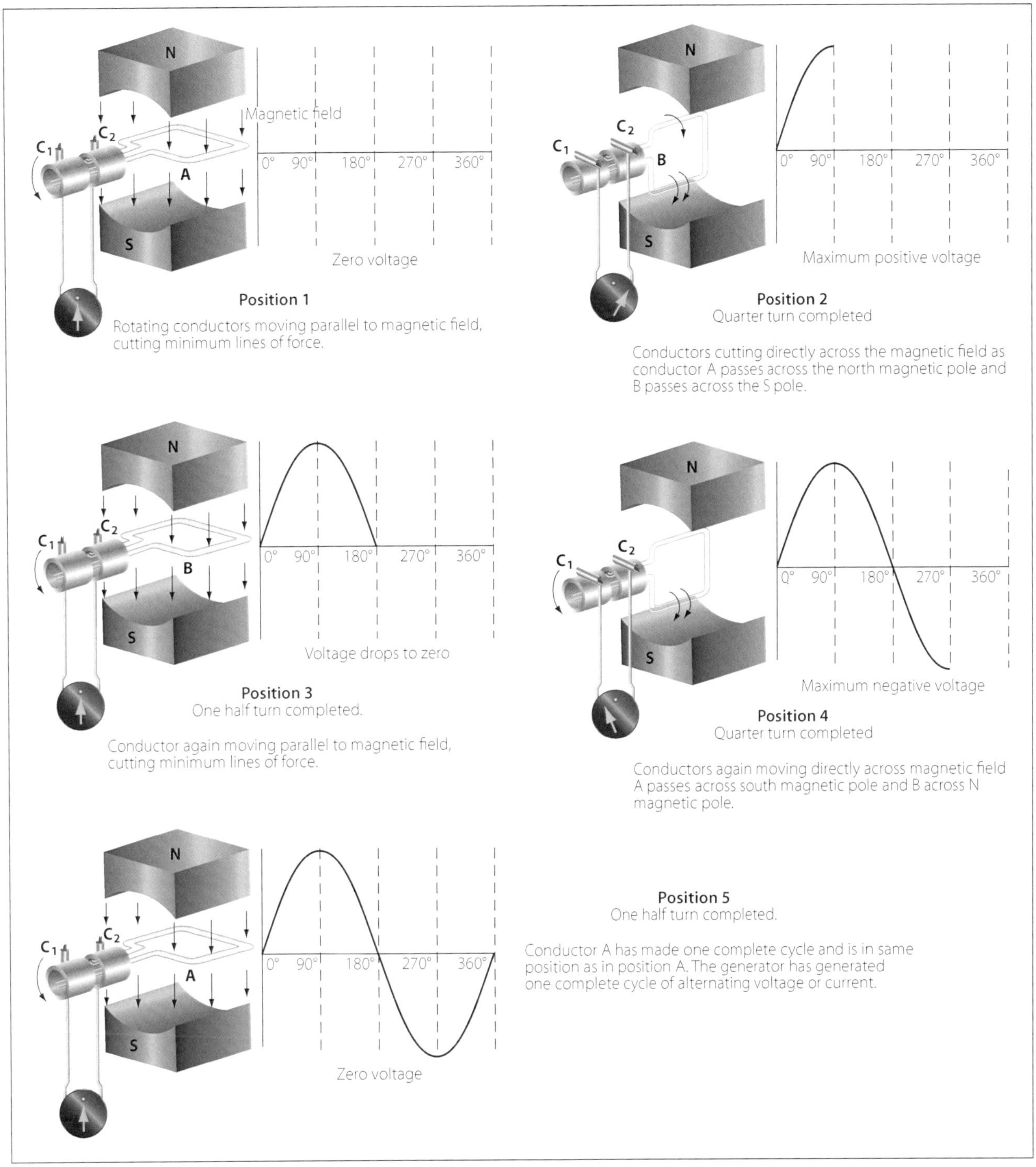

Figure 10-103. Generation of a sine wave.

Figure 10-103 illustrates the generation of alternating current with a simple loop conductor rotating in a magnetic field. As it is rotated in a counterclockwise direction, varying values of voltages are induced in it.

Position 1

The conductor A moves parallel to the lines of force. Since it cuts no lines of force, the induced voltage is zero. As the conductor advances from position 1 to position 2, the voltage induced gradually increases.

Position 2

The conductor is now moving perpendicular to the flux and cuts a maximum number of lines of force; therefore, a maximum voltage is induced. As the conductor moves beyond position 2, it cuts a decreasing amount of flux at each instant, and the induced voltage decreases.

Position 3

At this point, the conductor has made one-half of a revolution and again moves parallel to the lines of force, and no voltage is induced in the conductor. As the A conductor passes position 3, the direction of induced voltage now reverses since the A conductor is moving downward, cutting flux in the opposite direction. As the A conductor moves across the south pole, the induced voltage gradually increases in a negative direction, until it reaches position 4.

Position 4

Like position 2, the conductor is again moving perpendicular to the flux and generates a maximum negative voltage. From position 4 to 5, the induced voltage gradually decreases until the voltage is zero, and the conductor and wave are ready to start another cycle.

Position 5

The curve shown at position 5 is called a sine wave. It represents the polarity and the magnitude of the instantaneous values of the voltages generated. The horizontal base line is divided into degrees, or time, and the vertical distance above or below the base line represents the value of voltage at each particular point in the rotation of the loop.

Cycle and Frequency

Cycle Defined

A cycle is a repetition of a pattern. Whenever a voltage or current passes through a series of changes, returns to the starting point, and then again starts the same series of changes, the series is called a cycle. The cycle is represented by the symbol of a wavy line in a circle (~). In the cycle of voltage shown in Figure 10-104, the voltage increases from zero to a maximum positive value, decreases to zero; then increases to a maximum negative value, and again decreases to zero. At this point, it is ready to go through the same series of changes. There are two alternations in a complete cycle: the positive alternation and the negative. Each is half a cycle.

Frequency Defined

The frequency is the number of cycles of alternating current per second (1 second). The standard unit of frequency measurement is the hertz (Hz). [Figure 10-105]

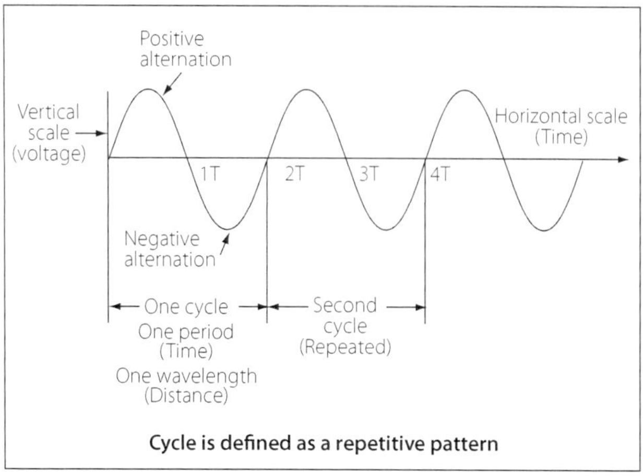

Figure 10-104. Cycle of voltage.

In a generator, the voltage and current pass through a complete cycle of values each time a coil or conductor passes under a north and south pole of the magnet. The number of cycles for each revolution of the coil or conductor is equal to the number of pairs of poles. The frequency, then, is equal to the number of cycles in one revolution multiplied by the number of revolutions per second. Expressed in equation form,

$$F = \frac{\text{Number of poles}}{2} \times \frac{\text{rpm}}{60}$$

where $^P\!/_2$ is the number of pairs of poles, and rpm/60 the number of revolutions per second. If in a 2 pole generator, the conductor is turning at 3,600 rpm, the revolutions per second are

$$\text{rps} = \frac{3600}{60} = 60 \text{ revolutions per second}$$

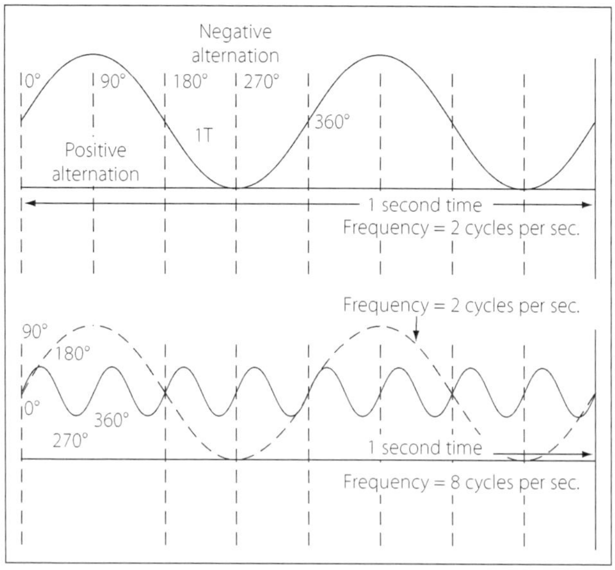

Figure 10-105. Frequency in cycles per second.

Since there are 2 poles, $^{P}/_{2}$ is 1, and the frequency is 60 cycles per second (cps). In a 4 pole generator with an armature speed of 1,800 rpm, substitute in the equation,

$$F = \frac{P}{2} \times \frac{rpm}{60} \text{ as follows}$$

$$F = \frac{4}{2} \times \frac{1800}{60}$$

$$F = 2 \times 30$$

$$F = 60 \text{ cps}$$

Period Defined

The time required for a sine wave to complete one full cycle is called a period. [Figure 10-104] The period of a sine wave is inversely proportional to the frequency. That is to say that the higher the frequency, the shorter the period will be. The mathematical relationship between frequency and period is given as:

$$\text{Period is } t = \frac{1}{f}$$

$$\text{Frequency is } f = \frac{1}{t}$$

Wavelength Defined

The distance that a waveform travels during a period is commonly referred to as a wavelength and is indicated by the Greek letter lambda (λ). The measurement of wavelength is taken from one point on the waveform to a corresponding point on the next waveform. [Figure 10-104]

Phase Relationships

In addition to frequency and cycle characteristics, alternating voltage and current also have a relationship called "phase." In a circuit that is fed (supplied) by one alternator, there must be a certain phase relationship between voltage and current if the circuit is to function efficiently. In a system fed by two or more alternators, not only must there be a certain phase relationship between voltage and current of one alternator, but there must be a phase relationship between the individual voltages and the individual currents. Also, two separate circuits can be compared by comparing the phase characteristics of one to the phase characteristics of the other.

In Phase Condition

Figure 10-106A, shows a voltage signal and a current signal superimposed on the same time axis. Notice that when the voltage increases in the positive alternation that the current also increases. When the voltage reaches it peak value, so does the current. Both waveforms then reverse and decrease back to a zero magnitude, then proceed in the same manner in the negative direction as they did in the positive direction. When two waves, such as these in Figure 10-106A, are exactly in step with each other, they are said to be in phase. To be in phase, the two waveforms must go through their maximum and minimum points at the same time and in the same direction.

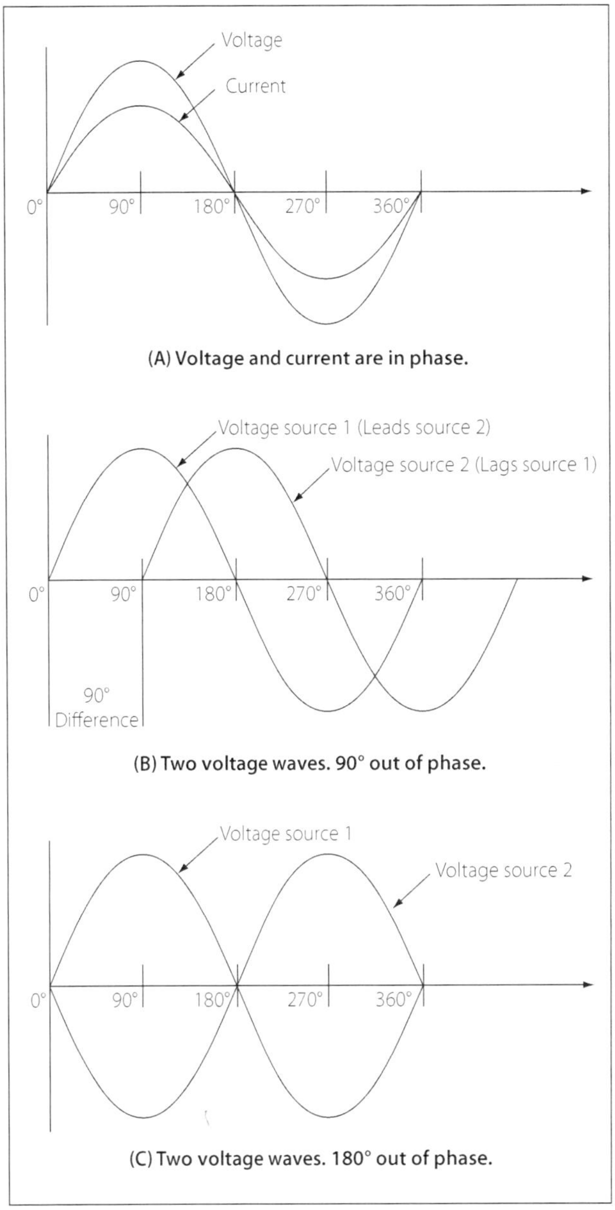

Figure 10-106. In phase and out of phase conditions.

Out of Phase Condition

When two waveforms go through their maximum and minimum points at different times, a phase difference will exist between the two. In this case, the two waveforms are said to be out of phase with each other. The terms lead and lag are often used to describe the phase difference between waveforms. The waveform that reaches its maximum or minimum value first is said to lead the other waveform. Figure 10-106B shows this relationship. Voltage source one starts to rise at the 0° position and voltage source two starts to rise at the 90° position. Because voltage source one begins its rise earlier in time (90°) in relation to the second voltage source, it is said to be leading the second source. On the other hand, the second source is said to be lagging the first source. When a waveform is said to be leading or lagging, the difference in degrees is usually stated. If the two waveforms differ by 360°, they are said to be in phase with each other. If there is a 180° difference between the two signals, then they are still out of phase even though they are both reaching their minimum and maximum values at the same time. [Figure 10-106]

A practical note of caution: *When encountering an aircraft that has two or more AC busses in use, it is possible that they may be split and not synchronized to be in phase with each other. When two signals that are not locked in phase are mixed, much damage can occur to aircraft systems or avionics.*

Values of Alternating Current

There are three values of alternating current, which are instantaneous, peak, and effective (root mean square, RMS).

Instantaneous Value

An instantaneous value of voltage or current is the induced voltage or current flowing at any instant during a cycle. The sine wave represents a series of these values. The instantaneous value of the voltage varies from zero at 0° to maximum at 90°, back to zero at 180°, to maximum in the opposite direction at 270°, and to zero again at 360°. Any point on the sine wave is considered the instantaneous value of voltage.

Peak Value

The peak value is the largest instantaneous value. The largest single positive value occurs when the sine wave of voltage is at 90°, and the largest single negative value occurs when it is at 270°. Maximum value is 1.41 times the effective value. These are called peak values.

Effective Value

The effective value is also known as the RMS value or root mean square, which refers to the mathematical process by which the value is derived. Most AC voltmeters will display the effective or RMS value when used. The effective value is less than the maximum value, being equal to .707 times the maximum value.

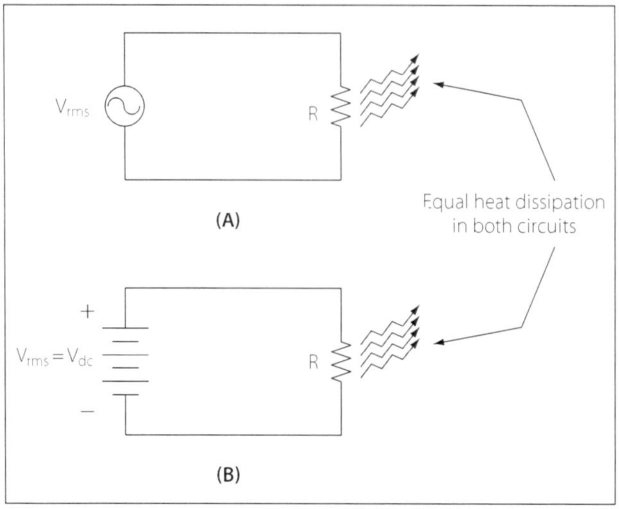

Figure 10-107. Sine wave effective value.

The effective value of a sine wave is actually a measure of the heating effect of the sine wave. Figure 10-107 illustrates what happens when a resistor is connected across an AC voltage source. In illustration A, a certain amount of heat is generated by the power in the resistor. Illustration B shows the same resistor now inserted into a DC voltage source. The value of the DC voltage source can now be adjusted so that the resistor dissipates the same amount of heat as it did when it was in the AC circuit. The RMS or effective value of a sine wave is equal to the DC voltage that produces the same amount of heat as the sinusoidal voltage.

The peak value of a sine wave can be converted to the corresponding RMS value using the following relationship.

$$Vrms = (\sqrt{0.5}) \times Vp$$

$$Vrms = 0.707 \times Vp$$

This can be applied to either voltage or current.

Algebraically rearranging the formula and solving for Vp can also determine the peak voltage. The resulting formula is:

$$Vp = 1.414 \times Vrms$$

Thus, the 110 volt value given for alternating current supplied to homes is only 0.707 of the maximum voltage of this supply. The maximum voltage is approximately 155 volts (110 × 1.41 = 155 volts maximum).

In the study of alternating current, any values given for current or voltage are assumed to be effective values unless otherwise specified, and in practice, only the effective values of voltage and current are used. Similarly, alternating current voltmeters and ammeters measure the effective value.

Capacitance

Another important property in AC circuits, besides resistance and inductance, is capacitance. While inductance is represented in a circuit by a coil, capacitance is represented by a capacitor. In its most basic form the capacitor is constructed of two parallel plates separated by a nonconductor, called a dielectric. In an electrical circuit, a capacitor serves as a reservoir or storehouse for electricity.

Capacitors in Direct Current

When a capacitor is connected across a source of direct current, such as a storage battery in the circuit shown in Figure 10-108A, and the switch is then closed, the plate marked B becomes positively charged, and the A plate negatively charged. Current flows in the external circuit during the time the electrons are moving from B to A. The current flow in the circuit is at a maximum the instant the switch is closed, but continually decreases thereafter until it reaches zero. The current becomes zero as soon as the difference in voltage of A and B becomes the same as the voltage of the battery. If the switch is opened as shown in Figure 10-108B, the plates remain charged. Once the capacitor is shorted, it will discharge quickly as shown Figure 10-108C.

It should be clear that during the time the capacitor is being charged or discharged, there is current in the circuit, even though the circuit is broken by the gap between the capacitor plates. Current is present only during the time of charge and discharge, and this period of time is usually short.

The RC Time Constant

The time required for a capacitor to attain a full charge is proportional to the capacitance and the resistance of the circuit. The resistance of the circuit introduces the element of time into the charging and discharging of a capacitor.

When a capacitior charges or discharges through a resistance, a certain amount of time is required for a full charge or discharge. The voltage across the capacitor will not change instantaneously. The rate of charging or discharging is determined by the time constant of the circuit. The time constant of a series RC (resistor/capacitor) circuit is a time interval that equals the product of the resistance in ohms and the capacitance in farad and is symbolized by the greek letter tau (τ).

$$\tau = RC$$

The time in the formula is that required to charge to 63% of the voltage of the source. The time required to bring the charge to about 99% of the source voltage is approximately 5 τ. Figure 10-109 illustrates this relationship of a time constant characteristics of charging.

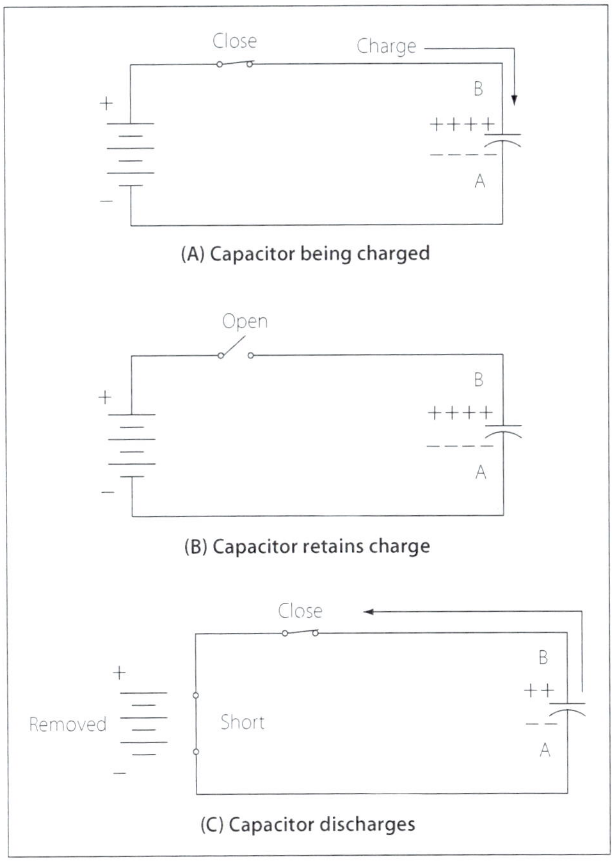

Figure 10-108. Capacitors in direct current.

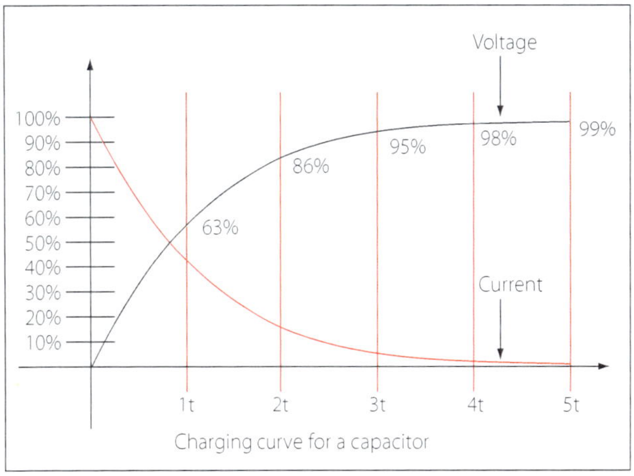

Figure 10-109. Capacitance discharge curve.

The measure of a capacitor's ability to store charge is its capacitance. The symbol used for capacitance is the letter C.

As can be seen from the time constant illustration there can be no continuous movement of direct current through a capacitor. A good capacitor will block direct current and will pass the effects of pulsing DC or alternating current.

Units of Capacitance

Electrical charge, which is symbolized by the letter Q, is measured in units of coulombs. The coulomb is given by the letter C, as with capacitance. Unfortunately this can be confusing. One coulomb of charge is defined as a charge having 6.28×10^{18} electrons. The basic unit of capacitance is the farad and is given by the letter f. By definition, one farad is one coulomb of charge stored with one volt across the plates of the capacitor. The general formula for capacitance in terms of charge and voltage is:

Where $C = \frac{Q}{E}$

C = Capacitance measured in farads.
E = Applied voltage measured in volts.
Q = Charge measured in coulombs.

In practical terms, one farad is a large amount of capacitance. Typically, in electronics, much smaller units are used. The two more common smaller units are the microfarad (μF), which is 10^{-6} farad and the picofarad (pF), which is 10^{-12} farad.

Voltage Rating of a Capacitor

Capacitors have their limits as to how much voltage can be applied across the plates. The aircraft technician must be aware of the voltage rating, which specifies the maximum DC voltage that can be applied without the risk of damage to the device. This voltage rating is typically called the breakdown voltage, the working voltage, or simply the voltage rating. If the voltage applied across the plates is too great, the dielectric will break down and arcing will occur between the plates. The capacitor is then short circuited, and the possible flow of direct current through it can cause damage to other parts of the equipment.

A capacitor that can be safely charged to 500 volts DC cannot be safely subjected to AC or pulsating DC whose effective values are 500 volts. An alternating voltage of 500 volts (RMS) has a peak voltage of 707 volts, and a capacitor to which it is applied should have a working voltage of at least 750 volts. The capacitor should be selected so that its working voltage is at least 50 percent greater than the highest voltage to be applied.

The voltage rating of the capacitor is a factor in determining the actual capacitance because capacitance decreases as the thickness of the dielectric increases. A high voltage capacitor that has a thick dielectric must have a larger plate area in order to have the same capacitance as a similar low voltage capacitor having a thin dielectric.

Factors Affecting Capacitance

1. The capacitance of parallel plates is directly proportional to their area. A larger plate area produces a larger capacitance and a smaller area produces less capacitance. If we double the area of the plates, there is room for twice as much charge. The charge that a capacitor can hold at a given potential difference is doubled, and since C = Q/E, the capacitance is doubled.
2. The capacitance of parallel plates is inversely proportional to their spacing.
3. The dielectric material affects the capacitance of parallel plates. The dielectric constant of a vacuum is defined as 1, and that of air is very close to 1. These values are used as a reference, and all other materials have values specified in relation to air (vacuum).

The strength of some commonly used dielectric materials is listed in Figure 10-110. The voltage rating also depends on frequency because the losses, and the resultant heating effect, increase as the frequency increases.

Dielectric	K	Dielectric Strength (volts per .001 inch)
Air	1.0	80
Paper		
(1) Paraffined	2.2	1,200
(2) Beeswaxed	3.1	1,800
Glass	4.2	200
Castor Oil	4.7	380
Bakelite	6.0	500
Mica	6.0	2,000
Fiber	6.5	50

Figure 10-110. Strength of some dielectric materials.

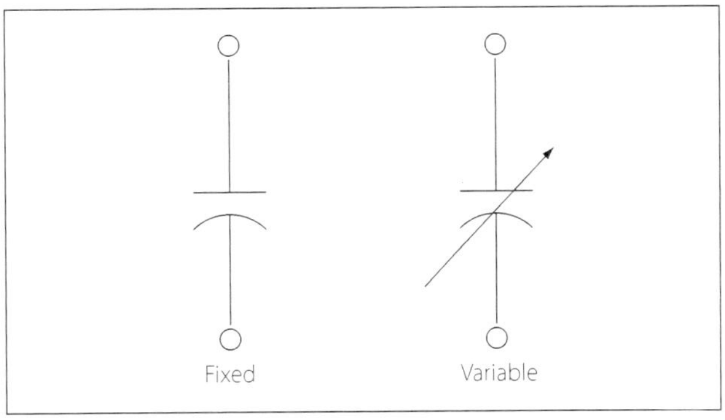

Figure 10-111. Schematic symbols for a fixed and variable capacitor.

Types of Capacitors

Capacitors come in all shapes and sizes and are usually marked with their value in farads. They may also be divided into two groups: fixed and variable. The fixed capacitors, which have approximately constant capacitance, may then be further divided according to the type of dielectric used. Some varieties are: paper, oil, mica, electrolytic and ceramic capacitors. Figure 10-111 shows the schematic symbols for a fixed and variable capacitor.

Fixed Capacitors

Mica Capacitors

The fixed mica capacitor is made of metal foil plates that are separated by sheets of mica, which form the dielectric. The whole assembly is covered in molded plastic, which keeps out moisture. Mica is an excellent dielectric and will withstand higher voltages than paper without allowing arcing between the plates. Common values of mica capacitors range from approximately 50 micromicrofarads, to about 0.02 microfarads.

Ceramic

The ceramic capacitor is constructed with materials, such as titanium acid barium for a dielectric. Internally these capacitors are not constructed as a coil, so they are well suited for use in high frequency applications. They are shaped like a disk, available in very small capacitance values and very small sizes. This type is fairly small, inexpensive, and reliable. Both the ceramic and the electrolytic are the most widely available and used capacitor.

Electrolytic

Two kinds of electrolytic capacitors are in use: (1) wet electrolytic and (2) dry electrolytic.

The wet electrolytic capacitor is designed of two metal plates separated by an electrolyte with an electrolyte dielectric, which is basically conductive salt in solvent. For capacitances greater than a few microfarads, the plate areas of paper or mica capacitors must become very large; thus, electrolytic capacitors are usually used instead. These units provide large capacitance in small physical sizes. Their values range from 1 to about 1,500 microfarads. Unlike the other types, electrolytic capacitors are generally polarized, with the positive lead marked with a "+" and the negative lead marked with a "–" and should only be subjected to direct voltage or pulsating direct voltage only.

The electrolyte in contact with the negative terminal, either in paste or liquid form, comprises the negative electrode. The dielectric is an exceedingly thin film of oxide deposited on the positive electrode of the capacitor. The positive electrode, which is an aluminum sheet, is folded to achieve maximum area. The capacitor is subjected to a forming process during manufacture, in which current is passed through it. The flow of current results in the deposit of the thin coating of oxide on the aluminum plate.

The close spacing of the negative and positive electrodes gives rise to the comparatively high capacitance value, but allows greater possibility of voltage breakdown and leakage of electrons from one electrode to the other.

The electrolyte of the dry electrolytic unit is a paste contained in a separator made of an absorbent material, such as gauze or paper. The separator not only holds the electrolyte in place but also prevents it from short circuiting the plates. Dry electrolytic capacitors are made in both cylindrical and rectangular block form and may be contained either within cardboard or metal covers. Since the electrolyte cannot spill, the dry capacitor may be mounted in any convenient position. Electrolytic capacitors are shown in Figure 10-112.

Tantalum

Similar to the electrolytic, these capacitors are constructed with a material called tantalum, which is used for the electrodes. They are superior to electrolytic capacitors, having better temperature and frequency characteristics. When tantalum powder is baked in order to solidify it, a crack forms inside. This crack is used to store an electrical charge. Like electrolytic capacitors, the tantalum capacitors are also polarized and are indicated with the "+" and "–" symbols.

Polyester Film

In this capacitor, a thin polyester film is used as a dielectric. These components are inexpensive, tem-

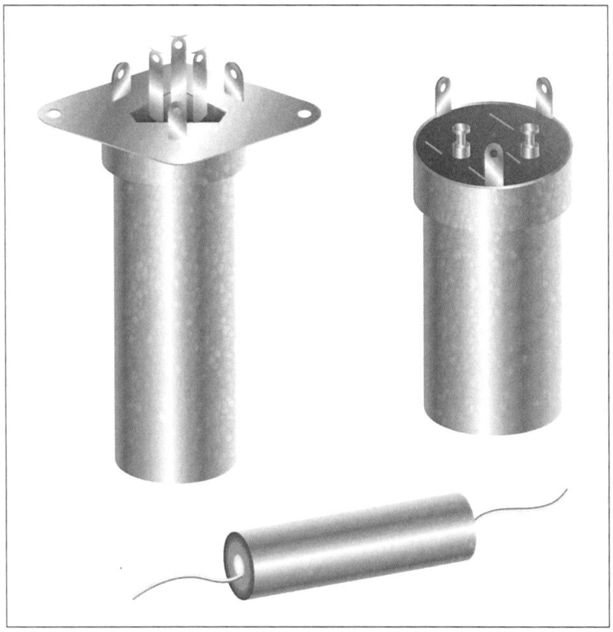

Figure 10-112. Electrolytic capacitors.

perature stable, and widely used. Tolerance is approximately 5 – 10 percent. It can be quite large depending on capacity or rated voltage.

Oil Capacitors

In radio and radar transmitters, voltages high enough to cause arcing, or breakdown, of paper dielectrics are often used. Consequently, in these applications capacitors that use oil or oil impregnated paper for the dielectric material are preferred. Capacitors of this type are considerably more expensive than ordinary paper capacitors, and their use is generally restricted to radio and radar transmitting equipment. [Figure 10-113]

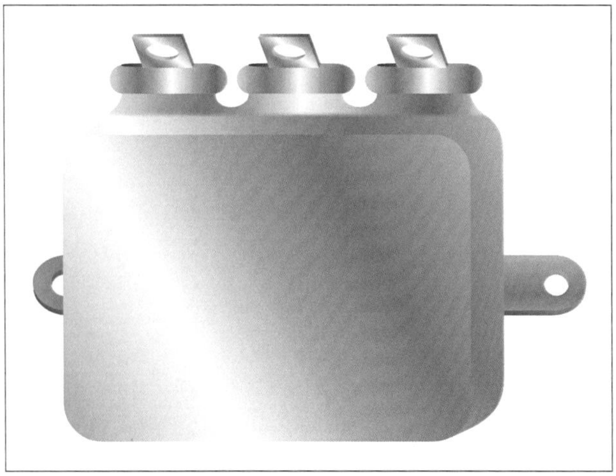

Figure 10-113. Oil capacitor.

Variable Capacitors

Variable capacitors are mostly used in radio tuning circuits, and they are sometimes called "tuning capacitors." They have very small capacitance values, typically between 100pF and 500pF.

Trimmers

The trimmer is actually an adjustable or variable capacitor, which uses ceramic or plastic as a dielectric. Most of them are color coded to easily recognize their tunable size. The ceramic type has the value printed on them. Colors are: yellow (5pF), blue (7pF), white (10pF), green (30pF), and brown (60pf).

Varactors

A voltage-variable capacitor or varactor is also known as a variable capacitance diode or a varicap. This device utilizes the variation of the barrier width in a reversed-biased diode. Because the barrier width of a diode acts as a non-conductor, a diode forms a capacitor when reversed biased. Essentially the N-type material becomes one plate and the junctions are the dielectric. If the reversed-bias voltage is increased, then the barrier width widens, effectively separating the two capacitor plates and reducing the capacitance.

Capacitors in Series

When capacitors are placed in series, the effective plate separation is increased and the total capacitance is less than that of the smallest capacitor. Additionally, the series combination is capable of withstanding a higher total potential difference than any of the individual capacitors. Figure 10-114 is a simple series circuit. The bottom plate of C_1 and the top plate of C_2 will be charged by electrostatic induction. The capacitors

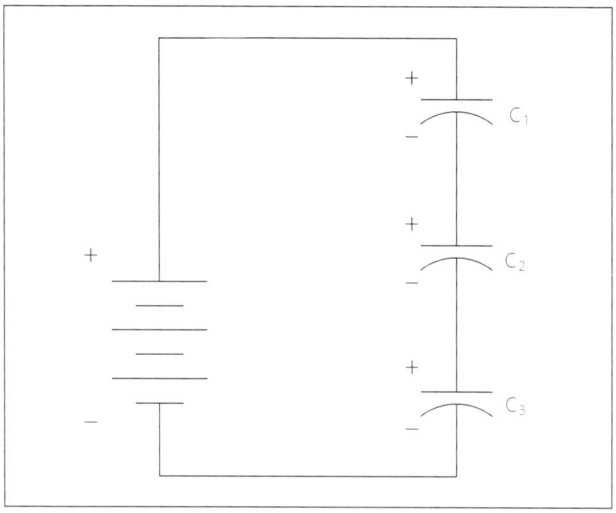

Figure 10-114. Simple series circuit.

charge as current is established through the circuit. Since this is a series circuit, the current must be the same at all points. Since the current is the rate of flow of charge, the amount of charge (Q) stored by each capacitor is equal to the total charge.

$$Q_T = Q_1 + Q_2 + Q_3$$

According to Kirchhoff's voltage law, the sum of the voltages across the charged capacitors must equal the total voltage, E_T. This is expressed as:

$$E_T = E_1 + E_2 + E_3$$

Equation E = Q/C can now be substituted into the voltage equation where we now get:

$$\frac{Q_T}{C_T} = \frac{Q_1}{C_1} + \frac{Q_2}{C_2} + \frac{Q_3}{C_3}$$

Since the charge on all capacitors is equal, the Q terms can be factored out, leaving us with the equation:

$$\frac{1}{C_T} = \frac{1}{C_1} + \frac{1}{C_2} + \frac{1}{C_3}$$

Consider the following example:

If $C_1 = 10\mu F$, $C_2 = 5\mu F$ and $C_3 = 8\mu F$

$$\text{Then } \frac{1}{C_T} = \frac{1}{10\mu F} + \frac{1}{5\mu F} + \frac{1}{8\mu F}$$

$$C_T = \frac{1}{0.425\mu F} = 2.35\mu F$$

Capacitors in Parallel

When capacitors are connected in parallel, the effective plate area increases, and the total capacitance is the sum of the individual capacitances. Figure 10-115 shows a simplified parallel circuit. The total charging current from the source divides at the junction of the parallel branches. There is a separate charging current through each branch so that a different charge can be stored by each capacitor. Using Kirchhoff's current law, the sum of all of the charging currents is then equal to the total current. The sum of the charges (Q) on the capacitors is equal to the total charge. The voltages (E) across all of the parallel branches are equal. With all of this in mind, a general equation for capacitors in parallel can be determined as:

$$Q_T = Q_1 + Q_2 + Q_3$$

Because Q = CE: $C_TE_T = C_1E_1 + C_2E_2 + C_3E_3$

Voltages can be factored out because:

$$E_T = E_1 + E_2 + E_3$$

Leaving us with the equation for capacitors in parallel:

$$C_T = C_1 + C_2 + C_3$$

Consider the following example:

If $C_1 = 330\mu F$, $C_2 = 220\mu F$

Then $C_T = 330\mu F + 220\mu F = 550\mu F$

Capacitors in Alternating Current

If a source of alternating current is substituted for the battery, the capacitor acts quite differently than it does with direct current. When an alternating current is applied in the circuit, the charge on the plates constantly changes. [Figure 10-116] This means that electricity must flow first from Y clockwise around to X, then from X counterclockwise around to Y, then from Y clockwise around to X, and so on. Although no current flows through the insulator between the plates of the capacitor, it constantly flows in the remainder of the circuit between X and Y. In a circuit in which there is only capacitance, current leads the applied voltage as contrasted with a circuit in which there is inductance, where the current lags the voltage.

Capacitive Reactance Xc

The effectiveness of a capacitor in allowing an AC flow to pass depends upon the capacitance of the circuit and the applied frequency. To what degree a

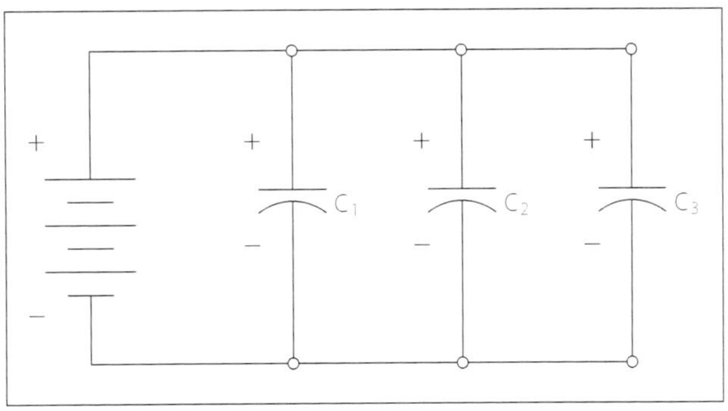

Figure 10-115. Simplified parallel circuit.

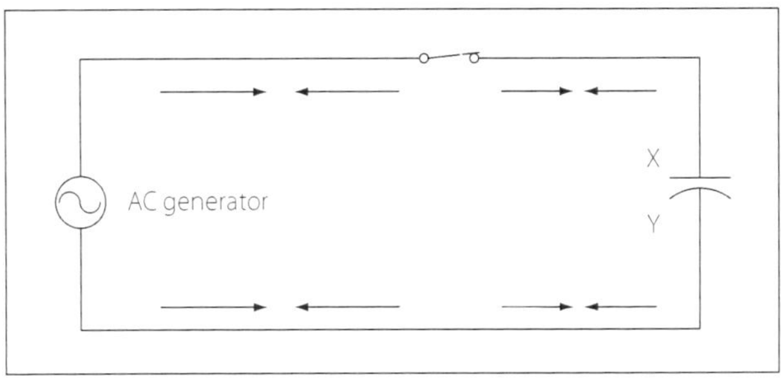

Figure 10-116. Capacitor in an AC circuit.

capacitor allows an AC flow to pass depends largely upon the capacitive value of the capacitor given in farads (f). The greater the capacitance of the capacitor, the greater the number of electrons, measured in Coulombs, necessary to bring the capacitor to a fully charged state. Once the capacitor approaches or actually reaches a fully charged condition, the polarity of the capacitor will oppose the polarity of the applied voltage, essentially acting then as an open circuit. To further illustrate this characteristic and how it manifests itself in an AC circuit, consider the following. If a capacitor has a large capacitive value, meaning that it requires a relatively large number of electrons to bring it to a fully charged state, then a rather high frequency current can alternate through the capacitor without the capacitor ever reaching a full charge. In this case, if the frequency is high enough and the capacitance large enough that there is never enough time for the capacitor to ever reach a full charge, it is possible that the capacitor may offer very little or no resistance to the current. However, the smaller the capacitance, the fewer electrons are required to bring it up to a full charge and it is more likely that the capacitor will build up enough of an opposing charge that it can present a great deal of resistance to the current if not to the point of behaving like an open circuit. In between these two extreme conditions lies a continuum of possibilities of current opposition depending on the combination of alpplied frequency and the selected capacitance. Current in an AC circuit can be controlled by changing the circuit capacitance in a similar manner that resistance can control the current. The actual AC reactance Xc, which just like resistance, is measured in ohms (Ω). Capacitive reactance Xc is determined by the following:

$$Xc = \frac{1}{2\pi fC}$$

Where Xc = Capacitive Reactance
f = frequency in cps
C = capacity in farads
$2\pi = 6.28$

Sample Problem:
A series circuit is assumed in which the impressed voltage is 110 volts at 60 cps, and the capacitance of a condenser is 80 Mf. Find the capacitive reactance and the current flow.

Solution:
To find capacitive reactance, the equation Xc = 1/(2 p f C) is used. First, the capacitance, 80 Mf, is changed to farads by dividing 80 by 1,000,000, since 1 million microfarads is equal to 1 farad. This quotient equals 0.000080 farad. This is substituted in the equation and

$$Xc = \frac{1}{6.28 \times 60 \times 0.000080}$$

$$Xc = 33.2 \text{ ohms reactance}$$

Once the reactance has been determined, ohm's law can then be used in the same manner as it is used in DC circuits to determine the current.

$$\text{Current} = \frac{\text{Voltage}}{\text{Capacitive reactance}}, \text{ or}$$

$$I = \frac{E}{Xc},$$

Find the current flow:

$$I = \frac{E}{Xc}$$

$$I = \frac{110}{33.2}$$

$$I = 3.31 \text{ amperes}$$

Capacitive Reactances in Series and in Parallel

When capacitors are connected in series, the total reactance is equal to the sum of the individual reactances. Thus,

$$Xct = (Xc)_1 + (Xc)_2$$

The total reactance of capacitors connected in parallel is found in the same way total resistance is computed in a parallel circuit:

$$(Xc)t = \frac{1}{\frac{1}{(Xc)_1} + \frac{1}{(Xc)_2} + \frac{1}{(Xc)_3}}$$

Phase of Current and Voltage in Reactive Circuits

Unlike a purely resistive circuit, the capacitive and inductive reactance has a significant effect on the phase relationship between the applied AC voltage and the corresponding current in the circuit.

In review, when current and voltage pass through zero and reach maximum value at the same time, the current and voltage are said to be in phase. [Figure 10-117A] If the current and voltage pass through zero and reach the maximum values at different times, the current and voltage are said to be out of phase. In a circuit containing only inductance, the current reaches a maximum value later than the voltage, lagging the voltage by 90°, or one-fourth cycle. [Figure 10-117B]

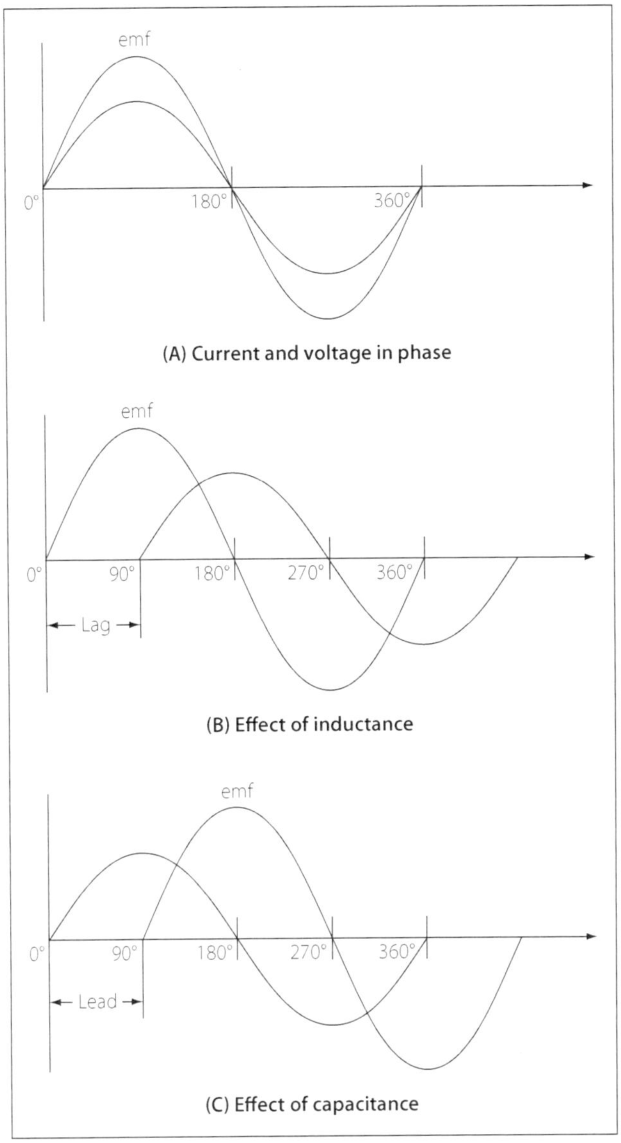

Figure 10-117. Phase of current and voltage.

In a circuit containing only capacitance, the current reaches its maximum value ahead of the voltage and the current leads the voltage by 90°, or one-fourth cycle. [Figure 10-117C] The amount the current lags or leads the voltage in a circuit depends on the relative amounts of resistance, inductance, and capacitance in the circuit.

Inductance

Characteristics of Inductance

Michael Faraday discovered that by moving a magnet through a coil of wire, a voltage was induced across the coil. If a complete circuit was provided, then a current was also induced. The amount of induced voltage is directly proportional to the rate of change of the magnetic field with respect to the coil. The simplest of experiments can prove that when a bar magnet is moved through a coil of wire, a voltage is induced and can be measured on a voltmeter. This is commonly known as Faraday's Law or the law of electromagnetic induction, which states:

The induced emf or electromagnetic force in a closed loop of wire is proportional to the rate of change of the magnetic flux through a coil of wire.

Conversely, current flowing through a coil of wire produces a magnetic field. When this wire is formed into a coil, it then becomes a basic inductor. The magnetic lines of force around each loop or turn in the coil effectively add to the lines of force around the adjoining loops. This forms a strong magnetic field within and around the coil. Figure 10-118A, illustrates this idea of a coil of wire strengthening a magnetic field. The magnetic lines of force around adjacent loops are deflected into an outer path when the loops are brought close together. This happens because the magnetic lines of force between adjacent loops are in opposition with each other. The total magnetic field for the two loops

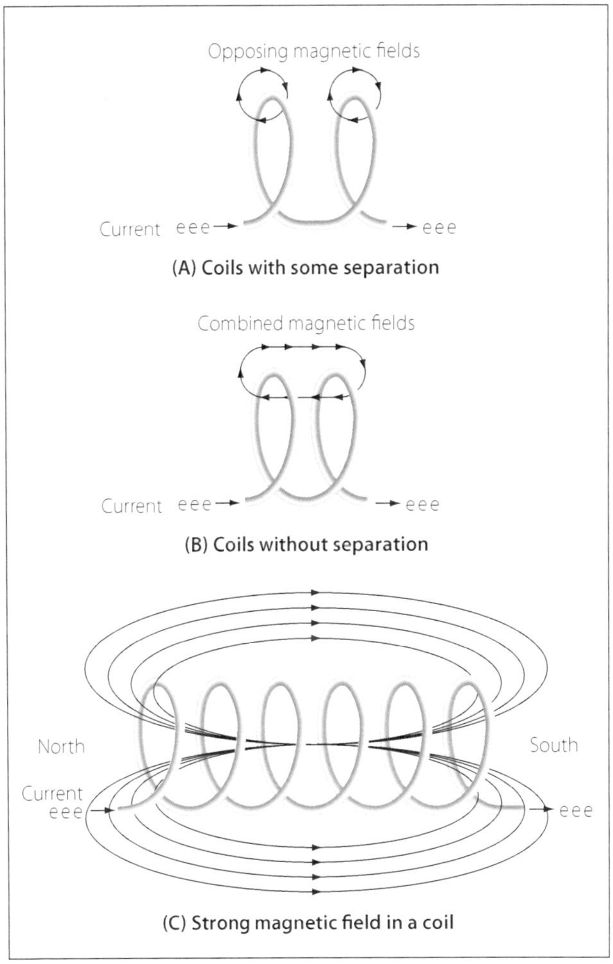

Figure 10-118. Many loops of a coil.

is shown in Figure 10-118B. As more loops are added close together, the strength of the magnetic field will increase. Figure 10-118C illustrates the combined effects of many loops of a coil. The result is a strong electromagnet.

The primary aspect of the operation of a coil is its property to oppose any change in current through it. This property is called inductance. When current flows through any conductor, a magnetic field starts to expand from the center of the wire. As the lines of magnetic force grow outward through the conductor, they induce an emf in the conductor itself. The induced voltage is always in the direction opposite to the direction of the current flow. The effects of this countering emf are to oppose the immediate establishment of the maximum current. This effect is only a temporary condition. Once the current reaches a steady value in the conductor, the lines of magnetic force will no longer be expanding and the countering emf will no longer be present.

At the starting instant, the countering emf nearly equals the applied voltage, resulting in a small current flow. However, as the lines of force move outward, the number of lines cutting the conductor per second becomes progressively smaller, resulting in a diminished counter emf. Eventually, the counter emf drops to zero and the only voltage in the circuit is the applied voltage and the current is at its maximum value.

The RL Time Constant

Because the inductors basic action is to oppose a change in its current, it then follows that the current cannot change instantaneously in the inductor. A certain time is required for the current to make a change from one value to another. The rate at which the current changes is determined by a time constant represented by the greek letter tau (τ). The time constant for the RL circuit is:

$$\tau = \frac{L}{R}$$

Where τ = seconds
L = inductance (H)
R = Resistance (Ω)

In a series RL circuit, the current will increase to 63% of its full value in 1 time constant after the circuit is closed. This build up of course is similar to the build up of voltage in a capacitor when charging an RC circuit. Both follow an exponential curve and reach 99% value after the 5th time constant. Figure 10-119 illustrates this characteristic.

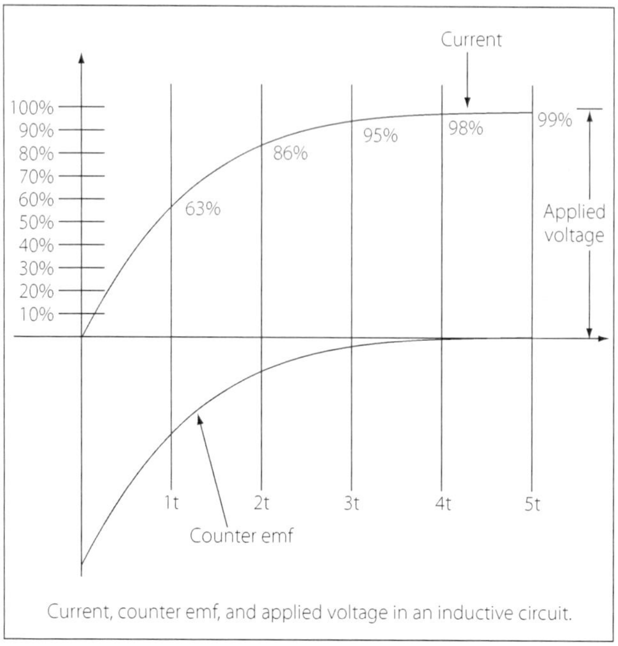

Figure 10-119. Inductor curve.

Physical Parameters

Some of the physical factors that affect inductance are:

1. The number of turns: Doubling the number of turns in a coil will produce a field twice as strong, if the same current is used. As a general rule, the inductance varies as the square of the number of turns.
2. The cross-sectional area of the coil: The inductance of a coil increases directly as the cross-sectional area of the core increases. Doubling the radius of a coil increases the inductance by a factor of four.
3. The length of a coil: Doubling the length of a coil, while keeping the same number of turns, halves the value of inductance.
4. The core material around which the coil is formed: Coils are wound on either magnetic or nonmagnetic materials. Some nonmagnetic materials include air, copper, plastic, and glass. Magnetic materials include nickel, iron, steel, or cobalt, which have a permeability that provides a better path for the magnetic lines of force and permit a stronger magnetic field.

Self-Inductance

The characteristic of self-inductance was summarized by German physicist Heinrich Lenz in 1833 and gives the direction of the induced electromotive force (emf) resulting from electromagnetic induction. This is commonly known as Lenz's Law, which states:

The emf induced in an electric circuit always acts in such a direction that the current it drives around a closed circuit produces a magnetic field which opposes the change in magnetic flux.

Self inductance is the generation of a voltage in an electric circuit by a changing current in the same circuit. Even a straight piece of wire will have some degree of inductance because current in a conductor produces a magnetic field. When the current in a conductor changes direction, there will be a corresponding change in the polarity of the magnetic field around the conductor. Therefore, a changing current produces a changing magnetic field around the wire. To further intensify the magnetic field, the wire can be rolled into a coil, which is called an inductor. The changing magnetic field around the inductor induces a voltage across the coil. This induced electromotive force is called self-inductance and tends to oppose any change in current within the circuit. This property is usually called inductance and symbolized with the letter L.

Types of Inductors

Inductors used in radio can range from a straight wire at UHF to large chokes and transformers used for filtering the ripple from the output of power supplies and in audio amplifiers. Figure 10-69 shows the schematic symbols for common inductors. Values of inductors range from nano-henries to tens of henries.

Inductors are classified by the type of core and the method of winding them. The number of turns in the inductor winding and the core material determine the capacity of the inductor. Cores made of dielectric material like ceramics, wood, paper provide small amounts of stored energy while cores made of ferrite substances have a much higher degree of stored energy. The core material is usually the most important aspect of the inductors construction. The conductors typically used in the construction of an inductor offer little resistance to the flow of current. However, with the introduction of a core, resistance is introduced in the circuit and the current now builds up in the windings until the resistance of the core is overcome. This buildup is stored as magnetic energy in the core. Depending on the core resistance, the buildup soon reaches a point of magnetic saturation and it can be released when necessary. The most common core materials are: Air, solid ferrite, powdered ferrite, steel, toroid and ferrite toroid.

Units of Inductance

The henry is the basic unit of inductance and is symbolized with the letter H. An electric circuit has an inductance of one henry when current changing at the rate of one ampere per second induces a voltage of one volt into the circuit. In many practical applications, millihenries (mH) and microhenries (μH) are more common units. The typical symbol for an inductor is shown in Figure 10-120.

Inductors in Series

If we connect two inductors in series as shown in Figure 10-121, the same current flows through both inductors and, therefore, both will be subject to the same rate of change of current. When inductors are connected in series, the total inductance L_T, is the sum of the individual inductors. The general equation for n number of inductors in series is:

$$L_T = L_1 + L_2 + L_3 + \ldots L_N$$

Inductors in Parallel

When two inductors are connected in parallel as shown in Figure 10-122, each must have the same potential difference between the terminals. When inductors are connected in parallel, the total inductance is less than

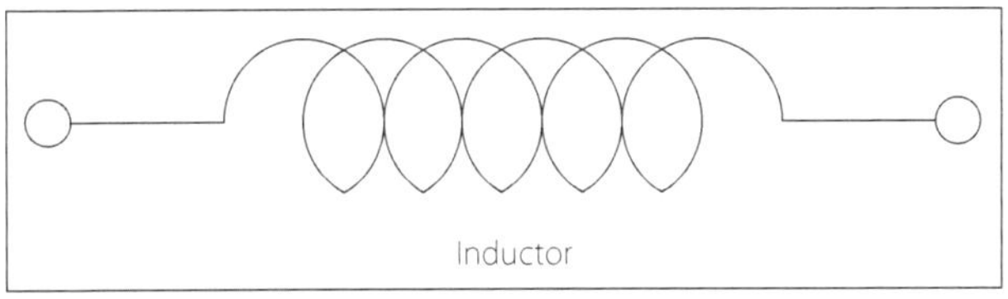

Figure 10-120. Typical symbol for an inductor.

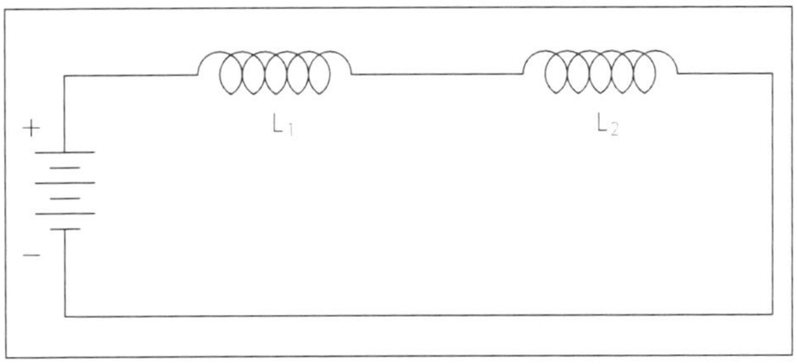

Figure 10-121. Two inductors in series.

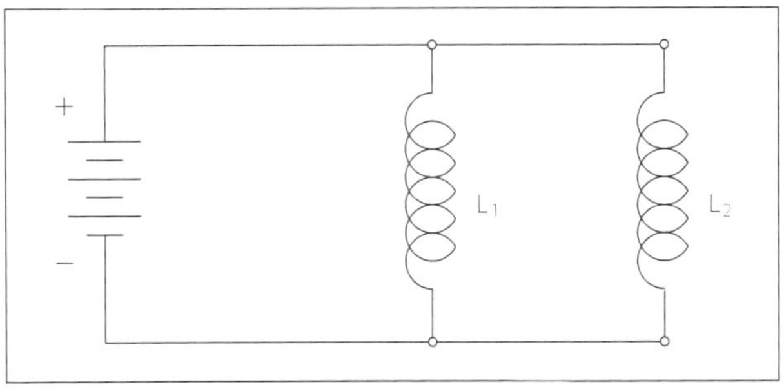

Figure 10-122. Two inductors in parallel.

the smallest inductance. The general equation for n number of inductors in parallel is:

$$L_T = \frac{1}{\frac{1}{L_1} + \frac{1}{L_2} + \frac{1}{L_3} + \dots \frac{1}{L_N}}$$

A simple example would be:

$L_1 = 10mH, L_2 = 5mH, L_3 = 2mH$

$$L_T = \frac{1}{\frac{1}{10mH} + \frac{1}{5mH} + \frac{1}{2mH}}$$

$$L_T = \frac{1}{0.8mH}$$

$$L_T = 1.25mH$$

Inductive Reactance

Alternating current is in a constant state of change; the effects of the magnetic fields are a continuously inducted voltage opposition to the current in the circuit. This opposition is called inductive reactance, symbolized by X_L, and is measured in ohms just as resistance is measured. Inductance is the property of a circuit to oppose any change in current and is measured in henries. Inductive reactance is a measure of how much the countering emf in the circuit will oppose current variations.

The inductive reactance of a component is directly proportional to the inductance of the component and the applied frequency to the circuit. By increasing either the inductance or applied frequency, the inductive reactance will likewise increase and present more opposition to current in the circuit. This relationship is given as:

$X_L = 2\pi fL$

Where: X_L = inductive reactance in ohms
f = frequency in cycles per second
$\pi = 3.1416$

In Figure 10-123, an AC series circuit is shown in which the inductance is 0.146 henry and the voltage is 110 volts at a frequency of 60 cycles per second. Inductive reactance is determined by the following method.

$X_L = 2\pi \times f \times L$

$X_L = 6.28 \times 60 \times 0.146$

To find current:

In any circuit where there is only resistance, the expression for the relationship of voltage and current is given by Ohm's law: I = E/R. Similarly, when there is inductance in an AC circuit, the relationship between voltage and current can be expressed as:

$$\text{Current} = \frac{\text{Voltage}}{\text{Reactance}} \text{ or } I = \frac{E}{X_L}$$

Where:
X_L = inductive reactance of the circuit in ohms.

$$I = \frac{E}{X_L}$$

$$I = \frac{110}{55}$$

I = 2 amperes

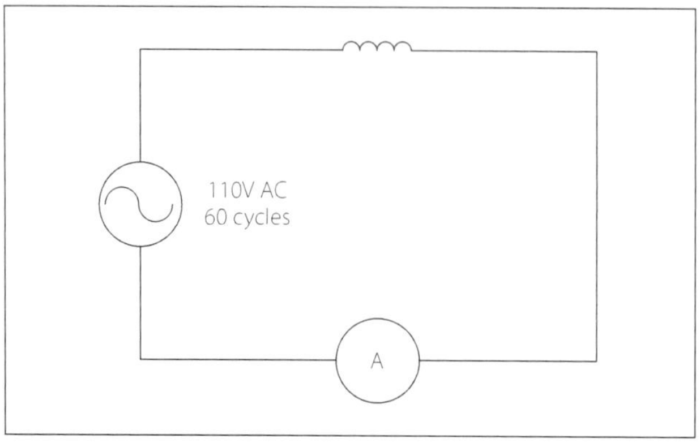

Figure 10-123. AC circuit containing inductance.

In AC series circuits, inductive reactances are added like resistances in series in a DC circuit. [Figure 10-124] Thus, the total reactance in the illustrated circuit equals the sum of the individual reactances.

The total reactance of inductors connected in parallel is found the same way as the total resistance in a parallel circuit. [Figure 10-125] Thus, the total reactance of inductances connected in parallel, as shown, is expressed as

$$(X_L)T = \frac{1}{\frac{1}{(X_L)_1} + \frac{1}{(X_L)_2} + \frac{1}{(X_L)_3}}$$

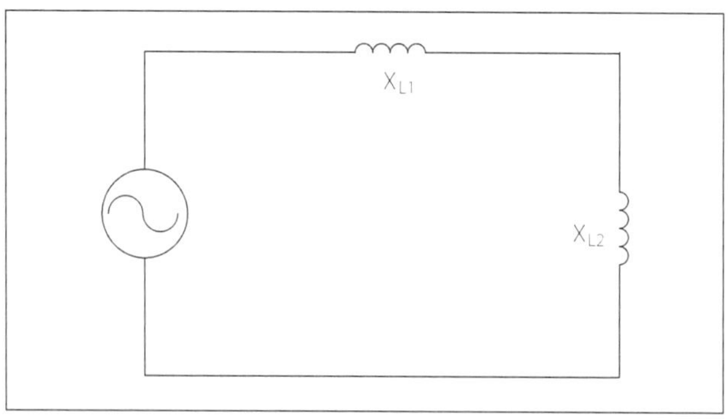

Figure 10-124. Inductances in series.

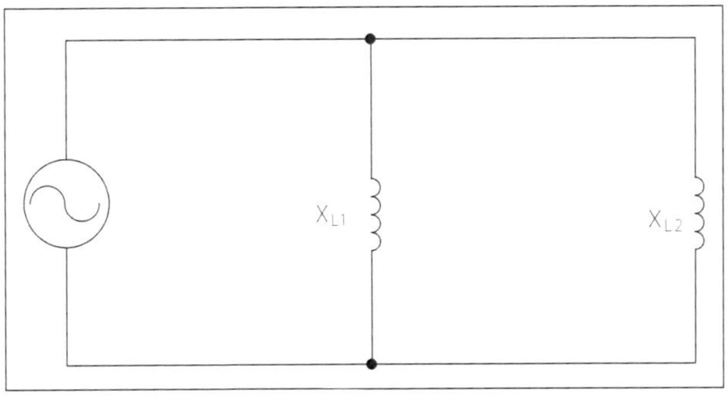

Figure 10-125. Inductances in parallel.

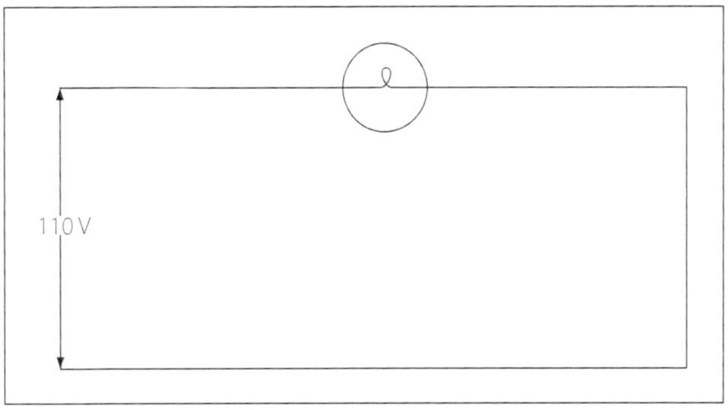

Figure 10-126. Applying DC and AC to a circuit.

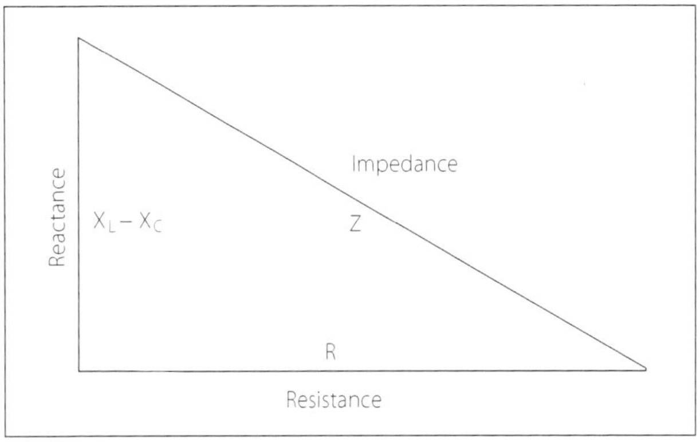

Figure 10-127. Impedance triangle.

AC Circuits

Ohm's Law for AC Circuits

The rules and equations for DC circuits apply to AC circuits only when the circuits contain resistance alone, as in the case of lamps and heating elements. In order to use effective values of voltage and current in AC circuits, the effect of inductance and capacitance with resistance must be considered.

The combined effects of resistance, inductive reactance, and capacitive reactance make up the total opposition to current flow in an AC circuit. This total opposition is called impedance and is represented by the letter Z. The unit for the measurement of impedance is the ohm.

Series AC Circuits

If an AC circuit consists of resistance only, the value of the impedance is the same as the resistance, and Ohm's law for an AC circuit, $I = E/Z$, is exactly the same as for a DC circuit. In Figure 10-126 a series circuit containing a lamp with 11 ohms resistance connected across a source is illustrated. To find how much current will flow if 110 volts DC is applied and how much current will flow if 110 volts AC are applied, the following examples are solved:

$$I = \frac{E}{R} \qquad I = \frac{E}{Z} \text{ (where } Z = R\text{)}$$

$$I = \frac{110V}{11W} \qquad I = \frac{110V}{11W}$$

$$I = 10 \text{ amperes DC} \qquad I = 10 \text{ amperes AC}$$

When AC circuits contain resistance and either inductance or capacitance, the impedance, Z, is not the same as the resistance, R. The impedance of a circuit is the circuit's total opposition to the flow of current. In an AC circuit, this opposition consists of resistance and reactance, either inductive or capacitive or elements of both.

Resistance and reactance cannot be added directly, but they can be considered as two forces acting at right angles to each other. Thus, the relation between resistance, reactance, and impedance may be illustrated by a right triangle. [Figure 10-127]

Since these quantities may be related to the sides of a right triangle, the formula for finding the impedance, or total opposition to current flow in an AC circuit, can be found by using the law of right triangles. This theorem, called the Pythagorean theorem, applies to any right triangle. It states that the square of the hypotenuse is equal to the sum of the squares of the other two sides. Thus, the value of any side of a right triangle can be found if the other two sides are known. If an AC circuit contains resistance and inductance, as shown in Figure 10-128, the relation between the sides can be stated as:

$$Z^2 = R^2 + X_L^2$$

The square root of both sides of the equation gives

$$Z = \sqrt{R^2 + X_L^2}$$

This formula can be used to determine the impedance when the values of inductive reactance and resistance are known. It can be modified to solve for impedance in

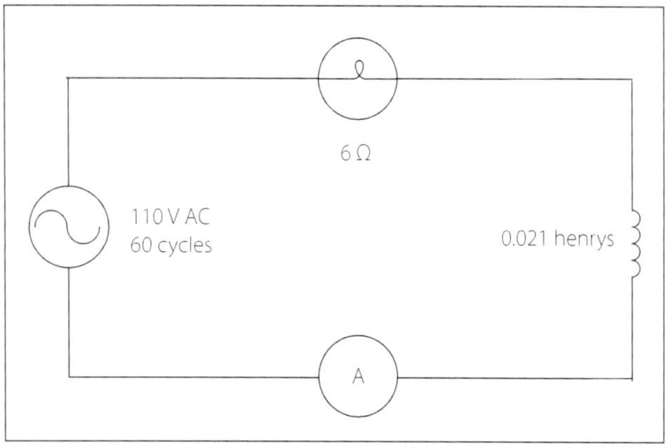

Figure 10-128. A circuit containing resistance and inductance.

circuits containing capacitive reactance and resistance by substituting X_C in the formula in place of X_L. In circuits containing resistance with both inductive and capacitive reactance, the reactances can be combined, but because their effects in the circuit are exactly opposite, they are combined by subtraction:

$X = X_L - X_C$ or $X = X_C - X_L$ (the smaller number is always subtracted from the larger).

In Figure 10-128, a series circuit consisting of resistance and inductance connected in series is connected to a source of 110 volts at 60 cycles per second. The resistive element is a lamp with 6 ohms resistance, and the inductive element is a coil with an inductance of 0.021 henry. What is the value of the impedance and the current through the lamp and the coil?

Solution:
First, the inductive reactance of the coil is computed:

$$X_L = 2\pi \times f \times L$$
$$X_L = 6.28 \times 60 \times 0.021$$
$$X_L = 8 \text{ ohms inductive reactance}$$

Next, the total impedance is computed:

$$Z = \sqrt{R^2 + X_L{}^2}$$
$$Z = \sqrt{6^2 + 8^2}$$
$$Z = \sqrt{36 + 64}$$
$$Z = \sqrt{100}$$
$$Z = 10 \text{ ohms impedance}$$

Then the current flow,

$$I = \frac{E}{Z}$$
$$I = \frac{110}{10}$$
$$I = 11 \text{ amperes current}$$

The voltage drop across the resistance (E^R) is

$$E^R = I \times R$$
$$E^R = 11 \times 6 = 66 \text{ volts}$$

The voltage drop across the inductance (E_{X_L}) is

$$E_{X_L} = I \times X_L$$
$$E_{X_L} = 11 \times 8 = 88 \text{ volts}$$

The sum of the two voltages is greater than the impressed voltage. This results from the fact that the two voltages are out of phase and, as such, represent the maximum voltage. If the voltage in the circuit is measured by a voltmeter, it will be approximately 110 volts, the impressed voltage. This can be proved by the equation,

$$E = \sqrt{(E_R)^2 + (E_{X_L})^2}$$
$$E = \sqrt{66^2 + 88^2}$$
$$E = \sqrt{4356 + 7744}$$
$$E = \sqrt{12100}$$
$$E = 110 \text{ volts}$$

In Figure 10-129, a series circuit is illustrated in which a capacitor of 200 µf is connected in series with a 10 ohm lamp. What is the value of the impedance, the current flow, and the voltage drop across the lamp?

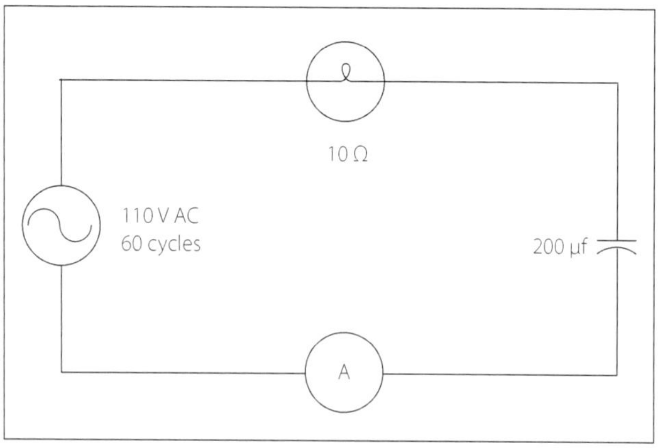

Figure 10-129. A circuit containing resistance and capacitance.

Solution:
First the capacitance is changed from microfarads to farads. Since 1 million microfarads equal 1 farad, then

$$200 \ \mu f. = \frac{200}{1,000,000} = 0.000200 \text{ farads}$$

$$X_C = \frac{1}{2 \pi f C}$$

$$X_C = \frac{1}{6.28 \times 60 \times 0.00200 \text{ farads}}$$

$$X_C = \frac{1}{0.07536}$$

X_C = 13 ohms capacitive reactance

To find the impedance,

$$Z = \sqrt{R^2 + X_C{}^2}$$
$$Z = \sqrt{10^2 + 13^2}$$
$$Z = \sqrt{100 + 169}$$
$$Z = \sqrt{269}$$
Z = 16.4 ohms capacitive reactance

To find the current,

$$I = \frac{E}{Z}$$
$$I = \frac{110}{16.4}$$
I = 6.7 amperes

The voltage drop across the lamp (E^R) is

$$E^R = 6.7 \times 10$$
$$E^R = 67 \text{ volts}$$

The voltage drop across the capacitor (E_{X_C}) is

$$E_{X_C} = I \times X_C$$
$$E_{X_C} = 6.7 \times 13$$
$$E_{X_C} = 86.1 \text{ volts}$$

The sum of these two voltages does not equal the applied voltage, since the current leads the voltage. To find the applied voltage,

the formula $E_T = \sqrt{(E_R)^2 + (E_{X_C})^2}$ is used.

$$E_T = \sqrt{67^2 + 86.1^2}$$
$$E_T = \sqrt{4489 + 7413}$$
$$E_T = \sqrt{11902}$$
$$E_T = 110 \text{ volts}$$

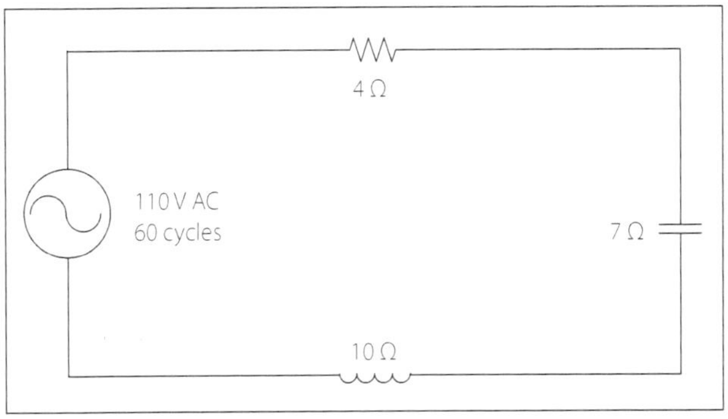

Figure 10-130. A circuit containing resistance, inductance, and capacitance.

When the circuit contains resistance, inductance, and capacitance, the equation

$$Z = \sqrt{R^2 + (X_L - X_C)^2}$$

is used to find the impedance.

Example: What is the impedance of a series circuit, consisting of a capacitor with a reactance of 7 ohms, an inductor with a reactance of 10 ohms, and a resistor with a resistance of 4 ohms? [Figure 10-130]

Solution:

$$Z = \sqrt{R^2 + (X_L - X_C)^2}$$
$$Z = \sqrt{4^2 + (10 - 7)^2}$$
$$Z = \sqrt{4^2 + 3^2}$$
$$Z = \sqrt{25}$$
Z = 5 ohms

Assuming that the reactance of the capacitor is 10 ohms and the reactance of the inductor is 7 ohms, then X_C is greater than X_L. Thus,

$$Z = \sqrt{R^2 + (X_L - X_C)^2}$$
$$Z = \sqrt{4^2 + (7 - 10)^2}$$
$$Z = \sqrt{4^2 + (-3)^2}$$
$$Z = \sqrt{16 + 9}$$
$$Z = \sqrt{25}$$
Z = 5 ohms

Parallel AC Circuits

The methods used in solving parallel AC circuit problems are basically the same as those used for series AC circuits. Out of phase voltages and currents can be added by using the law of right triangles. However, in solving circuit problems, the currents through the branches are added since the voltage drops across the various branches are the same and are equal to

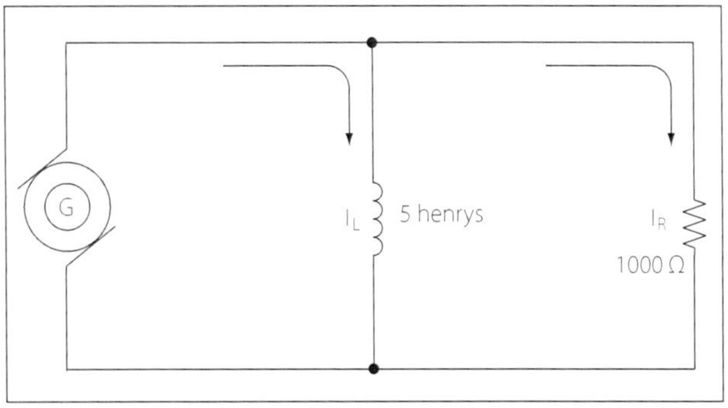

Figure 10-131. AC parallel circuit containing inductance and resistance.

the applied voltage. In Figure 10-131, a parallel AC circuit containing an inductance and a resistance is shown schematically. The current flowing through the inductance, I_L, is 0.0584 ampere, and the current flowing through the resistance is 0.11 ampere. What is the total current in the circuit?

Solution:

$$I_T = \sqrt{I_L^2 + I_R^2}$$
$$= \sqrt{(0.0584)^2 + (0.11)^2}$$
$$= \sqrt{0.0155}$$
$$= 0.1245 \text{ ampere}$$

Since inductive reactance causes voltage to lead the current, the total current, which contains a component of inductive current, lags the applied voltage. If the current and voltages are plotted, the angle between the two, called the phase angle, illustrates the amount the current lags the voltage.

In Figure 10-132, a 110-volt generator is connected to a load consisting of a 2 μf capacitance and a 10,000-ohm resistance in parallel. What is the value of the impedance and total current flow?

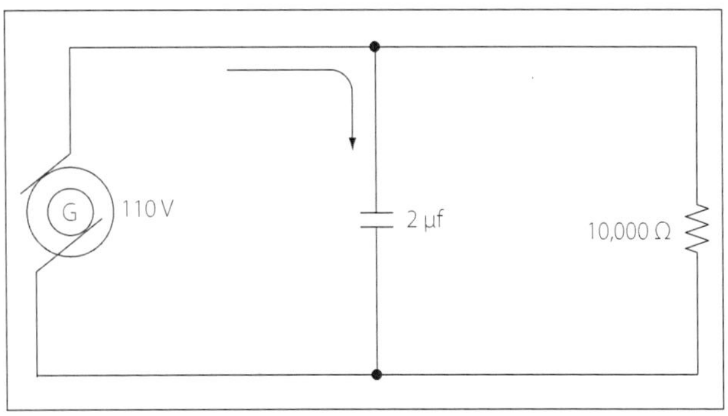

Figure 10-132. A parallel AC circuit containing capacitance and resistance.

Solution:

First, find the capacitive reactance of the circuit:

$$X_C = \frac{1}{2\,\pi\,f\,C}$$

Changing 2 μf to farads and entering the values into the formula given:

$$= \frac{1}{2 \times 3.14 \times 60 \times 0.000002}$$
$$= \frac{1}{0.00075360} \text{ or } \frac{10{,}000}{7.536}$$
$$= 1{,}327\ X_C \text{ capacitive reactance.}$$

To find the impedance, the impedance formula used in a series AC circuit must be modified to fit the parallel circuit:

$$Z = \frac{R_{X_C}}{\sqrt{R^2 + X_C^2}}$$
$$= \frac{10{,}000 \times 1327}{\sqrt{(10{,}000)^2 + (1327)^2}}$$
$$= 0.1315 \text{ W (approx.)}$$

To find the current through the capacitance:

$$I_C = \frac{E}{X_C}$$
$$I_C = \frac{110}{1327}$$
$$= 0.0829 \text{ ampere}$$

To find the current flowing through the resistance:

$$I_R = \frac{E}{R}$$
$$= \frac{110}{10{,}000}$$
$$= 0.0829 \text{ ampere}$$

To find the current flowing through the resistance:

$$I_R = \frac{E}{R}$$
$$= \frac{110}{10{,}000}$$
$$= 0.011 \text{ ampere}$$

To find the total current in the circuit:

$$I_T^2 = \sqrt{I_R^2 + I_C^2}$$
$$I_T = \sqrt{I_L^2 + I_R^2}$$
$$= 0.0836 \text{ ampere (approx.)}$$

Resonance

It has been shown that both inductive reactance ($X_L = 2 \pi f L$) and capacitive reactance:

$$X_C = \frac{1}{2 \pi f C}$$

...are functions of an alternating current frequency. Decreasing the frequency decreases the ohmic value of the inductive reactance, but a decrease in frequency increases the capacitive reactance. At some particular frequency, known as the resonant frequency, the reactive effects of a capacitor and an inductor will be equal. Since these effects are the opposite of one another, they will cancel, leaving only the ohmic value of the resistance to oppose current flow in a circuit. If the value of resistance is small or consists only of the resistance in the conductors, the value of current flow can become very high.

In a circuit where the inductor and capacitor are in series, and the frequency is the resonant frequency, or frequency of resonance, the circuit is said to be "in resonance" and is referred to as a series resonant circuit. The symbol for resonant frequency is Fn.

If, at the frequency of resonance, the inductive reactance is equal to the capacitive reactance, then:

$$X_L = Xc, \text{ or}$$

$$2 \pi f L = \frac{1}{2 \pi f C}$$

Dividing both sides by 2 fL,

$$Fn2 = \frac{1}{(2 \pi) 2 LC}$$

Extracting the square root of both sides gives:

$$Fn = \frac{1}{2 \pi \sqrt{LC}}$$

Where Fn is the resonant frequency in cycles per second, C is the capacitance in farads, and L is the inductance in henries. With this formula, the frequency at which a capacitor and inductor will be resonant can be determined.

To find the inductive reactance of a circuit use:

$$X_L = 2 (\pi) f l$$

The impedance formula used in a series AC circuit must be modified to fit a parallel circuit.

$$Z = \frac{X_L}{\sqrt{R^2 = X_L^2}}$$

To find the parallel networks of inductance and capacitive reactors, use:

$$X = \frac{X_L + X_C}{\sqrt{X_L + X_C}}$$

To find the parallel networks with resistance capacitive and inductance, use:

$$Z = \frac{R X_L X_C}{\sqrt{X_L^2 X_C^2 + (RX_L - RX_C)^2}}$$

Since at the resonant frequency X_L cancels X_C, the current can become very large, depending on the amount of resistance. In such cases, the voltage drop across the inductor or capacitor will often be higher than the applied voltage.

In a parallel resonant circuit, the reactances are equal and equal currents will flow through the coil and the capacitor. [Figure 10-133]

Since the inductive reactance causes the current through the coil to lag the voltage by 90°, and the capacitive reactance causes the current through the capacitor to lead the voltage by 90°, the two currents are 180° out of phase. The canceling effect of such currents would mean that no current would flow from the generator and the parallel combination of the inductor, and the capacitor would appear as infinite impedance. In practice, no such circuit is possible, since some value of resistance is always present, and the parallel circuit, sometimes called a tank circuit, acts as very high impedance. It is also called an antiresonant circuit, since its effect in a circuit is opposite to that of a series resonant circuit, in which the impedance is very low.

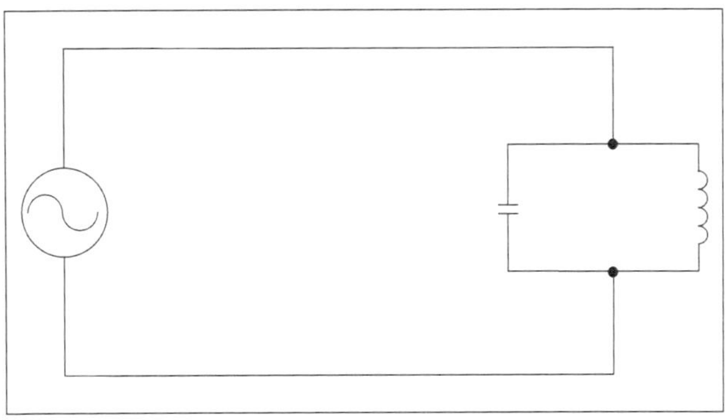

Figure 10-133. A parallel resonant circuit.

Power in AC Circuits

In a DC circuit, power is obtained by the equation, P = EI, (watts equal volts times amperes). Thus, if 1 ampere of current flows in a circuit at a pressure of 200 volts, the power is 200 watts. The product of the volts and the amperes is the true power in the circuit.

True Power Defined

The power dissipated in the resistance of a circuit, or the power actually used in the circuit.

In an AC circuit, a voltmeter indicates the effective voltage and an ammeter indicates the effective current. The product of these two readings is called the apparent power.

Apparent Power Defined

That power apparently available for use in an AC circuit containing a reactive component. It is the product of effective voltage times the effective current, expressed in volt-amperes. It must be multiplied by the power factor to obtain true power available.

Only when the AC circuit is made up of pure resistance is the apparent power equal to the true power. [Figure 10-134] When there is capacitance or inductance in the circuit, the current and voltage are not exactly in phase, and the true power is less than the apparent power. The true power is obtained by a wattmeter reading. The ratio of the true power to the apparent power is called the power factor and is usually expressed in percent. In equation form, the relationship is:

$$\text{Power Factor (PF)} = \frac{100 \times \text{Watts (True Power)}}{\text{Volts} \times \text{Amperes (Apparent Power)}}$$

Example: A 220-volt AC motor takes 50 amperes from the line, but a wattmeter in the line shows that only 9,350 watts are taken by the motor. What are the apparent power and the power factor?

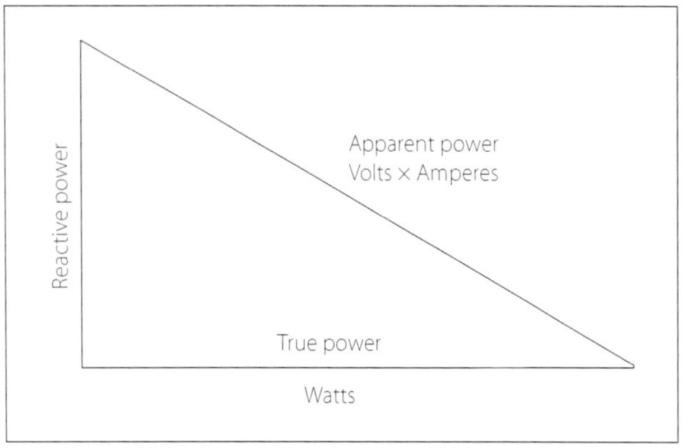

Figure 10-134. Power relations in AC circuit.

Solution:

Apparent power = Volts × Amperes

Apparent power = 220 × 50 = 11,000 watts or volt-amperes.

$$\text{(PF)} = \frac{\text{Watts (True Power)} \times 100}{\text{VA (Apparent Power)}}$$

$$\text{(PF)} = \frac{9{,}350 \times 100}{11{,}000}$$

$$\text{(PF)} = 85, \text{ or } 85\%$$

Transformers

A transformer changes electrical energy of a given voltage into electrical energy at a different voltage level. It consists of two coils that are not electrically connected, but are arranged so that the magnetic field surrounding one coil cuts through the other coil. When an alternating voltage is applied to (across) one coil, the varying magnetic field set up around that coil creates an alternating voltage in the other coil by mutual induction. A transformer can also be used with pulsating DC, but a pure DC voltage cannot be used, since only a varying voltage creates the varying magnetic field that is the basis of the mutual induction process.

A transformer consists of three basic parts. [Figure 10-135] These are an iron core which provides a circuit of low reluctance for magnetic lines of force, a primary winding which receives the electrical energy from the source of applied voltage, and a secondary winding which receives electrical energy by induction from the primary coil.

The primary and secondary of this closed core transformer are wound on a closed core to obtain maximum inductive effect between the two coils.

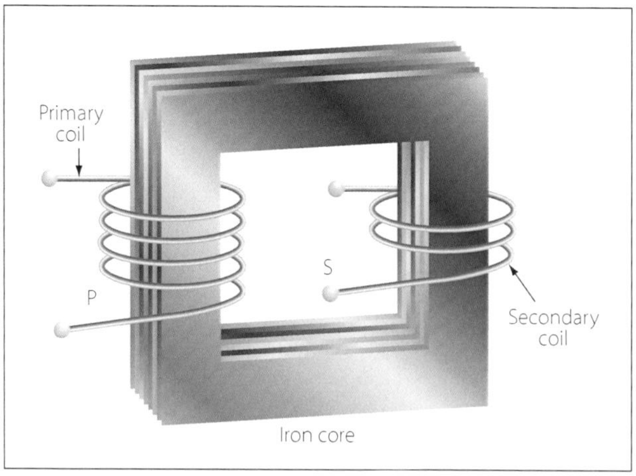

Figure 10-135. An iron-core transformer.

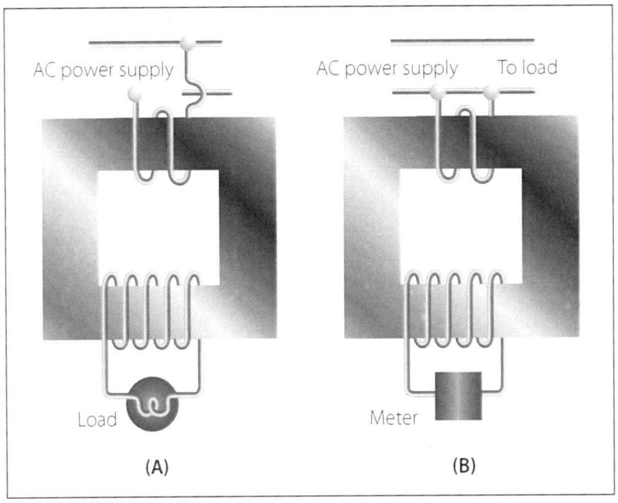

Figure 10-136. Voltage and current transformers.

There are two classes of transformers: (1) voltage transformers used for stepping up or stepping down voltages, and (2) current transformers used in instrument circuits.

In voltage transformers, the primary coils are connected in parallel across the supply voltage as shown in Figure 10-136A. The primary windings of current transformers are connected in series in the primary circuit [Figure 10-136B]. Of the two types, the voltage transformer is the more common.

There are many types of voltage transformers. Most of these are either step-up or step-down transformers. The factor that determines whether a transformer is a step-up, or step-down type is the "turns" ratio. The turns ratio is the ratio of the number of turns in the primary winding to the number of turns in the secondary winding. For example, the turns ratio of the step-down transformer shown in Figure 10-137A is 5 to 1, since there are five times as many turns in the primary as in the secondary. The step-up transformer shown in Figure 10-137B has a 1 to 4 turns ratio.

The ratio of the transformer input voltage to the output voltage is the same as the turns ratio if the transformer is 100 percent efficient. Thus, when 10 volts are applied to the primary of the transformer shown in Figure 10-137A, two volts are induced in the secondary. If 10 volts are applied to the primary of the transformer in Figure 10-137B, the output voltage across the terminals of the secondary will be 40 volts.

No transformer can be constructed that is 100 percent efficient, although iron core transformers can approach this figure. This is because all the magnetic lines of force set up in the primary do not cut across the turns

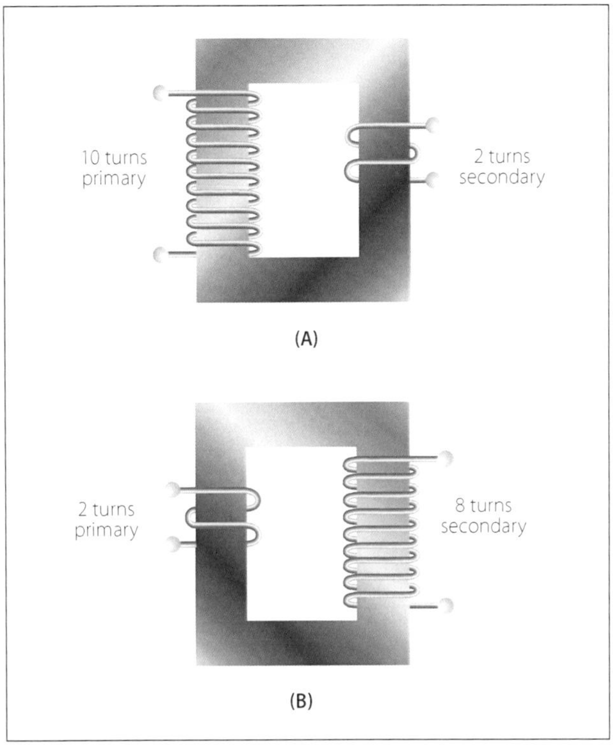

Figure 10-137. A step-down and a step-up transformer.

of the secondary coil. A certain amount of the magnetic flux, called leakage flux, leaks out of the magnetic circuit. The measure of how well the flux of the primary is coupled into the secondary is called the "coefficient of coupling." For example, if it is assumed that the primary of a transformer develops 10,000 lines of force and only 9,000 cut across the secondary, the coefficient of coupling would be 0.9 or, stated another way, the transformer would be 90 percent efficient.

When an AC voltage is connected across the primary terminals of a transformer, an alternating current will flow and self induce a voltage in the primary coil that is opposite and nearly equal to the applied voltage. The difference between these two voltages allows just enough current in the primary to magnetize its core. This is called the exciting, or magnetizing, current. The magnetic field caused by this exciting current cuts across the secondary coil and induces a voltage by mutual induction.

If a load is connected across the secondary coil, the load current flowing through the secondary coil will produce a magnetic field which will tend to neutralize the magnetic field produced by the primary current. This will reduce the self-induced (opposition) voltage in the primary coil and allow more primary current to flow. The primary current increases as the secondary load current increases, and decreases as the secondary load current decreases. When the secondary load is removed, the primary current is again reduced to the small exciting current sufficient only to magnetize the iron core of the transformer.

If a transformer steps up the voltage, it will step down the current by the same ratio. This should be evident if the power formula is considered, for the power (I × E) of the output (secondary) electrical energy is the same as the input (primary) power minus that energy loss in the transforming process. Thus, if 10 volts and 4 amps (40 watts of power) are used in the primary to produce a magnetic field, there will be 40 watts of power developed in the secondary (disregarding any loss). If the transformer has a step-up ratio of 4 to 1, the voltage across the secondary will be 40 volts and the current will be 1 amp. The voltage is 4 times greater and the current is one-fourth the primary circuit value, but the power (I × E value) is the same.

When the turns ratio and the input voltage are known, the output voltage can be determined as follows:

$$\frac{E_2}{E_1} = \frac{N_2}{N_1}$$

Where E is the voltage of the primary, E_2 is the output voltage of the secondary, and N_1 and N_2 are the number of turns of the primary and secondary, respectively.

Transposing the equation to find the output voltage gives:

$$E_2 = \frac{E_1 N_2}{N_1}$$

The most commonly used types of voltage transformers are as follows:

1. Power transformers are used to step up or step down voltages and current in many types of power supplies. They range in size from the small power transformer shown in Figure 10-138 used in a radio receiver to the large transformers used to step down high power line voltage to the 110–120 volt level used in homes.
   Figure 10-139 shows the schematic symbol for an iron core transformer. In this case, the secondary is made up of three separate windings. Each winding supplies a different circuit with a specific voltage, which saves the weight, space, and expense of three separate transformers. Each secondary has a midpoint connection, called a "center tap," which provides a selection of half the voltage across the whole winding. The leads from the various windings are color coded by the manufacturer, as labeled in Figure 10-139. This is a standard color code, but other codes or numbers may be used.
2. Audio transformers resemble power transformers. They have only one secondary and are designed to operate over the range of audio frequencies (20 to 20,000 cps).
3. RF transformers are designed to operate in equipment that functions in the radio range of frequencies. The symbol for the RF transformer is the same as for an RF choke coil. It has an air core as shown in Figure 10-140.
4. Autotransformers are normally used in power circuits; however, they may be designed for other uses. Two different symbols for autotransformers used in power or audio circuits are shown in Figure 10-141. If used in an RF communication or navigation circuit (Figure 10-141B), it is the

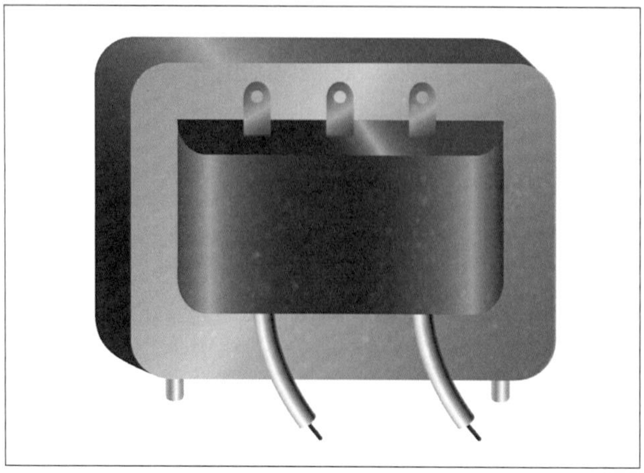

Figure 10-138. Power supply transformer.

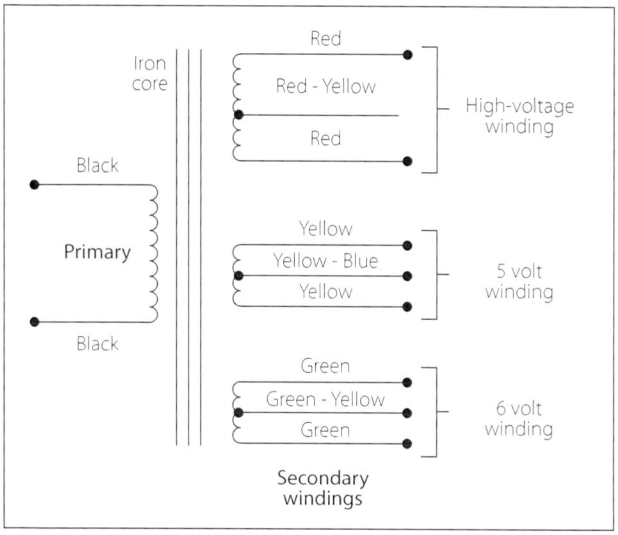

Figure 10-139. Schematic symbol for an iron-core power transformer.

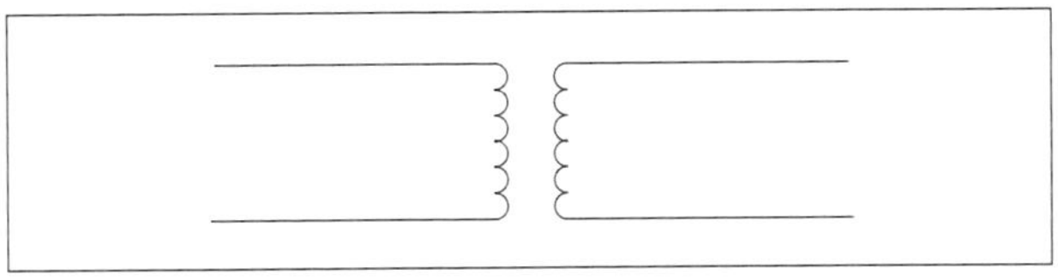

Figure 10-140. An air-core transformer.

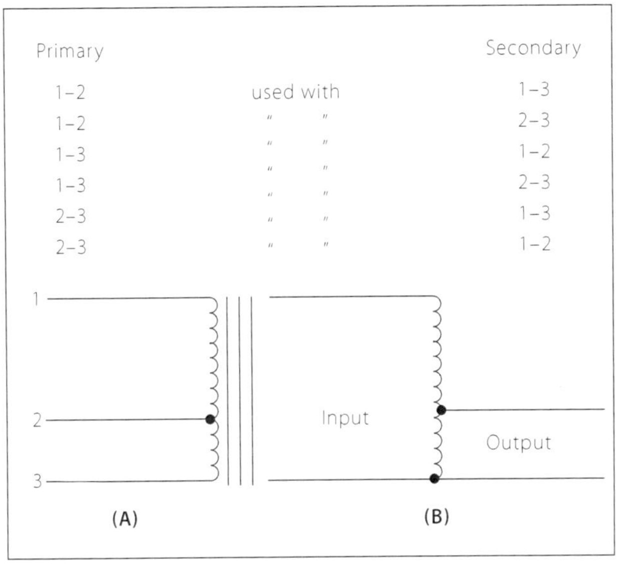

Figure 10-141. Autotransformers.

same, except there is no symbol for an iron core. The autotransformer uses part of a winding as a primary; and, depending on whether it is step up or step down, it uses all or part of the same winding as the secondary. For example, the autotransformer shown in Figure 10-141A could use the following possible choices for primary and secondary terminals.

Current Transformers

Current transformers are used in AC power supply systems to sense generator line current and to provide a current, proportional to the line current, for circuit protection and control devices.

The current transformer is a ring-type transformer using a current carrying power lead as a primary (either the power lead or the ground lead of the AC generator). The current in the primary induces a current in the secondary by magnetic induction.

The sides of all current transformers are marked "H1" and "H2" on the unit base. The transformers must be installed with the "H1" side toward the generator in the circuit in order to have proper polarity. The secondary of the transformer should never be left open while the system is being operated; to do so could cause dangerously high voltages, and could overheat the transformer. Therefore, the transformer output connections should always be connected with a jumper when the transformer is not being used but is left in the system.

Transformer Losses

In addition to the power loss caused by imperfect coupling, transformers are subject to "copper" and "iron" losses. The resistance of the conductor comprising the turns of the coil causes copper loss. The iron losses are of two types called hysteresis loss and eddy current loss. Hysteresis loss is the electrical energy required to magnetize the transformer core, first in one direction and then in the other, in step with the applied alternating voltage. Eddy current loss is caused by electric currents (eddy currents) induced in the transformer core by the varying magnetic fields. To reduce eddy current losses, cores are made of laminations coated with an insulation, which reduces the circulation of induced currents.

Power in Transformers

Since a transformer does not add any electricity to the circuit but merely changes or transforms the electricity that already exists in the circuit from one voltage to another, the total amount of energy in a circuit must remain the same. If it were possible to construct a perfect transformer, there would be no loss of power in it; power would be transferred undiminished from one voltage to another.

Since power is the product of volts times amperes, an increase in voltage by the transformer must result in a decrease in current and vice versa. There cannot be more power in the secondary side of a transformer than there is in the primary. The product of amperes times volts remains the same.

The transmission of power over long distances is accomplished by using transformers. At the power source, the voltage is stepped up in order to reduce the line loss during transmission. At the point of utilization, the voltage is stepped down, since it is not feasible to use high voltage to operate motors, lights, or other electrical appliances.

DC Measuring Instruments

Understanding the functional design and operation of electrical measuring instruments is very important, since they are used in repairing, maintaining, and troubleshooting electrical circuits. The best and most expensive measuring instrument is of no use unless the technician knows what is being measured and what each reading indicates. The purpose of the meter is to measure quantities existing in a circuit. For this reason, when a meter is connected to a circuit, it must not change the characteristics of that circuit.

Meters are either self-excited or externally excited. Those that are self-excited operate from a power source within the meter. Externally excited meters get their power source from the circuit that they are connected to. The most common analog meters in use today are the voltmeter, ammeter, and ohmmeter. All of which operate on the principles of electromagnetism. The fundamental principle behind the operation of the meter is the interaction between magnetic fields created by a current gathered from the circuit in some manner. This interaction is between the magnetic fields of a permanent magnet and the coils of a rotating magnet. The greater the current through the coils of the rotating magnet, the stronger the magnetic field produced. A stronger field produces greater rotation of the coil. While some meters can be used for both DC and AC circuit measurement, only those used as DC instruments are discussed in this section. The meters used for AC, or for both AC and DC, are discussed in the study of AC theory and circuitry.

D'Arsonval Meter Movement

This basic DC type of meter movement—first employed by the French scientist, d'Arsonval in making electrical measurement—is a current measuring device, which is used in the ammeter, voltmeter, and ohmmeter. The pointer is deflected in proportion to the amount of current through the coil. Basically, both the ammeter and the voltmeter are current measuring instruments, the principal difference being the method in which they are connected in a circuit. While an ohmmeter is also basically a current measuring instrument, it differs from the ammeter and voltmeter in that it provides its own source (self-excited) of power and contains other auxiliary circuits.

Current Sensitivity and Resistance

The current sensitivity of a meter movement is the amount of current required to drive the meter movement to a full-scale deflection. A simple example would be a meter movement that has 1mA sensitivity. What this indicates is that meter movement will require 1mA of current to move the needle to a full-scale indication. Likewise a half scale deflection will require only 0.5mA of current. Additionally, what is called movement resistance is the actual DC resistance of the wire used to construct the meter coil.

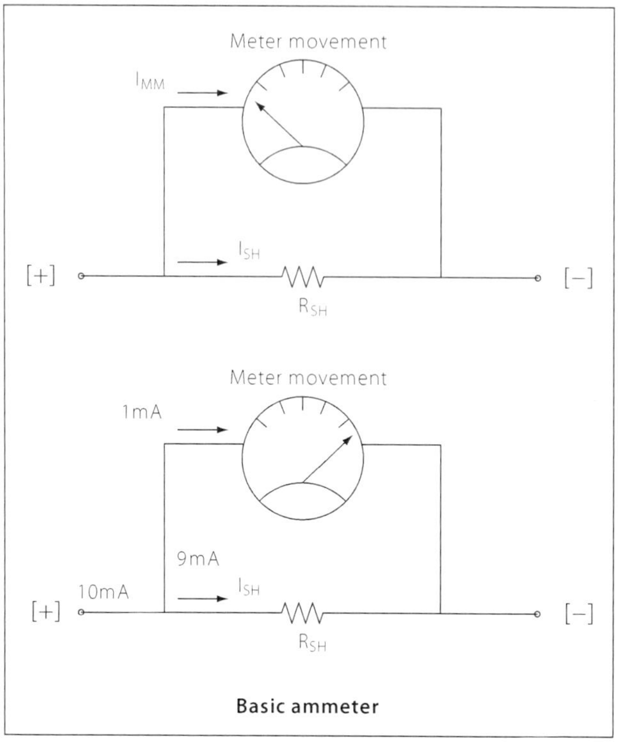

Figure 10-142. Basic meter drawing.

In a standard d'Arsonval meter movement may have a current sensitivity of 1mA and a resistance of 50Ω. If the meter is going to be used to measure more than 1mA then additional circuitry will be required to accomplish the task. This additional circuitry is a simple shunt resistor. The purpose of the shunt resistor is to bypass current that exceeds the 1mA limitation of the meter movement. To illustrate this, assume that the 1mA meter in question is needed to measure 10mA. The shunt resistor used should carry 9mA while the remaining 1mA is allowed to pass through the meter. [Figure 10-142]

To determine the proper shunt resistance for this situation:

R_{SH} = Shunt resistance
R_M = Meter resistance = 50Ω

Because the shunt resistance and the 50Ω meter resistance are in parallel, the voltage drop across both of them is the same.

$E_{SH} = E_M$

Using Ohm's law, this relationship can be rewritten as:

$$E_{SH} = I_{SH} \times R_{SH}$$
$$E_M = I_M \times R_M$$

$$I_{SH} \times R_{SH} = I_M \times R_M$$

Simply solve for R_{SH}

$$R_{SH} = \frac{I_M \times R_M}{I_{SH}}$$

Substituting the values

$$R_{SH} = \frac{1mA \times 50\Omega}{9mA} = 5.56\Omega$$

Damping

To make meter readings quickly and accurately, it is desirable that the moving pointer overshoot its proper position only a small amount and come to rest after not more than one or two small oscillations. The term "damping" is applied to methods used to bring the pointer of an electrical meter to rest after it has been set in motion. Damping may be accomplished by electrical means, by mechanical means, or by a combination of both.

Electrical Damping

A common method of damping by electrical means is to wind the moving coil on an aluminum frame. As the coil moves in the field of the permanent magnet, eddy currents are set up in the aluminum frame. The magnetic field produced by the eddy currents opposes the motion of the coil. The pointer will therefore swing more slowly to its proper position and come to rest quickly with very little oscillation.

Mechanical Damping

Air damping is a common method of damping by mechanical means. As shown in Figure 10-143, a vane is attached to the shaft of the moving element and enclosed in an air chamber. The movement of the shaft is retarded because of the resistance that the air offers to the vane. Effective damping is achieved if the vane nearly touches the walls of the chamber.

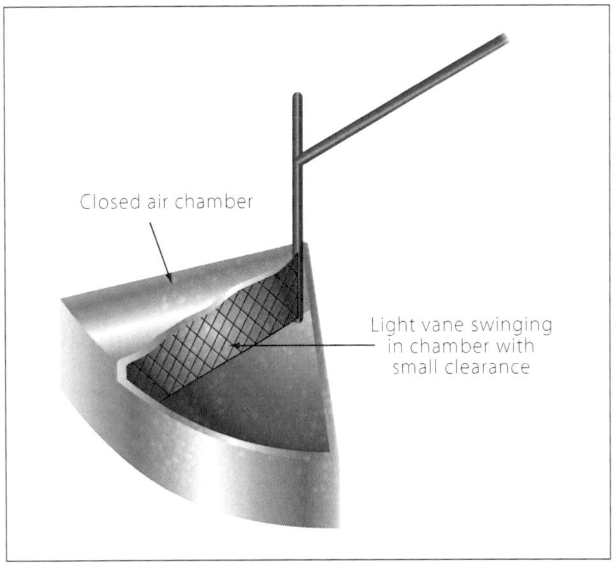

Figure 10-143. Air damping.

A Basic Multirange Ammeter

Building upon the basic meter previously discussed is the more complex and useful multirange meter, which is more practical. The basic idea of a multirange ammeter is to make the meter usable over a wide range of voltages. In order to accomplish this, each range must utilize a different shunt resistance. The example give in this text is that of a two-range meter. However, once the basics of a two range multirange ammeter are understood, the concepts can easily be transferred to the design of meters with many selectable ranges.

Figure 10-144 shows the schematic of an ammeter with two selectable ranges. This example builds upon the previous 10mA range meter by adding a 100mA range. With the switch selected to the 10mA range, the meter will indicate 10mA when the needle is deflected to full scale and will likewise indicate 100mA at full scale when selected to 100mA.

The value of the 100mA shunt resistor is determined the same way the 10mA shunt resistor was determined. Recall that the meter movement can only carry 1mA.

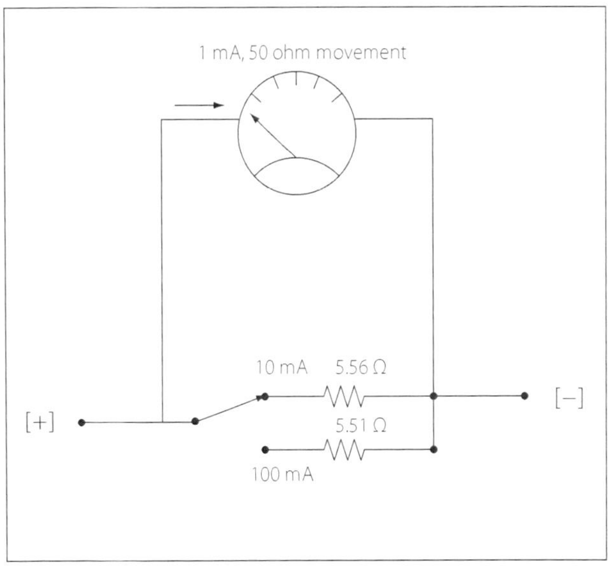

Figure 10-144. Ammeter with two ranges.

This means that in a 100mA range the remaining current of 99mA must pass through the shunt resistor.

$$R_{SH} = \frac{I_M \times R_M}{I_{SH}}$$

Substituting the values

$$R_{SH} = \frac{1mA \times 50\Omega}{99mA} = 0.51\Omega$$

Precautions

The precautions to observe when using an ammeter are summarized as follows:

1. Always connect an ammeter in series with the element through which the current flow is to be measured.
2. Never connect an ammeter across a source of voltage, such as a battery or generator. Remember that the resistance of an ammeter, particularly on the higher ranges, is extremely low and that any voltage, even a volt or so, can cause very high current to flow through the meter, causing damage to it.
3. Use a range large enough to keep the deflection less than full scale. Before measuring a current, form some idea of its magnitude. Then switch to a large enough scale or start with the highest range and work down until the appropriate scale is reached. The most accurate readings are obtained at approximately half-scale deflection. Many milliammeters have been ruined by attempts to measure amperes. Therefore, be sure to read the lettering either on the dial or on the switch positions and choose proper scale before connecting the meter in the circuit.
4. Observe proper polarity in connecting the meter in the circuit. Current must flow through the coil in a definite direction in order to move the indicator needle up scale. Current reversal because of incorrect connection in the circuit results in a reversed meter deflection and frequently causes bending of the meter needle. Avoid improper meter connections by observing the polarity markings on the meter.

The Voltmeter

The voltmeter uses the same type of meter movement as the ammeter but employs a different circuit external to the meter movement.

As shown before, the voltage drop across the meter coil is a function of current and the coil resistance. In another example, 50μA × 1000Ω = 50mV. In order for the meter to be used to measure voltages greater than 50mV, there must be added a series resistance to drop any excess voltage greater than that which the meter movement requires for a full scale deflection. The case of the voltmeter, this resistance is called multiplier resistance and will be designated as R_M. Figure 10-145 illustrates a basic voltmeter. This voltmeter only has one multiplier resistor for use in one range. In this example, the full scale reading will be 1 volt. R_M is determined in the follow way:

The meter movement drops 50mV at a full scale deflection of 50μA. The multiplying resistor RM must drop the remaining voltage of 1V − 50mV = 950 mV. Since RM is in series with the movement, it also carries 50μA at full scale.

$$RM = \frac{950\ mV}{50\mu A} = 19k\ \Omega$$

Therefore, for 1 volt full scale deflection, the total resistance of the voltmeter is 20 k Ω. That is, the multiplier resistance and the coil resistance.

Voltmeter Sensitivity

Voltmeter sensitivity is defined in terms of resistance per volt (Ω/V). The meter used in the previous example has a sensitivity of 20 k Ω and a full scale deflection of 1 volt.

Multiple Range Voltmeters

The simplified voltmeter in Figure 10-145 has only one range (1 volt), which means that it can measure voltages from 0 volts to 1 volt. In order for the meter to be more useful, additional multiplier resistors must be used. One resistor must be used for each desired range.

For a 50μA movement, the total resistance required is 20 k Ω for each volt of full scale reading. In other words, the sensitivity for a 50μA movement is always 20 k Ω regardless of the selected range. The full-scale meter current is 50μA at any range selection. To find the total meter resistance, multiply the sensitivity by the full scale voltage for that particular range. For

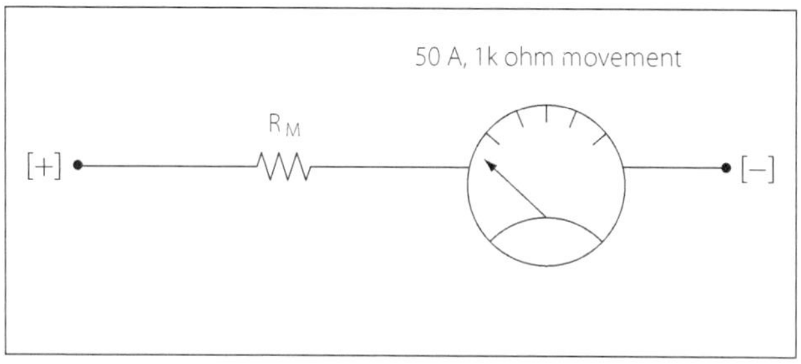

Figure 10-145. Basic voltmeter.

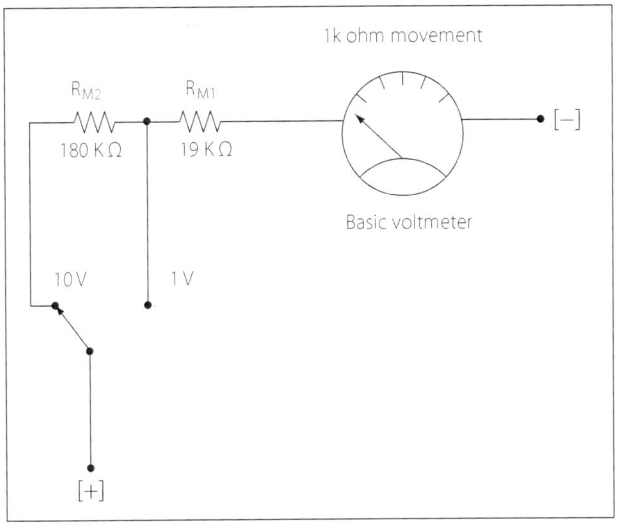

Figure 10-146. Two range voltmeter.

example for a 10 volt range, RT = (20k Ω/V)/(10 V) = 200k Ω. The total resistance for the 1 volt range is 20k Ω, so RM for a 10 V range will be 200k Ω – 20k Ω = 180k Ω. This two-range voltmeter is illustrated in Figure 10-146.

Voltmeter Circuit Connections

When voltmeters are used, they are connected in parallel with a circuit. If unsure about the voltage to be measured, take the first reading at the high value on the meter and then progressively move down through the range until a suitable read is obtained. Observe that the polarity is correct before connecting the meter to the circuit or damage will occur by driving the movement backwards.

Influence of the Voltmeter in the Circuit

When a voltmeter is connected across two points in a circuit, current will be shunted. If the voltmeter has low resistance, it will draw off a significant amount of current. This will lower the effective resistance of the circuit and change the voltage readings. When making a voltage measurement, use a high resistance voltmeter to prevent shunting of the circuit.

The Ohmmeter

The meter movement used for the ammeter and the voltmeter can also be used for the ohmmeter. The function of the ohmmeter is to measure resistance. A simplified one-stage ohmmeter is illustrated in Figure 10-147, which shows that the basic ohmmeter contains a battery and a variable resistor in series with the meter movement. To measure resistance, the leads of the meter are connected across an external resistance, which is to be measured. By doing this the ohmmeter circuit is completed. This connection allows the internal battery to produce a current through the movement coil, causing a deflection of the pointer proportional to the value of the external resistance being measured.

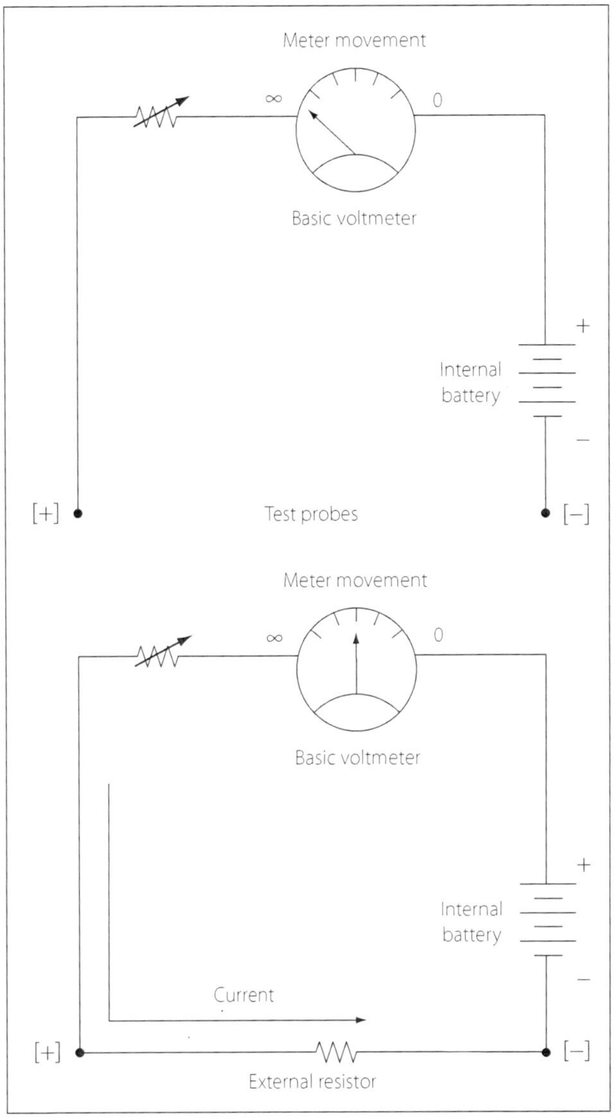

Figure 10-147. Basic ohmmeter.

Zero Adjustment

When the ohmmeter leads are open as shown in Figure 10-148, the meter is at a full scale deflection, indicating an infinite (∞) resistance or an open circuit. When the leads are shorted as shown in figure "zero adjust," the pointer will be at the full right hand position, indicating a short circuit or zero resistance. The purpose of the variable resistor in this figure is to adjust the current so that the pointer is at exactly zero when the leads are shorted. This is used to compensate for changes in the internal battery voltage due to aging.

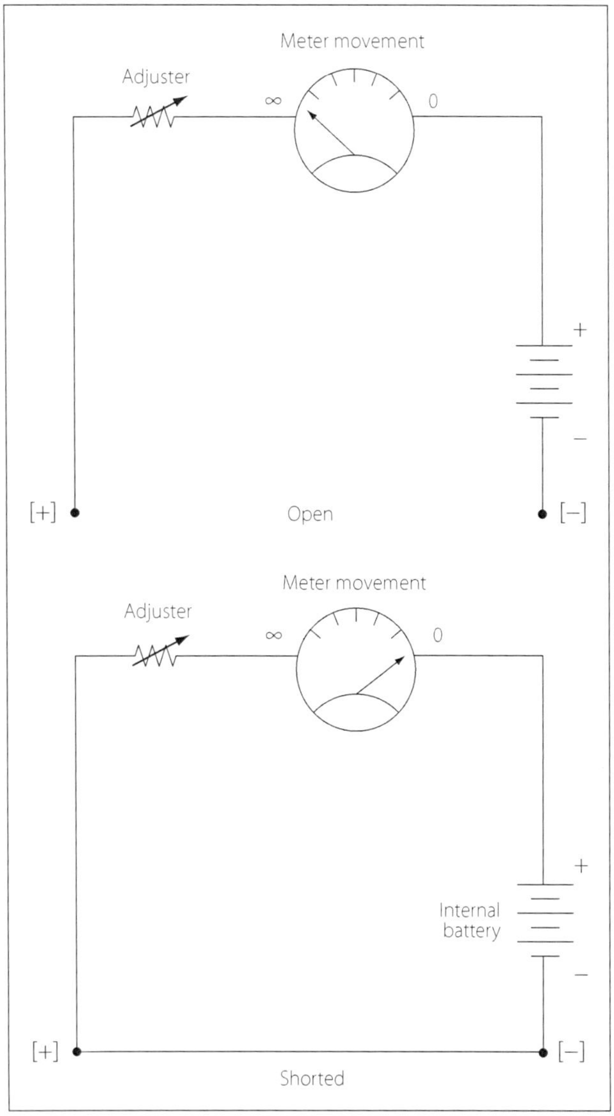

Figure 10-148. Zero adjustment.

Ohmmeter Scale

Figure 10-149 shows a typical analog ohmmeter scale. Between zero and infinity (∞), the scale is marked to indicate various resistor values. Because the values decrease from left to right, this scale is often called a back-off scale.

In the case of the example given, assume that a certain ohmmeter uses a 50μA, 1000Ω meter movement and has an internal 1.5 volt battery. A current of 50μA produces a full-scale deflection when the test leads are shorted. To have 50μA, the total ohmmeter resistance is 1.5 V/50μA = 30kΩ. Therefore, since the coil resistance is 1kΩ, the variable zero adjustment resistor must be set to 30kΩ – 1kΩ = 29kΩ.

Now consider that a 120kΩ resistor is connected to the ohmmeter leads. Combined with the 30kΩ internal resistance, the total R is 150kΩ. The current is 1.5 V/150kΩ = 10μA, which is 20% of the full scale current and which appears on the scale as shown in Figure 10-149.

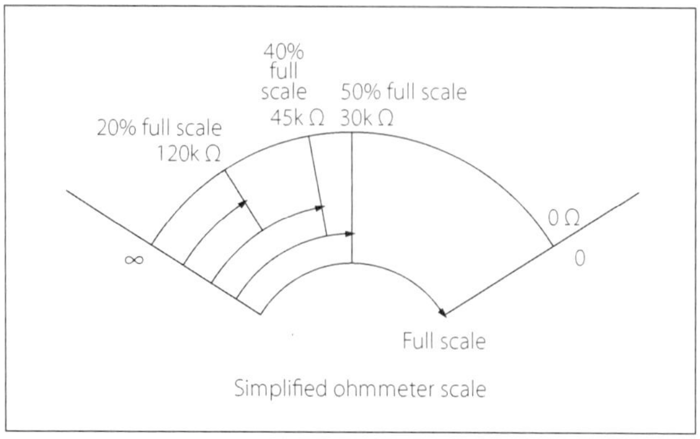

Figure 10-149. Ohm scale.

Now consider further that a 120kΩ resistor is connected to the ohmmeter leads. This will result in a current of 1.5V/75kΩ = 10μA, which is 40% of the full scale current and which is marked on the scale as shown. Additional calculations of this type show that the scale is nonlinear. It is more compressed toward the left side than the right side. The center scale point corresponds to the internal meter resistance of 30kΩ. The reason is as follows:

With 30kΩ connected to the leads, the current is 1.5V/60kΩ = 25μA, which is half of the full scale current of 50μA.

The Multirange Ohmmeter

A practical ohmmeter usually has several operational ranges. These typically are indicated by R × 1, R × 10, R × 100, R × 1k, R × 100k and R × 1M. These range selections are interpreted in a different manner than that of an ammeter or voltmeter. The reading on the ohmmeter scale is multiplied by the factor indicated by the range setting. For example, if the pointer is set on the scale and the range switch is set at R × 100, the actual resistance measurement is 20 × 100 or 2kΩ.

To measure small resistance values, the technician must use a higher ohmmeter current than is needed for measuring large resistance values. Shunt resistors are needed to provide multiple ranges on the ohmmeter to measure a range of resistance values from the very small to very large. For each range, a different value of shunt resistance is switched in. The shunt resistance increases for higher ohm ranges and is always equal to the center scale reading on any selected range. In some meters, a higher battery voltage is used for the

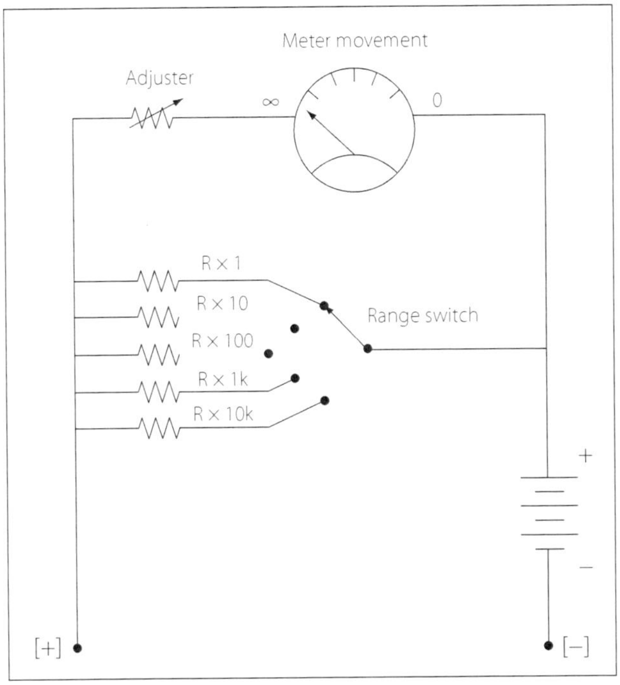

Figure 10-150. Multirange ohmmeter.

highest ohm range. A common circuit arrangement is shown in Figure 10-150.

Megger (Megohmmeter)

The megger, or megohmmeter, is a high range ohmmeter containing a hand-operated generator. It is used to measure insulation resistance and other high resistance values. It is also used for ground, continuity, and short circuit testing of electrical power systems. The chief advantage of the megger over an ohmmeter is its capacity to measure resistance with a high potential, or "breakdown" voltage. This type of testing ensures that insulation or a dielectric material will not short or leak under potential electrical stress.

The megger consists of two primary elements, both of which are provided with individual magnetic fields from a common permanent magnet: (1) a hand-driven DC generator, G, which supplies the necessary current for making the measurement, and (2) the instrument portion, which indicates the value of the resistance being measured. The instrument portion is of the opposed coil type. Coils A and B are mounted on the movable member with a fixed angular relationship to each other and are free to turn as a unit in a magnetic field. Coil B tends to move the pointer counterclockwise and coil A, clockwise. The coils are mounted on a light, movable frame that is pivoted in jewel bearings and free to move about axis 0. [Figure 10-151]

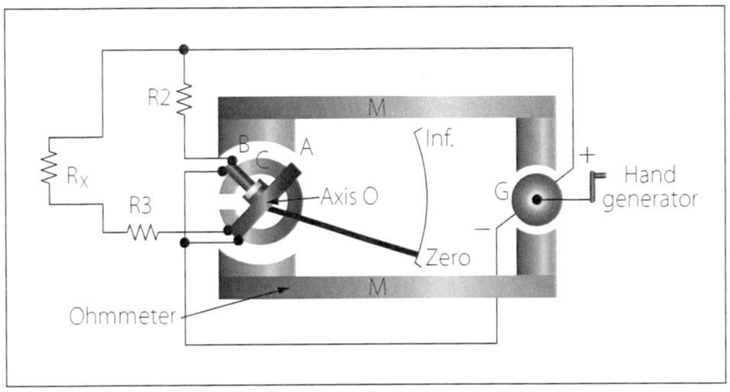

Figure 10-151. Simplified megger circuit.

Coil A is connected in series with R3 and the unknown resistance, R_X, to be measured. The series combination of coil A, R3, and R_X is connected between the + and – brushes of the DC generator. Coil B is connected in series with R2 and this combination is also connected across the generator. There are no restraining springs on the movable member of the instrument portion of the megger. When the generator is not in operation, the pointer floats freely and may come to rest at any position on the scale.

If the terminals are open circuited, no current flows in coil A, and the current in coil B alone controls the movement of the moving element. Coil B takes a position opposite the gap in the core (since the core cannot move and coil B can), and the pointer indicates infinity on the scale. When a resistance is connected between the terminals, current flows in coil A, tending to move the pointer clockwise. At the same time, coil B tends to move the pointer counterclockwise. Therefore, the moving element, composed of both coils and the pointer, comes to rest at a position at which the two forces are balanced. This position depends upon the value of the external resistance, which controls the relative magnitude of current of coil A. Because changes in voltage affect both coils A and B in the same proportion, the position of the moving element is independent of the voltage. If the terminals are short circuited, the pointer rests at zero because the current in A is relatively large. The instrument is not damaged under these circumstances because the current is limited by R3.

There are two types of hand-driven meggers: the variable type and the constant pressure type. The speed of the variable pressure megger is dependent on how fast the hand crank is turned. The constant pressure megger uses a centrifugal governor, or slip clutch. The governor becomes effective only when the megger is operated at a speed above its slip speed, at which speed its voltage remains constant.

AC Measuring Instruments

A DC meter, such as an ammeter, connected in an AC circuit will indicate zero, because the meter movements used in a d'Arsonval type movement is restricted to direct current. Since the field of a permanent magnet in the d'Arsonval type meter remains constant and in the same direction at all times, the moving coil follows the polarity of the current. The coil attempts to move in one direction during half of the AC cycle and in the reverse direction during the other half when the current reverses.

The current reverses direction too rapidly for the coil to follow, causing the coil to assume an average position. Since the current is equal and opposite during each half of the AC cycle, the direct current meter indicates zero, which is the average value. Thus, a meter with a permanent magnet cannot be used to measure alternating voltage and current. For AC measurements of current and voltage, additional circuitry is required. The additional circuitry has a rectifier, which converts AC to DC. There are two basic types of rectifiers: One is the half-wave rectifier and the other is the full-wave rectifier. Both of these are depicted in block diagram form in Figure 10-152.

Figure 10-152 also shows a simplified block diagram of an AC meter. In this depiction, the full-wave rectifier precedes the meter movement. The movement responds to the average value of the pulsating DC. The scale can then be calibrated to show anything the designer wants. In most cases, it will be root mean square (RMS) value or peak value.

Electrodynamometer Meter Movement

The electrodynamometer can be used to measure alternating or direct voltage and current. It operates on the same principles as the permanent magnet moving coil meter, except that the permanent magnet is replaced by an air core electromagnet. The field of the electrodynamometer is developed by the same current that flows through the moving coil. [Figure 10-153]

Because this movement contains no iron, the electrodynamometer can be used as a movement for both AC and DC instruments. Alternating current can be measured by connecting the stationary and moving coils in series. Whenever the current in the moving coil reverses, the magnetic field produced by the stationary coil reverses. Regardless of the direction of the current, the needle will move in a clockwise direction.

However, for either voltmeter or ammeter applications, the electrodynamometer is too expensive to economically compete with the d'Arsonval type movement.

Moving Iron Vane Meter

The moving iron vane meter is another basic type of meter. It can be used to measure either AC or DC. Unlike the d'Arsonval meter, which employs permanent magnets, it depends on induced magnetism for its operation. It utilizes the principle of repulsion between two concentric iron vanes, one fixed and one movable, placed inside a solenoid. A pointer is attached to the movable vane. [Figure 10-154]

When current flows through the coil, the two iron vanes become magnetized with north poles at their upper ends

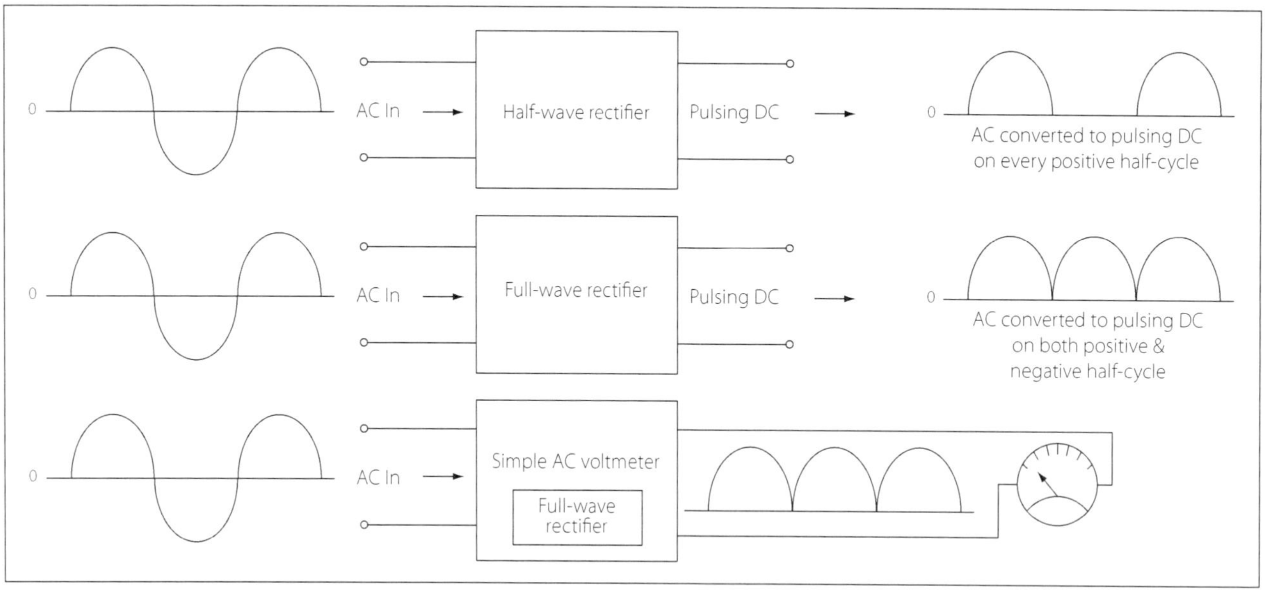

Figure 10-152. Simplified block diagram of AC meter.

and south poles at their lower ends for one direction of current through the coil. Because like poles repel, the unbalanced component of force, tangent to the movable element, causes it to turn against the force exerted by the springs.

The movable vane is rectangular in shape and the fixed vane is tapered. This design permits the use of a relatively uniform scale.

When no current flows through the coil, the movable vane is positioned so that it is opposite the larger portion of the tapered fixed vane, and the scale reading is zero. The amount of magnetization of the vanes depends on the strength of the field, which, in turn, depends on the amount of current flowing through the coil.

The force of repulsion is greater opposite the larger end of the fixed vane than it is nearer the smaller end. Therefore, the movable vane moves toward the smaller end through an angle that is proportional to the magnitude of the coil current. The movement ceases when the force of repulsion is balanced by the restraining force of the spring.

Because the repulsion is always in the same direction (toward the smaller end of the fixed vane), regardless of the direction of current flow through the coil, the moving iron vane instrument operates on either DC or AC circuits.

Mechanical damping in this type of instrument can be obtained by the use of an aluminum vane attached to the shaft so that, as the shaft moves, the vane moves in a restricted air space.

When the moving iron vane meter is used as an ammeter, the coil is wound with relatively few turns of large wire in order to carry the rated current.

When the moving iron vane meter is used as a voltmeter, the solenoid is wound with many turns of small wire. Portable voltmeters are made with self-contained series resistance for ranges up to 750 volts. Higher ranges are obtained by the use of additional external multipliers.

The moving iron vane instrument may be used to measure direct current but has an error due to residual magnetism in the vanes. Reversing the meter connections and averaging the readings may minimize the error. When used on AC circuits, the instrument has an accuracy of 0.5 percent. Because of its simplicity, relatively low cost, and the fact that no current is con-

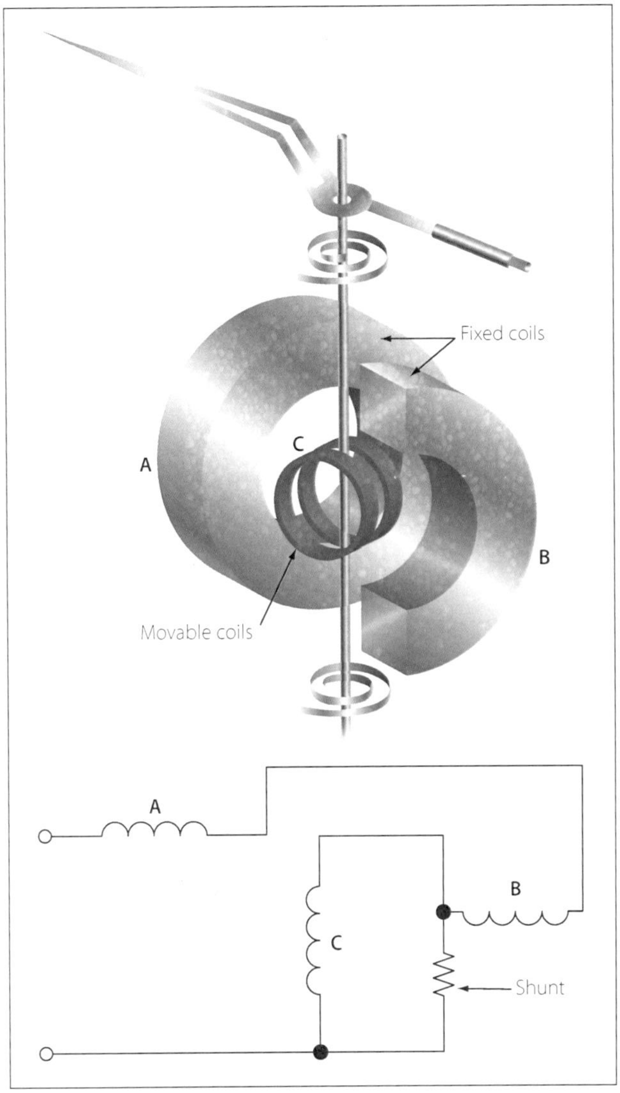

Figure 10-153. Simplified diagram of an electrodynamometer movement.

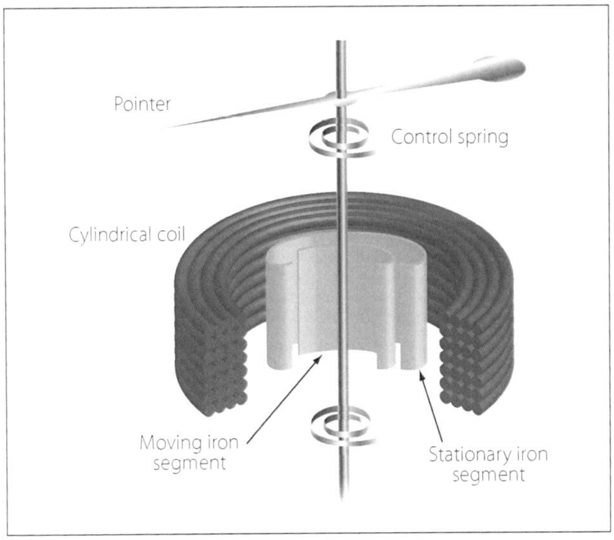

Figure 10-154. Moving iron vane meter.

ducted to the moving element, this type of movement is used extensively to measure current and voltage in AC power circuits. However, because the reluctance of the magnetic circuit is high, the moving iron vane meter requires much more power to produce full-scale deflection than is required by a d'Arsonval meter of the same range. Therefore, the moving iron vane meter is seldom used in high resistance low power circuits.

Inclined Coil Iron Vane Meter

The principle of the moving iron vane mechanism is applied to the inclined coil type of meter, which can be used to measure both AC and DC. The inclined coil, iron vane meter has a coil mounted at an angle to the shaft. Attached obliquely to the shaft, and located inside the coil, are two soft iron vanes. When no current flows through the coil, a control spring holds the pointer at zero, and the iron vanes lie in planes parallel to the plane of the coil. When current flows through the coil, the vanes tend to line up with magnetic lines passing through the center of the coil at right angles to the plane of the coil. Thus, the vanes rotate against the spring action to move the pointer over the scale.

The iron vanes tend to line up with the magnetic lines regardless of the direction of current flow through the coil. Therefore, the inclined coil, iron vane meter can be used to measure either alternating current or direct current. The aluminum disk and the drag magnets provide electromagnetic damping.

Like the moving iron vane meter, the inclined coil type requires a relatively large amount of current for full-scale deflection and is seldom used in high resistance low power circuits.

As in the moving iron vane instruments, the inclined coil instrument is wound with few turns of relatively large wire when used as an ammeter and with many turns of small wire when used as a voltmeter.

Varmeters

Multiplying the volts by the amperes in an AC circuit gives the apparent power: the combination of the true power (which does the work) and the reactive power (which does no work and is returned to the line). Reactive power is measured in units of vars (volt-amperes reactive) or kilovars (kilovolt-amperes reactive, abbreviated kVAR). When properly connected, wattmeters measure the reactive power. As such, they are called varmeters. Figure 10-155 shows a varmeter connected in an AC circuit.

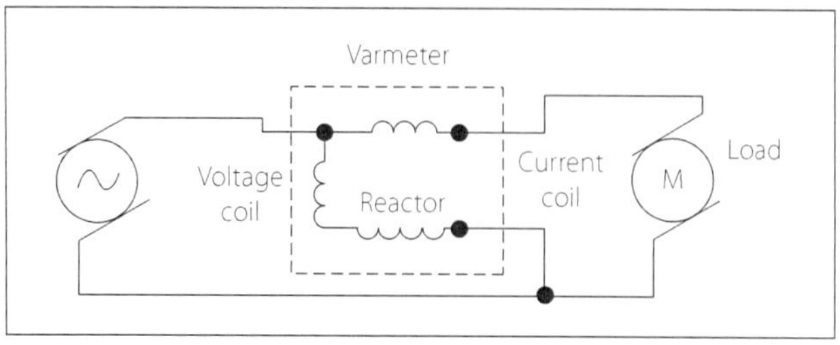

Figure 10-155. A varmeter connected in an AC circuit.

Wattmeter

Electric power is measured by means of a wattmeter. Because electric power is the product of current and voltage, a wattmeter must have two elements, one for current and the other for voltage. For this reason, wattmeters are usually of the electrodynamometer type. [Figure 10-156]

The movable coil with a series resistance forms the voltage element, and the stationary coils constitute the current element. The strength of the field around the potential coil depends on the amount of current that flows through it. The current, in turn, depends on the load voltage applied across the coil and the high resistance in series with it. The strength of the field around the current coils depends on the amount of current flowing through the load. Thus, the meter deflection is proportional to the product of the voltage across the potential coil and the current through the current coils. The effect is almost the same (if the

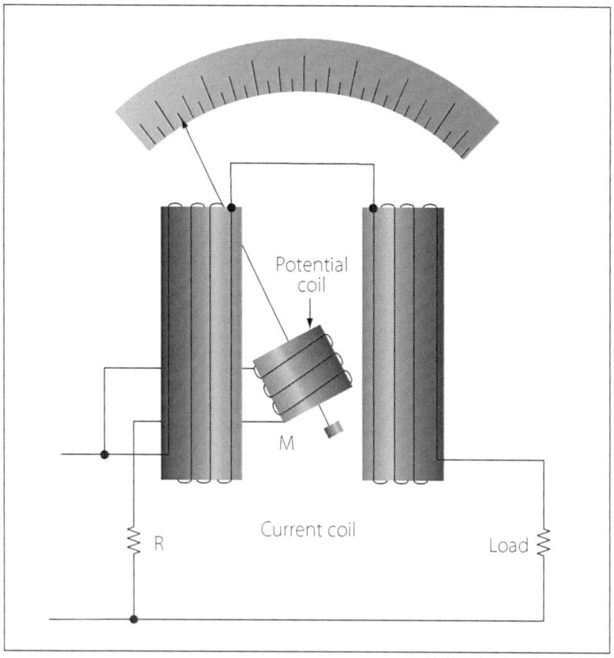

Figure 10-156. Simplified electrodynamometer wattmeter circuit.

scale is properly calibrated) as if the voltage applied across the load and the current through the load were multiplied together.

If the current in the line is reversed, the direction of current in both coils and the potential coil is reversed, the net result is that the pointer continues to read up scale. Therefore, this type of wattmeter can be used to measure either AC or DC power.

Frequency Measurement/Oscilloscope

The oscilloscope is by far one of the more useful electronic measurements available. The viewing capabilities of the oscilloscope make it possible to see and quantify various waveform characteristics such as phase relationships, amplitudes, and durations. While oscilloscopes come in a variety of configurations and presentations, the basic operation is typically the same. Most oscilloscopes in general bench or shop applications use a cathode-ray tube (CRT), which is the device or screen that displays the waveforms.

The CRT is a vacuum instrument that contains an electron gun, which emits a very narrow and focused beam of electrons. A phosphorescent coat applied to the back of the screen forms the screen. The beam is electronically aimed and accelerated so that the electron beam strikes the screen. When the electron beam strikes the screen, light is emitted at the point of impact.

Figure 10-157 shows the basic components of the CRT with a block diagram. The heated cathode emits electrons. The magnitude of voltage on the control grid determines the actual flow of electrons and thus controls the intensity of the electron beam. The acceleration anodes increase the speed of the electrons, and the focusing anode narrows the beam down to a fine point. The surface of the screen is also an anode and will assist in the acceleration of the electron beam.

The purpose of the vertical and horizontal deflection plates is to bend the electron beam and position it to a specific point of the screen. Figure 10-158 illustrates how the deflection plates are used to position the beam on the screen. By providing a neutral or zero voltage to a deflection plate, the electron beam will be unaffected. By applying a negative voltage to a plate, the electron beam with be repelled and driven away from the plate. Finally, by applying a positive voltage, the electron beam will be drawing to the plate. Figure 10-158 provides a few possible plate voltage combinations and the resultant beam position.

Horizontal Deflection

To get a visual representation of the input signal, an internally generated saw-tooth voltage is generated and then applied to the horizontal deflection plates. Figure 10-159 illustrates that the saw-tooth is a pattern of voltage applied, which begins at a negative voltage and increases at a constant rate to a positive voltage. This applied varying voltage will draw or trace the electron beam from the far left of the screen to the far right side of the screen. The resulting display is a straight line, if the sweep rate is fast enough. This saw-tooth applied voltage is a repetitive signal so that the beam is repeatedly swept across the tube. The rate at which the saw-tooth voltage goes from negative to positive is determined by the frequency. This rate then establishes the sweep rate of the beam. When the saw-tooth reaches the end of its sweep from left to right, the beam then rapidly returns to the left side and is ready to make another sweep. During this time, the electron beam is stopped or blanked out and does

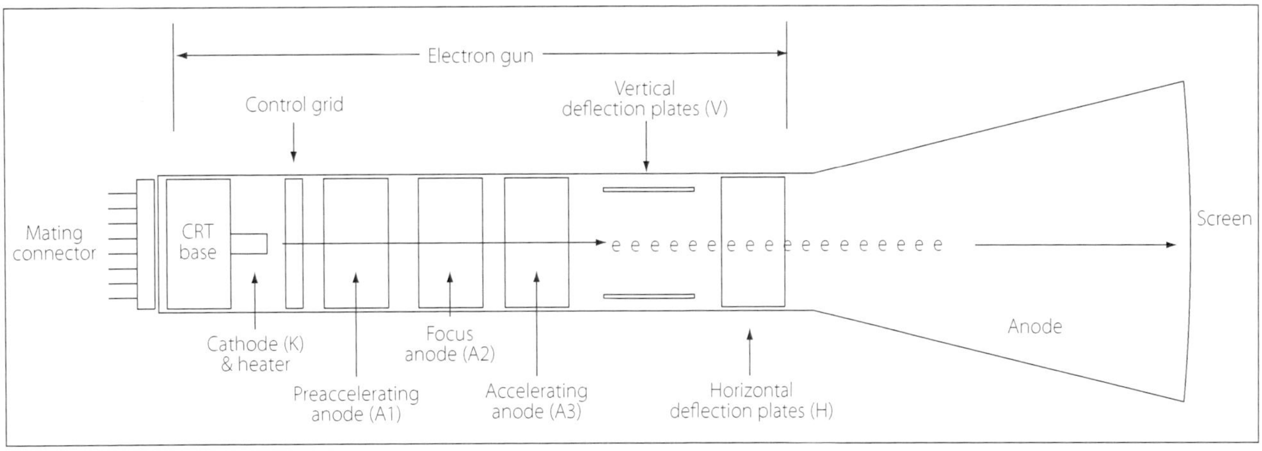

Figure 10-157. Basic components of the CRT with a block diagram.

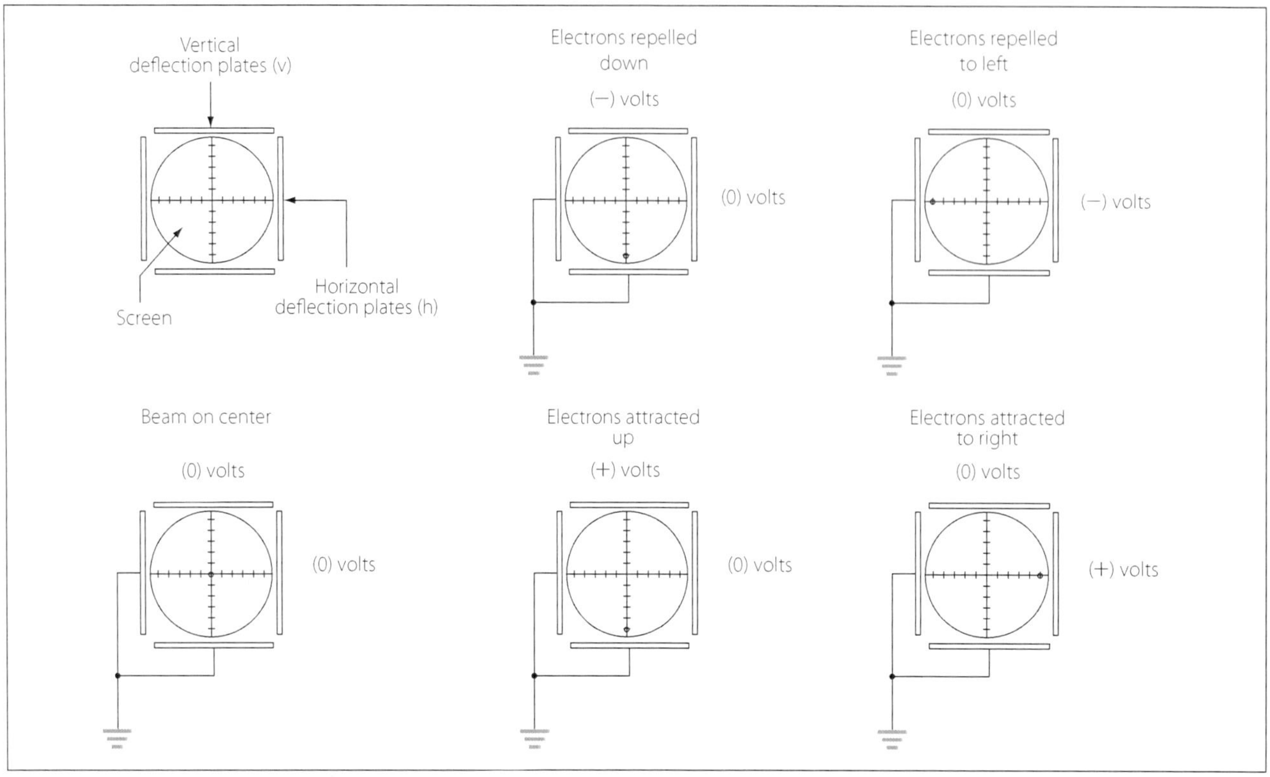

Figure 10-158. Possible plate voltage combinations and the resultant beam position.

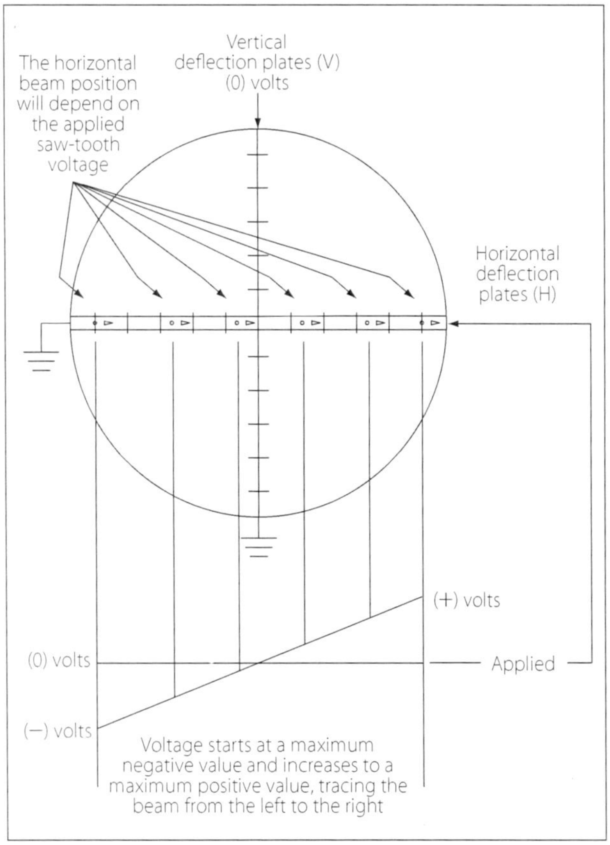

Figure 10-159. Saw-tooth applied voltage.

not produce any kind of a trace. This period of time is called flyback.

Vertical Deflection

If this same signal were applied to the vertical plates, it would also produce a vertical line by causing the beam to trace from the down position to the up position.

Tracing a Sine Wave

Reproducing the sine wave on the oscilloscope combines both the vertical and horizontal deflection patterns. [Figure 10-160] If the sine wave voltage signal is applied across the vertical deflection plates, the result will be the vertical beam oscillation up and down on the screen. The amount that the beam moves above the centerline will depend on the peak value of the voltage.

While the beam is being swept from the left to the right by the horizontal plates, the sine wave voltage is being applied to the vertical plates, causing the form of the input signal to be traced out on the screen.

Control Features on an Oscilloscope

While there are many different styles of oscilloscopes, which range from the simple to the complex, they all have some controls in common. Apart from the screen

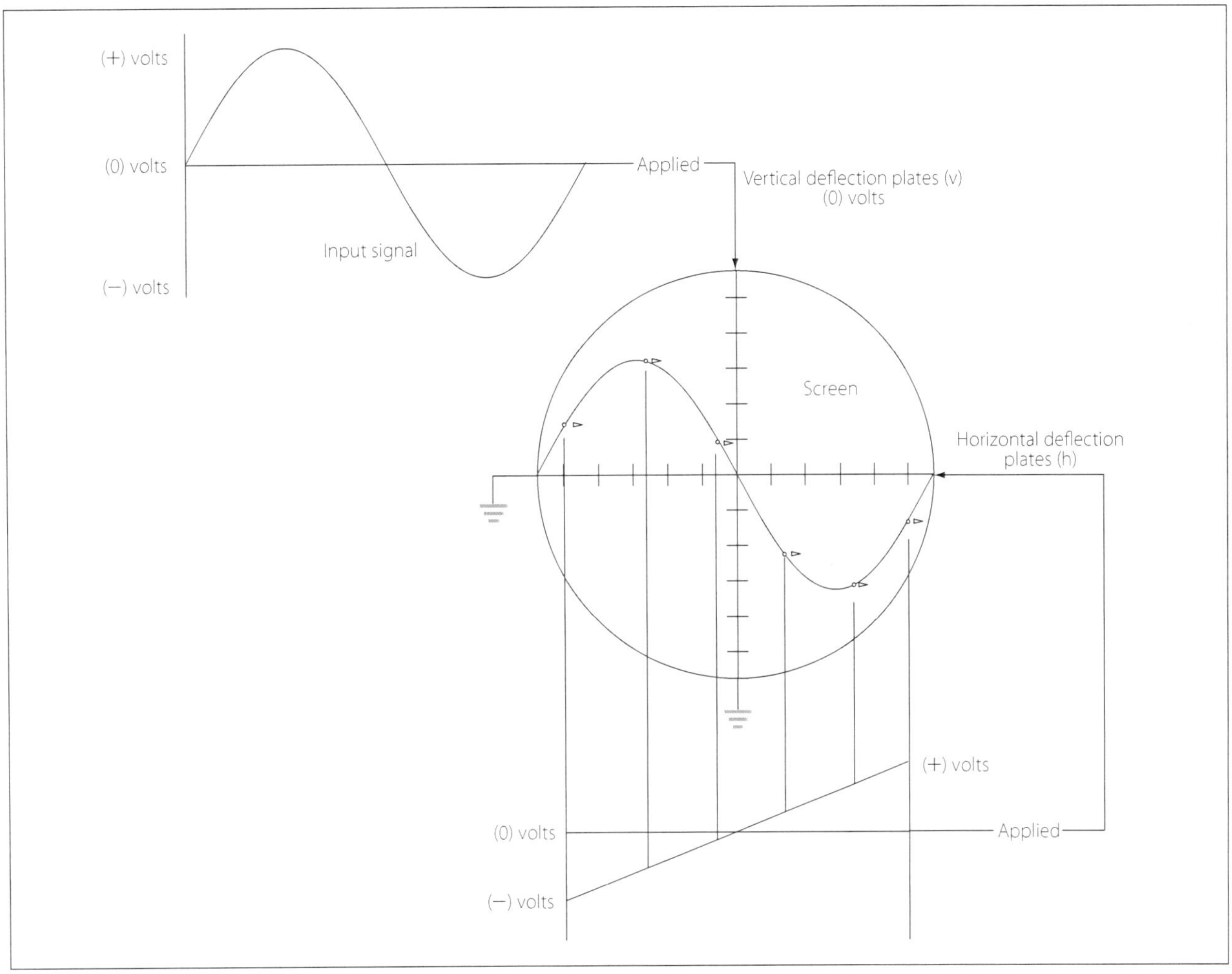

Figure 10-160. Sine wave voltage signal.

and the ON/OFF switch, some of these controls are listed below.

Horizontal Position: allows for the adjustment of the neutral horizontal position of the beam. Use this control to reposition the waveform display in order to have a better view of the wave or to take measurements.

Vertical Position: moves the traced image up or down allowing better observations and measurements.

Focus: controls the electron beam as it is aimed and converges on the screen. When the beam is in sharp focus, it is narrowed down to a very fine point and does not have a fuzzy appearance.

Intensity: essentially the brightness of the trace. Controlling the flow of electrons onto the screen varies the intensity. Do not keep the intensity too high for extended testing or when the beam is motionless and forms a dot on the screen. This can damage the screen.

Seconds/Division: a time-based control, which sets the horizontal sweep rate. Basically, the switch is used to select the time interval that each division on the horizontal scale will represent. These divisions can be seconds, milliseconds or even microseconds. A simple example would be if the technician had the seconds/division control set to 10 µS. If this technician is viewing a waveform that has a period of 4 divisions on the screen, then the period would be 40 µS. The frequency of this waveform can then be determined by taking the inverse of the period. In this case, 1/40 µS will equal a frequency of 25 kHz.

Volts/Division: used to select the voltage interval that each division on the vertical scale will represent. For example, suppose each vertical division was set to equal 10 mV. If a waveform was measured and had a peak value of 4 divisions, then the peak value in voltage would be 40 mV.

Trigger: The trigger control provides synchronization between the saw-tooth horizontal sweep and the applied signal on the vertical plates. The benefit is that the waveform on the screen appears to be stationary and fixed and not drifting across the screen. A triggering circuit is used to initiate the start of a sweep rather than the fixed saw-tooth sweep rate. In a typical oscilloscope, this triggering signal comes from the input signal itself at a selected point during the signal's cycle. The horizontal signal goes through one sweep, retraces back to the left side and waits there until it is triggered again by the input signal to start another sweep.

Flat Panel Color Displays for Oscilloscopes

While the standard CRT design of oscilloscope is still in service, the technology of display and control has evolved into use of the flat panel monitors. Furthermore, the newer oscilloscopes can even be integrated with the common personal computer (PC). This level of integration offers many diagnostic options unheard of only a few years ago. Some of the features of this technology include easy data capture, data transfer, documentation, and data analysis.

Digital Multimeter

Traditionally, the meters that technicians have used have been the analog voltmeter, ammeter, and the ohmmeter. These have usually been combined into the same instrument and called a multimeter or a VOM (volt-ohm-milliammeter). This approach has been both convenient and economical. Digital multimeters (DMM) and digital voltmeters (DVM) have now become more common due to their ease of use. These meters are easier to read and provide greater accuracy when compared to the older analog units with needle movement. The multimeter's single-coil movement requires a number of scales, which are not always easy to read accurately. In addition, the loading characteristics due to the internal resistance sometimes affect the circuit and the measurements. Not only does the DVM offer greater accuracy and less ambiguity, but also higher input resistance, which has less of a loading effect and influence on a circuit.

Basic Circuit Analysis and Troubleshooting

Troubleshooting is the systematic process of recognizing the symptoms of a problem, identifying the possible cause, and locating the failed component or conductor in the circuit. To be proficient at troubleshooting, the technician must understand how the circuit operates and know how to properly use the test equipment. There are many ways in which a system can fail and to cover all of the possibilities is beyond the scope of this text. However, there are some basic concepts that will enable the technician to handle many of the common faults encountered in the aircraft.

Before starting a discussion on basic circuits and troubleshooting, the following definitions are given.

- *Short circuit*—an unintentional low resistance path between two components in a circuit or between a component/conductor and ground. It usually creates high current flow, which will burn out or cause damage to the circuit conductor or components.
- *Open circuit*—a circuit that is not a complete or continuous path. An open circuit represents an infinitely large resistance. Switches are common devices used to open and close a circuit. Sometimes a circuit will open due to a component failure, such as a light bulb or a burned out resistor.
- *Continuity*—the state of being continuous, uninterrupted or connected together; the opposite of a circuit that is not broken or does not have an open.
- *Discontinuity*—the opposite of continuity, indicating that a circuit is broken or not continuous.

Voltage Measurement

Voltage is measured across a component with a voltmeter or the voltmeter position on a multimeter. Usually, there is a DC and an AC selection on the meter. Before the meter is used for measurements, make sure that the meter is selected for the correct type of voltage. When placing the probes across a component to take a measurement, take care to ensure that the polarity is correct. [Figure 10-161] Standard practice is for the red meter lead to be installed in the positive (+) jack and the black meter lead to be installed in the negative meter jack (−). Then when placing the probes across or in parallel with a component to measure the voltage, the leads should match the polarity of the component. The red lead shall be on the positive side of the component and the black on the negative side, which will prevent damage to the meter or incorrect readings.

All meters have some resistance and will shunt some of the current. This has the effect of changing the characteristic of the circuit because of this change in current. This is typically more of a concern with older analog type meters. If there are any questions about the magnitude of the voltage across a component, then the meter should be set to measure on the highest voltage range. This will prevent the meter from "pegging" and possible damage. The range should then be selected

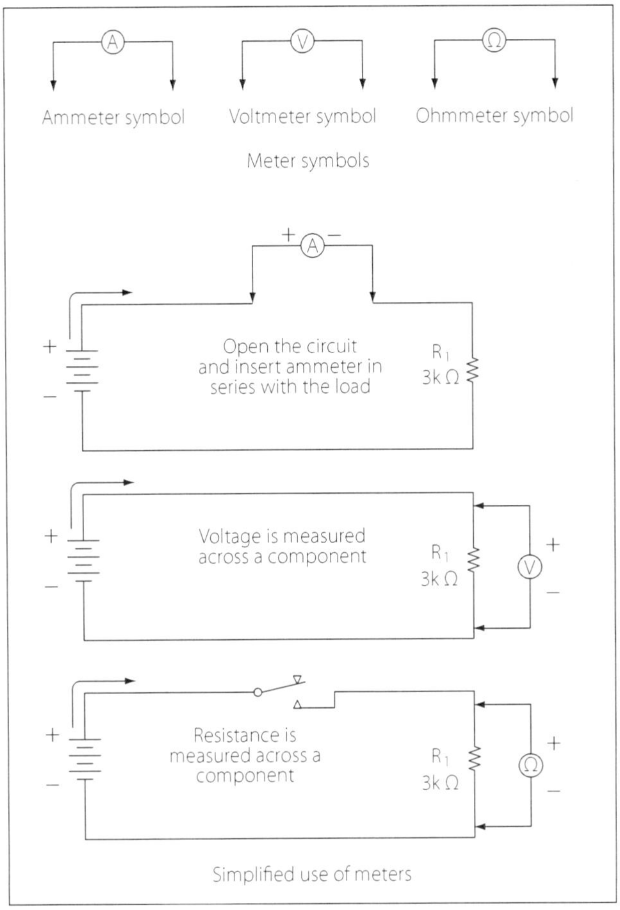

Figure 10-161. Current measurement.

to low values until the measured voltage is read at the mid-scale deflection. Readings taken at mid-scale are the most accurate.

Current Measurement

Current is measured with the ammeter connected in the current path by opening or breaking the circuit and inserting the meter in series as shown in Figure 10-161. Standard practice is for the red meter lead to be installed in the positive (+) jack and the black meter lead to be installed in the negative meter jack (−). The positive side of the meter is connected towards the positive voltage source. Ideally, the meter should not alter the current and influence the circuit and the measurements. However, the meter does have some effect because of its internal resistance that is connected with the rest of the circuit in series. The resistance is rather small and for most practical purposes, this can be neglected.

Checking Resistance in a Circuit

The ohmmeter is used to measure the resistance. In its more basic form, the ohmmeter consists of a variable resistor in series with a meter movement and a voltage source. The meter must first be adjusted before use.

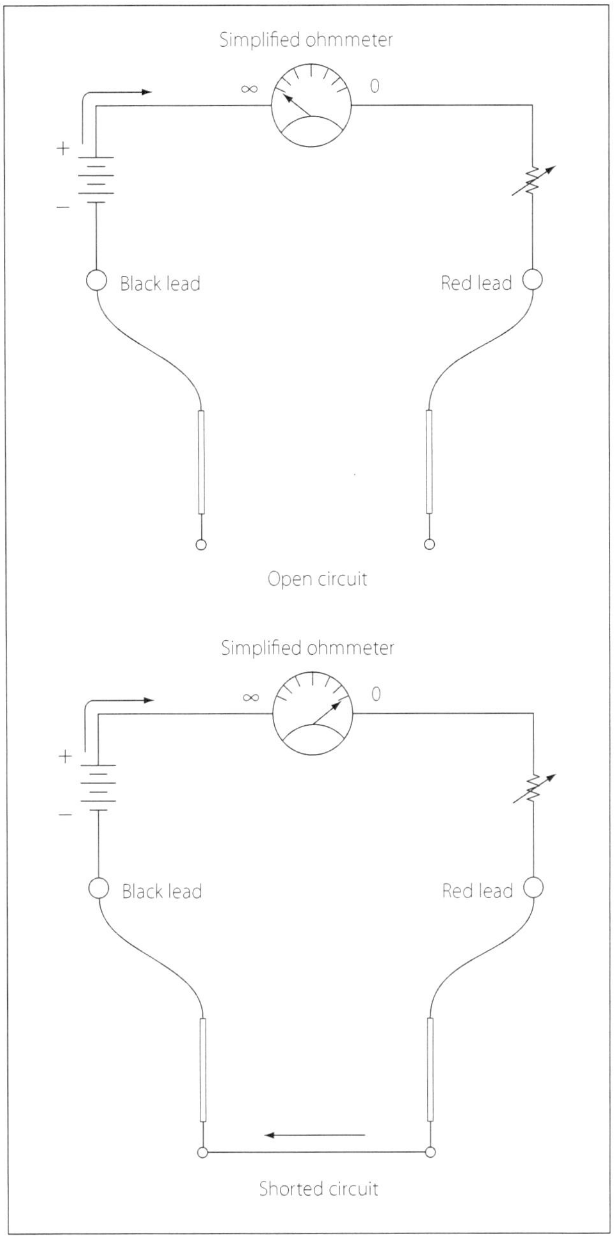

Figure 10-162. Meter configurations during adjustments.

Refer to Figure 10-162 for meter configurations during adjustments. When the meter leads are not connected (open), the needle will point to the full left-hand position, indicating infinite resistance or and open circuit. With the lead placed together, the circuit is shorted as shown with the meter needle to the full right-hand position. When a connection is made, the internal battery is allowed to produce a current through the movement coil, causing a deflection of the needle in proportion to the value of the external resistance. In this case, the resistance is zero because the leads are shorted.

The purpose of the variable resistor in the meter is to adjust the current so that the pointer will read exactly zero when the leads are shorted. This is needed because as the battery continues to be used, the voltage will change, thus requiring an adjustment. The meter should be "zeroed" before each use.

To check the value of a resistor, the resistor must be disconnected from the circuit. This will prevent any possible damage to the ohmmeter, and it will prevent the possibility of any inaccurate readings due to the circuit being in parallel with the resistor in question. [Figure 10-163]

Continuity Checks

In many cases, the ohmmeter is not used for measuring the resistance of a component but to simply check the integrity of a connection from one portion of a circuit to another. If there is a good connection, then the ohmmeter will read a near zero resistance or a short. If the circuit is open or has a very poor connection at some point like an over-crimped pin in a connector, then the ohmmeter will read infinity or some very high resistance. Keep in mind that while any measurement is being taken, contact with the circuit or probes should be avoided. Contact can introduce another parallel path and provide misleading indications.

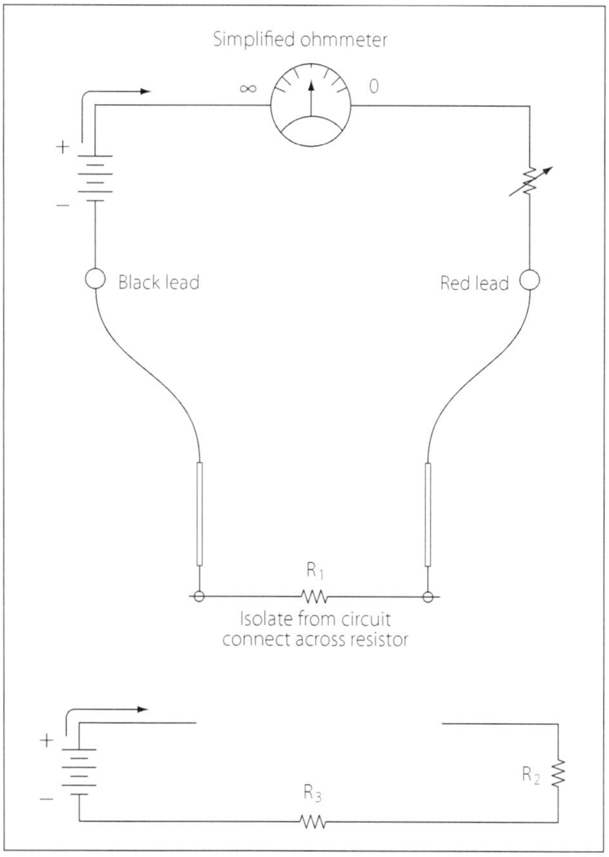

Figure 10-163. Meter adjustment.

Capacitance Measurement

Figure 10-164 illustrates a basic test of a capacitor with an ohmmeter. There are usually two common modes of fail for a capacitor. One is a complete failure characterized by short circuit through the capacitor due to the dielectric breaking down or an open circuit. The more insidious failure occurs due to degradation, which is a gradual deterioration of the capacitor's characteristics.

If a problem is suspected, remove the capacitor from the circuit and check with an ohmmeter. The first step

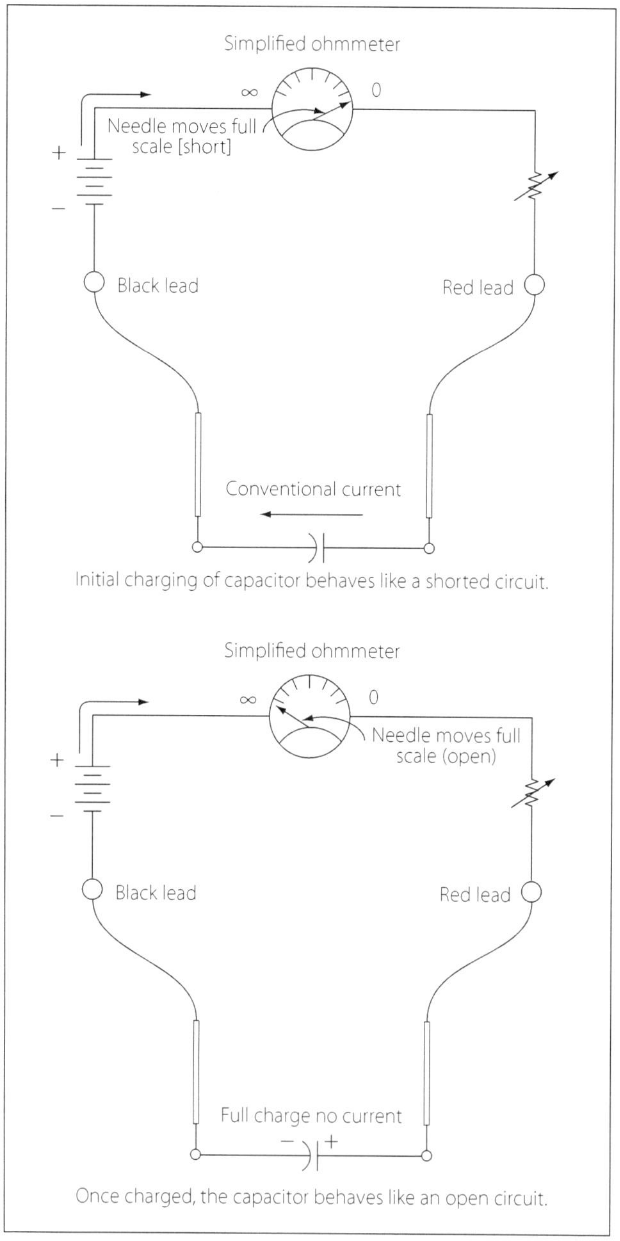

Figure 10-164. Basic test of a capacitor with an ohmmeter.

is to short the two leads of the capacitor to ensure that it is entirely discharged. Next, connect the two leads as shown in Figure 10-164 across the capacitor and observe the needle movement. At first, the needle should indicate a short circuit. Then as the capacitor begins to charge, the needle should move to the left or infinity and eventually indicate an open circuit. The capacitor takes its charge from the internal battery of the ohmmeter. The greater the capacitance, the longer it will take to charge. If the capacitor is shorted, then the needle will remain at a very low or shorted resistance. If there is some internal deterioration of the dielectric, then the needle will never reach a high resistance but some intermediate value, indicating a current.

Inductance Measurement

The common mode of failure in an inductor is an open. To check the integrity of an inductor, it must be removed from the circuit and tested as an isolated component just like the capacitor. If there is an open in the inductor, a simple check with an ohmmeter will show it as an open circuit with infinite resistance. If in fact the inductor is in good condition, then the ohmmeter will indicate the resistance of the coil.

On occasions, the inductor will fail due to overheating. When the inductor is overheated, it is possible for the insulation covering the wire in the coil to melt, causing a short. The effects of a shorted coil are that of reducing the number of turns. At this point, further testing of the inductor must be done with test equipment not covered in this text.

Troubleshooting the Open Faults in Series Circuit

One of the most common modes of failure is the "open" circuit. A component, such as a resistor, can overheat due to the power rating being exceeded. Other more frustrating problems can happen when a "cold" solder joint cracks leaving a wire disconnected from a relay or connector. This type of damage can occur during routine maintenance after a technician has accessed an area for inspections. In many cases, there is no visual indication that a failure has occurred, and the soon-to-be-frustrated technician is unaware that there is a problem until power is reapplied to the aircraft in the final days leading up to aircraft delivery and scheduled operations.

The first example is a simplified diagram shown in Figures 10-165 through 10-167. The circuit depicted in Figure 10-165 is designed to cause current to flow through a lamp, but because of the open resistor, the lamp will not light. To locate this open, a voltmeter or an ohmmeter should be used.

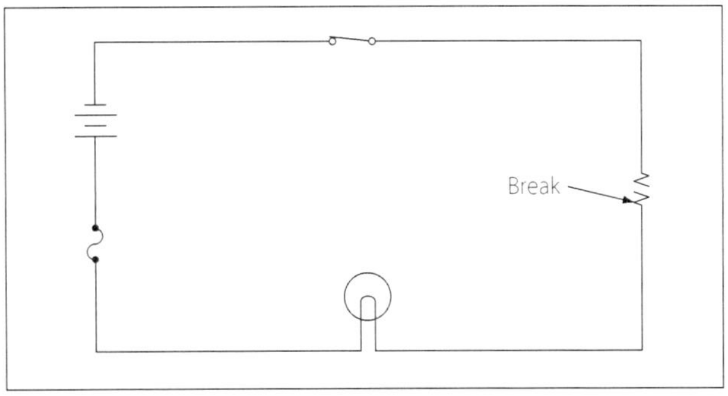

Figure 10-165. An open circuit.

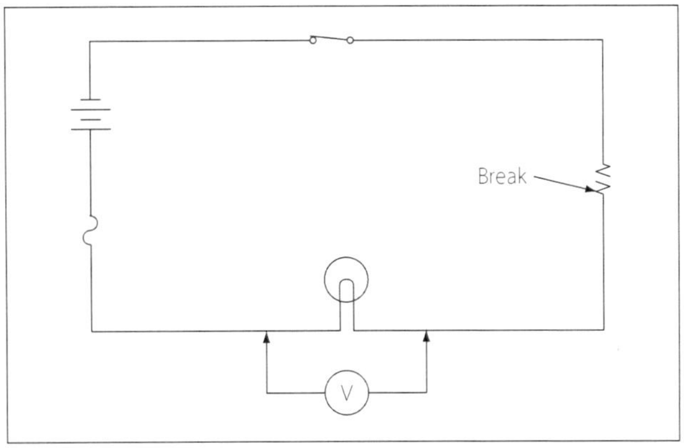

Figure 10-166. Voltmeter across a lamp in an open circuit.

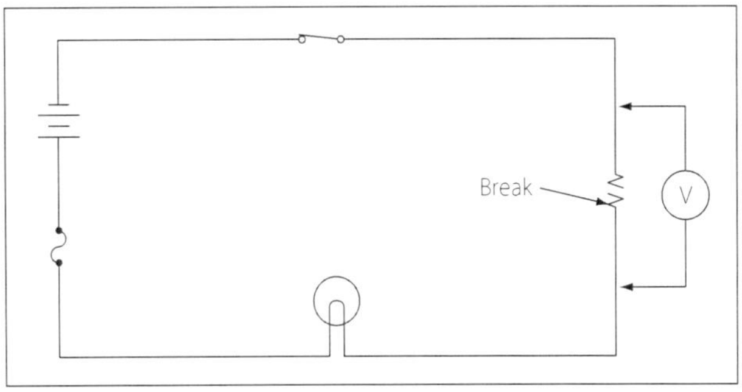

Figure 10-167. Voltmeter across a resistor in an open circuit.

Tracing Opens with the Voltmeter

A general procedure to follow in this case is to measure the voltage drop across each component in the circuit, keeping in mind the following points. If there is an open in a series circuit, then the voltage drops on sides of the component. In this case, the total voltage must appear across the open resistor as per Kirchhoff's voltage law.

If a voltmeter is connected across the lamp, as shown in Figure 10-166, the voltmeter will read zero. Since no current can flow in the circuit because of the open resistor, there is no voltage drop across the lamp indicating that the lamp is good.

Next, the voltmeter is connected across the open resistor, as shown in Figure 10-167. The voltmeter has closed the circuit by shunting (paralleling) the burned out resistor, allowing current to flow. Current will flow from the negative terminal of the battery, through the switch, through the voltmeter and the lamp, back to the positive terminal of the battery. However, the resistance of the voltmeter is so high that only a very small current flows in the circuit. The current is too small to light the lamp, but the voltmeter will read the battery voltage.

Tracing Opens with the Ohmmeter

A simplified circuit as shown in Figures 10-168 and 10-169 illustrates how to locate an open in a series circuit using the ohmmeter. A general rule to keep in mind when troubleshooting with an ohmmeter is: when an ohmmeter is properly connected across a circuit component and a resistance reading is obtained, the component has continuity and is not open.

When an ohmmeter is used, the circuit component to be tested must be isolated and the power source removed from the circuit. In this case, as shown in Figure 10-168, these requirements can be met by opening the circuit switch. The ohmmeter is zeroed and across all good components will be zero. The voltage drop across the open component will equal the total voltage across the series combination. This condition happens because the open component will prevent current to pass through the series circuit. With there being no current, there can be no voltage drop across any of the good components. Because the current is zero, it can be determined by Ohm's law that E = IR = 0 volts across a component. The voltage is the same on both places across (in parallel with) the lamp. In this testing configuration, some value of resistance is read indicating that the lamp is in good condition and is not the source of the open in the circuit.

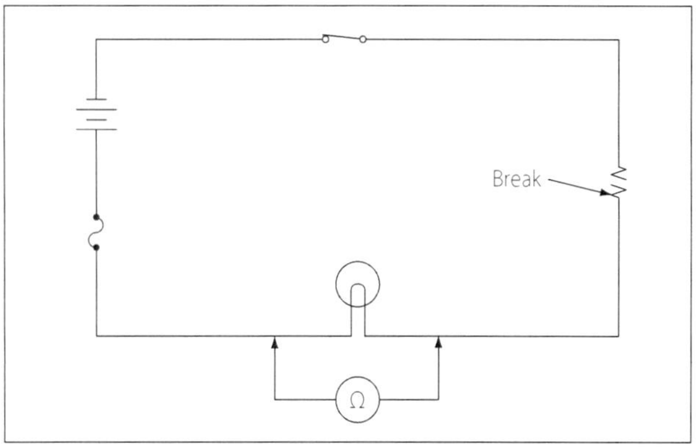

Figure 10-168. Using an ohmmeter to check a circuit component.

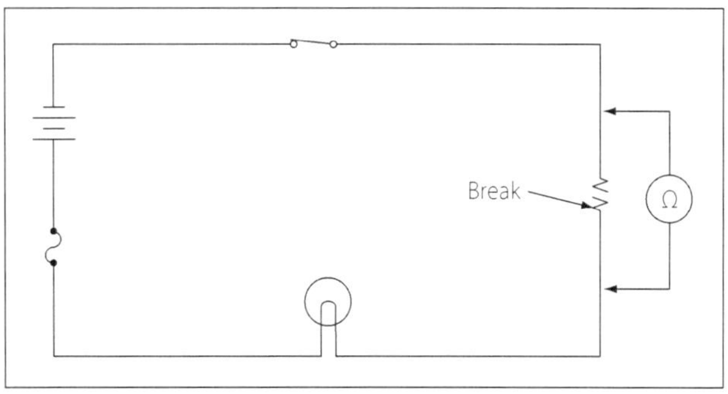

Figure 10-169. Using an ohmmeter to locate an open in a circuit component.

Now the technician should move to the resistor and place the ohmmeter probe across it as shown in Figure 10-169. When the ohmmeter is connected across the open resistor, it indicates infinite resistance, or a discontinuity. Thus, the circuit open has now been located.

Troubleshooting the Shorting Faults in Series Circuit

An open fault can cause a component or system not to work, which can be critical and hazardous. A shorting fault can potentially be more of a severe nature than the open type of fault. A short circuit, or "short," will cause the opposite effect. A short across a series circuit produces a greater than normal current flow. Faults of this type can develop slowly when a wire bundle is not properly secured and is allowed to chafe against the airframe structure or other systems such as hydraulic lines. Shorts can also occur due to a careless technician using incorrect hardware when installing an interior. If screws that are too long are used to install trim, it is possible to penetrate a wire bundle immediately causing numerous shorts. Worse yet, are the shorts that are not immediately seen but "latent" and do not show symptoms until the aircraft is in service. Another point to keep in mind is when closing panels. Wires can become pinched between the panel and the airframe causing either a short or a latent, intermittent short. The simplified circuit, shown in Figures 10-170 through 10-172, and Figure 10-173 will be used to illustrate troubleshooting a short in a series circuit.

In Figure 10-170, a circuit is designed to light a lamp. A resistor is connected in the circuit to limit current flow. If the resistor is shorted, as shown in the illustration, the current flow will increase and the lamp will become brighter. If the applied voltage were high enough, the lamp would burn out, but in this case the fuse would protect the lamp by opening first.

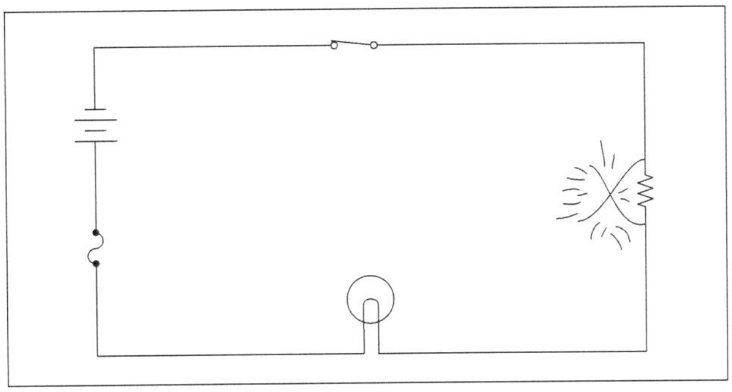

Figure 10-170. A shorted resistor.

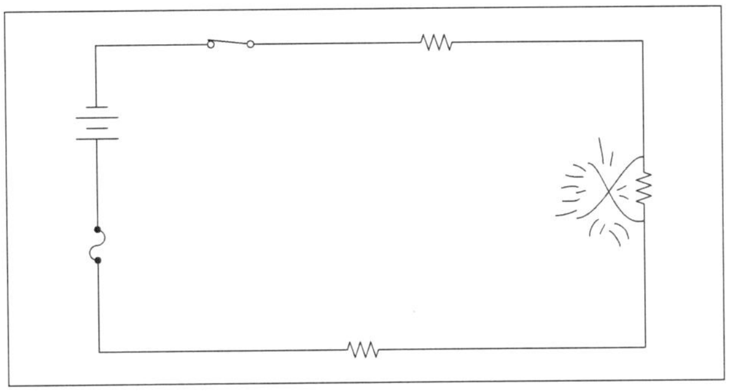

Figure 10-171. A short that does not open the circuit.

Usually a short circuit will produce an open circuit by either blowing (opening) the fuse or burning out a circuit component. But in some circuits, such as that illustrated in Figure 10-171, there may be additional resistors which will not allow one shorted resistor to increase the current flow enough to blow the fuse or burn out a component. Thus, with one resistor shorted out, the circuit will still function since the power dissipated by the other resistors does not exceed the rating of the fuse.

Tracing Shorts with the Ohmmeter

The shorted resistor shown in Figure 10-172 can be located with an ohmmeter. First the switch is opened to isolate the circuit components. In Figure 10-172, this circuit is shown with an ohmmeter connected across each of the resistors. Only the ohmmeter connected across the shorted resistor shows a zero reading, indicating that this resistor is shorted.

Tracing Shorts with the Voltmeter

To locate the shorted resistor while the circuit is functioning, a voltmeter can be used. Figure 10-173 illustrates that when a voltmeter is connected across any of the resistors, which are not shorted, a portion of the applied voltage will be indicated on the voltmeter scale. When it is connected across the shorted resistor, the voltmeter will read zero.

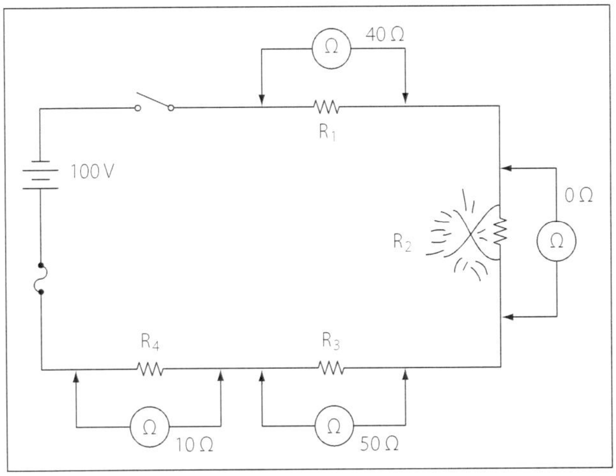

Figure 10-172. Using an ohmmeter to locate a shorted resistor.

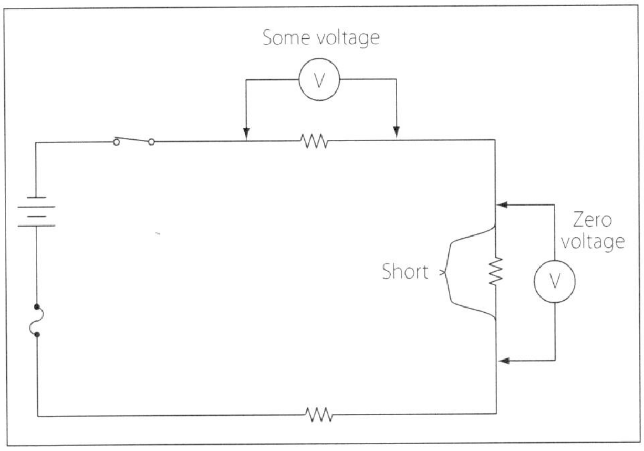

Figure 10-173. Voltmeter connected across resistors.

Troubleshooting the Open Faults in Parallel Circuit

The procedures used in troubleshooting a parallel circuit are sometimes different from those used in a series circuit. Unlike a series circuit, a parallel circuit has more than one path in which current flows. A voltmeter cannot be used, since, when it is placed across an open resistor, it will read the voltage drop in a parallel branch. But an ammeter or the modified use of an ohmmeter can be employed to detect an open branch in a parallel circuit.

If the open resistor shown in Figure 10-174 was not visually apparent, the circuit might appear to be functioning properly, because current would continue to flow in the other two branches of the circuit. To determine that the circuit is not operating properly, a determination must be made as to how the circuit should behave when working properly. First, the total resistance, total current, and the branch currents of the circuit should be calculated as if there were no open

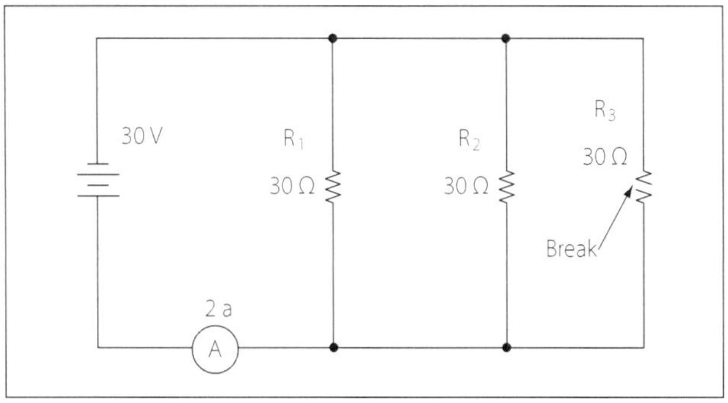

Figure 10-174. Finding an open branch in a parallel circuit.

in the circuit. In this case, the total resistance can be simply determined by:

$$R_T = \frac{R}{N}$$

Where R_T is the total circuit resistance
N is the number of resistors
R is the resistor value

$$R_T = \frac{30\ \Omega}{3} = 10\ \Omega$$

The total current of the circuit can now be determined by using Ohm's law:

$$I_T = \frac{E_S}{R_T}$$

Where I_T is the total current
E_S is the source voltage across the parallel branch
R_T is the total resistance of the parallel branch

$$I_T = \frac{30\ \text{v}}{10\ \Omega} = 3 \text{ amperes (total current)}$$

Each branch current should be determined in a similar manner. For the first branch, the current is:

$$I_1 = \frac{E_S}{R_1}$$

Where I_1 is the current in the first branch
E_S is the source voltage across the parallel branch
R_1 is the resistance of the first branch

$$I_1 = \frac{30\ \text{v}}{30\ \Omega} = 1 \text{ amperes}$$

Because the other two branches are of the same resistive value, then the current in each of those branches will be 1 ampere also. Adding up the amperes in each branch confirms the initial calculation of total current being 3 amperes.

Tracing an Open with an Ammeter

If the technician now places an ammeter in the circuit, the total current would be indicated as 2 amperes as show in Figure 10-174 instead of the calculated 3 amperes. Since 1 ampere of current should be flowing through each branch, it is obvious that one branch is open. If the ammeter is then connected into the branches, one after another, the open branch will eventually be located by a zero ammeter reading.

Tracing an Open with an Ohmmeter

A modified use of the ohmmeter can also locate this type of open. If the ohmmeter is connected across the open resistor, as shown in Figure 10-175, an erroneous reading of continuity would be obtained. Even though the circuit switch is open, the open resistor is still in parallel with R_1 and R_2, and the ohmmeter would indicate the open resistor had a resistance of 15 ohms, the equivalent resistance of the parallel combination of R_1 and R_2.

Therefore, it is necessary to open the circuit as shown in Figure 10-176 in order to check the resistance of R_3. In this way, the resistor is not shunted (paralleled) by R_1 and R_2. The reading on the ohmmeter will now

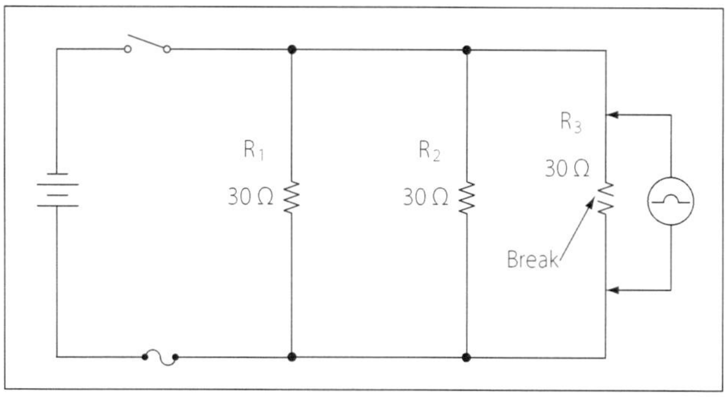

Figure 10-175. A misleading ohmmeter indication.

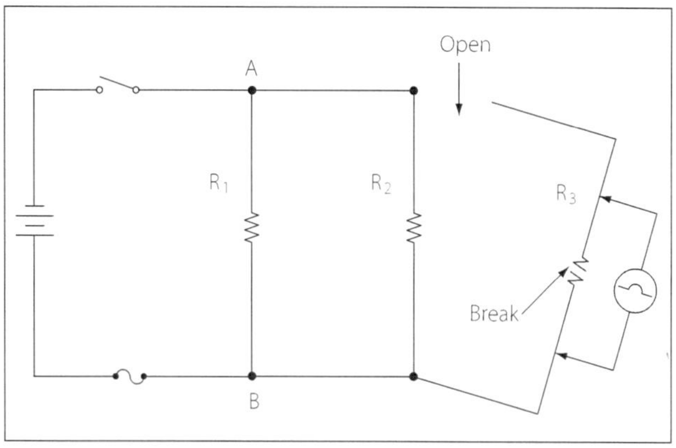

Figure 10-176. Opening a branch circuit to obtain an accurate ohmmeter reading.

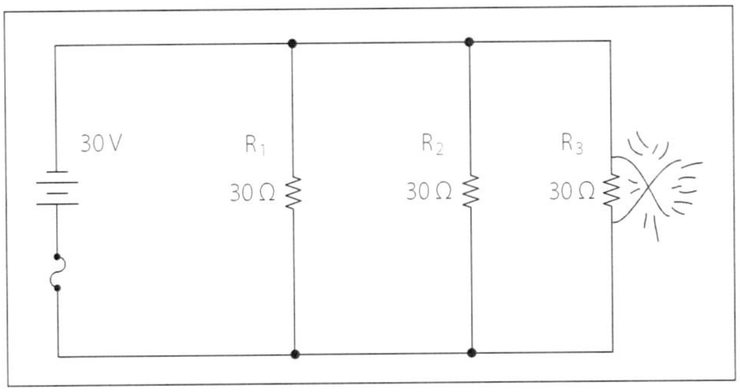

Figue 10-177. A shorted component causes the fuse to open.

indicate infinite resistance, which means the open component has been isolated.

Troubleshooting the Shorting Faults in Parallel Circuit

As in a series circuit, a short in a parallel circuit will usually cause an open circuit by blowing the fuse. But, unlike a series circuit, one shorted component in a parallel circuit will stop current flow by causing the fuse to open. Refer to the circuit in Figure 10-177. If resistor R_3 is shorted, a path of almost zero resistance will be offered the current, and all the circuit current will flow through the branch containing the shorted resistor. Since this is practically the same as connecting a wire between the terminals of the battery, the current will rise to an excessive value, and the fuse will open. Since the fuse opens almost as soon as a resistor shorts out, there is no time to perform a current or voltage check. Thus, troubleshooting a parallel DC circuit for a shorted component should be accomplished with an ohmmeter. But, as in the case of checking for an open resistor in a parallel circuit, a shorted resistor can be detected with an ohmmeter only if one end of the shorted resistor is disconnected and isolated from the rest of the circuit.

Troubleshooting the Shorting Faults in Series-Parallel Circuit

Logic in Tracing an Open

Troubleshooting a series-parallel resistive circuit involves locating malfunctions similar to those found in a series or a parallel circuit. Figures 10-178 through 10-180 illustrate three points of failure in a series-parallel circuit and their generalized effects.

1. In the circuit shown in Figure 10-178, an open has occurred in the series portion of the circuit. When the open occurs anywhere in the series portion of a series-parallel circuit, current flow in the entire circuit will stop. In this case, the circuit will not function, and the lamp, L_1, will not be lit.

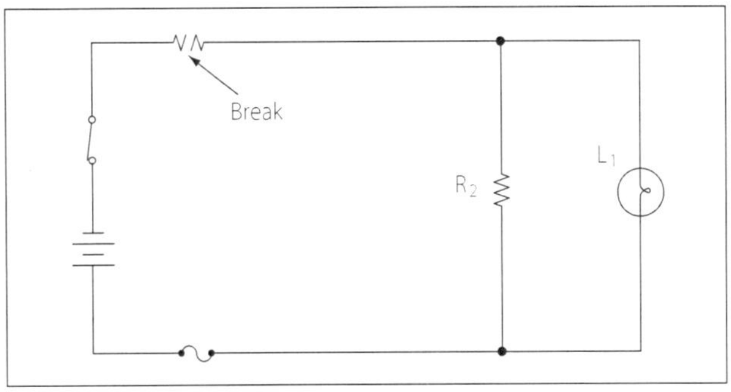

Figure 10-178. An open in the series portion of a series-parallel circuit.

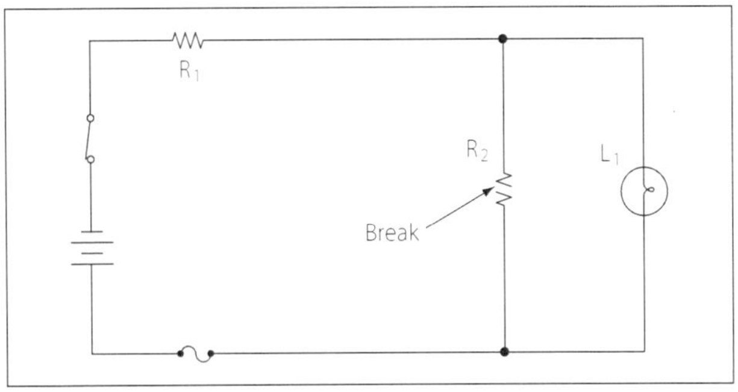

Figure 10-179. An open in the parallel portion of a series-parallel circuit.

2. If the open occurs in the parallel portion of a series-parallel circuit, as shown in Figure 10-179, part of the circuit will continue to function. In this case, the lamp will continue to burn, but its brightness will diminish, since the total resistance of the circuit has increased and the total current has decreased.
3. If the open occurs in the branch containing the lamp, as shown in Figure 10-180, the circuit will continue to function with increased resistance and decreased current, but the lamp will not light.

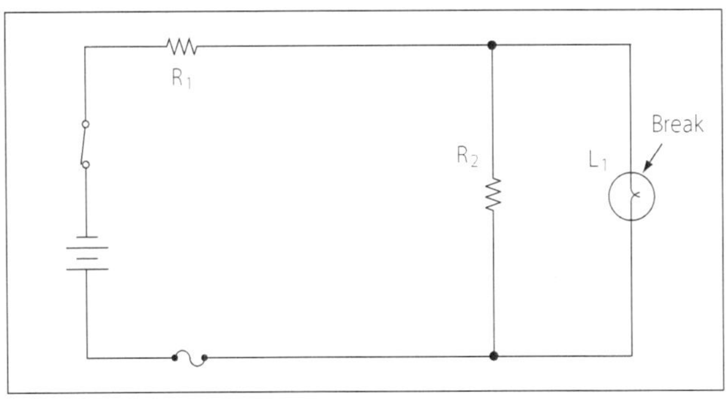

Figure 10-180. An open lamp in a series-parallel circuit.

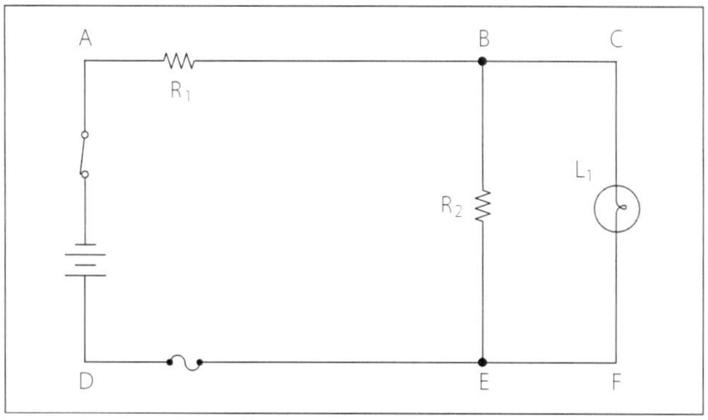

Figure 10-181. Using the voltmeter to troubleshoot a series-parallel circuit.

Tracing Opens with the Voltmeter

To explain how the voltmeter and ohmmeter can be used to troubleshoot series-parallel circuits, the circuit shown in Figure 10-181 has been labeled at various points. A point-to-point description is listed below with expected results:

1. By connecting a voltmeter between points A and D, the battery and switch can be checked for opens.
2. By connecting the voltmeter between points A and B, the voltage drop across R_1 can be checked. This voltage drop is a portion of the applied voltage.
3. If R_1 is open, the reading between B and D will be zero.
4. By connecting a voltmeter between A and E, the continuity of the conductor between the positive terminal of the battery and point E, as well as the fuse, can be checked. If the conductor or fuse is open, the voltmeter will read zero.
5. If the lamp is burning, it is obvious that no open exists in the branch containing the lamp, and the voltmeter could be used to detect an open in the branch containing R_2 by removing lamp, L_1, from the circuit.

Troubleshooting the series portion of a series-parallel circuit presents no difficulties, but in the parallel portion of the circuit, misleading readings can be obtained.

Batteries

Primary Cell

The dry cell is the most common type of primary-cell battery and is similar in its characteristics to that of an electrolytic cell. This type of a battery is basically designed with a metal electrode or graphite rod acting as the cathode (+) terminal, immersed in an electrolytic paste. This electrode⁄electrolytic build-up is then encased in a metal container, usually made of zinc, which itself acts as the anode (−) terminal. When the battery is in a discharge condition an electrochemical reaction takes place resulting in one of the metals being consumed. Because of this consumption, the charging process is not reversible. Attempting to reverse the chemical reaction in a primary cell by way of recharging is usually dangerous and can lead to a battery explosion.

These batteries are commonly used to power items such as flashlights. The most common primary cells today are found in alkaline batteries, silver-oxide and lithium batteries. The earlier carbon-zinc cells, with a carbon post as cathode and a zinc shell as anode were once prevalent but are not as common.

Secondary Cell

A secondary cell is any kind of electrolytic cell in which the electrochemical reaction that releases energy is reversible. The lead-acid car battery is a secondary-cell battery. The electrolyte is sulphuric acid (battery acid), the positive electrode is lead peroxide, and the negative electrode is lead. A typical lead-acid battery consists of six lead-acid cells in a case. Each cell produces 2 volts, so the whole battery produces a total of 12 volts.

Other commonly used secondary cell chemistry types are nickel cadmium (NiCd), nickel metal hydride (NiMH), lithium ion (Li-ion), and Lithium ion polymer (Li-ion polymer).

Lead-acid batteries used in aircraft are similar to automobile batteries. The lead acid battery is made up of a series of identical cells each containing sets of positive and negative plates. Figure 10-182 illustrates each cell contains positive plates of lead dioxide (PbO_2), negative plates of spongy lead, and electrolyte (sulfuric acid and water). A practical cell is constructed with many more plates than just two in order to get the required current output. All positive plates are connected together as well as all the negatives. Because each positive plate is always positioned between two negative plates, there are always one or more negative plates than positive plates.

Between the plates are porous separators that keep the positive and negative plates from touching each other and shorting out the cell. The separators have vertical ribs on the side facing the positive plate. This construction permits the electrolyte to circulate freely around the plates. In addition, it provides a path for sediment to settle to the bottom of the cell.

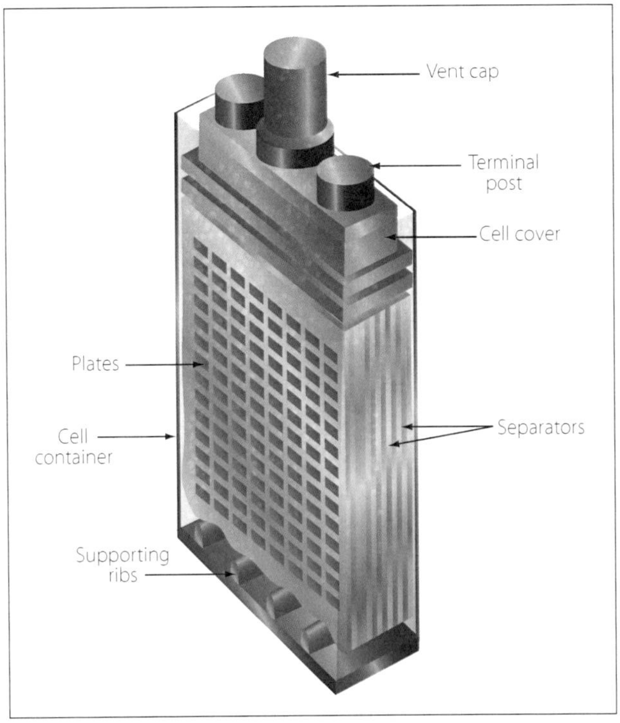

Figure 10-182. Lead-acid cell construction.

Each cell is seated in a hard rubber casing through the top of which are terminal posts and a hole into which is screwed a nonspill vent cap. The hole provides access for testing the strength of the electrolyte and adding water. The vent plug permits gases to escape from the cell with a minimum of leakage of electrolyte, regardless of the position the airplane might assume. Figure 10-183 shows the construction of the vent plug. In level flight, the lead weight permits venting of gases through a small hole. In inverted flight, this hole is covered by the lead weight.

The individual cells of the battery are connected in series by means of cell straps. [Figure 10-184] The complete assembly is enclosed in an acid resisting metal container (battery box), which serves as electrical shielding and mechanical protection. The battery box has a removable top. It also has a vent tube nipple at each end. When the battery is installed in an airplane, a vent tube is attached to each nipple. One tube is the intake tube and is exposed to the slipstream. The other is the exhaust vent tube and is attached to the battery drain sump, which is a glass jar containing a felt pad moistened with a concentrated solution of sodium bicarbonate (baking soda). With this arrangement, the airstream is directed through the battery case where battery gases are picked up, neutralized in the sump, and then expelled overboard without damage to the airplane.

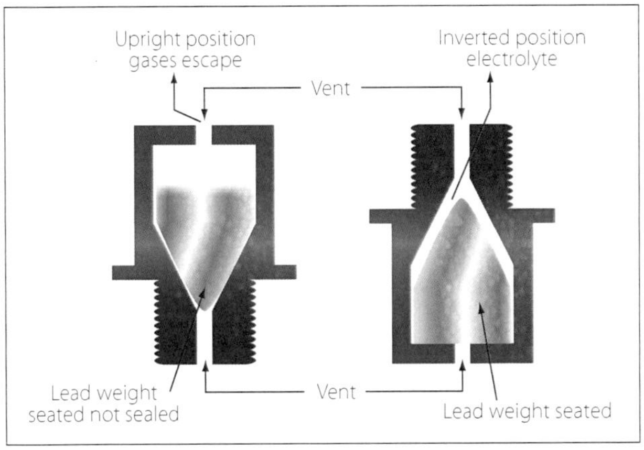

Figure 10-183. Nonspill battery vent plug.

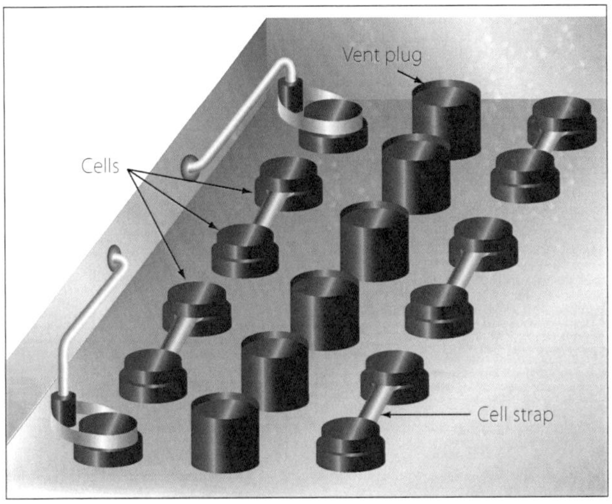

Figure 10-184. Connection of storage battery.

To facilitate installation and removal of the battery in some aircraft, a quick disconnect assembly is used to connect the power leads to the battery. This assembly attaches the battery leads in the aircraft to a receptacle mounted on the side of the battery. [Figure 10-185] The receptacle covers the battery terminal posts and prevents accidental shorting during the installation and removal of the battery. The plug consists of a socket and a handwheel with a course pitch thread. It can be readily connected to the receptacle by the handwheel. Another advantage of this assembly is that the plug can be installed in only one position, eliminating the possibility of reversing the battery leads.

The voltage of lead acid cell is approximately 2 volts in order to attain the voltage required for the application. Each cell is then connected in series with heavy gage metal straps to form a battery. In a typical battery, such as that used in a aircraft for starting, the voltage required is 12 or 24 volts. This voltage is achieved by

Figure 10-185. A battery quick-disconnect assembly.

connecting six cells or twelve cells respectively together in series and enclosing them in one plastic box.

Each cell containing the plates are filled with an electrolyte composed of sulphuric acid and distilled water with a specific gravity of 1.270 at 60 °F. This solution contains positive hydrogen ions and negative sulfate (SO_4) ions that are free to combine with other ions and form a new chemical compound. When the cell is discharged, electrons leave the negative plate and flow to the positive plates where they cause the lead dioxide (PbO_2) to break down into negative oxygen ions and positive lead ions. The negative oxygen ions join with positive hydrogen ions from the sulfuric acid and form water (H_2O). The negative sulfate ions join with the lead ions in both plates and form lead sulfate ($PbSO_4$). After the discharge, the specific gravity changes to about 1.150.

Battery Ratings

The voltage of a battery is determined by the number of cells connected in series to form the battery. Although the voltage of one lead-acid cell just removed from a charger is approximately 2.2 volts, a lead-acid cell is normally rated at approximately 2 volts. A battery rated at 12 volts consists of 6 lead-acid cells connected in series, and a battery rated at 24 volts is composed of 12 cells.

The most common battery rating is the amp-hour rating. This is a unit of measurement for battery capacity. It is determined by multiplying a current flow in amperes by the time in hours that the battery is being discharged.

A battery with a capacity of 1 amp-hour should be able to continuously supply a current of 1 amp to a load for exactly 1 hour, or 2 amps for 1/2 hour, or 1/3 amp for 3 hours, etc., before becoming completely discharged. Actually, the ampere-hour output of a particular battery depends on the rate at which it is discharged. Heavy discharge current heats the battery and decreases its efficiency and total ampere-hour output. For airplane batteries, a period of 5 hours has been established as the discharge time in rating battery capacity. However, this time of 5 hours is only a basis for rating and does not necessarily mean the length of time during which the battery is expected to furnish current. Under actual service conditions, the battery can be completely discharged within a few minutes, or it may never be discharged if the generator provides sufficient charge.

The ampere-hour capacity of a battery depends upon its total effective plate area. Connecting batteries in parallel increases ampere-hour capacity. Connecting batteries in series increases the total voltage but not the ampere-hour capacity.

Life Cycle of a Battery

Battery life cycle is defined as the number of complete charge/discharge cycles a battery can perform before its normal charge capacity falls below 80% of its initial rated capacity. Battery life can vary anywhere from 500 to 1,300 cycles. Various factors can cause deterioration of a battery and shorten its service life. The first is over-discharging, which causes excess sulphation; second, too-rapid charging or discharging which can result in overheating of the plates and shedding of active material. The accumulation of shed material, in turn, causes shorting of the plates and results in internal discharge. A battery that remains in a low or discharged condition for a long period of time may be permanently damaged. The deterioration can continue to a point where cell capacity can drop to 80% after 1,000 cycles. In a lot of cases the cell can continue working to nearly 2,000 cycles but with a diminished capacity of 60% of its original state.

Lead-Acid Battery Testing Methods

The state of charge of a storage battery depends upon the condition of its active materials, primarily the plates. However, the state of charge of a battery is indicated by the density of the electrolyte and is checked by a hydrometer, an instrument that measures the specific gravity (weight as compared with water) of liquids.

The most commonly used hydrometer consists of a small sealed glass tube weighted at its lower end so it

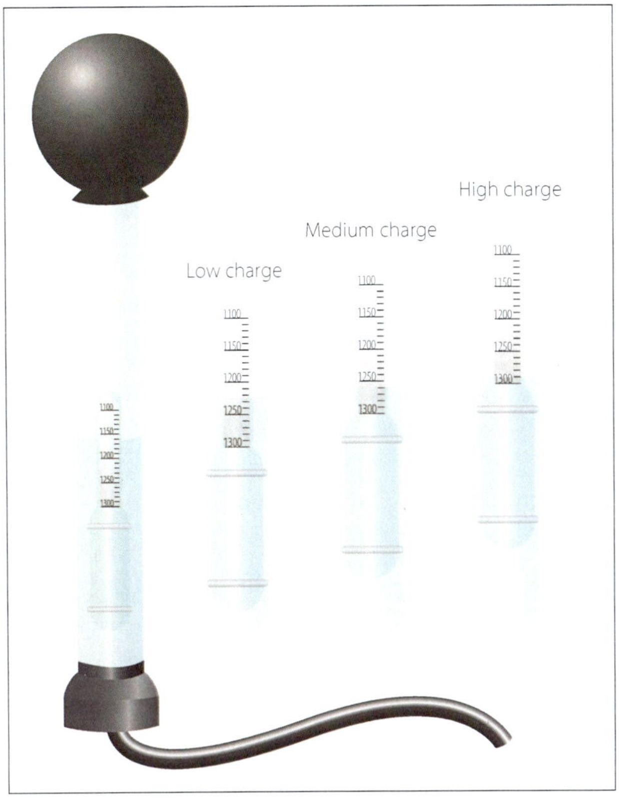

Figure 10-186. Hydrometer (specific gravity readings).

will float upright. [Figure 10-186] Within the narrow stem of the tube is a paper scale with a range of 1.100 to 1.300. When a hydrometer is used, a quantity of electrolyte sufficient to float the hydrometer is drawn up into the syringe. The depth to which the hydrometer sinks into the electrolyte is determined by the density of the electrolyte, and the scale value indicated at the level of the electrolyte is its specific gravity. The more dense the electrolyte, the higher the hydrometer will float; therefore, the highest number on the scale (1.300) is at the lower end of the hydrometer scale.

In a new, fully charged aircraft storage battery, the electrolyte is approximately 30 percent acid and 70 percent water (by volume) and is 1.300 times as heavy as pure water. During discharge, the solution (electrolyte) becomes less dense and its specific gravity drops below 1.300. A specific gravity reading between 1.300 and 1.275 indicates a high state of charge; between 1.275 and 1.240, a medium state of charge; and between 1.240 and 1.200, a low state of charge. Aircraft batteries are generally of small capacity but are subject to heavy loads. The values specified for state of charge are therefore rather high. Hydrometer tests are made periodically on all storage batteries installed in aircraft. An aircraft battery in a low state of charge may have perhaps 50 percent charge remaining, but is nevertheless considered low in the face of heavy demands that would soon exhaust it. A battery in such a state of charge is considered in need of immediate recharging.

When a battery is tested using a hydrometer, the temperature of the electrolyte must be taken into consideration. The specific gravity readings on the hydrometer will vary from the actual specific gravity as the temperature changes. No correction is necessary when the temperature is between 70 °F and 90 °F, since the variation is not great enough to consider. When temperatures are greater than 90 °F or less than 70 °F, it is necessary to apply a correction factor. Some hydrometers are equipped with a correction scale inside the tube. With other hydrometers, it is necessary to refer to a chart provided by the manufacturer. In both cases, the corrections should be added to, or subtracted from the reading shown on the hydrometer.

The specific gravity of a cell is reliable only if nothing has been added to the electrolyte except occasional small amounts of distilled water to replace that lost as a result of normal evaporation. Always take hydrometer readings before adding distilled water, never after. This is necessary to allow time for the water to mix thoroughly with the electrolyte and to avoid drawing up into the hydrometer syringe a sample that does not represent the true strength of the solution.

Exercise extreme care when making the hydrometer test of a lead-acid cell. Handle the electrolyte carefully because sulfuric acid will burn clothing and skin. If the acid does contact the skin, wash the area thoroughly with water and then apply bicarbonate of soda.

Lead-Acid Battery Charging Methods

Passing direct current through the battery in a direction opposite to that of the discharge current may charge a storage battery. Because of the internal resistance (IR) in the battery, the voltage of the external charging source must be greater than the open circuit voltage. For example, the open circuit voltage of a fully charged 12 cell, lead-acid battery is approximately 26.4 volts (12×2.2 volts), but approximately 28 volts are required to charge it. This larger voltage is needed for charging because of the voltage drop in the battery caused by the internal resistance. Hence, the charging voltage of a lead-acid battery must equal the open circuit voltage plus the IR drop within the battery (product of the charging current and the internal resistance).

Batteries are charged by either the constant voltage or constant current method. In the constant voltage method (Figure 10-187A), a motor generator set with

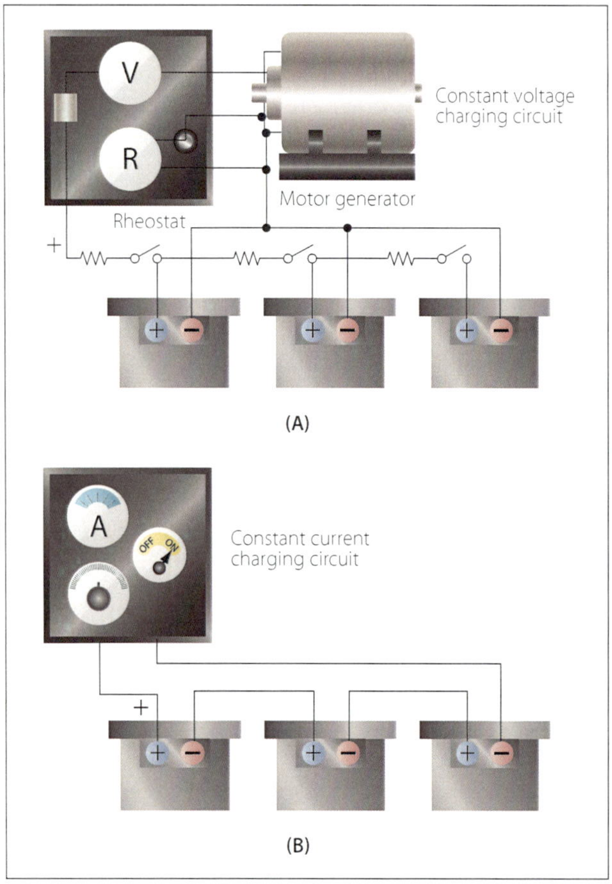

Figure 10-187. Battery charging methods.

a constant, regulated voltage forces the current through the battery. In this method, the current at the start of the process is high but automatically tapers off, reaching a value of approximately 1 ampere when the battery is fully charged. The constant voltage method requires less time and supervision than does the constant current method.

In the constant current method (Figure 10-187B), the current remains almost constant during the entire charging process.

This method requires a longer time to charge a battery fully and, toward the end of the process, presents the danger of overcharging, if care is not exercised.

In the aircraft, the storage battery is charged by direct current from the aircraft generator system. This method of charging is the constant voltage method, since the generator voltage is held constant by use of a voltage regulator.

When a storage battery is being charged, it generates a certain amount of hydrogen and oxygen. Since this is an explosive mixture, it is important to take steps to prevent ignition of the gas mixture. Loosen the vent caps and leave in place. Do not permit open flames, sparks, or other sources of ignition in the vicinity. Before disconnecting or connecting a battery to the charge, always turn off the power by means of a remote switch.

Nickel-Cadmium Batteries

Chemistry and Construction

Active materials in nickel-cadmium cells (Ni-Cad) are nickel hydrate (NiOOH) in the charged positive plate (Anode) and sponge cadmium (Cd) in the charged negative plate (Cathode). The electrolyte is a potassium hydroxide (KOH) solution in concentration of 20–34 percent by weight pure KOH in distilled water.

Sintered nickel-cadmium cells have relatively thin sintered nickel matrices forming a plate grid structure. The grid structure is highly porous and is impregnated with the active positive material (nickel-hydroxide) and the negative material (cadmium-hydroxide). The plates are then formed by sintering nickel powder to fine-mesh wire screen. In other variations of the process the active material in the sintered matrix is converted chemically, or thermally, to an active state and then formed. In general, there are many steps to these cycles of impregnation and formation. Thin sintered plate cells are ideally suited for very high rate charge and discharge service. Pocket plate nickel-cadmium cells have the positive or negative active material, pressed into pockets of perforated nickel plated steel plates or into tubes. The active material is trapped securely in contact with a metal current collector so active material shedding is largely eliminated. Plate designs vary in thickness depending upon cycling service requirements. The typical open circuit cell voltage of a nickel-cadmium battery is about 1.25 volts.

Operation of Nickel-Cadmium Cells

When a charging current is applied to a nickel-cadmium battery, the negative plates lose oxygen and begin forming metallic cadmium. The active material of the positive plates, nickel-hydroxide, becomes more highly oxidized. This process continues while the charging current is applied or until all the oxygen is removed from the negative plates and only cadmium remains.

Toward the end of the charging cycle, the cells emit gas. This will also occur if the cells are overcharged. This gas is caused by decomposition of the water in the electrolyte into hydrogen at the negative plates and oxygen at the positive plates. The voltage used during charging, as well as the temperature, determines when gassing will occur. To completely charge a nickel-cadmium battery, some gassing, however slight, must take place; thus, some water will be used.

The chemical action is reversed during discharge. The positive plates slowly give up oxygen, which is regained by the negative plates. This process results in the conversion of the chemical energy into electrical energy. During discharge, the plates absorb a quantity of the electrolyte. On recharge, the level of the electrolyte rises and, at full charge, the electrolyte will be at its highest level. Therefore, water should be added only when the battery is fully charged.

The nickel-cadmium battery is usually interchangeable with the lead-acid type. When replacing a lead-acid battery with a nickel-cadmium battery, the battery compartment must be clean, dry, and free of all traces of acid from the old battery. The compartment must be washed out and neutralized with ammonia or boric acid solution, allowed to dry thoroughly, and then painted with an alkali resisting varnish.

The pad in the battery sump jar should be saturated with a three percent (by weight) solution of boric acid and water before connecting the battery vent system.

General Maintenance and Safety Precautions

Refer to the battery manufacturer for detailed service instructions. Below are general recommendations for maintenance and safety precautions. For vented nickel-cadmium cells, the general maintenance requirements are:

1. Hydrate cells to supply water lost during overcharging.
2. Maintain inter-cell connectors at proper torque values.
3. Keep cell tops and exposed sides clean and dry.

Electrolyte spillage can form grounding paths. White moss around vent cap seals is potassium carbonate (K_2CO_3). Clean up these surfaces with distilled water and dry. While handling the caustic potassium hydroxide electrolyte, wear safety goggles to protect the eyes. The technician should also wear plastic gloves and an apron to protect skin and clothes. In case of spillage on hands or clothes, neutralize the alkali immediately with vinegar or dilute boric acid solution (one pound per gallon of water); then rinse with clear water.

During overcharging conditions, explosive mixtures of hydrogen and oxygen develop in nickel-cadmium cells. When this occurs, the cell relief valves vent these gases to the atmosphere, creating a potentially explosive hazard. Additionally, room ventilation should be such as to prevent a hydrogen build up in closed spaces from exceeding one percent by volume. Explosions can occur at concentrations above four percent by volume in air.

Sealed Lead Acid Batteries

In many applications, sealed lead acid (SLA) batteries are gaining in use over the Ni-Cad batteries. One leading characteristic of Ni-Cad batteries is that they perform well in low voltage, full-discharge, high cycle applications. However, they do not perform as well in extended standby applications, such as auxiliary or as emergency battery packs used to power inertial reference units or stand-by equipment (attitude gyro).

It is typical during the servicing of a Ni-Cad battery to match as many as twenty individual cells in order to prevent unbalance and thus cell reversal during end of discharge. When a Ni-Cad does reverse, very high pressure and heat can result. The result is often pressure seal rupture, and in the worst case, a cell explosion. With SLA batteries, cell matching is inherent in each battery. Ni-Cads also have an undesirable characteristic caused by constant overcharge and infrequent discharges, as in standby applications. It is technically known as "voltage depression" and commonly but erroneously called "memory effect." This characteristic is only detectable when a full discharge is attempted. Thus, it is possible to believe a full charge exists, while in fact it does not. SLA batteries do not have this characteristic voltage depression (memory) phenomenon, and therefore do not require scheduled deep cycle maintenance as do Ni-Cads.

The Ni-Cad emergency battery pack requires relatively complicated test equipment due to the complex characteristics of the Ni-Cad. Sealed lead acid batteries do not have these temperamental characteristics and therefore it is not necessary to purchase special battery maintenance equipment. Some manufactures of SLA batteries have included in the battery packs a means by which the battery can be tested while still installed on the aircraft. Ni-Cads must have a scheduled energy test performed on the bench due to the inability to measure their energy level on the aircraft, and because of their notable "memory" shortcoming.

The SLA battery can be designed to alert the technician if a battery is failing. Furthermore, it may be possible to test the failure detection circuits by activating a Built in Test (BITE) button. This practice significantly reduces FAA paperwork and maintenance workload.

Inverters

An inverter is used in some aircraft systems to convert a portion of the aircraft's DC power to AC. This AC is used mainly for instruments, radio, radar, lighting, and other accessories. These inverters are usually built to supply current at a frequency of 400 cps, but some are designed to provide more than one voltage; for example, 26 volt AC in one winding and 115 volts in another.

There are two basic types of inverters: the rotary and the static. Either type can be single phase or multiphase. The multiphase inverter is lighter for the same power rating than the single phase, but there are complications in distributing multiphase power and in keeping the loads balanced.

Rotary Inverters

There are many sizes, types, and configurations of rotary inverters. Such inverters are essentially AC generators and DC motors in one housing. The generator field, or armature, and the motor field, or armature, are mounted on a common shaft that will rotate within the housing. One common type of rotary inverter is the permanent magnet inverter.

Permanent Magnet Rotary Inverter

A permanent magnet inverter is composed of a DC motor and a permanent magnet AC generator assembly. Each has a separate stator mounted within a common housing. The motor armature is mounted on a rotor and connected to the DC supply through a commutator and brush assembly. The motor field windings are mounted on the housing and connected directly to the DC supply. A permanent magnet rotor is mounted at the

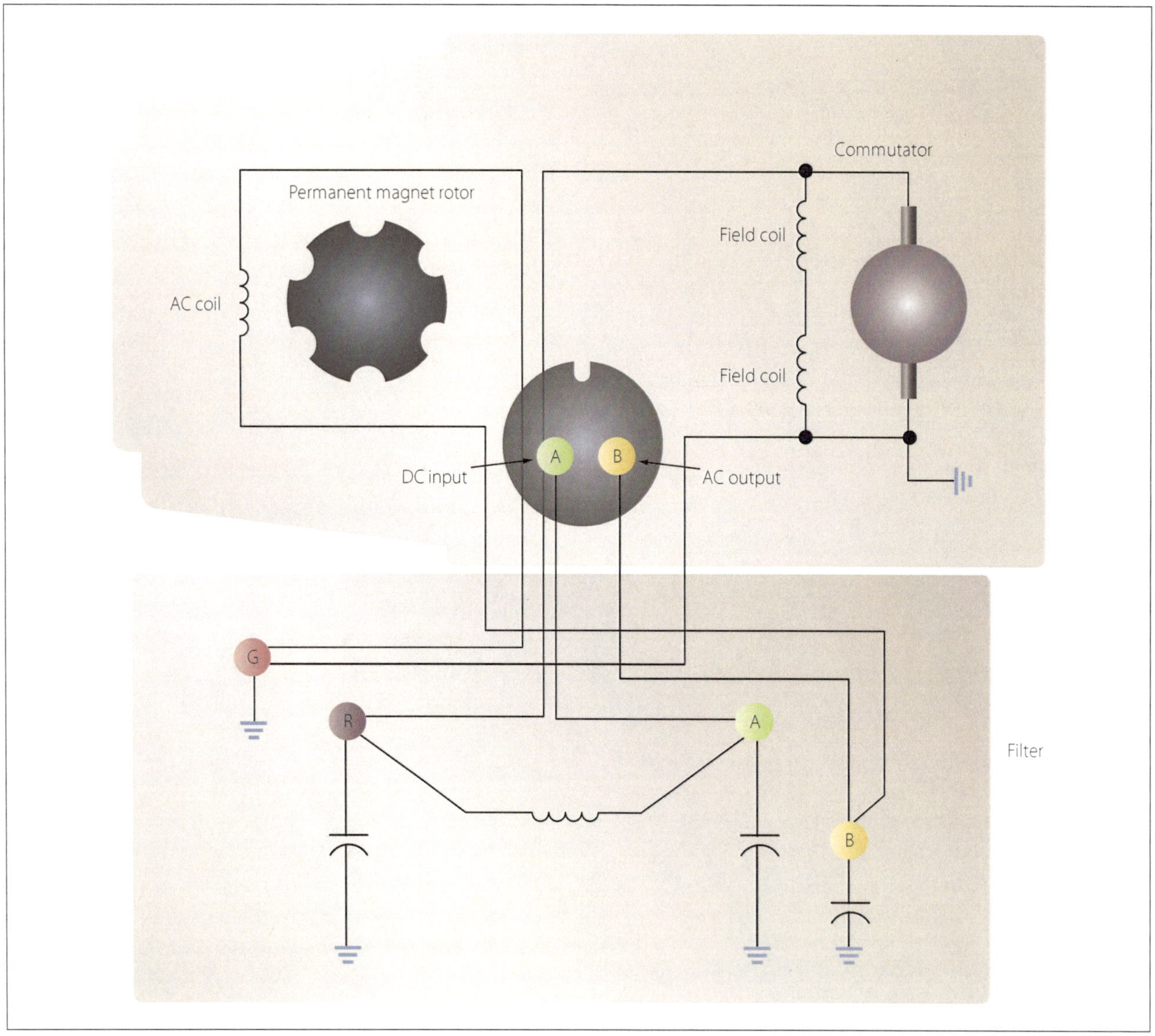

Figure 10-188. Internal wiring diagram of single-phase permanent magnet rotary inverter.

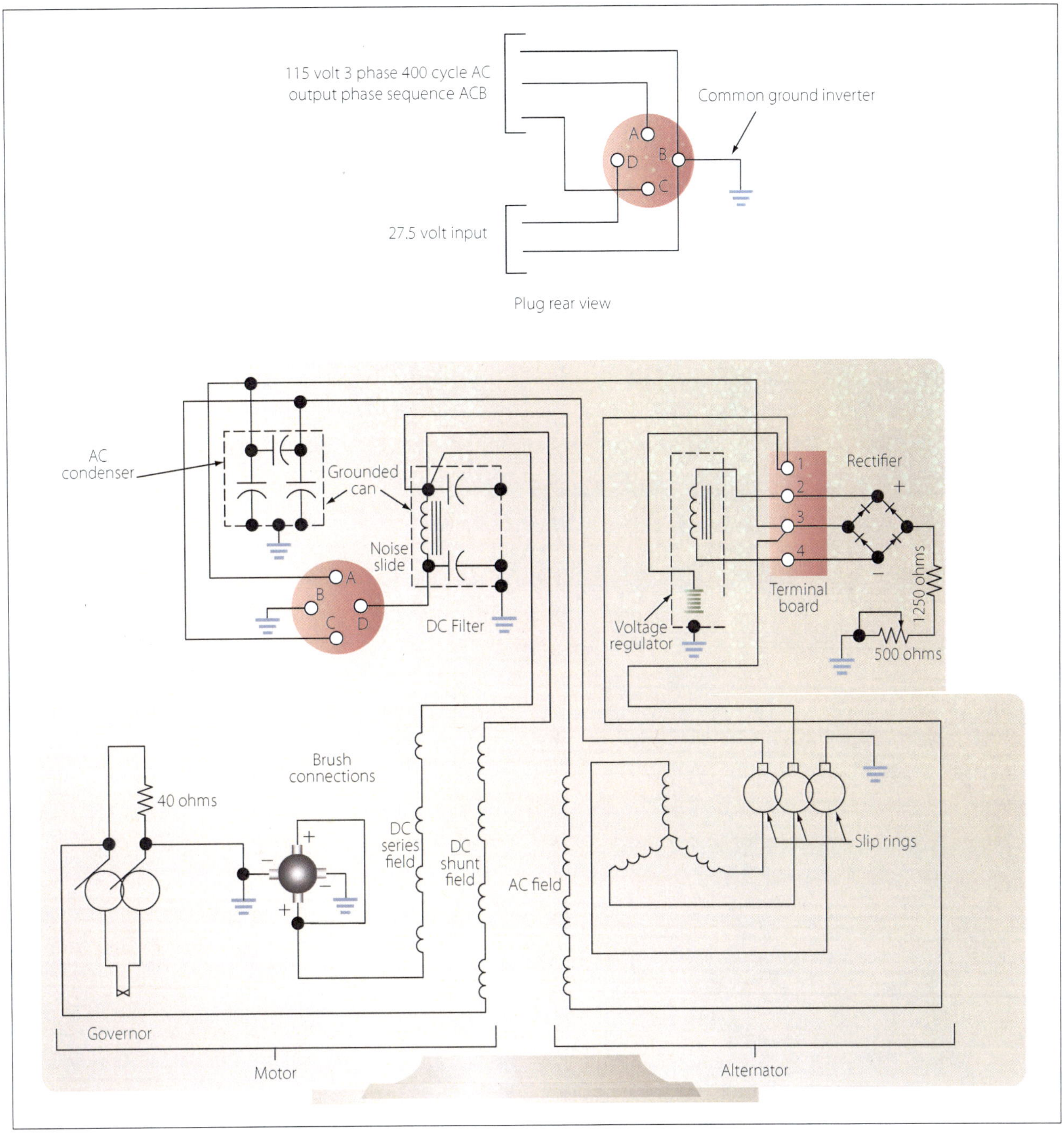

Figure 10-189. Internal wiring diagram of three-phase, revolving armature.

opposite end of the same shaft as the motor armature, and the stator windings are mounted on the housing, allowing AC to be taken from the inverter without the use of brushes. Figure 10-188 shows an internal wiring diagram for this type of rotary inverter. The generator rotor has six poles, magnetized to provide alternate north and south poles about its circumference.

When the motor field and armature are excited, the rotor will begin to turn. As the rotor turns, the permanent magnet will rotate within the AC stator coils, and the magnetic flux developed by the permanent magnets will be cut by the conductors in the AC stator coils. An AC voltage will be produced in the windings whose polarity will change as each pole passes the windings.

This type inverter may be made multiphase by placing more AC stator coils in the housing in order to shift the phase the proper amount in each coil.

As the name of the rotary inverter indicates, it has a revolving armature in the AC generator section. The illustration in Figure 10-189 shows the diagram of a revolving armature, three phase inverter.

The DC motor in this inverter is a four pole, compound wound motor. The four field coils consist of many turns of fine wire, with a few turns of heavy wire placed on top. The fine wire is the shunt field, connected to the DC source through a filter and to ground through a centrifugal governor. The heavy wire is the series field, which is connected in series with the motor armature. The centrifugal governor controls the speed by shunting a resistor that is in series with the shunt field when the motor reaches a certain speed.

The alternator is a three-phase, four-pole, star-connected AC generator. The DC input is supplied to the generator field coils and connected to ground through a voltage regulator. The output is taken off the armature through three slip rings to provide three-phase power.

The inverter would be a single-phase inverter if it had a single armature winding and one slip ring.

The frequency of this type unit is determined by the speed of the motor and the number of generator poles.

Inductor-Type Rotary Inverter

Inductor-type inverters use a rotor made of soft iron laminations with grooves cut laterally across the surface to provide poles that correspond to the number of stator poles, as illustrated in Figure 10-190. The field coils are wound on one set of stationary poles and the AC armature coils on the other set of stationary poles. When DC is applied to the field coils, a magnetic field is produced. The rotor turns within the field coils and, as the poles on the rotor align with the stationary poles, a low reluctance path for flux is established from the field pole through the rotor poles to the AC armature pole and through the housing back to the field pole. In this circumstance, there will be a large amount of magnetic flux linking the AC coils.

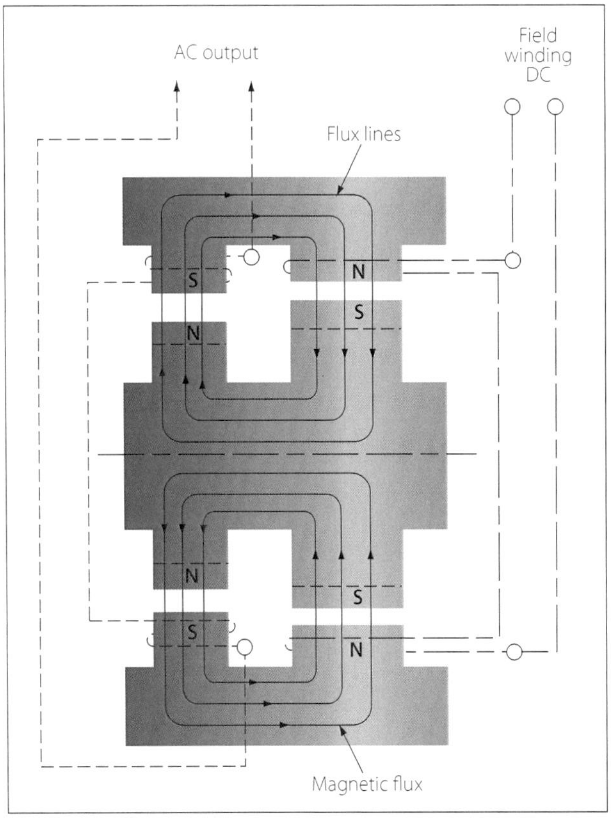

Figure 10-190. Diagram of basic inductor-type inverter.

When the rotor poles are between the stationary poles, there is a high reluctance path for flux, consisting mainly of air; then, there will be a small amount of magnetic flux linking the AC coils. This increase and decrease in flux density in the stator induces an alternating current in the AC coils.

The number of poles and the speed of the motor determine the frequency of this type of inverter. The DC stator field current controls the voltage. A cutaway view of an inductor-type rotary inverter is shown in Figure 10-191.

Figure 10-192 is a simplified diagram of a typical aircraft AC power distribution system, utilizing a main and a standby rotary inverter system.

Static Inverters

In many applications where continuous DC voltage must be converted to alternating voltage, static inverters are used in place of rotary inverters or motor generator sets. The rapid progress made by the semiconductor industry is extending the range of applications of such equipment into voltage and power ranges that would have been impractical a few years ago. Some such applications are power supplies for frequency sensitive military and commercial AC equipment, aircraft emergency AC systems, and conversion of wide frequency range power to precise frequency power.

The use of static inverters in small aircraft also has increased rapidly in the last few years, and the technology has advanced to the point that static inverters are available for any requirement filled by rotary inverters. For example, 250 VA emergency AC supplies operated from aircraft batteries are in production, as are 2,500 VA main AC supplies operated from a varying frequency generator supply. This type of equipment has certain advantages for aircraft applications, particularly the absence of moving parts and the adaptability to conduction cooling.

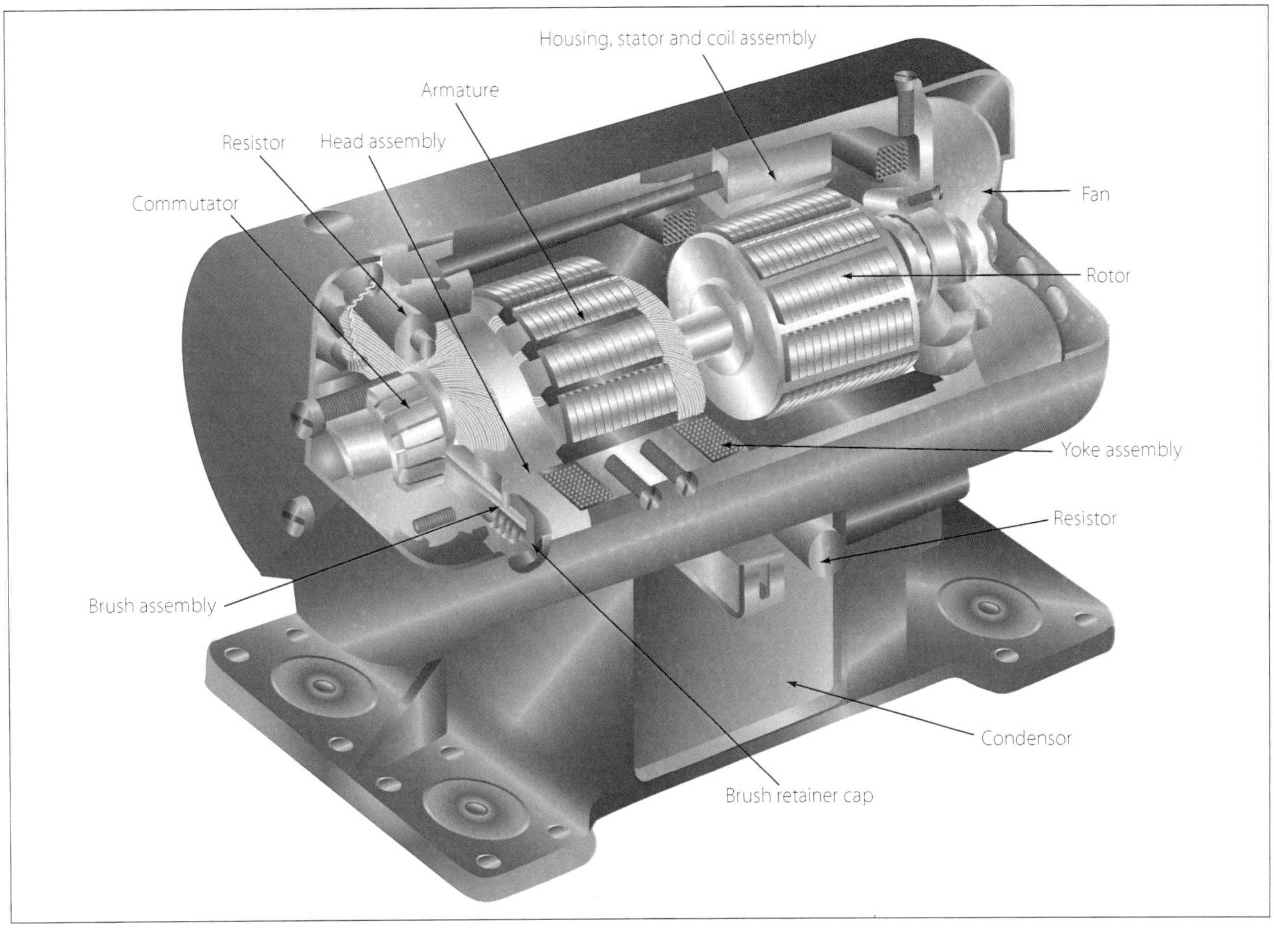

Figure 10-191. Cutaway view of inductor-type rotary inverter.

Static inverters, referred to as solid-state inverters, are manufactured in a wide range of types and models, which can be classified by the shape of the AC output waveform and the power output capabilities. One of the most commonly used static inverters produces a regulated sine wave output. A block diagram of a typical regulated sine wave static inverter is shown in Figure 10-193. This inverter converts a low DC voltage into higher AC voltage. The AC output voltage is held to a very small voltage tolerance, a typical variation of less than 1 percent with a full input load change. Output taps are normally provided to permit selection of various voltages; for example, taps may be provided for a 105, 115, and 125 volt AC outputs. Frequency regulation is typically within a range of one cycle for a 0–100 percent load change.

Variations of this type of static inverter are available, many of which provide a square wave output.

Since static inverters use solid-state components, they are considerably smaller, more compact, and much lighter in weight than rotary inverters. Depending on the output power rating required, static inverters that are no larger than a typical airspeed indicator can be used in aircraft systems. Some of the features of static inverters are:

1. High efficiency.
2. Low maintenance, long life.
3. No warmup period required.
4. Capable of starting under load.
5. Extremely quiet operation.
6. Fast response to load changes.

Static inverters are commonly used to provide power for such frequency sensitive instruments as the attitude gyro and directional gyro. They also provide power for autosyn and magnesyn indicators and transmitters, rate gyros, radar, and other airborne applications. Figure 10-194 is a schematic of a typical small jet aircraft auxiliary battery system. It shows the battery as input to the inverter, and the output inverter circuits to various subsystems.

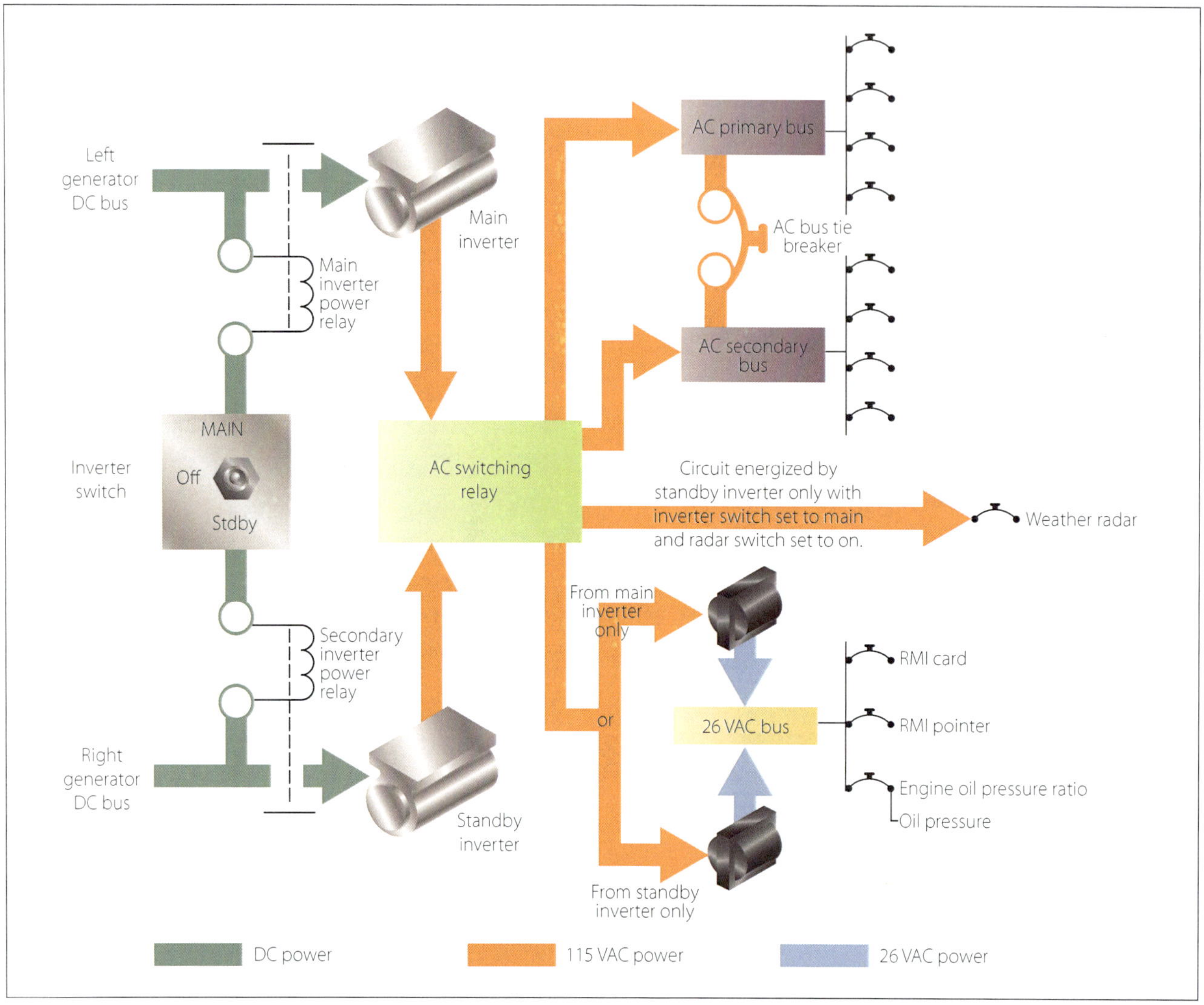

Figure 10-192. A typical aircraft AC power distribution system using main and standby rotary inverters.

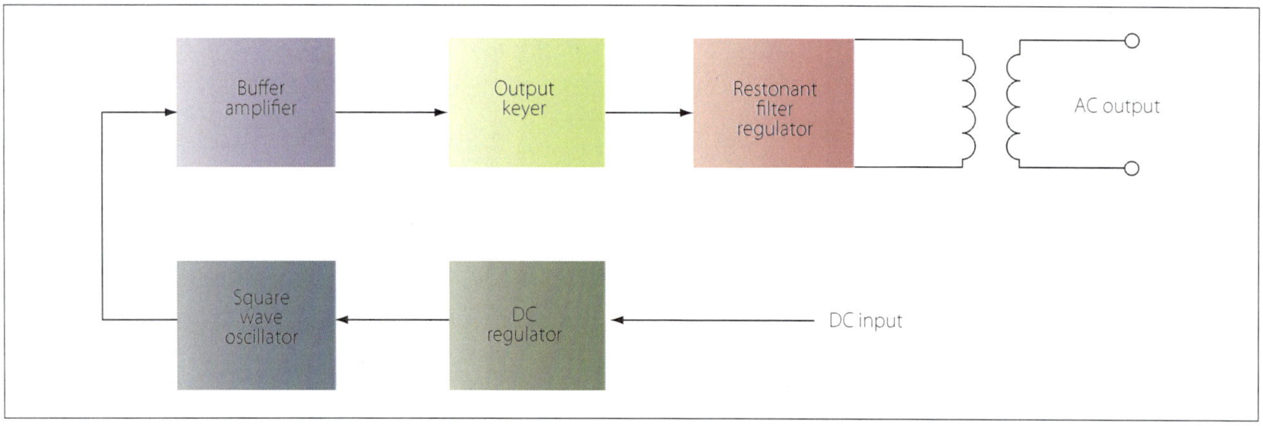

Figure 10-193. Regulated sine wave static inverter.

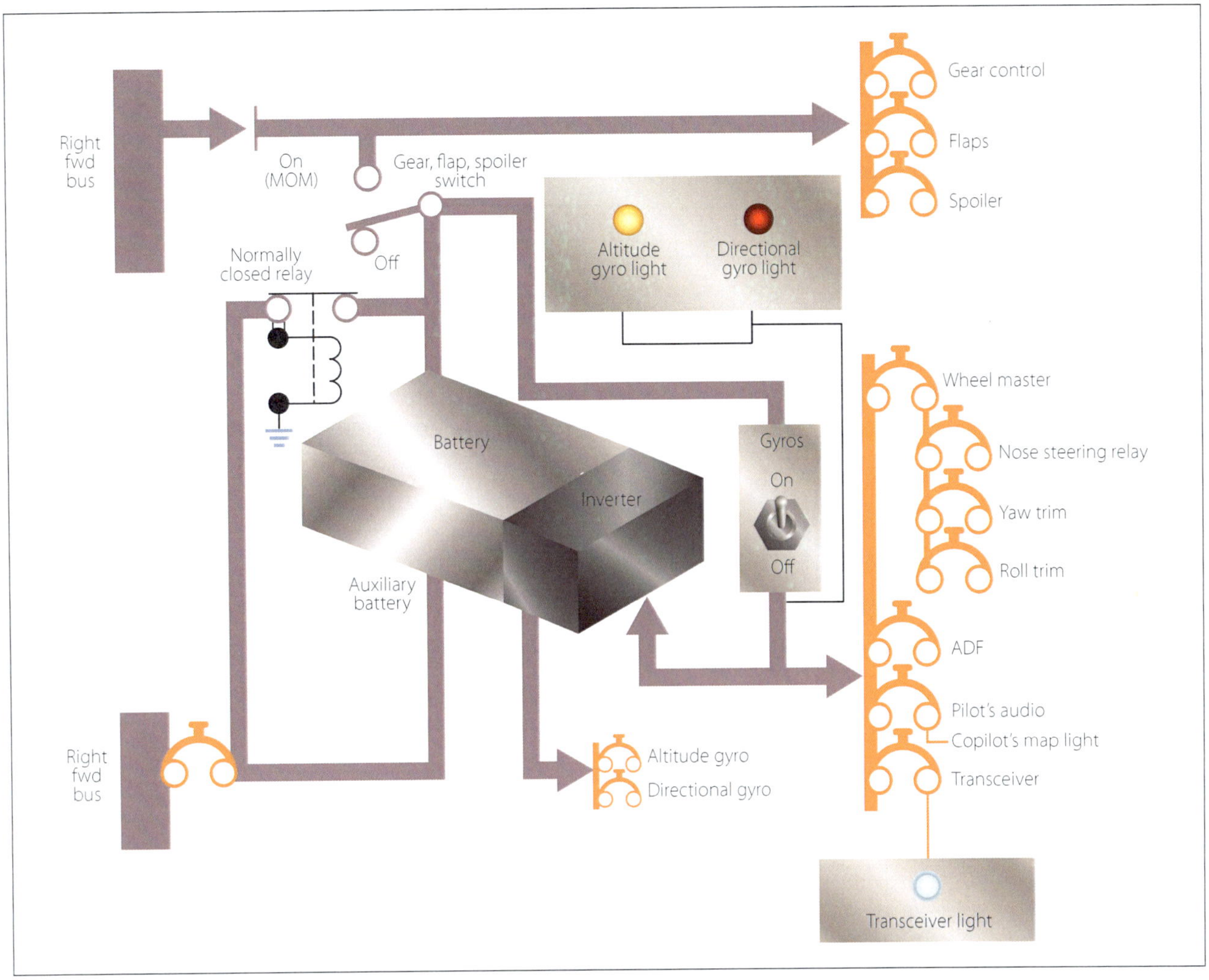

Figure 10-194. Auxiliary battery system using static inverter.

Semiconductors

To understand why solid-state devices function as they do, it is necessary to examine the composition and nature of semiconductors. The two most common materials used for semiconductors are germanium and silicon. The essential characteristic of these elements is that each atom has four valence electrons to share with adjacent atoms in forming bonds. While both elements are used in semiconductor construction, silicon is preferred in most modern applications due to its ability to operate over a wider range of temperatures. The nature of a bond between two silicon atoms is such that each atom provides one electron to share with the other. The two electrons shared are in fact shared equally between the two atoms. This form of sharing is known as a covalent bond. Such bonds are very stable, and hold the two atoms together very tightly, requiring much energy to break this bond. [Figure 10-195] In this case, all of the outer electrons are used to make covalent bonds with other silicon atoms. In this condition, because all of the outer shell atoms are used, silicon takes on the characteristic of a good insulator, due to the fact that there are no open positions available for electrons to migrate through the orbits.

For the silicon crystal to conduct electricity there must be some means available to allow some electrons to move from place to place within the crystal, regardless of the covalent bonds present between the atoms. One way to accomplish this is to introduce an impurity such as arsenic or phosphorus into the crystal structure, which will either provide an extra electron or create a vacant position in the outer shell for electrons to pass though. The method used to create this condition is called doping.

Doping

Doping is the process by which small amounts of additives called impurities are added to the semiconductor material to increase their current flow by adding a few electrons or a few holes. Once the material is doped,

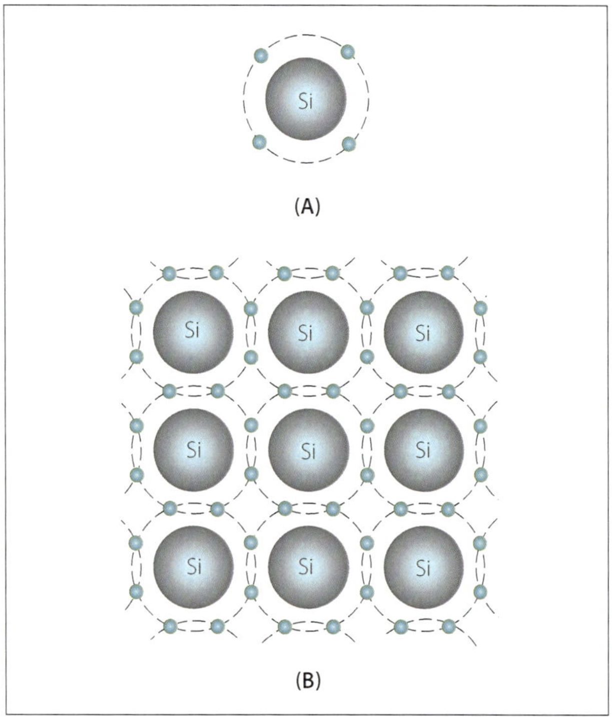

Figure 10-195. Valance electrons.

it then falls into one of two categories: the N-type semiconductor and the P-type semiconductor.

An N-type semiconductor material is one that is doped with an N-type or a donor impurity. Elements such as phosphorus, arsenic and antimony are added as impurities and have five outer electrons to share with other atoms. This will cause the semiconductor material to have an excess electron. Due to the surplus of electrons, the electrons are then considered the majority current carriers. This electron can easily be moved with only a small applied electrical voltage. Current flow in an N-type silicon material is similar to conduction in a copper wire. That is, with voltage applied across the material, electrons will move through the crystal towards the positive terminal just like current flows in a copper wire.

A P-type semiconductor is one that is doped with a P-type or an acceptor impurity. Elements such as boron, aluminum, and gallium have only three electrons in the valence shell to share with the silicon atom. Those three electrons will form covalent bonds with adjacent silicon atoms. However, the expected fourth bond cannot be formed and a complete connection is impossible here, leaving a "hole" in the structure of the crystal. There is an empty place where an electron would naturally go, and often an electron will move into that space to fill it. However, the electron filling the hole left a covalent bond behind to fill this empty space, which leaves another hole behind as it moves. Another electron may then move into that particular hole, leaving another hole behind. As this progression continues, holes appear to move as positive charges throughout the crystal. This type of semiconductor material is designated P-type silicon material. Figure 10-196 shows the progression of a hole moving through a number of atoms. Notice that the hole illustrated at the far left of the top depiction of Figure 10-196 attracts the next valance electron into the vacancy, which then produces another vacancy called a hole in the next position to the right. Once again this vacancy attracts the next valance electron. This exchange of holes and electrons continues to progress, and can be viewed in one of two ways: electron movement or hole movement. For electron movement, illustrated by the top depiction of Figure 10-196, the electron is shown as moving from the right to the left through a series of holes. In the second depiction in Figure 10-196, the motion of the vacated hole can be seen as migration from the left to the right, called hole movement. The valence electron in the structure will progress along a path detailed by the arrows. Holes, however, move along a path opposite that of the electrons.

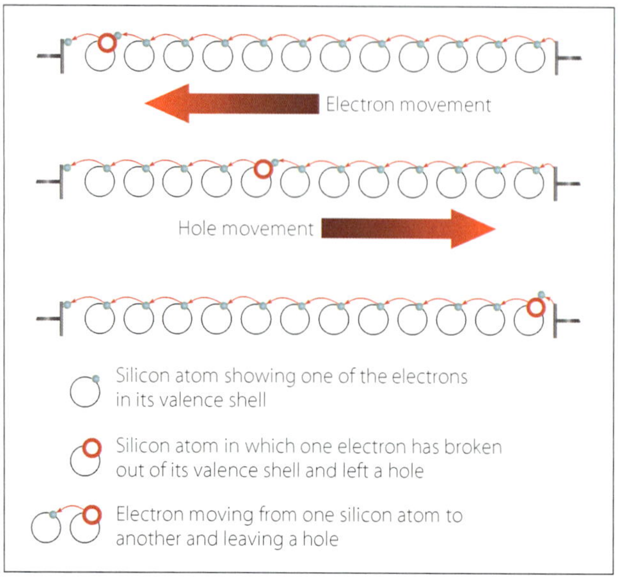

Figure 10-196. A hole moving through atoms.

PN Junctions and the Basic Diode

A single type of semiconductor material by itself is not very useful. Useful applications are developed only when a single component contains both P-type and N-type materials. The semiconductor diode is also known as a PN junction diode. This is a two-element semiconductor device that makes use of the rectifying properties of a PN junction to convert alternating current into direct current by permitting current flow in one direction only.

Figure 10-197 illustrates the electrical characteristics of an unbiased diode, which means that no external voltage is applied. The P-side in the illustration is shown to have many holes, while the N-side shows many electrons. The electrons on the N-side tend to diffuse out in all directions. When an electron enters the P region, it becomes a minority carrier. By definition, a minority carrier is an electron or hole, whichever is the less dominant carrier in a semiconductor device. In P-type materials, electrons are the minority carrier and in N-type material, the hole is considered the minority carrier. With so many holes around the electron, the electron will soon drop into a hole. When this occurs, the hole then disappears, and the conduction band electron becomes a valence electron.

Each time an electron crosses the PN junction, it creates a pair of ions. In Figure 10-197 this is shown in the area outlined by the dash lines. The circled plus signs and the circled negative signs are the positive and negative ions, respectively. These ions are fixed in the crystal and do not move around like electrons or holes in the conduction band. Thus, the depletion zone constitutes a layer of a fixed charge. An electrostatic field, represented by a small battery in Figure 10-195, is

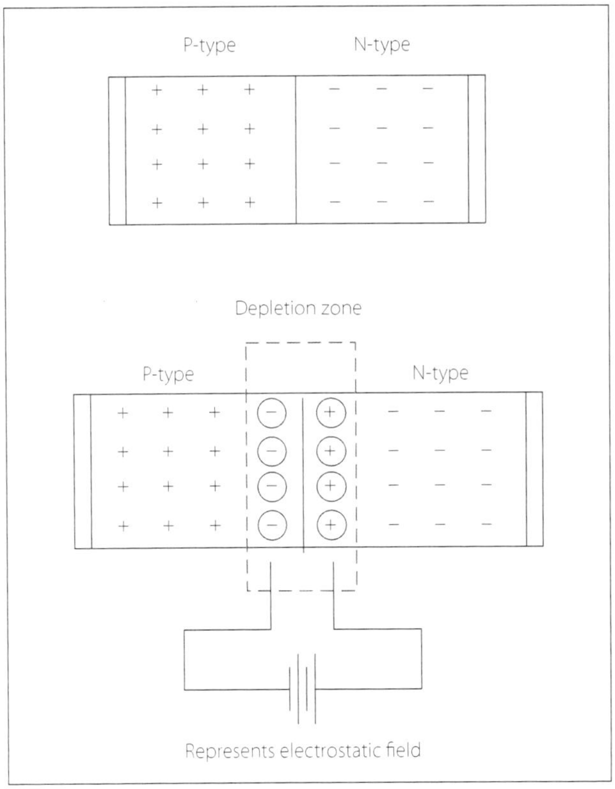

Figure 10-197. Depletion region.

established across the junction between the oppositely charged ions.

The junction barrier is an electrostatic field, which has been created by the joining of a section of N-type and P-type material. Because holes and electrons must overcome this field to cross the junction, the electrostatic field is usually called a barrier. Because there is a lack or depletion of free electrons and holes in the area around the barrier, this area is called the depletion region. [Figure 10-197] As the diffusion of electrons and holes across the junction continue, the strength of the electrostatic field will increase until it is strong enough to prevent electrons or holes from crossing over. At this point, a state of equilibrium exists, and there is no further movement across the junction. The electrostatic field created at the junction by the ions in the depletion zone is called a barrier.

Forward Biased Diode

Figure 10-198 illustrates a forward biased PN junction. When an external voltage is applied to a PN junction, it as called bias. In a forward biased PN junction or diode, the negative voltage source is connected to the N-type material and the positive voltage source is connected to the P-type material. In this configuration, the current can easily flow. If a battery is used to bias the PN junction and it is connected in such a way that the applied voltage opposes the junction field, it will have the effect of reducing the junction barrier and consequently aid in the current flow through the junction.

The electrons move toward the junction and the right end of the diode becomes slightly positive. This occurs because electrons at the right end of the diode move toward the junction and leave positively charged atoms behind. The positively charged atoms then pull

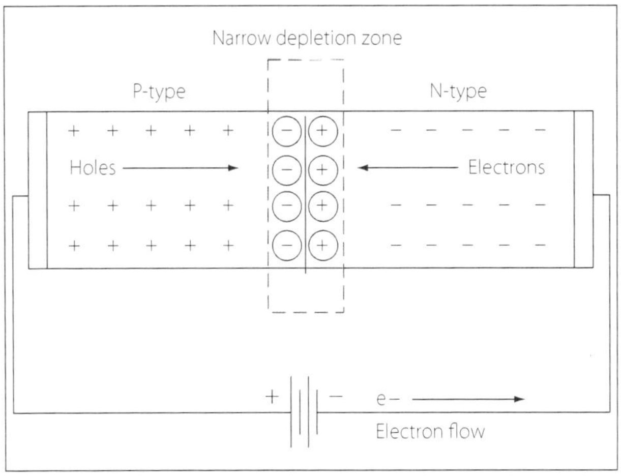

Figure 10-198. Forward biased PN junction.

electrons into the diode from the negative terminal of the battery.

When electrons on the N-type side approach the junction, they recombine with holes. Basically, electrons are flowing into the right end of the diode, while the bulk of the electrons in the N-type material move toward the junctions. The left edge of this moving front of electrons disappears by dropping into holes at the junction. In this way, there is a continuous current of electrons from the battery moving toward the junction.

When the electrons hit the junction, they then become valence electrons. Once a valence electron, they can then move through the holes in the P-type material. When the valence electrons move through the P-type material from the right to the left, a similar movement is occurring with the holes by moving from the left side of the P-type material to the right. Once the valence electron reaches the end of the diode, it will then flow back into the positive terminal of the battery.

In summary:

1. Electron leaves negative terminal of the battery and enters the right end (N-type material) of the diode.
2. Electron then travels through the N-type material.
3. The electron nears the junction and recombines and becomes a valence electron.
4. The electron now travels through the P-type material as a valence electron.
5. The electron then leaves the diode and flows back to the positive terminal of the battery.

Reverse Biased Diode

When the battery is turned around as shown in Figure 10-199, then the diode is reverse biased and current will not flow. The most noticeable effect seen in this illustration is the widened depletion zone.

The applied battery voltage is in the same direction as the depletion zone field. Because of this, holes and electrons tend to move away from the junction. Simply stated, the negative terminal attracts the holes away from the junction, and the positive terminal attracts the electrons away from the barrier. Therefore, the result is a wider depletion zone. This action increases the barrier width because there are more negative ions on the P-side of the junction and more positive ions on the N-side of the junction. This increase in the number of ions at the junction prevents current flow across the barrier by the majority carriers.

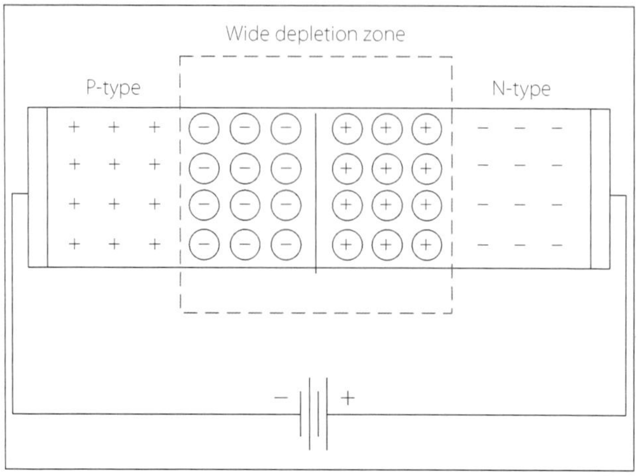

Figure 10-199. Reversed diode.

To summarize, the important thing to remember is that these PN junction diodes will offer very little resistance to current in a forward biased diode. Maximum resistance will happen when the diode is reversed biased. Figure 10-200 shows a graph of the current characteristics of a diode that is biased in both directions.

Rectifiers

Many devices in an aircraft require high amperage, low voltage DC for operation. This power may be furnished by DC engine driven generators, motor generator sets, vacuum tube rectifiers, or dry disk or solid-state rectifiers.

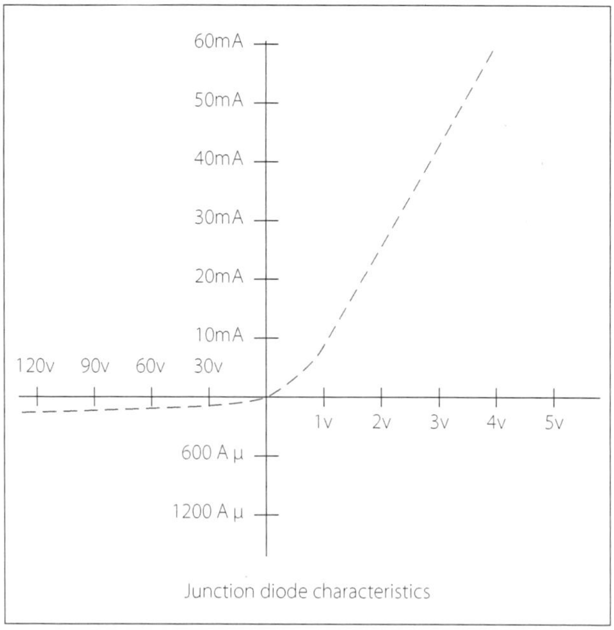

Figure 10-200. Diode biased in both directions.

In aircraft with AC systems, a special DC generator is not desirable since it would be necessary for the engine accessory section to drive an additional piece of equipment. Motor generator sets, consisting of air-cooled AC motors that drive DC generators, eliminate this objection because they operate directly off the AC power system. Vacuum tube or various types of solid-state rectifiers provide a simple and efficient method of obtaining high voltage DC at low amperage. Dry disk and solid-state rectifiers, on the other hand, are an excellent source of high amperage at low voltage.

A rectifier is a device that transforms alternating current into direct current by limiting or regulating the direction of current flow. The principal types of rectifiers are dry disk and solid state. Solid-state, or semiconductor, rectifiers have replaced virtually all other types; and, since dry disk and motor generators are largely limited to older model aircraft, the major part of the study of rectifiers is devoted to solid-state devices used for rectification.

The two methods discussed in this text are the half-wave rectifier and the full-wave rectifier.

Half-Wave Rectifier

Figure 10-201 illustrates the basic concept of a half-wave rectifier. When an AC signal is on a positive swing as shown in illustration A of the input signal, the polarities across the diode and the load resistor will also be positive. In this case, the diode is forward biased and can be replaced with a short circuit as shown in the illustration. The positive portion of the input signal will then appear across the load resistor with no loss in potential across the series diode.

Illustration B now shows the input signal being reversed. Note that the polarities across the diode and the load resistor are also reversed. In this case, the diode is now reverse biased and can be replaced with an equivalent open circuit. The current in the circuit is now 0 amperes and the voltage drop over the load resistor is 0 volts. The resulting waveform for a complete sinusoidal input can be seen at the far right of Figure 10-201. The output waveform is a reproduction of the input waveform minus the negative voltage swing of the wave. For this reason, this type of rectifier is called a half-wave rectifier.

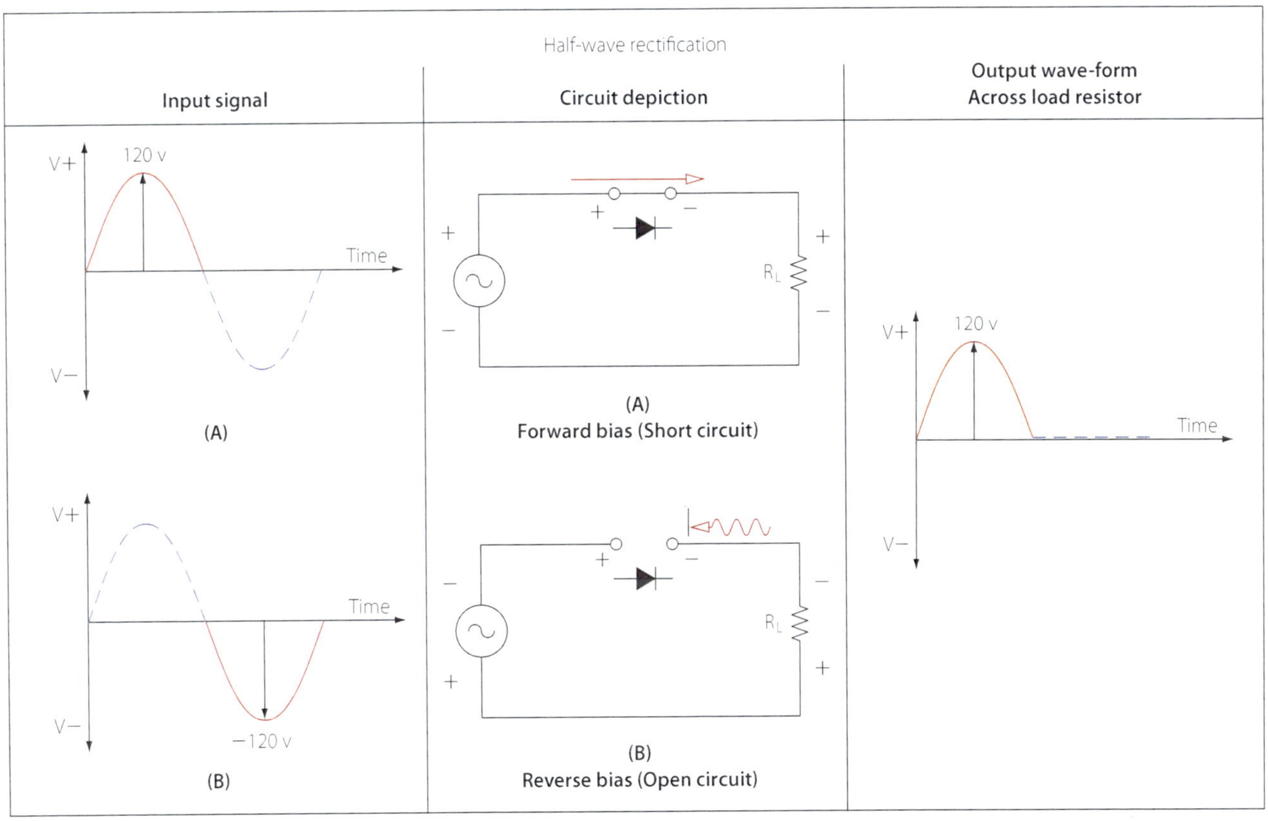

Figure 10-201. Basic concept of half-wave rectifier.

Full-Wave Rectifier

Figure 10-202 illustrates a more common use of the diode as a rectifier. This type of a rectifier is called a full-wave bridge rectifier. The term "full-wave" indicates that the output is a continuous sequence of pulses rather than having gaps that appear in the half-wave rectifier.

Illustration C shows the initial condition, during which, a positive portion of the input signal is applied to the network. Note the polarities across the diodes. Diodes D2 and D4 are reverse biased and can be replaced with an open circuit. Diodes D1 and D3 are forward biased and act as an open circuit. The current path through the diodes is clear to see, and the resulting waveform is developed across the load resistor.

During the negative portion of the applied signal, the diodes will reverse their polarity and bias states. The result is a network shown in illustration D. Current now passes through diodes D4 and D2, which are forward biased, while diodes D1 and D3 are essentially open circuits due to being reverse biased. Note that during both alternations of the input waveform, the current will pass through the load resistor in the same direction. This results in the negative swing of the waveform being flipped up to the positive side of the time line.

Dry Disk

Dry disk rectifiers operate on the principle that electric current flows through a junction of two dissimilar conducting materials more readily in one direction than it does in the opposite direction. This is true because the resistance to current flow in one direction is low, while in the other direction it is high. Depending on the materials used, several amperes may flow in the direction of low resistance but only a few milliamperes in the direction of high resistance.

Three types of dry disk rectifiers may be found in aircraft: the copper oxide rectifier, the selenium rectifier, and the magnesium copper-sulfide rectifier. The copper oxide rectifier consists of a copper disk upon which a layer of copper oxide has been formed by heating. [Figure 10-203] It may also consist of a chemical copper oxide preparation spread evenly over the copper surface. Metal plates, usually lead plates, are pressed against the two opposite faces of the disk to form a good contact. Current flow is from the copper to the copper oxide.

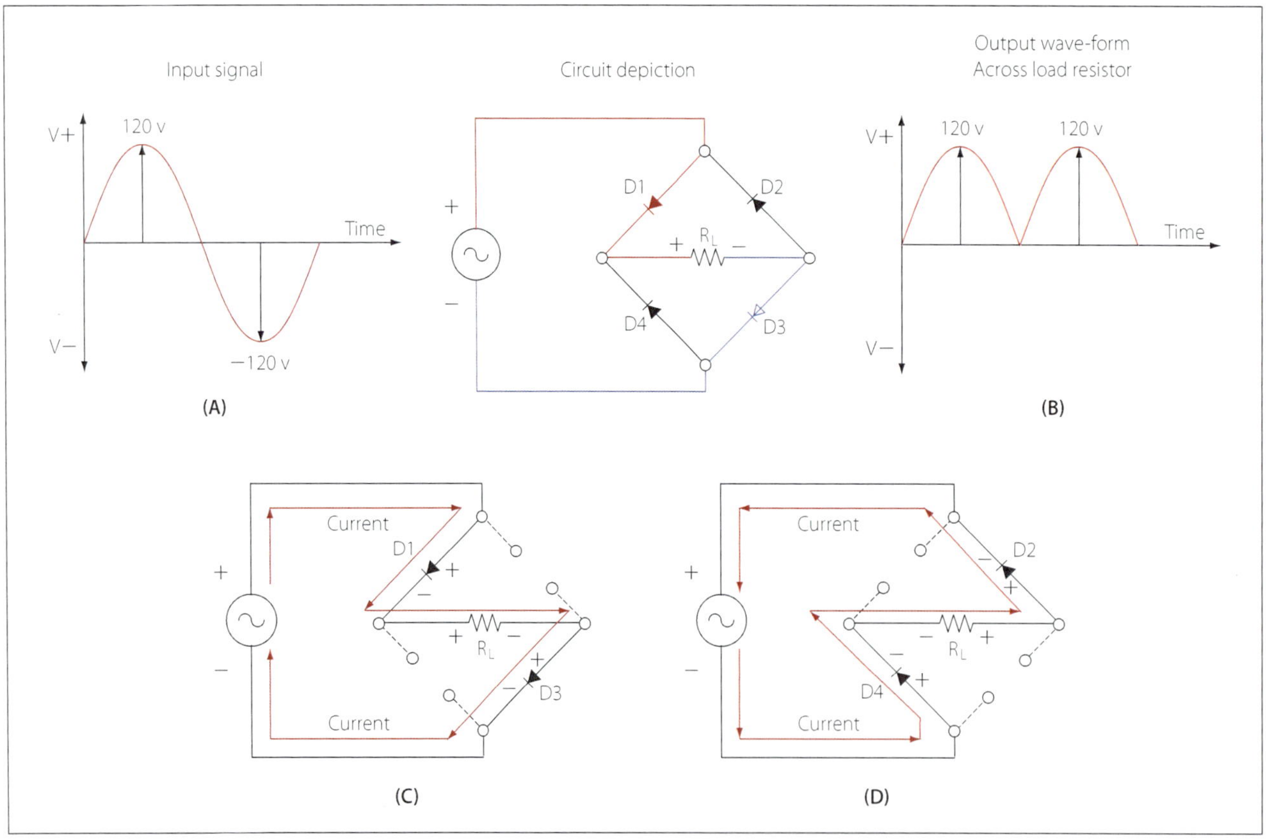

Figure 10-202. Full-wave bridge rectifier.

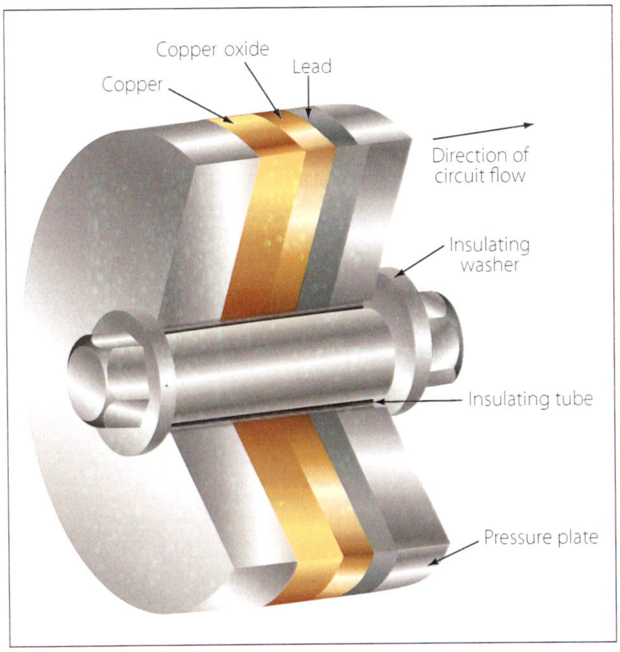

Figure 10-203. Copper oxide dry disk rectifier.

The selenium rectifier consists of an iron disk, similar to a washer, with one side coated with selenium. Its operation is similar to that of the copper oxide rectifier. Current flows from the selenium to the iron.

The magnesium copper-sulfide rectifier is made of washer-shaped magnesium disks coated with a layer of copper sulfide. The disks are arranged similarly to the other types. Current flows from the magnesium to the copper sulfide.

Types of Diodes

Today there are many varieties of diodes, which can be grouped into one of several basic categories.

Power Rectifier Diodes

The rectifier diode is usually used in applications that require high current, such as power supplies. The range in which the diode can handle current can vary anywhere from one ampere to hundreds of amperes. One common example of diodes is the series of diodes, part numbers 1N4001 to 1N4007. The "1N" indicates that there is only one PN junction, or that the device is a diode. The average current carrying range for these rectifier diodes is about one ampere and have a peak inverse voltage between 50 volts to1,000 volts. Larger rectifier diodes can carry currents up to 300 amperes when forward biased and have a peak inverse voltage of 600 volts. A recognizable feature of the larger rectifier diodes is that they are encased in metal in order to provide a heat sink. Figure 10-204 illustrates a line drawing of some general purpose diodes.

Zener Diodes

Zener diodes (sometimes called "breakdown diodes") are designed so that they will break down (allow current to pass) when the circuit potential is equal to or in

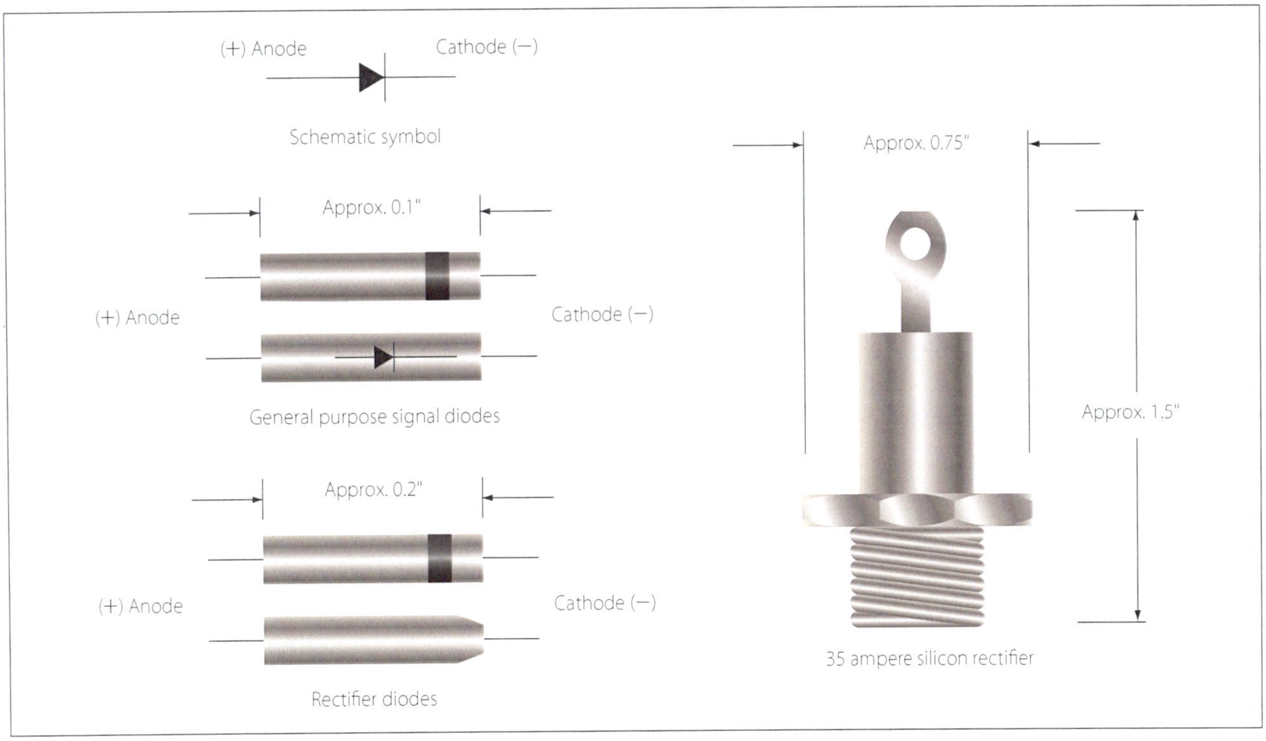

Figure 10-204. General purpose diodes.

excess of the desired reverse bias voltage. The range of reverse bias breakdown-voltages commonly found can range from 2 volts to 200 volts depending on design. Once a specific reverse bias voltage has been reached, the diode will conduct and behave like a constant voltage source. Within the normal operating range, the zener will function as a voltage regulator, waveform clipper, and other related functions. Below the desired voltage, the zener blocks the circuit like any other diode biased in the reverse direction. Because the zener diode allows free flow in one direction when it is used in an AC circuit, two diodes connected in opposite directions must be used. This takes care of both alternations of current. Power ratings of these devices range from about 250 milliwatts to 50 watts.

Special Purpose Diodes

The unique characteristics of semiconductor material have allowed for the development of many specialized types of diodes. A short description of some of the more common diode types is given for general familiarization. Figure 10-205 illustrates the schematic symbols for some of the special purpose diodes.

Light-Emitting Diode (LED)

In a forward biased diode, electrons cross the junction and fall into holes. As the electrons fall into the valence band, they radiate energy. In a rectifier diode, this energy is dissipated as heat. However, in the light-emitting diode (LED), the energy is dissipated as light. By using elements, such as gallium, arsenic, and phosphorous, an LED can be designed to radiate colors, such as red, green, yellow, blue and inferred light. LEDs that are designed for the visible light portion of the spectrum are useful for instruments, indicators, and even cabin lighting. The advantages of the LED over the incandescent lamps are longer life, lower voltage, faster on and off operations, and less heat.

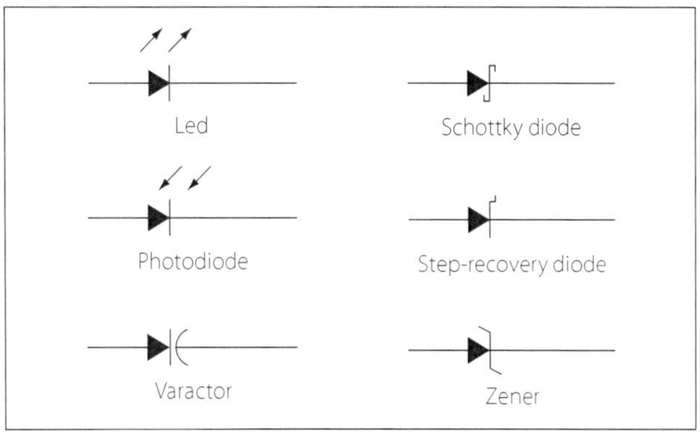

Figure 10-205. Schematic symbols for special purpose diodes.

Liquid Crystal Displays (LCD)

The liquid crystal display (LCD) has an advantage over the LED in that it requires less power to operate. Where LEDs commonly operate in the milliwatt range, the LCD operates in the microwatt range. The liquid crystal is encapsulated between two glass plates. When voltage is not applied to the LCD, the display will be clear. However, when a voltage is applied, the result is a change in the orientation of the atoms of the crystals. The incident light will then be reflected in a different direction. A frosted appearance will result in the regions that have voltage applied and will permit distinguishing of numeric values.

Photodiode

Thermal energy produces minority carriers in a diode. The higher the temperature, the greater the current in a reverse current diode. Light energy can also produce minority carriers. By using a small window to expose the PN junction, a photodiode can be built. When light fall upon the junction of a reverse- biased photodiode, electrons-hole pairs are created inside the depletion layer. The stronger the light, the greater the number of light-produced carriers, which in turn causes a greater magnitude of reverse-current. Because of this characteristic, the photodiode can be used in light detecting circuits.

Varactors

The varactor is simply a variable-capacitance diode. The reverse voltage applied controls the variable-capacitance of the diode. The transitional capacitance will decrease as the reverse voltage is increasingly applied. In many applications, the varactor has replaced the old mechanically tuned capacitors. Varactors can be placed in parallel with an inductor and provide a resonant tank circuit for a tuning circuit. By simply varying the reverse voltage across the varactor, the resonant frequency of the circuit can be adjusted.

Schottky Diodes

Schottky diodes are designed to have a metal, such as gold, silver, or platinum, on one side of the junction and doped silicon, usually an N-type, on the other side of the junction. This type of a diode is considered a unipoler device because free electrons are the majority carrier on both sides of the junction. The Schottky diode has no depletion zone or charge storage, which means that the switching time can be as high as 300 MHz. This characteristic exceeds that of the bipolar diode.

Diode Identification

Figure 10-204 illustrates a number of methods employed for identifying diodes. Typically manufacturers place some form of an identifier on the diode to indicate which end is the anode and which end is the cathode. Dots, bands, colored bands, the letter 'k' or unusual shapes indicate the cathode end of the diode.

Introduction to Transistors

The transistor is a three-terminal device primarily used to amplify signals and control current within a circuit. [Figure 10-206] The basic two-junction semiconductor must have one type of region sandwiched between two of the other type. The three regions in a transistor are the collector (C), which is moderately doped, the emitter (E), which is heavily doped and the base (B) significantly less in its doping. The alternating layers of semiconductor material type provide the common commercial name for each type of transistor. The interface between the layers is called a junction. Selenium and germanium diodes previously discussed are examples of junction diodes. Note that the sandwiched layer or base is significantly thinner than the collector or the emitter. In general this permits a "punching through" action for the carriers passing between the collector and emitter terminals.

Classification

The transistors are classified as either NPN or PNP according to the arrangement of their N and P-materials. The NPN transistor is formed by introducing a thin region of P-material between two regions of N-type material. The opposite is true for the PNP configuration.

The two basic types of transistors along with their circuit symbols are shown in Figure 10-207. Note that the two symbols are different. The horizontal line represents the base, and two angular lines represent the emitter and collector. The angular line with the arrow on it is the emitter, while the line without is the collector. The direction of the arrow on the emitter determines whether or not the transistor is a PNP or an NPN type. If the arrow is pointing in, the transistor is a PNP. On the other hand, if the arrow is pointing out, then it is an NPN type.

Transistor Theory

As discussed in the section on diodes, the movement of the electrons and holes can be considered current. Electron current moves in one direction, while hole current travels in the opposite direction. In transistors, both electrons and holes act as carriers of current.

A forward biased PN junction is comparable to a low-resistance circuit element because it passes a high current for a given voltage. On the other hand, a reverse-biased PN junction is comparable to a high-resistance circuit element. By using Ohm's law formula for power ($P = I^2R$) and assuming current is held constant through both junctions, it can be concluded that the power developed across the high resistance junction is greater than that developed across a low resistance junction. Therefore, if a crystal were to contain two PN junctions, one forward biased and the

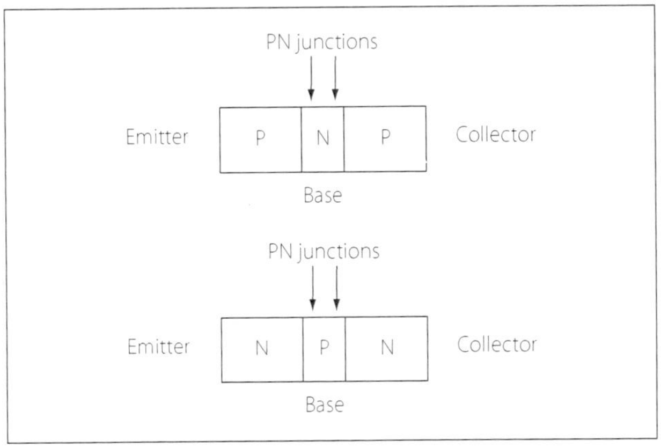

Figure 10-206. Transistor.

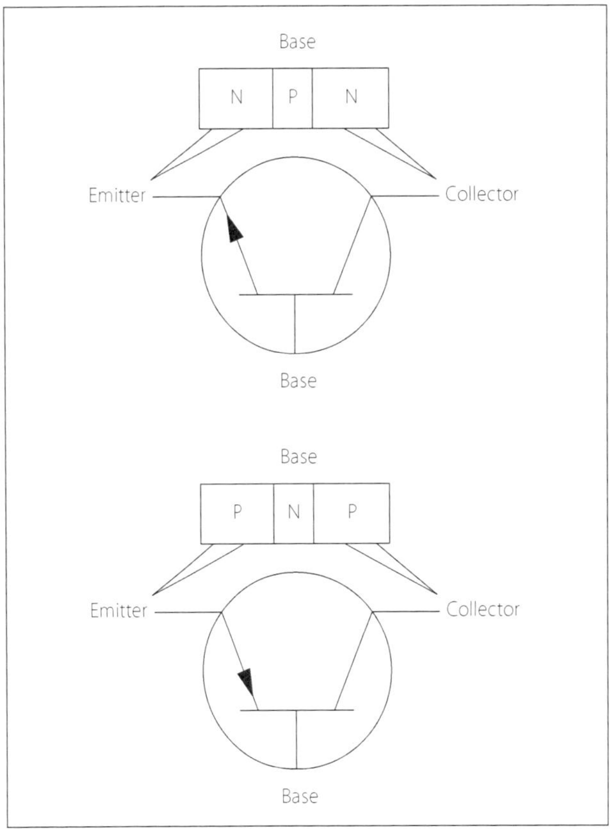

Figure 10-207. Two basic transistors with circuit symbols.

other reverse biased, and a low-power signal injected into the forward biased junction, a high power signal could be produced at the reverse-biased junction.

To use the transistor as an amplifier, some sort of external bias voltage must modify each of the junctions. The first PN junction (emitter-base) is biased in the forward direction. This produces a low resistance. The second junction, which is the collector-base junction, is reverse biased to produce a high resistance. Figure 10-208 illustrates the proper biasing of an NPN transistor.

With the emitter-base junction biased in the forward direction, electrons leave the negative terminal of the battery and enter the N-material. These electrons pass easily through the emitter, cross over the junction, and combine with the hole in the P-material in the base. For each electron that fills a hole in the P-material, another electron will leave the P-material, which creates a new hole and enters the positive terminal of the battery.

The second PN junction, which is the base-collector junction, is reverse biased. This will prevent the majority carriers from crossing the junction, thus creating a high resistance circuit. It is worth noting that there still is a small current passing through the reversed PN junction in the form of minority carriers—that is, electrons in the P-material and holes in the N-material. The minority carriers play a significant part in the operation of the NPN transistor.

Figure 10-209 illustrates the basic interaction of the NPN junction. There are two batteries in the circuit used to bias the NPN transistor. Vbb is considered the base voltage supply, rated in this illustration at 1 volt, and the battery voltage Vcc, rated at 6 volts, is called the collector voltage supply.

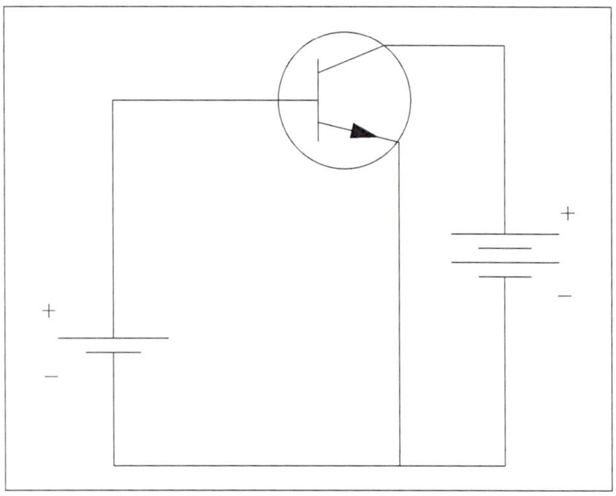

Figure 10-208. NPN transistor.

Current within the external circuit is simply the movement of free electrons originating at the negative terminal of the battery and flowing to the N-material. This is shown in Figure 10-209 as Ie or emitter-current.

As the electrons enter the N-material, they become the majority carrier and move through the N-material to the emitter-base PN junction. This emitter-base junction is forward biased at about 0.65 to 0.7 volts positive with respect to the emitter and presents no resistance to the flow of electrons from the emitter into the base, which is composed of P-material. As these electrons move into the base, they will drop into available holes. For every electron that drops into a hole, another electron exits the base by way of the base lead and becomes the base current or Ib. Of course, when one electron leaves the base, a new hole is formed. From the standpoint of the collector, these electrons that drop into holes are lost and of no use. To reduce this loss of electrons, the transistor is designed so that the base is very thin in relation to the emitter and collector, and the base is lightly doped.

Most of the electrons that move into the base will fall under the influence of the reverse bias of the collector. While collector-base junction is reverse biased with respect to the majority carriers, it behaves as if it is forward biased to the electrons or minority carriers in this case. The electrons are accelerated through the collector-base junction and into the collector. The collector is comprised of the N-type material; therefore, the electrons once again become the majority carrier. Moving easily through the collector, the electrons return to the positive terminal of the collector supply battery Vcc, which is shown in Figure 10-209 as Ic.

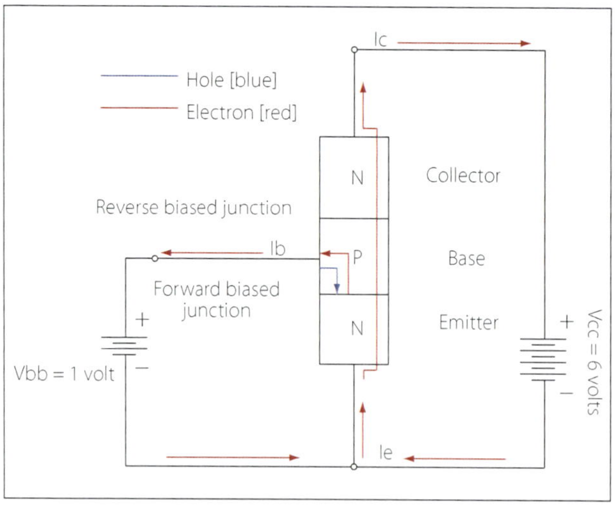

Figure 10-209. NPN Junction.

Because of the way this device operates to transfer current (and its internal resistances) from the original conduction path to another, its name is a combination of the words "transfer" and "resistor"—transistor.

PNP Transistor Operation

The PNP transistor generally works the same way as the NPN transistor. The primary difference is that the emitter, base, and collector materials are made of different material that the NPN. The majority and minority current carriers are the opposite in the PNP to that of the NPN. In the case of the PNP, the majority carriers are the holes instead of the electrons in the NPN transistor. To properly bias the PNP, the polarity of the bias network must be reversed.

Identification of Transistors

Figure 10-210 illustrates some of the more common transistor lead identifications. The methods of identifying leads will vary due to a lack of a standard and will require verification using manufacturer information to properly identify. However, a short description of the common methods is discussed below.

Illustration D in Figure 10-210 shows an oval-shaped transistor. The collector lead in this case is identified by the wide space between it and the lead for the base. The final lead at the far left is the emitter. In many cases, colored dots indicate the collector lead, and short leads relative to the other leads indicate the emitter. In a conventional power diode as seen in illustration E of Figure 10-210, the collector lead is usually a part of the mounting bases, while the emitter and collector are leads or tines protruding from the mounting surface.

Field Effect Transistors

Another transistor design that has become more important than the bipolar transistor is the field-effect transistor or FET. The primary difference between the bipolar transistor and the FET is that the bipolar transistor has two PN junctions and is a current controlled device, while the FET has only one PN junction and is a voltage controlled device. Within the FET family, there are two general categories of components. One category is called the junction FET (JFET), which has only one PN junction. The other category is known as the enhancement-type or metal-oxide JET (MOSFET).

Figure 10-211 shows the basic construction of the JFET and the schematic symbol. In this Figure, it can

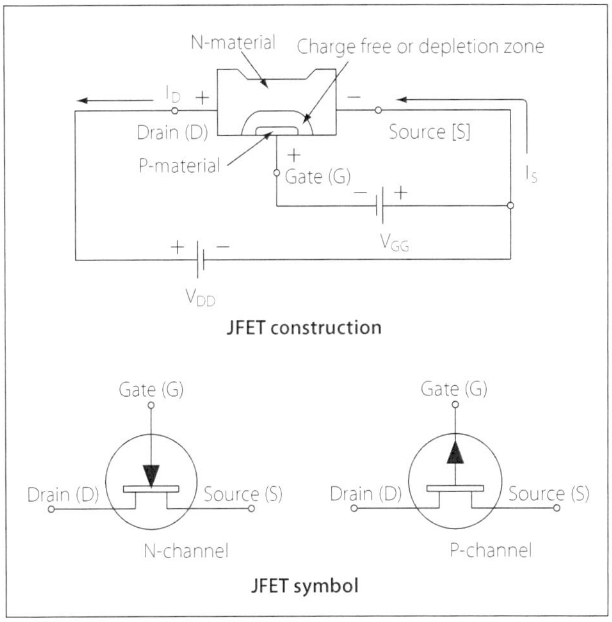

Figure 2-211. JFET and the schematic symbol.

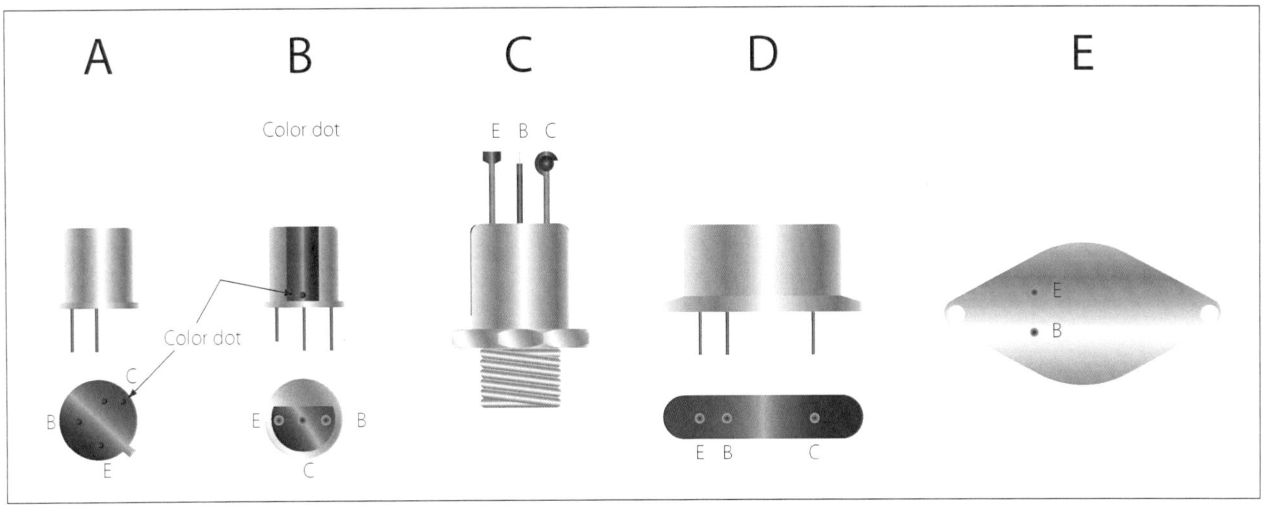

Figure 10-210. Common transistor lead identifications.

be seen that the drain (D) and source (S) are connected to an N-type material, and the gate (G) is connected to the P-type material. With gate voltage Vgg set to 0 volts and drain voltage Vdd set to some positive voltage, a current will flow between the source and the drain, through a narrow band of N-material. If then, Vgg is adjusted to some negative voltage, the PN junction will be reverse biased, and a depletion zone (no charge carriers) will be established at the PN junction. By reducing the region of noncarriers, it will have the effect of reducing the dimensions of the N-channel, resulting in a reduction of source to drain current.

Because JFETs are voltage-controlled devices, they have some advantages over the bipolar transistor. One such advantage is that because the gate is reverse biased, the circuit that it is connected to sees the gate as a very high resistance. This means that the JFET has less of an insertion influence in the circuit. The high resistance also means that less current will be used.

Like many other solid-state devices, careless handling and static electricity can damage the JFET. Technicians should take all precautions to prevent such damage.

Metal-Oxide-Semiconductor FET (MOSFET)

Figure 10-212 illustrates the general construction and the schematic symbol of the MOSFET transistor. The biasing arrangement for the MOSFET is essentially the same as that for the JFET. The term "enhancement" comes from the idea that when there is no bias voltage applied to the gate (G), then there is no channel for current conduction between the source (S) and the drain (D). By applying a greater voltage on the gate (G), the P-channel will begin to materialize and grow in size. Once this occurs, the source (S) to drain (D) current Id will increase. The schematic symbol reflects this characteristic by using a broken line to indicate that the channel does not exist without a gate bias.

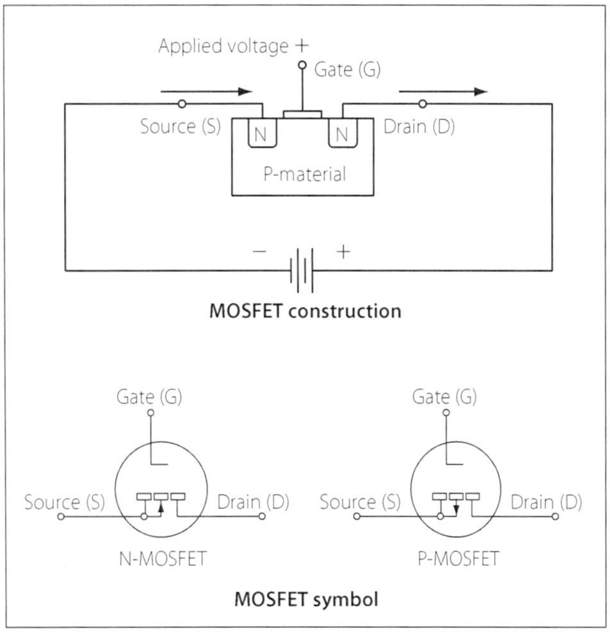

Figure 10-212. General construction and schematic symbol of MOSFET transistor.

Common Transistor Configurations

A transistor may be connected in one of three different configurations. The three basic configurations are: common-emitter (CE), common-base (CB), and common-collector (CC). The term "common" is used to indicate which element of the transistor is common to both the input and the output. Each configuration has its own characteristics, which makes each configuration suitable for particular applications. A way to determine what configuration you may find in a circuit is to first determine which of the three transistor elements is used for the input signal. Then, determine the element used for the output signal. At that point, the remaining element, (base, emitter, or collector) will be the common element to both the input and output, and thus you determine the configuration.

Common-Emitter Configuration

This is the configuration most commonly used in amplifier circuits because they provide good gains for voltage, current, and power. The input signal is applied to the base-emitter junction, which is forward biased (low resistance), and the output signal is taken off the collector-emitter junction, which is reverse biased (high resistance). Then the emitter is the common element to both input and output circuits. [Figure 10-213]

When the transistor is connected in a common-emitter configuration, the input signal is injected between the base and emitter, which is a low resistance, low-current circuit. As the input signal goes positive, it causes the base to go positive relative to the emitter. This causes a decrease in the forward bias, which in turn reduces the collector current I_C and increases the collector voltage (E_C being more negative). During the negative portion of the input signal, the voltage on the base is driven more negative relative to the emitter. This increases the forward bias and allows an increase in collector current I_C and a decrease in collector voltage (E_C being less negative and going positive). The collector current, which flows through the reverse-biased junction, also flows through a high resistance load resulting in a high level of amplification.

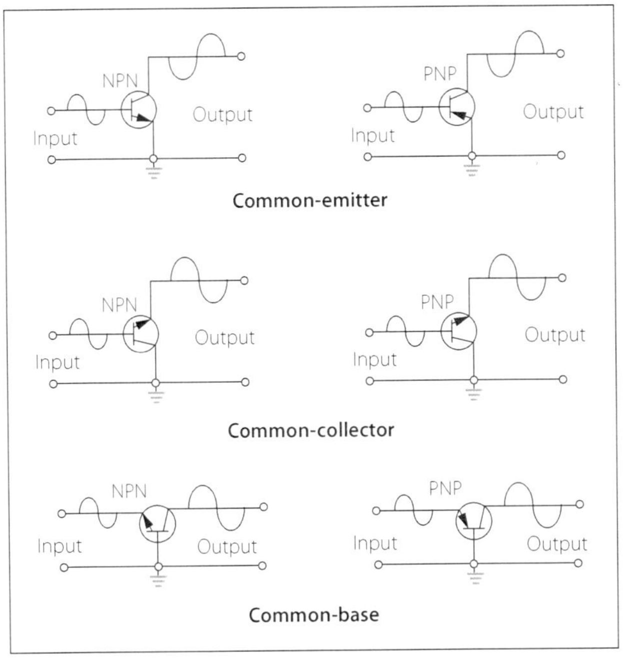

Figure 10-213. Transistor configurations (common-emitter, common-collector, common-base).

Because the input signal to the common-emitter goes positive when the output goes negative, the two signals are 180° out of phase. This is the only configuration that provides a phase reversal. The common-emitter is the most popular of the three configurations because it has the best combination of current and voltage gain. Gain is a term used to indicate the magnitude of amplification. Each transistor configuration has its unique gain characteristics even though the same transistors are used.

Common-Collector Configuration

This transistor configuration is usually used for impedance matching. It is also used as a current driver due to its high current gain. It is also very useful in switching circuits since it has the ability to pass signals in either direction. [Figure 10-213]

In the common-collector circuit, the input signal is applied to the base, and the output signal is taken from the emitter, leaving the collector as the common point between the input and the output. The input resistance of the CC circuit is high, while the output resistance is low. The current gain is higher than that in the common-emitter, but it has a lower power gain than either the common-emitter or common-base configuration. Just like the common-base configuration, the output signal of the common-collector circuit is in phase with the input signal. The common-collector is typically referred to as an emitter-follower because the output developed on the emitter follows the input signal applied to the base.

Common-Base Configuration

The primary use of this configuration is for impedance matching because it has low input impedance and high output resistance. Two factors, however, limit the usefulness of this circuit application. First is the low input resistance and second is its lack of current, which is always below 1. Since the CB configuration will give voltage amplification, there are some applications for this circuit, such as microphone amplifiers. [Figure 10-213]

In the common-base circuit, the input signal is applied to the emitter and the output signal is taken from the collector. In this case, both the input and the output have the base as a common element. When an input signal is applied to the emitter, it causes the emitter-base junction to react in the same manner as that in the common-emitter circuit. When an input adds to the bias, it will increase the transistor current; conversely, when the signal opposes the bias, the current in the transistor decreases.

The signal adds to the forward bias, since it is applied to the emitter, causing the collector current to increase. This increase in I_C results in a greater voltage drop across the load resistor RL, thus lowering the collector voltage E_C. The collector voltage, in becoming less negative, will swing in a positive direction and is therefore in phase with the incoming positive signal.

Vacuum Tubes

The use of vacuum tubes in aircraft electrical and electronic systems has rapidly declined due to the many advantages of using transistors. However, some systems still employ vacuum tubes in special applications, and possibly some older model aircraft still in service are equipped with devices that use vacuum tubes. While these components may still be in service, their infrequent occurrence does not warrant a detailed discussion.

Originally, vacuum tubes were developed for radio work. They are used in radio transmitters as amplifiers for controlling voltage and current, as oscillators for generating audio and radio frequency signals, and as rectifiers for converting alternating current into direct current. While there are many types of vacuum tubes for a variety of applications, the most common types fall into one of the following families: (1) diode, (2) triode, (3) tetrode, and (4) pentode. Each of these vacuum tube types operates on the following fundamental principles.

When a piece of metal is heated, the speed of the electrons in the metal is increased. If the metal is heated to a high enough temperature, the electrons are accelerated to the point where some of them actually leave the surface of the metal. In a vacuum tube, electrons are supplied by a piece of metal called a cathode, which is heated by an electric current. Within limits, the hotter the cathode, the greater the number of electrons it will give off or emit.

To increase the number of electrons emitted, the cathode is usually coated with special chemical compounds. If an external field does not draw the emitted electrons away, they form about the cathode into a negatively charged cloud called the space charge. The accumulation of negative electrons near the emitter repels others coming from the emitter. The emitter, if insulated, becomes positive because of the loss of electrons. This establishes an electrostatic field between the cloud of negative electrons and the now positive cathode. A balance is reached when only enough electrons flow from the cathode to the area surrounding it to supply the loss caused by diffusion of the space charge.

Filtering

One of the more common uses of the capacitor and inductor that the technician may find in the field is that of the filter.

Filtering Characteristics of Capacitors

The nature of capacitance opposes a voltage change across its terminal by storing energy in its electrostatic field. Whenever the voltage tends to rise, the capacitor converts this voltage change to stored energy. When the voltage tends to fall, the capacitor converts this stored energy back to voltage. The use of a capacitor for filtering the output of a rectifier is illustrated in Figure 10-214. The rectifier is shown as a block, and the capacitor C_1 is connected in parallel with the load R_1.

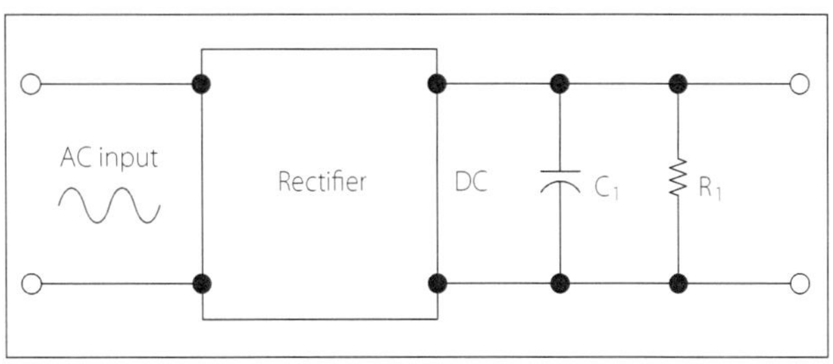

Figure 10-214. A capacitor used as a filter.

The capacitor C_1 is chosen to offer very low impedance to the AC ripple frequency and very high impedance to the DC component. The ripple voltage is therefore bypassed to ground through the low impedance path of the capacitor, while the DC voltage is applied unchanged to the load. The effect of the capacitor on the output of the rectifier can be seen in the waveshapes shown in Figure 10-215. Dotted lines show the rectifier output, while the solid lines show the effect of the capacitor. In this example, full-wave rectifier outputs are shown. The capacitor C_1 charges when the rectifier voltage output tends to increase and discharges when the voltage output tends to decrease. In this manner, the voltage across the load R_1 is kept fairly constant.

Filtering Characteristics of Inductors

The inductance provided by an inductor may be used as a filter, because it opposes a change in current through it by storing energy in its electromagnetic field. Whenever the current increases, the stored energy in the electromagnetic field increases. When the current through the inductor decreases, the inductor supplies the energy back into the circuit in order to maintain the existing flow of current. The use of an inductor for filtering the output of a rectifier is shown in Figure 10-216. Note that in this network the inductor L_1 is in series with the load R_1.

The inductance L_1 is selected to offer high impedance to the AC ripple voltage and low impedance to the DC component. The result is a very large voltage drop across the inductor and a very small voltage drop across the load R_1. For the DC component, however, a

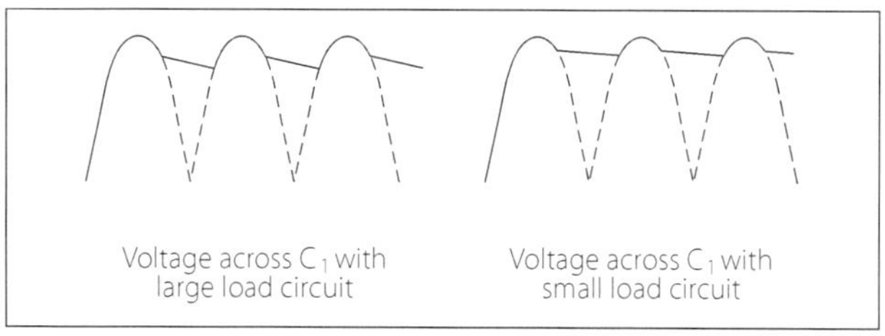

Figure 10-215. Half-wave and full-wave rectifier outputs using capacitor filter.

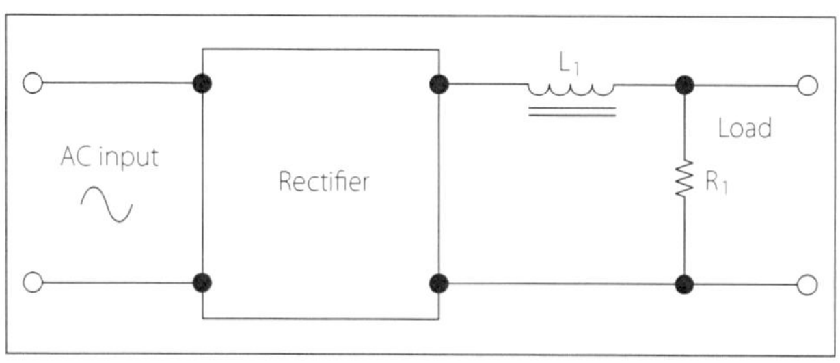

Figure 10-216. An inductor used as a filter.

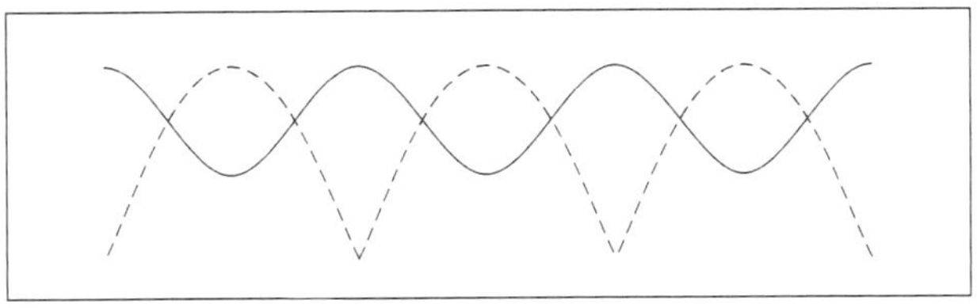

Figure 10-217. Output of an inductor filter rectifier.

very small voltage drop occurs across the inductor and a very large voltage drop across the load. The effect of an inductor on the output of a full-wave rectifier in the output waveshape is shown in Figure 10-217.

Common Filter Configurations

Capacitors and inductors are combined in various ways to provide more satisfactory filtering than can be obtained with a single capacitor or inductor. These are referred to collectively as "LC filters." Several combinations are shown schematically in Figure 10-218. Note that the L, or inverted L-type, and the T-type filter sections resemble schematically the corresponding letters of the alphabet. The pi-type filter section resembles the Greek letter pi (π) schematically.

All the filter sections shown are similar in that the inductances are in series and the capacitances are in parallel with the load. The inductances must, therefore, offer very high impedance and the capacitors very low impedance to the ripple frequency. Since the ripple frequency is comparatively low, the inductances are iron core coils having large values of inductance (several henries). Because they offer such high impedance to the ripple frequency, these coils are called chokes. The capacitors must also be large (several microfarads) to offer very little opposition to the ripple frequency. Because the voltage across the capacitor is DC, electrolytic capacitors are frequently used as filter capacitors. Always observe the correct polarity in connecting electrolytic capacitors.

LC filters are also classified according to the position of the capacitor and inductor. A capacitor input filter is one in which the capacitor is connected directly across the output terminals of the rectifier. A choke input filter is one in which a choke precedes the filter capacitor.

If it is necessary to increase the applied voltage to more than a single rectifier can tolerate, the usual solution is to stack them. These rectifiers are similar to resistors added in series. Each resistor will drop a portion of the applied voltage rather than the total voltage. The same theory applies to rectifiers added in series, or stacked. Series stacking increases the voltage rating. If, for example, a rectifier will be destroyed with an applied voltage exceeding 50 volts, and it is to be used in a circuit with an applied voltage of 150 volts, stacking of diodes can be employed. The result is shown in Figure 10-219.

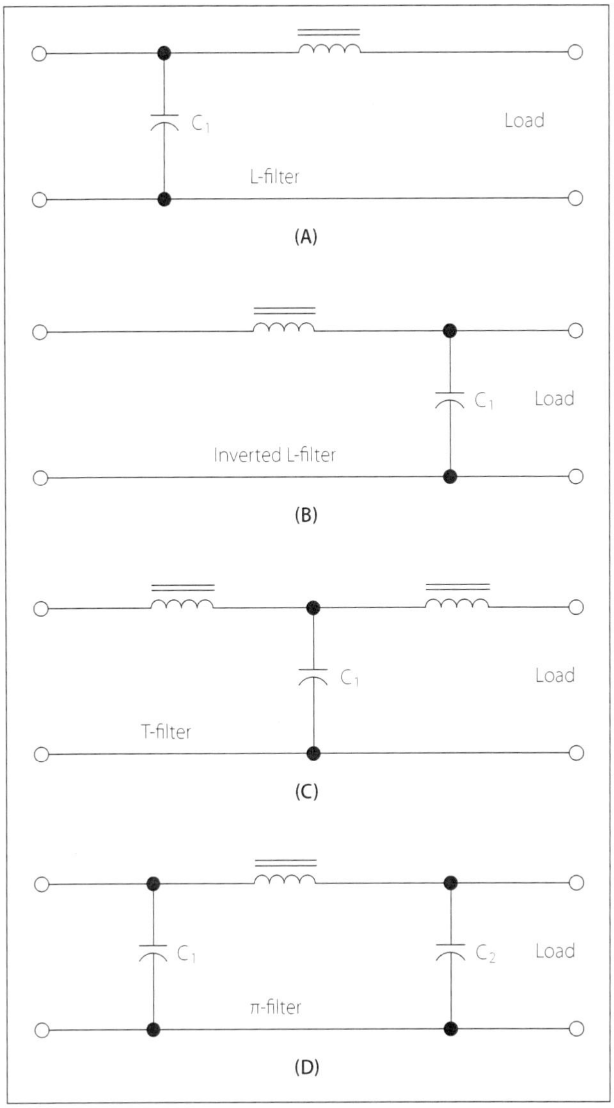

Figure 10-218. LC filters.

Basic LC Filters

Analog filters are circuits that perform signal processing functions, specifically intended to remove unwanted signal components such as ripple and enhance desired signals. The simplest analog filters are based on combinations of inductors and capacitors. The four basic categories of filters discussed are: low-pass, high-pass, band-pass and band-stop. All these types are collectively known as passive filters, because they do not depend on any external power source.

The operation of a filter relies on the characteristic of variable inductive and capacitive reactance based on

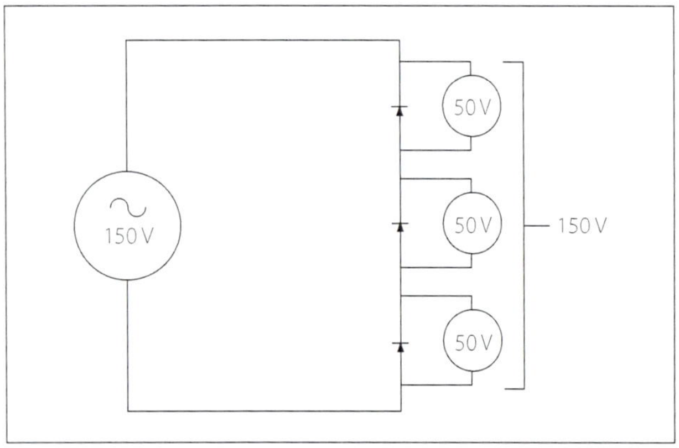

Figure 10-219. Stacking diodes in a circuit.

the applied frequency. In review, the inductor will block high-frequency signals (high reactance) and conduct low-frequency signals (low reactance), while capacitors do the reverse. A filter in which the signal passes through an inductor, or in which a capacitor provides a path to earth, presents less attenuation (reduction) to a low-frequency signal than to a high-frequency signal and is considered a low-pass filter. If the signal passes through a capacitor, or has a path to ground through an inductor, then the filter presents less attenuation to high-frequency signals than low-frequency signals and is then considered a high-pass filter. Typically after an AC signal is rectified the pulses of voltage are changed to usable form of DC by way of filtering.

Low-Pass Filter

A low-pass filter is a filter that passes low frequencies well, but attenuates (reduces) higher frequencies. The so-called cutoff frequency divides the range of frequencies that are passed and the range of frequencies that are stopped. In other words, the frequency components higher than the cutoff frequency will be stopped by a low-pass filter.

The actual amount of attenuation for each frequency varies by filter design.

An Inductive low-pass filter inserts an inductor in series with the load, where a capacitive low-pass filter inserts a resistor in series and a capacitor in parallel with the load. The former filter design tries to "block" the unwanted frequency signal while the latter tries to short it out. Figure 10-220 illustrates this type of circuit and the frequency/current flow response.

High-Pass Filter

A high-pass filter (HPF) is a filter that passes high frequencies well, but attenuates (reduces) frequencies lower than the cutoff frequency. The actual amount of attenuation for each frequency varies once again depending on filter design. In some cases it is called a low-cut filter. A high-pass filter is essentially the opposite of a low-pass filter.

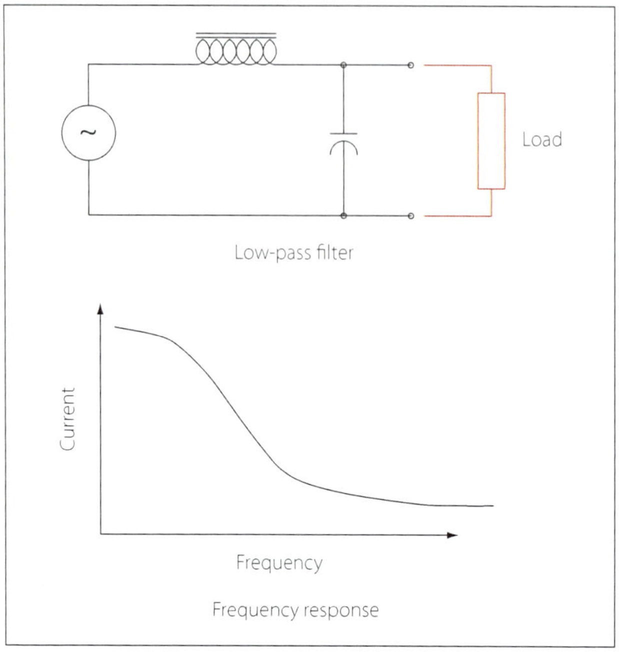

Figure 10-220. Low-pass filter.

It is useful as a filter to block any unwanted low frequency components of a signal while passing the desired higher frequencies. Figure 10-221 illustrates this type of circuit and the frequency/current flow response.

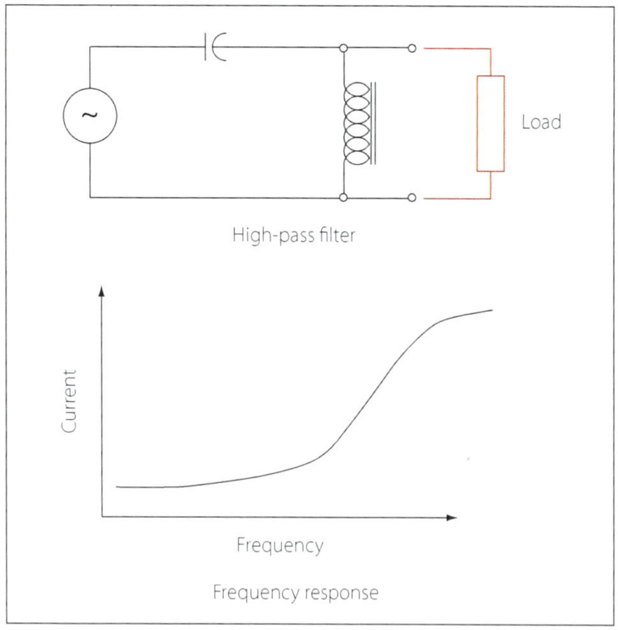

Figure 10-221. High-pass filter.

Band-Pass Filter

A band-pass filter is basically a combination of a high-pass and a low-pass. There are some applications where a particular range of frequencies need to be singled out or filtered from a wider range of frequencies. Band-pass filter circuits are designed to accomplish this task by combining the properties of low-pass and high-pass into a single filter. Figure 10-222 illustrates this type of circuit and the frequency/current flow response.

Band-Stop Filter

In signal processing, a band-stop filter or band-rejection filter is a filter that passes most frequencies unaltered, but attenuates those in a range to very low levels. It is the opposite of a band-pass filter. A notch filter is a band-stop filter with a narrow stopband (high Q factor). Notch filters are used in live sound reproduction (Public Address systems, also known as PA systems) and in instrument amplifier (especially amplifiers or preamplifiers for acoustic instruments such as acoustic guitar, mandolin, bass instrument amplifier, etc.) to reduce or prevent feedback, while having little noticeable effect on the rest of the frequency spectrum. Other names include "band limit filter," "T-notch filter," "band-elimination filter," and "band-rejection filter."

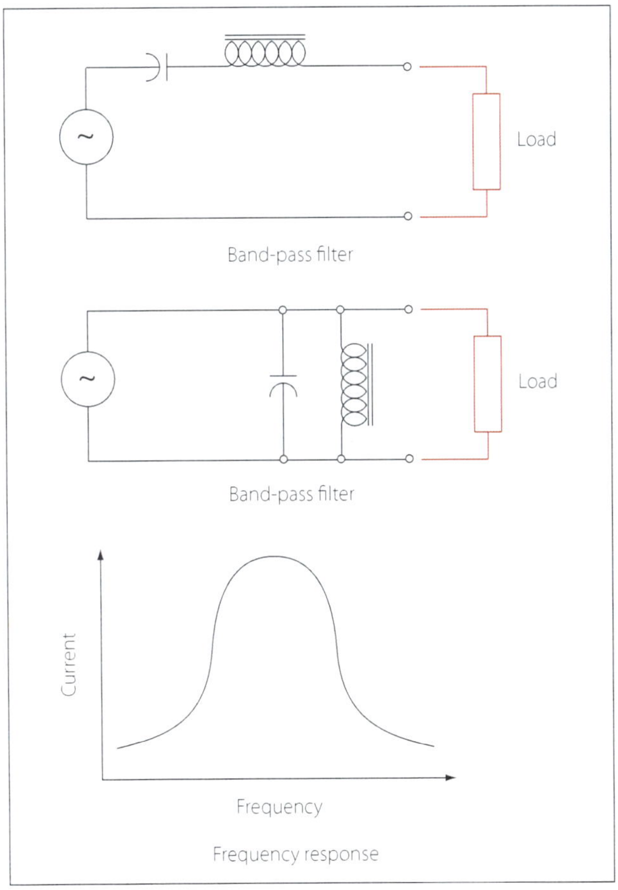

Figure 10-222. Band-pass filter.

Typically, the width of the stop-band is less than 1 to 2 decades (that is, the highest frequency attenuated is less than 10 to 100 times the lowest frequency attenuated). In the audio band, a notch filter uses high and low frequencies that may be only semitones apart.

A band-stop filter is the general case. A notch filter is a specific type of band-stop filter with a very narrow range.

Also called band-elimination, band-reject, or notch filters, this kind of filter passes all frequencies above and below a particular range set by the component values. Not surprisingly, it can be made out of a low-pass and a high-pass filter, just like the band-pass design, except that this time we connect the two filter sections in parallel with each other instead of in series. Figure 10-223 illustrates this type of circuit and the frequency/current flow response.

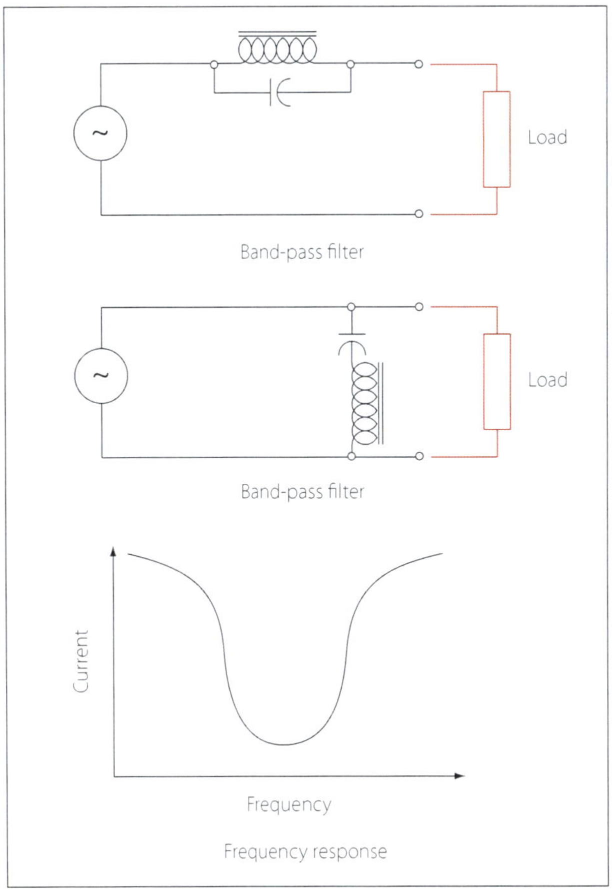

Figure 10-223. Band-stop filter.

Amplifier Circuits

An amplifier is a device that enables an input signal to control an output signal. The output signal will have some or all of the characteristics of the input signal but will generally be a greater magnitude than the input signal in terms of voltage, current, or power. Gain is the basic function of all amplifiers. Because of this gain, we can expect the output signal to be greater than the input signal. If for example we have an input signal of 1 volt and an output signal of 10 volts, then the gain factor can be determined by:

Gain = Signal out/Signal in
Gain = 10V/1V = 10

Voltage gain is usually used to describe the operation of a small gain amplifier. In this type of an amplifier, the output signal voltage is larger than the input signal voltage. Power gain, on the other hand, is usually used to describe the operation of large signal amplifiers. In the case of power gain amplifiers, the gain is not based on voltage but on watts. A power amplifier is an amplifier in which the output signal power is greater than the input signal power. Most power amplifiers are used as the final stage of amplification and drive the output device. The output device could be a cockpit or cabin speaker, an indicator, or antenna. Whatever the device, the power to make it work comes from the final stage of amplification. Drivers for autopilot servos are sometimes contained in line replaceable units (LRUs) called autopilot amplifiers. These units take the low signal commands from the flight guidance system and amplify the signals to a level usable for driving the servo motors.

Classification

The classification of a transistor amplifier circuit is determined by the percentage of the time that the current flows through the output circuit in relation to the input signal. There are four classifications of operation: A, AB, B, and C. Each class of operation has a certain use and characteristic. No individual class of amplifiers is considered the "best." The best use of an amplifier is a matter of proper selection for the particular operation desired.

Class A

Figure 10-224, shows a simplified Class A amplifier circuit. In the Class A operation, the current in the transistor flows for 100 percent or 360° of the input signal. Class A operation is the least efficient class of operation but provides the best fidelity. Fidelity simply means that the output signal is a good reproduction of the input signal in all respects other than the amplitude, which is amplified. In some cases, there may be some phase shifting between the input signal and the output signal. Typically, the phase difference is 180°. If the output signal is not a good reproduction of the input signal, then the signal is said to be distorted. Distortion is any undesired change to the signal from the input to the output.

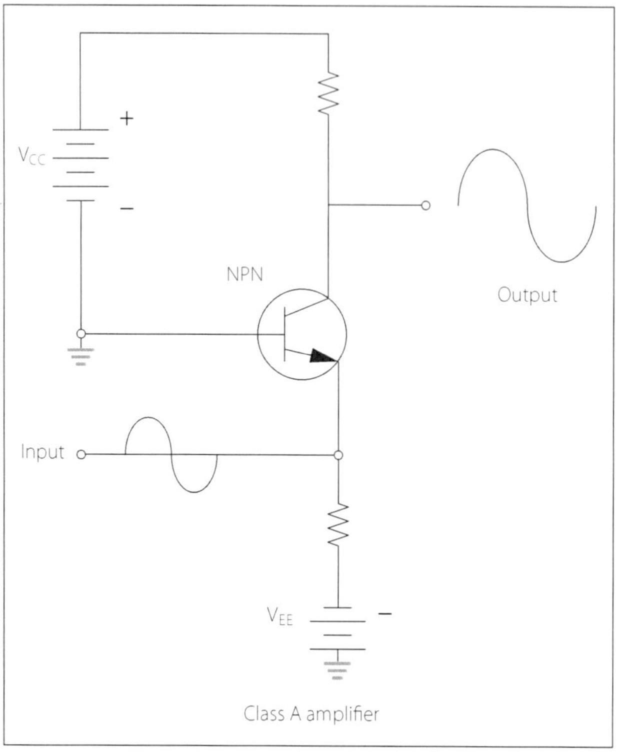

Figure 10-224. Simplified Class A amplifier circuit.

The efficiency of an amplifier refers to the amount of power delivered to the output compared to the power supplied to the circuit. Every device in the circuit consumes power in order to operate. If the amplifier operates for 360° of input signal, then it is using more power than if it was using only 180° of input signal. The more power consumed by the amplifier, the less there is available for the output signal. Usually the Class A amplifier is used where efficiency is of little concern and where fidelity in reproduction is desired.

Class AB

Figure 10-225, shows a simplified Class AB amplifier circuit. In the Class AB operation, the transistor current flows for more than 50 percent but less than 100 percent of the input signal. Unlike the Class A amplifier, the output signal is distorted. A portion of the output circuit appears to be truncated. This is due to the lack of current through the transistor during this point of operation. When the emitter in this case becomes positive enough, the transistor cannot conduct because the

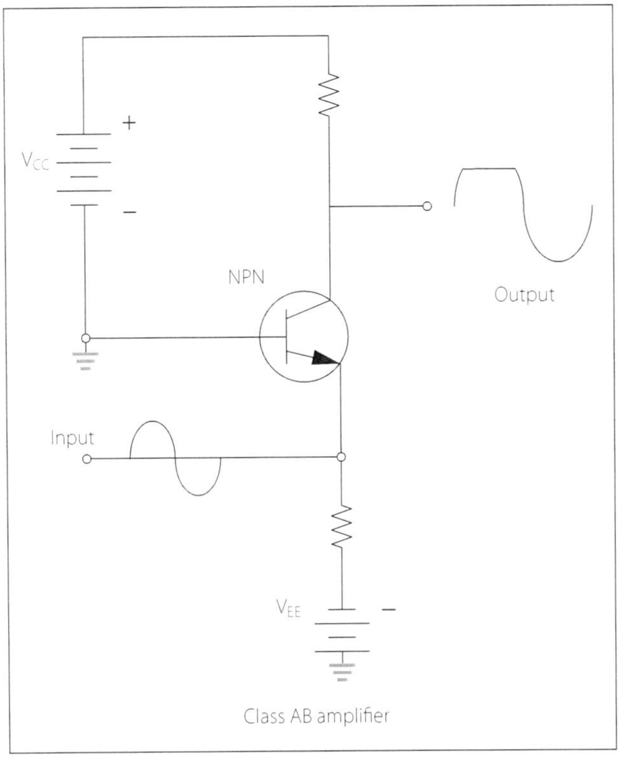

Figure 10-225. Simplified Class AB amplifier circuit.

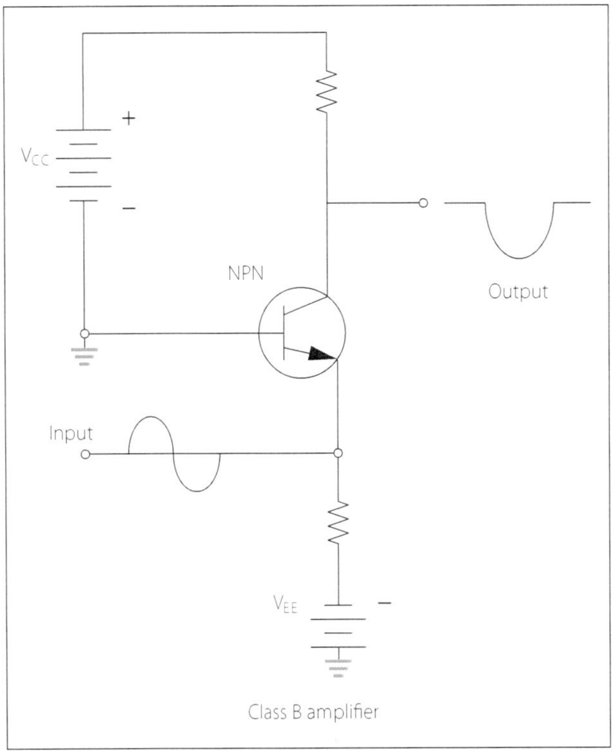

Figure 10-226. Simplified Class B amplifier circuit.

base to emitter junction is no longer forward biased. The input signal going positive beyond this point will not produce any further output and the output will remain level.

The Class AB amplifier has a better efficiency and a poorer fidelity than the Class A amplifier. These amplifiers are used when an exact reproduction of the input is not required but both the positive and negative portions of the input signals need to be available on the output.

Class B

Figure 10-226, shows a simplified Class B amplifier circuit. In Class B operation, the transistor current flows for only 50 percent of the input signal. In this illustration, the base-emitter bias will not allow the transistor to conduct whenever the input signal is greater than zero. In this case, only the negative portion of the input signal will be reproduced. Unlike the rectifier, the Class B amplifier will not only reproduce half of the input signal, but it will also amplify it. Class B amplifiers are twice as efficient as the Class A amplifier because the amplifying device only uses power for half of the input signal.

Class C

Figure 10-227, shows a simplified Class C amplifier circuit. In Class C operations, transistor current flows for less than 50 percent of the input signal. This class of operation is the most efficient. Because the transistor does not conduct except during a small portion of the input signal, this is the most efficient class of amplifier. The distortion of the Class C amplifier is greater (poor

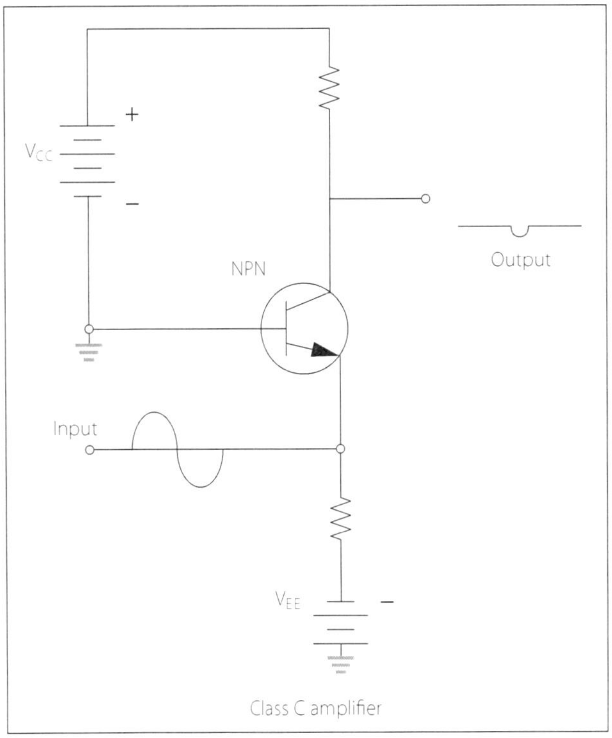

Figure 10-227. Simplified Class C amplifier circuit.

fidelity) than the Class A, AB, and B amplifiers because a small portion of the input signal is reproduced on the output. Class C amplifiers are used when the output signal is used for only small portions of time.

Methods of Coupling

Coupling is used to transfer a signal from one stage on an amplifier to another stage. Regardless of whether an amplifier is a single stage or one in a series of stages, there must be a method for the signal to enter and leave the circuit. Coupling is the process of transferring the energy between circuits. There are a number of ways for making this transfer and to discuss these methods in detail goes beyond the scope of this text. However, four methods are listed below with a brief description of their operation.

Direct Coupling

Direct coupling is the connection of the output of one stage directly to the input of the next stage. Direct coupling provides a good frequency response because no frequency-sensitive components such as capacitors and inductors are used. Yet this method is not used very often due to the complex power supply requirements and the impedance matching problems.

RC Coupling

RC coupling is the most common method of coupling and uses a coupling capacitor and signal developing resistors. Figure 10-228, shows a simplified RC coupling circuit. In this circuit, R1 acts as a load resistor for Q1 and develops the output signal for that stage. The capacitor C1 blocks the DC bias signal and passes the AC output signal. R2 then becomes the load over which the passes AC signal is developed as an input to the base of Q2. This arrangement allows for the bias voltage of each stage to be blocked, while the AC signal is passed to the next stage.

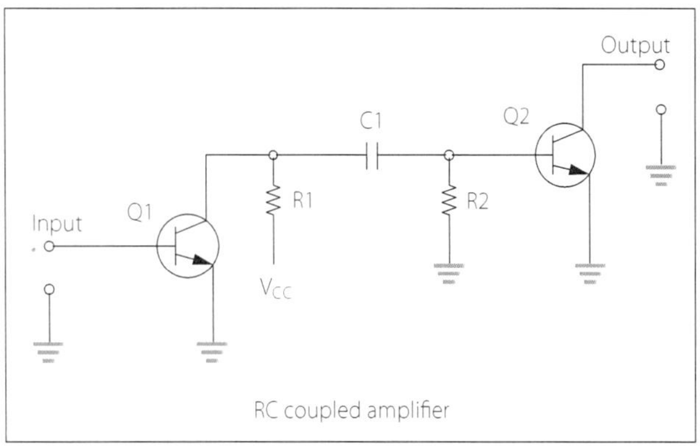

Figure 10-228. Simplified RC coupling circuit.

Impedance Coupling

Impedance coupling uses a coil as a load for the first stage but otherwise functions just as an RC coupling. Figure 10-229 shows a simplified impedance coupling circuit. This method is similar to the RC coupling method. The difference is that R1 is replaced with inductor L1 as the output load. The amount of signal developed on the output load depends on the inductive reactance of the coil. In order for the inductive reactance to be high, the inductance must be large; the frequency must be high or both. Therefore, load inductors should have relatively large amounts of inductance and are most effective at high frequencies.

Transformer Coupling

Transformer coupling uses a transformer to couple the signal from one stage to the next. Figure 10-230, shows a simplified transformer coupling circuit. The transformer action of T1 couples the signal from the first stage to the second stage. The primary coil of T1 acts as a load for the output of the first stage while the secondary coil acts as the developing impedance for the second stage Q2. Transformer coupling is very efficient and the transformer can aid in impedance matching.

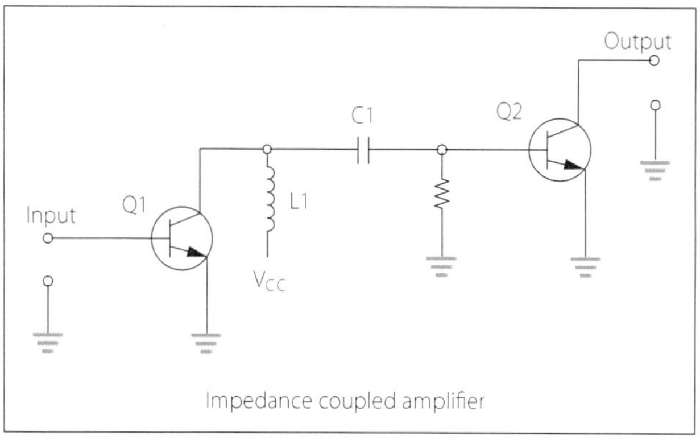

Figure 10-229. Simplified impedance coupling circuit.

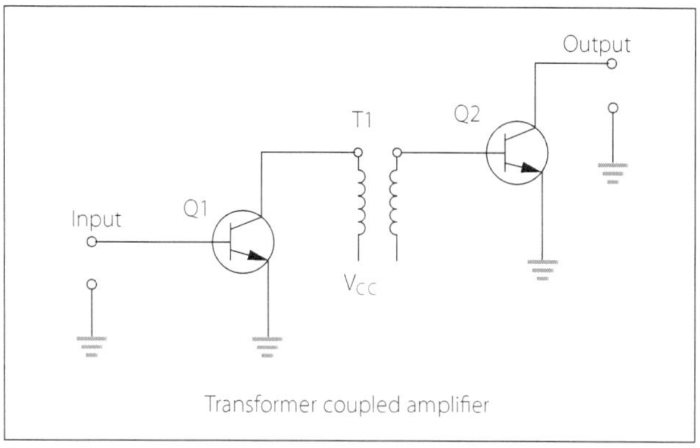

Figure 10-230. Simplified transformer coupling circuit.

Feedback

Feedback occurs when a small portion of the output signal is sent back to the input signal to the amplifier. There are two types of feedback in amplifiers:

1. Positive (regenerative)
2. Negative (degenerative)

The main difference between these two signals is whether the feedback signal adds to the input signal or if the feedback signal diminishes the input signal.

When the feedback is positive, the signal being returned to the input is in phase with the input signal and thus interferes constructively. Figure 10-231 illustrates this concept applied in the amplifier circuit through a block diagram. Notice that the feedback signal is in phase with the input signal, which will regenerate the input signal. This results in an output signal with amplitude greater than would have been without the constructive, positive feedback. This type of positive feedback is what causes an audio system to squeal.

Figure 10-231 also illustrates with a block diagram how negative or degenerative feedback occurs. In this case, the feedback signal is out of phase with the input signal. This causes destructive interference and degenerates the input signal. The result is a lower amplitude output signal than would have occurred without the feedback.

Operational Amplifiers

An operational amplifier (OP AMP) is designed to be used with other circuit components and performs either computing functions or filtering. Operational amplifiers are usually high-gain amplifiers with the amount of gain governed by the amount of feedback.

Operational amplifiers were originally developed for analog computers and used to perform mathematical functions. Today many devices use the operational amplifier for DC amplifiers, AC amplifiers, comparators, oscillators, and filter circuits. The widespread use is due to the fact that the OP AMP is a versatile device, small, and inexpensive. Built into the integrated chip, the operational amp is used as a basic building block of larger circuits.

Figure 10-232 shows the schematic symbol for the operational amplifier. There are two inputs to the

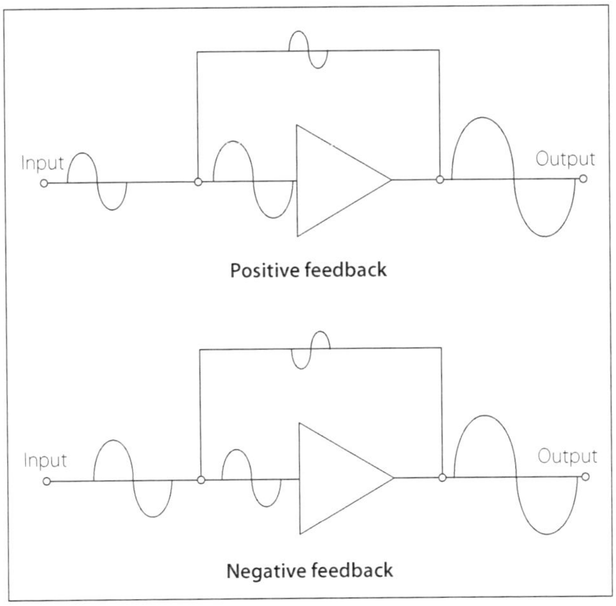

Figure 10-231. Feedback.

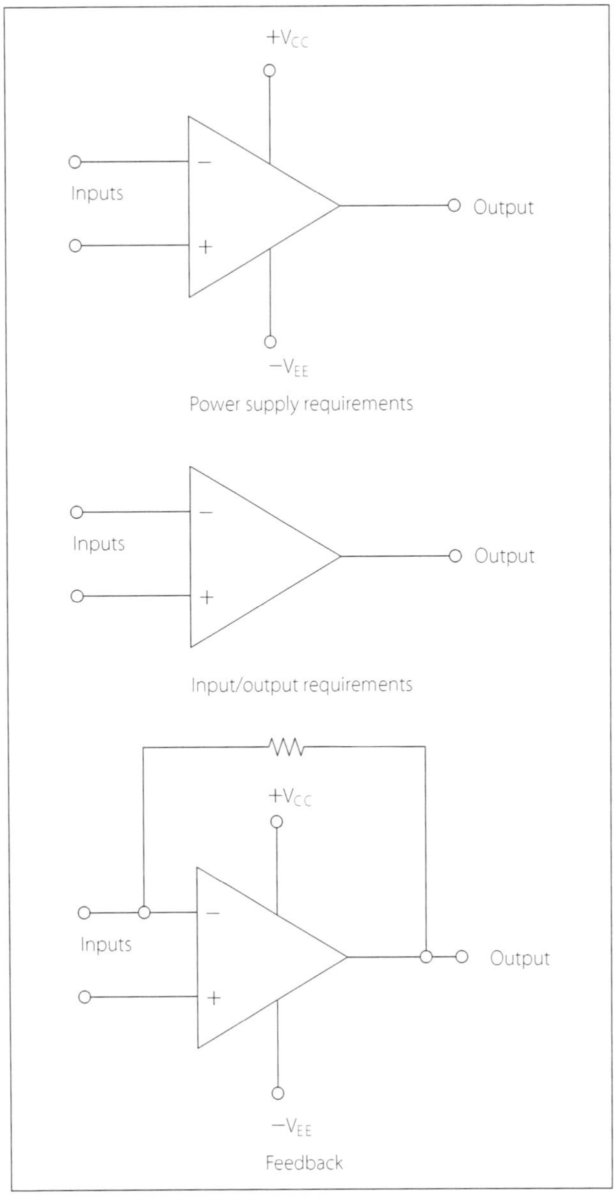

Figure 10-232. Schematic symbol for the operational amplifier.

operational amplifier, inverting (−) and non-inverting (+), and there is one output. The polarity of a signal applied to the inverting input (−) will be reversed at the output. A signal applied to the non-inverting (+) input will retain its polarity on the output. To be classified as an operational amplifier, the circuit must have certain characteristics:

1. Very high gain.
2. Very high input impedance.
3. Very high output impedance.

This type of a circuit can be made up of discrete components, such as resistors and transistors. However, the most common form of an operational amplifier is found in the integrated circuit. This integrated circuit or chip will contain the various stages of the operational amplifier and can be treated as if it were a single stage.

Applications

The number of applications for OP AMPs is too numerous to detail in this text. However, the technician will occasionally come across these devices in modern aircraft and should be able to recognize their general purpose in a circuit. Some of the basic applications are:

1. Go/No Go detectors.
2. Square wave circuits.
3. Non-inverting amplifier.
4. Inverting amplifier.
5. Half-wave rectifier.

Magnetic Amplifiers

Magnetic amplifiers do not amplify magnetism but use electromagnetism to amplify a signal. Essentially, the magnetic amplifier is a power amplifier with a very limited frequency response. The frequency range most commonly associated with the magnetic amplifier is 100 hertz and less, which places it in the audio range. As a technical point, the magnetic amplifier is a low-frequency amplifier.

Advantages of the magnetic amplifier are:

1. Very high efficiency, on the order of approximately 90 percent.
2. High reliability.
3. Very rugged, able to withstand vibrations, moisture, and overloads.
4. No warm-up time.

Some of the disadvantages of the magnetic amplifier are:

1. Incapacity to handle low voltage signals.
2. Not usable in high-frequency applications.
3. Time delay associated with magnetic affects.
4. Poor fidelity.

The basic operating principles of the magnetic amplifier are fairly simple. Keep in mind that all amplifiers are current control devices. In this particular case, power that is delivered to the load is controlled by a variable inductance.

If an AC voltage is applied to the primary winding of an iron core transformer, the iron core will be magnetized and demagnetized at the same frequency as that of the applied voltage. This, in turn, will induce a voltage in the transformers secondary winding. The output voltage across the terminals of the secondary will depend on the relationship of the number of turns in the primary and the secondary of the transformer.

The iron core of the transformer has a saturation point after which the application of a greater magnetic force will produce no change in the intensity of magnetization. Hence, there will be no change in transformer output, even if the input is greatly increased. The magnetic amplifier circuit in Figure 10-233 will be used to explain how a simple magnetic amplifier functions.

1. Assume that there is 1 ampere of current in coil A, which has 10 turns of wire. If coil B has 10 turns

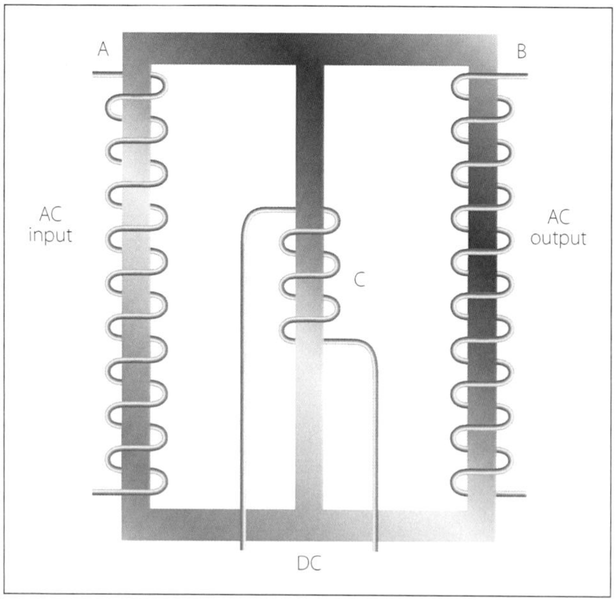

Figure 10-233. Magnetic amplifier circuit.

of wire, an output of 1 ampere will be obtained if coil B is properly loaded.

2. By applying direct current to coil C, the core of the magnetic amplifier coil can be further magnetized. Assume that coil C has the proper number of turns and, upon the application of 30 milliamperes, that the core is magnetized to the point where 1 ampere on coil A results in only 0.24 ampere output from coil B.
3. By making the DC input to coil C continuously variable from 0 to 30 milliamperes and maintaining an input of 1 ampere on coil A, it is possible to control the output of coil B to any point between 0.24 ampere and 1 ampere in this example.

The term "amplifier" is used for this arrangement because, by use of a few milliamperes, control of an output of 1 or more amperes is obtained.

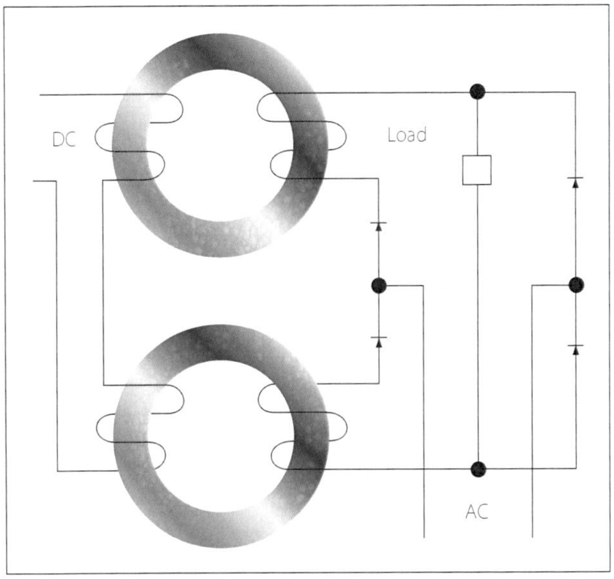

Figure 10-235. Self-saturating, full-wave magnetic amplifier.

Saturable-Core Reactor

The same procedure can be used with the circuit shown in Figure 10-234. A saturable-core reactor is a magnetic-core coil whose reactance is controlled by changing the permeability of the core. Varying the unidirectional flux controls the permeability of the core.

By controlling the extent of magnetization of the iron ring, it is possible to control the amount of current flowing to the load, since the amount of magnetization controls the impedance of the AC input winding. This type of magnetic amplifier is called a simple saturable reactor circuit.

Adding a rectifier to such a circuit would remove half the cycle of the AC input and permit a direct current to flow to the load. The amount of DC flowing in the load circuit is controlled by a DC control winding (sometimes referred to as bias). This type of magnetic amplifier is referred to as being self-saturating.

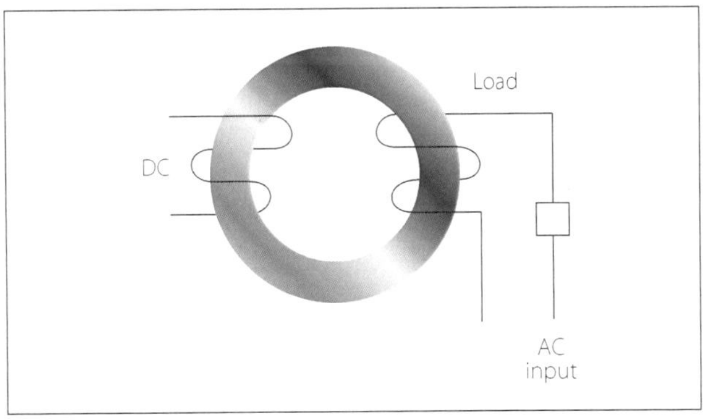

Figure 10-234. Saturable CORE reactor circuit.

To use the full AC input power, a circuit such as that shown in Figure 10-235 may be used. This circuit uses a full-wave bridge rectifier. The load will receive a controlled direct current by using the full AC input. This type of circuit is known as a self-saturating, full-wave magnetic amplifier.

In Figure 10-236 it is assumed that the DC control winding is supplied by a variable source, such as a sensing circuit. To control such a source and use its variations to control the AC output, it is necessary to include another DC winding that has a constant value. This winding, referred to as the reference winding, magnetizes the magnetic core in one direction.

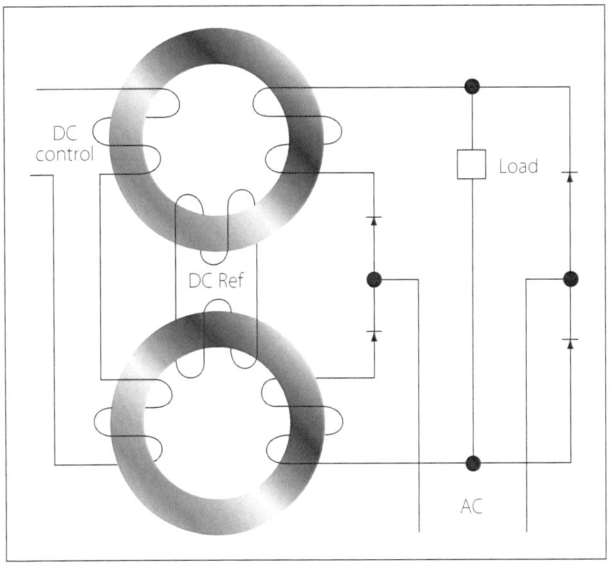

Figure 10-236. Basic preamplifier circuit.

The DC control winding, acting in opposition to the reference winding, either increases (degenerative) or decreases (regenerative) the magnetization of the core to change the amount of current flowing through the load. This is essentially a basic preamplifier.

Logic Circuits

Logic is considered the science of reasoning—the development of a reasonable conclusion based on known information. Human reasoning tells us that certain propositions are true if certain conditions or premises are true. An annunciator being lit in the master warning panel is an example of a proposition, which is either true or false. For example, predetermined and designed conditions must be met in order for an annunciator in a master warning panel to be lit. A "LOW HYDRAULIC PRESS" annunciator may have a simple set of conditions that will cause it to be illuminated. If the conditions are met, such as a hydraulic reservoir that is low on fluid, causing the line press to be low, then the logic is true and the annunciator will light. Several propositions, when combined will form a logical function. In the example above, the "LOW HYDRAULIC PRESS" annunciator will be on if the LED is not burned out AND the hydraulic press is low OR if the LED is not burned out AND the annunciator test is being asserted.

This section on logic circuits only serves as an introduction to the basic concepts. The technician will encounter many situations or problems in everyday life that can be expressed in some form of a logical function. Many problems and situations can be condensed down to simple yes/no or true/false statements, which if logically ordered can filter a problem down to a reasonable answer. The digital logic circuits are well suited for this task and have been employed in today's integrated circuits found in virtually all of the devices that we take for granted in modern aircraft. These logical circuits are used to carry out the logical functions for such things as navigation and communications. There are several fundamental elements that form the building blocks of the complex digital systems found in line replaceable units (LRUs) and avionics card cages. The following is a very basic outline of what those elements are and what logic conditions they will process. It is far beyond the scope of this text to cover digital logic systems due to the vast body of knowledge that it represents. However, this serves as an introduction and in some limited cases will be useful in reading system block diagrams that use logic symbols to aid the technician in understanding how a given circuit operates.

Logic Polarity

Electrical pulses can represent two logic conditions and any two differing voltages can be used for this purpose. For example, a positive voltage pulse could represent a true or 1 condition and a negative voltage pulse could then represent a false or 0 logic condition. The condition in which the voltage changes to represent a true or 1 logic is known as the logic polarity. Logic circuits are usually divided into two broad classes known as positive and negative polarity. The voltage levels used and a statement indicating the use of positive or negative logic will usually be specified in the logic diagrams provided by the original equipment manufacturers (OEMs).

Positive

When a signal that activates a circuit to a 1, true or high condition, has an electrical level that is relatively more positive that the other stated, then the logic polarity is said to be positive. An example would be:

Active State: 1 or True = +5 VDC
0 or False = −5 VDC

Negative

When the signal that actives a circuit to a 1, true or high condition, has an electrical level that is relatively more negative than the other stated, then the logic polarity is said to be negative. An example would be:

Active State: 1 or True = 0 VDC
0 or False = +5 VDC

Pulse Structure

Figure 10-237 illustrates the positive and negative pulse in an idealized form. In both forms, the pulse is

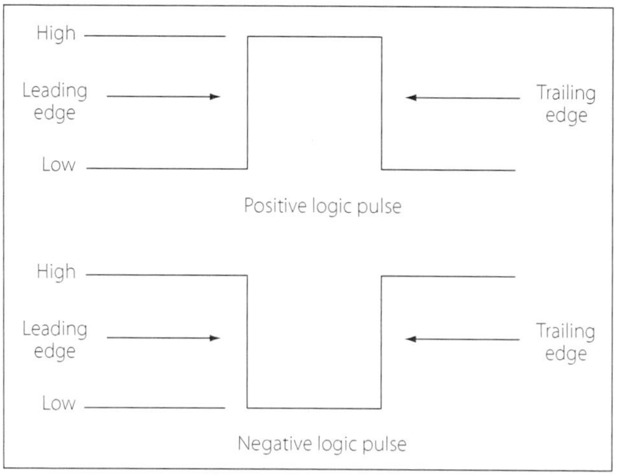

Figure 10-237. Positive and negative pulse in an idealized form.

composed of two edges—one being the leading edge and the other the trailing edge. In the case of the positive pulse logic, the positive transition from a lower state to a higher state is the leading edge and the trailing edge is the opposite. In the case of the negative logic pulse, the negative transition from a higher state to a lower state is the leading edge while the rise from the lower state back to the higher state is the trailing edge. Figure 10-237 is considered an ideal pulse because the rise and fall times are instantaneous. In reality, these changes take time, although in actual practice, the rise and fall can be assumed as instantaneous. Figure 10-238 shows the non-ideal pulse and its characteristics. The time required for a pulse to go from a low state to a high state is called the rise time, and the time required for the pulse to return to zero is called the fall time. It is common practice to measure the rise and fall time between 10 percent amplitude and 90 percent amplitude. The reason for taking the measurements in these points is due to the non-linear shape of the pulse in the first 10 percent and final 90 percent of the rise and fall amplitudes. The pulse width is defined as the duration of the pulse. To be more specific, it is the time between the 50 percent amplitude point on both the pulse rise and fall.

Basic Logic Circuits

Boolean logic is a symbolic system used in representing the truth value of statements. It is employed in the binary system used by digital computers primarily because the only truth values (true and false) can be represented by the binary digits 1 and 0. A circuit in computer memory can be open or closed, depending on the value assigned to it. The fundamental operations of Boolean logic, often called Boolean operators, are "and," "or," and "not"; combinations of these make up 13 other Boolean operators. Six of these operators are discussed.

The Inverter Logic

The inverter circuit performs a basic logic function called inversion. The purpose of the inverter is to convert one logic state into the opposite state. In terms of a binary digit, this would be like converting a 1 to a 0 or a 0 to a 1. When a high voltage is applied to the inverter input, low voltage will be the output. When a low voltage is applied to the input, a high voltage will be on the output. This operation can be put into what is known as a logic or truth table. Figure 10-240 shows the possible logic states for this gate. The standard logic symbol is shown in Figure 10-239. This is the common symbol for an amplifier with a small circle on the output. This type of logic can also be considered a NOT gate.

The AND Gate

The AND gate is made up of two or more inputs and a single output. The logic symbol is shown in Figure 10-241. Inputs are on the left and the output is on the right in each of the depictions. Gates with two, three, and four inputs are shown; however, any number of inputs can be used in the AND logic as long as the number is greater than one. The operation of the AND gate is such that the output is high only when all of the inputs are high. If any of the inputs are low, the output will also be low. Therefore, the basic purpose of an AND gate is to determine when certain conditions have been met at the same time. A high level on all inputs will produce a high level on the output. Figure 10-242 shows a simplified diagram of the AND logic with two switches and a light bulb. Notice that both switches need to be closed in order for the light bulb to turn on. Any other combination of switch positions will be an open circuit and the light will not turn on. An example of AND logic could possibly be engage logic, found in an autopilot. In this case, the autopilot would not be allowed to be engaged unless certain

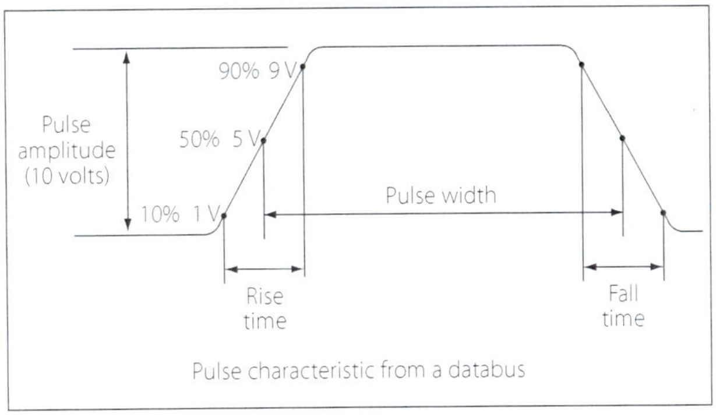

Figure 10-238. Non-ideal pulse and its characteristics.

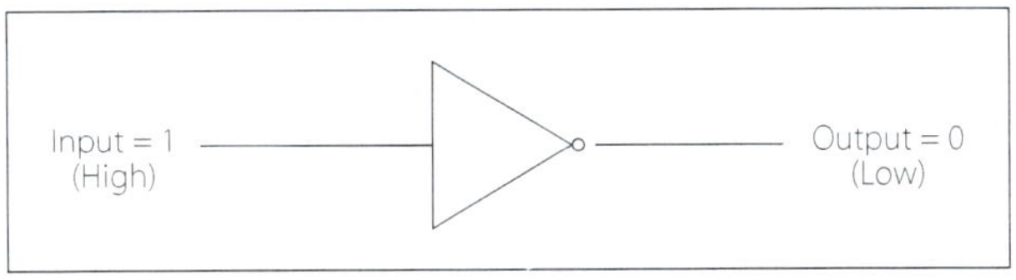

Figure 10-239. Standard logic symbol.

Input	Output
High	Low
Low	High

Figure 10-240. Possible logic states.

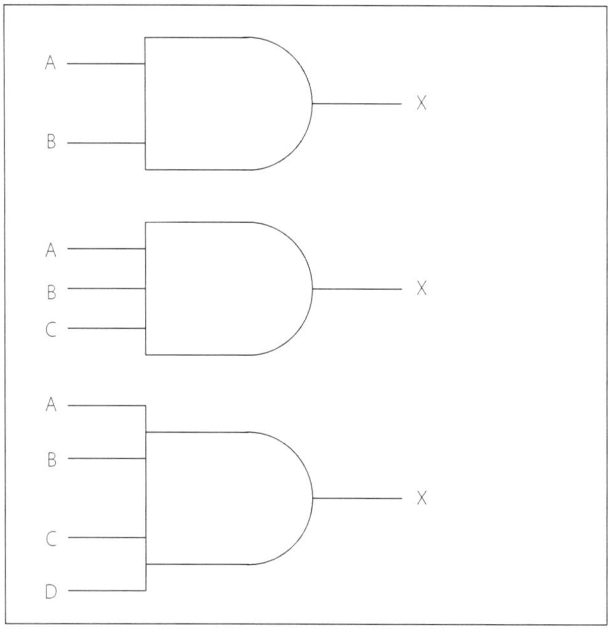

Figure 10-241. AND gate logic symbol.

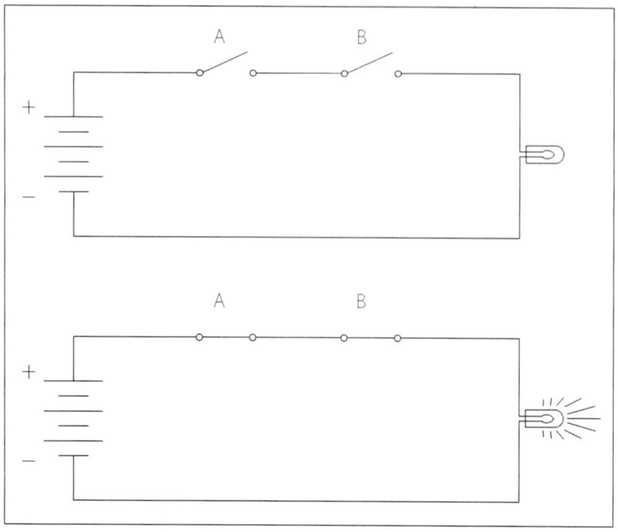

Figure 10-242. Simplified diagram of the AND logic.

conditions are first met. Such conditions could be: Vertical gyro is valid AND directional gyro is valid AND all autopilot control knobs are in detents AND servo circuits are operational. Only when these conditions are met will the autopilot be engaged. Figure 10-243 shows the logic of this system found in the aircraft wiring diagrams.

The OR Gate

The OR gate has two or more inputs and one output, and is normally represented by the standard logic symbol and truth table as shown in Figure 10-244. In this Figure, notice that the OR gate can have any number of inputs as long as it is greater than one. The operation of the OR gate is such that a high on any one of the inputs will produce a high on the output. The only time that a low is produced on the output is if there are no high levels on any input. Figure 10-245

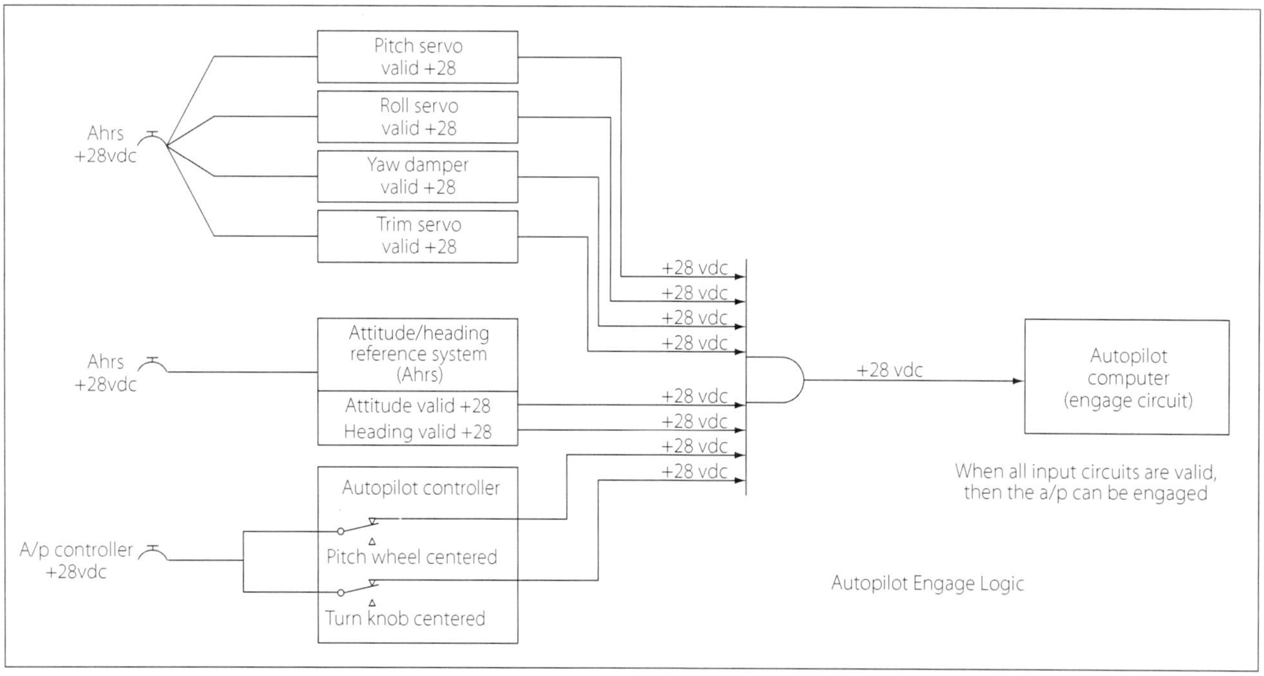

Figure 10-243. AND logic of system found in the aircraft wiring diagrams.

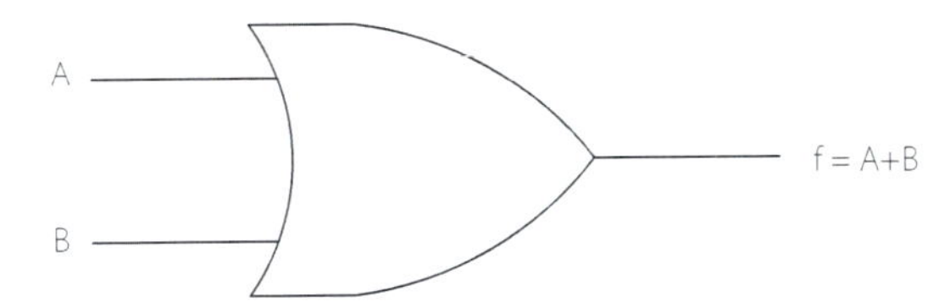

OR gate input/output

A	B	f
0	0	0
1	0	1
0	1	1
1	1	1

Truth table

Figure 10-244. OR gate.

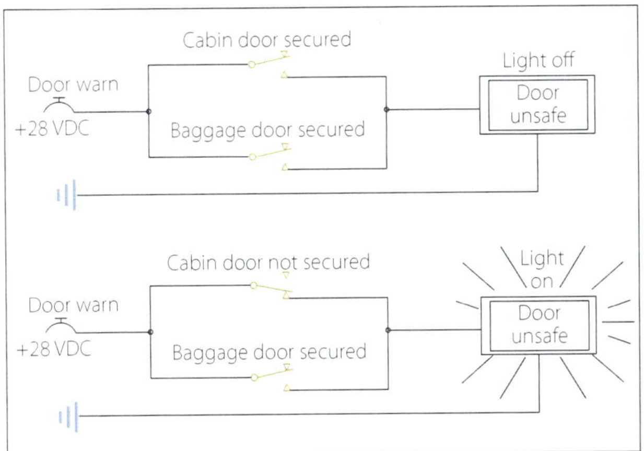

Figure 10-245. Smplified circuit that illustrates OR logic.

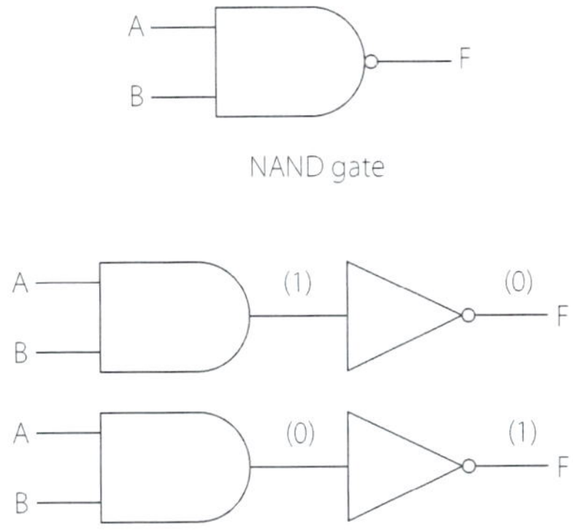

Equivalent circuit
NAND gate input/output

Figure 10-246. Standard logic symbol for two input NAND gate.

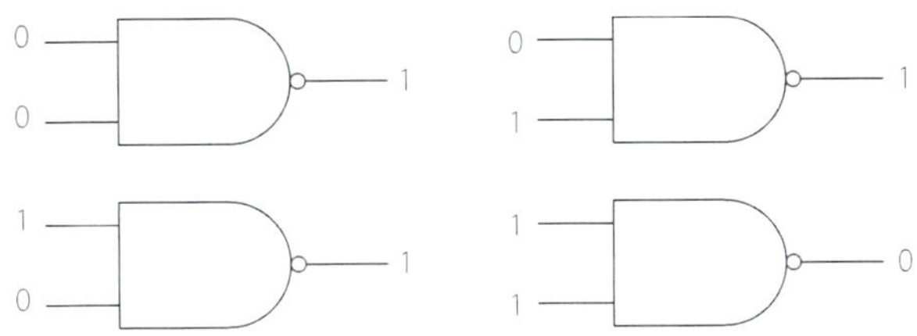

NAND gate input/output

A	B	f
0	0	1
1	0	1
0	1	1
1	1	0

NAND gate truth table

Figure 10-247. Two input NAND gate and corresponding truth table.

is a simplified circuit that illustrates the OR logic. The example used is a "DOOR UNSAFE" annunciator. Let's say in this case that the plane has one cabin door and a baggage door. In order for the annunciator light on the master warning panel to extinguish, both doors must be closed and locked. If any one of the doors is not secured properly, the baggage door OR the cabin door, then the "DOOR UNSAFE" annunciator will illuminate. In this case, two switches are in parallel with each other. If either one of the two switches is closed, the light bulb will light up. The lamp will be off only when both switches are open.

The NAND Gate

The term NAND is a combination of the NOT-AND gate and indicates an AND function with an inverted output. A standard logic symbol for a two input NAND gate is shown in Figure 10-246. Notice that an equivalent AND gate with an inverter is also shown. The logical operation of the NAND gate is such that a low output occurs only if all inputs are high. If any of the inputs are low, the output will be high. An example of a two input NAND gate and its corresponding truth table are shown in Figure 10-247.

The NOR Gate

The term NOR is a combination of the NOT and OR and indicates an OR function with an inverted output. The standard logic symbol for a two-inputs NOR gate is shown in Figure 10-248. Notice that an equivalent AND gate with an inverter is also shown. The logical

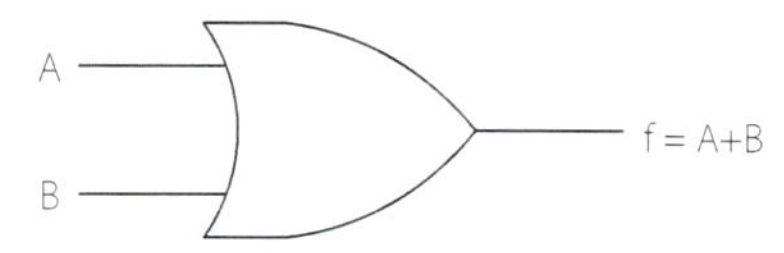

OR gate input/output

A	B	f
0	0	0
1	0	1
0	1	1
1	1	1

Truth table

Figure 10-248. Standard logic symbol for two inputs NOR gate.

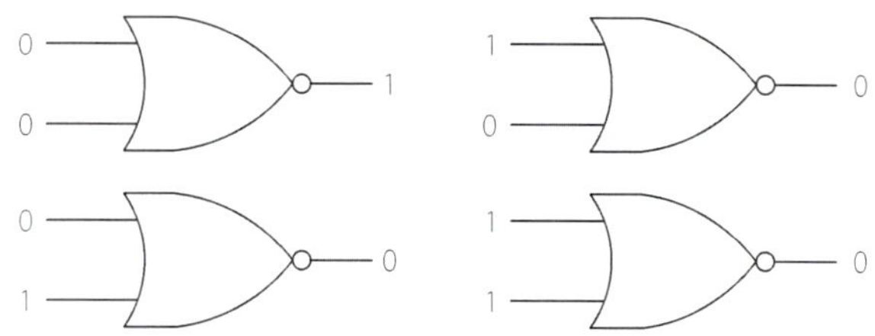

NOR gate input/output

A	B	f
0	0	1
0	1	0
1	0	0
1	1	0

NOR gate truth table

Figure 10-249. Logical operation of two-input NOR gate and truth table.

operation of the NOR gate is such that a low output happens when any of its inputs are high. Only when all of its inputs are low is the output high. The logic of this gate produces resultant outputs that are the opposite of the OR gate. In the NOR gate, the low output is the active output level. Figure 10-249 illustrates the logical operation of a two-input NOR gate for all of its possible combinations and the truth table.

Exclusive OR Gate

The exclusive OR gate is a modified OR gate that produces a 1 output when only one of the inputs is a 1. The abbreviation often used is X-OR. It is different from the standard OR gate in that when both inputs are a 1, then the output remains at a 0. The standard symbol and truth table for the X-OR gate are shown in Figure 10-250.

Exclusive NOR Gate

The exclusive NOR (X-NOR) gate is nothing more than an X-OR gate with an inverted output. It produces a 1 output when all inputs are 1s and also when all inputs are 0s. The standard symbol is shown in Figure 10-251.

The Integrated Circuit

All of the logic functions so far discussed plus many other components are available in some form of an integrated circuit. The digital systems found in today's aircraft owe their existence to a large extent to the design of the integrated circuit (IC). In most cases, the IC has an advantage over the use of discrete components in that they are smaller, consume less power, are very reliable, and are inexpensive. The most noticeable characteristic of the IC is its size and in comparison

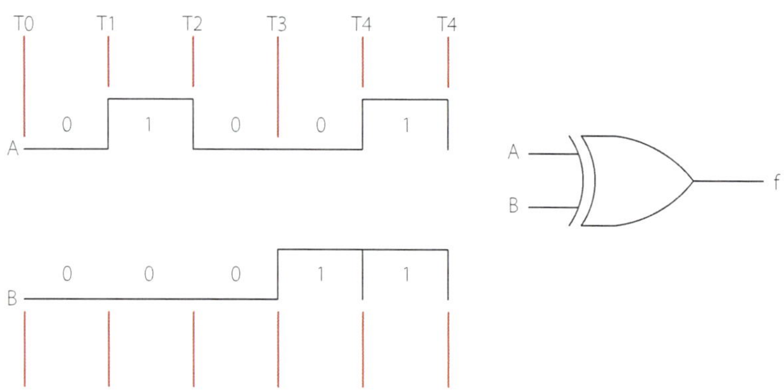

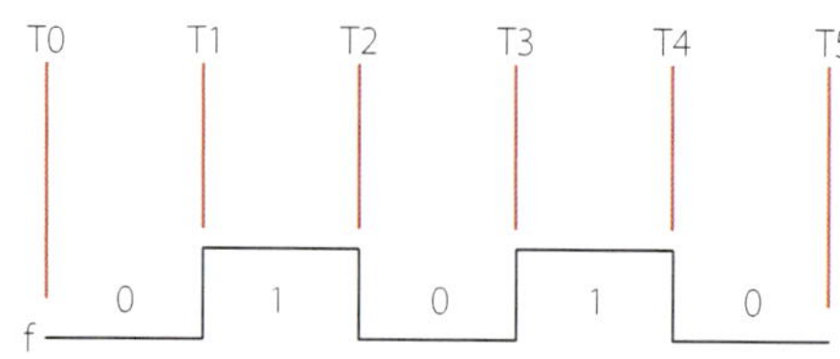

A	B	f
0	0	0
0	1	1
1	0	1
1	1	0

Truth table

X-OR gate input/output

Figure 10-250. Standard symbol and truth table for X-OR gate.

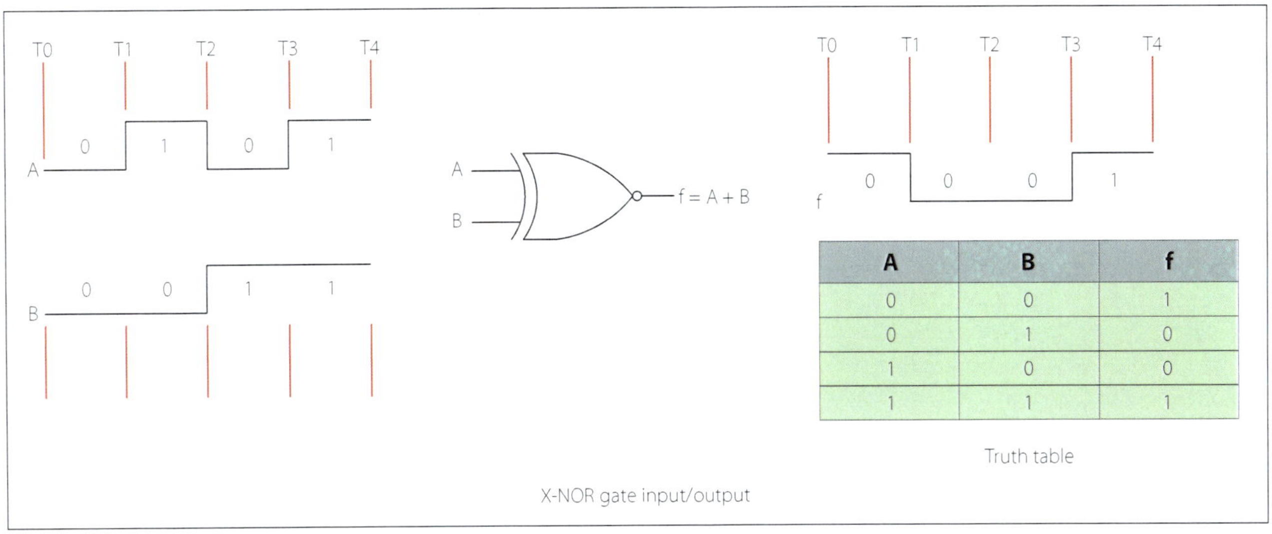

A	B	f
0	0	1
0	1	0
1	0	0
1	1	1

Truth table

Figure 10-251. Standard Symbol for X-NOR gate.

to the discrete semiconductor component, can easily be on the order of thousands of times smaller. [Figure 10-252]

A monolithic integrated circuit is an electronic circuit that is constructed entirely on a single chip or wafer of semiconductor material. All of the discrete components, such as resistors, transistors, diodes, and capacitors, can be constructed on these small pieces of semiconductor material and are an integral part of the chip. There are a number of levels of integration. Those levels are: small-scale integration, medium-scale integration, large-scale integration, and microprocessors. The small-scale integration is considered the least complex design of the digital ICs. These ICs contain the basic components such as the AND, OR, NOT, NOR and NAND gates. Figure 10-253 illustrates the schematic form of this type of circuit. The medium-scale integration can contain the same components as found in the small-scale design but in larger numbers ranging from 12 to 100. The medium-scale designs are house circuits that are more complex, such as encoders, decoders, registers, counters, multiplexers, smaller memories, and arithmetic circuits. Figure 10-254 illustrates the schematic form of this type of circuit. The large-scale integrated circuits contain even more logic

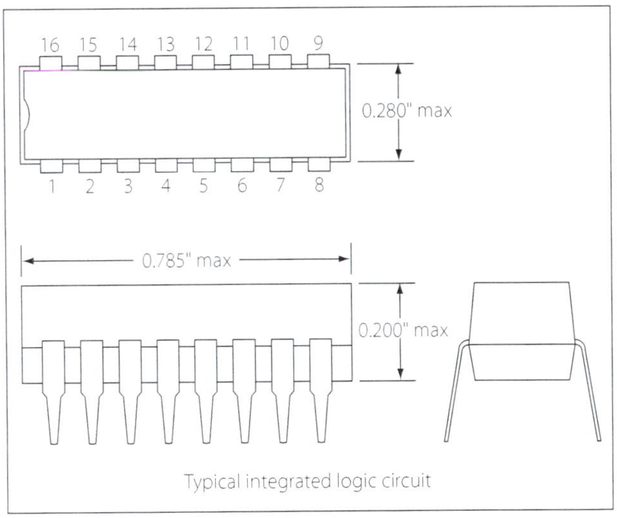

Figure 10-252. Integrated circuit.

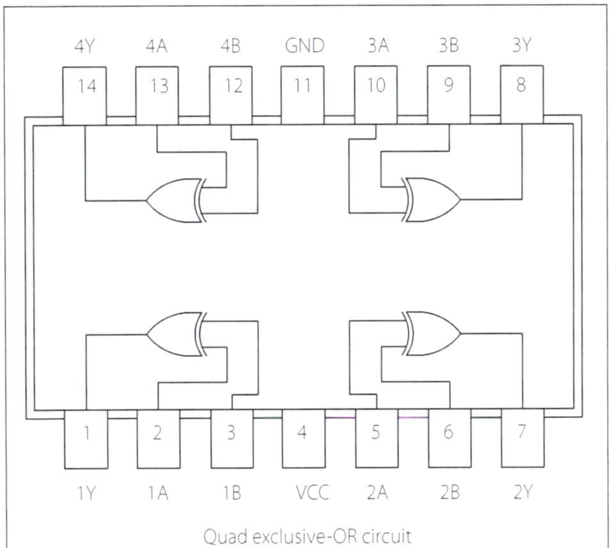

Input		Output
A	B	Y
0	0	0
0	1	1
1	0	1
1	1	0

Figure 10-253. Small-scale integration schematic form.

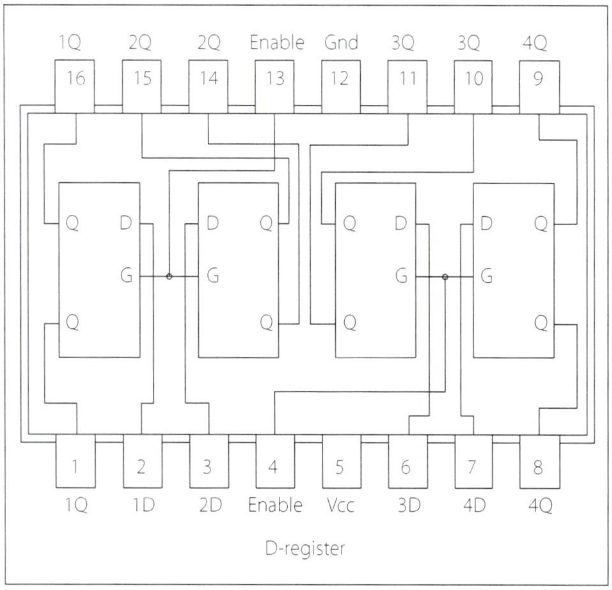

Figure 10-254. Medium-scale integration schematic form.

gates, larger memories than the medium-scale circuits, and in some cases microprocessors.

Microprocessors

The microprocessor is a device that can be programmed to perform arithmetic and logical operations and other functions in a preordered sequence. The microprocessor is usually used as the central processing unit (CPU) in today's computer systems when it is connected to other components, such as memory chips and input/output circuits. The basic arrangement and design of the circuits residing in the microprocessor is called the architecture.

DC Generators

Theory of Operation

In the study of alternating current, basic generator principles were introduced to explain the generation of an AC voltage by a coil rotating in a magnetic field. Since this is the basis for all generator operation, it is necessary to review the principles of generation of electrical energy.

When lines of magnetic force are cut by a conductor passing through them, voltage is induced in the conductor. The strength of the induced voltage is dependent upon the speed of the conductor and the strength of the magnetic field. If the ends of the conductor are connected to form a complete circuit, a current is induced in the conductor. The conductor and the magnetic field make up an elementary generator.

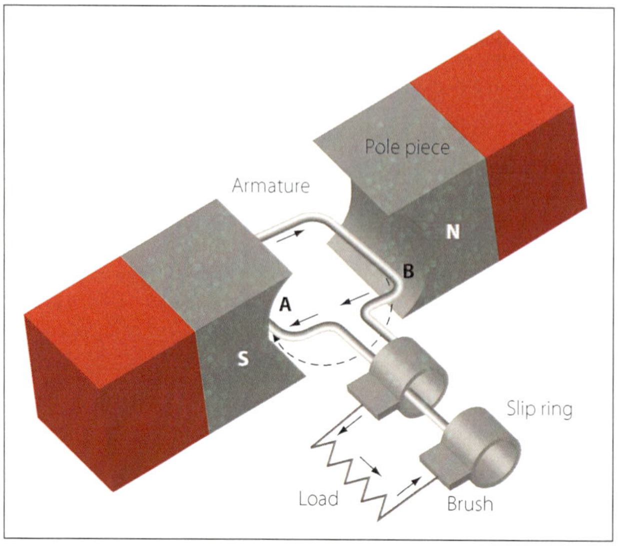

Figure 10-255. Inducing maximum voltage in an elementary generator.

This simple generator is illustrated in Figure 10-255, together with the components of an external generator circuit which collect and use the energy produced by the simple generator. The loop of wire (A and B of Figure 10-255) is arranged to rotate in a magnetic field. When the plane of the loop of wire is parallel to the magnetic lines of force, the voltage induced in the loop causes a current to flow in the direction indicated by the arrows in Figure 10-255. The voltage induced at this position is maximum, since the wires are cutting the lines of force at right angles and are thus cutting more lines of force per second than in any other position relative to the magnetic field. As the loop approaches the vertical position shown in Figure 10-256, the induced voltage

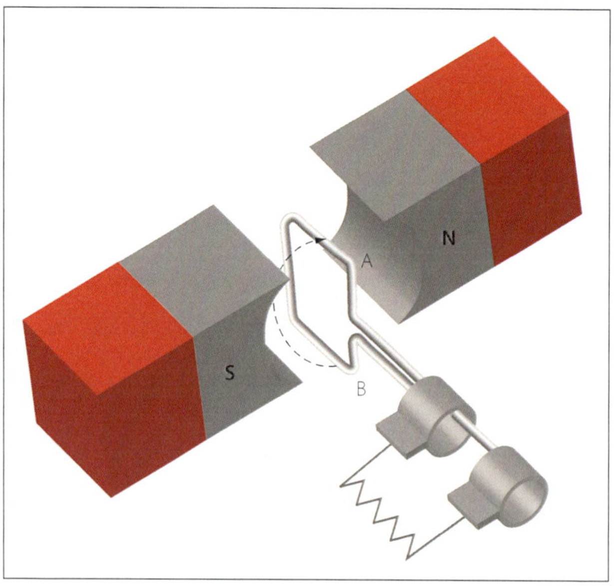

Figure 10-256. Inducing minimum voltage in an elementary generator.

decreases because both sides of the loop (A and B) are approximately parallel to the lines of force and the rate of cutting is reduced. When the loop is vertical, no lines of force are cut since the wires are momentarily traveling parallel to the magnetic lines of force, and there is no induced voltage. As the rotation of the loop continues, the number of lines of force cut increases until the loop has rotated an additional 90° to a horizontal plane. As shown in Figure 10-257, the number of lines of force cut and the induced voltage once again are maximum. The direction of cutting, however, is in the opposite direction to that occurring in Figures 10-255 and10-256, so the direction (polarity) of the induced voltage is reversed. As rotation of the loop continues, the number of lines of force having been cut again decreases, and the induced voltage becomes zero at the position shown in Figure 10-258, since the wires A and B are again parallel to the magnetic lines of force.

If the voltage induced throughout the entire 360° of rotation is plotted, the curve shown in Figure 10-259 results. This voltage is called an alternating voltage because of its reversal from positive to negative value—first in one direction and then in the other.

To use the voltage generated in the loop for producing a current flow in an external circuit, some means must be provided to connect the loop of wire in series with the external circuit. Such an electrical connection can be effected by opening the loop of wire and connecting its two ends to two metal rings, called slip rings, against which two metal or carbon brushes ride. The brushes are connected to the external circuit. By replacing the slip rings of the basic AC generator with two half cylinders, called a commutator, a basic DC generator is obtained. [Figure 10-260] In this illustration, the black side of the coil is connected to the black segment, and the white side of the coil to the white segment. The segments are insulated from each other. The two stationary brushes are placed on opposite sides of the commutator and are so mounted that each brush contacts each segment of the commutator as the latter revolves simultaneously with the loop. The rotating parts of a DC generator (coil and commutator) are called an armature. The generation of an emf by the loop rotating in the magnetic field is the same for both AC and DC generators, but the action of the commutator produces a DC voltage.

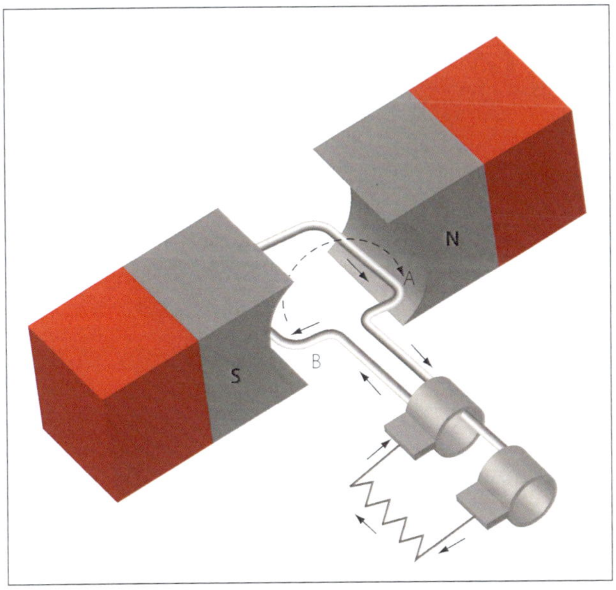

Figure 10-257. Inducing maximum voltage in the opposite direction.

Generation of a DC Voltage

Figure 10-261 illustrates in an elementary, step-by-step manner, how a DC voltage is generated. This is accomplished by showing a single wire loop rotating through a series of positions within a magnetic field.

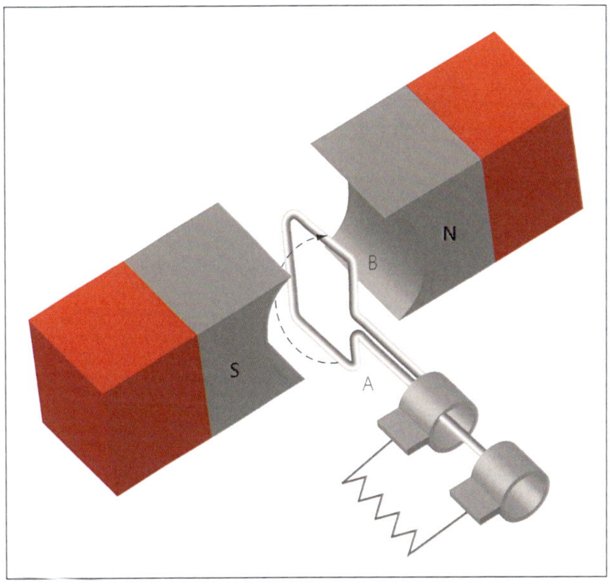

Figure 10-258. Inducing a minimum voltage in the opposite direction.

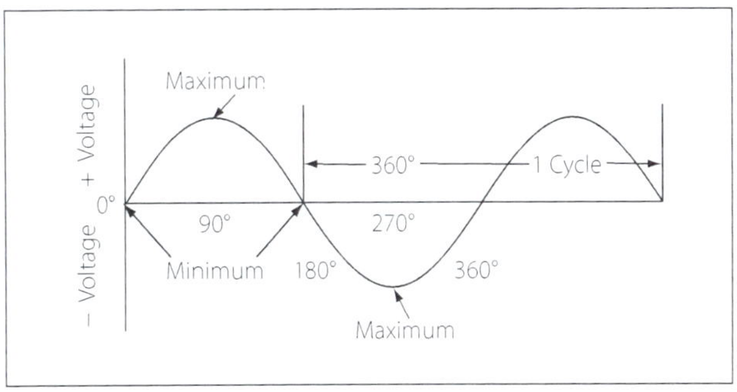

Figure 10-259. Output of an elementary generator.

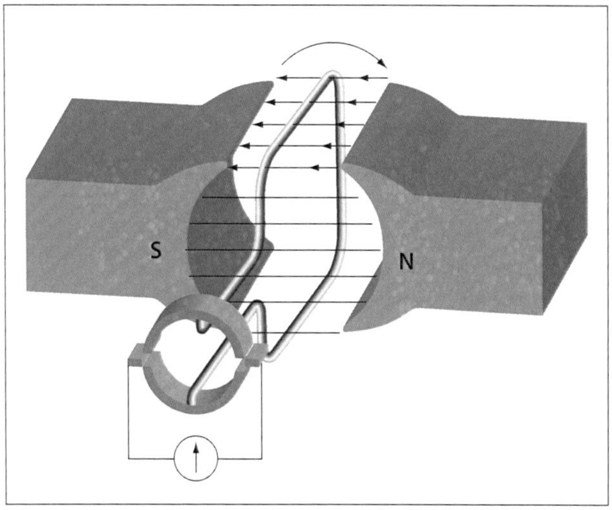

Figure 10-260. Basic DC generator.

Position A

The loop starts in position A and is rotating clockwise. However, no lines of force are cut by the coil sides, which means that no emf is generated. The black brush is shown coming into contact with the black segment of the commutator, and the white brush is just coming into contact with the white segment.

Position B

In position B, the flux is now being cut at a maximum rate, which means that the induced emf is maximum. At this time, the black brush is contacting the black segment, and the white brush is contacting the white segment. The deflection of the meter is toward the right, indicating the polarity of the output voltage.

Position C

At position C, the loop has completed 180° of rotation. Like position A, no flux lines are being cut and the output voltage is zero. The important condition to observe at position C is the action of the segments and brushes. The black brush at the 180° angle is contacting both black and white segments on one side of the commutator, and the white brush is contacting both segments on the other side of the commutator. After the loop rotates slightly past the 180° point, the black brush is contacting only the white segment, and the white brush is contacting only the black segment.

Because of this switching of commutator elements, the black brush is always in contact with the coil side moving downward, and the white brush is always in contact with the coil side moving upward. Though the current

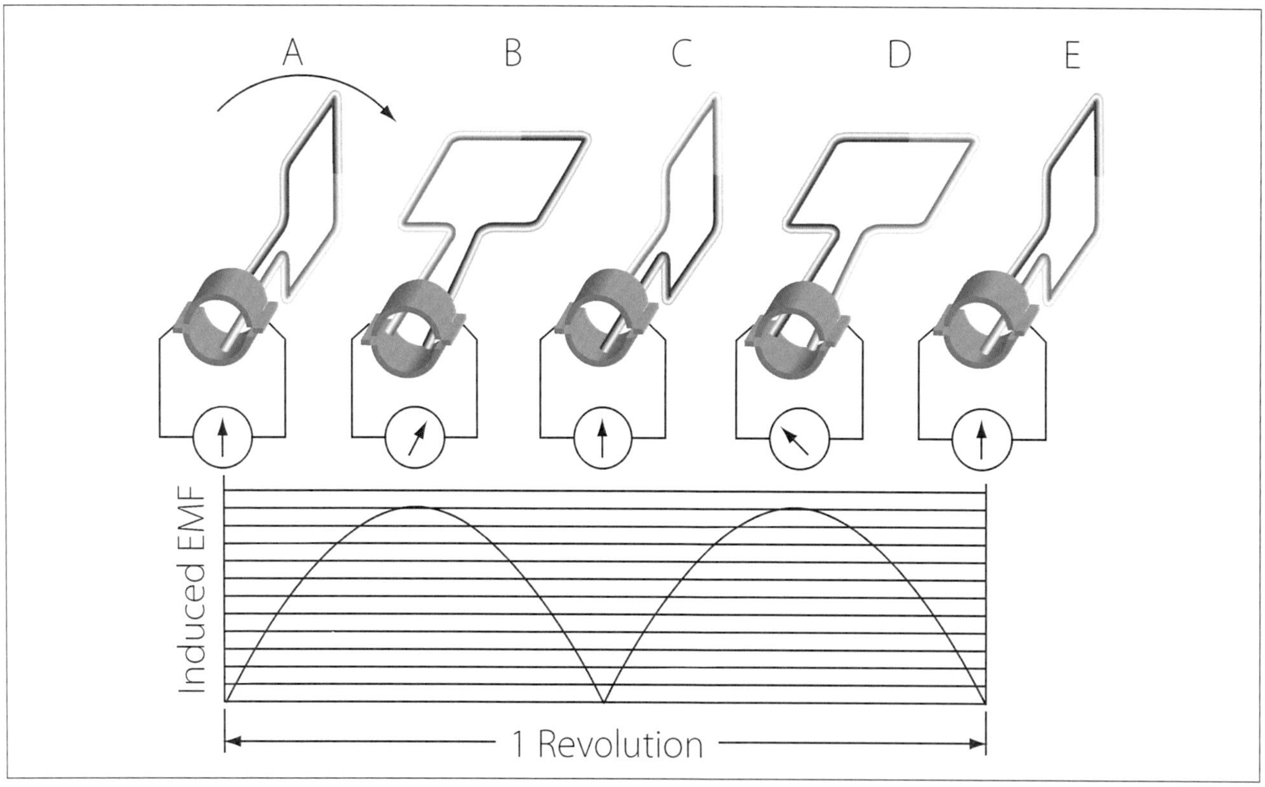

Figure 10-261. Operation of a basic DC generator.

actually reverses its direction in the loop in exactly the same way as in the AC generator, commutator action causes the current to flow always in the same direction through the external circuit or meter.

Position D

At position D, commutator action reverses the current in the external circuit, and the second half cycle has the same waveform as the first half cycle. The process of commutation is sometimes called rectification, since rectification is the converting of an AC voltage to a DC voltage.

The Neutral Plane

At the instant that each brush is contacting two segments on the commutator (positions A, C, and E in Figure 10-261), a direct short circuit is produced. If an emf were generated in the loop at this time, a high current would flow in the circuit, causing an arc and thus damaging the commutator. For this reason, the brushes must be placed in the exact position where the short will occur when the generated emf is zero. This position is called the neutral plane. If the brushes are installed properly, no sparking will occur between the brushes and the commutator. Sparking is an indication of improper brush placement, which is the main cause of improper commutation.

The voltage generated by the basic DC generator in Figure 10-261 varies from zero to its maximum value twice for each revolution of the loop. This variation of DC voltage is called "ripple," and may be reduced by using more loops, or coils, as shown in A of Figure 10-262. As the number of loops is increased, the variation between maximum and minimum values of voltage is reduced (view B of Figure 10-262), and the output voltage of the generator approaches a steady DC value. In view A of Figure 10-262 the number of commutator segments is increased in direct proportion to the number of loops; that is, there are two segments for one loop, four segments for two loops, and eight segments for four loops.

The voltage induced in a single turn loop is small. Increasing the number of loops does not increase the maximum value of generated voltage, but increasing the number of turns in each loop will increase this value. Within narrow limits, the output voltage of a DC generator is determined by the product of the number of turns per loop, the total flux per pair of poles in the machine, and the speed of rotation of the armature.

An AC generator, or alternator, and a DC generator are identical as far as the method of generating volt-

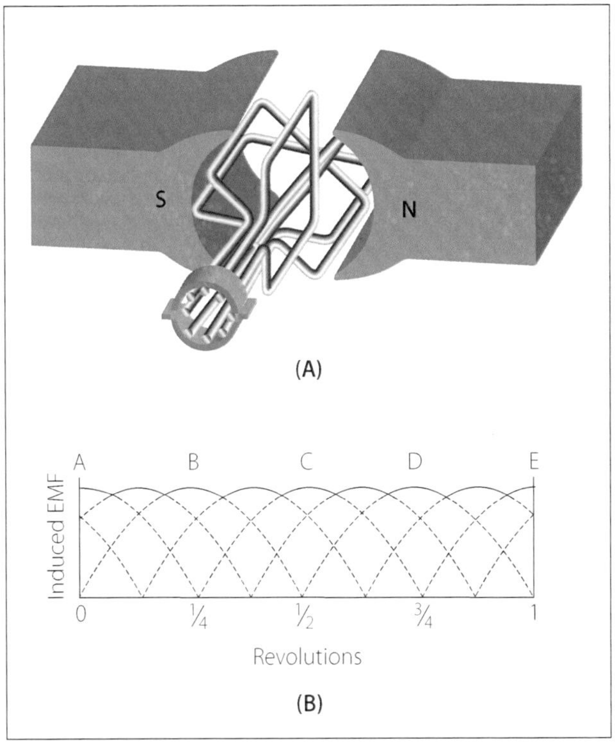

Figure 10-262. Increasing the number of coils reduces the ripple in the voltage.

age in the rotating loop is concerned. However, if the current is taken from the loop by slip rings, it is an alternating current, and the generator is called an AC generator, or alternator. If the current is collected by a commutator, it is direct current, and the generator is called a DC generator.

Construction Features of DC Generators

Generators used on aircraft may differ somewhat in design, since various manufacturers make them. All, however, are of the same general construction and operate similarly. The major parts, or assemblies, of a DC generator are a field frame (or yoke), a rotating armature, and a brush assembly. The parts of a typical aircraft generator are shown in Figure 10-263.

Field Frame

The field frame is also called the yoke, which is the foundation or frame for the generator. The frame has two functions: It completes the magnetic circuit between the poles and acts as a mechanical support for the other parts of the generator. In View A of Figure 10-264, the frame for a two-pole generator is shown in a cross-sectional view. A four-pole generator frame is shown in View B of Figure 10-264.

In small generators, the frame is made of one piece of iron, but in larger generators, it is usually made up of

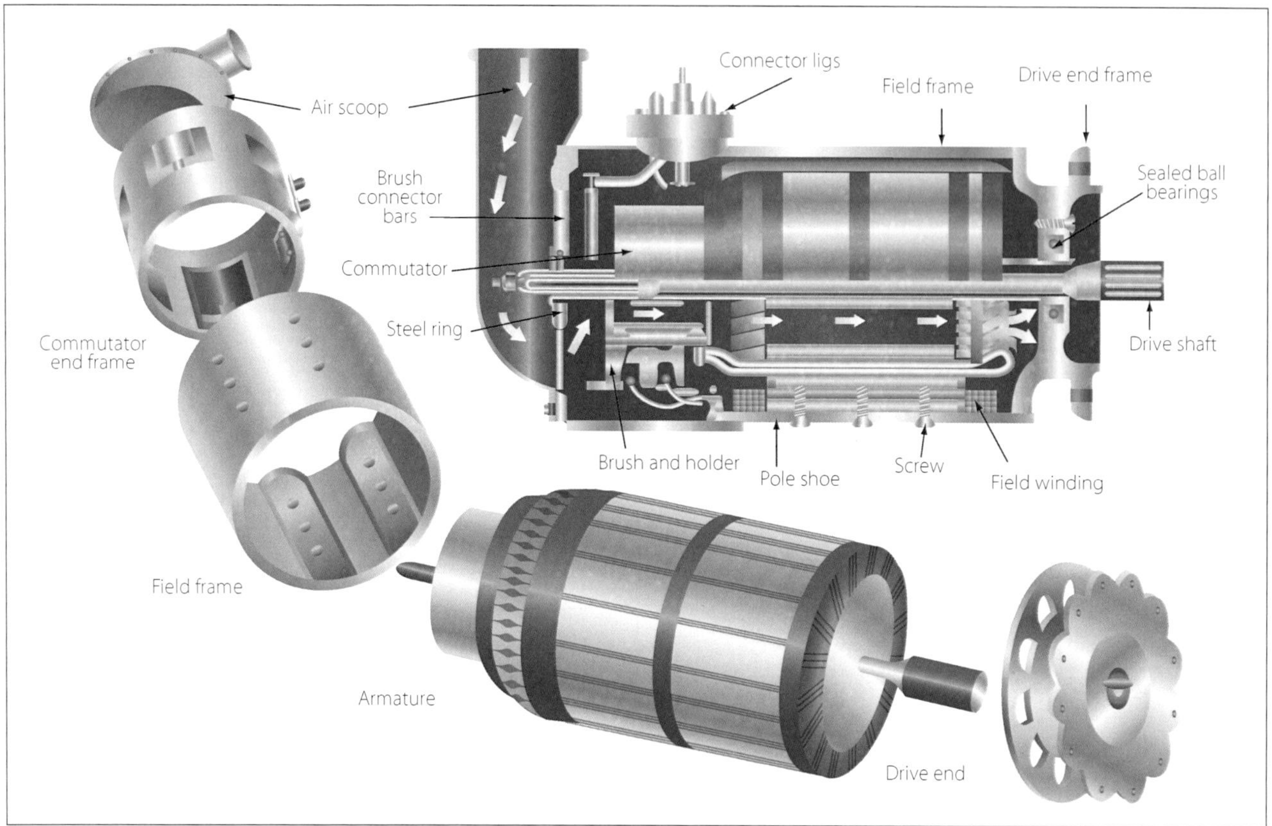

Figure 10-263. Typical 24-volt aircraft generator.

two parts bolted together. The frame has high magnetic properties and, together with the pole pieces, forms the major part of the magnetic circuit. The field poles, shown in Figure 10-264, are bolted to the inside of the frame and form a core on which the field coil windings are mounted.

The poles are usually laminated to reduce eddy current losses and serve the same purpose as the iron core of an electromagnet; that is, they concentrate the lines of force produced by the field coils. The entire frame including field poles, is made from high quality magnetic iron or sheet steel.

A practical DC generator uses electromagnets instead of permanent magnets. To produce a magnetic field of the necessary strength with permanent magnets would greatly increase the physical size of the generator.

The field coils are made up of many turns of insulated wire and are usually wound on a form that fits over the iron core of the pole to which it is securely fastened. [Figure 10-265] The exciting current, which is used to produce the magnetic field and which flows through the field coils, is obtained from an external source or from the generated DC of the machine. No electrical connection exists between the windings of the field coils and the pole pieces.

Most field coils are connected so that the poles show alternate polarity. Since there is always one north pole for each south pole, there must always be an even number of poles in any generator.

Note that the pole pieces in Figure 10-264 project from the frame. Because air offers a great amount of reluctance to the magnetic field, this design reduces the length of the air gap between the poles and the rotating armature and increases the efficiency of the generator. When the pole pieces are made to project as shown in Figure 10-264, they are called salient poles.

Armature

The armature assembly of a generator consists of many armature coils wound on an iron core, a commutator, and associated mechanical parts. These additional loops of wire are actually called windings and are evenly spaced around the armature so that the distance between each winding is the same. Mounted on a shaft, it rotates through the magnetic field produced by the field coils. The core of the armature acts as an iron conductor in the magnetic field and, for this reason, is laminated to prevent the circulation of eddy currents.

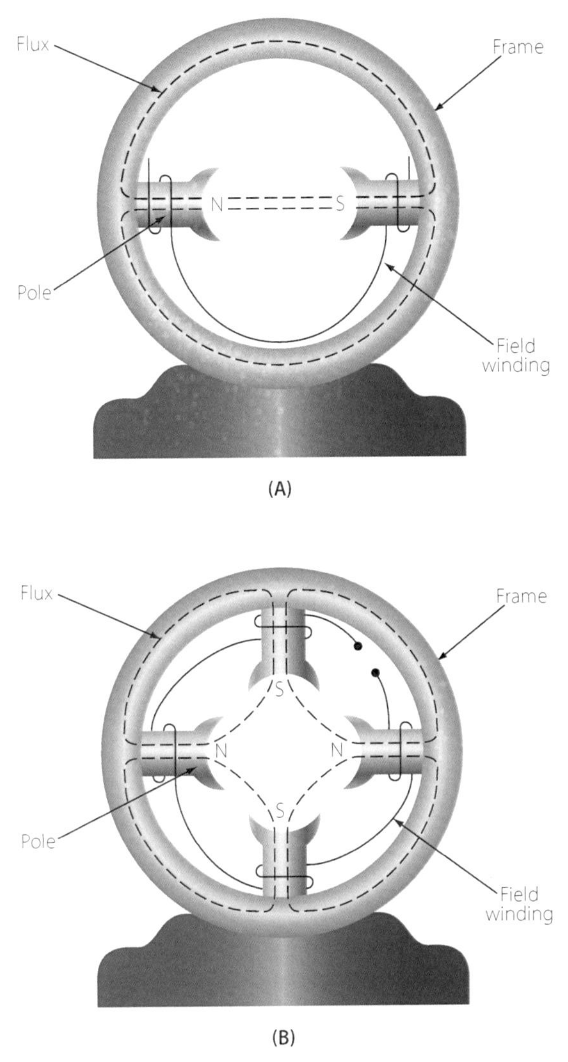

Figure 10-264. A two-pole and a four-pole frame assembly.

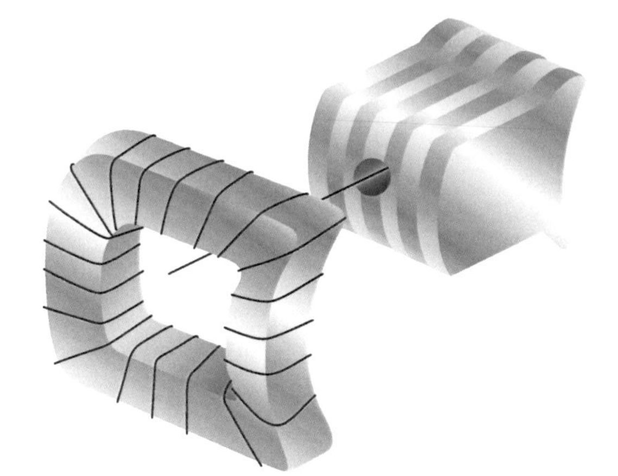

Figure 10-265. A field coil removed from a field pole.

Gramme-Ring Armature

There are two general kinds of armatures: the ring and the drum. Figure 10-266 shows a ring-type armature made up of an iron core, an eight-section winding, and an eight-segment commutator. The disadvantage of this arrangement is that the windings, located on the inner side of the iron ring, cut few lines of flux. As a result, they have very little voltage induced in them. For this reason, the Gramme-ring armature is not widely used.

Drum-Type Armature

A drum-type armature is shown in Figure 10-267. The armature core is in the shape of a drum and has slots cut into it where the armature windings are placed. The advantage is that each winding completely surrounds the core so that the entire length of the conductor cuts through the magnetic flux. The total induced voltage in this arrangement is far greater than that of the Gramme-ring type armature.

Drum-type armatures are usually constructed in one of two methods, each method having its own advantage. The two types of winding methods are the lap winding and the wave winding. Lap windings are used in generators that are designed for high current. The windings are connected in parallel paths and for this reason

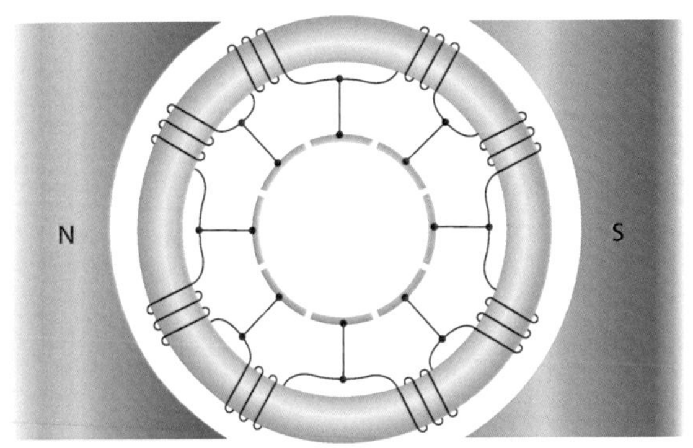

Figure 10-266. An eight-section, ring-type armature.

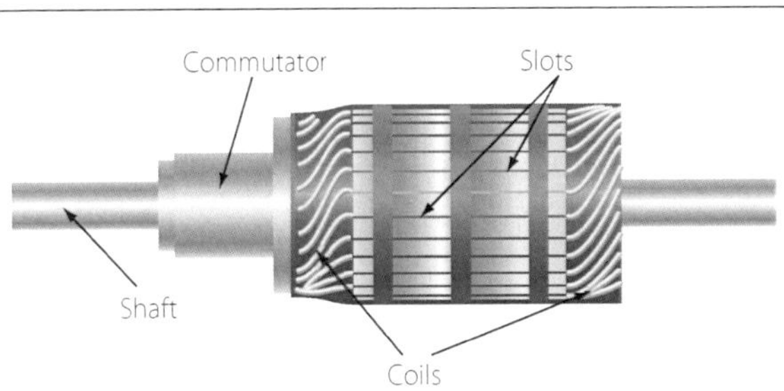

Figure 10-267. A drum-type armature.

require several brushes. The wave winding is used in generators that are designed for high voltage outputs. The two ends of each coil are connected to commutator segments separated by the distance between poles. This results in a series arrangement of the coils and is additive of all the induced voltages.

Commutators

Figure 10-268 shows a cross-sectional view of a typical commutator. The commutator is located at the end of an armature and consists of wedge shaped segments of hard drawn copper, insulated from each other by thin sheets of mica. The segments are held in place by steel V-rings or clamping flanges fitted with bolts. Rings of mica insulate the segments from the flanges. The raised portion of each segment is called a riser, and the leads from the armature coils are soldered to the risers. When the segments have no risers, the leads are soldered to short slits in the ends of the segments.

The brushes ride on the surface of the commutator, forming the electrical contact between the armature coils and the external circuit. A flexible, braided copper conductor, commonly called a pigtail, connects each brush to the external circuit. The brushes, usually made of high-grade carbon and held in place by brush holders insulated from the frame, are free to slide up and down in their holders in order to follow any irregularities in the surface of the commutator. The brushes are usually adjustable so that the pressure of the brushes on the commutator can be varied and the position of the brushes with respect to the segments can be adjusted.

The constant making and breaking of connections to the coils in which a voltage is being induced necessitates the use of material for brushes, which has a definite contact resistance. Also, this material must be such that the friction between the commutator and the brush is low, to prevent excessive wear. For these reasons, the material commonly used for brushes is high-grade carbon. The carbon must be soft enough to prevent undue wear of the commutator and yet hard enough to provide reasonable brush life. Since the

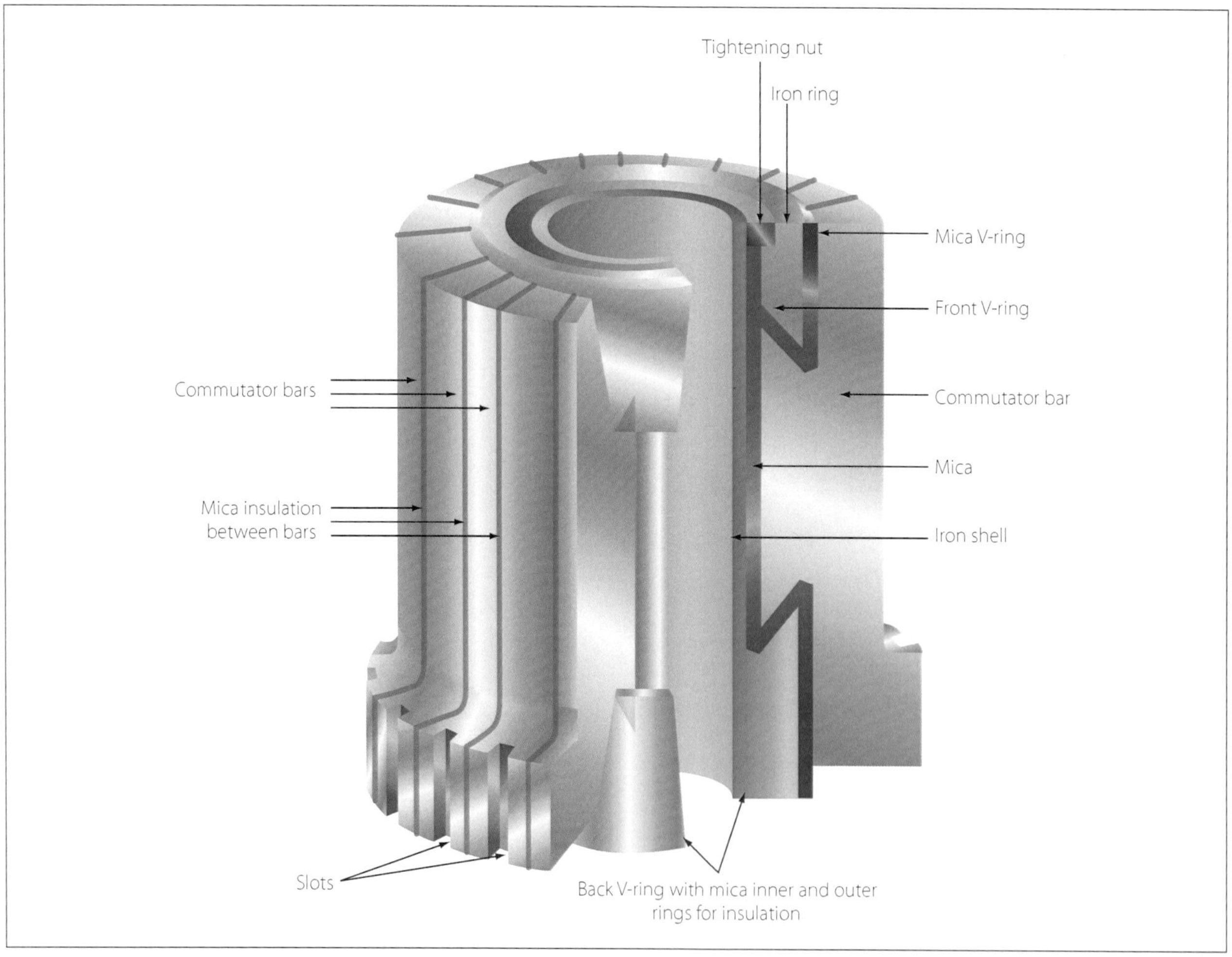

Figure 10-268. Commutator with portion removed to show construction.

contact resistance of carbon is fairly high, the brush must be quite large to provide a large area of contact. The commutator surface is highly polished to reduce friction as much as possible. Oil or grease must never be used on a commutator, and extreme care must be used when cleaning it to avoid marring or scratching the surface.

Armature Reaction

Current flowing through the armature sets up electromagnetic fields in the windings. These new fields tend to distort or bend the magnetic flux between the poles of the generator from a straight-line path. Since armature current increases with load, the distortion becomes greater with an increase in load. This distortion of the magnetic field is called armature reaction. [Figure 10-269]

Armature windings of a generator are spaced so that, during rotation of the armature, there are certain positions when the brushes contact two adjacent segments, thereby shorting the armature windings to these segments. When the magnetic field is not distorted, there is usually no voltage being induced in the shorted windings, and therefore no harmful results occur from the shorting of the windings. However, when the field is distorted, a voltage is induced in these shorted windings, and sparking takes place between the brushes and the commutator segments. Consequently, the commutator becomes pitted, the wear on the brushes becomes excessive, and the output of the generator is reduced. To correct this condition, the brushes are set so that the plane of the coils, which are shorted by the brushes, is perpendicular to the distorted magnetic field, which is accomplished by moving the brushes forward in the direction of rotation. This operation is called shifting the brushes to the neutral plane, or plane of commutation. The neutral plane is the position where the plane of the two opposite coils is perpendicular to the magnetic field in the generator. On a few generators, the brushes can be shifted manually ahead of the normal neutral plane to the neutral plane caused by field distortion. On nonadjustable brush generators, the manufacturer sets the brushes for minimum sparking.

Compensating windings or interpoles may be used to counteract some of the effects of field distortion, since shifting the brushes is inconvenient and unsatisfactory, especially when the speed and load of the generator are changing constantly.

Compensating Windings

The compensating windings consist of a series of coils embedded in slots in the pole faces. These coils are also

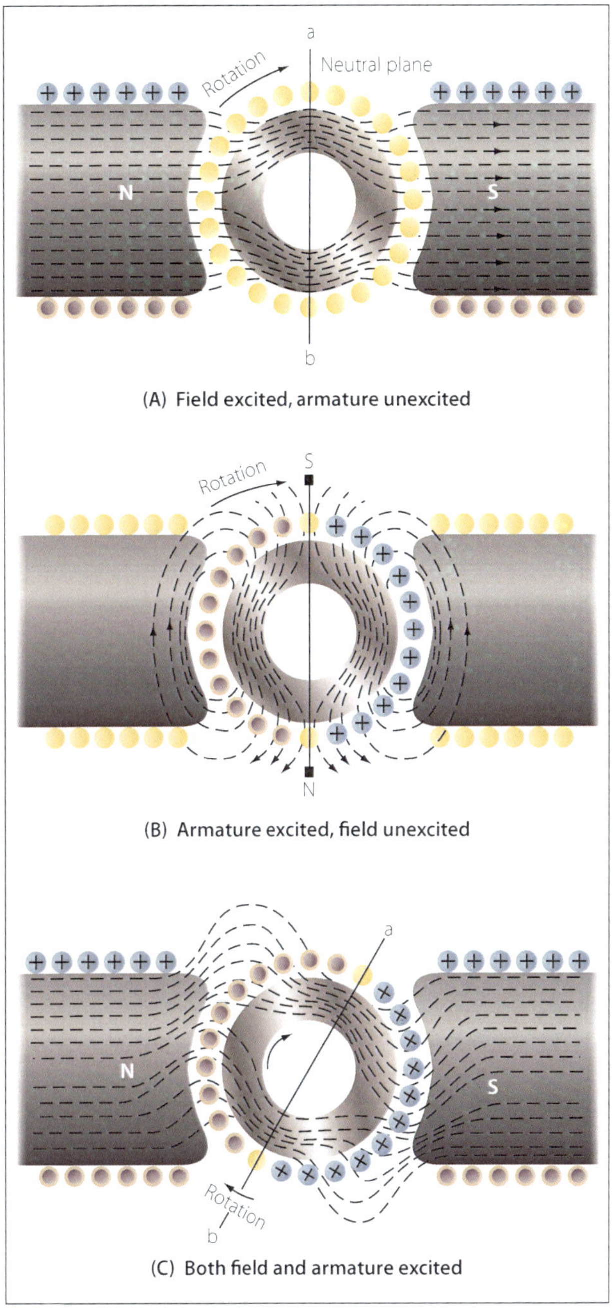

Figure 10-269. Armature reaction.

connected in series with the armature. Consequently, this series connection with the armature produces a magnetic field in the compensating windings that varies directly with the armature current. The compensating windings are wound in such a manner that the magnetic field produced by them will counteract the magnetic field produced by the armature. As a result, the neutral plane will remain stationary any magnitude of armature current. With this design, once the brushes are set correctly, they do not need to be moved again. Figure 10-270A illustrates how the windings are set into the pole faces.

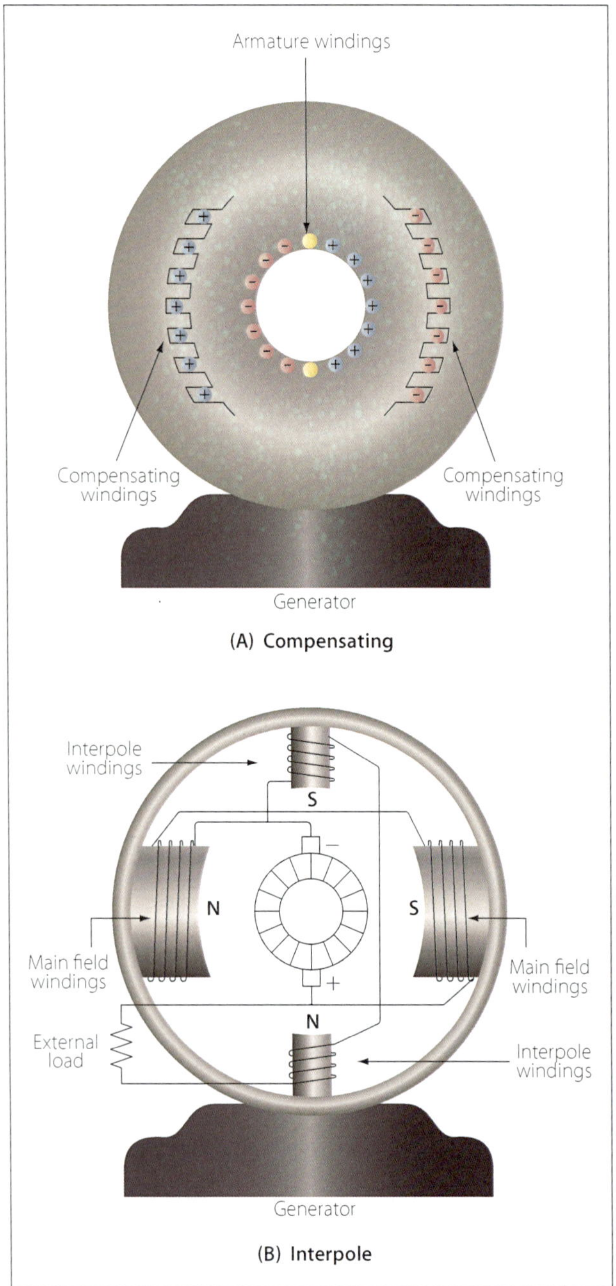

Figure 10-270. Simple two-pole generator with two interpoles.

Interpoles

An interpole is a pole placed between the main poles of a generator. An example of interpole placement is shown in Figure 10-270B. This is a simple two-pole generator with two interpoles.

An interpole has the same polarity as the next main pole in the direction of rotation. The magnetic flux produced by an interpole causes the current in the armature to change direction as an armature winding passes under it. This cancels the electromagnetic fields about the armature windings. The magnetic strength of the interpoles varies with the load on the generator; and since field distortion varies with the load, the magnetic field of the interpoles counteracts the effects of the field set up around the armature windings and minimizes field distortion. Thus, the interpole tends to keep the neutral plane in the same position for all loads on the generator; therefore, field distortion is reduced by the interpoles, and the efficiency, output, and service life of the brushes are improved.

Types of DC Generators

There are three types of DC generators: series wound, shunt wound, and shunt series or compound wound. The difference in type depends on the relationship of the field winding to the external circuit.

Series Wound DC Generators

The field winding of a series generator is connected in series with the external circuit, called the load. [Figure 10-271] The field coils are composed of a few turns of large wire; the magnetic field strength depends more on the current flow rather than the number of turns in the coil. Series generators have very poor voltage regulation under changing load, since the greater the current through the field coils to the external circuit, the greater the induced emf and the greater the terminal or output voltage. Therefore, when the load is increased, the voltage increases; likewise, when the load is decreased, the

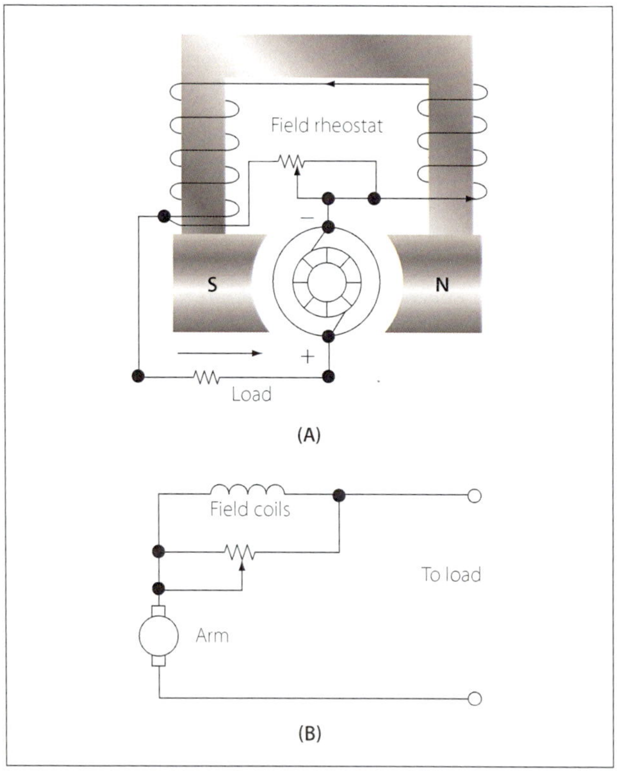

Figure 10-271. Diagram and schematic of a series wound generator.

voltage decreases. The output voltage of a series wound generator may be controlled by a rheostat in parallel with the field windings, as shown in Figure 10-271A. Since the series wound generator has such poor regulation, it is never employed as an airplane generator. Generators in airplanes have field windings, which are connected either in shunt or in compound.

Shunt Wound DC Generators

A generator having a field winding connected in parallel with the external circuit is called a shunt generator, as shown in views A and B of Figure 10-272. The field coils of a shunt generator contain many turns of small wire; the magnetic strength is derived from the large number of turns rather than the current strength through the coils. If a constant voltage is desired, the shunt wound generator is not suitable for rapidly fluctuating loads. Any increase in load causes a decrease in the terminal or output voltage, and any decrease in load causes an increase in terminal voltage; since the armature and the load are connected in series, all current flowing in the external circuit passes through the armature winding. Because of the resistance in the armature winding, there is a voltage drop (IR drop = current × resistance). As the load increases, the armature current increases and the IR drop in the armature increases. The voltage delivered to the terminals is the difference between the induced voltage and the voltage drop; therefore, there is a decrease in terminal voltage. This decrease in voltage causes a decrease in field strength, because the current in the field coils decreases in proportion to the decrease in terminal voltage; with a weaker field, the voltage is further decreased. When the load decreases, the output voltage increases accordingly, and a larger current flows in the windings. This action is cumulative, so the output voltage continues to rise to a point called field saturation, after which there is no further increase in output voltage.

The terminal voltage of a shunt generator can be controlled by means of a rheostat inserted in series with the field windings as shown in Figure 10-272A. As the resistance is increased, the field current is reduced; consequently, the generated voltage is reduced also. For a given setting of the field rheostat, the terminal voltage at the armature brushes will be approximately equal to the generated voltage minus the IR drop produced by the load current in the armature; thus, the voltage at the terminals of the generator will drop as the load is applied. Certain voltage sensitive devices are available which automatically adjust the field rheostat to compensate for variations in load. When

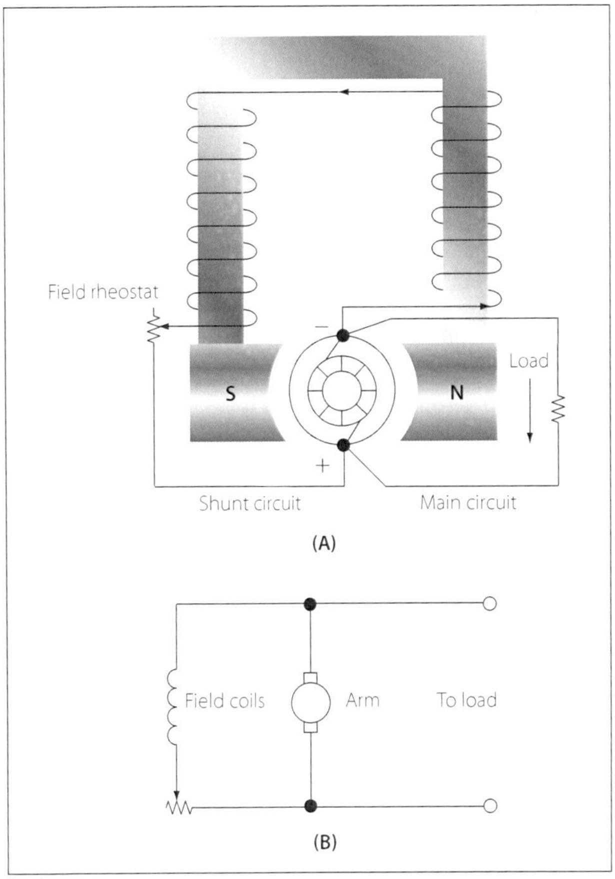

Figure 10-272. Shunt wound generator.

these devices are used, the terminal voltage remains essentially constant.

Compound Wound DC Generators

A compound wound generator combines a series winding and a shunt winding in such a way that the characteristics of each are used to advantage. The series field coils are made of a relatively small number of turns of large copper conductor, either circular or rectangular in cross section, and are connected in series with the armature circuit. These coils are mounted on the same poles on which the shunt field coils are mounted and, therefore, contribute a magnetomotive force which influences the main field flux of the generator. A diagrammatic and a schematic illustration of a compound wound generator is shown in A and B of Figure 10-273.

If the ampere turns of the series field act in the same direction as those of the shunt field, the combined magnetomotive force is equal to the sum of the series and shunt field components. Load is added to a compound generator in the same manner in which load is added to a shunt generator, by increasing the number of parallel paths across the generator terminals. Thus,

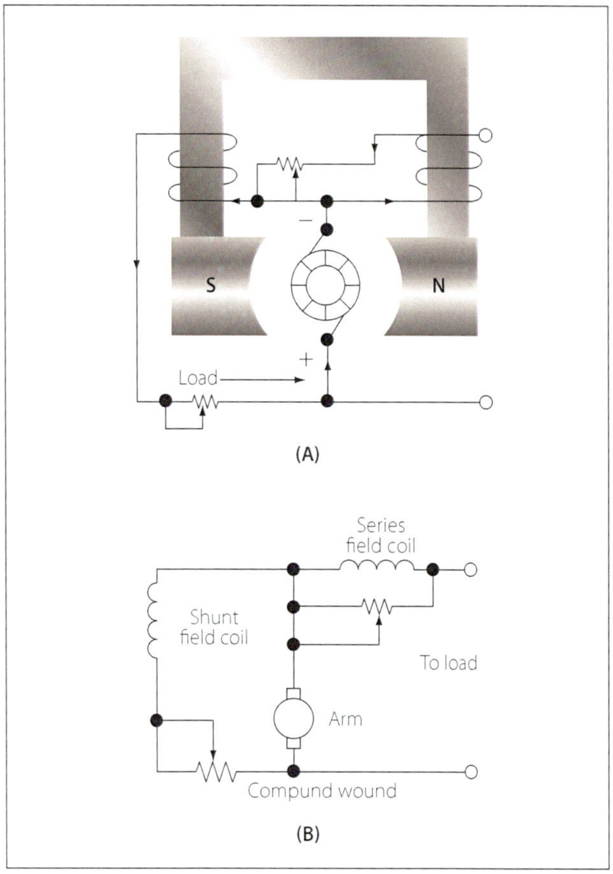

Figure 10-273. Compound wound generator.

the decrease in total load resistance with added load is accompanied by an increase in armature circuit and series field circuit current. The effect of the additive series field is that of increased field flux with increased load. The extent of the increased field flux depends on the degree of saturation of the field as determined by the shunt field current. Thus, the terminal voltage of the generator may increase or decrease with load, depending on the influence of the series field coils. This influence is referred to as the degree of compounding. A flat compound generator is one in which the no load and full load voltages have the same value; whereas an under compound generator has a full load voltage less than the no load value, and an over compound generator has a full load voltage which is higher than the no load value. Changes in terminal voltage with increasing load depend upon the degree of compounding.

If the series field aids the shunt field, the generator is said to be cumulative compounded.

If the series field opposes the shunt field, the machine is said to be differentially compounded, or is called a differential generator. Compound generators are usually designed to be overcompounded. This feature permits varied degrees of compounding by connecting a variable shunt across the series field. Such a shunt is sometimes called a diverter. Compound generators are used where voltage regulation is of prime importance.

Differential generators have somewhat the same characteristics as series generators in that they are essentially constant current generators. However, they generate rated voltage at no load, the voltage dropping materially as the load current increases. Constant current generators are ideally suited as power sources for electric arc welders and are used almost universally in electric arc welding.

If the shunt field of a compound generator is connected across both the armature and the series field, it is known as a long shunt connection, but if the shunt field is connected across the armature alone, it is called a short shunt connection. These connections produce essentially the same generator characteristics.

A summary of the characteristics of the various types of generators discussed is shown graphically in Figure 10-274.

Generator Ratings

A generator is rated in power output. Since a generator is designed to operate at a specified voltage, the rating usually is given as the number of amperes the generator can safely supply at its rated voltage.

Generator rating and performance data are stamped on the nameplate attached to the generator. When replacing a generator, it is important to choose one of the proper rating.

The rotation of generators is termed either clockwise or counterclockwise, as viewed from the driven end. Usually, the direction of rotation is stamped on the data

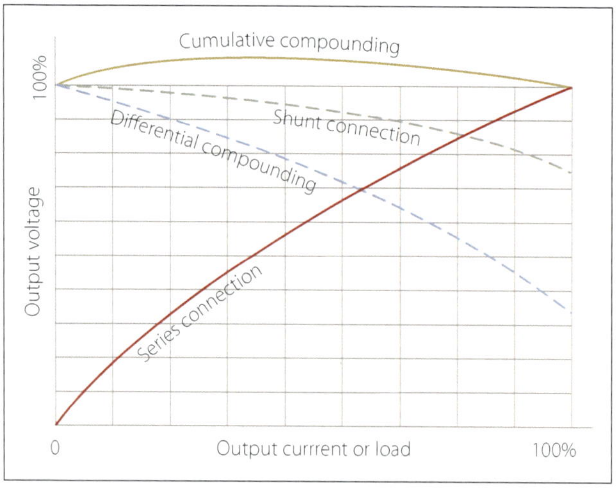

Figure 10-274. Generator characteristics.

plate. If no direction is stamped on the plate, the rotation may be marked by an arrow on the cover plate of the brush housing. It is important that a generator with the correct direction of rotation be used; otherwise, the voltage will be reversed.

The speed of an aircraft engine varies from idle rpm to takeoff rpm; however, during the major portion of a flight, it is at a constant cruising speed. The generator drive is usually geared to revolve the generator between 1-1/8 and 1-1/2 times the engine crankshaft speed. Most aircraft generators have a speed at which they begin to produce their normal voltage. Termed the "coming in" speed, it is usually about 1,500 rpm.

Generator Terminals

On most large 24-volt generators, electrical connections are made to terminals marked B, A, and E. The positive armature lead in the generator connects to the B terminal. The negative armature lead connects to the E terminal. The positive end of the shunt field winding connects to terminal A, and the opposite end connects to the negative terminal brush. Terminal A receives current from the negative generator brush through the shunt field winding. This current passes through the voltage regulator and back to the armature through the positive brush. Load current, which leaves the armature through the negative brushes, comes out of the E lead and passes through the load before returning to the armature through the positive brushes.

DC Generator Maintenance

Inspection

The following information about the inspection and maintenance of DC generator systems is general in nature because of the large number of differing aircraft generator systems. These procedures are for familiarization only. Always follow the applicable manufacturer's instructions for a given generator system.

In general, the inspection of the generator installed in the aircraft should include the following items:

1. Security of generator mounting.
2. Condition of electrical connections.
3. Dirt and oil in the generator. If oil is present, check engine oil seal. Blow out dirt with compressed air.
4. Condition of generator brushes.
5. Generator operation.
6. Voltage regulator operation.

Condition of Generator Brushes

Sparking of brushes quickly reduces the effective brush area in contact with the commutator bars. The degree of such sparking should be determined. Excessive wear warrants a detailed inspection.

The following information pertains to brush seating, brush pressure, high mica condition, and brush wear. Manufacturers usually recommend the following procedures to seat brushes that do not make good contact with slip rings or commutators.

Lift the brush sufficiently to permit the insertion of a strip of No. 000, or finer, sandpaper under the brush, rough side out. [Figure 10-275] Pull the sandpaper in the direction of armature rotation, being careful to keep the ends of the sandpaper as close to the slip ring or commutator surface as possible in order to avoid rounding the edges of the brush.

When pulling the sandpaper back to the starting point, raise the brush so it does not ride on the sandpaper. Sand the brush only in the direction of rotation.

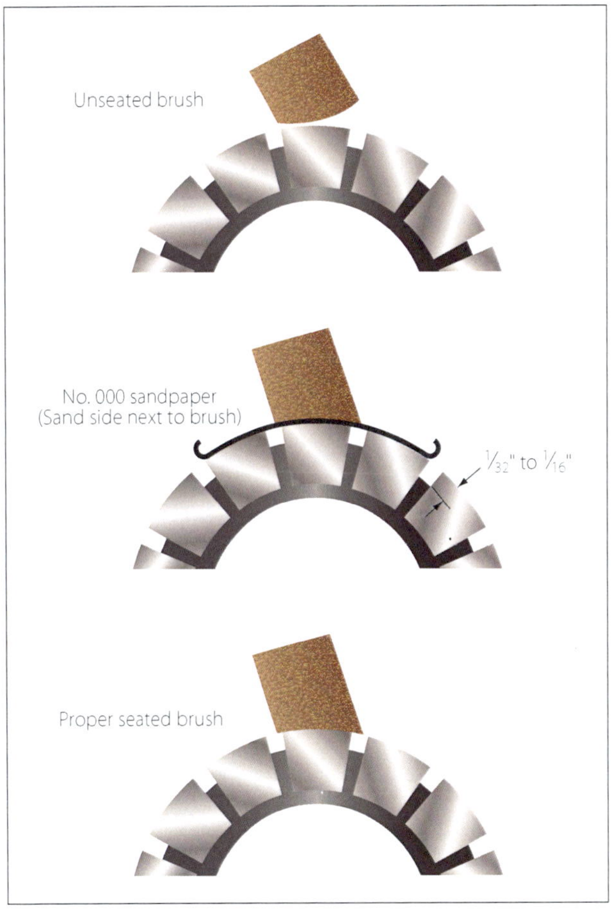

Figure 10-275. Seating brushes with sandpaper.

After the generator has run for a short period, brushes should be inspected to make sure that pieces of sand have not become embedded in the brush and are collecting copper.

Under no circumstances should emery cloth or similar abrasives be used for seating brushes (or smoothing commutators), since they contain conductive materials that will cause arcing between brushes and commutator bars.

Excessive pressure will cause rapid wear of brushes. Too little pressure, however, will allow "bouncing" of the brushes, resulting in burned and pitted surfaces.

A carbon, graphite, or light metalized brush should exert a pressure of $1\frac{1}{2}$ to $2\frac{1}{2}$ psi on the commutator. The pressure recommended by the manufacturer should be checked by the use of a spring scale graduated in ounces. Brush spring tension is usually adjusted between 32 to 36 ounces; however, the tension may differ slightly for each specific generator.

When a spring scale is used, the measurement of the pressure that a brush exerts on the commutator is read directly on the scale. The scale is applied at the point of contact between the spring arm and the top of the brush, with the brush installed in the guide. The scale is drawn up until the arm just lifts off the brush surface. At this instant, the force on the scale should be read.

Flexible low resistance pigtails are provided on most heavy current carrying brushes, and their connections should be securely made and checked at frequent intervals. The pigtails should never be permitted to alter or restrict the free motion of the brush.

The purpose of the pigtail is to conduct the current, rather than subjecting the brush spring to currents that would alter its spring action by overheating. The pigtails also eliminate any possible sparking to the brush guides caused by the movement of the brushes within the holder, thus minimizing side wear of the brush.

Carbon dust resulting from brush sanding should be thoroughly cleaned from all parts of the generators after a sanding operation. Such carbon dust has been the cause of several serious fires as well as costly damage to the generator.

Operation over extended periods of time often results in the mica insulation between commutator bars protruding above the surface of the bars. This condition is called "high mica" and interferes with the contact of the brushes to the commutator. Whenever this condition exists, or if the armature has been turned on a lathe, carefully undercut the mica insulation to a depth equal to the width of the mica, or approximately 0.020 inch.

Each brush should be a specified length to work properly. If a brush is too short, the contact it makes with the commutator will be faulty, which can also reduce the spring force holding the brush in place. Most manufacturers specify the amount of wear permissible from a new brush length. When a brush has worn to the minimum length permissible, it must be replaced.

Some special generator brushes should not be replaced because of a slight grooving on the face of the brush. These grooves are normal and will appear in AC and DC generator brushes which are installed in some models of aircraft generators. These brushes have two cores made of a harder material with a higher expansion rate than the material used in the main body of the brush. Usually, the main body of the brush face rides on the commutator. However, at certain temperatures, the cores extend and wear through any film on the commutator.

DC Motors

Most devices in an airplane, from the starter to the automatic pilot, depend upon mechanical energy furnished by direct current motors. A direct current motor is a rotating machine, which transforms direct current energy into mechanical energy. It consists of two principal parts—a field assembly and an armature assembly. The armature is the rotating part in which current carrying wires are acted upon by the magnetic field.

Whenever a current carrying wire is placed in the field of a magnet, a force acts on the wire. The force is not one of attraction or repulsion; however, it is at right angles to the wire and also at right angles to the magnetic field set up by the magnet. The action of the force upon a current carrying wire placed in a magnetic field is shown in Figure 10-276. A wire is located between two permanent magnets. The lines of force in the magnetic field are from the north pole to the south pole. When no current flows, as in Figure 10-276A, no force is exerted on the wire, but when current flows through the wire, a magnetic field is set up about it, as shown in Figure 10-276B. The direction of the field depends on the direction of current flow. Current in one direction creates a clockwise field about the wire, and current in the other direction, a counterclockwise field.

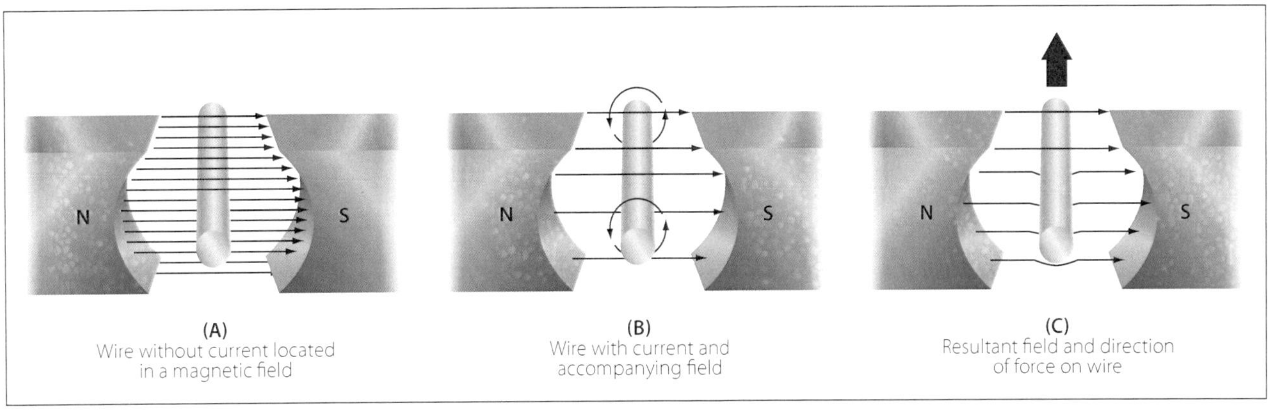

Figure 10-276. Force on a current carrying wire.

Since the current carrying wire produces a magnetic field, a reaction occurs between the field about the wire and the magnetic field between the magnets. When the current flows in a direction to create a counterclockwise magnetic field about the wire, this field and the field between the magnets add or reinforce at the bottom of the wire because the lines of force are in the same direction. At the top of the wire, they subtract or neutralize, since the lines of force in the two fields are opposite in direction. Thus, the resulting field at the bottom is strong and the one at the top is weak. Consequently, the wire is pushed upward as shown in Figure 10-276C. The wire is always pushed away from the side where the field is strongest.

If current flow through the wire were reversed in direction, the two fields would add at the top and subtract at the bottom. Since a wire is always pushed away from the strong field, the wire would be pushed down.

Force between Parallel Conductors

Two wires carrying current in the vicinity of one another exert a force on each other because of their magnetic fields. An end view of two conductors is shown in Figure 10-277. In A, electron flow in both conductors is toward the reader, and the magnetic fields are clockwise around the conductors. Between the wires, the fields cancel because the directions of the two fields oppose each other. The wires are forced in the direction of the weaker field, toward each other. This force is one of attraction. In B, the electron flow in the two wires is in opposite directions.

The magnetic fields are, therefore, clockwise in one and counterclockwise in the other, as shown. The fields reinforce each other between the wires, and the wires are forced in the direction of the weaker field, away from each other. This force is one of repulsion.

To summarize: conductors carrying current in the same direction tend to be drawn together; conductors carrying current in opposite directions tend to be repelled from each other.

Developing Torque

If a coil in which current is flowing is placed in a magnetic field, a force is produced which will cause the coil to rotate. In the coil shown in Figure 10-278, current flows inward on side A and outward on side B. The magnetic field about B is clockwise and that about A, counterclockwise. As previously explained, a force will

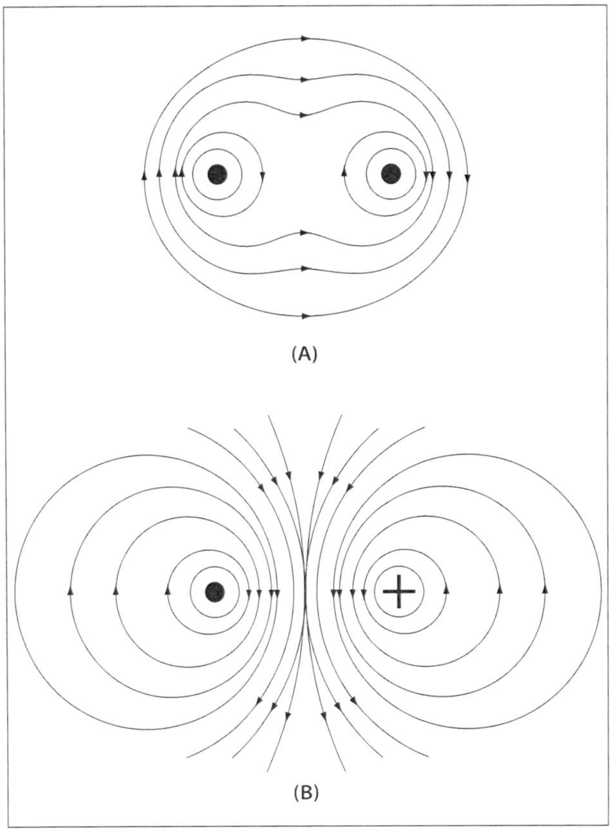

Figure 10-277. Fields surrounding parallel conductors.

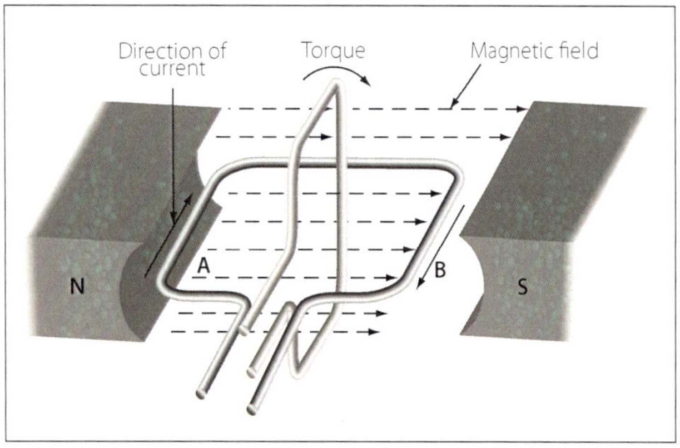

Figure 10-278. Developing a torque.

develop which pushes side B downward. At the same time, the field of the magnets and the field about A, in which the current is inward, will add at the bottom and subtract at the top. Therefore, A will move upward. The coil will thus rotate until its plane is perpendicular to the magnetic lines between the north and south poles of the magnet, as indicated in Figure 10-278 by the white coil at right angles to the black coil.

The tendency of a force to produce rotation is called torque. When the steering wheel of a car is turned, torque is applied. The engine of an airplane gives torque to the propeller. Torque is developed also by the reacting magnetic fields about the current carrying coil just described. This is the torque, which turns the coil.

The right-hand motor rule can be used to determine the direction a current carrying wire will move in a magnetic field. As illustrated in Figure 10-279, if the index finger of the right hand is pointed in the direction of the magnetic field and the second finger in the direction of current flow, the thumb will indicate the direction the current carrying wire will move.

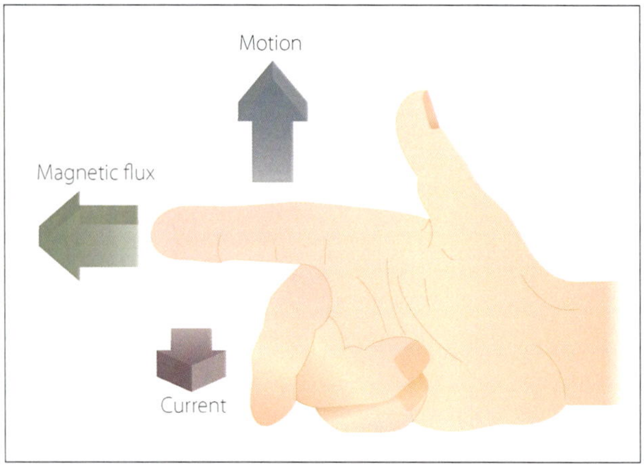

Figure 10-279. Right-hand motor rule.

The amount of torque developed in a coil depends upon several factors: the strength of the magnetic field, the number of turns in the coil, and the position of the coil in the field. Magnets are made of special steel that produces a strong field. Since there is torque acting on each turn, the greater the number of turns on the coil, the greater the torque. In a coil carrying a steady current located in a uniform magnetic field, the torque will vary at successive positions of rotation, as shown in Figure 10-280. When the plane of the coil is parallel to the lines of force, the torque is zero. When its plane cuts the lines of force at right angles, the torque is 100 percent. At intermediate positions, the torque ranges between zero and 100 percent.

Basic DC Motor

A coil of wire through which the current flows will rotate when placed in a magnetic field. This is the technical basis governing the construction of a DC motor. Figure 10-281 shows a coil mounted in a magnetic field in which it can rotate. However, if the connecting wires from the battery were permanently fastened to the terminals of the coil and there was a flow of current, the coil would rotate only until it lined itself up with the magnetic field. Then, it would stop, because the torque at that point would be zero.

A motor, of course, must continue rotating. It is therefore necessary to design a device that will reverse the current in the coil just at the time the coil becomes parallel to the lines of force. This will create torque again and cause the coil to rotate. If the current reversing device is set up to reverse the current each time the coil is about to stop, the coil can be made to continue rotating as long as desired.

One method of doing this is to connect the circuit so that, as the coil rotates, each contact slides off the terminal to which it connects and slides onto the

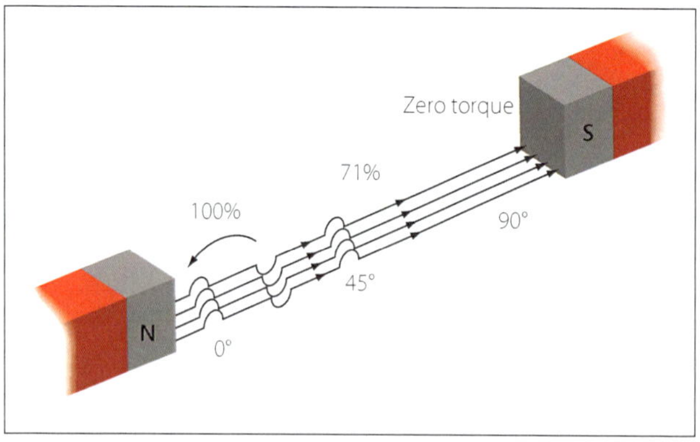

Figure 10-280. Torque on a coil at various angles of rotation.

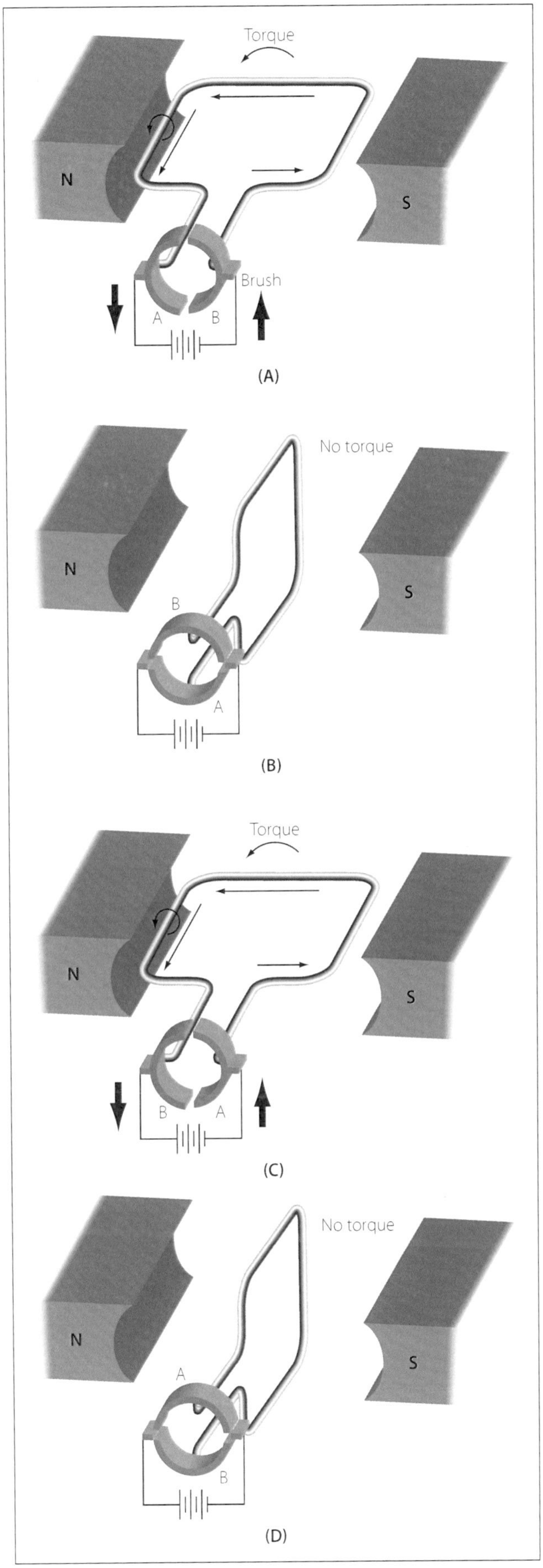

Figure 10-281. Basic DC motor operation.

terminal of opposite polarity. In other words, the coil contacts switch terminals continuously as the coil rotates, preserving the torque and keeping the coil rotating. In Figure 10-281, the coil terminal segments are labeled A and B. As the coil rotates, the segments slide onto and past the fixed terminals or brushes. With this arrangement, the direction of current in the side of the coil next to the north-seeking pole flows toward the reader, and the force acting on that side of the coil turns it downward. The part of the motor, which changes the current from one wire to another, is called the commutator.

Position A

When the coil is positioned as shown in Figure 10-281A, current will flow from the negative terminal of the battery to the negative (−) brush, to segment B of the commutator, through the loop to segment A of the commutator, to the positive (+) brush, and then, back to the positive terminal of the battery. By using the right-hand motor rule, it is seen that the coil will rotate counterclockwise. The torque at this position of the coil is maximum, since the greatest number of lines of force is being cut by the coil.

Position B

When the coil has rotated 90° to the position shown in Figure 10-281B, segments A and B of the commutator no longer make contact with the battery circuit and no current can flow through the coil. At this position, the torque has reached a minimum value, since a minimum number of lines of force are being cut. However, the momentum of the coil carries it beyond this position until the segments again make contact with the brushes, and current again enters the coil; this time, though, it enters through segment A and leaves through segment B. However, since the positions of segments A and B have also been reversed, the effect of the current is as before, the torque acts in the same direction, and the coil continues its counterclockwise rotation.

Position C

On passing through the position shown in Figure 10-281C, the torque again reaches maximum.

Position D

Continued rotation carries the coil again to a position of minimum torque, as in Figure 10-281D. At this position, the brushes no longer carry current, but once more the momentum rotates the coil to the point where current enters through segment B and leaves through A. Further rotation brings the coil to the starting point and, thus, one revolution is completed.

The switching of the coil terminals from the positive to the negative brushes occurs twice per revolution of the coil.

The torque in a motor containing only a single coil is neither continuous nor very effective, for there are two positions where there is actually no torque at all. To overcome this, a practical DC motor contains a large number of coils wound on the armature. These coils are so spaced that, for any position of the armature, there will be coils near the poles of the magnet. This makes the torque both continuous and strong. The commutator, likewise, contains a large number of segments instead of only two.

The armature in a practical motor is not placed between the poles of a permanent magnet but between those of an electromagnet, since a much stronger magnetic field can be furnished. The core is usually made of a mild or annealed steel, which can be magnetized strongly by induction. The current magnetizing the electromagnet is from the same source that supplies the current to the armature.

DC Motor Construction

The major parts in a practical motor are the armature assembly, the field assembly, the brush assembly, and the end frame. [Figure 10-282]

Armature Assembly

The armature assembly contains a laminated, soft iron core, coils, and a commutator, all mounted on a rotatable steel shaft. Laminations made of stacks of soft iron, insulated from each other, form the armature core. Solid iron is not used, since a solid iron core revolving in the magnetic field would heat and use energy needlessly. The armature windings are insulated copper wire, which are inserted in slots insulated with fiber paper (fish paper) to protect the windings. The ends of the windings are connected to the commutator segments. Wedges or steel bands hold the windings in place to prevent them from flying out of the slots when the armature is rotating at high speeds. The commutator consists of a large number of copper segments insulated from each other and the armature shaft by pieces of mica. Insulated wedge rings hold the segments in place.

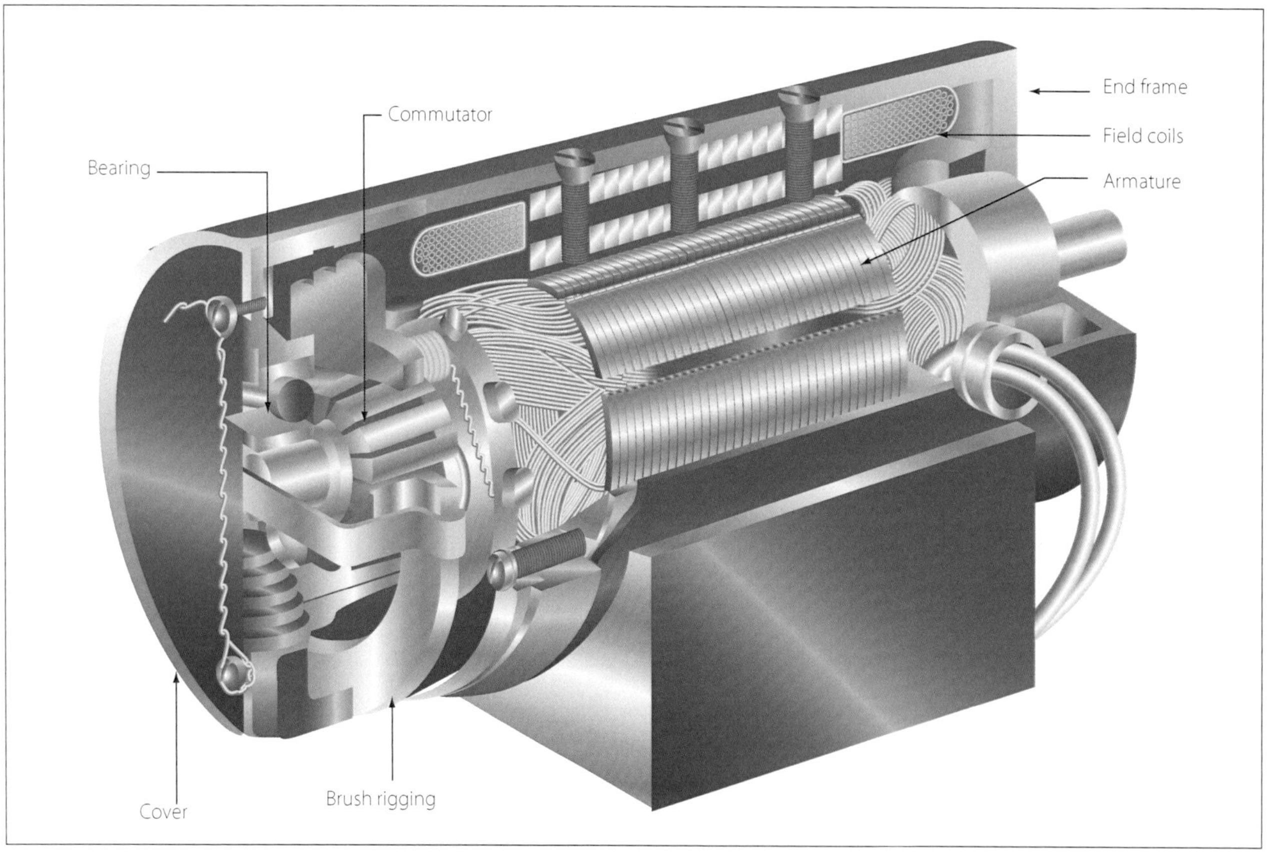

Figure 10-282. Cutaway view of practical DC motor.

Field Assembly

The field assembly consists of the field frame, the pole pieces, and the field coils. The field frame is located along the inner wall of the motor housing. It contains laminated soft steel pole pieces on which the field coils are wound. A coil, consisting of several turns of insulated wire, fits over each pole piece and, together with the pole, constitutes a field pole. Some motors have as few as two poles, others as many as eight.

Brush Assembly

The brush assembly consists of the brushes and their holders. The brushes are usually small blocks of graphitic carbon, since this material has a long service life and also causes minimum wear to the commutator. The holders permit some play in the brushes so they can follow any irregularities in the surface of the commutator and make good contact. Springs hold the brushes firmly against the commutator. A commutator and two types of brushes are shown in Figure 10-283.

End Frame

The end frame is the part of the motor opposite the commutator. Usually, the end frame is designed so that it can be connected to the unit to be driven. The bearing for the drive end is also located in the end frame. Sometimes the end frame is made a part of the unit driven by the motor. When this is done, the bearing on the drive end may be located in any one of a number of places.

Types of DC Motors

There are three basic types of DC motors: (1) series motors, (2) shunt motors, and (3) compound motors. They differ largely in the method in which their field and armature coils are connected.

Series DC Motor

In the series motor, the field windings, consisting of a relatively few turns of heavy wire, are connected in series with the armature winding. Both a diagrammatic and a schematic illustration of a series motor are shown in Figure 10-284. The same current flowing through the field winding also flows through the armature winding. Any increase in current, therefore, strengthens the magnetism of both the field and the armature.

Because of the low resistance in the windings, the series motor is able to draw a large current in starting. This starting current, in passing through both the field and armature windings, produces a high starting torque, which is the series motor's principal advantage.

The speed of a series motor is dependent upon the load. Any change in load is accompanied by a substantial change in speed. A series motor will run at high speed when it has a light load and at low speed with a heavy load. If the load is removed entirely, the motor may operate at such a high speed that the armature will fly apart. If high starting torque is needed under heavy load conditions, series motors have many applications. Series motors are often used in aircraft as engine start-

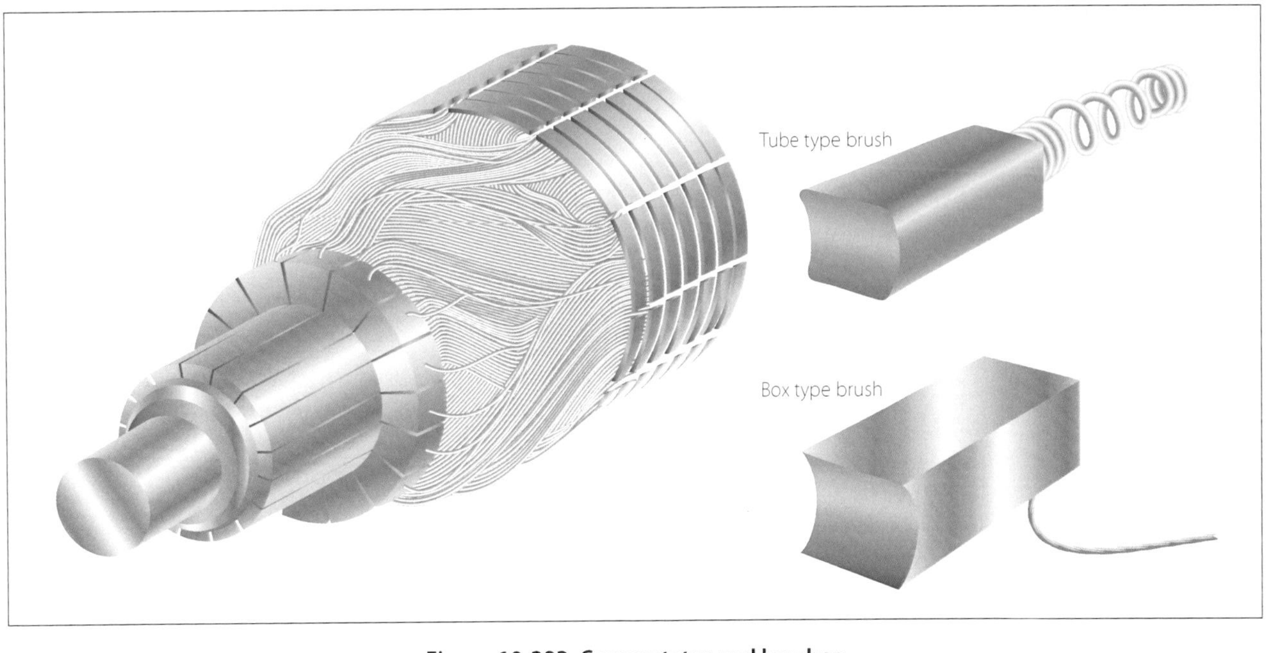

Figure 10-283. Commutator and brushes.

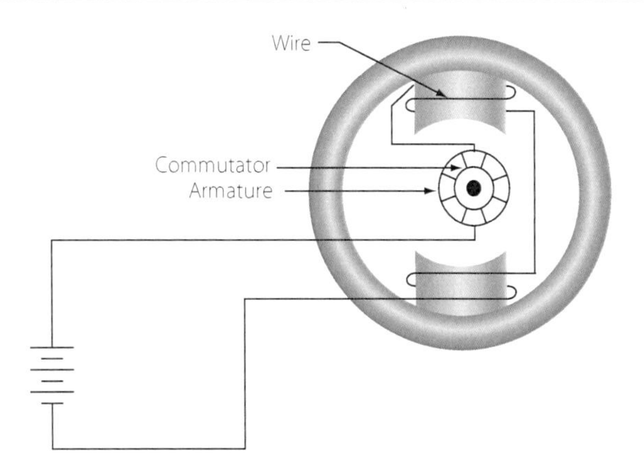

(A) Diagrammatic

(B) Schematic

Figure 10-284. Series motor.

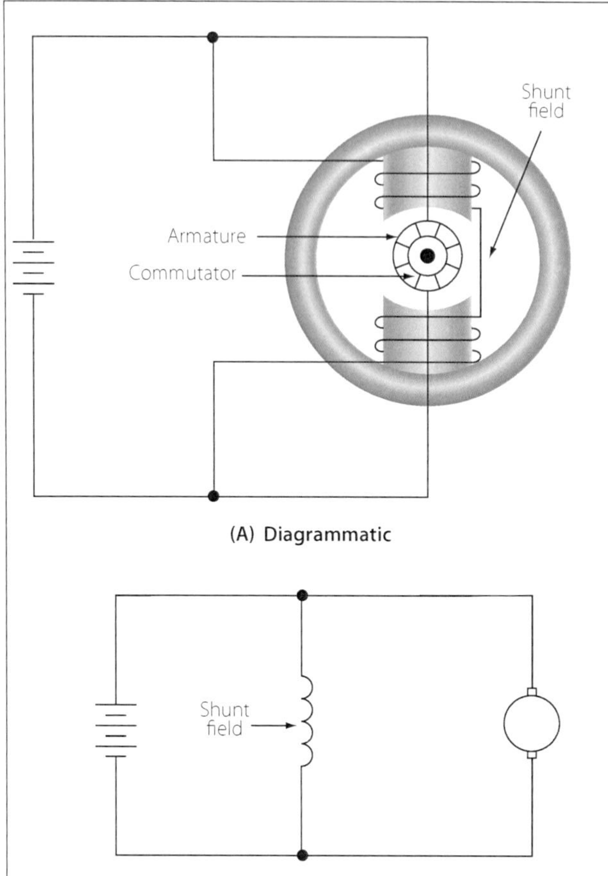

(A) Diagrammatic

(B) Schematic

Figure 10-285. Shunt motor.

ers and for raising and lowering landing gears, cowl flaps, and wing flaps.

Shunt DC Motor

In the shunt motor, the field winding is connected in parallel or in shunt with the armature winding. [Figure 10-285] The resistance in the field winding is high. Since the field winding is connected directly across the power supply, the current through the field is constant. The field current does not vary with motor speed, as in the series motor and, therefore, the torque of the shunt motor will vary only with the current through the armature. The torque developed at starting is less than that developed by a series motor of equal size.

The speed of the shunt motor varies very little with changes in load. When all load is removed, it assumes a speed slightly higher than the loaded speed. This motor is particularly suitable for use when constant speed is desired and when high starting torque is not needed.

Compound DC Motor

The compound motor is a combination of the series and shunt motors. There are two windings in the field: a shunt winding and a series winding. A schematic of a compound motor is shown in Figure 10-286. The shunt winding is composed of many turns of fine wire and is connected in parallel with the armature winding. The series winding consists of a few turns of large wire and is connected in series with the armature winding. The starting torque is higher than in the shunt motor but lower than in the series motor. Variation of speed with load is less than in a series wound motor but greater than in a shunt motor. The compound motor is used whenever the combined characteristics of the series and shunt motors are desired.

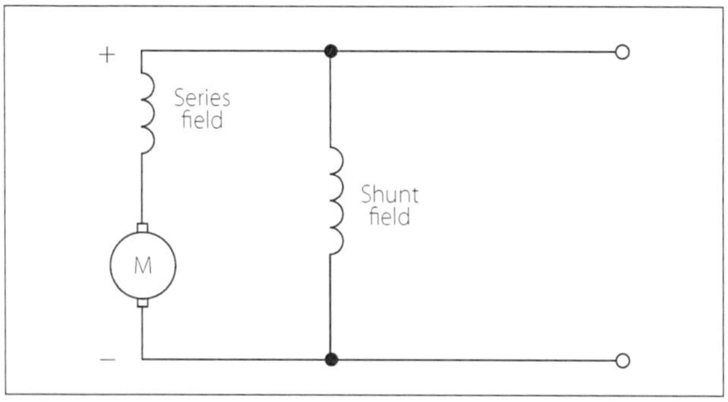

Figure 10-286. Compound motor.

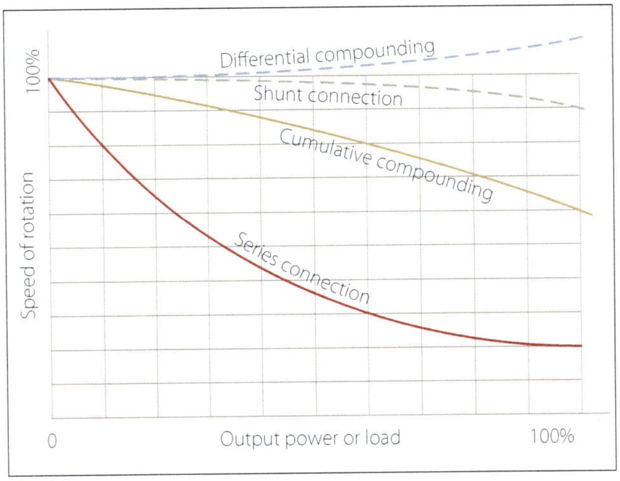

Figure 10-287. Load characteristics of DC motors.

Like the compound generator, the compound motor has both series and shunt field windings. The series winding may either aid the shunt wind (cumulative compound) or oppose the shunt winding (differential compound). The starting and load characteristics of the cumulative compound motor are somewhere between those of the series and those of the shunt motor.

Because of the series field, the cumulative compound motor has a higher starting torque than a shunt motor. Cumulative compound motors are used in driving machines, which are subject to sudden changes in load. They are also used where a high starting torque is desired, but a series motor cannot be used easily.

In the differential compound motor, an increase in load creates an increase in current and a decrease in total flux in this type of motor. These two tend to offset each other and the result is a practically constant speed. However, since an increase in load tends to decrease the field strength, the speed characteristic becomes unstable. Rarely is this type of motor used in aircraft systems.

A graph of the variation in speed with changes of load of the various types of DC motors is shown in Figure 10-287.

Counter Electromotive Force (emf)

The armature resistance of a small, 28-volt DC motor is extremely low, about 0.1 ohm. When the armature is connected across the 28-volt source, current through the armature will apparently be

$$I = \frac{E}{R} = \frac{28}{0.1} = 280 \text{ amperes}$$

This high valve of current flow is not only impracticable but also unreasonable, especially when the current drain, during normal operation of a motor, is found to be about 4 amperes. This is because the current through a motor armature during operation is determined by more factors than ohmic resistance.

When the armature in a motor rotates in a magnetic field, a voltage is induced in its windings. This voltage is called the back or counter emf (electromotive force) and is opposite in direction to the voltage applied to the motor from the external source.

Counter emf opposes the current, which causes the armature to rotate. The current flowing through the armature, therefore, decreases as the counter emf increases. The faster the armature rotates, the greater the counter emf. For this reason, a motor connected to a battery may draw a fairly high current on starting, but as the armature speed increases, the current flowing through the armature decreases. At rated speed, the counter emf may be only a few volts less than the battery voltage. Then, if the load on the motor is increased, the motor will slow down, less counter emf will be generated, and the current drawn from the external source will increase. In a shunt motor, the counter emf affects only the current in the armature, since the field is connected in parallel across the power source. As the motor slows down and the counter emf decreases, more current flows through the armature, but the magnetism in the field is unchanged. When the series motor slows down, the counter emf decreases and more current flows through the field and the armature, thereby strengthening their magnetic fields. Because of these characteristics, it is more difficult to stall a series motor than a shunt motor.

Types of Duty

Electric motors are called upon to operate under various conditions. Some motors are used for intermittent operation; others operate continuously. Motors built for intermittent duty can be operated for short periods only and, then, must be allowed to cool before being operated again. If such a motor is operated for long periods under full load, the motor will be overheated. Motors built for continuous duty may be operated at rated power for long periods.

Reversing Motor Direction

By reversing the direction of current flow in either the armature or the field windings, the direction of a motor's rotation may be reversed. This will reverse the magnetism of either the armature or the magnetic field in which the armature rotates. If the wires connecting the motor to an external source are interchanged, the direction of rotation will not be reversed, since changing these wires reverses the magnetism of both field and armature and leaves the torque in the same direction as before.

One method for reversing direction of rotation employs two field windings wound in opposite directions on the same pole. This type of motor is called a split field motor. Figure 10-288 shows a series motor with a split field winding. The single pole, double throw switch makes it possible to direct current through either of the two windings. When the switch is placed in the lower position, current flows through the lower field winding, creating a north pole at the lower field winding and at the lower pole piece, and a south pole at the upper pole piece. When the switch is placed in the upper position, current flows through the upper field winding, the magnetism of the field is reversed, and the armature rotates in the opposite direction. Some split field motors are built with two separate field windings wound on alternate poles. The armature in such a motor, a four pole reversible motor, rotates in one direction when current flows through the windings of one set of opposite pole pieces, and in the opposite direction when current flows through the other set of windings.

Another method of direction reversal, called the switch method, employs a double pole, double throw switch which changes the direction of current flow in either the armature or the field. In the illustration of the switch method shown in Figure 10-289, current direction may be reversed through the field but not through the armature. When the switch is thrown to the "up" position, current flows through the field winding to establish a north pole at the right side of the motor and a south pole at the left side of the motor. When the switch is thrown to the "down" position, this polarity is reversed and the armature rotates in the opposite direction.

Motor Speed

Motor speed can be controlled by varying the current in the field windings. When the amount of current flowing through the field windings is increased, the field strength increases, but the motor slows down since a greater amount of counter emf is generated in the armature windings. When the field current is decreased, the field strength decreases, and the motor speeds up because the counter emf is reduced. A motor in which speed can be controlled is called a variable speed motor. It may be either a shunt or series motor.

In the shunt motor, speed is controlled by a rheostat in series with the field windings. [Figure 10-290] The speed depends on the amount of current that flows through the rheostat to the field windings. To increase the motor speed, the resistance in the rheostat is increased, which decreases the field current. As a result,

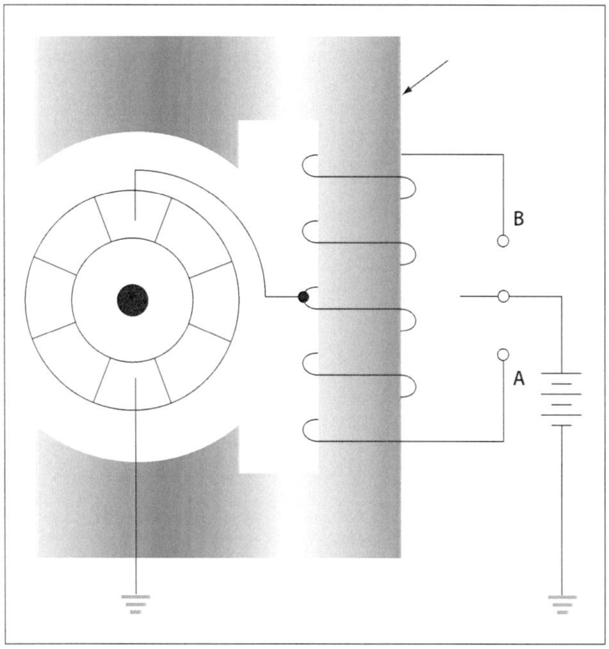

Figure 10-288. Split field series motor.

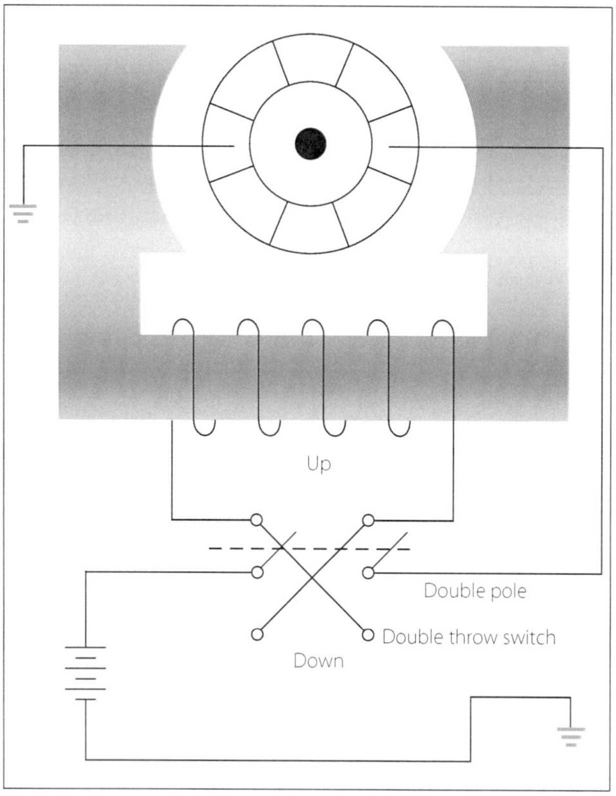

Figure 10-289. Switch method of reversing motor direction.

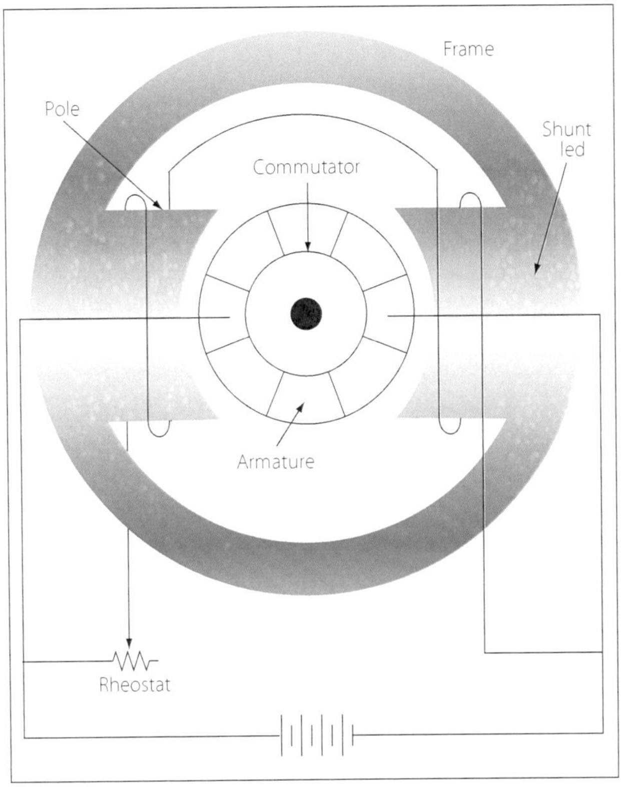

Figure 10-290. Shunt motor with variable speed control.

there is a decrease in the strength of the magnetic field and in the counter emf. This momentarily increases the armature current and the torque. The motor will then automatically speed up until the counter emf increases and causes the armature current to decrease to its former value. When this occurs, the motor will operate at a higher fixed speed than before.

To decrease the motor speed, the resistance of the rheostat is decreased. More current flows through the field windings and increases the strength of the field; then, the counter emf increases momentarily and decreases the armature current. As a result, the torque decreases and the motor slows down until the counter emf decreases to its former value; then the motor operates at a lower fixed speed than before.

In the series motor, the rheostat speed control is connected either in parallel or in series with the motor field, or in parallel with the armature. When the rheostat is set for maximum resistance, the motor speed is increased in the parallel armature connection by a decrease in current. When the rheostat resistance is maximum in the series connection, motor speed is reduced by a reduction in voltage across the motor. For above normal speed operation, the rheostat is in parallel with the series field. Part of the series field current is bypassed and the motor speeds up. [Figure 10-291]

Energy Losses in DC Motors

Losses occur when electrical energy is converted to mechanical energy (in the motor), or mechanical energy is converted to electrical energy (in the generator). For the machine to be efficient, these losses must be kept to a minimum. Some losses are electrical; others are mechanical. Electrical losses are classified

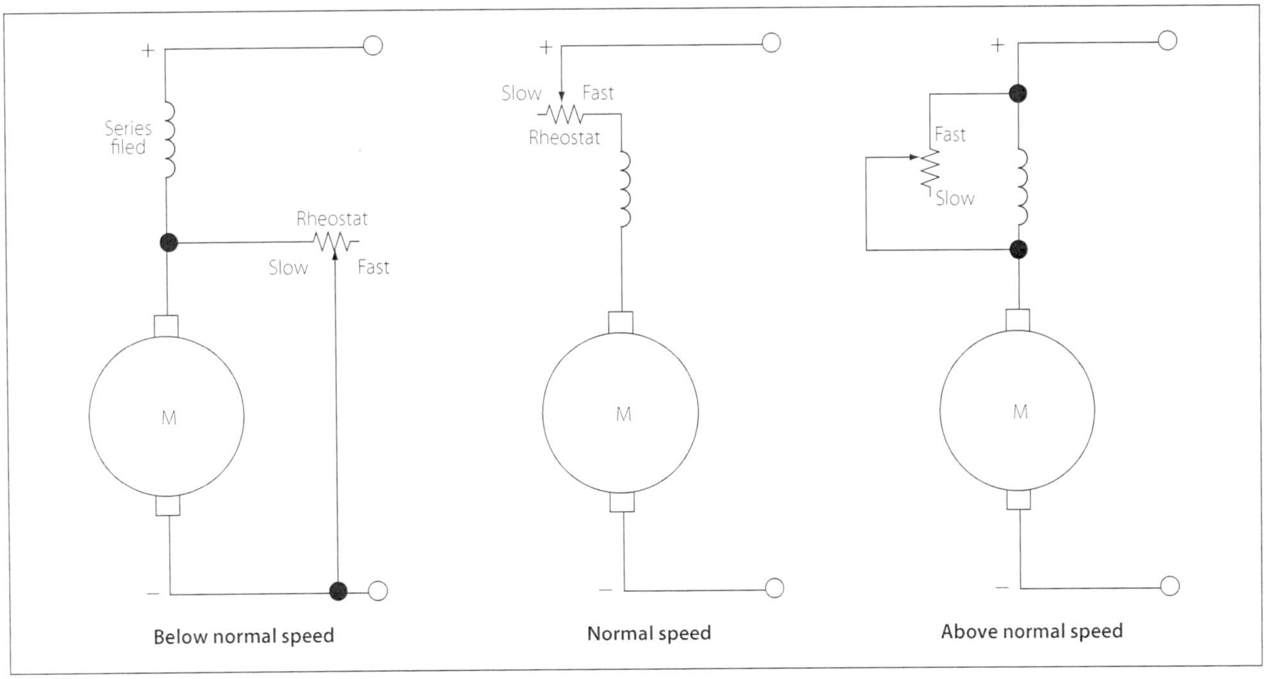

Figure 10-291. Controlling the speed of a series DC motor.

as copper losses and iron losses; mechanical losses occur in overcoming the friction of various parts of the machine.

Copper losses occur when electrons are forced through the copper windings of the armature and the field. These losses are proportional to the square of the current. They are sometimes called I^2R losses, since they are due to the power dissipated in the form of heat in the resistance of the field and armature windings.

Iron losses are subdivided in hysteresis and eddy current losses. Hysteresis losses are caused by the armature revolving in an alternating magnetic field. It, therefore, becomes magnetized first in one direction and then in the other. The residual magnetism of the iron or steel of which the armature is made causes these losses. Since the field magnets are always magnetized in one direction (DC field), they have no hysteresis losses.

Eddy current losses occur because the iron core of the armature is a conductor revolving in a magnetic field. This sets up an emf across portions of the core, causing currents to flow within the core. These currents heat the core and, if they become excessive, may damage the windings. As far as the output is concerned, the power consumed by eddy currents is a loss. To reduce eddy currents to a minimum, a laminated core usually is used. A laminated core is made of thin sheets of iron electrically insulated from each other. The insulation between laminations reduces eddy currents, because it is "transverse" to the direction in which these currents tend to flow. However, it has no effect on the magnetic circuit. The thinner the laminations, the more effectively this method reduces eddy current losses.

Inspection and Maintenance of DC Motors

Use the following procedures to make inspection and maintenance checks:

1. Check the operation of the unit driven by the motor in accordance with the instructions covering the specific installation.
2. Check all wiring, connections, terminals, fuses, and switches for general condition and security.
3. Keep motors clean and mounting bolts tight.
4. Check brushes for condition, length, and spring tension. Minimum brush lengths, correct spring tension, and procedures for replacing brushes are given in the applicable manufacturer's instructions.
5. Inspect commutator for cleanness, pitting, scoring, roughness, corrosion or burning. Check for high mica (if the copper wears down below the mica, the mica will insulate the brushes from the commutator). Clean dirty commutators with a cloth moistened with the recommended cleaning solvent. Polish rough or corroded commutators with fine sandpaper (000 or finer) and blow out with compressed air. Never use emery paper since it contains metallic particles which may cause shorts. Replace the motor if the commutator is burned, badly pitted, grooved, or worn to the extent that the mica insulation is flush with the commutator surface.
6. Inspect all exposed wiring for evidence of overheating. Replace the motor if the insulation on leads or windings is burned, cracked, or brittle.
7. Lubricate only if called for by the manufacturer's instructions covering the motor. Most motors used in today's airplanes require no lubrication between overhauls.
8. Adjust and lubricate the gearbox, or unit which the motor drives, in accordance with the applicable manufacturer's instructions covering the unit.

When trouble develops in a DC motor system, check first to determine the source of the trouble. Replace the motor only when the trouble is due to a defect in the motor itself. In most cases, the failure of a motor to operate is caused by a defect in the external electrical circuit, or by mechanical failure in the mechanism driven by the motor.

Check the external electrical circuit for loose or dirty connections and for improper connection of wiring. Look for open circuits, grounds, and shorts by following the applicable manufacturer's circuit testing procedure. If the fuse is not blown, failure of the motor to operate is usually due to an open circuit. A blown fuse usually indicates an accidental ground or short circuit. A low battery usually causes the chattering of the relay switch, which controls the motor. When the battery is low, the open circuit voltage of the battery is sufficient to close the relay, but with the heavy current draw of the motor, the voltage drops below the level required to hold the relay closed. When the relay opens, the voltage in the battery increases enough to close the relay again. This cycle repeats and causes chattering, which is very harmful to the relay switch, due to the heavy current causing an arc, which will burn the contacts.

Check the unit driven by the motor for failure of the unit or drive mechanism. If the motor has failed as a result of a failure in the driven unit, the fault must be corrected before installing a new motor.

If it has been determined that the fault is in the motor itself (by checking for correct voltage at the motor terminals and for failure of the driven unit), inspect the commutator and brushes. A dirty commutator or defective or binding brushes may result in poor contact between brushes and commutator. Clean the commutator, brushes, and brush holders with a cloth moistened with the recommended cleaning solvent. If brushes are damaged or worn to the specified minimum length, install new brushes in accordance with the applicable manufacturer's instructions covering the motor. If the motor still fails to operate, replace it with a serviceable motor.

AC Motors

Because of their advantages, many types of aircraft motors are designed to operate on alternating current. In general, AC motors are less expensive than comparable DC motors. In many instances, AC motors do not use brushes and commutators so sparking at the brushes is avoided. AC motors are reliable and require little maintenance. They are also well suited for constant speed applications and certain types are manufactured that have, within limits, variable speed characteristics. Alternating current motors are designed to operate on polyphase or single phase lines and at several voltage ratings.

The speed of rotation of an AC motor depends upon the number of poles and the frequency of the electrical source of power:

$$\text{rpm} = \frac{120 \times \text{Frequency}}{\text{Number of poles}}$$

Since airplane electrical systems typically operate at 400 cycles, an electric motor at this frequency operates at about seven times the speed of a 60 cycle commercial motor with the same number of poles. Because of this high speed of rotation, 400-cycle AC motors are suitable for operating small high-speed rotors, through reduction gears, in lifting and moving heavy loads, such as the wing flaps, the retractable landing gear, and the starting of engines. The 400-cycle induction type motor operates at speeds ranging from 6,000 rpm to 24,000 rpm. Alternating current motors are rated in horsepower output, operating voltage, full load current, speed, number of phases, and frequency. Whether the motors operate continuously or intermittently (for short intervals) is also considered in the rating.

Types of AC Motors

There are two general types of AC motors used in aircraft systems: induction motors and synchronous motors. Either type may be single phase, two phase, or three phase. Three phase induction motors are used where large amounts of power are required. They operate such devices as starters, flaps, landing gears, and hydraulic pumps. Single phase induction motors are used to operate devices such as surface locks, intercooler shutters, and oil shutoff valves in which the power requirement is low. Three phase synchronous motors operate at constant synchronous speeds and are commonly used to operate flux gate compasses and propeller synchronizer systems. Single phase synchronous motors are common sources of power to operate electric clocks and other small precision equipment. They require some auxiliary method to bring them up to synchronous speeds; that is, to start them. Usually the starting winding consists of an auxiliary stator winding.

Three Phase Induction Motor

The three phase AC induction motor is also called a squirrel cage motor. Both single phase and three phase motors operate on the principle of a rotating magnetic field. A horseshoe magnet held over a compass needle is a simple illustration of the principle of the rotating field. The needle will take a position parallel to the magnetic flux passing between the two poles of the magnet. If the magnet is rotated, the compass needle will follow. A rotating magnetic field can be produced by a two or three phase current flowing through two or more groups of coils wound on inwardly projecting poles of an iron frame. The coils on each group of poles are wound alternately in opposite directions to produce opposite polarity, and each group is connected to a separate phase of voltage. The operating principle depends on a revolving, or rotating, magnetic field to produce torque. The key to understanding the induction motor is a thorough understanding of the rotating magnetic field.

Rotating Magnetic Field

The field structure shown in Figure 10-292A has poles whose windings are energized by three AC voltages, a, b, and c. These voltages have equal magnitude but differ in phase, as shown in Figure 10-292B: at the instant of time shown as 0, the resultant magnetic field produced by the application of the three voltages has its greatest intensity in a direction extending from pole 1 to pole 4. Under this condition, pole 1 can be considered as a north pole and pole 4 as a south pole. At the instant of time shown as 1, the resultant magnetic field

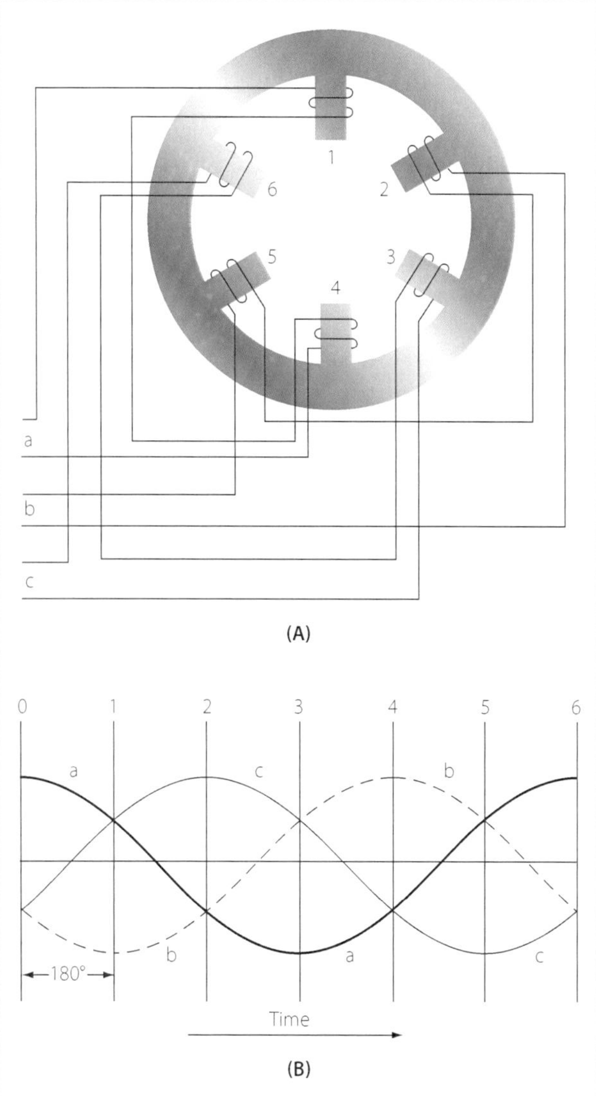

Figure 10-292. Rotating magnetic field developed by application of three phase voltages.

will have its greatest intensity in the direction extending from pole 2 to pole 5; in this case, pole 2 can be considered as a north pole and pole 5 as a south pole. Thus, between instant 0 and instant 1, the magnetic field has rotated clockwise. At instant 2, the resultant magnetic field has its greatest intensity in the direction from pole 3 to pole 6, and the resultant magnetic field has continued to rotate clockwise. At instant 3, poles 4 and 1 can be considered as north and south poles, respectively, and the field has rotated still farther. At later instants of time, the resultant magnetic field rotates to other positions while traveling in a clockwise direction, a single revolution of the field occurring in one cycle. If the exciting voltages have a frequency of 60 cps, the magnetic field makes 60 revolutions per second, or 3,600 rpm. This speed is known as the synchronous speed of the rotating field.

Construction of Induction Motor

The stationary portion of an induction motor is called a stator, and the rotating member is called a rotor. Instead of salient poles in the stator, as shown in A of Figure 10-292, distributed windings are used; these windings are placed in slots around the periphery of the stator. It is usually impossible to determine the number of poles in an induction motor by visual inspection, but the information can be obtained from the nameplate of the motor. The nameplate usually gives the number of poles and the speed at which the motor is designed to run. This rated, or nonsynchronous, speed is slightly less than the synchronous speed. To determine the number of poles per phase on the motor, divide 120 times the frequency by the rated speed. Written as an equation, it is:

$$P = \frac{120 \times f}{N}$$

Where: P is the number of poles per phase,
f is the frequency in cps,
N is the rated speed in rpm, and
120 is a constant.

The result will be very nearly equal to the number of poles per phase. For example, consider a 60 cycle, three phase motor with a rated speed of 1,750 rpm. In this case:

$$P = \frac{120 \times 60}{1{,}750} = \frac{7{,}200}{1{,}750} = 4.1$$

Therefore, the motor has four poles per phase. If the number of poles per phase is given on the nameplate, the synchronous speed can be determined by dividing 120 times the frequency by the number of poles per phase. In the example used above, the synchronous speed is equal to 7,200 divided by 4, or 1,800 rpm.

The rotor of an induction motor consists of an iron core having longitudinal slots around its circumference in which heavy copper or aluminum bars are embedded. These bars are welded to a heavy ring of high conductivity on either end. The composite structure is sometimes called a squirrel cage, and motors containing such a rotor are called squirrel cage induction motors. [Figure 10-293]

Induction Motor Slip

When the rotor of an induction motor is subjected to the revolving magnetic field produced by the stator windings, a voltage is induced in the longitudinal bars. The induced voltage causes a current to flow through

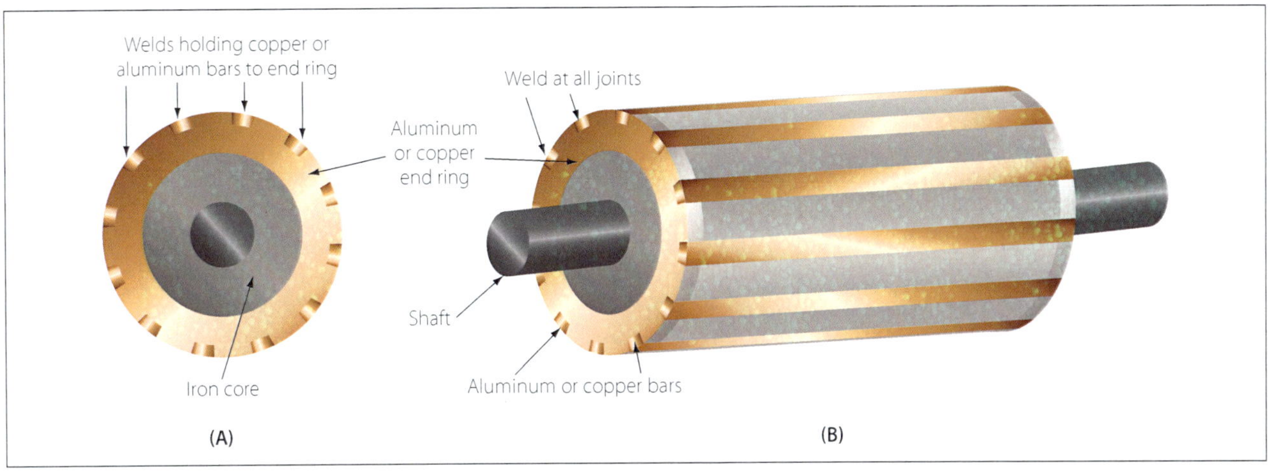

Figure 10-293. Squirrel cage rotor for an AC induction motor.

the bars. This current, in turn, produces its own magnetic field, which combines with the revolving field so that the rotor assumes a position in which the induced voltage is minimized. As a result, the rotor revolves at very nearly the synchronous speed of the stator field, the difference in speed being just sufficient enough to induce the proper amount of current in the rotor to overcome the mechanical and electrical losses in the rotor. If the rotor were to turn at the same speed as the rotating field, the rotor conductors would not be cut by any magnetic lines of force, no emf would be induced in them, no current could flow, and there would be no torque. The rotor would then slow down. For this reason, there must always be a difference in speed between the rotor and the rotating field. This difference in speed is called slip and is expressed as a percentage of the synchronous speed. For example, if the rotor turns at 1,750 rpm and the synchronous speed is 1,800 rpm, the difference in speed is 50 rpm. The slip is then equal to 50/1,800 or 2.78 percent.

Single Phase Induction Motor

The previous discussion has applied only to polyphase motors. A single-phase motor has only one stator winding. This winding generates a field, which merely pulsates, instead of rotating. When the rotor is stationary, the expanding and collapsing stator field induces currents in the rotor. These currents generate a rotor field opposite in polarity to that of the stator. The opposition of the field exerts a turning force on the upper and lower parts of the rotor trying to turn it 180° from its position. Since these forces are exerted through the center of the rotor, the turning force is equal in each direction. As a result, the rotor does not turn. If the rotor has started turning, it will continue to rotate in the direction in which it is started, since the turning force in that direction is aided by the momentum of the rotor.

Shaded Pole Induction Motor

The first effort in the development of a self-starting, single-phase motor was the shaded pole induction motor. [Figure 10-294] This motor has salient poles, a portion of each pole being encircled by a heavy copper ring. The presence of the ring causes the magnetic field through the ringed portion of the pole face to lag appreciably behind that through the other part of the pole face. The net effect is the production of a slight component of rotation of the field, sufficient to cause the rotor to revolve. As the rotor accelerates, the torque increases until the rated speed is obtained. Such

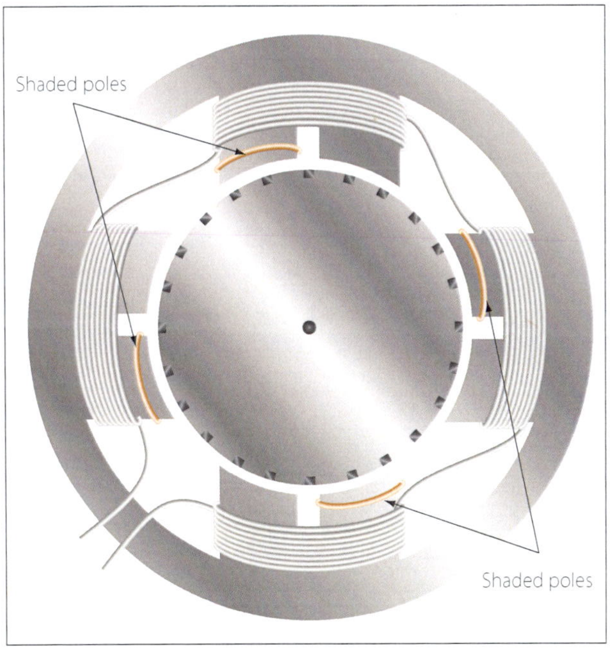

Figure 10-294. Shaded pole induction motor.

motors have low starting torque and find their greatest application in small fan motors where the initial torque required is low.

In Figure 10-295, a diagram of a pole and the rotor is shown. The poles of the shaded pole motor resemble those of a DC motor.

A low resistance, short-circuited coil or copper band is placed across one tip of each small pole, from which, the motor gets the name of shaded pole. The rotor of this motor is the squirrel cage type. As the current increases in the stator winding, the flux increases. A portion of this flux cuts the low resistance shading coil. This induces a current in the shading coil, and by Lenz's law, the current sets up a flux that opposes the flux inducing the current. Hence, most of the flux passes through the unshaded portion of the poles, as shown in Figure 10-295.

When the current in the winding and the main flux reaches a maximum, the rate of change is zero; thus, no emf is induced in the shading coil. A little later, the shading coil current, which causes the induced emf to lag, reaches zero, and there is no opposing flux. Therefore, the main field flux passes through the shaded portion of the field pole. The main field flux, which is now decreasing, induces a current in the shading coil. This current sets up a flux that opposes the decrease of the main field flux in the shaded portion of the pole. The effect is to concentrate the lines of force in the shaded portion of the pole face. In effect, the shading coil retards, in time phase, the portion of the flux passing through the shaded part of the pole. This lag in time phase of the flux in the shaded tip causes the flux to produce the effect of sweeping across the face of the pole, from left to right in the direction of the shaded tip. This behaves like a very weak rotating magnetic field, and sufficient torque is produced to start a small motor. The starting torque of the shaded pole motor is exceedingly weak, and the power factor is low. Consequently, it is built in sizes suitable for driving such devices as small fans.

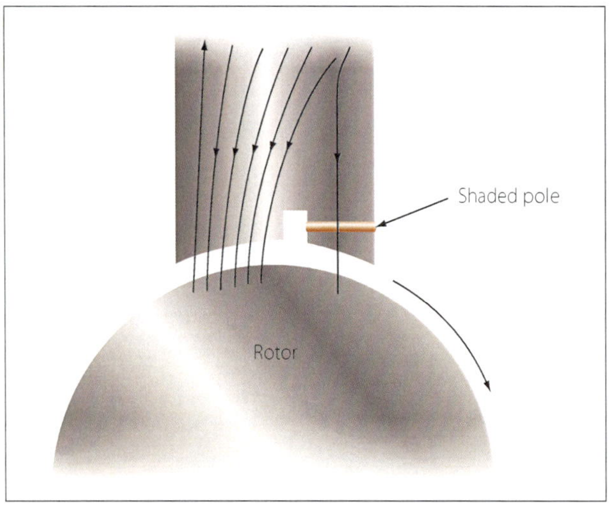

Figure 10-295. Diagram of a shaded pole motor.

Split Phase Motor

There are various types of self-starting motors, known as split phase motors. Such motors have a starting winding displaced 90 electrical degrees from the main or running winding. In some types, the starting winding has a fairly high resistance, which causes the current in this winding to be out of phase with the current in the running winding. This condition produces, in effect, a rotating field and the rotor revolves. A centrifugal switch disconnects the starting winding automatically, after the rotor has attained approximately 25 percent of its rated speed.

Capacitor Start Motor

With the development of high capacity electrolytic capacitors, a variation of the split phase motor, known as the capacitor start motor, has been made. Nearly all fractional horsepower motors in use today on refrigerators and other similar appliances are of this type. [Figure 10-296] In this adaptation, the starting winding and running winding have the same size and resistance value. The phase shift between currents of the two windings is obtained by using capacitors connected in series with the starting winding.

Capacitor start motors have a starting torque comparable to their torque at rated speed and can be used in applications where the initial load is heavy. Again, a centrifugal switch is required for disconnecting the starting winding when the rotor speed is approximately 25 percent of the rated speed.

Although some single phase induction motors are rated as high as 2 horsepower (hp), the major field of application is 1 hp, or less, at a voltage rating of 115 volts for the smaller sizes and 110 to 220 volts for one-fourth hp and up. For even larger power ratings, polyphase motors generally are used, since they have excellent starting torque characteristics.

Direction of Rotation of Induction Motors

The direction of rotation of a three phase induction motor can be changed by simply reversing two of the leads to the motor. The same effect can be obtained in a two phase motor by reversing connections to one phase. In a single phase motor, reversing connections to the starting winding will reverse the direction of rotation.

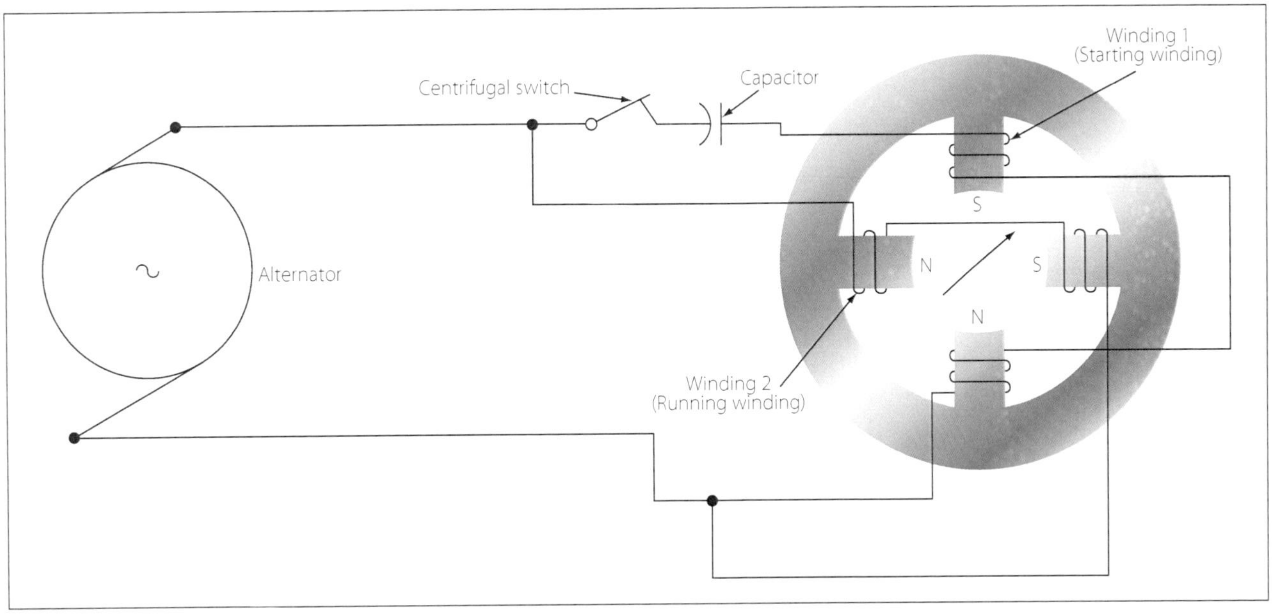

Figure 10-296. Single phase motor with capacitor starting winding.

Most single phase motors designed for general application have provision for readily reversing connections to the starting winding. Nothing can be done to a shaded pole motor to reverse the direction of rotation because the direction is determined by the physical location of the copper shading ring. If, after starting, one connection to a three phase motor is broken, the motor will continue to run but will deliver only one-third the rated power. Also, a two phase motor will run at one-half its rated power if one phase is disconnected. Neither motor will start under these abnormal conditions.

Synchronous Motor

The synchronous motor is one of the principal types of AC motors. Like the induction motor, the synchronous motor makes use of a rotating magnetic field. Unlike the induction motor, however, the torque developed does not depend on the induction of currents in the rotor. Briefly, the principle of operation of the synchronous motor is as follows: A multiphase source of AC is applied to the stator windings, and a rotating magnetic field is produced. A direct current is applied to the rotor winding, and another magnetic field is produced. The synchronous motor is so designed and constructed that these two fields react to each other in such a manner that the rotor is dragged along and rotates at the same speed as the rotating magnetic field produced by the stator windings.

An understanding of the operation of the synchronous motor can be obtained by considering the simple motor of Figure 10-297. Assume that poles A and B are being rotated clockwise by some mechanical means in order to produce a rotating magnetic field, they induce poles of opposite polarity in the soft iron rotor, and forces of attraction exist between corresponding north and south poles.

Consequently, as poles A and B rotate, the rotor is dragged along at the same speed. However, if a load is applied to the rotor shaft, the rotor axis will momentarily fall behind that of the rotating field but, thereafter, will continue to rotate with the field at the same speed, as long as the load remains constant. If the load is too large, the rotor will pull out of synchronism with the rotating field and, as a result, will no longer rotate with the field at the same speed. Thus the motor is said to be overloaded.

Such a simple motor as that shown in Figure 10-297 is never used. The idea of using some mechanical means of rotating the poles is impractical because another motor would be required to perform this work. Also, such an arrangement is unnecessary because a rotating magnetic field can be produced electrically by using phased AC voltages. In this respect, the synchronous motor is similar to the induction motor.

The synchronous motor consists of a stator field winding similar to that of an induction motor. The stator winding produces a rotating magnetic field. The rotor may be a permanent magnet, as in small single phase synchronous motors used for clocks and other small precision equipment, or it may be an electromagnet, energized from a DC source of power and fed through slip rings into the rotor field coils, as in an alternator.

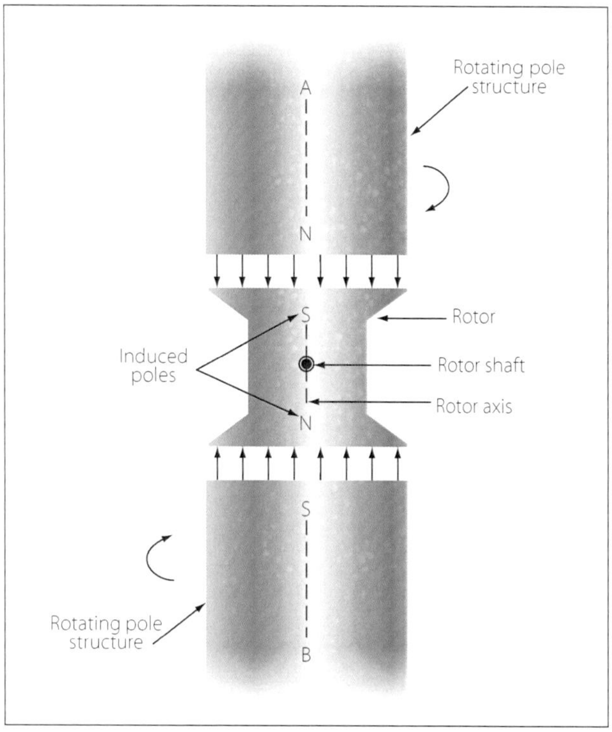

Figure 10-297. Illustrating the operation of a synchronous motor.

In fact, an alternator may be operated either as an alternator or a synchronous motor.

Since a synchronous motor has little starting torque, some means must be provided to bring it up to synchronous speed. The most common method is to start the motor at no load, allow it to reach full speed, and then energize the magnetic field. The magnetic field of the rotor locks with the magnetic field of the stator and the motor operates at synchronous speed.

The magnitude of the induced poles in the rotor shown in Figure 10-298 is so small that sufficient torque cannot be developed for most practical loads. To avoid such a limitation on motor operation, a winding is placed on the rotor and energized with DC. A rheostat placed in series with the DC source provides the operator of the machine with a means of varying the strength of the rotor poles, thus placing the motor under control for varying loads.

The synchronous motor is not a self-starting motor. The rotor is heavy and, from a dead stop, it is impossible to bring the rotor into magnetic lock with the rotating magnetic field. For this reason, all synchronous motors have some kind of starting device. One type

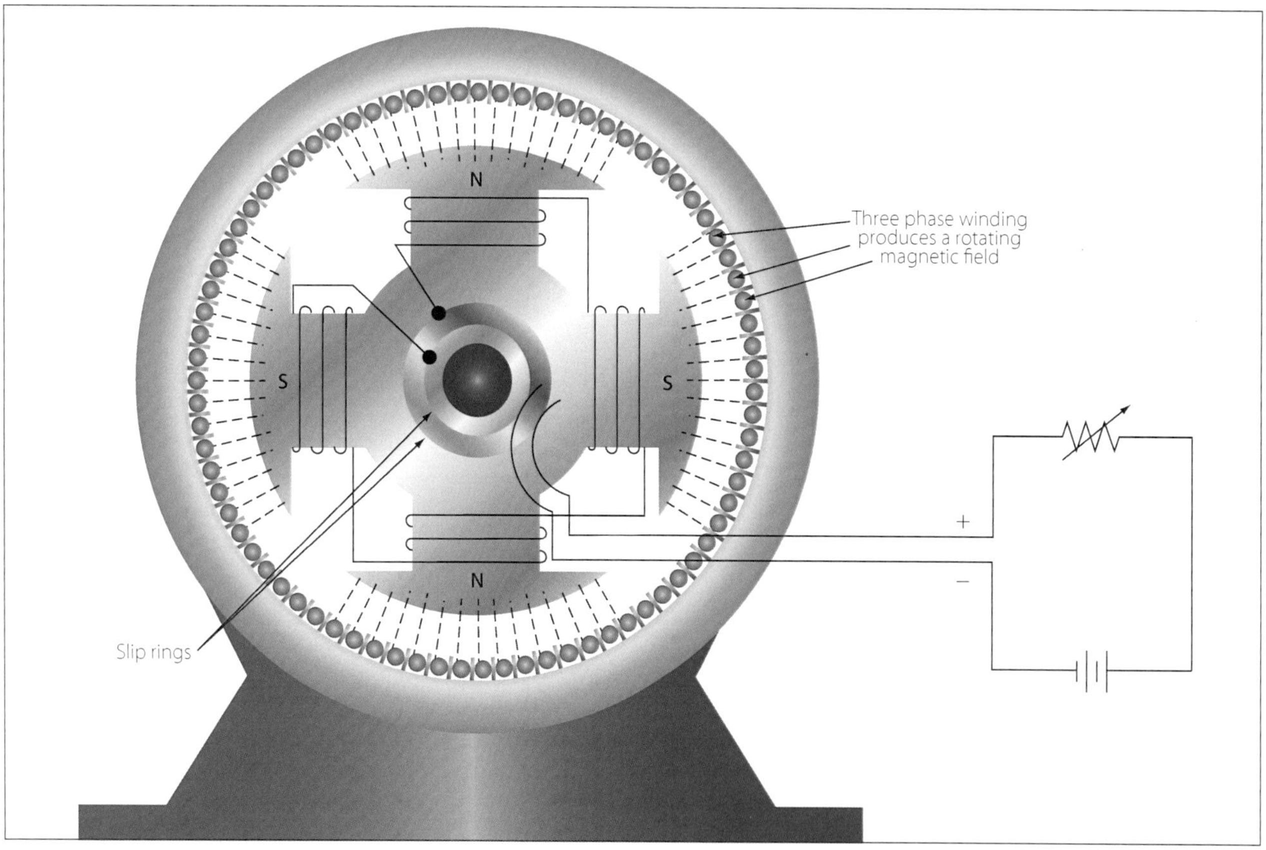

Figure 10-298. Synchronous motor.

of simple starter is another motor, either AC or DC, which brings the rotor up to approximately 90 percent of its synchronous speed. The starting motor is then disconnected, and the rotor locks in step with the rotating field. Another starting method is a second winding of the squirrel cage type on the rotor. This induction winding brings the rotor almost to synchronous speed, and when the DC is connected to the rotor windings, the rotor pulls into step with the field. The latter method is the more commonly used.

AC Series Motor

An alternating current series motor is a single phase motor, but is not an induction or synchronous motor. It resembles a DC motor in that it has brushes and a commutator. The AC series motor will operate on either AC or DC circuits. It will be recalled that the direction of rotation of a DC series motor is independent of the polarity of the applied voltage, provided the field and armature connections remain unchanged. Hence, if a DC series motor is connected to an AC source, a torque will be developed which tends to rotate the armature in one direction. However, a DC series motor does not operate satisfactorily from an AC supply for the following reasons:

1. The alternating flux sets up large eddy current and hysteresis losses in the unlaminated portions of the magnetic circuit and causes excessive heating and reduced efficiency.
2. The self induction of the field and armature windings causes a low power factor.
3. The alternating field flux establishes large currents in the coils, which are short circuited by the brushes; this action causes excessive sparking at the commutator.

To design a series motor for satisfactory operation on AC, the following changes are made:

1. The eddy current losses are reduced by laminating the field poles, frame and armature.
2. Hysteresis losses are minimized by using high permeability, transformer-type, silicon steel laminations.
3. The reactance of the field windings is kept satisfactorily low by using shallow pole pieces, few turns of wire, low frequency (usually 25 cycles for large motors), low flux density, and low reluctance (a short air gap).
4. The reactance of the armature is reduced by using a compensating winding embedded in the pole pieces. If the compensating winding is connected in series with the armature, as shown in Figure 10-299, the armature is conductively compensated.

If the compensating winding is designed as shown in Figure 10-300, the armature is inductively compensated. If the motor is designed for operation on both DC and AC circuits, the compensating winding is connected in series with the armature. The axis of the compensating winding is displaced from the main field axis by an angle of 90°. This arrangement is similar to the compensating winding used in some DC motors and generators to overcome armature reaction. The compensating winding establishes a counter magnetomotive force, neutralizing the effect of the armature magnetomotive force, preventing distortion of the main field flux, and reducing the armature reactance. The inductively compensated armature acts like the primary of a transformer, the secondary

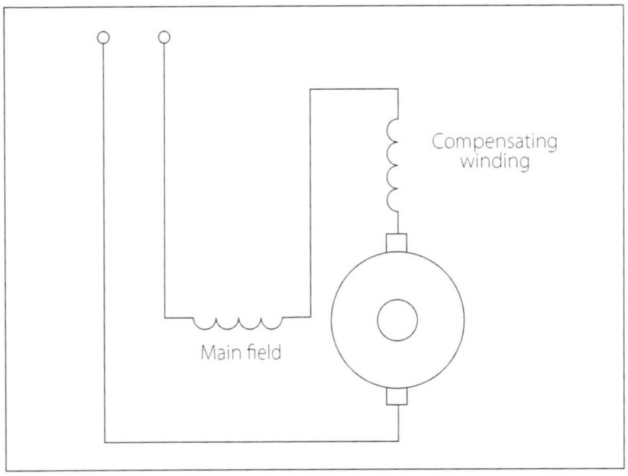

Figure 10-299. Conductivety compensated armature of AC series motor.

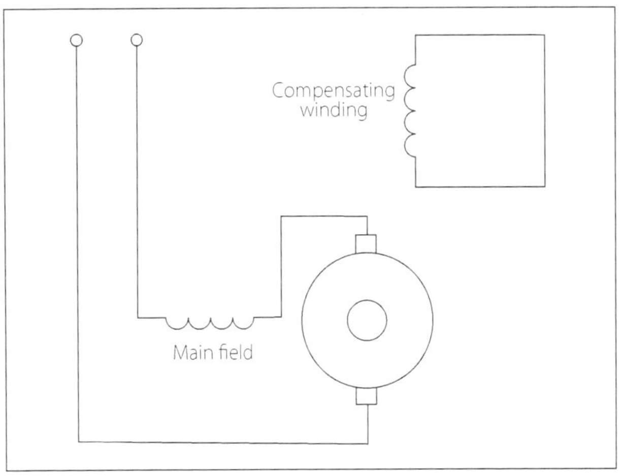

Figure 10-300. Inductively compensated armature of AC series motor.

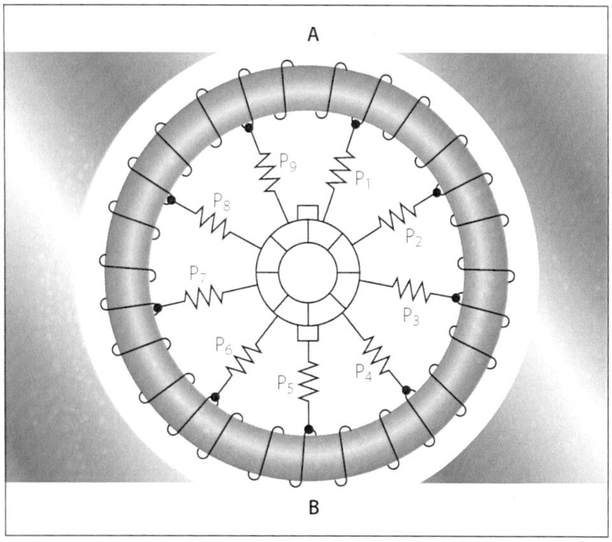

Figure 10-301. Preventive coils in AC series motor.

of which is the shorted compensating winding. The shorted secondary receives an induced voltage by the action of the alternating armature flux, and the resulting current flowing through the turns of the compensating winding establishes the opposing magnetomotive force, neutralizing the armature reactance.

5. Sparking at the commutator is reduced by the use of preventive leads P_1, P_2, P_3, and so forth, as shown in Figure 10-301, where a ring armature is shown for simplicity. When coils at A and B are shorted by the brushes, the induced current is limited by the relatively high resistance of the leads. Sparking at the brushes is also reduced by using armature coils having only a single turn and multipolar fields. High torque is obtained by having a large number of armature conductors and a large diameter armature. Thus, the commutator has a large number of very thin commutator bars and the armature voltage is limited to about 250 volts.

Fractional horsepower AC series motors are called universal motors. They do not have compensating windings or preventive leads. They are used extensively to operate fans and portable tools, such as drills, grinders, and saws.

Maintenance of AC Motors

The inspection and maintenance of AC motors is very simple. The bearings may or may not need frequent lubrication. If they are the sealed type, lubricated at the factory, they require no further attention. Be sure the coils are kept dry and free from oil or other abuse. The temperature of a motor is usually its only limiting operating factor. A good rule of thumb is that a temperature too hot for the hand is too high for safety. Next to the temperature, the sound of a motor or generator is the best trouble indicator. When operating properly, it should hum evenly. If it is overloaded it will "grunt." A three phase motor with one lead disconnected will refuse to turn and will "growl." A knocking sound generally indicates a loose armature coil, a shaft out of alignment, or armature dragging because of worn bearings.

In all cases, the inspection and maintenance of all AC motors should be performed in accordance with the applicable manufacturer's instructions.

Alternators

Basic Alternators and Classifications

An electrical generator is a machine, which converts mechanical energy into electrical energy by electromagnetic induction. A generator which produces alternating current is referred to as an AC generator and, through combination of the words "alternating" and "generator," the word "alternator" has come into widespread use. In some areas, the word "alternator" is applied only to small AC generators. This text treats the two terms synonymously and uses the term "alternator" to distinguish between AC and DC generators.

The major difference between an alternator and a DC generator is the method of connection to the external circuit; that is, the alternator is connected to the external circuit by slip rings, but the DC generator is connected by a commutator.

Method of Excitation

One means of classification is by the type of excitation system used. In alternators used on aircraft, excitation can be affected by one of the following methods:

1. *A direct connected, direct current generator.* This system consists of a DC generator fixed on the same shaft with the AC generator. A variation of this system is a type of alternator which uses DC from the battery for excitation, after which the alternator is self-excited.
2. *By transformation and rectification from the AC system.* This method depends on residual magnetism for initial AC voltage buildup, after which the field is supplied with rectified voltage from the AC generator.
3. *Integrated brushless type.* This arrangement has a direct current generator on the same shaft with

an alternating current generator. The excitation circuit is completed through silicon rectifiers rather than a commutator and brushes. The rectifiers are mounted on the generator shaft and their output is fed directly to the alternating current generator's main rotating field.

Number of Phases

Another method of classification is by the number of phases of output voltage. Alternating current generators may be single phase, two phase, three phase, or even six phase and more. In the electrical systems of aircraft, the three phase alternator is by far the most common.

Armature or Field Rotation

Still another means of classification is by the type of stator and rotor used. From this standpoint, there are two types of alternators: the revolving armature type and the revolving field type. The revolving armature alternator is similar in construction to the DC generator, in that the armature rotates through a stationary magnetic field. The revolving armature alternator is found only in alternators of low power rating and generally is not used. In the DC generator, the emf generated in the armature windings is converted into a unidirectional voltage (DC) by means of the commutator. In the revolving armature type of alternator, the generated AC voltage is applied unchanged to the load by means of slip rings and brushes.

The revolving field type of alternator has a stationary armature winding (stator) and a rotating field winding (rotor). [Figure 10-302] The advantage of having a stationary armature winding is that the armature can be connected directly to the load without having sliding contacts in the load circuit. A rotating armature would require slip rings and brushes to conduct the load current from the armature to the external circuit. Slip rings have a relatively short service life and arc over is a continual hazard; therefore, high voltage alternators are usually of the stationary armature, rotating field type. The voltage and current supplied to the rotating field are relatively small, and slip rings and brushes for this circuit are adequate. The direct connection to the armature circuit makes possible the use of large cross-section conductors, adequately insulated for high voltage. Since the rotating field alternator is used almost universally in aircraft systems, this type will be explained in detail, as a single phase, two phase, and three phase alternator.

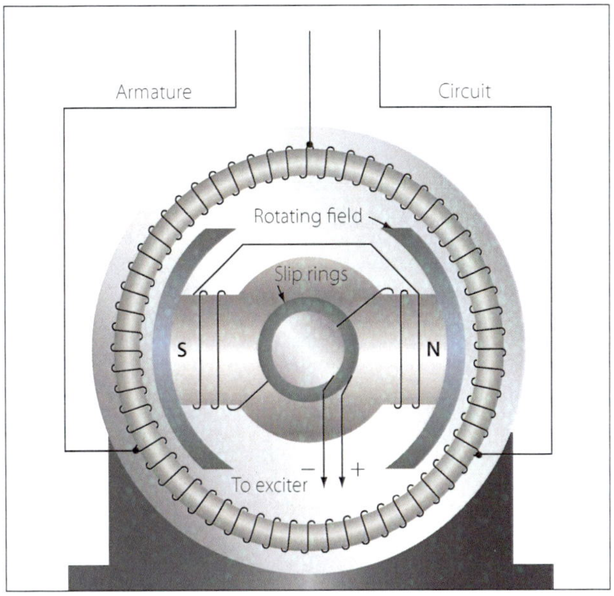

Figure 10-302. Alternator with stationary armature and rotating field.

Single Phase Alternator

Since the emf induced in the armature of a generator is alternating, the same sort of winding can be used on an alternator as on a DC generator. This type of alternator is known as a single phase alternator, but since the power delivered by a single phase circuit is pulsating, this type of circuit is objectionable in many applications.

A single phase alternator has a stator made up of a number of windings in series, forming a single circuit in which an output voltage is generated. Figure 10-303 illustrates a schematic diagram of a single phase alternator having four poles. The stator has four polar groups evenly spaced around the stator frame. The rotor has four poles, with adjacent poles of opposite polarity. As the rotor revolves, AC voltages are induced in the stator windings. Since one rotor pole is in the same position relative to a stator winding as any other rotor pole, all stator polar groups are cut by equal numbers of magnetic lines of force at any time.

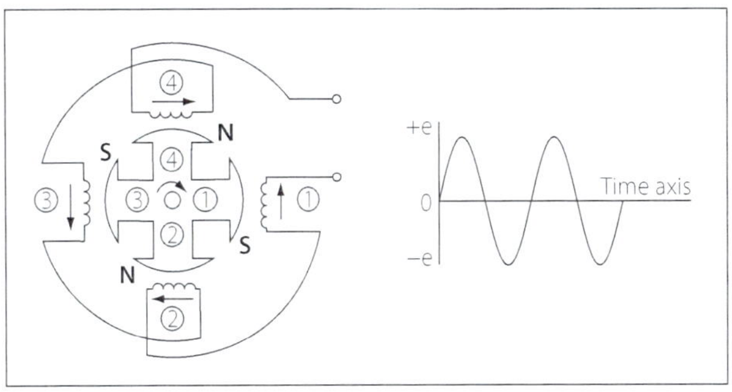

Figure 10-303. Single phase alternator.

As a result, the voltages induced in all the windings have the same amplitude, or value, at any given instant. The four stator windings are connected to each other so that the AC voltages are in phase, or "series adding." Assume that rotor pole 1, a south pole, induces a voltage in the direction indicated by the arrow in stator winding 1. Since rotor pole 2 is a north pole, it will induce a voltage in the opposite direction in stator coil 2 with respect to that in coil 1. For the two induced voltages to be in series addition, the two coils are connected as shown in the diagram. Applying the same reasoning, the voltage induced in stator coil 3 (clockwise rotation of the field) is the same direction (counterclockwise) as the voltage induced in coil 1. Similarly, the direction of the voltage induced in winding 4 is opposite to the direction of the voltage induced in coil 1. All four stator coil groups are connected in series so that the voltages induced in each winding add to give a total voltage that is four times the voltage in any one winding.

Two Phase Alternator

Two phase alternators have two or more single phase windings spaced symmetrically around the stator. In a two phase alternator, there are two single phase windings spaced physically so that the AC voltage induced in one is 90° out of phase with the voltage induced in the other. The windings are electrically separate from each other. When one winding is being cut by maximum flux, the other is being cut by no flux. This condition establishes a 90° relation between the two phases.

Three Phase Alternator

A three phase, or polyphase circuit, is used in most aircraft alternators, instead of a single or two phase alternator. The three phase alternator has three single phase windings spaced so that the voltage induced in each winding is 120° out of phase with the voltages in the other two windings. A schematic diagram of a three phase stator showing all the coils becomes complex and difficult to see what is actually happening.

A simplified schematic diagram, showing each of three phases, is illustrated in Figure 10-304. The rotor is omitted for simplicity. The waveforms of voltage are shown to the right of the schematic. The three voltages are 120° apart and are similar to the voltages which would be generated by three single phase alternators whose voltages are out of phase by angles of 120°. The three phases are independent of each other.

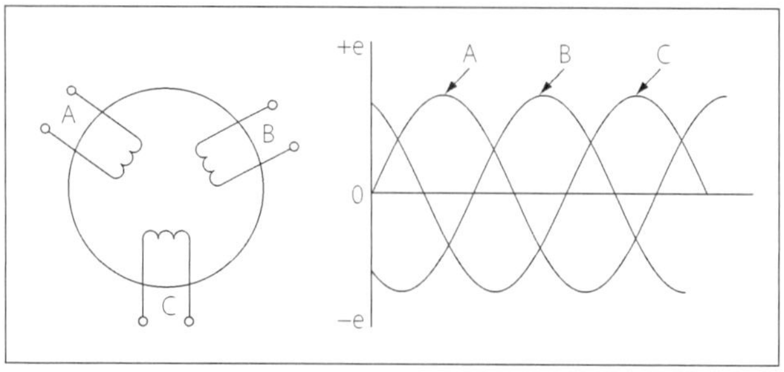

Figure 10-304. Simplified schematic of three phase alternator with output waveforms.

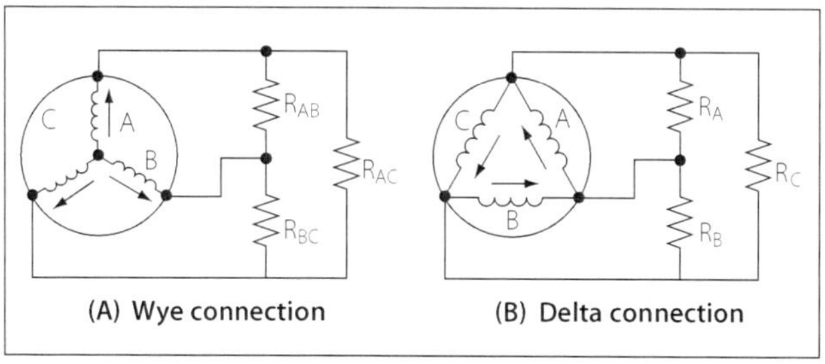

Figure 10-305. Wye- and delta-connected alternators.

Wye Connection (Three Phase)

Rather than have six leads from the three phase alternator, one of the leads from each phase may be connected to form a common junction. The stator is then called wye or star connected. The common lead may or may not be brought out of the alternator. If it is brought out, it is called the neutral lead. The simplified schematic (Figure 10-305A) shows a wye-connected stator with the common lead not brought out. Each load is connected across two phases in series. Thus, RAB is connected across phases A and B in series; RAC is connected across phases A and C in series; and RBC is connected across phases B and C in series. Therefore, the voltage across each load is larger than the voltage across a single phase. The total voltage, or line voltage, across any two phases is the vector sum of the individual phase voltages. For balanced conditions, the line voltage is 1.73 times the phase voltage. Since there is only one path for current in a line wire and the phase to which it is connected, the line current is equal to the phase current.

Delta Connection (Three Phase)

A three phase stator can also be connected so that the phases are connected end to end as shown in Figure 10-305B. This arrangement is called a delta connection. In a delta connection, the voltages are equal to the phase voltages; the line currents are equal to the vector sum of the phase currents; and the line current is equal to 1.73 times the phase current, when the loads are balanced. For equal loads (equal output), the delta connection supplies increased line current at a value of line voltage equal to phase voltage, and the wye connection supplies increased line voltage at a value of line current equal to phase current.

Alternator Rectifier Unit

A type of alternator used in the electrical system of many aircraft weighing less than 12,500 pounds is shown in Figure 10-306. This type of power source is sometimes called a DC generator, since it is used in DC systems. Although its output is a DC voltage, it is an alternator rectifier unit. This type of alternator rectifier is a self-excited unit but does not contain a permanent magnet. The excitation for starting is obtained from the battery; immediately after starting, the unit is self-exciting. Cooling air for the alternator is conducted into the unit by a blast air tube on the air inlet cover.

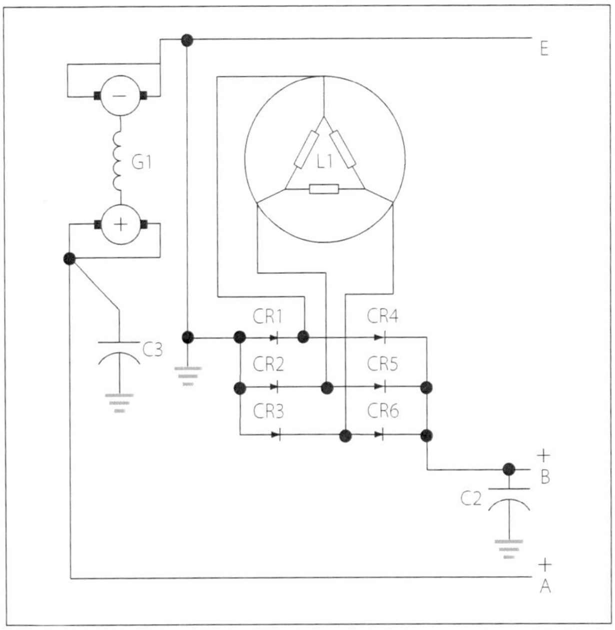

Figure 10-307. Wiring diagram of alternator-rectifier unit.

The alternator is directly coupled to the aircraft engine by means of a flexible drive coupling. The output of the alternator portion of the unit is three phase alternating current, derived from a three phase, delta connected system incorporating a three phases, full-wave bridge rectifier. [Figure 10-307] This unit operates in a speed range from 2,100 to 9,000 rpm, with a DC output voltage of 26–29 volts and 125 amperes.

Brushless Alternator

This design is more efficient because there are no brushes to wear down or to arc at high altitudes. This generator consists of a pilot exciter, an exciter, and the main generator system. The need for brushes is eliminated by using an integral exciter with a rotating armature that has its AC output rectified for the main AC field, which is also of the rotating type. A brushless alternator is illustrated in Figure 10-308.

The pilot exciter is an 8 pole, 8,000 rpm, 533 cps, AC generator. The pilot exciter field is mounted on the main generator rotor shaft and is connected in series with the main generator field. The pilot exciter armature is mounted on the main generator stator. The AC output of the pilot exciter is supplied to the voltage regulator, where it is rectified and controlled, and is

Figure 10-306. Exploded view of alternator rectifier.

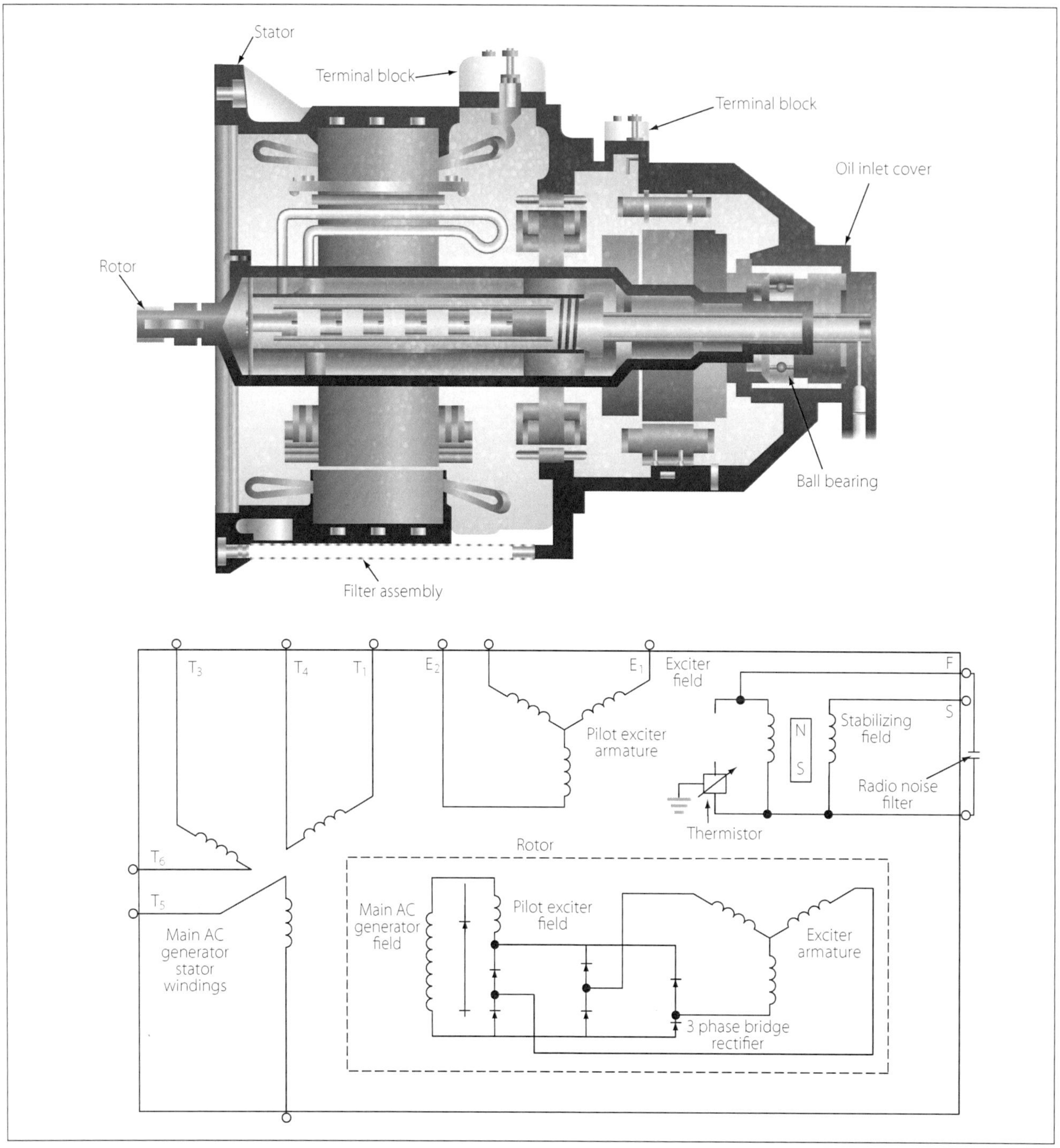

Figure 10-308. A typical brushless alternator.

then impressed on the exciter field winding to furnish excitation for the generator.

The exciter is a small AC generator with its field mounted on the main generator stator and its three phase armature mounted on the generator rotor shaft. Included in the exciter field are permanent magnets mounted on the main generator stator between the exciter poles.

The exciter field resistance is temperature compensated by a thermistor. This aids regulation by keeping a nearly constant resistance at the regulator output terminals. The exciter output is rectified and impressed on the main generator field and the pilot exciter field. The exciter stator has a stabilizing field, which is used to improve stability and to prevent voltage regulator over-corrections for changes in generator output voltage.

The AC generator shown in Figure 10-308 is a 6 pole, 8,000 rpm unit having a rating of 31.5 kilovoltamperes (kVA), 115/200 volts, 400 cps. This generator is three phase, 4 wire, wye connected with grounded neutrals. By using an integral AC exciter, the necessity for brushes within the generator has been eliminated. The AC output of the rotating exciter armature is fed directly into the three phase, full-wave, rectifier bridge located inside the rotor shaft, which uses high temperature silicon rectifiers. The DC output from the rectifier bridge is fed to the main AC generator rotating field.

Voltage regulation is accomplished by varying the strength of the AC exciter stationary fields. Polarity reversals of the AC generator are eliminated and radio noise is minimized by the absence of the brushes. A noise filter mounted on the alternator further reduces any existing radio noise. The rotating pole structure of the generator is laminated from steel punchings, containing all six poles and a connecting hub section. This provides optimum magnetic and mechanical properties.

Some alternators are cooled by circulating oil through steel tubes. The oil used for cooling is supplied from the constant speed drive assembly. Ports located in the flange connecting the generator and drive assemblies make oil flow between the constant speed drive and the generator possible.

Voltage is built up by using permanent magnet interpoles in the exciter stator. The permanent magnets assure a voltage buildup, precluding the necessity of field flashing. The rotor of the alternator may be removed without causing loss of the alternator's residual magnetism.

Alternator Rating

The maximum current that can be supplied by an alternator depends upon the maximum heating loss (I^2R power loss) that can be sustained in the armature and the maximum heating loss that can be sustained in the field. The armature current of an alternator varies with the load. This action is similar to that of DC generators. In AC generators, however, lagging power factor loads tend to demagnetize the field of an alternator, and terminal voltage is maintained only by increasing DC field current. For this reason, alternating current generators are usually rated according to kVA, power factor, phases, voltage, and frequency. One generator, for example, may be rated at 40 kVA, 208 volts, 400 cycles, three phase, at 75 percent power factor. The kVA indicates the apparent power. This is the kVA output, or the relationship between the current and voltage at which the generator is intended to operate. The power factor is the expression of the ratio between the apparent power (volt-amperes) and the true or effective power (watts). The number of phases is the number of independent voltages generated. Three phase generators generate three voltages 120 electrical degrees apart.

Alternator Frequency

The frequency of the alternator voltage depends upon the speed of rotation of the rotor and the number of poles. The faster the speed, the higher the frequency will be; the lower the speed, the lower the frequency becomes. The more poles on the rotor, the higher the frequency will be for a given speed. When a rotor has rotated through an angle so that two adjacent rotor poles (a north and a south pole) have passed one winding, the voltage induced in that winding will have varied through one complete cycle. For a given frequency, the greater the number of pairs of poles, the lower the speed of rotation will be. A two-pole alternator rotates at twice the speed of a four-pole alternator for the same frequency of generated voltage. The frequency of the alternator in cycles per minute is related to the number of poles and the speed, as expressed by the equation

$$F = \frac{P}{2} \times \frac{N}{60} = \frac{PN}{120}$$

where P is the number of poles and N the speed in rpm. For example, a two pole, 3,600 rpm alternator has a frequency of

$$\frac{2 \times 3{,}600}{120} = 60 \text{ cps}$$

A four pole, 1,800 rpm alternator has the same frequency; a six pole, 500 rpm alternator has a frequency of

$$\frac{6 \times 500}{120} = 25 \text{ cps}$$

A 12 pole, 4,000 rpm alternator has a frequency of

$$\frac{2 \times 4{,}000}{120} = 400 \text{ cps}$$

Alternator Maintenance

Maintenance and inspection of alternator systems is similar to that of DC systems. Check the exciter brushes for wear and surfacing. On most large aircraft with two or four alternator systems, each power panel has three signal lights, one connected to each phase of the power bus, so the lamp will light when the panel power is on. The individual buses throughout the airplane can be checked by operating equipment from that particular bus. Consult the manufacturer's instructions on operation of equipment for the method of testing each bus.

Alternator test stands are used for testing alternators and constant speed drives in a repair facility. They are capable of supplying power to constant speed drive units at input speeds varying from 2,400 rpm to 9,000 rpm.

A typical test stand motor uses 220/440 volt, 60 cycle, three phase power. Blowers for ventilation, oil coolers, and necessary meters and switches are integral parts of the test stand. A load bank supplies test circuits. An AC motor generator set for ground testing is shown in Figure 10-309.

A typical, portable, AC electrical system test set is an analyzer, consisting of a multirange ohmmeter, a multirange combination AC DC voltmeter, an ammeter with a clip-on current transformer, a vibrating reed type frequency meter, and an unmounted continuity light.

A portable load bank unit furnishes a load similar to that on the airplane for testing alternators, either while mounted in the airplane or on the shop test stand. A complete unit consists of resistive and reactive loads controlled by selector switches and test meters mounted on a control panel. This load unit is compact and convenient, eliminating the difficulty of operating large loads on the airplane while testing and adjusting the alternators and control equipment.

Proper maintenance of an alternator requires that the unit be kept clean and that all electrical connections are tight and in good repair. If the alternator fails to build up voltage as designated by applicable manufacturer's technical instructions, test the voltmeter first by checking the voltages of other alternators, or by checking the voltage in the suspected alternator with another voltmeter and comparing the results. If the voltmeter is satisfactory, check the wiring, the brushes, and the drive unit for faults. If this inspection fails to reveal the trouble, the exciter may have lost its residual magnetism. Residual magnetism is restored to the exciter by flashing the field. Follow the applicable manufacturer's instructions when flashing the exciter field. If, after flashing the field, no voltage is indicated, replace the alternator, since it is probably faulty.

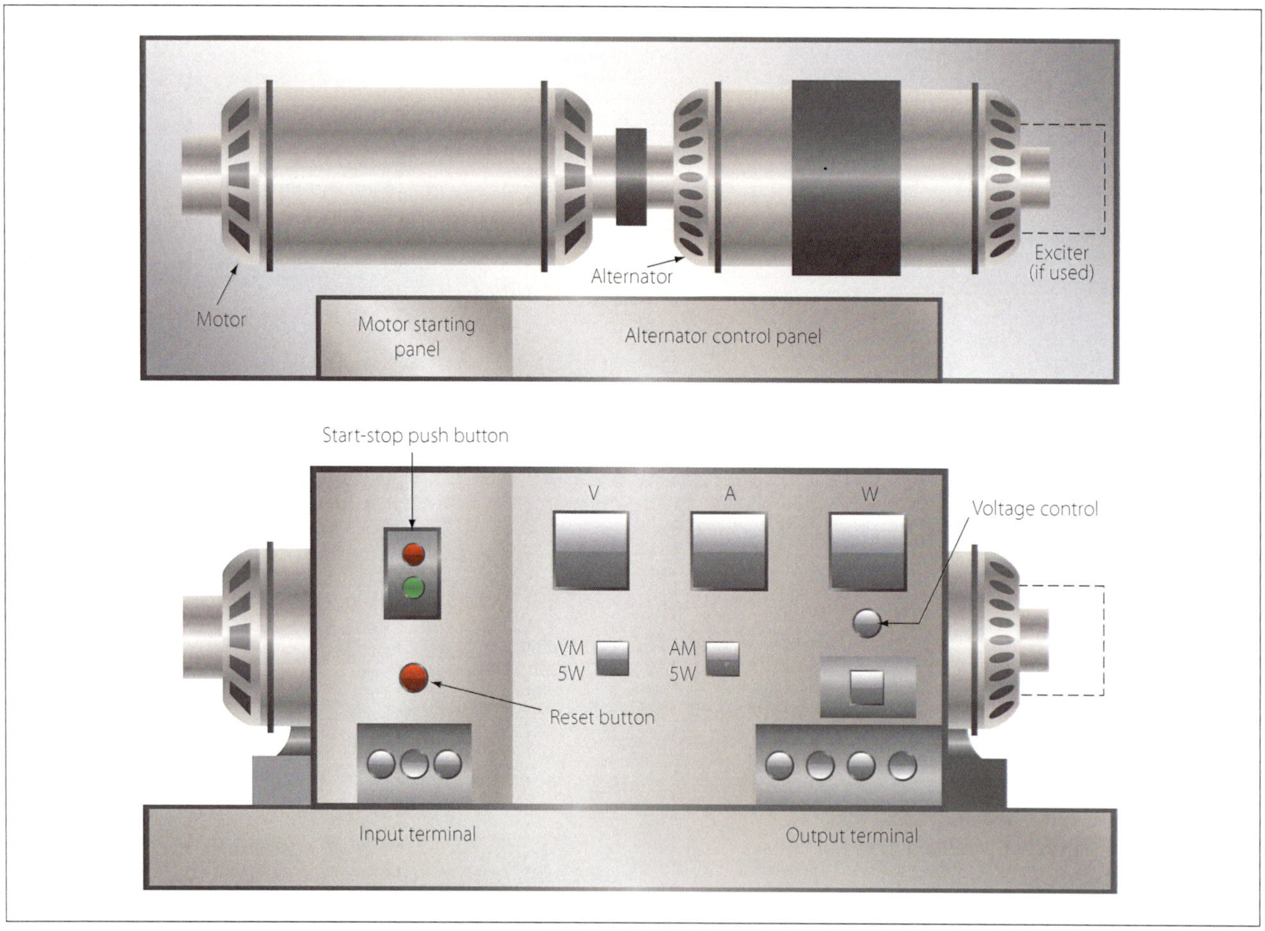

Figure 10-309. AC motor generator set for ground testing.

Clean the alternator exterior with an approved fluid; smooth a rough or pitted exciter commutator or slip ring with 000 sandpaper; then clean and polish with a clean, dry cloth. Check the brushes periodically for length and general condition. Consult the applicable manufacturer's instructions on the specific alternator to obtain information on the correct brushes.

Regulation of Generator Voltage

Efficient operation of electrical equipment in an airplane depends on a constant voltage supply from the generator. Among the factors, which determine the voltage output of a generator, only one, the strength of the field current, can be conveniently controlled. To illustrate this control, refer to the diagram in Figure 10-310, showing a simple generator with a rheostat in the field circuit. If the rheostat is set to increase the resistance in the field circuit, less current flows through the field winding and the strength of the magnetic field in which the armature rotates decreases. Consequently, the voltage output of the generator decreases. If the resistance in the field circuit is decreased with the rheostat, more current flows through the field windings, the magnetic field becomes stronger, and the generator produces a greater voltage.

Voltage Regulation with a Vibrating-Type Regulator

Refer to Figure 10-311. With the generator running at normal speed and switch K open, the field rheostat is adjusted so that the terminal voltage is about 60 percent of normal. Solenoid S is weak and contact B is held closed by the spring. When K is closed, a short circuit is placed across the field rheostat. This action causes the field current to increase and the terminal voltage to rise.

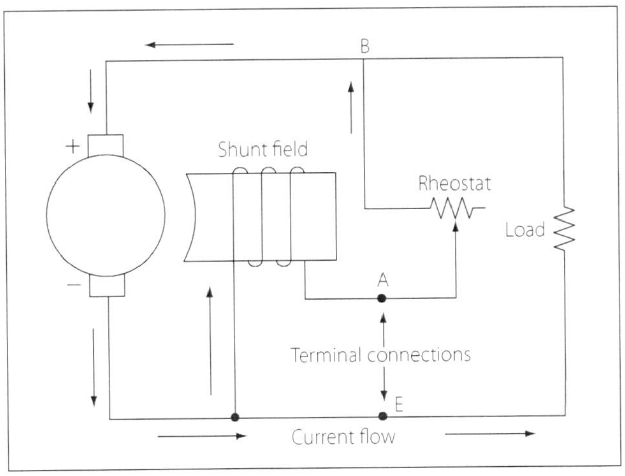

Figure 10-310. Regulation of generator voltage by field rheostat.

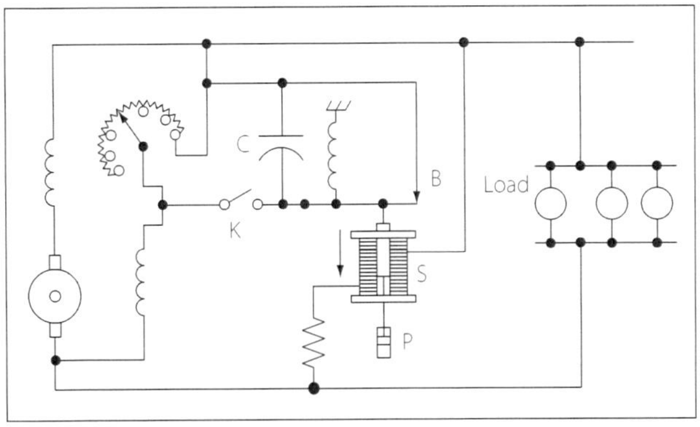

Figure 10-311. Vibrating-type voltage regulator.

When the terminal voltage rises above a certain critical value, the solenoid downward pull exceeds the spring tension and contact B opens, thus reinserting the field rheostat in the field circuit and reducing the field current and terminal voltage.

When the terminal voltage falls below a certain critical voltage, the solenoid armature contact B is closed again by the spring, the field rheostat is now shorted, and the terminal voltage starts to rise. The cycle repeats with a rapid, continuous action. Thus, an average voltage is maintained with or without load change.

The dashpot P provides smoother operation by acting as a damper to prevent hunting. The capacitor C across contact B eliminates sparking. Added load causes the field rheostat to be shorted for a longer period of time and, thus, the solenoid armature vibrates more slowly. If the load is reduced and the terminal voltage rises, the armature vibrates more rapidly and the regulator holds the terminal voltage to a steady value for any change in load, from no load to full load, on the generator.

Vibrating-type regulators cannot be used with generators, which require a high field current, since the contacts will pit, or burn. Heavy-duty generator systems require a different type of regulator, such as the carbon pile voltage regulator.

Three Unit Regulators

Many light aircraft employ a three unit regulator for their generator systems. This type of regulator includes a current limiter and a reverse current cutout in addition to a voltage regulator.

The action of the voltage regulator unit is similar to the vibrating-type regulator described earlier. The second of the three units is a current regulator to limit the output current of the generator. The third unit is a reverse current cutout that disconnects the battery from the generator. If the battery is not disconnected, it will discharge through the generator armature when the generator voltage falls below that of the battery, thus driving the generator as a motor. This action is called "motoring" the generator and, unless it is prevented, it will discharge the battery in a short time.

The operation of a three unit regulator is described in the following paragraphs. [Figure 10-312]

The action of vibrating contact C1 in the voltage regulator unit causes an intermittent short circuit between points R1 and L2. When the generator is not operating, spring S1 holds C1 closed; C2 is also closed by S2. The shunt field is connected directly across the armature.

When the generator is started, its terminal voltage will rise as the generator comes up to speed, and the armature will supply the field with current through closed contacts C2 and C1.

As the terminal voltage rises, the current flow through L1 increases and the iron core becomes more strongly magnetized. At a certain speed and voltage, when the magnetic attraction on the movable arm becomes strong enough to overcome the tension of spring S1, contact points C1 are separated. The field current now flows through R1 and L2. Because resistance is added to the field circuit, the field is momentarily weakened and the rise in terminal voltage is checked. Also, since the L2 winding is opposed to the L1 winding, the magnetic pull of L1 against S1 is partially neutralized, and spring S1 closes contact C1. Therefore, R1 and L2 are again shorted out of the circuit, and the field current again increases; the output voltage increases, and C1 is opened because of the action of L1. The cycle is rapid and occurs many times per second. The terminal voltage of the generator varies slightly, but rapidly, above

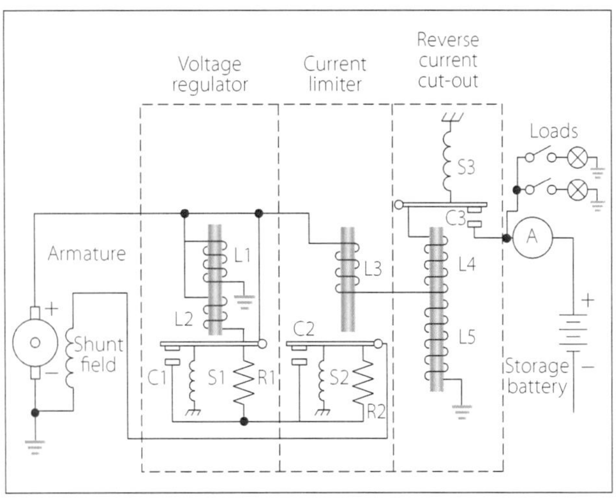

Figure 10-312. Three unit regulator for variable speed generators.

and below an average value determined by the tension of spring S1, which may be adjusted.

The purpose of the vibrator-type current limiter is to limit the output current of the generator automatically to its maximum rated value in order to protect the generator. As shown in Figure 10-312, L3 is in series with the main line and load. Thus, the amount of current flowing in the line determines when C2 will be opened and R2 placed in series with the generator field. By contrast, the voltage regulator is actuated by line voltage, whereas the current limiter is actuated by line current. Spring S2 holds contact C2 closed until the current through the main line and L3 exceeds a certain value, as determined by the tension of spring S2, and causes C2 to be opened. The increase in current is due to an increase in load. This action inserts R2 into the field circuit of the generator and decreases the field current and the generated voltage. When the generated voltage is decreased, the generator current is reduced. The core of L3 is partly demagnetized and the spring closes the contact points. This causes the generator voltage and current to rise until the current reaches a value sufficient to start the cycle again. A certain minimum value of load current is necessary to cause the current limiter to vibrate.

The purpose of the reverse current cutout relay is to automatically disconnect the battery from the generator when the generator voltage is less than the battery voltage. If this device were not used in the generator circuit, the battery would discharge through the generator. This would tend to make the generator operate as a motor, but because the generator is coupled to the engine, it could not rotate such a heavy load. Under this condition, the generator windings may be severely damaged by excessive current.

There are two windings, L4 and L5, on the soft iron core. The current winding, L4, consisting of a few turns of heavy wire, is in series with the line and carries the entire line current. The voltage winding, L5, consisting of a large number of turns of fine wire, is shunted across the generator terminals.

When the generator is not operating, the contacts, C3 are held open by the spring S3. As the generator voltage builds up, L5 magnetizes the iron core. When the current (as a result of the generated voltage) produces sufficient magnetism in the iron core, contact C3 is closed, as shown. The battery then receives a charging current. The coil spring, S3, is so adjusted that the voltage winding will not close the contact points until the voltage of the generator is in excess of the normal voltage of the battery. The charging current passing through L4 aids the current in L5 to hold the contacts tightly closed. Unlike C1 and C2, contact C3 does not vibrate. When the generator slows down or, for any other cause, the generator voltage decreases to a certain value below that of the battery, the current reverses through L4 and the ampere turns of L4 oppose those of L5. Thus, a momentary discharge current from the battery reduces the magnetism of the core and C3 is opened, preventing the battery from discharging into the generator and motoring it. C3 will not close again until the generator terminal voltage exceeds that of the battery by a predetermined value.

Differential Relay Switch

Aircraft electrical systems normally use some type of reverse current relay switch, which acts not only as a reverse current relay cutout but also serves as a remote control switch by which the generator can be disconnected from the electrical system at any time. One type of reverse current relay switch operates on the voltage level of the generator, but the type most commonly used on large aircraft is the differential relay switch, which is controlled by the difference in voltage between the battery bus and the generator.

The differential type relay switch connects the generator to the main bus bar in the electrical system when the generator voltage output exceeds the bus voltage by 0.35 to 0.65 volt. It disconnects the generator when a nominal reverse current flows from the bus to the generator. The differential relays on all the generators of a multiengine aircraft do not close when the electrical load is light. For example, in an aircraft having a load of 50 amperes, only two or three relays may close. If a heavy load is applied, the equalizing circuit will lower the voltage of the generators already on the bus and, at the same time, raise the voltage of the remaining generators, allowing their relays to close. If the generators have been paralleled properly, all the relays stay closed until the generator control switch is turned off or until the engine speed falls below the minimum needed to maintain generator output voltage.

The differential generator control relay shown in Figure 10-313 is made up of two relays and a coil-operated contactor. One relay is the voltage relay and the other is the differential relay. Both relays include permanent magnets, which pivot between the pole pieces of temporary magnets wound with relay coils. Voltages of one polarity set up fields about the temporary magnets with polarities that cause the permanent magnet to move in the direction necessary to close the relay contacts; voltages of the opposite polarity establish fields that cause the relay contacts to open. The differential relay has two coils wound on the same core. The coil-operated contactor, called the main contactor, consists of movable contacts that are operated by a coil with a movable iron core.

Closing the generator switch on the control panel connects the generator output to the voltage relay coil. When generator voltage reaches 22 volts, current flows through the coil and closes the contacts of the voltage relay. This action completes a circuit from the generator to the battery through the differential coil.

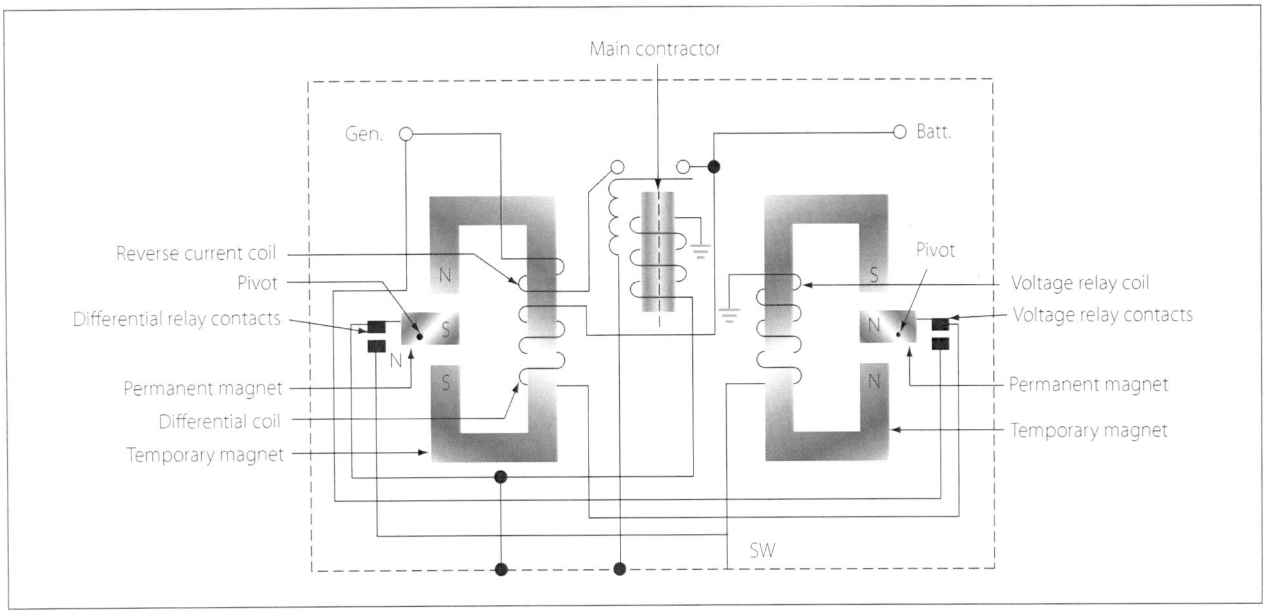

Figure 10-313. Differential generator control relay.

When the generator voltage exceeds the bus voltage by 0.35 volt, current will flow through the differential coil, the differential relay contact will close and, thus, complete the main contractor coil circuit. The contacts of the main contactor close and connect the generator to the bus.

When the generator voltage drops below the bus (or battery) voltage, a reverse current weakens the magnetic field about the temporary magnet of the differential relay. The weakened field permits a spring to open the differential relay contacts, breaking the circuit to the coil of the main contactor relay, opening its contacts, and disconnecting the generator from the bus. The generator battery circuit may also be broken by opening the cockpit control switch, which opens the contacts of the voltage relay, causing the differential relay coil to be de-energized.

Overvoltage and Field Control Relays

Two other items used with generator control circuits are the overvoltage control and the field control relay.

As its name implies, the overvoltage control protects the system when excessive voltage exists. The overvoltage relay is closed when the generator output reaches 32 volts and completes a circuit to the trip coil of the field control relay. The closing of the field control relay trip circuit opens the shunt field circuit and completes it through a resistor, causing generator voltage to drop; also, the generator switch circuit and the equalizer circuit (multiengine aircraft) are opened. An indicator light circuit is completed, warning that an overvoltage condition exists. A "reset" position of the cockpit switch is used to complete a reset coil circuit in the field control relay, returning the relay to its normal position.

Generator Control Units (GCU)

Basic Functions of a Generator Control Unit

The generator control unit (GCU) is more commonly found on turbine power aircraft. The most basic generator control units perform a number of functions related to the regulation, sensing, and protection of the DC generation system.

Voltage Regulation

The most basic of the GCU functions is that of voltage regulation. Regulation of any kind requires the regulation unit to take a sample of an output and to compare that sample with a controlled reference. If the sample taken falls outside of the limits set by the reference, then the regulation unit must provide an adjustment to the unit generating the output so as to diminish or increase the output levels. In the case of the GCU, the output voltage from a generator is sensed by the GCU and compared to a reference voltage. If there is any difference between the two, the error is usually amplified and then sent back to the field excitation control portion of the circuit. The field excitation control then makes voltage/excitation adjustments in the field winding of the generator in order to bring the output voltage back into required bus tolerances.

Overvoltage Protection

Like the voltage regulation feature of the GCU, the overvoltage protection system compares the sampled voltage to reference voltage. The output of the overvoltage protection circuit is used to open the relay that controls the output for the field excitation. These types of faults can occur for a number of reasons. The most common, however, is the failure of the voltage regulation circuit in the GCU.

Parallel Generator Operations

The paralleling feature of the GCU allows for two or more GCU/generator systems to work in a shared effort to provide current to the aircraft electrical system. Comparing voltages between the equalizer bus and the interpole/compensator voltage, and amplifying the differences accomplishes the control of this system. The difference is then sent to the voltage regulation circuit, where adjustments are then made in the regulation output. These adjustments will continue until all of the busses are equalized in their load sharing.

Over-Excitation Protection

When a GCU in a paralleled system fails, a situation can occur where one of the generators becomes overexcited and tries to carry more than its share of the load, if not all of the loads. When this condition is sensed on the equalizing bus, the faulted generation control system will shut down by receiving a de-excitation signal. This signal is then transmitted to the overvoltage circuit, and then opens the field excitation output circuit.

Differential Voltage

When the GCU allows the logic output to close the generator line contactor, the generator voltage must be within a close tolerance of the load bus. If the output is not within the specified tolerance, then the contactor is not allowed to connect the generator to the bus.

Reverse Current Sensing

If the generator is unable to maintain the required voltage level, it will eventually begin to draw current instead of providing it. In this case, the faulty generator will be seen as a load to the other generators and will need to be removed from the bus. Once the generator is off-line, it will not be permitted to be reconnected to the bus until such time that the generator faults are cleared and the generator is capable of providing a current to the bus. In most cases, the differential voltage circuit and the reverse current sensing circuit are one in the same.

Alternator Constant Speed Drive System

Alternators are not always connected directly to the airplane engine like DC generators. Since the various electrical devices operating on AC supplied by alternators are designed to operate at a certain voltage and at a specified frequency, the speed of the alternators must be constant; however, the speed of an airplane engine varies. Therefore, the engine, through a constant speed drive installed between the engine and the alternator, drives some alternators.

A typical hydraulic-type drive is shown in Figure 10-314. The following discussion of a constant speed drive system will be based on such a drive, found on large multiengine aircraft.

The constant speed drive is a hydraulic transmission, which may be controlled either electrically or mechanically.

The constant speed drive assembly is designed to deliver an output of 6,000 rpm, provided the input remains between 2,800 and 9,000 rpm. If the input, which is determined by engine speed, is below 6,000 rpm, the drive increases the speed in order to furnish the desired output. This stepping up of speed is known as overdrive.

In overdrive, an automobile engine will operate at about the same rpm at 60 mph as it does in conventional drive at 49 mph. In aircraft, this principle is applied in the same manner. The constant speed drive enables the alternator to produce the same frequency at slightly above engine idle rpm as it would at takeoff or cruising rpm.

With the input speed to the drive set at 6,000 rpm, the output speed will be the same. This is known as straight drive and might be compared to an automobile in high gear. However, when the input speed is greater than 6,000 rpm, it must be reduced to provide an output of 6,000 rpm. This is called underdrive, which is comparable to an automobile in low gear. Thus, the large

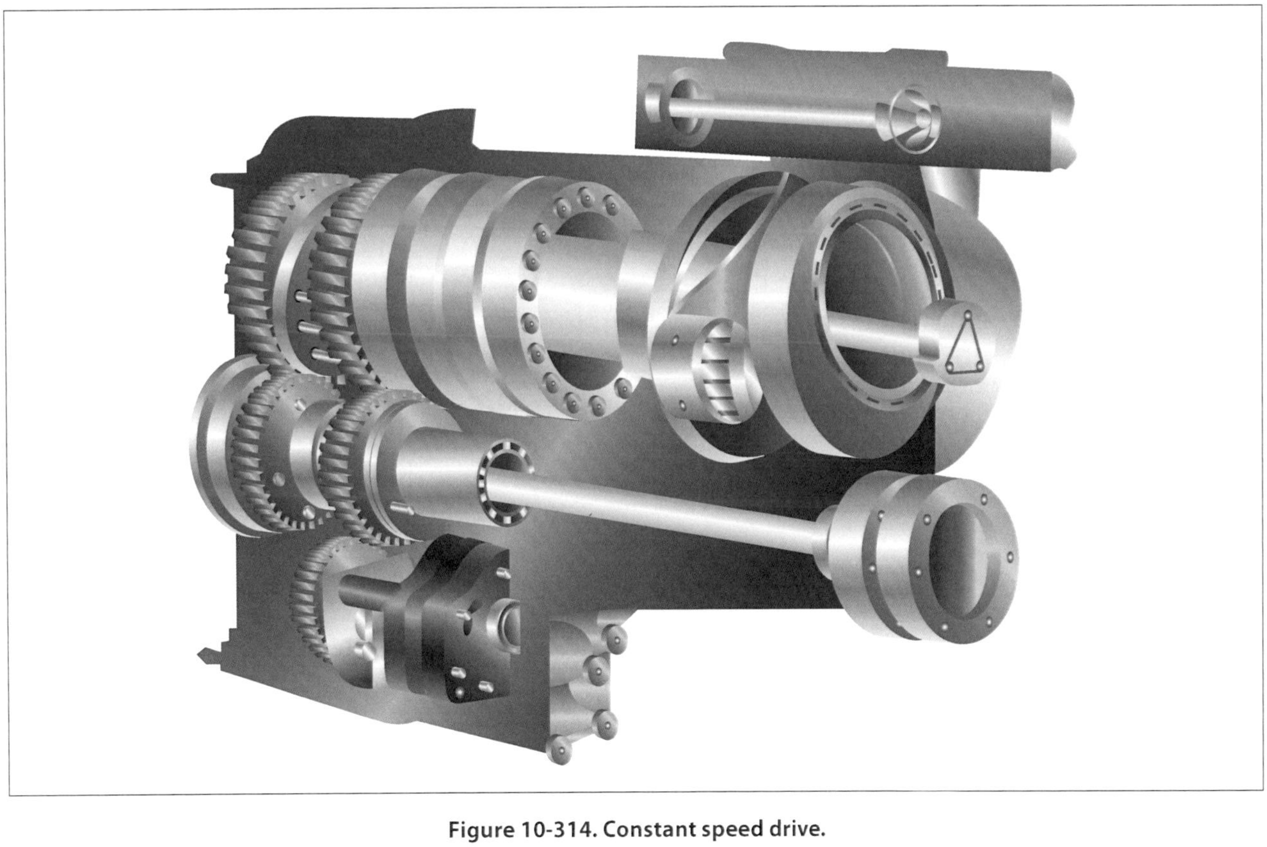

Figure 10-314. Constant speed drive.

input, caused by high engine rpm, is reduced to give the desired alternator speed.

As a result of this control by the constant speed drive, the frequency output of the generator varies from 420 cps at no load to 400 cps under full load.

This, in brief, is the function of the constant speed drive assembly. Before discussing the various units and circuits, the overall operation of the transmission should be discussed as follows.

Hydraulic Transmission

The transmission is mounted between the generator and the aircraft engine. Its name denotes that hydraulic oil is used, although some transmissions may use engine oil. Refer to the cutaway view of such a transmission in Figure 10-315. The input shaft D is driven from the drive shaft on the accessory section of the engine. The output drive F, on the opposite end of the transmission, engages the drive shaft of the generator.

The input shaft is geared to the rotating cylinder block gear, which it drives, as well as to the makeup and scavenger gear pumps E.

The makeup (charge) pump delivers oil (300 psi) to the pump and motor cylinder block, to the governor system, and to the pressurized case, whereas the scavenger pump returns the oil to the external reservoir.

The rotating cylinder assembly B consists of the pump and motor cylinder blocks, which are bolted to opposite

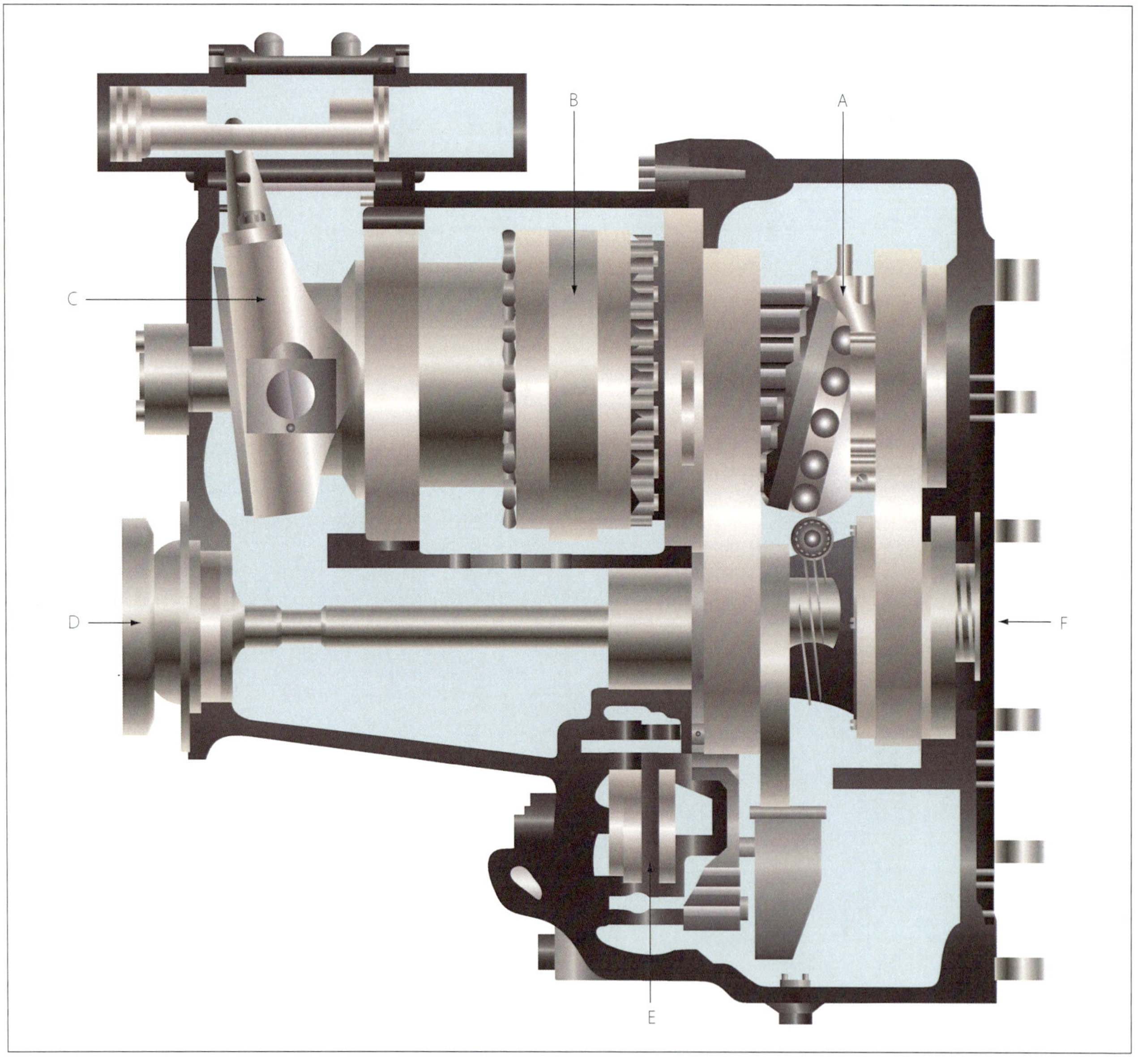

Figure 10-315. Cutaway of a hydraulic transmission.

sides of a port plate. The two other major parts are the motor wobbler A and the pump wobbler C. The governor system is the unit at the top of the left side in the illustration.

The cylinder assembly has two primary units. The block assembly of one of the units, the pump, contains 14 cylinders, each of which has a piston and pushrod. Charge pressure from the makeup pump is applied to each piston in order to force it outward against the pushrod. It, in turn, is pushed against the pump wobble plate.

If the plate remained as shown in Figure 10-316A, each of the 14 cylinders would have equal pressure, and all pistons would be in the same relative position in their respective cylinders. But with the plate tilted, the top portion moves outward and the lower portion inward, as shown in Figure 10-316B. As a result, more oil enters the interior of the upper cylinder, but oil will be forced from the cylinder of the bottom piston.

If the pump block were rotated while the plate remained stationary, the top piston would be forced inward because of the angle of the plate. This action would cause the oil confined within the cylinder to be subjected to increased pressure great enough to force it into the motor cylinder block assembly.

Before explaining what the high-pressure oil in the motor unit will do, it is necessary to know something about this part of the rotating cylinder block assembly. The motor block assembly has 16 cylinders, each with its piston and pushrod. These are constantly receiving charge pressure of 300 psi. The position of the piston depends upon the point at which each pushrod touches the motor wobble plate. These rods cause the wobble plate to rotate by the pressure they exert against its sloping surface.

The piston and pushrod of the motor are pushed outward as oil is forced through the motor valve plate from the pump cylinder. The pushrods are forced against the motor wobble plate, which is free to rotate but cannot change the angle at which it is set. Since the pushrods cannot move sideways, the pressure exerted against the motor wobble plate's sloping face causes it to rotate.

In the actual transmission, there is an adjustable wobble plate. The control cylinder assembly determines the tilt of the pump wobble plate. For example, it is set at an angle which causes the motor cylinders to turn the motor wobble plate faster than the motor assembly, if the transmission is in overdrive. The greater pressure in the pump and motor cylinders produces the result described.

With the transmission in underdrive, the angle is arranged so there is a reduction in pumping action. The subsequent slippage between the pushrods and motor wobble plate reduces the output speed of the transmission. When the pump wobble plate is not at an angle, the pumping action will be at a minimum and the transmission will have what is known as hydraulic lock. For this condition, the input and output speed will

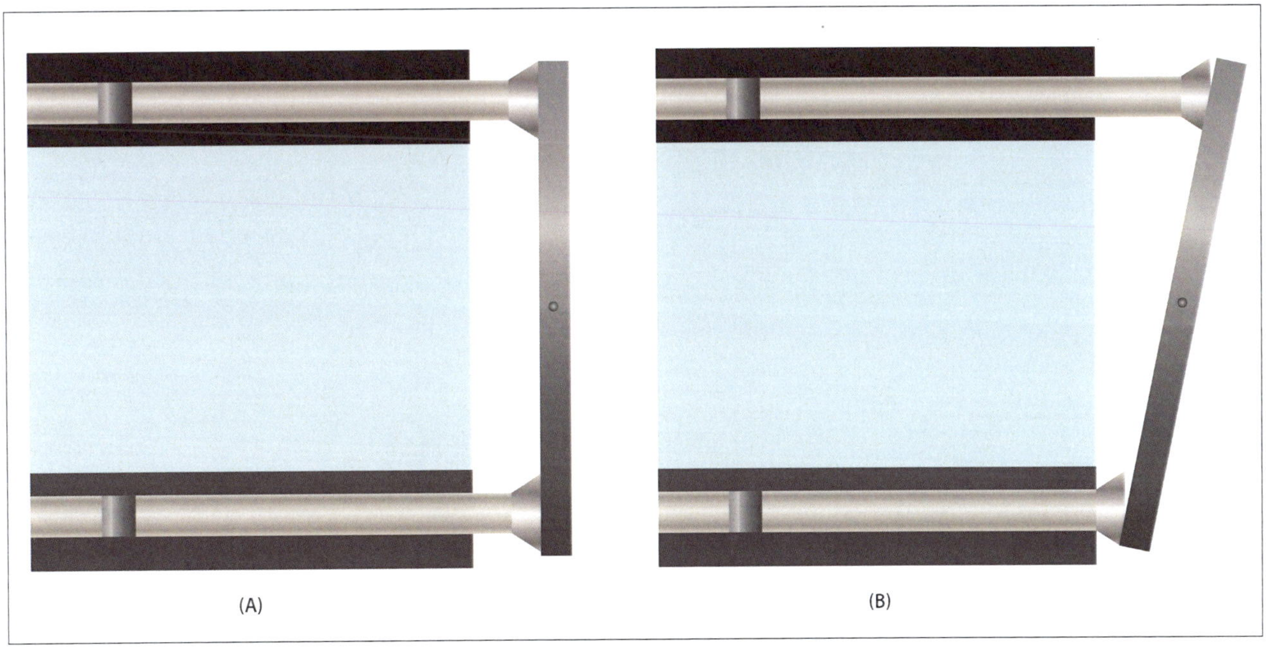

Figure 10-316. Wobble plate position.

be about the same, and the transmission is considered to be in straight drive.

To prevent the oil temperature from becoming excessively high within the cylinder block, the makeup pressure pump forces oil through the center of this block and the pressure relief valve. From this valve, the oil flows into the bottom of the transmission case. A scavenger pump removes the oil from the transmission case and circulates it through the oil cooler and filters before returning it to the reservoir. At the start of the cycle, oil is drawn from the reservoir, passed through a filter, and forced into the cylinder block by the makeup pressure pump.

The clutch, located in the output gear and clutch assembly, is an overrunning one way, sprag-type device. Its purpose is to ratchet if the alternator becomes motorized; otherwise, the alternator might turn the engine. Furthermore, the clutch provides a positive connection when the transmission is driving the alternator.

There is another unit of the drive that must be discussed—the governor system. The governor system, which consists of a hydraulic cylinder with a piston, is electrically controlled. Its duty is to regulate oil pressure flowing to the control cylinder assembly. [Figure 10-317]

The center of the system's hydraulic cylinder is slotted so the arm of the pump wobble plate can be connected to the piston. As oil pressure moves the piston, the pump wobble plate is placed in either overspeed, underspeed, or straight drive.

Figure 10-318 shows the electrical circuit used to govern the speed of the transmission. First, the main points of the complete electrical control circuit will be discussed. [Figures 10-318 and 10-319] Then, for simplification, two portions, the overspeed circuit and the load division circuit, will be considered as individual circuits.

Note, then, in Figure 10-318, that the circuit has a valve and solenoid assembly O and a control cylinder E, and that it contains such units as the tachometer generator D, the rectifier C, and adjustable resistor B, rheostat A, and the control coil Q.

Since it is driven by a drive gear in the transmission, the tachometer (often called tach) generator, a three phase unit, has a voltage proportional to the speed of the output drive. The rectifier changes its voltage from AC to DC. After rectification, the current flows through the resistor, rheostat, and valve and solenoid. Each of these units is connected in series. [Figure 10-319]

Under normal operating conditions, the output of the tach generator causes just enough current to enter the valve and solenoid coil to set up a magnetic field of sufficient strength to balance the spring force in the valve. When the alternator speed increases as the result of a decrease in load, the tach generator output also increases. Because of the greater output, the coil in the solenoid is sufficiently strengthened to overcome the spring force. Thus, the valve moves and, as a result, oil pressure enters the reduced speed side of the control cylinder.

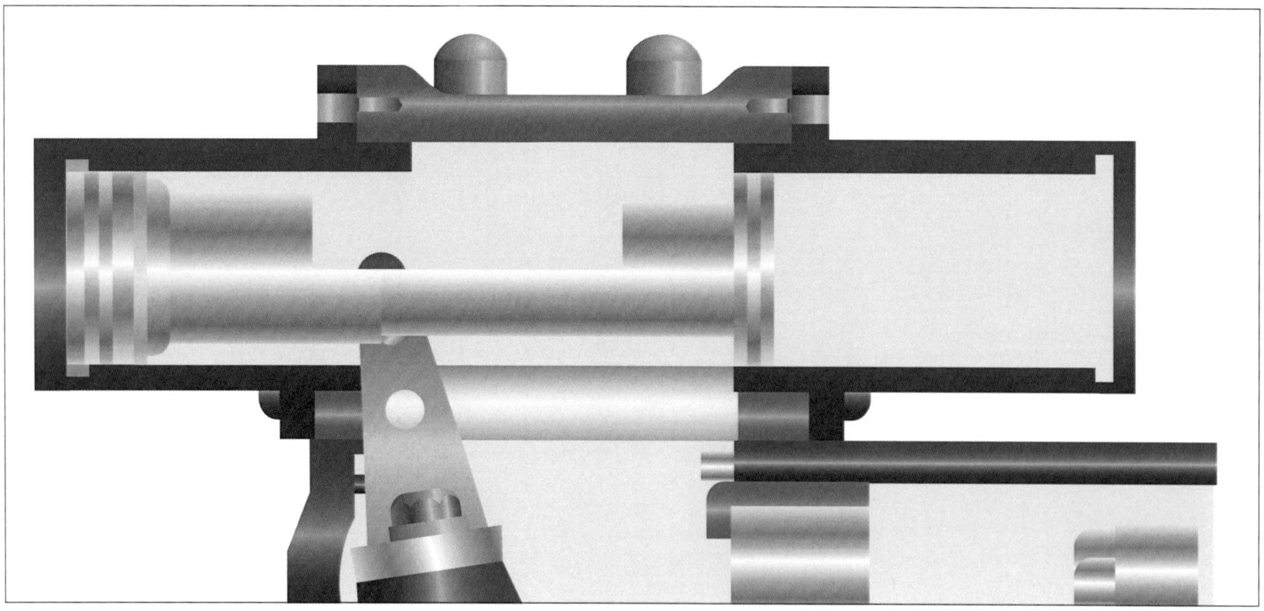

Figure 10-317. Control cylinder.

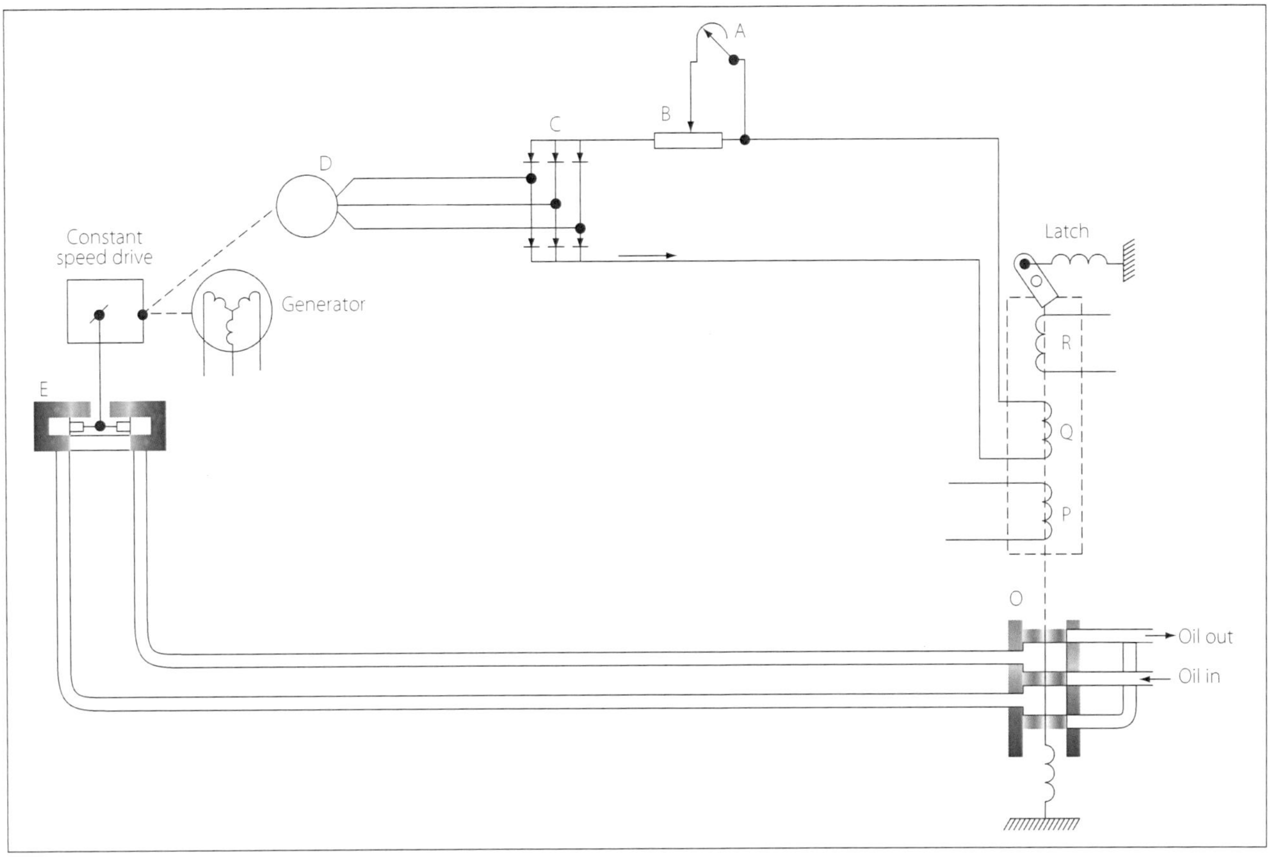

Figure 10-318. Electrical hydraulic control circuit.

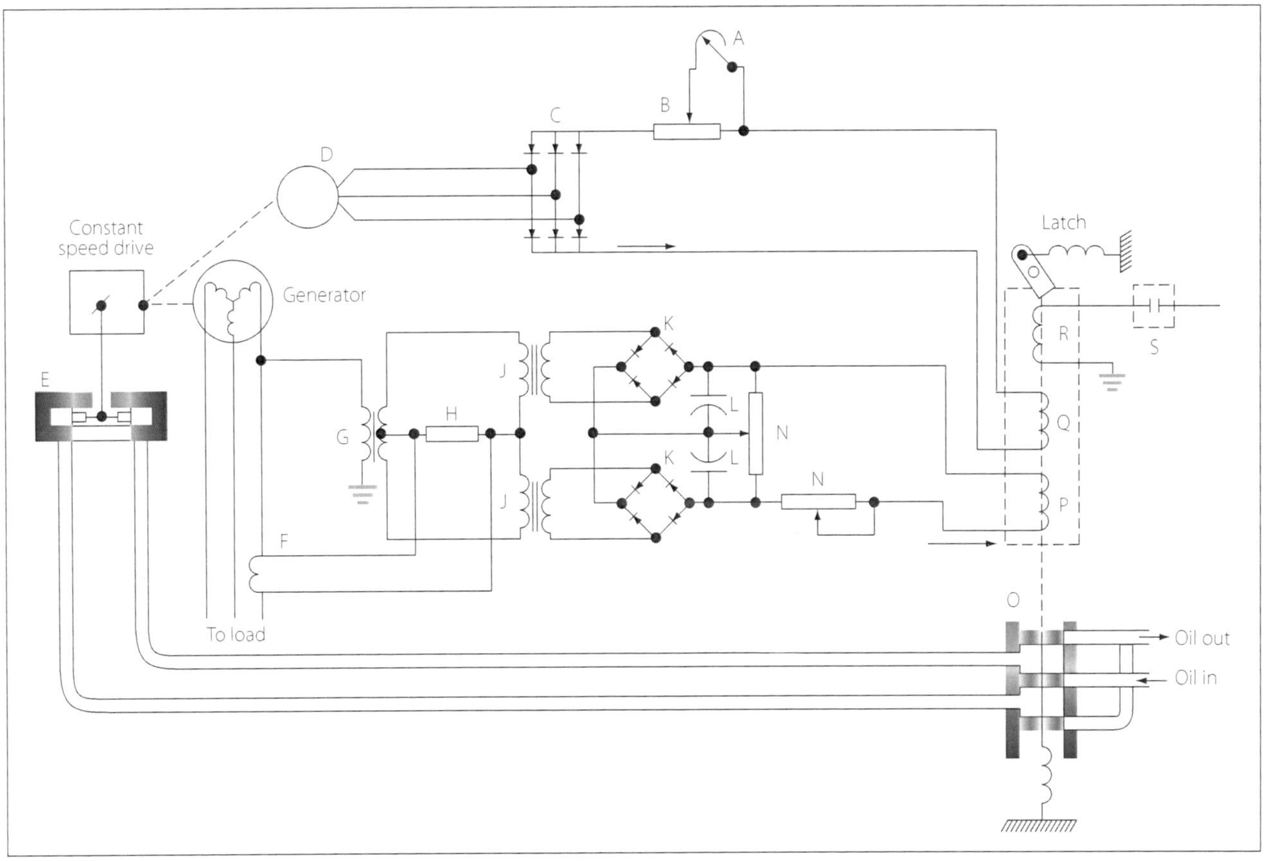

Figure 10-319. Speed control circuit.

In turn, the pressure moves the piston, causing the angle of the pump wobble plate to be reduced. The oil on the other side of the piston is forced back through the valve into the system return. Since the angle of the pump wobble plate is smaller, there is less pumping action in the transmission. The result is decreased output speed. To complete the cycle, the procedure is reversed.

With the output speed reduction, tach generator output decreases; consequently, the flow of current to the solenoid diminishes. Therefore, the magnetic field of the solenoid becomes so weak that the spring is able to overcome it and reposition the valve.

If a heavy load is put on the AC generator, its speed decreases. The generator is not driven directly by the engine; the hydraulic drive will allow slippage. This decrease will cause the output of the tach generator to taper off and, as a result, weaken the magnetic field of the solenoid coil. The spring in the solenoid will move the valve and allow oil pressure to enter the increase side of the control cylinder and the output speed of the transmission will be raised.

There are still two important circuits, which must be discussed: the overspeed circuit and the load division circuit. The generator is prevented from overspeeding by a centrifugal switch (S in Figure 10-320) and the overspeed solenoid coil R, which is located in the solenoid and valve assembly. The centrifugal switch is on the transmission and is driven through the same gear arrangement as the tach generator.

The aircraft DC system furnishes the power to operate the overspeed coil in the solenoid and coil assembly. If the output speed of the transmission reaches a speed of 7,000 to 7,500 rpm, the centrifugal switch closes the DC circuit and energizes the overspeed solenoid. This component then moves the valve and engages the latch that holds the valve in the underdrive position. To release the latch, energize the underdrive release solenoid.

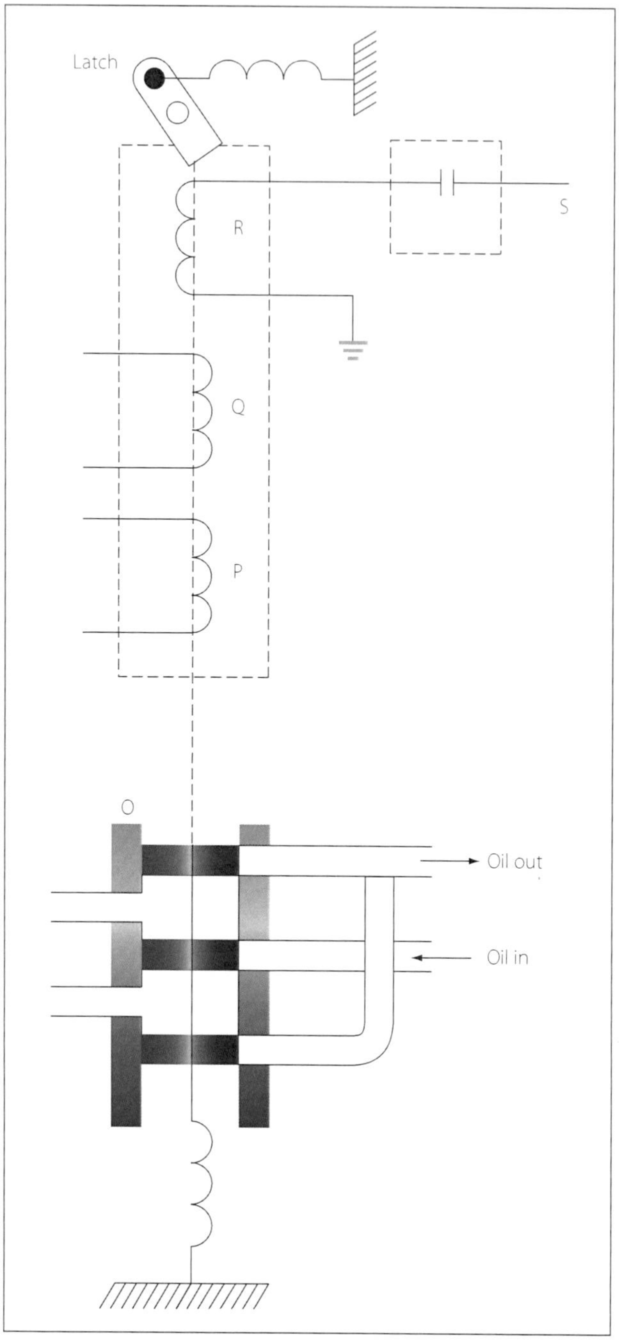

Figure 10-320. Overspeed circuit.

The load division circuit's function is to equalize the loads placed on each of the alternators, which is necessary to assure that each alternator assumes its share; otherwise, one alternator might be overloaded while another would be carrying only a small load.

In Figure 10-321, one phase of the alternator provides power for the primary in transformer G, whose secondary supplies power to the primaries of two other transformers, J1 and J2. Rectifiers K then change the output of the transformer secondaries from AC to DC.

The function of the two capacitors, L, is to smooth out the DC pulsations.

The output of the current transformer F depends upon the amount of current flowing in the line of one phase. In this way, it measures the real load of the generator. The output voltage of the current transformer is applied across resistor H. This voltage will be added vectorially to the voltage applied to the upper winding of transformer J by the output of transformer F. At the same time as it adds vectorially to the upper winding of

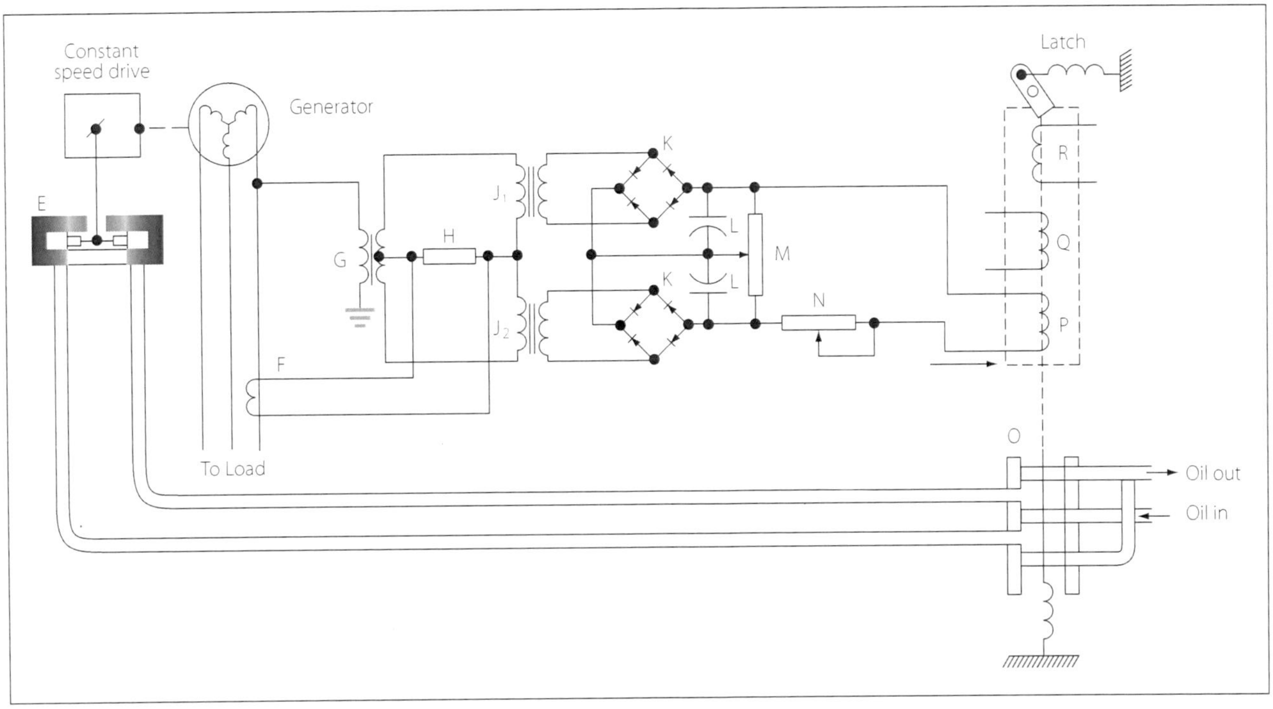

Figure 10-321. Droop circuit.

transformer J, it subtracts vectorially from the voltage applied to the lower winding of J.

This voltage addition and subtraction depends on the real load of the generator. The amount of real load determines the phase angle and the amount of voltage impressed across resistor H. The greater the real load, the greater the voltage across H, and hence, the greater the difference between the voltages applied to the two primaries of transformer J. The unequal voltages applied to resistor M by the secondaries of transformer J cause a current flow through the control coil P.

The control coil is wound so that its voltage supplements the voltage for the control coil in the valve and solenoid assembly. The resulting increased voltage moves the valve and slows down the generator's speed. Why should the speed be decreased if the load has been increased? Actually, systems using only one generator would not have decreased speed, but for those having two or more generators, a decrease is necessary to equalize the loads.

The load division circuit is employed only when two or more generators supply power. In such systems, the control coils are connected in parallel. If the source voltage for one of these becomes higher than the others, it determines the direction of current flow throughout the entire load division circuit. As explained before, the real load on the generator determines the amount of voltage on the control coil; therefore, the generator with the highest real load has the highest voltage.

As shown in Figure 10-322, current through No. 1 control coil, where the largest load exists, aids the control coil of the valve and solenoid, thereby slowing down the generator. (The source voltage of the control coils is represented by battery symbols in the illustration.) The current in the remaining control coils opposes the control coil of the valve and solenoid, in order to increase the speed of the other generators so the load will be more evenly distributed.

On some drives, instead of an electrically controlled governor, a flyweight-type governor is employed, which consists of a recess-type revolving valve driven by the output shaft of the drive, flyweights, two coil springs, and a nonrotating valve stem. Centrifugal force, acting on the governor flyweights, causes them

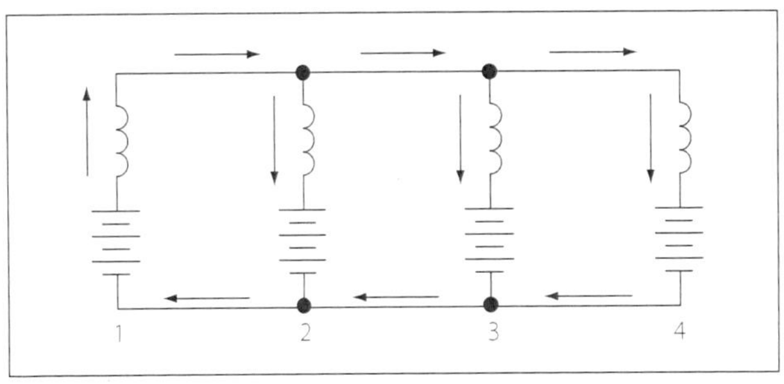

Figure 10-322. Relative direction of current in droop coil circuit with unequal loads.

to move outward, lifting the valve stem against the opposition of a coil spring.

The valve stem position controls the directing of oil to the two oil outlines. If the output speed tends to exceed 6,000 rpm, the flyweights will lift the valve stem to direct more oil to the side of the control piston, causing the piston to move in a direction to reduce the pump wobble plate angle. If the speed drops below 6,000 rpm, oil is directed to the control piston so that it moves to increase the wobble plate angle.

Overspeed protection is installed in the governor. The drive starts in the underdrive position. The governor coil springs are fully extended and the valve stem is held at the limit of its downward travel. In this condition, pressure is directed to the side of the control piston giving minimum wobble plate angle. The maximum angle side of the control piston is open to the hollow stem. As the input speed increases, the flyweights start to move outward to overcome the spring bias. This action lifts the valve stem and starts directing oil to the maximum side of the control piston, while the minimum side is opened to the hollow stem.

At about 6,000 rpm, the stem is positioned to stop drainage of either side, and the two pressures seek a balance point as the flyweight force is balanced against the spring bias. Thus, a mechanical failure in the governor will cause an underdrive condition. The flyweight's force is always tending to move the valve stem to the decrease speed position so that, if the coil spring breaks and the stem moves to the extreme position in that direction, output speed is reduced. If the input to the governor fails, the spring will force the stem all the way to the start position to obtain minimum output speed.

An adjustment screw on the end of the governor regulates the output speed of the constant speed drive. This adjustment increases or decreases the compression of a coil spring, opposing the action of the flyweights. The adjustment screws turn in an indented collar, which provides a means of making speed adjustments in known increments. Each "click" provides a small change in generator frequency.

Voltage Regulation of Alternators

The problem of voltage regulation in an AC system does not differ basically from that in a DC system. In each case, the function of the regulator system is to control voltage, maintain a balance of circulating current throughout the system, and eliminate sudden changes in voltage (anti-hunting) when a load is applied to the system. However, there is one important difference between the regulator system of DC generators and alternators operated in a parallel configuration. The load carried by any particular DC generator in either a two or four generator system depends on its voltage as compared with the bus voltage, while the division of load between alternators depends upon the adjustments of their speed governors, which are controlled by the frequency and droop circuits discussed in the previous section on alternator constant-speed drive systems.

When AC generators are operated in parallel, frequency and voltage must both be equal. Where a synchronizing force is required to equalize only the voltage between DC generators, synchronizing forces are required to equalize both voltage and speed (frequency) between AC generators. On a comparative basis, the synchronizing forces for AC generators are much greater than for DC generators. When AC generators are of sufficient size and are operating at unequal frequencies and terminal voltages, serious damage may result if they are suddenly connected to each other through a common bus. To avoid this, the generators must be synchronized as closely as possible before connecting them together.

Regulating the voltage output of a DC exciter, which supplies current to the alternator rotor field, best controls the output voltage of an alternator. This is accomplished by the regulation of a 28-volt system connected in the field circuit of the exciter. A regulator controls the exciter field current and thus regulates the exciter output voltage applied to the alternator field.

Alternator Transistorized Regulators

Many aircraft alternator systems use a transistorized voltage regulator to control the alternator output. Before studying this section, a review of transistor principles may be helpful.

A transistorized voltage regulator consists mainly of transistors, diodes, resistors, capacitors, and, usually, a thermistor. In operation, current flows through a diode and transistor path to the generator field. When the proper voltage level is reached, the regulating components cause the transistor to cut off conduction to control the alternator field strength. The regulator operating range is usually adjustable through a narrow range. The thermistor provides temperature compensation for the circuitry. The transistorized voltage regulator shown in Figure 10-323 will be referred to in explaining the operation of this type of regulator.

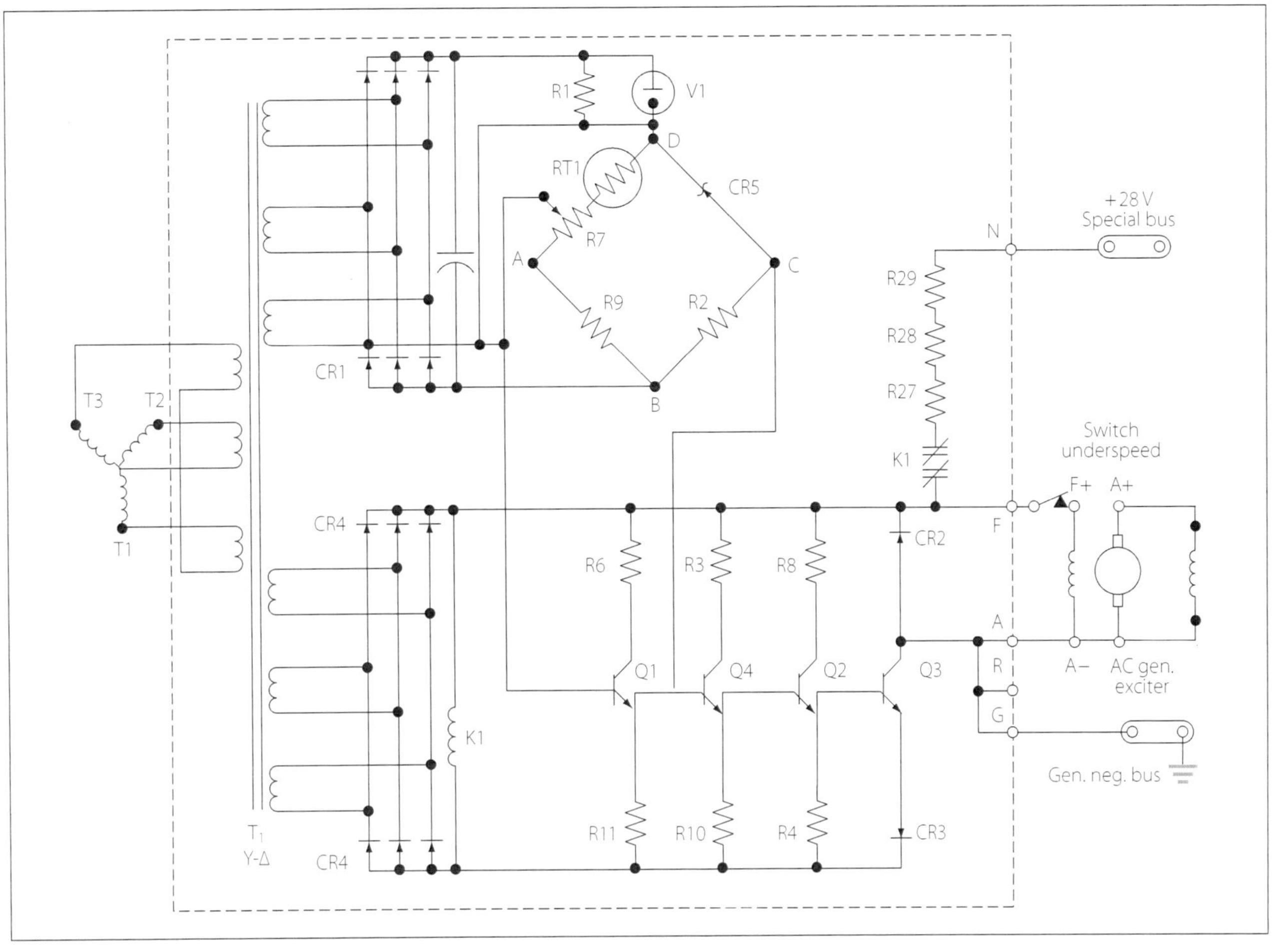

Figure 10-323. Transistorized voltage regulator.

The AC output of the generator is fed to the voltage regulator, where it is compared to a reference voltage, and the difference is applied to the control amplifier section of the regulator. If the output is too low, field strength of the AC exciter generator is increased by the circuitry in the regulator. If the output is too high, the field strength is reduced.

The power supply for the bridge circuit is CR1, which provides full-wave rectification of the three phase output from transformer T1. The DC output voltages of CR1 are proportional to the average phase voltages. Power is supplied from the negative end of the power supply through point B, R2, point C, zener diode (CR5), point D, and to the parallel hookup of V1 and R1. Takeoff point C of the bridge is located between resistor R2 and the zener diode. In the other leg of the reference bridge, resistors R9, R7, and the temperature compensating resistor RT1 are connected in series with V1 and R1 through points B, A, and D. The output of this leg of the bridge is at the wiper arm of R7.

As generator voltage changes occur, for example, if the voltage lowers, the voltage across R1 and V1 (once V2 starts conducting) will remain constant. The total voltage change will occur across the bridge circuit. Since the voltage across the zener diode remains constant (once it starts conducting), the total voltage change occurring in that leg of the bridge will be across resistor R2. In the other leg of the bridge, the voltage change across the resistors will be proportional to their resistance values. Therefore, the voltage change across R2 will be greater than the voltage change across R9 to wiper arm of R7. If the generator output voltage drops, point C will be negative with respect to the wiper arm of R7. Conversely, if the generator voltage output increases, the polarity of the voltage between the two points will be reversed.

The bridge output, taken between points C and A, is connected between the emitter and the base of transistor Q1. With the generator output voltage low, the voltage from the bridge will be negative to the emitter and positive to the base. This is a forward bias signal to the transistor, and the emitter to collector current will therefore increase. With the increase of current, the voltage across emitter resistor R11 will increase.

This, in turn, will apply a positive signal to the base of transistor Q4, increasing its emitter to collector current and increasing the voltage drop across the emitter resistor R10.

This will give a positive bias to the base of Q2, which will increase its emitter to collector current and increase the voltage drop across its emitter resistor R4. This positive signal will control output transistor Q3. The positive signal on the base of Q3 will increase the emitter to collector current.

The control field of the exciter generator is in the collector circuit. Increasing the output of the exciter generator will increase the field strength of the AC generator, which will increase the generator output.

To prevent exciting the generator when the frequency is at a low value, there is an underspeed switch located near the F+ terminal. When the generator reaches a suitable operating frequency, the switch will close and allow the generator to be excited.

Another item of interest is the line containing resistors R27, R28, and R29 in series with the normally closed contacts of the K1 relay. The operating coil of this relay is found in the lower left-hand part of the schematic. Relay K1 is connected across the power supply (CR4) for the transistor amplifier. When the generator is started, electrical energy is supplied from the 28-volt DC bus to the exciter generator field, to "flash the field" for initial excitation. When the field of the exciter generator has been energized, the AC generator starts to produce, and as it builds up, relay K1 is energized, opening the "field flash" circuit.

Chapter 11
Safety, Ground Operations, & Servicing

Aircraft maintenance technicians devote a portion of their aviation career to ground handling and operating aircraft. Technicians also need to be proficient in operating ground support equipment. The complexity of support equipment and the hazards involved in the ground handling of aircraft require that maintenance technicians possess a detailed knowledge of safety procedures used in aircraft servicing, taxiing, runup, and in the use of ground support equipment. The information provided in this chapter is intended as a general guide for safely servicing and operating aircraft.

Human factors should be introduced to aircraft maintenance personnel to make them aware of how it affects the maintenance performed. Although there are many factors involved when dealing with human performance, several areas can be considered. Some of these include fatigue, deadline pressure, stress, distractions, poor communication skills, complacency, and lack of information. Maintenance technicians should be aware of how human factors can affect their performance and safety while performing maintenance practices.

Shop Safety

Keeping hangars, shop, and the flight line orderly and clean is essential to safety and efficient maintenance. The highest standards of orderly work arrangements and cleanliness should be observed during the maintenance of aircraft.

Where continuous work shifts are established, the outgoing shift should remove and properly store personal tools, rollaway boxes, all workstands, maintenance stands, hoses, electrical cords, hoists, crates, and boxes that were needed for the work to be accomplished.

Signs should be posted to indicate dangerous equipment or hazardous conditions. There should also be signs that provide the location of first aid and fire equipment.

Safety lanes, pedestrian walkways, and fire lanes should be painted around the perimeter inside the hangars. This is a safety measure to prevent accidents and to keep pedestrian traffic out of work areas.

Safety is everyone's business, and communication is key to ensuring everyone's safety. Technicians and supervisors should watch for their own safety and for the safety of others working around them. If other personnel are conducting their actions in an unsafe manner, communicate with them, reminding them of their safety and that of others around them.

Electrical Safety

Physiological Safety

Working with electrical equipment poses certain physiological safety hazards. It is known that when electricity is applied to the human body, it can create severe burns in the area of entrance to and at the point of exit from the body. In addition, the nervous system is affected and can be damaged or destroyed.

To safely deal with electricity, the technician must have a working knowledge of the principles of electricity, and a healthy respect for its capability to do both work and damage.

Wearing or use of proper safety equipment can provide a psychological assurance at the same time it physically protects the user. The use of rubber gloves, safety glasses, rubber or grounded safety mats, and other safety equipment contributes to the physiological safety of the technician working on or with electrical equipment.

Two factors that affect safety when dealing with electricity are fear and overconfidence. These two factors are major causes of accidents involving electricity. While both a certain amount of respect for electrical equipment is healthy and a certain level of confidence is necessary, extremes of either can be deadly.

Lack of respect is often due to lack of knowledge. Personnel who attempt to work with electrical equipment and have no knowledge of the principles of electricity lack the skills to deal with electrical equipment safely.

Overconfidence leads to risk taking. The technician who does not respect the capabilities of electricity will, sooner or later, become a victim of electricity's awesome power.

Fire Safety

Anytime current flows, whether during generation or transmission, a byproduct of that flow is heat. The greater the current flow, the greater the amount of heat created. When this heat becomes too great, protective coatings on wiring and other electrical devices can melt, causing shorting, which leads to more current flow and greater heat. This heat can become so great that metals can melt, liquids vaporize, and flammable substances ignite.

An important factor in preventing electrical fires is to keep the area around electrical work or electrical equipment clean, uncluttered, and free of all unnecessary flammable substances.

Ensure that all power cords, wires, and lines are free of kinks and bends which can damage the wire. Never place wires or cords where they will be walked on or run over by other equipment. When several wires inside a power cord are broken, the current passing through the remaining wires increases. This generates more heat than the insulation coatings on the wire are designed to withstand and can lead to a fire.

Closely monitor the condition of electrical equipment. Repair or replace damaged equipment before further use.

Safety Around Compressed Gases

Compressed air, like electricity, is an excellent tool as long as it is under control. A typical nitrogen bottle set is shown in Figure 11-1.

The following "do's and don'ts" apply when working with or around compressed gases:

- Inspect air hoses frequently for breaks and worn spots. Unsafe hoses should be replaced immediately.
- Keep all connections in a "no-leak condition."
- Maintain in-line oilers, if installed, in operating condition.
- The system should have water sumps installed and should be drained at regular intervals.

Figure 11-1. A typical nitrogen bottle.

- Air used for paint spraying should be filtered to remove oil and water.
- Never use compressed air to clean hands or clothing. Pressure can force debris into the flesh leading to infection.
- Never spray compressed air in the area of other personnel.
- Air hoses should be straightened, coiled, and properly stored when not in use.

Many accidents involving compressed gases occur during aircraft tire mounting. To prevent possible personal injury, use tire dollies and other appropriate lifting and mounting devices in mounting or removing heavy aircraft tires.

When inflating tires on any type of aircraft wheels, always use tire cage guards. Because of possible personal injury, extreme caution is required to avoid overinflation of high pressure tires. Use pressure regulators on high pressure air bottles to eliminate the possibility of overinflation of tires. Tire cages need not be used when adjusting pressure in tires installed on aircraft.

Safety Around Hazardous Materials

Material safety diamonds are very important with regard to shop safety. These forms and labels are a simple and quick way to determine the risk and, if used properly with the tags, will indicate what personal safety equipment to use with the hazardous material.

Figure 11-2. A risk diamond.

The most observable portion of the Material Safety Data Sheet (MSDS) label is the risk diamond. It is a four color segmented diamond that represents Flammability (Red), Reactivity (Yellow), Health (Blue), and special Hazard (White). In the Flammability, Reactivity, and Health blocks, there should be a number from 0 to 4. Zero represents little or no hazard to the user; 4 means that the material is very hazardous. The special hazard segment contains a word or abbreviation to represent the special hazard. Some examples are: RAD for radiation, ALK for alkali materials, Acid for acidic materials, and CARC for carcinogenic materials. The letter W with a line through it stands for high reactivity to water. [Figure 11-2]

The Material Safety Data Sheet (MSDS) is a more detailed version of the chemical safety issues. They all have the same information requirements, but the exact location of the information on the sheet varies by MSDS manufacturer. These forms have the detailed breakdown of the chemicals, including formulas and action to take if personnel come into contact with the chemical(s). The U.S. Department of Labor Occupational Safety and Health Administration (OSHA) requires certain information be on every MSDS.

These forms are necessary for a safe shop that meets all the requirements of the governing safety body, the U.S. Department of Labor Occupational Safety and Health Administration (OSHA).

Safety Around Machine Tools

Hazards in a shop's operation increase when the operation of lathes, drill presses, grinders, and other types of machines are used. Each machine has its own set of safety practices. The following discussions regarding precautions should be followed to avoid injury.

The drill press can be used to bore and ream holes, to do facing, milling, and other similar types of operations. The following precautions can reduce the chance of injury:

- Wear eye protection.
- Securely clamp all work.
- Set the proper RPM for the material used.
- Do not allow the spindle to feed beyond its limit of travel while drilling.
- Stop the machine before adjusting work or attempting to remove jammed work.
- Clean the area when finished.

Lathes are used in turning work of a cylindrical nature. This work may be performed on the inside or outside of the cylinder. The work is secured in the chuck to provide the rotary motion, and the forming is done by contact with a securely mounted tool. The following precautions can reduce the chance of injury:

- Wear eye protection.
- Use sharp cutting tools.
- Allow the chuck to stop on its own. Do not attempt to stop the chuck by hand pressure.
- Examine tools and work for cracks or defects before starting the work.
- Do not set tools on the lathe. Tools may be caught by the work and thrown.
- Before measuring the work, allow it to stop in the lathe.

Milling machines are used to shape or dress; cut gear teeth, slots, or key ways; and similar work. The following precautions can reduce the chance of injury:

- Wear eye protection.
- Clean the work bed prior to work.
- Secure the work to the bed to prevent movement during milling.
- Select the proper tools for the job.
- Do not change the feed speed while working.

- Lower the table before moving under or away from the work.
- Ensure all clamps and bolts will pass under the arbor.

Grinders are used to sharpen tools, dress metal, and perform other operations involving the removal of small amounts of metal. The following precautions can reduce the chance of injury:

- Wear eye protection even if the grinder has a shield.
- Inspect the grinding wheel for defects prior to use.
- Do not force grinding wheels onto the spindle. They fit snugly, but do not require force to install them. *Placing side pressure on a wheel could cause it to explode.*
- Check the wheel flanges and compression washer. They should be one-third the diameter of the wheel.
- Do not stand in the arc of the grinding wheel while operating, in case the wheel explodes.

Welding should be performed only in designated areas. Any part to be welded should be removed from the aircraft, if possible. Repair would then be accomplished in the welding shop under a controlled environment.

A welding shop should be equipped with proper tables, ventilation, tool storage, and fire prevention and extinguishing equipment.

Welding on an aircraft should be performed outside, if possible. If welding in the hangar is necessary, observe these precautions:

- During welding operations, there should be no open fuel tanks, and no work on fuel systems should be in progress.
- No painting should be in progress.
- No aircraft are to be within 35 feet of the welding operation.
- No flammable material should be in the area around the welding operation.
- Only qualified welders should be permitted to do the work.
- The welding area should be roped off and placarded.
- Fire extinguishing equipment of a minimum rating of 20B should be in the immediate area with 80B rated equipment as a backup. These ratings will be explained later in this chapter.
- There should be trained fire watches in the area around the welding operation.
- Aircraft being welded should be in towable condition, with a tug attached, and the aircraft parking brakes released. A qualified operator should be on the tug, and mechanics available to assist in the towing operation should it become necessary to tow the aircraft. If the aircraft is in the hangar, the hangar doors should be opened.

Flight Line Safety

Hearing Protection

The flight line is a place of dangerous activity. Technicians who perform maintenance on the flight line must constantly be aware of what is going on around them.

The noise on a flight line comes from many places. Aircraft are only one source of noise. There are auxiliary-power units (APUs), fuel trucks, baggage handling equipment, and so forth. Each has its own frequency of sound. Combined all together, the ramp or flight line can cause hearing loss.

There are many types of hearing protection available. Hearing protection can be external or internal. The external protection is the earmuff/headphone type. The internal type fit into the auditory canal. Both types will reduce the sound level reaching the eardrum and reduce the chances of hearing loss.

Hearing protection should also be used when working with pneumatic drills, rivet guns, or other loud or noisy tools or machinery. Because of their high frequency, even short duration exposure to these sounds can cause a hearing loss. Continued exposure *will* cause hearing loss.

Foreign Object Damage (FOD)

FOD is any damage caused by any loose object to aircraft, personnel, or equipment. These loose objects can be anything from broken runway concrete to shop towels to safety wire.

To control FOD, keep ramp and operation areas clean, have a tool control program, and provide convenient receptacles for used hardware, shop towels, and other consumables.

The modern gas turbine engine will create a low pressure area in front of the engine that will cause any loose object to be drawn into the engine. The exhaust of these engines can propel loose objects great distances with enough force to damage anything that is hit.

The importance of an FOD program cannot be overstressed when a technician considers the cost of engines, components, or the cost of a human life. Never leave tools or other items around the intake of a turbine engine.

Safety Around Airplanes

As with the previously mentioned items, it is important to be aware of propellers. Do not assume the pilot of a taxiing aircraft can see you. Technicians must stay where the pilot can see them while on the ramp area. Turbine engine intakes and exhaust can also be very hazardous areas. There should be no smoking or open flames anywhere near an aircraft in operation. Be aware of aircraft fluids that can be detrimental to skin. When operating support equipment around aircraft, be sure to allow space between it and the aircraft and secure it so it cannot roll into the aircraft. All items in the area of operating aircraft must be stowed properly.

Safety Around Helicopters

Every type of helicopter has its own differences. These differences must be learned to avoid damaging the helicopter or injuring the technician.

When approaching a helicopter while the blades are turning, observe the rotor head and blades to see if they are level. This will allow maximum clearance as you approach the helicopter. Observe the following:

- Approach the helicopter in view of the pilot.
- Never approach a helicopter carrying anything with a vertical height that the blades could hit. This could cause blade damage and injury to the person.
- Never approach a single-rotor helicopter from the rear. The tail rotor is invisible when operating.
- Never go from one side of the helicopter to the other by going around the tail. Always go around the nose of the helicopter.

When securing the rotor on helicopters with elastometric bearings, check the maintenance manual for the proper method. Using the wrong method could damage the bearing.

Fire Safety

Performing maintenance on aircraft and their components requires the use of electrical tools which can produce sparks, along with heat-producing tools and equipment, flammable and explosive liquids, and gases. As a result, a high potential exists for fire to occur. Measures must be taken to prevent a fire from occurring and to also have a plan for extinguishing it.

The key to fire safety is knowledge of what causes fire, how to prevent it, and how to put it out. This knowledge must be instilled in each technician emphasized by their supervisors through sound safety programs, and occasionally practiced. Airport or other local fire departments can normally be called upon to assist in training personnel and helping to establish fire safety programs for the hangar, shops, and flight line.

Fire Protection

Requirements for Fire To Occur

Three things are required for a fire: (1) fuel—something that will, in the presence of heat, combine with oxygen, thereby releasing more heat and as a result reduces itself to other chemical compounds; (2) heat—accelerates the combining of oxygen with fuel, in turn releasing more heat; and (3) oxygen—the element which combines chemically with another substance through the process of oxidation. Rapid oxidation, accompanied by a noticeable release of heat and light, is called combustion or burning. [Figure 11-3] Remove any one of these things and the fire extinguishes.

Classification of Fires

For commercial purposes, the National Fire Protection Association (NFPA) has classified fires into three basic types: Class A, Class B, and Class C.

1. Class A fires occur in ordinary combustible materials, such as wood, cloth, paper, upholstery materials, and so forth.
2. Class B fires occur in flammable petroleum products of other flammable or combustible liquids, greases, solvents, paints, and so forth.
3. Class C fires occur involve energized electrical wiring and equipment.

Figure 11-3. Three elements of fire.

A fourth class of fire, with which the technician should be familiar, the Class D fire, is defined as fire in flammable metal. Class D fires are not commercially considered by the National Fire Protection Association to be a basic type or category of fire since they are caused by a Class A, B, or C fire. Usually Class D fires involve magnesium in the shop or in aircraft wheels and brakes, or are the result of improper or poorly conducted welding operations.

Any one of these types of fires can occur during maintenance on or around, or operations involving, aircraft. There is a particular type extinguisher which is most effective for each type of fire.

Types and Operation of Shop and Flight Line Fire Extinguishers

Water extinguishers are the best type to use on Class A fires. Water has two effects on fire: it deprives fire of oxygen and cools the material being burned.

Since most petroleum products float on water, water-type fire extinguishers are not recommended for Class B fires.

Extreme caution must be used when fighting electrical fires with water-type extinguishers. Not only must all electrical power be removed or shut off to the burning area, but residual electricity in capacitors, coils, and so forth must be considered to prevent severe injury, and possibly death from electrical shock.

Never use water-type fire extinguishers on Class D fires. Because metals burn at extremely high temperatures, the cooling effect of water causes an explosive expansion of the metal.

Water fire extinguishers are operated in a variety of ways. Some are hand pumped, while some are pressurized. The pressurized types of extinguishers may have a gas charge stored in the container with the water, or it may contain a "soda-acid" container where acid is spilled into a container of soda inside the extinguisher. The chemical reaction of the soda and the acid causes pressure to build inside the fire extinguisher, forcing the water out.

Carbon dioxide (CO_2) extinguishers are used for Class A, B, and C fires, extinguishing the fire by depriving it of oxygen. [Figure 11-4] Additionally, like water-type extinguishers, CO_2 cools the burning material.

Never use CO_2 on Class D fires. As with water extinguishers, the cooling effect of CO_2 on the hot metal can cause explosive expansion of the metal.

Figure 11-4. Carbon dioxide fire extinguisher.

When using CO_2 fire extinguishers, all parts of the extinguisher can become extremely cold, and remain so for a short time after operation. Wear protective equipment or take other precautions to prevent cold injury (such as frostbite) from occurring.

Extreme caution must be used when operating CO_2 fire extinguishers in closed or confined areas. Not only can the fire be deprived of oxygen, but so too can the operator.

CO_2 fire extinguishers generally use the self-expelling method of operation. This means that the CO_2 has sufficient pressure at normal operating pressure to expel itself. This pressure is held inside the container by some type of seal or frangible disk, which is broken or punctured by a firing mechanism, usually a pin. This means that once the seal or disk is broken, pressure in the container is released, and the fire extinguisher is spent, requiring replacement. [Figure 11-5]

Halogenated hydrocarbon extinguishers are most effective on Class B and C fires. They can be used on Class A and D fires but they are less effective. Halogenated hydrocarbon, (commonly called Freon™ by the industry), are numbered according to chemical formulas with Halon™ numbers.

Carbon tetrachloride (Halon 104), chemical formula CCl_4, has an Underwriters Laboratory (UL) toxicity rating of 3. As such, it is extremely toxic. [Figure 11-6]

Extinguishing Materials	Classes of fire				Self-Generating	Self-Expelling	Cartridge of N_2 Cylinder	Stored Pressure	Pump	Hand
	A	B	B	D						
Water and antifreeze	X						X	X	X	X
Soda-acid (water)	X				X					
Wetting agent	X						X			
Foam	X	X			X					
Loaded stream	X	X+					X	X		
Multipurpose dry chemical	X+	X	X				X	X		
Carbon dioxide		X+	X			X				
Dry chemical		X	X				X	X		
Bromotrifluoromethane — Halon 1301		X	X			X				
Bromochlorodifluoromethane — Halon 1211		X	X					X		
Dry powder (metal fires)				X			X			X

+ Smaller sizes of these extinguishers are not recognized for use on these classes of fires.

Figure 11-5. Extinguisher operation and methods of expelling.

Hydrochloric acid vapor, chlorine and phosgene gas are produced whenever carbon tetrachloride is used on ordinary fires. The amount of phosgene gas is increased whenever carbon tetrachloride is brought in direct contact with hot metal, certain chemicals, or continuing electrical arcs. It is not approved for any fire extinguishing use. Old containers of Halon 104 found in or around shops or hangars should be disposed of in accordance with Environmental Protection Agency (EPA) regulations and local laws and ordinances.

Group	Definition	Examples
6 (Least toxic)	Gases or vapors which in concentrations up to 20% by volume for durations of exposure of up to approximately 2 hours do not appear to produce injury.	Bromotrifluoromethane (Halon 1301)
5a	Gases or vapors much less toxic than Group 4 but more toxic than Group 6.	Carbon dioxide
4	Gases or vapors which in concentrations of the order of 2 to 2½% for durations of exposure of up to approximately 2 hours are lethal or produce serious injury.	Dibromodifluormethane (Halon 1202)
3	Gases or vapors which in concentrations of the order of 2 to 2½% for durations of exposure of the order of 1 hour are lethal or produce serious injury.	Bromochloromethane (Halon 1011) Carbon tetrachloride (Halon 104)
2	Gases or vapors which in concentrations of approximately ½ to 1% for durations of exposure of up to approximately ½ hour are lethal or produce serious injury.	Methyl bromide (Halon 1001)

Figure 11-6. Toxicity table.

Methyl bromide (Halon 1001), chemical formula CH_3Br, is a liquefied gas with a UL toxicity rating of 2. Very toxic, it is corrosive to aluminum alloys, magnesium, and zinc. Halon 1001 is not recommended for aircraft use.

Chlorobromomethane (Halon 1011), chemical formula CH_2ClBr, is a liquefied gas with a UL toxicity rating of 3. Like methyl bromide, Halon 1011 is not recommended for aircraft use.

Dibromodifluoromethane (Halon 1202), chemical formula CBr_2F_2, has a UL toxicity rating of 4. Halon 1202 is not recommended for aircraft use.

Bromochlorodifluoromethane (Halon 1211), chemical formula $CBrClF_2$, is a liquefied gas with a UL toxicity rating of 5. It is colorless, noncorrosive and evaporates rapidly leaving no residue. It does not freeze or cause cold burns, and will not harm fabrics, metals, or other materials it contacts. Halon 1211 acts rapidly on fires by producing a heavy blanketing mist that eliminates oxygen from the fire source. But more importantly, it interferes chemically with the combustion process of the fire. It has outstanding properties in preventing reflash after the fire has been extinguished.

Bromotrifluoromethane (Halon 1301), chemical formula CF_3Br, is also a liquefied gas with a UL toxicity rating of 6. It has all the characteristics of Halon 1211. The significant difference between the two is: Halon 1211 forms a spray similar to CO_2, while Halon 1301 has a vapor spray that is more difficult to direct.

Note: The Environmental Protection Agency (EPA) has restricted Halon to its 1986 production level due to its effect on the ozone layer.

Dry powder extinguishers, while effective on Class B and C fires, are the best for use on Class D fires.

The method of operation of dry powder fire extinguishers varies from gas cartridge charges, or stored pressure within the container which forces the powder charge out of the container, to tossing the powder on the fire by hand, by scooping pails or buckets of the powder from large containers or barrels.

Dry powder is not recommended for aircraft use (except on metal fires as a fire extinguisher) because the leftover chemical residues and dust often make cleanup difficult, and can damage electronic or other delicate equipment.

Inspection of Fire Extinguishers

Fire extinguishers should be checked periodically utilizing a checklist. If a checklist is unavailable, check the following as a minimum:

- Proper location of appropriate extinguisher
- Safety seals unbroken
- All external dirt and rust removed
- Gauge or indicator in operable range
- Proper weight
- No nozzle obstruction
- No obvious damage

Airport or other local fire departments can usually help in preparing and often can provide extinguisher checklists. In addition, these fire departments can be helpful in answering questions and assisting in obtaining repairs to or replacement of fire extinguishers.

Identifying Fire Extinguishers

Fire extinguishers should be marked to indicate suitability for a particular class of fire. The markings on Figure 11-7 should be placed on the fire extinguisher and in a conspicuous place in the vicinity of the fire extinguisher. When the location is marked, however, extreme care must be taken to ensure that the fire extinguisher kept at that location is in fact the type depicted by the marking. In other words, if a location is marked for a Class B fire extinguisher, ensure that the fire extinguisher in that location is in fact suitable for Class B fires.

Markings should be applied by decalcomanias (decals), painting, or similar methods. They should be legible and as durable as necessary for the location. For example, markings used outside need to be more durable than those in the hangar or office spaces.

Where markings are applied to the extinguisher, they should be located on the front of the shell (if one is installed) above or below the extinguisher nameplate. Markings should be large enough and in a form that is easily seen and identifiable by the average person with average eyesight at a distance of at least 3 feet.

Where markings are applied to wall panels, and so forth, in the vicinity of extinguishers, they should be large enough and in a form that is easily seen and identifiable by the average person with average eyesight, at a distance of at least 25 feet. [Figure 11-8]

Figure 11-7. Typical extinguisher markings.

Using Fire Extinguishers

When using a fire extinguisher, make sure you have the correct type for the fire. Most extinguishers have a pin to pull that will allow the handle to activate the agent. Stand back 8 feet and aim at the base of the fire or flames. Squeeze the lever and sweep side to side until the fire is extinguished.

Tiedown Procedures

Preparation of Aircraft

Aircraft should be tied down after each flight to prevent damage from sudden storms. The direction in which

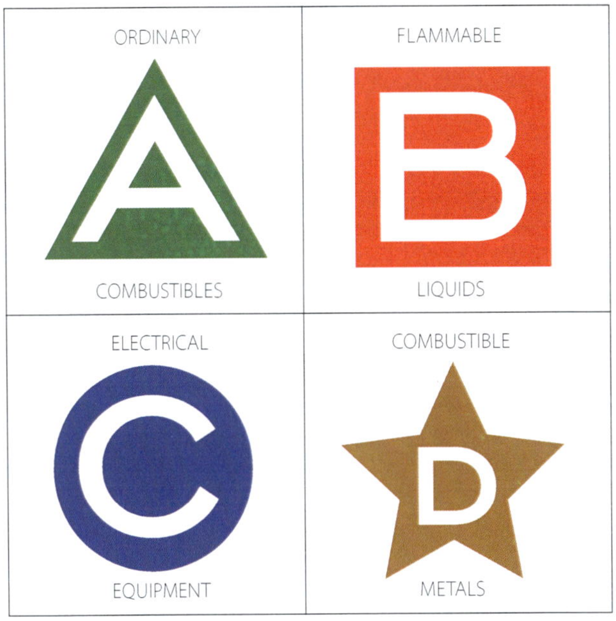

Figure 11-8. Identification of fire extinguisher type location.

aircraft are to be parked and tied down is determined by prevailing or forecast wind direction.

Aircraft should be headed as nearly as possible into the wind, depending on the locations of the parking area's fixed tiedown points. Spacing of tiedowns should allow for ample wingtip clearance. [Figure 11-9] After the aircraft is properly located, lock the nosewheel or the tailwheel in the fore-and-aft position.

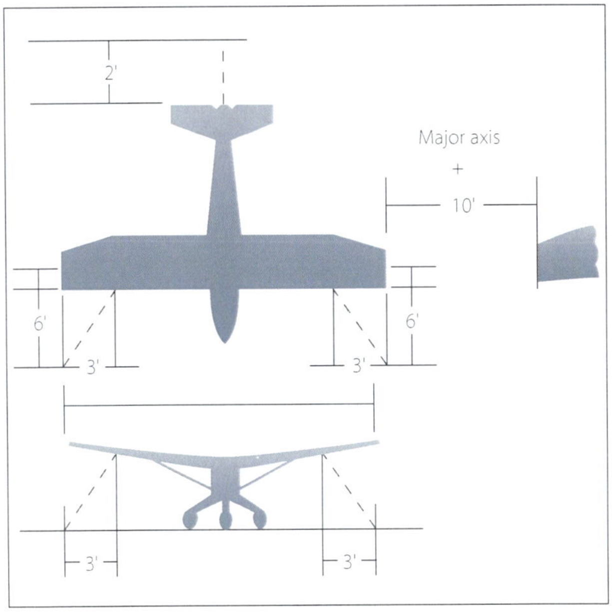

Figure 11-9. Diagram of tiedown dimensions.

Tiedown Procedures for Land Planes

Securing Light Aircraft

Light aircraft are most often secured with ropes tied only at the aircraft tiedown rings provided for securing purposes. Rope should never be tied to a lift strut, since this practice can bend a strut if the rope slips to a point where there is no slack. Manila rope shrinks when wet; about 1 inch (1") of slack should be provided for movement. Too much slack allows the aircraft to jerk against the ropes. Tight tiedown ropes put inverted flight stresses on the aircraft, many of which are not designed to take such loads.

A tiedown rope holds no better than the knot. Anti-slip knots such as the bowline are quickly tied and are easy to untie. [Figure 11-10] Aircraft not equipped with tiedown fittings should be secured in accordance with the manufacturer's instructions. Ropes should be tied to outer ends of struts on high-wing monoplanes, and suitable rings should be provided where structural conditions permit, if the manufacturer has not already provided them.

Securing Heavy Aircraft

The normal tiedown procedure for heavy aircraft can be accomplished with rope or cable tiedown. The number of such tiedowns should be governed by anticipated weather conditions.

Most heavy aircraft are equipped with surface control locks, which should be engaged or installed when the aircraft is secured. Since the method of locking controls will vary on different type aircraft, check the manufacturer's instructions for proper installation or engaging procedures. If high winds are anticipated, control surface battens can also be installed to prevent damage. Figure 11-11 illustrates four common tiedown points on heavy aircraft.

The normal tiedown procedure for heavy aircraft should generally include the following:

1. Head airplane into prevailing wind whenever possible.
2. Install control locks, all covers and guards.
3. Chock all wheels fore and aft. [Figure 11-12]
4. Attach tiedown reels to airplane tiedown loops and to tiedown anchors or tiedown stakes. Use tiedown stakes for temporary tiedown only. If tiedown reels are not available, 1/4" wire cable or $1\frac{1}{2}$" manila line may be used.

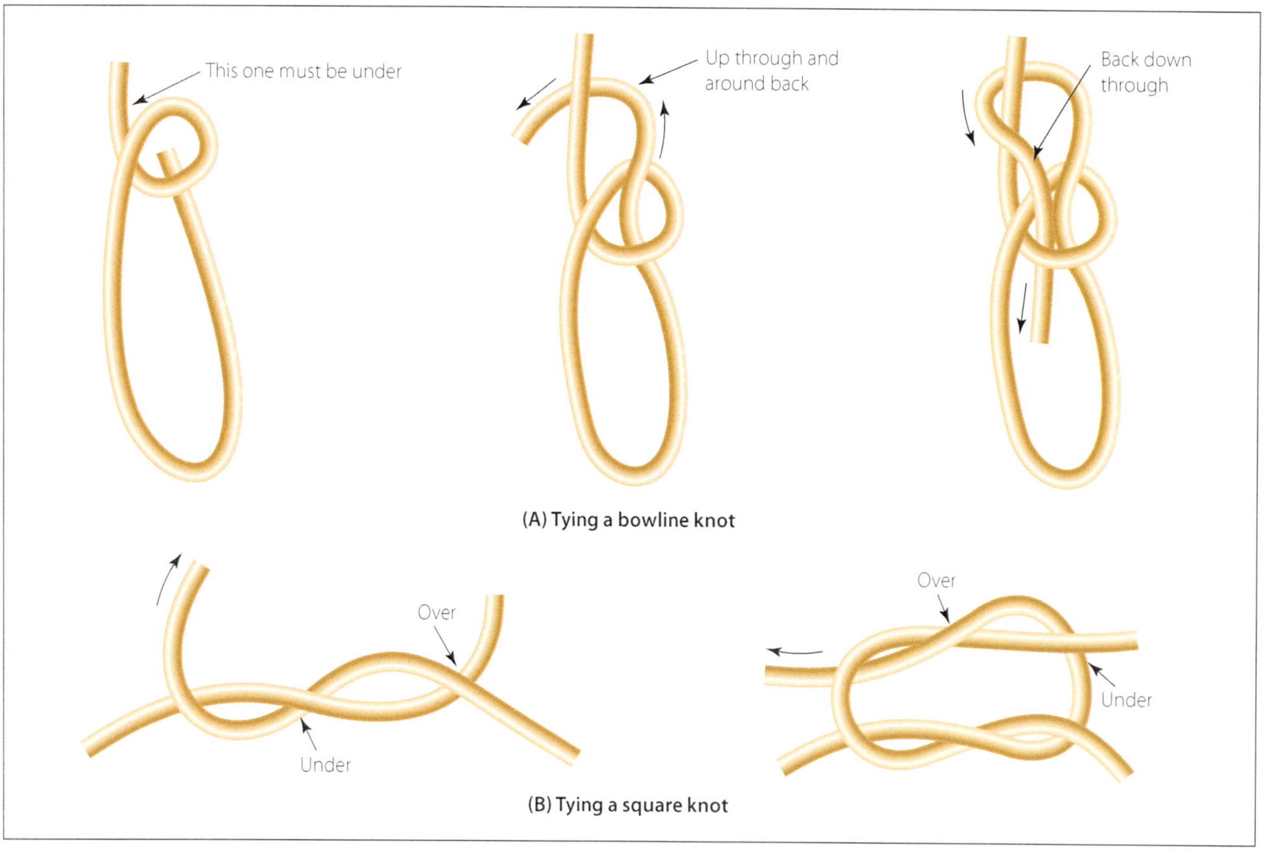

Figure 11-10. Knots commonly used for aircraft tiedown.

Figure 11-11. Common tiedown points.

Tiedown Procedures for Seaplanes

Seaplanes can be moored to a buoy, weather permitting, or tied to a dock. Weather causes wave action, and waves cause the seaplane to bob and roll. This bobbing and rolling while tied to a dock can cause damage.

When warning of an impending storm is received and it is not possible to fly the aircraft out of the storm area, some compartments of the seaplane can be flooded, partially sinking the aircraft. In addition, the aircraft should be tied down securely to anchors. Seaplanes tied down on land have been saved from high-wind damage by filling the floats with water in addition to tying the aircraft down in the usual manner.

During heavy weather, if possible, remove the seaplane from the water and tie down in the same manner as a land plane. If this is not possible, the seaplane could be anchored in a sheltered area away from wind and waves.

Tiedown Procedures for Ski Planes

Ski planes are tied down, if the securing means are available, in the same manner as land planes.

Ski-equipped airplanes can be secured on ice or in snow by using a device called a dead-man. A dead-man is any item at hand (such as a piece of pipe, log, and so forth) that a rope is attached to and buried in a snow or ice trench. Using caution to keep the free end of the rope dry and unfrozen, snow is packed in the trench. If available, pour water into the trench; when it is frozen, tie down the aircraft with the free end of the rope.

Figure 11-12. Wheels chocked fore and aft.

Operators of ski-equipped aircraft sometimes pack soft snow around the skis, pour water on the snow, and permit the skis to freeze to the ice. This, in addition to the usual tiedown procedures, aids in preventing damage from windstorms. Caution must be used when moving an aircraft that has been secured in this manner to ensure that a ski is not still frozen to the ground. Otherwise, damage to the aircraft or skis can occur.

Tiedown Procedures for Helicopters

Helicopters, like other aircraft are secured to prevent structural damage, which can occur from high-velocity surface winds.

Helicopters should be secured in hangars, when possible. If not, they should be tied down securely. Helicopters that are tied down can usually sustain winds up to approximately 65 mph. If at all possible, helicopters should be evacuated to a safe area if tornadoes or hurricanes are anticipated.

For added protection, helicopters should be moved to a clear area so that they will not be damaged by flying objects or falling limbs from surrounding trees.

If high winds are anticipated with the helicopter parked in the open, the main rotor blades should be tied down. Detailed instructions for securing and mooring each type of helicopter can be found in the applicable maintenance manual. [Figure 11-13] Methods of securing helicopters vary with weather conditions, the length of time the aircraft is expected to remain on the ground,

Figure 11-13. Example of mooring of a helicopter.

and location and characteristics of the aircraft. Wheel chocks, control locks, rope tiedowns, mooring covers, tip socks, tiedown assemblies, parking brakes, and rotor brakes are used to secure helicopters.

Typical mooring procedures are as follows:

1. Face the helicopter in the direction from which the highest forecast wind or gusts are anticipated.
2. Spot the helicopter slightly more than one rotor-span distance from other aircraft.
3. Place wheel chocks ahead of and behind all wheels (where applicable). On helicopters equipped with skids, retract the ground handling wheels, lower the helicopter to rest on the skids, and install wheel position lockpins or remove the ground handling wheels. Ground handling wheels should be secured inside the aircraft or inside the hangar or storage buildings. Do not leave them unsecured on the flight line.
4. Align the blades and install tiedown assemblies as prescribed by the helicopter manufacturer. [Figure 11-14] Tie straps snugly without strain, and during wet weather, provide some slack to avoid the possibility of the straps shrinking, causing undue stresses on the aircraft and/or its rotor system(s).
5. Fasten the tiedown ropes or cables to the forward and aft landing gear cross tubes and secure to ground stakes or tiedown rings.

Procedures for Securing Weight-Shift Control Aircraft

There are many types of weight-shift control aircraft—engine powered and nonpowered. These types of aircraft are very suitable to wind damage. The wings can be secured in a similar manner as a conventional aircraft in light winds. But in high winds, the mast can be disconnected from the wing and the wing placed close to the ground and secured. This type of aircraft can also be partially disassembled or moved into a hangar for protection.

Procedures for Securing Powered Parachutes

Powered parachutes should have the parachute packed in a bag to prevent the chute from filling with air from the wind and dragging the seat and engine. The engine and seat can also be secured if needed.

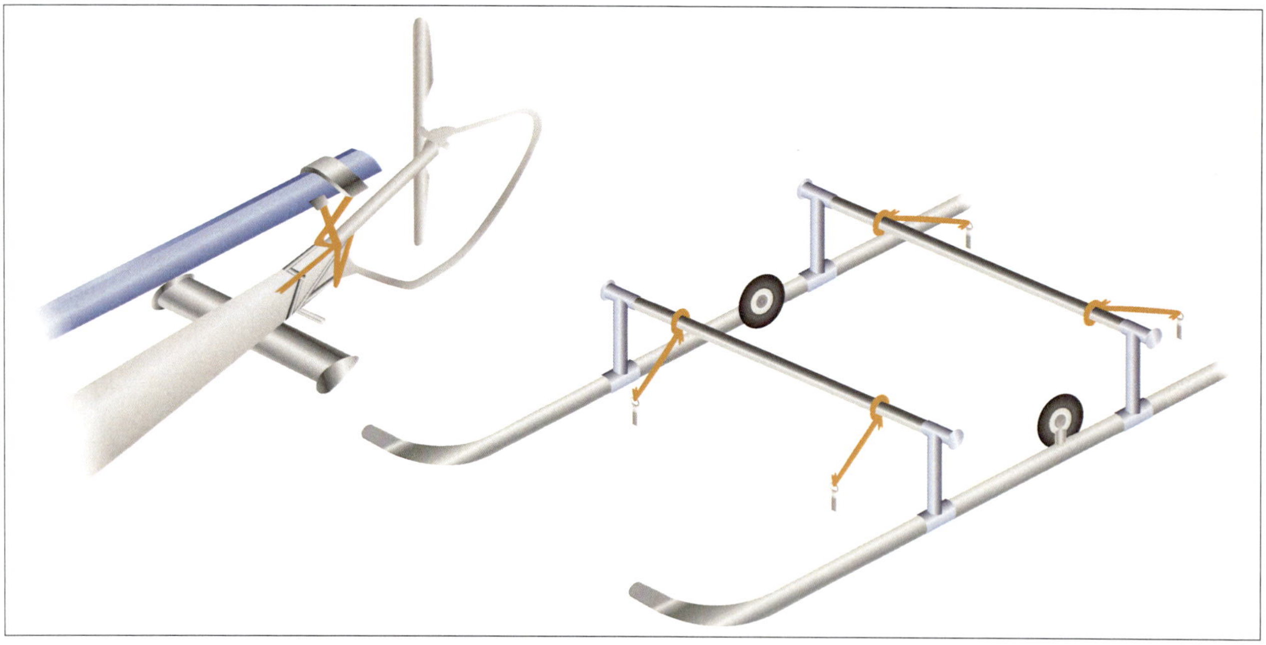

Figure 11-14. Securing helicopter blades and fuselage.

Ground Movement of Aircraft

Engine Starting and Operation

The following instructions cover the starting procedures for reciprocating, turboprop, turbofan, and auxiliary power units (APU). These procedures are presented only as a general guide for familiarization with typical procedures and methods. Detailed instructions for starting a specific type of engine can be found in the manufacturer's instruction book.

Before starting an aircraft engine:

1. Position the aircraft to head into the prevailing wind to ensure adequate airflow over the engine for cooling purposes.
2. Make sure that no property damage or personal injury will occur from the propeller blast or jet exhaust.
3. If external electrical power is used for starting, ensure that it can be removed safely and it is sufficient for the total starting sequence.
4. During any and all starting procedures, a "fireguard" equipped with a suitable fire extinguisher shall be stationed in an appropriate place. A fireguard is someone familiar with aircraft starting procedures. The fire extinguisher should be a CO_2 extinguisher of at least 5-pound capacity. The appropriate place is adjacent to the outboard side of the engine, in view of the pilot, and also where he or she can observe the engine/aircraft for indication of starting problems.
5. If the aircraft is turbine engine powered, the area in front of the jet inlet must be kept clear of personnel, property, and/or debris (FOD).
6. These "before starting" procedures apply to all aircraft powerplants.
7. Follow manufacturer's checklists for start procedures and shutdown procedures.

Reciprocating Engines

The following procedures are typical of those used to start reciprocating engines. There are, however, wide variations in the procedures for the many reciprocating engines. No attempt should be made to use the methods presented here for actually starting an engine. Instead, always refer to the procedures contained in the applicable manufacturer's instructions.

Reciprocating engines are capable of starting in fairly low temperatures without the use of engine heating or oil dilution, depending on the grade of oil used. The various covers (wing, tail, cockpit, wheel, and so forth) protecting the aircraft must be removed before attempting to turn the engine. External sources of electrical power should be used when starting engines equipped with electric starters, if possible or needed. This eliminates an excessive burden on the aircraft battery. All unnecessary electrical equipment should be left off until the generators are furnishing electrical power to the aircraft power bus.

Before starting a radial engine that has been shut down for more than 30 minutes, check the ignition switch for off; turn the propeller three or four complete revolutions by hand to detect a hydraulic lock, if one is present. Any liquid present in a cylinder is indicated by the abnormal effort required to rotate the propeller, or by the propeller stopping abruptly during rotation. Never use force to turn the propeller when a hydraulic lock is detected. Sufficient force can be exerted on the crankshaft to bend or break a connecting rod if a lock is present.

To eliminate a lock, remove either the front or rear spark plug from the lower cylinders and pull the propeller through. Never attempt to clear the hydraulic lock by pulling the propeller through in the direction opposite to normal rotation. This tends to inject the liquid from the cylinder into the intake pipe. The liquid will be drawn back into the cylinder with the possibility of complete or partial lock occurring on the subsequent start.

To start the engine, proceed as follows:

1. Turn the auxiliary fuel pump on, if aircraft is so equipped.
2. Place the mixture control to the position recommended for the engine and carburetor combination being started. As a general rule, the mixture control should be in the "idle cut-off" position for fuel injection and in the "full rich" position for float-type carburetors. Many light aircraft are equipped with a mixture control pull rod which has no detent intermediate positions. When such controls are pushed in flush with the instrument panel, the mixture is set in the "full rich" position. Conversely, when the control rod is pulled all the way out, the carburetor is in the "idle cut-off" or "full lean" position. The operator can select unmarked intermediate positions between these two extremes to achieve any desired mixture setting.
3. Open the throttle to a position that will provide 1,000 to 1,200 rpm (approximately 1/8 to 1/2 inch from the "closed" position).

4. Leave the pre-heat or alternate air (carburetor air) control in the "cold" position to prevent damage and fire in case of backfire. These auxiliary heating devices should be used after the engine warms up. They improve fuel vaporization, prevent fouling of the spark plugs, ice formation, and eliminate icing in the induction system.
5. Move the primer switch to "on" intermittently (press to prime by pushing in on the ignition switch during the starting cycle), or prime with one to three strokes of priming pump, depending on how the aircraft is equipped. The colder the weather, the more priming will be needed.
6. Energize the starter after the propeller has made at least two complete revolutions, and turn the ignition switch on. On engines equipped with an induction vibrator (shower of sparks, magneto incorporates a retard breaker assembly), turn the switch to the "both" position and energize the starter by turning the switch to the "start" position. After the engine starts, release the starter switch to the "both" position. When starting an engine that uses an impulse coupling magneto, turn the ignition switch to the "left" position. Place the start switch to the "start" position; when the engine starts, release the start switch. Do not crank the engine continuously with the starter for more than 1 minute. Allow a 3- to 5-minute period for cooling the starter (starter duty cycle) between successive attempts. Otherwise, the starter may be burned out due to overheating.
7. After the engine is operating smoothly, move the mixture control to the "full rich" position if started in the "idle cutoff" position. Carbureted engines will already be in the rich mixture position. Check for oil pressure.

Instruments for monitoring the engine during operation include a tachometer for rpm, manifold pressure gauge, oil pressure gauge, oil temperature gauge, cylinder head temperature gauge, exhaust gas temperature gauge and fuel flow gauge.

Hand Cranking Engines

If the aircraft has no self-starter, the engine must be started by turning the propeller by hand (hand propping the propeller). The person who is turning the propeller calls: "Fuel on, switch off, throttle closed, brakes on." The person operating the engine will check these items and repeat the phrase. The switch and throttle must not be touched again until the person swinging the prop calls "contact." The operator will repeat "contact" and then turn on the switch. Never turn on the switch and then call "contact."

A few simple precautions will help to avoid accidents when hand propping the engine. While touching a propeller, always assume that the ignition is on. The switches which control the magnetos operate on the principle of short-circuiting the current to turn the ignition off. If the switch is faulty, it can be in the "off" position and still permit current to flow in the magneto primary circuit. This condition could allow the engine to start when the switch is off.

Be sure the ground is firm. Slippery grass, mud, grease, or loose gravel can lead to a fall into or under the propeller. Never allow any portion of your body to get in the way of the propeller. This applies even though the engine is not being cranked.

Stand close enough to the propeller to be able to step away as it is pulled down. Stepping away after cranking is a safeguard in case the brakes fail. Do not stand in a position that requires leaning toward the propeller to reach it. This throws the body off balance and could cause you to fall into the blades when the engine starts.

In swinging the prop, always move the blade downward by pushing with the palms of the hands. Do not grip the blade with the fingers curled over the edge, since "kickback" may break them or draw your body in the blade path.

Excessive throttle opening after the engine has fired is the principal cause of backfiring during starting. Gradual opening of the throttle, while the engine is cold, will reduce the potential for backfiring. Slow, smooth movement of the throttle will assure correct engine operation.

Avoid overpriming the engine before it is turned over by the starter. This can result in fires, scored or scuffed cylinders and pistons, and, in some cases, engine failures due to hydraulic lock. If the engine is inadvertently flooded or overprimed, turn the ignition switch off and move the throttle to the "full open" position. To rid the engine of the excess fuel, turn it over by hand or by the starter. If excessive force is needed to turn over the engine, stop immediately. Do not force rotation of the engine. If in doubt, remove the lower cylinder spark plugs.

Immediately after the engine starts, check the oil pressure indicator. If oil pressure does not show within 30 seconds, stop the engine and determine the trouble.

If oil pressure is indicated, adjust the throttle to the aircraft manufacturer's specified rpm for engine warm-up. Warm-up rpm will usually be in the 1,000 to 1,300 rpm range.

Most aircraft reciprocating engines are air cooled and depend on the forward speed of the aircraft to maintain proper cooling. Therefore, particular care is necessary when operating these engines on the ground.

During all ground running, operate the engine with the propeller in full low pitch and headed into the wind with the cowling installed to provide the best degree of engine cooling. Closely monitor the engine instruments at all times. Do not close the cowl flaps for engine warm-up; they should be in the open position while operating on the ground. When warming up the engine, ensure that personnel, ground equipment that may be damaged, or other aircraft are not in the propeller wash.

Extinguishing Engine Fires

In all cases, a fireguard should stand by with a CO_2 fire extinguisher while the aircraft engine is being started. This is a necessary precaution against fire during the starting procedure. The fireguard should be familiar with the induction system of the engine so that in case of fire he or she can direct the CO_2 into the air intake of the engine to extinguish it. A fire could also occur in the exhaust system of the engine from liquid fuel being ignited in the cylinder and expelled during the normal rotation of the engine.

If an engine fire develops during the starting procedure, continue cranking to start the engine and blow out the fire. If the engine does not start and the fire continues to burn, discontinue the start attempt. The fireguard should extinguish the fire using the available equipment. The fireguard must observe all safety practices at all times while standing by during the starting procedure.

Turboprop Engines

The starting of any turbine engine consists of three steps that must be carried out in the correct sequence. The starter turns the main compressor to provide airflow though the engine. At the correct speed (which provides enough airflow) the igniters are turned on and provide a hot spark to light the fuel which is engaged next. As the engine accelerates, it will reach a self-sustaining speed and the starter is disengaged.

The various covers protecting the aircraft must be removed. Carefully inspect the engine exhaust areas for the presence of fuel or oil. Make a close visual inspection of all accessible parts of the engines and engine controls, followed by an inspection of all nacelle areas to determine that all inspection and access plates are secured. Check sumps for water. Inspect air inlet areas for general condition and foreign material. Check the compressor for free rotation, when the installation permits, by reaching in and turning the blades by hand.

The following procedures are typical of those used to start turboprop engines. There are, however, wide variations in the procedures applicable to the many turboprop engines, and no attempt should be made to use these procedures in the actual starting of a turboprop engine. These procedures are presented only as a general guide for familiarization with typical procedures and methods. For starting of all turboprop engines, refer to the detailed procedures contained in the applicable manufacturer's instructions or their approved equivalent.

Turboprop engines are usually fixed turbine or free turbine. The propeller is connected to the engine directly in a fixed turbine, which results in the propeller being turned as the engine starts. This provides extra drag which must be overcome during starting. If the propeller is not at the "start" position, difficulty may be encountered in making a start due to high loads. Because of this, the propeller is in flat pitch at shut down and subsequently in flat pitch during start.

The free turbine engine has no mechanical connection between the gas generator and the power turbine which is connected to the propeller. In this type of engine, the propeller remains in the feather position during starting and starts to turn only as the gas generator accelerates.

Instrumentation for turbine engines varies according to the type of turbine engine. Turboprop engines use the normal instruments—oil pressure, oil temperature, interturbine temperature (ITT) and fuel flow. They also use instruments to measure gas generator speed, propeller speed, and torque produced by the propeller. [Figure 11-15] A typical turboprop uses a set of engine controls, such as power levelers (throttle), propeller levers, and condition levers. [Figure 11-16]

The first step in starting a turbine engine is to provide an adequate source of power for the starter. On smaller turbine engines, the starter is an electric motor which turns the engine through electrical power. Larger engines needed a much more powerful starter. Electric motors would be limited by current flow and weight. Air turbine starters were developed which were lighter and produced sufficient power to turn the engine at

Figure 11-15. Typical examples of turboprop instruments.

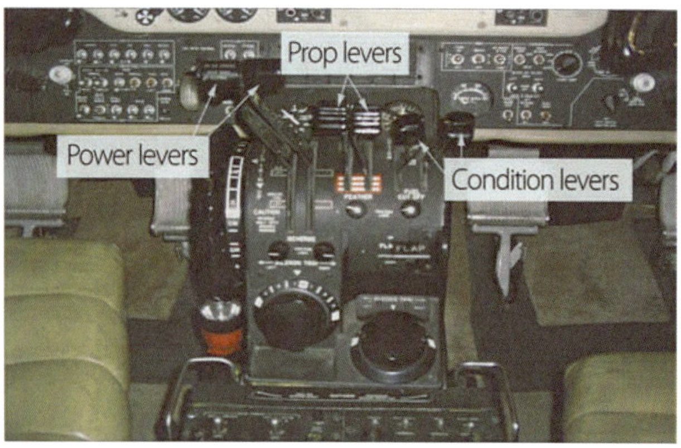

Figure 11-16. Engine controls of a turboprop aircraft.

the correct speed for starting. Where an air turbine starter is used, the starting air supply may be obtained from an auxiliary power unit onboard the aircraft, an external source (ground air cart), or an engine cross-bleed operation. In some limited cases, a low-pressure large-volume tank can provide the air for starting an engine. Many smaller turboprop engines are started using the starter/generator, which is both the engine starter and the generator.

While starting an engine, always observe the following:

- Always observe the starter duty cycle. Otherwise, the starter can overheat and be damaged.
- Assure that there is enough air pressure or electrical capacity before attempting a start.
- Do not perform a ground start if turbine inlet temperature (residual temperature) is above that specified by the manufacturer.
- Provide fuel under low pressure to the engine's fuel pump.

Turboprop Starting Procedures

To start an engine on the ground, perform the following operations:

1. Turn the aircraft boost pumps on.
2. Make sure that the power lever is in the "start" position.
3. Place the start switch in the "start" position. (This will start the engine turning.)
4. Place the ignition switch on. (On some engines, the ignition is activated by moving the fuel lever.)
5. The fuel is now turned on. This is accomplished by moving the condition lever to the "on" position.
6. Monitor the engine lights of the exhaust temperature. If it exceeds the limits, the engine should be shut down.
7. Check the oil pressure and temperature.
8. After the engine reaches a self-sustaining speed, the starter is disengaged.
9. The engine should continue to accelerate up to idle.
10. Maintain the power lever at the "start" position until the specified minimum oil temperature is reached.
11. Disconnect the ground power supply, if used.

If any of the following conditions occur during the starting sequence, turn off the fuel and ignition switch, discontinue the start immediately, make an investigation, and record the findings.

- Turbine inlet temperature exceeds the specified maximum. Record the observed peak temperature.

- Acceleration time from start of propeller rotation to stabilized rpm exceeds the specified time.
- There is no oil pressure indication at 5,000 rpm for either the reduction gear or the power unit.
- Torching (visible burning in the exhaust nozzle).
- The engine fails to ignite by 4,500 rpm or maximum motoring rpm.
- Abnormal vibration is noted or compressor surge occurs (indicated by backfiring).
- Fire warning bell rings. (This may be due to either an engine fire or overheat.)

Turbofan Engines

Unlike reciprocation engine aircraft, the turbine-powered aircraft does not require a preflight run-up unless it is necessary to investigate a suspected malfunction.

Before starting, all protective covers and air inlet duct covers should be removed. If possible, the aircraft should be headed into the wind to obtain better cooling, faster starting, and smoother engine performance. It is especially important that the aircraft be headed into the wind if the engine is to be trimmed.

The run-up area around the aircraft should be cleared of both personnel and loose equipment. The turbofan engine intake and exhaust hazard areas are illustrated in Figure 11-17. Exercise care to ensure that the run-up area is clear of all items, such as nuts, bolts, rocks, shop towels, or other loose debris (FOD). Many very serious accidents have occurred involving personnel in the vicinity of turbine engine air inlets. Use extreme caution when starting turbine aircraft.

Check the aircraft fuel sumps for water or ice, and inspect the engine air inlet for general condition and the presence of foreign objects. Visually inspect the fan blades, forward compressor blades, and the compressor inlet guide vanes for nicks and other damage. If possible, check the fan blades for free rotation by turning the fan blades by hand. All engine controls should be operated, and engine instruments and warning lights should be checked for proper operation.

Starting a Turbofan Engine

The following procedures are typical of those used to start many turbine engines. There are, however, wide variations in the starting procedures used for turbine engines, and no attempt should be made to use these procedures in the actual starting of an engine. These procedures are presented only as a general guide for familiarization with typical procedures and methods. In the starting of all turbine engines, refer to the detailed procedures contained in the applicable manufacturer's instructions or their approved equivalent.

Most turbofan engines can be started by either air turbine or electrical starters. Air-turbine starters use compressed air from an external source as discussed earlier. Fuel is turned on either by moving the start lever to "idle/start" position or by opening a fuel shutoff valve. If an air turbine starter is used, the engine should "light off" within a predetermined time after the fuel is turned on. This time interval, if exceeded, indicates a malfunction has occurred and the start should be discontinued.

Most turbofan engine controls consist of a power lever, reversing levers, and start levers. Newer aircraft have replaced the start levers with a fuel switch. [Figure 11-18] Turbofan engines also use all the normal instruments speeds, (percent of total rpm) exhaust gas temperature, fuel flow, oil pressure, and temperature. An instrument that measures the amount of thrust being delivered is the engine pressure ratio. This measures the ratio between the inlet pressure to the outlet pressure of the turbine.

The following procedures are useful only as a general guide, and are included to show the sequence of events in starting a turbofan engine.

1. If the engine is so equipped, place the power lever in the "idle" position.
2. Turn the fuel boost pump(s) switch on.
3. A fuel inlet pressure indicator reading ensures fuel is being delivered to engine fuel pump inlet.
4. Turn engine starter switch on; note that the engine rotates to a preset limit; check for oil pressure.
5. Turn ignition switch on. (This is usually accomplished by moving the start lever toward the "on" position. A micro switch connected to the leveler turns on the ignition.)
6. Move the start lever to "idle" or "start" position; this will start fuel flow into the engine.
7. Engine start (light off) is indicated by a rise in exhaust gas temperature.
8. If a two spool engine, check rotation of fan or N1.
9. Check for proper oil pressure.
10. Turn engine starter switch off at proper speeds.
11. After engine stabilizes at idle, ensure that none of the engine limits are exceeded.

Newer aircraft will drop off the starter automatically.

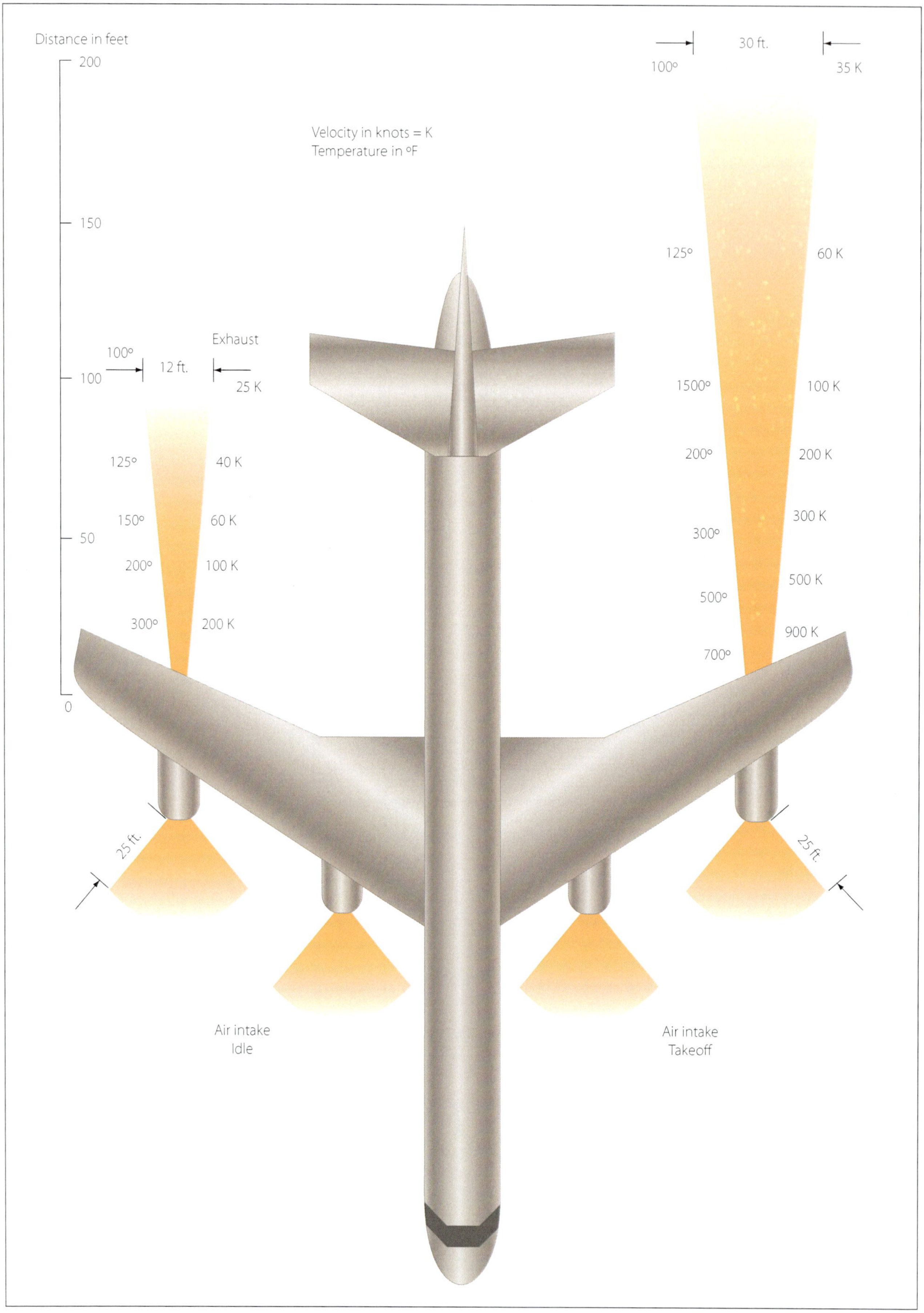

Figure 11-17. Engine intake and exhaust hazard areas.

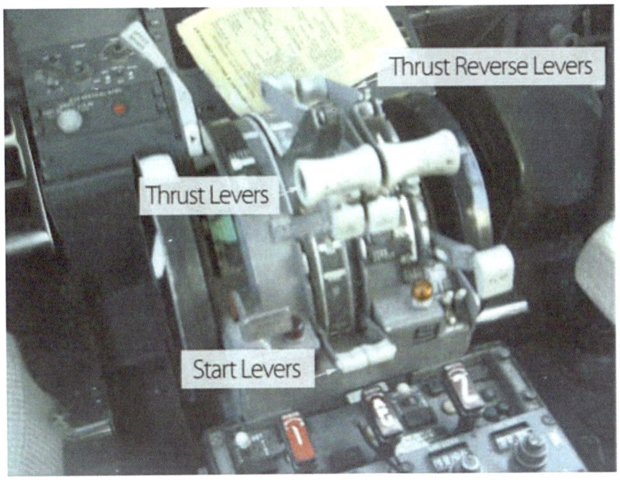

Figure 11-18. Turbofan engine control levers.

Auxiliary Power Units (APUs)

APUs are generally smaller turbine engines that provide compressed air for starting engines, cabin heating and cooling, and electrical power while on the ground. Their operation is normally simple. By turning a switch on and up to the start position (spring loaded to on position), the engine will start automatically. During start, the exhaust gas temperature must be monitored. APUs are at idle at 100 percent rpm with no load. After the engine reaches its operating rpm, it can be used for cooling or heating the cabin and for electrical power. It is normally used to start the main engines.

Unsatisfactory Turbine Engine Starts

Hot Start

A hot start occurs when the engine starts, but the exhaust gas temperature exceeds specified limits. This is usually caused by an excessively rich fuel/air mixture entering the combustion chamber. This condition can be caused by either too much fuel or not enough airflow. The fuel to the engine should be shut off immediately.

False or Hung Start

False or hung starts occur when the engine starts normally, but the rpm remains at some low value rather than increasing to the normal starting rpm. This is often the result of insufficient power to the starter, or the starter cutting off before the engine starts self-accelerating. In this case, the engine should be shut down.

Engine Will Not Start

The engine will not start within the prescribed time limit. It can be caused by lack of fuel to the engine, insufficient or no electrical power to the exciter in the ignition system, or incorrect fuel mixer. If the engine fails to start within the prescribed time, it should be shut down.

In all cases of unsatisfactory starts the fuel and ignition should be turned off. Continue rotating the compressor for approximately 15 seconds to remove accumulated fuel from the engine. If unable to motor (rotate) the engine, allow a 30-second fuel draining period before attempting another start.

Towing of Aircraft

Movement of large aircraft on an airport and about the flight line and hangar is usually accomplished by towing with a tow tractor (sometimes called a "tug"). [Figure 11-19] In the case of small aircraft, some moving is accomplished by hand, by pushing on the correct areas of the aircraft. Aircraft may also be taxied about the flight line, but usually only by certain qualified persons.

Towing aircraft can be a hazardous operation, causing damage to the aircraft and injury to personnel, if done recklessly or carelessly. The following paragraphs outline the general procedure for towing aircraft; however, specific instructions for each model of aircraft are detailed in the manufacturer's maintenance instructions and should be followed in all instances.

Before the aircraft to be towed is moved, a qualified person must be in the cockpit to operate the brakes in case the tow bar should fail or become unhooked. The aircraft can then be stopped, preventing possible damage.

Some types of tow bars available for general use can be used for many types of towing operations. [Figure 11-20] These bars are designed with sufficient tensile strength

Figure 11-19. Example of a tow tractor.

Figure 11-20. Tow bar for a large aircraft.

to pull most aircraft, but are not intended to be subjected to torsional or twisting loads. Many have small wheels that permit them to be drawn behind the towing vehicle going to or from an aircraft. When the bar is attached to the aircraft, inspect all the engaging devices for damage or malfunction before moving the aircraft.

Some tow bars are designed for towing various types of aircraft; however, other special types can be used on a particular aircraft only. Such bars are usually designed and built by the aircraft manufacturer.

When towing the aircraft, the towing vehicle speed must be reasonable, and all persons involved in the operation must be alert. When the aircraft is stopped, do not rely upon the brakes of the towing vehicle alone to stop the aircraft. The person in the cockpit should coordinate the use of the aircraft brakes with those of the towing vehicle. A typical smaller aircraft tow tractor (or tug) is shown in Figure 11-21.

Figure 11-21. Typical smaller aircraft tow tractor.

The attachment of the tow bar varies on different types of aircraft. Aircraft equipped with tailwheels are generally towed forward by attaching the tow bar to the main landing gear. In most cases, it is permissible to tow the aircraft in reverse by attaching the tow bar to the tailwheel axle. Any time an aircraft equipped with a tailwheel is towed, the tailwheel must be unlocked or the tailwheel locking mechanism will be damaged or broken. Aircraft equipped with tricycle landing gear are generally towed forward by attaching a tow bar to the axle of the nosewheel. They may also be towed forward or backward by attaching a towing bridle or specially designed towing bar to the towing lugs on the main landing gear. When an aircraft is towed in this manner, a steering bar is attached to the nosewheel to steer the aircraft.

The following towing and parking procedures are typical of one type of operation. They are examples, and not necessarily suited to every type of operation. Aircraft ground-handling personnel should be thoroughly familiar with all procedures pertaining to the types of aircraft being towed and local operation standards governing ground handling of aircraft. Only competent persons properly checked out should direct an aircraft towing team.

1. The towing vehicle driver is responsible for operating the vehicle in a safe manner and obeying emergency stop instructions given by any team member.
2. The person in charge should assign team personnel as wing walkers. A wing walker should be stationed at each wingtip in such a position that he or she can ensure adequate clearance of any obstruction in the path of the aircraft. A tail walker should be assigned when sharp turns are to be made, or when the aircraft is to be backed into position.
3. A qualified person should occupy the pilot's seat of the towed aircraft to observe and operate the brakes as required. When necessary, another qualified person is stationed to watch and maintain aircraft hydraulic system pressure.
4. The person in charge of the towing operation should verify that, on aircraft with a steerable nosewheel, the locking scissors are set to full swivel for towing. The locking device must be reset after the tow bar has been removed from the aircraft. Persons stationed in the aircraft should not attempt to steer or turn the nosewheel when the tow bar is attached to the aircraft.
5. Under no circumstances should anyone be permitted to walk or to ride between the nosewheel of an aircraft and the towing vehicle, nor ride on the outside of a moving aircraft or on the towing vehicle. In the interest of safety, no attempt to board or leave a moving aircraft or towing vehicle should be permitted.
6. The towing speed of the aircraft should not exceed that of the walking team members. The aircraft's engines usually are not operated when the aircraft is being towed into position.
7. The aircraft brake system should be charged before each towing operation. Aircraft with faulty brakes should be towed into position only for repair of brake systems, and then only with personnel standing by ready with chocks for emergency use. Chocks must be immediately available in case of an emergency throughout any towing operation.
8. To avoid possible personal injury and aircraft damage during towing operations, entrance doors should be closed, ladders retracted, and gear downlocks installed.
9. Prior to towing any aircraft, check all tires and landing gear struts for proper inflation. (Inflation of landing gear struts of aircraft in overhaul and storage is excluded.)
10. When moving aircraft, do not start and stop suddenly. For added safety, aircraft brakes must never be applied during towing except in emergencies, and then only upon command by one of the tow team members.
11. Aircraft should be parked in specified areas only. Generally, the distance between rows of parked aircraft should be great enough to allow immediate access of emergency vehicles in case of fire, as well as free movement of equipment and materials.
12. Wheel chocks should be placed fore and aft of the main landing gear of the parked aircraft.
13. Internal or external control locks (gust locks or blocks) should be used while the aircraft is parked.
14. Prior to any movement of aircraft across runways or taxiways, contact the airport control tower on the appropriate frequency for clearance to proceed.
15. An aircraft should not be parked in a hangar without immediately being statically grounded.

Taxiing Aircraft

As a general rule, only rated pilots and qualified airframe and powerplant technicians are authorized to start, run up, and taxi aircraft. All taxiing operations should be performed in accordance with applicable local regulations. Figure 11-22 contains the standard taxi light signals used by control towers to control and expedite the taxiing of aircraft. The following section provides detailed instructions on taxi signals and related taxi instructions.

Taxi Signals

Many ground accidents have occurred as a result of improper technique in taxiing aircraft. Although the pilot is ultimately responsible for the aircraft until the

Lights	Meaning
Flashing green	Cleared to taxi
Steady red	Stop
Flashing red	Taxi clear of runway in use
Flashing white	Return to starting point
Alternating red and green	Exercise extreme caution

Figure 11-22. Standard taxi light signals.

engine is stopped, a taxi signalman can assist the pilot around the flight line. In some aircraft configurations, the pilot's vision is obstructed while on the ground. The pilot cannot see obstructions close to the wheels or under the wings, and has little idea of what is behind the aircraft. Consequently, the pilot depends upon the taxi signalman for directions. Figure 11-23 shows a taxi signalman indicating his readiness to assume guidance of the aircraft by extending both arms at full length above his head, palms facing each other.

The standard position for a signalman is slightly ahead of and in line with the aircraft's left wingtip. As the signalman faces the aircraft, the nose of the aircraft is on the left. [Figure 11-24] The signalman must stay far enough ahead of the wingtip to remain in the pilot's field of vision. It is a good practice to perform a foolproof test to be sure the pilot can see all signals.

Figure 11-23. The taxi signalman.

If the signalman can see the pilot's eyes, the pilot can see the signals.

Figure 11-24 shows the standard aircraft taxiing signals published in the FAA Aeronautical Information Manual (AIM). It should be emphasized that there are other standard signals, such as those published by the Armed Forces. In addition, operation conditions in many areas may call for a modified set of taxi signals. The signals shown in Figure 11-24 represent a minimum number of the most commonly used signals. Whether this set of signals or a modified set is used is not the most important consideration, as long as each flight operational center uses a suitable, agreed-upon set of signals.

Figure 11-25 illustrates some of the most commonly used helicopter operating signals.

The taxi signals to be used should be studied until the taxi signalman can execute them clearly and precisely. The signals must be given in such a way that the pilot cannot confuse their meaning. Remember that the pilot receiving the signals is always some distance away, and must often look out and down from a difficult angle. Thus, the signalman's hands should be kept well separated, and signals should be over-exaggerated rather than risk making indistinct signals. If there is any doubt about a signal, or if the pilot does not appear to be following the signals, use the "stop" sign and begin the series of signals again.

The signalman should always try to give the pilot an indication of the approximate area in which the aircraft is to be parked. The signalman should glance behind himself or herself often when walking backward to prevent backing into a propeller or tripping over a chock, fire bottle, tiedown line, or other obstruction.

Taxi signals are usually given at night with the aid of illuminated wands attached to flashlights. [Figure 11-26] Night signals are made in the same manner as day signals with the exception of the stop signal. The stop signal used at night is the "emergence stop" signal. This signal is made by crossing the wands to form a lighted "X" above and in front of the head.

Servicing Aircraft

Servicing Aircraft Air/Nitrogen, Oil, and Fluids

Checking or servicing aircraft fluids is an important maintenance function. Before servicing any aircraft, consult the specific aircraft maintenance manual to determine the proper type of servicing equipment and procedures. In general, aircraft engine oil is checked with a dipstick or a sight gauge. There are markings

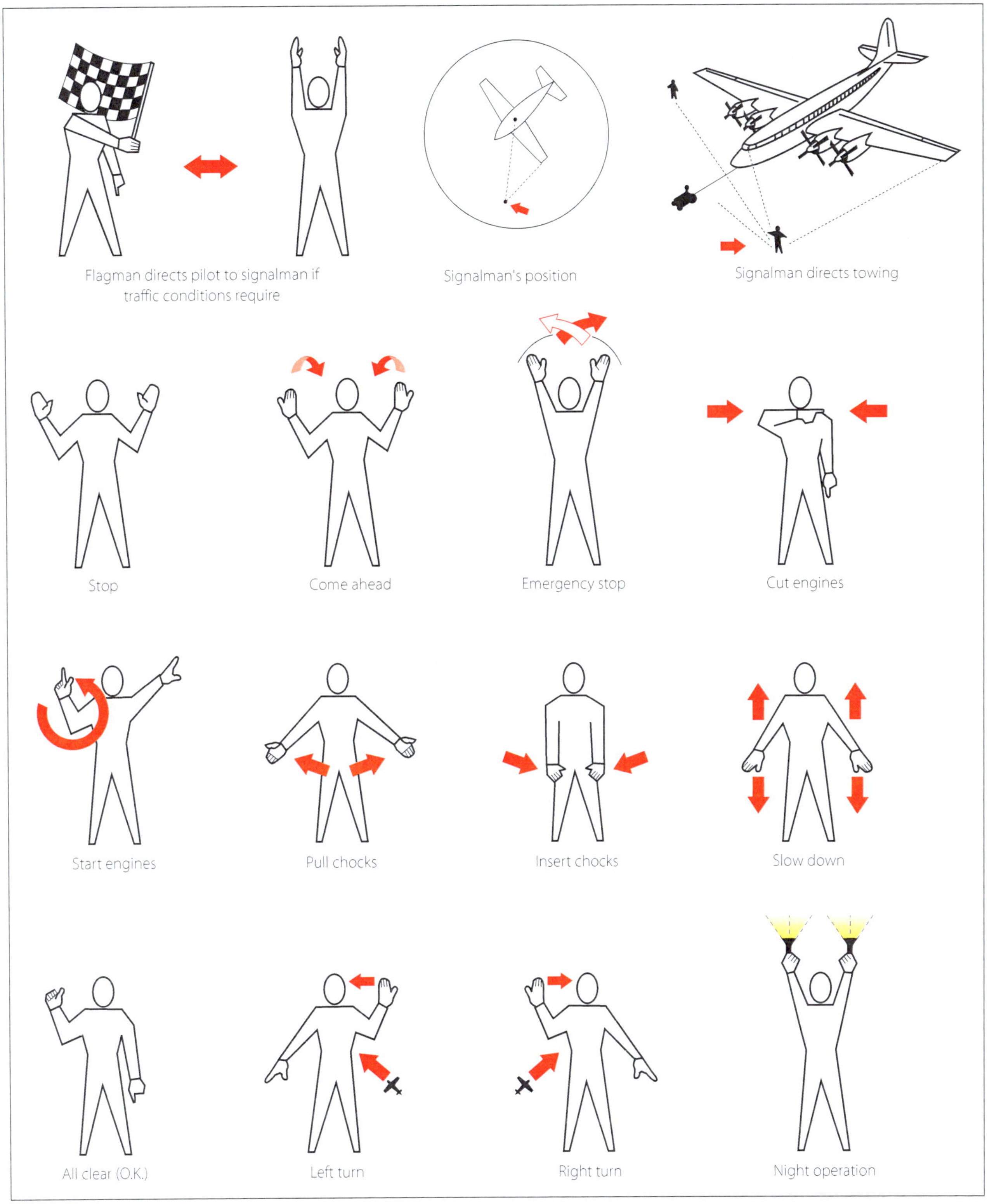

Figure 11-24. Standard FAA hand taxi signals.

on the stick or around the sight gauge to determine the correct level. Reciprocating engines should be checked after the engine has been inactive, while the turbine engine should be checked just after shutdown. Dry sump oil systems tend to hide oil that has seeped from the oil tank into the gear case of the engine. This oil will not show up on the dipstick until the engine has been started or motored. If serviced before this oil is pumped back into the tank, the engine will be overfilled. Never overfill the oil tank. Oil will foam as it is circulated through the engine. The expansion space in the oil tank allows for this foaming (oil mix-

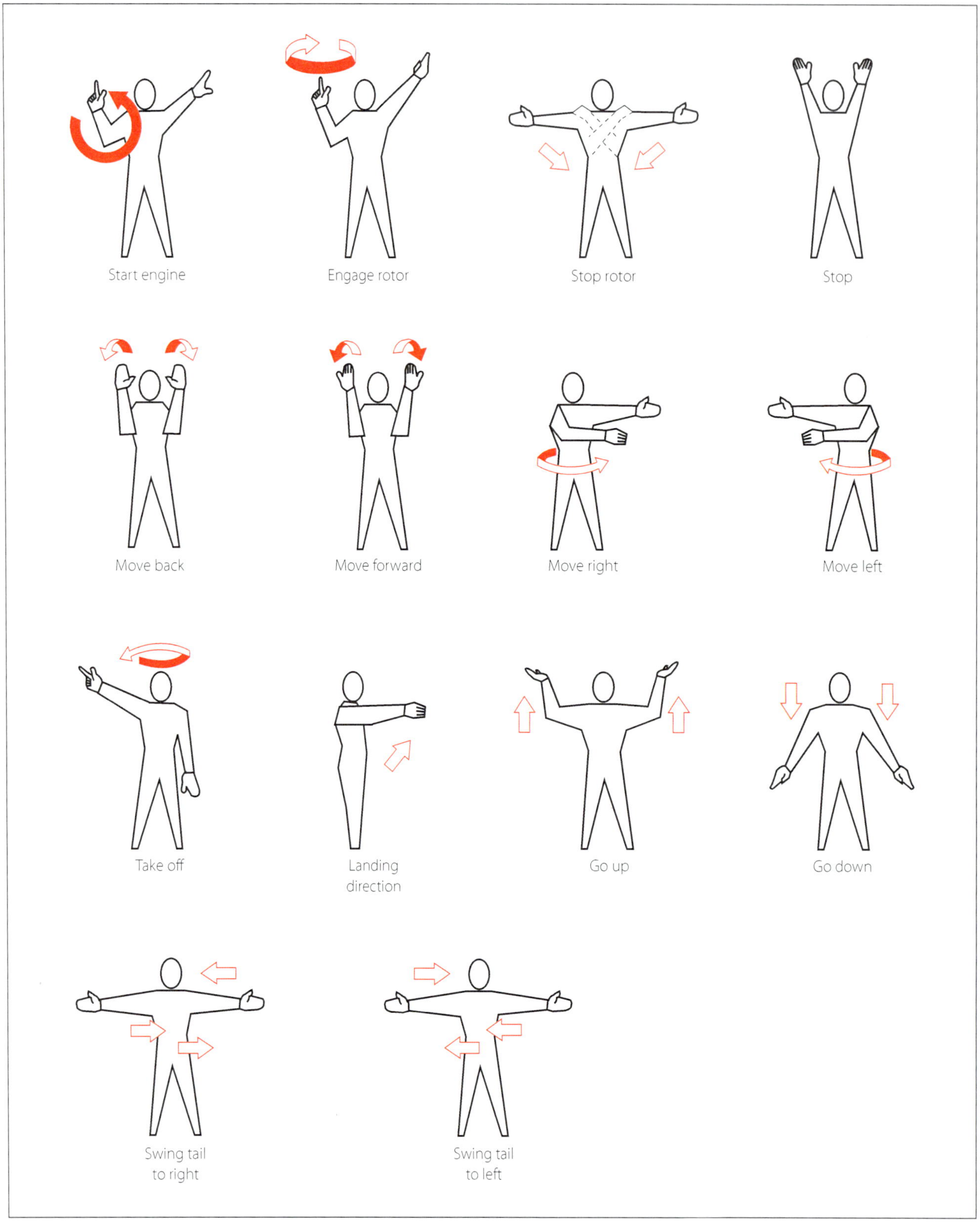

Figure 11-25. Helicopter operating signals.

ing with air). Also the correct type of oil must be used for the appropriate engine being serviced. Hydraulic fluid, fuel, and oil, if spilled on clothes or skin, must be removed as soon as possible because of fire danger and health reasons.

When servicing a hydraulic reservoir the correct fluid must be used. Normally, this can be determined by the container or by color. Some reservoirs are pressurized by air which must be bled off before servicing. Efforts must be made to prevent any type of contamination dur-

Figure 11-26. Night operations with wands.

ing servicing. Also, if changing hydraulic filters, assure that the pressure is off the system before removing the filters. After servicing the filters (if large amounts of fluids were lost) or system quantity, air should be purged and the system checked for leaks. While servicing tires or struts with high pressure nitrogen, the technician must use caution while performing maintenance. Clean areas before connecting filling hose and do not overinflate.

Ground Support Equipment

Electric Ground Power Units

Ground support electrical auxiliary power units vary widely in size and type. However, they can be generally classified by towed, stationary, or self-propelled items of equipment. Some units are mainly for in-hangar use during maintenance. Others are designed for use on the flight line either at a stationary gate area or towed from aircraft to aircraft. The stationary type can be powered from the electrical service of the facility. The movable type ground power unit (GPU) generally has an onboard engine that turns a generator to produce power. Some smaller units use a series of batteries.

The towed power units vary in size and range of available power.

The smallest units are simply high-capacity batteries used to start light aircraft. These units are normally mounted on wheels or skids and are equipped with an extra-long electrical line terminated in a suitable plug-in adapter.

Larger units are equipped with generators. Providing a wider range of output power, these power units are normally designed to supply constant-current, variable-voltage DC electrical power for starting turbine aircraft engines, and constant-voltage DC for starting reciprocating aircraft engines. Normally somewhat top-heavy, large towed power units should be towed at restricted speeds, and sharp turns should be avoided. An example of a large power unit is shown in Figure 11-27.

Self-propelled power units are normally more expensive than the towed units and in most instances supply a wider range of output voltages and frequencies. Another example, the stationary power unit shown in Figure 11-28 is capable of supplying DC power in varying amounts, as well as 115/200-volt, 3-phase, 400-cycle AC power continuously for 5 minutes.

Figure 11-27. A mobile electrical power unit.

Figure 11-28. A stationary electrical power unit.

When using ground electrical power units, it is important to position the unit to prevent collision with the aircraft being serviced, or others nearby, in the event the brakes on the unit fail. It should be parked so that the service cable is extended to near its full length away from the aircraft being serviced, but not so far that the cable is stretched or undue stress is placed on the aircraft electrical receptacle.

Observe all electrical safety precautions when servicing an aircraft. Additionally, never move a power unit when service cables are attached to an aircraft or when the generator system is engaged.

Hydraulic Ground Power Units

Portable hydraulic test stands are manufactured in many sizes and cost ranges. [Figure 11-29] Some have a limited range of operation, while others can be used to perform all the system tests that fixed shop test stands are designed to perform. Hydraulic power units, sometimes called a hydraulic mule, provide hydraulic pressure to operate the aircraft systems during maintenance. They can be used to:

- Drain the aircraft hydraulic systems.
- Filter the aircraft system hydraulic fluid.
- Refill the aircraft system with clean fluid.
- Check the aircraft hydraulic systems for operation and leaks.

This type of portable hydraulic test unit is usually an electrically powered unit. It uses a hydraulic system capable of delivering a variable volume of fluid from zero to approximately 24 gallons per minute at variable pressures up to 3,000 psi.

Figure 11-29. A portable hydraulic power unit.

Operating at pressures of 3,000 psi or more, extreme caution must be used when operating hydraulic power units. At 3,000 psi, a small stream from a leak can cut like a sharp knife. Therefore, inspect lines used with the system for cuts, frays, or any other damage, and keep them free of kinks and twists. When not in use, hydraulic power unit lines should be stored (preferably wound on a reel) and kept clean, dry, and free of contaminants.

Ground Support Air Units

Air carts are used to provide low pressure (up to 50 psi high volume flow) air which can be used for starting the engines, and heating and cooling the aircraft on the ground (using the onboard aircraft systems). It generally consists of an APU built into the cart that provides bleed air from the APU's compressor for operating aircraft systems or starting engines. [Figure 11-30]

Ground Air Heating and Air Conditioning

Most airport gates have facilities that can provide heated or cooled air. The units that cool or heat the air are permanent installations, which connect to the aircraft by a large hose that connects to the aircraft's ventilation system. Portable heating and air conditioning units can also be moved close to the aircraft and connected by a duct, which provides air to keep the cabin temperature comfortable.

Oxygen Servicing Equipment

Before servicing any aircraft, consult the specific aircraft maintenance manual to determine the proper type of servicing equipment to be used.

Two personnel are required to service an aircraft with gaseous oxygen. One person should be stationed at the

Figure 11-30. A portable compressed air cart.

control valves of the servicing equipment and one person stationed where he or she can observe the pressure in the aircraft system. Communication between the two people is required in case of an emergency.

Aircraft should not be serviced with oxygen during fueling, defueling, or other maintenance work, which could provide a source of ignition. Oxygen servicing of aircraft should be accomplished outside hangars.

Oxygen used on aircraft is available in two types: gaseous and liquid. The type to use on any specific aircraft depends on the type of equipment in the aircraft. Gaseous oxygen is stored in large steel cylinders, while liquid oxygen (commonly referred to as LOX) is stored and converted into a usable gas in a liquid oxygen converter.

Oxygen is commercially available in three general types: aviator's breathing, industrial, and medical. Only oxygen marked "Aviator's Breathing Oxygen" which meets Federal Specification BB-0-925A, Grade A, or its equivalent should be used in aircraft breathing oxygen systems. Industrial oxygen may contain impurities, which could cause the pilot, crew, and/or passengers to become sick. Medical oxygen, although pure, contains water, which can freeze in the cold temperatures found at the altitudes where oxygen is necessary.

Oxygen Hazards

Gaseous oxygen is chemically stable and is nonflammable. However, combustible materials ignite more rapidly and burn with greater intensity in an oxygen-rich atmosphere. In addition, oxygen combines with oil, grease, or bituminous material to form a highly explosive mixture, which is sensitive to compression or impact. Physical damage to, or failure of oxygen containers, valves, or plumbing can result in an explosive rupture, with extreme danger to life and property. It is imperative that the highest standard of cleanliness be observed in handling oxygen and that only qualified and authorized persons be permitted to service aircraft gaseous oxygen systems.

In addition to aggravating the fire hazard, because of its low temperature (it boils at −297 °F), liquid oxygen causes severe "burns" (frostbite) if it comes in contact with the skin.

Fuel Servicing Of Aircraft

Types of Fuel and Identification

Two types of aviation fuel in general use are aviation gasoline, also known as AVGAS, and turbine fuel, also known as JET A fuel.

Color	Grade
Red	80
Blue	100
Green	100LL
Purple	115

Aviation gasoline (AVGAS) is used in reciprocating-engine aircraft. Currently, there are three grades of fuel in general use: 80/87, 100/130, and 100LL (low lead). A fourth grade, 115/145, is in limited use in the large reciprocating-engine aircraft. The two numbers indicate the lean mixture and rich mixture octane rating numbers of the specific fuel. In other words, with 80/87 aviation gasoline, the 80 is the lean mixture rating and 87 is the rich mixture rating number. To avoid confusing the types of AVGAS, it is generally identified as grade 80, 100, 100LL, or 115. AVGAS can also be identified by a color code. The color of the fuel should match the color band on piping and fueling equipment.

Turbine fuel/jet fuel is used to power turbojet and turbo-shaft engines. Three types of turbine fuel generally used in civilian aviation are JET A and JET A-1, which are made from kerosene and JET B, which is a blend of kerosene and aviation gasoline. While jet fuel is identified by the color black on piping and fueling equipment, the actual color of jet fuel can be clear or straw colored.

Never mix AVGAS and turbine fuel. Adding jet fuel to AVGAS will cause a decrease in the power developed by the engine and could cause damage to the engine (through detonation) and loss of life. Adding AVGAS to jet fuel, although allowed, can cause lead deposits in the turbine engine and can lead to reduced service life.

Contamination Control

Contamination is anything in the fuel that is not supposed to be there. The types of contamination found in aviation fuel include water, solids, and microbial growths.

The control of contamination in aviation fuel is extremely important, since contamination can lead to engine failure, or stoppage and the loss of life. The best method of controlling contamination is to prevent its introduction into the fuel system. Some forms of contamination will still occur inside the fuel system.

Either way, the filter, separators, and screens should remove most of the contamination.

Water in aviation fuels will generally take two forms: dissolved (vapor) and free water.

The dissolved water is not a major problem until, as the temperature lowers, it becomes free water. This then poses a problem if ice crystals form, clogging filters and other small orifices.

Free water can appear as water slugs or entrained water. Water slugs are concentrations of water. This is the water that is drained after fueling an aircraft.

Entrained water is suspended water droplets. These droplets may not be visible to the eye, but will give the fuel a cloudy look. The entrained water will settle out in time.

Solid contaminants are insoluble in fuel. The more common types are rust, dirt, sand, gasket material, lint, and fragments of shop towels. The close tolerances of fuel controls and other fuel-related mechanisms can be damaged or blocked by particles as small as one-twentieth the diameter of a human hair.

Microbiological growths are a problem in jet fuel. There are a number of varieties of micro-organisms that can live in the free water in jet fuel. Some variations of these organisms are airborne, others live in the soil. The aircraft fuel system becomes susceptible to the introduction of these organisms each time the aircraft is fueled.

Favorable conditions for the growth of micro-organisms in the fuel are warm temperatures and the presence of iron oxide and mineral salts in the water.

The effects of micro-organisms are:

- Formation of slime or sludge that can foul filters, separators, or fuel controls.
- Emulsification of the fuel.
- Corrosive compounds that can attack the fuel tank's structure. In the case of a wet wing tank, the tank is made from the aircraft's structure. They can also have offensive odors.

The best way to prevent microbial growth is to keep the fuel dry.

Fueling Hazards

The volatility of aviation fuels creates a fire hazard that has plagued aviators and aviation engine designers since the beginning of powered flight. Volatility is the ability of a liquid to change into a gas at a relatively low temperature. In its liquid state, aviation fuel will not burn. It is, therefore, the vapors, or gaseous state to which the liquid fuel changes that is not only useful in powering the aircraft, but also a fire hazard.

Static electricity is a byproduct of one substance rubbing against another. Fuel flowing through a fuel line causes a certain amount of static electricity. The greatest static electricity concern around aircraft is that during flight, the aircraft moving through the air causes static electricity to build in the airframe. If that static electricity is not dissipated prior to refueling, the static electricity in the airframe will try to return to ground through the fuel line from the servicing unit. The spark caused by the static electricity can ignite any vaporized fuel.

Breathing the vapors from fuel can be harmful and should be limited. Any fuel spilled on the clothing or skin must be removed as soon as possible.

Fueling Procedures

The proper fueling of an aircraft is the responsibility of the owner/operator. This does not, however, relieve the person doing the fueling of the responsibility to use the correct type of fuel and safe fueling procedures.

There are two basic procedures when fueling an aircraft. Smaller aircraft are fueled by the over-the-wing method. This method uses the fuel hose to fill through fueling ports on the top of the wing. The method used for larger aircraft is the single point fueling system. This type of fueling system uses receptacles in the bottom leading edge of the wing, which is used to fill all the tanks from this one point. This decreases the time it takes to refuel the aircraft, limits contamination, and reduces the chance of static electricity igniting the fuel. Most pressure fueling systems consist of a pressure fueling hose and a panel of controls and gauges that permit one person to fuel or defuel any or all fuel tanks of an aircraft. Each tank can be filled to a predetermined level. These procedures are illustrated in Figures 11-31 and 11-32.

Prior to fueling, the person fueling should check the following:

1. Ensure all aircraft electrical systems and electronic devices, including radar, are turned off.
2. Do not carry anything in the shirt pockets. These items could fall into the fuel tanks.

Figure 11-31. Refueling an aircraft by the over-the-wing method.

3. Ensure no flame-producing devices are carried by anyone engaged in the fueling operation. A moment of carelessness could cause an accident.
4. Ensure that the proper type and grade of fuel is used. Do not mix AVGAS and JET fuel.
5. Ensure that all the sumps have been drained.
6. Wear eye protection. Although generally not as critical as eye protection, other forms of protection, such as rubber gloves and aprons, can protect the skin from the effects of spilled or splashed fuel.
7. Do not fuel aircraft if there is danger of other aircraft in the vicinity blowing dirt in the direction of the aircraft being fueled. Blown dirt, dust, or

other contaminants can enter an open fuel tank, contaminating the entire contents of the tank.

8. Do not fuel an aircraft when there is lightning within 5 miles.
9. Do not fuel an aircraft within 500 feet of operating ground radar.

When using mobile fueling equipment:

1. Approach the aircraft with caution, positioning the fuel truck so that if it is necessary to depart quickly, no backing will be needed.
2. Set the hand brake of the fuel truck, and chock the wheels to prevent rolling.
3. Ground the aircraft and then ground the truck. Next, ground or bond them together by running a connecting wire between the aircraft and the fuel truck. This may be done by three separate ground wires or by a "Y" cable from the fuel truck.
4. Ensure that the grounds are in contact with bare metal or are in the proper grounding points on the aircraft. Do not use the engine exhaust or propeller as grounding points. Damage to the propeller can result, and there is no way of quickly ensuring a positive bond between the engine and the airframe.
5. Ground the nozzle to the aircraft, then open the fuel tank.
6. Protect the wing and any other item on the aircraft from damage caused by spilled fuel or careless handling of the nozzle, hose, or grounding wires.
7. Check the fuel cap for proper installation and security before leaving the aircraft.
8. Remove the grounding wires in the reverse order. If the aircraft is not going to be flown or moved soon, the aircraft ground wire can be left attached.

When fueling from pits or cabinets, follow the same procedures as when using a truck. Pits or cabinets are usually designed with permanent grounding, eliminating the need to ground the equipment. However, the aircraft still must be grounded, and then the equipment must be grounded to the aircraft as it was with mobile equipment.

Defueling

Defueling procedures differ with different types of aircraft. Before defueling an aircraft, check the maintenance/service manual for specific procedures and cautions.

Defueling can be accomplished by gravity defueling or by pumping the fuel out of the tanks. When the gravity method is used, it is necessary to have a method of collecting the fuel. When the pumping method is used, care must be taken not to damage the tanks, and the removed fuel should not be mixed with good fuel.

General precautions when defueling are:

- Ground the aircraft and defueling equipment.
- Turn off all electrical and electronic equipment.
- Have the correct type of fire extinguisher available.
- Wear eye protection.

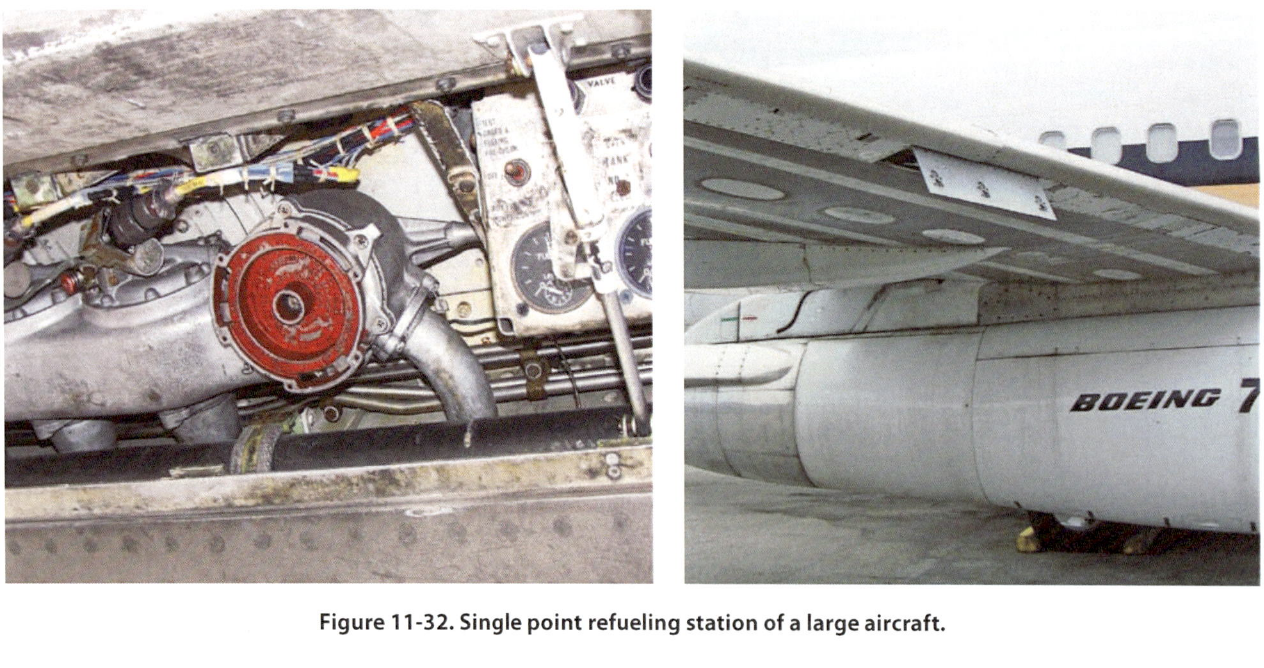

Figure 11-32. Single point refueling station of a large aircraft.

Publications, Forms, & Records

Overview — Title 14 of the Code of Federal Regulations (14 CFR)

Figure 12-1 reflects the changes in aviation related regulations, which have occurred during the time frame indicated in the left column. Just as aircraft continue to evolve with ever improving technology, so do the regulations, publications, forms, and records required to design, build, and maintain them.

The Federal Aviation Administration (FAA) regulations that govern today's aircraft are found in Title 14 of the Code of Federal Regulations (14 CFR). [Figure 12-2] There are 68 regulations organized into three volumes under Title 14, Aeronautics and Space. A fourth volume deals with the Department of Transportation, and the fifth volume is focused on NASA.

These 68 regulations can be separated into the following three categories:

1. Administrative
2. Airworthiness Certification
3. Airworthiness Operation

Since 1958, these rules have typically been referred to as "FARs," short for Federal Aviation Regulations. However, another set of regulations (Title 48) is titled Federal Acquisitions Regulations," and this has led

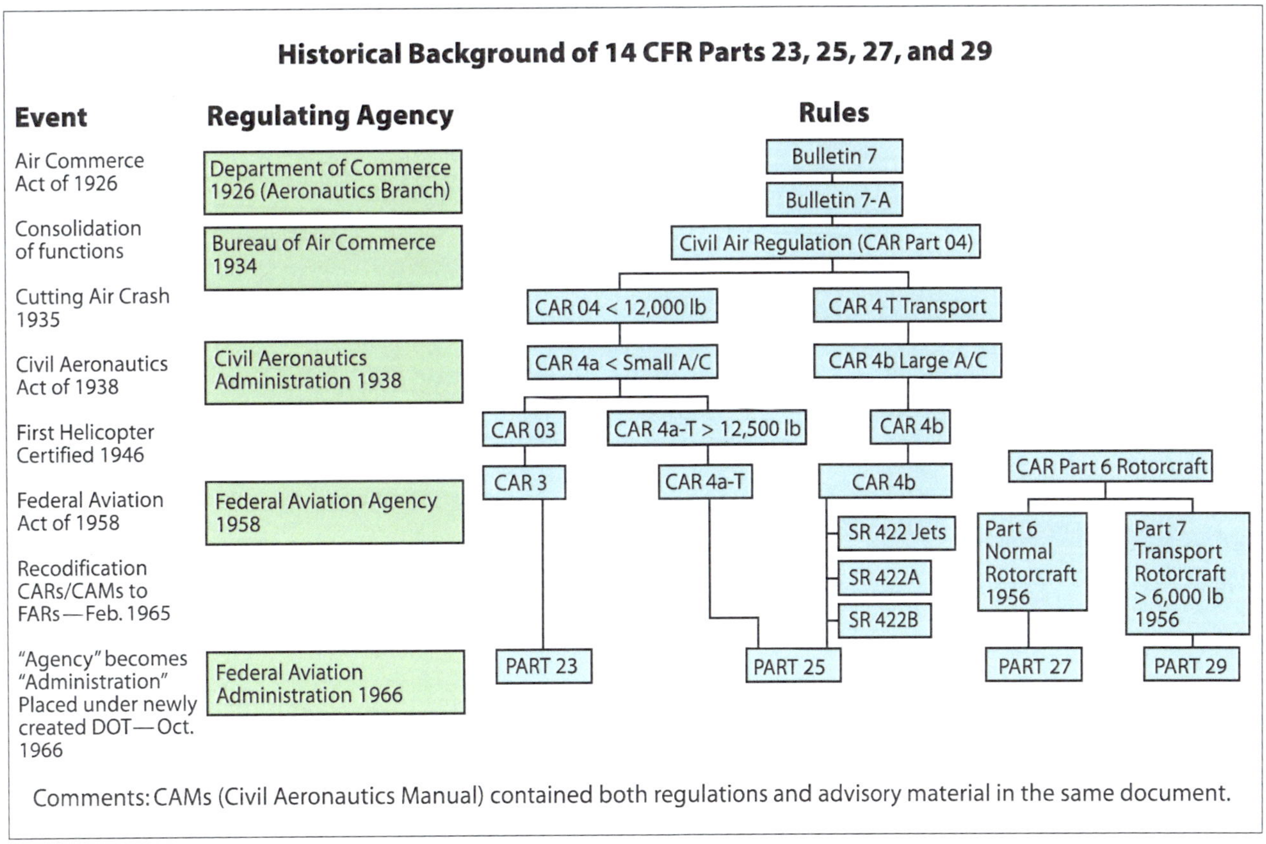

Figure 12-1. FAA historical background of aviation related regulations.

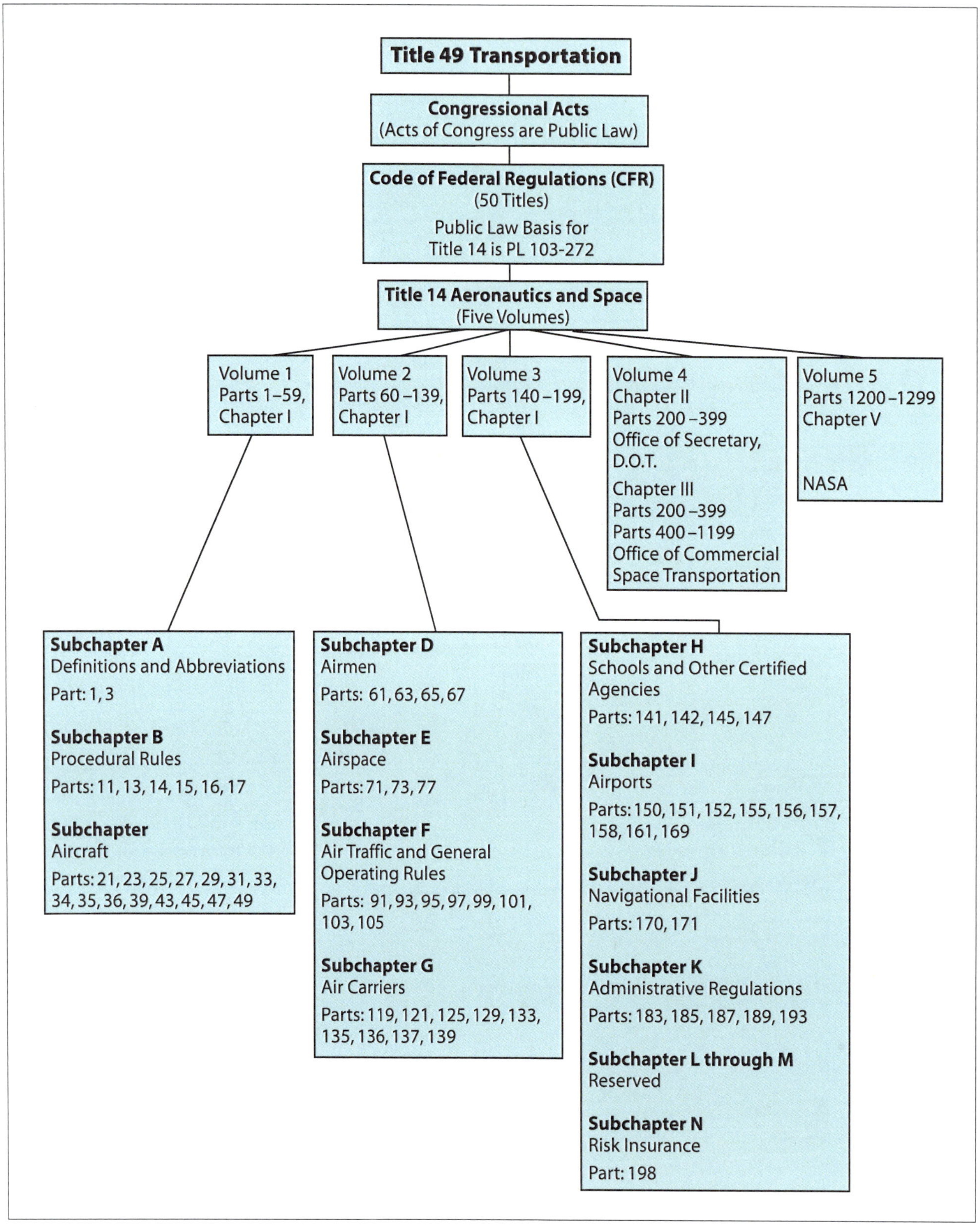

Figure 12-2. Title 14 of the Code of Federal Regulations.

Special Federal Aviation Regulations		
SFAR No.	**Title**	**Appears in 14 CFR**
13	—	Part 25
14	—	Part 121
23	—	Part 23
36	—	Parts 121, 135, 145
50–2	Special Flight Rules in the Vicinity of the Grand Canyon National Park, AZ	Parts 91, 121, 135
60	Air Traffic Control System Emergency Operation	Parts 91, 93
71	Special Operating Rules for Air Tour Operators in the State of Hawaii	Parts 121, 135
73	Robinson R–22/R–44 Special Training and Experience Requirements	Part 61
77	Prohibition Against Certain Flights Within the Territory and Airspace of Iraq	Part 91
79	Prohibition Against Certain Flights Within the Flight Information Region (FIR) of the Democratic People's Republic of Korea (DPRK)	Part 91
80	Alternative Communications Dispatching Procedures	Part 121
87	Prohibition Against Certain Flights Within the Territory and Airspace of Ethiopia	Part 91
88	Fuel Tank System Fault Tolerance Evaluation Requirements	Part 21
92–5	Flightcrew Compartment Access and Door Designs	Part 121
93	Temporary Extension of Time to Allow for Certain Training and Testing	Parts 61, 63, 121, 135, 142
97	Special Operating Rules for the Conduct of Instrument Flight Rules (IFR) Area Navigation (RNAV) Operations using Global Positioning Systems (GPS) in Alaska	Parts 71, 91, 95, 121, 125, 129, 135
98	Construction or Alteration in the Vicinity of the Private Residence of the President of the United States	Part 77
100–1	Relief for U.S. Military and Civilian Personnel Who Are Assigned Outside the United States in Support of U.S. Armed Forces Operations	Parts 61, 63, 65
103	Process for Requesting Waiver of Mandatory Separation Age for Certain Federal Aviation Administration (FAA) Air Traffic Control Specialists in Flight Service Stations, Enroute or Terminal Facilities, and the David J. Hurley Air Traffic Control System Command Center	Part 65
104	Prohibition Against Certain Flights by Syrian Air Carriers to the United States	Part 91
105	Operating Limitations for Unscheduled Operations at Chicago's O'Hare International Airport	Part 93
106	Rules for Use of Portable Oxygen Concentrator Systems Onboard Aircraft	Parts 121, 125, 135
107	Prohibition Against Certain Flights Within the Territory and Airspace of Somalia	Part 91

This chart is current as of the book publishing date. Please visit www.faa.gov for a current list of SFARs.

Figure 12-3. Appendix 3 of Advisory Circular (AC) 00-44, Status of Federal Aviation Regulations.

to confusion with the use of the acronym "FAR." Therefore, the FAA began to refer to specific regulations by the term "14 CFR part XX." Most regulations and the sections within are odd numbered because the FAA realized in 1958 when the Civil Aeronautics Regulations were recodified into the Federal Aviation Regulations, that it would be necessary to add regulations later.

Over the years, the FAA has sometimes seen the need to issue Special Federal Aviation Regulations (SFAR). [Figure 12-3] These are frequently focused very specifically on a unique situation, and are usually given a limited length of time for effectiveness. Note that the SFAR number is purely a sequential number and has no relevance to the regulation it is addressing or attached to.

The remainder of this text focuses only on those regulations relative to airworthiness certification. There are 20 of these listed in Figure 12-4, and they are shown graphically in Figure 12-5, along with five

Part 1:	Definitions and Abbreviations
Part 13:	Investigative and Enforcement Procedures
Part 21:	Certification Procedures for Products and Parts
Part 23:	Airworthiness Standards: Normal, Utility, Acrobatic, and Commuter Category Airplanes
Part 25:	Airworthiness Standards: Transport Category Airplanes
Part 33:	Airworthiness Standards: Aircraft Engines
Part 34:	Fuel Venting and Exhaust Emission Requirements for Turbine Engine Powered Airplanes
Part 35:	Airworthiness Standards: Propellers
Part 39:	Airworthiness Directives
Part 43:	Maintenance, Preventive Maintenance, Rebuilding, and Alteration
Part 45:	Identification and Registration Marking
Part 47:	Aircraft Registration
Part 65:	Certification: Airmen Other Than Flight Crewmembers
Part 91:	General Operating and Flight Rules
Part 121:	Operating Requirements: Domestic, Flag, and Supplemental Operations
Part 125:	Certification and Operations: Airplanes Having a Seating Capacity of 20 or More Passengers or a Maximum Payload Capacity of 6,000 Pounds or More; and Rules Governing Persons on Board Such Aircraft
Part 135:	Operating Requirements: Commuter and On Demand Operations and Rules Governing Persons on Board Such Aircraft
Part 145:	Repair Stations
Part 183:	Representatives of the Administrator

Figure 12-4. List of FAA regulations relative to airworthiness certification.

additional regulations that will not be addressed in this chapter. (Part 31—Airworthiness Standards, Manned Free Balloons; Part 34—Fuel Venting and Exhaust Emission Requirements for Turbine Engine Powered Airplanes; Part 36—Noise Standards: Aircraft Type and Airworthiness Certification; Part 61—Certification: Pilots, Flight Instructors, and Ground Instructors; and Part 63—Certification of Flight Crew Members Other Than Pilots.) A significant benefit of this chart is the visual effect showing the interaction of the regulation with other regulations and the placement of the regulation relative to its impact on airworthiness. It is fundamentally important that the definition of the term "airworthy" be clearly understood.

Only recently did the FAA actually define the term "airworthy" in a regulation. (Refer to the 14 CFR part 3 excerpt following this paragraph.) Prior to this definition in part 3, the term could be implied from reading part 21, section (§) 21.183. The term was defined in other non-regulatory FAA publications, and could also be implied from the text found in block 5 of FAA Form 8100-2, Standard Airworthiness Certificate. This certificate is required to be visibly placed on board each civil aircraft. (Refer to "Forms" presented later in this chapter.)

> **Part 3 General Requirements**
>
> **Sec. 3.5 Statements about products, parts, appliances and materials.**
>
> **(a) Definitions.** The following terms will have the stated meanings when used in this section:
>
> ***Airworthy*** means the aircraft conforms to its type design and is in a condition for safe operation.
>
> ***Product*** means an aircraft, aircraft engine, or aircraft propeller.

There are three primary regulations that govern the airworthiness of an aircraft. The "Big Three" are:

1. 14 CFR Part 21—Certification Procedures for Products and Parts
2. 14 CFR Part 43—Maintenance, Preventive Maintenance, Rebuilding, and Alterations
3. 14 CFR Part 91—General Operating and Flight Rules

Note that the chart in Figure 12-5 shows most of the other airworthiness certification regulations link to one of these regulations. Note also the solid line located in the center of the chart. This line represents the separation between an aircraft or aircraft parts obtaining Original Airworthiness (left side of the line) and Recurrent Airworthiness (right side of the line).

Although the history section that opens this chapter discusses "The FAA" as if it was a single unit, it is important to understand that there are various subgroups within the FAA, which have differing responsibilities of oversight in the aviation industry. These may vary by organizational chart or geographic location.

The maintenance technician will interact mostly with FAA personnel from the Flight Standard Service (AFS) but may also have some interaction with FAA personnel from the Aircraft Certification Service (AIR).

Since regulations change over the years, it should be noted that not every aircraft currently flying meets the current design regulations as printed this year. When regulations are revised, they are printed in the Federal Register and released with an "Amendment number" that ties them to the regulation being revised. Aircraft are required to meet only the specifications in force at the time the aircraft is built. (*Note:* The preceding statement does not apply to the mandatory requirements imposed by Airworthiness Directives (AD), which usually have a compliance date included in the text of the AD note.)

Maintenance Related Regulations

14 CFR Part 1 — Definitions and Abbreviations

This section is a very comprehensive, but certainly not all inclusive, list of definitions that both pilots and mechanics should become familiar with. Many regulations often provide additional definitions, which are unique to their use and interpretation in that specific part.

The Abbreviations and Symbols section (§1.2) tends to be highly focused on those abbreviations related to flight.

14 CFR Part 21 — Certification Procedures for Products and Parts

This regulation (the first of the "Big Three" identified earlier) identifies the requirements of and the procedures for obtaining type certificates, supplemental type certificates, production certificates, airworthiness certificates, and import and export approvals. Some of the other major areas covered in this part are the authorization procedures for obtaining a delegation option authorization, becoming a designated alteration station (DAS), or obtaining a Part Manufacture Approval (PMA) or an authorization related to producing a Technical Standard Order (TSO) part are also included in this regulation. Note that most of the part 21 triangle lies to the left of the vertical line reflecting its significance in the original airworthiness phase. [Figure 12-5] One of the most important sections of this regulation is §21.50, Instructions for continued airworthiness and manufacturer's maintenance manuals having airworthiness limitations sections. When an aircraft is delivered new from the manufacturer, it comes with maintenance manuals that define the inspection and maintenance actions necessary to maintain the aircraft in airworthy condition. Also any STC modification that was developed after 1981 must have, as part of the STC documentation, a complete set of instructions for continued airworthiness (ICA). This ICA contains inspection and maintenance information intended to be used by the technician in maintaining that part of the aircraft which has been altered since it was new. This ICA is comprised of 16 specific subjects. [Figure 12-6.] An ICA developed in accordance with this checklist should be acceptable to the ASI reviewing major alteration.

14 CFR Part 23 — Airworthiness Standards: Normal, Utility, Acrobatic, and Commuter Category Airplanes

Aircraft certificated under part 23 represent the greatest portion of what the industry refers to as "general aviation." These aircraft vary from the small two-place piston engine, propeller-driven trainers that are frequently used for flight training, to turbine powered corporate jets used to transport business executives. Seating capacity is limited to nine or less on all aircraft except the commuter aircraft where the maximum passenger seating is 19, excluding the pilot and copilot seats.

This part specifies the airworthiness standards that must be met in order for a manufacturer to receive a type certificate and for the aircraft to receive an airworthiness certificate. Part 23 aircraft are those aircraft that have a maximum certificated takeoff weight of 12,500 pounds or less, except for those aircraft in the commuter category. The maximum certificated takeoff weight limit rises to 19,000 pounds or less for these aircraft.

Part 23 has seven subparts, six of them providing detailed criteria for the design of these aircraft. The first, Subpart A, defines the applicability of this regulation. The others are:

- Subpart B — Flight
- Subpart C — Structure
- Subpart D — Design and Construction
- Subpart E — Powerplant
- Subpart F — Equipment
- Subpart G — Operating Limitations and Information

Within each of these subparts are numerous sections that specify details, such as center of gravity, gust load factors, removable fasteners, the shape of certain cockpit controls, engine and propeller requirements, fuel tank markings, cockpit instrumentation mark-

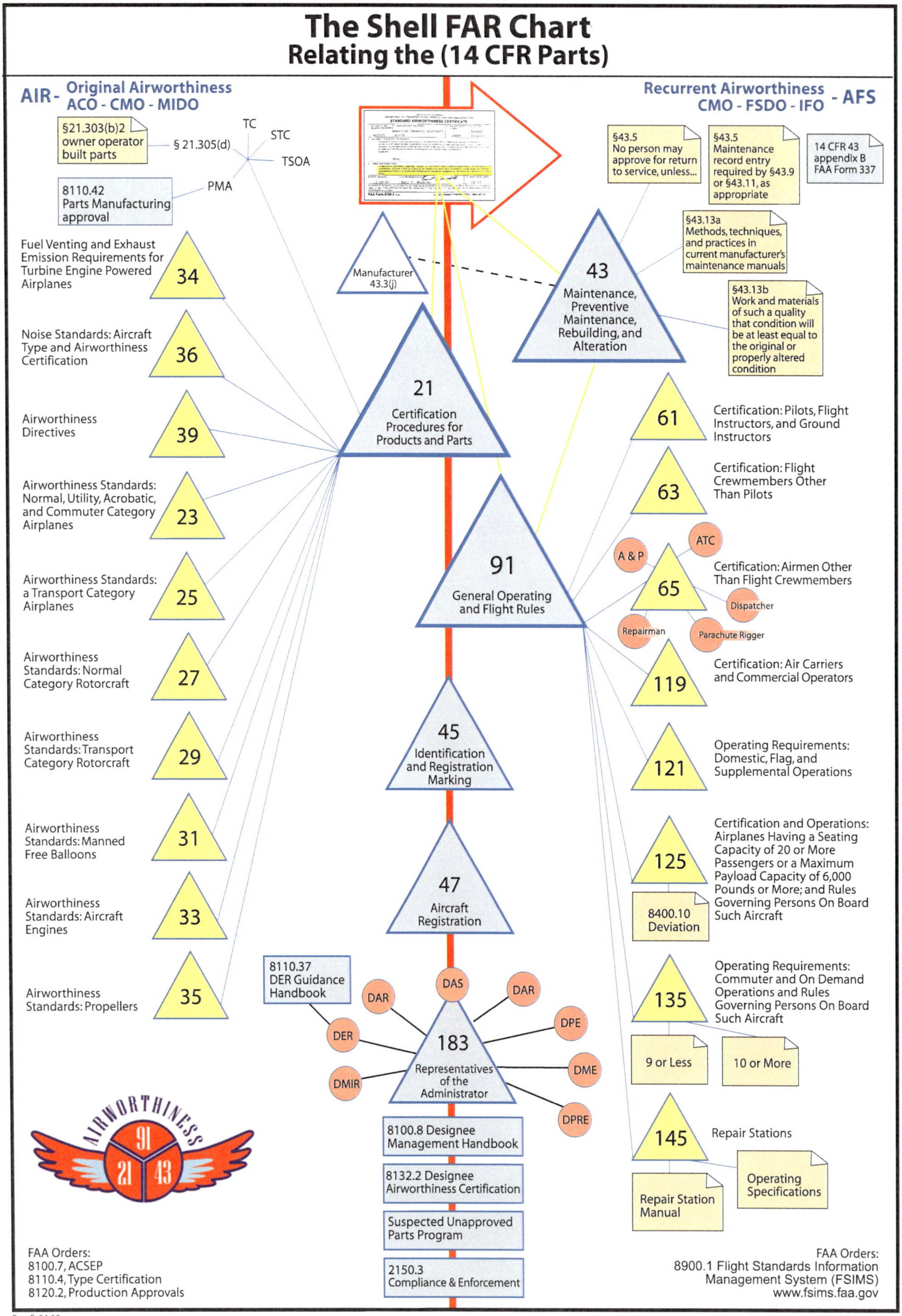

Rev. C: 04-08

Figure 12-5. Graphic chart of FAA regulations.

UNITED STATES OF AMERICA
DEPARTMENT OF TRANSPORTATION—FEDERAL AVIATION ADMINISTRATION
STANDARD AIRWORTHINESS CERTIFICATE

1 NATIONALITY AND REGISTRATION MARKS	2 MANUFACTURER AND MODEL	3 AIRCRAFT SERIAL NUMBER	4 CATEGORY
N620FT	American General Aircraft AG-5B	10009	Normal Utility

5 AUTHORITY AND BASIS FOR ISSUANCE
This airworthiness certificate is issued pursuant to the Federal Aviation Act of 1958 and certifies that as of the date of issuance the aircraft to which issued has been inspected and found to conform to the type certificate therefor, to be in condition for safe operation, and has been shown to meet the requirements of the applicable comprehensive and detailed airworthiness code as provided by Annex 8 to the Convention on International Civil Aviation, except as noted herein.
Exceptions

None

6 TERMS AND CONDITIONS
Unless sooner surrendered, suspended, revoked, or a termination date is otherwise established by the Administrator, this airworthiness certificate is effective as long as the maintenance, preventative maintenance, and alterations are performed in accordance with Parts 21, 43, and 91 of the Federal Aviation Regulations, as appropriate, and the aircraft is registered in the United States.

DATE OF ISSUANCE	FAA REPRESENTATIVE	DESIGNATION NUMBER
12-20-90	Emil F. Miehlke	CE-51

Any alteration, reproduction, or misuse of this certificate may be punishable by a fine not exceeding $1,000 or imprisonment not exceeding 3 years, or both. THIS CERTIFICATE MUST BE DISPLAYED IN THE AIRCRAFT IN ACCORDANCE WITH APPLICABLE FEDERAL AVIATION REGULATIONS.

FAA Form 8100-2 (8-82) U.S. GOVERNMENT PRINTING OFFICE: 1988-0-562-105

Administrative Regulations
Subchapters A and B

14 CFR Part 1: Definitions and Abbreviations
- 1.1 General Definitions
- 1.2 Abbreviations and Symbols
- 1.3 Rules of Construction

14 CFR Part 11: General Rulemaking Procedures
- Subpart A – Rulemaking Procedures
- Subpart B – Paperwork Reduction Act Control Numbers

14 CFR Part 13: Investigative and Enforcement Procedures
- Subpart A – Investigative Procedures
- Subpart B – Administrative Actions
- Subpart C – Legal Enforcement Actions
- Subpart D – Rules and Practice for FAA Hearings
- Subpart E – Orders of Compliance Under the Hazardous Materials Transportation Act
- Subpart F – Formal Fact-Finding Investigation Under an Order of Investigation
- Subpart G – Rules of Practice in FAA Civil Penalty Actions
- Subpart H – Civil Monetary Penalty Inflation Adjustment
- Subpart I – Flight Operational Quality Assurance Programs

Acceptable Data: methods, techniques, and practices used in the absence of current manufacturer's maintenance instructions or Instructions for Continued Airworthiness [see 14 CFR §43.13(a) and AC 43.13-1B(1)]

Airworthiness: that condition of an aircraft, airframe, engine, propeller, appliance, or component part where it conforms to its **approved type design** and is in a condition for safe operation [see 14 CFR §3.5(a)]

Approved Data: unless used with reference to another person or entity, means data approved by the Administrator [required for major repairs or alterations, see 14 CFR §65.95(a), 14 CFR §145.201(c)(2)]

Repair: a procedure, executed in accordance with data, to restore an aircraft, airframe, engine, propeller, appliance, or component part to at least equal its original or properly altered condition.

This chart is intended to be used as a template for discussions relating to the regulatory issues surrounding the certification of airworthiness. It is by no means all inclusive, but intended to be used as an educational tool in conjunction with complete and current regulatory and related materials.

When considering the airworthiness status of an aircraft, airframe, aircraft engine, propeller, appliance, or component part, first consider how it originally crossed the fence [red vertical line, from left to right] and gained its airworthiness identity. For used components/materials, you must then determine how that airworthiness status has been maintained or restored.

Created by C. Rayner Hutchinson III, Aviation Consulting Crayneriii@aol.com © 1999–2007
From a concept developed by Wyman Shell, Manager FAA MIDO-VNY

Figure 12-5. Graphic chart of FAA regulations. (continued)

ing and placards, cabin aisle width, and flammability resistant standards.

14 CFR Part 25—Airworthiness Standards: Transport Category Airplanes

The standards in this part apply to large aircraft with a maximum certificated takeoff weight of more than 12,500 pounds. This segment of aviation is usually referred to as "commercial aviation" and includes most of the aircraft seen at a large passenger airport, except for the commuter aircraft included in part 23. However, the ability to carry passengers is not a requirement for aircraft certified to part 25. Many of these aircraft are also used to transport cargo. This chapter is subdivided into the same design subpart categories and the same sequence as the requirements specified in part 23.

14 CFR Part 27—Airworthiness Standards: Normal Category Rotorcraft

This regulation deals with the small rotor wing aircraft and is consistent to part 23 with limiting the passenger seating to nine or less. However, the maximum certificated weight is limited to 7,000 pounds. It contains the same six design subparts identified in part 23, which provides the details for developing the aircraft.

Item	Subject
1.	**Introduction:** This section briefly describes the aircraft, engine, propeller, or component that has been altered. Include any other information on the content, scope, purpose, arrangement, applicability, definitions, abbreviations, precautions, units of measurement, referenced publications, and distribution of the ICA as applicable.
2.	**Description:** Of the major alteration, its functions, including an explanation of its interface with other systems, if any.
3.	**Control, operation information:** Or special procedures, if any.
4.	**Servicing information:** Such as types of fluids used, servicing points, and location of access panels, as appropriate.
5.	**Maintenance instructions:** Such as recommended inspection/maintenance periods in which each of the major alteration components is inspected, cleaned, lubricated, adjusted, and tested, including applicable wear tolerances and work recommended at each scheduled maintenance period. This section can refer to the manufacturer's instructions for the equipment installed where appropriate (e.g., functional checks, repairs, inspections). It should also include any special notes, cautions, or warnings, as applicable.
6.	**Troubleshooting information:** Information describing probable malfunctions, how to recognize those malfunctions, and the remedial actions to be taken.
7.	**Removal and replacement information:** This section describes the order and method of removing and replacing products, parts, and any necessary precautions. This section should also describe or refer to manufacturer's instructions to make required tests, trim checks, alignment, calibrations, center of gravity changes, lifting or shoring, etc., if any.
8.	**Diagrams:** Of access plates and information, if needed to gain access for inspection.
9.	**Special inspection requirements:** Such as X-ray, ultrasonic testing, or magnetic particle inspection, if required.
10.	**Application of protective treatments:** To the affected area after inspection and/or maintenance, if any.
11.	**Data:** Relative to structural fasteners such as type, torque, and installation requirements, if any.
12.	**List of special tools:** Special tools that are required, if any.
13.	**For commuter category aircraft:** The following additional information must be furnished, as applicable: A. Electrical loads. B. Methods of balancing flight controls. C. Identification of primary and secondary structures. D. Special repair methods applicable to the aircraft.
14.	**Recommended overhaul periods:** Are required to be noted on the ICA when an overhaul period has been set by the manufacturer of a component or equipment. If there is no overhaul period, the ICA should state for item 14: "No additional overhaul time limitations."
15.	**Airworthiness limitation section:** Include any "approved" airworthiness limitations identified by the manufacturer of FAA Type Certificate holding Office (e.g., an STC incorporated in a larger field-approved major alteration may have an airworthiness limitation). The FAA inspector shall not establish, alter or cancel airworthiness limitations without coordinating with the appropriate FAA Type Certificate Holding Office. If there are no changes to the airworthiness limitations, the ICA should state for item 15: "No additional airworthiness limitations" or "Not Applicable."
16.	**Revision:** This section should include information on how to revise the ICA. For example, a letter will be submitted to the local FSDO with a copy of the revised FAA Form 337 and revised ICA. The FAA inspector accepts the change by signing block 3 and including the following statement: "The attached revised/new Instructions for Continued Airworthiness (date___) for the above aircraft or component major alteration have been accepted by the FAA, superseding the Instructions for Continued Airworthiness (date____)." Once the revision has been accepted, a maintenance record entry will be made, identifying the revision, its location, and date of the Form 337.

Figure 12-6. ICA Checklist.

14 CFR Part 29—Airworthiness Standards: Transport Category Rotorcraft

This part (section) specifies those standards applicable to rotorcraft with a maximum certified weight greater than 7,000 pounds. However, it also includes additional parameters based upon seating capacity and an additional weight limit. Those parameters are passenger seating, (nine or less, ten or more) and whether the rotorcraft is over or under a maximum weight of 20,000 pounds. The same six design subparts exist in this part as are in parts 23, 25, and 27.

14 CFR Part 33—Airworthiness Standards: Aircraft Engines

Each of the four preceding 14 CFR regulations require that the engine used in the aircraft must be "type certificated." Part 33 details the requirements for both reciprocating and turbine style aircraft engines. It not only specifies the design and construction requirements, but also the "block test" requirements that subject the engine to extremely demanding testing in order to prove its capability of enduring the stresses of powering the aircraft.

14 CFR Part 35—Airworthiness Standards: Propellers

Just as each engine used on an aircraft must have a type certificate, the propeller must also be type certificated. This part is arranged the same way that part 33 is, in that subpart B specifies design and construction and subpart C covers tests and inspections.

14 CFR Part 39—Airworthiness Directives

In spite of all the emphasis on proper design and certification testing, sometimes the actual day-to-day use of the aircraft causes unanticipated wear or failure to occur. When that happens, if the FAA determines that the wear or failure represents an unsafe condition, and that the condition is likely to exist in other products of the same type of design, it will issue an AD. Actual AD notes are not included in part 39, but rather are printed in the Federal Register and are linked to this part as amendments to part 39, §39.13. AD notes are legally enforceable rules that apply to aircraft, aircraft engines, propellers, and appliances.

14 CFR Part 43—Maintenance, Preventive Maintenance, Rebuilding, and Alteration

This regulation represents the heart of aviation maintenance, and is the second of the "Big Three" regulations previously identified. The 13 rules and six appendices contained within part 43 provide the standard for maintaining all 185,000 civilian aircraft currently registered in the United States. Note that the entire part 43 triangle lies to the right of the vertical line, indicating its relationship to maintaining continued airworthiness. A more detailed explanation of this regulation is presented later in this text. [Figure 12-5]

14 CFR Part 45—Identification and Registration Marking

Part 45 identifies the requirements for the identification of aircraft, engines, propellers, certain replacement and modification parts, and the nationality and registration marking required on U.S.-registered aircraft. All type-certificated products must have the following information on a fireproof dataplate or similar approved fireproof method.

1. Builder's name
2. Model designation
3. Builder's serial number
4. Type certificate number (if any)
5. Production certificate number (if any)
6. For aircraft engines, the established rating.
7. Reference to compliance or exemption to14 CFR Part 34, Fuel Venting and Exhaust Emission Requirements for Turbine Engine Powered Airplanes
8. Any other information that the FAA determines to be appropriate

Replacement and modification parts, which are produced in accordance with a parts manufacturing approval (PMA) (14 CFR §21.303), must have the following information permanently and legibly marked:

1. The letters "FAA-PMA."
2. The name, symbol, or trademark of the holder of the PMA.
3. The part number.
4. The name and model designation for each type-certificated product on which it can be installed.

If a part has a specified replacement time, inspection interval, or other related procedure specification in the maintenance manual or Instruction for Continued Airworthiness (ICA), that part must have a part number and a serial number (or the equivalent of each).

The manufacturer of a life-limited part must either provide marking instructions for that part, or state that the part cannot be marked without a compromise to its integrity.

Exceptions are made for the identification of parts that are too small to be practical to mark the required data.

Nationality and registration marks (commonly known as the N-number for U.S.-registered aircraft) can vary in size, depending on the year that the aircraft was built, and whether or not the aircraft has been repainted. The most common size is at least 12 inches in height. Small aircraft built at least 30 years ago or replicas of these, or experimental exhibition or amateur-built aircraft may use letters at least 2 inches in height. Only a few aircraft are authorized to display registration markings of at least 3 inches. Note that this regulation sits directly on the vertical line in Figure 12-5 indicating that it applies to both original and recurrent airworthiness. [Figure 12-5]

14 CFR Part 47—Aircraft Registration

This regulation provides the requirements for registering aircraft. It includes procedures for both owner and dealer registration of aircraft.

14 CFR Part 65—Certification: Airmen Other Than Flight Crewmembers

Note: SFAR 100-1. Relief for U.S. Military and Civilian Personnel who are assigned outside the United States in support of U.S. Armed Forces Operations is a good example of the specific nature and limited time frame that are part of an SFAR.

Pilots, flight instructors, and ground instructors are certificated under part 61; flight crew other than pilots are certificated under part 63. However, many other people are also required to be certificated by the FAA for the U.S. aviation fleet to operate smoothly and efficiently. Part 65 addresses many of those other persons.

- Subpart B—Air Traffic Control Tower Operators
- Subpart C—Aircraft Dispatchers
- Subpart D—Mechanics
- Subpart E—Repairmen
- Subpart F—Parachute Riggers

A more detailed discussion of this chapter with a special emphasis on mechanics is included in Chapter 13, The Mechanic Certificate.

14 CFR Part 91—General Operating and Flight Rules

This is the final regulation of the "Big Three" identified earlier in this chapter. Notice that most of this icon is located to the right of the central vertical line in Figure 12-5, which visually indicates its "operational" involvement or "recurrent airworthiness." (Although it is an operational regulation that is focused toward the owner, operator, and/or pilot of the aircraft, the maintenance technician must have an awareness of this regulation.) Two examples of these maintenance related issues are:

1. §91.207—Emergency locator transmitters

   Paragraph (c)(2)—battery replacement interval and requirement for a logbook entry indicating the expiration date of the new battery.

2. §91.213—Inoperative Instruments and Equipment

   Paragraph (a)(2)—a letter of authorization from the Flight Standards District Office (FSDO), authorizing the operation of the aircraft under a minimum equipment list (MEL) constitutes a supplemental type certificate (STC) and must be carried in the aircraft during flight.

Subpart E—Maintenance, Preventive Maintenance, and Alterations (sections 91.401 through 91.421) is the section of most interest to the technician. He or she should be familiar with it, because it does carry some (indirect) responsibility for the technician. Note that the part 91 icon in Figure 12-5 has a direct line to part 43. This is because §91.403(b) states "No person may perform maintenance, preventive maintenance, or alterations on an aircraft other than as prescribed in this subpart and other applicable regulations, including part 43 of this chapter."

A more complete discussion of this regulation, especially Subpart E—Maintenance, Preventive Maintenance, and Alterations is presented later in this chapter.

14 CFR Part 119—Certification: Air Carriers and Commercial Operators

In order to better understand the next three regulations discussed here (parts 121, 125, and 135) a brief overview of part 119 will be beneficial. See Figure 12-7, 14 CFR part 119 Applicability of Regulations. There are more than 50 Advisory Circulars in the 120 series alone, providing additional non-regulatory information concerning the variety of procedures involved with these operations. There are basically three different

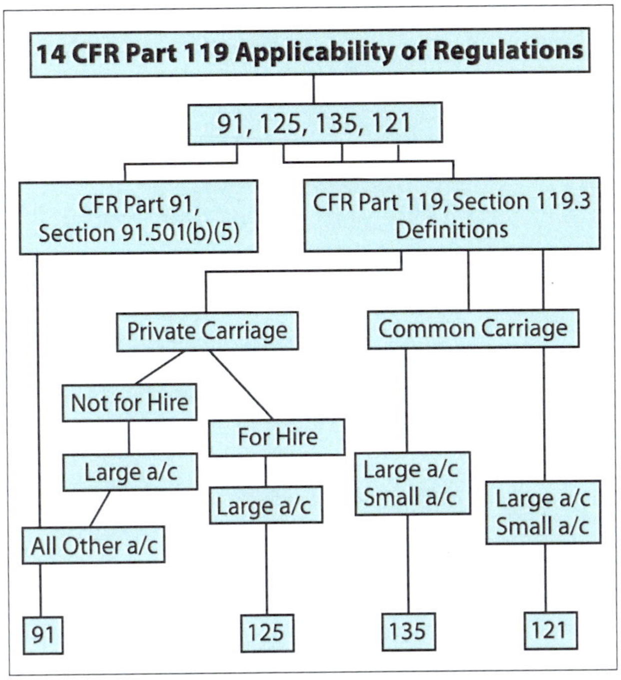

Figure 12-7. Applicability of Regulations.

criteria which must be analyzed in order to properly determine which regulation applies. These are:

1. Is the service provided for private carriage or common carriage?
2. Is the aircraft for hire or is it not for hire?
3. Is it a large or small aircraft?

AC 120-12, as revised, provides the following definition regarding the above criteria:

"A carrier becomes a common carrier when it "holds itself out" to the public, or to a segment of the public, as willing to furnish transportation within the limits of its facilities to any person who wants it. There are four elements in defining a common carrier; (1) a holding out of a willingness to (2) transport persons or property (3) from place to place (4) for compensation. This "holding out" which makes a person a common carrier can be done in many ways and it does not matter how it is done. Signs and advertising are the most direct means of "holding out" but are not the only ones… Carriage for hire which does not involve "holding out" is private carriage. Private carriers for hire are sometimes called "contract carriers," but the term is borrowed from the Interstate Commerce Act and legally inaccurate when used in connection with the Federal Aviation Act. Private carriage for hire is carriage for one or several selected customers, generally on a long term basis. The number of contracts must not be too great; otherwise, it implies a willingness to make a contract with anybody. A carrier operating pursuant to 18 to 24 contracts has been held to be a common carrier because it held itself out to serve the public generally to the extent of its facilities. Private carriage has been found in cases where three contracts have been the sole basis of the operator's business."

Operations that constitute common carriage are required to be conducted under 14 CFR part 121 or 135. Private carriage may be conducted under part 91 or 125.

The term "for hire" is not defined in any of the FAA documents, but is generally understood to mean that compensation for both direct and indirect expenses associated with the flight, as well as a profit margin for the operator are collected from the person or persons benefiting from the flight operation.

The determination of whether the aircraft is large or small is based upon the definition provided in 14 CFR part 1 (i.e., if the aircraft has maximum certificated takeoff weight of 12,500 pounds or more, it is a large aircraft. All aircraft less than 12,500 maximum certificated takeoff weight are considered to be small aircraft).

It may also help the reader understand when 121, 125, and 135 regulations apply, by taking a brief look at a list of flight operations to which part 119 does not apply.

1. Student instruction
2. Nonstop sightseeing flights with less than 30 seats and less than 25 nautical miles from the departure airport
3. Ferry or training flights
4. Crop dusting or other agricultural operations
5. Banner towing
6. Aerial photography or surveying
7. Fire fighting
8. Powerline or pipeline patrol
9. Parachute operations on nonstop flights within 25 nautical miles form the departure airport
10. Fractional ownership in accordance with part 91, subpart K

14 CFR Part 121 — Operating Requirements: Domestic, Flag, and Supplemental Operations

Part 121 establishes the operational rules for air carriers flying for compensation or hire. A domestic operation is any scheduled operation (within the 48 contiguous states, the District of Columbia, or any territory or possession) being conducted with either a turbo-jet aircraft, an airplane having 10 or more passenger seats, or a payload capacity greater than 7,500 pounds.

A "flag" operation means any scheduled operation (operating in Alaska or Hawaii to any point outside of those states, or to any territory or possession of the United States, or from any point outside the United States to any point outside the United States) being conducted with either a turbo-jet aircraft, an airplane having 10 or more passenger seats, or a payload capacity greater than 7,500 pounds.

"Supplemental" operation means any common carriage operation conducted with airplanes having more than 30 passenger seats (if less than 30, the airplane must also be listed on the operations specifications of domestic and flag carriers), with a payload capacity of more than 7,500 pounds.

Part 121 operators are required by part 119 to have the following personnel:

- Director of Safety
- Director of Operations
- Director of Maintenance
- Chief Pilot
- Chief Inspector

There are 23 subparts and 15 appendices in this regulation; however, only two subparts (J and L) are of concern for the mechanic. Subpart J—Maintenance, Preventive Maintenance, and Alterations, identifies Special Airworthiness Requirements, which deals with many of the mechanical aspects of a passenger or cargo aircraft. Subpart L—Maintenance, Preventive Maintenance, and Alterations, requires that a part 121 operator must have an operational manual which contains the following information:

- Organizational chart.
- List of persons who may perform required inspection.
- Company maintenance, preventive maintenance, or alterations.
- A system to both preserve and retrieve maintenance and inspection related information.

There is also a section (§121.368) added in 2002 that establishes the requirement for conducting inspections on aging aircraft. Additional §121.370 lists various transport category commercial aircraft that require FAA-ACO approval to operate beyond a specified number of flights. Both of these regulations came about following FAA and NTSB review of various commercial airline accidents.

14 CFR Part 125—Certification and Operations: Airplanes Having a Seating Capacity of 20 or More Passengers or a Maximum Payload Capacity of 6,000 Pounds or More; and Rules Governing Persons on Board Such Aircraft

This regulation applies to private and noncommon carriage when such operations are conducted in airplanes having 20 or more seats (excluding crewmembers) or having a payload capacity of 6,000 pounds or more. There must also be "operations specifications" issued to the operator which include the following information:

- Kinds of operations authorized
- Types of aircraft and registration numbers of the airplanes authorized for use
- Approval of the provisions of the operator's manual relating to airplane inspections, together with the necessary conditions and limitations
- Registration numbers of the airplanes that are to be inspected under an approved airplane inspection program (AAIP) under §125.247
- Procedures for the control of weight and balance of airplanes
- Any other item that the administrator determines is necessary

Just as in part 121, subpart (E) identifies special airworthiness requirements dealing mostly with the mechanical devices of the aircraft.

14 CFR Part 135—Operating Requirements: Commuter and On Demand Operations and Rules Governing Persons on Board Such Aircraft

As the title of this section states, this regulation is applicable to short distance commercial aircraft operations or "commuters" and nonscheduled carriers that operate "on demand." These aircraft are frequently referred to as air taxi or air charter aircraft.

Preceding the regulation are several SFARs. One of these, SFAR 36, is of interest to the technician because, although it is linked here directly to part 135, it is also linked in the text to parts 121 and 145. The significance of SFAR 36 is that it allows for a company with sufficient engineering and certification personnel to perform major repairs on products it is authorized to work on, without having the technical data approved by the FAA Administrator.

Aircraft operated under part 135 must be operated and maintained in accordance with the certificate holder's operations manual. This manual, when accepted by the FAA, specifies how the flight crew, ground personnel, and maintenance technicians will conduct their operations.

A pivotal portion of this regulation is §135.411, Applicability, which is the first section in subpart J. This section specifies that having a type certificated passenger seating configuration of nine or less may be maintained in accordance with the maintenance manual provided by the aircraft manufacturer. Those aircraft having a type certificated passenger seating configuration of 10 or more seats must be maintained in accordance with a maintenance manual written by the air carrier, which must then be submitted to the FAA for review and approval. The requirements for the maintenance manual are specified in §135.427. Sections 135.415 through 135.417, and 135.423 through 135.443 specify additional maintenance requirements. It should be noted that sections (§§)135.415, 135.416 and 135.417 are applicable regardless of the number of seats in the aircraft.

A major change in the "nine or less" aircraft maintenance requirements occurred in February of 2005 when §135.422 regarding aging aircraft was incorporated into part 135.

This new subpart (note the even number) to 14 CFR part 135 specifically prohibits a certificate holder from operating certain aircraft unless the Administrator has completed the aging aircraft inspection and records review. This inspection requires the certificate holder to show the FAA that the maintenance of age-sensitive parts and components has been adequate to ensure safety.

This section applies only to multiengine aircraft in scheduled operation, with nine or fewer passenger seats. It does *not* apply to aircraft operating wholly, or even partly, in Alaska.

The required record review start date varies depending on the age of the aircraft. However, once initiated, the repetitive inspection intervals are not to exceed seven years.

The certificate holder must make both aircraft and records available to the FAA for inspection and review. The certificate holder must notify the Administrator at least 60 days in advance of the availability of the aircraft and the records for review.

The records must include the following information:

1. Total years in service of the airplane
2. Total time in service of the airframe
3. Date of the last inspection and records review required by this section
4. Current status of life-limited parts
5. Time since the last overhaul of all structural components required to be overhauled on a specific time basis
6. Current inspection status of the airplane, including the time since the last inspection required by the inspection program under which the airplane is maintained
7. Current status of applicable airworthiness directives, including the date and methods of compliance, and, if the AD involves recurring action, the time and date when the next action is required
8. A list of major structural alterations
9. A report of major structural repairs and the current inspection status of those repairs

14 CFR Part 145—Repair Stations

This regulation underwent a major rewrite released in 2004 and was the most comprehensive change in nearly 20 years.

It may be of interest to note that one does not need to hold an airframe and powerplant (A&P) certificate to be employed at a repair station. The repair station may employ both certified technicians (repairman—in accordance with the requirements of part 65, subpart E) and noncertified personnel. All work that is signed off is done so using the repair station certificate number, and must be done only by persons authorized by part 65 to approve an article for return to service. Just as other certificate holders must have an operations manual, the repair station must have a "repair station manual" which must contain:

- An organizational chart.
- Procedures for maintaining rosters.
- Description of housing, facilities, and equipment.
- Procedures for revising the capability list and conducting a self-evaluation (audit).
- Procedures for revising the training program.
- Procedures governing work done at another location.

- Procedures for working on air carrier aircraft.
- Description of the required records and record keeping.
- Procedures for revising the repair station manual.
- Description of the system to identify and control the sections of the manual.

All records from repair station maintenance activity must be kept a minimum of 2 years.

Domestic repair station certificates are effective until they are surrendered, suspended, or revoked. The certificates of foreign repair stations expire, usually after 1 or 2 years, and must be renewed.

14 CFR Part 147—Aviation Maintenance Technician Schools

Part 147 defines the requirements for obtaining a maintenance training certificate. This certificate may be for either airframe, or powerplant, or a combination of the two. The minimum number of curriculum hours for conducting either airframe or powerplant training independently is 1,150.

If both airframe and powerplant ratings are offered, the combined total curriculum hours are 1,900. This is because of the 1,150 hours specified to obtain either the Airframe or the Powerplant rating; 400 hours are devoted to general studies. Only one set of general studies hours is applicable to the combined total. Therefore, 400 hours can be subtracted from the implied total of 2,300 hours (1,150 × 2) to obtain the reduced figure of 1,900 hours. Requirements are detailed as follows:

- Appendix A lists Curriculum Requirements
- Appendix B lists General Curriculum Subjects
- Appendix C lists Airframe Curricular Subjects
- Appendix D lists Powerplant Curriculum Subjects

14 CFR Part 183—Representatives of the Administrator

As the aviation industry grows and the design, manufacture, and testing of aircraft gets more complex, the FAA faces both budget constraints and personnel shortages. As early as 1962, the FAA began a program to allow private sector persons in various areas of industry to be "designees" or "representatives of the FAA Administrator." These people are *not* FAA employees, but rather are designated by the FAA to act on their behalf. Most people are aware that regular doctors may serve as "aviation medical examiners," skilled pilots can become "pilot examiners," and frequently maintenance school instructors serve as "technical person examiners" to administer the oral and practical portion of the FAA testing. Other lesser known designees are the designated engineering representatives (DER), the designated manufacturing inspection representatives (DMIR), and the designated airworthiness representatives (DAR).

- DERs approve data based upon their engineering training and their knowledge of FAA regulations.
- DMIRs make conformity inspections *only* at their employer. They are similar to "designated repairmen" because they are only authorized to inspect parts at their employers' facility.
- DARs conduct conformity inspections whenever they are authorized to do so by the FAA. They can do this for either manufacturing or maintenance facilities depending on their specific authorization.

Detailed Explanation of Primary Regulations (Parts 43 and 91)

14 CFR Part 43—Maintenance, Preventive Maintenance, Rebuilding, and Alteration

§43.1 Applicability

Paragraph (a) states quite clearly that aircraft (whether U.S.- *or* foreign-registered operating under part 121 or 135) and component parts thereof, must be maintained in accordance with the rules set forth in this part. However, although paragraph (b) states quite clearly the type of aircraft for which this part does *not* apply, it seems to have led to considerable confusion within the aviation industry. If an aircraft is flying with a Special Airworthiness—Experimental certificate (FAA Form 8130-7, Special Airworthiness Certificate—pink color certificate) and that is *the only* airworthiness certificate this aircraft has ever had, then part 43 does not apply.

Conversely, sometimes during maintenance (especially Supplemental Type Certificate (STC) modification—the STC will be addressed later in this chapter) it becomes necessary to temporarily place the aircraft into Special Airworthiness—Experimental. This is done to show compliance with federal regulations. These aircraft must still be maintained in accordance with part 43 because the aircraft had a different kind of airworthiness (in this example a Standard) prior to being issued the Special.

§43.2 *Records of overhaul and rebuilding*

These terms are not defined in part 1, and are given full explanation in this subpart. Each term states that it may not be used to describe work done on an aircraft, airframe, aircraft engine, propeller, appliance, or component part unless that item has been:

- Disassembled,
- Cleaned,
- Inspected,
- Repaired as Necessary,
- Reassembled, and
- Tested,

The key difference between the two terms is in determining how the item will be tested. If it will be "tested in accordance with approved standards acceptable to the Administration which have been developed and documented by the manufacturer, the item is said to be overhauled." This is basically another way of describing "service limits," which is a term frequently used to describe manufacturer specified acceptable limits for used parts. A "rebuilt" item, on the other hand, must be tested to the same tolerances and limits as a new item.

§43.3 *Persons authorized to perform maintenance, preventive maintenance, rebuilding, and alterations*

There are nine different persons who may perform maintenance: (Reminder: Per part 1, the FAA definition of a person is "an individual, firm, partnership, corporation, association, joint-stock association, or governmental entity. It includes a trustee, receiver, assignee, or similar representative of any of them.")

1. Certified mechanic per part 65.
2. Certified repairman per part 65.
3. Person working under the supervision of a certified mechanic or repairman.
4. Holder of repair station certificate.
5. Holder of an air carrier certificate.
6. Holder of pilot certificate (other than a sport pilot certificate) may perform preventive maintenance on an aircraft he or she owns or operates. This does not apply to aircraft operated under part 121, 129, or 135. (*Note:* Preventive maintenance is identified in appendix A, paragraph C of this part.)
7. Pilot of a rotorcraft (when operated under part 135 and in remote areas) may perform specific preventive maintenance actions. These actions may only be accomplished under the following conditions:
    - The mechanical difficulty or malfunction occurred enroute to or in the remote area.
    - The pilot has been satisfactorily trained and is authorized in writing by the certificate holder to perform the required maintenance.
    - There is no certificated mechanic available.
    - The certificate holder has procedures to evaluate the work performed when a decision for airworthiness is required.
    - The work done is listed in paragraph (c) or Appendix A of this chapter.
8. Holder of part 135 certificate may allow pilots of aircraft with nine or less passenger seats to remove and reinstall cabin seats and stretchers, and cabin mounted medical oxygen bottles.

These actions may only be accomplished under the following conditions:

- The pilot has been satisfactorily trained and is authorized in writing by the certificate holder to perform the required maintenance
- The certificate holder has written procedures available to the pilot to evaluate the work performed

9. Manufacturer may inspect and rebuild any item it has manufactured.

§43.5 *Approval for return to service after maintenance, preventive maintenance, rebuilding, or alteration*

Returning an aircraft component of an aircraft to service after maintenance, preventive maintenance, rebuilding or alteration must be done by creating an appropriate maintenance record entry as required by either §43.9 or 43.11. This may include the use of FAA Form 337, Major Repair and Alteration, if the maintenance action was a major repair or a major alteration.

Whenever a maintenance action is being planned, it is critical that the technician understands exactly:

1. What he/she is going to do,
2. How that work is classified by the FAA, and
3. What type of documentation will be required to support this activity.

The first question to consider is "Is this a repair or an alteration?" This should be a relatively simply decision since a repair basically returns the aircraft to its previous or unaltered condition (i.e., replacing magnetos, an exhaust system, tires, or brakes). Even replacing an entire engine (although it is a big job) is still a repair if it is the one properly specified for that aircraft. An alteration, on the other hand, *always changes or modifies* the aircraft from its previous state (i.e., installing winglets, new avionics, or an engine that is NOT listed in the aircraft TCDS). Please refer to part 1 and part 43, appendix A, for additional clarification and examples.

The second question to consider is whether or not the work that will be performed constitutes a major or a minor maintenance action. A "major" action is typically one that might appreciably affect weight, balance, structural strength, performance, powerplant operation, flight characteristics, or other qualities affecting airworthiness, and that are not done according to accepted practices or cannot be done by elementary operations. (The reader is again requested to refer to part 1 and part 43, appendix A, for additional clarification and examples.) This is a much more complex question, but it is extremely important as it drives the final question concerning the substantiating documentation.

The third question deals with the type of documentation required to substantiate the work performed. Minor repairs and alterations need only to refer to "acceptable" data, such as manufacturers' maintenance manuals or AC 43.13-1, the maintenance action can simply be recorded in the maintenance record as a logbook entry. Major repairs and alterations require "approved data" Some examples of approved data are: AD Notes, STCs, TCDS's, DAS specific authorizations, DER specific delegations, and FAA-approved manufacturer Service Bulletins.

Sometimes the repair or alteration being performed does not have previously approved data. In that case the technician may request that the FAA accomplish a "Field Approval." In this procedure, the technician completes the front side of Form 337 through block 6 (leaving block 3 open for later FAA approval) and then indicates in block 8 on the back what work is to be done, and what the substantiating reference data is. Form 337 is then submitted to the local FAA FSDO office for review and approval by an ASI. If necessary, this ASI may seek input from other ASIs or FAA specialists to assist in the review of the data. If the data is found to comply with FAA regulations, the ASI will enter one of the following statements in block 3, depending on whether the ASI has performed a review of the data only, or has physically inspected the aircraft:

- "The technical data identified herein has been found to comply with applicable airworthiness requirements and is hereby approved for use only on the above described aircraft, subject to conformity inspection by a person authorized in part 43, section 43.7."

or

- "The alteration or repair identified herein complies with the applicable airworthiness requirements and is approved for use only on the above described aircraft, subject to conformity inspection by a person authorized in part 43, section 43.7."

§43.7 Persons authorized to approve aircraft, airframes, aircraft engines, propellers, appliances, or component parts for return to service after maintenance, preventive maintenance, rebuilding, or alteration

There are seven different persons listed in this section who may sign return to service (RTS) documentation:

1. Certified mechanic or holder of an inspection authorization.
2. Holder of a repair station certificate.
3. Manufacturer.
4. Holder of an air carrier certificate.
5. Certificated private pilot.
6. Repairman certificated with a maintenance rating for light sport aircraft only.
7. Certificated sport pilot for preventive maintenance on an aircraft owned and or operated by him or her.

Note that although a certificated repairman is authorized to work on a product undergoing maintenance, preventive maintenance, rebuilding, or alterations (refer to §43.3), he or she is *not* authorized to approve that product for return to service. He or she must make the appropriate maintenance record entry per the requirements of §43.9 or 43.11.

§43.9 Content, form and disposition of maintenance, preventive maintenance, rebuilding, and alteration records (except inspections performed in accordance with parts 91 and 125, and §§135.411(a)(1) and 135.419 of this chapter).

The first observation is that this section specifically excludes inspection entries (those are covered in

§43.11). This section deals exclusively with maintenance record entries.

The next observation should be that the list of maintenance actions includes "preventive maintenance." As stated in the explanation of §43.3, a certificated pilot is authorized to perform preventive maintenance on the aircraft he or she owns or operates. Therefore, remember that the pilot *must* make a record entry of the preventive maintenance he or she has accomplished.

There are three distinct issues to be addressed in the maintenance entry, and they answer the questions of "what? when? and who?"

What — a description of the work performed.

When — the date the work was completed.

Who — the name of the person who did the work if other than the person who approves the Return to Service

— the signature, certificate number, and type of certificate of the person who is approving the work for return to service.

Note: Frequently, logbooks have a statement entered that ends something like this: "... and is hereby returned to service. Joe Fixer A&P, Certificate #123456789." As this section of the regulation currently reads, that part of the record entry is not required, since §43.9 clearly states that "the signature constitutes the approval for return to service only for the work performed," which also leads to another point. The technician is only signing off the work he or she has done. We will see later in the explanation of §43.11 that an inspection write-up usually carries a broader scope of responsibility. This section is very clear that the entry completed in accordance with this section *only* holds the technician responsible for the service maintenance action he or she entered.

If the maintenance accomplished was a major repair or alteration (both of which require supporting approved data), the work must be documented on FAA Form 337. If the maintenance action was a major repair and it was done by a certificated repair station, a signed copy of the completed customer work order accompanied by a signed maintenance release may be used in lieu of the FAA Form 337.

§43.10 Disposition of life-limited aircraft parts

(Note the even number again; this regulation became part of 14 CFR part 43 in 2002.)

This section presents two terms not previously defined in 14 CFR:

1. Life-limited part means any part for which a mandatory replacement limit has been specified.
2. Life status means the accumulated cycles, hours, or any other mandatory limit of a life-limited part.

This section then goes on to specify what to do with parts that are temporarily removed from and then reinstalled on a type-certificated product.

1. With parts that are removed from a type certified product and not immediately reinstalled.
2. How to transfer life-limited parts from one type-certificated product to another.

When a life-limited part is removed, the person removing it from the type-certificated product must control the part and assure proper tracking of the life-limiting factor. This is to prevent the installation of the part *after* it has reached its life limit.

There are seven possible methods the technician or repair facility may choose from to comply with this requirement.

1. Recordkeeping.
2. Tagging.
3. Non-permanent marking.
4. Permanent marking.
5. Segregation.
6. Mutilation.
7. Any other method approved or accepted by the FAA.

When a life-limited part is transferred, the information concerning the life status of that part must be transferred with it. (Although regulations already did exist which required the tracking of life-limited parts when they were installed on an aircraft, this regulation was generated to govern the disposition of such parts when they were removed from the aircraft.)

§43.11 Content, form, and disposition of records for inspections conducted under parts 91 and 125, and §135.411(a)(1) and 135.419 of this chapter.

Although this section deals exclusively with inspection record entries, the requirements are similar to §43.9 in that information of what, when, and who is required.

1. **What**—type of inspection, including a brief description.
2. **When**—date of the inspection and the total time in service
3. **Who**—the signature, certificate number, and kind of certificate of the person approving *or disapproving* the return to service.

Since this is an inspection write-up and not a maintenance entry, it is quite possible that the inspecting technician could reject or disapprove the item being inspected for the return to service. When that situation occurs, the regulation states in paragraph (b) that a list of discrepancies must be given to the owner. A reference to this list and its delivery to the aircraft owner must be reflected in the record entry. Although the regulation neither specifies how those discrepancies can be cleared, nor who may do them, it should be noted that any appropriately rated repair station or certificated technician can perform the required maintenance actions. When they are completed and the proper maintenance record entries are generated in accordance with §43.9, the aircraft is approved for return to service. It is neither necessary to have an additional inspection, nor is it necessary to contact the disapproving inspector.

If the aircraft is on a progressive inspection program, the inspection statement changes slightly from the statement referenced earlier by adding the reference to both a "routine inspection" and a "detailed inspection." (Refer to explanatory text of §43.15 for a definition of these terms.)

Inspections accomplished in accordance with other inspection program requirements must identify that particular program and which part of the program the inspection completed.

§43.12 Maintenance records: Falsification, reproduction, or alteration

The aviation community relies heavily on trust and honesty in both oral and written communication. The maintenance log entries described in §§43.9 and 43.11 provide the documentation trail relied upon by aircraft owners, pilots, and technicians regarding the aircraft's maintenance history. Falsification of these records is potentially dangerous to the personnel who rely on the accuracy of these records.

This section identifies that fraudulent entries are unacceptable. If someone commits such an act, that action is the basis for suspension or revocation of the appropriate certificate, authorization, or approval. A technician who is encouraged by his or her employer, or by anyone else, to falsify records in any way should remember this comment: "Companies come and go, but my signature lasts a lifetime. I will not use it inappropriately."

§43.13 Performance rules (general)

This section deals with the specific requirements for conducting maintenance. (*Note:* This section is probably the best to reflect the relationship between the FAA's numbering of ACs and the regulations they are related to.) Paragraph 3 on the cover page of AC 43.13-1B, Acceptable Methods, Techniques, and Practices—Aircraft Inspection and Repair (dated 9/8/98) states:

"3. Reference: Title 14 of the Code of Federal Regulations part 43, §43.13(a) states that each person performing maintenance, alteration, or preventive maintenance on an aircraft, engine, propeller, or appliance shall use the methods, techniques, and practices prescribed in the current manufacturer's maintenance manual or Instructions for Continued Airworthiness prepared by the manufacturer, or other methods techniques or practices acceptable to the Administrator, except as noted in §43.16." [Figure 12-8]

Although not all ACs are this directly linked, there is a definite relationship. (Refer to the text in this chapter on advisory circulars for additional information.)

Aircraft maintenance technicians are highly skilled personnel because aviation maintenance work requires a great attention to detail. The complexity of technology on today's aircraft demands a significant level of communication to properly accomplish maintenance, preventive maintenance, rebuilding, or alteration. This communication frequently comes in written form, i.e., manufacturer's maintenance manuals or instructions for continued airworthiness. If neither of these documents provides the guidance the technician needs, it is possible that the information found in AC 43.13 may be sufficient.

In addition to the documentation, the technician must also use the proper tools, equipment, and test apparatus

U.S. Department of Transportation

Federal Aviation Administration

Advisory Circular

TITLE 14 OF THE CODE OF FEDERAL REGULATIONS (14 CFR) GUIDANCE MATERIAL

Subject: ACCEPTABLE METHODS, TECHNIQUES, AND PRACTICES—AIRCRAFT INSPECTION AND REPAIR	**Date:** 9/8/98 **Initiated by:** AFS-640	**AC No:** 43.13-1B **Change: 1**

1. **PURPOSE.** This advisory circular (AC) contains methods, techniques, and practices acceptable to the Administrator for the inspection and repair of nonpressurized areas of civil aircraft, only when there are no manufacturer repair or maintenance instructions. This data generally pertains to minor repairs. The repairs identified in this AC may only be used as a basis for FAA approval for major repairs. The repair data may also be used as approved data, and the AC chapter, page, and paragraph listed in block 8 of FAA form 337 when:

   a. the user has determined that it is appropriate to the product being repaired;

   b. it is directly applicable to the repair being made; and

   c. it is not contrary to manufacturer's data.

2. **CANCELLATION.** The AC 43.13-1A dated 1988 is canceled.

3. **REFERENCE:** Title 14 of the Code of Federal Regulations part 43, section 43.13(a) states that each person performing maintenance, alteration, or preventive maintenance on an aircraft, engine, propeller, or appliance shall use the methods, techniques, and practices prescribed in the current manufacturer's maintenance manual or Instructions for Continued Airworthiness prepared by its manufacturer, or other methods, techniques, or practices acceptable to the Administrator, except as noted in section 43.16. FAA inspectors are prepared to answer questions that may arise in this regard. Persons engaged in the inspection and repair of civil aircraft should be familiar with 14 CFR part 43, Maintenance, Preventive Maintenance, Rebuilding, and Alteration, and part 65, Subparts A, D, and E of Certification: Airmen Other Than Flight Crewmembers, and the applicable airworthiness requirements under which the aircraft was type certificated.

4. **ACKNOWLEDGMENTS.** The FAA would like to thank the following persons and organization for their assistance in producing AC 43.13-1B: Richard Finch, Richard Fischer, Michael Grimes, Ray Stits, William A. Watkins, and the SAE, Aerospace Electronics and Electrical Systems Division. Acknowledgment is also extended to all in the aviation community who commented on the document.

5. **COMMENTS INVITED.** Comments regarding this AC should be directed to DOT/FAA; ATTN: Airworthiness Programs Branch, AFS-610; PO Box 25082; Oklahoma City, OK 73125

Acting Deputy Director, Flight Standards Service

Figure 12-8. AC 43.13-1B excerpt.

that will ensure that the work complies with accepted industry practices. If the test equipment specified by the manufacturer is not available, equipment that is determined to be equivalent and acceptable to the Administrator may be used. The technician should be cautious, however, as "proving" the equivalence of test equipment may not be as simple as it seems.

Air carriers (commercial — "scheduled" airlines operating under part 121 and the "commuter/on demand" aircraft operating under part 135, and foreign air carriers and operators of U.S.-registered aircraft under part 129) may use the maintenance manual required by the operations specifications to comply with the requirements of this section. The operator must provide a continuous airworthiness maintenance and inspection program acceptable to the Administrator.

§43.15 Additional Performance Rules for Inspections

This section presents a general comment concerning the responsibility of conducting an inspection and then provides details of three separate conditions.

1. If a rotorcraft is being inspected, specific items, such as rotor transmissions and drive shafts, must be inspected.
2. When performing an annual or 100-hour inspection, a checklist must be used. This checklist may be a personal one or one from the manufacturer; either way, it must include the scope and detail of the inspection in appendix D. Specific engine performance is also required to be tested (or monitored) as part of RTS for an annual or 100-hour inspection. This applies whether the aircraft is reciprocating or turbine powered.
3. If a progressive inspection is being conducted, it must be preceded by a complete aircraft inspection. (*Note:* A progressive inspection is the result of breaking down the large task of conducting a major inspection into smaller tasks which can be accomplished periodically without taking the aircraft out of service for an extended period of time.) Two new definitions are also presented here, "routine" and "detailed." A routine inspection is a visual examination or check of the item, but no disassembly is required. A detailed inspection is a thorough examination of the item including disassembly. The overhaul of a component is considered to be a detail inspection.

Any inspection work that is conducted away from the normal work location must be performed as if it were occurring at the maintenance facility, including using the same forms and procedures.

§43.16 Airworthiness limitations

The technician performing inspection or maintenance actions on an aircraft must be certain he/she has all appropriate data available. This includes manufacturers' maintenance manuals, operations specifications approved by the FAA under part 121, 123, or 135, or an inspection program approved under §91.409. Instructions for Continued Airworthiness, as required by §21.50 must also be consulted when available. (Since 1998 the FAA has required ICAs to be generated for all major alterations which are accomplished by the field approval process.) This section specifies that the technician is responsible to perform inspection or maintenance in accordance with all the preceding instructions.

§43.17 Maintenance, preventive maintenance, or alterations performed on U.S. aeronautical products by certain Canadian persons

This section was significantly revised in 2005 as the result of a Bilateral Aviation Safety Agreement (BASA) between the United States and Canada. The two countries have enjoyed a long and professional relationship with respect to reciprocal aviation maintenance activity. This section of part 43 both defines some terms and gives specific limitations as to what an Aviation Maintenance Engineer (AME is the Canadian equivalent to the U.S. A&P) may do to maintain U.S.-registered aircraft located in Canada. It also provides similar limitations for an Approved Maintenance Organization. (AMO is the Canadian equivalent to the U.S.-certified repair stations.)

Appendix A — Major Alterations, Major Repairs, and Preventive Maintenance

This appendix provides a comprehensive, but not exclusive, list of the subjects stated. Paragraph (a) is titled Major Alteration, and is further subdivided as follows:

- Airframe
- Powerplant
- Propeller
- Appliance

That same subdivision is used in paragraph (b), which is titled Major Repairs.

Paragraph (c) is titled Preventive Maintenance and identifies those maintenance actions, which are defined as preventive maintenance (provided the maintenance does not involve complex assembly operations). Preventive maintenance work may be accomplished by the holder of at least a private pilot certificate provided he or she is the owner or operator of that aircraft, and it is not operated under part 121, 129, or 135.

Appendix B — Recording of Major Repairs and Major Alterations

In most cases when a major repair or alteration is accomplished, FAA Form 337, Major Repair or Alteration, is completed at least in duplicate with the original going to the aircraft owner and a copy sent to the FAA Aircraft Registration Branch in Oklahoma City where all civil aircraft information is compiled and retained. *Note:* Historically the second copy was sent to the local FAA Flight Standards District Office (FSDO) within 48 hours after RTS. This copy is reviewed by an ASI and then forwarded by the FSDO to FAA records on OK City. However, in the fall of 2005 the FAA made a significant change to this submittal process and now requires the technician to submit the Form 337 directly to the Aircraft Registration Branch in Oklahoma City. (Although a third copy is not required, it makes good business sense for the technician or certified repair station to keep a copy of the work that was accomplished.)

However, if a certificated repair station completes a major repair, it may provide the customer with a signed copy of the work order and a maintenance release signed by an authorized representative of the repair station, instead of the FAA Form 337.

If the major repair or alteration was done by an authorized Canadian maintenance person (AME or AMO), the copy normally provided to the FAA-FSDO is sent directly to the FAA Aircraft Registration Branch.

Finally, if extended range tanks are installed in either passenger or cargo compartments, the technician must generate a third FAA Form 337 for the modification. This copy must be placed and retained in the aircraft. (Refer to §91.417(d).)

Appendix C is reserved for future use and therefore currently contains no information.

Appendix D — Scope and Detail of Items To Be Included in Annual and 100-Hour Inspections

Some important items to consider in this appendix are:

1. The list of items and areas to be inspected are exactly the same for an annual as a 100 hour. (The only difference is who is authorized to approve the aircraft for return to service following the inspection.) Refer to §65.95(a)(2), which states that an IA must perform an annual inspection.
2. The aircraft and engine must be cleaned prior to conducting the inspection.
3. Any miscellaneous item not covered in the detailed list provided must also be inspected for improper installation and operation.
4. There are eight specific areas identified for detail inspection. They are the fuselage hull group, cabin/cockpit group, engine/nacelle group, landing gear group, wing/center section group, empennage assembly, propeller group, and the radio group.

Appendix E — Altimeter System Test and Inspection

This is commonly referred to as "the 411 test." (Refer to §91.411, which requires that no person may operate an aircraft in controlled airspace under IFR unless the aircraft has had this test completed successfully within the preceding 24 months.) This section, which requires detailed testing of the static pressure system, the altimeter, and the automatic pressure altitude reporting equipment, also requires that the test information be recorded in the maintenance logs *and* on the altimeter.

Appendix F — ATC Transponder Tests and Inspections

This is commonly referred to as "the 413 test." (Refer to §91.413, which requires that no person may use a transponder unless it has had this test completed successfully within the preceding 24 months.) This section specifies an extremely complex set of tests, which may be accomplished either as a bench test, or by using portable test equipment. Major categories of the testing required are radio reply frequency, suppression, receiver sensitivity, radio frequency peak output power, and mode S (when applicable). Upon completion of testing, proper entries must be made in the maintenance record.

14 CFR Part 91 — General Operating and Flight Rules

Subpart A — General

As mentioned in the brief overview of the regulation portion earlier in this chapter, this part is actually addressing the operation of the aircraft. For example, §91.7(a) states "no person may operate a civil aircraft unless it is in an airworthy condition." We learned earlier that this term means that the aircraft conforms to its approved type design and is in condition for safe operation. When the pilot performs a preflight inspection, he or she is making a determination concerning the "condition for safe operation." The pilot does not usually determine "conformity to type design" unless he or she performs a review of the maintenance records. However, since that is fundamental to the definition of airworthy, it is still part of their responsibility. Therefore, a professional and ethical technician will want to help the customer understand his or her responsibilities in maintaining and documenting the airworthiness of the aircraft.

Subpart E — Maintenance, Preventive Maintenance, and Alterations

§91.401 — Applicability

Although this subpart describes in general the rules regarding maintenance, preventive maintenance, and alteration, certain sections do not apply if the aircraft is operated in accordance with part 121, 125, 129, or 135.

§91.403 — General

The owner/operator holds the primary responsibility for maintaining the aircraft in airworthy condition. This includes compliance with all applicable airworthiness directives and is the reason that the FAA sends new AD notes to the registered owners of the affected aircraft. All maintenance performed must be accomplished in accordance with part 43. Compliance with the appropriate manufacturer maintenance manuals and Instructions for Continued Airworthiness is also required. Mandatory replacement times, inspection intervals, and related procedures as specified in the FAA-approved operations specifications must also be complied with.

§91.405 — Maintenance required

The owner/operator is required to have the appropriate inspections made, and to have discrepancies repaired in accordance with part 43. He or she is also required to ensure that the appropriate entries have been made in the maintenance records. Any inoperative instruments or equipment must be properly placarded as inoperative.

§91.407 — Operation after maintenance, preventive maintenance, rebuilding or alteration

Whenever the aircraft has undergone maintenance, preventive maintenance, rebuilding or alteration, it must have been approved for return to service and a proper entry made in the maintenance records. If the maintenance that was done could have appreciably changed the flight characteristics, an appropriately rated pilot must perform an operational flight check of the aircraft and must make an entry of the flight in the maintenance records. If ground testing and inspection can show conclusively that the maintenance has not adversely affected the flight characteristics, no flight test is required.

§91.409 — Inspections

This section identifies various types of inspection applicable to the civilian aircraft fleet. Paragraph (a) defines the requirement for an annual inspection. However, there are certain exceptions to this regulation:

1. Aircraft operating under a special flight permit.
2. Aircraft operating with a current experimental certificate.
3. Aircraft operating with a light sport aircraft certificate.
4. Aircraft operating with a provisional certificate.
5. Aircraft operating in accordance with the operational specification of a part 125 or 135 certificate holder.
6. Aircraft operating under a progressive inspection program.

Annual inspections are usually the inspection method associated with small "general aviation" aircraft. If this same aircraft is used for hire (including flight instruction for hire), then the aircraft must also be inspected every 100 hours of time in service. This requirement for a 100-hour inspection to be conducted on an aircraft may be exceeded by as much as 10 hours if the aircraft is en route to reach a facility that will be conducting the inspection. Any time accrued between 100 and 110 hours is subtracted from the hours remaining before the *next* 100-hour inspection.

Since aircraft used for hire only generate revenue when they are flying, any time that the aircraft is "down for inspection" can result in a loss of income for the owner/operator. Therefore, the FAA has made provision to minimize the impact of the 100-hour and annual inspection requirement. The owner/operator may petition the local FSDO for approval of a progressive inspection program. This program breaks the complete inspection of the aircraft into smaller, less time-consuming steps. (Refer to part 43, appendix D.) This inspection may be either performed or supervised by a technician holding an Inspection Authorization (IA). The program must ensure at all times that the aircraft is airworthy. The owner/operator must submit an inspection schedule with his or her application to the FAA. This schedule must identify the time intervals (hours or days) when routine and detailed inspections are to be accomplished. (Refer to §43.15.) Just as with the 100-hour inspection, a 10-hour maximum extension of a specified inspection interval is allowed if the aircraft is en route. A change in the inspection interval is also allowed for changes in service experience.

If the progressive inspection is discontinued, the aircraft will again be subject to the traditional annual and 100-hour inspections.

Other inspection programs, which may be applicable to other aircraft, are a continuous airworthiness inspection program and an approved aircraft inspection program (AAIP). The former program is applicable to either a part 121 or 135 carrier, but the latter program is limited to part 135 operators only. Finally, the owner/operator may use either a current inspection program recommended by the aircraft manufacturer or one established by the owner/operator and approved by the local FSDO. Any subsequent changes to that program must also be approved by the local FSDO.

There may be an instance when the operator of an aircraft wishes to change from one type of inspection program to another. In that case the time in service, calendar times, or cycles of operation from the current program must be carried over to the subsequent program.

§91.410—Special maintenance program requirements

After the separation of an 18-foot section of the upper fuselage area of an Aloha Airlines Boeing 737 in 1988, Boeing, the FAA, and the NTSB carefully studied this catastrophic failure. It was determined that corrosion and fatigue failure had occurred due to the numerous pressurization cycles this aircraft had experienced during its use by Aloha Airlines. This 19-year-old aircraft had over 90,000 landing cycles on it. The subsequent lessons learned from this event have led the FAA to impose flight cycle life limits on various commercial aircraft. Those limits and the specific aircraft impacted are identified in paragraph (a) of this section. Paragraph (b) specifies requirements for inspection of the fuel tank system on all turbine powered transport category aircraft having a passenger capacity of 30 or more, or a maximum certificated payload capacity of 7,500 pounds or more, if they were type certificated after January 1, 1958. This requirement is a result of the in-flight breakup of a 25-year-old Boeing Model 747-100 series airplane on July 17, 1996, after takeoff from Kennedy International Airport in New York, resulting in 230 fatalities. The accident investigation conducted by the National Transportation Safety Board (NTSB) indicated that the center wing fuel tank exploded due to an unknown ignition source. This accident prompted the FAA to examine the underlying safety issues surrounding fuel tank explosions, the adequacy of the existing regulations, the service history of airplanes certificated to these regulations, and existing maintenance practices relative to the fuel tank system. This rule requires design approval holders of certain turbine-powered transport category airplanes, and of any subsequent modifications to these airplanes, to substantiate that the design of the fuel tank system precludes the existence of ignition sources within the airplane fuel tanks. It also requires developing and implementing maintenance and inspection instructions to assure the safety of the fuel tank system. For new type designs, this rule also requires demonstrating that ignition sources cannot be present in fuel tanks when failure conditions are considered, identifying any safety-critical maintenance actions, and incorporating a means either to minimize development of flammable vapors in fuel tanks or to prevent catastrophic damage if ignition does occur. These actions are based on accident investigations and adverse service experience, which have shown that unforeseen failure modes and lack of specific maintenance procedures on certain airplane fuel tank systems may result in degradation of design safety features intended to preclude ignition of vapors within the fuel tank.

§91.411—Altimeter system and altitude reporting equipment tests and inspections

Commonly referred to as "the 411 test." This section specifies the requirements for testing the static pressure system, each altimeter instrument, and each automatic pressure altitude reporting system every 24 calendar months. The static system must also be tested any time it has been "opened and closed" except for the normal use of the system drain, and alternate static system pressure valves. If the automatic pressure altitude reporting system of the air traffic control (ATC) transponder is either installed or subjected to maintenance actions, the system must also be tested per appendix E of part 43.

Due to the inherent design and accuracy of this system, only certain persons may perform these tests: the aircraft manufacturer, a properly rated repair station, or a certificated airframe mechanic. It should be noted that the airframe technician may *only* perform the inspection and test of the static pressure system. Calibration and maintenance of related instruments is specifically prohibited to the technician by the language of 14 CFR §65.81 and specifically allowed in 14 CFR §145.59 for repair stations holding an "instrument rating."

TSO'd items are considered to be "tested and inspected" as of the date they were manufactured. The maximum altitude for which the system was tested is the maximum altitude at which the aircraft can be flown IFR in controlled airspace.

§91.413—ATC transponder tests and inspections

This "413 test" is the other test required every 24 months. Whenever the ATC transponder is installed or has undergone maintenance, the complete system must be tested and inspected in accordance with appendix E of part 43. The transponder itself must be tested and inspected in accordance with appendix F of part 43. As with the 411 test, only certain persons are authorized to conduct the tests. They are: the manufacturer of the aircraft, a properly certificated repair station, or the holder of a continuous airworthiness maintenance program under part 121 or 135.

§91.415—Changes to aircraft inspection programs

If the FAA determines that the inspection program established and approved under either §91.409 or 91.1109 must be revised to ensure continued safety and adequacy of the program, the owner/operator must make the necessary changes as identified by the Administrator. If the owner/operator desires to contest this request, they must petition the FAA to reconsider their request to change the program within 30 days of receiving the change request from the FAA.

§91.417—Maintenance records

The understanding and implementation of this section is fundamental to the aircraft industry, in general, and the aircraft owner/operator, in specific. A professional maintenance technician should not only be knowledgeable of this section, but should also be able to help the owner/operator understand it. [Figure 12-9] This section identifies four types of records—two are quite specific (paragraphs (a) and (d)) and two are more general: (a)(1) and (a)(2). Paragraph (a) refers to the 411 and 413 testing, which requires testing every 24 months; therefore records must be kept for that length of time.

Paragraph (d) refers to the installation of fuel tanks in the cabin or cargo area. The FAA Form 337 authorizing this installation must be kept on board the aircraft all the time.

Note: Other than this paragraph, there is no requirement that the maintenance records of the aircraft be carried on the aircraft. In fact, there are very logical reasons that in most cases they shouldn't be. The two biggest concerns are damaged or lost records. It is much safer to retain the logs in a filing system in the office. It is also a very wise idea to have the logbook copied or scanned, and retained at a separate location should a catastrophic event (fire, flood, tornado, hurricane, and so forth) occur at the site the original records are retained.

Subparagraph (a)(1) then lists those records which are later defined in (b)(1) as being only retained for 1 year or until the work is repeated or superseded. Subparagraph (a)(2) specifies the records that are permanent records, and are identified in subparagraph (b)(2) as those which must be transferred with the aircraft. Refer to the chart for further clarification. [Figure 12-9]

Paragraph (c) requires that all of the maintenance records mandated by this section be made available upon request to the Administrator or any authorized representative of the National Transportation Safety Board. Furthermore, the owner/operator must provide the Form 337 required to be aboard the aircraft whenever additional fuel tanks are installed in either the passenger compartment or the baggage compartment (per paragraph (d)) to any law enforcement officer upon request.

§91.419—Transfer of maintenance records

When an aircraft is sold, it is logical that the records are transferred with it. They may be either in plain language or coded. However, the purchaser may elect to permit the seller to retain the actual records, but if that occurs, the purchaser (now the current owner/operator) must still make available upon request, these records to either the FAA or the NTSB.

§91.421—Rebuilt engine maintenance records

This section presents the term "zero time." Although not truly given as a definition, the wording of the regulation is very clear that an aircraft engine, when rebuilt by the engine manufacturer or an agency approved by the manufacturer, may be given a new maintenance record, showing no previous operating history. This new record must include a signed statement with the date it was rebuilt, any changes incorporated by compliance with AD notes, and compliance with any of the manufacturer's service bulletins.

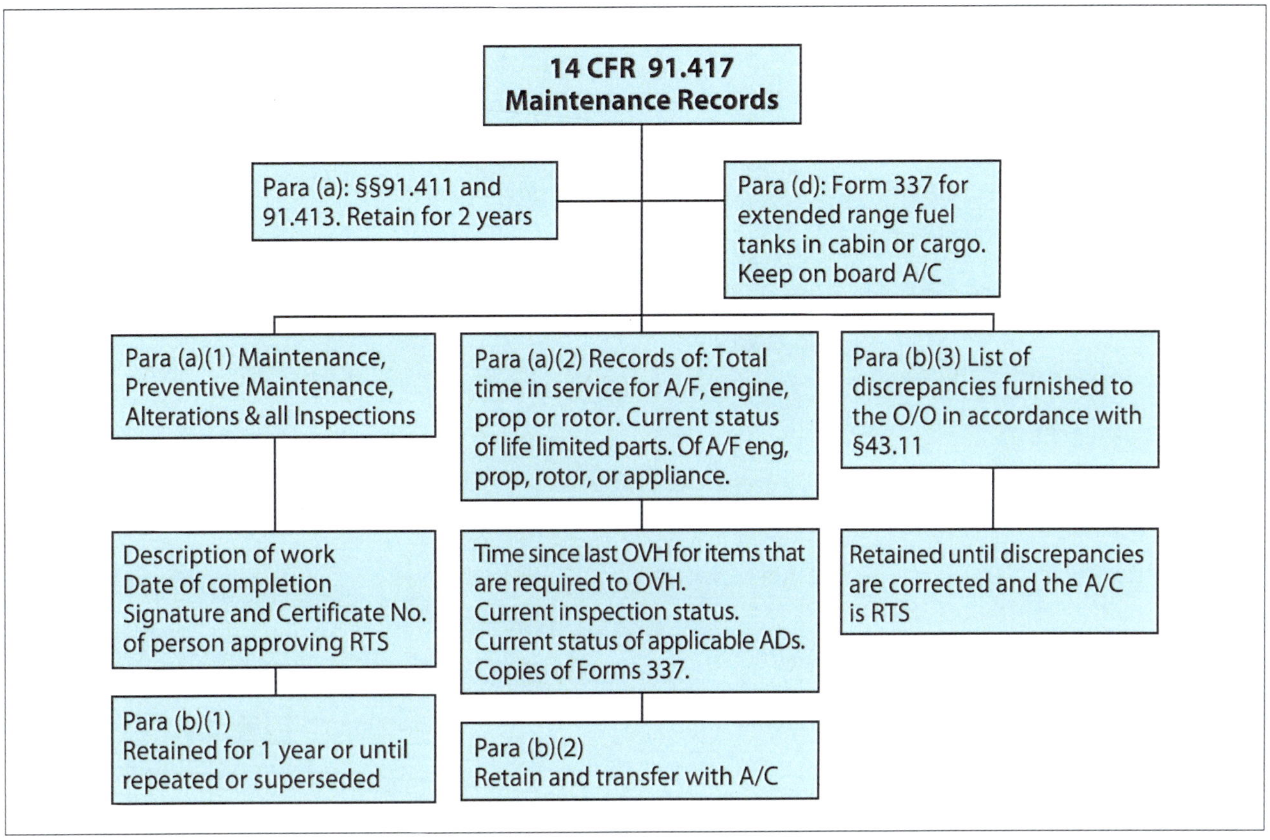

Figure 12-9. Maintenance records.

Civil Air Regulations (CARs)

Prior to 1926, access to flying was totally uncontrolled. No licensing or certification was required. By the middle of the 1920s, it became obvious that unregulated private and commercial flying was dangerous. There was a growing awareness and acceptance that regulation could improve safety and encourage growth in aviation. Therefore, in 1926 the aviation industry requested Congress to enact federal legislation to regulate civil aviation. Thus, the Air Commerce Act of 1926 provided for the:

1. Establishment of airways.
2. Development of aviation aids.
3. Investigation of aviation accidents.
4. Licensing of pilots.
5. Certification of aircraft.

The CARs were part of the original certification basis for aircraft first certified in the 1940s, 1950s, and 1960s by the CAA. Therefore, the CARs may still be needed as a reference for these older aircraft. [Figure 12-10]

CAR 3—Airplane Airworthiness—Normal, Utility, Acrobatic, and Restricted Purpose Categories

As the name implies, and Figure 12-1 indicates, this specific regulation is the basis for the current part 23 regulation. It has the following subpart categories:

A—Airworthiness Requirements

B—Flight Requirements—General

C—Strength Requirements—General

D—Design and Construction—General

E—Powerplant Installations—Reciprocating Engines

F—Equipment

Some examples of CAR 3 aircraft are Piper PA 22, PA 28, PA 32 and Cessna 182, 195, and 310.

Note: The "CAR" acronym actually has two interpretations: Civil Air Regulations and Canadian Aviation Regulations. The technician must clearly understand the difference, and recognize when one or the other is appropriate.

Predecessor Regulations to the Federal Aviation Regulations (14 CFR)	
	Aeronautical Bulletins
Aeronautics Bulletin No. 7A	Airworthiness Requirements for Aircraft
Aeronautics Bulletin No. 7H	Alteration and Repair of Aircraft
	Civil Air Regulations
CAR Part 1 Amdt. 1-9	Certification, Identification, and Marking of Aircraft and Related Products
CAR Part 3 Amdt. 1-14	Airplane Airworthiness—Normal, Utility, Acrobatic, and Restricted Purpose Categories
CAR Part 4a Amdt. 1	Airplane Airworthiness
CAR Part 4b Amdt. 1-15	Airplane Airwothiness: Transport Categories
CAR Part 6 Amdt. 1-7	Rotorcraft Airworthiness: Normal Category
CAR Part 7 Amdt. 1-5	Rotorcraft Airworthiness: Transport Categories
CAR Part 8	Aircraft Airworthiness: Restricted Category
CAR Part 9 Amdt. 1	Aircraft Airworthiness: Limited Category
CAR Part 10 Amdt. 1	Certification and Approval of Import Aircraft and Related Products
CAR Part 13 Amdt. 1-6	Aircraft Engine Airworthiness
CAR Part 14 Amdt. 1	Aircraft Propeller Airworthiness
CAR Part 18 Amdt. 1-3	Maintenance, Repair, and Alteration of Airframes, Powerplants, Propellers, and Appliances
	Civil Aeronautics Manual
CAM 1	Certification, Identification, and Marking of Aircraft and Related Products
CAM 3	Airplane Airworthiness: Normal, Utility, and Acrobatic Categories
CAM 4b	Airplane Airworthiness: Transport Category
CAM 6	Rotorcraft Airworthiness: Normal Category
CAM 7	Rotorcraft Airworthiness: Transport Category
CAM 8	Aircraft Airworthiness: Restricted Category
CAM 9	Aircraft Airworthiness: Limited Category
CAM 10	Certification and Approval of Imported Aircraft and Related Products
CAM 13	Aircraft Engine Airworthiness
CAM 14	Aircraft Propeller Airworthiness
CAM 18	Maintenance, Repair, and Alterations of Airframe, Powerplants, Propellers, and Appliances

Figure 12-10. CARs.

CAR 4a — Airplane Airworthiness

This regulation, which originated in 1936, was last amended on April 7, 1950. The subparts included in this regulation are:

A—Airworthiness Requirements

B—Definitions

C—Structural Loading Conditions, General Structural Requirements

D—Proof of Structure

E—Detail Design and Construction

F—Equipment

G—Powerplant Installation

H—Performance

I—Miscellaneous Requirements

Initially this regulation was the basis for establishing the design requirements for virtually all aircraft produced in the 1930s, 1940s, and 1950s.

Eventually CAR 3 evolved as the regulatory material specific to small aircraft, and CAR 4a and b focused on regulatory requirements for large aircraft. It is very important to review the type certificate data sheet for each aircraft. The Cessna 140 for example, was certified as a landplane under CAR 3, but under CAR 4a as a ski-plane or seaplane.

Another example of a more current and larger aircraft is the Gulfstream 1159 and 1159A. The former is certified under CAR 4b, but the latter is certified to 14 CFR part 25.

Suspected Unapproved Parts (SUPs)

There are four types of aircraft parts:

1. Good parts with good paperwork.
2. Good parts with bad paperwork.
3. Bad parts with "good" (bogus) paperwork.
4. Bad parts with bad paperwork.

The first of those listed represents properly authorized parts and when properly installed are approved parts, and the aircraft can be returned to service.

The last of those listed represent obviously unauthorized and unapproved parts. The technician should be alert for these, and must never install them on an aircraft.

The center two categories of parts represent suspected unapproved parts. If either the physical part *or* the paperwork associated with the part is questionable, it is best to contact the shop foreman, shift supervisor, or the assigned quality individual to discuss your concerns. Suspected unapproved parts (SUPs) should be segregated and quarantined until proper disposition can be determined. Contacting the manufacturer of the product is a good way to start gathering the facts concerning the product in question. (Refer to the current version of Advisory Circular (AC) 21-29, Detecting and Reporting Suspected Unapproved Parts, for additional information.) Contact the FAA—System Surveillance and Analysis Division (AIR-300), P.O. Box 17030, Washington, DC 20041; using FAA Form 8120-11, Suspected Unapproved Parts Report, or call the FAA Aviation Safety Hotline at 1-800-255-1111.

Other FAA Documents

Advisory Circulars (ACs)

The FAA issues advisory circulars to inform the aviation public in a systematic way of non-regulatory material. Unless incorporated into a regulation by reference, the contents of an advisory circular are not binding on the public. Advisory circulars are issued in a numbered-subject system corresponding to the subject areas of the Federal Aviation Regulations (Title 14 of the Code of Federal Regulations, Chapter I, Federal Aviation Administration) and Chapter III, Commercial Space Transportation, Federal Aviation Administration, Department of Transportation, Parts 400–450. An AC is issued to provide guidance and information in a designated subject area or to show a method acceptable to the Administrator for complying with a related federal aviation regulation.

Because of their close relationship to the CFRs, ACs are arranged in a numbered system which corresponds to the subject areas of the CFRs. In some series, consecutive numbers may be missing. These numbers were either assigned to advisory circulars still in preparation, which will be issued at a later date or were assigned to advisory circulars that have been canceled.

They are issued and updated irregularly with new revisions designated by letters of the alphabet. The suffix following an AC number indicates a revision to that AC and cancels the previous edition. For example, AC 00-6A replaces and cancels AC 00-6; AC 00-7D replaces and cancels AC 00-7C. ACs may be canceled without replacement, either by the issuing office, by another AC, or by another publication. [Figure 12-11]

Advisory Circular Numbering System

The Circular Numbering System

1. General. The advisory circular numbers relate to the 14 CFR subchapter titles and correspond to the Parts, and when appropriate, to the specific sections of 14 CFR.
2. General and specific subject numbers. The subject numbers and related subject areas are as follows:

General Subject Number (1)	Specific Subject Number (2)	Subject
00		**General**
	1	Definitions and Abbreviations
10		**Procedural Rules**
	11	General Rule-Making Procedures
	13	Investigation and Enforcement Procedures
20		**Aircraft**
	21	Certification Procedures for Products and Parts
	23	Airworthiness Standards: Normal, Utility, and Acrobatic Category Airplanes
	25	Airworthiness Standards: Transport Category Airplanes
	27	Airworthiness Standards: Normal Category Rotorcraft
	29	Airworthiness Standards: Transport Category Rotorcraft
	31	Airworthiness Standards: Manned Free Balloons
	33	Airworthiness Standards: Aircraft Engines
	34	Fuel Venting and Exhaust Emission Requirements for Turbine Engine Powered Airplanes
	35	Airworthiness Standards: Propellers
	36	Noise Standards: Aircraft Type and Airworthiness Certification
	39	Airworthiness Directives
	43	Maintenance, Preventive Maintenance, Rebuilding, and Alteration
	45	Identification and Registration Marking
	47	Aircraft Registration
	49	Recording of Aircraft Titles and Security Documents
60		**Airmen**
	61	Certification: Pilots and Flight Instructors
	63	Certification: Flight Crewmembers Other Than Pilots
	65	Certification: Airmen Other Than Flight Crewmembers
	67	Medical Standards and Certification
70		**Airspace**
	71	Designation of Federal Airways, Area Low Routes, Controlled Airspace, and Reporting Points
	73	Special Use Airspace
	75	Establishment of Jet Routes and Area High Routes
	77	Objects Affecting Navigable Airspace
90		**Air Traffic and General Operating Rules**
	91	General Operating and Flight Rules
	93	Special Air Traffic Rules and Airport Traffic Patterns
	95	IFR Altitudes
	97	Standard Instrument Approach Procedures
	99	Security Control of Air Traffic
	101	Moored Balloons, Kites, Unmanned Rockets, and Unmanned Free Balloons
	103	Ultralight Vehicles
	105	Parachute Jumping
	107	Airport Security
	108	Airplane Operators Security
	109	Indirect Air Carrier Security
119		**Certification: Air Carriers and Commercial Operators**
120		**Air Carriers, Air Travel Clubs, and Operators for Compensation or Hire: Certification and Operations**
	121	Certification and Operations: Domestic, Flag, and Supplemental Air Carriers and Commercial Operators of Large Aircraft

Figure 12-11. List of advisory circular numbers.

Advisory Circular Numbering System *(continued)*

General Subject Number (1)	Specific Subject Number (2)	Subject
120		*(Continued)*
	125	Certification and Operations: Airplanes Having a Seating Capacity of 20 or More Passengers or a Maximum Payload Capacity of 6,000 Pounds of More
	127	Certification and Operations of Scheduled Air Carriers with Helicopters
	129	Operations of Foreign Air Carriers
	133	Rotorcraft External-Load Operations
	135	Air Taxi Operators and Commercial Operators
	137	Agricultural Aircraft Operations
	139	Certification and Operations: Land Airports Serving CAB-Certificated Air Carriers
140		**Schools and Other Certificated Agencies**
	141	Pilot Schools
	143	Ground Instructors
	145	Repair Stations
	147	Aviation Maintenance Technician Schools
150		**Airport Noise Compatibility Planning**
	151	Federal Aid to Airports
	152	Airport Aid Program
	155	Release of Airport Property from Surplus Property Disposal Restrictions
	156	State Block Grant Pilot Program
	157	Notice of Construction, Alteration, Activation, and Deactivation of Airports
	158	Passenger Facilities Charges
	159	National Capital Airports
	159/10	Washington National Airport
	159/20	Dulles International Airport
	161	Notice and Approval of Airport Noise and Access Restrictions
	169	Expenditures of Federal Funds for Nonmilitary Airports or Air Navigational Facilities Thereon

General Subject Number (1)	Specific Subject Number (2)	Subject
170		**Navigational Facilities**
	170	Establishment and Discontinuance Criteria for Airport Traffic Control Tower Facilities
	171	Non-Federal Navigation Facilities
180		**Administrative Regulations**
	183	Representatives of the Administrator
	185	Testimony by Employees and Production of Records in Legal Proceedings
	187	Fees
	189	Use of Federal Aviation Administration Communication System
190		**Withholding Security Information**
	191	Withholding Security Information from Disclosure Under the Air Transportation Security Act of 1974
	198	Aviation Insurance Program
210		**Flight Information**
	211	Aeronautical Charts and Flight Information Publications
	212	Publication Specification: Charts and Publications
400		**Commercial Space Transportation**
	440	Financial Responsibility

1— Based on 14 CFR Subchapter Titles (excluding the 210 series)
2— Based on 14 CFR Part Titles (excluding the 210 series)

3. Within the General Subject Number Areas, specific selectivity in advisory circular mail lists is available corresponding to the applicable 14 CFR parts. For example: under the 60 general subject area, separate mail lists for advisory circulars issued in the 61, 63, 65, or 67 series are available. An AC numbered "60" goes to all numbers in the 60 series. When the volume of circulars in a series warrants a subsubject breakdown, the general number is followed by a slash and a subsubject number.
4. Individual circular identification numbers. Each circular has a subject number followed either by a dash and a consecutive number (135-15) or a period with a specific 14 CFR section number, followed by a dash and a consecutive number (135.169-2) identifying the individual circular. This consecutive number is not used again in the same subject series. Revised circulars have a letter (A, B, C, etc.) after the consecutive number to show complete revisions. Changes to circulars have Chg. 1, Chg. 2, Chg. 3, etc., after the identification number on pages that have been changed. The date on a revised page is changed to the date of the change transmittal.

Figure 12-11. List of advisory circular numbers. (continued)

Airworthiness Directives (ADs)

In accordance with part 39 of Title 14 of the Code of Federal Regulations, the FAA issues airworthiness directives in response to deficiencies and/or unsafe conditions found in aircraft, engines, propellers, or other aircraft parts. ADs, as they are known in the industry, require that the relevant problem must be corrected on all aircraft or aircraft parts using the same design. ADs are initiated as either proposed, corrective, or final (telegraphic) via the Federal Register. The Federal Register is the official daily publication of the United States Government. It is the printed method of informing the public of laws that are enacted or will be enacted. Electronic versions of ADs are available from the Federal Register and from the Regulatory and Guidance Library. You can search by manufacturer, model, or AD number. All ADs are "incorporated by reference" into part 39, and are considered final. ADs must be followed to remain in compliance with the FAA. Once an AD has been issued, a person/company is authorized to use the affected aircraft or part only if it has been corrected in accordance with the AD.

AD Content

Generally, ADs include:

- A description of the unsafe condition.
- The product to which the AD applies.
- The required corrective action or operating limitations, or both.
- The AD effective date.
- A compliance time.
- Where to go for more information.
- Information on alternative methods of compliance with the requirements of the AD.

AD Number

ADs have a three-part number designator. The first part is the calendar year of issuance. The second part is the biweekly period of the year when the number is assigned. The third part is issued sequentially within each biweekly period.

Applicability and Compliance

The AD subject line specifically identifies the type certificate holder of the aircraft or products affected by the AD. The specific models affected, and any special considerations, such as specific installed part numbers or modifications, are listed in the AD applicability section. In order to find all applicable ADs for a specific product, you must search for ADs on the product, aircraft, engine(s), propeller, or any installed appliance. If there are multiple series under the aircraft or engine model, you must also search for ADs applicable to the model as well as the specific series of that model. The final determination of ADs applicable to a particular product can only be made by a thorough examination of the ADs and the product logbooks. No person may operate a product to which an AD applies, except in accordance with the requirements of the AD. Furthermore, the owner or operator of an aircraft is required by 14 CFR §91.403 to maintain the aircraft in compliance with all ADs. The AD specifies a compliance time that relates to the effective date of the AD. That compliance time determines when the actions are required.

Alternative Method of Compliance

Different approaches or techniques that are not specified in an AD can, after FAA approval, be used to correct an unsafe condition on an aircraft or aircraft product. Although the alternative was not known at the time the AD was issued, an alternative method may be acceptable to accomplish the intent of the AD. A compliance time that differs from the requirements of the AD can also be approved if the revised time period and approved alternative method provides an acceptable level of safety as the requirements of the AD.

Aircraft Listings

This document contains the specifications and data for certificated aircraft when the number of registered aircraft is 50 or fewer.

Aircraft Specifications

"Specifications" were originated during implementation of the Air Commerce Act of 1926. Specifications are FAA recordkeeping documents issued for both type certificated and non-type-certificated products which have been found eligible for U.S. airworthiness certification. Although they are no longer issued, specifications remain in effect and will be further amended. Specifications covering type-certificated products may be converted to type certificate data sheets at the option of the type certificate holder. However, to do so requires the type certificate holder to provide an equipment list. A specification is *not* part of a type certificate. Specifications are subdivided into five major groups as follows:

1. Group I—Type Certificate Aircraft, Engines and Propellers. Covering standard, restricted, and limited types issued for domestic, foreign, and military surplus products.

2. Group II—Aircraft, Engine, and Propeller Approvals. Covering domestic, foreign, and military surplus products constructed or modified between October 1, 1927, and August 22, 1938, all of which have met minimum airworthiness requirements without formal type certification. Such products are eligible for standard airworthiness certification as though they are type-certificated products.
3. Group III—Aircraft, Engine, and Propeller Approvals. Covering domestic products manufactured prior to October 1, 1927, foreign products manufactured prior to June 20, 1931, and certain military surplus engines and propellers all of which have met minimum airworthiness requirements of the Air commerce Act of 1926 and implementing Air Commerce Regulations without formal type certification. Such products are eligible for standard airworthiness certification as though they are type-certificated products.
4. Group IV—Engine Ratings. Covering unapproved engines rated for maximum power and speed only, their use being limited to specific aircraft with maximum gross weights less than 1,000 pounds. Such engines are not eligible for independent airworthiness certification. These ratings are no longer issued.
5. Group V—Engine Approvals. Covering military surplus engines meeting Civil Air Regulation (CAR) 13 design requirements without formal type certification. Such engines are eligible for airworthiness certification as though they are type-certificated engines.

Aviation Maintenance Alerts (AC 43-16)

This is a monthly electronic publication of the FAA that provides information concerning malfunction or defects observed by technicians throughout the aviation industry. The data is supplied to the FAA on FAA Form 8010-4, Malfunction or Defect Report, which may be submitted to the FAA either electronically via the Internet, or by downloading a paper copy and completing it manually. After review, the FAA may print and format the inspection information into the AC. The professional maintenance technician will make full use of the information and will review it regularly when conducting an inspection. The monthly material is separated as necessary into airplanes, helicopters, powerplants, propeller, and accessory categories and then presented alphabetically in the category. Below is a sample Alert from the January 2005 issue. Notice the reference to the ATA code after the title. Refer to "Non-FAA Documents" presented later in this chapter for an explanation of these codes.

> **Cessna; 182T; Locked Rudder Trim Wheel; ATA 2721**
>
> A transit customer required help as his rudder trim was stuck in the full-right position. The attending mechanics found the trim indicator pin had jumped free of its positioning track and locked the trim wheel. After adjustment, the system was cycled to full extremes several times. The submitter notes they could replicate the "jammed" trim condition with extreme R/H trim. This aircraft was the second 182 observed by these mechanics having this particular problem.
>
> Part total time: 31.9 hours.

Supplemental Type Certificates (STC)

A supplemental type certificate (STC) is a document issued by the FAA approving a product (aircraft, engine, or propeller) modification. [Figure 12-12] The STC defines the product design change, states how the modification affects the existing type design, and lists serial number effectivity. It also identifies the certification basis listing specific regulatory compliance for the design change. Information contained in the certification basis is helpful for those applicants proposing subsequent product modifications and evaluating certification basis compatibility with other STC modifications. Refer to Figure 12-13 for a listing of how TCs and STCs are numbered.

Possession of the STC document does not constitute rights to the design data or installation of the modification. The STC and its supporting data (drawings, instructions, specifications, and so forth) are the property of the STC holder. You must contact the STC holder to obtain rights for the use of the STC.

Type Certificate Data Sheets (TCDS)

The TCDS is a formal description of the aircraft, engine, or propeller. It lists limitations and information required for type certification including airspeed limits, weight limits, thrust limitations, and so forth.

Type certificate data sheets and specifications set forth essential factors and other conditions which are necessary for U.S. airworthiness certification. Aircraft, engines, and propellers which conform to a U.S. type certificate (TC) are eligible for U.S. airworthiness certification when found to be in a condition for safe

United States of America

Department of Transportation—Federal Aviation Administration

Supplemental Type Certificate

Number SA7855SW 150-308 ARH

This certificate, issued to Sarnes Aircraft Company
1100 Airport Drive
Bethany, Oklahoma 73008

certifies that the change in the type design for the following product with the limitations and conditions therefor as specified hereon meets the airworthiness requirements of Part 23 of the Federal Aviation Regulations.

Original Product—Type Certificate Number: A12SO
Make: DEMANDER
Model: 114

Description of Type Design Change:

Installation of McCauley B3D32C419/82NHA-5 Propeller on Demander Model 114 airplane in accordance with Sarnes Aircraft Co., Installation Instructions dated July 13, 1990, Revision A dated August 22, 1990, or later FAA approved revision.

Limitations and Conditions:

FAA approved Sarnes Aircraft Co. Flight Manual Supplement dated August 23, 1990, must accompany this modification. Compatibility of this modification with previously installed equipment must be determined by installer.

This certificate and the supporting data which is the basis for approval shall remain in effect until surrendered, suspended, revoked, or a termination date is otherwise established by the Administrator of the Federal Aviation Administration.

Date of application: February 9, 1990
Date reissued:
Date of issuance: August 23, 1990
Date amended: Amd. 1, September 10, 1990

By direction of the Administrator

Jane Doe
(Signature)

Jane Doe
Airplane Certification Office
(Title)

FEDERAL AVIATION ADMINISTRATION

Any alteration of this certificate is punishable by a fine of not exceeding $1,000, or imprisonment not exceeding 3 years, or both.

This certificate may be transferred in accordance with FAR 21.47.

FAA Form 8110-2 (10-68)

12-33

Figure 12-12. Supplemental type certificate.

Forms and Guidance for Certification Projects

Type Certificate Numbering System (5 Parts)

1. **S** indicates a supplemental type certificate (blank if a type certificate)
2. **X** = An alpha digit to identify the type of product; that is,

   A = Small Airplane
   B = Balloon
   E = Engine
   G = Glider
   P = Propeller
   R = Rotorcraft
   S = Airship
   T = Transport Airplane

3. **nnnnn** = Assigned number automatically generated by ACOS; for example 00146
4. **YY** = Two alpha digits to identify the issuing Aircraft Certification Office (ACO)

   AC = ASW-150, Ft. Worth Airplane Certification Office
   AK = ACE-115N, Anchorage Aircraft Certification Office
   AT = ACE-15A, Atlanta Aircraft Certification Office
   BO = ANE-150, Boston Aircraft Certification Office
   CH = ACE-115C, Chicago Aircraft Certification Office
   DE = ANM-100D, Denver Aircraft Certification Office
   EN = ANE-140, Engine Certification Office, Boston
   LA = ANM-100L, Los Angeles Aircraft Certification Office
   MC = ACE-100M, Military Certification Office
   NY = ANE-170, New York Aircraft Certification Office
   RC = ASW-170, Ft. Worth Rotorcraft Aircraft Certification Office
   SE = ANM-100S, Seattle Aircraft Certification Office
   SC = ASW-190, Ft. Worth Special Certification Office
   WI = ACE-115W, Wichita Aircraft Certification Office

5. **D** = Indicates Designated Alteration Station (DAS) issued STC (blank if not DAS issued)

As an example, SA00125AT would be an STC issued by the Atlanta ACO on a small airplane with the assigned number 00125.

Figure 12-13. Numbering system for TCs and STCs.

operation and ownership requisites are fulfilled. [Figure 12-14]

Type certificate data sheets were originated and first published in January 1958. 14 CFR part 21, §21.41 indicates they are part of the type certificate. As such, a type certificate data sheet is evidence the product has been type certificated. Generally, type certificate data sheets are compiled from details supplied by the type certificate holder; however, the FAA may request and incorporate additional details when conditions warrant. [Figure 12-15]

Under federal law, no civil aircraft registered in the United States can operate without a valid airworthiness certificate. This certificate must be approved and issued by the FAA; and it will only be issued if the aircraft and its engines, propellers, and appliances are found to be airworthy and meet the requirements of an FAA-approved type certificate. The FAA issues a type certificate when a new aircraft, engine, propeller, and so forth, is found to meet safety standards set forth by the FAA. The TCDS lists the specifications, conditions, and limitations under which airworthiness requirements were met for the specified product, such as engine make and model, fuel type, engine limits, airspeed limits, maximum weight, minimum crew, and so forth. TCDS's are issued and revised as necessary to accommodate new models or other major changes in the certified product. TCDS's are categorized by TCD holder and product type.

The United States of America
Federal Aviation Administration
Department of Transportation

Type Certificate

Number A6WE

This certificate issued to McDONNELL DOUGLAS CORPORATION

certifies that the type design for the following product with the operating limitations and conditions therefor as specified in the Federal Aviation Regulations and the Type Certificate Data Sheet, meets the airworthiness requirements of Part 4b of the Federal Aviation Regulations.

DC-9-11, DC-9-12, DC-9-13, DC-9-14, DC-9-15,
DC-9-31, DC-9-15F, DC-9-32, DC-9-32F, DC-9-41,
DC-9-33F, DC-9-21, DC-9-51, DC-9-34F, DC-9-34,
DC-9-81, DC-9-82, DC-9-83

This certificate, and the Type Certificate Data Sheet which is a part hereof, shall remain in effect until surrendered, suspended, revoked, or a termination date is otherwise established by the Administrator of the Federal Aviation Administration.

This certificate consists of three pages.

Date of application: See Page 2.

Date of issuance: See Page 3.

By direction of the Administrator

(Signature) Keith D. Anderson

(Title) MANAGER, LOS ANGELES AIRCRAFT CERTIFICATION OFFICE

This certificate may be transferred if endorsed as provided on the reverse hereof.

Any alteration of this certificate and/or the Type Certificate Data Sheet is punishable by a fine of not exceeding $1,000, or imprisonment not exceeding 3 years, or both.

FAA FORM 8110—9 (7-67) SUPERSEDES FAA FORM 331

(21660NEW8.PG) 8-4

Figure 12-14. Type certificate.

Department of Transportation
Federal Aviation Administration

A6WE
Revision 26
McDonnell Douglas
DC-9-11, DC-9-12, DC-9-13, DC-9-14, DC-9-15, DC-9-15F, DC-9-21, DC-9-31, DC-9-32,DC-9-32 (VC-9C), DC-9-32F, DC-9-32F (C-9A, C-9B), DC-9-33F, DC-9-34, DC-9-34F, DC-9-41, DC-9-51, DC-9-81 (MD-81), DC-9-82 (MD-82), DC-9-83 (MD-83), DC-9-87 (MD-87), MD-88, MD-90, 717-200
November 1, 2001

Type Certificate Data Sheet No. A6WE

This Data sheet, which is part of Type Certificate No. A6WE, prescribes condition and limitations under which the product for which the Type Certificate was issued meets the airworthiness requirements of the Civil Air Regulations and Federal Aviation Regulations.

Type Certificate Holder: McDonnell Douglas Corporation
Long Beach, California

Type Certificate Ownership Record: McDonnell Douglas Corporation transferred ownership of the TC A6WE to The Boeing Company on January 30, 1998. MDC is a subsidiary of The Boeing Company.

The DC-9, -10, -20, -30, -40, and -50 Series; and DC-9-81/-82/-83/-87, MD-88, MD-90, and 717 are defined as follows:

DC-9-10 Series DC-9-11, DC-9-12, DC-9-13, DC-9-14, DC-9-15, DC-9-15F
DC-9-20 Series............................DC-9-21
DC-9-30 Series............................DC-9-31, DC-9-32, DC-9-32 (VC-9C), DC-9-32F, DC-9-32F (C-9A, C-9B), DC-9-33F, DC-9-34, and DC-9-34F
DC-9-40 Series............................DC-9-41
DC-9-50 Series............................DC-9-51
DC-9-81/-82/-83/-87 ....DC-9-81 (MD-81), DC-9-82 (MD-82), DC-9-83 (MD-83), DC-9-87 (MD-87)
MD-88..MD-88
MD-90..MD-90-30
717..717-200

Page No.	**1**	**2**	**3**	**4**	**5**	**6**	**7**	**8**	**9**	**10**
Rev. No.	26	26	26	26	26	26	26	26	26	26
Page No.	**11**	**12**	**13**	**14**	**15**	**16**	**17**	**18**	**19**	**20**
Rev. No.	26	26	26	26	26	26	26	26	26	26
Page No.	**21**	**22**	**23**	**24**	**25**	**26**	**27**	**28**	**29**	**30**
Rev. No.	26	26	26	26	26	26	26	26	26	26
Page No.	**31**	**32**	**33**	**34**	**35**	**36**	**37**	**38**	**39**	**40**
Rev. No.	26	26	26	26	26	26	26	26	26	26
Page No.	**41**	**42**	**43**	**44**	**45**	**46**	**47**	**48**	**49**	**50**
Rev. No.	26	26	26	26	26	26	26	26	26	26
Page No.	**51**	**52**	**53**	**54**	**55**	**56**	**57**	**58**	**59**	**60**
Rev. No.	26	26	26	26	26	26	26	26	26	26
Page No.	**61**	**62**	**63**	**64**	**65**	**66**	**67**	**68**	**69**	**70**
Rev. No.	26	26	26	26	26	26	26	26	26	26
Page No.	**71**	**72**	**73**	**74**	**75**	**76**	**77**	**78**	**79**	**80**
Rev. No.	26	26	26	26	26	26	26	26	26	26

Note: Revision 26 of the A6WE Type Certificate Data Sheet is formatted as the most recent change to FAA Type Certificate Data Sheet presentation. Revision 26 to the A6WE Type Certificate Data Sheet incorporates all previous A6WE revisions.

Figure 12-15. Type Certificate Data Sheet.

JASC/ATA 100 Code Listings					
11	Placards and Markings	34	Navigation	64	Tail Rotor
12	Servicing	35	Oxygen	65	Tail Rotor Drive
14	Hardware	36	Pneumatic	67	Rotors Flight Control
18	Helicopter Vibration	37	Vacuum	71	PowerPlant
21	Air Conditioning	38	Water/Waste	72	Turbine/Turboprop Engine
22	Auto Flight	45	Central Maintenance System	73	Engine Fuel and Control
23	Communications	49	Airborne Auxiliary Power	74	Ignition
24	Electrical Power	51	Standard Practices/Structures	75	Air
25	Equipment/Furnishings	52	Doors	76	Engine Controls
26	Fire Protection	53	Fuselage	77	Engine Indicating
27	Flight Controls	54	Nacelles/Pylons	78	Engine Exhaust
28	Fuel	55	Stabilizers	79	Engine Oil
29	Hydraulic Power	56	Windows	80	Starting
30	Ice and Rain Protection	57	Wings	81	Turbocharging
31	Instruments	61	Propellers/Propulsors	82	Water Injection
32	Landing Gear	62	Main Rotor	83	Accessory Gearboxes
33	Lights	63	Main Rotor Drive	85	Reciprocating Engine

Figure 12-16. Maintenance classification.

Non-FAA Documents

Air Transport Association (ATA) 100

To standardize the technical data and maintenance activities on large and therefore complex aircraft, the ATA has established a classification of maintenance related actions. These are arranged with sequential numbers assigned to ATA chapters. These chapters are consistent regardless of which large aircraft is being worked on. [Figure 12-16]

Manufacturers' Published Data

The original equipment manufacturer (OEM) is usually the best source of information for the operation of and maintenance on a particular product. Whether the product is a complete aircraft (Cessna 172, Boeing 777, or Airbus A380), an engine, or a component of the engine (i.e., thrust reverser, hydraulic pump, generator, and so forth), the manufacturer is required by 14 CFR part 21, §21.5 to provide a current approved airplane or rotorcraft flight manual and (if applicable) a rotorcraft maintenance manual. If the product is a TC'd or STC'd item, §21.50 requires the holder of the design approval to provide one set of complete Instructions for Continued Airworthiness (ICAs). Additional requirements for ICAs are specified in §§23.1529, 25.1529, 27.1529 and 29.1529. These sections further refer the reader to appendix G (for parts 23 and 25) and appendix A (for parts 27 and 29). Regardless of which appendix is referred, the requirements in the appendix for the ICA are as follows:

General: The aircraft ICA must contain instructions for continued airworthiness for each engine, propeller, or appliance and the interface of those appliances and products with the aircraft.

Format: The ICA must be in the form of a manual or manuals appropriate to the data being provided.

Content: The manual contents must be in English and must include:

- Introductory information including an explanation of the airplane's features and data as necessary to perform maintenance or preventive maintenance.

- A description of the aircraft and its systems, including engine, propeller, and appliances.
- Basic operating information describing how the aircraft and its components are controlled.
- Servicing information with such detail as servicing parts, task capacities, types of fluid to be used, applicable pressures for the various systems, access panels for inspection and servicing, lubrication points, and types of lubricants to be used.

The maintenance instructions must include the following data:

- Recommended schedule for cleaning, inspecting, adjusting, testing, and lubricating the various parts.
- Applicable wear tolerances.
- Recommended overhaul periods.
- Details for an inspection program that identifies both the frequency and the extent of the inspections necessary to provide for continued airworthiness.
- Troubleshooting information.
- The order and method for proper removal and replacement of parts.
- Procedures for system testing during ground operations.
- Diagrams for structural access plates.
- Details for application of special inspection techniques.
- Information concerning the application of protective treatments after inspection.
- Information relative to the structural fasteners.
- List of any special tools needed.

Airworthiness Limitations

The ICA must contain a separate and clearly distinguishable section titled "Airworthiness Limitations." Within this section are mandatory replacement times, structural inspection interval, and related inspection procedures.

All of this is quite understandable to be included in the initial release of documents when the aircraft is delivered. However, over the course of the life of an aircraft, various modifications can and often do occur. Whether these are as simple as a new cabin to galley sliding door, or as complex as a navigation related STC, any major alteration requires that this type of maintenance data be provided to the owner, so that subsequent maintenance, inspection, and repair can be properly accomplished. As aircraft and their systems become more and more complex, and society continues its preoccupation with litigation for every incident, it is imperative that the technician have the right information, that it is current, and that he or she has the proper tools, including those required for any special inspection, and correct replacement parts. If any one these items is required, and the technician does not have it accessible, he or she is in violation of 14 CFR §§65.81(b), 43.13(a), and 43.16 if he or she attempts to return the aircraft to service.

Manufacturers may provide this required information in a variety of different manuals:

- Operating Instructions—The Airplane Flight Manual (AFM) or the Pilot's Operating Handbook (POH) provides the pilot with the necessary and proper information to properly operate the aircraft. These manuals are usually listed in the aircraft type certificate data sheet, and therefore are a required item for the aircraft to be considered airworthy.
- Maintenance Manuals—These manuals are often referred to as AMM (Aircraft Maintenance Manual) or CMM (Component Maintenance Manual).

The AMM is focused on the entire aircraft and therefore provides the "big picture" for the maintenance technician. It provides information concerning the maintenance (including troubleshooting and repair) of the aircraft and systems on the aircraft.

The CMM, on the other hand, is focused on a specific item or component such as hydraulic pump, generator, or thrust reverser. It will provide the bench mechanic with detail troubleshooting information and will usually serve also as an overhaul manual giving details for disassembly, cleaning, inspection, repair as necessary, reassembly, and testing in accordance with approved standards and technical data accepted by the Administrator. (Refer to 14 CFR §43.2(a).) When maintenance is done according to the CMM, the technician should always include the appropriate references in the maintenance record entry required by §43.9 or 43.11.

Service Bulletins (SB)

Throughout the life of a product (whether TC'd or not), manufacturing defects, changes in service, or design improvements often occur. When that happens, the OEM will frequently use an SB to distribute the information to the operator of the aircraft. SBs are good information and should be strongly considered by the owner for implementation to the aircraft. However, SBs are not required *unless* they are referred to in an AD note or if compliance is required as a part of the authorized inspection program. (Refer to §39.27.)

Structural Repair Manual (SRM)

As the name implies, this manual carries detail information for the technician concerning an aircraft's primary and secondary structure, criteria for evaluating the severity of the detected damage, determining the feasibility of a repair, and alignment/inspection information. This manual is usually a separate manual for large aircraft. On small aircraft, this information is often included in the AMM.

Forms

Airworthiness Certificates

In addition to the registration certificate, which indicates the ownership of an aircraft, an airworthiness certificate indicates the airworthiness of the aircraft. Advisory Circular (AC) 21-12, Application for U.S. Airworthiness Certificate, FAA Form 8130-6, is a comprehensive guide for the completion of the application form for this certificate. There are two certificates: standard and special. FAA Form 8100-2, Standard Airworthiness Certificate, may be issued to allow operation of a type-certificated aircraft in one or more of the following categories: [Figure 12-17]

- Normal
- Utility
- Acrobatic
- Commuter
- Transport
- Manned Free Balloon
- Special Classes

FAA Form 8130-7, Special Airworthiness Certificate, may be issued to authorize the operation of an aircraft in the following categories: [Figure 12-18]

- Primary
- Restricted
- Multiple
- Limited
- Light-Sport
- Experimental
- Special Flight Permit
- Provisional

Airworthiness certificates may be issued by either FAA personnel or FAA designees. (Refer to 14 CFR §§183.31 and 183.33.) The certificate must not only be

UNITED STATES OF AMERICA
DEPARTMENT OF TRANSPORTATION—FEDERAL AVIATION ADMINISTRATION
STANDARD AIRWORTHINESS CERTIFICATE

1 NATIONALITY AND REGISTRATION MARKS	2 MANUFACTURER AND MODEL	3 AIRCRAFT SERIAL NUMBER	4 CATEGORY

5 AUTHORITY AND BASIS FOR ISSUANCE
This airworthiness certificate is issued pursuant to the Federal Aviation Act of 1958 and certifies that, as of the date of issuance, the aircraft to which issued has been inspected and found to conform to the type certificate therefor, to be in condition for safe operation, and has been shown to meet the requireme[illegible] comprehensive and detailed airworthiness code as provided by Annex 8 to the [illegible] except as noted herein
Exceptions:

SAMPLE

6 TERMS AND CONDITIONS
Unless sooner surrendered, suspended, revoked, or a termination date is otherwise established by the Administrator, this airworthiness certificate is effective as long as the maintenance, preventative maintenance, and alterations are performed in accordance with Parts 21, 43, and 91 of the Federal Aviation Regulations, as appropriate, and the aircraft is registered in the United States

DATE OF ISSUANCE	FAA REPRESENTATIVE	DESIGNATION NUMBER

Any alteration, reproduction, or misuse of this certificate may be punishable by a fine not exceeding $1,000, or imprisonment not exceeding 3 years, or both. THIS CERTIFICATE MUST BE DISPLAYED IN THE AIRCRAFT IN ACCORDANCE WITH APPLICABLE FEDERAL AVIATION REGULATIONS.

FAA Form 8100-2 (8-82) *U.S. G.P.O. 1995 669-242

Figure 12-17. FAA Form 8100-2, Standard Airworthiness Certificate.

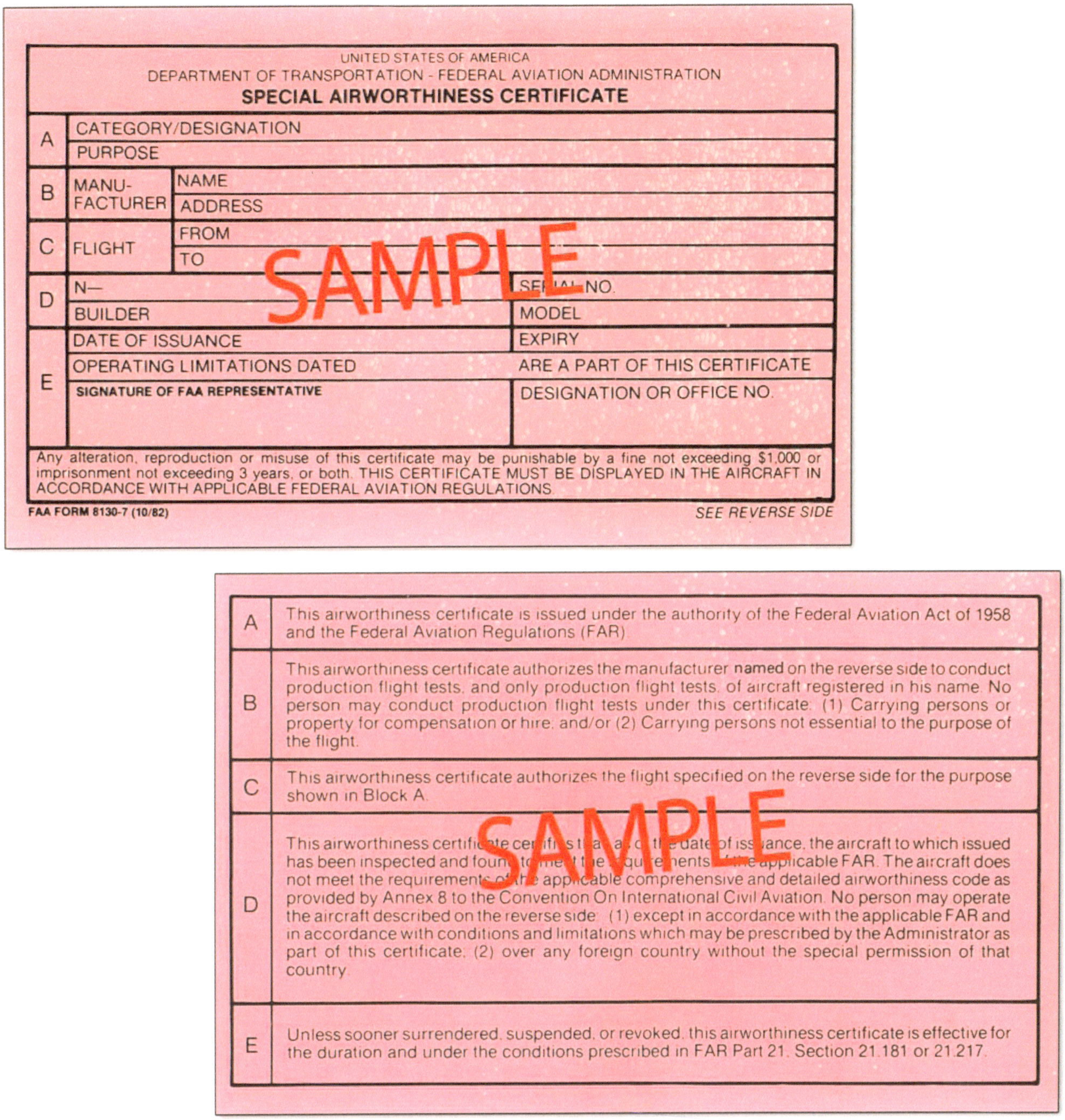

UNITED STATES OF AMERICA
DEPARTMENT OF TRANSPORTATION - FEDERAL AVIATION ADMINISTRATION
SPECIAL AIRWORTHINESS CERTIFICATE

A	CATEGORY/DESIGNATION		
	PURPOSE		
B	MANU-FACTURER	NAME	
		ADDRESS	
C	FLIGHT	FROM	
		TO	
D	N—		SERIAL NO.
	BUILDER		MODEL
E	DATE OF ISSUANCE		EXPIRY
	OPERATING LIMITATIONS DATED		ARE A PART OF THIS CERTIFICATE
	SIGNATURE OF FAA REPRESENTATIVE		DESIGNATION OR OFFICE NO.

Any alteration, reproduction or misuse of this certificate may be punishable by a fine not exceeding $1,000 or imprisonment not exceeding 3 years, or both. THIS CERTIFICATE MUST BE DISPLAYED IN THE AIRCRAFT IN ACCORDANCE WITH APPLICABLE FEDERAL AVIATION REGULATIONS.

FAA FORM 8130-7 (10/82) SEE REVERSE SIDE

SAMPLE

A	This airworthiness certificate is issued under the authority of the Federal Aviation Act of 1958 and the Federal Aviation Regulations (FAR).
B	This airworthiness certificate authorizes the manufacturer named on the reverse side to conduct production flight tests, and only production flight tests, of aircraft registered in his name. No person may conduct production flight tests under this certificate: (1) Carrying persons or property for compensation or hire; and/or (2) Carrying persons not essential to the purpose of the flight.
C	This airworthiness certificate authorizes the flight specified on the reverse side for the purpose shown in Block A.
D	This airworthiness certificate certifies that as of the date of issuance, the aircraft to which issued has been inspected and found to meet the requirements of the applicable FAR. The aircraft does not meet the requirements of the applicable comprehensive and detailed airworthiness code as provided by Annex 8 to the Convention On International Civil Aviation. No person may operate the aircraft described on the reverse side: (1) except in accordance with the applicable FAR and in accordance with conditions and limitations which may be prescribed by the Administrator as part of this certificate; (2) over any foreign country without the special permission of that country.
E	Unless sooner surrendered, suspended, or revoked, this airworthiness certificate is effective for the duration and under the conditions prescribed in FAR Part 21, Section 21.181 or 21.217.

SAMPLE

Figure 12-18. FAA Form 8130-7, Special Airworthiness Certificate.

on board the aircraft (14 CFR §91.203(a)(1)), but must also be "displayed at the cabin or cockpit entrance so that it is legible to the passengers or crew" (§91.203(b)). Since the ability to obtain this certificate is based upon the requirement to inspect the aircraft to determine that it conforms to type design and is in condition for safe operation, it can also be revoked by the FAA if either of those two requirements ceases to exist.

Aircraft Registration

Aircraft must be registered in the United States if the aircraft is not registered under the laws of a foreign country and is owned by either a citizen of the United States, a foreign citizen lawfully admitted to the United States, or a corporation organized in and doing business under U.S. laws and primarily based in the United States. This registration is accomplished by the use of FAA Form 8050-1, Aircraft Registration Application. This is one of only a few FAA forms that are pre-printed, and cannot be obtained by downloading and computer printing. This form consists of three sheets, colored white, green, and pink. The white original and the green copy must be submitted to the FAA Aircraft Registration Branch (AFS-750). The pink copy serves as the temporary authority to operate the aircraft until the official registration is received from AFS-750, and is valid for a maximum of 90 days. This form may be obtained from either the local Flight Standards District Office (FSDO) or by contacting: FAA Aircraft Registration Branch (AFS-750), P.O. Box 25504, Oklahoma City, OK 73125-0504.

In addition to the completed application form, the owner must also submit evidence of his or her ownership (such as a bill of sale) and a registration fee. A successful review of the application will result in the issuance of AC Form 8050-3, Certificate of Aircraft Registration. (Note the AC prefix.)

14 CFR §91.203(a)(2) requires that either the pink copy of the application or the actual certificate of registration be on board the aircraft during its operation.

If the registration is ever lost or damaged, it may be replaced by contacting the FAA Aircraft Registration Branch and providing them with the aircraft specific data, including make, model, N-number, and serial number. A replacement certificate fee and an explanation of the reason for the replacement certificate are also required.

Radio Station License

A radio station license is required if the aircraft is equipped with radios, and the aircraft is planned to be flown outside the boundaries of the United States. A radio station license is *not* required for aircraft that are operated domestically. (A major change occurred on February 8, 1996, when the telecommunications Act of 1996 was signed into law.)

The Federal Communications Commission (FCC) formerly required that any communication transmitter installed in aircraft be licensed. These FCC licenses were valid for 5 years. Note that this is *not* an FAA requirement. FAA inspectors who conducted ramp inspections and detected an expired radio station license were not required to notify the FCC, nor could they issue a violation to the owner/operator. Simply informing the operator of the expired radio station license was their only responsibility.

FSGA 96-06, a Flight Standards Information Bulletin (FSIB) for General Aviation (FSGA) titled "Elimination of Aircraft Radio Station Licenses" became effective on July 8, 1996. Although that FSIB had an effectivity of only 1 year, the elimination of the requirement for aircraft used only in domestic operations continues.

FAA Form 337 — Major Repair and Alteration

(Refer to current issue of AC 43.9-1, Instructions for Completion of FAA Form 337 (OMB No. 2120-0020), Major Repair and Alteration (Airframe, Powerplant, Propeller, or Appliance).) [Figure 12-19]

As the name clearly states, this form is to be used whenever major repairs or alterations are accomplished on an aircraft. (The only exception would be that 14 CFR part 43, appendix B, allows for a certificated repair station to return to service (RTS) an aircraft after a major *repair* by using a signed and dated work order and a signed maintenance release.)

Information in item 1 should be taken directly from the aircraft dataplate, except for the tail number which should be compared to the aircraft registration form.

Information in item 2 must reflect the name and address listed on AC Form 8050-3, Certificate of Registration.

Item 3 is used only when there is no existing approved data for the intended repair or alteration. In that case, the technician can request that the local FSDO Principal Maintenance Inspector (PMI) review the data and then grant a field approval which is shown by completing and signing this area. In many cases, this block will be blank because the technician has found, used, and made reference to data already approved by the FAA.

Item 4 should have "X" marked in either the Repair or the Alteration column.

Item 5 — If the repair or alteration is being done to the aircraft airframe, no entry is required since the data will be identical to that in item 1. However, if the repair or alteration is being done to an engine, a propeller, or other appliance, entries must include the appropriate make, model and serial number information.

Item 6 — Enter appropriate data as specified, and check the proper box in B. The technician is encouraged to carefully read the preprinted statement in subparagraph D prior to signing this section.

Item 7 must be completed by the IA or authorized individual from the repair station.

Item 8 (on the reverse side) is for the description of the work accomplished. It must include a reference to the approved data used to conduct the required maintenance.

US Department of Transportation
Federal Aviation Administration

MAJOR REPAIR AND ALTERATION
(Airframe, Powerplant, Propeller, or Appliance)

Form Approved OMB No. 2120-0020 11/30/2007	Electronic Tracking Number
For FAA Use Only	

INSTRUCTIONS: Print or type all entries. See Title 14 CFR §43.9, Part 43 Appendix B, and AC 43.9-1 (or subsequent revision thereof) for instructions and disposition of this form. This report is required by law (49 U.S.C. §44701). Failure to report can result in a civil penalty for each such violation. (49 U.S.C. §46301(a))

1. Aircraft	Nationality and Registration Mark	Serial No.	
	Make	Model	Series
2. Owner	Name *(As shown on registration certificate)*	Address *(As shown on registration certificate)* Address ______ City ______ State ______ Zip ______ Country ______	

3. For FAA Use Only

4. Type		**5. Unit Identification**			
Repair	Alteration	Unit	Make	Model	Serial No.
☐	☐	AIRFRAME	————	*(As described in Item 1 above)*	————
☐	☐	POWERPLANT			
☐	☐	PROPELLER			
☐	☐	APPLIANCE	Type Manufacturer		

6. Conformity Statement

A. Agency's Name and Address	B. Kind of Agency	
Name ______	☐ U. S. Certificated Mechanic	☐ Manufacturer
Address ______	☐ Foreign Certificated Mechanic	C. Certificate No.
City ______ State ______	☐ Certificated Repair Station	
Zip ______ Country ______	☐ Certificated Maintenance Organization	

D. I certify that the repair and/or alteration made to the unit(s) identified in item 5 above and described on the reverse or attachments hereto have been made in accordance with the requirements of Part 43 of the U.S. Federal Aviation Regulations and that the information furnished herein is true and correct to the best of my knowledge.

Extended range fuel per 14 CFR Part 43 App. B ☐	Signature/Date of Authorized Individual

7. Approval for Return to Service

Pursuant to the authority given persons specified below, the unit identified in item 5 was inspected in the manner prescribed by the Administrator of the Federal Aviation Administration and is ☐ Approved ☐ Rejected

BY	☐ FAA Flt. Standards Inspector	☐ Manufacturer	☐ Maintenance Organization	☐ Persons Approved by Canadian Department of Transport
	☐ FAA Designee	☐ Repair Station	☐ Inspection Authorization	Other *(Specify)*

Certificate or Designation No.	Signature/Date of Authorized Individual

FAA Form 337 (10-06)

Figure 12-19. FAA Form 337, Major Repair and Alteration.

NOTICE

Weight and balance or operating limitation changes shall be entered in the appropriate aircraft record. An alteration must be compatible with all previous alterations to assure continued conformity with the applicable airworthiness requirements.

8. Description of Work Accomplished
(If more space is required, attach additional sheets. Identify with aircraft nationality and registration mark and date work completed.)

Nationality and Registration Mark	Date

☐ Additional Sheets Are Attached

FAA Form 337 (10-06)

Figure 12-19. FAA Form 337, Major Repair and Alteration. (continued)

The form must be completed at least in duplicate, with the original provided to the owner/operator and a copy sent to the FAA within 48 hours of completing the maintenance and return to service. If the FAA Form 337 is used to document additional fuel tanks in the cabin or cargo, then an additional copy must be signed and in the aircraft at all times. Professional business sense would be to also make and retain a copy at the maintenance facility.

Records

Making Maintenance Record Entries

14 CFR §§43.9 and 43.11 require the technician to make appropriate entries of maintenance actions or inspection results in the aircraft maintenance record. 14 CFR §91.417 defines how long those records must be kept.

Whenever maintenance, preventive maintenance, rebuilding, or alteration work occurs on an aircraft, airframe, aircraft engine, propeller, appliance, or component part, a maintenance record entry must be created. The importance of compliance with this requirement cannot be overemphasized. Complete and organized maintenance logs for an aircraft can have significant (and usually positive) effect during the buy/sell negotiations of an aircraft. On the other hand, poorly organized and incomplete logs can have a detrimental effect upon the selling price of an aircraft.

Temporary Records — 14 CFR §91.417(a)(1) and (b)(1)

These are records that must be kept by the owner until the work is repeated, superseded, or 1 year has transpired since the work was performed. These are typically records referring to maintenance, preventive maintenance, alteration, and all inspections. They include a description of the work performed (or reference to the FAA-accepted data); the date of completion; and the name, signature and certificate number of the person doing the return to service (RTS).

Permanent Records — 14 CFR §91.417(a)(2) and (b)(2)

These records must be retained by the owner during the time he or she operates the aircraft. They are transferred with the aircraft at the time of sale. Typically, these are documents relating to total time in service, current status of life-limited parts, time since last overhaul, current inspection status, current status of applicable AD notes, and major alteration forms as required by 14 CFR §43.9.

Electronic Records

During the last 25 years, the field of aviation maintenance has seen a significant change in the documentation requirements for aircraft and related parts. Nowhere is that change seen as revolutionary as the introduction of electronic data and record retention. Just as the arrival of the personal computer placed the possibility of the power and versatility of a computer in the hands of the average person, it made it available to the maintenance technician. Initially some technicians developed their own programs for listing data (TCDS, AD notes, and so forth), but soon commercially available programs were developed. Basically these were developed by either one of the following two groups:

1. Computer literate persons who felt the aviation industry could benefit from the computer.
2. Aviation professionals who felt the aviation industry must benefit from the computer.

Some of those initial programs were either not very user friendly (if developed by computer wizards) or not very sophisticated (if developed by the maintenance technician).

Today there is a mixture of these various database programs. A review of the advertisement section in any current aviation maintenance magazine will offer the reader numerous options for electronic (computerized) maintenance records. Many of these programs offer a combination of the data research, such as ADs, SBs, STCs, and TCDS's, required to conduct proper maintenance and inspections and data recording (logbook entries, AD compliance history, length of component time in service, and so forth) desired to improve the efficiency of the technician.

Although some large shops and certified repair stations may have a separate group of people responsible for "records and research," the professional maintenance technician should be aware of the benefits of these systems. Some factors to consider when reviewing a system are:

- What is the typical aircraft size on which maintenance is being done? (i.e., less than 12,500 pounds, more than 12,500? Mixed?)
- Does the program have built-in templates for the aircraft you are working on?
- What FAA forms (if any) are available in the program?

- Does it have a user-friendly template to enter the data for the form, or must you enter directly onto the form?
- Can it calculate weight and balance data?
- Does it have adequate word search capabilities?
- Is it networkable?
- Are the updates sent via U.S. mail, or can you download from the Internet?
- What is the maximum number of aircraft that the system can handle?
- Can the system handle both single- and multiengine aircraft? Fixed and rotary wing? Piston and jet?
- Can an item removed from an aircraft be tracked?
- Is the data from this system exportable to other electronic formats?
- Can it forecast "items due" for maintenance or inspection?

Since no program can be considered the best, the technician should learn all he or she can about the numerous systems that exist. Exposure to the pros and cons of these different systems can be one of the benefits of attending various trade shows, maintenance seminars, or IA renewal sessions. Continuous learning and personal improvement should be the goal of every professional maintenance technician.

Light Sport Aircraft (LSA) Maintenance

The LSA category includes gliders, airplanes, gyroplanes, powered parachutes, weight-shift and lighter-than-air aircraft. There two general types of LSAs: Special (SLSA) and Experimental (ELSA). The SLSA are factory built and the ESLA are kit built. This new category of aircraft was added to the regulations in 2004. (Refer to 14 CFR §§21.190, 65.107, and 91.327, all dated July 27, 2004.)

Just as "industry standard" specifications have replaced many of the Military Standards to define products that are destined to be part of the Department of Defense inventory, so too have industry standards come into the FAA sights for documenting certain things. Quality is one example, where the Society of Automotive Engineers (SAE) has developed AS 9100 and AS 9110 as auditing standards for aerospace facilities and specifically repair stations. Likewise, ISO9001 is being adopted by the FAA as a system of measuring their performance. Therefore, it was logical that when the FAA looked to develop the standards for this newest category of aircraft, they again looked to industry, and this time it was the American Society for Testing and Materials (ASTM). The ASTM developed a comprehensive list of consensus standards for use by manufacturers, regulators, maintenance facilities, LSA owners, and service providers. It is very unique that these standards are the first ones in over 100 years to solely address the issue of recreational aircraft use. It is also the first complete set of industry consensus standards covering the design, manufacture, and use of recreational aircraft that was developed by a non-government agency. The ASTM committee that developed these LSA standards did so to ensure the quality of products and services to support both the national and the international regulatory structures for LSAs. Over 20 standards have been generated, and more are being developed to cover this diversity of aircraft. This text only incorporates a review of F2483-05, Standard Practice for Maintenance and the Development of Maintenance Manuals for Light Sport Aircraft (LSA). That 6-page document is comprised of the following 12 sections:

1. Scope
2. Referenced Documents
3. Terminology
4. Significance and Use
5. Aircraft Maintenance Manual
6. Line Maintenance, Repairs and Alterations
7. Heavy Maintenance, Repairs and Alterations
8. Overhaul
9. Major Repairs and Alterations
10. Task-Specific Training
11. Safety Directives
12. Keywords

The scope of that document is basically twofold:

- To provide guidelines for the qualification necessary to accomplish various levels of maintenance on LSA.
- To provide the content and structure of maintenance manuals for aircraft and their components that are operated as LSAs.

Some additional definitions from section 3, Terminology, that will help to better understand the LSA concepts are:

Annual Condition Inspection is defined as a detailed inspection accomplished once a year in accordance with instructions provided in the maintenance manual supplied with the LSA. The purpose of this inspection is to look for any wear, corrosion, or damage that would cause the LSA not to be in condition for safe operation.

Heavy maintenance is any maintenance, inspection, repair, or alteration a manufacturer has designated that requires specialized training, equipment, or facilities.

Line maintenance is any repair, maintenance, scheduled checks, servicing, inspections, or alterations not considered heavy maintenance that are approved by the manufacturer and is specified in the manufacturer's maintenance manual.

LSA repairman—inspection is a U.S. FAA-certified LSA repairman with an inspection rating per 14 CFR part 65. This person is authorized to perform the 100-hour/annual inspection of the aircraft, which he or she owns.

LSA repairman—maintenance is a U.S. FAA-certified LSA repairman with a maintenance rating per 14 CFR part 65. This person is allowed to perform the required maintenance and can also accomplish the 100-hour/annual inspection.

Major repair, alteration, or maintenance is any repair, alteration, or maintenance for which instructions to complete the task are *excluded* from the maintenance manual.

Minor repair, alteration, or maintenance is any repair, alteration, or maintenance for which instructions to complete the task are *included* in the maintenance manual.

The 100-hour inspection is the same as the annual inspection, except for the interval of time. The requirements for whether or not the 100-hour inspection is applicable are exactly the same as the criteria for the standard 100-hour/annual required of non-LSA aircraft.

Aircraft Maintenance Manual

Although these manuals do not require any FAA approval, the regulations do require that the manual be developed in accordance with industry standards. This ASTM sets that standard by requiring:

- General specifications including capacities, servicing, lubrication, and ground handling must be listed.
- An inspection checklist for the annual condition or 100-hour inspection.
- A description of and the instructions for the maintenance, repair, and overhaul of the LSA engine.
- A description of and the instructions for the maintenance, repair, and alteration of the aircraft's primary structure.

Other items for which maintenance procedures must be provided are:

- Fuel systems,
- Propeller,
- Utility system,
- Instruments and avionics,
- Electrical system,
- Structural repair, and
- Painting and coatings,

The Inspection, Repair, and Alterations section must specifically list any special tools and parts needed to complete the task, as well as the type of maintenance action (line, heavy, or overhaul) necessary to accomplish the activity. Directly associated with that information is the requirement to specify the level of certification needed to do the job, i.e., LSA repairman, A&P, or repair station. The manual may make reference to existing FAA advisory circulars.

Line Maintenance, Repairs, and Alterations

The minimum level of certification necessary to accomplish line maintenance is LSA inspection. Some typical tasks considered to be line maintenance are:

- 100-hour/annual condition inspection.
- Servicing of fluids.
- Removing and replacing components when instructions to do so are provided in the maintenance manual.
  - Batteries
  - Fuel pump
  - Exhaust
  - Spark plugs and wires
  - Floats and skis

- Repair or alteration of components when specific instructions are provided in the maintenance manual.
  - Patching a hole in the fabric
  - Installation of a strobe light kit

Heavy maintenance, repairs, and alterations can only be accomplished by either a certified mechanic (A or P or A&P) or an LSA repairman—maintenance who has received additional "task specific" training. Some examples of this would be the removal and replacement of the following: complete engine, cylinder, piston and valve assemblies; primary flight controls; and landing gear.

Heavy repair of components or structure can be accomplished when instructions are provided in the maintenance manual or other service directed instructions. A few examples of this activity are:

- Repainting of control surfaces.
- Structural repairs.
- Recovering of a dope and fabric.

Heavy alterations of components can be accomplished when instructions are provided in the maintenance manual or other service directed instructions. Examples of this activity are initial installation of skis and installation of new additional pitot static instruments.

Overhaul of components can be performed only by the manufacturer (or someone authorized to perform) of the LSA or the component to be overhauled. An overhaul manual is required and must be a separate manual from the manufacturer's maintenance manual. Items typically considered for overhaul are engines, carburetors, starters, generators, alternators, and instruments.

Major Repairs and Alterations

Another major difference between LSA maintenance and traditional aircraft maintenance is that FAA Form 337, Major Repair and Alteration, is *not required* to document major repairs and alterations. Instead, any major repair or alteration that is accomplished after the LSA has gone through production acceptance testing must be evaluated relative to the applicable ASTM requirements. After this evaluation has been accomplished (either by the manufacturer or an entity approved by them), a written affidavit must be provided attesting that the LSA still meets the requirements of the applicable ASTMs.

The manufacturer (or other approved entity) must provide written instructions defining the level of certification necessary to perform the maintenance, and also include any ground test or flight testing necessary to verify that the LSA complies with the original LSA acceptance test standards, and is in condition for safe operation. Proper documentation of this maintenance activity is required to be entered in the LSA records, and is also defined by the manufacturer.

Task specific training is not required to be FAA approved. This is solely the responsibility of the manufacturer. Some examples of this are: an engine manufacturer's overhaul school, or the EAA Sport Air fabric covering school.

Safety Directives are issued against an LSA or component, but are *not issued* by the FAA, but rather by the original aircraft manufacturer. (*Note:* if the LSA includes a product that is TC'd by the FAA, the manufacturer is required to issue a safety directive.)

Typical instructions within a Safety Directive are:

- List of tools required for the task.
- List of parts needed.
- Type of maintenance (line, heavy, overhaul).
- Level of certification needed.
- Detailed instructions and diagrams.
- Inspection and test methods.

Safety directives are considered to be mandatory except for experimental use LSAs.

Chapter 13 — The Mechanic Certificate

Overview of the Maintenance Technician

The Mechanic Certificate — Maintenance Technician Privileges and Limitations

Since part 65 was covered only briefly in Chapter 12, it was left for this chapter to develop it more completely. Therefore, this chapter discusses the Federal Aviation Administration (FAA) regulation governing the certification of airmen other than flight crew members. This chapter is based on the material contained in Title 14 of the Code of Federal Regulations (14 CFR) part 65, which has the following subparts:

- Subpart A — General
- Subpart B — Air Traffic Control Operators
- Subpart C — Aircraft Dispatchers
- Subpart D — Mechanics
- Subpart E — Repairmen
- Subpart F — Parachute Riggers

This chapter will only focus on the certification of maintenance technicians, and therefore subparts B, C, E, and F will not be addressed.

The FAA certifies two separate categories of maintenance technicians, mechanic and repairman.

The fundamental difference between these two is that the mechanic certificate is transportable, is issued to the technician based upon his or her training and knowledge, and is not dependent on the technician's location. Although the repairman certificate is also based upon the training and knowledge of the technician, it is specifically issued to that technician while he or she is employed at a distinct location of a specific company. This certificate carries a literal address where he or she is authorized to work using his or her repairman skills. When the technician is no longer employed there, the repairman certificate must be returned to the Flight Standards District Office (FSDO) that issued it.

Mechanic Certification — General (by 14 CFR Section)

65.3 Certification of Foreign Airmen Other Than Flight Crewmembers

Normally, the FAA issues these certificates only to U.S. citizens or resident aliens residing in the United States. However, on occasion if the FAA determines that the issuance of a certificate to a person located outside of the United States is necessary for the operation and continued airworthiness of a U.S.-registered civil aircraft, it will issue a certificate to that person, providing they meet the necessary requirements.

65.11 Application and Issue

Any person who meets the criteria for obtaining a mechanic certificate must apply by means of a form and in a manner prescribed by the Administrator. That form is FAA Form 8610-2, Airman Certificate and/or Rating Application. If a mechanic has had a certificate suspended, they may not apply for additional ratings during the time of suspension. A revocation of a mechanic certificate prevents that person from applying for a certificate within a period of 1 year after the revocation.

65.12 Offenses Involving Alcohol and Drugs

Any person, who has been convicted of violating federal or state statutes relating to drug offenses, can be denied their application for a certificate or rating up to 1 year after the date of conviction. The violation can be relating to any one or more of the following actions: growing, processing, manufacturing, selling, disposing, transporting, or importing narcotic drugs, marijuana, depressants, or stimulants. They may also face the suspension or revocation of any certificate that they currently hold.

65.13 Temporary Certificate

A qualified applicant who successfully passes all required tests (a minimum score of 70 percent is required) may be issued a temporary certificate, which is valid for not more than 120 days. During this time, the FAA will review the application and supplementary documentation, and will issue the official certificate and rating.

65.14 Security Disqualification

This section was added following the terrorist attacks of September 11, 2001. It basically states that anyone determined by the TSA to be a security threat will either have their application held if they are applying for a certificate, or have the certificate that they do hold revoked.

65.15 Duration of Certificate

Mechanics certificates are effective until they are surrendered, suspended, or revoked. The difference in these terms can be summarized briefly this way:

- Surrendered means given up voluntarily.
- Suspended means the FAA temporarily removed the certificate from the holder.
- Revoked means the FAA permanently removed the certificate from the holder.

65.16 Change of Name: Replacement of Lost or Destroyed Certificate

If the technician changes his or her name, or is seeking a replacement certificate, an application must be submitted to the FAA at:

Federal Aviation Administration
Airmen Certification Branch (AFS-760)
P.O. Box 25082
Oklahoma City, OK 73125

There is a charge for this service.

65.17 Test: General Procedure

The FAA has designated certain persons to administer the tests associated with obtaining a mechanic certificate. The minimum passing score for these tests is 70 percent.

65.18 Written Tests: Cheating or Other Unauthorized Content

If the mechanic or repairmen applicant is determined to be cheating, or otherwise involved in unauthorized conduct, they are not eligible for *any* certificate or rating under this chapter for a period of 1 year. Furthermore, current ratings the person already holds may also be suspended or revoked. Unacceptable conduct for written tests consists of:

- Copying or intentionally removing the test.
- Giving or receiving any part of a copy of the test.
- Giving or receiving help during the test taking period.
- Using any material or aid during the test taking period.
- Intentionally causing, assisting, or participating in any of the previous acts.

65.19 Retesting After Failure

Should the mechanic or repairman fail to achieve the required minimum passing grade, there are two options they may consider when desiring to apply for retesting:

- Wait a period of 30 days after the date of test failure.
- Seek additional instruction in the subject matter areas he or she failed, and provide a signed statement from the certificated technician providing the instruction.

65.20 Applications, Certificates, Logbooks, Reports, and Records: Falsification, Reproduction, or Alteration

In 14 CFR part 43, paragraphs 9 and 11, defines the requirements for a technician to make appropriate entries in the maintenance/inspection records for the work performed. This proper documentation is fundamental to safe and efficient operation of the U.S. civil aircraft fleet. Therefore, the FAA takes strong action against those who would participate in the falsification of those records. The following actions are the basis for suspending or revoking any certificate or rating held by the person invoked:

- Fraudulent or intentional false statement on an application.
- Fraudulent or intentional false statement in any logbook, record, or report required to show compliance with any certificate requirements.
- Any reproduction (for fraudulent purposes) of a certificate or rating.
- Any alteration of any certificate or rating under this part.

65.21 Change of Address

If the technician changes his or her address, the FAA (at the address shown below) must be notified in writing within 30 days after the change of permanent residence:

> Federal Aviation Administration
> Airmen Certification Branch (AFS-760)
> P.O. Box 25082
> Oklahoma City, OK 73125

65.23 Refusal to Submit to a Drug or Alcohol Test

Any technician who refuses to submit to a drug test, which is required by 14 CFR part 121, appendix I or J, is subject to denial by the FAA of any application for additional certification or ratings, and suspension or revocation of any existing certificate or rating he or she currently holds. Appendix I of part 121 is titled Drug Testing Program, and requires a urine sample from the employee. Appendix J is titled Alcohol Misuse Prevention Program and requires that the employee submit to a breath test. Each appendix contains a "Definitions" section and a section titled "Employees who must be tested." Persons involved with "Aircraft maintenance or preventative maintenance duties" are listed in both appendices. There are various types (or rather times) when testing is required.

- Pre-employment
- Periodic
- Random
- Post-accident
- Testing based upon reasonable cause
- Return to duty testing
- Follow-up testing

The numerous test methods and the harsh penalty imposed by the FAA on those who would involve themselves with these unauthorized substances, or abuse the allowable use of alcohol indicates the concern that the FAA has for the possible impairment of technicians. Aviation maintenance is a professional career choice that demands the highest caliber technical person to be capable of functioning at his or her maximum potential. There is no room in this profession for a person to be involved with substance abuse. By doing so, the technician not only endangers themselves, but their co-workers, and ultimately the customer who is expecting to have an airworthy aircraft delivered following a maintenance activity.

Mechanic Certification — Specific (by 14 CFR Section)

65.71 Eligibility Requirements: General

The requirements for obtaining a mechanic certificate are:

- Be at least 18 years of age.
- Be able to read, write, speak, and understand the English language. (Note: If the applicant does not meet this requirement and is employed outside the United States by a U.S. carrier, the certificate will be endorsed "valid only outside the United States.")
- Have passed all the required tests (written, oral, and practical) within the preceding 24 months from application.
- Possess and demonstrate the appropriate knowledge and skill for the certificate rating being sought.

If a technician has one of the ratings, and desires to add the other, he/she must meet the requirements set forth in section (§) 65.77, and take the written, oral, and practical tests within 24 months.

65.73 Ratings

The FAA recognizes two ratings, airframe and powerplant, which may be obtained by a person upon successful application and testing. These may be attained either individually, or as a combined certificate.

Any person holding an aircraft (A) or aircraft engine (E) certificate prior to June 15, 1952, and which was valid on that date, may exchange it for the corresponding current certificate. If both ratings were held, the A & E certificate may be exchanged for an A & P.

65.75 Knowledge Requirements

Any applicant meeting the experience requirements listed in 65.77 must pass a written test (minimum passing score of 70, reference §65.17) covering the construction and maintenance of aircraft. Applicable portions of 14 CFR 43 and 91 are also included in the testing. Basic principles for the installation and maintenance of propellers are included with the testing that is administered for the powerplant rating. Successful completion of the written test is required before the candidate may apply for the oral and practical tests identified in section 65.79.

65.77 Experience Requirements

Each mechanic applicant must have a certificate of completion from an FAA certified technician school (14 CFR part 147) or provide documented evidence of a minimum of 18 months practical experience related to either airframe or powerplant maintenance (30 months required if applying for certification for both airframe and powerplant).

65.79 Skill Requirements

Oral and practical tests to determine the applicant's basic knowledge and skills necessary for the certificate or rating sought are required to be completed after the applicant has successfully completed the written test. Minor repairs and minor alterations to propellers are required to be demonstrated as part of the powerplant rating.

65.80 Certificated Aviation Maintenance Technician School Students

Whenever satisfactory evidence is shown to the FAA that a student enrolled in an aviation maintenance training school certificated under part 147 is making satisfactory progress, that student may take the oral and practical tests required by paragraph 79, prior to completing the school (as required by paragraph 77) and prior to taking the written test required by paragraph 75.

65.81 General Privileges and Limitations

Once a technician becomes a certificated mechanic, he or she may perform or supervise the maintenance, preventive maintenance, or alterations of an aircraft or appliance (or part thereof) for which he or she is rated. However, he or she is *not* permitted to perform major repair or major alterations to propellers nor accomplish any repair to or alteration of instruments. These activities are reserved for certificated repairmen at an authorized repair station. Also, he or she may not supervise the maintenance, preventive maintenance, or alteration of any aircraft or appliance (or part thereof) for which he or she is rated, unless he or she has satisfactorily performed this work at an earlier date. This is where the benefit of keeping an on the job training (OJT) log cannot be overemphasized. Whether the technician attends a part 147 maintenance training school or receives the required number of months as practical experience, he or she has only scratched the surface of the tremendously complex world of aviation maintenance. The technician must either work with someone (like a shop mentor) or must perform the task satisfactorily for the FAA. The certified mechanic must have and be able to comprehend the maintenance manuals and/or instructions for continued airworthiness for the task he or she is accomplishing.

65.83 Recent Experience Requirements

In addition to having the proper documentation, the mechanic is required by this regulation to have recent and relevant work experience. Although, as it was stated earlier in this chapter, the A & P certificate is valid until it is surrendered, suspended, or revoked, it may not be exercised if the holder has not been actively working as a mechanic at least 6 of the preceding 24 months. This activity can be any one of the following, or any combination of them:

- Served as a mechanic under the certificate and rating
- Technically supervised other mechanics
- Supervised (in an executive capacity) the maintenance or alteration of an aircraft

65.85 Airframe Rating: Additional Privileges

A mechanic who holds an airframe rating may approve and return to service an airframe, an appliance, or any related part after he or she has performed, supervised, or inspected minor repairs or alterations. He or she may also perform the maintenance actions required for a major repair or alteration, and should initiate the appropriate form (FAA Form 337, Major Repair and Alteration) associated with that work. However, the return to service action must be accomplished by a certificated A & P technician holding an Inspection Authorization (IA). (Refer to 14 CFR §65.95.) The airframe mechanic is also authorized to perform the 100-hour inspection (if required per 14 CFR part 91 §91.409) on the airframe.

The FAA recently added a new category of aircraft called Light Sport. (Refer to 14 CFR part 21, §21.190.) A certificated Airframe technician can approve and return to service the airframe after performing and inspecting a major repair or major alteration. The work must have been done on products that are not produced under FAA approval (i.e., are not type certificated) and must have been performed in accordance with instructions developed by the manufacturer or person acceptable to the FAA.

65.87 Powerplant Rating: Additional Privileges

Similarly, a mechanic holding a powerplant rating has the same limitations imposed regarding the powerplant and propeller as the airframe technician has on the airframe rating. He or she may perform and return to service minor repairs or alterations. He or she may also accomplish the work activities required for a major repair or alteration, but the work must be signed off for return to service by an IA. The privilege of performing a 100-hour inspection (if required by part 91) on a powerplant or propeller is also authorized.

A certificated Powerplant technician can approve and return to service a Light Sport powerplant or propeller after performing and inspecting a major repair or major alteration. The work must have been done on products that are not produced under FAA approval (i.e., are not type certificated) and must have been performed in accordance with instructions developed by the manufacturer or person acceptable to the FAA.

65.89 Display of Certificate

Once a technician receives his or her mechanic certificate, the certificate must be kept in the immediate area where he or she normally conducts work and exercises the privileges of the certificate. When requested, the technician is required to present the certificate for inspection to the FAA, or any authorized representation from the National Transportation Safety Board (NTSB), or any federal, state, or local law enforcement officer.

Inspection Authorization (by 14 CFR Section)

65.91 Inspection Authorization

An A & P mechanic who has held his or her certificate for at least 3 years, and has been active for the last 2 years, may submit application using FAA Form 8610-1, Mechanic's Application for Inspection Authorization, to the FAA for consideration as an IA. In addition to the preceding time requirements, the IA candidate must have:

- A fixed base of operation where he or she can be located in person or by phone.
- Available equipment, facilities, and inspection data necessary to properly inspect the airframe, powerplants, propeller, or any related part or appliance he or she will be approving for return to service.

The applicant who meets all the above criteria must then pass a written (computerized) test to determine his or her ability to inspect the airworthiness of an aircraft following either a major repair or alteration action or the performance of an annual or progressive inspection.

The minimum passing score for the computer test is 70 percent. If the applicant fails the test, retesting cannot be attempted until a minimum of 90 days have elapsed from the failure date. Unlike the A & P test, there is no reduction in this time if the applicant receives additional training.

65.92 Inspection Authorization: Duration

An IA certificate expires on March 31 of each odd numbered year, but may only be exercised during the time the technician holds a currently effective mechanic certificate. The IA ceases to be effective if:

- The technician surrenders it, or it is suspended or revoked.
- The technician no longer has a fixed base of operations.
- The technician no longer has the required facilities equipment or inspection data available.

Whenever the certificate is suspended or revoked, the technician must return it to the Administrator when requested by the FAA to do so.

65.93 Inspection Authorization: Renewal

An IA certificate may be renewed in one of the following ways each year the technician is seeking renewal:

- The performance of at least one annual inspection for each 90 days the technician has held the IA rating.
- The performance of the inspections of at least two major repairs or alterations for each 90 days the technician has held the IA rating. (Note: The inspections can be counted regardless of the approval or disapproval of the work.)
- The performance (or supervision) and approval of at least one progressive inspection.
- The attendance and successful completion of a refresher course (acceptable to the Administrator) that is at least 8 hours of instruction. This can be either a single day seminar or a combination of individual classes acceptable to the Administrator. Some seminars are sponsored by the FAA FSDOs and are free; others are low cost. Private industry also frequently conducts one-day sessions and usually charge for their efforts. Regardless of who is conducting the seminar, it is usually an excellent way to accomplish renewal, learn about new issues, and develop a network among peers.

No renewal is required for someone who received the IA during the first quarter of the calendar year, since the regulation states that anyone holding an IA for less than 90 days need not meet the preceding renewal requirements.

It should be noted that regulations clearly state the number of annual inspections and major repair or alteration inspections required for renewal are *for* each 90-day period and not *in* each 90-day period. Therefore, an IA could actually go 11 months without performing any inspection activity relative to renewal. Then, in March he or she could conduct all four necessary annual inspections, or all eight 337-related inspections. The regulations do not provide for the mixing of any of these renewal activities (i.e., two annual inspections and four Major Repair and Alteration forms).

Another method of renewal is to meet with the FAA-assigned FSDO inspector who will determine that the applicant possesses current knowledge of the applicable regulations and standards. Although this is often considered the renewal method of last resort, it should not be considered a negative experience. If the IA has been performing his or her activities in a professional manner throughout the year, this session can be considered a professional follow-up or consultation. Proper IA-to-FSDO inspector interaction can be enhanced with such a meeting.

65.95 Inspection Authorization: Privileges and Limitations

The IA may perform an annual inspection, or perform or supervise a progressive inspection. He or she may also approve for return to service any aircraft related part or appliance which has undergone a major repair or alteration (except aircraft maintained in accordance with a continuous airworthiness program operated under part 121).

The IA must keep his or her certificate available for inspection by any one of the following persons:

- Aircraft owner
- A & P technician
- FAA Administrator
- Authorized representative of the NTSB
- Any federal, state, local, or law enforcement officer

If the holder of an IA moves his or her fixed base of operation, he or she must notify in writing the FSDO responsible for the location he or she is moving to, before beginning to exercise the privileges of an IA. Although it is not required, good business etiquette and professional responsibility would suggest that a similar letter be written to the responsible FAA principal maintenance inspector (PMI) at the FSDO in the area he or she is leaving.

Ethics

This is a tremendously broad and diverse area of study. It is also an area that is coming under more scrutiny by consumers, individual watchdog groups, and government review committees. Ethics, or more appropriately the lack of ethics, has caused the loss of millions of dollars through fraudulent accounting practices, shoddy workmanship, etc. This chapter examines some definitions of ethics and some examples of poor business ethics in order to raise the awareness of the technician to the importance of ethics.

The word "ethics" is actually a philosophical term that comes from the Greek word "ethos," which means character or custom. So, it is logical that a current definition of ethics is "the study of standards of conduct and moral judgment." Although situations involving questionable ethics can exist wherever and whenever business decisions are made, the scope of this discussion is limited to areas with which the technician will probably be associated.

A Scenario

The following incident illustrates one way that both personal ethics and technician knowledge of regulations can work together to provide him or her with the ability to make the right decision. Unfortunately, others in that shop did not appear as concerned as the technician sharing the incident.

A technician working for an airline was involved in a situation that required a repair or replacement of a fuselage ice shield. The computer inventory indicated that a replacement part was in stock, so the technician removed the damaged component. It was then found that the replacement part was not actually in stock. At this point, a crucial decison was to be made: Can the damaged item be reinstalled? The steps in properly documenting a maintenance event are to record the removal of the damaged part, then document the installation of an airworthy part. Once the technician has committed to removing the damaged part, it becomes unairworthy and cannot be reinstalled regardless of its deferability in the minimum equipment list (MEL).

The actual sequence of events is as follows:

Significant impact damage to the ice shield was observed and recorded.

The inspector reviewed and instructed the technician to replace the ice shield.

Availability of the replacement part was confirmed by computer.

The damaged part was removed, and the technician prepared the surface for the replacement part.

The new part was ordered from inventory, but the part was not in stock. (Inventory Error)

The inspector instructed the technician to reinstall the old one.

The technician refused.

The inspector instructed the technician to repair it.

The technician researched the structural repair manual (SRM) and found that the facility did not have the proper facility authorization to repair the damaged part.

The inspector told the technician to apply 5-minute epoxy to the area, sand it down, and paint it.

The technician walked away.

The inspector found someone else to compromise standards. The aircraft departed on time—illegal and unairworthy.

This happens more often than one would like, is probably overlooked by many people, and, unfortunately, might be considered standard operating procedure (SOP) for some maintenance facilities. It is the responsibility of the mechanic to follow regulations and to question the actions of his or her supervisors if the policy is to circumvent rules to make an on-time departure.

This incident provides some valuable insights into how day-to-day events can lead to pressure to produce and ultimately compromise the decision-making.

1. The incident occurred while working for a commercial airline. The pressure for getting the aircraft in the air is tremendous in this environment.
2. Inventory error added to the pressure. The damaged part had been removed because the technician had queried and believed a replacement part was immediately available.
3. The inspector was either unaware of regulatory requirements or simply did not care.
4. The second technician was either unaware of regulatory requirements or simply did not care.

Final Observation

The underlying company culture was apparently lacking concern for ethical decisions and regulatory compliance. An effective organizational culture should always encourage ethical behavior and discourage unethical behavior. This means that not only does the upper management of an organization say that they conduct themselves ethically, they must do it consistently; employees, customers, vendors, and even competitors should know this company has "high ethical standards."

This latter issue may sometimes have painful consequences, if the businesses are competing for a customer's business. The ethical company may estimate the maintenance activities to take 8 weeks and quotes that time frame to the customer. The unethical company may also know the work will take 8 weeks, but tells the customer only 6 weeks, hoping to get the job. Once the plane is "captured" and maintenance has begun, explanations and excuses extend the original time estimate of 6 weeks to the actual 8 weeks or longer. Although the customer would be disappointed in this situation, few customers would be able to remove an aircraft undergoing maintenance. This "bait and switch" tactic is often used by unscrupulous companies to get an aircraft into their shop no matter what it takes. Although the shop's retention of clients is frequently very low, there always seem to be new ones willing to accept a shorter-than-normal turnaround time quote. Often these same shops will underbid the job, and then continually add extra costs as the work progresses. The technician is encouraged to avoid employment at maintenance facilities that do not think twice about trying to deceive the customer.

Since companies are usually in business to make money, the "bottom line" mentality frequently drives management, and ultimately technician decisions. But short-term, quick-fix solutions that focus only on immediate financial success promote the idea that everything boils down to monetary gain. Ethical behavior is not about monetary gain.

made the error. The MEDA concept is based on the following three principles:

- Positive employee intent (In other words, maintenance technicians want to do the best job possible and do not make intentional errors.)
- Contribution of multiple factors (There is often a series of factors that contribute to an error.)
- Manageability of errors (Most of the factors that contribute to an error can be managed.)

When a company is willing to adopt these principles, then the MEDA process can be implemented to help the maintenance organization achieve the dual goals of identifying those factors that contribute to existing errors, and avoiding future errors. In creating this five-step process, Boeing initially worked with British Airways, Continental Airlines, United Airlines, a maintenance worker labor union, and the FAA. The five steps are:

1. **Event:** the maintenance organization must select which error that caused events will be investigated.
2. **Decision:** was the event maintenance related? If the answer is yes, then the MEDA investigation continues.
3. **Investigation:** using the MEDA results form, the operator conducts an investigation to record general information about the airplane—when the maintenance and the event occurred, what event initiated the investigation, the error that caused the event, the factors contributing to the error, and a list of possible presentation strategies.
4. **Prevention strategies:** the operator reviews, prioritizes, implements, and then tracks the process improvements (prevention strategies) in order to avoid or reduce the likelihood of similar errors in the future.
5. **Feedback:** the operator provides feedback to the maintenance workplace so technicians know that changes have been made to the maintenance system as a result of this MEDA process.

The implantation and continuous use of MEDA is a long-term commitment and not a "quick fix." However, airline operators and maintenance facilities frequently decide to use the MEDA approach to investigate serious, high visibility events which have caused significant cost to the company. The desire to do this is based upon the potential "payback" of such an investigation. This may ultimately be counterproductive because a highly visible event may not really be the best opportunity to investigate errors. Those involved in the process may be intimidated by the attention coming from upper management and various regulatory authorities.

By using the MEDA process properly, the organization can investigate the factors that contributed to an error, discover exactly what led to that error, and fix those factors. Successful implementation of MEDA will allow the organization to avoid rework, lost revenue, and potentially dangerous situations related to events caused by maintenance errors.

The "SHEL" model is another concept for investigating and evaluating maintenance errors. [Figure 13-3] As with other human factors tools, its goal is to determine not only what the problem is, but where and why it exists. SHEL was initiated by Professor Elwyn Edwards (Professor Emeritus, Aston University, Birmingham, U.K.) in 1972. It was later modified slightly by the late Capt. Frank Hawkins, a Human Factors consultant to KLM (Royal Dutch Airlines), in 1975. The acronym SHEL represents:

- **S**oftware
- **H**ardware
- **E**nvironment
- **L**iveware

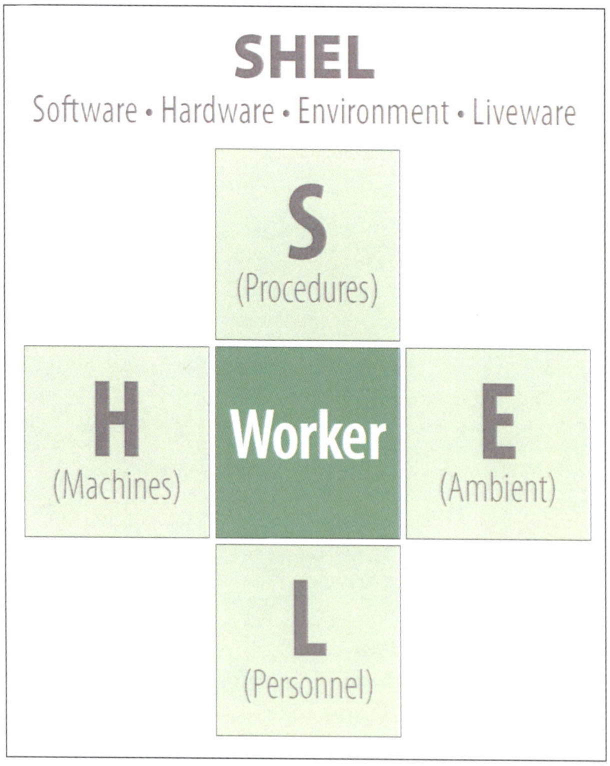

Figure 13-3. SHEL Model.

4. Test each option by asking such questions as:
   - —Is it legal?
   - —Is it right?
   - —Is it beneficial?
5. Make your decision.
6. Double check your decision by asking:
   - —How would I feel if my family found out about this?
   - —How would I feel if my decision is printed in the local newspaper?
7. Take action (Schermerhorn 1989).

Finally, the technician is encouraged to read the following code of ethics developed by Professional Aviation Maintenance Association (PAMA), Inc., and consider adopting it as their own.

"As a certified technician, my performance is a public service and, as such, I have a responsibility to the United States Government and its citizens. I must ensure that all citizens have confidence in my integrity, and that I will perform my work according to the highest principles of ethical conduct. Therefore, I swear that I shall hold in sacred trust the rights and privileges conferred upon me as a certified technician. The safety and lives of others are dependent on my skill and judgment; therefore, I shall never knowingly subject others to risks which I would not be willing to assume for myself, or those who are dear to me.

"As a certified technician, I am aware that it is not possible to have knowledge and skill in every aspect of aviation maintenance for every airplane, so I pledge that I will never undertake work or approve work which I believe to be beyond the limits of my knowledge. I shall not allow any superior to persuade me to approve aircraft or equipment as airworthy when there is doubt in my mind as to the validity of my action. Under no circumstances will I permit the offer of money or other personal favors to influence me to act contrary to my best judgment, nor to pass as airworthy aircraft or equipment about which I am in doubt.

"The responsibility that I have accepted as a certified technician demands that I exercise my judgment on the airworthiness of aircraft and equipment; therefore, I pledge unyielding adherence to these precepts for the advancement of aviation and for the dignity of my vocation."

Professionalism

The aviation technician is the central figure in aviation maintenance. From the day he or she sets out to become a certificated technician per 14 CFR part 65 to the day he or she retires or leaves the field, the technician must be fully qualified as an aviation professional. This means the technician successfully blends technical training with ethical thinking, and understands the ramifications of the various aspects of human factors. Although the word "professionalism" is widely used, it is rarely defined. In fact, no single definition can encompass all of the qualifications and considerations that must be present for true professionalism to exist. Though not all inclusive, the following list provides major considerations and qualifications that should be included in the definition of professionalism.

1. Professionalism exists only when a service is performed for someone or for the common good.
2. Professionalism is achieved only after extended training and preparation.
3. Professionalism is performance based on study and research.
4. Professionalism is reasoning logically and accurately.
5. Professionalism is making good judgment decisions.
6. Professionalism is not limiting actions and decisions to standard patterns and practices.
7. Professionalism demands a code of ethics.
8. Professionalism is being true to one's values and ethics and to those being served. Anything less than a sincere performance is quickly detected, and immediately destroys effectiveness.

Chapter 14

Human Factors

Introduction

Why are human conditions, such as fatigue, complacency, and stress, so important in aviation maintenance? These conditions, along with many others, are called human factors. Human factors directly cause or contribute to many aviation accidents. It is universally agreed that 80 percent of maintenance errors involve human factors. If they are not detected, they can cause events, worker injuries, wasted time, and even accidents. *[Figure 14-1]*

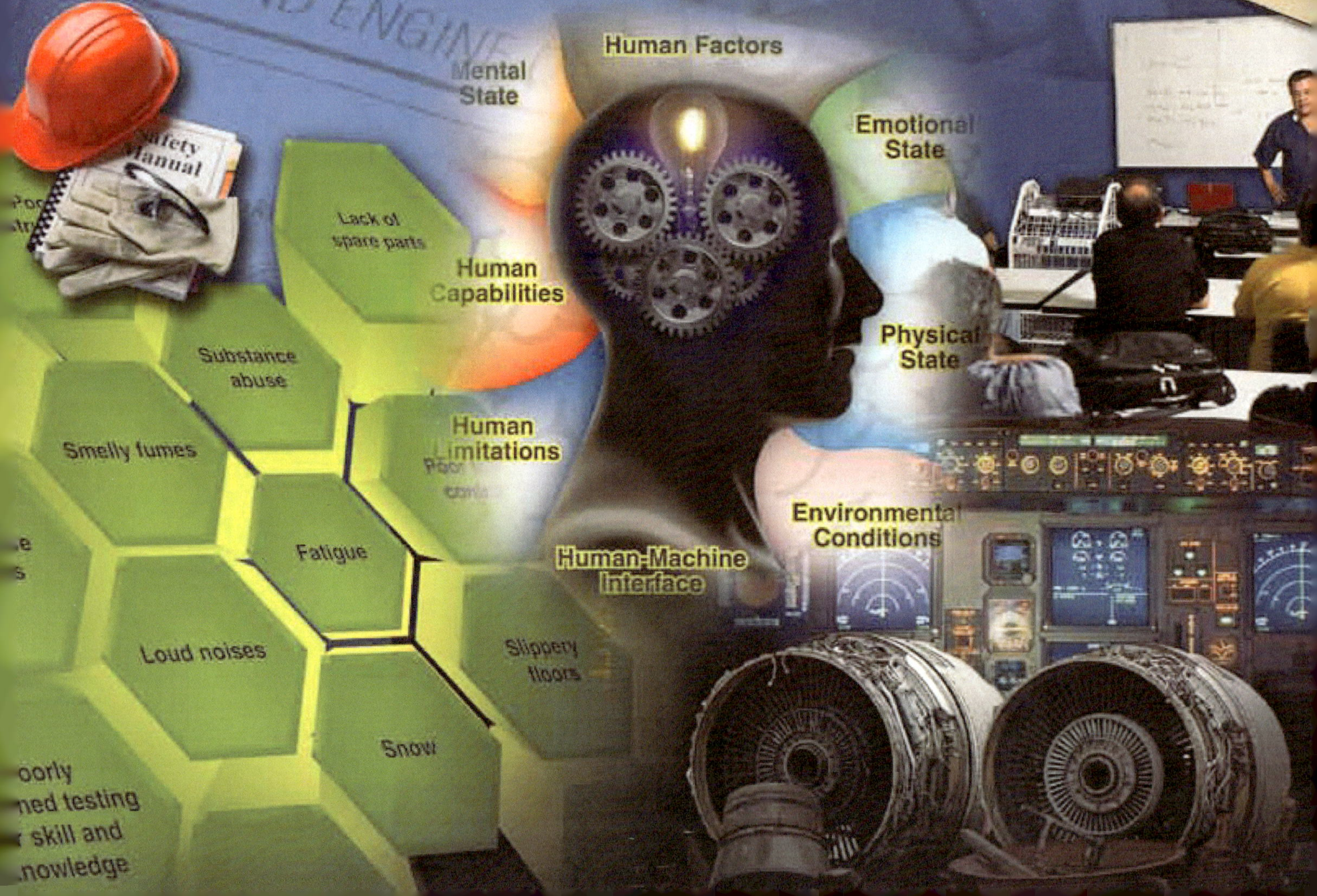

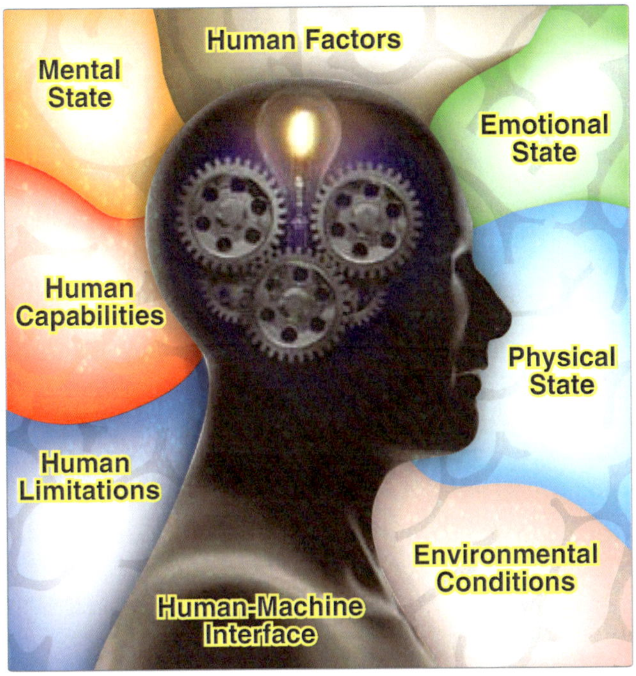

Figure 14-1. *Human factors and how they affect people are very important to aviation maintenance.*

Aviation safety relies heavily on maintenance. When it is not done correctly, it contributes to a significant proportion of aviation accidents and incidents. Some examples of maintenance errors are parts installed incorrectly, missing parts, and necessary checks not being performed. In comparison to many other threats to aviation safety, the mistakes of an aviation maintenance technician (AMT) can be more difficult to detect. Often times, these mistakes are present but not visible and have the potential to remain latent, affecting the safe operation of aircraft for longer periods of time.

AMTs are confronted with a set of human factors unique within aviation. Often times, they are working in the evening or early morning hours, in confined spaces, on platforms that are up high, and in a variety of adverse temperature/humidity conditions. The work can be physically strenuous, yet it also requires attention to detail. *[Figure 14-2]* Because of the nature of the maintenance tasks, AMTs commonly spend more time preparing for a task than actually carrying it out. Proper documentation of all maintenance work is a key element, and AMTs typically spend as much time updating maintenance logs as they do performing the work. *[Figure 14-3]*

Human factors awareness can lead to improved quality, an environment that ensures continuing worker and aircraft safety, and a more involved and responsible work force. More specifically, the reduction of even minor errors can provide measurable benefits including cost reductions, fewer missed

Figure 14-2. *Aviation maintenance technicians (AMTs) are confronted with many human factors due to their work environments.*

Figure 14-3. *AMT documenting repair work.*

deadlines, reduction in work related injuries, reduction of warranty claims, and reduction in more significant events that can be traced back to maintenance error. Within this chapter, the many aspects of human factors are discussed in relation to aviation maintenance. The most common human factors are introduced along with ways to mitigate the risk to stop them from developing into a problem. Several Federal Aviation Administration (FAA) human factor resources are provided to include the most direct link to aviation maintenance human factors which can be found at https://hfskyway.faa.gov.

What is Human Factors

The term human factors has grown increasingly popular as the commercial aviation industry realize that human error, rather than mechanical failure, underlies most aviation accidents and incidents. Human factors science or technologies are multidisciplinary fields incorporating contributions from psychology, engineering, industrial design, statistics, operations research, and anthropometry. It is a term that covers the science of understanding the properties of human capability, the application of this understanding to the design, development, and deployment of systems and services, and the art of ensuring successful application of human factor principles into the maintenance working environment.

The list of human factors that can affect aviation maintenance and work performance is broad. They encompass a wide range of challenges that influence people very differently as humans do not all have the same capabilities, strengths, weaknesses, or limitations. Unfortunately, aviation maintenance tasks that do not account for the vast amount of human limitations can result in technical error and injuries. *Figure 14-4* shows some of the human factors that affect AMTs. Some are more serious than others but, in most cases, when you combine three or four of the factors, they create a problem that contributes to an accident or incident.

Elements of Human Factors

Human factors are comprised of many disciplines. This section discusses ten of those disciplines: Clinical Psychology, Experimental Psychology, Anthropometrics, Computer Science, Cognitive Science, Safety Engineering, Medical Science, Organizational Psychology, Educational Psychology, and Industrial Engineering. *[Figure 14-5]*

The study and application of human factors is complex because there is not just one simple answer to fix or change how people are affected by certain conditions or situations. Aviation maintenance human factors research has the overall goal to identify and optimize the factors that affect human performance in maintenance and inspection. The focus initiates on the technician but extends to the entire engineering and technical organization. Research is optimized by incorporating the many disciplines that affect human factors and help to understand how people can work more efficiently and maintain work performance.

By understanding each of the disciplines and applying them to different situations or human behaviors, we can correctly recognize potential human factors and address them before they develop into a problem or create a chain of problems that result in an accident or incident.

Clinical Psychology

Clinical psychology includes the study and application of psychology for the purpose of understanding, preventing, and relieving psychologically-based distress or dysfunction and to promote subjective well-being and personal development. It focuses on the mental well-being of the individual. Clinical psychology can help individuals deal with stress, coping mechanisms for adverse situations, poor self image, and accepting criticism from coworkers.

Experimental Psychology

Experimental psychology includes the study of a variety of basic behavioral processes, often in a laboratory environment. These processes may include learning, sensation, perception, human performance, motivation, memory, language, thinking, and communication, as well as the physiological processes underlying behaviors, such as eating, reading, and problem solving. In an effort to test the efficiency of work policies and procedures, experimental studies help measure performance, productivity, and deficiencies.

Anthropometrics

Anthropometry is the study of the dimensions and abilities of the human body. This is essential to aviation maintenance due to the environment and spaces that AMTs have to work with.

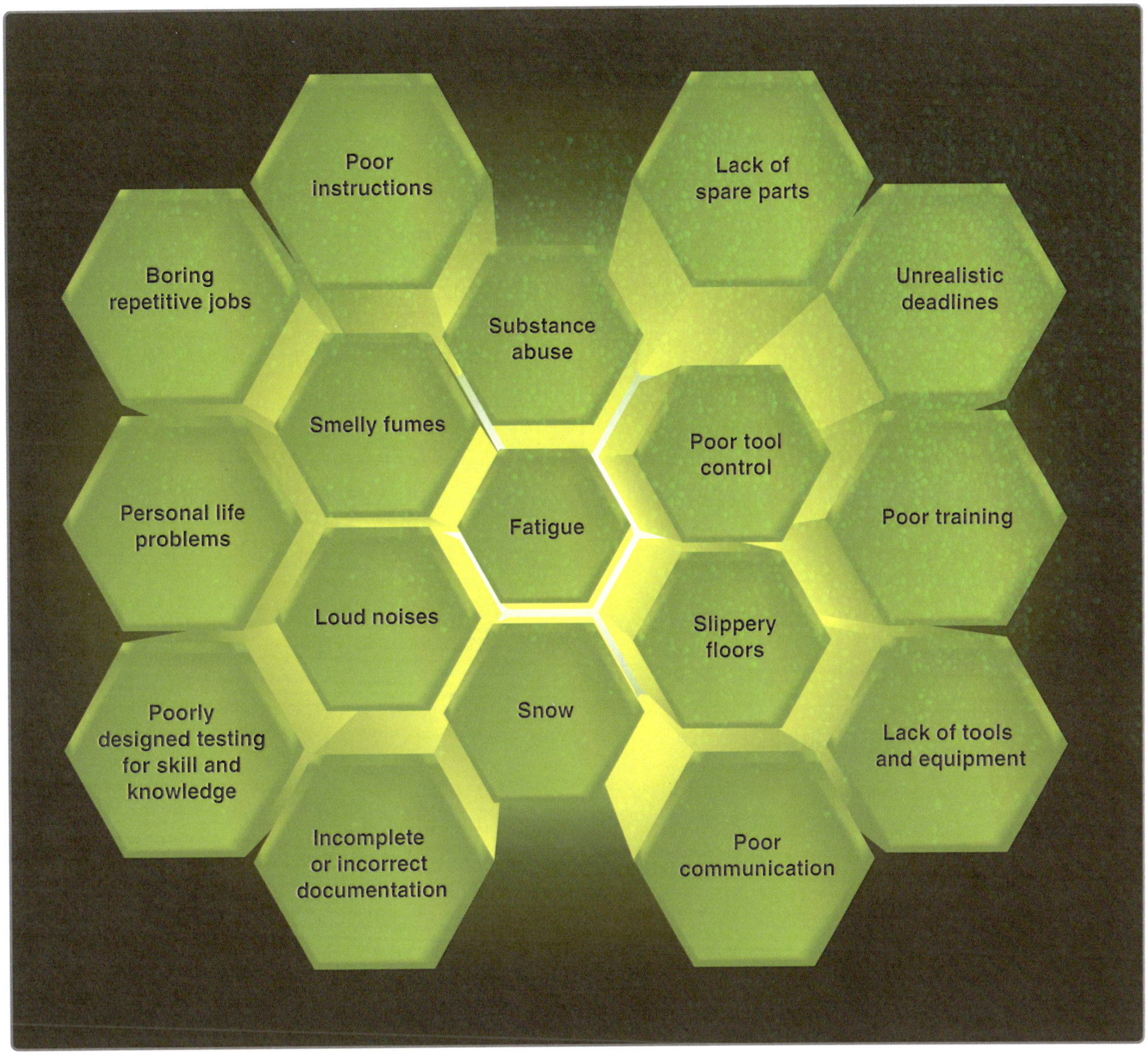

Figure 14-4. *A list of human factors that affect AMTs.*

For example, a man who is 6 feet 3 inches and weighs 230 pounds may be required to fit into a small crawl space of an aircraft to conduct a repair. Another example is the size and weight of equipment and tools. Men and women are generally on two different spectrums of height and weight. Although both are equally capable of completing the same task with a high level of proficiency, someone who is smaller may be able to perform more efficiently with tools and equipment that is tailored to their size. In other words, one size does not fit all and the term "average person" does not apply when employing such a diverse group of people.

Computer Science

The technical definition for computer science is the study of the theoretical foundations of information and computation and of practical techniques for their implementation and application in computer systems. How this relates to aviation maintenance is a lot simpler. As mentioned earlier, AMTs spend as much time documenting repairs as they do performing them. It is important that they have computer work stations that are comfortable and reliable. Software programs and computer-based test equipment should be easy to learn and use, and not intended only for those with a vast level of computer literacy.

Cognitive Science

Cognitive science is the interdisciplinary scientific study of minds as information processors. It includes research on how information is processed (in faculties such as perception, language, reasoning, and emotion), represented, and transformed in a nervous system or machine (e.g., computer). It spans many levels of analysis from low-level

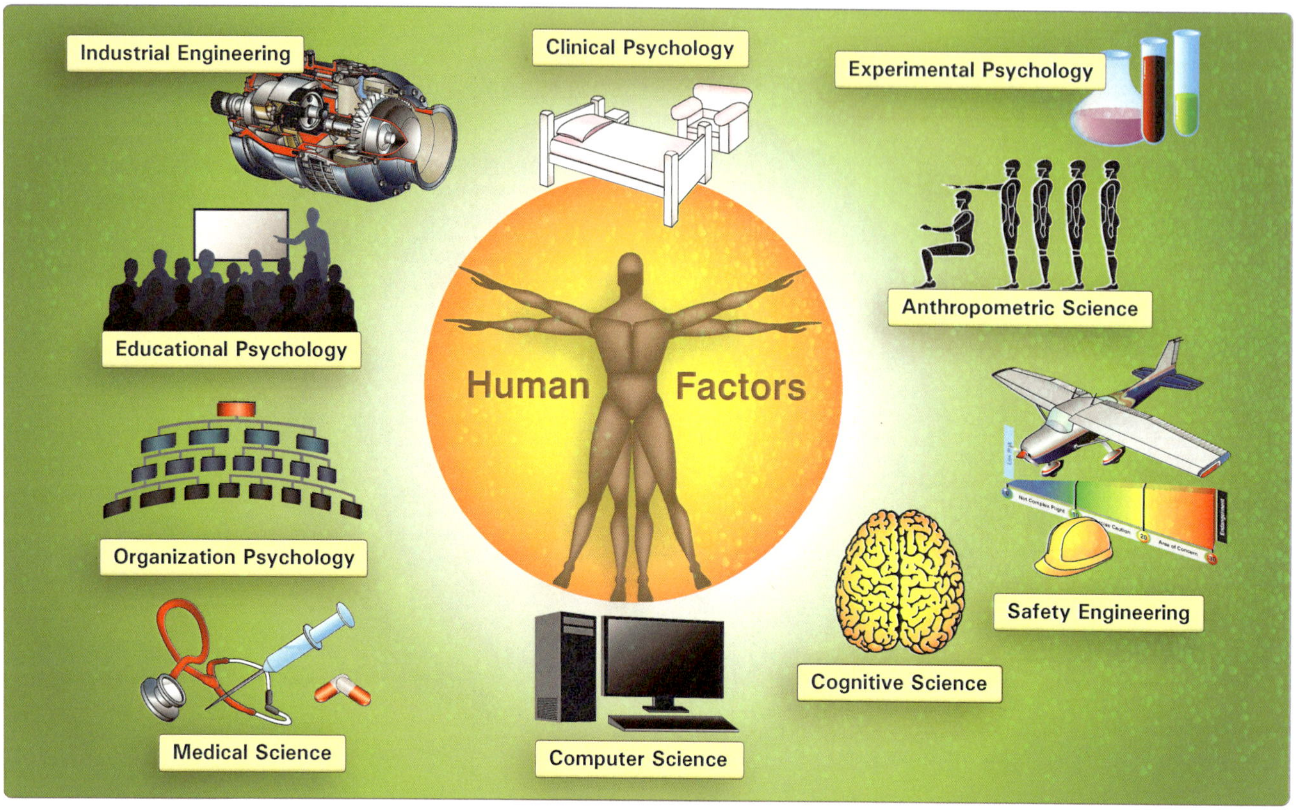

Figure 14-5. *Human factor disciplines.*

learning and decision mechanisms to high-level logic and planning. AMTs must possess a great ability to problem solve quickly and efficiently. They constantly have to troubleshoot a situation and quickly react to it. This can be a viscous cycle creating an enormous amount of stress. The discipline of cognitive science helps us understand how to better assist AMTs during situations that create high levels of stress so that their mental process does not get interrupted and effect their ability to work.

Safety Engineering

Safety engineering assures that a life-critical system behaves as needed even when the component fails. Ideally, safety engineers take an early design of a system, analyze it to find what faults can occur, and then propose safety requirements in design specifications up front and changes to existing systems to make the system safer. Safety cannot be stressed enough when it comes to aviation maintenance, and everyone deserves to work in a safe environment. Safety engineering plays a big role in the design of aviation maintenance facilities, storage containers for toxic materials, equipment used for heavy lifting, and floor designs to ensure no one slips, trips, or falls. In industrial work environments, the guidelines of the Occupational Safety and Health Administration (OSHA) are important.

Medical Science

Medicine is the science and art of healing. It encompasses a variety of health care practices evolved to maintain and restore health by the prevention and treatment of illness. Disposition and physical well-being are very important and directly correlated to human factors. Just like people come in many shapes and sizes, they also have very different reactions to situations due to body physiology, physical structures, and biomechanics.

Organizational Psychology

Organizational psychologists are concerned with relations between people and work. Their interests include organizational structure and organizational change, workers' productivity and job satisfaction, consumer behavior, and the selection, placement, training, and development of personnel. Understanding organizational psychology helps aviation maintenance supervisors learn about the points listed below that, if exercised, can enhance the work environment and productivity.

- Rewards and compensations for workers with good safety records.
- Motivated workers that want to do well and work safely.

- Unified work teams and groups that get along and work together to get the job done right.
- Treat all workers equally.

Educational Psychology

Educational psychologists study how people learn and design the methods and materials used to educate people of all ages. Everyone learns differently and at a different pace. Supervisors should design blocks of instruction that relate to a wide variety of learning styles.

Industrial Engineering

Industrial engineering is the organized approach to the study of work. It is important for supervisors to set reasonable work standards that can be met and exceeded. Unrealistic work standards create unnecessary stressors that cause mistakes. It is also beneficial to have an efficient facility layout so that there is room to work. Clean and uncluttered environments enhance work performance. Another aspect of industrial engineering that helps in the understanding of human factors is the statistical analysis of work performance. Concrete data of work performance, whether good or bad, can show the contributing factors that may have been present when the work was done.

History of Human Factors

Around 1487, Leonardo DiVinci began research in the area of anthropometrics. The Vitruvian Man, one of his most famous drawings, can be described as one of the earliest sources presenting guidelines for anthropometry. *[Figure 14-6]* Around the same time, he also began to study the flight of birds. He grasped that humans are too heavy and not strong enough to fly using wings simply attached to the arms. Therefore, he sketched a device in which the aviator lies down on a plank and works two large, membranous wings using hand levers, foot pedals, and a system of pulleys. *[Figure 14-7]* Today, anthropometry plays a considerable role in the fields of computer design, design for access and maintainability, simplicity of instructions, and ergonomics issues.

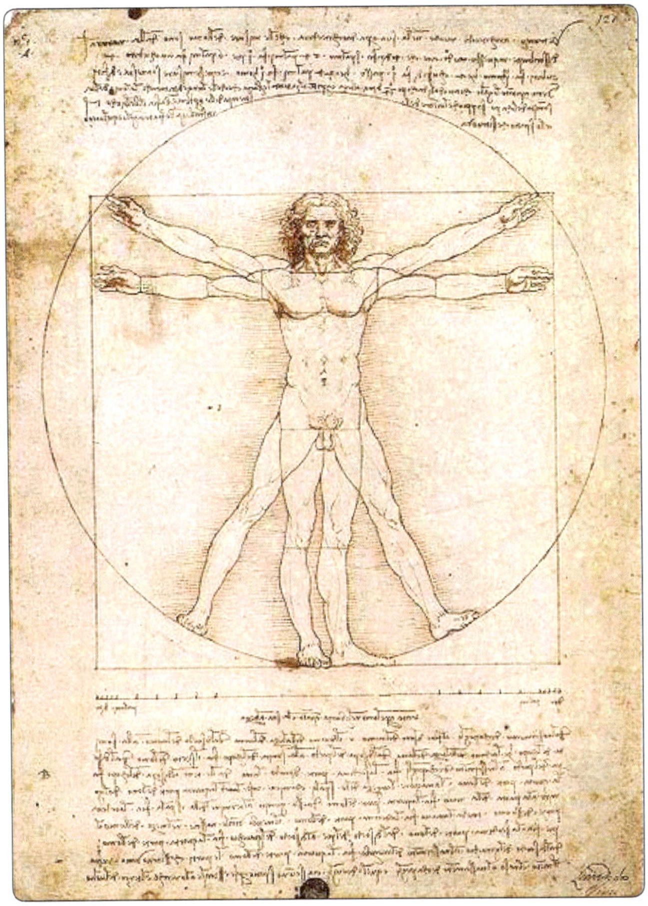

Figure 14-6. *Vitruvian Man, one of Leonardo DiVinci's famous drawings about anthropometry.*

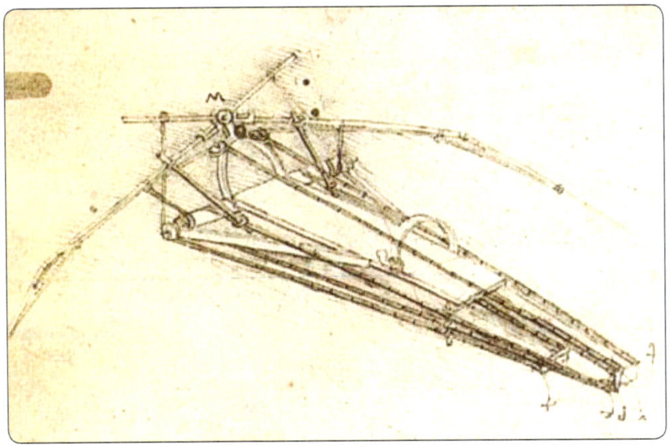

Figure 14-7. *Leonardo DiVinci's rendering of a flying device for man.*

In the early 1900s, industrial engineers Frank and Lillian Gilbreth were trying to reduce human error in medicine. *[Figures 14-8* and *14-9]* They developed the concept of using call backs when communicating in the operating room. For example, the doctor says "scalpel" and the nurse repeats "scalpel" and then hands it to the doctor. That is called the challenge-response system. Speaking out loud reinforces what tool is needed and provides the doctor with an opportunity to correct his/herself if that is not the necessary tool. This same verbal protocol is used in aviation today. Pilots are required to read back instructions or clearances given by air traffic control (ATC) to ensure that the pilot receives the correct instructions and gives ATC an opportunity to correct if the information is wrong. Frank and Lillian Gilbreth also are known for their research on fatigue.

Also in the early 1900s, Orville and Wilbur Wright were the first to fly a powered aircraft and also pioneered many human factors considerations. While others were trying to

Figure 14-8. *Frank Gilbreth – Industrial Engineer.*

Figure 14-9. *Lillian Gilbreth – Industrial Engineer.*

develop aircraft with a high degree of aerodynamic stability, the Wrights intentionally designed unstable aircraft with cerebralized control modeled after the flight of birds. Between 1901 and 1903, the brothers worked with large gliders at Kill Devil Hills, near Kitty Hawk, North Carolina, to develop the first practical human interactive controls for aircraft pitch, roll, and yaw. On December 17, 1903, they made four controlled powered flights over the dunes at Kitty Hawk with their Wright Flyer. *[Figure 14-10]* They later developed practical in-flight control of engine power, plus an angle of attack sensor and stick pusher that reduced pilot workload. The brothers' flight demonstrations in the United States and Europe during 1908-1909, awakened the world to the new age of controlled flight. Orville was the first aviator to use a seat belt and also introduced a rudder boost/trim control that gave the pilot greater control authority. The Wrights' flight training school in Dayton, Ohio included a flight simulator of their own design. The Wrights patented their practical airplane and flight control concepts, many of which are still in use today.

Figure 14-10. *The Wright Brothers on December 17, 1903, flying over the dunes at Kitty Hawk with their Wright Flyer.*

Prior to World War I, the only test of human to machine compatibility was that of trial and error. If the human functioned with the machine, he was accepted, if not he was rejected. There was a significant change in the concern for humans during the American Civil War. The U.S. Patent Office was concerned about whether the mass produced uniforms and new weapons could effectively be used by the infantry men.

Evolution of Maintenance Human Factors

With the onset of World War I (1914–1918), more sophisticated equipment was being developed and the inability of personnel to use such systems led to an increased interest in human capability. Up to this point, the focus of aviation psychology was on the pilot, but as time progressed, the focus shifted onto the aircraft. Of particular concern was the design of the controls and displays, the effects of altitude, and environmental factors on the pilot. The war also brought on the need for aeromedical research and the need for testing and measurement methods. By the end of World War I, two aeronautical labs were established, one at Brooks Air Force Base, Texas, and the other at Wright Field outside of Dayton, Ohio.

Another significant development was in the civilian sector, where the effects of illumination on worker productivity were examined. This led to the identification of the Hawthorne Effect, which suggested that motivational factors could significantly influence human performance.

With the onset of World War II (1939–1945), it was becoming increasingly harder to match individuals to pre-existing jobs. Now the design of equipment had to take into account human limitations and take advantage of human capabilities. This change took time as there was a lot of research to be done to determine the human capabilities and limitations that had to be accomplished. An example of this is the 1947 study done by Fitts and Jones, who studied the most effective configuration of control knobs to be used in aircraft flight decks. Much of this research transcended into other equipment with the aim of making the controls and displays easier for the operators to use.

In the initial 20 years after World War II, most human factors research was done by Alphonse Chapanis, Paul Fitts, and Arnold Small. The beginning of the Cold War led to a major expansion of Department of Defense supported research laboratories, and many of the labs established during the war started expanding. Most of the research following the war was military sponsored and large sums of money were granted to universities to conduct research. The scope of the research also broadened from small equipment to entire workstations and systems. In the civilian industry, the focus shifted from research to participation through advice to engineers in the design of equipment.

The Pear Model

There are many concepts related to the science and practice of human factors. However, from a practical standpoint, it is most helpful to have a unified view of the things we should be concerned about when considering aviation maintenance human factors. A good way to gain this understanding is by using a model. For more than a decade, the term "PEAR" has been used as a memory jogger, or mnemonic, to characterize human factors in aviation maintenance. PEAR prompts recall of the four important considerations for human factors programs, which are listed below.

- People who do the job.
- Environment in which they work.
- Actions they perform.
- Resources necessary to complete the job.

People

Aviation maintenance human factors programs focus on the people who perform the work and address physical, physiological, psychological, and psychosocial factors. *[Figure 14-11]* It must focus on individuals, their physical capabilities, and the factors that affect them. It also should consider their mental state, cognitive capacity, and conditions that may affect their interaction with others. In most cases, human factors programs are designed around the people in the company's existing workforce. You cannot apply identical strength, size, endurance, experience, motivation, and certification standards equally to all employees. The company must match the physical characteristics of each person to the tasks each performs.

The company must consider factors like each person's size, strength, age, eyesight, and more to ensure each person is physically capable of performing all the tasks making up the job. A good human factors program considers the limitations of humans and designs the job accordingly. An important

Figure 14-11. *People who do the job.*

element when incorporating human factors into job design is planned rest breaks. People can suffer physical and mental fatigue under many work conditions. Adequate breaks and rest periods ensure the strain of the task does not overload their capabilities. Another "People" consideration, which also is related to "E" for "Environment," is ensuring there is proper lighting for the task, especially for older workers. Annual vision testing and hearing exams are excellent proactive interventions to ensure optimal human physical performance.

Attention to the individual does not stop at physical abilities. A good human factors program must address physiological and psychological factors that affect performance. Companies should do their best to foster good physical and mental health. Offering educational programs on health and fitness is one way to encourage good health. Many companies have reduced sick leave and increased productivity by making healthy meals, snacks, and drinks available to their employees. Companies also should have programs to address issues associated with chemical dependence, including tobacco and alcohol. Another "People" issue involves teamwork and communication. Safe and efficient companies find ways to foster communication and cooperation among workers, managers, and owners. For example, workers should be rewarded for finding ways to improve the system, eliminate waste, and help ensure continuing safety.

Environment

There are at least two environments in aviation maintenance. There is the physical workplace on the ramp, in the hangar, or in the shop. In addition, there is the organizational environment that exists within the company. A human factors program must pay attention to both environments. *[Figure 14-12]*

Physical

The physical environment is obvious. It includes ranges of temperature, humidity, lighting, noise control, cleanliness, and workplace design. Companies must acknowledge these conditions and cooperate with the workforce to either accommodate or change the physical environment. It takes a corporate commitment to address the physical environment. This topic overlaps with the "Resources" component of PEAR when it comes to providing portable heaters, coolers, lighting, clothing, and workplace and task design.

Organizational

The second, less tangible, environment is the organizational one. The important factors in an organizational environment are typically related to cooperation, communication, shared values, mutual respect, and the culture of the company. An excellent organizational environment is promoted with leadership, communication, and shared goals associated with safety, profitability, and other key factors. The best companies guide and support their people and foster a culture of safety. A safe culture is one where there is a shared value and attitude toward safety. In a safe culture, each person understands their individual role is contributing to overall mission safety.

Actions

Successful human factors programs carefully analyze all the actions people must perform to complete a job efficiently and safely. Job task analysis (JTA) is the standard human factors approach to identify the knowledge, skills, and attitudes necessary to perform each task in a given job. The JTA helps identify what instructions, tools, and other resources are necessary. Adherence to the JTA helps ensure each worker is properly trained and each workplace has the necessary equipment and other resources to perform the job. Many regulatory authorities require the JTA serve as the basis for the company's general maintenance manual and training plan. Many human factors challenges associated with use of job cards and technical documentation fall under "Actions." A crystal clear understanding and documentation of actions ensures instructions and checklists are correct and useable. *[Figure 14-13]*

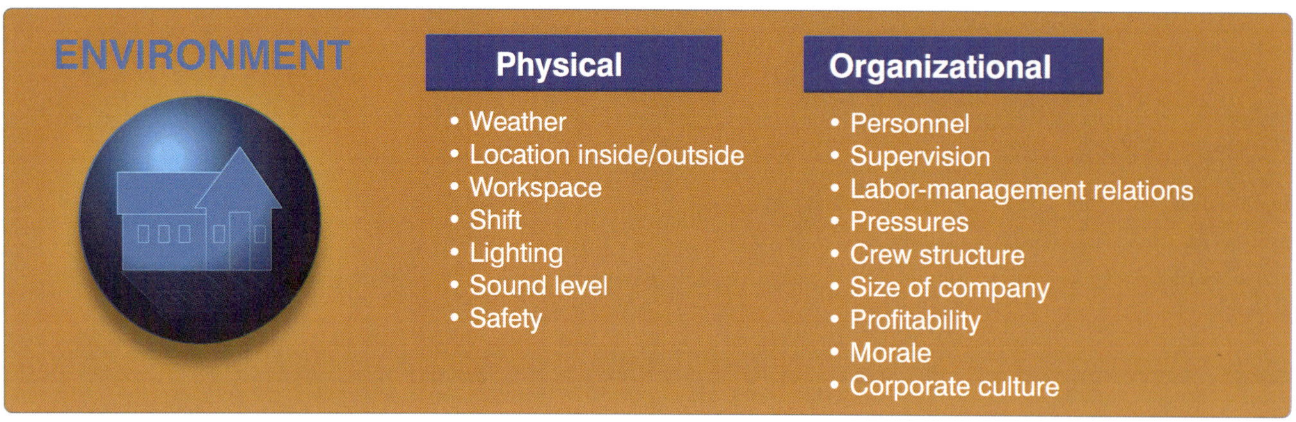

Figure 14-12. *Environment in which they work.*

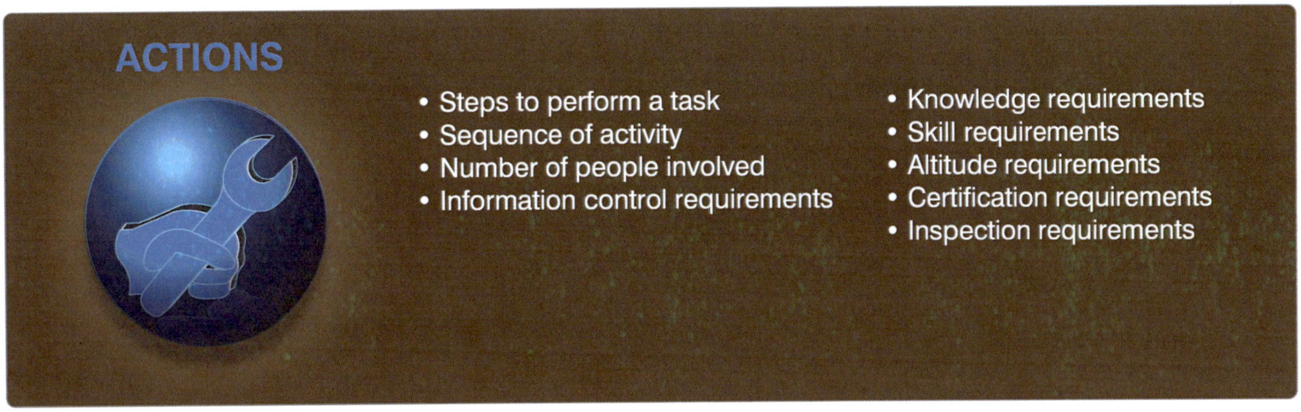

Figure 14-13. *Actions they perform.*

Resources

The final PEAR letter is "R" for "Resources." *[Figure 14-14]* Again, it is sometimes difficult to separate resources from the other elements of PEAR. In general, the characteristics of the people, environment, and actions dictate the resources. Many resources are tangible, such as lifts, tools, test equipment, computers, technical manuals, and so forth. Other resources are less tangible. Examples include the number and qualifications of staff to complete a job, the amount of time allocated, and the level of communication among the crew, supervisors, vendors, and others. Resources should be viewed (and defined) from a broad perspective. A resource is anything a technician (or anyone else) needs to get the job done. For example, protective clothing is a resource. A mobile phone can be a resource. Rivets can be resources. What is important to the "Resource" element in PEAR is focusing on identifying the need for additional resources.

Human Error

Human error is defined as a human action with unintended consequences. There is nothing inherently wrong or troublesome with error itself, but when you couple error with aviation maintenance and the negative consequences that it produces, it becomes extremely troublesome. Training, risk assessments, safety inspections, etc., should not be restricted to attempt to avoid errors but rather to make them visible and identify them before they produce damaging and regrettable consequences. Simply put, human error is not avoidable but it is manageable. *[Figure 14-15]*

Figure 14-15. *Safety awareness will help foresee and mitigate the risk of human error.*

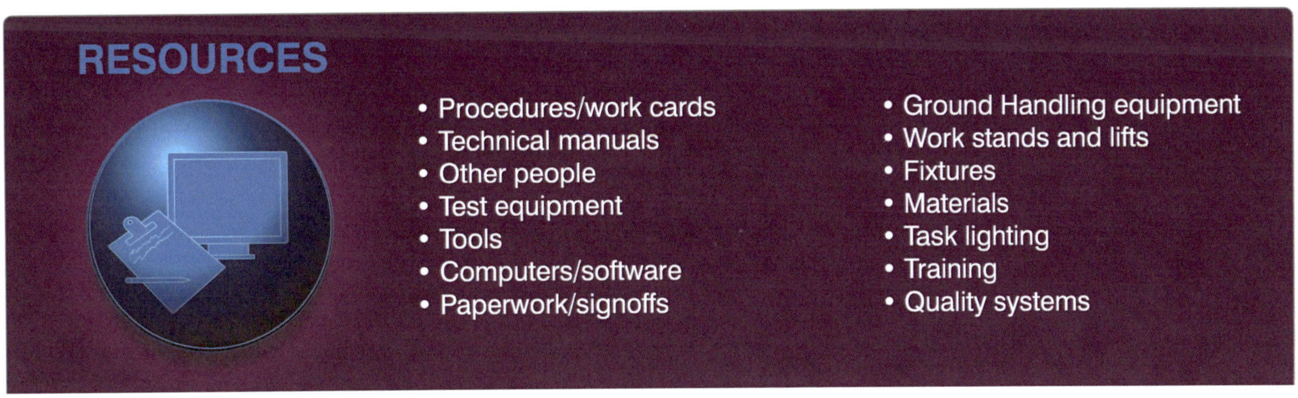

Figure 14-14. *Resources necessary to complete the job.*

Types of Errors

Unintentional

An unintentional error is an unintentional wandering or deviation from accuracy. This can include an error in your action (a slip), opinion, or judgment caused by poor reasoning, carelessness, or insufficient knowledge (a mistake). For example, an AMT reads the torque values from a job card and unintentionally transposed the number 26 to 62. He or she did not mean to make that error but unknowingly and unintentionally did. An example of an unintentional mistake would be selecting the wrong work card to conduct a specific repair or task. Again, not an intentional mistake but a mistake nonetheless.

Intentional

In aviation maintenance, an intentional error should really be considered a violation. If someone knowingly or intentionally chooses to do something wrong, it is a violation, which means that one has deviated from safe practices, procedures, standards, or regulations.

Kinds of Errors

Active and Latent

An active error is the specific individual activity that is an obvious event. A latent error is the company issues that lead up to the event. For example, an AMT climbs up a ladder to do a repair knowing that the ladder is broken. In this example, the active error was falling from the ladder. The latent error was the broken ladder that someone should have replaced.

The "Dirty Dozen"

Due to a large number of maintenance-related aviation accidents and incidents that occurred in the late 1980s and early 1990s, Transport Canada identified twelve human factors that degrade people's ability to perform effectively and safely, which could lead to maintenance errors. These twelve factors, known as the "dirty dozen," were eventually adopted by the aviation industry as a straight forward means to discuss human error in maintenance. It is important to know the dirty dozen, how to recognize their symptoms, and most importantly, know how to avoid or contain errors produced by the dirty dozen. Understanding the interaction between organizational, work group, and individual factors that may lead to errors and accidents, AMTs can learn to prevent or manage them proactively in the future.

Lack of Communication

Lack of communication is a key human factor that can result in suboptimal, incorrect, or faulty maintenance. *[Figure 14-16]* Communication occurs between the AMT and many people (i.e., management, pilots, parts suppliers, aircraft servicers). Each exchange holds the potential for misunderstanding or omission. But communication between AMTs may be the most important of all. Lack of communication between technicians could lead to a maintenance error and result in an aircraft accident. This is especially true during procedures where more than one technician performs the work on the aircraft. It is critical that accurate, complete information be exchanged to ensure that all work is completed without any step being omitted. Knowledge and speculation about a task must be clarified and not confused. Each step of the maintenance procedure must be performed according to approved instructions as though only a single technician did the work.

A common scenario where communication is critical and a lack thereof can cause problems, is during shift change in an airline or fixed base operator (FBO) operation. A partially completed job is transferred from the technician finishing his or her workday to the technician coming on duty. Many steps in a maintenance procedure are not able to be seen or verified once completed due to the installation of components hiding the work. No steps in the procedure can be omitted and some steps still to be performed may be contingent on the work already completed. The departing technician must thoroughly explain what has occurred so that the arriving technician can correctly complete the job. A recounting of critical steps and any difficulties encountered gives insight. A lack of communication at this juncture could result in the work being continued without certain required operations having been performed.

The approved steps of a maintenance procedure must be signed off by the technician doing the work as it is performed. Continuing a job that has been started by someone else should only occur after a face-to-face meeting of technicians. The applicable paperwork should be reviewed, the completed work discussed, and attention for the next step should be drawn. Absence of either a written or oral turnover serves as warning that an error could occur.

It is vital that work not be continued on a project without both oral and written communication between the technician who started the job and the technician continuing it. Work should always be done in accordance with the approved written procedure and all of the performed steps should bear the signature of the technician who accomplishes the work. If necessary, a phone call can be made to obtain an oral turnover when technicians cannot meet face-to-face at the work area. In general, the technician must see his or her role as part of a greater system focused on safe aircraft operation and must communicate well with all in that system to be effective.

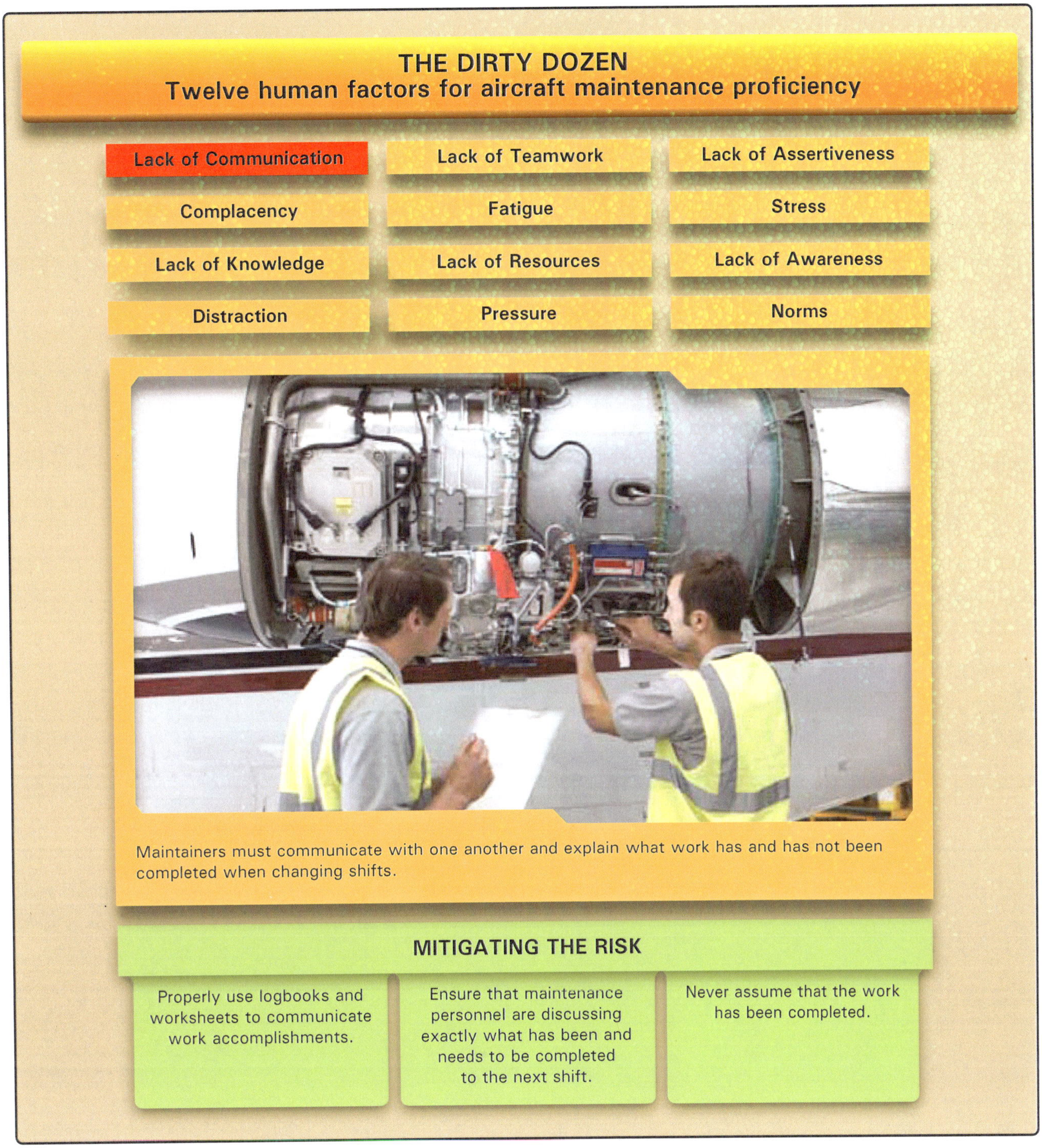

Figure 14-16. *Lack of communication.*

Complacency

Complacency is a human factor in aviation maintenance that typically develops over time. *[Figure 14-17]* As a technician gains knowledge and experience, a sense of self satisfaction and false confidence may occur. A repetitive task, especially an inspection item, may be overlooked or skipped because the technician has performed the task a number of times without ever finding a fault. The false assumption that inspection of the item is not important may be made. However, even if rare, a fault may exist. The consequences of the fault not being detected and corrected could cause an incident or accident. Routine tasks performed over and over allow time for the technician's mind to wander, which may also result in a required task not being performed.

When a technician finds him or herself performing work without documentation, or documenting work that was not performed, it is a sign that complacency may exist.

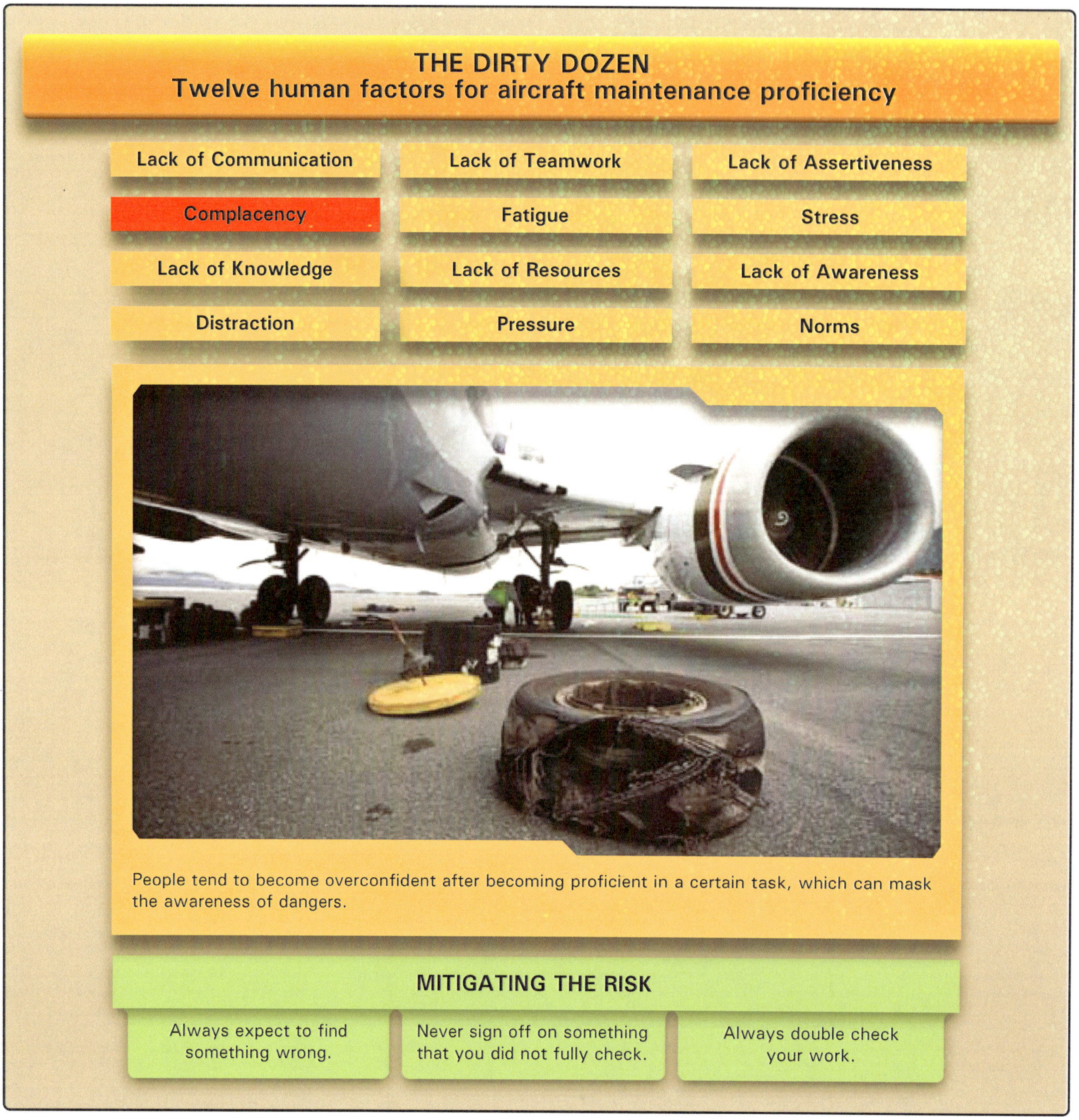

Figure 14-17. *Complacency.*

Approved, written maintenance procedures should be followed during all maintenance inspections and repairs. Executing the proper paperwork draws attention to a work item and reinforces its significance.

To combat complacency, a technician must train oneself to expect to find the fault that created the inspection item in the first place. He or she must stay mentally engaged in the task being performed. All inspection items must be treated with equal importance, and it must never be assumed that an item is acceptable when it has not been inspected. A technician should never sign for any work that has not been performed. Prior to the pen touching the paper for a signature, the technician should read the item before signing and confirm it has been performed.

Lack of Knowledge

A lack of knowledge when performing aircraft maintenance can result in a faulty repair that can have catastrophic results. *[Figure 14-18]* Differences in technology from aircraft to aircraft and updates to technology and procedures on a single aircraft also make it challenging to have the knowledge required to perform airworthy maintenance.

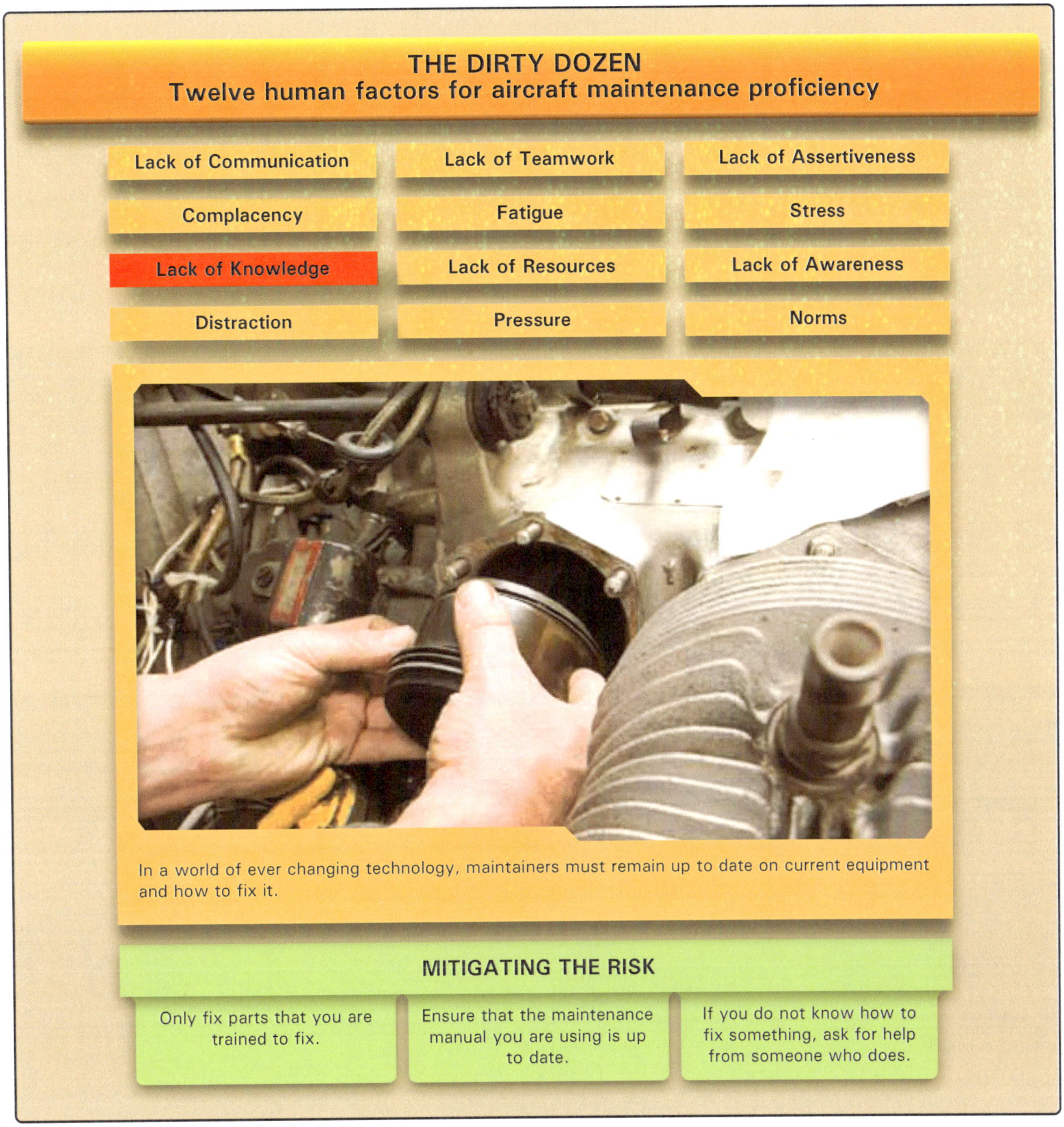

Figure 14-18. *Lack of knowledge.*

All maintenance must be performed to standards specified in approved instructions. These instructions are based on knowledge gained from the engineering and operation of the aircraft equipment. Technicians must be sure to use the latest applicable data and follow each step of the procedure as outlined. They must also be aware that differences exist in the design and maintenance procedures on different aircraft. It is important for technicians to obtain training on different types of aircraft. When in doubt, a technician with experience on the aircraft should be consulted. If one is not available, or the consulted technician is not familiar with the procedure, a manufacturer's technical representative should be contacted. It is better to delay a maintenance procedure than to do it incorrectly and cause an accident.

Distraction

A distraction while performing maintenance on an aircraft may disrupt the procedure. *[Figure 14-19]* When work resumes, it is possible that the technician skips over a detail that needs attention. It is estimated that 15 percent of maintenance related errors are caused by distractions.

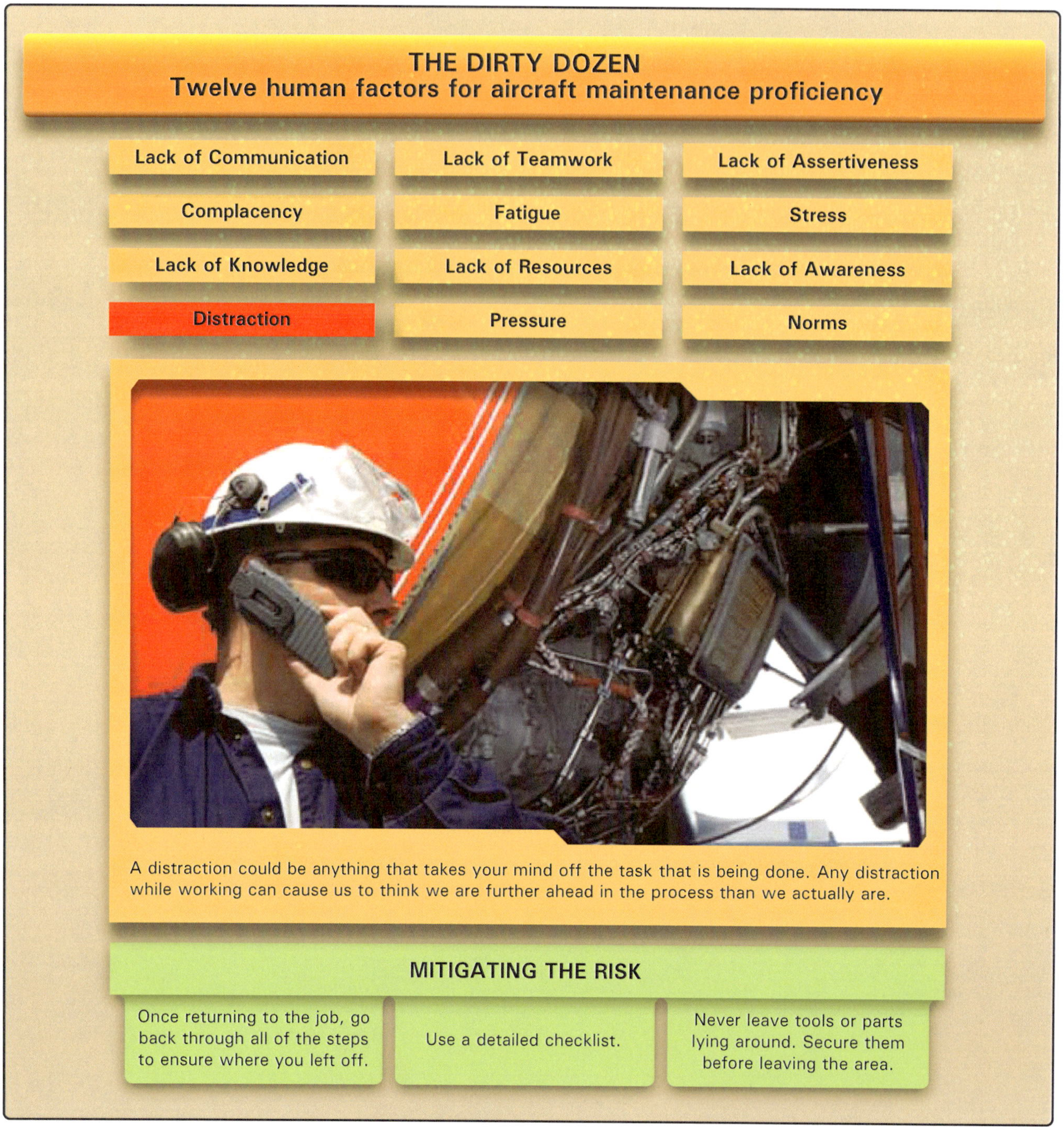

Figure 14-19. *Distraction.*

Distractions can be mental or physical in nature. They can occur when the work is located on the aircraft or in the hangar. They can also occur in the psyche of the technician independent of the work environment. Something as simple as a cell phone call or a new aircraft being pushed into the hangar can disrupt the technician's concentration on a job. Less visible is a difficult family or financial matter or other personal issues that may occupy the technicians thought process as work is performed. This can make performance of the required maintenance less effective.

Regardless of their nature, numerous distractions may occur during the course of maintaining an aircraft. The technician must recognize when attention to the job at hand is being diverted and assure that work continues correctly. A good practice is to go back three steps in the work procedure when one is distracted and resume the job from that point. Use of a detailed step-by-step written procedure and signing off each step only after it is completed also helps. Incomplete work can be marked or tagged, especially when the technician is pulled from the work by a distraction, and it is unknown when work will be resumed and by whom. Disconnect any connector and leave it plainly visible if an installation is not complete. There is a tendency to think a job is finished when a component is "hooked up." Similarly, when a step in the maintenance procedure is complete, be sure to immediately lock wire or torque the fasteners if required. This can be used as an indication that all is well up to that point in the procedure.

Lack of Teamwork

A lack of teamwork may also contribute to errors in aircraft maintenance. *[Figure 14-20]* Closely related to lack of communication, teamwork is required in aviation maintenance in many instances. Sharing of knowledge between technicians, coordinating maintenance functions, turning work over from shift to shift, and working with flight personnel to troubleshoot and test aircraft are all are executed better in an atmosphere of teamwork. Often associated with improved safety in the workplace, teamwork involves everyone understanding and agreeing on actions to be taken. A gear swing or other operational check involves all the members of a team working together. Multiple technicians contribute to the effort to ensure a single outcome. They communicate and look out for one

THE DIRTY DOZEN
Twelve human factors for aircraft maintenance proficiency

Lack of Communication	**Lack of Teamwork**	Lack of Assertiveness
Complacency	Fatigue	Stress
Lack of Knowledge	Lack of Resources	Lack of Awareness
Distraction	Pressure	Norms

Personality difference in the workplace must be left at the door. Organizations should emphasize that a lack of teamwork can ultimately effect the safety of maintenance work.

MITIGATING THE RISK

Ensure that lines of communication are open between personnel.	Discuss specific duties when jobs require more than one person to eliminate any questions.	Always look out for co-workers with safety in mind.

Figure 14-20. *Lack of teamwork.*

another as they do the job. A consensus is formed that the item is airworthy or not airworthy.

The technician primarily deals with the physical aspect of the aircraft and its airworthiness. Others in the organization perform their roles and the entire company functions as a team. Teams can win or lose depending on how well everyone in the organization works together toward a common objective. A lack of teamwork makes all jobs more difficult and, in maintenance, could result in a miscommunication that affects the airworthiness of the aircraft.

Fatigue

Fatigue is a major human factor that has contributed to many maintenance errors resulting in accidents. *[Figure 14-21]* Fatigue can be mental or physical in nature. Emotional fatigue also exists and effects mental and physical performance. A person is said to be fatigued when a reduction or impairment in any of the following occurs: cognitive ability, decision-making, reaction time, coordination, speed, strength, and balance. Fatigue reduces alertness and often reduces a person's ability to focus and hold attention on the task being performed.

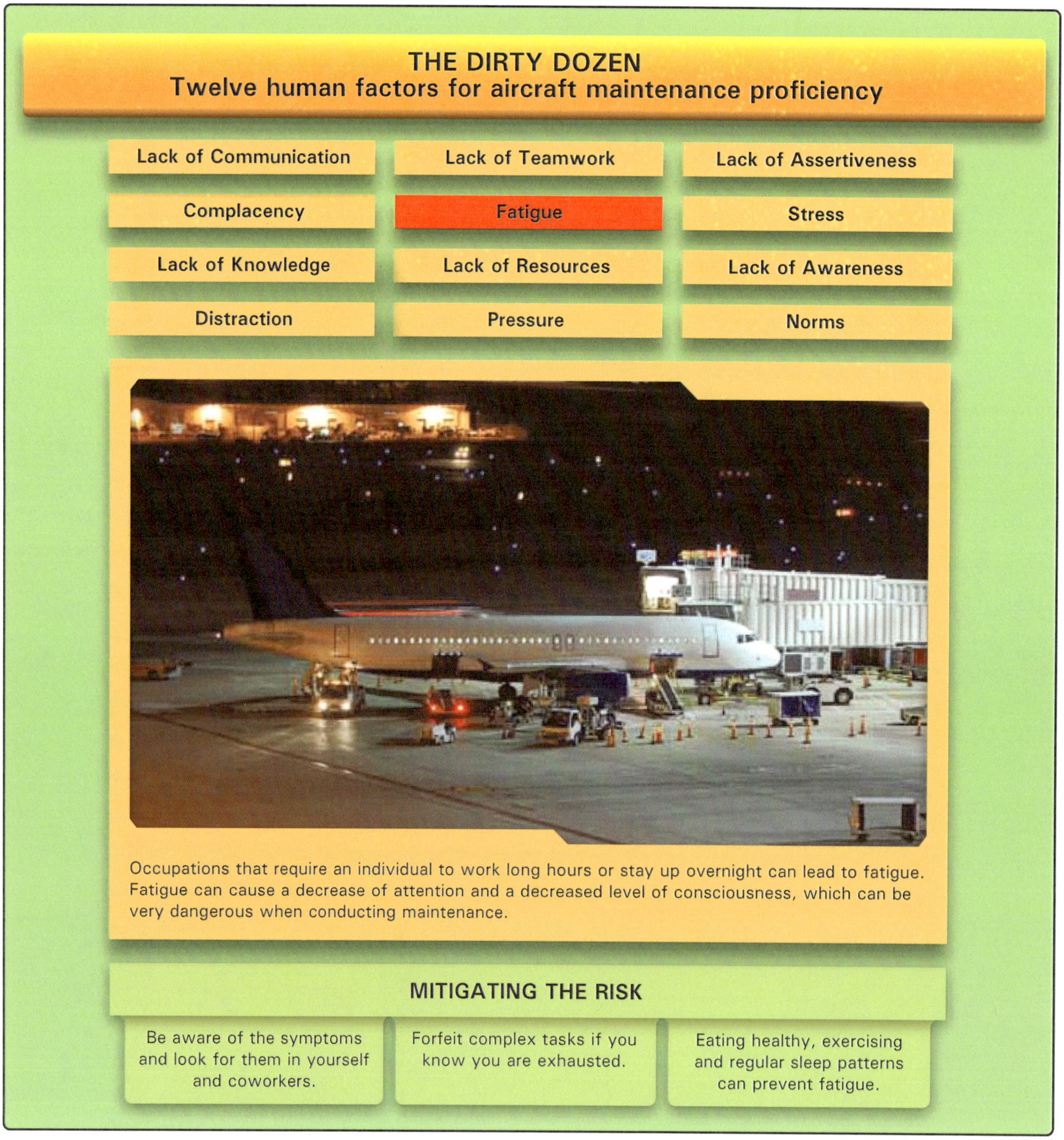

Figure 14-21. *Fatigue.*

Symptoms of fatigue may also include short-term memory problems, channeled concentration on unimportant issues while neglecting other factors that may be more important, and failure to maintain a situational overview. A fatigued person may be easily distracted or may be nearly impossible to distract. He or she may experience abnormal mood swings. Fatigue results in an increase in mistakes, poor judgment, and poor decisions or perhaps no decisions at all. A fatigued person may also lower his or her standards.

Tiredness is a symptom of fatigue. However, sometimes a fatigued person may feel wide awake and engaged in a task. The primary cause of fatigue is a lack of sleep. Good restful sleep, free from drugs or alcohol is a human necessity to prevent fatigue. Fatigue can also be caused by stress and overworking. A person's mental and physical state also naturally cycles through various levels of performance each day. Variables such as body temperature, blood pressure, heart rate, blood chemistry, alertness, and attention rise and fall in a pattern daily. This is known as one's circadian rhythm. *[Figure 14-22]* A person's ability to work (and rest) rises and falls during this cycle. Performance counter to circadian rhythm can be difficult. Until it becomes extreme, a person may be unaware that he or she is fatigued. It is easier recognized by another person or in the results of tasks being performed. This is particularly dangerous in aviation maintenance since the lives of people depend on maintenance procedures performed at a high level of proficiency. Working alone when fatigued is particularly dangerous.

The best remedy for fatigue is to get enough sleep on a regular basis. The technician must be aware of the amount and quality of sleep obtained. Caution or time off is justified when too little sleep has occurred and errors are probable during maintenance.

Countermeasures to fatigue are often used. Effectiveness can be short lived and many countermeasures may make fatigue worse. Caffeine is a common fatigue countermeasure. Pseudoephedrine found in sinus medicine and amphetamines are also used. While effective for short periods, a fatigued person remains fatigued and may have trouble getting the rest needed once off the job due to this drug use.

Suggestions to help mitigate the problems caused by fatigue include looking for symptoms of fatigue in one's self and in others. Have others check your work, even if an inspector sign off is not required. Avoid complex tasks during the bottom of your circadian rhythm. Sleep and exercise daily. Eight to nine hours of daily sleep are recommended to avoid fatigue. AMTs in airline operations are part of a system in which most maintenance is performed at night. Fleet aircraft are operated primarily during the daytime hours to generate company revenue. Therefore, shift work is required to maintain the fleet. It is already known that turning work over to other technicians during shift change is a problem that can lead to errors due to lack of communication. But shift work alone is a cause of fatigue that can degrade performance and also lead to errors. Shift work requires technicians to work during low cycles of their natural circadian rhythm. It also makes sleep more difficult when not on the job. Furthermore, regular night shift work makes one's body more sensitive to environmental disturbances. It can degrade performance, morale, and safety. It can also affect one's physical health. All of these can be reflected in degraded maintenance performance—a dangerous situation.

The technician must be aware that shift work is the norm in aviation. Avoidance of fatigue is part of the job. Title 14 of the Code of Federal Regulations (14 CFR) part 121, section

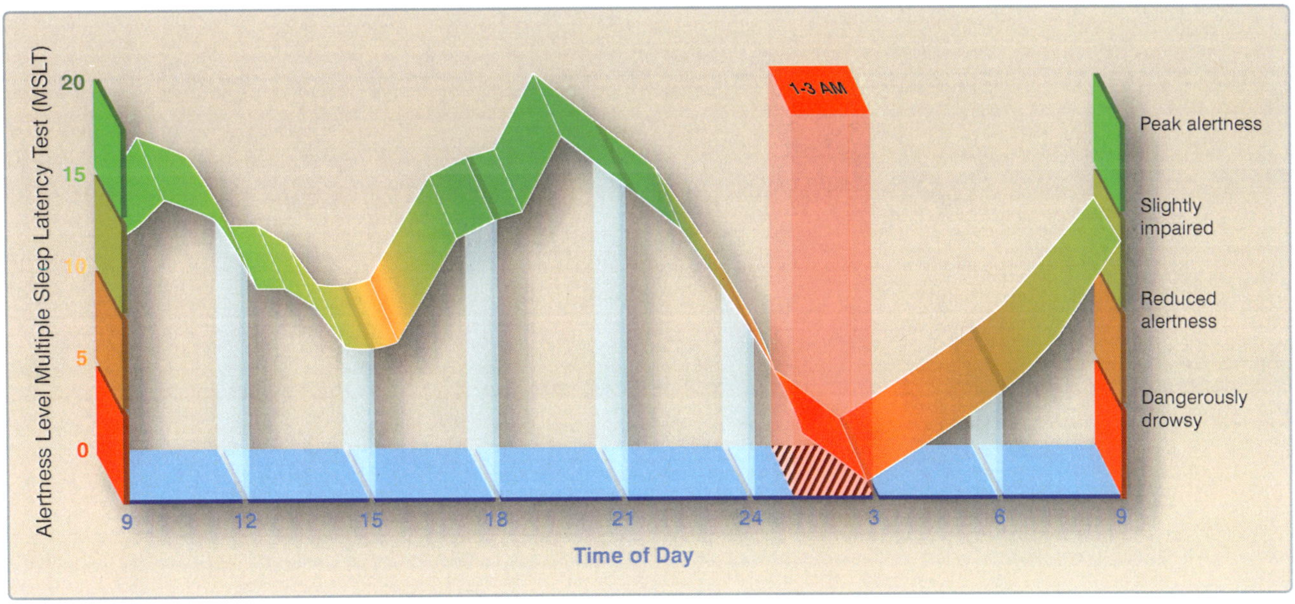

Figure 14-22. *Many human variables rise and fall daily due to one's natural circadian rhythm.*

377, only requires 24 hours time off during a week of work. Since this is obviously not enough, it is up to companies and technicians to regulate shift work and time off to reduce the potential for errors. Most importantly, each technician must monitor and control his or her sleep habits to avoid fatigue.

Lack of Resources

A lack of resources can interfere with one's ability to complete a task because there is a lack of supply and support. *[Figure 14-23]* Low quality products also affect one's ability to complete a task. Aviation maintenance demands proper tools and parts to maintain a fleet of aircraft. Any lack of resources to safely carry out a maintenance task can cause both non-fatal and fatal accidents. For example, if an aircraft is dispatched without a functioning system that is typically not needed for flight but suddenly becomes needed, this could create a problem.

Parts are not the only resources needed to do a job properly, but all too frequently parts become a critical issue. AMTs can

THE DIRTY DOZEN
Twelve human factors for aircraft maintenance proficiency

Lack of Communication	Lack of Teamwork	Lack of Assertiveness
Complacency	Fatigue	Stress
Lack of Knowledge	Lack of Resources	Lack of Awareness
Distraction	Pressure	Norms

When there is a lack of resources available to properly fix something, a decision should be made to cease maintenance until the proper parts are available.

MITIGATING THE RISK

Maintain a sufficient supply of parts and order any anticipated parts before they are required.	Never replace a part with one that is not compatible for the sake of getting the job done.	Preserve all equipment through proper maintenance.

Figure 14-23. *Lack of resources.*

try to be proactive by checking suspected areas or tasks that may require parts at the beginning of the inspection. Aircraft on ground (AOG) is a term in aviation maintenance indicating that a problem is serious enough to prevent an aircraft from flying. Generally, there is a rush to acquire the parts to put the aircraft back into service and prevent further delays or cancellations of the planned itinerary. AOG applies to any aviation materials or spare parts that are needed immediately for an aircraft to return to service. AOG suppliers refer qualified personnel and dispatch the parts required to repair the aircraft for an immediate return to service. AOG also is used to describe critical shipments for parts or materials for aircraft "out of service" (OTS) at a location.

If the status of an aircraft is AOG and materials required are not on hand, parts and personnel must be driven, flown, or sailed to the location of the grounded aircraft. Usually the problem is escalated through an internal AOG desk, then the manufacturer's AOG desk, and finally competitors' AOG desks. All major air carriers have an AOG desk that is manned 24 hours a day, 7 days a week by personnel trained in purchasing, hazardous materials shipping, and parts manufacturing and acquisition processes.

Within an organization, making sure that personnel have the correct tools for the job is just as important as having the proper parts when they are needed. Having the correct tools means not having to improvise. For example, an aircraft that had received a new interior needed to be weighed prior to being released to fly. Two days before the planned release, the aircraft was weighed without the proper electronic load cells placed between the aircraft jack and the aircraft. Because the correct equipment was not used, the aircraft slipped off of one of the load cells and the jack point creased the spar. The cost of improvising can be very steep. The right tools to do the job need to be used at all times, and if they are broken, out of calibration, or missing, they need to be repaired, calibrated, or returned as soon as possible.

Technical documentation is another critical resource that can lead to problems in aviation maintenance. When trying to find out more about the task at hand or how to troubleshoot and repair a system, often the information needed cannot be found because the manuals or diagrams are not available. If the information is not available, personnel should ask a supervisor or speak with a technical representative or technical publications department at the appropriate aircraft manufacturer. Most manuals are in a constant state of revision and, if organizations do not identify missing information in the manuals, then nothing is done to correct the documentation. Resources, such as publication departments and manufacturer's technical support, are available and should be used rather than ignoring the problem.

Another valuable resource that the maintenance department should rely on is the flight crew. Organizations should encourage open communication between the flight crews and the maintenance crews. The flight crew can provide valuable information when dealing with a defective part or problem. *Figure 14-24* shows a number of questions that flight crews can be asked to help resolve and understand maintenance issues.

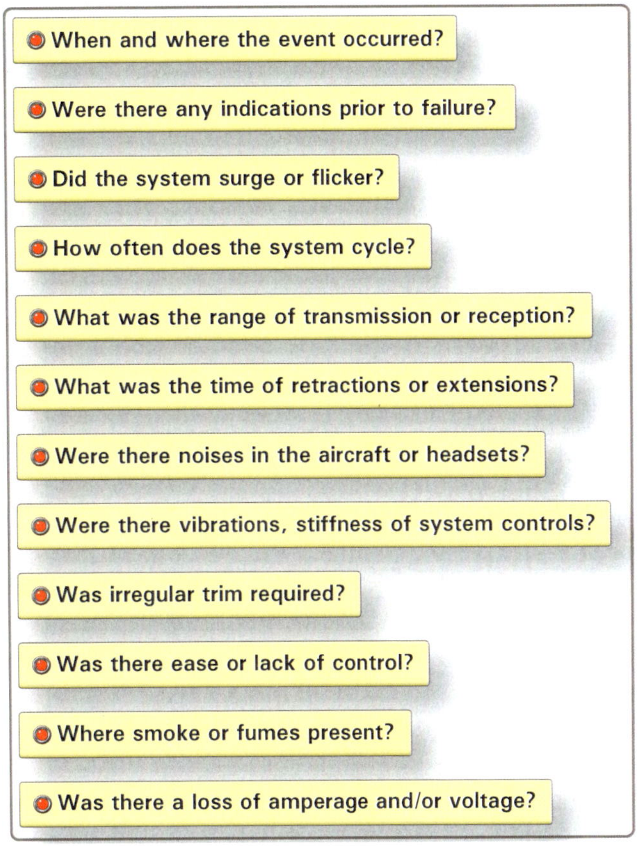

Figure 14-24. *Questions that technicians can ask flight crews in an effort to resolve and understand maintenance issues of resources.*

When the proper resources are available for the task at hand, there is a much higher probability that maintenance will do a better, more efficient job and higher likelihood that the job will be done correctly the first time. Organizations must learn to use all of the resources that are available and, if the correct resources are not available, make the necessary arrangements to get them in a timely manner. The end result saves time, money, and enables organizations to complete the task knowing the aircraft is airworthy.

Pressure

Aviation maintenance tasks require individuals to perform in an environment with constant pressure to do things better and faster without making mistakes and letting things fall through the cracks. Unfortunately, these types of job pressures can affect the capabilities of maintenance workers

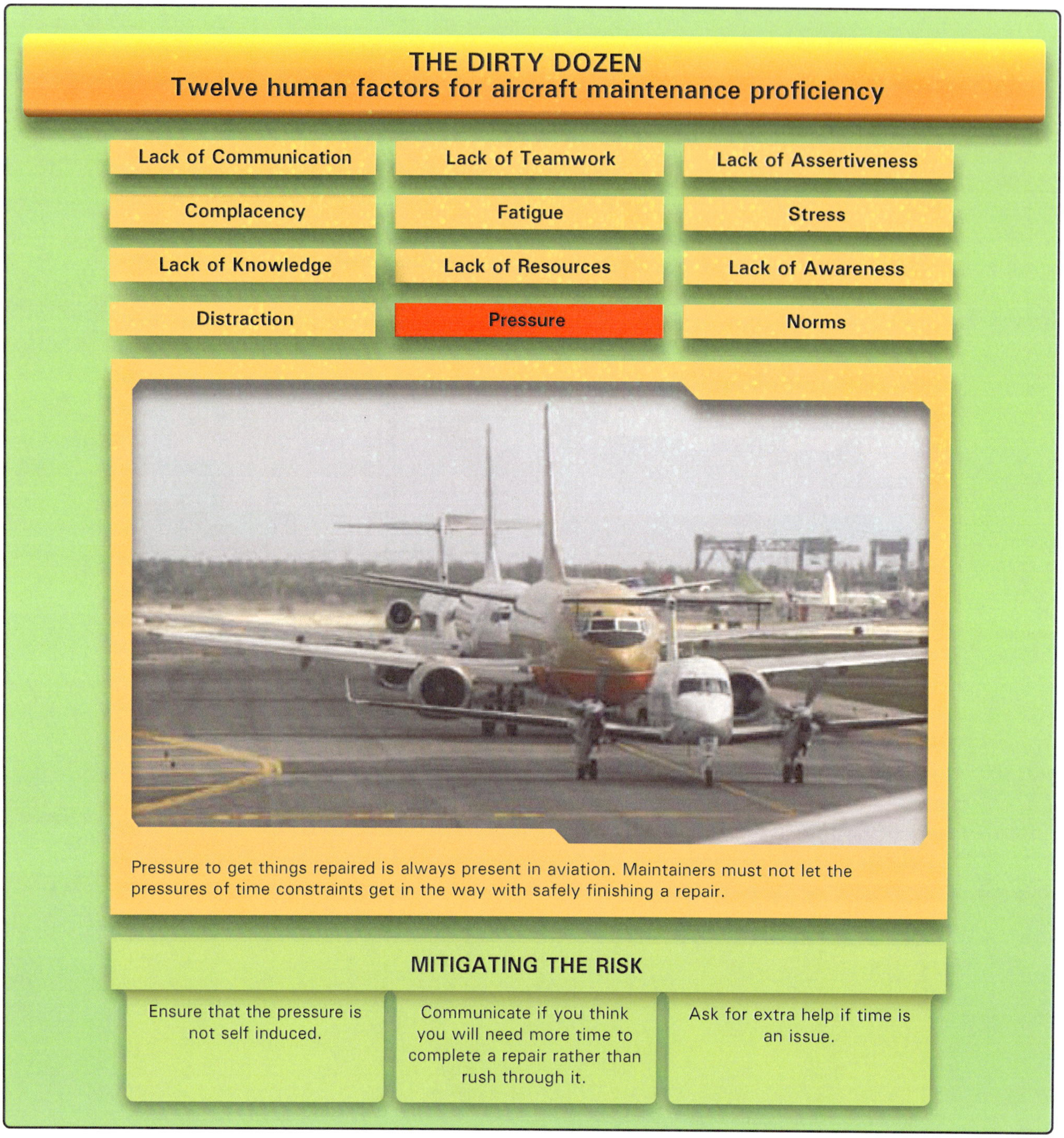

Figure 14-25. *Pressure.*

to get the job done right. *[Figure 14-25]* Airlines have strict financial guidelines, as well as tight flight schedules, that force mechanics to be under pressure to identify and repair mechanical problems quickly so that the airline industry can keep moving. Most important, aircraft mechanics are responsible for the overall safety of everyone who uses flying as a mode of transportation.

Organizations must be aware of the time pressures that are put on aircraft mechanics and help them manage all of the tasks that need to be completed so that all repairs, while done in a timely manner, are completed correctly with safety being the ultimate goal. Sacrificing quality and safety for the sake of time should not be tolerated or accepted. Likewise, AMTs need to recognize on their own when time pressures are clouding their judgments and causing them to make unnecessary mistakes. Self-induced pressures are those occasions where one takes ownership of a situation that was not of their doing.

In an effort to combat self-induced pressure, technicians should ask for help if they feel overwhelmed and under a time constraint to get a repair fixed. Another method is to have someone check the repair thoroughly to ensure that all maintenance tasks were completed correctly.

Lastly, if given a repair with a specific time limitation that you do not feel is realistic or compromises safety, bring it to the attention of the organization's management and openly discuss a different course of action.

Lack of Assertiveness

Assertiveness is the ability to express your feelings, opinions, beliefs, and needs in a positive, productive manner and should not be confused with being aggressive. *[Figure 14-26]* It is important for AMTs to be assertive when it pertains to aviation repair rather than choosing or not being allowed to voice their concerns and opinions. The direct result of not being assertive could ultimately cost people their lives. The following are examples of how a lack of assertiveness can be offset:

1. Address managers and supervisors directly by stating the problem.

   Example: "John, I have a concern with how this repair is being rushed."

2. Explain what the consequences will be.

   Example: "If we continue, the result will be that the part will break sooner rather than later."

3. Propose possible solutions to the problem.

   Example: "We could try doing things another way or you may want to try this way."

4. Always solicit feedback and include other opinions.

   Example: "John, what do you think?"

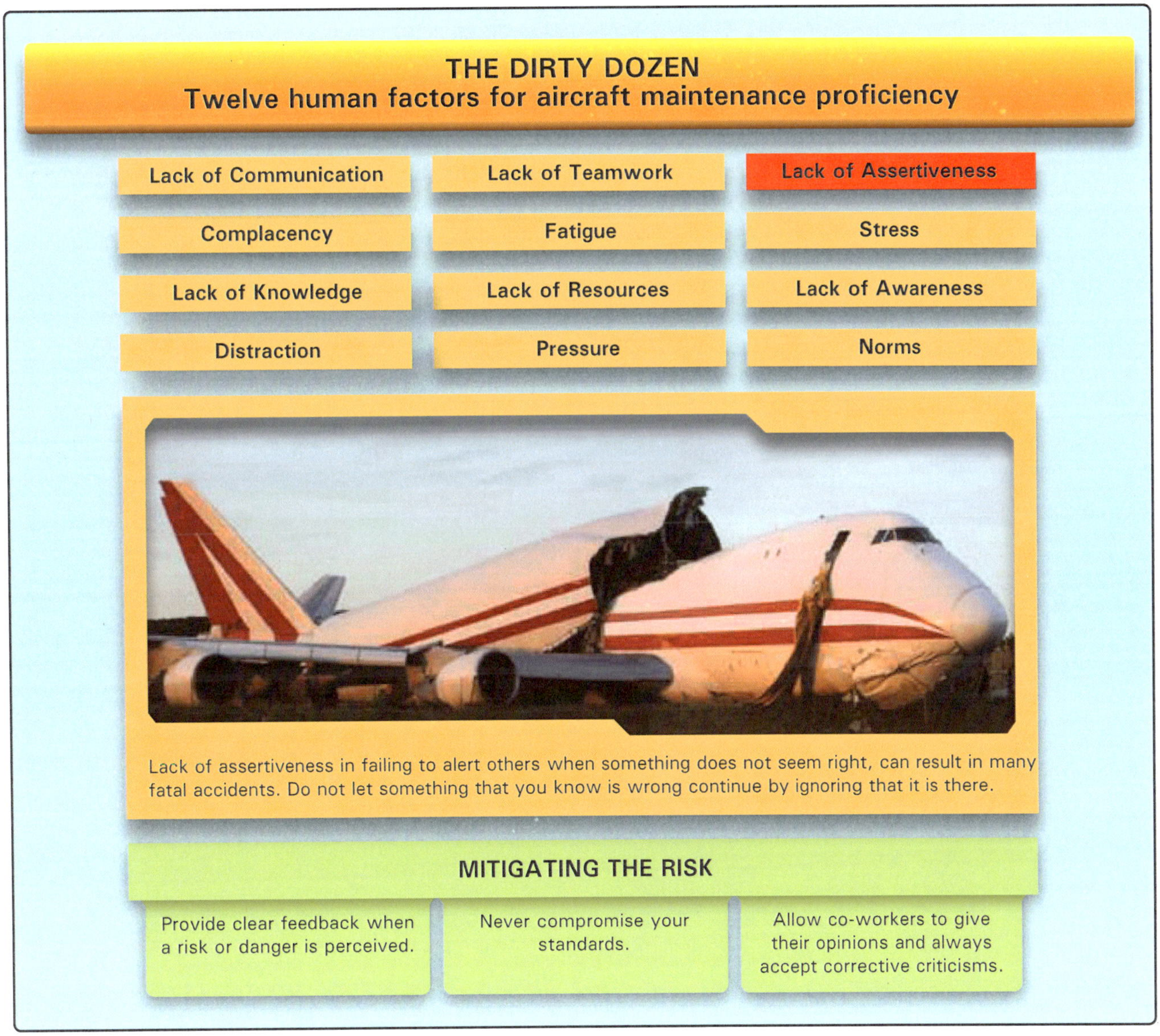

Figure 14-26. *Lack of assertiveness.*

When being assertive with co-workers or management, deal with one issue at a time rather than trying to tackle a number of problems at once. It is also important to have documentation and facts to back up your argument, which can give people a visual account of what you are trying to explain. A lack of assertiveness in failing to speak up when things do not seem right has resulted in many fatal accidents. This can easily be changed by promoting good communication between co-workers and having an open relationship with supervisors and management. Maintenance managers must be familiar with the behavior style of the people they supervise and learn to utilize their talents, experience, and wisdom. As the employees become aware of behavior styles and understand their own behavior, they see how they unwittingly contribute to some of their own problems and how they can make adjustments. Assertive behavior may not be a skill that comes naturally to every individual, but it is a critical skill to achieve effectiveness. AMTs should give supervisors and management the kind of feedback required to ensure that they will be able to assist the mechanic to do their job.

Stress

Aviation maintenance is a stressful task due to many factors. *[Figure 14-27]* Aircraft must be functional and flying in order for airlines to make money, which means that maintenance must be done within a short timeframe to avoid flight delays and cancellations. Fast-paced technology that is always changing can add stress to technicians. This demands that AMTs stay trained on the latest equipment. Other stressors include working in dark, tight spaces, lack of resources to get the repair done correctly, and long hours. The ultimate stress of aviation maintenance is knowing that the work they do, if not done correctly, could result in tragedy.

Everyone handles stress differently and particular situations can bring about different degrees of difficulty for different people. For example, working under a strict timeline can be a stressor for one person and normal for another. The causes of stress are referred to as stressors. They are categorized as physical, psychological, and physiological stressors. Following, is a list of each and how they may affect maintenance.

Physical Stressors

Physical stressors add to the personnel's workload and make it uncomfortable for him or her in their work environment.

- Temperature—high temperatures in the hanger increases perspiration and heart rate causing the body to overheat. Low temperatures can cause the body to feel cold, weak, and drowsy.
- Noise—hangers that have high noise levels (due to aircraft taking off and landing close by) can make it difficult for maintenance personnel to focus and concentrate.
- Lighting—poor lighting within a work space makes it difficult to read technical data and manuals. Likewise, working inside an aircraft with poor lighting increases the propensity to miss something or to repair something incorrectly.
- Confined spaces—small work spaces make it very difficult to perform tasks as technicians are often contorted into unusual positions for a long period of time.

Psychological Stressors

Psychological stressors relate to emotional factors, such as a death or illness in the family, business worries, poor interpersonal relationships with family, co-workers, supervisors, and financial worries.

- Work-related stressors—over anxiousness can hinder performance and speed while conducting maintenance if there is any apprehension about how to do a repair or concerns about getting it done on time.
- Financial problems—impending bankruptcy, recession, loans, and mortgages are a few examples of financial problems that can create stressors.
- Marital problems—divorce and strained relationships can interfere with one's ability to perform their job correctly.
- Interpersonal problems—problems with superiors and colleagues due to miscommunication or perceived competition and backstabbing can cause a hostile work environment.

Physiological Stressors

Physiological stressors include fatigue, poor physical condition, hunger, and disease.

- Poor physical condition—trying to work when ill or not feeling well can force the body to use more energy fighting the illness and less energy to perform vital tasks.
- Proper meals—not eating enough, or foods lacking the proper nutrition, can result in low energy and induce symptoms like headaches and shaking.
- Lack of sleep—fatigued, the maintainer is unable to perform to standard for long periods of time and can become sloppy with repairs and miss important mistakes.
- Conflicting shift schedules—the affect of changing sleep patterns on the body's circadian cycle can lead to a degradation of performance.

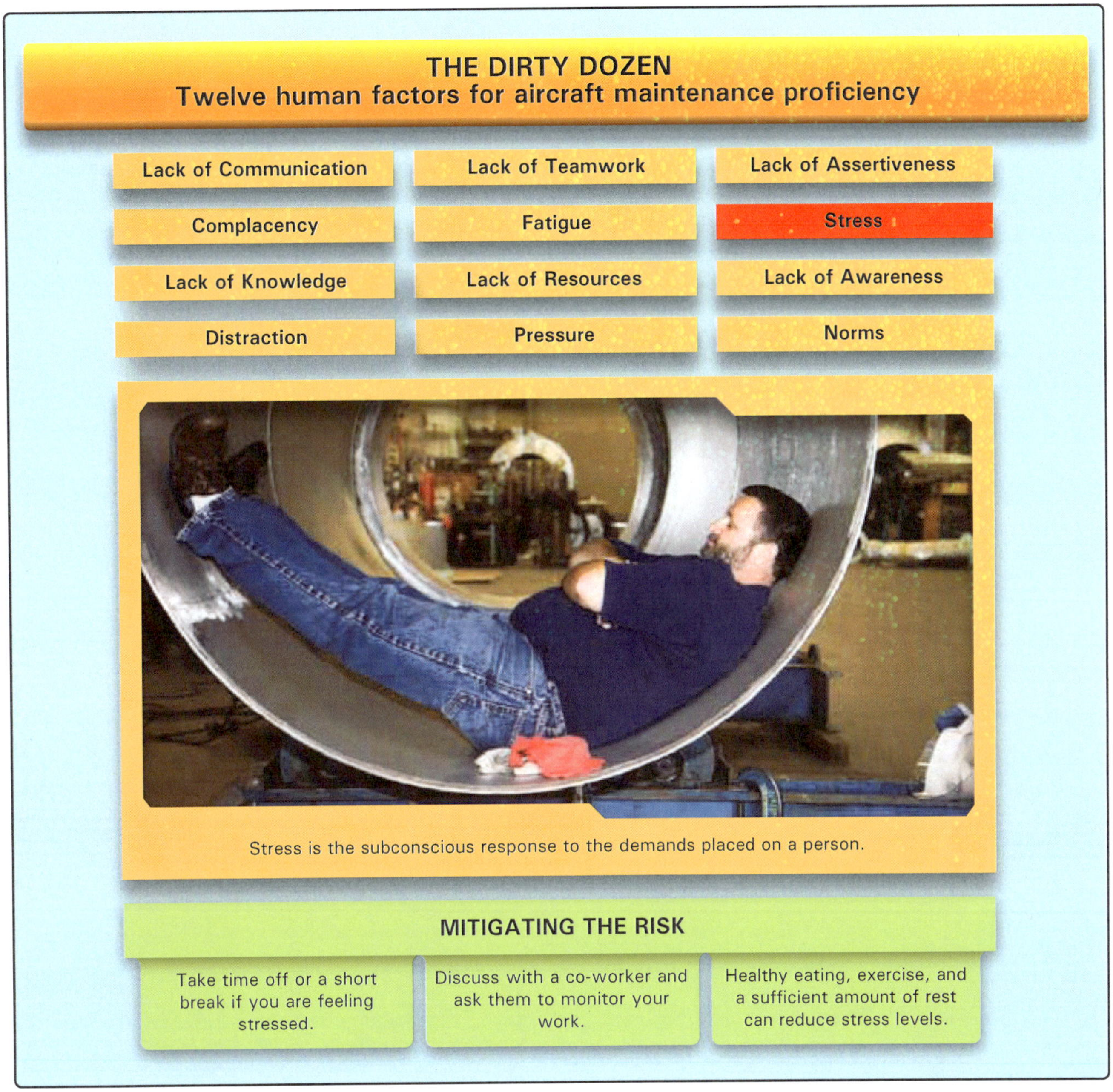

Figure 14-27. *Stress.*

People cope with stress in many different ways. Specialists say that the first step is to identify stressors and the symptoms that occur after exposure to those stressors. Other recommendations involve development or maintenance of a healthy lifestyle with adequate rest and exercise, a healthy diet, limited consumption of alcoholic drinks, and avoidance of tobacco products.

Lack of Awareness

Lack of awareness is defined as a failure to recognize all the consequences of an action or lack of foresight. *[Figure 14-28]* In aviation maintenance, it is not unusual to perform the same maintenance tasks repeatedly. After completing the same task multiple times, it is easy for technicians to become less vigilant and develop a lack of awareness for what they are doing and what is around them. Each time a task is completed it must be treated as if it were the first time.

Norms

Norms is short for "normal," or the way things are normally done. *[Figure 14-29]* They are unwritten rules that are followed or tolerated by most organizations. Negative norms can detract from the established safety standard and cause an accident to occur. Norms are usually developed to solve problems that have ambiguous solutions. When faced with an ambiguous situation, an individual may use another's

Figure 14-28. *Lack of awareness.*

behavior as a frame of reference around which to form his or her own reactions. As this process continues, group norms develop and stabilize. Newcomers to the situation are then accepted into the group based on adherence to norms. Very rarely do newcomers initiate change in a group with established norms.

Some norms are unsafe in that they are non-productive or detract from the productivity of the group. Taking shortcuts in aircraft maintenance, working from memory, or not following procedures are examples of unsafe norms. Newcomers are better able to identify these unsafe norms than long-standing members of the group. On the other hand, the newcomer's credibility depends on his or her assimilation into the group. The newcomer's assimilation, however, depends on adherence to the group norms. Everyone should be aware of the perceptiveness of newcomers in identifying unhealthy norms and develop a positive attitude toward the possibility that norms may need to be changed. Finally, as newcomers become assimilated into the group structure, they build credibility with others. Once this has been done, a relative newcomer may begin to institute change within the group. Unfortunately,

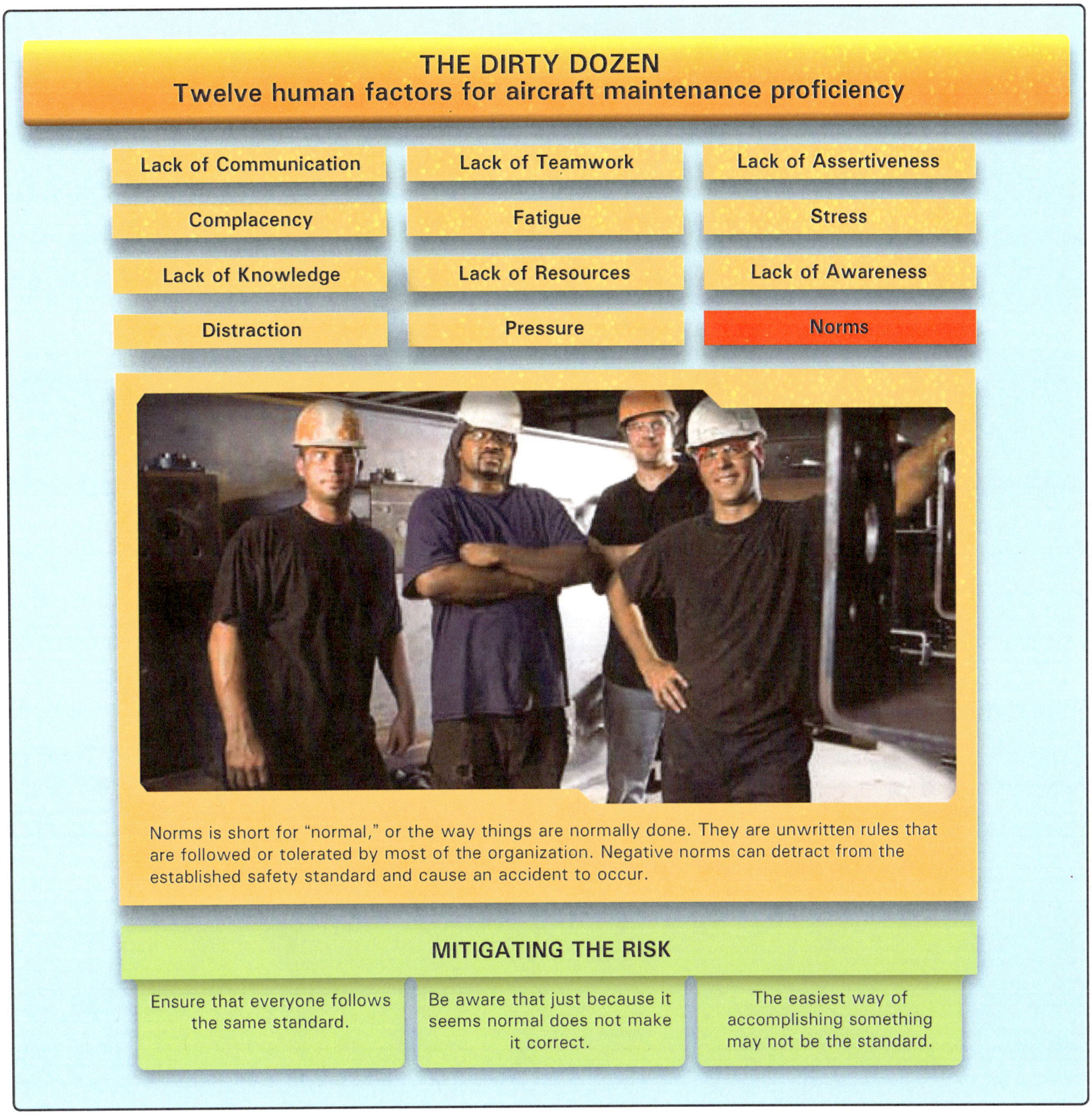

Figure 14-29. *Norms.*

such actions are often difficult to do and rely heavily on the group's perception of the newcomer's credibility.

Norms have been identified as one of the dirty dozen in aviation maintenance and a great deal of anecdotal evidence points to the use of unsafe norms on the line. The effect of unsafe norms may range from the relatively benign, such as determining accepted meeting times, to the inherently unsafe, such as signing off on incomplete maintenance tasks. Any behavior commonly accepted by the group, whether as a standard operating procedure (SOP) or not, can be a norm. Supervisors need to ensure that everyone adheres to the same standards and not tolerate unsafe norms. AMTs should pride themselves on following procedure, rather than unsafe norms that may have been adopted as regular practice.

Example of Common Maintenance Errors

In an effort to identify the most frequently occurring maintenance discrepancies, the United Kingdom Civil Aviation Authority (CAA) conducted in-depth studies of maintenance sites on aviation maintenance operations. The following list is what they found to be the most common occurring maintenance errors.

1. Incorrect installation of components.
2. Fitting of wrong parts.
3. Electrical wiring discrepancies to include crossing connections. *[Figure 14-30]*

Incident

On March 20, 2001 a Lufthansa Airbus A320 almost crashed shortly after takeoff because of reversed wiring in the captain's sidestick flight control. Quick action by the co-pilot, whose sidestick was not faulty, prevented a crash.

Cause

The investigation has focused on maintenance on the captain's controls carried out by Lufthansa Technik just before the flight. During the previous flight, a problem with one of the two elevator/aileron computers (ELAC) had occurred. An electrical pin in the connector was found to be damaged and was replaced. It has been confirmed that two pairs of pins inside the connector had accidentally been crossed during the repair. This changed the polarity in the sidestick and the respective control channels "bypassing" the control unit, which might have sensed the error and would have triggered a warning. Clues might have been seen on the electronic centralized aircraft monitor (ECAM) screen during the flight control checks, but often pilots only check for a deflection indication, not the direction. Before the aircraft left the hangar, a flight control check was performed by the mechanic, but only using the first officer's sidestick.

Figure 14-30. *A description of a Lufthansa Airbus A320 that almost crashed due to reversed wiring of the flight controls.*

4. Forgotten tools and parts.
5. Failure to lubricate. *[Figure 14-31]*

Accident

Alaska Airlines Flight 261, a McDonnell Douglas MD-83 aircraft, experienced a fatal accident on January 31, 2000, in the Pacific Ocean. The two pilots, three cabin crewmembers, and 83 passengers on board were killed and the aircraft was destroyed.

Cause

The subsequent investigation by the National Transportation Safety Board (NTSB) determined that inadequate maintenance led to excessive wear and catastrophic failure of a critical flight control system during flight. The probable cause was stated to be "a loss of airplane pitch control resulting from the in-flight failure of the horizontal stabilizer trim system jackscrew assembly's acme nut threads. The thread failure was caused by excessive wear resulting from Alaska Airlines insufficient lubrication of the jackscrew assembly."

Figure 14-31. *A description of Alaska Airlines Flight 261 that crashed due to insufficient lubrication of the jackscrew assembly.*

6. Failure to secure access panels, fairings, or cowlings.
7. Fuel or oil caps and fuel panels not secured.
8. Failure to remove lock pins. *[Figure 14-32]*

Figure 14-32. *Lock pins located on the wheels of an aircraft.*

All of the maintenance discrepancies listed above can be avoided if the proper procedures are followed on the job card that is being used. *[Figure 14-33]* Regardless of how many times the task has been completed, each time you pick up a job card, treat it like it is the first time you have ever completed the task with diligence and complete accuracy.

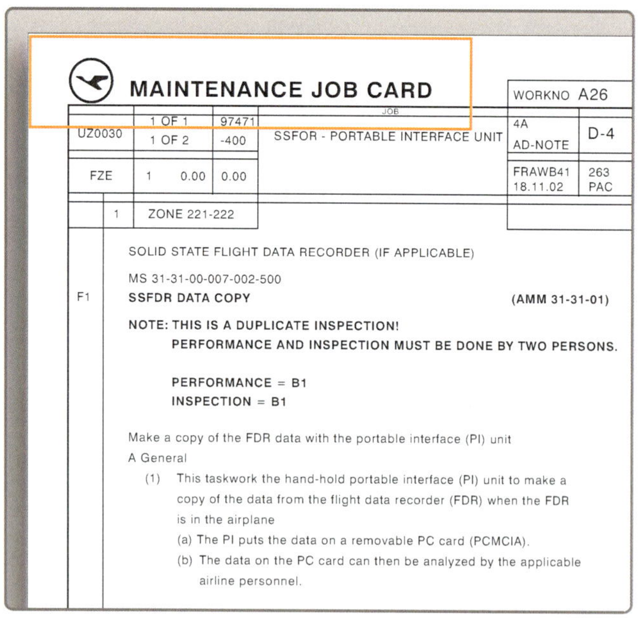

MAINTENANCE JOB CARD

				WORKNO A26
UZ0030	1 OF 1 / 1 OF 2	97471 / -400	SSFOR - PORTABLE INTERFACE UNIT	4A AD-NOTE / D-4
FZE	1 0.00	0.00		FRAWB41 18.11.02 / 263 PAC
1	ZONE 221-222			

SOLID STATE FLIGHT DATA RECORDER (IF APPLICABLE)

MS 31-31-00-007-002-500

F1 SSFDR DATA COPY (AMM 31-31-01)

NOTE: THIS IS A DUPLICATE INSPECTION!
PERFORMANCE AND INSPECTION MUST BE DONE BY TWO PERSONS.

PERFORMANCE = B1
INSPECTION = B1

Make a copy of the FDR data with the portable interface (PI) unit

A General

(1) This taskwork the hand-hold portable interface (PI) unit to make a copy of the data from the flight data recorder (FDR) when the FDR is in the airplane

(a) The PI puts the data on a removable PC card (PCMCIA).

(b) The data on the PC card can then be analyzed by the applicable airline personnel.

Figure 14-33. *A sample picture of a maintenance job card that explains the steps of each maintenance task.*

Historically, twenty percent of all accidents are caused by a machine failure, and eighty percent of all accidents are caused by human factors. *[Figure 14-34]* Originally focusing on the pilot community, human factors awareness has now spread into the training sphere of maintenance technicians. An in-depth review of an aviation incident reveals time and again that a series of human errors (known also as a chain of events) was allowed to build until the accident occurred. If the chain of events is broken at the maintenance level, the likelihood of the accident occurring can be drastically decreased. *Figure 14-35* is a list of maintenance related incidents/accidents and their causes. It is easy to see how many of the "Dirty Dozen" contributed to the cause or was considered a contributing factor.

Where to Get Information

Following, is a list of websites and references that are good sources of information on human factors.

Federal Aviation Administration (FAA)

There are a number of human factors resources within the FAA. The most direct link for aviation maintenance human factors is the FAA Human Factors website at https://hfskyway.faa.gov. It offers document access and services, including most of the FAA maintenance human factors documents dating back to the 1988 start of FAA's maintenance human factors research and development program. New documents include videos, PowerPoint presentations, and other media.

FAA's Maintenance Fatigue Section

The FAA has sponsored a multi-disciplinary subject matter expert work group involving industry, labor, research, and government to investigate the issues associated with maintenance fatigue, and the practical science based methods that can be used to manage fatigue risk. For more information, visit the website at www.mxfatigue.com.

FAA Safety Team

The FAA Safety Team has a dedicated website that provides up-to-date information safety concerns, upcoming seminars, featured courses and resources. For more information, visit the website at www.faasafety.gov.

Other Resources

Decoding Human Factors

Decoding Human Factors Aviation Safety Training, Consulting and Support provides a repository of human factors and safety management system educational material. The website also provides the latest news in human

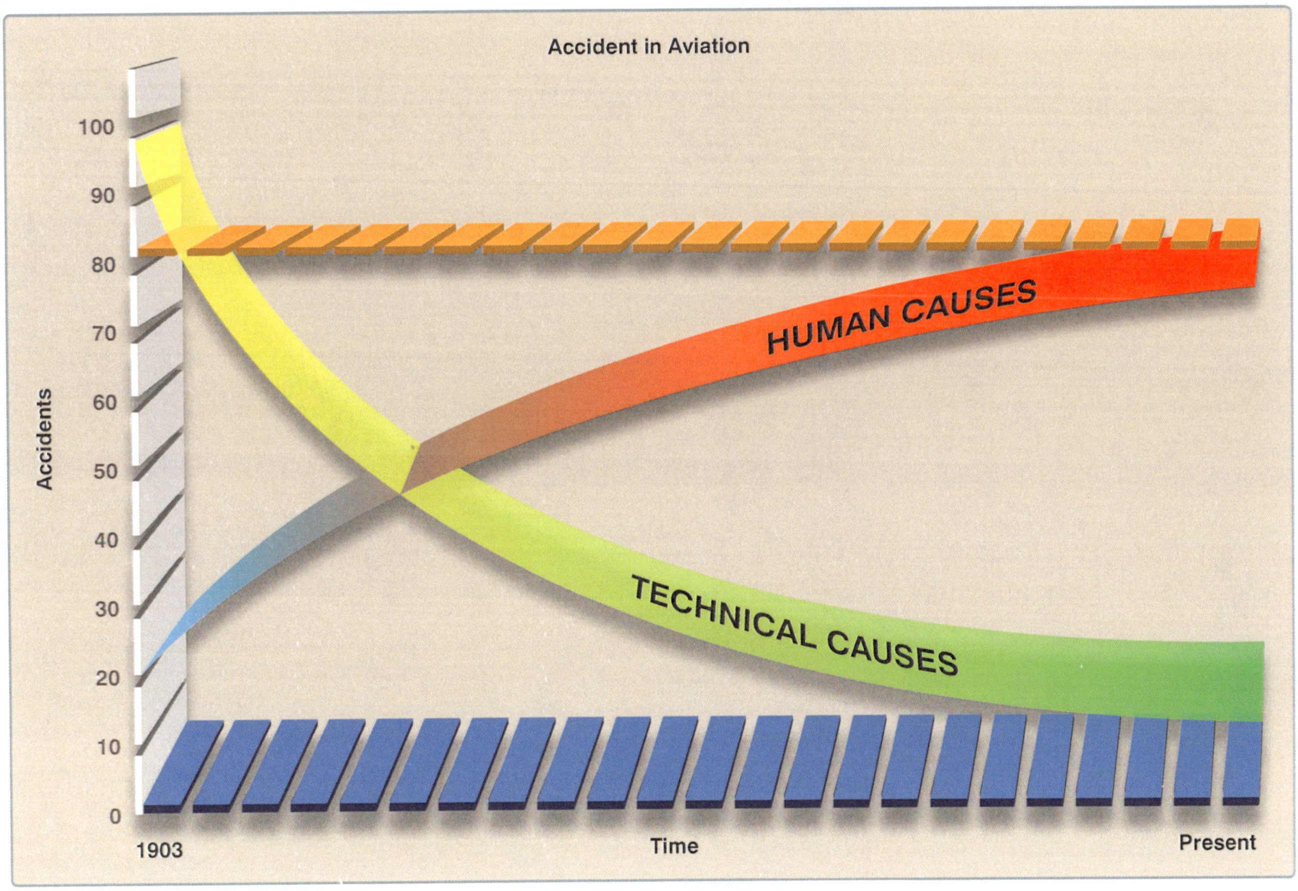

Figure 14-34. *Statistical graph showing that 80 percent of all aviation accidents are caused by human factors.*

INCIDENT

August 26, 1993, an Excalibur Airways Airbus 320 took off from London-Gatwick Airport (LGW) and exhibited an undemanded roll to the right on takeoff, a condition which persisted until the aircraft landed back at LGW 37 minutes later. Control of the aircraft required significant left sidestick at all times and the flight control system was degraded by the loss of spoiler control.

CAUSE

Technicians familiar with Boeing 757 flap change procedures lacked the knowledge required to correctly lock out the spoilers on the Airbus during the flap change work that was done the day before the flight. Turnover to technicians on the next shift compounded the problem. No mention of incorrect spoiler lockout procedure was given since it was assumed that the 320 was like the 757. The flap change was operationally checked, but the spoiler remained locked out incorrectly and was not detected by the flight crew during standard functional checks. The lack of knowledge on Airbus procedures was considered a primary cause of this incident.

INCIDENT

April 26, 2001, an Emery Worldwide Airlines DC-8-71F left main landing gear would not extend for landing.

CAUSE

Probable cause was failure of maintenance to install the correct hydraulic landing gear extension component and the failure of inspection to comply with post-maintenance test procedures. No injuries.

ACCIDENT

On May 25, 2002, China Airlines Flight 611 Boeing 747 broke into pieces in mid-air and crashed, killing all 225 people on board.

CAUSE

The accident was the result of metal fatigue caused by inadequate maintenance after a previous incident.

ACCIDENT

On August 26, 2003, a Colgan Air Beech 1900D crashed just after takeoff from Hyannis, Massachusetts. Both pilots were killed.

CAUSE

The improper replacement of the forward elevator trim cable and subsequent inadequate functional check of the maintenance performed that resulted in a reversal of the elevator trim system and a loss of control in flight. Factors were the flight crew's failure to follow the checklist procedures and the aircraft manufacturer's erroneous depiction of the elevator trim drum in the maintenance manual.

ACCIDENT

On September 28, 2007, American Airlines Flight 1400 DC-9 experienced an in-flight engine fire during departure climb from Lambert St. Louis International Airport (STL). During the return to STL, the nose landing gear failed to extend, and the flight crew executed a go-around, during which the crew extended the nose gear using the emergency procedure. The flight crew conducted an emergency landing, and the 2 flight crewmembers, 3 flight attendants, and 138 passengers deplaned on the runway. No occupant injuries were reported, but the airplane sustained substantial damage from the fire.

CAUSE

American Airlines' maintenance personnel's use of an inappropriate manual engine-start procedure, which led to the uncommanded opening of the left engine air turbine starter valve, and a subsequent left engine fire.

Figure 14-35. *A list of maintenance related incidents/accidents and their causes.*

factors, as well as the Aviation Human Factors Industry Newsletter. For more information, visit their website at www.decodinghumanfactors.com.

System Safety Services

The mission of System Safety Services is to assist clients in developing the best possible safety system to meet their needs. They have an experienced and professional team of individuals with years of experience in aviation and human factors. The website provides a lot of information on human factors to include articles, upcoming events, presentations, safety videos, training aids and workshops. For more information, visit their website at www.system-safety.com.

Human Factors and Ergonomics Society (HFES)

The Human Factors and Ergonomics Society (HFES) is the only organization in the United States dedicated specifically to the human factors profession. The HFES was formed in 1957 and typically maintains about 5,000 members. For more information, visit their website at www.hfes.org.

International Ergonomics Association (IEA)

The International Ergonomics Association (IEA) is a federation of over 40 ergonomics and human factors societies located all over the world. All members of the HFES are automatically also members of the IEA. The main contact point within the IEA is through the office of their Secretary General. For more information, visit their website at www.iea.cc.

GLOSSARY

Absolute humidity. The actual amount of the water vapor in a mixture of air and water.

Absolute pressure. Equal to gauge pressure plus atmospheric pressure. Also known as psia.

Absolute temperature. Temperature measured relative to absolute zero. Absolute temperature scales include Kelvin and Rankine.

Acceleration due to gravity. The acceleration of an object caused by gravity. On earth, it is measured as 32.2 feet per second per second (32.2 fps/s).

Addition. The process in which the value of one number is added to the value of another.

Advisory Circulars (AC). Issued to inform the aviation public in a systematic way of nonregulatory material. An AC is issued to provide guidance and information in a designated subject area or to show a method acceptable to the Administrator for complying with a related 14 CFR part.

Aircraft Specifications. FAA recordkeeping documents issued for both type-certificated and non-type-certificated products which have been found eligible for U.S. airworthiness certification.

Airfoil. Any device that creates a force, or lift, based on Bernoulli's principles or Newton's laws, when air is caused to flow over the surface of the device.

Airworthiness certificate. A document required to be onboard an aircraft that indicates the aircraft conforms to type design and is in condition for safe operation.

Airworthiness directive (AD). Issued by the FAA in response to deficiencies and/or unsafe conditions found in aircraft, engines, propellers, or other aircraft parts. Compliance with an AD is mandatory.

Alclad aluminum. Used to designate sheets that consist of an aluminum alloy core coated with a layer of pure aluminum to a depth of approximately $5\frac{1}{2}$ percent on each side.

Algebra. The branch of mathematics that uses letters or symbols to represent numbers in formulas and equations.

Allowance. The difference of the upper and lower variation of a part.

Alodizing. A simple chemical treatment for all aluminum alloys to increase their corrosion resistance and to improve their paint bonding qualities.

Alteration. A change or modification to an aircraft from its previous state

Alternating current. An electric current that reverses direction in a circuit at regular intervals.

Ammeter. An instrument for measuring electric current in amperes.

Ampere. A unit of measure of the rate of electron flow or current in an electrical conductor. One ampere of current represents one coulomb of electrical charge (6.24×10^{18} charge carriers) moving past a specific point in one second.

Annealing. The process of heating a metal to a prescribed temperature, holding it there for a specified length of time, and then cooling the metal back to room temperature.

Annual inspection. An inspection required by the FAA once every 12 calendar months if other suitable inspections do not occur within that timeframe. An A&P technician with inspection authorization must perform this inspection.

Anodizing. The most common surface treatment of nonclad aluminum alloy surfaces. The aluminum alloy sheet or casting is the positive pole in an electrolytic bath in which chromic acid or other oxidizing agent produces an aluminum oxide film on the metal surface. Aluminum oxide is naturally protective, and anodizing merely increases the thickness and density of the natural oxide film.

Apparent power. That power apparently available for use in an AC circuit containing a reactive component. It is the product of effective voltage times the effective current, expressed in volt-amperes.

Archimedes' principle. The buoyant force that a fluid exerts upon a submerged body is equal to the weight of the fluid the body displaces.

Area. A measurement of the amount of surface inside a two-dimensional object.

Arm. The horizontal distance that a part of the aircraft or a piece of equipment is located from the datum.

Armature. The rotating part of an electric generator or motor.

Aspect ratio. The relationship of the length (wingtip to wingtip), or span, of an airfoil to its width, or chord.

Assembly drawing. A description of an object made up of two or more parts.

Atom. The smallest particle composed of a nucleus that contains protons, neutrons, and electrons, which revolve around the nucleus.

Ballast. A weight installed or carried in an aircraft to move the center of gravity to a location within its allowable limits.

Base. In mathematics, used to refer to a particular mathematical object that is used as a building block. A base-*a* system is one that uses *a* as a new unit from which point counting starts again. (See decimal system.) In the mathematical expression a^n, read as "*a* to the nth power," *a* is the base.

Basic empty weight. Standard empty weight plus optional equipment.

Bernoulli's principle. Equivalent to the principle of conservation of energy, this principle states that the static pressure of a fluid (liquid or gas) decreases at points where the velocity of the fluid increases, provided no energy is added to or taken away from the fluid.

Binary number system. The binary number system is a number system that has only two digits, 0 (zero) and 1. Binary numbers are made from a series of zeros and ones. An example of an 8-bit binary number is 11010010. The prefix "bi" in the word binary is a Latin root for the word "two."

Block diagrams. Used to show a simplified relationship of a more complex system of components.

Borescope. A device that enables the inspector to see inside areas that could not otherwise be inspected without disassembly.

Boyle's law. States that the volume of an enclosed dry gas varies inversely with its absolute pressure, provided the temperature remains constant.

Break lines. Line on a drawing indicating that a portion of the object is not shown on the drawing.

British thermal unit (Btu). The amount of heat required to change the temperature of 1 pound of water by 1 degree Fahrenheit.

Buoyancy. The upward force that any fluid exerts on a body submerged in it.

Buttock line (BL). The longitudinal axis of the aircraft that serves as the reference location for positions to the left and right of center. The positions are usually dimensioned in inches.

Calorie. The amount of heat required to change the temperature of 1 gram of water by 1 degree Centigrade.

Camber. The curvature of a wing as viewed by cross section. A wing has upper camber on its top surface and lower camber on its bottom surface. The upper camber is more pronounced; the lower camber is comparatively flat. This causes the velocity of the airflow immediately above the wing to be much higher than that below the wing.

Capacitance (C). The property of an electric conductor that characterizes its ability to store an electric charge.

Capacitive reactance (X_C). The measure of a capacitor's opposition to alternating current.

Capacitor. An electrical component that stores an electric charge.

Case hardening. A process in which the surface of a metal is changed chemically by introducing a high carbide or nitride content. Case hardening produces a hard, wear-resistant surface, or case, over a strong, tough core.

Center of gravity (CG). The point about which the nose-heavy and tail-heavy moments are exactly equal in magnitude.

Center of gravity range. The center of gravity range for an aircraft is the limits within which the aircraft must balance. It is identified as a forward-most limit (arm) and an aft-most limit (arm).

Centrifugal force. The apparent force occurring in curvilinear motion acting to deflect objects outward from the axis of rotation. For instance, when pulling out of a dive, it is the force pushing the pilot down in his or her seat.

Centripetal force. The force in curvilinear motion acting toward the axis of rotation. For instance, when pulling out of a dive, it is the force that the seat exerts on the pilot to offset the centrifugal force.

Charles' Law. States that all gases expand and contract in direct proportion to the change in the absolute temperature, provided the pressure is held constant.

Chemical energy. Energy released from chemical reactions.

Circular magnetization. The induction of a magnetic field consisting of concentric circles of force about and within a part, which is achieved by passing electric current through the part.

Circumference (of a circle). The linear measurement of the distance around a circle. The circumference is calculated by multiplying the diameter of the circle by 3.1416.

Code of Federal Regulations (CFR). Established by law to provide for the safe and orderly conduct of flight operations and to prescribe airmen privileges and limitations.

Compression ratio. The ratio of the volume of a cylinder with the piston at the bottom of its stroke to the volume of the cylinder with the piston at the top of its stroke.

Computer aided design (CAD). Using a computer in the design of a product.

Computer aided design drafting (CADD). Using a computer in the design and drafting process.

Computer aided engineering (CAE). Using a computer in the engineering of a product.

Computer aided manufacturing (CAM). Using a computer in the manufacturing of a product.

Computer graphics. Drawing with the use of a computer.

Conduction. The transfer of heat which requires physical contact between an object that has a large amount of heat energy and one that has a smaller amount of heat energy.

Conductor. A material that will carry electric current.

Convection. The process by which heat is transferred by movement of a heated fluid (gas or liquid).

Corrosion. The deterioration of metal by chemical or electrochemical attack.

Cosine (cos). A trigonometric function comparing two sides of a right triangle as follows:

$$\text{Cos} = \frac{\text{adjacent side}}{\text{hypotenuse}}$$

Coulomb. A measure of electrical output. One coulomb is 6.24×10^{18} electrons.

Countersink. A tool that cuts a cone-shaped depression around a hole in order to allow a rivet or screw to set flush with the surface of the material.

Current. The flow of electrical charge.

Dalton's Law. States that a mixture of several gases which do not react chemically exerts a pressure equal to the sum of the pressures which the several gases would exert separately if each were allowed to occupy the entire space alone at the given temperature.

Datum. An imaginary vertical plane or line from which all measurements of arm are taken. The datum is established by the manufacturer. Once the datum has been selected, all moment arms and the location of CG range are measured from this point.

Debonding. Separation of the bond between the skin laminates and the core of a composite structure.

Decibels. The unit for measuring sound intensity. One decibel is the smallest change in sound intensity the human ear can detect.

Decimal system. The number system, also called the base-ten system, based on the number 10. Consisting of ten symbols, or digits (0, 1, 2, 3, 4, 5, 6, 7, 8, 9), the main principle is that 10 is considered as a new unit from which point counting starts again.

Degradation. The alteration of material properties (e.g., strength, modulus, coefficient of expansion) which may result from deviations in manufacturing or from repeated loading and/or environmental exposure.

Delamination. Separation of the bond between the individual plies of a laminated composite structure.

Denominator. The lower part of a fraction (represented by the letter D in $^N/_D$), the quantity by which the numerator is divided.

Density. The weight of a substance per unit volume.

Detail drawing. A description of a single part, given in such a manner as to describe by lines, notes, and symbols the specifications for size, shape, material, and methods of manufacture to be used in making the part.

Detailed inspection. A thorough examination of an item including disassembly. The overhaul of a component is considered to be a detailed inspection.

Detonation. Uncontrolled burning of fuel in the cylinder of a reciprocating engine. Detonation causes explosive burning of the fuel which creates an increased cylinder pressure, excessive cylinder head temperature, and decreased engine performance.

Dew point. The temperature to which humid air must be cooled at constant pressure to become saturated.

Dial indicator. Measures variations in a surface by using an accurately machined probe mechanically linked to a circular hand whose movement indicates thousandths of an inch, or is displayed on a liquid crystal display (LCD) screen.

Diameter (circle). The length of a line passing directly through the center of a circle. Twice the radius of the circle.

Die. Used for cutting external threads on round stock.

Difference. The answer to a subtraction problem.

Direct current (DC). Electricity that flows in one direction at all times.

Directional stability. Stability about the vertical axis of an aircraft, whereby an aircraft tends to return, on its own, to flight aligned with the relative wind when disturbed from the equilibrium state.

Discontinuity. An interruption in the normal physical structure or configuration of a part, such as a crack, forging lap, seam, inclusion, porosity, and the like. A discontinuity may or may not affect the usefulness of a part.

Dissimilar metal corrosion. Caused by contact between dissimilar metal parts in the presence of a conductor.

Dividend. In a division problem, the number to be divided by the divisor. In $6 \div 2 = 3$, the dividend is 6.

Division. The process of finding how many times one number (the divisor) is contained in another number (the dividend).

Divisor. In a division problem, the number by which dividend is to be divided. In $6 \div 2 = 3$, the divisor is 2.

Doping. The process by which small amounts of additives called impurities are added to the semiconductor material to increase their current flow by adding a few electrons or a few holes.

Dynamic stability. The property of an aircraft that causes it, when disturbed from straight-and level flight, to develop forces or moments that restore the original condition of straight and level.

Eddy current inspection. An inspection method where eddy currents are induced into the material to be tested. In aircraft manufacturing plants, eddy current is used to inspect castings, stampings, machine parts, forgings, and extrusions.

Electrical energy. Electrical energy is converted to heat energy when an electric current flows through any form of resistance such as an electric iron, electric light, or an electric blanket.

Electromotive force (EMF). The pressure or force that causes electrons to flow in an electrical circuit.

Electrostatic field. A field of force that exists around a charged body.

Empennage. The section of the airplane that consists of the vertical stabilizer, the horizontal stabilizer, and the associated control surfaces.

Empty-weight center of gravity range. The distance between the allowable forward and aft empty-weight CG limits.

Empty-weight center of gravity. The center of gravity of an aircraft when it contains only the items specified in the aircraft empty weight.

Empty weight. See standard empty weight.

Energy. The capacity of a physical system to perform work. There are two types of energy, kinetic and potential.

Exploded view drawing. A pictorial drawing of two or more parts that fit together as an assembly. The view shows the individual parts and their relative position to the other parts before they are assembled.

Exponent (power). A shorthand method of indicating how many times a number, called the base, is multiplied by itself. For example, in the number 4^3, the 3 is the power or exponent and 4 is the base. That is, 4^3 is equal to $4 \times 4 \times 4 = 64$.

Extension lines. Used to extend the line showing the side or edge of a figure for the purpose of placing a dimension to that side or edge.

FAA Form 337. This form must be completed when a major repair or alteration is accomplished.

Ferrous metals. Metals having iron as their principal constituent.

Force. The intensity of an impetus, or the intensity of an input.

Foreign object damage (FOD). Any damage caused by any loose object to aircraft, personnel, or equipment. These loose objects can be anything from broken runway concrete to shop towels and safety wire.

Fraction. A number written in the form $^N\!/_D$ in which N is the numerator and D is the denominator. For example, $^5\!/_{16}$ is a fraction.

Frequency. The number of cycles (on/off) completed per unit of time. Usually expressed in Hertz.

Fretting corrosion. Occurs when two mating surfaces, normally at rest with respect to one another, are subject to slight relative motion.

Friction. The opposition to movement between objects.

Fuel grade. The rating system used for aviation gasoline. It rates fuel according to its antidetonation characteristics.

Fuse. A protective device containing a special wire that melts when current exceeds the rated value for a definite period.

Fuselage stations (FS). Reference locations, usually given in inches, used to determine forward and aft positions on an aircraft. FS – 0 is the datum.

Heat. The total kinetic energy of the molecules of any substance.

Henry. The basic unit of inductance, symbolized with the letter H. An electric circuit has an inductance of one henry when current changing at the rate of one ampere per second induces a voltage of one volt into the circuit.

Hermaphrodite caliper. Generally used as a marking gauge in layout work. It should not be used for precision measurement.

Hidden lines. Indicates invisible edges or contours.

Horsepower. A measure of power equal to 550 foot-pounds per second or 33,000 foot-pounds per minute and 746 Watts.

Hot start. Occurs when the engine starts, but the exhaust gas temperature exceeds specified limits. This is usually caused by an excessively rich fuel/air mixture entering the combustion chamber.

Humidity. The amount of water vapor in the air.

Hung start. Occurs when the engine starts normally, but the rpm remains at some low value rather than increasing to the normal starting rpm. This is often the result of insufficient power to the starter, or the starter cutting off before the engine starts self-accelerating.

Hydrometer. A device used for measuring the specific gravity of liquids.

Hydrometer. An instrument for determining the specific gravity of liquids.

Hypotenuse. The side of a right triangle that is opposite the right angle. The hypotenuse is the longest side of a right triangle.

Improper fraction. A fraction with the numerator equal to or greater than the denominator.

Inside calipers. Calipers with outward curved legs for measuring inside diameters, such as diameters of holes.

Installation drawing. A drawing that includes all necessary information for a part or an assembly in the final installed position in the aircraft.

Intergranular corrosion. An attack along the grain boundaries of an alloy that commonly results from a lack of uniformity in the alloy structure.

Inductance. The ability of a coil or conductor to oppose a change in current flow.

Inductive reactance. The opposition to the flow of current which inductances put in a circuit.

Inductor. A coil of wire that produces inductance in an electrical circuit.

Insulator. A material that does not conduct electrical current very well or not at all. Examples are glass, ceramic, and plastic.

Ion. An atom or group of atoms in which the number of electrons is different from the number of protons. It is a positive ion if the number of electrons is less than the number of protons, and a negative ion if the number of electrons is greater than the number of protons.

Isometric drawings. A drawing that uses a combination of the views of an orthographic projection and tilts the object forward so that portions of all three views can be seen in one view.

Joule. The amount of work done by a force of one newton when it acts through a distance of one meter.

Kinetic energy. Energy due to motion, defined as one half mass times velocity squared.

Kirchhoff's Law (voltage). A basic law of electrical currents stating that the algebraic sum of the applied voltage and the voltage drop around any closed circuit is zero.

Lateral stability. The stability about the longitudinal axis of an aircraft; the rolling stability, or the ability of an airplane to return to level flight due to a disturbance that causes one of the wings to drop.

Lever. The simplest machine. There are three basic parts in all levers: the fulcrum "F," a force or effort "E," and a resistance "R."

Load cell. A component in an electronic weighing system that is placed between the jack and the jack pad on the aircraft. The load cell contains strain gauges whose resistance changes with the weight on the cell.

Longitudinal magnetization. The magnetic field is produced in a direction parallel to the long axis of the part. This is accomplished by placing the part in a solenoid excited by electric current.

Longitudinal stability. The tendency for an aircraft nose to pitch up or pitch down, rotating around the lateral axis (wingtip to wingtip).

Magnetic particle inspection. A method of detecting invisible cracks and other defects in ferromagnetic materials such as iron and steel. The inspection process consists of magnetizing the part and then applying ferromagnetic particles to the surface area to be inspected.

Maintenance. This includes inspection, overhaul, repair, preservation, and the replacement of parts, but excludes preventive maintenance.

Major alteration. An alteration not listed in the aircraft, aircraft engine, or propeller specifications: (1) that might appreciably affect weight, balance, structural strength, performance, powerplant operation, flight characteristics, or other qualities affecting airworthiness; or (2) that is not done according to accepted practices or cannot be done by elementary operations.

Major repairs. A repair that (1) if improperly done, might appreciably affect weight, balance, structural strength, performance, powerplant operation, flight characteristics, or other qualities affecting airworthiness, or (2) is not done according to accepted practices, or cannot be done by elementary operations.

Malfunction or Defect Report. A report (FAA Form 8010-4) providing the FAA and industry with a very essential service record of mechanical difficulties encountered in aircraft operations. Such reports contribute to the correction of conditions or situations which otherwise will continue to prove costly and/or adversely affect the airworthiness of aircraft.

Manufacturer's maintenance manual. A manual provided by an aircraft manufacturer that outlines the methods, techniques, and practices prescribed for each person performing maintenance, alteration, or preventive maintenance on an aircraft, engine, propeller, or appliance.

Mass. A measure of the quantity of matter in an object.

Matter. Any substance that has mass and takes up space.

Maximum landing weight. The heaviest weight an aircraft can have when it lands. For large wide body commercial airplanes, it can be 100,000 pounds less than maximum takeoff weight, or even more.

Maximum ramp weight. The heaviest weight to which an aircraft can be loaded while it is sitting on the ground, sometimes referred to as the maximum taxi weight.

Maximum takeoff weight. The heaviest weight an aircraft can have when it starts the takeoff roll. The difference between this weight and the maximum ramp weight would equal the weight of the fuel that would be consumed prior to takeoff.

Maximum weight. The maximum authorized weight of the aircraft and its contents, and is indicated in the Aircraft Specifications or Type Certificate Data Sheet.

Maximum zero fuel weight. The heaviest weight an aircraft can be loaded to without having any usable fuel in the fuel tanks. Any weight loaded above this value must be in the form of fuel.

Mean aerodynamic chord (MAC). The average distance from the leading edge to the trailing edge of the wing.

Mechanical advantage. A ratio of the resistance force to the effort force.

Mechanical energy. This includes all methods of producing increased motion of molecules such as friction, impact of bodies, or compression of gases.

METO horsepower. The maximum power allowed to be continuously produced by an engine. Takeoff power is usually limited to a given amount of time, such as 1 minute or 5 minutes.

Mixed number. A combination of a whole number and a fraction. For example, $5\,{}^{3}\!/_{8}$ is a mixed number.

Molecule. The smallest particle of an element or compound that retains the chemical properties of the element or compound.

Moment. In determining weight and balance, the moment is the product of a weight multiplied by its arm.

MS flareless fittings. Designed primarily for high-pressure (3,000 psi) hydraulic systems that may be subjected to severe vibration or fluctuating pressure. Using this type of fitting eliminates all tube flaring, yet provides a safe and strong, dependable tube connection. The fitting consists of three parts: a body, a sleeve, and a nut.

Multiplication. The process of repeated addition.

Negative number. A number that is less than zero.

Nomogram. A graph that usually consists of three sets of data. Knowledge of any two sets of data enables the reader to determine the third set.

Normalizing. The process of heating the part to the proper temperature, holding it at that temperature until it is uniformly heated, and then cooling it in still air.

Nuclear energy. Energy stored in the nucleus of atoms is released during the process of nuclear fission in a nuclear reactor or atomic explosion.

Numerator. The upper part of a fraction (represented by the letter N in ${}^{N}\!/_{D}$).

Oblique view. A view that is similar to an isometric view except with two of the three drawing axes always at right angles to each other.

Ohm's Law. Explains the relationship between voltage, current, and resistance in an electrical circuit, and states that current flow in an electrical circuit is directly proportional to the amount of voltage applied to the circuit.

Ohmmeter. A current measuring instrument that provides its own source (self-excited) of power.

Ohm. The standard unit used to measure resistance.

One-hundred-hour inspection. A complete inspection that is required for all aircraft operated for hire every 100 hours. An annual inspection must be conducted by an A&P mechanic with Inspection Authorization.

Operating center of gravity range. The center of gravity for an aircraft loaded and ready for flight.

Orthographic projection. A method of showing all six possible views of an object: front, top, bottom, rear, right side, and left side.

Outside calipers. Used for measuring outside dimensions, such as the diameter of a piece of round stock.

Parallel circuit. A circuit in which two or more electrical resistances or loads are connected across the same voltage source.

Pascal's Law. The law that states that pressure applied anywhere to a body of fluid causes a force to be transmitted equally in all directions; the force acts at right angles to any surface in contact with the fluid.

Percentage. Used to express a number as a fraction of 100. Using the percentage sign, %, 90 percent is expressed as 90%.

Permeability. Used to refer to the ease with which a magnetic flux can be established in a given magnetic circuit.

Perspective view. A drawing that shows a three-dimensional object (portraying height, width, and depth) as it appears to an observer. It most closely resembles the way an object would look in a photograph.

Phantom line. Composed of one long and two short evenly spaced dashes, indicates an alternate position of parts of the object or the relative position of a missing part.

Pictorial drawing. A drawing that is similar to a photograph. It shows an object as it appears to the eye, but it is not satisfactory for showing complex forms and shapes.

Pitch. Term used to describe the frequency of a sound.

Plumb bob. A heavy metal object, cylinder- or cone-shaped, with a sharp point at one end that is suspended by a string to produce a vertical reference line useful in aircraft measurements.

Positive number. A number that is greater than zero.

Potential difference. A difference in electrical pressure.

Potential energy. Energy that is stored.

Potentiometer. A variable tapped resistor that can be used as a voltage divider.

Power (exponent). A shorthand method of indicating how many times a number, called the base, is multiplied by itself. For example, in the number 4^3, 3 is the power, or exponent, and 4 is the base. That is, 4^3 is equal to $4 \times 4 \times 4 = 64$.

Power. Power is the time rate at which work is done or energy is transferred.

Powers of ten. Also called scientific notation. It is a shorthand method of depicting very large or very small numbers.

Pressure. The amount of force acting on a specific amount of surface area, typically measured in pounds per square inch or psi.

Preventive maintenance. Simple or minor preservation operations and the replacement of small standard parts not involving complex assembly operations.

Product. The result of multiplication.

Progressive inspection. Breaking down the large task of conducting a major inspection into smaller tasks which can be accomplished periodically without taking the aircraft out of service for an extended period of time.

Proportion. A proportion is a statement of equality between two or more ratios. The example of A is to B as C is to D can be represented A:B = C:D or A/B = C/D.

Pythagorean Theorem. An equation used to find the length of a third side of any right triangle when the lengths of two sides are known. The Pythagorean Theorem states that $a^2 + b^2 = c^2$. The square of the hypotenuse (side opposite the right angle) is equal to the sum of the squares of the other two sides (a and b).

Quenching. The rapid cooling of metal in the heat treatment process.

Quotient. The result of dividing two numbers.

Radiant energy. Electromagnetic waves of certain frequencies produce heat when they are absorbed by the bodies they strike such as x-rays, light rays, and infrared rays.

Radiation. The continuous emission of energy from the surface of all bodies.

Radical sign. The symbol √, used to indicate the root of a number.

Radiographic inspection. Inspection using radiography to locate defects or flaws in airframe structures or engines with little or no disassembly.

Radius (circle). Equal to one-half the diameter of a circle.

Ratio. The comparison of two numbers or quantities.

Reamers. Tools made of either carbon tool steel or high-speed steel that are used to smooth and enlarge holes to exact size.

Rectifier. A device for converting alternating current to direct current.

Relative humidity. The ratio of the amount of water vapor actually present in the atmosphere to the amount that would be present if the air were saturated at the prevailing temperature and pressure.

Remainder. The leftover number in the process of division.

Repair. The restoration of an aircraft component to its previous state.

Repair station. A maintenance facility certificated under 14 CFR part 145 to perform maintenance functions.

Resistance. The opposition a device or material offers to the flow or current.

Resonance. The increase in amplitude of vibrations of an electric or mechanical system exposed to a periodic force whose frequency is equal or very close to the natural frequency of the system.

Retentivity. The ability of a material to hold its magnetism after the magnetizing field has been removed.

Rheostat. A variable resistor used to vary the amount of current flowing in a circuit.

Root. A number that when multiplied by itself a specified number of times will produce a given number. The two most commonly used roots are the square root and the cube root.

Routine inspection. A visual examination or check of an item in which no disassembly is required.

Schematic diagram. A diagram that locates components with respect to each other within a system.

Scientific notation. Used as a type of shorthand to express very large or very small numbers. For example, to express 1,250,000,000,000 in scientific notation is 1.25×10^{12}.

Sea level pressure. The atmospheric pressure at sea level. Average sea level pressure is 29.92 inches of mercury, or 1013.25 millibars.

Sectional view. A view obtained by cutting away part of an object to show the shape and construction at the cutting plane.

Semiconductor. Any device based on either preferred conduction through a solid in one direction, as in rectifiers, or on a variation in conduction characteristics through a partially conductive material, as in a transistor.

Series circuit. The most basic electrical circuit in which there is only one possible path for current to flow. Current must pass through the circuit components, the battery and the resistor, one after the other, or "in series."

Series-parallel DC circuits. A grouping of parallel resistors connected in series with other resistors.

Signed numbers. A signed number can be either a positive or negative number. A positive number is a number that is greater than zero. A negative number is a number that is less than zero.

Sine. A trigonometric function comparing two sides of a right triangle as follows:

$$\text{Sine} = \frac{\text{opposite side}}{\text{hypotenuse}}$$

Sine wave. A continuous waveform with a constant frequency and amplitude.

Sketch. A simple rough drawing that is made rapidly and without much detail.

Slide caliper. Often used to measure the length of an object. It provides greater accuracy than a ruler.

Solenoid. A loop of wire, often wrapped around a metal core, which produces a magnetic field when an electrical current is passed through it.

Specific gravity. The ratio of the mass of a solid or liquid to the mass of an equal volume of water.

Specific heat. The quantity of heat necessary to increase the temperature of a unit of the mass of a substance 1 °C. The specific heat of a substance is the ratio of its specific heat capacity to the specific heat capacity of water.

Speed of sound. The speed of sound at sea level under standard temperature and pressure conditions is 1,108 feet per second or 658 knots.

Spirit level. A leveling instrument placed on or against a specified place on the aircraft. Spirit levels have vials that are full of liquid, except for a small air bubble. When the air bubble is centered between the two black lines, a level condition is indicated.

Square root. A non-negative number that must be multiplied by itself to equal a given number.

Standard empty weight. The weight of the airframe, engines, all permanently installed equipment, and unusable fuel. Depending upon the part of the Federal regulations under which the aircraft was certificated, either the undrainable oil or full reservoir of oil is included.

Standard weights. Values used in weight and balance calculations if specific weight for an item is unknown. The following are examples:

- Aviation gasoline — 6 pounds per gallon
- Crew and passengers — 170 pounds per person
- Lubricating oil — 7.5 pounds per gallon
- Turbine fuel — 6.7 pounds per gallon
- Water — 8.35 pounds per gallon

Static stability. The initial response that an airplane displays after its equilibrium is disrupted.

Strain. A deformity or change in an object due to stress.

Stress corrosion. Occurs as the result of the combined effect of sustained tensile stresses and a corrosive environment.

Stress. The internal resistance of an object to external forces attempting to strain or deform that object. Measured in pounds per square foot or pounds per square inch (psi).

Subtraction. The process where the value of one number is taken from the value of another.

Sum. The resulting answer in the addition process.

Supplemental Type Certificates (STC). A document issued by the FAA approving a product (aircraft, engine, or propeller) modification.

Surface corrosion. Caused by either direct chemical or electrochemical attack, it appears as a general roughening, etching, or pitting of the surface of a metal, frequently accompanied by a powdery deposit of corrosion products.

Swaged Fittings. These fittings create a permanent connection that is virtually maintenance free. Swaged fittings are used to join hydraulic lines in areas where routine disconnections are not required and are often used with titanium and corrosion resistant steel tubing.

Tangent (tan). A trigonometric function comparing two sides of a right triangle as follows:

$$\text{Tan} = \frac{\text{opposite side}}{\text{adjacent side}}$$

Tap. Instrument used to cut threads on the inside of a hole.

Tare weight. The weight of any chocks or devices used to hold an aircraft on scales when it is weighed. The tare weight must be subtracted from the scale reading to get the net weight of the aircraft.

Tempering. Process that reduces the brittleness imparted by hardening and produces definite physical properties within the steel. Tempering always follows, never precedes, the hardening operation.

Thermal expansion. The increase in size of a material as temperature increases.

Tolerance. The sum of the plus and minus allowance figures.

Torque. The tendency of a force to cause or change rotational motion of a body.

Transformer. A device that changes electrical energy of a given voltage into electrical energy at a different voltage level. It consists of two coils that are not electrically connected, but arranged so that the magnetic field surrounding one coil cuts through the other coil.

Transistor. A three-terminal device primarily used to amplify signals and control current within a circuit.

Trapezoid. A four-sided figure with one pair of parallel sides.

True power. The power dissipated in the resistance of a circuit, or the power actually used in the circuit.

Triangle. A three-sided figure in which the sum of the three angles equal 180°.

Trigonometry. The study of the relationships between the angles and sides of a triangle.

Type Certificate Data Sheet (TCDS). The FAA issues a type certificate when a new aircraft, engine, propeller, etc., is found to meet safety standards set forth by the FAA. The TCDS lists the specifications, conditions and limitations under which airworthiness requirements were met for the specified product, such as engine make and model, fuel type, engine limits, airspeed limits, maximum weight, minimum crew, etc.

Ultrasonic inspection. Uses high frequency sound energy to conduct examinations and make measurements. Ultrasonic inspection can be used for flaw detection/evaluation, dimensional measurements, and material characterization.

Useful load. Fuel, any other fluids that are not part of empty weight, passengers, baggage, pilot, copilot, and crewmembers. It is determined by subtracting the empty weight from the maximum allowable gross weight.

Vapor pressure. The portion of atmospheric pressure that is exerted by the moisture in the air (expressed in tenths of an inch of mercury).

Volt. The basic unit of electrical potential or electromotive force. A potential of one volt appears across a resistance of one ohm when a current of one ampere flows through that resistance.

Voltmeter. A current-measuring instrument, designed to indicate voltage by measuring the current flow through a resistance of known value.

Volume. The amount of space within a three-dimensional solid.

Waterline (WL). A horizontal reference plane used to locate vertical positions on an aircraft. Positions are usually given in inches above or below the waterline.

Watt. A unit of power equal to one joule per second.

Weighing points. Locations on an aircraft that the manufacturer designates for the placement of scales when weighing aircraft.

Weight. A measure of the pull of gravity acting on the mass of an object.

Whole numbers. The numbers: 0, 1, 2, 3, 4, 5, and so on.

Wiring diagrams. A diagram that shows the electrical wiring and circuitry, coded for identification, of all the electrical appliances and devices used on aircraft.

Work. The amount of energy transferred by a force.

Zero fuel weight. The weight of an aircraft without fuel.

Zone numbers. On drawings, these are similar to the numbers and letters printed on the borders of a map, used for locating a particular point in the drawing.

INDEX

C

D

E

F

G

H

I

N

O

P

Q

R

S

www.ingramcontent.com/pod-product-compliance
Lightning Source LLC
LaVergne TN
LVRC101545220926
848486LV00041B/381

* 9 7 8 1 7 8 2 6 6 0 8 6 6 *